# READINGS IN SOLID-STATE AND MATERIALS CHEMISTRY

## Selected Research Papers of C. N. R. Rao

on the Occasion of his 80th Birthday

# READINGS IN SOLID-STATE AND MATERIALS CHEMISTRY

## Selected Research Papers of C. N. R. Rao

on the Occasion of his 80th Birthday

C. N. R. Rao

*Jawaharlal Nehru Centre for Advanced Scientific Research, Bangalore, India*

World Scientific

NEW JERSEY • LONDON • SINGAPORE • BEIJING • SHANGHAI • HONG KONG • TAIPEI • CHENNAI

*Published by*

World Scientific Publishing Co. Pte. Ltd.
5 Toh Tuck Link, Singapore 596224
*USA office:* 27 Warren Street, Suite 401-402, Hackensack, NJ 07601
*UK office:* 57 Shelton Street, Covent Garden, London WC2H 9HE

**British Library Cataloguing-in-Publication Data**
A catalogue record for this book is available from the British Library.

**READINGS IN SOLID-STATE AND MATERIALS CHEMISTRY**
**Selected Research Papers of C N R Rao**



ISBN 978-981-4531-68-9

# PREFACE

I cannot believe that I will soon be 80 years old. I have been doing research in chemical sciences for 62 years of which I have been a faculty member for 54 years. I started working on the chemistry of solid materials over 50 years ago. There were not many practitioners in the subject at that time. Much of the preoccupation of those who worked in this area was on defects in solids and nonstoichiometry. J.S. Anderson (Oxford), who was a source of inspiration to many of us, was interested in the relation between nonstoichiometry and defects. Nonstoichiometry was a problem that plagued chemists for a long time since people were dismayed by compounds such as $Ti_4O_7$ and $Pr_6O_{11}$ with unusual cation : anion ratios. In the late 1960's, it was established that these problems arose due to certain novel structural manifestations, typical of them being crystallographic shear planes and defect complexes. The subject of solid state chemistry gradually grew over the years, with people working on the structure and properties of materials. In particular, research on transition metal oxides was pursued by a number of solid state chemists, and the effort in organic solid state was marginal.

Transition metal oxides are fascinating because of the very wide range of properties and structures exhibited by them. I got interested in this area around 1958, and one of the major researchers at that time was John Goodenough. Several important discoveries have been made in this area during that period. For example, metallic oxides such as $ReO_3$ were reported. $ReO_3$ looks like copper and conducts like copper. Oxides undergoing transitions from the insulating state to the metallic state were discovered, the most prominent amongst them being $V_2O_3$, showing a 10-million fold jump in resistivity around 150 K. My interest in such transitions was aroused by my association with J.M. Honig. $V_2O_3$ continues to be a matter of interest even today. Research on other types of materials such as chalcogenides and pnictides was also pursued, but oxides of different structures, specially those belonging to the perovskite family, received greater attention. I really enjoyed working in this area, and specially recall my association with Nevil Mott (Cambridge) whom I consider to be the father of modern solid state and materials science.

The biggest explosion in materials chemistry and physics occurred in late 1986 when high-temperature superconductivity was discovered in a lanthanum cuprate, a material which was a ceramic and on which a few of us had worked earlier. As stated in a report of the US National Academy of Sciences, this discovery changed the role of chemistry in the study of materials, and materials chemistry became a more significant part of materials science. It was around this time that chemists began to treat solid state chemistry as an integral and important part of mainstream chemistry.

In the last thirty years or so, the subject of solid state chemistry has been transformed into materials chemistry by absorbing various features of modern chemical science. The materials investigated by chemists are no longer limited to inorganic materials but include a variety of organic materials. Synthesis has become a major aspect of materials chemistry. Materials chemists employ a variety of chemical strategies, soft chemical approaches, in particular, for synthesis. Studies of structure, properties and phenomena, and structure-to-properties relations are important aspects of materials chemistry. Computation and theory are used extensively. Supramolecular strategies are employed to design materials. Materials chemistry is now truly interdisciplinary, not only within chemical science itself but also by having interfaces with biology and other subjects. It has immediate connections with physics, engineering and technology.

In this collection, I have included a set of my papers which represent some of the developments in solid state and materials chemistry. There is a section on oxidic materials which includes high-temperature superconductivity, colossal magnetoresistance, electronic phase separation and multiferroics. I could have included my papers on characterization methods and other areas as well as chalcogenides, nitrides and other materials, but have not done so because of limitation of space. There is a section dealing with open-framework and hybrid materials which make use of the rigidity of inorganic structures and the functionality and flexibility of the organic residues to generate novel structures and properties. I have had the pleasure of carrying out some collaborative work on hybrid frameworks with A.K. Cheetham. Nanomaterials, the new fascination, constitute a large section of the book. I have not been able to include many papers on organic materials since I have not worked in this area extensively.

The book is a selection of my papers that is intended to present the flavor of solid state and materials chemistry. I believe that I have succeeded in doing so. It is difficult to be truly up-to-date in such a field since new materials, methods and phenomena are being discovered constantly. Even as I write this preface, new results of vital importance are being obtained in my laboratory and elsewhere. I earnestly hope that the reader will find the papers in this volume interesting and representative of the subject. It has been a pleasure to see the subject grow, and to grow along with it. I have no doubt that materials chemistry will continue to develop, creating new horizons in the years to come.

I remember my teachers at school and elsewhere, my research guides in various places, specially at Purdue, for the inspiration and encouragement that they provided me. I have carried out almost all of my research in the Indian Institute of Science, Indian Institute of Technology Kanpur and Jawaharlal Nehru Centre of Advanced Scientific Research. I sincerely thank these institutions for their unstinted support. All through my professional career, I have been supported by various agencies in India through small and big grants. I owe much to my country, to my fine students and coworkers, and to my wonderful wife, Indu.

Bangalore
2013

C.N.R. Rao

# BRIEF BIODATA OF PROFESSOR C.N.R. RAO

Prof. C.N.R. Rao (born on 30 June 1934, Bangalore, India) is the National Research Professor as well as Honorary President and Linus Pauling Research Professor at the Jawaharlal Nehru Centre for Advanced Scientific Research. He is also an Honorary Professor at the Indian Institute of Science. His main research interests are in solid state and materials chemistry. He is an author of over 1400 research papers and 45 books. He received the M.Sc. degree from Banaras, Ph.D. from Purdue, D.Sc. from Mysore universities and has received honoris causa doctorate degrees from 60 universities including Purdue, Bordeaux, Banaras, Calcutta, Sri Venkateswara, Delhi, IIT Bombay, IIT Kharagpur, IISERs (Bhopal, Kolkata, Mohali, Pune), Northwestern, Notre Dame, Novosibirsk, Oxford, Stellenbosch, Grenoble, Uppsala, Wales, Wroclaw, Caen, Liverpool, St. Andrews and Desikottama from Visva-Bharati.

Prof. Rao is a member of many of the major science academies in the world including the Royal Society, London, the National Academy of Sciences, U.S.A., the Russian Academy of Sciences, French Academy of Sciences, Japan Academy as well as the American Philosophical Society. He is a Member of the Pontifical Academy of Sciences, Foreign Fellow of Academia Europaea and the Royal Society of Canada. He is on the editorial boards of several leading professional journals and is a distinguished visiting professor of the University of California and Cambridge University.

Among the various medals, honours and awards received by him, mention may be made of the Marlow Medal of the Faraday Society (1967), Bhatnagar Prize (1968), Padma Shri (1974), Centennial Foreign Fellowship of the American Chemical Society (1976), Royal Society of Chemistry (London) Medal (1981), Padma Vibhushan (1985), Honorary Fellowship of the Royal Society of Chemistry, London (1989), Hevrovsky Gold Medal of the Czechoslovak Academy (1989), Blackett Lectureship of the Royal Society (1991), Einstein Gold Medal of UNESCO (1996), Linnett Professorship of the University of Cambridge (1998), Centenary Medal of the Royal Society of Chemistry, London (2000), the Hughes Medal of the Royal Society, London, for original discovery in physical sciences (2000), Karnataka Ratna (2001) by the Karnataka Government, the Order of Scientific Merit (Grand-Cross) from the President of Brazil (2002), Gauss Professorship of Germany (2003) and the Somiya Award of the International Union of Materials Research (2004). He is the first recipient of the India Science Award by the Government of India and received the Dan David Prize for science in the future dimension for his research in Materials Science in 2005. He was named as Chemical Pioneer by the American Institute of Chemists (2005), Chevalier de la Légion d'Honneur by the President of the French Republic (2005) and received the Honorary Fellowship of the Institute of Physics, London (2006) and St. Catherine's College, Oxford (2007). He received the Nikkei Asia Prize for Science, Technology and Innovation (2008). He was awarded the Royal Medal by the Royal Society (2009) and the August-Wilhelm-von-Hoffmann Medal for his outstanding contributions to chemistry by the German Chemical Society (2010). He received the Ernesto Illy Trieste Science Prize for materials research in 2011 and was Albert Einstein Professor of the Chinese Academy of Sciences in 2012.

Prof. Rao is Chairman, Scientific Advisory Council to the Prime Minister, immediate past President of The Academy of Sciences for the Developing World (TWAS) and Member of the Atomic Energy Commission of India. He is Founder-President of both the Chemical Research Society of India and of the Materials Research Society of India. Prof. Rao was President of the Indian National Science Academy (1985–86), the Indian Academy of Sciences (1989–91) and the International Union of Pure and Applied Chemistry (1985–97). He was the Director of the Indian Institute of Science (1984–94), Chairman of the Science Advisory Council to Prime Minister Rajiv Gandhi (1985–89) and Chairman, Scientific Advisory Committee to the Union Cabinet (1997–98) and Albert Einstein Research Professor (1995–99).

# CONTENTS

# ACKNOWLEDGEMENTS

Sincere thanks are due to the following publishers for providing copyright permission to reproduce the articles listed below in this volume.

(1) **American Chemical Society**

Article Nos. 8, 12, 22, 24, 26, 27, 30, 31, 32, 39, 41, 42, 43, 46

(2) **Royal Society of Chemistry**

Article Nos. 1, 3, 6, 7, 9, 10, 13, 20, 28, 33, 34, 35, 37, 38, 49

(3) **The Royal Society, London**

Article Nos. 15, 16, 19, 23

(4) **Materials Research Society, USA**

Article No. 25

(5) **Wiley-VCH Verlag GmbH & Co., KGaA**

Article Nos. 21, 29, 36, 45, 47, 48

(6) **Elsevier**

Article Nos. 2, 5, 11, 14, 17, 18, 40, 44

(7) **CSIR, India**

Article No. 4

# I. SOME ARTICLES OF INTEREST

*Commentary by C.N.R. Rao*

MATERIALS CHEMISTRY as we understand it today is relatively of recent origin. A few decades ago, the subject generally included cement, steel, and a few other topics of an applied nature. In the last 30 years of so, the subject has been increasingly recognized as an important new direction in modern chemistry, having incorporated all the salient aspects of solid state chemistry as well.[1,2] While solid state chemistry may be considered as representing the chemical counterpart of solid state physics, materials chemistry deals with structure, response, and function and has the ultimate purpose of developing novel materials and of understanding structure-property relations as well as the phenomena related to a wide range of materials. The materials can be organic, inorganic, or biological and can be in any condensed state of matter. With this development, it becomes difficult to classify materials as in the earlier years when it was common to classify them as ceramics, metals, organics and so on. We now have organic metals, superconductors, and nonlinear materials. So, it is probably more convenient to classify materials based on properties or phenomena. For example, porous solids, superconductors and ferroics cover all types of chemical constituents. It is, however, common to distinguish molecular solids from extended solids, as they represent two limiting descriptions.

Materials chemistry contains all the elements of modern chemistry. These include synthesis, structure, dynamics, and properties. In synthesis, one employs all possible methods and conditions from high-temperature and high-pressure techniques to mild solution methods (chimie douce or soft chemistry).[3] Chemical methods generally tend to be more delicate, often yielding novel, metastable products. Kinetic rather than thermodynamic control of reactions favors the formation of such structures. Supramolecular organization provides new ways of designing materials.[4]

All available methods of diffraction, microscopy, and spectroscopy are used for structure elucidation in present-day materials chemistry.[1,2] For the detailed structure determination, even powders suffice for the most part because of the advances in diffraction profile analysis. These advances in structural tools enable more meaningful correlations of structure with properties and phenomena. Catalysis is becoming more of a science partly because of our ability to unravel the structures and surfaces of catalysts. More and more phase transitions of all varieties[5] are being investigated by chemists.

In this section, I have included a few of my papers dealing with synthesis, defects, certain properties of oxidic materials and a new family of emerging materials, the borocarbonitrides, besides a general article dealing with important directions in materials chemistry.

## References

1. C.N.R. Rao and J. Gopalakrishnan, *New Directions in Solid State Chemistry*, Cambridge University Press, Second Edition, 1997.
2. C.N.R. Rao (Ed.), *Chemistry of Advanced Materials*, IUPAC 21st Century Monograph, Blackwell, Oxford, 1993.
3. C.N.R. Rao, *Chemical Approaches to the Synthesis of Inorganic Materials*, John Wiley, New York, 1994.
4. W. Jones and C.N.R. Rao (Eds.), *Supramolecular Organization and Materials Design*, Cambridge University Press, 2002.
5. C.N.R. Rao and K.J. Rao, *Phase Transitions in Solids*, McGraw Hill, New York, 1978.
6. *Trends in the Chemistry of Materials*, Selected Research Papers of C.N.R. Rao, World Scientific, Singapore, 2009.

JOURNAL OF Materials CHEMISTRY Feature Article

# Novel materials, materials design and synthetic strategies: recent advances and new directions†

C. N. R. Rao‡

*Solid State and Structural Chemistry Unit, Indian Institute of Science, Bangalore 560 012, India and Chemistry & Physics of Materials Unit, Jawaharlal Nehru Centre for Advanced Scientific Research, Jakkur P.O., Bangalore 560 064, India*

*Received 5th May 1998, Accepted 12th June 1998*

**There have been major advances in solid state and materials chemistry in the last two decades and the subject is growing rapidly. In this account, a few of the important aspects of materials chemistry of interest to the author are presented. Accordingly, transition metal oxides, which constitute the most fascinating class of inorganic materials, receive greater attention. Metal–insulator transitions in oxides, high temperature superconductivity in cuprates and colossal magnetoresistance in manganates are discussed at some length and the outstanding problems indicated. We then discuss certain other important classes of materials which include molecular materials, biomolecular materials and porous solids. Recent developments in synthetic strategies for inorganic materials are reviewed. Some results on metal nanoparticles and nanotubes are briefly presented. The overview, which is essentially intended to provide a flavour of the subject and show how it works, lists references to many crucial reviews in the recent literature.**

I have been working in solid state and materials chemistry for over four decades. When I first got interested in the subject, it was not an integral part of main-stream chemistry. In the first book on solid state chemistry[1a] that I edited in 1970, it was stated in the preface, that 'Solid state chemistry has not become part of the formal training programmes in chemistry. Being one of the frontiers of chemistry, it has a tremendous future and undoubtedly demands the active involvement of many more chemists'. In 1986, in the first edition of the book *New Directions in Solid State Chemistry*,[1b] the preface states, 'At the present time, solid state chemistry is mainly concerned with the development of new methods of synthesis, new ways of identifying and characterizing materials and of describing their structure, and above all, with new strategies of tailor-making materials with desired and controllable properties, be they electronic, magnetic, dielectric, optical, absorptive or catalytic. It is heartening that the subject is increasingly coming to be recognized as an emerging area of chemical science'. In the last few years, solid state chemistry has given room to the broader area of materials chemistry. The subject has come of age and there are many reviews and books dealing with it.[2] I am delighted to have this opportunity to present an overview of the subject in this first materials chemistry discussion meeting. In so doing, I will try to present the highlights of some of the areas and refer to important reviews in the literature, but I cannot help dealing with those aspects which are of personal interest to me in somewhat greater detail. It is indeed impractical to cover every aspect of this vast subject in an article. Furthermore, all research in solid state chemistry is not necessarily related to materials development. Materials chemistry, as distinct from solid state chemistry, deals with structure, response and function, and has an ultimate technological objective. I illustrate this aspect of materials chemistry in Table 1.

**Table 1** A description of materials chemistry

| Constituent units | State[b] | Function | Advanced technology |
|---|---|---|---|
| atoms<br>molecules[a]<br>ions[a] | crystalline<br>(molecular, ionic, polymeric, metallic *etc.*)<br>non-crystalline (glasses)<br>clusters and nanomaterials<br>liquid crystalline | miniaturization<br>selectivity and recognition<br>transformation<br>transduction<br>transport<br>energy storage | nanolithography<br>microelectronics<br>magnets sensors and transducers<br>photonic devices<br>energy devices<br>porous solids and membranes<br>micromachines |

[a]Inorganic or organic. [b]In pure (monophasic) form or in the form of aggregates or composites.

If one were to list the most important discoveries in solid state and materials science in the last decade, it would include high-temperature cuprate superconductors (1986), fullerenes and related materials (1990), mesoporous silica (1992) and colossal magnetoresistance (CMR) in manganates (1993). Three of these deal with metal oxides. I have been involved in research on metal oxides for many years and it has been exciting to witness the increasing importance gained by these materials. I shall deal with certain aspects of the chemistry of transition metal oxides at some length. This discussion is presented as a personal account and would also serve as a case-study. I will indicate the developments in molecular materials, porous solids and biomaterials and highlight out the role of chemical synthesis in inorganic systems. I will end the article with a discussion of some of my recent interests in nanotubes and metal nanoparticles.

## Transition metal oxides

Transition metal oxides constitute the most fascinating class of materials, exhibiting a variety of structures and properties.[3] The metal–oxygen bond can vary anywhere between highly ionic to covalent or metallic. The unusual properties of transition metal oxides are clearly due to the unique nature of the outer d-electrons. The phenomenal range of electronic and magnetic properties exhibited by transition metal oxides is noteworthy. Thus, the electrical resistivity in oxide materials spans the extraordinary range of $10^{-10}$ to $10^{20}$ ohm cm. We have oxides with metallic properties (*e.g.* $RuO_2$, $ReO_3$, $LaNiO_3$) at one end of the range and oxides with highly insulating behaviour (*e.g.* $BaTiO_3$) at the other. There are also

† Basis of the presentation given at Materials Chemistry Discussion No. 1, 24–26 September 1998, ICMCB, University of Bordeaux, France.

‡ Also at the Materials Research Laboratory, University of California, Santa Barbara, CA 93106-5050, USA.

oxides that traverse both these regimes with changes in temperature, pressure, or composition (*e.g.* $V_2O_3$, $La_{1-x}Sr_xVO_3$). Interesting electronic properties also arise from charge density waves (*e.g.* $K_{0.3}MoO_3$), charge-ordering (*e.g.* $Fe_3O_4$) and defect ordering (*e.g.* $Ca_2Mn_2O_5$, $Ca_2Fe_2O_5$). Oxides with diverse magnetic properties anywhere from ferromagnetism (*e.g.* $CrO_2$, $La_{0.5}Sr_{0.5}MnO_3$) to antiferromagnetism (*e.g.* NiO, $LaCrO_3$) are known. Many oxides possess switchable orientation states as in ferroelectric (*e.g.* $BaTiO_3$, $KNbO_3$) and ferroelastic [*e.g.* $Gd_2(MoO_4)_3$] materials. Then, there are a variety of oxide bronzes showing a gamut of properties.[4] Superconductivity in transition metal oxides has been known for some time, but the highest $T_c$ reached was around 13 K; we now have oxides with $T_c$s in the region of 160 K. The discovery of high $T_c$ superconductors[5] focused worldwide scientific attention on the chemistry of metal oxides and at the same time revealed the inadequacy of our understanding of these materials.

The unusual properties of transition metal oxides that distinguish them from the metallic elements and alloys, covalent semiconductors and ionic insulators are due to several factors. (a) Oxides of d-block transition elements have narrow electronic bands, because of the small overlap between the metal d and the oxygen p orbitals. The bandwidths are typically of the order of 1–2 eV (rather than 5–15 eV as in most metals). (b) Electron correlation effects play an important role, as expected because of the narrow electronic bands. The local electronic structure can be described in terms of atomic-like states [*e.g.* $Cu^+(d^{10})$, $Cu^{2+}$ ($d^9$) and $Cu^{3+}$ ($d^8$) for Cu in CuO] as in the Heitler–London limit. (c) The polarizability of oxygen is also of importance. The divalent oxide ion, $O^{2-}$, does not exactly describe the state of oxygen and configurations such as $O^-$ have to be included, especially in the solid state. This gives rise to polaronic and bipolaronic effects. Species such as $O^-$, which are oxygen holes with a $p^5$ configuration instead of the filled $p^6$ configuration of $O^{2-}$, can be mobile and correlated. (d) Many transition metal oxides are not truly three-dimensional, but have low-dimensional features. For example, $La_2CuO_4$ and $La_2NiO_4$ with the $K_2NiF_4$ structure are two-dimensional compared to $LaCuO_3$ and $LaNiO_3$, which are three-dimensional perovskites. Because of the varied features of individual oxides, it has not been possible to establish satisfactory theoretical models for complex transition metal oxides. I started working on transition metal oxides in the late 1950s and my early preoccupation was with the structures, defects and phase transitions of oxides such as $TiO_2$, $Pr_6O_{11}$ and $Bi_2O_3$. Soon I became interested in the electronic and magnetic properties of oxides, especially those with the perovskite structure. The perovskite structure dominates all classes of materials, whether they are dielectrics or superconductors.

Transition metal oxides that transform from the metallic state to an insulating or semiconducting state are of great interest. It has always amazed me how a simple oxide like $V_2O_3$ conducts like copper at ordinary temperatures and becomes like wood on cooling it to 150 K. Interest in these transitions has blossomed over the years and the subject has itself become a frontier area of investigation.[6,7] Typical transitions of this type in oxide materials are: (a) pressure-induced transitions, as in NiO, in which the pressure increases the wavefunction overlap between neighbors to induce a change from localized to itinerant behavior of electrons; (b) transitions as in $Fe_3O_4$ involving charge-ordering; (c) transitions in $LaCoO_3$ that are initially induced because of the different spin configurations of the transition metal ion; electron transfer between the two spin states initiates a process that eventually renders the oxide metallic around 1200 K; (d) transitions as in EuO arising from the disappearance of spin polarization band-splitting effects when the ferromagnetic Curie temperature is reached; (e) compositionally induced transitions, as in $La_{1-x}Sr_xCoO_3$ and $LaNi_{1-x}Mn_xO_3$, in which changes of band structure in the vicinity of the Fermi level are brought about by a change in composition or are due to disorder-induced localization; (f) transitions in two-dimensional systems, such as $La_2NiO_4$, in which Ni—O—Ni interactions can only occur in the *ab* plane (unlike in the three-dimensional analogue in $LaNiO_3$) and (g) temperature-induced transitions in a large class of oxides such as $Ti_2O_3$, $VO_2$ and $V_2O_3$. Although I have only mentioned metal–insulator transitions in oxides, it should be noted that these transitions are found in a variety of other systems such as doped semiconductors, metal-ammonia solutions and expanded metals. Global aspects of metal–insulator transitions are truly fascinating.[6,7] In spite of considerable effort, we do not yet understand many aspects of these transitions, specially the 10-million fold jump in resistivity at the transition in $V_2O_3$.

The perovskite structure is ideal for 180° cation–anion–cation interactions and there are no cation–cation interactions. One can vary the covalency (or the interaction between the metal and oxygen orbitals) in this structure by varying the central cation and/or the transition metal ion and obtain a wide variation in properties. Interestingly, a high proportion of oxide materials of technological importance (*e.g.* ferroelectrics, superconductors, CMR materials) possess the perovskite structure. If we examine the electronic and magnetic properties of the family of perovskites of the type $LnMO_3$ (Ln, rare earth; M, first-row transition metal), we find a great variety. The titanates and nickelates possessing low-spin trivalent transition metal ions are metallic or nearly metallic. On the other hand, $LnCrO_3$, $LnMnO_3$ and $LnFeO_3$ are antiferromagnetic insulators. $LaCoO_3$ and other cobaltates are interesting, in that $Co^{3+}$ can be in the low-spin ($t_{2g}^6$) or the high spin ($t_{2g}^4e_g^2$) state. While at low temperatures, $LaCoO_3$ is a diamagnetic insulator, with increasing temperature, the high-spin population increases eventually resulting in a phase transition due to the ordering of the two spin states. At high temperatures (*ca.* 1000 K), the material becomes metallic, resulting from the electron transfer between different spin/oxidation states of cobalt. Substituting for Ln in $LnCoO_3$ with a divalent ion such as $Sr^{2+}$ or $Ba^{2+}$, as in $Ln_{1-x}Sr_xCoO_3$, progressively renders it metallic and ferromagnetic due to the fast electron transfer between the $Co^{3+}$ and $Co^{4+}$ states. We have investigated these cobaltates by a variety of methods including $^{57}Co$ Mössbauer spectroscopy, ferromagnetic resonance, photoemission spectroscopy and EXAFS. Electron transport properties of $LnCoO_3$ doped lightly with $Sr^{2+}$ and other similar oxide systems can be understood in terms of variable range hopping as in disordered systems. The metal–nonmetal transition in $Ln_{1-x}Sr_xCoO_3$ [Fig. 1(*a*)] is a situation similar to impurity doping.[8] Another system showing metal-nonmetal transitions is $LaNi_{1-x}M_xO_3$ (M = Mn, Co, *etc.*), where $LaNiO_3$ ($x$ = 0.0) is metallic [see Fig. 1(*b*) for the Mn system]. On increasing $x$, the material becomes an insulator.[8] Interestingly, in all such systems, the coefficient of resistivity changes sign across a

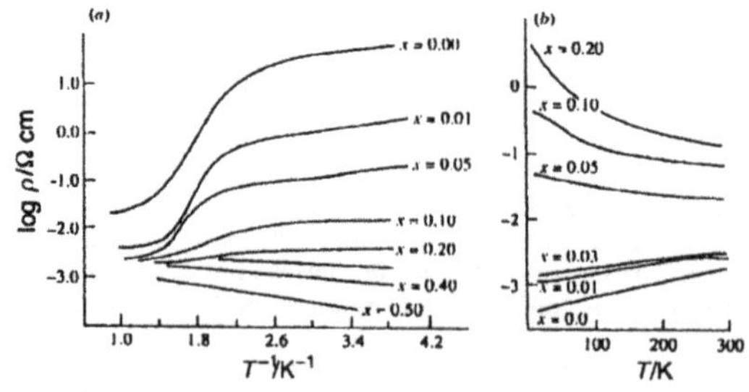


**Fig. 1** Compositionally controlled metal–insulator transitions in (*a*) $La_{1-x}Sr_xCoO_3$ and (*b*) $LaNi_{1-x}Mn_xO_3$ [from Rao (reproduced with permission from ref. 8)].

universal value of resistivity, corresponding to Mott's minimum metallic conductivity.

Studies of the properties of quasi two-dimensional oxides with the $K_2NiF_4$ structure were initiated in this laboratory as early as 1970. It was found that Cu in $Ln_2CuO_4$ had no magnetic moment and that $La_2CuO_4$ was antiferromagnetic. Magnetic and electrical properties of $La_2NiO_4$ and its oxygen-excess nonstoichiometry were investigated. Solid solutions of $La_2CuO_4$ (T) and $Nd_2CuO_4$ (T') showed an interesting phase diagram. When we wrote an overview article[9] on the structure and properties of two-dimensional oxides with the $K_2NiF_4$ structure in 1984, little did we realize then that $La_2CuO_4$ was going to become famous because of its high-temperature superconductivity. The effect of dimensionality on the electronic and magnetic properties of layered nickel oxides (*e.g.* $LaNiO_3$, $La_4Ni_3O_{10}$, $La_3Ni_2O_7$, $La_2NiO_4$ with decreasing dimensionality from 3 to 2) was examined some time ago. Materials of this type have the general formula $(LaO)(LaMO_3)_n$ where the (LaO) is the rock salt layer and the $LaMO_3$ is the perovskite layer.[10] When M = Ni, the $n=\infty$ member ($LaNiO_3$) is metallic, but the metallicity decreases as $n$ decreases. The $n=1$ material has the $K_2NiF_4$ (two-dimensional) structure. One of the more fascinating families of such materials prepared by us was of the type $(SrO)(La_{1-x}Sr_xMnO_3)_n$. These materials have now become familiar because of the colossal magnetoresistance recently discovered in them.[11] In $La_{2-x}Sr_xNiO_4$, the electronic and magnetic properties depend crucially on $x$ and on the oxygen stoichiometry.

Of the many other oxide systems we have investigated, I would like to briefly discuss oxides forming intergrowth structures. Several metal oxide systems exhibit chemically well defined recurrent intergrowth structures with large periodicities, rather than random solid solutions with variable composition. The ordered intergrowth structures themselves frequently show the presence of wrong sequences. High resolution electron microscopy (HREM) enables direct examination of the extent to which a particular ordered arrangement repeats itself, and the presence of different sequences of intergrowths, often of unit cell dimensions. Selected area electron diffraction, which forms an essential part of HREM, provides useful information (not generally provided by X-ray diffraction) regarding the presence of supercells due to the formation of intergrowth structures. Several systems forming ordered intergrowth structures have been discovered and they generally exhibit homology.[12]

Aurivillius described a family of oxides of the general formula $Bi_2A_{n-1}B_nO_{3n+3}$ where the perovskite slabs, $(A_{n-1}B_nO_{3n+1})^{2-}$, $n$ octahedra thick, are interleaved by $(Bi_2O_2)^{2+}$ layers. Typical members of this family are $Bi_2WO_6$ ($n=1$), $Bi_3Ti_{1.5}W_{0.5}O_9$ ($n=2$), $Bi_4Ti_3CrO_{12}$ ($n=3$) and $Bi_5Ti_3CrO_{15}$ ($n=4$). These oxides form intergrowth structure of the general formula $Bi_4A_{m+n-2}B_{m+n}O_{3(m+n)+6}$ involving alternate stacking of two Aurivillius oxides with different $n$ values (Fig. 2). The method of preparation simply involves heating a mixture of the component metal oxides around 1000 K. Ordered intergrowth structures with ($m$, $n$) values of (1,2), (2,3) and (3,4) have been fully characterized by X-ray diffraction and HREM. What is amazing is that such intergrowth structures with long-range order are indeed formed while either member ($m$ and $n$) can exist as a stable entity. These materials seem to be truly representative of recurrent intergrowth. The periodicity found in recurrent intergrowth solids formed by the Aurivillius family of oxides is indeed impressive.[12] $WO_3$ forms tetragonal, hexagonal or perovskite-type bronzes by interaction with alkali and other metals. The family of intergrowth tungsten bronzes (ITB) involving the intergrowth of $nWO_3$ slabs and one to three strips of the hexagonal tungsten bronze (HTB), first described by Kihlborg, is of relevance to the discussion here. In the intergrowth tungsten bronzes of the general formula $M_xWO_3$, $x$ is generally 0.1 or less and depends on whether the HTB strip is one or two tunnels wide. In the ITB phases of bismuth characterized by us some time ago, the HTB strips were always one tunnel wide. Displacement of adjacent tunnel rows due to the tilting of $WO_3$ octahedra often results in doubling of the long-period axis of the ITB. Evidence for the ordering of the intercalating bismuth atoms in the tunnels is found in terms of satellites around the superlattice spots in the electron diffraction patterns. The occurrence of recurrent intergrowth structures continues to fascinate chemists and crystallographers alike.

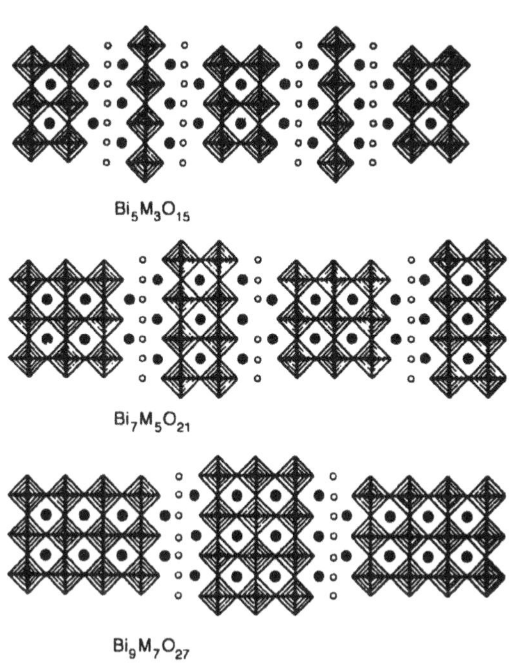


**Fig. 2** Different types of intergrowth structures formed by the Aurivillius family of bismuth oxides.

## Superconducting cuprates

Late in 1986 the solid state community received the big news of the discovery of a high temperature oxide superconductor.[5] The material that was found to be superconducting with a $T_c$ of *ca.* 30 K was based on the cuprate $La_2CuO_4$ (Fig. 3). In February 1987, we decided to look for an yttrium cuprate in the Y–Ba–Cu–O system as a candidate for high $T_c$ superconductivity. Wu *et al.*[13] announced an yttrium cuprate with superconductivity above liquid nitrogen temperature around March 1987. Our strategy to prepare such a material was different. Wu *et al.* obtained a mixture of black and green compounds in their effort to make an yttrium cuprate doped with Ba. The black compound was superconducting ($T_c \approx 90$ K), but they did not know its composition or structure. Since we knew that $Y_2CuO_4$ could not be made, we started examining the $Y_{3-x}Ba_{3+x}Cu_6O_{14}$ system. We obtained 90 K superconductivity in the $x=1$ composition corresponding to $YBa_2Cu_3O_7$. When we independently discovered 90 K superconductivity in March 1987, we knew the composition and the approximate structure.[14]

We worked on a variety of cuprates but unfortunately the area was so competitive and overcrowded that it became difficult to get due credit for much of the work. In spite of these limitations, we made several contributions to this area. These include, besides the independent synthesis and charac-

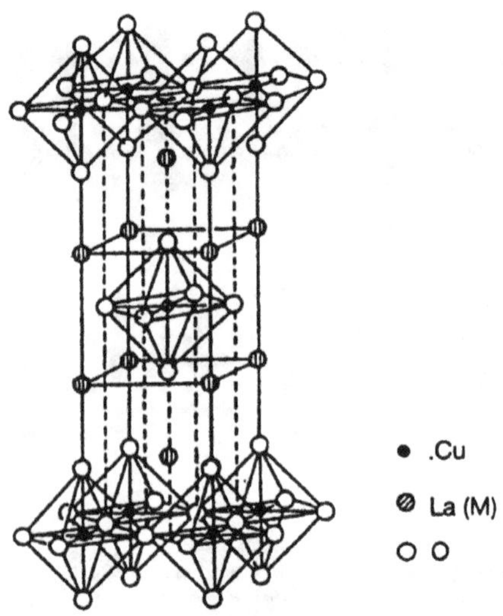


**Fig. 3** Structure of $La_2CuO_4$ and $La_{2-x}A_xCuO_4$ (A = Sr or Ba). The $CuO_6$ octahedra are Jahn–Teller distorted.

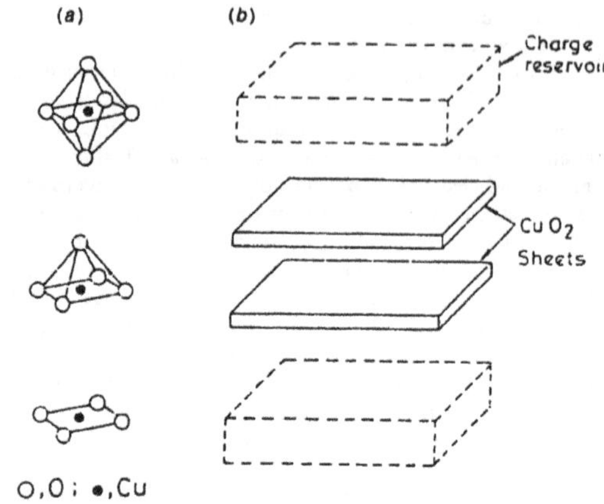


**Fig. 4** (*a*) Cu–O polyhedra found in superconducting cuprates. (*b*) Schematic representation of the cuprates.

terization of $YBa_2Cu_3O_7$, the first observation of non-resonant microwave absorption in high $T_c$ cuprates, metastability of $YBa_2Cu_3O_{7-\delta}$ (in the 60 K regime, $\delta$=0.3–0.4) and its transformation to a novel type of $YBa_2Cu_4O_8$, the importance of Cu–O charge transfer energy in relation to the superconductivity of cuprates and the nature of charge carriers, optimization of hole concentration in cuprates and the relation between $T_c$ and hole concentration. On the synthetic front, we made some contributions which we can consider to be Bangalore brand. For example, we could make a series of superconductors in the Tl–Ca–Sr–Cu–O system by substitution of Tl by Pb or Ca by a rare earth metal; a new family of superconductors of the type $TlSr_{n+1}Cu_nO_{2n+3}$ ($n$=1 and $n$=2) wherein Sr is partly substituted by a rare earth or Tl by Pb was discovered. Another noteworthy contribution was the synthesis of bismuth cuprate superconductors without superlattice modulation. This was significant since all the Bi cuprates showed such modulation. The modulation could be removed by suitable cation substitution (Pb/Bi) and the adjustment of oxygen stoichiometry. An interesting aspect was the synthesis and characterization of oxyanion derivatives of cuprate superconductors containing anions such as $CO_3^{2-}$, $SO_4^{2-}$, $PO_4^{3-}$ and $BO_3^{3-}$, as integral parts of the oxide framework. Besides the mercury cuprates with the highest $T_c$ to date, many interesting cuprates have been synthesized in the last 4–5 years.[15,16] These include copper oxycarbonates, copper oxyhalides, $Sr_2CuO_2F_{2+\delta}$ and ladder cuprates. Here, I must place on record the enormous contributions of Raveau and coworkers[17] to the synthesis and characterization of cuprate superonductors.

The several families of cuprate superconductors with holes as carriers all have the general features shown in Fig. 4. These features are common to the mercury cuprate showing the highest $T_c$ as well. Electron superconductors of the type $Nd_{2-x}M_xCuO_4$ (M = Ce, Th) with the T′ structure possess square-planar $CuO_4$ units instead of the elongated octahedra in the T structure. The hole superconductors generally have the square pyramidal units. Many of the cuprate families have an antiferromagnetic insulator member at one end of the composition. What is more interesting is that all the cuprates are at a metal–insulator boundary. Some of them undergo a metal–nonmetal transition as a function of composition (*e.g.* $Bi_2Ca_{1-x}Y_xSr_2Cu_2O_8$ and $TlY_{1-x}Ca_xSr_2Cu_2O_7$ with change in $x$). The essential feature of the cuprates is the presence of $CuO_2$ sheets with or without apical oxygens. The mobile charge carriers in the cuprates are in the $CuO_2$ sheets. All the cuprates have charge reservoirs as exemplified by the Cu–O chains in the 123 and 124 cuprates and the TlO, BiO and HgO layers in the other cuprates. That the $CuO_2$ sheets are the seat of high-temperature superconductivity is demonstrated by the fact that intercalation of iodine between BiO layers in the bismuth cuprates does not affect the superconducting transition temperature while introduction of fluorite layers between the $CuO_2$ sheets adversely affects superconductivity. In the different series of cuprates with varying number of $CuO_2$ sheets studied hitherto, the $T_c$ reaches a maximum when $n$=3 except in single thallium layer cuprates where the maximum is at $n$=4. The infinite layered cuprates, where the $CuO_2$ sheets are separated by alkaline earth and other cations, show $T_c$ values in the 40–110 K range. Superconductivity in these materials appears to be due to the presence of Sr–O defect layers corresponding to the insertion of $Sr_3O_{2\pm x}$ blocks. We shall briefly present the common features of cuprate superconductors[18] and examine some of the outstanding problems.

The chemistry of the majority of the cuprate superconductors is governed by oxygen holes. (Rouxel[19] has carried out fine work with sulfur holes in metal sulfides). The excess positive charge in cuprates can be represented in terms of the formal valence of copper, which in the absence of holes will be +2 in the $CuO_2$ sheets. In hole superconducting cuprates, it is generally around +2.2. In electron superconductors, it would be less than +2 as expected. The actual concentration of holes, $n_h$, in the $CuO_2$ sheets in $La_{2-x}A_xCuO_4$, $YBa_2Cu_3O_7$ and Bi cuprates is readily determined by redox titrations. In the 123 cuprates, the concentration of mobile holes in the $CuO_2$ sheets can be delineated from that in the Cu–O chains. Determination of $n_h$ in thallium cuprates poses some problems, but in single Tl–O layer cuprates, chemical methods have been developed to obtain reasonable estimates. Generally, $T_c$ in a given family of cuprates reaches a maximum value at an optimal value of $n_h$ as shown in Fig. 5; the maximum is around $n_h \approx 0.2$ in most cuprates. Notice that the points in the underdoped region in Fig. 5 fall close to a straight line, but deviations occur in the overdoped region. Single layer thallium cuprates also show this behaviour. In $Tl_{1-y}Pb_yY_{1-x}Ca_xSr_2Cu_2O_7$ where the substitution of $Tl^{3+}$ by $Pb^{4+}$ has an effect opposite to that due to the substitution of $Y^{3+}$ by $Ca^{2+}$, the $T_c$ becomes a maximum at an optimal value of $(x-y)$, which is a measure of the hole concentration. By suitably manipulating $x$ and $y$, the $T_c$ of this system can be increased from 85–90 K to 110 K.

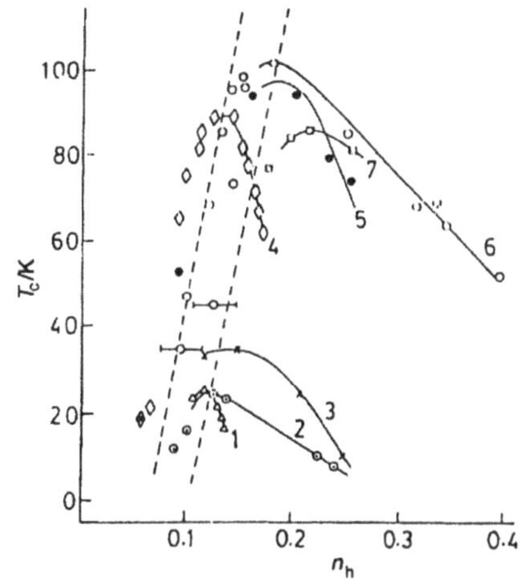


**Fig. 5** Variation of $T_c$ with the hole concentration, $n_h$: 1,2,6 and 7, Bi cuprates; 5, 123 cuprates; 3, $La_{2-x}Sr_xCuO_4$; 4, Tl cuprates (from Rao and Ganguli[18a]).

It is possible to obtain useful correlations between $T_c$ and the Cu–O distances, both in-plane and apical. The Cu and oxygen Madelung energies also provide good correlations.[18a] The $T_c$s vary sensibly with the Cu–O charge-transfer energy as determined by core-level intensities. In spite of such relations with structural parameters, we are far from understanding the high $T_c$ cuprates, particularly in the normal state.[18] The doped and undoped cuprates are both strongly correlated and we do not fully understand the strong correlation in these materials. Many puzzles remain. (i) For example, we do not fully understand the global composition–temperature phase diagram and the apparent lack of symmetry between hole and electron superconductors. (ii) The cuprates are unusual metals in-plane and insulators perpendicular to it; on cooling, they become three-dimensional superconductors. The anomalous temperature variation of resistivity in cuprates is noteworthy (Fig. 6), particularly the extraordinary linear variation of the *ab* plane metallic resistivity. (iii) The ac optical conductivity data do not conform the behavior expected of metals. (iv) The absence of splitting in the energy of states coupled by bilayer tunneling (as found from ARPES) is surprising. All the above observations cannot be reconciled with the picture of an anisotropic metal. (v) The linear resistivity (*ab* plane) with respect to temperature, with zero-intercept and nearly constant slope (independent of $T_c$) cannot be explained easily. The anisotropy is large and temperature dependent. The Hall resistivity varies inversely with temperature. (vi) Photoemission spectra suggest that the quasiparticles are not well defined in the normal state. (vii) Above all, there appears to be a characteristic low energy scale in cuprates. These pseudogap effects are indeed unique to cuprates. (viii) The superconducting state also has some unusual features, such as anisotropic order parameter, and well defined quasiparticles. Even a doped ladder cuprate has been rendered superconducting. Clearly, the status of our theoretical understanding of cuprate superconductors is far from satisfactory. It appears that cuprates constitute an experimentally overdetermined system. In the meantime, if one discovers a room-temperature superconductor, it will be an experimentalist's joy, but a theoretician's nightmare. It is noteworthy that the high $T_c$ cuprates are finding applications in electronic devices, although their use in fabricating generators and magnets may take time.

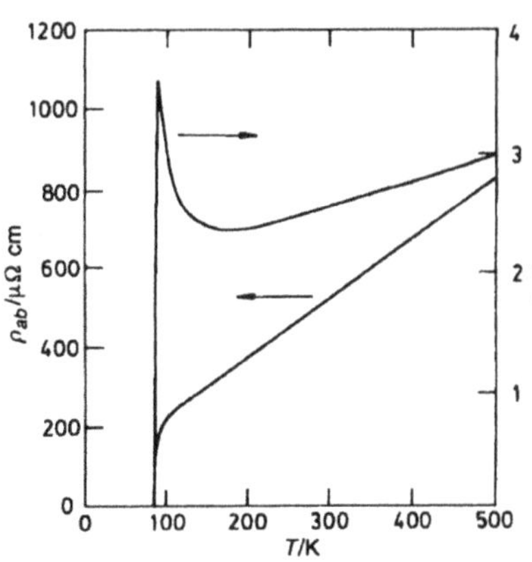


**Fig. 6** Temperature variation of resistivity of $Bi_2Sr_2CaCu_2O_8$ single crystal along the *ab* plane and *c*-axis [data from Ong *et al.* as quoted by Ramakrishnan[18c] reproduced with permission from ref. 18(*c*)].

## Colossal magnetoresistance, charge ordering and related properties of rare earth manganates

Giant magnetoresistance was known to occur in bilayer and granular metallic materials. Perovskite manganates of the general formula $Ln_{1-x}A_xMnO_3$ (Ln = rare earth, A = alkaline earth) have created wide interest because they exhibit colossal magnetoresistance (CMR). These oxides become ferromagnetic at an optimal value of $x$ (or $Mn^{4+}$ content) and undergo an insulator–metal (I–M) transition around the ferromagnetic $T_c$. These properties are attributed to double exchange associated with electron hopping from $Mn^{3+}$ to $Mn^{4+}$. The double-exchange which favours itinerant electron behavior is opposed by the Jahn–Teller distortion due to the presence of $Mn^{3+}$. The manganates show charge ordering especially when the average size of the A-site cations is small. Charge ordering also competes with double exchange and favours insulating behavior. Thus, the manganates exhibit a variety of properties associated with spin, charge as well as orbital ordering. In what follows, we discuss some important aspects of colossal magnetoresistance, charge ordering and related properties of the manganates.[8,11,20]

$LaMnO_3$ is an insulator with an orthorhombically distorted perovskite structure ($b>a>c\sqrt{2}$) due to Jahn–Teller distortion. Typically, as prepared samples of $LaMnO_3$ contain some $Mn^{4+}$. $LaMnO_3$ with a small proportion of $Mn^{4+}$ (<5%) becomes antiferromagnetically ordered at low temperatures ($T_N \approx 150$ K). When the $La^{3+}$ in $LaMnO_3$ is progressively substituted by a divalent cation as in $La_{1-x}A_xMnO_3$ (A = Ca, Sr or Ba), it becomes ferromagnetic with a well defined Curie temperature, $T_c$, and metallic below $T_c$. The saturation moment is typically *ca.* 3.8 $\mu_B$ which is close to the theoretical estimate based on localized spin-only moments, suggesting that the carriers are spin-polarized. Below $T_c$, the manganates exhibit metal-like conductivity. Fig. 7 illustrates how the ferromagnetism and the I–M transition occur around $T_c$ in a typical $La_{1-x}Ca_xMnO_3$ composition. The simultaneous observation of itinerant electron behavior and ferromagnetism in the manganates is explained by Zener's double-exchange mechanism. The basic process in this mechanism is the hopping of a d hole from $Mn^{4+}$ ($d^3$, $t_{2g}^3$, $S=3/2$) to $Mn^{3+}$ ($d^4$, $t_{2g}^3e_g^1$, $S=2$) *via* the oxygen, so that the $Mn^{4+}$ and $Mn^{3+}$ ions change places. The $t_{2g}^3$ electrons of the $Mn^{3+}$ ion are localized on the Mn site giving rise to a local spin of 3/2, but the $e_g$ state, which is hybridized with the oxygen 2p state, can be localized or itinerant. There is a strong Hund's rule interaction between the $e_g$ and the $t_{2g}^3$ electrons. Goodenough pointed out that the ferromagnetism is governed not only by double exchange,

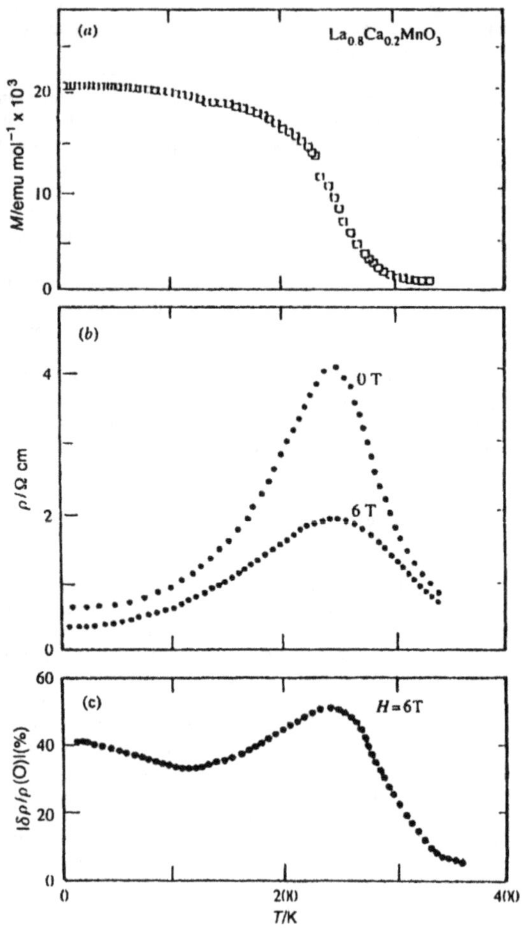


**Fig. 7** Temperature variation of (*a*) the magnetization, (*b*) the resistivity (at 0 and 6 T) and (*c*) the magnetoresistance in $La_{0.8}Ca_{0.2}MnO_3$ [from Rao (reproduced with permission from ref. 8)].

but also by the nature of the superexchange interactions. It should be noted that the $Mn^{3+}$-O-$Mn^{4+}$ superexchange interaction is ferromagnetic while both the $Mn^{3+}$-O-$Mn^{3+}$ and $Mn^{4+}$-O-$Mn^{4+}$ interactions are antiferromagnetic.

The orthorhombic distortion of $LaMnO_3$ decreases as La is progressively substituted by a divalent ion and until it becomes rhombohedral or pseudocubic. The Mn-O-Mn angle approaches 180° as the distortion decreases. When the size of the A-site cations is small, orthorhombic distortion is favoured. The key factor responsible for the distortion in $LaMnO_3$ is the Jahn-Teller distortion due to $Mn^{3+}$ ions. Accordingly, $LaMnO_3$ has three Mn-O distances (1.91, 1.96 and 2.19 Å). Across the I-M transition, the Jahn-Teller distortion decreases in $Ln_{1-x}A_xMnO_3$, the distortion being greater in the insulating regime. The structural parameters, specifically the oxygen thermal parameters, show a significant change across the M-I transition. Clearly, $Mn^{4+}$ ions decrease the Jahn-Teller distortion, increase the itinerancy of electrons through double exchange and increase the tolerance factor of the perovskite structure. Temperature-composition phase diagrams for $La_{1-x}A_xMnO_3$ have been worked out with increasing $x$ or $Mn^{4+}$ content. A composition of $x \approx 0.3$ is ideal to observe a good I-M transition and negative CMR. The magnitude of CMR is maximum at the I-M transition (Fig. 7) and scales with the field-induced magnetization. CMR is found in cation-deficient $LaMnO_3$ containing $Mn^{4+}$ as well, and the features of the magnetic and metal-insulator transitions are similar to those in Fig. 7.

CMR is found over a wide range of compositions of $La_{1-x}Ca_xMnO_3$ ($x<0.5$). However, this material shows significant lattice effects in particular for low $x$. EXAFS studies reveal significant changes in the local structure in the $x=0.3$ composition in the 80–300 K range which are attributed to the formation of small polarons due to the Jahn-Teller distortion when $T>T_c$. Lattice polaron formation and local structural distortion have important implications on the electron transport (high resistivity) in these materials. $La_{1-x}Sr_xMnO_3$ is somewhat different from $La_{1-x}Ca_xMnO_3$, the A-site ion size being larger. The ferromagnetic metallic regime in $La_{1-x}Sr_xMnO_3$ extends over larger values of $x$ and the Curie temperatures are higher. Magnetization in the manganates decreases as $T^2$ and the resistivity increases as $T^2$. In $La_{1-x}Sr_xMnO_3$ ($x=0.17$), the local spin moments and charge carriers couple strongly to changes in the structure and the structure can be switched by the application of a magnetic field depending on the temperature. The magnetic field which causes this crossover from the orthorhombic to the rhombohedral structure is rather low (*ca.* 2 T), showing thereby that a very small energy difference exists between these two structures.

Hydrostatic pressure stabilizes the ferromagnetic metallic state in $La_{1-x}A_xMnO_3$ (*i.e.* increases $T_c$). A similar effect can be brought about by increasing the radius of the A-site cation. There is a direct relationship between $T_c$ and the average radius of the A-site cation, $\langle r_A \rangle$, the $T_c$ increasing with $\langle r_A \rangle$. A small $\langle r_A \rangle$ gives rise to a distortion of the $MnO_6$ octahedra by bending the Mn-O-Mn bond and in turn causing the narrowing of the $e_g$ bandwidth to a greater extent. Accordingly, yttrium substitution in $La_{1-x}A_xMnO_3$ decreases $T_c$, and enhances the magnitudes of resistivity as well as magnetoresistance. By plotting the $T_c$ against $\langle r_A \rangle$, one obtains a phase diagram separating the ferromagnetic metal and the paramagnetic insulator regimes. In Fig. 8, we show a phase diagram obtained in this manner. In the lower left-hand corner, we see the ferromagnetic insulator regime. CMR generally decreases with increase in $\langle r_A \rangle$, just as the peak resistivity at the I-M transition.

The effect of particle size on the electron transport and magnetic properties of polycrystalline $La_{0.7}Ca_{0.3}MnO_3$ has been investigated. Although $T_c$ decreases with decreasing particle size, the magnetoresistance (MR) is insensitive to particle size. The I-M transition becomes broader when the particle size is small, but the %MR is nearly constant over a

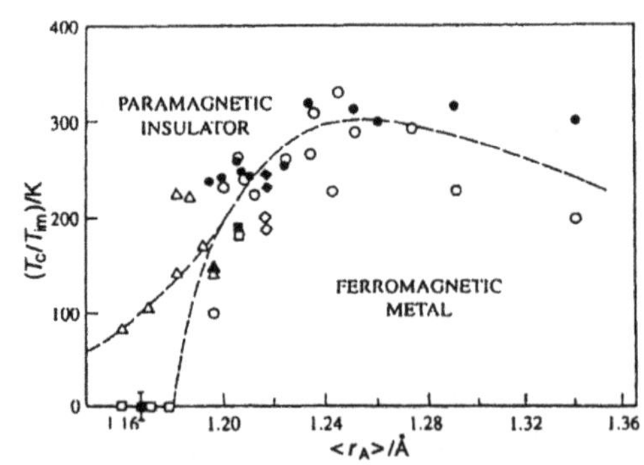


**Fig. 8** Variation of the ferromagnetic $T_c$ or the metal-insulator transition temperature, $T_{im}$, in $Ln_{1-x}A_xMnO_3$ with weighted average radius of the A site cations, $\langle r_A \rangle$ [from Rao (reproduced with permission from ref. 8)].

wide temperature range. The low-temperature MR is significantly dependent on the grain size. For larger grain sizes, the MR at the lowest temperature decreases eventually becoming zero as in good single crystals. Samples of manganates with different particle sizes prepared with different heat treatments often show large differences between the ferromagnetic $T_c$ and the temperature of the I–M transition. Grain boundaries appear to contribute substantially to magnetoresistance in polycrystalline materials, in particular at $T \ll T_c$.

The effects of dimensionality on CMR and related properties have been studied by examining the $(SrO)(La_{1-x}Sr_xMnO_3)_n$ family. The $n=1$ member which is two dimensional with the $K_2NiF_4$ structure, is an insulator, while the $n=\infty$ member is the three-dimensional perovskite showing CMR and other properties. The $n=2$ member exhibits the sharpest I–M transition and high MR. In $La_{1.4}Sr_{1.6}Mn_2O_7$, the current perpendicular to the $MnO_2$ planes is large, particularly close to $T_c$, and the barrier to transport provided by the intervening insulating layers is removed by the magnetic field as shown by Tokura *et al.* Interlayer transport seems to involve tunneling. We see some comparable features between the layered cuprate superconductors and the layered manganates. The big difference is that electrons are spin polarized in the latter.

Studies of the electrical, magnetic and other properties of $Ln_{1-x}A_xMnO_3$ systems have revealed certain unusual features with respect to the charge carriers in these oxides. The manganates exhibit high resistivities even in the metallic state, particularly at low temperatures. The values of resistivities are considerably higher than Mott's value of maximum metallic resistivity in many of these materials, the resistivity reaching constant values as high as $10^3$–$10^4$ ohm cm at low temperatures. A clear picture of the nature of carriers responsible for the high resistivity is yet to emerge. The cause of the observed temperature dependence of resistivity at low temperatures is also not entirely clear. Another unusual feature of the manganates is related to the occurrence of large MR in a small field at relatively high temperatures. The scale of Zeeman energy associated with a field of 1 T in these spin systems is $\approx 5$ K. This is much too small compared to the scale of thermal energy (200–300 K) where one sees such a large MR.

That double exchange is necessary for CMR in the manganates is shown by a study of the $LnMn_{1-x}Cr_xO_3$ system where $Cr^{3+}$ replaces $Mn^{4+}$. In fact, double exchange probably suffices to explain most properties of $La_{1-x}Sr_xMnO_3$. The basic question that remains is why these manganates are insulating when $T>T_c$. The anomalous resistance has been explained by considering electron–phonon interactions arising from the Jahn–Teller splitting of the Mn d-levels. Jahn–Teller distortion localizes conduction electrons as polarons, especially in narrow band materials like $La_{1-x}Ca_xMnO_3$. As the temperature decreases, the transfer integral $t_{ij}$ increases and the ratio of the Jahn–Teller energy to $t_{ij}$ decreases. The coupling varies with temperature and composition, and increases across the metal–insulator transition. The large effects of local lattice distortion and crystal structure symmetry on electron transport and magnetism show that one cannot explain the unusual behavior without taking the lattice into account, especially in systems such as $La_{1-x}Ca_xMnO_3$. The observed $T_c$ (*ca.* 250–400 K) in most manganates is considerably lower than that estimated from the measured spin-wave stiffness constant implying that the ferromagnetic exchange arising from the double exchange mechanism is not the only factor that affects $T_c$. The electrical resistivity of the rare earth manganates across the I–M transition is indeed difficult to understand especially at low temperatures based on double exchange or electron–phonon interaction alone. We also do not understand why CMR actually occurs in these oxides.[20] There is however every expectation that CMR will be exploited in the recording and sensor industries. We need to explore pure as well as composite materials showing CMR at ambient temperatures at low magnetic fields. The exploration should not be limited to rare earth manganates since many other inorganic materials, including $Fe_{1-x}Cu_xCr_2S_4$ and $EuO_x$, exhibit large MR effects.

## Marginal metallicity in oxide materials

We had earlier looked briefly at the metal–insulator transitions in oxides of the type $La_{1-x}Sr_xCoO_3$ or $LaNi_{1-x}Mn_xO_3$ where the temperature coefficient of resistance changes sign across a universal value of conductivity, *viz.* Mott's minimum metallic conductivity. These oxides not only obey the minimum metallic conductivity criterion arising from consideration of disorder, but also the $n_c^{1/3}a \approx 0.25$ criterion ($n_c$ = critical carrier concentration at the transition) arising from electron interactions. The same situation holds for the cuprate superconductors which follow both these criteria. The rare earth manganates, $Ln_{1-x}A_xMnO_3$, exhibiting colossal magnetoresistance, on the other hand, are quite different in that many of them exhibit high residual resistivities far above Mott's value. The nature of carriers in the manganates would be much more difficult to describe than those in cuprate superconductors, which are themselves far from being understood. What is to be noted is that three of the most fascinating phenomena in materials, namely insulator–metal transitions, high temperature superconductivity and CMR are exhibited by oxide materials which are marginally metallic. Marginal metallicity appears to be a universal feature,[21] which requires further probing. It is interesting to speculate whether a universal minimum electron diffusivity due to localized states is present in these materials, besides the extended states.

## Charge ordering in the manganates

Charge ordering is not new to transition metal oxides. $Fe_3O_4$ undergoes the Verwey transition with a resistivity anomaly due to charge ordering at 120 K and ferrimagnetic ordering at 860 K. Charge ordering of the $Mn^{3+}$ and $Mn^{4+}$ ions occurs in the perovskite manganates. $Ln_{1-x}A_xMnO_3$ is more interesting and is associated with dramatic changes in properties.[8,11,20] Charge ordering in the manganates is strongly dependent on the average radius of the A-site cations, $\langle r_A \rangle$. The ferromagnetic $T_c$ increases with increase in $\langle r_A \rangle$ while charge ordering is favoured by small $\langle r_A \rangle$. Charge ordering and double exchange are thus competing interactions. We shall examine charge ordering in $Ln_{0.5}A_{0.5}MnO_3$ with equal proportions of $Mn^{3+}$ and $Mn^{4+}$.

When $\langle r_A \rangle$ is sufficiently large, as in $La_{0.5}Sr_{0.5}MnO_3$ ($\langle r_A \rangle = 1.26$ Å), the manganates become ferromagnetic and undergo an insulator–metal transition around $T_c$ as described earlier. $Nd_{0.5}Sr_{0.5}MnO_3$ with a slightly smaller $\langle r_A \rangle$ (1.236 Å), is ferromagnetic and metallic at 250 K ($T_c$), but transforms to a charge-ordered (CO), antiferromagnetic state at 150 K as shown by Tokura.[20] The charge-ordering transition temperature, $T_{CO}$, is the same as $T_N$. There is a significant specific heat anomaly at $T_{CO}$ ($T_N$) and the resistivity increases markedly due to the metal–insulator (FM–CO) transition. This situation is to be contrasted with that of $Nd_{0.5}Ca_{0.5}MnO_3$ ($\langle r_A \rangle = 1.17$ Å) which exhibits a $T_{CO}$ of 250 K in the paramagnetic state exhibiting a small heat capacity anomaly. $Nd_{0.5}Ca_{0.5}MnO_3$ is an insulator at all temperatures. $Y_{0.5}Ca_{0.5}MnO_3$ with a $\langle r_A \rangle$ of 1.13 Å also gets charge ordered in the paramagnetic state ($T_{CO}$, 240 K) and becomes antiferromagnetic at 140 K ($T_N$). In Fig. 9 we show a simple diagram where we delineate the different types of behavior of the manganates depending on the $\langle r_A \rangle$. The charge ordered state of $Nd_{0.5}Sr_{0.5}MnO_3$ is melted by the application of a magnetic field, rendering the material metallic. On the other hand, a magnetic field of 6 T has no effect on the charge-ordered insulating state of $Y_{0.5}Ca_{0.5}MnO_3$. The charge ordered states in $Nd_{0.5}Sr_{0.5}MnO_3$ and $Nd_{0.5}(Y_{0.5})Ca_{0.5}MnO_3$ are

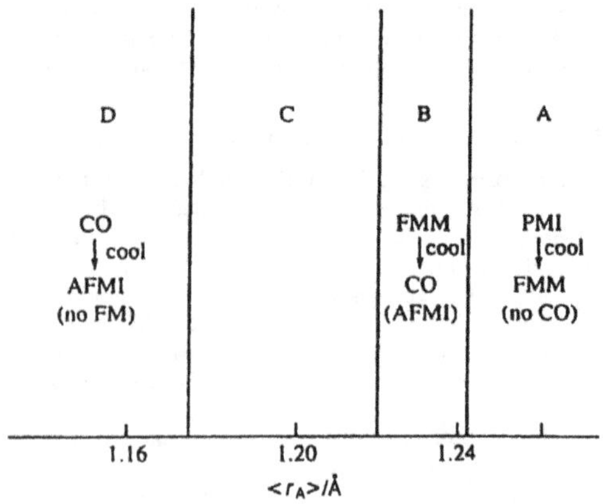


**Fig. 9** Schematic diagram showing the different types of behavior in $La_{0.5}A_{0.5}MnO_3$. Region A corresponds to manganates showing CMR while region B corresponds to manganates of the type $Nd_{0.5}Sr_{0.5}MnO_3$ transforming from the ferromagnetic metallic (FMM) state to a charge-ordered (CO) state on cooling. Region D corresponds to manganates where the CO state is generally found in the paramagnetic insulating (PMI) state (*e.g.* $Nd_{0.5}Ca_{0.5}MnO_3$ and $Y_{0.5}Ca_{0.5}MnO_3$). Region C is expected to show complex behavior. AFMI stands for the antiferromagnetic insulating state [from Rao (reproduced with permission from ref. 8)].

clearly of different kinds, the difference being caused by the difference in $<r_A>$.

The occurrence of two types of CO states can be understood qualitatively in terms of the variation of the exchange couplings, $J_{FM}$ and $J_{AFM}$, and the single-ion Jahn–Teller energy ($E_{JT}$) with $<r_A>$. While $J_{FM}$ and $J_{AFM}$ are expected to decrease with a decrease in $<r_A>$, albeit with different slopes, $E_{JT}$ would be insensitive to $<r_A>$. In the small $<r_A>$ regime, a cooperative Jahn–Teller effect involving long-range elastic strain would dominate charge ordering, while at moderate values of $<r_A>$ (when $J_{AFM}>E_{JT}$, $J_{FM}$), the $e_g$ electrons which are localized magnetically lower the configuration energy by charge ordering. Such a CO state, as exemplified by $Nd_{0.5}Sr_{0.5}MnO_3$, would be sensitive to magnetic fields unlike the manganates such as $Y_{0.5}Ca_{0.5}MnO_3$ in the small $<r_A>$ regime. In $Nd_{0.5}Sr_{0.5}MnO_3$, the gain in Zeeman energy resulting from the application of a magnetic field stabilizes the ferromagnetic metallic state over the antiferromagnetic insulating state. The general phase diagram in the $T$–$<r_A>$ plane has been constructed by Kumar and Rao based on these considerations.[8] Although we understand the gross features of charge-ordering and associated properties in manganates, we do not have a proper theory or model to explain all the varied aspects of these fascinating materials.

The high sensitivity of charge ordering in manganates to $<r_A>$ cannot be understood on the basis of the variation of the Mn–O–Mn angle and the Mn–O distance. This is because these geometrical changes do not give rise to sufficiently large changes in bandwidth as are necessary to explain the significant changes in the charge ordering temperature and the Curie temperature with $<r_A>$. The study of the complex regime (region C in Fig. 9), has yielded fascinating results. Thus, $Nd_{0.25}La_{0.25}Ca_{0.5}MnO_3$ ($<r_A>=1.19$ Å) undergoes a reentrant transition from an incipient charge-ordered state to a FMM state on cooling, driven by a first order phase transition.[22] The charge-ordering gap collapses at the transition. Charge ordering in the manganates needs to be pursued further, especially regarding the marked effects of electric fields and chemical substitution. The role of the A-site ion mismatch on CO and FM states also needs attention.

## Porous solids

Porous inorganic materials have many applications in catalytic and separation technologies. The subject has grown explosively with 8000 or more literature citations. The synthesis of microporous solids with connectivities and pore chemistry different from zeolitic materials has attracted considerable attention. A variety of such open framework structures, in particular Al and Ga phosphates as well as many other metal phosphates prepared in the presence of structure directing agents, have been characterized. The work of Cheetham, Ferey, Flanigen, Stucky and Thomas in this direction is noteworthy.[23a] Zeolites and zeolite-like materials[23,24] with micropores in the 5–20 Å range have contributed much to the advances in catalysis. There have been many break-throughs in the design and synthesis of these molecular seives with well defined crystalline architecture and there is a continuing quest for extra-large pore molecular sieves.[24] Several types of new materials including tunnel and layer structures of porous manganese oxides have been reported.[25] The progress has resulted mainly from our understanding of the role of structure-directing agents. The discovery of mesoporous silica (pore diameter 20–100 Å) by Mobil chemists added a new dimension to the study of porous solids. The synthesis of mesoporous materials also makes use of structure-directing surfactants[24,26] (cationic, anionic, and neutral) and a variety of mesoporous oxidic materials (*e.g.* $ZrO_2$, $TiO_2$, $AlPO_4$, aluminoborates), have been prepared and characterized.[26,27] In Fig. 10, the X-ray diffraction patterns of the three forms of $CaAlBO_4$ are shown.

An appropriate integration of hydrogen bond interactions at the inorganic–organic interface and the use of sol–gel and emulsion chemistry has enabled the synthesis of a large class of porous materials.[28a] Today, we have a better understanding of the structure, topology and phase transitions of mesoporous solids. Thus, block copolymers have been used to prepare mesoporous materials with large pore sizes[28a] (>30 nm). There is also some understanding of the lamellar–hexagonal–cubic phase transitions in mesoporous oxides[28b] (Fig. 11). Derivatized mesoporous materials have been explored for potential applications in catalysis and other areas.[29a] It has been found recently that transition metal complexes encapsu-

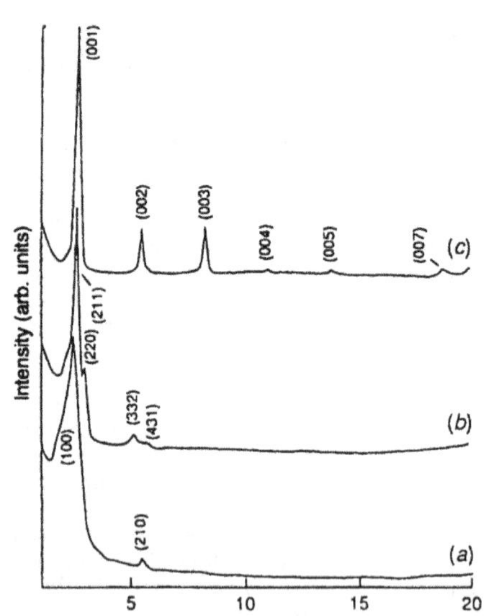


**Fig. 10** X-Ray diffraction patterns of (*a*) hexagonal, (*b*) cubic and (*c*) lamellar forms of mesoporous calcium aluminoborate (from Ayyappan and Rao[27]).

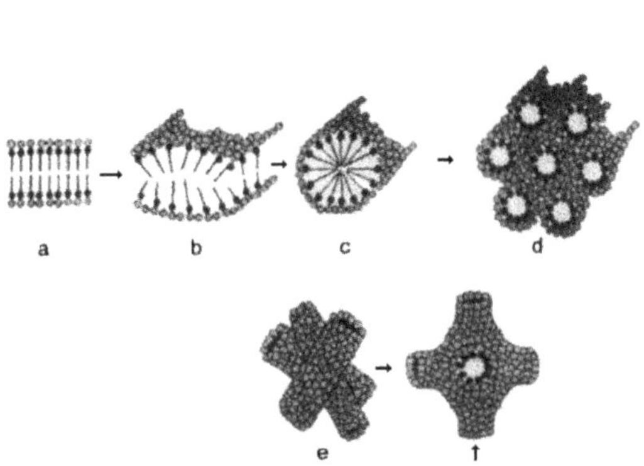


**Fig. 11** Phase transitions in mesoporous solids: a–d, lamellar–hexagonal; e–f, hexagonal–cubic. The circular objects around the surfactant assemblies are the metal-oxo species (from Neeraj and Rao[28b]).

lated in cubic mesoporous phases show high catalytic activity in oxidation reactions.[29b]

Organic inclusion compounds and clathrates have been known for a long time. While these compounds are still being investigated, there have been efforts to synthesise novel organic structures by supramolecular means.[30] A recent example in this direction is the noncovalent synthesis of a novel channel structure formed by trithiocyanuric acid and bipyridine.[30b] The channels can accommodate benzene, xylenes and other molecules (Fig. 12) and the process is reversible.

## Synthesis of inorganic materials

Although rational design and synthesis have remained primary objectives of materials chemistry, we are far from achieving this goal universally. There are many recent instances where rational synthesis has worked (*e.g.*, porous solids), but by and large the success has been limited. Thus, one may obtain the right structure and composition, but not the properties. There are very few instances where both the structure and properties of the material obtained are as desired. The synthesis of $ZrP_{2-x}V_xO_7$ solid solutions showing zero or negative thermal expansion is one such example.[31] The synthesis of $HgBa_2Ca_2Cu_3O_8$ with the highest superconducting $T_c$ (*ca.* 160 K under pressure) is another example.[15,32] The preparation of modulation free bismuth cuprate superconductors by the substitution of Bi by Pb is also a good example.[33] One may

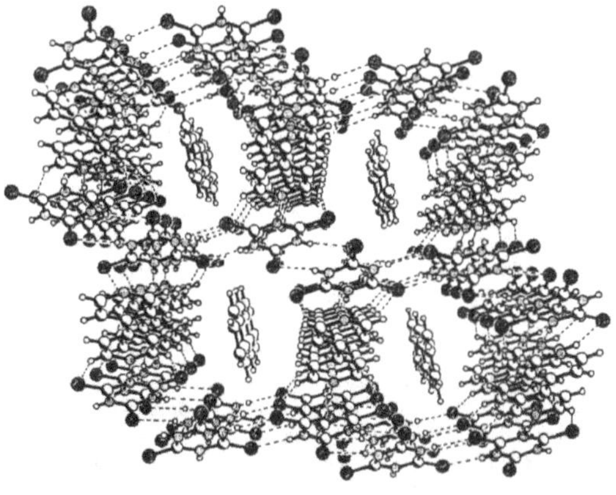

**Fig. 12** Noncovalent synthesis of a layered network with large cavity based on hydrogen bonding between trithiocyanuric acid and bipyridine. Three-dimensional channels with benzene molecules can be seen [from Pedireddi *et al.* (reproduced with permission from ref. 30(*b*))].

also cite examples of the synthesis of ionic conductors and other materials. Most inorganic materials are still discovered accidentally, the cases of high $T_c$ cuprates and CMR manganates being well known. What is noteworthy is the development of new strategies for synthesis, particularly those involving soft chemical routes. The subject has been reviewed adequately in the recent literature.[16,34,35]

The methods of soft chemistry include sol–gel, electrochemical, hydrothermal, intercalation and ion-exchange processes. Many of these methods are employed routinely for the synthesis of ceramic materials.[36,37] There have been recent reviews of the electrochemical methods,[38] intercalation reactions,[39] and the sol–gel technique.[40] The sol–gel method has been particularly effective with wide-ranging applications in ceramics, catalysts, porous solids and composites[41] and has given rise to fine precursor chemistry. Hydrothermal synthesis has been employed for the synthesis of oxidic materials under mild conditions[42] and most of the porous solids and open-framework structures using organic templates are prepared hydrothermally.[43] The advent of supramolecular chemistry has started to make an impact on synthesis,[16] mesoporous solids being well known examples.[43]

Many of the traditional methods continue to be exploited to synthesise novel materials. In the area of cuprates, the synthesis of a superconducting ladder cuprate and of carbonato- and halocuprates is noteworthy. High pressure methods have been particularly useful in the synthesis of certain cuprates and other materials.[36,44] The combustion method has come of age for the synthesis of oxidic and other materials.[45] Microwave synthesis is becoming popular[46] while sonochemistry has begun to be exploited.[47]

In the oxide literature, there are several reviews specific to the different families, such as layered transition metal oxides[48] and metal phosphonates.[49] Synthesis of metal nitrides has been discussed by Di Salvo.[50] Precursor synthesis of oxides, chalcogenides and other materials is being pursued with vigour.[16,36] Thus, single-molecule precursors of chalcogenide containing compound semiconductors have been developed.[51] Kanatzidis[52] has reviewed the application of molten polychalcophosphate fluxes for the synthesis of complex metal thiophosphates and selenophosphates.

In our own work over the years,[53] we have found the use of precursor carbonates to be effective in the synthesis of many oxide materials. Electrochemical techniques have been useful in oxidation reactions (*e.g.*, preparation of $SrCoO_3$, ferromagnetic $LaMnO_3$ containing *ca.* 30% $Mn^{4+}$, $La_2NiO_{4.25}$, $La_2CuO_{4+\delta}$). Intercalation and deintercalation of amines can be used to produce novel phases of oxides such as $WO_3$ and $MoO_3$. Topochemical oxidation, reduction and dehydration reactions also give rise to novel oxide phases, which cannot be prepared otherwise. Thus, $La_2Ni_2O_5$ can only be made by the reduction of $LaNiO_3$. Special mention must be made of the simple chemical technique of nebulized spray pyrolysis of solutions of organometallic precursors, to obtain epitaxial films of complex oxide materials; this technique can also be used to prepare nanoscale oxide powders.[54] Epitaxial films of PZT, $LaMnO_3$, $LaNiO_3$ and other oxide materials have been prepared by this method. Good films of copper and other metallic systems have also been obtained.

## Molecular materials

Molecular materials, especially organic ones, have gained prominence in the last few years, with practical applications as optical and electronic materials already becoming possible. This rich area of research has benefitted from attempts to carry out rational design based on crystal engineering, including supramolecular chemistry[55,56] and the interest to investigate the structure and chemistry of the organic solid state.[57] The conventional areas of polymer research have become more

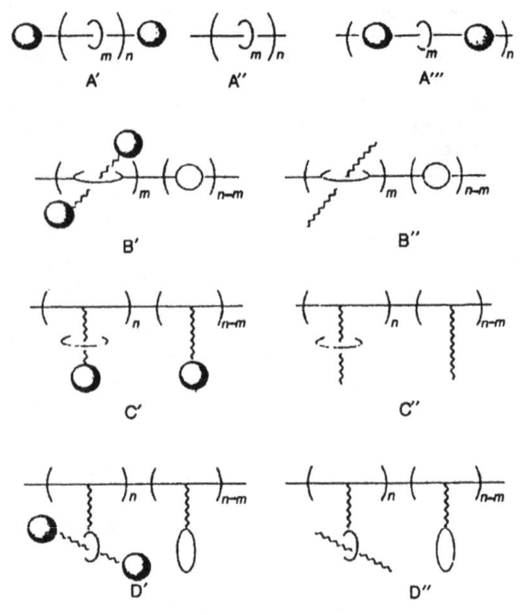


**Fig. 13** Various types of main chain and side chain polyrotaxanes: blocking group (shaded circle); hollow circle (cyclic component); ellipses (cyclics threaded by linear species) [from Gong and Gibson (reproduced with permission from ref. 58)].

focussed on developing new synthetic methods and controlling polymerization. Since architecture and functionality at a molecular level control many of the properties of polymers, there are efforts to synthesise polymers with well defined topologies. Polymers involving rotaxanes, catenanes, rods, dendrimers and hyperbranched systems[58,59] are some recent examples of such efforts. Recognizing that hydrogen bonding is the driving force for the formation of crown ether based polyrotaxanes, mechanically linked polymers have been prepared, with the possibility of molecular shuttles at polymeric level (Fig. 13). Standard synthetic routes for dendrimers (Fig. 14) and hyperbranched polymers suggest many new possibilities.

Molecular magnetic materials constitute an interesting area of research, although practical applications may not be feasible in the short term. Synthesis and design of molecular ferromagnets, ferrimagnets and weak ferromagnets, providing permanent magnetization with fairly high $T_c$ values are the main objectives of this effort. While purely organic ferromagnets have only reached modest $T_c$s below 2 K, materials incorporating transition metals have shown promise [*e.g.* $V(TCNE)_x$]. Molecular magnetic materials generally make use of free radicals such as nitroxides, high-spin metal complex magnetic clusters or a combination of radicals and metal complexes.[60,61] There has been a renaissance in spin-crossover materials with emphasis on active elements for memory devices and light-induced spin changes.[62] The study of diradicals and radical pairs has been strengthened by our understanding of how to stabilize the triplet state.[61] $C_{60}$–TDAE ($T_c \approx 16$ K) and related compounds have attracted some attention (TDEA = tetrakisdimethylaminoethylene). BEDT-TTF salts with magnetic anions have been discovered recently.[63] Organometallic materials based on neutral nitroxide radicals and charge-transfer (CT) complexes derived from the radicals have been examined in some detail.[64]

Semiconducting, metallic and superconducting molecular materials have been investigated by several workers in the last two decades. New types of TTF type molecules, transition metal complexes with elongated $\pi$ ligands, neutral radicals and Te-containing $\pi$ donors have been synthesized and an organic superconductor with an organic anion, $\beta''$-$(ET)_2SF_5CH_2CF_2SO_3$, has been discovered.[65] Cation radical salts of BEDT-TTF and similar donors as well as BEDT-TTF salts are still a fertile ground for the production of new materials.[63,65] Synthesis of linearly extended TTF analogues[66] as well as of new TCNQ and DCNQ acceptors[67] have been reported. The use of organic semiconductors as active layers in thin films holds promise, α-sexithiophene being an example.[68,69]

High performance photonic devices have been fabricated from conjugated polymers such as poly(*p*-phenylenevinylene), polyaniline and polythiophene.[70] The devices include diodes, light-emitting diodes, photodiodes, field effect transistors, polymer grid triodes, light emitting electrochemical cells, optocouplers and lasers. The performance levels of many of these organic devices have reached those of the inorganic counterparts. The high photoluminescence efficiency and large cross section for stimulated emission of semiconducting polymers persist upto high concentrations (unlike dyes). By combination with InGaN, hybrid lasers can now be fabricated. It is gratifying that plastic electronics is moving rapidly from fundamental research to industrial applications.

The area of molecular nonlinear optics has been rejuvenated by efforts to investigate three-dimensional multipolar systems, functionalized polymers as optoelectronic materials, near infrared optical parametric oscillators and related aspects.[71] There have been some advances in chromophore design for second-order nonlinear optical materials;[72] these include one-dimensional CT molecules, octopolar compounds and organometallics. Some of the polydiacetylenes and poly(*p*-phenylenevinylene)s appear to possess the required properties for use as third-order nonlinear optical materials for photonic switching.[73]

Increasing interest in molecular materials has made an impact on the direction of organic chemistry as well, giving rise to large-scale activity in this area. The discovery of fullerenes and fullerene-based materials has contributed to some interesting developments.[74–76] Organic–inorganic hybrids offer many interesting possibilities, with the composites exhibiting properties and phenomena not found in either component.[77a] Two types of materials have been classified, organic–inorganic units held together by weak bonds and organometallic polymers.[77b] Silicon based hybrids and related materials have been reviewed by Livage.[40] In Fig. 15, we show some examples of polyhedral oligomeric silsesquioxanes used as building blocks for the sol–gel synthesis of hybrid materials. It is indeed gratifying that organic molecular devices have become a reality. The concepts and directions suggested by Whitesides[78a] have been useful in designing devices. Special mention must be made of the devices developed by Willner[78b] making use of self-organized thiols and other molecular systems. In this connection, the device-related studies of functional polymers by Wegner and others are also relevant.

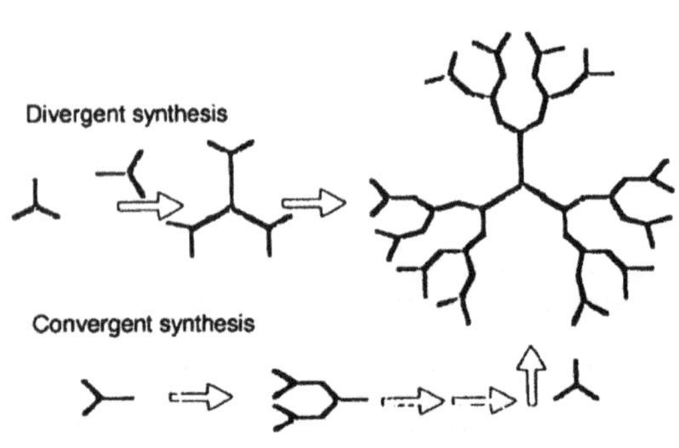


**Fig. 14** Synthetic routes to dendrimers [from Hobson and Harrison (reproduced with permission from ref. 59)].

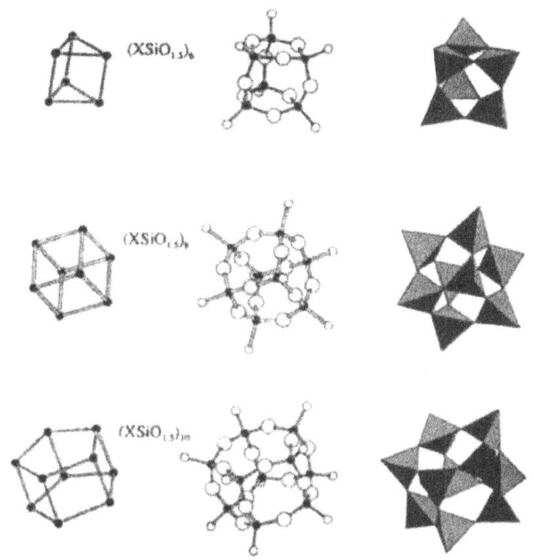

**Fig. 15** Examples of silsesquioxanes used as building blocks for the sol–gel synthesis of hybrid materials [from Livage (reproduced with permission from ref. 60)].

## Biomolecular materials

Biomolecular materials constitute a class of materials at the interface between biology and materials science. They not only provide an insight to Nature's ways, but also suggest new methods of synthesis and design of materials. Thus we can learn much from a study of the principles of structure–function relations in mineralized biological materials, the various forms of calcium carbonate forming the basis for many things that happen.[79,80] Some of the materials such as shells can, in principle, be considered to be part of traditional ceramic materials. The synthesis of mesoporous silica and related materials also owes its origins to biomineralization. Biomineralization has inspired the synthesis of nanocomposites, nanoparticles and nanoparticulate films.[81,82] Collagen fibrillar structure in mineralized and nonmineralized tissue, in particular the ultrastructure orientation of mineral crystals in bone relative to that of the collagen fibril, is an important problem receiving attention.[83]

Some of the recent developments in biomolecular materials include designed drug delivery systems consisting of self-assemblies of lipid and biocompatible lipids,[84] protein-based biological motors,[85] formation of new types of liposomes (*e.g.* spherical liposomes) and emulsions, tissue engineering, biogels, higher-order self-assemblies in biomaterials, and new methods of biopolymer synthesis.[86] Assembly of soft biofunctional and biocompatible interfaces of solids by the deposition of ultrathin soft polymer films, supported lipid–protein bilayers or membranes separated from solids by soft polymer cushions are other areas of vital interest.[87] The ability of polymer molecules attached at one end to a surface to prevent or enhance protein adsorption is being examined.[88]

Glasses and ceramics can bond to living tissues if there is bioactive layer. The development of a bioactive hydroxyapatite layer *in vivo* at body temperature is therefore an important problem.[89] Materials with the highest level of bioactivity develop a silica layer that enhances the formation of such a layer. Such sol–gel processes are used to produce bioactive coatings, powders and substrates which allow molecular control over the incorporation and behavior of proteins and cells with applications as sensors and implants. Sol–gel encapsulation of biomolecules within silica matrices has encompassed enzymes, metalloproteins, photoactive biomolecules and even whole cells.[78] Recently, interfacing electronic materials with lipids and proteins has shown promise in several fields, including biosensors.[90]

## Nanomaterials

The synthesis, characterization and properties of nanomaterials have become very active areas of research in the last few years. In particular, nanostructured materials assembled by means of supramolecular organization offer many exciting possibilities. These include self-assembled monolayers and multilayers with different functionalities, intercalation in preassembled layered hosts and inorganic three-dimensional networks. The reader is referred to the special issue of *Chemistry of Materials*[91] for an overview of present day interests. There are many recent reviews on the varied aspects of nanomaterials. The work of Alivisatos[92] on the structural transitions, elec-

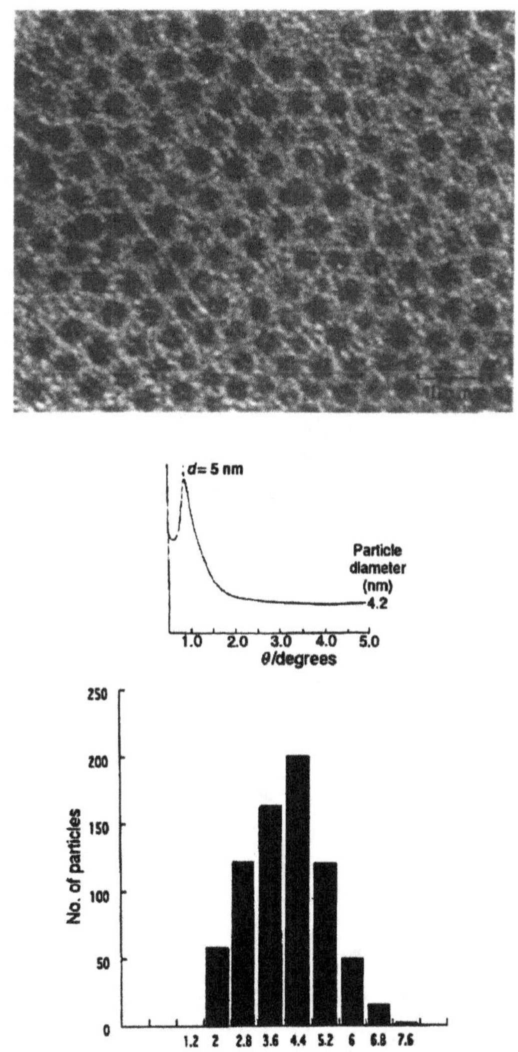


**Fig. 16** TEM image of a nanocrystalline array of thiol-derivatized Au particles [from Sarathy *et al.* (reproduced with permission from ref. 96(*b*))].

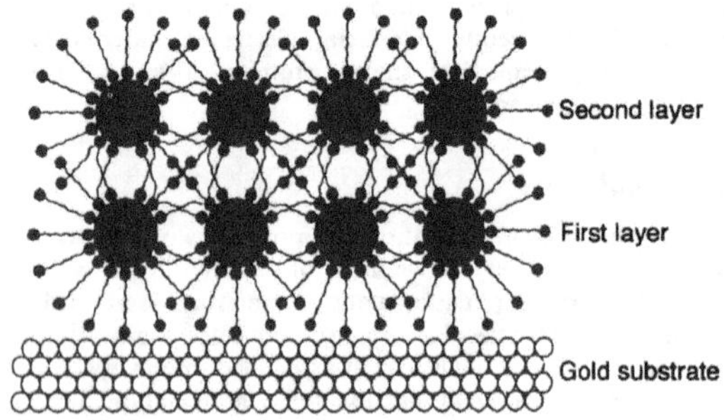


**Fig. 17** Schematic representation of successive deposition of layers of metal (Pt or Au) nanoparticles and dithiols.

tronic properties and related aspects of nanoparticles of CdS, InAs and such materials is particularly noteworthy.

We have been interested in investigating the size-dependent electronic structure and reactivity of metal clusters deposited on solid substrates. Thus, we have shown that when the cluster size is small ($\lesssim$1 nm), an energy gap opens up.[93a] Bimetallic clusters show additive effects due to alloying and cluster size[93b] in their electronic properties. Small metal clusters of Cu, Ni and Pd show enhanced chemical reactivity with respect to CO and other molecules.[94] Metal clusters and colloids, especially those with protective ligands, have been reviewed in relation to nanomaterials.[95] We have recently developed methods of preparing nanoparticles of various metals as well as nanocrystalline arrays of thiolized nanoparticles of Au, Ag and Pt.[96] In Fig. 16, we show the TEM image of thiol-derivatized Au nanoparticles forming a nanocrystalline array. More interestingly, by using dithiols, it has been possible to accomplish layer-by-layer deposition of dithiol–metal nanoparticle films (Fig. 17). This is somewhat similar to the layer-by-layer self assembly of polyelectrolyte–inorganic nanoparticle sandwich films.[97] Such superlattices involving vertical organization of arrays of metal quantum dots may have novel properties.

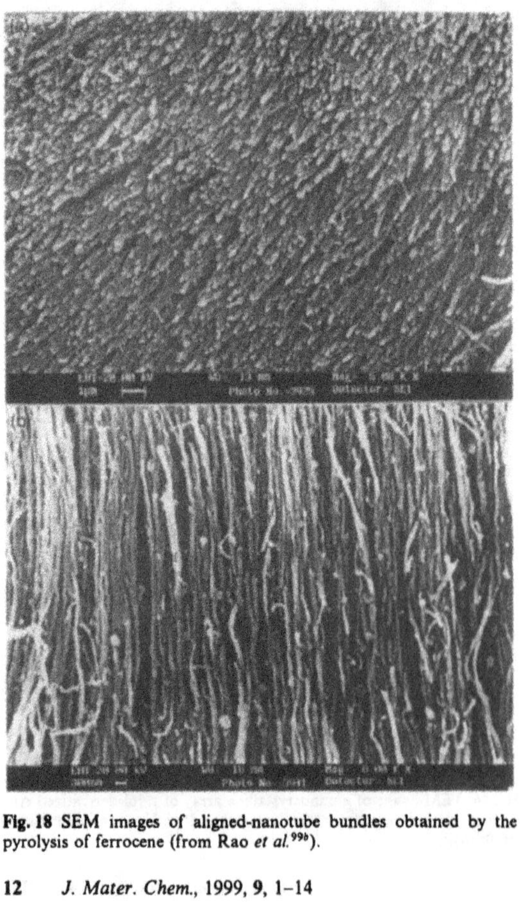

**Fig. 18** SEM images of aligned-nanotube bundles obtained by the pyrolysis of ferrocene (from Rao *et al.*[99b]).

Unprecedented interest has developed in carbon science ever since the discovery of fullerenes and nanotubes. Solid state properties of $C_{60}$ and $C_{70}$ have been of interest to the author since 1991. Some of the aspects investigated in this laboratory include orientational phase transitions, amorphization and polymerization under pressure, molecular magnetism of the TDAE derivative *etc.*[74] We have been working on the synthesis and characterization of carbon nanotubes for some time.[74] Opening, filling, closing and functionalizing carbon nanotubes have been accomplished.[98] Since metal particles are essential as catalysts to prepare nanotubes by the pyrolysis of hydrocarbons, we have employed organometallic precursors to generate nanotubes.[99a] Single-wall nanotubes have been obtained by the pyrolysis of metallocene or $Fe(CO)_5$–hydrocarbon mixtures under controlled conditions. It has also been possible to obtain copious quantities of aligned-nanotube bundles by the pyrolysis of ferrocene (Fig. 18) precursors. C–N and B–C–N nanotubes have been prepared by precursor pyrolysis as well.[99a] By using acid-treated carbon nanotubes as templates, ceramic oxide nanotubes have been prepared.[100] The procedure involves coating the acid-treated nanotubes by metal alkoxides and such precursors, and burning off the carbon.

## Concluding remarks

In a brief overview of this type, it has not been possible to do justice to the work of many leading practitioners of materials chemistry or cover all the known classes of materials and phenomena. Thus I have not covered chalcogenides, nitrides and the like. An area that has not been discussed is noncrystalline materials, especially glasses, which have been of vital interest to materials chemists for many years. The subject is growing rapidly because of the increasing number of applications. Amorphous carbon, amorphous silicon, fluoride glasses, superionic glasses and metallic glasses are some of the amorphous materials of great interest.[101] The glass transition continues to arouse interest.[102] It is an unusual phase transition indeed. The subject of phase transitions itself is of great importance in materials chemistry and a proper understanding of the subject is essential to appreciate many of the materials properties.[103] Ionic conductors, energy storage materials and dielectrics constitute important areas of vital technological relevance. Recent developments in solid state electrochemistry are nicely covered in the book edited by Bruce.[104] An aspect of materials chemistry that is being increasingly exploited is computer simulation and modelling of structures, surfaces, processes and mechanisms.[105] Simulation and modelling techniques are particularly useful in understanding phenomena and structures in situations where experimentation is difficult (*e.g.* molecular processes in zeolites). Another development that deserves notice is the atomic layer-by-layer synthesis of inorganic materials,[106] which suggests the possibility of making use of oxides and other materials in integrated circuits. In spite of the brevity, I trust that the article has succeeded in communicating the nature of present-day materials chemistry. I believe that the references to many of the recent reviews will be useful to students, teachers and practitioners of the subject.

## References

1 (*a*) *Modern Aspects of Solid State Chemistry*, ed. C. N. R. Rao, Plenum Press, New York, 1970; (*b*) C. N. R. Rao and

J. Gopalakrishnan, *New Directions in Solid State Chemistry*, first edition, Cambridge University Press, 1986 (second edition 1997).
2 (*a*) A. R. West, *Solid State Chemistry and Applications*, Wiley, Chichester, 1984; (*b*) *Solid State Chemistry*, ed. A. K. Cheetham and P. Day, Clarendon Press, Oxford 1987, 1992; (*c*) *Chemistry of Advanced Materials*, ed. C. N. R. Rao, Blackwell, Oxford, 1994; (*d*) L. V. Interrante and M. J. Hampden-Smith, *Chemistry of Advanced Materials*, Wiley-VCH, 1998.
3 (*a*) J. B. Goodenough, *Prog. Solid State Chem.*, 1971, **5**, 149; (*b*) C. N. R. Rao and B. Raveau, *Transition Metal Oxides*, second edition, Wiley-VCH, New York, 1998.
4 M. Greenblatt, *Chem. Rev.*, 1988, **88**, 31.
5 J. G. Bednorz and K. A. Müller, *Z. Phys. B*, 1986, **64**, 189.
6 (*a*) *Metal-insulator Transitions*, ed. P. P. Edwards and C. N. R. Rao, Taylor and Francis, London, 1985; (*b*) *Metal-insulator Transition Revisited*, ed. P. P. Edwards and C. N. R. Rao, Taylor and Francis, London, 1995.
7 (*a*) J. M. Honig and L. L. Van Zandt, *Annu. Rev. Mater. Sci.*, 1975, **5**, 225; (*b*) J. M. Honig and J. Spalek, *Proc. Ind. Nat. Sci. Acad. A*, 1986, **52**, 232.
8 C. N. R. Rao, *Philos. Trans. R. Soc. London*, 1998, **356**, 23.
9 P. Ganguly and C. N. R. Rao, *J. Solid State Chem.*, 1984, **53**, 193.
10 C. N. R. Rao, P. Ganguly, K. K. Singh and R. A. Mohanram, *J. Solid State Chem.*, 1988, **72**, 14.
11 C. N. R. Rao and A. K. Cheetham, *Adv. Mater.*, 1997, **9**, 1009 and references therein.
12 C. N. R. Rao and J. M. Thomas, *Acc. Chem. Res.*, 1985, **18**, 113.
13 M. K. Wu *et al.*, *Phys. Rev. Lett.*, 1987, **58**, 908.
14 C. N. R. Rao *et al.*, *Nature*, 1987, **326**, 856.
15 M. Hervieu, *Curr. Opin. Solid State Mater. Sci.*, 1996, **1**, 29.
16 J. Gopalakrishnan, N. S. P. Bhuvanesh and K. K. Rangan, *Curr. Opin. Solid State Mater. Sci.*, 1996, **1**, 285.
17 B. Raveau, C. Michel, M. Hervieu and D. Groult, *Crystal Chemistry of High $T_c$ Superconducting Copper Oxides*, Springer, Berlin, 1991.
18 (*a*) C. N. R. Rao and A. K. Ganguli, *Chem. Soc. Rev.*, 1995, **24**, 1; (*b*) C. N. R. Rao, *Philos. Trans. R. Soc. London*, 1991, **336**, 595; (*c*) T. V. Ramakrishnan, in *Critical Problems in Physics*, Princeton University Press, 1996.
19 J. Rouxel, *Curr. Sci. (India)*, 1997, **73**, 31.
20 (*a*) *Colossal Magnetoresistance, Charge-ordering and Related Aspects of Manganese Oxides*, ed. C. N. R. Rao and B. Raveau, World Scientific, Singapore, 1998; (*b*) Y. Tokura in ref. 20(a).
21 C. N. R. Rao, *Chem. Commun.*, 1996, 2217.
22 A. Arulraj, A. Biswas, A. K. Raychaudhuri, C. N. R. Rao, P. M. Woodward, T. Vogt, D. E. Cox and A. K. Cheetham, *Phys. Rev. B*, 1998, **57**, R8115.
23 (*a*) S. Natarajan, M. Eswaramoorthy, A. K. Cheetham and C. N. R. Rao, *Chem. Commun.*, 1998, 1561 and references therein; (*b*) S. I. Zones and M. E. Davis, *Curr. Opin. Solid State Mater. Sci.*, 1996, **1**, 107; (*c*) J. M. Thomas, *Angew. Chem., Int. Ed. Engl.*, 1994, **33**, 913; (*d*) G. Ozin, *Adv. Mater.*, 1992, **4**, 612.
24 (*a*) C. J. Brinker, *Curr. Opin. Solid State Mater. Sci.*, 1996, **1**, 795; (*b*) M. E. Davis, *Chem. Eur. J.*, 1997, **3**, 1745.
25 S. L. Suib, *Curr. Opin. Solid State Mater Sci.*, 1998, **3**, 63.
26 J. S. Beck and J. C. Vartuli, *Curr. Opin. Solid State Mater. Sci.*, 1996, **1**, 76.
27 S. Ayyappan and C. N. R. Rao, *Chem. Commun.*, 1997, 575 and references therein.
28 (*a*) D. Zhao, P. Yang, Q. Huo, B. F. Chmelka and G. D. Stucky,*Curr. Opin. Solid State Mater. Sci.*, 1998, **3**, 111; (*b*) Neeraj and C. N. R. Rao, *J. Mater. Chem.*, 1998, **8**, 1631.
29 (*a*) T. Maschmeyer, *Curr. Opin. Solid State Mater. Sci.*, 1998, **3**, 71; (*b*) M. Eswaramoorthy, Neeraj and C. N. R. Rao,*Chem. Commun.*, 1998, 615.
30 (*a*) O. M. Yagi, G. Li and H. Li, *Nature*, 1995, **378**, 703; (*b*) V. R. Pedireddi, S. Chatterji, A. Ranganathan and C. N. R. Rao, *J. Am. Chem. Soc.*, 1997, **119**, 10867.
31 A. W. Sleight, *Endeavour*, 1995, **19**, 64.
32 E. V. Antipov, S. M. Loureno, C. Chaillout, J. J. Capponi, P. Bordet, J. L. Tholence, S. N. Putlin and M. Marezio, *Physica C*, 1993, **215**, 1.
33 V. Manivannan, J. Gopalakrishnan and C. N. R. Rao, *Phys. Rev. B*, 1991, **43**, 8686.
34 J. Gopalakrishnan, *Chem. Mater.*, 1995, **7**, 1265.
35 *Soft Chemistry Routes to New Materials—Chemie Douce*, ed. J. Rouxel, M. Tournoux and R. Brec, Trans. Tech. Publications, Aedermannsdorf, Switzerland, 1994.
36 (*a*) *Preparative Solid State Chemistry*, ed. P. Hagenmuller, Academic Press, New York, 1997, p. 2; (*b*) C. N. R. Rao, *Chemical Approaches to the Synthesis of Inorganic Materials*, John Wiley, Chichester, 1994; (*c*) J. D. Corbett, in *Solid State Chemistry—Techniques*, ed. A. K. Cheetham and P. Day, Clarendon Press, Oxford, 1987.
37 D. Segal, *J. Mater. Chem.*, 1997, **7**, 1297.
38 J. C. Grenier, M. Pouchard and A. Wattiaux, *Curr. Opin. Solid State Mater. Sci.*, 1996, **1**, 233.
39 G. Ouvrad and D. Guyomard, *Curr. Opin. Solid State Mater. Sci.*, 1996, **1**, 260.
40 J. Livage, *Curr. Opin. Solid State Mater. Sci.*, 1997, **2**, 132.
41 See special issue of *Chem. Mater.*, 1997, **9**, 2247–2670.
42 M. S. Whittingham, *Curr. Opin. Solid State Mater. Sci.*, 1996, **1**, 227.
43 M. E. Davis and I. E. Maxwell, *Curr. Opin. Solid State Mater. Sci.*, 1996, **1**, 55.
44 M. Takano and A. Onodera, *Curr. Opin. Solid State Mater. Sci.*, 1997, **2**, 166.
45 K. C. Patil, S. T. Aruna and S. Ekambaram, *Curr. Opin. Solid State Mater. Sci.*, 1997,**2**, 158.
46 (*a*) K. J. Rao and P. D. Ramesh, *Bull Mater. Sci.*, 1995, **18**, 447; (*b*) D. M. P. Mingos, *Chem. Soc. Rev.*, 1998, **27**, 213.
47 D. Peters, *J. Mater. Chem.*, 1996, **6**, 1605.
48 P. A. Salvador, T. O. Mason, M. E. Hagerman and K. R. Poeppelmeier, in *Chemistry of Advanced Materials: An Overview*, ed L. V. Interrante and M. J. Hampden-Smith, Wiley—VCH, New York, 1998.
49 A. Clearfield, *Curr. Opin. Solid State Mater. Sci.*, 1996, **1**, 268.
50 F. J. Di Salvo, *Curr. Opin. Solid State Mater. Sci.*, 1996, **1**, 241.
51 P. O'Brien and R. Nomura, *J. Mater. Chem.*, 1995, **5**, 1761.
52 M. G. Kanatzidis, *Curr. Opin. Solid State Mater. Sci.*, 1997, **2**, 139.
53 C. N. R. Rao, *Pure Appl. Chem.*, 1994, **66**, 1765; 1997, **69**, 199.
54 (*a*) H. N. Aiyer, A. R. Raju, G. N. Subbanna and C. N. R. Rao, *Chem. Mater.*, 1997, **9**, 755; (*b*) P. Murugavel, M. Kalaiselvam, A. R. Raju and C. N. R. Rao, *J. Mater. Chem.*, 1997, **7**, 1433.
55 A. Gavezzotti, *Curr. Opin. Solid State Mater. Sci.*, 1996, **1**, 501.
56 G. R. Desiraju, *Curr. Opin. Solid State Mater. Sci.*, 1997, **2**, 451.
57 See special issue of *Chem. Mater.* (dedicated to M.C. Etter), 1994, **6**, 1087–1461.
58 C. Gong and H. W. Gibson, *Curr. Opin. Solid State Mater. Sci.*, 1997, **2**, 647.
59 L. C. Hobson and R. M. Harrison, *Curr. Opin. Solid State Mater. Sci.*, 1997, **2**, 683.
60 D. Gatteschi, *Curr. Opin. Solid State Mater. Sci.*, 1996, **1**, 192.
61 K. Matsuda and H. Iwamura, *Curr. Opin. Solid State Mater. Sci.*, 1997, **2**, 446.
62 O. Kahn, *Curr. Opin. Solid State Mater. Sci.*, 1996, **1**, 547.
63 P. Day and M. Kurmod, *J. Mater. Chem.*, 1997, **7**, 1291.
64 S. Nakatsuji and H. Auzai, *J. Mater. Chem.*, 1997, **7**, 2161.
65 H. Kobayasbi, *Curr. Opin. Solid State Mater. Sci.*, 1997, **2**, 440.
66 J. Roncali, *J. Mater. Chem.*, 1997, **7**, 2307.
67 N. Martin, J. Segura and C. Sevane, *J. Mater. Chem.*, 1997, **7**, 1661.
68 F. Garnier, *Curr. Opin. Solid State Mater. Sci.*, 1997, **2**, 455.
69 H. E. Katz, *J. Mater. Chem.*, 1997, **7**, 369.
70 A. J. Heeger and M. A. Diaz-Garcia, *Curr. Opin. Solid State Mater. Sci.*, 1998, **3**, 16.
71 J. Zyss and J-F, Nicoud, *Curr. Opin. Solid State Mater. Sci.*, 1996, **1**, 533.
72 T. Verbiest, S. Houbrechts, M. Kauranen, K. Clays and A. Persons, *J. Mater. Chem.*, 1997, **7**, 2175.
73 B. Luther-Davies and M. Samic, *Curr. Opin. Solid State Mater. Sci.*, 1997, **2**, 213.
74 C. N. R. Rao, R. Seshadri, R. Sen and A. Govindaraj, *Mater. Sci. Eng. Rep.*, 1995, **R15**, 209; also see *Curr. Opin. Solid State Mater. Sci.*, 1996, **1**, 279.
75 K. Prassides, *Curr. Opin. Solid State Mater. Sci.*, 1997, **2**, 433.
76 M. Prato, *J. Mater. Chem.*, 1997, **7**, 1097.
77 (*a*) Y. Chujo, *Curr. Opin. Solid State Mater. Sci.*, 1996, **1**, 806; (*b*) P. Judeinstein and C. Sanchez, *J. Mater. Chem.*, 1996, **6**, 511.
78 (*a*) See for example G. M. Whitesides, *Acc. Chem. Res.*, 1995, **28**, 37, 219; (*b*) I. Willner, *Acc. Chem. Res.*, 1997, **30**, 347.
79 *Biomimetic Materials Chemistry*, ed. S. Mann, VCH Publishers, Weinheim, 1996.
80 S. Weiner and L. Addadi, *J. Mater. Chem*, 1997, **7**, 689.
81 S. Manne and L. A. Aksay, *Curr. Opin. Solid State Mater. Sci.*, 1997, **2**, 358.
82 J.H. Fendler, *Curr. Opin. Solid State Mater. Sci.*, 1997, **2**, 365.
83 A. Veis, *Curr. Opin. Solid State Mater. Sci.*, 1997, **2**, 370.
84 D. D. Lasic and D. Papahadzopouls, *Curr. Opin. Solid State Mater. Sci.*, 1996, **1**, 392.
85 F. Gittes and C. F. Schmidt, *Curr. Opin. Solid State Mater. Sci.*, 1996, **1**, 412.

86 C.R. Safinya and L. Addadi, *Curr. Opin. Solid State Mater. Sci.*, 1996, **1**, 387.
87 J. Rädler and E. Sackmann, *Curr. Opin. Solid State Mater. Sci.*, 1997, **2**, 330.
88 I. Szleifer, *Curr. Opin. Solid State Mater. Sci.*, 1997, **2**, 337.
89 L. L. Hench, *Curr. Opin. Solid State Mater. Sci.*, 1997, **2**, 604.
90 J. A. Zasadzinski, *Curr. Opin. Solid State Mater. Sci.*, 1997, **2**, 345.
91 See special issue of *Chem. Mater.*, 1996, **8**, 1569–2193.
92 P. Alivisatos, *J. Phys. Chem.*, 1996, **100**, 13226.
93 (*a*) C. P. Vinod, G. U. Kulkarni and C. N. R. Rao, *Chem. Phys. Lett.*, 1998, **289**, 329. (*b*) K. R. Harikumar, S. Ghosh and C. N. R. Rao, *J. Phys. Chem.*, 1997, **101**, 536.
94 A. K. Santra, S. Ghosh and C. N. R. Rao, *Langmuir*, 1994,**10**, 3937.
95 G. Schmid and G. L. Hornyak, *Curr. Opin. Solid State Mater. Sci.*, 1997, **2**, 204.
96 (*a*) S. Ayyappan, R. S. Gopalan, G. N. Subbanna and C. N. R. Rao, *J. Mater. Res.*, 1997, **12**, 398; (*b*) K. V. Sarathy, G. Raina, R. T. Yadav, G. U. Kulkarni and C. N. R. Rao, *J. Phys. Chem.*, 1997, **101**, 9876.
97 J. H. Fendler, *Chem. Mater.*, 1996, **8**, 1616.
98 B. C. Satishkumar, A. Govindaraj and C. N. R. Rao, *J. Phys. B*, 1996, **29**, 4925.
99 (*a*) R. Sen, A. Govindaraj and C. N. R. Rao, *Chem. Mater.*, 1997, **9**, 2078; *Chem. Phys. Lett.*, 1998, **287**, 671; (*b*) C. N. R. Rao, R. Sen, B. C. Satishkumar and A. Govindaraj, *Chem. Commun.*, 1998, 1525.
100 C. N. R. Rao, B. C. Satishkumar and A. Govindaraj, *Chem. Commun.*, 1997, 1581.
101 S. R. Elliott and R. Street, *Curr. Opin. Solid State Mater. Sci.*, 1996, **1**, 555; 1997, **2**, 397.
102 C. A. Angell, *Curr. Opin. Solid State Mater. Sci.*, 1996, **1**, 578.
103 C. N. R. Rao, *Acc. Chem. Res.*, 1984, **17**, 83; see also *J. Mol. Struct.*, 1993, **292**, 229.
104 *Solid State Electrochemistry*, ed. P. G. Bruce, Cambridge University Press, 1995.
105 M. Stoneham and M. L. Klein, *Curr. Opin. Solid State Mater. Sci.*, 1996, **1**, 817; M. Stoneham and S. Panteldes, *Curr. Opin. Solid State Mater. Sci.*, 1997, **2**, 6.
106 A. Gupta, *Curr. Opin. Solid State Mater. Sci.*, 1997, **2**, 23.


*Paper 8/04467H*

Reprinted from the *Journal of Inorganic and Nuclear Chemistry*, Vol. 28, pp. 2167–2173 (1966), Copyright © 1966, with kind permission from Elsevier Science Ltd (Pergamon imprint), The Boulevard, Langford Lane, Kidlington OX5 1GB, UK.

# SOLVOLYTIC DISPROPORTIONATION OF NON-STOICHIOMETRIC OXIDES OF PRASEODYMIUM AND TERBIUM*

R. L. N. SASTRY†

Department of Inorganic and Physical Chemistry, Indian Institute of Science, Bangalore, India

and

P. N. MEHROTRA and C. N. R. RAO‡

Department of Chemistry, Indian Institute of Technology, Kanpur, India

**Abstract**—The solvolytic disproportionation of non-stoichiometric $PrO_x$ and $TbO_x$ in acid solutions to produce higher oxides has been investigated. Some new non-stoichiometric phases have been reported. A number of interesting features of the non-stoichiometric rare earth oxides have been discussed and the need for a satisfactory structural model has been pointed out.

THERE has been considerable interest in recent years in the study of non-stoichiometric rare earth oxides, particularly those of praseodymium.[1–3] The phase equilibria in the praseodymium–oxygen system have been investigated in detail and several well-defined phases of narrow composition range have been reported.[3,4] An interesting result from the studies on the phase equilibria of the praseodymium oxide system is the coexistence of a homologous series[2,5] of the formula $2Pr_2O_3{\cdot}mPrO_2$ $(0 \leqslant m \leqslant 8)$ or $Pr_nO_{2n-2}(4 \leqslant n \leqslant 12)$. Such a series reminds one of the MAGNELI structures in the molybdenum–oxygen, tungsten–oxygen and titanium–oxygen systems.[6,7]

An early study of the solubility of praseodymia, $Pr_6O_{11}$, in acids was made by PRANDTL and HÜTTNER[8] who found that a residue of $Pr_4O_8{\cdot}H_2O$ remained in acid solution. EYRING§ had observed that old samples of $Pr_6O_{11}$ exposed to the atmosphere showed small traces of $PrO_2$. Normally, $PrO_2$ is made by the treatment of the lower oxides of praseodymium with oxygen at a high pressure and temperature[9–11] and $TbO_2$ is made by the treatment of the lower oxides with atomic oxygen.[12] It was

* Partly presented in the Third Rare Earth Research Conference, 1963.
† Presently at Defence Metallurgical laboratory, Hyderabad, India.
‡ To whom all the correspondence should be addressed.
§ Personal communication to Professor G. BRAUER.

(1) L. EYRING and B. HOLMBERG, *Adv. Chem. Ser.* **39**, 46 (1963).
(2) L. EYRING and N. C. BAENZIGER, *J. appl. Phys.* (*Suppl.*) **33**, 28 (1962).
(3) J. M. HONIG, A. F. CLIFFORD and P. A. FAETH, *Inorg. Chem.* **2**, 791 (1963).
(4) R. E. FERGUSON, E. D. GUTH and L. EYRING, *J. Am. chem. Soc.* **76**, 3890 (1954).
(5) C. N. R. RAO, P. A. FAETH and J. M. HONIG. Unpublished results.
(6) A. D. WADSLEY, *Rev. pure appl. Chem.* **5**, 165 (1959); *Adv. Chem. Ser.* **39**, 23 (1963).
(7) S. ANDERSON, A. SUNDHOLM and A. MAGNELI, *Acta chem. scand.* **13**, 989 (1959).
(8) W. PRANDTL and K. HÜTTNER, *Z. anorg. allg. Chem.* **149**, 235 (1925).
(9) J. D. MCCULLOUGH, *J. Am. chem. Soc.* **72**, 1386 (1950).
(10) S. W. RABIDEAU, *J. chem. phys.* **19**, 874 (1951).
(11) S. W. RABIDEAU and G. GLOCKER, *J. Am. chem. Soc.* **73**, 488 (1951).
(12) D. M. GRUEN, W. C. KOEHLER and J. J. KATZ, *J. Am. chem. Soc.* **73**, 1475 (1951).

therefore considered important to investigate the quantitative solvolytic behaviour of non-stoichiometric praseodymium and terbium oxides in acid solutions. If the dioxides of praseodymium and terbium were to remain as residues in acid solutions, the situation would be similar to the leaching of PbO from $Pb_3O_4$ in dilute nitric acid leaving a residue of $PbO_2$[13] or that of MnO from $Mn_2O_3$ leaving a residue of $MnO_2$.[13]

## EXPERIMENTAL

Praseodymium oxide, $Pr_6O_{11}$, (Lindsay, 99·99% Pure) ignited to constant weight was used for these studies. Determination of active oxygen (Pr IV) by iodimetry as well as by use of arsenic (III) as reducing agent,[14] confirmed that the composition was exactly $PrO_{1 \cdot 83}$ (or $Pr_6O_{11}$). The X-ray powder pattern when indexed on the basis of an f.c.c. lattice gave a lattice constant of 5·468 ± 0·002 Å which agrees with the literature value.[1] A sample of $PrO_{1 \cdot 72}$ was made according to the procedure of KUNTZ and EYRING[15] by heating $Pr_6O_{11}$ in vacuum for 2 hr at 490°C and the composition was verified by active oxygen analysis.[14] This oxide was rhombohedral as reported by EYRING and HOLMBERG.[1] An authentic sample of $PrO_2$ was made by heating $Pr_6O_{11}$ in oxygen at 5 atm and at 360°C for 3 days. The lattice constant of this fluorite structure was 5·393 ± 0·001 Å. Lindsay 99·9% pure terbium oxide was ignited in air to constant weight and its composition was found to be $TbO_{1 \cdot 81}$ by active oxygen estimation. The lattice constant of the oxide was 5·295 ± 0·01 Å when the important lines were indexed on the basis of a cubic lattice; to be exact this phase should be considered as triclinic[1] with the angles showing very small deviations from 90°.

X-ray powder pattterns were taken on a GE unit or a Reich-Seifert unit employing filtered Cu or Mo radiation. In the X-ray determination of the lattice constants of cubic substances, absorption correction was applied in relevant cases using the Nelson–Riley function. Thermogravimetric analysis was carried out emplying a balance described earlier.[16] Differential thermal analysis was carried out with an apparatus[17] similar to that described by NORTON.[18]

The procedure for studying the solvolysis was as follows. About 300 mg of the non-stoichiometric oxide was added to a known quantity of the acid contained in a stopped flask and maintained at a particular temperature and for a fixed period with frequent agitation. At the end of the period, the residue was washed with acid of the same concentration as used in the experiment and several times with water. The residue was dried to constant weight at 110°C and analysed for active oxygen.[14] The quantity of oxide dissolved in the acid was determined by drying and ignition to the stable non-stoichiometric oxide. The kinetics of solvolysis were studied by first equilibrating stoppered flasks containing a known amount of acid in a thermostat and then adding $Pr_6O_{11}$. The flasks were frequently agitated and at the end of the required period, an aliquot portion of the acid was pipetted out into a crucible and the amount dissolved determined by drying and ignition to $Pr_6O_{11}$.

## RESULTS

### *Solvolysis of non-stoichiometric praseodymium oxides*

The results of the solvolysis of $Pr_6O_{11}$ ($PrO_{1 \cdot 83}$) in different acid solutions are presented in Table 1. The residue was found to be $PrO_2$ by active oxygen analysis in almost all cases except when acetic acid concentration was very high. The residue of $PrO_2$ was finely divided, but crystalline, as shown by the X-ray powder patterns.

(13) J. R. PARTINGTON, *Textbook of Inorganic Chemistry*, Macmillan, London (1953).

(14) G. L. BARTHAUER and D. W. PEARCE. *Ind. Engng. Chem. Analyt. Edn* **18**, 479 (1946); J. S. ANDERSON and M. STERNS, *J. inorg. nucl. Chem.* **11**, 272 (1959).

(15) U. E. KUNTZ and L. EYRING, in *Kinetics of High Temperature Processes* (Edited by W. D. KINGERY) Technology Press of Massachusetts Institute of Technology and J. Wiley, New York (1960).

(16) A. R. V. MURTHY, R. M. MALLYA and D. S. BHARDWAJ, *Chemy Ind.* **16**, 300 (1956).

(17) N. S. K. PRASAD and C. C. PATEL, *J. Indian Inst. Sci.* **36A**, 23 (1954).

(18) F. H. NORTON, *J. Am. chem. Soc.* **22**, 54 (1939).

TABLE 1.—LEACHING OF $PrO_{1\cdot83}$ BY ACIDS*

| Conc. of acid | Amount dissolved as $Pr_6O_{11}$ | Residue (wt.%) | Active oxygen in residue (%) | Composition of the residue: $x$ in $PrO_x$ | Lattice constant from X-ray data (Å) |
|---|---|---|---|---|---|
| 0·01 or 0·001 HCl | — | — | 4·63 | 2·00 | 5·393 ± 0·002 |
| 1·6 N HAc | 32·1 | 63·8 | 4·63 | 2·00 | 5·393 ± 0·002 |
| 4 N HAc | 32·2 | 63·5 | 4·62 | 1·98 | 5·394 ± 0·004 |
| 8 N HAc | 32·9 | 63·8 | 4·63 | 2·00 | 5·393 ± 0·004 |
| 12 N HAc | 32·5 | 63·6 | 4·62 | 1·98 | 5·399 ± 0·004 |
| 16 N HAc | — | — | 3·64 | 1·92 | 5·430 ± 0·006 |
| Glacial HAc | 2·7 | 96·4 | 3·52 | 1·88 | 5·465 ± 0·004 |

* When acetic acid solutions were employed the period was 5 days; in the case of HCl solutions the period was 2 hr, Temp. was 25 ± 2°C.

In order to verify the applicability of this method of leaching out $Pr^{3+}$ in acid solutions to the other non-stoichiometric oxides, $PrO_{1\cdot72}$ was prepared. It is known from studies of phase equilibria[3] that this composition is the best defined non-stoichiometric oxide of praseodymium other than $PrO_{1\cdot83}$. In the case of $PrO_{1\cdot72}$ also, the residue from 8N acetic acid leaching was found to be $PrO_2$.

The $PrO_2$ obtained by leaching of $PrO_{1\cdot83}$ gave good sharp X-ray patterns. A well crystallized single phase alone was present. It is not likely that any minor phase was present and the iodometric estimation of active oxygen agreed closely with the composition $PrO_2$. The X-ray lines fitted a fluorite lattice and the lattice constants of the various samples correspond closely to the literature value[1] (5·393 ± 0·001 Å). The $PrO_2$ obtained from solvolysis was subjected to thermogravimetric (TGA) and differential thermal analysis (DTA). The TGA curve showed that it starts losing weight from 110°C till about 320°C when more rapid weight loss occurrs. The loss below 320°C is due to moisture and chemisorbed acetate as pointed out by CLIFFORD *et al.*[19] and the rapid weight loss occurring between 320 and 400°C corresponds to the reaction; $PrO_2 \rightarrow PrO_{1\cdot83}$. The residue from TGA was found to be $Pr_6O_{11}$ by X-ray powder diffraction and chemical analysis. The authentic sample of $PrO_2$ prepared from $Pr_6O_{11}$ by treatment with oxygen under high pressure showed oxygen loss between 360 and 420°C. The DTA curve of $PrO_2$ from solvolysis showed a small endothermic peak at 400°C due to decomposition to $Pr_6O_{11}$. The slow decomposition due to chemisorbed acetate noticed in TGA was not observed in the DTA.

The residues from glacial and 16 N acetic acids gave lower active oxygen contents than $PrO_2$ and compositions are in the region of $PrO_{1\cdot90}$ (Table 1). This composition is not known to be a stable phase.[3,20] The X-ray powder patterns of these residues yielded lattice constants which are in between those of $Pr_6O_{11}$ and $PrO_2$. It is possible that these are mixtures of $Pr_6O_{11}$ and $PrO_2$; the proportion should be $Pr_6O_{11}$: $2PrO_2$ to give $PrO_{1\cdot88}$. Whether these are unique phases occurring in the miscibility range of $PrO_{1\cdot83}$ to $PrO_2$ is difficult to ascertain from the present study.

It can be seen from Table 1 that the residue in dilute acetic acid (~64 per cent) is roughly the same in all cases. If $Pr_6O_{11}$ were to be written as $2Pr_2O_3{\cdot}8PrO_2$ according

[19] A. F. CLIFFORD, P. A. FAETH and K. J. HUGHES, paper presented in Third Rare Earth Conf. (1963).
[20] C. L. SIEGLAFF and L. EYRING, *J. Am. chem. Soc.* **79**, 3024 (1957).

to the homologous series, and if all the $Pr_2O_3$ went into solution, the residue of $PrO_2$ would be 67·7 per cent. In fact the quantity dissolved was about 33 per cent. The agreement in the percentage of the $PrO_2$ residue is not very good due to incomplete recovery caused by the fine particle size. These results seem to establish the quantitative nature of the separation of $PrO_2$ from $Pr_6O_{11}$ in dilute acetic acid solution and are in accordance with the stoichiometry demanded by the homologous series. Apparently $PrO_2$ is insoluble in acetic acid solutions as was confirmed with the authentic sample.

When $Pr_6O_{11}$ was hydrolysed by boiling water for 24 hr the residue consisted of $PrO_2$ and $Pr(OH)_3$. All the lines in the X-ray pattern of the residue could be fitted into lattices of either $PrO_2$ or $Pr(OH)_3$.

$$Pr_6O_{11} + 3H_2O \rightarrow 2Pr(OH)_3 + 4PrO_2$$

These results agree with those of BRAUER and PFEIFFER.[21]

*Solvolysis of non-stoichiometric terbium oxides*

It was found that terbia was nearly insoluble in acetic acid of any concentration at room temperature (20–30°C). Various concentrations of hydrochloric acid were therefore employed and the results are presented in Table 2. Decinormal hydrochloric acid was found to be useful in obtaining $TbO_2$ from $TbO_{1\cdot 81}$. Boiling $TbO_{1\cdot 81}$ with a 1 : 1 acetic and hydrochloric acid mixture also gives $TbO_2$ as residue.[22] Unlike $Pr_6O_{11}$, the residue obtained does not exactly conform to that expected from the formulation according to the homologous series. The amount of residue also varies with time. It is possible that $TbO_2$ dissolves sufficiently in 0·1 N HCl solution. The composition of $TbO_{1\cdot 83}$ ($Tb_6O_{11}$) obtained in one of the experiments has not been clearly defined in the literature and it is likely that this is the analog of $Pr_6O_{11}$. The ratio of the cell edges of $Tb_6O_{11}$ and $Pr_6O_{11}$ is nearly the same as the ratio of the cell edges of $Tb_2O_3$(C) and $Pr_2O_3$(C).

TABLE 2.—SOLVOLYSIS OF TERBIUM OXIDE, $TbO_{1\cdot 81}$

| Medium | Residue (%) | Active oxygen (%) | Composition of the residue x in TbO- | Lattice constant from X-ray data (Å) |
|---|---|---|---|---|
| 8 N HAc (5 days) | 99·0 | 2·39–2·56 | 1·78–1·80 | — |
| 0·01 N HCl | 92·0 | 2·47–2·64 | 1·79–1·81 | 5·29 ± 0·01 |
| 0·1 N HCl (8 hr) | — | 2·81 | 1·83 | 5·26 ± 0·01 |
| 0·1 N HCl (24 hr or 5 days) | 29·4 | 4·00–4·19 | 1·98–2·00 | 5·200 ± 0·001 |
| 1·0 N hydrochloric acid | 0·0 | — | — | — |

The residue of $TbO_2$ obtained from 0·1 N hydrochloric acid was bright red in colour and gave a good X-ray powder pattern. The data fit a f.c.c. cell fairly well and lines from the back reflection region which were resolved into $\alpha_1$ and $\alpha_2$ components gave a lattice constant of 5·219 ± 0·0005 Å. An examination of the pattern

[21] G. BRAUER and B. PFEIFFER, *J. less-common Metals* **5**, 171 (1963).
[22] G. BRAUER and B. PFEIFFER, *Angew. Chem.* (*Int.*) **1**, 551 (1962).

failed to reveal any splitting or broadening of the lines due to a possible distortion of the cubic cell. $TbO_2$ looses oxygen to give $TbO_{1\cdot8}$ around 340°C.

The patterns of non-stoichiometric praseodymium and terbium oxides as well as the dioxides contained many lines other than those belonging to the unit cells already described. It was possible to index some of these lines on the basis of a unit cell of the C-type sesquioxides. In the case of $TbO_2$ the extra lines were fitted into a b.c.c. lattice with a cube edge of $10{\cdot}73 \pm 0{\cdot}03$ Å. The cell dimension agrees with the literature value[1] of $10{\cdot}7281 \pm 0{\cdot}0005$ Å for C-$Tb_2O_3$.

*Kinetics of solvolyis of* $Pr_6O_{11}$

Rates of dissolution of $Pr_6O_{11}$ in 8 N acetic acid solution were found to follow the first-order rate law with an energy of activation of 4 to 5 kcal. This value of the energy of activation could be reproduced although the values of the rate constants varied from sample to sample. For one sample, the rate constants were $1{\cdot}20 \times 10^{-3}$ and $1{\cdot}58 \times 10^{-3}$ min$^{-1}$ at 29 and 40°C respectively, while for another sample the values were $2{\cdot}9 \times 10^{-4}$ and $3{\cdot}5 \times 10^{-4}$ min$^{-1}$ at 40° and 50°C respectively. The low value of the energy of activation is significant and together with the first-order dependence of the reaction, it indicates that the process is akin to a simple dissolution. An attempt to study the rate of dissolution of a 1:4 mixture of $Pr_2O_3$ and $PrO_2$ (corresponds to $Pr_6O_{11}$) was unsuccessful since the dissolution of $Pr_2O_3$ was complete in about 5 min at 20°C. Further, X-ray patterns of $Pr_2O_3$ and $PrO_2$ mixed in the proper proportions to form $PrO_{1\cdot71}$ and $PrO_{1\cdot833}$ showed only lines belonging to $Pr_2O_3$ and $PrO_2$.

In order to examine the nature of the intermediate products, samples of residues from solvolysis at 40°C were separated after intervals of 15 min and 1 hr and their powder patterns were taken. The pattern of the 1 hr sample turned out to be that of $PrO_2$ ($a = 5{\cdot}400 \pm 0{\cdot}005$ Å) with several additional lines which belonged neither to a fluorite type lattice nor to $Pr(OH)_3$. The pattern of the specimen after 15 and 25 min showed larger *d*-spacings and the fluorite lines could be accounted for by cells with lattice constants of $5{\cdot}460 \pm 0{\cdot}005$ and $5{\cdot}435 \pm 0{\cdot}003$ Å respectively. While even these patterns did not show any $Pr_6O_{11}$ lines, they showed the same additional lines as the 1-hr specimen. The absence of $Pr_6O_{11}$ lines after only 15 min of solvolysis is surprising. A fluorite cell with $a = 5{\cdot}435$ Å which is between the size of the $Pr_6O_{11}$ and $PrO_2$ cells could mean a solid solution. Using a smooth curve obtained by plotting the unit cell lengths of all the cubic and rhombohedral phases of praseodymium oxides from the literature[1] vs. the composition (pseudo cell dimensions were used where necessary), it was found that $a = 5{\cdot}435$ Å corresponds to a composition of $PrO_{1\cdot92}$. A similar composition with about the same lattice constant was also obtained by solvolysis of $Pr_6O_{11}$ with 16 N acetic acid (Table 1). A miscibility gap is known to exist between $Pr_6O_{11}$ and $PrO_2$ and how such a uniphasic composition in this region could result is not clear; but it is possible that in the medium of a solution of praseodymium ions, such solid solutions could form. This would also explain the absence of $Pr_6O_{11}$ lines and is a strong argument in favour of solid solution formation.

## DISCUSSION

The formation of $MnO_2$ and $PbO_2$ by the action of acids on the mixed valence oxides $Mn_2O_3$ and $Pb_3O_4$ respectively is well known.[13] $Pb_3O_4$ consists of metal ions in non-equivalent positions[23] and if these correspond to different valence states, those in the lower valence state (being the more basic) are preferentially attacked by acids; the crystalline sub-particles that are left behind would then rearrange into a stable structure. In the non-stoichiometric oxides of praseodymium, however, the $Pr^{3+}$ and $Pr^{4+}$ ions are supposed to be crystallographically indistinguishable.[24] It is, therefore, difficult to offer an unified explanation for the solvolysis of non-stoichiometric rare earth oxides and the mixed valence oxides of lead and manganese, based on the existing knowledge of the crystallography of these systems. Several interesting features of these non-stoichiometric oxides are apparent.

(1) The X-ray analysis of the quenched samples of the various non-stoichiometric praseodymium and terbium oxides does not seem to be satisfactory; for example, several extra lines are found in the patterns, which cannot be assigned to the main phase.[2,5] These extra lines are not due to an impurity of sesquioxide, but are likely to arise from the disordered nature of these oxides. An X-ray pattern of an intimate mixture of $Pr_2O_3$ and $PrO_2$ in the proportion 1:4 (giving a composition of $Pr_6O_{11}$) was not identical to the pattern of $Pr_6O_{11}$ and clearly showed the distinct lines of both the components.

(2) An examination of the thermodynamic data[25–27] on the oxides of praseodymium and terbium oxides has shown that the heat of formation of the non-stoichiometric oxides is equal to the sum of the heats of formation of the stoichiometric oxides taken in proper proportion (following the homologous series $2Pr_2O_3 \cdot m\, PrO_2$). This is similar to the linearity of thermodynamic data in the stoichiometric range found by ARIYA and MOROZOWA[28] in the Ti–O system.

| Composition | $\Delta H^\circ_{298.16}$ kcal/g atom of Pr, observed | $\Delta H^\circ_{298.16}$ kcal/g atom of Pr, derived from the values of C—$Pr_2O_3$ and $PrO_2$ |
|---|---|---|
| $PrO_{1.83}$ | $-227{\cdot}6 \pm 0{\cdot}8$ | $-228{\cdot}1$ |
| $PrO_{1.72}$ | $-224{\cdot}0 \pm 0{\cdot}9$ | $-224{\cdot}6$* |

* Value calculated for $PrO_{1.714}$

This striking additivity of the heats of formation following the homologous series can only mean that energetically, there is little or no difference between a non-stoichiometric rare earth oxide and a corresponding mixture of the stoichiometric oxides.

(3) Recent studies on the phase equilibria in the praseodymium–oxygen system have shown very large hysteresis effects,[29] particularly in the monophasic regions. Observation of hysteresis effects may mean that one is not dealing with true equilibrium systems and throws some doubt on the X-ray analysis of quenched samples.

[23] S. T. GROSS, *J. Am. chem. Soc.* **63,** 1168 (1941); **65,** 1107 (1943); M. STRAUMANIS, *Z. phys. Chem.* **52B,** 127 (1942) A. BYSTROM and A. WESTGREN, *Ark. Kemi.* **B16,** 14 (1943).
[24] L. EYRING, in *Rare Earth Research*, Gordon and Breach, New York (1962).
[25] L. EYRING, H. R. LOHR and B. B. CUNNINGHAM, *J. Am. chem. Soc.* **74,** 1186 (1962).
[26] C. T. STUBBLEFIELD, H. EICK and L. EYRING, *J. Am. chem. Soc.* **78,** 3018 (1956).
[27] C. T. STUBBLEFIELD, H. EICK and L. EYRING, *J. Am. chem. Soc.* **78,** 3877 (1965).
[28] S. M. ARIYA and M. P. MOROZOWA, *Russ. J. gen. Chem.* **28,** 2647 (1958).
[29] P. A. FAETH and A. F. CLIFFORD, *J. phys. Chem.* **67,** 1453 (1963).

(4) From the available thermodynamic data[3,30] it is found that the entropy change, $\Delta S°$, for the formation of $PrO_{1 \cdot 714}$ from $Pr_2O_3$ is considerably higher than the normal differential molar standard entropy change for such oxidation reactions. The higher value for $\Delta S°$ would also indicate that one may not be dealing with an equilibrium situation. However, assuming a random distribution of 4 moles of $Pr^{3+}$, 3 moles of $Pr^{4+}$, 2 moles of vacancies and 12 moles of $O^{2-}$ in $PrO_{1 \cdot 714}$ ($2Pr_2O_3 \cdot 3PrO_2$), one can account for a major portion of the discrepancies in $\Delta S°$.

(5) Preliminary ESR studies have indicated that the signals for the ions in the non-stoichiometric and stoichiometric oxides are not different and magnetic susceptibilities[31] of non-stoichiometric praseodymium oxides can be interpreted in terms of the additive susceptibilities of $Pr_2O_3$ and $PrO_2$ in suitable proportions.

From the discussion above, it becomes clear that the structures of the non-stoichiometric praseodymium and terbium oxides need clarification. The crystallography of these systems should account for the number of extra weak lines in the X-ray pattern, the additivity of heats of formation, the ready separation of the dioxides in acid solutions and the existence of the homologous series. A plausible model seems to be one involving a "string" as the structural entity (analogous to the shear plane of $Ti_nO_{2n-1}^{(6)}$) generating a series of $LnO_{2n-2}$ from the parent $LnO_2$.[32] In this model, cubic $Ln_2O_3$ can be pictured as containing non-intersecting linear strings of $LnO_6$ octahedra ($LnO_6V_2$ cubes where V stands for oxygen ion vacancy) arranged in four directions. $LnO_{1 \cdot 71}$ (or $Ln_7O_{12}$) consists of parallel infinite strings of $RO_6V_2$ cubes surrounded by sheaths of 7 co-ordinated cations. The composition of the phase is determined by the concentration of strings and the symmetry by the geometrical arrangement in the (111) directions. Thus, any non-stoichiometric phase $Ln_2O_{2n-2}$ or $LnO_{2(1-1/n)}$ will consist of $1/n$-th of the cations in the string with each of them having $RO_6V_2$ co-ordination. This model seems to be able to explain most of the unusual features of non-stoichiometric rare earth oxides described earlier.

*Acknowledgement*—Authors thanks are due to Dr. B. G. HYDE, University of Western Australia, Perth, for providing them with a preprint of his paper and to Dr. M. R. A. RAO for his interest.

(30) A. KUZNETSOV and T. N. REZUKHINA, *Zh. fiz. Khim.* **136**, 1364 (1962).

(31) S. KERN, *J. chem. Phys.* **40**, 208 (1964).

(32) B. G. HYDE. D. J. M. BEVAN and L. EYRING. Unpublished results, to be presented in Melbourne *Conf. Non-stoichiometric Compounds*, 1965; J. O. SAWYER, B. G. HYDE and L. EYRING, Coll. Int. C.N.R.S. *Bull. Soc. chim. Fr.* 1190 (1965).

Reprinted with permission from the *Transactions of the Faraday Society*,
No. 563, Vol. 65, Part 11, pp. 3088–3092 (1969),


# Thermal and Particle Size Effects in Magnesium Oxide

By M. Natarajan, T. S. Sarma, J. C. Ahluwalia and C. N. R. Rao*

Department of Chemistry, Indian Institute of Technology, Kanpur, India

*Received 9th December*, 1968

X-ray line widths, surface areas and heats of solution of MgO samples prepared by vacuum dehydration of $Mg(OH)_2$ at different temperatures in the range 350-1 200°C have been studied. While the surface energy of particles shows a steady decrease with the temperature of preparation, the heats of solution shows an abrupt decrease over a narrow temperature range. The results have been interpreted in terms of pseudocrystallization of the highly disordered oxide followed by sintering; the pseudocrystallization of colloidal oxides appear to be similar to glass transitions.

---

Giauque [1] showed that the heats of solution of finely divided and crystalline samples of MgO are different and expressed his findings in terms of the thermochemical equation:

$$MgO_{(\text{crystalline})} = MgO_{(\text{finely divided})}, \Delta H = 800 \text{ cal mol}^{-1}. \quad (1)$$

Giauque,[1] and Taylor and Wells,[2] suggested that this heat-content anomaly could be due to particle-size effects since the particle size of MgO increases appreciably with the temperature of preparation; however, these workers had no quantitative measure of the particle sizes of their samples. Since crystalline MgO samples are generally prepared by the decomposition of the hydroxide at relatively high temperatures compared to the finely divided samples, Rao and Pitzer,[3] as well as Rao, Yoganarasimhan and Lewis [4] have attributed the major part of such heat-content anomalies to pseudo-crystallization involving the removal of internal defects. Crystallization of amorphous oxides is associated with large evolution of heat.[5, 6] The increased energy content of finely divided MgO was shown to be primarily due to crystallite size effects by Thomas and Baker,[7] who measured the crystallite sizes (from X-ray line widths) of samples prepared by the dehydration of brucite at different temperatures. These workers showed that the energy contents of the crystallites calculated by the Weissenbach-relation [8] agree well with the experimental values of Livey and coworkers [9] for the heats of solution. There is evidence that microstresses in MgO samples may also contribute to the energy content of small particles produced by the decomposition of various magnesium compounds.[10]

While the reduction in surface area (or increase in crystallite size) is expected to produce energy changes continuously over a wide temperature range,[7, 11], the pseudo-crystallization reaction would occur sharply in a narrow temperature range.[3, 4] Even though a smooth variation of heats of solution of MgO with the temperature of preparation of oxide has been suggested,[7, 9] a close examination of the experimental data revealed that the data could as well be shown to indicate a " transition " around 600°C. We have investigated thermal and particle size effects on MgO in order to examine the relative importance of these different factors in causing increased energy contents in finely divided MgO samples. For this purpose, we have prepared several

* to whom all the correspondence should be addressed.

samples of MgO by the dehydration of $Mg(OH)_2$ at different temperatures between 350 and 1 200°C and measured their X-ray line profiles, surface areas and heats of solution in 0.1 NHCl.

## EXPERIMENTAL

Magnesium hydroxide precipitated by the dropwise addition of 6N ammonia solution to a solution of $MgCl_2$ was allowed to settle overnight, filtered, washed and dried in an air oven at ~110°C for 8 h. The hydroxide was heated at 350°C *in vacuo* ($10^{-5}$ torr) for ~200 h to produce the finely divided MgO. 2-g portions of the oxide thus prepared were heated at 450, 550, 600, 700, 800, 950 and 1 200°C *in vacuo* for 3 h and then sealed in Pyrex bulbs. The samples thus prepared contained no water (or hydroxide).

X-ray diffraction patterns [7, 12] of the MgO samples were recorded employing a North American Philips X-ray diffractometer with Copper $K$-$\alpha$ radiation. The crystal structure and lattice dimensions of all the MgO samples were the same, but the line widths varied with the temperature of preparation. Line broadening $\beta$ due to the particle size is related to the mean particle size $\varepsilon_{hkl}$ for the particular reflection ($hkl$) by the equation,[13]

$$\beta = \lambda K/\varepsilon_{hkl} \cos \theta, \tag{2}$$

where $\theta$ is the Bragg angle and $K$ is the shape factor normally assumed to be close to unity. In order to allow for the instrumental broadening $b$, a sample of MgO heated to 1 200°C was taken as the reference [14] since the half-line width of the sample was comparable to that of standard unstrained quartz. By making use of $\beta$ values after applying the Warren correction,[13] the $\varepsilon_{hkl}$ values for the samples of MgO were obtained. The $\varepsilon_{hkl}$ values have an uncertainty of $\pm 10$ %.

B.E.T. surface areas of the samples were measured from nitrogen adsorption at low temperatures employing a volumetric apparatus. The molecular area of 16.2 Å$^2$ was assumed for nitrogen.

The submarine calorimeter, used for the measurement of heats of solution, was essentially similar to that described in the literature.[15, 16] The calorimeter was calibrated by measuring the heats of solution of KCl; good agreement (within 0.2 %) was obtained with the standard values obtained by Gunn.[17] Outgassed samples of MgO in sealed bulbs were opened under the acid solution in the calorimeter. The heats of solution of the MgO samples in 0.1 N HCl were determined at 35.0 ±0.10°C. The heat of solution of the 700°C sample of MgO was determined at 25°C as 35.07 ±0.05 kcal mol$^{-1}$; the value at 35°C was 33.11 ±0.05 kcal mol$^{-1}$. Further, the heats of solution of two sintered MgO samples with particle sizes of $\leqslant 37\ \mu$ and $\geqslant 125\ \mu$ were identical (35.81 ±0.05 kcal mol$^{-1}$) at 25°C.

## RESULTS AND DISCUSSION

The experimentally-determined values of the crystallite size $\varepsilon_{200}$, the surface area $A_s$, and the heats of solution $\Delta H_s$, are given in table 1; $\Delta H_s$ shows little variation up to ~450°C but suddenly decreases in the temperature range 450-650°C. The results differ from those of Livey and coworkers [1] who found a progressive decrease of $\Delta H_s$ with the temperature of preparation. The excess energy in small particles of MgO can be expressed in terms of $\Delta(\Delta H_s)$ (with respect to the sample at 950°C); the $\Delta(\Delta H_s)$ values are plotted against temperature in fig. 1. The figure clearly shows the sharpness of the energy decrease in MgO particles between 450 and 650°C.

The excess energy $\Delta E$ of MgO particles was calculated by the relation of Weissenbach [8]:

$$\Delta E = 111/n + 1\,035/n^2 - 1\,944/n^3 + 1\,350/n^4, \tag{3}$$

where $n$ is the number of atomic diameters on the particle edge and is approximately given by $\varepsilon_{hkl}/2$. This relation assumes that MgO powder consists of cube-shaped particles of uniform size bounded by (100) faces. The values of $\Delta E$ are given in table 1 and the $\Delta(\Delta E)$ values (with respect to the 950°C sample) are shown in fig. 1.

Giauque [1] attempted to estimate the surface energy of finely divided MgO by applying the thermodynamic relation,

$$\Delta G = \Delta H - T\Delta S = 2\sigma^F V/r, \tag{4}$$

where $\Delta G$ is for reaction (1), $r$ is the radius of the particle, and $V$ is the molar volume (11.10 cm$^3$ mol$^{-1}$). Giauque [1] could not, however, obtain an estimate of the surface free energy $\sigma^F$, due to the lack of particle size data. Assuming that for very finely divided MgO, $r \approx \varepsilon_{200}/2$, $\Delta H = \Delta(\Delta H_s)$ in table 1, and that $\Delta S$ is about 0.3 cal. deg.$^{-1}$ mol$^{-1}$, $\sigma^F$ can be estimated to be approximately 1 850 ergs cm$^{-2}$ for the 350°C sample.

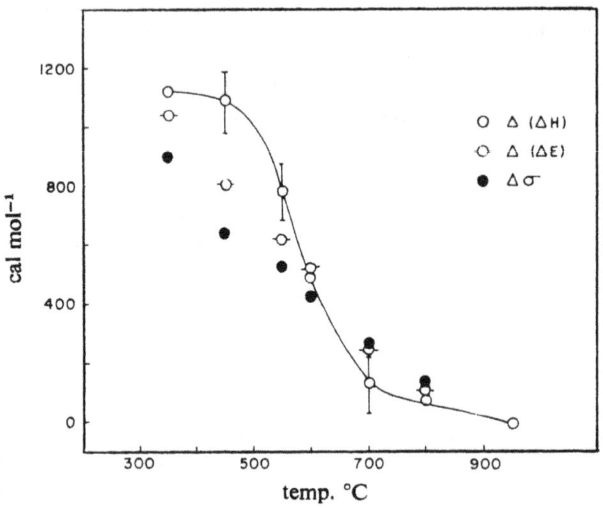


FIG. 1.—Variation of $\Delta(\Delta H_s)$ (with respect to the 950°C sample) obtained from experimental heats of solution of MgO samples prepared by the decomposition of the hydroxide at different temperatures. Values of $\Delta(\Delta E)$ and $\Delta\sigma$ (with respect to the 950°C sample) are also shown in the figure for comparison.

TABLE 1. CRYSTALLITE SIZES, SURFACE AREAS AND HEATS OF SOLUTION OF MgO

| temp. °C | $\varepsilon_{(200)}$, Å (a) | $(\Delta E)$ cal mol$^{-1}$ | $\Delta(\Delta E)$ cal mol$^{-1}$ | $A_s$ (b) m$^2$ g$^{-1}$ | $\sigma$ (c) cal mol$^{-1}$ | $\Delta H_s$ (d) kcal mol$^{-1}$ | $\Delta(\Delta H_s)$ cal mol$^{-1}$ |
|---|---|---|---|---|---|---|---|
| 350 | 210 | 1 150 | 1 040 | 90 | 1 176 | 34.11 | 1 120 |
| 450 | 260 | 910 | 800 | 70 | 919 | 34.08 | 1 090 |
| 550 | 320 | 730 | 620 | 62 | 814 | 33.78 | 790 |
| 600 | 370 | 630 | 520 | 53 | 696 | 33.48 | 490 |
| 700 | 510 | 360 | 250 | 42 | 552 | 33.12 | 130 |
| 800 | 920 | 220 | 110 | 32 | 420 | 33.07 | 80 |
| 950 | 2200 | 110 | 0 | 22 | 280 | 32.99 | 0 |

(a) the 1 200°C sample was taken as the reference for instrumental broadening.
(b) $A_s$ of 1 200°C sample was 10 m$^2$ g$^{-1}$.
(c) calculated from Lennard-Jones value [18] of the surface energy.
(d) at 35°C.

By making use of the $\Delta(\Delta H_s)$ value of the 350°C sample and the surface areas of the 350 and 1 200°C samples, the surface enthalpy, $\sigma^H$, of the MgO was calculated to be ~1 470 ergs cm$^{-2}$. The surface energy was estimated to be ~1 600 ergs cm$^{-2}$ by employing the $\Delta(\Delta E)$ value at 350°C. These values of $\sigma^H$ and $\sigma$ compare favourably with the value of 1 362 ergs $cm^{-2}$ proposed by Lennard-Jones and Taylor.[18] The $\sigma$ values of the various MgO samples in table 1 were calculated by assuming the surface energy of Lennard-Jones and Taylor.[18] The $\Delta\sigma$ values (with respect to the 950°C sample) are plotted in fig. 1. These values are close to the $\Delta E$ from Weissenbach's relation up to about 600°C; apparently the Weissenbach relation is fairly satisfactory in estimating surface energies.

An examination of the data in table 1 and fig. 1 indicates that the surface energy ($\Delta(\Delta E)$ as well as $\Delta\sigma$) decreases steadily as a function of temperature unlike $\Delta(\Delta H_s)$, which shows an abrupt decrease in a narrow temperature range (450-600°C). Beyond 600°C, the $\Delta\sigma$, $\Delta(\Delta E)$ and $\Delta(\Delta H_s)$ are about the same within experimental error of the last quantity (fig. 1). Apparently, there are other factors besides surface area or particle size responsible for the observed variation of $\Delta(\Delta H_s)$. While lattice microstresses [10] may contribute to some extent to the abrupt decrease in $\Delta(\Delta H_s)$ with temperature, the major contribution may be due to a pseudocrystallization or the so-called disorder-order (or defect-annealing) reaction.[3, 4] The magnitude of the heat evolution in such a reaction would be much smaller than that in the usual crystallization reactions of amorphous oxides.[5]

The $\Delta(\Delta H_s)$ curve in fig. 1 is reminiscent of a glass transition. In finely divided solids prepared by precipitation or by low-temperature decomposition of compounds, crystal imperfections may be " frozen in " as suggested by Kobayashi,[19] and Rao and coworkers.[4] A more plausible explanation for the excess energy of the finely divided particles of MgO is on the basis of surface and bulk defects discussed by Nelson, Tench and coworkers [20-22] in their studies of irradiated MgO. These workers have shown that the concentration of point defects on, or close, to the surface ($S$-centres) as well as that of the bulk centres ($F$ and $V$) may decrease with decrease in particle size. Small amounts of impurities present in MgO may also affect the properties of the surface, but their role is not clear.

The energy difference $\Delta(\Delta H_s)-\Delta\sigma$, or $\Delta(\Delta H_s)-\Delta(\Delta E)$, attributed to pseudocrystallization or defect annealing decreases sharply between 450 and 600°C. Above 600°C the increase in crystallite size becomes more marked (table 1) probably due to the onset of sintering. These results are in general agreement with the surface area studies of Razouk and Mikhail [11] who have suggested that three processes may occur when MgO is heated: (i) recrystallization, (ii) crystal growth, and (iii) sintering. These processes apparently take place at different temperatures at different rates.

The authors thank the U.S. National Bureau of Standards for a research grant (G-51) under their Special International Programmes.

*Note added in proof:* The present results on the anomalous behaviour of finely divided MgO are indirectly supported by studies of Anderson and Horlock (*Trans. Faraday Soc.*, 1969, **65**, 251) on the heats of adsorption on microcrystalline samples.

1 W. F. Giauque, *J. Amer. Chem. Soc.*, 1949, **71**, 3192.
2 K. Taylor and L. S. Wells, *Bur. Stand. J. Res.*, 1938, **21**, 133.
3 C. N. R. Rao and K. S. Pitzer, *J. Phys. Chem.*, 1960, **64**, 282.
4 C. N. R. Rao, S. R. Yoganarasimhan and M. P. Lewis, *Can. J. Chem.*, 1960, **38**, 2359.
5 M. Natarajan, G. V. Chandrashekar, and C. N. R. Rao, *J. Chem. Eng. Data*, 1968, **13**, 235.
6 R. Kelly and C. Jech, *Proc. Brit. Ceram. Soc.*, 1967, **9**, 243.
7 D. K. Thomas and T. W. Baker, *Proc. Phys. Soc.*, 1959, **74**, 673.
8 B. Weissenbach, *Radex Rundschau*, 1951, **6**, 257.
9 D. T. Livey, B. M. Walklyn, M. Hewitt and P. Murray, *Trans. Brit. Cer. Soc.*, 1957, **56**, 217.
10 Z. Librant and R. Pampuch, *J. Amer. Ceram. Soc.*, 1968, **51**, 109.
11 R. I. Razouk and R. Sh. Mikail, *J. Phys. Chem.*, 1959, **63**, 1050.
12 H. E. Swanson and E. Tatge, *N.B.S. Circ.* 539, 1953.

[13] H. P. Klug and L. E. Alexander, *X-ray Diffraction Procedures* (John Wiley and Sons, Inc., New York, 1954).
[14] L. S. Birks and H. Friedman, *J. Appl. Phys.*, 1946, **17,** 687.
[15] J. C. Ahluwalia and J. W. Cobble, *J. Amer. Chem. Soc.*, 1964, **86,** 5377.
[16] S. Subramanian and J. C. Ahluwalia, *J. Phys. Chem.*, 1968, **72,** 2525.
[17] S. R. Gunn, *J. Phys. Chem.*, 1965, **69,** 2902.
[18] J. E. Lennard-Jones and G. I. Taylor, *Proc. Roy. Soc. A*, 1925, **109,** 476.
[19] K. Kobayashi, *Sci. Reports Tohoku Univ.*, 1951, **35,** 173, 283.
[20] R. L. Nelson, A. J. Tench and B. J. Harmsworth, *Trans. Faraday Soc.*, 1967, **63,** 1427.
[21] R. L. Nelson and A. J. Tench, *Trans. Faraday Soc*,, 1967, **63,** 3039.
[22] R. L. Nelson, J. W. Hale, B. J. Harmsworth and A. J. Tench, *Trans. Faraday Soc.*, 1968, **64,** 2521.

PRINTED IN GREAT BRITAIN AT
THE UNIVERSITY PRESS
ABERDEEN

Indian Journal of Chemistry
Vol. 23A, April 1984, pp. 265-284

*Advances in Contemporary Research*

# Superstructures, Ordered Defects & Nonstoichiometry in Metal Oxides of Perovskite & Related Structures†

C N R RAO*‡
Jawaharlal Nehru Visiting Professor, Department of Physical Chemistry, University of Cambridge, Cambridge CB2 1EP, UK
and
J GOPALAKRISHNAN & K VIDYASAGAR
Solid State and Structural Chemistry Unit, Indian Institute of Science, Bangalore 560012

*Received* 14 *February* 1984

## 1. Introduction

Nonstoichiometry has been a problem of vital interest from the early beginnings of chemistry[1]. Nonstoichiometry was until recently explained in terms of equilibria involving point defects (vacancies and interstitials)[2,3], often invoking interaction of defects[4]. As more and more grossly nonstoichiometric compounds belonging to different families and possessing a variety of novel structural features were discovered, a new look at this old problem became necessary[5]. On the basis of careful structural studies, it was soon realized that the concept of point defects was most inadequate to explain nonstoichiometry in many of the metal oxides and other materials. Point defects may explain properties of solids with very small deviations in stoichiometry, but for most compounds where deviations in the cation or the anion site occupancy is sizeable (a small fraction of a percent or more), alternative models based on experimental findings have become possible. The models involve ordering of point derects or new structural features which eliminate point defects[6-9]. Ordering of point defects is found in simple solids such as TiO (where 15% of both the cations and the anions are absent) and FeO where there is a relatively small cation deficiency. The formation of crystallographic shear planes[6,7] provides an important means of eliminating point defects and is found in oxides based on $ReO_3$ and $TiO_2$ structures, giving rise to homologous series of oxides such as $Ti_nO_{2n-1}$, $W_nO_{3n-1}$ and $W_nO_{3n-2}$. It is interesting that the oxide qf titanium with the composition $TiO_{1.9995}$ shows evidence of crystallographic shear planes rather than point defects. The subject of nonstoichiometry and defect structures has been reviewed widely in the last few years[8-11]. The main conclusion is that isolated point defects are rarely of relevance to the chemistry of a large variety of nonstoichiometric inorganic compounds. Instead, the new structural principles of solid state inorganic chemistry enable us to rationalize the structures of nonstoichiometric compounds giving rise to new families of inorganic compounds and thereby transforming a subject of curiosity to a select bunch of chemists into one belonging to main stream chemistry.

In this article, we concern ourselves with the defect structures of metal oxides of the general formula $ABO_3$ possessing the perovskite structure, the topic being of great significance not only because of the wide variety of oxides crystallizing in structures related to the perovskite, but also because of the interesting electronic, magnetic, dielectric, catalytic and related properties exhibited by these metal oxides[12,13]. A number of perovskite oxides exhibit nonstoichiometry. Defects in these oxides can arise from cation deficiency (in A or B site), oxygen deficiency or oxygen excess. Such vacancies often give rise to ordered superstructures which are readily studied by electron diffraction and high resolution electron microscopy (HREM). X-ray diffraction is inadequate to examine vacancy ordering and aspects related to local structure such as stacking sequences and disorder. Electron diffraction, is eminently suited to the study of superstructures formed by vacancy ordering. Electron diffraction forms an integral part of HREM which yields images of lattices in direct space and these images enable us to ùnravel the intricate details of local structure at the atomic or the unit cell level. We shall be discussing the results of HREM of defect perovskites throughout this article.

## 2. Metal oxides with the perovskite structure

The ideal perovskite ($ABO_3$) structure which is a simple cubic structure with the space group Pm3m, provides the basis for the structures of a large variety of inorganic solids. The perovskits structure is conventionally described as consisting of a $BO_3$ array

†Contribution No. 246 from the Solid State and Structural Chemistry Unit.

‡Permanent address: Indian Institute of Science, Bangalore 560012.

formed by corner-sharing $BO_6$ octahedra. The large A-cations occupy twelve-coordinated body-centre sites in the $BO_3$ network (Fig. 1a). The structure can be thought of as consisting of alternating $BO_2$ and AO layers stacked one over the other in the [001] direction (Fig. 1b). The geometrical (size) requirement for the perovskite structure is expressed in terms of Goldschmidt's tolerance factor,

$$t=(r_A+r_O)/\sqrt{2}(r_B+r_O)$$

where $r_A$, $r_B$ and $r_O$ are the ionic radii of A, B and oxide ions respectively. For the ideal structure, $t=1$, but the perovskite structure occurs over a range of $t$ with a lower limit of ~0.7. The perovskite structure also requires that A and B ions be stable in 12 and 6 (octahedral) coordinations respectively. In oxides, this limits the radii of A and B ions to $r_A>0.9$ and $r_B>0.51$ Å. For lower values of $t$, the cubic structure is distorted to optimize A–O bond lengths. When $0.75<t<0.90$, the $BO_6$ octahedra are tilted cooperatively to give an enlarged orthorhombic unit cell (Fig. 1c). For $t$ slightly less than unity, a rhombohedral distortion occurs as in $LaAlO_3$. In general, distortions from the ideal structure can be traced to cation displacement or tilting of the $BO_6$ octahedral or to a combination of both. Different kinds of octahedral tiltings have been classified and discussed[14,15].

Katz and Ward[16] have given an alternative description of the perovskite structure in terms of close packing of A and O ions. This description is more general and helps to understand the occurrence of the hexagonal polytypes of perovskites. An $ABO_3$ perovskite structure in Katz and Ward model is thought to consist of close-packed $AO_3$ layers (Fig. 2a) stacked one over the other; the B-cations occupy octahedral holes that are exclusively surrounded by oxygens. When the stacking of the $AO_3$ layers is all-cubic (ccc....), the B-cation octahedra share all their corners, giving rise to the perovskite structure (3C) (Fig. 2b). When the stacking of $AO_3$ is all-hexagonal (hhh....), $BO_6$ octahedra share opposite faces forming infinite chains parallel to the hexagonal $c$-axis (Fig. 2c). Mixed stacking sequences of cubic and hexagonal stacking arrangements give rise to polytypes with larger periodicities. In Fig. 2d-f, we have shown the arrangements, cchcch (6H), chch (4H) and $(chh)_3$ (9H). The hexagonal stacking of $AO_3$ layers is favoured when the A-cation is large ($t>1.0$). Hexagonal stacking results in considerable loss of the Madelung energy as a consequence of B-cation repulsion between face-shared octahedra. Several factors such as B–B bonding, B–O covalency and B-site vacancy which compensate for the loss in the Madelung energy, stabilize hexagonal stacking in preference to cubic stacking.

### 3. Cation-deficient perovskites

*A-site vacancies*: Since the $BO_3$ array in the perovskite structure forms a stable network, the large A-cations at twelve-coordinated sites can be missing either partially or wholly. The tungsten bronzes, $A_xWO_3$, which constitute an important family of nonstoichiometric oxides, may be regarded as A-site deficient perovskites[17]. Of the three different structures known for tungsten bronzes, hexagonal, tetragonal and cubic (Fig. 3), only the cubic one is related to the perovskite structure consisting of A-site vacancies. The cubic bronzes are referred to as perovskite tungsten bronzes (PTB) and two types of PTBs can be distinguished: those with small values of $x(x \ll 1.0)$ and others with large values of $x \leqslant 1.0$. PTBs with a large $x$-value show true bronze characteristics, viz. metallic behaviour and chemical inertness. An important question with regard to the non-stoichiometry of these phases is whether the A-site atoms and vacancies are ordered or not. An early neutron diffraction study[18] of $Na_{0.75}WO_3$ showed that sodium was ordered in six out of eight A-sites in a doubled perovskite cell. But the problem of ordering of A-side vacancies in metallic PTBs is not fully resolved.

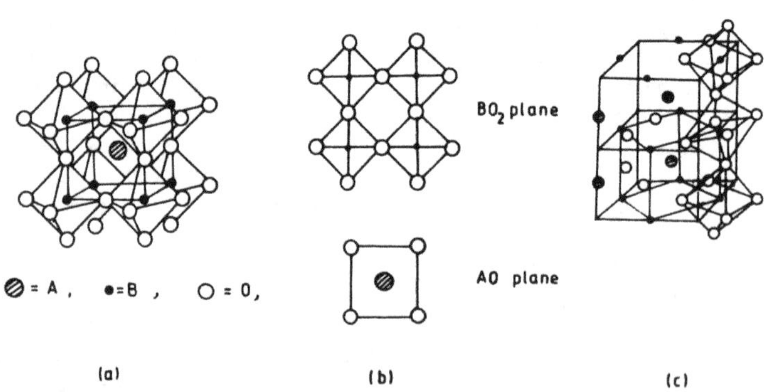


Fig. 1—(a) Ideal perovskite structure; (b) layer sequence in the perovskite structure parallel to (001); and (c) $GdFeO_3$ structure.

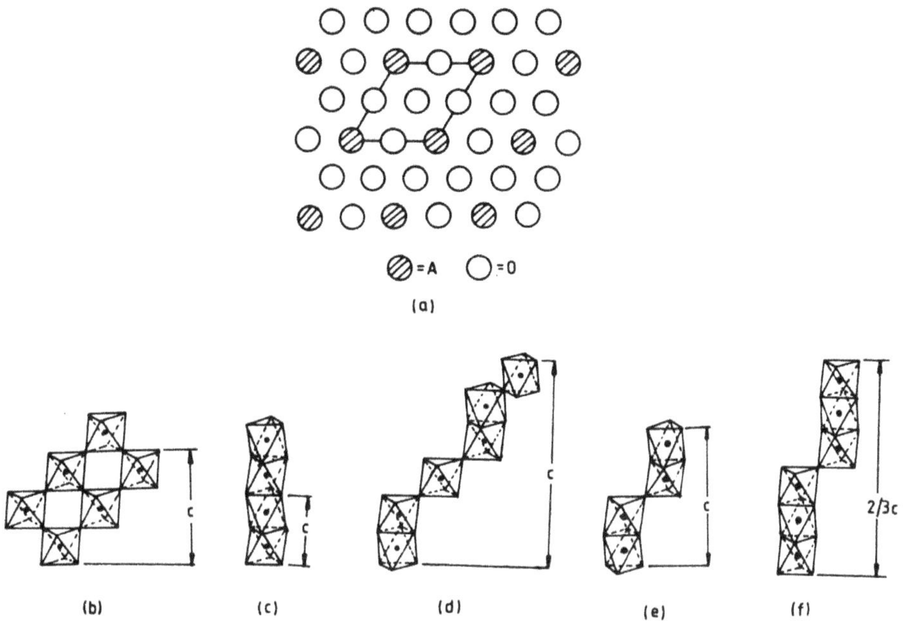


Fig. 2—(a) Close-packed $AO_3$ layer in perovskites; (b)-(f) $BO_6$ octahedra in different perovskite polytypes: (b) 3C; (c) 2H; (d) 6H; (e) 4H; and (f) 9R.

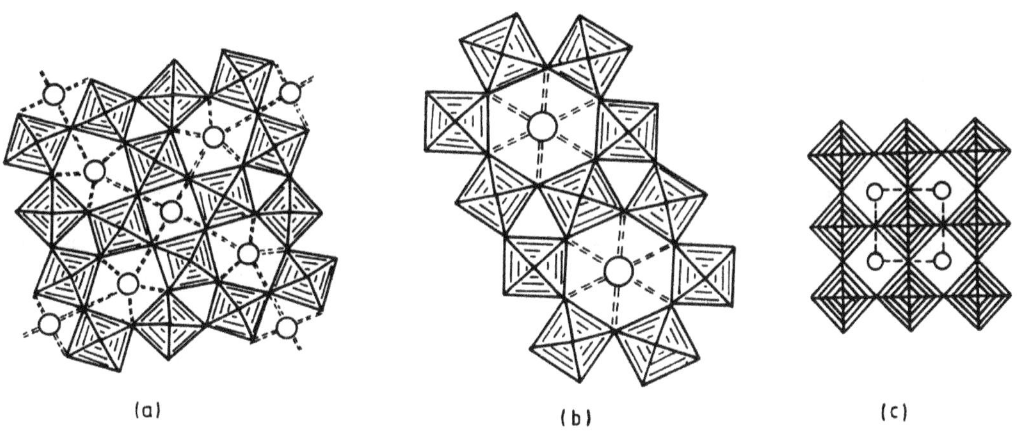


Fig. 3 Tungsten bronze structures: (a) tetragonal tungsten bronze (TTB); (b) hexagonal tungsten bronze (HTB); and (c) perovskite tungsten bronze (PTB).

In this context, it is relevant to note the difference between essentially ionic and essentially metallic solids as regards their nonstoichiometric behaviour[19]. In metallic systems, point defects would essentially be screened from one another by the conduction electrons; long-range defect interactions would be small and it is therefore possible that isolated random point defects or weakly interacting point defect clusters are present. This is probably the reason why in metallic PTBs no significant long-range ordering of vacancies is seen. But this is not the case in nonmetallic PTBs where one would expect A-site vacancy ordering.

Besides PTBs, A-site defective perovskite oxides are known to be formed when B = Ti, Nb, Ta and so on[7,13]. Such compounds exhibit metallic properties and perovskite structures when the B atom occurs in a low oxidation state. Compositions such as $A_{0.5}NbO_3$ (A = Ba, Pb etc.) where niobium is in the highest oxidation state adopt non-perovskite network structures. An interesting example[20,21] of a A-site defective perovskite is $Cu_{0.5}TaO_3$ which crystallizes in a pseudocubic perovskite structure. The unit cell is orthorhombic with $a = 7.523$, $b = 7.525$ and $c = 7.520$ Å and eight formula units per cell. Tantalum atoms form a $TaO_3$ framework as in a cubic perovskite, while the copper atoms are ordered at the A-sites. Three of them are located in the middle of the cube edges, while the fourth is statistically distributed over the three face centres. Twelve oxygens around copper are arranged as three mutually perpendicular squares of different size: Cu – O bond lengths are 2.763 and 3.923 Å.

Besides the perovskite, tetragonal and hexagonal tungsten bronzes, a new variety of tungsten bronzes formed by the intergrowth of $WO_3$ slabs (*n*-octahedra wide) with strips of hexagonal tungsten bronze (HTB), one to three tunnels wide, is noteworthy[22]. These intergrowth tungsten bronzes (ITB) can be fairly ordered as shown in Plate 1 or disordered as shown in Plate 2. The width of the HTB strips in Plate 1 is two tunnels wide and this is in fact the most stable configuration of alkali metal intergrowth tungsten bronzes. Recently we have synthesized[23] intergrowth tungsten bronzes of bismuth where the HTB strip is always one tunnel wide (Plate 3). What is interesting in the $Bi_xWO_3$ system is that when $x \leq 0.02$, a PTB phase seems to be formed, the ITBs being found only when $x > 0.02$. Furthermore, the ITB phase does not give place to the HTB at large $x$; the maximum value of $x$ in this system is $\sim 0.1$.

$WO_3$ and $ReO_3$ may be regarded as the limiting cases of A-site vacancy perovskites. Both the oxides possess corner-linked framework of the octahedra, but unlike $ReO_3$, $WO_3$ is never cubic. It shows several polymorphic transitions starting from the low temperature triclinic structure to more symmetric forms with increasing temperature. The transitions arise from temperature-dependent displacements of the tungsten atom from the centre of the $WO_6$ octahedron[24].

*B-site vacancies*: B-site vacancies in perovskite oxides are energetically not favoured because of the large formal charge and the small size of B-site cations. If vacancies are to occur at the B-sites, there must be other compensating factors such as B – O covalency and B – B interaction. While the covalency of the B – O bond would increase with increasing charge and decreasing size of the B-cation, the B – B interaction would be favoured by hexagonal rather than cubic stacking of $AO_3$ layers. Thus one would expect B-site vacancies to occur more frequently in perovskite oxides of highly charged B-cations possessing hexagonal polytypic structures. B-site vacancy perovskites are indeed far more common with hexagonal or mixed hexagonal-cubic $AO_3$ stackings than with exclusive cubic stacking of $AO_3$ layers. Rauser and Kemmler-Sack[25] have listed cubic perovskites exhibiting B-site vacancy. Some of the examples of B-site vacancy ordered perovskites are: $Ba_2Sm_{2/3}UO_6$ (5), $Ba_2Ce^{4+}_{3/4}Sb^{5+}O_6$ (7), $Ba_2Ce^{4+}Sb^{5+}_{4/5}O_6$ (9) and $Ba_2Sm^{3+}U^{6+}_{5/6}O_6$ (11), where the numbers in parentheses represent the number of B-cations per vacancy. Ordering of these vacancies is revealed by the formation of superstructures.

In Table 1, are listed some of the well-characterized B-site vacancy hexagonal perovskites. Normally, B-site vacancies are ordered between h-h layers where the $BO_6$ octahedra share faces. This is consistent with Pauling's rules for the sharing of coordination polyhedra, according to which corner-sharing is more favourable than face-sharing of polyhedra in essentially ionic structures[37]. Accordingly, $Ba_5Ta_4O_{15}$ adopts a five-layer (hhccc) sequence[30] wherein the octahedral site between h-h layers is vacant (Fig. 4a). The arrangement gives rise to isolated clusters of four octahedra that share each two opposite corners, $[Ta_4O_{15}]^{10-}$. The structure of $Ba_3Re_2O_9$ is similar[26] consisting of the $(hhc)_3$ layer sequence wherein the $ReO_6$ octahedra share corners, and the vacancy occurs at octahedral sites between h-h layers (Fig. 4b). Somewhat exceptional to this general trend[27,28,31] are

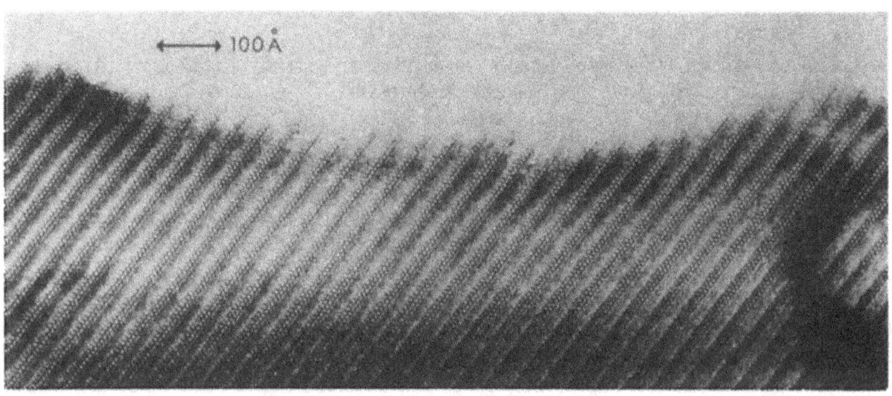


Plate 1—Lattice image of a highly ordered $Cs_xWO_3$ ITB (from Ref. 22). The narrow HTB strip can be readily distinguished from the $WO_3$ slabs. The ITB phase corresponds to the (1,7) system containing two tunnel-wide HTB strips separated by 7 $WO_6$ octahedra.

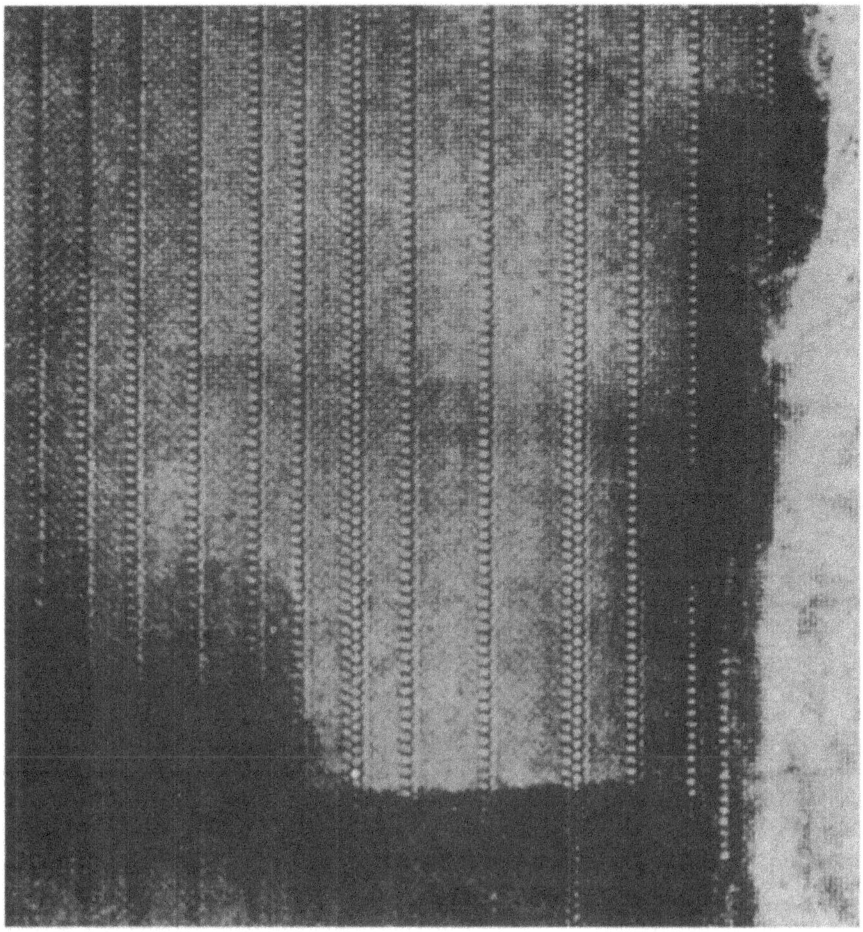

Plate 2—Lattice image of a disordered $Cs_xWO_3$ ITB showing different $WO_3$ slab widths (from ref. 22).

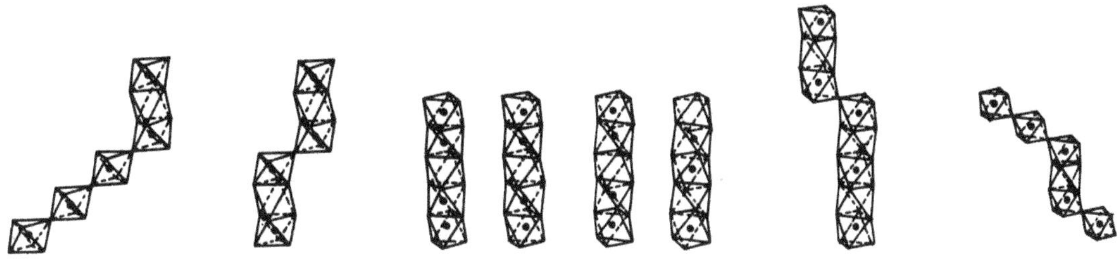


Fig. 4—Schematic illustration of linkage of $BO_6$ octahedra in hexagonal perovskites with B-site vacancy ordering: (a) $Ba_5Ta_4O_{15}$; (b) $Ba_3Re_2O_9$; (c) $Ba_3Te_2O_9$; (d) $Ba_3W_2O_9$; (e) $Ba_8Re_2W_3O_{24}$; and (f) $Ba_9Nb_6WO_{27}$.

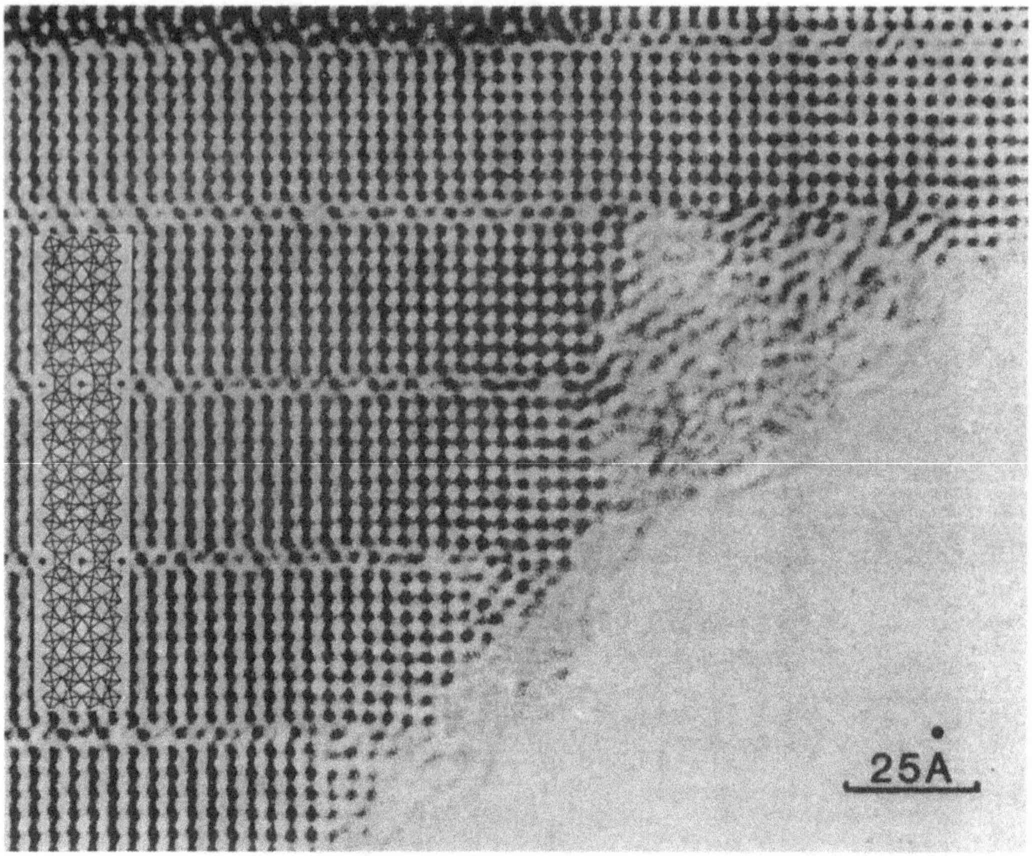


Plate 3—HREM (500kV) image of the ordered ITB phase of $Bi_{0.1}WO_3$ (nominal composition) showing Bi atoms in the tunnels (from ref. 23a).

$Ba_3W_2O_9$ and $Ba_3Te_2O_9$. Both these compounds adopt an all-hexagonal $BaO_3$ layer sequence. In $Ba_3W_2O_9$, two-thirds of the octahedral sites in every layer are filled by tungsten, while in $Ba_3Te_2O_9$, octahedral sites in two adjacent layers are fully occupied, every third layer being completely vacant. Both the orderings give rise to $[B_2O_9]^{6-}$ (B = W, Te) groups comprising two $BO_6$ octahedra sharing a common face (confacial bioctahedra) as shown in Figs. 4c and 4d. In both the structures, B-atoms are displaced considerably towards the vacancies, resulting in a minimization of B – B repulsion.

Oxide perovskites consisting of ordered B-site vacancies exhibit novel luminescence properties. $Ba_3W_2O_9$, for instance, shows an efficient blue (460 nm) photoluminescence[38] below 150K, unlike $BaWO_4$ and $Ba_3WO_6$ which consist of isolated $WO_4$ tetrahedra and $WO_6$ octahedra respectively. It has been suggested that clustering of $WO_6$ octahedra in $Ba_3W_2O_9$ is responsible for the efficient luminescence. More interestingly, B-site vacancy ordered perovskites show different colour emissions, when they are doped with different rare-earth activators. For instance, when $Ba_{3-x}Sr_xLaScW_2O_{12}$, 12-layered (hhcc)$_3$ perovskite, is doped with $Eu^{3+}$ and $Tb^{3+}$ simultaneously, a green terbium emission ($Tb^{3+}: {}^5D_4 \rightarrow {}^7F_5$) at 547 nm and a red europium emission ($Eu^{3+}: {}^5D_0 \rightarrow {}^7F_2$) at 615 nm are seen. The former shows up when the phosphor is excited in the $WO_6$ charge-transfer band at 300 nm and the latter emission is seen when excitation occurs at 340 nm corresponding to charge-transfer from oxygen to $Eu^{3+}$.Similarly $Sr_3La_2W_2O_{12}$ doped with $Eu^{3+}$ and $Tm^{3+}$ shows red and blue emissions when excited suitably, and the same host doped with $Er^{3+}$ and $Tm^{3+}$ shows green and blue emissions[39,40].

The family of bismuth oxides of the general formula $Bi_mB_{m-1}O_{3m}$ (B = Ti, Nb and W), first described by Aurivillius[41], may be regarded as B-cation deficient perovskites. Typical members of this family are $Bi_2WO_6$, $Bi_3TiNbO_9$ and $Bi_4Ti_3O_{12}$. These phases adopt layer structures consisting of PbO-like $(Bi_2O_2)^{2+}$ layers which alternate in the orthorhombic

Table 1—Hexagonal Perovskite Oxides Exhibiting B-Site Vacancy-ordering

| Compound | Structural details | Reference |
|---|---|---|
| $Ba_3Re_2O_9$ | $(hhc)_3$: 9H. $ReO_6$ octahedra are corner-shared. Vacancy is found between h-h layers. | 26 |
| $Ba_3W_2O_9$ | $(h)_6$: 6H. Isostructural with $Cs_3Tl_2Cl_9$. Clusters of $(W_2O_9)$ formed by two $WO_6$ octahedra sharing a common face. Two-thirds of octahedral sites in every layer are filled. | 27 |
| $Ba_3Te_2O_9$ | $(h)_6$: 6H. Isostructural with $Cs_3Fe_2F_9$. Clusters of $(Te_2O_9)$ similar to $(W_2O_9)$ clusters in $Ba_3W_2O_9$. Octahedral sites in two adjacent layers are completely filled by Te followed by an empty layer of octahedral sites. | 28 |
| $Ba_5M_4O_{15}$ (M = Nb, Ta) | (hhccc): 5H. B-site vacancy is ordered between two h layers. | 29,30 |
| $Ba_3LaM_3O_{12}$ $Ba_4M_2WO_{12}$ (M = Nb, Ta) | $(hhcc)_3$: 12H. B-site vacancy occurs at the centre of three face-connected octahedra. | 31-33 |
| $Ba_6B_2^{III}W_3O_{18}$ ($B^{III}$ = Gd-Lu, Y) | $(hhcccc)_3$: 18H. Structure consists of three octahedra sharing faces which alternate with three octahedra sharing corners. B-site vacancy is ordered at the central octahedral site of the face-sharing groups of octahedra. | 34 |
| $Ba_8Re_2W_3O_{24}$ | $(hhhhchhc)_3$: 24H. Structure consists of groups of three and five face-shared octahedra which are linked through corners (Fig. 4e). B-site vacancy is ordered in alternate face-shared octahedral sites. | 35 |
| $Ba_9Nb_6WO_{27}$ | $(hhcchhcc)_3$: 27H. Structure consists of groups of three face-shared octahedra which are linked to one another through one and two corner-sharing octahedra alternatively (Fig. 4f). Central B-site in the group of three face-sharing octahedra is vacant. | 36 |

c-direction with perovskite-like slabs of the formula $(A_{n-1}B_nO_{3n-1})^{2-}$ where $n=m-1$. $Bi_2WO_6$ may also be considered to be the first member of the homologous series $Bi_2W_nO_{3n+3}$; the $n=2$ member of this series, $Bi_2W_2O_9$, is known but the $n=3$ member only occurs as isolated strips of intergrowth[42] in $Bi_2W_2O_9$.

## 4. Anion-deficient perovskites and vacancy-ordered structures

Anion vacancy nonstoichiometry in perovskite oxides is more common than that involving vacancies in the cation sublattice. Anion vacancy nonstoichiometry is known with both $ABO_3$ and $BO_3$ frameworks. Nonstoichiometric tungsten trioxide, $WO_{3-x}$, is a typical example of the latter. Anion deficiency in $WO_3$ is accommodated by crystallographic shear (CS). In Fig. 5 is shown the formation of CS in $WO_{3-x}$ schematically. When $WO_3$ becomes oxygen-deficient, oxygen vacancies are accumulated on specific planes (Fig. 5a) and this is accompanied by shearing of the lower part of the crystal relative to the upper part in the direction shown in Fig. 5a. This results in the oxygen atoms, marked by asterisk, being pushed to the position of oxygen vacancies. Vacancies are thus eliminated. Inspection of the sheared structure (Fig. 5b) reveals that cation coordination is restorted to six but the octahedra are now linked through edges instead of corners in the shear plane. This remarkable process of CS involving the elimination of anion vacancies results in the formation of the homologous series of phases, $W_nO_{3n-2}$, in the $WO_{3-x}$ system, when the shear plane on $\{103\}$ recurs periodically.

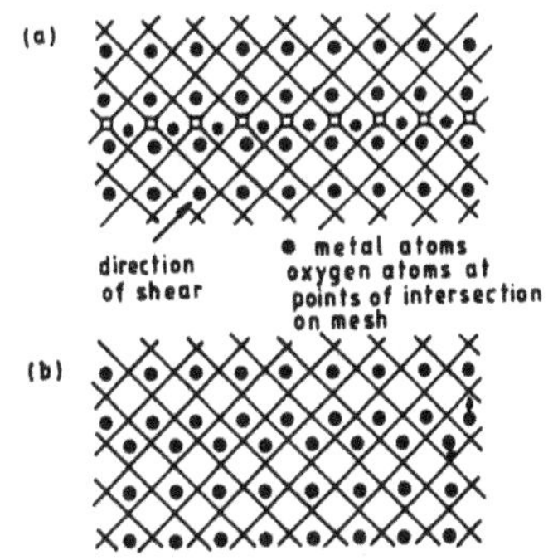


Fig. 5—Schematic representation of a shear plane (CS) formation: (a) idealized $WO_3$ structure showing anion vacancies aligned on a plane; and (b) sheared structure after vacancies are eliminated (following ref. 43).

An isolated CS plane is referred to as a Wadsley defect and a random array of CS planes is considered to constitute planar (extended) defects which are entirely different from point defects. It is obvious that when CS planes occur at regular intervals, the composition of the crystal is stoichiometric, whereas a random array of CS planes results in nonstoichiometric compositions. While we have invoked anion vacancies which are later annihilated in our description of CS plane formation, we must point out that vacancies are not essential precursors for the formation of CS planes. Accommodating anion-deficient nonstoichiometry through CS mechanism is a special feature restricted to $d^0$ metal oxides such as $WO_3$, $Nb_2O_5$ and $TiO_2$ which exhibit soft phonon modes. Soft phonon modes in metal oxides arise from soft metal-orxygen potentials which permit large cation relaxation. The latter

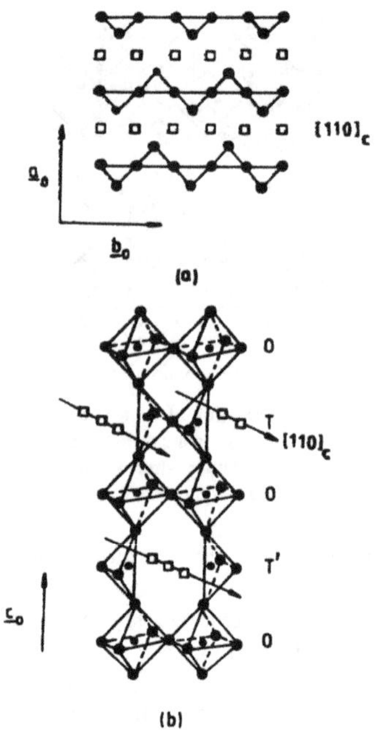


Fig. 6—Structure of brownmillerite: (a) vacancy ordering in the *a-b* plane; and (b) alternating sequence of octahedral and tetrahedral layers in the *c*-direction.

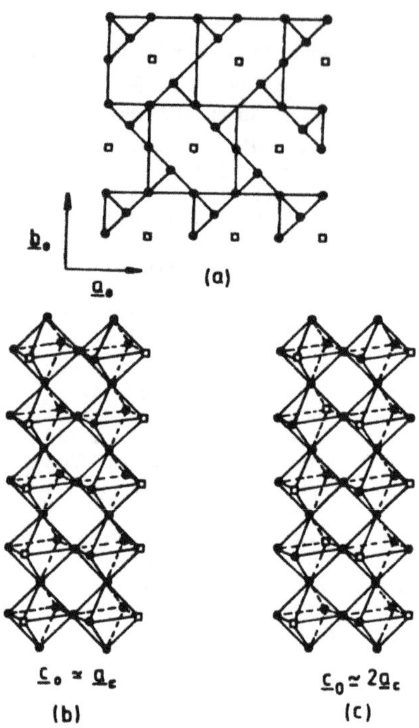


Fig. 7—Structures of $Ca_2Mn_2O_5$ and $Ca_2Co_2O_5$: (a) vacancy ordering in the *a-b* plane; vacancy ordering parallel to the *c*-direction in $Ca_2Mn_2O_5$; and $Ca_2Co_2O_5$ is shown in (b) and (c).

appears to be an essential prerequisite for CS plane formation[43]. CS planes are not formed for instance in $SnO_{2-x}$ and $TeO_{3-x}$ although both $SnO_2$ and $TeO_3$ crystallize in rutile and $WO_3$ related structures respectively.

Anion-deficient nonstoichiometry in $ABO_{3-x}$ perovskites is not accommodated by the CS mechanism. The reason probably is that the constant A/B ratio required by the composition of perovskites prevents formation of CS planes. Defect-ordering in $ABO_{3-x}$ oxides involves a conservative mechanism in the sense that the vacancies are assimilated into the structure resulting in large supercells of the basic perovskite structure. The type of superstructure formed depends however on the identity of the B-cation.

One of the best-characterized perovskite oxides with ordering of anion vacancies[44,45] is the brownmillerite structure exhibited by $Ca_2Fe_2O_5$ and $Ca_2FeAlO_5$. The compositions could be considered as anion-deficient perovskites with one-sixths of anion sites being vacant. The orthorhombic unit cell of brownmillerite structure ($a = 5.425$, $b = 5.598$ and $c = 14.768$ Å for $Ca_2Fe_2O_5$) arises because of vacancy ordering and is related to the cubic perovskite as $a \simeq \sqrt{2}a_c$, $b \simeq \sqrt{2}a_c$, $c \simeq 4a_c$. Oxygen vacancies are ordered in alternate (001) $BO_2$ planes of the cubic perovskite structure such that alternate [110] rows of oxygens are missing (Fig. 6a). The layer sequence along the *c*-axis is therefore $CaO-FeO_2-CaO-FeO$... A slight shift of the iron atoms in the FeO layers leads to a tetrahedral coordination of iron in this layer resulting in alternate sheets of $FeO_6$ octahedra and $FeO_4$ tetrahedra (OTOT'....) in the *c*-direction (Fig. 6b). We have recently prepared a new oxide, $Ca_2FeCoO_5$, possessing the brownmillerite structure[46] wherein $Fe^{3+}$ and $Co^{3+}$ occupy tetrahedral and octahedral sites respectively.

The compounds $Ca_2Mn_2O_5$ and $Ca_2Co_2O_5$ which have compositions similar to brownmillerite have recently been reported[46-48], but their structures involve different modes of vacancy ordering. $Ca_2Mn_2O_5$ adopts an orthorhombic structure with $a = 5.432$, $b = 10.242$ and $c = 3.742$ Å, the relationship with the cubic perovskite structure being $a \simeq \sqrt{2}a_c$, $b \simeq \sqrt{2}a_c$ and $c \simeq a_c$. Oxygen vacancies in $Ca_2Mn_2O_5$ are ordered in every (001) $BO_2$ plane of the cubic perovskite such that one-half of the oxygen atoms in alternate [110] rows are removed (Fig. 7). The composition of the plane is now $MnO_{1.5}$. If these are stacked in the sequence $CaO-MnO_{1.5}-CaO-MnO_{1.5}$..., the structure of $Ca_2Mn_2O_5$ is obtained. An important feature of the structure which distinguishes it from brownmillerite is that manganese is square-pyramidally coordinated. It has been justifiably argued[47] that the $3d^4$-configuration of $Mn^{3+}$ favours square-pyramidal coordination rather than tetrahedral coordination, thereby determining the anion vacancy ordering in $Ca_2Mn_2O_5$. $Ca_2Co_2O_5$ also crystallizes in an orthorhombic structure with $a = 11.12$, $b = 10.74$ and $c = 7.48$ Å. Comparison of these parameters with those of $Ca_2Mn_2O_5$ reveals a doubling of the $a$ and $c$ axes in $Ca_2Co_2O_5$. We propose that vacancy-ordering in $Ca_2Co_2O_5$ is similar to that in $Ca_2Mn_2O_5$, but the doubling of the $c$ axis is likely to be due to the ordering scheme shown in Fig. 7. $Sr_2Co_2O_5$, on the other hand, crystallizes in an anion-deficient 2H perovskite structure at low temperatures and transforms to the brownmillerite structure[49,50] above 900°C. $SrCoO_{3-x}$ $(O < x < 0.5)$ prepared under high oxygen pressures (50-2600 bars) adopts a cubic perovskite structure. The oxygen deficiency influences the unit cell parameter and the Curie temperature[51].

The compounds $SrTiO_{2.5}$, $SrVO_{2.5}$ and $KTiO_{2.5}$ are a few other anion vacancy perovskites[7] with the composition $ABO_{2.5}$. $KTiO_{2.5}$ shows little similarity to the perovskites, each titanium having a trigonal bipyramid oxygen coordination. X-ray diffraction studies of $SrTiO_{2.5}$ and $SrVO_{2.5}$ show cubic perovskite unit cells with $a \simeq 3.9$ Å, implying that anion vacancies are random. An electron diffraction study[52] of $SrTiO_{2.5}$ has revealed a six-fold superlattice ordering in the [111] direction. Subsequent work by Tofield[53] however did not substantiate the results. It is noteworthy that both $SrTiO_{3-x}$ and $SrVO_{3-x}$ exhibit metallic properties indicating that $d$-electrons associated with B-cations are itinerant. This accounts for the difference in the type of vacancy-ordering in these two oxides from the ordering in $CaMO_{2.5}$ (M = Mn, Fe, Co) where the $d$-electrons are localized.

In many of the $ABO_{3-x}$ perovskites, non-stoichiometry spans the composition range $0 < x \leq 0.5$ and we have so far discussed only the structures of the end members with $x = 0.5$. It is of interest to see the type of structures prevailing at intermediate compositions. In $SrFeO_{3-x}$, MacChesney et al.[54] reported a cubic structure when $0.0 < x < 0.12$, a tetragonally distorted perovskite structure when $0.15 < x < 0.28$ and a mixture of the perovskite and brownmillerite phases when $0.28 < x < 0.5$. $SrFeO_{2.75}$ was investigated by Tofield et al.[55] who found a supercell related to the perovskite as $a \simeq 2\sqrt{2}a_c$, $b \simeq 2a_c$ and $c \simeq 2\sqrt{2}a_c$ in the electron diffraction patterns. The vacancy-ordering suggested for this phase is related to that of brownmillerite; every other oxygen atom is lost in alternate $[110]_c$ strings so as to give five-coordination to half the iron atoms, the other half retaining octahedral coordination as in perovskite. In $Sr_xNd_{1-x}FeO_{3-y}$, superstructures involving random doubling of one of the perovskite cell axes (in $a$, $b$ or $c$ directions) has been found[56]. This has been considered to be due to the existence of a microdomain texture wherein the domains intergrow randomly in three dimensions.

Reller et al.[57,58] have investigated in great detail the anion-deficient $CaMnO_{3-x}$ system over the range $0 < x \leq 0.5$. Five distinct compositions, $CaMnO_{2.5}$, $CaMnO_{2.556}$, $CaMnO_{2.667}$, $CaMnO_{2.75}$ and $CaMnO_{2.80}$, with ordered oxygen vacancies have been identified and in all these structural features of the parent perovskite are preserved. The structure of $CaMnO_{2.5}$ possessing square-pyramidal coordination of $Mn^{3+}$ was discussed earlier. In the other compositions, the proportion of square-pyramids relative to octahedra decreases as we go from $CaMnO_{2.5}$ to $CaMnO_{3.0}$. These structures may be pictured as superlattice repeats of the parent undistorted perovskite, but with the superlattice mesh rotated by an angle $R$ in the (001) plane. In Fig. 8, is presented a comparison of the schematic arrangement of $MnO_6$ octahedra in an idealized cubic perovskite $CaMnO_3$ with the observed arrangement of the slightly distorted $MnO_6$ octahedra in $CaMnO_3$. In Fig. 9, the arrangements of $MnO_6$ octahedra and $MnO_5$ square-pyramids in $CaMnO_{2.8}$ are shown. The superlattice in $CaMnO_{2.8}$ is $\sqrt{5} \times \sqrt{5}$, but rotated by 26.5° and this structure represented in Fig. 10 gives tetragonal unit cell with $a = b = 8.34$ Å and $c = 7.46$ Å. In order to check this structure, lattice images have been computed and compared with the HREM image (Plate 4); the observed agreement tends to support the structural model. The structure of $CaMnO_{2.75}$ can be represented as $\sqrt{2} \times 2\sqrt{2}\,R\,45°$ or

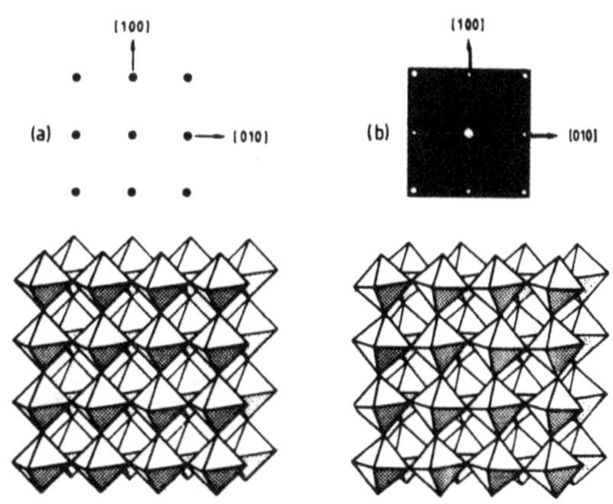


Fig. 8—(a) [001] zone axis electron diffraction pattern and schematic arrangement of $MnO_6$ octahedra in the idealized cubic perovskite, $CaMnO_3$ $(a = b = c = 3.73$ Å) (b) observed [001] zone axis electron diffraction and schematic arrangement of the slightly distorted $MnO_6$ octahedra of $CaMnO_3$ with $a = b = c = 7.46$ Å (after ref. 57).

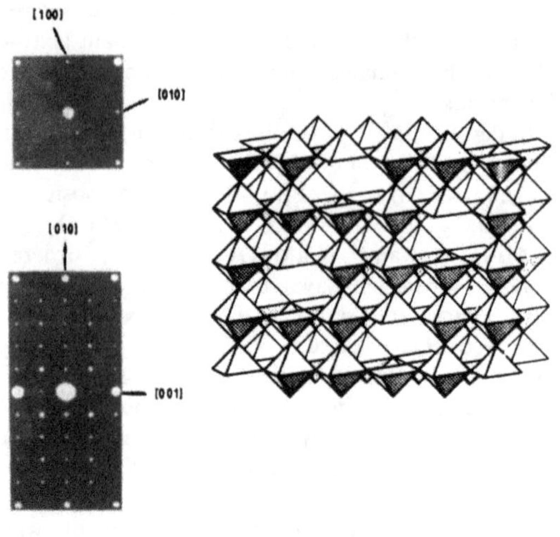


Fig. 9—Observed [001] and [100] zone axes electron diffraction patterns and the schematic arrangement of $MnO_6$ octahedra and $MnO_5$ square pyramids of $CaMnO_{2.8}$ with $a=b=8.34$ Å and $c=7.46$ Å (after ref. 57).

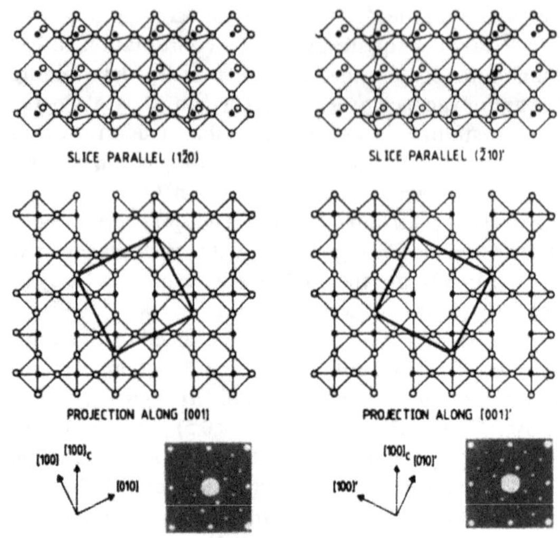


Fig. 10—Schematic representation of the structure and [001] zone axis electron diffraction patterns of $CaMnO_{2.8}$: Full circles, oxygen; closed circles, Mn; dotted broken circles, oxygen vacancies, Ca atoms are not shown for simplicity (after ref. 57).

$\sqrt{2}\times 4\sqrt{2}\,R$ 45°, while that of $CaMnO_{2.667}$ is $\sqrt{2}\times 3\sqrt{2}\,R$ 45°. Four ordered structures of $CaMnO_{2.5}$ have been observed, but the most common one is the $\sqrt{2}\times 2\sqrt{2}\,R$ 45° phase described by Poeppelmeier *et al*[47]. The relation between the defect structures of the members of the $CaMnO_{3-x}$ system and their catalytic activity has been commented upon by Reller *et al*[58].

Vacancy ordering in anion deficient $LaNiO_3$ was investigated by Gai and Rao[59]. Thermogravimetry in air and oxygen provided evidence for the formation of a series of phases pf the general formula $La_nNi_nO_{3n-1}$ with $n=7$, 9, 13 and 30 (Fig. 11). An electron diffraction study of the anion deficient samples

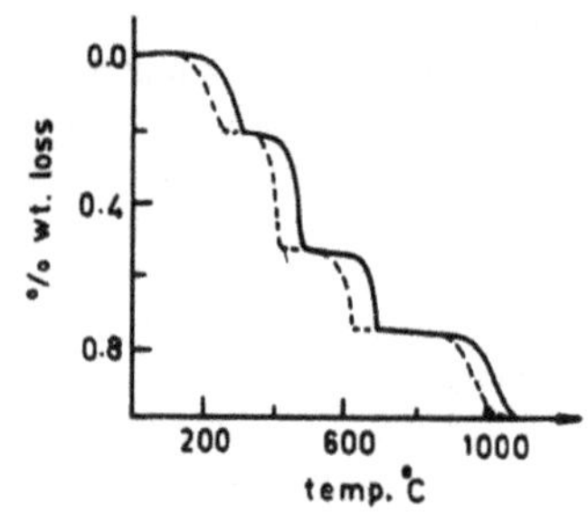


Fig. 11—Thermogram of $LaNiO_3$ showing stages and the relative ease with which oxygen is lost (after ref. 59).

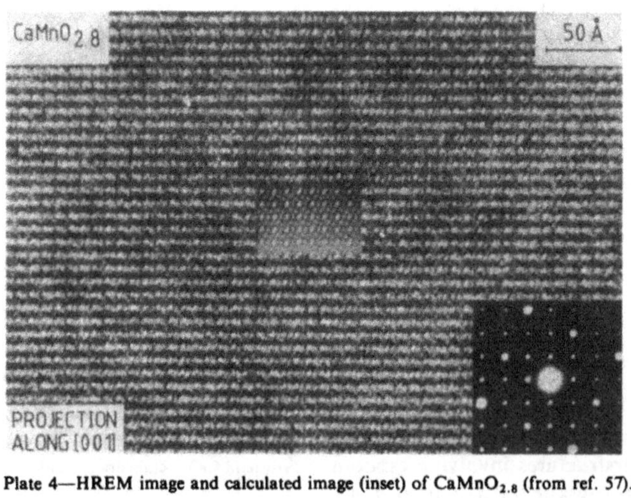


**Plate 4—HREM image and calculated image (inset) of $CaMnO_{2.8}$ (from ref. 57).**

suggested the formation of superstructures due to vacancy ordering. An orthorhombic or body-centred tetragonal cell with $a \simeq c \simeq 2\sqrt{2a_c}$ and $b \simeq 2a_c$ has been identified by electron diffraction. Crespin *et al.*[60] have recently reported that $La_2Ni_2O_5$ and $LaNiO_2$ are obtained by hydrogen reduction of $LaNiO_3$ around 550 and 670K respectively. The X-ray diffraction pattern of $La_2Ni_2O_5$, which corresponds to the $n = 2$ member of the $La_nNi_nO_{3n-1}$ series proposed by Gai and Rao[59], has been indexed on a monoclinic system with $a = 11.068$, $b = 11.168$, $c = 7.824$ Å and $\beta = 92.21°$ and a brownmillerite-type ordering of anion vacancies has been suggested. Since tetrahedral coordination of $Ni^{2+}$ is unlikely to be present in oxides, we feel that anion vacancy ordering in $La_2Ni_2O_5$ may involve octahedral and square-planar coordination around nickel. Regarding $LaNiO_2$, there is much more doubt; it is rather strange that an oxide with nickel in the 1+ state is stable in air.

Stabilities of $LaBO_3$ (B = V, Cr, Mn, Fe, Co and Ni) perovskites at 1270K in reducing atmosphere have been investigated by Nakamura *et al*[61] by thermogravimetry. The results reveal that the stabilities depend on the identity of the B atom, varying in the order, $LaVO_3 \sim LaCrO_3 > LaFeO_3 > LaMnO_3 > LaCoO_3 > LaNiO_3$. Under the experimental conditions employed in this study, $LaBO_{3-x}$ were however not obtained.

Grenier *et al.*[45,62] have investigated $Ca_xLa_{1-x}FeO_{3-y}$ (compositions intermediate between $Ca_2Fe_2O_5$ and $LaFeO_3$) and $CaTi_{1-x}Fe_xO_{3-y}$ (compositions intermediate between $Ca_2Fe_2O_5$ and $CaTiO_3$) systems in an attempt to characterize anion vacancy ordering at compositions intermediate between brownmillerite and perovskite. Their work has revealed the existence of a homologous series of phases of the general formula $A_nB_nO_{3n-1}$ (A = La, Ca; B = Ti, Fe). The structure of these phases consists of a succession of $(n-1)$ perovskite-like $BO_6$ sheets (O layers) and a sheet of $MO_4$ tetrahedra (T layers) in the $c$-direction. When $n = 2$ we have the brownmillerite phase where the sequence of octahedral and tetrahedral layers is ....OTOT′OTOT′.... where T and T′ stand for two different orientations of tetrahedra in the structure (Fig. 6). When $n = \infty$, there will be no tetrahedral layers and the sequence is ....OOOO.... corresponding to the perovskite structure. Other members of this series are $Ca_2LaFe_3O_8$, $Ca_3Fe_2TiO_8$ ($n = 3$) and $Ca_4Fe_2Ti_2O_{11}$ ($n = 4$). The layer sequences in the $n = 3$ and $n = 4$ members are ....OOT OOT.... and ....OOOT OOOT.... respectively. An interesting member of this series is $Ca_4YFe_5O_{13}$ which corresponds to $n = 2.5$ in the series. The structure of this oxide, determined by high resolution (1 MeV) electron

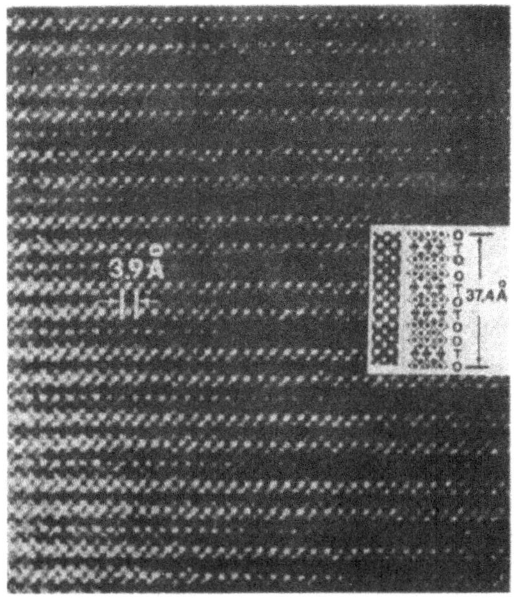


Plate 5—HREM image of $Ca_4YFe_5O_{13}$ showing ordered intergrowth of $n = 2$ and $n = 3$ units of the $A_nB_nO_{3n-1}$ series (from ref. 63).

microscopy[63], consists of a regular intergrowth of $n = 2$ and $n = 3$ units of the $A_nB_nO_{3n-1}$ series, the polyhedral layer sequence being ....OTOOTOTOOT.... (Plate 5). Other ordered intergrowth phases reported are $Ca_7Fe_6TiO_{18}$ ($n = 2.33$) and $Ca_5Fe_4TiO_{13}$ ($n = 2.5$).

Recent studies[64-66] have shown the non-stoichiometric behaviour in such Fe containing systems to be highly complex, the type of long-range ordering (or the absence of it) being dependent on sample preparation conditions. Intermediate compositions in the $Ca_xLa_{1-x}FeO_{3-y}$ ($2/3 < x < 1$) system exhibit two different types of behaviour. When the preparations are carried out in a non-oxidizing atmosphere at relatively low temperatures (1100°C), the material exhibits a disordered intergrowth structure comprising fragments of $Ca_2LaFe_3O_8$ ($n = 3$) and $Ca_2Fe_2O_5$ (Plate 6). When the samples are prepared at 1400°C in air, they become oxygen-rich and exhibit a totally different structural behaviour. Powder X-ray diffraction patterns of the latter samples show deceptively simple cubic perovskite structures with $a \simeq 3.85$ Å. Electron diffraction and HREM reveal the structures to be more complex consisting of microdomains of brownmillerite or $Ca_2LaFe_3O_8$ which are intergrown three-dimensionally so as to give an arrangement that appears cubic perovskite-like on the average. The dimension of individual domains is of the order $\sim 10^6$ Å$^3$ and decreases with increasing oxygen content. It is significant that even at the highest

INDIAN J. CHEM., VOL. 23A, APRIL 1984

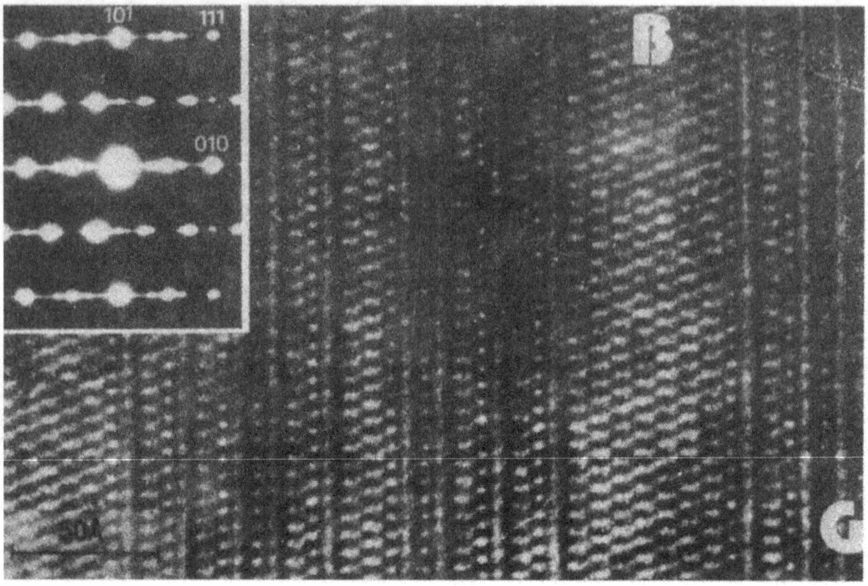


**Plate 6—Lattice image of $Ca_xLa_{1-x}FeO_{3-y}$ showing disordered intergrowth. B stands for brownmillerite structure and G stands for $Ca_2LaFe_3O_8$ structure (from ref. 65).**

temperatures of preparation the system does not exhibit classical nonstoichiometric behaviour consisting of randomly distributed point defects; apparent point defects are always structurally ordered giving microdomain texture of the material.

Anion-deficient hexagonal perovskites exhibit interesting structural behaviour involving polytypism. $BaMO_{3-x}$ (M = Mn, Fe, Co etc.) are prototype examples of this type of nonstoichiometric system. The nonstoichiometry is accommodated by these systems by adjusting the relative proportion of cubic and hexagonal stacking of $AO_{3-x}$ layers. HREM provides direct verification of the stacking arrangement of the $BaO_3$ layers in $BaMO_3$ polytypes[67]; a typical image of a 4H polytype of $BaCrO_3$ is shown in Plate 7. 6H-$BaCrO_3$, 9R-$BaRuO_3$ and 4H-$BaRuO_3$ have also been identified by HREM. By employing HREM, the structure of $BaIrO_3$ has been solved[68]; this material is found to possess a 9H structure (rather than the 9*R* structure of $BaRuO_3$) admixed with 4H sequences.

The results of Negas and Roth[69] on the $BaMnO_{3-x}$ system are presented in Table 2 to illustrate the structural complexity of this system. We see that with decreasing oxygen content, cubic stacking increases in the structure. The structural changes accompanying oxygen nonstoichiometry are however not true polytypic transformations. The $BaFeO_{3-x}$ system[70,71] is equally complex and shows a different structural behaviour: $BaFeO_{2.5}$ (triclinic, brownmillerite type), $BaFeO_{2.62-2.64}$ (rhombohedral), $BaFeO_{2.67}$ (triclinic), $BaFeO_{2.75-2.81}$ (tetragonal) and $BaFeO_{2.69-2.95}$ (6H hexagonal). A 12H phase has also been described but its composition is not known definitely. A structural investigation[72] of 6H $BaFeO_{2.79}$ shows that oxygen vacancies are unevenly distributed between the cubic and hexagonal layers which are arranged in the sequence $(cch)_2$. The hexagonal layers have the composition $BaO_{2.5}$ and the cubic layers $BaO_{2.835}$. The oxygen vacancy ordering in $BaO_{2.5}$ layer appears to be similar to the ordering in $SrO_{2.5}$ layers of $SrFeO_{2.5}$ where one [110] string of oxygen in every four is removed converting part of face-shared $FeO_6$ octahedra into corner-shared $FeO_4$ tetrahedra. The structure of 6H $BaFeO_{2.79}$ is to be contrasted with that of the manganese system. In 4H $Ba_{0.5}Sr_{0.5}MnO_{2.85}$, for instance, all the vacancies are found in the

Table 2—Phase Relations of $BaMnO_{3-x}$ System in Air*

| $BaMnO_{3-x}$ | Temp °C | Layer sequence | % Cubic stacking |
|---|---|---|---|
| $x=0$ | 1150 | 2H | 0 |
| $0<x<0.02$ | 1150-1300 | 15H | 20 |
| $0.03<x<0.05$ | 1300-1350 | 8H | 25 |
| $0.10<x<0.15$ | 1350-1475 | 6H | $33\frac{1}{3}$ |
| $0.175<x<0.20$ | 1475-1550 | 10H | 40 |
| $x \sim 0.25$ | 1550 | 4H | 50 |

*After ref. 69.

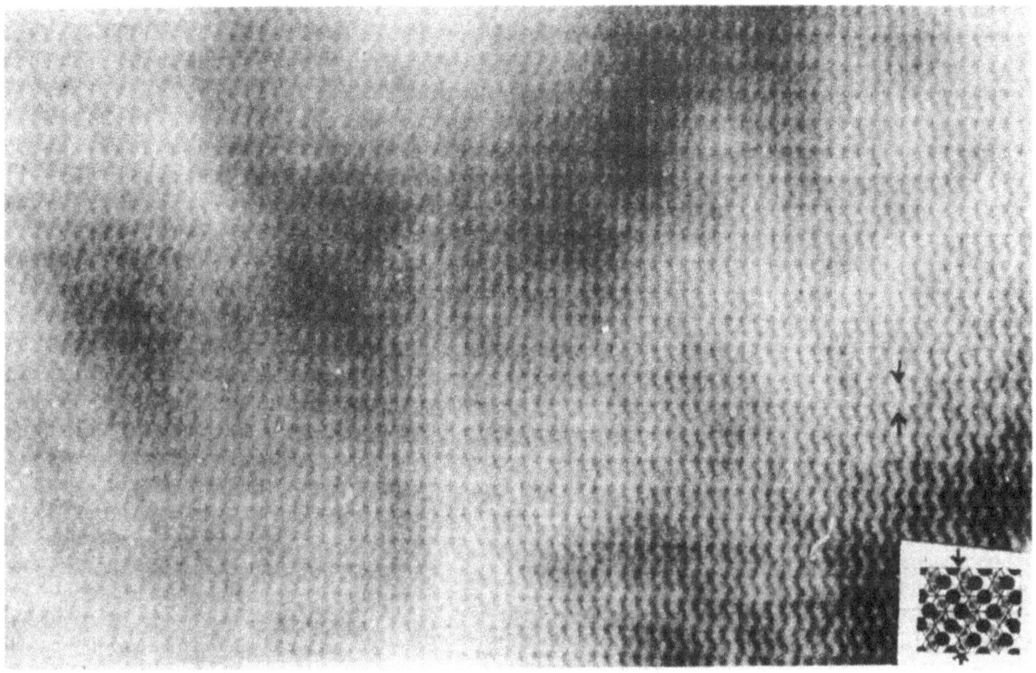

**Plate 7—Lattice image of 4H-$BaCrO_3$ showing corrugative pattern of $BaO_3$ due to chch stacking (from ref. 67). A projection drawing is shown in the inset.**

hexagonal layers, converting pairs of face-sharing octahedra in the parent 4H structure into edge-sharing trigonal bipyramides[73,74]. The difference in structural behaviour between the iron and manganese systems has been attributed to the difference in coordination tendencies of $Fe^{3+}$ and $Mn^{3+}$ high-spin $Mn^{3+}$ ($d^4$) prefers trigonal bipyramidal or square-pyramidal environment while high-spin $Fe^{3+}$ ($d^5$) accepts both octahedral and tetrahedral coordination.

Structures of the members of the $BaCoO_{3-x}$ system are interesting. Zanne *et al.*[75] found the following phases in this system at 1178 K: 2H $BaCoO_{3.0-2.85}$, 7H $BaCoO_{2.575-2.520}$, 12H $BaCoO_{2.49-2.43}$, 15H $BaCoO_{2.23-2.10}$ and an orthorhombic $BaCoO_{2.07}$. Among the various phases reported, a 12H phase is well-established in this system. The structure of 12H $BaCoO_{2.6}$ has been determined by HREM and powder neutron diffraction[76,77]. The layer sequence of this phase is $(ccchhh)_2$. Oxygen vacancies are non-random in this structure being accommodated by replacing one-sixths of $BaO_3$ layers by $BaO_2$ layers (Fig. 12). As a consequence, one-third of the cobalt atoms are tetrahedrally coordinated by oxygen. The $CoO_4$ tetrahedra are linked through corners to strings of four face-shared octahedra that contain the remaining cobalt. Assuming that tetrahedral-site cobalts are 4+ and octahedral-site cobalts are 3+, one arrives at the ideal composition of the 12H phase as $BaCoO_{2.67}$. Replacement of $BaO_3$ by $BaO_2$ layers leading to tetrahedral coordination of B-metal atoms is a novel way of accommodating oxygen deficiency in $ABO_3$ perovskites, which may be prevalent in other systems as well. $BaVO_{3-x}$ which crystallizes in a structure closely related to $Ba_3V_2O_8$ may belong to this category[78].

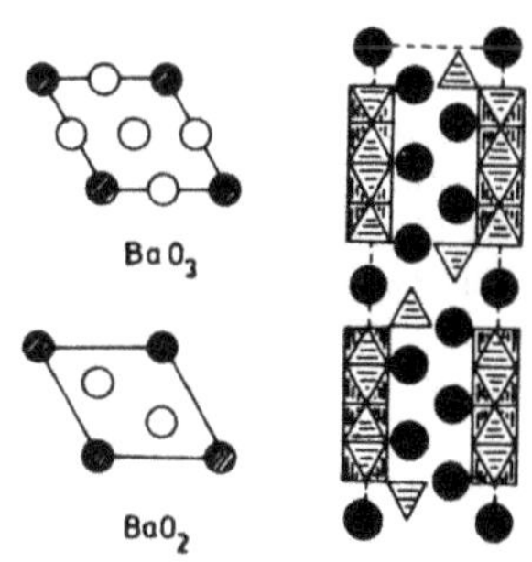


Fig. 12—Structure of $BaCoO_{2.61}$ (a) $BaO_3$ and $BaO_2$ layers and (b) (11.0) projection of the structure showing linkage of $CoO_4$ tetrahedra and $CoO_6$ octahedra (after ref. 77).

An interesting case[79] of a perovskite with high oxygen-deficiency ($x>0.5$) where the B-metal is present in three different coordination environments is $La_3Ba_3Cu_6O_{18-y}$. Phases with varying $y$ adopt a tetragonal cell related to the perovskite structure: $a_T \simeq \sqrt{2}a_c$ and $C_T \simeq 3a_c$ where $a_c$ is the cubic perovskite parameter. The detailed structure of $La_3Ba_3Cu_6O_{14.1}$ has been determined. Oxygen vacancies are ordered in the AO layers of the perovskite structure and the layer sequence in the ordered structure along [001] is $Cu_2O_4 - A_2O \quad - Cu_2O_4 - A_2 \quad _2 - Cu_2O_4 - A_2O$ .... This type of ordering of anion vacancies gives rise to three different coordination environments for copper: square-planar, square-pyramidal and distorted octahedral (Fig. 13). It has been suggested that $Cu^{3+}$ ions occupy octahedral sites in this structure. It is clear from the foregoing examples that vacancy-ordering in anion-deficient $BaMO_{3-x}$ or $Ba_{1-y}Sr_yMO_{3-x}$ systems where M = Mn, Fe, Co or Cu is dictated by the coordination preference of the transition metal ion.

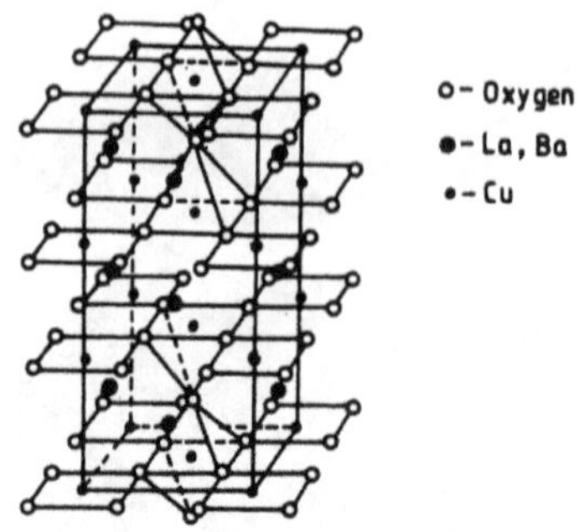


Fig. 13—Structure of $La_3Ba_3Cu_6O_{14.1}$ showing vacancy ordering and different coordination polyhedra around copper (after ref. 79).

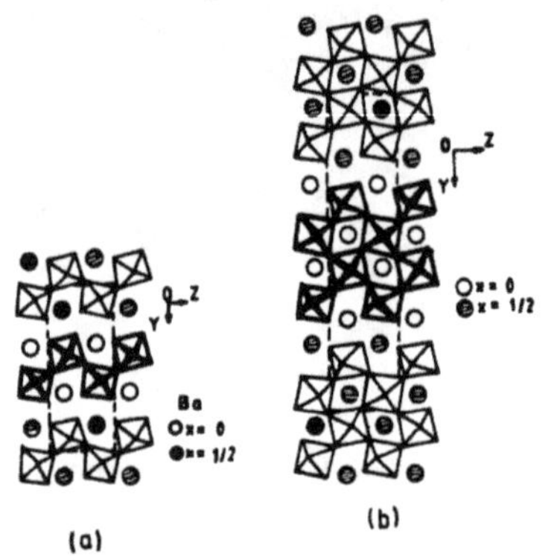


Fig. 14—Structure of $A_nB_nO_{3n+2}$ layered perovskites: (a) $n = 2$ member and (b) $n = 4$ member (after ref. 84).

### 5. Anion-excess nonstoichiometry

Anion-excess nonstoichiometry in perovskite oxides is not as common as anion-deficient nonstoichiometry probably because introduction of interstitial oxygen in perovskite structure is energetically unfavourable. There are a few systems which show apparent oxygen-excess nonstoichiometry: $LaMnO_{3+x}$, $Ba_{1-x}La_xTiO_{3+x/2}$ and $EuTiO_{3+x}$. Tofield and Scott[80] investigated oxidative nonstoichiometry in $LaMnO_{3+x}$ by neutron diffraction. These workers suggest three possible models to accommodate oxygen excess in the perovskite structure: (i) interstitial oxygen at ($\frac{1}{2}$00) face or ($\frac{1}{4}\frac{1}{4}\frac{1}{4}$) position of the cubic perovskite structure, (ii) cation vacancies at A and/or B sites leaving a perfect oxygen sublattice and (iii) formation of new oxidized phases. In $LaMnO_{3.12}$, diffraction results show that oxygen excess is accommodated by vacancies at A and B sites with partial elimination of La (as $La_2O_3$), the composition of the perovskite being $La_{0.94}\ _{0.06}Mn_{0.98}\ _{0.02}O_3$. Similar behaviour[81] is exhibited by $La_2TiCoO_{6.35}$, the composition being a mixture of a perovskite phase $La_{1.77}Ti^{4+}Co^{3+}_{0.7}Co^{2+}_{0.3}O_6$ and $La_2O_3$.

Where there is large oxygen excess, new phases possessing perovskite-related structures are formed. One of the ways of accommodating oxygen excess while retaining the features of the perovskite structure is to slice the cubic perovskite parallel to (110) face to give slabs of composition $(A_{n-1}B_nO_{3n+2})_\alpha$ and stack them one over the other by adding a layer of A-atoms in between. The process gives rise to a homologous series of general formula[11] $A_nB_nO_{3n+2}$ (Fig. 14). Important examples of this family are $Ca_2Nb_2O_7$ (ref. 82) and $La_2Ti_2O_7$ (ref. 83) which are $n = 4$ members in the series. They may be regarded as oxidation products of $CaNbO_3$ and $LaTiO_3$ perovskites. Members of this family with $n$-values ranging from 1 to 6 have been characterized[84].

### 6. Nonstoichiometry in oxides of structures related to perovskites

The first family of oxides that we consider here will be that related to the $K_2NiF_4$ structure. These oxides of the general formula $A_2BO_4$ contain alternating $ABO_3$ perovskite layers and AO rock-salt layers. Structural chemistry of these oxides has been reviewed recently[85]. Oxygen-excess nonstoichiometry is found in $La_2NiO_4$ where intergrowth phases containing $La_3Ni_2O_7$, $La_4Ni_3O_{10}$ etc. are found when the deviation is large[86]. Oxygen-deficient nonstoichiometry is found in $Ca_2MnO_4$ which can be reduced topotactically to $Ca_2MnO_{3.5}$ (Ref. 47). The unit cell dimensions are $a = 5.30$, $b = 10.05$ and $c = 12.24$Å, the relationship with the tetragonal $K_2NiF_4$ structure being $a \simeq \sqrt{2}a_T$, $b \simeq 2\sqrt{2}a_T$ and $c \simeq c_T$. The anion-vacancies are ordered along alternate $[110]_c$ rows as in $CaMnO_{2.5}$ to give sheets of $(MnO_5)$ square-pyramids. Vidyasagar *et al.*[46] have recently characterized an oxide of the formula $Ca_2FeO_{3.5}$ with $a = 14.79$, $b = 13.71$ and $c = 12.19$Å. The structure of this oxide is different from that of $Ca_2MnO_{3.5}$. It is likely that the ordering of anion vacancies in $Ca_2FeO_{3.5}$ is similar to that in the brownmillerite structure consisting of

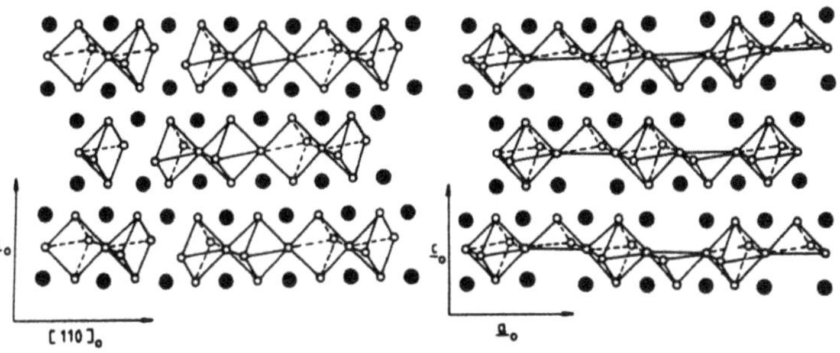

(a) (b)

Fig. 15—Vacancy ordering in (a) $Ca_2MnO_{3.5}$ and (b) $Ca_2FeO_{3.5}$.

alternating rows of tetrahedra and octahedra in the perovskite-like layers of the $K_2NiF_4$ structure (Fig. 15).

Oxide pyrochlores, $A_2B_2O_6O'$, can tolerate vacancies at the A and O′ sites giving phases $A_2B_2O_6\square$ ($ABO_3$) and $A\square B_2O_6\square$ ($AB_2O_6$)[87]. A small number of $A_2B_2O_6$ compounds adopt the pyrochlore structure in preference to the perovskite structure when the A and B ions are highly polarizable but not too electropositive. Typical of these oxides are $Tl_2B_2O_6$ (B = Nb, Ta or U) and $A_2Sb_2O_6$. In the A – Sb – O system[88] (A = Na, K or Ag), a series of non-stoichiometric compounds $ASb_yO_z$, with the maximum ratio of A : Sb depending on A, is known. Some of these compounds are superionic conductors. The defect pyrochlores of the composition $AB_2O_6$ will not be discussed here since they are not of direct relevance to this article.

## 7. Coherent intergrowth phases

In the last few years, the phenomenon of coherent intergrowth, wherein two or more phases possessing a common plane of nearly identical features intergrow randomly or in an ordered fashion, has emerged to become an important aspect of solid state inorganic chemistry. A variety of oxide systems belonging to the perovskite family show disordered intergrowth; disordered intergrowth of two or more phases necessarily involves nonstoichiometry if the parent phases have different compositions. Recurrent ordered intergrowth is found in some systems, $Ca_4YFe_5O_{13}$ being a typical example (Plate 5).

The $A_nB'_nO_{3n+2}$ family shows an extraordinarily large number of ordered intergrowth phases when *n* is between 4 and 4.5 in the Na – Ca – Nb – O system[84]; the oxide with $n = 4.5$ itself is a result of regular intergrowth between $n = 4$ and $n = 5$ compounds. When the *n* varies between 4.0 and 4.5 in these intergrowth phases, *x* in $ABO_{3+x}$ varies between 3.444 and 3.500, but the large number of ordered phases found in this narrow range is truly remarkable. A typical ordered intergrowth phase (with a medium periodicity in this family) has the sequence 444544445 44454444445 (where 4 and 5 represent compositions corresponding to $n = 4$ and $n = 5$. In Plate 8 the HREM image of a member of this family is shown to illustrate the ordered recurrent intergrowth. Unlike the $A_nB_nO_{3n+2}$ family, the $A_{n+1}B_{3n+1}$ family obtained by cutting the perovskite structure into slabs along (100) planes does not form ordered intergrowth phases; instead disordered intergrowth is commonly observed as týified by the Sr – Ti – O system[89]. In Plate 9 HREM image of $Sr_3Ti_2O_7$ containing lamellae of several other members of this series is shown. The La – Ni – O system also shows disordered intergrowth of different members of the $A_{n+1}B_nO_{3n+1}$ family. Aurivillius phases on the other hand form intergrowth phases of the general formula $(Bi_2O_2)^{2+}$ $(A_{n-1}B_nO_{3n+1})^{2-}$ $(Bi_2O_2)^{2+}$ (A

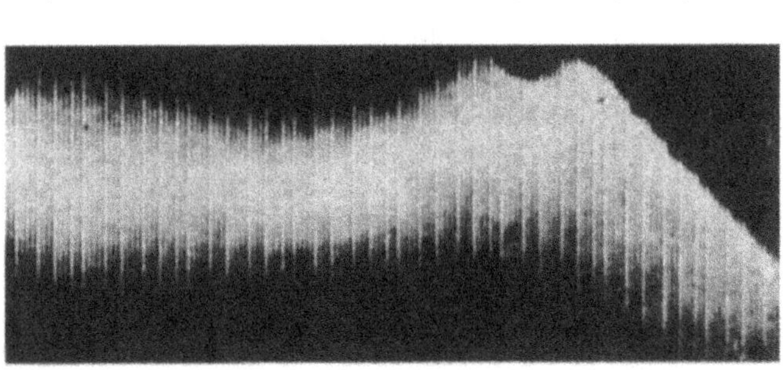

Plate 8—Lattice image of $A_nB_nO_{3n+2}$ oxides in the Na – Ca – Nb – O system showing ordered recurrent intergrowth (from ref. 84).

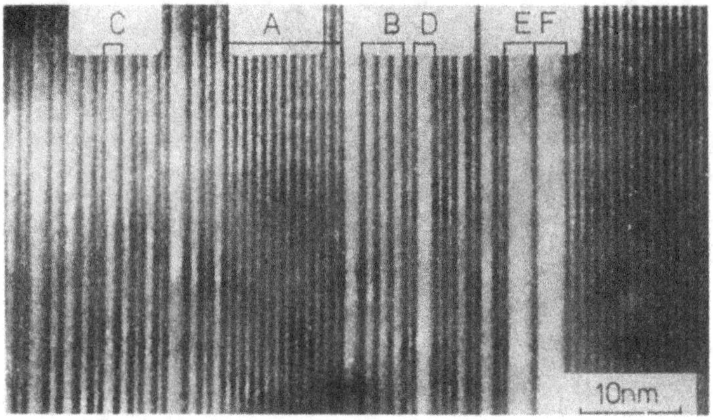


Plate 9—Lattice image of an oxide in the Sr – Ti – O system showing disordered intergrowth among different members of the $Sr_{n+1}Ti_nO_{3n+1}$ family. A, $n=2$; B, $n=3$; C, $n=4$; D, $n=5$; E, $n=7$; and F, $n=8$ (from ref. 89).

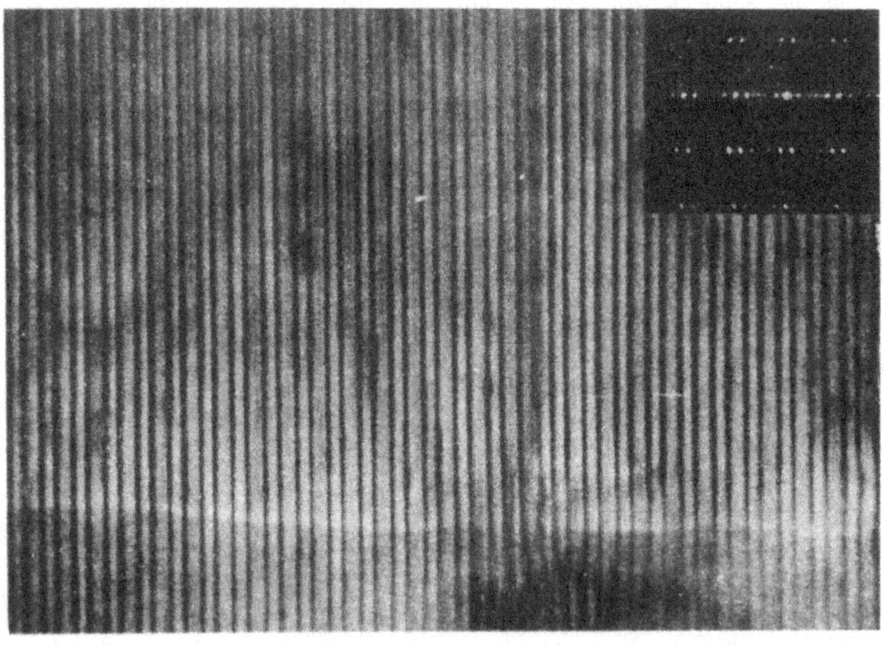

Plate 10—One-dimensional lattice image of $Bi_9Ti_6CrO_2$- showing ordered intergrowth between $n=3$ and $n=4$ members of the Aurivillius family.

$_{n}B_{n-1}O_{3n-4})^{2-}$. For example, $Bi_4Ti_3O_{12}$ ($n = 3$) forms a regular ordered intergrowth phase with $n = 4$ member $Bi_5Ti_3CrO_{15}$; in Plate 10 is shown a one-dimensional image of this compound to indicate the remarkable order in the intergrowth phase[90]. The image of a similar layered intergrowth phase between $Bi_4Ti_3O_{12}$ and $BaBi_4Ti_4O_{15}$ is shown in Plate 11. Long period Aurivillius phases show disordered intergrowth[91] as illustrated in Plate 12.

Barium ferrites of the general formula $M_pY_q$ formed between $BaFe_{12}O_{19}$ (M. phase) and $Ba_2Me_2Fe_{12}O_{22}$ with Me = Zn, Co, Ni etc. (Y-phase) may also be considered here since the structures of these oxides consist of layers of barium and oxygen in a ratio of 1 : 3 just as in the {111} sheet of a cubic barium perovskite. This family of intergrowth phases starting from $M_2Y$ to $M_8Y_{12}$ with $c$-dimension periodicities going upto several hundred angstroms is found within the narrow

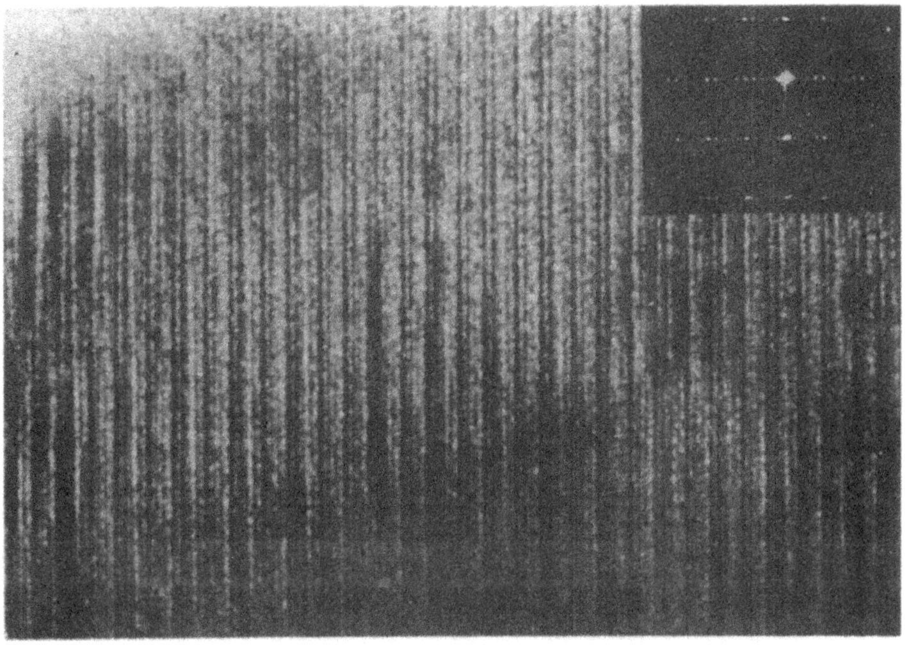

Plate 11—One-dimensional lattice image of $BaBi_8Ti$-$O_2$- showing ordered intergrowth of $n = 3$ and $n = 4$ members of the Aurivillius family.

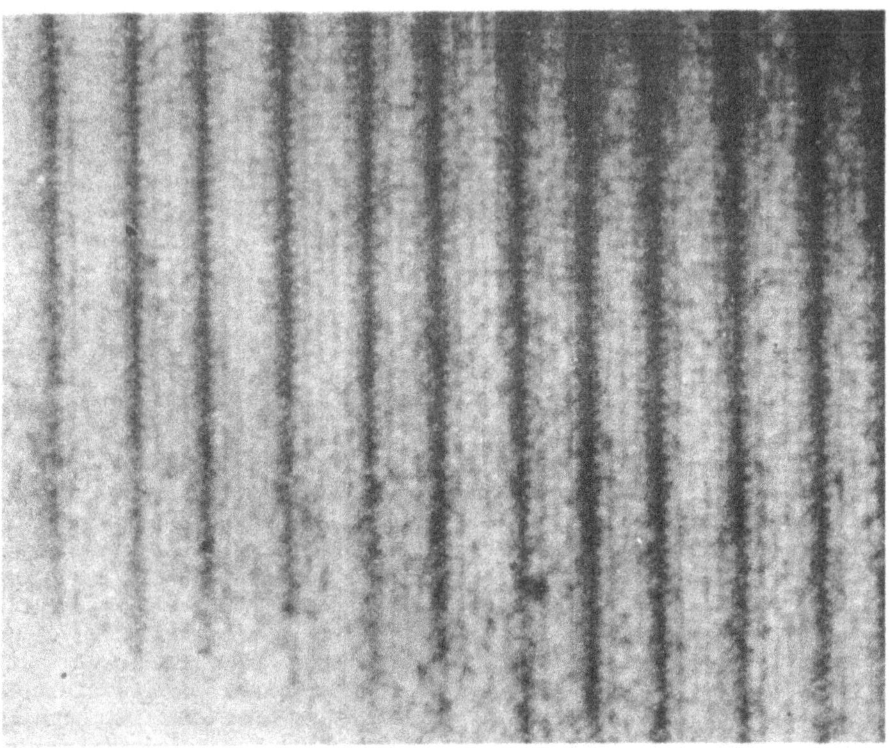

Plate 12—Disordered intergrowth of Aurivillius phases as illustrated in the case of $Bi_9Ti_3Fe_5O_2$- (from ref. 91).

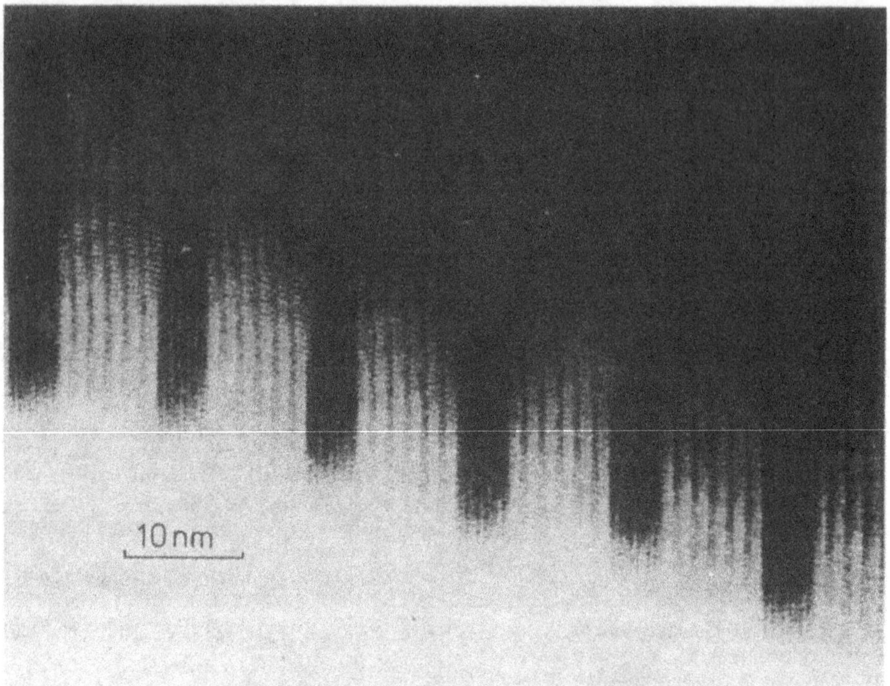


Plate 13—Lattice image of $MYMY_6$ intergrowth phase in the barium hexaferrite system (from ref. 92).

oxygen to metal ratio of 1.39179 and 1.42857. The large period intergrowth phases show considerable disorder. In Plate 13 we show the lattice image of a regular ordered intergrowth compound[92] belonging to this family.

It is truly remarkable that in none of the ordered or disordered intergrowth phases do we see any evidence of vacancies or other point defects. Nonstoichiometry is however present in all systems showing disordered intergrowth of two or more phases.

## 8. Concluding remarks

Defects in perovskite oxides can be due to cation vacancies (A or B site), anion vacancies or anion excess. Cation-deficient oxides such as $A_xWO_3$ give rise to oxide bronze structures, $WO_3$ itself representing the limiting case of the A-site deficient oxide; A-site vacancies are seldom ordered in these metallic systems. B-site vacancies are favoured in hexagonal perovskites and ordering of these vacancies gives rise to superstructures in some of the oxides.

Oxygen bacancies are commonly encountered in oxide perovskites. In $ABO_{3-x}$, unlike in $WO_{3-x}$ and $TiO_{2-x}$, crystallographic shear planes are not found. Instead, a variety of superstructures are seen due to the ordering of vacancies. The brownmillerite phase of $Ca_2Fe_2O_5$ and oxides of the $CaMnO_{3-x}$ family are good examples of such vacancy-ordered structures. Complex intergrowth phases (ordered as well as disordered) involving brownmillerite and other related phases are commonly found in some of the anion-deficient oxides. Some of them also show polytypism due to different modes of stacking of the hexagonal and cubic layers.

Anion excess is found in some perovskites, though not commonly. $LaMnO_{3+x}$ is an example of an anion-excess perovskite with cation vacancies, while $LaTiO_{3.5}$ is a case of anion-excess perovskite with a layered structure. Anion excess also results in the formation of new structures, an example being the $A_nB_nO_{3n+2}$ system of oxides.

Ordering of point defects such as vacancies does not appear to be favoured in oxides with itinerant electrons (e.g. $SrTiO_{3-x}$). It is not clear that this is universal since oxides such as TiO exhibit vacancy ordering.

In spite of several investigations of defect perovskite systems in the last few years, there is much scope for study in this interesting aspect of solid state inorganic chemistry. The nature of defects as well as the ordering of defects in many systems are yet to be studied. For example, not much is known about the defect structures of $LaNiO_3$ and $LaCoO_3$, both of which are interesting materials. Relations between defect structures and properties, especially catalytic propoerties, of perovskite oxides can only be established based on systematic investigations of both structure and reactivity.

**Acknowledgement**

The authors thank the UGC, New Delhi and the Department of Science and Technology, Government of India, for supporting this work.

**References**

1 Rao C N R, *Chemica Scripta*, **19** (1982) 124.
2 Kroger F A, *The chemistry of imperfect crystals* (North-Holland, Amsterdam) 1964.
3 Libowitz G G, *Prog Solid State Chem*, Vol. 2, edited by H Reiss (Pergamon Press, Oxford) 1965.
4 Anderson J S, *Proc Roy Soc London*, **185** (1946) 67.
5 Anderson J S, *Modern aspects of solid state chemistry*, edited by C N R Rao (Plenum Press, New York) 1970.
6 Magneli A, *Arkiv Kemi*, **2** (1950) 513; see also *Acta Cryst*, **6** (1953) 495.
7 Wadsley A D, *Non-stoichiometric compounds*, edited by L Mandelcorn (Academic Press, New York) 1964.
8 Anderson J S, *Defects and transport in oxides*, edited by M S Seltzer & R I Jaffe (Plenum Press, New York) 1974.
9 *The chemistry of extended defects in nonmetallic solids*, edited by L Eyring & M O'Keeffe (North-Holland, Amsterdam) 1970.
10 *Nonstoichiometric oxides*, edited by O T Sorensen (Academic Press, New York) 1981.
11 Tilley R J D, *Chemical physics of solids and their surfaces*, Vol. 8, edited by M W Roberts & J M Thomas (Royal Society of Chemistry, London) 1980.
12 Rao C N R & Subba Rao G V, *Physica Status Solid*, **A1** (1970) 597.
13 Goodenough J B & Longo J M, *Landolt-Bornstein Tabellen*, New Series, Group III/Vol. 4a (Springer Verlag, Berlin) 1970.
14 Glazer A M & Megaw H D, *Phil Mag*, **25** (1972) 1119.
15 Glazer A M, *Acta Cryst*, **B28** (1972) 3384.
16 Katz L & Ward R, *Inorg Chem*, **3** (1964) 205.
17 Ekstrom T & Tilley R J D, *Chemica Scripta*, **16** (1980) 1.
18 Atoji M & Rundle R E, *J chem Phys*, **32** (1960) 627.
19 Anderson J S, *Surface and defect properties of solids*, Vol. 1, edited by M W Roberts & J M Thomas (The Chemical Society, London) 1972.
20 Longo J M & Sleight A W, *Materials Res Bull*, **10** (1975) 1273.
21 Vincent H, Bochu B, Aubert J J, Joubert C C & Marezio M, *J Solid State Chem*, **24** (1978) 245.
22 Kihlborg L, *Chemica Scripta*, **14** (1978) 187.
23 (a) Jefferson D A, Uppal M K, Smith D J, Gopalakrishnan J, Ramanan A & Rao C N R, *Materials Res Bull*, (1984) in print.
(b) Ramanan A, Gopalakrishnan J, Uppal M K, Jefferson D A & Rao C N R, *Proc Roy Soc London*, under publication.
24 Sundberg M, *Chem Commun*, (Univ Stockholm), **5** (1981).
25 Rauser G & Kemmler-Sack S, *J Solid State Chem*, **33** (1980) 135.
26 Calvo C, Ng N N & Chamberland B L, *Inorg Chem*, **17** (1978) 699.
27 Poeppelmeier K R, Jacobson A J & Longo J M, *Materials Res Bull*, **15** (1980) 339.
28 Jacobson A J, Scanlon J C, Poeppelmeier K R, Longo J M & Cox D E, *Materials Res Bull*, **16** (1981) 359.
29 Galasso F & Katz L, *Acta Cryst*, **14** (1961) 647.
30 Shannon J & Katz L, *Acta Cryst*, **B26** (1970) 102.
31 Kemmler-Sack S, *Z Anorg Allgem Chem*, **454** (1979) 63.
32 Kemmler-Sack S, *Z Anorg Allgem Chem*, **457** (1979) 157.
33 Kemmler-Sack S, *Z Anorg Allgem Chem*, **461** (1980) 151.
34 Schittenheim H J, Fadini A & Kemmler-Sack S, *Z Anorg Allgem Chem*, **457** (1979) 149.
35 Kemmler-Sack S & Treiber U, *Z Anorg Allgem Chem*, **451** (1979) 129.
36 Kemmler-Sack S & Treiber U, *Z Anorg Allgem Chem*, **462** (1980) 166.
37 Pauling L, *The nature of the chemical bond* (Cornell Univ. Press, New York) (1960) 559.
38 Blasse G & Dirksen G J, *J Solid State Chem*, **36** (1981) 124.
39 Brown R & Kemmler-Sack S, *Naturwissenchaften*, **70** (1983) 463.
40 Kemmler-Sack S & Ehmann A, *Naturwissenchaften*, **70** (1983) 250.
41 Aurivillius B, *Ark Kemi*, **1** (1949) 463, 499; Ark. Kemi, **2** (1950) 519.
42 Jefferson D A, Gopalakrishnan J & Ramanan A, *Materials Res Bull*, **17** (1982) 269.
43 Catlow C R A & James R, *Chemical physics of solids and their surfaces*, Vol. 8, edited by M W Roberts & J M Thomas (The Royal Society of Chemistry, London) 1980, 108.
44 Berggren J, *Acta chem Scand*, **25** (1971) 3616.
45 Grenier J C, Pouchard M & Hagenmuller P, *Structure and bonding*, **47** (1981) 1.
46 Vidyasagar K, Gopalakrishnan J & Rao C N R, *Inorg Chem*, (1984) in press.
47 Poeppelmeier K R, Leonowicz M E & Longo J M, *J Solid State Chem*, **44** (1982) 89.
48 Poéppelmeier K R, Leonowicz M E, Scanlon J C, Longo J M & Yelon W B, *J Solid State Chem*, **45** (1982) 71.
49 Takeda T & Watanabe H, *J Phys Soc Japan*, **33** (1972) 973.
50 Grenier J C, Ghodbane S, Demazeau G, Pouchard M & Hagenmuller P, *Materials Res Bull*, **14** (1979) 831.
51 Taguchi H, Shimada M & Koizumi M, *J Solid State Chem*, **29** (1979) 221.
52 Alario-Franco M A & Regi M V, *Nature*, **270** (1977) 706.
53 Tofield B C, *Nature*, **272** (1978) 713.
54 MacChesney J B, Sherwood R C & Potter J F, *J chem Phys*, **43** (1965) 1907.
55 Tofield B C, Greaves C & Fender B E F, *Materials Res Bull*, **10** (1975) 737.
56 Alario-Franco M A, Joubert J C & Levy J P, *Materials Res Bull*, **17** (1982) 733.
57 Reller A, Jefferson D A, Thomas J M & Uppal M K, *J phys Chem*, **87** (1983) 913.
58 Reller A, Thomas J M, Jefferson D A & Uppal M K, *Proc Roy Soc*, (1984) in print.
59 Gai P L & Rao C N R, *Z Naturforsch*, **30A** (1975) 1092.
60 Crespin M, Levitz P & Gatineau L, *J chem Soc Faraday* II, **79** (1983) 1181.
61 Nakamura T, Petzow G & Gauckler L J, *Materials Res Bull*, **14** (1979) 649.
62 Grenier J C, Darriet J, Pouchard M & Hagenmuller P, *Materials Res Bull*, **11** (1976) 1219.
63 Bando Y, Sekikawa Y, Nakamara H & Matsui Y, *Acta Cryst*, **A37** (1981) 723.
64 Alario-Franco M A, Henche M J R, Regi M V, Calbet J M G, Grenier J C, Wattiaux A & Hagenmuller P, *J Solid State Chem*, **46** (1983) 23.
65 Alario-Franco M A, Calbet J M G, Regi M V & Grenier J C, *J Solid State Chem*, **49** (1983) 219.
66 Calbet J M G, Regi M V, Alario-Franco M A & Grenier J C, *Materials Res Bull*, **18** (1983) 285.

67 Gai P L & Rao C N R, *Pramana*, **5** (1975) 274.
68 Gai P L, Jacobson A J & Rao C N R, *Inorg Chem*, **15** (1976) 480.
69 Negas T & Roth R S, *J Solid State Chem*, **3** (1971) 323.
70 Mori S, *J phys Soc Japan*, **28** (1970) 44.
71 Zanne M & Gleitzer C, *Bull Soc Chim Fr*, (1971) 1567.
72 Jacobson A J, *Acta Cryst*, **B32** (1976) 1087.
73 Jacobson A J & Horrox A J W, *Acta Cryst*, **B32** (1976) 1003.
74 Hutchison J L & Jacobson A J, *J Solid State Chem*, **20** (1977) 417.
75 Zanne M, Courtois A & Gleitzer C, *Bull Soc Chim Fr*, (1972) 4470.
76 Jacobson A J & Hutchison J L, *Chem Commun*, (1976) 116.
77 Jacobson A J & Hutchison J L, *J Solid State Chem*, **35** (1980) 334.
78 Palanisamy T, Gopalakrishnan J & Sastri M V C, *Z Anorg Allgem Chem*, **415** (1975) 275.
79 Er-Rahko L, Michel C, Provost J & Raveau B, *J Solid State Chem*, **37** (1981) 151.
80 Tofield B C & Scott W R, *J Solid State Chem*, **10** (1974) 183.
81 Ramadass N, Gopalakrishnan J & Sastri M V C, *J inorg nucl Chem*, **40** (1978) 1453.
82 Hervieu M, Studer F & Raveau B, *J Solid State Chem*, **22** (1977) 273.
83 Scheunemann K & Muller-Buschbaum H K, *J inorg nucl Chem*, **37** (1975) 1879, 2261.
84 Portier R, Carpy A, Fayard M & Galy J, *Phys Stat Solidi*, **A30** (1975) 683.
85 Ganguly P & Rao C N R, *J Solid State Chem*, (1984) in print.
86 Drennan J, Travares C P & Steele B C H, *Materials Res Bull*, **17** (1982) 621.
87 Subramanian M A, Aravamudan G & Subba Rao G V, *Prog Solid State Chem*, **15** (1983) 55.
88 Piffard Y & Tournox M, *Acta Cryst*, **35B** (1979) 1450 and references therein.
89 Tilley R J D, *J Solid State Chem*, **21** (1977) 293.
90 Ramanan A, Gopalakrishnan J, Smith D J, Jefferson D A & Rao C N R, unpublished results.
91 Hutchison J L, Anderson J S & Rao C N R, *Proc Roy Soc London*, **355** (1977) 301.
92 Anderson J S & Hutchison J L, *Contemp Phys*, **16** (1975) 443.

Materials Science and Engineering, B18 (1993) 1–21

## Critical Review

# Chemical synthesis of solid inorganic materials*

C. N. R. Rao

*CSIR Centre of Excellence in Chemistry and Solid State and Structural Chemistry Unit, Indian Institute of Science, Bangalore 560 012 (India)*

(Received August 19, 1992; accepted in revised form September 1, 1992)

**Abstract**

Chemical methods of synthesis of materials play a crucial role in designing and discovering new materials and also in providing better and less cumbersome methods for preparing known materials. In this article, we shall discuss the chemical synthesis of inorganic solids, in particular oxidic materials. We shall first briefly examine the different classes of chemical reactions generally employed for synthesis and then discuss the various methods used along with several case studies and examples. In addition to the traditional ceramic method, the topics discussed include the combustion method (self-propagating high-temperature synthesis), the precursor method, topochemical routes, intercalation compounds, the ion-exchange method, the sol-gel process, the alkali-flux method, electrochemical methods, the pyrosol process and high pressure methods. The last topic includes hydrothermal synthesis of zeolitic materials. Intergrowth structures and superconducting cuprates are discussed in separate sections. It is hoped that the article will provide a useful survey of chemical methods of synthesis of inorganic materials and will serve as a ready reference to practitioners of the subject.

## 1. Introduction

There is much chemical ingenuity in the synthesis of solid materials [1–6] and this aspect of materials science is being increasingly recognized as a vital component of the subject. While tailor-making materials of the desired structure and properties remains the main goal of solid state chemistry and materials science, it is not always possible. One can evolve a rational approach to the synthesis of solids [7], but there is always an element of surprise which is encountered not uncommonly. A well known example of an oxide discovered serendipitously is $NaMo_4O_6$ containing condensed $Mo_6$ octahedral metal clusters [8]. This was discovered during attempts to prepare the lithium analogue of $NaZn_2Mo_3O_8$. Another serendipitous discovery was that of the phosphorus–tungsten bronze $Rb_xP_8W_{32}O_{112}$ formed by the reaction of phosphorus present in the silica of the ampoule during the preparation of rubidium–tungsten bronze [9]. Since the material could not be prepared in a platinum crucible, it was suspected that a constituent of the silica ampoule must have been incorporated. This discovery led to the synthesis of several phosphorus–tungsten bronzes of the type $A_xP_4O_8(WO_3)_{2m}$. Chevrel phases of the type $A_xMo_6S_8$ (A ≡ Cu, Pb, La etc.) were also discovered accidentally [10].

Rational synthesis of materials requires an understanding of the principles of crystal chemistry, and of thermodynamics, phase equilibria and reaction kinetics. There are many examples of rational synthesis. A good example is SIALON [11] where aluminium and oxygen were partly substituted for silicon and nitrogen in $Si_3N_4$. The fast $Na^+$ ion conductor NASICON, $Na_3Zr_2PSi_2O_{12}$, was synthesized with a clear understanding of the coordination preferences of cations and the nature of oxide networks formed by them [12]. The zero-expansion ceramic $Ca_{0.5}Ti_2P_3O_{12}$ possessing the NASICON framework was later synthesized based on the idea that the property of zero-expansion would be exhibited by two or three coordination polyhedra linked in such a manner as to leave substantial empty space in the network [7]. Another example of rational synthesis is that of silicate-based porous materials, making use of organic templates to predetermine the pore or cage geometries [13]. A microporous phosphate of the formula $(Me_4N)_{1.3}(H_3O)_{0.7}Mo_4O_8(PO_4)_2\cdot 2H_2O$ where the tetramethylammonium ions fill the voids in the three-

*Contribution No. 870 from Solid State & Structural Chemistry Unit.

0921-5107/93/$6.00

dimensional structure made up of $Mo_4O_8$ cubes and $PO_4$ tetrahedra, has been prepared in this manner [14].

A large variety of inorganic solid materials has been prepared in recent years by the traditional ceramic method, which involves mixing and grinding powders of the constituent oxides, carbonates and such compounds and heating them at high temperatures with intermediate grinding when necessary. A wide range of conditions, often bordering on the extreme, such as very high temperatures or pressures, very low oxygen fugacities and rapid quenching have been employed in materials synthesis. The low-temperature chemical routes, however, are of greater interest. The trend nowadays is to avoid brute-force methods in order to have better control of the structure, stoichiometry and phase purity. Noteworthy chemical methods of synthesis include the precursor method, coprecipitation and soft-chemistry routes, the combustion method, the sol-gel method, topochemical methods and high-pressure methods. In this article, we shall discuss the synthesis of inorganic solids by chemical methods with several examples, especially oxide materials including superconducting cuprates synthesized by these means. For the purpose of brevity we shall cite only some of the key references which guide the reader to the original literature.

## 2. Common reactions encountered in the synthesis of inorganic solids

Various types of chemical reactions have been used for the synthesis of solid materials. Corbett [3] has written an excellent article on the subject. Some of the common reactions employed for the synthesis of inorganic solids are listed below:

(1) decomposition

$$A(s) \rightarrow B(s) + C(g)$$

(2) combination

$$A(s) + B(g) \rightarrow C(s)$$

(3) metathetic (combining (1) and (2) above)

$$A(s) + B(g) \rightarrow C(s) + D(g)$$

(4) addition

$$A(s) + B(s) \rightarrow C(s)$$

$$A(s) + B(l) \rightarrow C(s)$$

$$A(g) + B(g) \rightarrow C(s)$$

(5) exchange

$$AX(s) + BY(s) \rightarrow AY(s) + BX(s)$$

$$AX(s) + BY(g) \rightarrow AY(s) + BX(g)$$

Typical examples of the above simple reactions include the following:

(1) $CaCO_3(s) \rightarrow CaO(s) + CO_2(g)$

$M_mO_n(s) \rightarrow M_mO_{n-\delta}(s) + \frac{\delta}{2} O_2(g)$ (M = metal)

(2) $YBa_2Cu_3O_6(s) + O_2(g) \rightarrow YBa_2Cu_3O_7(s)$

(3) $Pr_6O_{11}(s) + 2H_2(g) \rightarrow 3Pr_2O_3(s) + 2H_2O(g)$

(4) $ZnO(s) + Fe_2O_3(s) \rightarrow ZnFe_2O_4(s)$

$BaO(s) + TiO_2(s) \rightarrow BaTiO_3(s)$

$3SiCl_4(g) + 4NH_3(g) \rightarrow Si_3N_4(s) + 12HCl(g)$

$GaMe_3(g) + 4NH_3(g) \rightarrow GaAs(s) + 3CH_4(g)$

(5) $ZnS(s) + CdO(s) \rightarrow CdS(s) + ZnO(s)$

$MnCl_2(s) + 2HBr \rightarrow MnBr_2(s) + 2HCl$

More complex reactions involving more than one type of reaction are also commonly employed in solid state synthesis. For example, in the preparation of complex oxides it is common to carry out thermal decomposition of a compound followed by oxidation (in air or $O_2$) essentially in one step:

$$2Ca_{0.5}Mn_{0.5}CO_3(s) + \tfrac{1}{2}O_2(g) \rightarrow CaMnO_3(s) + 2CO_2(g)$$

Vapour phase reactions also yield solid products in many instances. In chemical vapour transport reactions, a gaseous reagent acts as a carrier to transport a solid by transforming it into the vapour state:

$$MgO(s) + Cr_2O_3(s) \xrightarrow{O_2} MgCr_2O_4(s)$$

Typical transport reaction equilibria include the following:

$$ZnS(s) + I_2 \rightleftharpoons ZnI_2 + \tfrac{1}{2}S_2$$

$$TaOCl_2(s) + TaCl_5 \rightleftharpoons TaOCl_3 + TaCl_4$$

$$Nb_2O_5(s) + 3NbCl_5 \rightleftharpoons 5NbOCl_3$$

$$GaAs(s) + HCl \rightleftharpoons GaCl + \tfrac{1}{2}H_2 + As$$

In the last reaction, the starting reactants would be $AsCl_3$, gallium and $H_2$. Table 1 lists several examples of chemical transport systems.

Specific reagents and reaction conditions are employed to carry out various processes such as reduction, oxidation and halogenation in the synthesis of solids. For example, reduction of oxides is carried out in an atmosphere of (flowing) pure or dilute hydrogen (*e.g.* $N_2$–$H_2$ mixtures) or sometimes in an atmosphere of CO or CO–$CO_2$ mixtures. Reduction of oxides for the purpose of lowering the oxygen content is also achieved by heating oxides in argon or nitrogen; appli-

TABLE 1. Examples of chemical transport

| Solid | Transporting agent | Solid | Transporting agent |
|---|---|---|---|
| $Nb_2O_5$ | $Cl_2$, $NbCl_5$ | CrOCl | $Cl_2$ |
| $TiO_2$ | $I_2+S_2$ | $FeWO_4$ | $Cl_2$ |
| $IrO_2$ | $O_2$ | $MgFe_2O_4$ | HCl |
| $WO_3$ | $H_2O$ | $CaNb_2O_6$ | $Cl_2$, HCl |
| $NbS_2$ | S | ZrOS | $I_2$ |
| $TaS_3$ | S | $LaTe_2$ | $I_2$ |
| $MnGeO_3$ | HCl | $V_nO_{2n-1}$ | $TeCl_4$ |
| $MgTiO_3$ | $Cl_2$ | $NbS_2Cl_2$ | $NbCl_5$ |

cation of vacuum at an appropriate temperature (vacuum annealing or decomposition at low pressures) is also used. The obvious means of reducing solid compounds by hydrogen is employed not only for reducing oxides, but also halides and other compounds. Thermal decomposition of metal halides also yields lower halides:

$$M_2O_3(s)+H_2(g) \longrightarrow 2MO(s)+H_2O(g) \quad (e.g. M\equiv Fe)$$

$$ABO_3(s)+H_2(g) \longrightarrow ABO_{2.5}(s)+\tfrac{1}{2}H_2O(g) \quad (e.g.\ LaCoO_3)$$

$$MCl_3(s)+H_2(g) \longrightarrow MCl_2(s)+HCl(g) \quad (e.g.\ M\equiv Fe)$$

$$MX_3 \xrightarrow{heat} MX_2+\tfrac{1}{2}X_2 \quad (e.g.\ M\equiv Cr)$$

Reduction of oxides can be accomplished by reacting with elemental carbon or with a metal. Reduction of halides is also carried out by metals:

$$2MCl_3+M \longrightarrow 3MCl_2 \quad (e.g.\ M\equiv Nd, Fe)$$

$$3MCl_4+M'(s) \longrightarrow 3MCl_3(s)+M'Cl_3(g) \quad (e.g.\ M\equiv Hf, M'\equiv Al)$$

$$M_2O_5+3M \longrightarrow 5MO \quad (e.g.\ M\equiv Nb)$$

Metal oxychlorides are obtained by heating oxides with $Cl_2$ (LaOCl from $La_2O_3$). Fluorination is generally carried out using elemental fluorine. There are examples where oxides are reacted with a fluoride such as $BaF_2$ to attain partial fluorination.

## 3. Ceramic method

The most common method of preparing solid materials is by reaction of the component materials in the solid state at elevated temperatures. Several oxides, sulphides, phosphides, etc., have been prepared by this method. Knowledge of the phase diagram is generally helpful in fixing the desired composition and conditions for synthesis. Some caution is necessary in choosing the container; platinum, silica and alumina containers are generally used for the synthesis of metal oxides, while graphite containers are employed for sulphides and other chalcogenides as well as pnictides. If one of the constituents is volatile or sensitive to the atmosphere, the reaction is carried out in sealed evacuated capsules. Most ceramic preparations require relatively high temperatures which are generally attained by resistance heating. Electric arc and skull techniques give temperatures up to 3300 K while high-power $CO_2$ lasers give temperatures up to 4300 K.

The ceramic method suffers from several disadvantages. When no melt is formed during the reaction, the entire reaction has to occur in the solid state, initially by a phase boundary reaction at the points of contact between the components and later by diffusion of the constituents through the product phase. As the reaction progresses, diffusion paths become increasingly longer and the reaction rate slower. The product interface between the reacting particles acts as a barrier. The reaction can be speeded up to some extent by intermittent grinding between heating cycles. There is no simple way of monitoring the progress of the reaction in the ceramic method. It is only by trial and error (by carrying out X-ray diffraction and other measurements periodically) that one decides on appropriate conditions that lead to completion of the reaction. Because of this difficulty, one frequently ends up with mixtures of reactants and products. Separation of the desired product from these mixtures is generally difficult, if not impossible. It is sometimes difficult to obtain a compositionally homogeneous product by the ceramic technique, even when the reaction proceeds almost to completion.

In spite of such limitations, ceramic techniques have been widely used for the synthesis of solid materials. Mention must be made, among others, of the use of this technique for the synthesis of rare earth monochalcogenides such as SmS and SmSe. The method involves heating the elements, first at lower temperatures (870–1170 K) in evacuated silica tubes; the contents are then homogenized, sealed in tantalum tubes and heated to around 2300 K by passing a high current through the tube [15].

Various modifications of the ceramic technique have been employed to overcome some of the limitations. One of these relates to decreasing the diffusion path lengths. In a polycrystalline mixture of reactants, the individual particles are approximately 10 $\mu$m in size, representing diffusion distances of roughly 10 000 unit cells. By using freeze-drying, spray-drying, coprecipitation, and sol-gel and other techniques, it is possible to reduce the particle size to a few hundred ångströms and thus effect a more intimate mixing of the reactants. In spray-drying, suitable constituents dissolved in a solvent are sprayed in the form of fine droplets into a hot chamber. The solvent evaporates instantaneously, leaving behind an intimate mixture of reactants, which on heating at elevated temperatures gives the product. In freeze-drying, reactants in a common solvent are frozen by immersing in liquid nitrogen and the solvent is removed at low pressures. In coprecipitation, the required metal cations taken as soluble salts (*e.g.* nitrates) are coprecipitated from a common medium, usually as hydroxides, carbonates, oxalates, formates or citrates. In actual practice, one takes oxides or carbonates of the relevant metals, digests them with an acid (usually $HNO_3$) and then the precipitating reagent is added to the solution. The precipitate obtained after drying is heated to the required temperature in a desired atmosphere to produce the final product. For example, tetraethylammonium oxalate has been used to prepare superconducting $YBa_2Cu_4O_8$. The decomposition temperatures of the precipitates are generally lower than the temperatures employed in the ceramic method.

## 4. Combustion method

Combustion synthesis or self-propagating high-temperature synthesis is a versatile method used for the synthesis of a variety of solids. The method makes use of a highly exothermic reaction between the reactants to produce a flame due to spontaneous combustion which then yields the desired product or its precursor in finely divided form. Borides, carbides, oxides, chalcogenides and other metal derivatives have been prepared by this method and the topic has been reviewed recently by Merzhanov [16]. In order for combustion to occur, one has to ensure that the initial mixture of reactants is highly dispersed and has high chemical energy. For example, one may add a fuel and an oxidizer when preparing oxides by the combustion method, both these additives being removed during combustion to yield only the product or its precursor. Thus, one can take a mixture of nitrates (oxidizer) of the desired metals along with a fuel (*e.g.* hydrazine, glycine or urea) in solution, evaporate the solution to dryness and heat the resulting solid to around 423 K to obtain spontaneous combustion, yielding an oxidic product in fine particulate form. Even if the desired product is not formed immediately after combustion, the fine particulate nature of the product facilitates its formation on further heating.

In order to carry out combustion synthesis, the powdered mixture of reactants (0.1–100 μm particle size) is generally placed in an appropriate gas medium which favours an exothermic reaction on ignition. The combustion temperature is anywhere between 1500 and 3500 K, depending on the reaction. Reaction times are very short since the desired product results soon after the combustion. A gas medium is not always necessary. This is so in the synthesis of borides, silicides and carbides where the elements are quite stable at high temperatures (*e.g.* $Ti + 2B \rightarrow TiB_2$). Combustion in a nitrogen atmosphere yields nitrides. Nitrides of various metals have been prepared in this manner. Azides have been used as sources of nitrogen.

The following are some typical combustion reactions:

$$MoO_3 + 2SiO_2 + 7Mg \longrightarrow MoSi_2 + 7MgO$$

$$WO_3 + C + 2Al \longrightarrow WC + Al_2O_3$$

$$TiO_2 + B_2O_3 + 5Mg \longrightarrow TiB_2 + 5MgO$$

$$Ta \xrightarrow{N_2} Ta_2N \xrightarrow[\text{after burning}]{N_2} TaN$$

Recently, $MoS_2$ and other refractories have been prepared starting from halides [17].

Use of the combustion method in an atmosphere of air or oxygen to prepare complex metal oxides seems obvious. In the last three to four years, a variety of oxides have been prepared using nitrate mixtures with a fuel such as glycine or urea. It seems that almost any ternary or quaternary oxide can be prepared by this method. All the superconducting cuprates have been prepared by this method, although the resulting products in fine particulate form have to be heated to an appropriate high temperature in a desired atmosphere to obtain the final cuprate [18]. Table 2 lists typical materials prepared by the combustion method.

## 5. Precursor method

As mentioned earlier, diffusion distances for the reacting cations are rather large in the ceramic method. Diffusion distances are markedly reduced by incorporating the cations in the same solid precursor. Synthesis

TABLE 2. Typical materials prepared by the combustion method

| | |
|---|---|
| Oxides | $BaTiO_3$, $LiNbO_3$, $PbMoO_4$, $Bi_4Ti_3O_{12}$, $BaFe_{12}O_{19}$, $YBa_2Cu_3O_7$ |
| Carbides | TiC, $Mo_2C$, NbC |
| Borides | $TiB_2$, $CrB_2$, $MoB_2$, FeB |
| Silicides | $MoSi_2$, $TiSi_2$, $ZrSi_2$ |
| Phosphides | NbP, MnP, TiP |
| Chalcogenides | $WS_2$, $MoS_2$, $MoSe_2$, $TaS_2$, $LaTaS_3$ |
| Hydrides | $TiH_2$, $NdH_2$ |

of complex oxides by the decomposition of compound precursors has been known for some time. For example, thermal decomposition of $LaCo(CN)_6{\cdot}5H_2O$ and $LaFe(CN)_6{\cdot}6H_2O$ in air readily yields $LaCoO_3$ and $LaFeO_3$ respectively. $BaTiO_3$ can be prepared by the thermal decomposition of $Ba[TiO(C_2O_4)_2]$, while $LiCrO_2$ can be prepared from $Li[Cr(C_2O_4)_2(H_2O)_2]$. Ferrite spinels of the general formula $MFe_2O_4$ ($M \equiv$ Mg, Mn, Ni, Co) are prepared by the thermal decomposition of acetate precursors of the type $M_3Fe_6(CH_3COO)_{17}O_3OH{\cdot}12C_5H_5N$. Chromites of type $MCr_2O_4$ are obtained by the decomposition of $(NH_4)_2M(CrO_4)_2{\cdot}6H_2O$.

Carbonates of metals such as calcium, magnesium, manganese, iron, cobalt, zinc and cadmium are all isostructural, possessing the calcite structure. We can therefore prepare a large number of carbonate solid solutions containing two or more cations in different proportions [19, 20]. The rhombohedral $a_R$ of the carbonate solid solutions varies systematically with the weighted mean cation radius (Fig. 1). Carbonate solid solutions are ideal precursors for the synthesis of monoxide solid solutions of rock-salt structure. For example, the carbonates are decomposed in vacuum or in flowing dry nitrogen, to obtain monoxides of the type $Mn_{1-x}M_xO$ ($M \equiv$ Mg, Ca, Co or Cd) of rock-salt structure. Oxide solid solutions of $M \equiv$ Mg, Ca and Co would require temperatures of 770–970 K for formation, while those containing cadmium are formed at still lower temperatures. The facile formation of rock-salt oxides by the decomposition of carbonates of calcite structure is due to the close (possible topotactic) relationship between the structures of calcite and rock-salt. The monoxide solid solutions can be used as precursors for preparing spinels and other complex oxides.

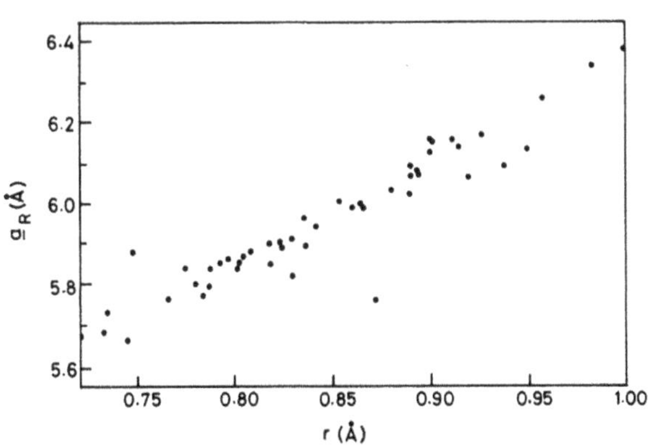


Fig. 1. Plot of the rhombohedral lattice parameters $a_R$ of a variety of binary and ternary carbonates of calcite structure (e.g. Ca–M, Ca–M–M, Mg–M, M–M where M,M ≡ Mn, Fe, Co, Cd, etc.) against the mean cation radius.

Besides monoxide solid solutions, a number of ternary and quaternary oxides of novel structures can be prepared by decomposing carbonate precursors containing the different cations in the required proportions. Thus, one can prepare $Ca_2Fe_2O_5$ and $CaFe_2O_4$ by heating the corresponding carbonate solid solutions in air at 1070 and 1270 K respectively for about 1 h. $CaFe_2O_5$ is a defect perovskite with ordered oxide ion vacancies and has the well known brown-millerite structure (Fig. 2) with the $Fe^{3+}$ ions in alternate octahedral (O) and tetrahedral (T) sites. Two new oxides of similar compositions, $Ca_2Co_2O_5$ and $CaCo_2O_4$, have been prepared by decomposing the appropriate carbonate precursors in oxygen atmosphere around 940 K. Unlike $Ca_2Fe_2O_5$, in $Ca_2Mn_2O_5$, anion-vacancy ordering in the perovskite structure gives a square-pyramidal coordination (SP) around the transition metal ion (Fig. 2). One can also synthesize quaternary oxides, $Ca_2FeCoO_5$, $Ca_2Fe_{1.6}Mn_{0.4}O_5$, $Ca_3Fe_2MnO_8$ etc. belonging to the $A_nB_nO_{3n-1}$ family, by the carbonate precursor route. In the Ca–Fe–O system, there are several other oxides such as $CaFe_4O_7$, $CaFe_{12}O_{19}$ and $CaFe_2O_4(FeO)_n$ ($n = 1, 2, 3$) which can, in principle, be synthesized by starting from the appropriate carbonate solid solutions and decomposing them in a proper atmosphere.

A good example of multistep solid state synthesis achieved starting from carbonate solid solution precursors is provided by the $Ca_2Fe_{2-x}Mn_xO_5$ series of oxides. The structures of both the end members, $Ca_2Fe_2O_5$ and $Ca_2Mn_2O_5$, are derived from that of the perovskite (Fig. 2). Solid solutions between the two oxides would be expected to show oxygen vacancy

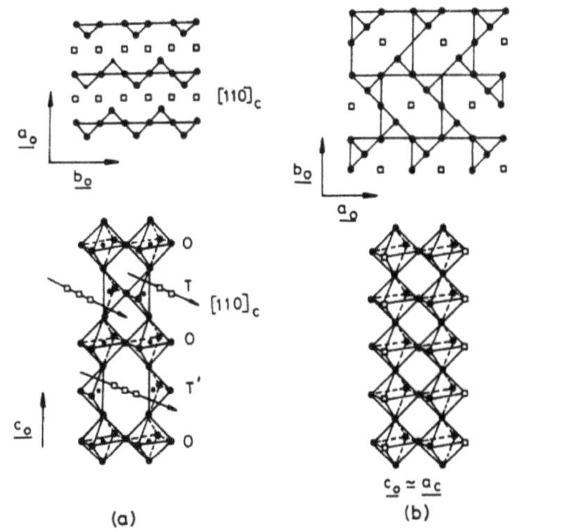


Fig.2. Structures of (a) $Ca_2Fe_2O_5$ (brown-millerite) and (b) $Ca_2Mn_2O_5$. Oxygen vacancy ordering in the $a$–$b$ plane is also shown.

ordered superstructures with $Fe^{3+}$ in octahedral (O) and tetrahedral (T) coordinations and $Mn^{3+}$ in square-pyramidal (SP) coordination, but they cannot be prepared by the ceramic method. These solid solutions have indeed been prepared starting from the carbonate solid solutions, $Ca_2Fe_{2-x}Mn_x(CO_3)_4$. The carbonates decompose in air at around 1200–1350 K to give perovskite-like oxides, $Ca_2Fe_{2-x}Mn_xO_{6-y}$ ($y<1$). The compositions of the perovskites obtained with $x=2/3$ and 1 are $Ca_3Fe_2MnO_8$ and $Ca_3Fe_{1.5}Mn_{1.5}O_{8.1}$. X-ray and electron diffraction patterns show that they are members of the $A_nB_nO_{3n-1}$ homologous series with anion-vacancy ordered superstructures with $n=3$ ($A_3B_3O_{8+x}$). Careful reduction of $Ca_3Fe_2MnO_8$ in dilute hydrogen gives $Ca_3Fe_{4/3}Mn_{2/3}O_5 = Ca_3Fe_2MnO_{7.5}$ (Fig. 3). During this step, only $Mn^{4+}$ in the parent oxides is topochemically reduced to $Mn^{3+}$, and $Fe^{3+}$ remains unreduced. The most probable superstructure of $Ca_3Fe_2MnO_{7.5}$ involves SP, O and T polyhedra along the *b*-direction. On heating in vacuum at 1140 K, however, it transforms to the more stable brown-millerite structure with only O and T coordinations. In Fig. 4 we show typical oxides prepared from precursor solid solutions to illustrate the usefulness of the method.

A number of ternary and quaternary metal oxides of perovskite and related structures can be prepared by employing hydroxide, nitrate and cyanide solid solution precursors as well [20]. For example, hydroxide solid solutions of the general formula $Ln_{1-x}M_x(OH)_3$ where Ln ≡ La or Nd and M ≡ Al, Cr, Fe, Co or Ni) and $La_{1-x-y}M'_xM''_y(OH)_3$ where (where M' ≡ Ni and M'' ≡ Co or Cu) crystallizing in the rare earth trihydroxide structure are decomposed at relatively low temperatures (around 870 K) to yield $LaNiO_3$, $NdNiO_3$, $LaNi_{1-x}Co_xO_3$, $LaNi_{1-x}Cu_xO_3$ etc.

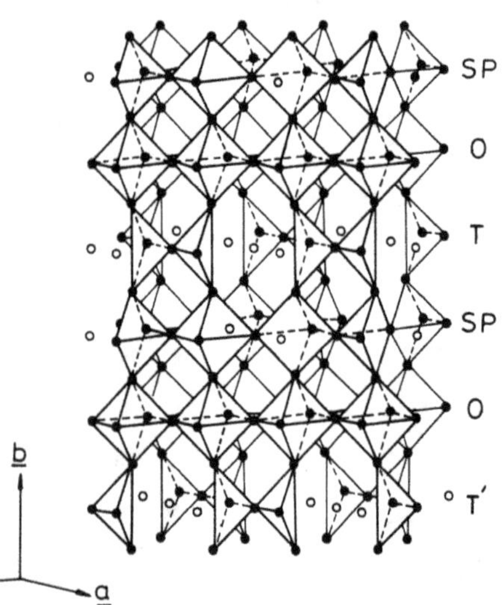


Fig. 3. $Ca_3Fe_2MnO_{7.5}$ obtained by the topotactic reduction of $Ca_3Fe_2MnO_8$. The latter is prepared by decomposition of the precursor carbonate, $Ca_2Fe_{4/3}Mn_{2/3}(CO_3)_4$.

Making use of the fact that anhydrous alkaline earth nitrates $A(NO_3)_2$ (A ≡ Ca, Sr, Ba) and $Pb(NO_3)_2$ are isostructural, nitrate solid solutions of the formula $A_{1-x}Pb_x(NO_3)_2$ have been used as precursors for the preparation of ternary oxides such as $BaPbO_3$, $Ba_2PbO_4$, and $Sr_2PbO_4$. Quaternary oxides of type $LaFe_{0.5}Co_{0.5}O_3$ and $La_{0.5}Nd_{0.5}CoO_3$ which cannot be readily prepared by the ceramic method have been obtained by the decomposition of $LaFe_{0.5}Co_{0.5}(CN)_6{\cdot}5H_2O$ and $La_{0.5}Nd_{0.5}Co(CN)_6{\cdot}5H_2O$ respectively [20]. A hyponitrite precursor has been used to prepare superconducting $YBa_3Cu_3O_7$ free from $BaCO_3$ impurity [21].

Chevrel compounds of general formula $A_xMO_6S_8$ with A ≡ Cu, Pb, La etc. (Fig. 5) are generally prepared by the ceramic method. A novel precursor compound has been employed [22] to obtain these compounds by a one-step reduction as given by the reaction:

$$2A_x(NH_4)_yMo_3S_9 + 10H_2 \rightarrow A_{2x}Mo_6S_8 + 10H_2S + 2yNH_3 + yH_2$$

Ammonium thiomolybdate, $(NH_4)_2MoS_4$, was reacted with the metal chloride ($AX_n$) to obtain the precursor compound.

Precursor solid solutions or compounds can be used to prepare metal alloys. Thus, Mo–W alloys have been prepared [23] by the hydrogen reduction of $(NH_4)_6[Mo_{7-x}W_xO_{24}]$. Metal alloys can be used as precursors to obtain the desired oxides on treatment with oxygen under appropriate conditions. For example, an Eu–Ba–Cu alloy has been oxidized at 1170 K to obtain superconducting $EuBa_2Cu_3O_7$ [24].

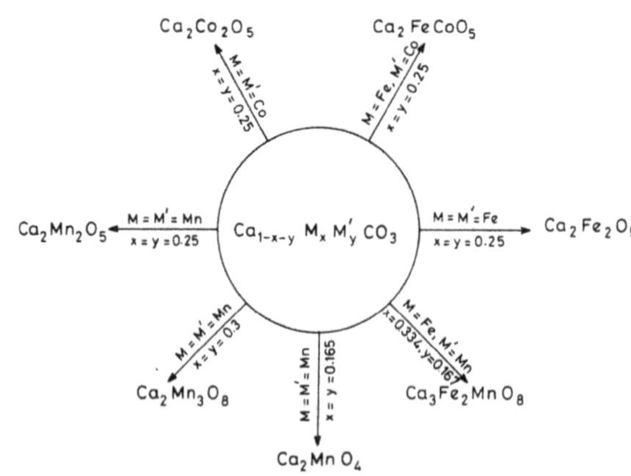


Fig. 4. Some of the complex oxides prepared by the decomposition of carbonate precursors.

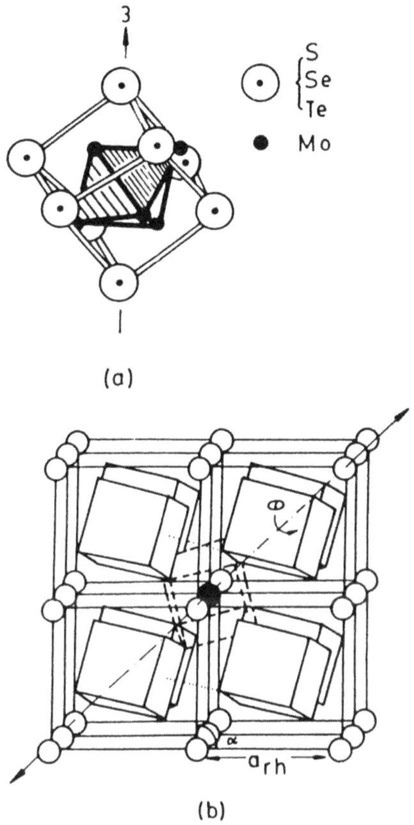


Fig. 5. Structure of Chevrel compounds, $A_xMo_6Ch_8$ (Ch ≡ S, Se, Te): (a) $Mo_6Ch_8$ building block, (b) structure of Chevrel compounds where the A cation is shown by the closed circle.

Organometallic precursors are used in the synthesis of a variety of semiconducting materials and superconducting cuprates. Organoaluminium silicate precursors have been used for the synthesis of aluminosilicates [25], while polymeric methylsilylamines have been used to obtain $SiC$–$Si_3N_4$ fibres [26]. Silicon nitride can also be made using organometallic precursors.

## 6. Topochemical reactions

A solid state reaction is said to be topochemically controlled when the reactivity is controlled by the crystal structure rather than by the chemical nature of the constituents. The products obtained in many solid state decompositions are determined by topochemical factors, especially when the reaction occurs within the solid without the separation of a new phase [27, 28]. In topotactic solid state reactions, the atomic arrangement in the reactant crystal remains largely unaffected during the course of the reaction, except for changes in dimension in one or more directions. Dehydration of $WO_3 \cdot H_2O$ or $MoO_3 \cdot H_2O$ to give $WO_3$ or $MoO_3$ is one such reaction [29]. Dehydration of many other hydrates such as $VOPO_4 \cdot 2H_2O$ and $HMoO_2PO_4 \cdot H_2O$ is also found to be topotactic. $\gamma$-FeOOH transforms to $\gamma$-$Fe_2O_3$ by treatment with an organic base. Intercalation reactions are generally topotactic in nature. Decomposition of $WO_3$, $MoO_3$ or $TiO_2$ to give lower mixed valent oxides (Magneli phases) is accommodated by the collapse of the structure in specific crystallographic directions. Decomposition of $V_2O_5$ to form $V_6O_{13}$ is a similar reaction. The reduction of NiO to nickel metal proceeds topochemically [30].

### 6.1. Dehydration of $Mo_{1-x}W_xO_3 \cdot H_2O$

Many of the modern developments in solid state chemistry owe much to the investigations carried out on $MoO_3$ and $WO_3$, the crystallographic shear planes being a major discovery. $WO_3$ crystallizes in an $ReO_3$-like structure, but $MoO_3$ possesses a layered structure (Fig. 6). $MoO_3$ can be stabilized in the $WO_3$ structure by partly substituting tungsten for molybdenum. $Mo_{1-x}W_xO_3$ solid solutions can be prepared by the ceramic method (by heating $MoO_3$ and $WO_3$ in sealed tubes at around 870 K) or by the thermal decomposition of mixed ammonium metallates. These methods, however, do not always yield monophasic products owing to the difference in volatilities of $MoO_3$ and $WO_3$. We therefore sought to prepare $Mo_{1-x}W_xO_3$ by topochemical dehydration of the hydrates [31], the process being very gentle. $MoO_3 \cdot H_2O$ and $WO_3 \cdot H_2O$ are isostructural and the solid solutions between the two hydrates are prepared readily by adding a solution of $MoO_3$ and $WO_3$ in ammonia to hot 6 M $HNO_3$. The hydrates $Mo_{1-x}W_xO_3 \cdot H_2O$ crystallize in the same structure as $MoO_3 \cdot H_2O$ and $WO_3 \cdot H_2O$ with a monoclinic unit cell. The hydrate solid solutions undergo dehydration under mild conditions (around 500 K) yielding $Mo_{1-x}W_xO_3$ which crystallizes in the $ReO_3$-related structure of $WO_3$. The dehydration of these hydrates has been studied by *in situ* electron diffraction where the decomposition occurs owing to beam heating [31]. Electron diffraction patterns clearly show how $WO_3 \cdot H_2O$ transforms to $WO_3$ topotactically with the required orientational relationships. The mixed hydrates, $Mo_{1-x}W_xO_3 \cdot H_2O$, also form under topotactic dehydration with similar orientational relations. What is more interesting is that the dehydration of $MoO_3 \cdot H_2O$ under electron beam heating gives $MoO_3$ in the $ReO_3$ structure, instead of the expected layered structure [31]. The $ReO_3$ structure of $MoO_3$ is metastable and is produced only by topotactic dehydration under mild conditions. We believe that the preparation

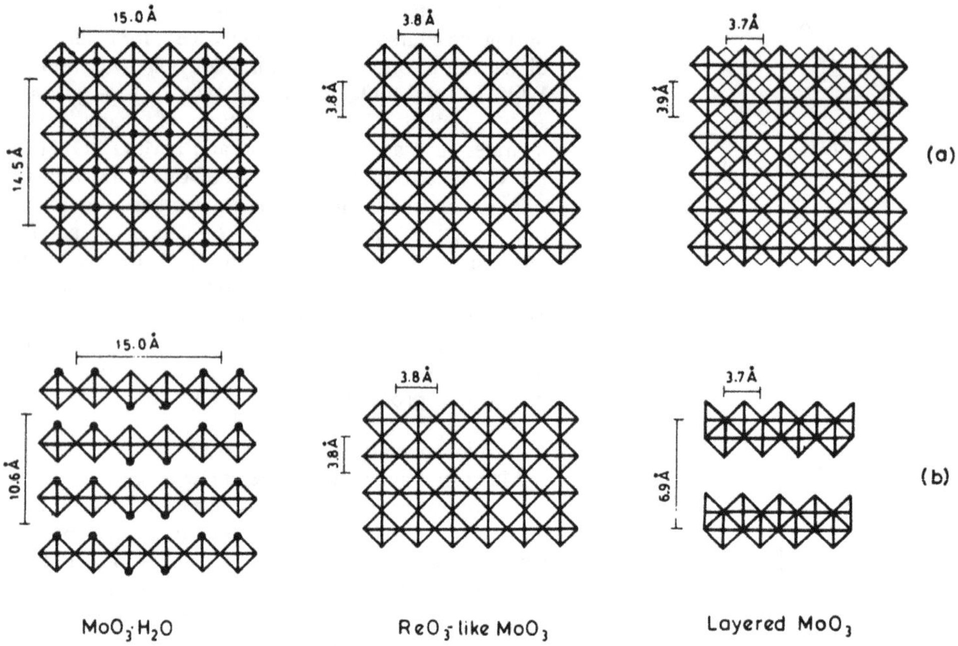


Fig. 6. Schematic representation of $MoO_3{\cdot}H_2O$ (or $WO_3{\cdot}H_2O$), $ReO_3$-like $MoO_3$ (or $WO_3$) and the layered structure of $MoO_3$.

of $ReO_3$-like $MoO_3$ by mild chemical processing is significant. Bulk quantities of $MoO_3$ in the $ReO_3$ structure have been prepared by mild dehydration of the hydrate [32].

### 6.2. *Reduction of perovskite oxides*

Reduction of $ABO_3$ perovskites to give $A_2B_2O_5$ and such defect oxides is found to be topochemical in many instances (*e.g.* $CaMnO_3$). We shall examine the reduction of $LaNiO_3$ and $LaCoO_3$ which crystallize in the rhombohedral perovskite structure. The occurrence of the $La_nNi_nO_{3n-1}$ homologous series was proposed some time ago [33] on the basis of a thermogravimetric study of the decomposition of $LaNiO_3$. However, it was not known whether a similar series exists in the case of $LaCoO_3$. Controlled reduction of $LaNiO_3$ and $LaCoO_3$ in hydrogen shows the formation of $La_2Ni_2O_5$ and $La_2Co_2O_5$ representing the $n=2$ members of the homologous series $LaB_nO_{3n-1}$ (B$\equiv$Co or Ni) [34]. $La_2Ni_2O_5$ can be prepared by the reduction of $LaNiO_3$ at 600 K in pure or dilute hydrogen, but $La_2Co_2O_5$ can only be prepared by the reduction of $LaCoO_3$ in dilute hydrogen at 670 K. Both the oxides can be oxidized back to the parent perovskites at low temperatures. Neither $La_2Ni_2O_5$ nor $La_2Co_2O_5$ can be prepared by the solid state reaction of $La_2O_3$ and the transition metal oxide. X-ray and electron diffraction data reveal that $La_2Co_2O_5$ adopts the brown-millerite structure. The X-ray diffraction pattern of $La_2Ni_2O_5$ is different from that of $La_2Co_2O_5$, but could be indexed on a tetragonal cell related to cubic perovskite. The formation of $La_2Co_2O_5$ and $La_2Ni_2O_5$ by the reduction of $LaCoO_3$ and $LaNiO_3$ respectively is certainly topochemical.

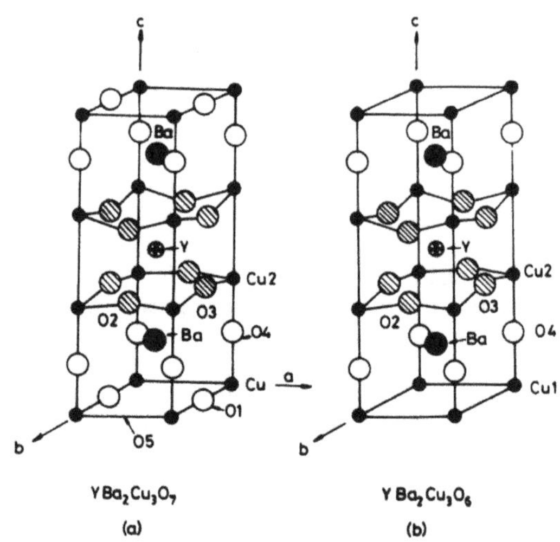


Fig. 7. Structures of $YBa_2Cu_3O_7$ and $YBa_2Cu_3O_6$.

It should be noted that reduction of the high-temperature superconductor $YBa_2Cu_3O_7$ to $YBa_2Cu_3O_6$ (Fig. 7) is a topochemical process.

## 7. Intercalation compounds

Intercalation reactions of solids involve the insertion of a guest species (ion or molecule) into a solid host lattice without any major rearrangement of the solid structure:

$$x(\text{guest}) + \square_x[\text{host}] \rightleftharpoons (\text{guest})_x[\text{host}]$$

where $\square$ stands for a vacant lattice site. Redox intercalation reactions (*e.g.* $Li_xTiS_2$ where the lithium metal reduces the $TiS_2$ layers) can be written as

$$x(\text{guest})_x^{+} + xe^{-} + \square_x[\text{host}] \rightleftharpoons (\text{guest})_x[\text{host}]$$

A variety of layered structures act as hosts. The general feature of these structures is that the interlayer interactions are weak while the intralayer bonding is strong. Intercalation compounds show interesting phase relations, staging (Fig. 8) being an important feature in some of them. Higher stages correspond to lower guest concentrations. Intercalation reactions have been reviewed by many authors [2, 35, 36] and we will present the essential features with a few typical examples.

Alkali metal intercalation involving a redox reaction is readily carried out electrochemically by using the host ($MCh_2$ dichalcogenide) as the cathode, the alkali metal as the anode and the non-aqueous solution of the alkali metal salt as the electrolyte:

Na/NaI–propylenecarbonate/$MCh_2$ (Ch ≡ S, Se, Te)

Li/$LiClO_4$–dioxolane/$MCh_2$

The reaction is spontaneous if a reverse potential is applied or the cell is short circuited. Low alkali metal concentrations are obtained by using solutions of salts such as sodium or potassium naphthalide in THF or *n*-butyllithium in hexane:

$$xC_4H_9Li + TiS_2 \longrightarrow Li_xTiS_2 + \frac{x}{2} C_8H_{18}$$

Alkali metal intercalation in dichalcogenides is also achieved by direct reaction of the elements at around 1070 K (*e.g.* $A_xMCh_2$ where M ≡ V, Nb or Ta) in sealed tubes. Alkali metal intercalation compounds with dichalcogenides form hydrated phases, $A_x(H_2O)_y$-$MCh_2$, just like some of the layered oxides (*e.g.* $VOPO_4{\cdot}2H_2O$, $MoO_3{\cdot}2H_2O$). $NH_3$ is intercalated to dichalcogenides by direct reaction (by distilling liquid $NH_3$ into the dichalcogenide). Intercalation of organic compounds in dichalcogenides is carried out by thermal reaction at temperatures up to 470 K.

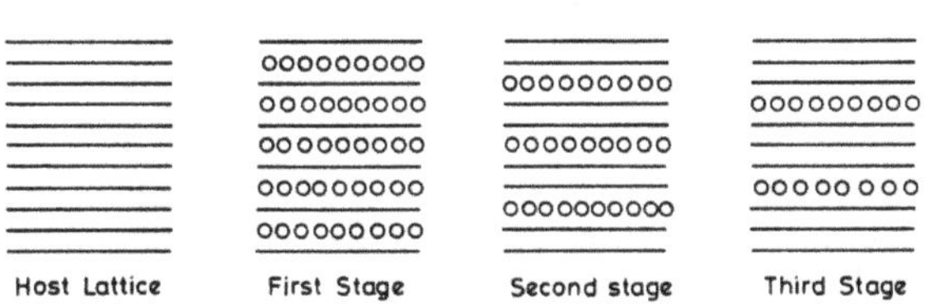


Fig. 8. Schematic diagram of staging in intercalation compounds. Guest molecules are represented by circles in between the layers (shown by lines).

Metal phosphorus trisulphides undergo redox intercalation reactions just like the dichalcogenides and also ion exchange reactions, giving $(\text{guest})_x^{+}$-$[M_{1-x/2}\square_{x/2}PS_3]$. Metal oxyhalides (*e.g.* FeOCl) show intercalation reactions similar to the dichalcogenides. In addition, they undergo irreversible substitution reactions where the halogens of the top layer are substituted by other groups such as $NHCH_3$ and $CH_3$. Layered transition metal oxides such as $MoO_3$, $V_2O_5$, $MOPO_4$ and $MoAsO_4$ (M ≡ V, Nb, Ta) show reduction reactions similar to the dichalcogenides. Layered oxides of type $AMO_2$ and $HTiNbO_5$ and $H_2Ti_4O_9$ undergo ion-exchange and oxidative deintercalation reactions.

Ready deintercalation of lithium from $LiMO_2$ enables these materials to be used as cathodes in lithium cells. Delithiation occurs not only by electrochemical methods, but also by reaction with $I_2$ or $Br_2$ in solution phase. Lithium insertion in close-packed oxides such as $TiO_2$, $Fe_2O_3$ and $Mn_3O_4$ results in interesting structural changes. Delithiation gives rise to oxides in unusual metastable structures (*e.g.* $VO_2$ obtained from delithiation of $LiVO_2$). Delithiation of $LiVS_2$ gives $VS_2$ which cannot otherwise be prepared. Table 3 lists the important hosts and guests in intercalation compounds. In Table 4, we list lithium intercalated compounds to show the variety in this system.

Intercalation of sodium and potassium differs from that of lithium. In layered $A_xMX_n$, lithium is always octahedrally coordinated, while sodium and potassium occupy octahedral or trigonal prismatic sites; octahedral coordination is favoured by large values of $x$

TABLE 3. Examples of hosts and guests in intercalation compounds

| Hosts | Guests |
|---|---|
| *Neutral layers* | |
| Graphite | $FeCl_3$, K, $Br_2$ |
| $MCh_2$ (M ≡ Ti, Zr, Nb, Ta, etc. Ch ≡ S, Se, Te) | Li, Na, $NH_3$, organic amines, $CoCP_2$ |
| $MPCh_3$ (M ≡ Mg, V, Fe, Zn, etc.) | Li, $CoCp_2$ |
| $MoO_3$, $V_2O_5$ | H, alkali metal |
| $MOPO_4$ and $MOAsO_4$ (M ≡ V, Nb, Ta) | $H_2O$, pyridine, Li |
| MOCl and MOBr (M ≡ V, Fe, etc.) | Li, $FeCp_2$ |
| *Negatively charged layers* | |
| (A)$MX_2$ (M ≡ Ti, V, Cr, Fe, X ≡ O, S) | A=Group IA (Li, Na... |
| Layered silicates and clays | Organic compounds |
| $M(HPO_4)_2$ (M ≡ Ti, Zr...) | |
| $K_2Ti_4O_9$ | |

TABLE 4. Intercalation compounds of lithium

| Host | Description | Reference |
|---|---|---|
| $TiS_2$ | $Li_xTiS_2$, $0<x\leqslant 1$ | 37 |
| $VS_2$ | $Li_xVS_2$, $0<x\leqslant 1$. Phases obtained by deintercalation of Li from $LiVS_2$ using $I_2$–$CH_3CN$. Three different phase regions: $0.25\leqslant x\leqslant 0.33$, $0.48\leqslant x\leqslant 0.62$ and $0.85\leqslant x\leqslant 1$ apart from $VS_2$ | 38 |
| $NbS_2$(3R) | $Li_{0.5}NbS_2$ and $Li_{0.70}NbS_2$ | 37 |
| $MoS_3$ | $Li_xMoS_3$, $0<x\leqslant 4$ | 39 |
| $MO_2$ | $Li_xMO_2$, $x\geqslant 1$ (M ≡ Mo, Ru, Os or Ir), $MO_2$ of rutile structure | 40 |
| $TiO_2$ (anatase) | $Li_xTiO_2$, $0<x\leqslant 0.7$. $Li_{0.5}TiO_2$ transforms irreversibly to $LiTi_2O_4$ spinel at 770 K | 41 |
| $CoO_2$ | $Li_xCoO_2$, $0<x<1$. Phases obtained by electrochemical delithiation of $LiCoO_2$ | 42 |
| $VO_2$ | (a) $Li_xVO_2$, $0<x<1$. Phases obtained by delithiation of $LiVO_2$ using $Br_2$–$CHCl_3$ | 43 |
| | (b) $Li_xVO_2$, $0<x<2/3$. Lithiation using *n*-butyl lithium | 44 |
| $Fe_2O_3$ | $Li_xFe_2O_3$, $0<x<2$. Anion array transforms from h.c.p. to c.c.p. on lithiation | 45 |
| $Fe_3O_4$ | $Li_xFe_3O_4$, $0<x<2$. $Fe_2O_4$ subarray of the spinel structure remains intact | 45 |
| $Mn_3O_4$ | $Li_xMn_3O_4$, $0<x<1.2$. Li insertion suppresses tetragonal distortion of $Mn_3O_4$ | 45 |
| $MoO_3$ | $Li_xMoO_3$, $0<x<1.55$ | 46 |
| $V_2O_5$ | $Li_xV_2O_5$, $0<x<1.1$. Intercalation of Li using LiI | 47 |
| $ReO_3$ | $Li_xReO_3$, $0<x<2$. Three phases $0<x\leqslant 0.35$, $x=1$ and $1.8\leqslant x\leqslant 2$ | 48 |

and low formal oxidation states of M. For smaller $x$ and higher oxidation states of M, the coordination of sodium and potassium is trigonal prismatic. Intercalated caesium in $MX_n$ is always trigonal prismatic. Intercalation of sodium and potassium in layered $MX_2$ oxides and sulphides results in structural transformations involving a change in the sequence of anion layer stacking.

Tungsten and molybdenum bronzes, $A_xWO_3$ and $A_xMoO_3$ (A ≡ K, Rb, Cs) are generally prepared by reaction of the alkali metals with the host oxide. Electrochemical methods are also employed for these preparations. A novel reaction that has been employed to prepare bronzes which are otherwise difficult to obtain involves the reaction of oxide host with anhydrous alkali iodides [49]:

$$Mo_{1-x}W_xO_3 + y(AI) \longrightarrow A_yMo_{1-x}W_xO_3 + \frac{y}{2} I_2$$

Atomic hydrogen has been inserted into many binary and ternary oxides. Recently, iodine has been intercalated into the superconducting cuprate, $Bi_2CaSr_2Cu_2O_8$, causing an expansion of the $c$-parameter of the unit cell, without destroying the superconductivity [50].

Chevrel phase compounds, $A_xMo_6Ch_8$ (Ch ≡ S, Se or Te), may also be considered as intercalation compounds. $Mo_6S_8$ can be prepared by acid-leaching copper from $Cu_xMo_6S_8$.

## 8. Ion exchange method

Ion-exchange in fast-ion conductors such as $\beta$-alumina is well known. The exchange can be carried out both in aqueous and molten salt conditions. Thus, sodium $\beta$-alumina has been exchanged with $H_3O^+$, $NH_4^+$ and other monovalent and divalent cations, giving rise to different $\beta$-aluminas [51]. Ion-exchange in inorganic solids is a general phenomenon, not restricted to fast ion conductors alone. The kinetic and thermodynamic aspects of ion-exchange in inorganic solids were examined by England *et al.* [52]. Their results reveal that ion-exchange is a phenomenon that occurs even when the diffusion coefficients are as small as approximately $10^{-12}$ cm$^2$ s$^{-1}$, at temperatures far below the sintering temperatures of solids. Ion-exchange occurs at a considerable rate in stoichiometric solids as well. Mobile ion vacancies introduced by non-stoichiometry or doping seem to be unnecessary for exchange to occur. Since the exchange occurs topochemically, it enables the preparation of metastable phases that are inaccessible by high-temperature reactions.

England *et al.* [52] have shown that a variety of metal oxides having layered, tunnel or close-packed structures can be ion exchanged in aqueous solutions or molten salt media to produce new phases. Typical examples are

$$\alpha\text{-NaCrO}_2 \xrightarrow[570\text{ K, }24\text{ h}]{\text{LiNO}_3} \alpha\text{-LiCrO}_2$$

$$\text{KAlO}_2 \xrightarrow{\text{AgNO}_3(\text{l})} \beta\text{-AgAlO}_2$$

$$\alpha\text{-LiFeO}_2 \xrightarrow{\text{CuCl(l)}} \text{CuFeO}_2$$

The structure of the framework is largely retained during ion-exchange except for minor changes to accommodate the structural preferences of the incoming ion. Thus, when $\alpha$-$LiFeO_2$ is converted to $CuFeO_2$ by exchange with molten CuCl, the structure changes from that of $\alpha$-$NaCrO_2$ to that of delafossite to provide a linear anion coordination for $Cu^+$. Similarly, when $KAlO_2$ is converted to $\beta$-$AgAlO_2$ by ion-exchange, there is a change in structure from cristobalite type to ordered wurtzite type. The change probably occurs to provide a tetrahedral coordination for $Ag^+$.

An interesting ion-exchange reaction is the conversion of $LiNbO_3$ and $LiTaO_3$ to $HNbO_3$ and $HTaO_3$ respectively, by treatment with hot aqueous acid [53]. The exchange of $Li^+$ by protons is accompanied by a topotactic transformation of the rhombohedral $LiNbO_3$ structure to the cubic perovskite structure of $HNbO_3$. The mechanism suggested for the transformation is the reverse of the transformation of cubic $ReO_3$ to rhombohedral $LiReO_3$ and $Li_2ReO_3$ [50], involving a twisting of the octahedra along the [111] cubic direction so as to convert the 12-coordinated perovskite tunnel sites to two 6-coordinated sites in the rhombohedral structure. An interesting structural change accompanying ion-exchange is found in $Na_{0.7}CoO_2$ [54] where the anion layer sequence is ABBAA; cobalt ions occur in alternate interlayer octahedral sites and sodium ions in trigonal prismatic coordination in between the $CoO_2$ units. When this material is ion exchanged with LiCl, a metastable form of $LiCoO_2$ with the layer sequence ABCBA is obtained. The phase transforms irreversibly to the stable $LiCoO_2$ (ABCABC) at around 520 K.

A variety of inorganic solids has been exchanged with protons to give new phases, some of which exhibit high protonic conduction [2]. Typical of these are $HTaWO_6{\cdot}H_2O$, $HMO_3{\cdot}xH_2O$ ($M \equiv$ Sb, Nb, Ta), pyrochlores and $HTiNbO_5$ [2]. Ion-exchange has also been reported in metal sulphides. For example, $KFeS_2$ undergoes topochemical exchange of potassium in aqueous solutions with alkaline earth metal cations to give new phases in which the $[FeS_{4/2}]$ tetrahedral chain is preserved [55].

## 9. Sol-gel method

The sol-gel method has emerged to become an important means of preparing inorganic oxides in recent years. It is a wet chemical method and a multistep process involving both chemical and physical processes such as hydrolysis, polymerization, drying and densification. The name "sol-gel" is given to the process because of the distinctive increase in viscosity which occurs at a particular point in the sequence of steps. A sudden increase in viscosity is the common feature in sol-gel processing, indicating the onset of gel formation. In the sol-gel process, synthesis of inorganic oxides is achieved from inorganic or organometallic precursors (generally metal alkoxides). Most of the sol-gel literature deals with synthesis from alkoxides. The important features of the sol-gel techniques are better homogeneity compared with the traditional ceramic method, high purity, lower processing temperature, more uniform phase distribution in multicomponent systems, better size and morphological control, the possibility of preparing new crystalline and non-crystalline materials, and lastly easy preparation of thin films and coatings. The sol-gel method is widely used in ceramic technology and the subject has been adequately reviewed [56, 57]. We shall deal with the topic briefly for the purpose of completeness. The important steps in sol-gel synthesis are as follows.

*Hydrolysis.* The process of hydrolysis may start with a mixture of a metal alkoxide and water in a solvent (usually alcohol) at the ambient or a slightly elevated temperature. Acid or base catalysts are added to speed up the reaction.

*Polymerization.* This step involves condensation of adjacent molecules wherein $H_2O$ and ROH are eliminated and metal oxide linkages are formed. Polymeric networks grow to colloidal dimensions in the liquid (sol) state.

*Gelation.* In this step, the polymeric networks link up to form a three-dimensional network throughout the liquid. The system becomes somewhat rigid, characteristic of a gel. The solvent as well as water and alcohol remain inside the pores of the gel. Aggregation of smaller polymeric units to the main network continues progressively on aging the gel.

*Drying.* Here, water and alcohol are removed at a moderate temperature (less than 470 K), leaving a hydroxylated metal oxide with residual organic content. If the objective is to prepare a high surface area of aerogel powder with low bulk density, the solvent is removed supercritically.

*Dehydration.* This step is carried out between 670 and 1070 K to drive off the organic residues and chemically bound water, yielding a glassy metal oxide with up to 20%–30% microporosity.

*Densification.* Temperatures in excess of 1270 K are used to form the dense oxide product.

The various steps in the sol-gel technique described above may or may not be strictly followed in practice. Thus, many complex metal oxides are prepared by a modified sol-gel route without actually preparing metal alkoxides. For example, a transition metal salt solution is converted into a gel by the addition of an appropriate organic reagent. In the case of cuprate superconductors, an equimolar proportion of citric acid is added to the solution of metal nitrates, followed by ethylenediamine until the solution attains a pH of 6–6.5. The blue sol is concentrated to obtain the gel. The xerogel is obtained by heating at approximately 420 K. The xerogel is decomposed at an appropriate temperature to obtain the cuprate.

The sol-gel technique has been used to prepare submicrometre metal oxide powders [58] with a narrow particle size distribution and unique particle shapes (*e.g.* $Al_2O_3$, $TiO_2$, $ZrO_2$, $Fe_2O_3$). Uniform $SiO_2$ spheres have been grown from aqueous solutions of colloidal $SiO_2$ [59]. Small metal clusters (*e.g.* nickel, copper, gold) have been prepared by *in situ* chemical reduction of metal salts [60]. Metal-ceramic composites (*e.g.* $Ni$–$Al_2O_3$, $Pt$–$ZrO_2$) can also be prepared in this manner. By employing several variants of the basic sol-gel technique, a number of multicomponent oxide systems have been prepared. Typical of these are, $SiO_2$–$B_2O_3$, $SiO_2$–$TiO_2$, $SiO$–$ZrO_2$, $SiO_2$–$Al_2O_3$, $ThO_2$–$UO_2$ [56, 57]. A variety of ternary and still more complex oxides such as $PbTiO_3$, $PbTi_{1-x}Zr_xO_3$ and NASICON have been prepared by this technique [56, 57]. In the last five years, different types of cuprate superconductors have been prepared by this method [61]. These include $YBa_2Cu_3O_7$, $YBa_2Cu_4O_8$, $Bi_2CaSr_2Cu_2O_8$ and $Pb_2Sr_2Ca_{1-x}Y_xCu_3O_8$.

## 10. Alkali flux method

The use of strong alkaline media, either in the form of solid fluxes or molten (or aqueous) solutions, has enabled the synthesis of novel oxides. The alkali flux method stabilizes higher oxidation states of the metal by providing an oxidizing atmosphere. Alkali carbonate fluxes have traditionally been used to prepare transition metal oxides such as $LaNiO_3$. A good example of an oxide synthesized in a strongly alkaline medium is the pyrochlore, $Pb_2(Ru_{2-x}Pb_x)O_{7-y}$ where Pb is in the 4+ state [62]; this oxide is a bifunctional electrocatalyst. The procedure for preparation involves bubbling oxygen through a solution of lead and rubidium salts in strong KOH at 320 K. The so-called alkaline hypochlorite method is used in many instances. For example, $La_4NiO_{10}$ was prepared by bubbling $Cl_2$ gas through a NaOH solution of lanthanum and nickel nitrates.

Superconducting $La_2CuO_{4+\delta}$ has been prepared by reacting a mixture of $La_2O_3$ and CuO in molten KOH–NaOH around 520 K [63]. $YBa_2Cu_4O_8$ has been prepared by using a $Na_2CO_3$–$K_2CO_3$ flux in a flowing oxygen atmosphere [64]. KOH melt has been used to prepare superconducting $Ba_{1-x}K_xBiO_3$ [65].

## 11. Electrochemical method

Electrochemical methods have been employed to advantage for the synthesis of many solid materials [2, 3, 66, 67]. Typical of the materials prepared in this manner are metal borides, carbides, silicides, oxides and sulphides. Vanadate spinels of formula $MV_2O_4$ as well as tungsten bronzes $A_xWO_3$, have been prepared by the electrochemical route. Tungsten bronzes are obtained at the cathode when current is passed through two inert electrodes immersed in a molten solution of the alkali metal tungstate, $A_2WO_4$ and $WO_3$; oxygen is liberated at the anode [68]. Blue molybdenum bronzes have been prepared by fused salt electrolysis [69]. Monosulphides of uranium, gadolinium, thorium and other metals are obtained from a solution of the normal valent metal sulphide and chloride in an NaCl–KCl eutectic. $LaB_6$ is prepared by taking $La_2O_3$ and $B_2O_3$ in an $LiBO_2$–LiF melt and using gold electrodes. Crystalline transition metal phosphides are prepared from solutions of oxides with alkali metal phosphates and halides.

As mentioned earlier, intercalation of alkali metals in host solids in readily accomplished electrochemically. It is easy to see how both intercalation (reduction of the host) and deintercalation (oxidation of the host) are processes suited for this method. Thus, lithium intercalation is carried out using a lithium anode and a lithium salt in a non-aqueous solvent

$$MS_2(s) + xLi^+ + xe^- \rightleftharpoons Li_xMS_2(s)$$

Superconducting $Ba_{1-x}K_xBiO_3$ has been prepared electrochemically.

Although the electrochemical method is old, the processes involved in the synthesis of various solids are not entirely understood. Generally one uses solvents whose decomposition potentials are high (*e.g.* alkali metal phosphates, borates, fluorides, *etc.*). Changes in melt composition could cause limitations in certain instances. There is considerable scope for investigating the chemistry in the electrochemical synthesis of solids.

## 12. High pressure methods

The use of high pressures for solid state synthesis has become increasingly common in recent years. With the development of high-pressure technology, commercial equipment permitting simultaneous use of both high-pressure and high-temperature conditions has become available. Reviews of various experimental aspects in this area have been published in the literature [2, 3, 70–72]. For the 1–10 kbar pressure range, the hydrothermal method is often employed. In this method, the reaction is carried out either in an open or a closed system. In the open system, the solid is in direct contact with the reacting gas ($F_2$, $O_2$ or $N_2$) which also serves as a pressure intensifier. A gold container is generally used in this type of synthesis. This method has been used for the synthesis of transition metal compounds such as $RhO_2$, $PtO_2$ and $Na_2NiF_6$ where the transition metal is in a high oxidation state. Hydrothermal high pressure synthesis under closed system conditions has been employed for the preparation of higher-valence metal oxides. An internal oxidant such as $KClO_3$ is added to the reactants, which on decomposition under reaction conditions provides the necessary oxygen pressure. For example, pyrochlores of palladium(IV) and platinum(IV), $Ln_2M_2O_7$, (Ln = rare earth) have been prepared by this method (970 K, 3 kbar). $(H_3O)Zr_2(PO_4)_3$ and a family of zero thermal expansion ceramics (*e.g.* $Ca_{0.5}Ti_2P_3O_{12}$) have also been prepared hydrothermally [73, 74]. Another good example is the synthesis of borates of aluminium, yttrium and such metals wherein the sesquioxides are reacted with boric acid [72]. Oxyfluorides have been prepared in HF medium [75].

Zeolites are generally prepared under hydrothermal conditions in the presence of alkali [13, 76, 77]. The alkali, the silica component and the source of aluminium are mixed in appropriate proportions and heated. The reactant mixture forms a hydrous gel which is then allowed to crystallize under pressure for several hours to several weeks between 330 and 470 K. In a typical synthesis, $Al_2O_3{\cdot}3H_2O$ dissolved in concentrated NaOH solution (20 N) is mixed with a 1 N solution of $Na_2SiO_3{\cdot}9H_2O$ to obtain a gel (of composition $2.1Na_2O{\cdot}Al_2O_3{\cdot}2.1SiO_2{\cdot}60H_2O$) which is then crystallized to give zeolite A. The $Na_2O$–$SiO_2$–$Al_2O_3$–$H_2O$ system yields a large number of materials with the zeolitic framework. Under alkaline conditions, aluminium is present as $Al(OH)_4$ anions. The $OH^-$ ions act as a mineralizing catalyst while the cations present in the reactant mixture determine the kinds of zeolite formed. Besides water, some inorganic salts are also encapsulated in some zeolites. Several zeolite structures are found in the $K_2O$–$SiO_2$–$Al_2O_3$–$H_2O$ system as well. $Li_2O$, however, does not give rise to many microporous materials. Group IIA cations yield several zeolitic products.

Zeolitization in the presence of organic bases is useful for synthesizing silica-rich zeolites. Silicalite with a tetrahedral framework enclosing a three-dimensional system of channels (defined by 10 rings wide enough to absorb molecules up to 0.6 nm in diameter) has been synthesized by the reaction of tetrapropylammonium (TPA) hydroxide and a reactive form of silica between 370 and 470 K. The precursor crystals have the composition $(TPA)_2 0.48SiO_2{\cdot}H_2O$ and the organic cation is removed by chemical reaction or thermal decomposition to yield microporous silicalite which may be considered to be a new polymorph of $SiO_2$ [77]. The clathrasil (silica analogue of a gas hydrate), dodecasil-1H, is prepared from an aqueous solution of tetramethoxysilane and $N(CH_3)_4OH$; after the addition of aminoadamantane, the solution is treated hydrothermally under nitrogen for four days at 470 K [78].

The use of template cations has enabled the synthesis of a variety of zeolite materials. Cations such as $(NMe_4)^+$ fit snugly into the cages (*e.g.* sodalite cages of sodalite and SAPO or gmelinite cages of zeolites omega). Neutral organic amines have also been used (*e.g.* in the synthesis of ZSM-5). Many new microporous materials, including those based on $AlPO_4$ (analogue of $SiO_2$), gallosilicates and aluminogerminates (analogues of aluminosilicates), have been prepared. $AlPO_4$-based materials are prepared by the crystallization of gels formed by adding an organic template to a mixture of active alumina, $H_3PO_4$ and water at a pH of 5–8 around 470 K.

Pressures in the range 10–150 kbar are commonly used for solid-state synthesis. In the piston-cylinder apparatus consisting of a tungsten carbide chamber and a piston assembly, the sample is contained in a suitable metal capsule surrounded by a pressure-transducer (pyrophyllite). Pressure is generated by moving the piston through the blind hole in the cylinder. A microfurnace made of graphite or molybdenum is incorporated in the design. Pressures up to 50 kbar and temperatures up to 1800 K are readily reached in a volume of 0.1 $cm^3$ using this design. In the anvil apparatus, first designed by Bridgman, the sample is subjected to pressure by simply squeezing it between two opposed anvils. Although pressures of around 200 kbar and temperatures up to 1300 K are reached in this technique, it is not popular for solid-state synthesis since only milligram quantities can be handled. An extension of the opposed anvil principle is the tetrahedral anvil design, where four massively supported anvils disposed tetrahedrally ram towards the centre where the sample is located in a pyrophyllite medium together with a heating arrangement. The multi-anvil design has been extended to cubic geometry, where six anvils act on the faces of a pyrophyllite cube located at the centre.

The belt apparatus provides the best high-pressure–high-temperature combination for solid-state synthesis. This apparatus, which was used for the synthesis of diamonds some years ago is a combination of the piston-cylinder and the opposed anvil designs. The apparatus consists of two conical pistons made of tungsten carbide, which ram through a specially shaped chamber from opposite directions. The chamber and pistons are laterally supported by several steel rings making it possible routinely to reach fairly high pressures (around 150 kbar) and high temperatures (approximately 2300 K). In the belt apparatus, the sample is contained in a noble metal capsule (a BN or MgO container is used for chalcogenides) and surrounded by pyrophyllite and a graphite sleeve, the latter serving as an internal heater. In a typical high-pressure run, the sample is loaded, the pressure raised to the desired value and then the temperature increased. After holding the pressure for about 30 min, the sample is quenched ($400\ K\ s^{-1}$) while still under pressure. The pressure is released after the sample has cooled at room temperature.

High-pressure methods have been used for the synthesis of several materials that cannot possibly be made otherwise. In general, the formation of a new compound from its components requires that the new composition have a lower free energy than the sum of the free energies of the components. Pressure can aid in the lowering of free energy in different ways [70]. (a) Pressure delocalizes outer d electrons in transition-metal compounds by increasing the magnitude of coupling between the d electrons on neighbouring cations, thereby lowering the free energy. A typical example is the synthesis of $ACrO_3$ (A≡Ca, Sr, Pb) perovskites and $CrO_2$. (b) Pressure stabilizes higher-valence states of transition metals, thus promoting the formation of a new phase. For example, in the Ca–Fe–O system only $CaFeO_{2.5}$ (brown-millerite) is stable under ambient pressures. Under high oxygen pressures, iron is oxidized to the $4^+$ state and hence $CaFeO_3$ with the perovskite structure is formed. (c) Pressure can suppress the ferroelectric displacement of cations, thereby adding the synthesis of new phases. The synthesis of $A_xMoO_3$ bronzes, for example, requires populating the empty d orbitals centred on molybdenum; at ambient pressures, $MoO_3$ is stabilized by a ferroelectric distortion of $MoO_6$ octahedra up to the melting point. (d) Pressure alters site-preference energies of cations, and facilitates the formation of new phases. For example, it is not possible to synthesize $A^{2+}Mn^{4+}O_3$ (A≡Mg, Co, Zn) ilmenites because of the strong tetrahedral site preference of the divalent cations. One therefore obtains a mixture of $A[AMn]O_4$(spinel) + $MnO_2$(rutile) under atmospheric pressure instead of monophasic $AMnO_3$. However, the latter is formed at high pressures with a corundum-type structure in which both the A and Mn ions are in octahedral coordination. (e) Pressure can suppress the $6s^2$ core polarization in oxides containing isoelectronic $Tl^+$, $Pb^{2+}$, $Bi^{3+}$ cations. For example, perovskite-type $PbSnO_3$ cannot be made at atmospheric pressure because the mixture of $PbO + SnO_2$ is more stable than the perovskite.

Stabilization of unusual oxidation states and spin states of transition metals is of considerable interest (*e.g.* $La_2Pd_2O_7$). Such stabilization can be rationalized by making use of correlations of structural factors with the electronic configuration. Six-coordinated high-spin iron(IV) has been stabilized in $La_{1.5}Sr_{0.5}Li_{0.5}Fe_{0.5}O_4$ which has the $K_2NiF_4$ structure [79]. The elongated $FeO_6$ octahedra and the presence of ionic Li–O bonds resulting from the $K_2NiF_4$ structure favour the high-spin iron(IV) state. The lithium and iron ions in this oxide are ordered in the *a–b*-plane, as evidenced by the supercell spots in the electron diffraction pattern. Such an oxide is prepared under oxidizing conditions. $CaFeO_3$ and $SrFeO_3$ prepared under oxygen pressure also contain octahedral iron(IV). While iron(IV) in $SrFeO_3$ is in the high-spin state with the $e_g$ electron in the narrow $\sigma^*$ band down to 4 K, iron(IV) in $CaFeO_3$ disproportionates to iron(III) and iron(V)below 290 K [80]. $La_2LiFeO_6$ prepared under high oxygen pressure has the perovskite structure with the iron in the pentavalent state [79].

Nickel in the 3+ state is present in the perovskite $LaNiO_3$ which can be prepared at atmospheric pressure; other rare earth nickelates have been prepared at high oxygen pressures. Recently $NdNiO_3$ has been prepared by the sol-gel and other chemical routes [20, 81]. $MNiO_{3-x}$ (M≡Ba or Sr) prepared under high pressure contains nickel(IV) [82]. In $La_2Li_{0.5}Co_{0.5}O_4$, there is evidence for the transformation of the low-spin cobalt(III) to the intermediate-spin as well as high-spin states. The lithium and cobalt ions are ordered in the *a–b*-plane of this oxide of $K_2NiF_4$ structure; the highly elongated $CoO_6$ octahedra seem to stabilize the intermediate-spin state. Oxides in perovskite and $K_2NiF_4$ structures with trivalent copper have been prepared under high oxygen pressure [79]. High $F_2$ pressure has been employed to prepare $Cs_2NiF_6$ and other fluorides

[83]. Monel autoclaves have been used in such reactions of $F_2$.

As noted earlier, solid-state reactions are generally slow under ordinary pressures even when the product is thermodynamically stable. Pressure has a marked effect on the kinetics of the reaction, reducing the reaction times considerably, and at the same time giving more homogeneous and crystalline products. For instance, $LnFeO_3$, $LnRhO_3$ and $LnNiO_3$ (Ln = rare earth) are prepared in a matter of hours under high-pressure–high-temperatue conditions, whereas at ambient pressure the reactions require several days ($LnFeO_3$ and $LnRhO_3$) or they do not occur at all ($LnNiO_3$). Thus $LnFeO_3$ is formed in 30 min at 50 kbar [84]. In several $(AX)(ABX_3)_n$ series of compounds, the end members $ABX_3$ and $A_2BX_4$, having the perovskite and $K_2NiF_4$ structures respectively, are formed at atmospheric pressures, but not the intermediate phases such as $A_3B_2X_7$ and $A_4B_3X_{10}$. Pressure facilitates the synthesis of such solids. $Sr_3Ru_2O_7$ and $Sr_4Ru_3O_{10}$ are formed in 15 min at 20 kbar and 1300 K. $TaS_3$, $NbSe_3$ and such solids are prepared in 30 min at 2 GPa and 970 K.

High-pressure methods have been employed in the synthesis of novel superconducting cuprates. A rudimentary example is the preparation of oxygen-excess $La_2CuO_4$ under high oxygen pressure. A more interesting example is the synthesis of the next homologue with two $CuO_2$ layers. $La_2Ca_{1-x}Sr_xCu_2O_6$ which had earlier been found to be an insulator was rendered superconducting by heating it under oxygen pressure [85]. $YBa_2Cu_4O_8$ was first prepared under high oxygen pressure, but this was soon found unnecessary [86]. However, superconducting cuprates with infinite $CuO_2$ layers of the type $Ca_{1-x}Sr_xCuO_2$ or $Sr_{1-x}Nd_xCuO_2$ can only be prepared under high hydrostatic pressures which help to give materials with shorter Cu-O bonds [87, 88]. It should be noted that $Ca_{1-x}Sr_xCuO_2$ prepared at ambient pressure is insulating [89].

## 13. The pyrosol process: a novel chemical method for depositing films

Pyrolysis of sprays is a well known method for depositing films. Thus, one can obtain films of oxidic materials such as CoO, ZnO and $YBa_2Cu_3O_7$ by the spray pyrolysis of solutions containing salts (*e.g.* nitrates) of the cations. A novel improvement in this technique is the pyrosol process involving the transport and subsequent pyrolysis of a spray generated by an ultrasonic atomizer [90]. When a high frequency (100 kHz–10 MHz range) ultrasonic beam is directed at a gas–liquid interface, a geyser is formed and the height of the geyser is proportional to the acoustic intensity. Its formation is accompanied by the generation of a spray, resulting from the vibrations at the liquid surface and cavitation at the liquid–gas interface. The quantity of spray is a function of the intensity. Ultrasonic atomization is accomplished using an appropriate transducer made of PZT located at the bottom of the liquid container. A 500–1000 kHz transducer is generally adequate. The atomized spray which goes up in a column fixed to the liquid container is deposited onto a suitable solid substrate and then heat treated to obtain the film of the material concerned. The flow rate of the spray is controlled by the flow rate of air or any other gas. The liquid is heated to some extent, but its vaporization should be avoided.

The source liquid contains the relevant cations in the form of salts dissolved in an organic solvent. Organometallic compounds are often used (e.g. acetates, alkoxides, acetylacetonates, etc.). Proper gas flow is crucial to obtain satisfactory conditions for a good liquid spray. The pyrosol process is in some way between chemical vapour deposition and spray pyrolysis, but the choice of source compounds for the pyrosol process is much larger than that available for chemical vapour deposition. Films of a variety of materials have been obtained by the pyrosol method. The thickness of the films can be anywhere between a few hundred ångströms to a few micrometres. Table 5 lists typical materials prepared by this method [90]. Films of superconducting cuprates such as $YBa_2Cu_3O_7$ have also been prepared by the pyrosol process. Epitaxy has been observed in films deposited onto single-crystal substrates.

## 14. Intergrowth substrates

Several metal oxide systems exhibit chemically well defined recurrent intergrowth structures with large periodicities, rather than random solid solutions with variable composition. However, the ordered intergrowth structures themselves frequently show the presence of wrong sequences. The presence of wrong sequences or lamellae is best revealed by a technique that is more suited to the study of local structure. High resolution electron microscopy (HREM) enables direct examination of the extent to which a particular ordered arrangement repeats itself, the presence of different sequences of intergrowths, often of unit cell dimensions. Selected area electron diffraction, which forms an essential part of HREM, provides useful information (not generally provided by X-ray diffraction) regarding the presence of supercells due to the formation of intergrowth structures. Many systems forming ordered intergrowth structures have been discovered recently [91, 92]. These systems generally exhibit

TABLE 5. Typical films prepared by the pyrosol process

| Material | Compound used | Solvent | Gas | Substrate[a] |
|---|---|---|---|---|
| Pt | Pt acetylacetonate | Acetylacetone | Air | Glass, $Al_2O_3$, Si (670 K) |
| $In_2O_3$ | In acetylacetonate | Acetylacetone | Air | Glass, $Al_2O_3$, Si (770 K) |
| $SnO_2$ | $SnCl_4$ | Methanol | Air, $N_2$ | Glass, $Al_2O_3$ (670 K) |
| $CdIn_2O_4$ | In acetylacetonate + Cd acetate | Acetylacetone, methanol | Air | Glass, $Al_2O_3$ (710 K) |
| $TiO_2$ | Butyl-orthotitanate | Butanol, acetylacetone | Air, $N_2$ | Glass, steel (670 K) |
| $\gamma$-$Fe_2O_3$ | Fe acetylacetonate | Butanol | Argon | Glass (760 K) |
| $(Ni,Zn)Fe_2O_4$ | Ni, Zn, Fe acetylacetonates | Butanol | Air | Glass (770 K) |
| $Al_2O_3$ | Al isopropoxide | Butanol | Air | Glass (920 K) |

[a]Approximate substrate temperature is shown.

TABLE 6. Recurrent ordered intergrowth structures forming homologous series

(1) Barium ferrites
- (i) $M_pY_q$ where $M = BaFe_{12}O_{19}$ and $Y = Ba_2Me_2Fe_{12}O_{22}$ (Me ≡ Zn, Ni etc.); $Ba_{2n+p}Me_{2n}Fe_{12(n+p)}O_{22n+19p}$ with $n = 1$–47 and $p = 2$–12
- (ii) $MS_n$ where $S = Me_2Fe_4O_8$ with $n = 1, 2, 3, 4$

(2) Perovskites
- (i) $Bi_4A_{m+n-2}B_{m+n}O_{3(m+n)+6}$ formed by Aurivillius oxides of the type $Bi_2A_{n-1}B_nO_{3n+3}$
- (ii) $A_nB_nO_{3n+2}$ such as $(Na,Ca)_nNb_nO_{3n+2}$ with $n = 4$–4.5
- (iii) $A_{n+1}B_nO_{3n+1}$ as exemplified by Sr–Ti–O and La–Ni–O systems

(3) Tungsten bronzes
- (i) $A_xWO_3$ (ITB) with A = alkali metal, Bi etc. ($0.0 < x < 0.1$)
- (ii) $A_xM_xW_{1-x}O_3$ (bronzoids) with M ≡ V, Nb etc. ($0.0 < x < 0.1$)
- (iii) $A_xP_4O_8(WO_3)_{2m}$ with $m = 4$–10
- (iv) $K_xP_2O_4(WO_3)_{2m}$ with $m = 2$
- (v) $P_4O_8(WO_3)_{4m}$
- (vi) $ACu_3M_7O_{21}$ with M ≡ Ta, Nb

(4) Siliconiobates
- (i) $(A_3M_6Si_4O_{26})_n(A_3Nb_{8-x}M_xO_{21})$ with A ≡ Ba, Sr; M ≡ Ta, Nb
- (ii) $(Ba_3Nb_6Si_4O_{26})_nA_3Nb_xM_xO_{21}$ with A ≡ K or Ba; M ≡ Ti, Ni, Zn, etc.

(5) Others
- (i) $(ATi_6O_{13})_n(ATi_4O_9)_m$ with A ≡ Na, A ≡ Ba
- (ii) $La_2O_3$–$ThO_2$ system

homology. In Table 6, various known intergrowth structures are listed.

If the $ABO_3$ perovskite structure is cut parallel to the (110) plane, slabs of the composition $A_{n-1}B_nO_{3n+2}$ are obtained; if these slabs are stacked, an extra sheet of A is introduced, giving rise to the family of oxides of the general formula $A_nB_nO_{3n+2}$. Typical members of this family are $Ca_2Nb_2O_7$ ($n = 4$), $NaCa_4Nb_5O_{17}$ ($n = 5$) and $Na_2Ca_4Nb_6O_{20}$ ($n = 6$). HREM and X-ray studies show that an ordered intergrowth structure with $n = 4.5$ with the composition $NaCa_8Nb_9O_{31}$ corresponds to alternate stacking of $n = 4$ and $n = 5$ lamellae. What is curious is that $NaCa_8Nb_9O_{31}$ is prepared by the standard procedure of heating the mixture of component oxides and yet shows such extraordinary periodicity. Between $n = 4$ and 4.5, a large number of ordered solids are found with the $b$ parameter of the unit cell ranging anywhere from 58.6 Å in the $n = 4.5$ compound to a few thousand ångströms in longer period structures. These solids seem to belong to a class of infinitely adaptive structures [2].

Aurivillius described the family of oxides of the general formula $Bi_2A_{n-1}B_nO_{3n+3}$ where the perovskite slabs, $(A_{n-1}B_nO_{3n+1})^{2-}$, $n$ octahedra thick, are interleaved by $(Bi_2O_2)^{2+}$ layers. Typical members of this family are $Bi_2WO_6$ ($n = 1$), $Bi_3Ti_{1.5}W_{0.5}O_9$ ($n = 2$), $Bi_4Ti_3CrO_{12}$ ($n = 3$) and $Bi_5Ti_3CrO_{15}$ ($n = 4$). These oxides form intergrowth structure of the general formula $Bi_4A_{m+n-2}B_{m+n}O_{3(m+n)+6}$ involving alternate stacking of two Aurivillius oxides with different $n$ values (Fig. 9). The method of preparation simply involves heating a mixture of the component metal oxides at around 1000 K. Ordered intergrowth structures with $(m,n)$ values of (1,2), (2,3) and (3,4) have been fully characterized by X-ray diffraction and HREM (see Fig. 10). What is amazing is that such intergrowth structures with long-range order are indeed formed while either member ($m$ and $n$) can exist as a stable entity. These materials seem to be truly representative of recurrent intergrowth. The periodicity found in recurrent intergrowth solids formed by the Aurivillius family of oxides is indeed impressive.

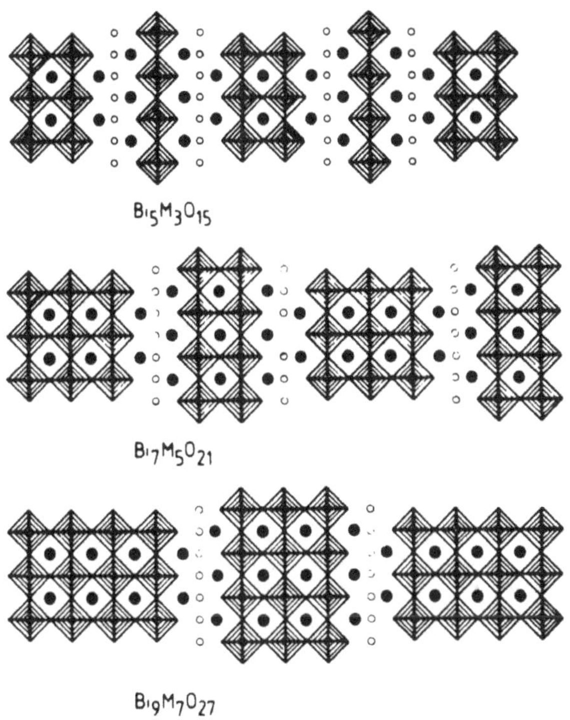


Fig. 9. Different types of intergrowth structures formed by the Aurivillius family of bismuth oxides.

$WO_3$ forms tetragonal, hexagonal or perovskite-type bronzes by interaction with alkali and other metals. The family of intergrowth tungsten bronzes (ITB) involving the intergrowth of $nWO_3$ slabs and one to three strips of the hexagonal tungsten bronze (HTB) is of relevance to our discussion here. In these intergrowth tungsten bronzes of the general formula $M_xWO_3$, $x$ is generally 0.1 or less and depending on whether the HTB strip is one or two-tunnel wide, ITBs are classified as belonging to $(0,n)$ and $(1,n)$ series (Fig. 11). HTB strips of two-tunnel width seem to be most stable in ITBs and many ordered sequences of the $(0,n)$ and the $(1,n)$ series have been identified. ITB phases of bismuth were characterized some time ago in this laboratory. In this system, the HTB strips are always one-tunnel wide (Fig. 12). Displacement of adjacent tunnel rows due to the tilting of $WO_3$ octahedra often results in doubling of the long-period axis of the ITB. Evidence for the ordering of the intercalating bismuth atoms in the tunnels has been found in terms of satellites around the superlattice spots in the electron diffraction patterns. Among the other systems exhibiting ordered intergrowth, mention must be made of hexagonal barium ferrites $M_pY_q$ ($M = BaFe_{12}O_{19}$ and $Y = Ba_2Me_{12}O_{22}$, where Me is Zn, Ni, Mg, *etc.*). A large number of intergrowth structures of this family have been identified.

## 15. Superconducting cuprates

Bednorz and Muller [93] discovered high $T_c$ superconductivity (approximately 30 K) in $La_{2-x}Ba_xCuO_4$.

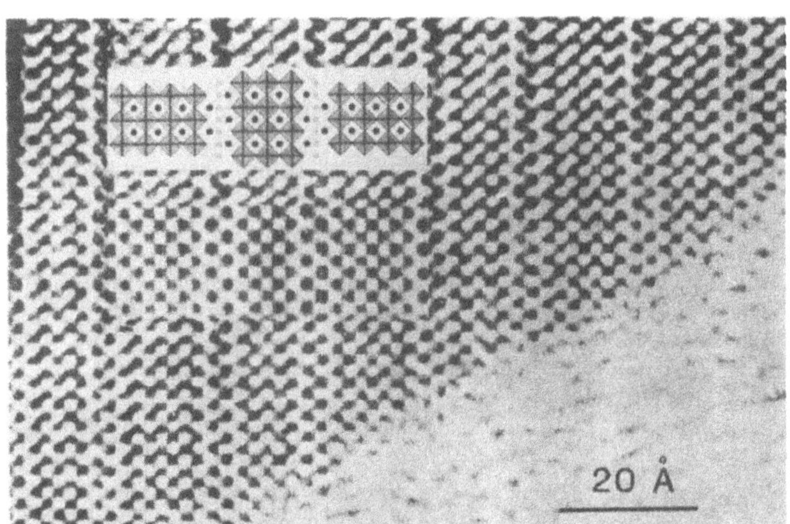


Fig. 10. High-resolution electron micrograph of a (3,4) intergrowth structure involving the Aurivillius phases $Bi_4Ti_3O_{12}$ ($n=3$) and $Bi_5Ti_3CrO_{15}$ ($n=4$).

Fig. 11. Schematic drawing of (1,4), (1,5) and (1,6) intergrowth tungsten bronzes. Hexagonal tunnels of HTB strips separate the $WO_3$ slabs shown in polyhedral form (after Kihlborg [92]).

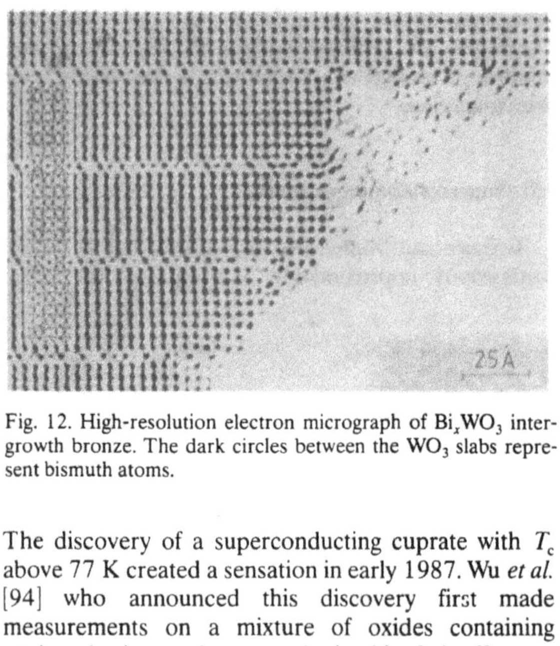

Fig. 12. High-resolution electron micrograph of $Bi_xWO_3$ intergrowth bronze. The dark circles between the $WO_3$ slabs represent bismuth atoms.

The discovery of a superconducting cuprate with $T_c$ above 77 K created a sensation in early 1987. Wu *et al.* [94] who announced this discovery first made measurements on a mixture of oxides containing yttrium, barium and copper obtained in their efforts to obtain the yttrium-analogue of $La_{2-x}Ba_xCuO_4$. In this laboratory, we worked independently on the Y–Ba–Cu–O system on the basis of solid state chemistry [95]. We knew that $Y_2CuO_4$ could not be made and that substituting yttrium by barium in this cuprate was not the way to proceed. We therefore tried to make $Y_3Ba_3Cu_6O_{14}$ by analogy with the known $La_3Ba_3Cu_6O_{14}$ and varied the yttrium barium ratio as in $Y_{3-x}Ba_{3+x}Cu_6O_{14}$. By putting $x=1$, we obtained $YBa_2Cu_3O_7$ ($T_c \approx 90$ K). We knew the structure had to be that of a defect perovskite from the beginning, because of the route we adopted for the synthesis.

We shall briefly examine some preparative aspects of the various types of cuprate superconductors [61, 96, 97]. The cuprates are ordinarily made by the traditional ceramic method (mix, grind and heat), which involves thoroughly mixing the various oxides and/or carbonates (or any other salt) in the desired proportion and heating the mixture (preferably in pellet form) at a high temperature. The mixture is ground again after some time and reheated until the desired product is formed, as indicated by X-ray diffraction. This method may not always yield the product with the desired structure, purity or oxygen stoichiometry. Variants of this method are often employed. For example, decomposing a mixture of nitrates has been found to yield a better product in the case of the 123 compounds by some workers; others prefer to use $BaO_2$ in place of $BaCO_3$ for the synthesis.

One of the problems with the bismuth cuprates is the difficulty of obtaining phasic purity (minimizing intergrowth of the different layered phases). The glass or the melt route has been employed to obtain better samples. This method involves preparing a glass by quenching the melt; the glass is then crystallized by heating it above the crystallization temperature. Thallium cuprates are best prepared in sealed tubes (gold or silver). Heating $Tl_2O_3$ with a matrix of the other oxides (already heated to 1100–1200 K) in a sealed tube is preferred by some workers. It is important that thallium cuprates are not prepared in open furnaces since $Tl_2O_3$ (which readily sublimes) is highly toxic. In order to obtain superconducting compositions corresponding to a particular copper content (number of $CuO_2$ sheets) by the ceramic method, one often has to start with various arbitrary compositions, especially in the case of the thallium cuprates. The real composition of a bismuth or a thallium cuprate superconductor is not likely to be anywhere near the starting composition. The actual composition has to be determined by analytical electron microscopy and other methods.

Heating oxidic materials under high oxygen pressures or in flowing oxygen often becomes necessary to attain the desired oxygen stoichiometry. Thus, $La_2CuO_4$ and $La_2Ca_{1-x}Sr_xCu_2O_6$ heated under high oxygen pressures become superconducting with $T_c$ values of 40 and 60 K respectively. In the case of the 123 compounds, one of the problems is that they lose oxygen easily. Note that superconducting $LnBa_2Cu_3O_7$ (Ln ≡ Y, rare earth) is orthorhombic while insulating $LnBa_2Cu_3O_{6.4}$ is tetragonal. It therefore becomes necessary to heat the material in an oxygen atmosphere below the orthorhombic–tetragonal transition temperature. Oxygen stoichiometry is, however, not a

TABLE 7. Superconducting cuprates

| Cuprate | Approximate $T_c$ (K) | Methods of synthesis[a] |
|---|---|---|
| $L_{2-x}Sr_x(Ba_x)CuO_4$ | 35 | Ceramic*, sol-gel, combustion |
| $La_2Ca_{1-x}Sr_xCu_2O_6$ | 60 | Ceramic (high $O_2$ pressure)* |
| $La_2CuO_{4+\delta}$ | 35 | Ceramic (high $O_2$ pressure)*, alkali-flux |
| $YBa_2Cu_3O_7$[b] | 90 | Ceramic (flowing $O_2$)*, sol-gel*, coprecipitation*, combustion |
| $YBa_2Cu_4O_8$[b] | 80 | Ceramic (high $O_2$ pressure), ceramic (with $Na_2O_2$)*, sol-gel*, coprecipitation* |
| $Bi_2CaSr_2Cu_2O_8$ | 90 | Ceramic (air-quench)*, sol-gel*, combustion, melt (glass) route* |
| $Bi_2Ca_2Sr_2Cu_3O_{10}$ | 110 | Ceramic*, sol-gel, melt route |
| $TlCaBa_2Cu_2O_{6+\delta}$ | 90 | Ceramic (sealed Ag/Au tube)* |
| $TlCa_2Ba_2Cu_3O_{8+\delta}$ | 115 | Ceramic (sealed Ag/Au tube)* |
| $Tl_2CaBa_2Cu_2O_8$ | 110 | Ceramic (sealed Ag/Au tube)* |
| $Tl_2Ca_2Ba_2Cu_3O_{10}$ | 125 | Ceramic (sealed Ag/Au tube)* |
| $Tl_{0.5}Pb_{0.5}CaSr_2Cu_2O_{6+\delta}$ | 110 | Ceramic (sealed Ag/Au tube)* |
| $Pb_2Sr_2Ca_{1-x}Y_xCu_3O_8$ | 70 | Ceramic (low $O_2$ partial pressure), sol-gel (low $O_2$ partial pressure)* |
| $Nd_{2-x}Ce_xCuO_4$ | 30 | Ceramic (low $O_2$ partial pressure), coprecipitation (low $O_2$ partial pressure)* |
| $Ca_{1-x}Sr_xCuO_2$ | 40–110 | Ceramic (high pressure)* |
| $Sr_{1-x}Nd_xCuO_2$ | 40–110 | Ceramic (high pressure)* |

[a]Recommended methods are indicated by asterisks.
[b]Other rare earth compounds of this type are also prepared by similar methods.

problem with the bismuth cuprates. The 124 superconductors were first prepared under high oxygen pressures, but it was later found that heating the oxide or nitrate mixture in the presence of $Na_2O_2$ in flowing oxygen is sufficient to obtain 124 compounds. Superconducting lead cuprates, however, can only be prepared in the presence of very little oxygen ($N_2$ with a small percentage of $O_2$). In the case of the electron superconductor $Nd_{2-x}Ce_xCuO_4$, it is necessary to heat the material in an oxygen-deficient atmosphere; otherwise, the electron given by cerium will merely result in an oxygen-excess material. It may be best to prepare $Nd_{2-x}Ce_xCuO_4$ by a suitable method (say decomposition of mixed oxalates or nitrates) and then reduce it with hydrogen.

The sol-gel method has been conveniently employed for the synthesis of 123 compounds such as $YBa_2Cu_3O_7$ and the bismuth cuprates. Materials prepared by such low-temperature methods have to be annealed or heated under suitable conditions to obtain the desired oxygen stoichiometry as well as the characteristic high $T_c$ value. 124 cuprates, lead cuprates and even thallium cuprates have been made by the sol-gel method; the first two are particularly difficult to make by the ceramic method. Coprecipitation of all the cations in the form of a sparingly soluble salt such as carbonate in a proper medium (*e.g.* using tetraethylammonium oxalate), followed by thermal decomposition of the dried precipitate has been employed by many workers to prepare cuprates.

Several other novel strategies have been employed for the synthesis of superconducting cuprates; some of them were mentioned earlier while discussing the various methods. Especially noteworthy are the use of the combustion method and the alkali-flux method for cuprate synthesis. Superconducting infinite-layered cuprates seem to be possible only when prepared under high pressures because of bonding (structural) considerations [87, 88]. In Table 7 we list the various cuprate superconductors along with their properties and the preferred methods of synthesis.

Strategies where structure and bonding considerations are involved in the synthesis are generally more interesting. One such example is the synthesis of modulation-free superconducting bismuth cuprates [98]. Superconducting bismuth cuprates such as $Bi_2CaSr_2Cu_2O_8$ exhibit superlattice modulation. Since such modulation had something to do with the oxygen content in the Bi–O layers and lattice mismatch, $Bi^{3+}$ was partly substituted by $Pb^{2+}$ to eliminate the modulation, without losing the superconductivity.

## 16. Concluding remarks

Chemical synthesis of materials has become an extensive area with newer types of materials being prepared frequently by novel methods. We have reviewed some of the important methods and materials here and there are many more. For example, we have not discussed metal cluster or metal-rich compounds and pnictides; there are several interesting halides and chalcogenides as well. We have only touched on the chemistry involved with chemical transport. The ingenuity with which properties of oxides and other materials are modified drastically by appropriate substitutions or by modification of the structure also forms a part of synthetic strategies, just as the preparation of materials in different forms (such as fine powders) involves chemical inputs. While we have not brought in all these aspects of materials synthesis in this article, it is hoped that what has been covered gives the flavour of this vital area and helps to underscore the important role of chemistry in the synthesis of materials.

## Acknowledgment

The author thanks the Indo-French Centre for Pure and Applied Research and CSIR (India) for support.

## References


1 P. Hagenmuller (ed.), *Preparative Methods in Solid State Chemistry*, Academic Press, New York, 1972.
2 C. N. R. Rao and J. Gopalakrishnan, *New directions in Solid State Chemistry*, Cambridge University Press, Cambridge, 1986.
3 J. D. Corbett, in A. K. Cheetham and P. Day (eds.), *Solid State Chemistry — Techniques*, Clarendon Press, Oxford, 1987, p. 1.
4 F. J. Di Salvo, *Science, 247* (1990) 647.
5 A. K. Cheetham and P. Day (eds.), *Solid State Chemistry — Compounds*, Clarendon Press, Oxford, 1992.
6 C. N. R. Rao (ed.), *Chemistry of Advanced Materials*, IUPAC 21st Century Monograph Series, Blackwell, Oxford, 1992.
7 R. Roy, *Solid State Ionics, 32–33* (1989) 3.
8 C. Torardi and R. E. McCarley, *J. Am. Chem. Soc., 101* (1979) 3963.
9 J. P. Giroult, M. Goreaud, P. H. Labbe and B. Raveau, *Acta Crystallogr. B, 36* (1980) 2570.
10 K. Yvon, in E. Kaldis (ed.), *Current Topics in Materials Science, Vol. 3*, North-Holland, Amsterdam, 1979.
11 K. H. Jack, *Mater. Res. Bull., 13* (1978) 1327.
12 J. B. Goodenough, H. Y. P. Hong and J. A. Kafalas, *Mater. Res. Bull., 11* (1976) 203.
13 J. M. Newsam, in A. K. Cheetham and P. Day (eds.), *Solid State Chemistry — Compounds*, Clarendon Press, Oxford, 1992, p. 234.
14 R. C. Haushalter, K. G. Strohmaeu and F. W. Lai, *Science, 246* (1989) 1289.
15 A. Jayaraman, P. D. Dernier and L. D. Longinolti, *High Temp. High Pressures, 7* (1975) 1.
16 A. G. Merzhanov, in C. N. R. Rao (ed.), *Chemistry of Advanced Materials*, IUPAC 21st Century Monograph Series, Blackwell, Oxford, 1992, p. 19.
17 P. R. Bonneau, R. F. Jarvis, Jr., and R. B. Kaner, *Nature, 349* (1991) 510.
18 R. Mahesh, V. A. Pavate, Om Parkash and C. N. R. Rao, *Supercond. Sci. Technol., 5* (1992) 174.
19 K. Vidyasagar, J. Gopalakrishnan and C. N. R. Rao, *Inorg. Chem., 23* (1984) 1206.
20 C. N. R. Rao and J. Gopalakrishnan, *Acc. Chem. Res., 20* (1987) 20.
21 H. S. Horowitz, S. J. McLain and A. W. Sleight, *Science, 243* (1989) 66.
22 K. S. Nanjundaswamy, N. Y. Vasantacharya, J. Gopalakrishnan and C. N. R. Rao, *Inorg. Chem., 26* (1987) 4286.
23 A. K. Cheetham, *Nature, 228* (1980) 469.
24 K. Matsuzaki, A. Inoue, H. Kimura, K. Aoki and T. Masumoto, *Jpn. J. Appl. Phys., 26* (1987) L1310.
25 L. V. Interrante and A. G. Williams, *Polym. Prep. (Am. Chem. Soc. Div. Polym. Chem.), 25* (1984) 13.
26 D. Seyferth and G. H. Wiseman, *J. Ceram. Soc., 67* (1984) C132.
27 J. M. Thomas, *Philos. Trans. R. Soc. London, Ser. A, 277* (1974) 251.
28 G. W. Brindley, *Prog. Ceram. Sci., 3* (1963).
29 J. R. Gunter, *J. Solid State Chem., 5* (1972) 354.
30 A. Revcolevachi and G. Dhalenne, *Nature, 316* (1985) 335.
31 L. Ganapathi, A. Ramanan, J. Gopalakrishnan and C. N. R. Rao, *J. Chem. Soc. Chem. Commun.*, (1986) 62.
32 E. M. McCarron, *J. Chem. Soc. Chem. Commun.*, (1986) 336.
33 P. L. Gai and C. N. R. Rao, *Z. Naturfosch. A, 30* (1975) 1092.
34 K. Vidyasagar, A. Reller, J. Gopalakrishnan and C. N. R. Rao, *J. Chem. Soc. Chem. Commun.*, (1985) 7.
35 M. S. Whittingham and A. J. Jacobson (eds.), *Intercalation Chemistry*, Academic Press, New York, 1982.
36 A. J. Jacobson, in A. K. Cheetham and P. Day (eds.), *Solid State Chemistry — Compounds*, Clarendon Press, Oxford, 1992 and references listed therein, p. 182.
37 M. S. Whittingham, *Prog. Solid State Chem., 12* (1978) 41.
38 D. W. Murphy, C. Cros, F. J. Di Salvo and J. V. Waszezak, *Inorg. Chem., 16* (1977) 3027.
39 A. J. Jacobson, R. R. Chianelli, S. M. Rich and M. S. Whittingham, *Mater. Res. Bull., 14* (1979) 1437.
40 D. W. Murphy, F. J. Di Salvo, J. N. Carides and J. V. Waszezak, *Mater. Res. Bull., 13* (1978) 1395.
41 D. W. Murphy, M. Greenblatt, S. M. Zahurak, R. J. Cava, J. V. Waszezak and R. S. Hutton, *Rev. Chim. Miner., 19* (1982) 441.
42 K. Mizushima, P. C. Jones, P. J. Wiseman and J. B. Goodenough, *Mater. Res. Bull., 15* (1980) 783.
43 K. Vidyasagar and J. Gopalakrishnan, *J. Solid State Chem., 42* (1982) 217.
44 D. W. Murphy and P. A. Christian, *Science, 205* (1979) 651.
45 M. M. Thackeray, W. I. F. David and J. B. Goodenough, *Mater. Res. Bull., 17* (1982) 785; *18* (1983) 461.
46 D. G. Dickens and M. F. Pye, in M. S. Whittingham and A. J. Jacobson (eds.), *Intercalation Chemistry*, Academic Press, New York, 1982, p. 539.
47 P. G. Dickens, S. J. French, A. T. Hight and M. F. Pye, *Mater. Res. Bull., 14* (1979) 1295.
48 R. J. Cava, A. Santoro, D. W. Murphy, S. Zahurak and R. S. Roth, *J. Solid State Chem., 42* (1982) 251.
49 A. K. Ganguli, J. Gopalakrishnan and C. N. R. Rao, *J. Solid State Chem., 74* (1988) 228.
50 X. D. Xiang, S. McKernan and W. A. Vareka, *Nature, 348* (1990) 145.
51 B. C. Tofield, in M. S. Whittingham and A. J. Jacobson (eds.), *Intercalation Chemistry*, Academic Press, New York, 1982, p. 181.
52 W. A. England, J. B. Goodenough and P. J. Wiseman, *J. Solid State Chem., 49* (1983) 289.

53 C. E. Rice and J. L. Jackel, *J. Solid State Chem., 41* (1982) 308.
54 C. Delmas, J. J. Braconnier and P. Hagenmuller, *Mater. Res. Bull., 17* (1982) 117.
55 H. Boller, *Monatsh. Chem., 109* (1978) 975.
56 D. R. Uhlmann, B. J. Zelinski and G. E. Wnek, in C. J. Brinker, D. E. Clark and D. R. Ulrich (eds.), *Better Ceramics through Chemistry*, North-Holland, New York, 1984, p. 59.
57 B. J. Zelinski, C. J. Brinker, D. E. Clark and D. R. Ulrich (eds.), *Better Ceramics through Chemistry IV*, MRS symposium, vol. 180, Materials Research Society, Pittsburgh, PA, 1990, and earlier volumes.
58 E. Matijevic, in L. L. Hench and D. Ulrich (eds.), *Ultrastructure Processing of Ceramics, Glasses and Composites*, Wiley, New York, 1984, p. 123.
59 R. K. Iler, *The Chemistry of Silica*, Wiley, New York, 1979.
60 E. Matijevic, *Discuss. Faraday Soc., 92* (1991) 229.
61 C. N. R. Rao, R. Nagarajan and R. Vijayaraghawam, *Supercond. Sci. Technol.*, (1992) in press.
62 H. S. Horowitz, J. M. Longo and J. T. Lewandowski, *Mater. Res. Bull., 16* (1981) 489.
63 W. K. Ham, G. F. Holland and A. M. Stacy, *J. Am. Chem. Soc., 110* (1988) 5214.
64 R. J. Cava, J. J. Krajewski and W. F. Peck, *Nature, 338* (1989) 328.
65 L. F. Schneemeyer, J. K. Thomas and T. Siegrist, *Nature, 335* (1988) 421.
66 A. Wold and D. Bellawance, in P. Hagenmuller (ed.), *Preparative Methods in Solid State Chemistry*, Academic Press, new York, 1972, p. 279.
67 R. S. Feigelson, *Adv. Chem. Sci., 186* (1980) 243.
68 M. S. Whittingham and R. A. Huggins, in R. S. Roth and S. J. Schneider, Jr., (eds.), *Solid State Chemistry*, National Bureau of Standards, US Department of Commerce, Washington, DC, 1972, p. 133.
69 E. Banks and A. Wold, in C. N. R. Rao (ed.), *Solid State Chemistry*, Marcel Dekker, New York, 1974.
70 J. B. Goodenough, J. A. Kafalas and J. M. Longo, in P. Hagenmuller (ed.), *Preparative Methods in Solid State Chemistry*, Academic Press, New York, 1972, p. 2.
71 C. W. F. T. Pistorius, *Prog. Solid State Chem., 11* (1976) 1.
72 J. C. Joubert and J. Chenavas, in N. B. Hannay (ed.), *Treatise on Solid State Chemistry*, Vol. 5, Plenum, New York, 1975, p. 463.
73 R. Roy, D. K. Agarwal, J. Alamo and R. A. Roy, *Mater. Res. Bull., 19* (1984) 471.
74 M. A. Subramanian, B. D. Roberts and A. Clearfield, *Mater. Res. Bull., 19* (1984) 1471.
75 A. W. Sleight, *Inorg. Chem., 8* (1969) 1764.
76 R. M. Barrer, *Zeolites, 1* (1981) 130.
77 E. M. Flanigen, J. M. Bennett, R. W. Grose, J. P. Cohen, R. L. Patton, R. M. Kirchner and J. V. Smith, *Nature, 271* (1978) 512.
78 E. J. J. Groenen, N. C. M. Alma, A. G. T. M. Bastein, G. R. Hays, R. Huis and A. G. T. G. Kortbeck, *J. Chem. Soc. Chem. Commun.*, (1983) 1360.
79 G. Demazeau, B. Buffat, M. Pouchard and P. Hagenmuller, *J. Solid State Chem., 45* (1982) 881; *Z. Inorg. Allg. Chem., 491* (1982) 60.
80 M. Takano and Y. Takeda, *Bull. Inst. Chem. Res., (Kyoto Univ., Jpn.), 61* (1983) 406.
81 J. K. Vassilou, M. Hornbostel, R. Ziebarth and F. J. Di Salvo, *J. Solid State Chem., 81* (1989) 208.
82 Y. Takeda, F. Kanamaru, M. Shimada and M. Koizumi, *Acta Crystallogr. B, 32* (1976) 2464.
83 P. Hagenmuller (ed.), *Solid Inorganic Fluorides*, Academic Press, New York, 1985.
84 F. Sugawara, Y. Syono and S. Akimoto, *Mater. Res. Bull., 3* (1968) 529.
85 R. J. Cava, B. Battlogg, L. F. Schneemeyer *et al.*, *Nature, 345* (1990) 602.
86 C. N. R. Rao, G. N. Subbanna, R. Nagarajan *et al.*, *J. Solid State Chem., 88* (1990) 163.
87 M. Azuma, M. Tanako *et al.*, *Nature, 356* (1992) 775.
88 M. Takano, Z. Hiroi, M. Azuma and Y. Takeda, in C. N. R. Rao (ed.), *Chemistry of High-Temperature Superconductors*, World Scientific, Singapore, 1992, p. 243.
89 T. Siegerist, S. M. Zahurak, D. W. Murphy and R. S. Roth, *Nature, 334* (1988) 155.
90 M. Langlett and J. C. Jobert, in C. N. R. Rao (ed.), *Chemistry of Advanced Materials*, IUPAC 21st Century Monograph Series, Blackwell, Oxford, 1992, p. 55.
91 C. N. R. Rao, *Bull. Mater. Sci., 7* (1985) 155; C. N. R. Rao and J. M. Thomas, *Acc. Chem. Res., 18* (1985) 113.
92 L. Kihlborg (ed.), Direct imaging of atoms in crystals and molecules, *Nobel Symposium, 47* (1979) 101.
93 J. G. Bednorz and K. A. Muller, *Z. Phys. B, 64* (1986) 189.
94 M. K. Wu, J. R. Ashburn, C. J. Torng, P. H. Hor, R.-L. Meng, L. L. Gao, Z. H. Huang, Y. Q. Wang and C. W. Chu, *Phys. Rev. Lett., 58* (1987) 908.
95 C. N. R. Rao, P. Ganguly, R. A. Mohan Ram, A. K. Raychaudhuri and K. Sreedhar, *Nature, 326* (1987) 856.
96 C. N. R. Rao, *Philos. Trans. R. Soc. London, 336* (1991) 596.
97 C. N. R. Rao (ed.), *Chemistry of High-Temperature Superconductors*, World Scientific, Singapore, 1992.
98 V. Manivannan, J. Gopalakrishnan and C. N. R. Rao, *Phys. Rev. B, 43* (1991) 8686.

ChemComm FEATURE ARTICLE

# Virtues of marginally metallic oxides

**C. N. R. Rao**

*CSIR Centre of Excellence in Chemistry & Solid State and Structural Chemistry Unit, Indian Institute of Science, Bangalore 560 012, India and Materials Research Laboratory, University of California, Santa Barbara, CA 93106, USA*

**Transition-metal oxides at the metal–insulator boundary, especially those belonging to the perovskite family, exhibit fascinating phenomena such as insulator–metal transitions controlled by composition, high-temperature superconductivity and giant magnetoresistance (GMR). Interestingly, many of these marginally metallic oxides obey the established criteria for metallicity and have a finite density of states at the Fermi level. The perovskite manganates exhibiting GMR, on the other hand, are unusual in that they possess very high resistivities in the 'metallic' state and show no significant density of states at the Fermi level. Marginal metallicity in oxide systems is a problem of great complexity and contemporary interest and its understanding is of crucial significance to the diverse phenomena exhibited by these materials.**

Properties of conventional metals such as gold and copper are familiar to most of us. One of these properties is the low electrical resistivity of metals. Some transition-metal compounds, in particular the oxides, exhibit low electrical resistivities just like conventional metals. The resistivity behaviour of the oxides such as $ReO_3$ and $RuO_2$ is indeed comparable to that of copper (Fig. 1). Metal oxides, however, exhibit a very wide range of electrical resistivity, anywhere between $10^{20}$ to $10^{-10}$ ohm cm, probably representing the widest range in any physical property known to date.[1] Besides the oxides with metallic resistivities and those with insulating properties, there are oxides which exhibit transitions from the insulating state to the metallic state. The insulator-to-metal (I–M) transition in $V_2O_3$ is well known for the unparalleled 10 or 100 million-fold jump in resistivity at 150 K; it is as if wood becomes copper at the transition temperature. The I–M transition in oxides can be brought about by changing the temperature, pressure or composition.[1,2] The perovskite oxides which exhibit compositionally controlled I–M transitions are especially noteworthy.[3] $La_{1-x}Sr_xCoO_3$ is an example of an oxide system which becomes metallic at a particular Sr concentration, the parent $LaCoO_3$ ($x$ = 0.0) being an insulator at ordinary temperatures.

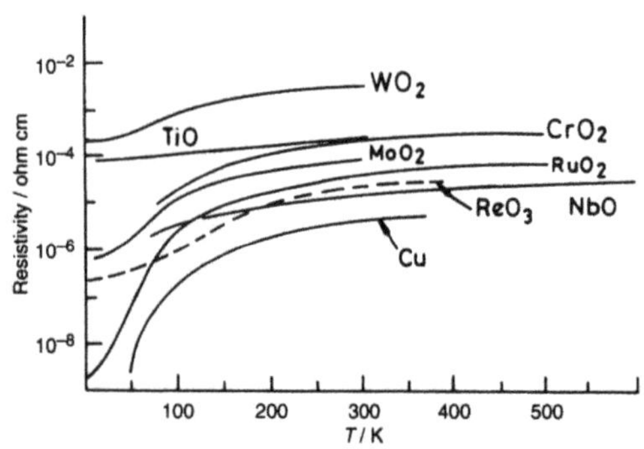


**Fig 1** Resistivities of highly conducting (metallic) oxides

$LaNi_{1-x}Mn_xO_3$ is an example of an oxide system which becomes insulating as $x$ is increased, $LaNiO_3$ ($x$ = 0.0) being metallic. I shall discuss briefly the nature of such metal oxides to point out how the I–M transition occurs across a state which can be considered as barely metallic and yet obey the known criteria for metallicity.†

The cuprates which exhibit high-temperature superconductivity[4,5] are marginally metallic in the normal (non-superconducting) state and exhibit certain anomalous properties. What is even more interesting is that perovskite manganates of the general formula $La_{1-x}A_xMnO_3$ (Ln = rare earth, A = divalent ion such as Ca, Sr, Ba or Pb) exhibit high resistivities in what is considered to be the metallic state and do not seem to obey the criteria for metallicity. Marginal metallicity seems to be generally associated with all the oxides exhibiting giant magnetoresistance.[6,7] Thus, marginal metallicity of metal oxides is closely linked to three of the most important aspects of present-day solid-state science, namely insulator–metal transitions, superconductivity and giant magnetoresistance. In Fig. 2 we compare the temperature variation

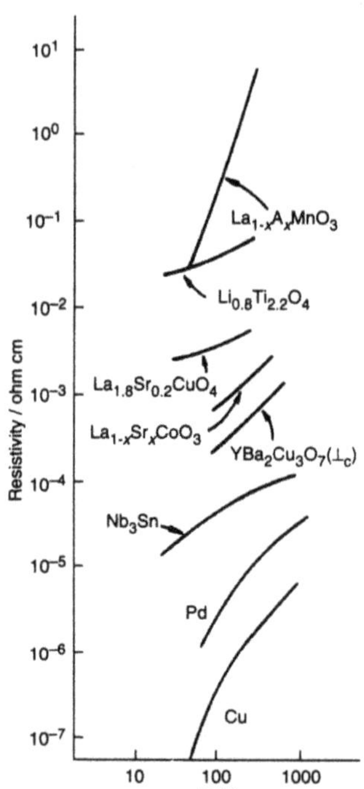


**Fig 2** Resistivities of some of the oxides exhibiting marginal metallicity. Data on metallic $Nb_3Sn$, Pd and Cu are shown for comparison.

of resistivity of cuprates and other oxides to give an indication of the resistivity ranges encountered in marginally metallic oxides. In this article, I examine the nature of marginal metallicity encountered in transition-metal oxide systems after a brief discussion of the criteria for metallicity.

## Operational criteria for metallicity

An important criterion for metallicity is the existence of the Fermi surface or the presence of a finite density of states at the Fermi level.† Two of the successful criteria for metallicity derived from theories of electron transport in solids are due to Mott.[8] Several years ago, Mott[9] pointed out how a material would be metallic at high charge carrier (electron) densities or small interparticle distances; in this state, the material would have a finite electrical conductivity at $T = 0$ K. On the other hand, at low carrier densities or large interparticle distances, the material would be an insulator with zero conductivity at 0 K. A discontinuous transition from the metallic state to the insulating state (Mott transition) can occur at a critical distance or a critical carrier concentration. At the transition, $n_c^{1/3}a_H \approx 0.25$, where $n_c$ is the critical carrier concentration and $a_H$ is the Bohr orbit radius of an isolated centre. A large number of systems, including doped semiconductors and metal–ammonia solutions, obey this relation.[2] The metallicity criterion as defined by this expression is essentially a manifestation of electron interactions or Coulomb correlation resulting in an energy gap at low densities.

Another criterion for metallicity arises from electron localization induced by disorder proposed by Anderson.[10] Electrons diffuse when the disorder is small, but at a critical disorder they do not diffuse (giving rise to zero conductivity). A transition from the metallic to the insulating state occurs as the disorder increases. Mott[11] proposed that there exists a minimum metallic conductivity, $\sigma_{min}$, or maximum metallic resistivity, $\rho_{max}$, for which the material may still be viewed as being metallic, prior to the localization of electrons due to disorder. The $\sigma_{min}$ is given by $\rho^2/h\pi^3 l$, where $l$ is the mean free path of the electrons. Mott's $\sigma_{min}$ criterion assumes that the disorder-driven metal–insulator transition is discontinuous, but the scaling theory[12] predicts it to be continuous. It is, however, found that $\sigma_{min}$ or $\rho_{max}$ is a useful experimental criterion and represents the value of conductivity where the temperature coefficient of resistance changes sign from metal-like (+) to insulator-like (−) behaviour.[2,3] The value of Mott's $\rho_{max}$ is around 1–10 mohm cm in oxide systems.

A useful way of describing and classifying transition-metal oxides in terms of their electronic properties is that due to Zaanen, Sawatzky and Allen (ZSA)[13] who consider the intra-atomic Coulomb strength, $U$, the ligand–metal charge-transfer energy, $\Delta$, and the metal (d)–oxygen(p) hybridization strength, $t_{pd}$, to provide a electronic phase diagram (with $U/t_{pd}$ plotted against $\Delta/t_{pd}$). This diagram has been modified by Sarma to account for magnetic properties and doping effects. Such a modified phase diagram[14] is shown in Fig. 3. In this phase diagram, region D corresponds to d-band metals with $U < \Delta$, and $U < W$ where $W$ is the d-band width and region C represents mixed-valent or p-type metals with $U > \Delta$. Mott insulators with $\Delta > U > W$ wherein $U$ essentially determines the band gap are in region B. Charge-transfer insulators with $U > \Delta$ wherein the band gap is mainly controlled by $\Delta$ are in region A. Covalent insulators in region E are those where $t_{pd}$ plays a crucial role; this region had not been identified in the ZSA phase diagram. Amongst the binary oxides, NiO is a charge-transfer insulator (region A) while $ReO_3$ is a d-band metal (D). Amongst the ternary oxides, $LaTiO_3$ is best described as a Mott insulator (B) while $LaNiO_3$ is a metal in the C region. Both $LaCoO_3$ and $LaMnO_3$ have mixed-valent ground states, but $LaCoO_3$ is closer to A than B, probably on the borderline, while $LaMnO_3$ is closer to B than A. $La_2CuO_4$, which is the parent cuprate for the numerous superconductor families, is described as a charge-transfer insulator with $\Delta \ll t_{pd}$.

## Compositionally controlled metal–insulator transitions in perovskite oxides

In Fig. 4 we show typical compositionally controlled I–M transitions in two perovskite oxide systems. In $La_{1-x}Sr_xCoO_3$, the material becomes metallic as $x$ is increased beyond a critical composition. A similar behaviour is seen in $La_{1-x}Sr_xVO_3$. Metallicity arises in these oxides because of fast hopping of

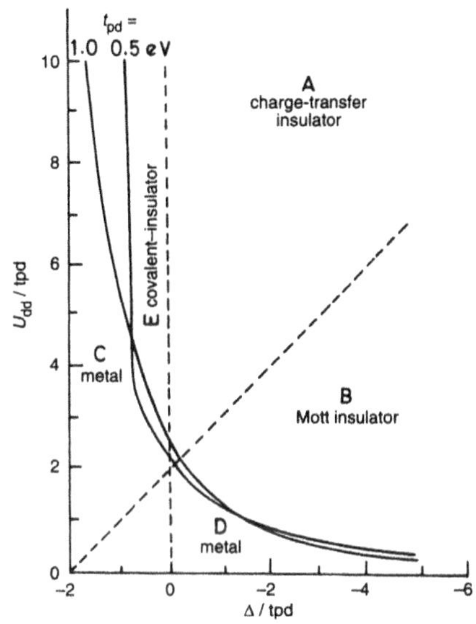


**Fig. 3** ZSA phase diagram as modified by Sarma.[14] The solid lines separate the metallic and insulating regimes.

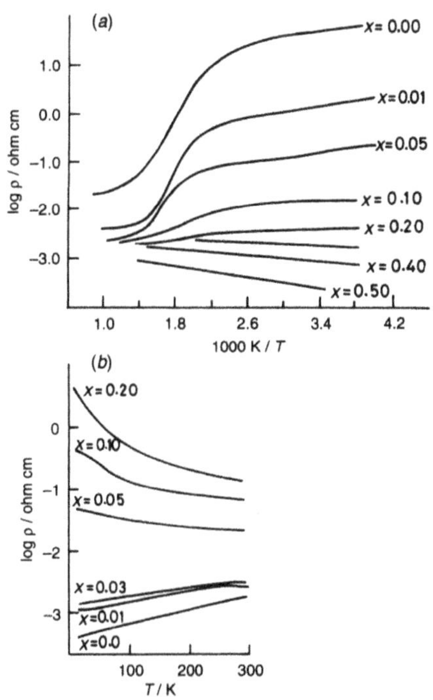


**Fig. 4** Compositionally controlled insulator metal transitions in (*a*) $La_{1-x}Sr_xCoO_3$ and (*b*) $LaNi_{1-x}Mn_xO_3$ (from the author's laboratory)

electrons between the 3+ and 4+ states of the transition metal. It is noteworthy that the temperature coefficient of resistance changes sign around a resistivity value close to Mott's maximum resistivity, $\rho_{max}$. In $LaNi_{1-x}Mn_xO_3$ where the material becomes insulating with increasing $x$, the change from the metallic to the insulating state occurs again around Mott's $\rho_{max}$. The value of $\rho_{max}$ or $\sigma_{min}$ is known to scale with the critical carrier concentration,[15] $n_c$, as shown in Fig. 5 where log $\rho_{max}$ is plotted against log $n_c$ for a variety of systems including doped semiconductors. The points corresponding to the metal oxides exhibiting compositionally controlled metal–insulator transitions fall in line with the other electronic materials at the threshold of metallicity. It is noteworthy that these oxides obey the $\rho_{max}$ or the $\sigma_{min}$ criterion as well as the $n_c^{1/3}a_H \approx 0.25$ criterion. Although the oxides obey the $\sigma_{min}$ criterion, it is possible that some of them may show a departure from this behaviour and become insulating at very low temperatures (say < 50 K), with the resistivity increasing with decreasing temperature[16] (*i.e.* conductivity is zero at 0 K). If so, $\sigma_{min}$ serves as the high-temperature limit for the conductivity of these marginal metals.

Photoelectron spectroscopic studies[14] of $La_{1-x}Sr_xCoO_3$ in the valence band region show the presence of a finite density of states at the Fermi level in these marginally metallic oxides (Fig. 6). The oxygen K-absorption spectra of $La_{1-x}Sr_xCoO_3$ show a feature increasing in intensity and width with increasing $x$, suggesting a progressive creation of doped hole states with substantial oxygen p-character as expected of a charge-transfer insulator. These states overlap the top of the valence band above a critical concentration. A finite density of states at the Fermi level is seen in $LaNi_{1-x}Mn_xO_3$ as well, even slightly beyond the critical $x$ value above which the material is insulating.

### Cuprate superconductors

Most of the cuprates which exhibit high-temperature superconductivity exhibit metallic resistivity (with the resistivity decreasing with temperature) before they become superconducting (Fig. 7). The values of resistivities in the metallic regime before they become superconducting is in the range 2–5 mohm cm, not far from Mott's $\rho_{max}$. Interestingly, the cuprates conform to the $n_c^{1/3}a_H \approx 0.25$ relation (Fig. 8) and also follow the linear log $\rho_{max}$–log $n_c$ plot (Fig. 5). Although these observations would suggest that the cuprates are similar to $La_{1-x}Sr_xVO_3$ and such oxides showing I–M transitions, they exhibit certain anomalous properties[18] in the metallic state, one of them being the linear variation of resistivity with temperature in the metallic state (see Fig. 7). High-energy spectroscopic studies[19] of the cuprates show that the oxygen 2p contribution in the hole doped states near the Fermi level is significant. The cuprates have strong correlation effects and there is as yet no unanimous view as to the best way of describing the marginally metallic state.

A novel feature of the cuprates is that they exhibit compositionally controlled metal–insulator transitions in the normal state ($T > T_c$) as can be seen in Fig. 9(*a*), (*b*). It is indeed curious that the insulating state transforms directly to the superconducting state in some compositions [Fig. 9(*b*)]. Such an insulator–metal transition is found in $La_{2-x}Sr_xCuO_4$ as well. The question that arises is whether there can be insulating and superconducting ground states without a metallic ground state in between. A careful comparison[20] shows that the Hall coefficients and other properties of such cuprate compositions are different from those of oxide systems such as $La_{1-x}Sr_xTiO_3$ which exhibit insulator–metal transitions. Clearly, the superconducting cuprates are associated with a unique kind of a marginal metallic state.

### Perovskite manganates exhibiting GMR

Of all the oxide systems considered here, the manganates of the general formula $Ln_{1-x}A_xMnO_3$ (Ln = La, Pr, Nd; A = Ca, Sr, Ba, Pb) possessing the perovskite structure are most unusual in

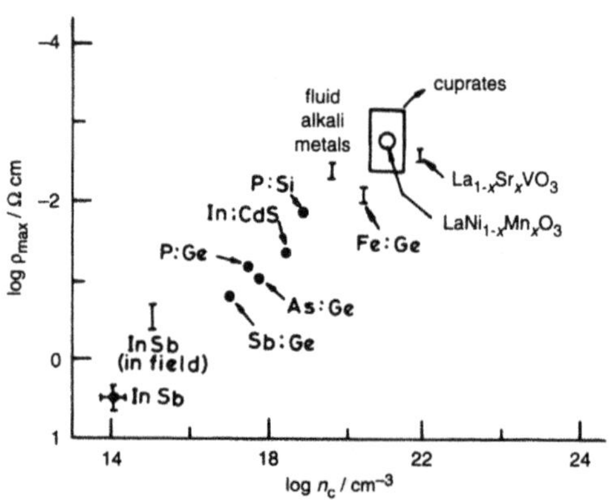


**Fig. 5** A log–log plot of the maximum metallic resistivity, $\rho_{max}$, against the critical carrier density, $n_c$, at the insulator–metal transition

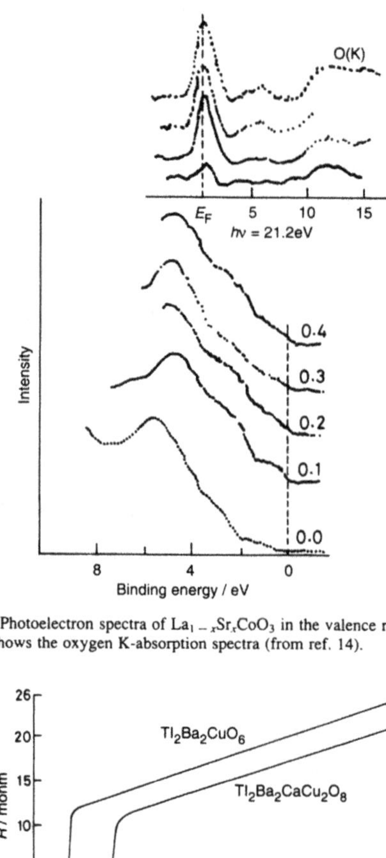


**Fig. 6** Photoelectron spectra of $La_{1-x}Sr_xCoO_3$ in the valence region. The inset shows the oxygen K-absorption spectra (from ref. 14).

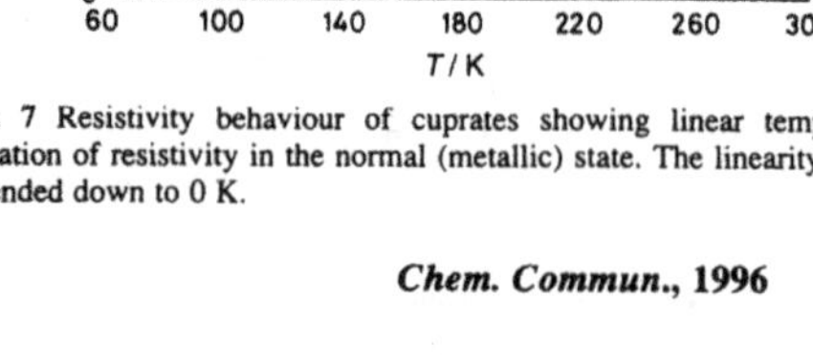


**Fig. 7** Resistivity behaviour of cuprates showing linear temperature variation of resistivity in the normal (metallic) state. The linearity can be extended down to 0 K.

that they possess high resistivities in the metallic state which cannot be accounted for or explained on the basis of the criteria discussed earlier. Let us briefly examine the nature of the metallic state in these materials. In $La_{1-x}A_xMnO_3$, the $Mn^{4+}$ content increases with $x$ (when $x = 0.0$, Mn will be nominally in the 3+ state).‡ The $Mn^{3+}$–O–$Mn^{4+}$ interaction is ferromagnetic, unlike the $Mn^{3+}$–O–$Mn^{3+}$ and $Mn^{4+}$–O–$Mn^{4+}$ interactions which are antiferromagnetic. The manganates become ferromagnetic above a critical concentration of $Mn^{4+}$ (or value of $x$) and simultaneously exhibit metallic conductivity.§ The primary process involved is the hopping of a charge carrier as described by,

$$Mn^{3+}(i)\text{–O–}Mn^{4+}(j) \rightarrow Mn^{4+}(i)\text{–O–}Mn^{3+}(j)$$

where $i$ and $j$ are nearest neighbours. Zener[22] suggested that a paramagnetic to ferromagnetic transition should occur with a transition temperature $T_c$ according to the relation $k_BT_c \approx x_htZ$, where $x_h$ is the $Mn^{4+}$(hole) concentration, $t$ is the hopping amplitude and $Z$ is the number of nearest neighbours. These materials exhibit a transition from an insulating to a metallic state as the temperature is lowered, the transition occurring close to the ferromagnetic Curie temperature, $T_c$. The material is insulating at $T > T_c$ because thermal fluctuations of the magnetic moments impede the motion of holes. The I–M transition in an $La_{1-x}Ca_xMnO_3$ composition is shown in Fig. 10. The resistivity peak in this figure arises from the I–M transition, with the temperature variation of resistivity being metal-like at $T < T_c$ ($T_{IM}$) and insulator-like when $T > T_c$ ($T_{IM}$).

When a magnetic field is applied to such a material the resistivity decreases enormously[23] (Fig. 10). This phenomenon, called giant magnetoresistance (GMR), is of great importance in magnetic recording and other technological applications. GMR-related aspects of the manganates are not within the scope of this article¶ and I shall restrict myself to the metallic state of these materials.

The resistivity of the manganates at the I–M transition is generally high, anywhere from 10 mohm cm to several ohm cm (see Fig. 2). No metal can have such resistivities, the observed values being much higher than Mott's $\rho_{max}$. Furthermore, the resistivity in the metallic regime ($T < T_c$) in some of the manganates is comparable to or even higher than that in the insulating state ($T > T_c$).‖ Residual resistivities at low temperatures (*ca.* 4 K) are also very large (several mohm cm), larger than in disordered metals. While the effects of the lattice and of magnetic polarons may contribute to such high resistivities, we note that electron correlation would play a major role in a Mott insulator such as $LaMnO_3$. What is curious is that photoelectron spectra in the valence-band region[14] of $La_{1-x}Sr_xMnO_3$ show the absence of any significant density of states in the 'metallic' compositions at ordinary temperatures as can be seen from Fig. 11. The oxygen K-absorption spectra show a progressive formation of hole states with the increase in $x$ *ca.* 1 eV above the Fermi level (Fig. 11).

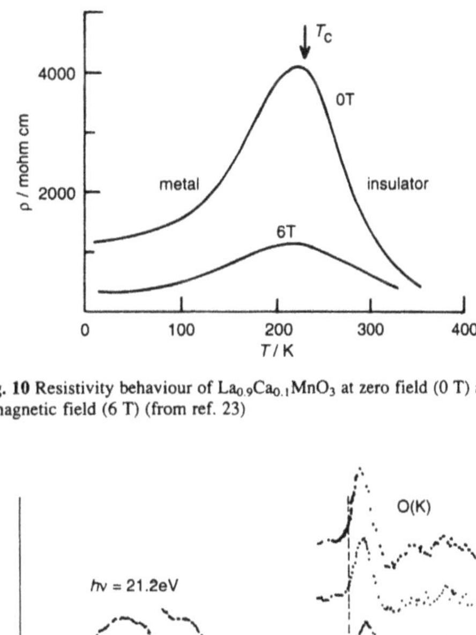

Fig. 10 Resistivity behaviour of $La_{0.9}Ca_{0.1}MnO_3$ at zero field (0 T) and in a magnetic field (6 T) (from ref. 23)

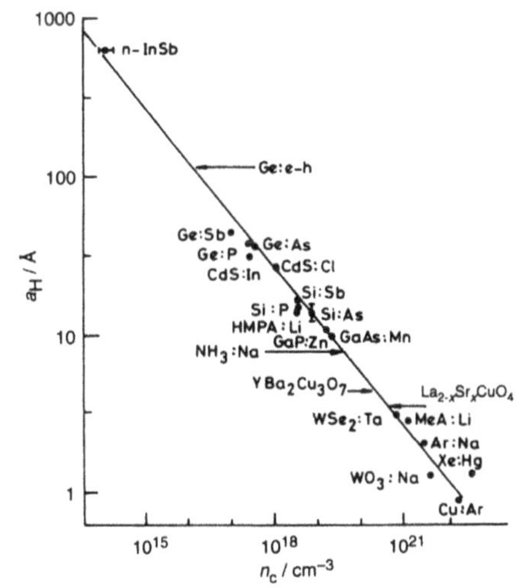

Fig. 8 Relation between the critical carrier concentration, $n_c$, and the Bohr radius, $a_H$. The solid line represents $n_c^{1/3}a_H$ = 0.26 (from ref. 17).

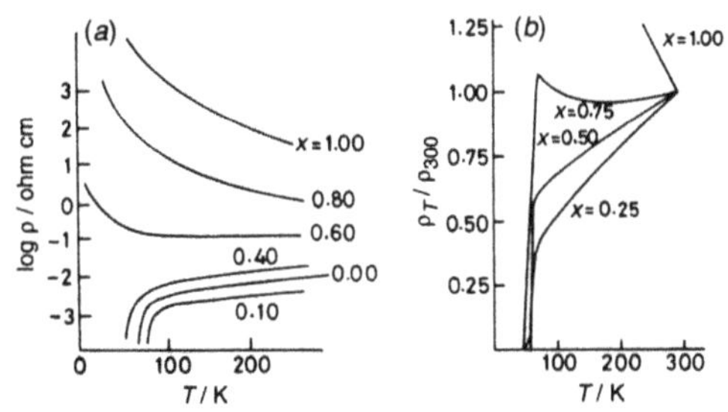

Fig. 9 Temperature variation of the resistivity of (*a*) superconducting $Bi_2Ca_{1-x}Y_xSr_2Cu_2O_{8+\delta}$ and (*b*) $TlCa_{1-x}Nd_xSr_2CuO_6$ (from the author's laboratory)

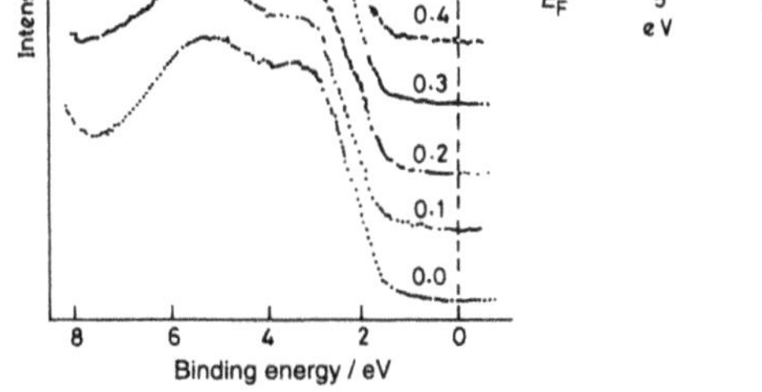

Fig. 11 Photoelectron spectra of $La_{1-x}Sr_xMnO_3$ in the valence region. Inset shows the oxygen K-absorption spectra (from ref. 14).

## Concluding remarks

I have discussed three types of complex metal oxides which are marginally metallic. The first category of oxides exhibiting compositionally controlled insulator–metal transitions have a finite density of states at the Fermi level in the metallic state and obey the known criteria for metallicity. The second category involves the superconducting cuprates. They are marginally metallic in the normal state, obey the criteria for metallicity and undergo unusual insulator–metal and insulator–superconductor transitions, besides exhibiting certain anomalous properties in the metallic state. The first two categories of oxides seem to obey both the criteria of Mott arising from considerations of electron interactions (correlation) and of disorder. It is obvious that both these factors should be considered in describing these oxides. The importance of disorder in these oxides is also indicated by the fact that the phenomena are dependent on a critical concentration of one of the component species. Correlation is certainly important in all the systems considering that the parent oxides, $LaCoO_3$, $LaMnO_3$ and $La_2CuO_4$, all have mixed-valent ground states and fall in the category of Mott or charge-transfer insulators (Fig. 3). Yet, those who work on electron correlation models of oxides ignore disorder and *vice versa*. Unlike the two categories of oxides mentioned above, the manganates exhibiting giant magnetoresistance have high resistivities beyond the range of all known metallic oxides and show no evidence for any significant density of states at the Fermi level in the so-called 'metallic' state. The metallicity of the manganates is clearly of an entirely different category, not encountered hitherto in any other oxide system.

Clearly, there is need for models which include both electron interactions and disorder to describe marginally metallic oxides. This is not an easy task and requires new ideas. The problem does not end here. Besides electron interactions and disorder, other factors such as electron–lattice interactions, magnetic effects and the effect of finite temperature would have to be taken into account depending on the situation as shown schematically in Fig. 12. The theoretical complexity of the marginally metallic state is truly formidable, a proper description of either one of the factors alone (*e.g.* pure correlation or pure disorder) being fraught with many difficulties. Some workers have considered simple models which essentially assume the coexistence of localized and itinerant electrons to describe insulator–metal transitions[25,26] and superconductivity[26] in oxides. For example, Burdett[26] makes use of the interaction or crossing of two diabatic potential-energy curves for the insulating and metallic states to describe the two phenomena. Although educative, such models do not throw light on the nature of marginal metallicity. An effort to combine both electron interactions and disorder has been made recently by Logan *et al.*[27] Whether the metallic state has a universal minimum (electron) diffusity and hence minimum conductivity[28] due to localized states (in addition to extended electron states) is another question that needs to be explored.

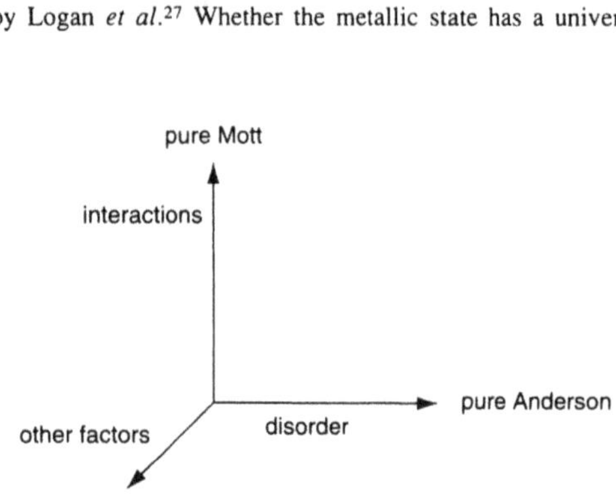


**Fig. 12** Complexity of the problem of marginal metallicity (adapted from ref. 27). The oxides discussed in this article fall somewhere in the three-dimensional space indicated here. The 'other factors' include electron–lattice interaction, magnetic polaron and finite temperature effects.

### Acknowledgements

The author thanks the Science Office of the European Union and the Department of Science and Technology, Government of India, for support of this research.

**C. N. R. Rao is Albert Einstein Research Professor at the Indian Institute of Science and President of the Jawaharlal Nehru Centre for Advanced Scientific Research, Bangalore. He was born in Bangalore and educated in Mysore, Banaras and Purdue Universities. He was Visiting Professor at the Oxford and Cambridge Universities and is Honorary Professor at the University of Wales, Cardiff. He is a Fellow of the Royal Society, Foreign Associate of the US National Academy of Sciences and Member of the Pontifical Academy of Sciences. He is the recipient of the Marlow Medal and the Solid State Chemistry Medal of the Royal Society of Chemistry, of which he is an honorary fellow.**

### Footnotes

† An important experimental criterion for metallicity is the presence of a finite density of states at the Fermi level, which is readily established by photoelectron spectroscopy in the valence region and other high-energy spectroscopic techniques. Two of the most useful operational criteria for metallicity are those due to Mott discussed in the next section.

‡ $LaMnO_3$ as prepared in the laboratory by the solid-state reaction of the oxides and carbonates of the component metals at high temperatures, generally contains around 10% $Mn^{4+}$. The origin of $Mn^{4+}$ is the presence of cation vacancies in both the A(La) and B(Mn) sites in roughly equal proportions. It cannot be due to oxygen excess since the perovskite structure cannot accommodate excess oxygen.[21]

§For small $x$, the material is an antiferromagnetic insulator; the same is true for large $x$ ($x \geqslant 0.5$).

¶ The ferromagnetic $T_c$ in $Ln_{1-x}A_xMnO_3$ increases markedly with the increase in the average radius of the A-site cations, $<r_A>$; accordingly, $T_{IM}$ also increases with $<r_A>$. The magnitude of GMR as well as the peak resistivity at the I–M transition decrease with the increase in $<r_A>$. Increasing $<r_A>$ in these perovskites is analogous to increasing pressure.[24] It should also be noted that GMR is not confined to perovskites or manganates.[7]

‖ The magnitude of GMR itself increases with an increase in the peak resistivity (at $T_{IM}$). In other words, one needs a bad metal to observe good GMR. Furthermore, even formally antiferromagnetic compositions ($x > 0.5$) show GMR because of the presence of ferromagnetic clusters.

### References

1 C. N. R. Rao and B. Raveau, *Transition Metal Oxides*, VCH, Cambridge, 1995; C. N. R. Rao, *Annu. Rev. Phys. Chem.*, 1989, **40**, 291.
2 P. P. Edwards, T. V. Ramakrishnan and C. N. R. Rao, *J. Phys. Chem.*, 1995, **99**, 5228.
3 C. N. R. Rao and P. Ganguly, in *The metallic and the nonmetallic states of matter*, ed. P. P. Edwards and C. N. R. Rao, Taylor and Francis, London, 1985.
4 C. N. R. Rao, *Philos. Trans. R. Soc. London, A*, 1991, **336**, 595.
5 C. N. R. Rao and A. K. Ganguli, *Chem. Soc. Rev.*, 1995, **24**, 1; *Acta Crystallogr., Sect. B*, 1995, **51**, 604.
6 C. N. R. Rao and A. K. Cheetham, *Science.*, 1996, **272**, 369.
7 G. Briceno, H. Chang, X. Sun, P. G. Schultz and X. D. Xiang, *Science*, 1995, **270**, 273; T. Shimakawa, Y. Kubo and T. Manako, *Nature (London)*, 1996, **379**, 53.
8 N. F. Mott, *Metal–insulator transitions*, 2nd edn., Taylor and Francis, London, 1990.
9 N. F. Mott, *Philos. Mag.*, 1961, **6**, 287.
10 P. W. Anderson, *Phys. Rev.*, 1958, **109**, 1492.
11 N. F. Mott, *Philos. Mag.*, 1972, **26**, 1015.
12 E. Abrahams, P. W. Anderson, D. C. Liccardello and T. V. Ramakrishnan, *Phys. Rev. Lett.*, 1979, **42**, 693.
13 J. Zaanen, G. A. Sawatzky and J. W. Allen, *Phys. Rev. Lett.*, 1985, **55**, 418.

14 D. D. Sarma, in *Metal–insulator transitions revisited*, ed. P. P. Edwards and C. N. R. Rao, Taylor and Francis, London, 1995.
15 N. F. Mott, *Proc. R. Soc. London, A*, 1982, **382**, 1.
16 A. K. Raychaudhuri, *Phys. Rev. B*, 1991, **44**, 8572.
17 G. A. Thomas, *J. Phys. Chem.*, 1983, **88**, 3749; P. P. Edwards and M. J. Sienko, *Phys. Rev. B*, 1978, **17**, 2575.
18 T. V. Ramakrishnan and C. N. R. Rao, *J. Phys. Chem.*, 1989, **93**, 4414.
19 D. D. Sarma, in *Chemistry of high temperature superconductors*, ed. C. N. R. Rao, World Scientific, Singapore, 1991.
20 Y. Iye, in *Metal–insulator transitions revisited*, ed. P. P. Edwards and C. N. R. Rao, Taylor and Francis, London, 1995.
21 M. Hervieu, R. Mahesh, N. Rangavittal and C. N. R. Rao, *Eur. J. Solid State Inorg. Chem.*, 1995, **32**, 79.
22 C. Zener, *Phys. Rev.*, 1951, **82**, 403.
23 It is possible to increase the $Mn^{4+}$ content in the parent $LaMnO_3$ and render it ferromagnetic. $LaMnO_3$ with $\geqslant$20% $Mn^{4+}$ shows the I–M transition and GMR: R. Mahesh, K. R. Kannan and C. N. R. Rao, *J. Solid State Chem.*, 1995, **114**, 294; R. Mahendiran, R. Mahesh, S. K. Tewari, N. Rangavittal, A. K. Raychaudhuri, T. V. Ramakrishnan and C. N. R. Rao, *Phys. Rev. B*, 1995, **53**, 3348.
24 R. Mahesh, R. Mahendiran, A. K. Raychaudhuri and C. N. R. Rao, *J. Solid State Chem.*, 1995, **120**, 204.
25 J. M. Honig and J. Spalek, in *Advances in Solid State Chemistry*, ed. C. N. R. Rao, Indian National Science Academy, New Delhi, 1986.
26 J. K. Burdett, *Acc. Chem. Res.*, 1995, **28**, 227.
27 D. E. Logan, Y. H. Szczechand and M. A. Tusch, in *Metal–insulator transitions revisited*, ed. P. P. Edwards and C. N. R. Rao, Taylor and Francis, London, 1995.
28 N. Kumar and A. M. Jayannavar, in *Metal–insulator transitions revisited*, ed. P. P. Edwards and C. N. R. Rao, Taylor and Francis, London, 1995.

*Received, 21st March 1996; 6/01957I*

Journal of
Materials Chemistry A

RSCPublishing

FEATURE ARTICLE

# Borocarbonitrides, $B_xC_yN_z$

**Cite this:** *J. Mater. Chem. A*, 2013, **1**, 5806

Nitesh kumar,[a] Kota Moses,[b] K. Pramoda,[a] Sharmila N. Shirodkar,[a] Abhishek Kumar Mishra,[a] Umesh V. Waghmare,[a] A. Sundaresan[a] and C. N. R. Rao*[a]

Received 3rd December 2012
Accepted 1st February 2013

DOI: 10.1039/c3ta01345f

**www.rsc.org/MaterialsA**

Various forms of carbon, especially the nanocarbons, have received considerable attention in recent years. There has also been some effort to investigate borocarbonitrides, $B_xC_yN_z$, comprising besides carbon, the two elements on either side. Although uniformly homogeneous compositions of borocarbonitrides may be difficult to generate, there have been attempts to prepare them by solid state as well as gas phase reactions. Some of the products so obtained show evidence for the presence of BCN networks. Then, there are composites (G–BN) containing hexagonal BN (h-BN) and graphene (G) domains, $G_{1-x}(BN)_x$, in varying proportions. Nanotubes of $B_xC_yN_z$ have been reported by several workers. The borocarbonitrides exhibit some interesting electronic and gas adsorption properties. Thus, some of the preparations show selective $CO_2$ adsorption. They also exhibit excellent characteristics for supercapacitor applications. In order to understand the nature of these understudied materials, it is necessary to examine the results from first-principles calculations. These calculations throw light on the variation in the band gap of G–BN with the concentration of h-BN, for different geometries of the domains and their boundaries. The possibility of formation of Stone–Wales (SW) defects at the interfaces of graphene and h-BN has been studied and the estimates of the formation energies of SW defects at the interfaces are ~4 to 6 eV. The presence of such defects at the interfaces influences the electronic structure near the band gap and the associated properties. For example, adsorption of $CH_4$ and $CO_2$ occurs with significantly stronger binding at the interfacial defects.

## 1 Introduction

The new forms of carbon have aroused great interest in recent years. Besides the zero-dimensional fullerenes[1] and one-dimensional nanotubes,[2,3] two-dimensional graphene[4–7] has opened up new directions of research with regard to the properties of low-dimensional materials.[8,9] Hexagonal boron nitride (h-BN) which has a structure similar to that of graphite can exist in the form of nanotubes.[10–13] Recently, single-layer and few-layer graphene-like structures of BN have been prepared and characterized.[14,15] BN is an insulator while graphene is a gapless conductor. Graphene also exhibits other novel properties such as high surface area and has also been doped with boron and nitrogen.[16] It would, therefore, be most pertinent to explore the properties of borocarbonitrides, $B_xC_yN_z$, containing the

[a]*Chemistry and Physics of Materials Unit, New Chemistry Unit, Theoretical Sciences Unit and International Centre for Materials Science, Jawaharlal Nehru Centre for Advanced Scientific Research, Jakkur P.O., Bangalore-560064, India. E-mail: cnrrao@jncasr.ac.in*

[b]*Materials Research Centre, Indian Institute of Science, Bangalore-560012, India*

*Nitesh Kumar obtained his B.Sc. degree from Delhi University and his M.S. degree from JNCASR and is pursuing research for a Ph. D. degree at JNCASR.*

*Kota Moses obtained his M.Sc. degree from the University of Hyderabad and is pursuing research for a Ph. D. degree in the Indian Institute of Science.*

three elements in different proportions. There have been some studies on compositions formed between boron and carbon as well as carbon and nitrogen in the literature and it is only recently that materials containing boron, carbon and nitrogen have received attention.[17] The borocarbonitrides are not only of academic interest but would be expected to exhibit novel properties.[18,19] $B_xC_yN_z$ nanotubes of different structures and compositions have been reported in the literature including those containing separated layers of BN and carbon or an outer $BC_2N$ layer with inner graphitic layers.[20–24] A mesoporous borocarbonitride has been prepared by the substitution reaction in mesoporous carbon at 1550 °C using boron trioxide as the boron source; reaction of carbon nitride with a borane complex also yields a borocarbonitride.[25,26] Pyrolysis of the $BH_3$–trimethylamine adduct is reported to yield $B_xC_yN_z$ nanotubes.[12] Multiwalled $B_5CN_5$ nanotubes have been obtained by chemical vapour deposition.[27] $B_xC_yN_z$ type materials have been prepared by the nitridation of boric acid and carbonization of saccharose in molten urea.[28] Urea gives $NH_3$ on heating and acts as a good nitrogen source.[29] Accordingly, $BC_4N$ nanotubes have been prepared by the high-temperature reaction of amorphous carbon nanotubes with boric acid and urea[30] while microspheres of a similar composition have been produced by the reaction of low-surface area carbon spheres with urea and boric acid.[31] Such products of solid state synthesis seem to give rise to borocarbonitrides with well-defined compositions, but they are unlikely to possess entirely homogeneous compositions of B, C and N atoms throughout the structure. Spectroscopic and theoretical studies suggest these compositions to contain graphene-like networks, without long-range registry along the *c*-direction. Such $B_xC_yN_z$ species would therefore be comparable to the few-layer graphenes wherein the hexagonal networks are formed with B–C, B–N, C–N and C–C bonds but with no B–B or N–N bonds. Gas phase synthesis of $B_xC_yN_z$ by the reaction of a hydrocarbon such as $CH_4$ with ammonia and $BCl_3$ was carried out sometime ago by Bartlett and coworkers.[32] Recently, a similar reaction of $BBr_3$, $NH_3$ and ethylene has been found to give borocarbonitrides.[33] Composites containing hybrid structures of BN and graphene have also been prepared.[34,35] Some of the composites contain stacks of graphene and BN or involve deposition of graphene on BN surfaces.[36–39]

In this article, we discuss the results of recent studies directed towards the preparation and characterization of borocarbonitrides containing hexagonal BCN networks albeit with no long range order, and also of the composites of BN with graphene (Fig. 1). We examine properties of these materials, especially their high surface areas and the selective adsorption of $CO_2$ and $CH_4$. We also discuss supercapacitor applications of borocarbonitrides.

In order to understand the structures and properties of borocarbonitrides adequately, it is essential to resort to first-principles calculations. We present the results of calculations

*S. N. Shirodkar obtained her M.Sc. degree from the University of Pune, and is pursuing research for a Ph. D. degree at JNCASR.*

*A. Sundaresan obtained his Ph. D. degree from IIT Bombay and did post-doctoral work at Caen, Grenoble and Tsukuba. He is an associate professor of chemistry of materials at JNCASR and is a recipient of medals of the Materials and Chemical Research Societies of India.*

*U. V. Waghmare obtained his Ph. D. degree from Yale University and did Post-doctoral work at Harvard. He is now a professor of theoretical sciences at JNCASR. He is a fellow of the Indian Academy of Sciences and is a recipient of the Shanti Swarup Bhatnagar Prize in Physics.*

*C. N. R. Rao is National Professor and Linus Pauling Research Professor at JNCASR and is an Honorary Fellow of RSC. He is a member of several science academies including the Royal Society. He is the recipient of Hughes and Royal Medals of the Royal Society, Dan David Prize as well as Illy Trieste Science Prize for materials research and the August von Wilhelm Hoffmann medal of the German Chemical Society.*

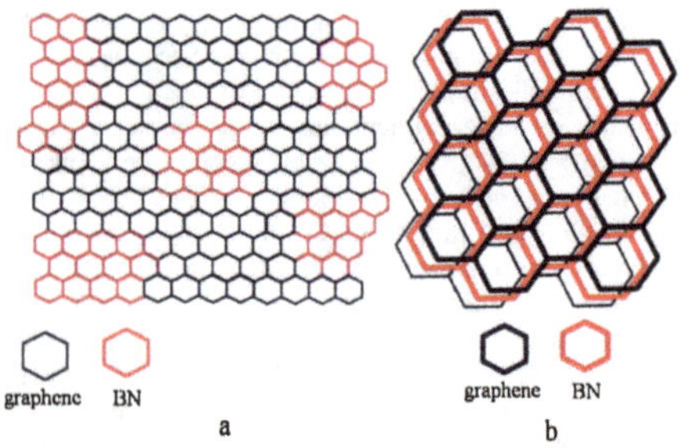


**Fig. 1** (a) Schematic of graphene–BN layers and (b) stacking of graphene and BN layers.

on BCN networks as well as BN–graphene composites (G–BN). A solid solution of hexagonal BN and graphene containing domains of BN in a matrix of graphene would be expected to show interesting changes in electronic properties,[40] with the properties depending on the size of the BN domains and the type of interface (zigzag or armchair) with the graphene matrix. A zigzag interface of a BN–graphene composite would have edge states at the graphene edge, whereas the armchair interfaces would be devoid of such states. Due to the electric potential of the BN domain, the edge states can give rise to half metallicity in graphene[41] and the type of interface therefore is an important factor. A systematic study of the variation of the band gap of graphene doped with varying concentrations of h-BN for different types of interfaces (zigzag and armchair) has been carried out. The energetics of domain formation of BN in graphene and the structural changes accompanying Stone–Wales (SW) defects at the interfaces have also been examined in these calculations. It should be noted that despite several theoretical studies of solid solutions of BN and graphene,[41–43] the nature and role of defects have not received adequate attention. We briefly discuss the adsorption of gas molecules on G–BN interfaces, with and without the SW defects.

## 2 Synthesis and characterization of borocarbonitrides

Borocarbonitrides which are likely to possess hexagonal BCN networks would be of great chemical interest. In this section, we shall discuss the synthesis of these materials.

### 2.1 Solid state synthesis based on urea

In an effort to generate borocarbonitrides, possibly containing hexagonal $B_xC_yN_z$ networks, an intimate mixture of carbon (high surface-area activated charcoal), boric acid and urea has been employed as the starting material.[44] In a typical synthesis, different ratios of the three starting materials are mixed in water and kept at 80 °C in order to obtain a slurry. The slurry is then heated at 900–1000 °C in a nitrogen atmosphere followed by a heat treatment in an ammonia atmosphere at around 900 °C. The reactions involved in the synthesis may be represented by the following chemical equations:

$$2H_3BO_3 \xrightarrow{\Delta} B_2O_3 + 3H_2O \tag{1}$$

$$CO(NH_2)_2 \xrightarrow{\Delta} NH_3 + HNCO \tag{2}$$

$$B_2O_3 + 2NH_3 + 2C \xrightarrow{\Delta} 2BCN + 3H_2O \tag{3}$$

$$B_2O_3 + 2NH_3 + 4C \xrightarrow{\Delta} 2BC_2N + 3H_2O \tag{4}$$

The compositions of the products can be varied by changing the molar ratio of the reactants as well as the temperature of the reaction. These products have been characterized by X-ray photoelectron spectroscopy (XPS) to obtain their compositions and to establish the type of bonds present. Fig. 2 shows typical core-level spectra of two compositions BCN-1 and BCN-3. By taking account of the capture cross-sections, the core level spectra of B, C and N atoms gave the compositions of BCN-1, BCN-2 and BCN-3 to be $BC_{1.6}N$, $BC_{1.7}N_{0.6}$ and $BC_{1.9}N$ respectively. The B 1s feature of these compositions can be deconvoluted into two peaks arising from the contribution from B–N and B–C bonds centred at 190.8 eV and 192.7 eV respectively as shown in Fig. 2.

The C 1s feature of this composition can be deconvoluted into three peaks corresponding to C–B, C–C and C–N bonds at 284.6 eV, 285.8 eV and 288.6 eV, respectively. The N 1s feature can be deconvoluted into N–B and N–C contributions centred at 398.4 eV and 400.3 eV, respectively (Fig. 2).[27] The XPS data thus suggest the presence of different bonding environments of the B, C and N atoms and hence the presence of random BCN networks. It seems that such networks are present along with

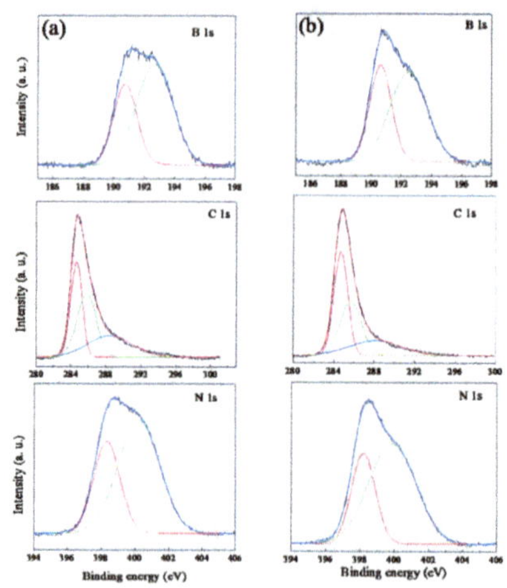


**Fig. 2** Core level XP spectra of B, C and N atoms for (a) $BC_{1.9}N$, BCN-3 and (b) $BC_{1.6}N$, BCN-1.

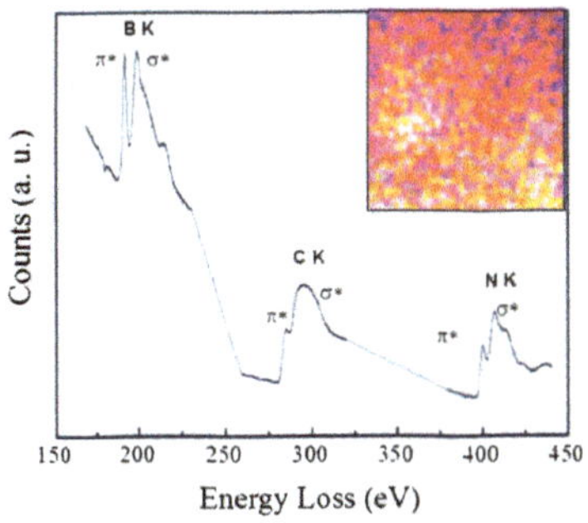


**Fig. 3** EEL spectra of BCN-2 with the inset showing elemental mapping with B, C and N atoms as red, yellow and purple dots, respectively.

domains of BN and carbon (Fig. 1). Fig. 3 shows the electron energy loss (EEL) spectra of BCN-2 where we see splitting of the $\pi^*$ and $\sigma^*$ in B 1s, C 1s and N 1s levels, indicating $sp^2$ hybridization of these atoms as expected in graphene-like structures. In this figure, we also show the EELS maps to demonstrate how the three elements B, C and N are distributed fairly uniformly. The $^{13}C$ NMR spectra showed a shift of the band from 118.1 ppm (in the starting carbon source) to 122.2 ppm in the BCN samples. The $^{11}B$ NMR spectra of the BCN samples showed features around 20 and 50 ppm due to B–N and B–C type species, respectively.

Fig. 4 shows atomic force microscopic images of the three compositions BCN-1, BCN-2 and BCN-3 along with their height profiles. The height profiles correspond to 2–6 layers of $B_xC_yN_z$ with lateral dimensions in the 70–150 nm range irrespective of the compositions. High resolution transmission electron microscope (HREM) images (Fig. 4) of these compositions show layered structures with an interlayer separation of ~0.36 nm.

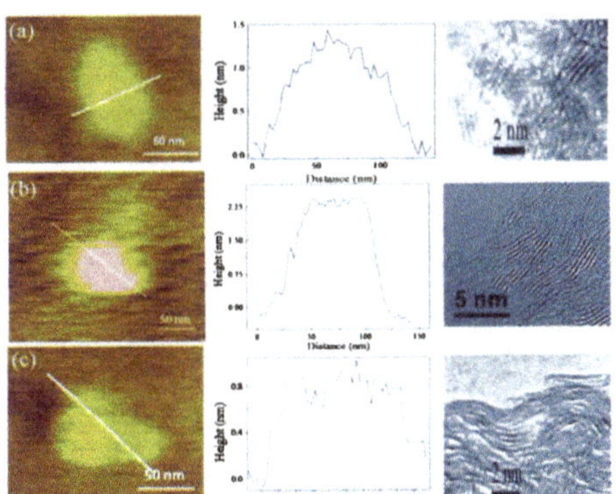


**Fig. 4** AFM images with the corresponding height profiles and HREM images of (a) BCN-3, (b) BCN-2 and (c) BCN-1 (reproduced from ref. 44).

X-ray diffraction patterns for BCN-1, BCN-2 and BCN-3 contain broad features corresponding to the (002) reflection and convoluted (100) and (101) reflections. Broadening of these reflections arises partly because of the presence of only a few layers. The Raman spectra (Fig. 5) show two bands where the ~1300 $cm^{-1}$ band corresponds to the in-plane B–N vibrational mode, probably in admixture with the D band of graphene, while the band at around1600 $cm^{-1}$ is due to the $E_{2g}$ band of graphene-like domains.

The borocarbonitrides obtained by the urea route using activated charcoal generally possess high surface areas consistent with the presence of two-dimensional layers. Nitrogen adsorption–desorption curves of BCN-1 and BCN-3 at 77 K are shown in Fig. 6. Specific surface areas calculated by the Brunauer–Emmett–Teller (BET) method for BCN-1 and BCN-3 are 1509 and 1991 $m^2\ g^{-1}$ respectively. The isotherms show type IV behavior with H3 kind of hysteresis loops involved due to capillary condensation of nitrogen in the mesopores.

**Nanotubes.** Borocarbonitrides obtained by using high surface area carbon possess layered structures, but have no well-defined morphology. Borocarbonitride nanotubes have been prepared by the template assisted method.[30] In this method, amorphous carbon nanotubes (a-CNTs) are prepared by a polycarbonate membrane with a pore diameter of 220 nm. a-CNTs are soaked in a viscous mixture of boric acid and urea followed by drying and heating in a nitrogen atmosphere at 970 °C and

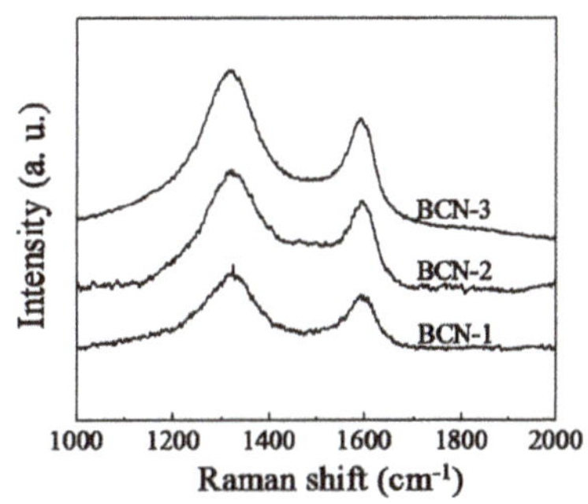


**Fig. 5** (a) Raman spectra of borocarbonitrides prepared by the urea route (reproduced from ref. 33).

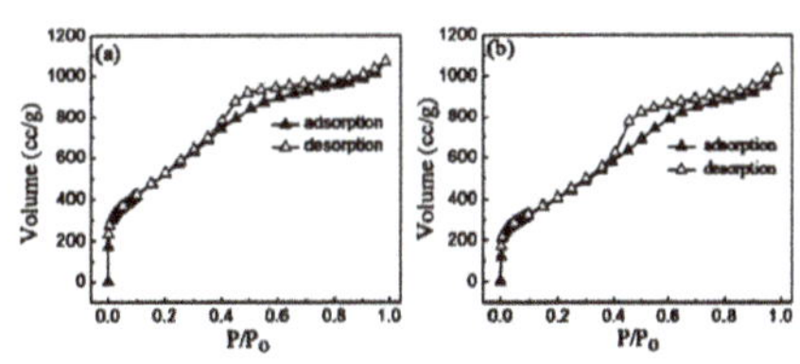


**Fig. 6** $N_2$ adsorption–desorption isotherms of (a) BCN-3 and (b) BCN-1 at 77 K and 1 atm (reproduced from ref. 44).

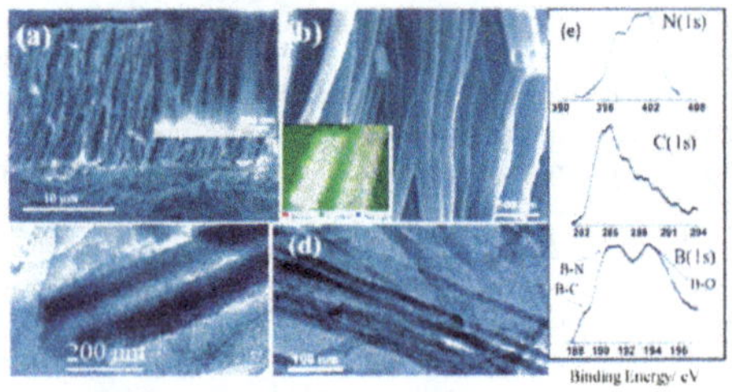

Fig. 7 (a) FESEM images of $B_xC_yN_z$ nanotube brushes with the average diameter of a single tube being around 170 nm. The inset shows a FESEM image of $B_xC_yN_z$ nanotube brushes of 40 nm diameter. (b) Higher magnification FESEM images of BCN nanotube brushes with the inset showing elemental mapping (with red, green and blue dots as B, C and N atoms respectively). TEM image of a $B_xC_yN_z$ nanotube (c) 170 nm diameter, (d) 40 nm diameter. (e) XP spectra of $BC_4N$ nanotubes (reproduced from ref. 30).

then at 1050 °C in an ammonia atmosphere. Fig. 7a shows vertically aligned a-CNTs. Borocarbonitride nanotubes are shown in higher magnification FESEM images in Fig. 7b where the opening of the tubes can be easily seen. Fig. 7c and d show TEM images of nanotubes of different diameters. XPS analysis carried out on the nanotubes (Fig. 7e) gave a composition of $BC_4N$. The composition was also confirmed by EEL spectroscopy. The EELS map shows that the B, C and N atoms are uniformly distributed. The $B_xC_yN_z$ nanotubes are highly insulating with the resistance falling in the MΩ range.

**Microspheres.** The urea route has also been used to obtain microspheres of borocarbonitrides. The carbon source was amorphous carbon spheres prepared by a hydrothermal method using glucose as the carbon source. FESEM images of carbon spheres used and borocarbonitride microspheres are shown in Fig. 8a–d. XPS and EELS studies of the microspheres gave a composition of $BC_4N$.

A typical EEL spectrum of borocarbonitride microspheres is shown in Fig. 8e along with an elemental EELS map (Fig. 8f) indicating that B, C, and N atoms are present uniformly throughout the sample. We see that the B 1s, C 1s and N 1s levels appear as doublets. Like the borocarbonitride nanotubes, the microspheres are electrically insulating. More significantly, they are stable in air up to 1000 °C without any weight loss as seen in TGA. The $BC_4N$ nanotubes are also stable up to 900 °C. The increased thermal stability over graphene and other pure carbon structures shows that the nanotubes and microspheres contain BCN networks and not mixtures of BN and carbon. Fig. 9 shows the nitrogen adsorption–desorption curves at 77 K of the $BC_4N$ nanotubes and microspheres along with the pore-size distributions. Both the materials contain micropores as well as mesopores. The BET surface areas of the $BC_4N$ nanotubes and microspheres are 356 and 428 $m^2\,g^{-1}$ respectively.

### 2.2 Gas phase synthesis

Chemical vapor deposition techniques have been used to prepare graphene and boron nitride. A common practice is to use a substrate like Si and SiC or transition metal foils such as Cu and Ru, *etc.* Ci *et al.*[35] have attempted a synthesis of $B_xC_yN_z$ by the gas phase reaction of $CH_4$ and $NH_3$–$BH_3$ wherein the product always had a B : N ratio of unity. The films obtained generally contained 2–3 layers. TEM studies (Fig. 10) revealed turbostatic stacking without any long range ordering resulting from AB type stacking. EEL spectra (Fig. 11) showed evidence for $sp^2$ hybridization of the three atoms. The optical gap of BCN was 4.48 eV and decreased with the increase in carbon content.

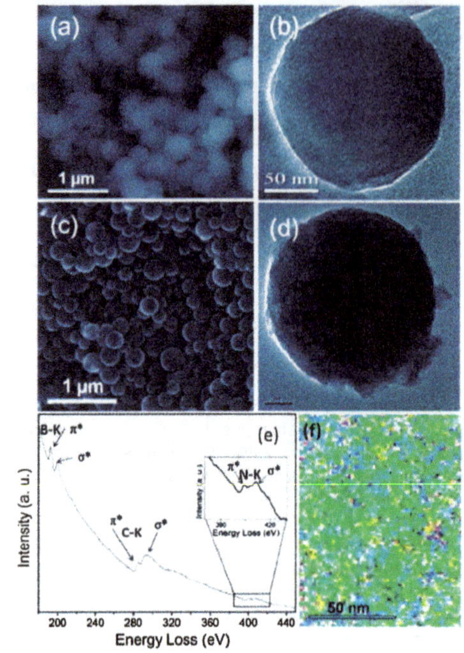


Fig. 8 (a) FESEM image and (b) TEM image of carbon spheres. (c) FESEM image and (d) TEM image of $B_xC_yN_z$ spheres. (e) EEL spectrum (inset shows core edge fine structure of nitrogen) and (f) EELS elemental map (boron (red), carbon (green), and nitrogen (blue)) (reproduced from ref. 31).

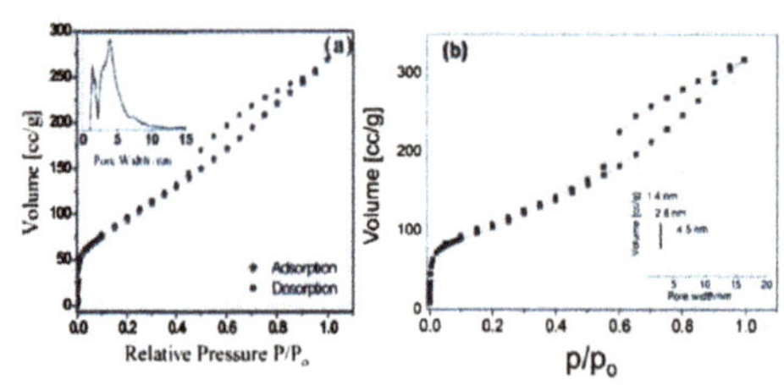


Fig. 9 $N_2$ adsorption–desorption isotherms of (a) $BC_4N$ nanotubes and (b) $BC_4N$ microspheres along with their pore-size distributions (reproduced from ref. 31).

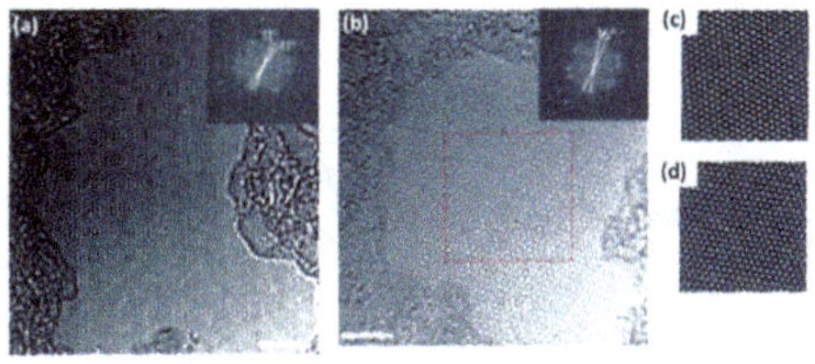


**Fig. 10** (a) An atomic-scale Moiré pattern of a region of h-BNC; the inset FFT indicates a three-layer stacked region with a rotational angle of 11 °C. (b) A Moiré pattern from a different region; the FFT in the inset reveals a two-layer packing with a relative rotational angle of 16°. (c and d) Two individual atomic layers reconstructed by masking the FFT pattern from the area in (b) (reproduced from ref. 35).

Synthesis of different $B_xC_yN_z$ compositions can be achieved by the pyrolysis of precursors such as pyridine–$BH_3$ and trimethylamine–$BH_3$. By this means, it should be possible to obtain well-defined compositions of borocarbonitrides. High surface area ($\sim$850 $m^2\,g^{-1}$) $BC_4N$ has been obtained by the pyrolysis of a boron imidazolate. This material exhibits good hydrogen adsorption and has one-dimensional morphology.[45] Goto *et al.*[46] obtained $BC_{1.5}N_{0.4}$ and $BC_{0.9}N_{0.4}$ by the pyrolysis of a pyrrolidine–borane complex while Sauter *et al.*[47] obtained $[B_2(NCN)_3$ with a hexagonal planar atomic arrangement by the pyrolysis of poly(borosesquicarbodiimide). Thermolysis of the ionic liquid *N,N′*-ethylmethylimidazoliumtetracyanoborate yields a composition close to $BC_3N$.[48]

Direct synthesis of such compounds by gas phase reactions has been carried out without involving substrates. A borocarbonitride was first synthesized by this means by Bartlett and co-workers[32] using $BCl_3$, acetylene and ammonia as boron, carbon and nitrogen sources, respectively. They obtained a product of the composition $B_{0.35}C_{0.30}N_{0.35}$ ($\sim$BCN), but the detailed characterization of the product was not carried out. We have prepared $B_xC_yN_z$ by the gas phase reaction of $BBr_3$, methane and ammonia at 950 °C.[33] The XP spectra of the product reveal a composition of $BC_{1.6}N$. The core-level spectra of B, C and N could be fitted using more than one Gaussian indicating different types of bonding environments around these atoms. The B 1s feature centred around 191.5 eV can be deconvoluted into peaks at 190.5 and 190.0 eV corresponding to B–C and B–N bonds, respectively. The C 1s band can be deconvoluted into three peaks at 284, 286 and 290 eV corresponding to B–C, C–C and C–N bonds, respectively. Similarly, the N 1s feature can be deconvoluted into two peaks at 397.5 and 400 eV for B–N and C–N bonds, respectively. In order to verify the composition obtained by XPS, EELS measurements (Fig. 12) show the K-shell ionization edges of boron, carbon and nitrogen and yield a composition of $BC_2N$. The inset of Fig. 12 shows the EEL spectrum of another region of the sample with the composition BCN. The peaks for B, C and N atoms are split into two peaks corresponding to the $\pi^*$ and $\sigma^*$ levels confirming that all the three atoms are $sp^2$ hybridized. The EELS map indicates uniform distribution of B, C, and N atoms.

In Fig. 13a and b we show TEM and HREM images of $BC_{1.6}N$, respectively. The images reveal disc-type morphology with an outer rim. A layered structure can be inferred from the HREM

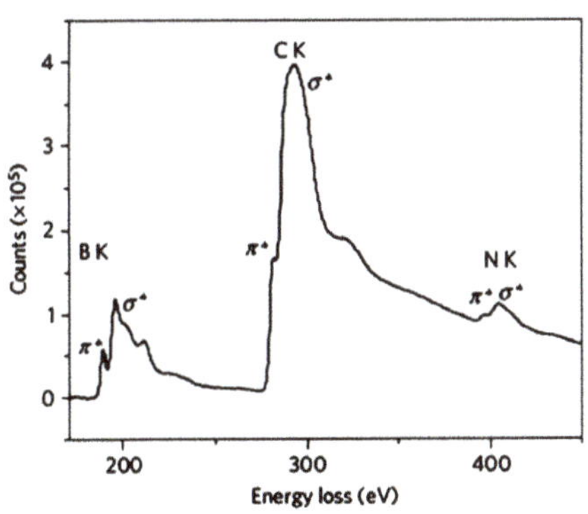


**Fig. 11** K-shell excitations of B, C and N in the EEL spectrum of a BNC film (reproduced from ref. 35).

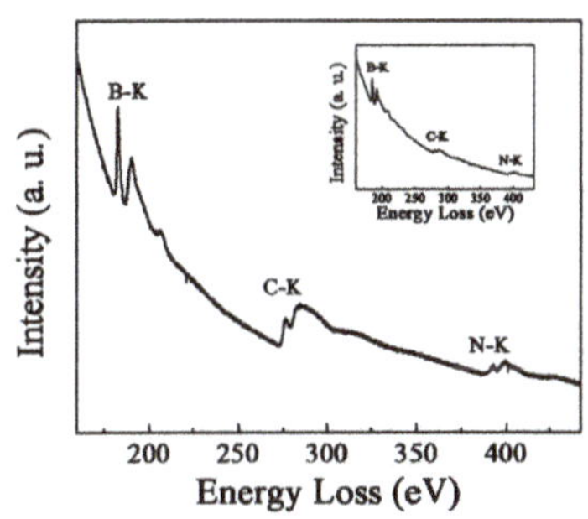


**Fig. 12** K-shell ionization edges of B, C and N of $BC_2N$ with the inset showing a different region of the sample with a composition BCN (reproduced from ref. 33).

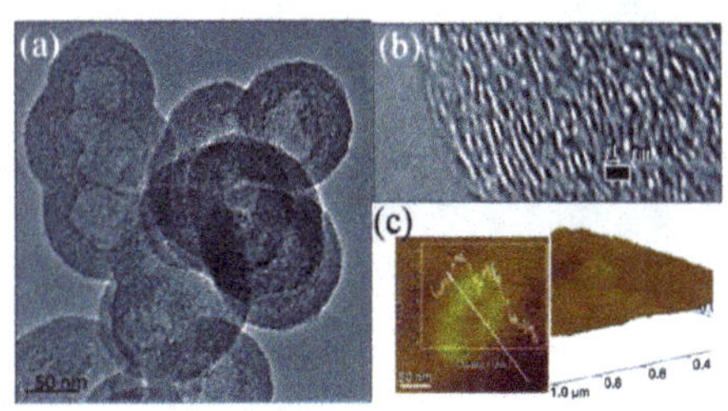


**Fig. 13** (a) TEM image of borocarbonitride nanopans, (b) HREM of the rim portion of the borocarbonitride, (c) AFM image with the corresponding height profile (reproduced from ref. 33).

image of the rim with an interlayer separation of ~0.34 nm. The XRD pattern of $BC_{1.6}N$ showed broad (002) and (101) reflections as expected from such a nanostructure. The AFM image along with the height profile as shown in Fig. 13c indicates a dip at the centre of the disc suggesting more appropriately a pan-like structure with a lateral dimension of ~100 nm. The height of the centre of the pan corresponds to the monolayer thickness of $B_xC_yN_z$. Thermogravimetric analysis showed the decomposition temperature of the borocarbonitride to be 740 °C, only slightly higher than graphene.

### 2.3 Graphene–BN composites

Composites obtained by alternate stacking of graphene and BN would be of value in electronic devices. Liu *et al.*[36] have carried out a CVD synthesis of BN–graphene stacks by employing a two-step process. The stacked structure consists of two to few layers as revealed by TEM (Fig. 14). The Raman spectrum shows G and 2D bands of graphene and the $E_{2g}$ band of BN. This spectrum is different from that of borocarbonitrides with BCN networks where one observes a broad D band. Depth profile XPS analysis shows that the stacked structures also indicate the presence of BN and graphene. Yan *et al.*[38] have attempted to grow graphene on BN by using solid carbon sources.

Random stacking of BN and graphene can be accomplished by mixing solutions or dispersions of the two in suitable solvents.[39] Such composites would contain graphene and BN sheets and exhibit additive Raman spectra. In Fig. 15a we show the Raman spectrum of such a composite of BN and graphene (RGO) exhibiting the band due to BN appearing next to the D-band of graphene. A composite of single-layer graphene with BN would only show the band corresponding to BN, without any interference from the D-band of graphene. In Fig. 16a, we show

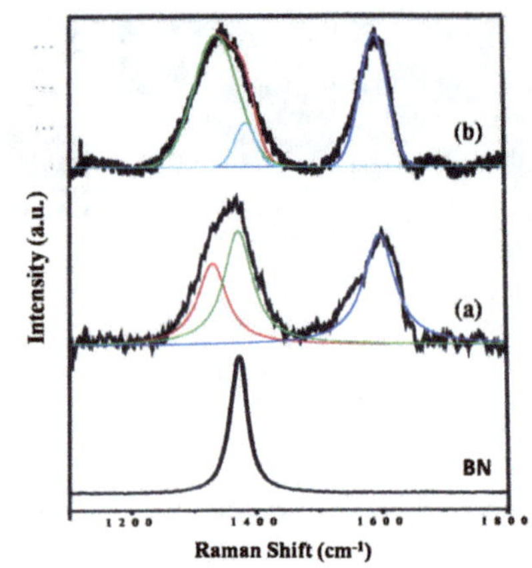


Fig. 15 Raman spectra of BN, (a) a physical mixture of BN and graphene (RGO) and (b) covalently attached BN and graphene.

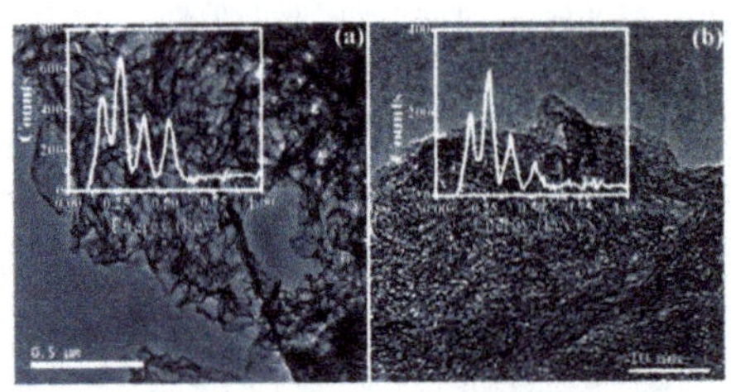

Fig. 16 TEM images and the corresponding EDAX spectra of (a) a physical mixture of BN and graphene (RGO) and (b) covalently attached BN and graphene.

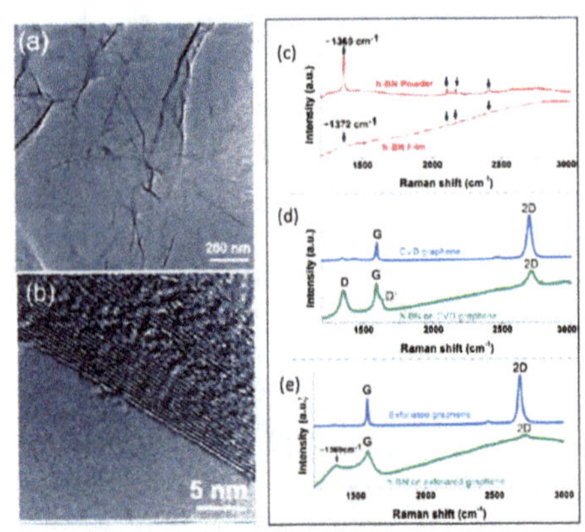


Fig. 14 (a) TEM image showing the morphology of a graphene–BN film. (b) HREM image of the film edge showing the number of layers and thickness of the film. Raman spectra of (c) BN powder and film with a band at ~1370 $cm^{-1}$ (additional peaks are indicated by arrows), (d) graphene and h-BN on CVD graphene and (e) mechanically exfoliated graphene and graphene after the growth of BN (reproduced from ref. 36).

a TEM image and EDAX data of a sample of the composite. There is much scope for studying such composites, especially those obtained at the liquid–liquid interface (Fig. 17).[49]

Graphene chemically bonded to BN has been prepared by using the amino group on the BN surface ($BN-NH_2$) and the carboxyl groups on the oxidized surface of graphene (COOH–G).

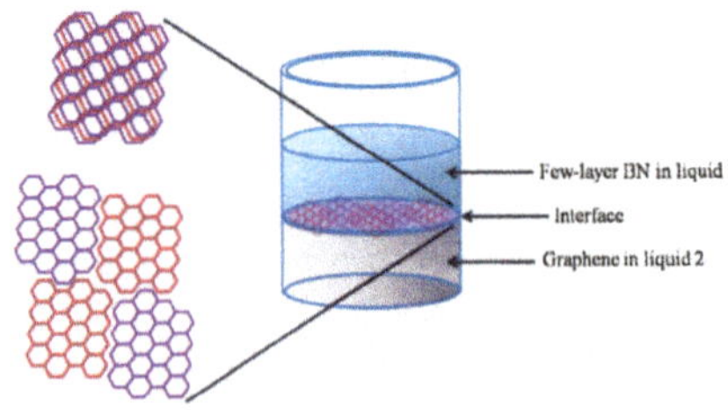


Fig. 17 Schematic of the formation of graphene–BN composites at the liquid–liquid interface.

$$\text{BN-NH}_2 + \text{HOOC-G} \xrightarrow{\text{EDC+DMAP}} \text{BNNHCOG}$$

Here, EDC (1-ethyl-3-(3-dimethylaminopropyl)carbodiimide) is the coupling agent and 4-dimethylaminopyridine (DMAP) acts as the catalyst. The Raman spectrum of such an adduct shows a band due to BN next to the D-band of graphene (Fig. 15b). The infrared spectrum shows the B–N stretching band, and CO and N–H stretching bands of the amide linkage. In Fig. 16b, we show a TEM image of the covalently bonded composite along with the EDAX data. A better way of preparing covalently linked BN-G composites would be to react $BN\text{-}NH_2$ with ClCO–G to create the amide linkage.

## 3 Magnetic and electrical properties

Graphene shows magnetic hysteresis at room temperature. Such high temperature ferromagnetism in graphene and in many inorganic nanomaterials[50,51] has been of much interest in the last few years. Several investigations of graphene have shown that the magnetism is not due to impurities. Thus, interaction with electron donor and acceptor molecules as well as hydrogenation affect the magnetic properties. Graphene-like BN sheets exhibit magnetic hysteresis with the saturation magnetization increasing with the decreasing number of layers.[51] It would, therefore, be not surprising to find room-temperature ferromagnetism in borocarbonitrides. We have studied magnetic properties of a borocarbonitride with the composition $BC_{1.9}N$ prepared by the urea route with a BET surface area of 1990 $m^2\,g^{-1}$ Fig. 18a shows magnetization data under zero field cooled (ZFC) and field cooled conditions (FC) at 500 Oe. The data clearly show divergence between the ZFC and FC curves up to room temperature. The curves overlap at a higher field of 1 tesla as shown in the inset of Fig. 18a. The behavior resembles that of magnetically frustrated systems. Room temperature hysteresis data in Fig. 18b give 0.014 emu $g^{-1}$ as the saturation magnetization ($M_S$) with a coercive field ($H_C$) value of ~120 Oe. The inset of Fig. 18b shows the hysteresis loop at 5 K with $M_S$ and $H_C$ values of 0.092 emu $g^{-1}$ and 300 Oe, respectively. Borocarbonitride prepared by the gas phase route with the composition $BC_{1.6}N$ also shows similar behavior with a lower magnetization. Its ZFC and FC curves at 500 Oe are shown in Fig. 18c. Divergence between the two curves prevails until 390 K. The room temperature hysteresis loop (Fig. 18d) gives a $M_S$ of $2.2 \times 10^{-3}$ emu $g^{-1}$ and a $H_C$ of 65 Oe. The inset of Fig. 18d shows the hysteresis loop at 5 K.

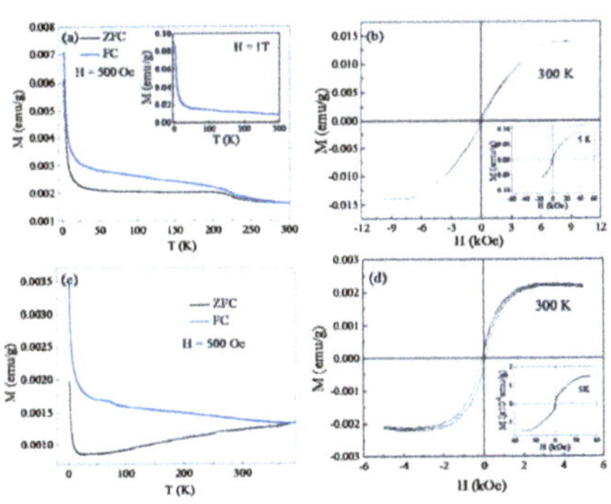


**Fig. 18** (a) ZFC–FC magnetization curves for high surface area borocarbonitride (BCN-3) at 500 Oe with the inset showing the same at 1 T. (b) Room temperature magnetic hysteresis loop of BCN-3; the inset shows the loop at 5 K. (c) ZFC–FC magnetization curves of gas phase synthesized borocarbonitride at 500 Oe. (d) Room temperature hysteresis with the inset showing hysteresis at 5 K.

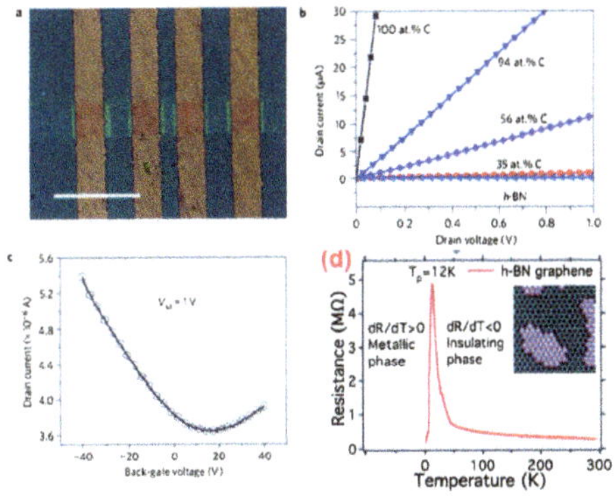


**Fig. 19** (a) Optical image of a four-terminal BNC ribbon field-effect transistor. Scale bar: 30 μm. (b) Current–voltage (*I*–*V*) characteristics at room temperature of as-grown BCN with different percentages of carbon measured at room temperature. (c) Drain current as a function of the voltage applied to the back gate for a 7 μm-wide BCN ribbon with 40% carbon. The drain–source voltage is fixed to 1 V. (d) Temperature dependence of the resistance of BCN with 90 at% carbon under zero magnetic field in a cooling run. The inset shows atomic models for hybridized h-BN–graphene (reproduced from ref. 35 and 40).

Many of the borocarbonitride compositions prepared by the gas phase or solid state routes are generally found to be insulating with large band gaps. Studies on a device made of $B_xC_yN_z$ from the gas phase route show increasing conductivity with increase in carbon content (Fig. 19a and b). At 40% C, the $B_xC_yN_z$ field-effect transistor exhibits ambipolar behavior (Fig. 19c). $B_xC_yN_z$ with 90% C exhibits an insulator-metal transition (Fig. 19d). The transition is considered to arise from the occurrence of both percolation *via* graphene networks and hopping between the edge states of BN. $BC_3N$ obtained by thermolysis of an ionic liquid is reported to be a p-type semiconductor with potential application as an electrode material.[48a] Preliminary studies of Bhattacharyya *et al.* on high surface area borocarbonitrides prepared by the urea route indicate that these materials can be used in electrodes of Li batteries.[48b]

## 4 Gas adsorption properties

Apart from metal organic frameworks (MOFs), carbon-based materials such as carbon nanotubes, activated carbons and graphene have attracted interest as adsorbents for gases for the purpose of storage as well as removal of greenhouse gases. We

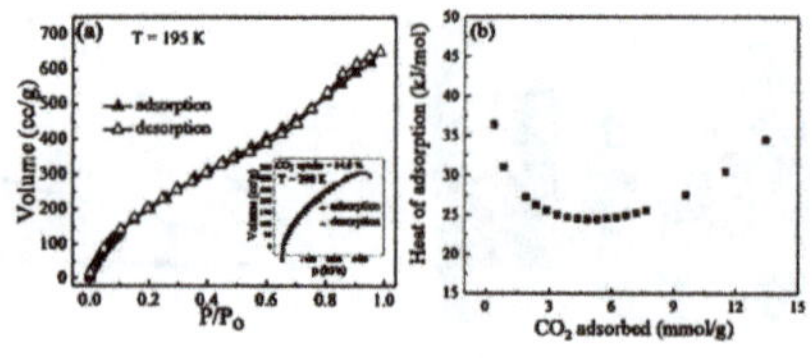


Fig. 20 (a) $CO_2$ adsorption–desorption isotherms at 195 K and 1 atm for BCN-3 with the inset showing the adsorption–desorption data at 298 K and 5 MPa (reproduced from ref. 44). (b) Isosteric heat of adsorption of $CO_2$ on BCN-2.

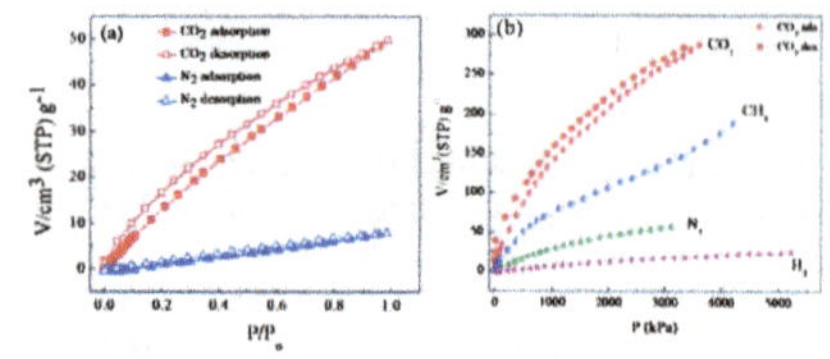

Fig. 22 (a) Comparison of $CO_2$ and $N_2$ uptakes of BCN-2 at 293 K and 1 atm. (b) High pressure adsorption isotherms of $CO_2$, $CH_4$, $N_2$ and $H_2$ at 293 K for BCN-2.

have studied the gas adsorption properties of borocarbonitrides prepared by the urea route. Among them, $B_xC_yN_z$ synthesized using activated charcoal, urea and boric acid exhibit high BET surface areas. $CO_2$ adsorption–desorption experiments carried out at 195 K and 1 atm show a remarkable $CO_2$ uptake of 128 wt% for BCN-3 which has the highest surface area of 1990 $m^2 g^{-1}$. A typical $CO_2$ adsorption–desorption isotherm is shown in Fig. 20a. At room temperature and 5 MPa, BCN-3 shows a $CO_2$ uptake of 64 wt % as shown in the inset of Fig. 20a. $CO_2$ molecules are found to be physisorbed with isosteric heat of adsorption ($Q_{st}$) of 36.4 kJ $mol^{-1}$ at 1 atm. With an onset value of 36.4 kJ $mol^{-1}$, $Q_{st}$ decreases initially with the increase in $CO_2$ loading but increases again to attain a value comparable to the onset as shown in Fig. 20b. The $CH_4$ adsorption–desorption isotherm at 298 K and 5 MPa for BCN-3 is shown in Fig. 21. It shows an uptake of 15 wt% which increases to 17 wt% at 273 K (inset of Fig. 21). The percentage uptake of $CH_4$ over borocarbonitrides was found to increase exponentially with the increase in surface area. The adsorption of $CO_2$ is found to be almost 10 times more selective over $N_2$ at 1 atm and 293 K as can be calculated from the adsorption–desorption data shown in Fig. 22a. In order to compare the uptake of different gases, adsorption isotherms for $CO_2$, $CH_4$, $N_2$ and $H_2$ at 293 K and 5000 kPa in the same scale are shown in Fig. 22b. Selectivities of $CO_2$ over $H_2$, $N_2$ and $CH_4$ are 28 : 1, 8 : 1 and 4 : 1, respectively. A remarkable selectivity of 68 : 1 is observed for $CH_4$ over $H_2$ at 293 K and 4000 kPa. $CO_2$ adsorption measurements carried out on microspheres of $BC_4N$ show 40 wt% uptake at 195 K and 1 atm whereas 9 wt% at room temperature and 4 MPa. The material also shows 1.1 wt% uptake of $H_2$ at 77 K and 5 MPa. Nanotubes of borocarbonitrides of the same composition prepared by the urea route show $CO_2$ uptake of 23.5 wt% at 195 K and 1 atm and $H_2$ uptake of 0.4 wt% at 77 K and 1 atm.

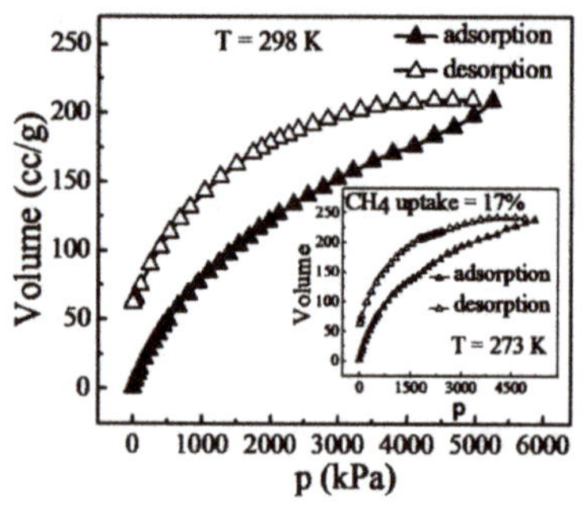


Fig. 21 Methane adsorption–desorption isotherms at 298 K and 5 MPa for BCN-3 with the inset showing the data at 273 K and 5 Mpa (reproduced from ref. 44).

## 5 Supercapacitor applications

Borocarbonitride nanotubes obtained by the pyrolysis of melamine diborate around 1000 °C, with a surface area of 347 $m^2 g^{-1}$, are reported to exhibit a high specific capacitance of 321 F $g^{-1}$ in aqueous medium, when the nanotubes were aligned.[52] The non-aligned nanotubes showed a specific capacitance of 167.3 F $g^{-1}$. These measurements were based on a 3-electrode system.

High surface-area borocarbonitrides prepared by the urea route have been examined for supercapacitor applications.[53]

Measurements with a 2-electrode system on a borocarbonitride with a surface area of 1280 $m^2 g^{-1}$ have yielded interesting results (Fig. 23). $BC_{4.5}N$ shows a specific capacitance of 178 F $g^{-1}$ at 5 mV $s^{-1}$ in aqueous medium (0–1 V range). In an ionic liquid where the voltage range is 0–2.5 V, the specific capacitance value of $BC_{4.5}N$ was 240 F $g^{-1}$ at 5 mV $s^{-1}$. The galvanostatic charge–discharge measurements give a capacitance value of 140 F $g^{-1}$ at a current density of 100 mA $g^{-1}$. In aqueous medium, $BC_{4.5}N$ exhibits supercapacitor characteristics superior to those of nitrogen-doped graphene, or reduced graphene oxide.

## 6 Results of first-principles calculations

First-principles calculations have been carried out based on the density functional theory (DFT) as implemented in the

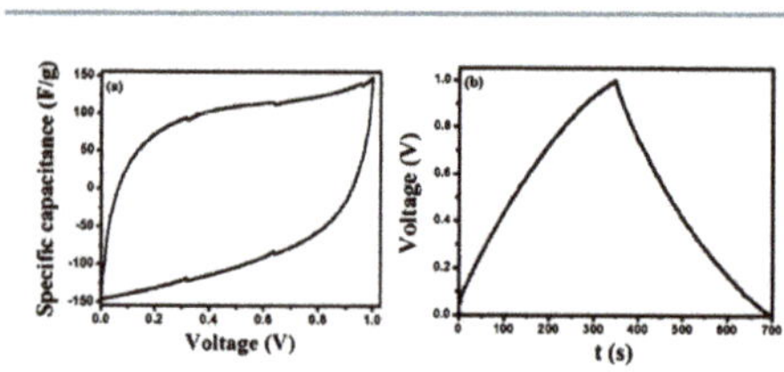


Fig. 23 (a) Cyclic voltammograms of $BC_{4.5}N$ at a scan rate of 100 mV $s^{-1}$; (b) galvanostatic charge–discharge curves for $BC_{4.5}N$ electrodes at a current density of 100 mA $g^{-1}$.

 

Quantum ESPRESSO code,[54] by employing the Local Density Approximation (LDA) with Caperley Alder functional for exchange correlation energy.[55] In these calculations interactions between ionic cores and valence electrons are represented by ultrasoft pseudopotentials[56] and interaction between the periodic images of the sheets is made negligible with an inclusion of a vacuum of 10 Å in the $z$ direction. An energy cut off of 30 Ry is used for the plane wave basis in representing wave functions and a grid with a cut off of 180 Ry in representing the density. The structures are relaxed till the magnitude of Hellman–Feynman forces on the ions become smaller than 0.02 eV $Å^{-1}$. A solid solution of graphene and BN has broadly two types of interfaces between the BN and graphene domains: armchair (AC) and zigzag (ZZ). AC interfaces are non-magnetic (NM) but ZZ interfaces in nanoribbons are anti-ferromagnetic (AFM) half-metals when the number of zigzag chains of BN are ≥6 and the number of zigzag chains of C are ≤8.[41] To study the effects of SW defects on both magnetic and non-magnetic configurations, two supercells with varying number of graphene and BN stripes can be considered. We refer to the NM supercell as SC1 and the magnetic supercell as SC2. Since the defect states form the highest occupied molecular orbital (HOMO) and the lowest unoccupied molecular orbital (LUMO), and get localized on the SW defect, we expect the defect sites to adsorb gases effectively. We study the influence of such interfacial SW defect sites on the adsorption of $CO_2$ and $CH_4$.

For the AC interface, the unit cell of SC1 consists of 3 AC chains of graphene and 3 AC chains of BN along the $y$ direction. Supercells SC1 and SC2 are constructed by repeating the unit cell 4 times and 5 times along the $x$ axis, respectively. Brillouin zone integrations were carried out with a 3 × 3 × 1 mesh of $k$-points for both SC1 and SC2. Similarly, for the ZZ interface, the unit cell of SC1 consists of 4 ZZ chains of graphene and 4 ZZ chains of BN (both with ZZ edges) and SC2 consists of 5 ZZ chains of graphene and 9 ZZ chains of BN, both along the $y$ direction. Supercells SC1 and SC2 are constructed by repeating the unit cell 6 times and 5 times along the $x$ axis, respectively. Brillouin zone integrations are carried out with a 3 × 3 × 1 and 12 × 5 × 1 mesh of $k$-points for SC1 and SC2, respectively. Since magnetic calculations critically depend on the $k$-point sampling, a large number of $k$-points are considered for SC2, which shows a stable AFM state without the SW defect.

We examine the energetics of different configurations of G-BN with random distribution of B, C and N atoms by considering a 2 × 2 supercell with 8 atoms. Here, integrations over its Brillouin zones (BZ) were sampled with a Monkhorst-Pack scheme[57] with an 8 × 8 × 1 mesh of $k$ points, using two different supercells for the study of the band gap. A 24 atom supercell of graphene was constructed for studying the variation of band gap with varying h-BN doping concentration. In the first, a 1 × 6 × 1 rectangular supercell of BN makes a zigzag interface with graphene (SZ), and in the second, a 3 × 2 × 1 rectangular supercell of BN contains graphene domains (SD). The Brillouin Zone (BZ) integrations were carried out on a 12 × 2 × 1 and 6 × 4 × 1 mesh of $k$-points for SZ and SD, respectively.

In simulations of gas adsorption, we include semiempirical dispersive van der Waals interactions (vdW) according to Grimme,[58] along with a generalized gradient approximation (GGA)[59] to exchange the correlation energy of electrons and ultrasoft pseudopotentials[56] to represent interaction between ionic cores and valence electrons. Kohn–Sham wave functions were represented with a plane wave basis with an energy cut off of 25 Ry and charge density cut off of 150 Ry. Supercell SC1 with an AC interface was considered for this study. Integration over the Brillouin zone was sampled with a Monkhorst–Pack scheme[57] with a 3 × 3 × 1 mesh of $k$ points.

### 6.1 Chemical ordering of boron, nitrogen and carbon atoms in graphene

Monte Carlo simulations and recent experiments predict formation of BN domains in graphene.[35,60–64] When the ratio of the number of boron to nitrogen atoms is unity (B/N = 1), the materials show segregation of BN islands in graphene (for carbon rich samples) or graphene islands in the BN matrix (for BN rich samples).[62] In nitrogen-rich samples of BN doped graphene, the carbon atoms appear to get dispersed into the BN lattice,[62] favored by the energetically favourable C–N bonds.

We first analyze different configurations of G-BN with a hexagonal supercell containing $C_4B_2N_2$ with pseudo-random distribution of B, C and N atoms (Fig. 24). Among these, configuration (a) is the most stable one, and is characterized by the largest number of B–N and C–C bonds. Configuration (b) is the second lowest in energy, but higher than configuration (a) by 2.3 eV, and this is because it contains more C–N and B–N bonds than configuration (a). Configurations (c) and (d) are even higher in energy and the chemical cause for the same is the presence of N–N and B–B bonds which are energetically least favorable. As a result, configurations (c) and (d) are 3.5 eV and 4.2 eV higher than configuration (a), respectively. These results are consistent with the hierarchy of energetic stability of the bonds, B–N > C–C > C–N > C–B > B–B > N–N,[64] and suggest that formation of structures containing B–B and N–N bonds is unlikely. We find that configuration (a) (with pseudo-random distribution of B and N in graphene, see Fig. 24a) is higher in

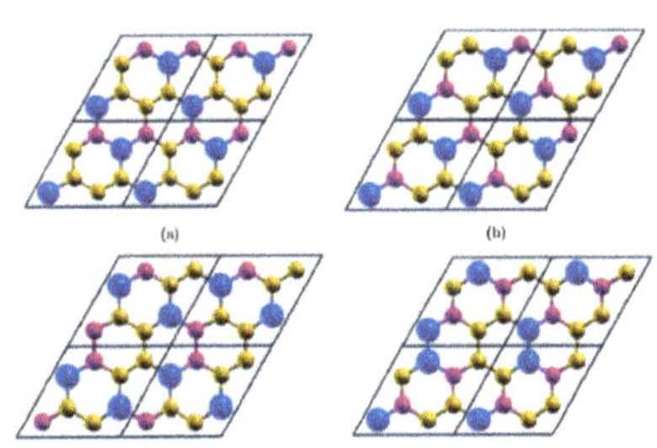

Fig. 24 Different configurations of $C_4B_2N_2$, with decreasing stability from configuration (a) to (d). (a) The most stable configuration has 3 B–N, 3 C–C, 3 C–N and 3 C–B bonds. Configuration (b) has 2 B–N, 2 C–C, 4 C–N and 4 C–B bonds. Configuration (c) has 2 B–N, 3 C–C, 2 C–N, 4 C–B and 1 N–N bonds, and configuration (d) has 2 B–N, 3 C–C, 4 C–N, 2 C–B and 1 B–B bonds. C atoms are denoted by yellow, B by blue and N by pink.

energy by ~1.8 eV than the configuration with ordered BN and graphene domains (of the same stoichiometry, $C_4B_2N_2$, see below). Thus formation of BN domains is thermodynamically preferred over the random and homogeneous distribution of B and N atoms in the graphene matrix.[33,60] Depending on the kinetics involved in the experimental synthesis of G–BN however, some degree of randomness in chemical ordering of B and N is expected.

Two types of chemically ordered periodic supercells can be considered to simulate $G_{1-x}(BN)_x$. In the first one, the zigzag chains of BN are replaced by zigzag chains of graphene chains in the SZ configuration (Fig. 25d). The chemical ordering in the SZ-system is such that the zigzag chains of BN are adjacent to one another, amounting to simulation of a zigzag stripe of graphene in h-BN. In the second case, we replace a pair of boron and nitrogen atoms with two carbon atoms at a time in the SD configuration such that they form segregated carbon islands[62] (where the largest domain is 5 Å × 6 Å in size) in the BN matrix, maintaining the stoichiometry (B/N = 1). From the cohesive energies of SZ and SD configurations for different compositions (Table 1), we conclude that those with zigzag interfaces between BN and graphene are energetically more favourable compared to domains of carbon atoms in a h-BN sheet, since the number of energetically less favourable C–N and C–B bonds[64] are more in the SD configurations. Secondly, the armchair interface is even lower in energy than the zigzag interface between BN and graphene, since the interface energy of an armchair edge is 0.39 eV Å$^{-1}$ compared to 0.58 eV Å$^{-1}$ of the zigzag edge. In the experimental synthesis of $G_xBN_{(1-x)}$ however, the randomness in the positions of substituted atoms and associated entropy are expected to favor formation of graphene domains over zigzag stripes.[62]

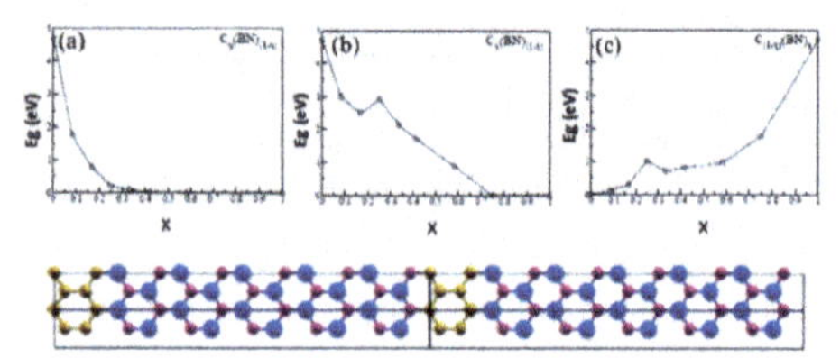

Fig. 25 Variation of the band gap ($E_g$) of h-BN doped with graphene for different doping fractions ($x$). (a) and (b) are for SZ and SD configurations, respectively. Figure (c) denotes the variation of the band gap of graphene doped with BN in the SD configuration, and (d) shows the SZ supercell periodically repeated along $x$ and $y$ directions for $C_xBN_{1-x}$ (the black curve denotes a single SZ supercell). Definitive experimental $E_g$ values are not available.

Table 1 Cohesive energies of SZ and SD for different compositions (with GGA estimation in parenthesis)

| | Energy (eV per atom) | |
|---|---|---|
| $x$ | SZ | SD |
| 0 | −9.45 (−8.66) | −9.45 (−8.66) |
| 0.083 | −9.45 | −9.41 |
| 0.166 | −9.48 (−8.68) | −9.41 (−8.60) |
| 0.25 | −9.52 | −9.45 |
| 0.333 | −9.56 (−8.74) | −9.45 (−8.64) |
| 0.417 | −9.60 | −9.50 |
| 1.0 | −9.97 (−9.10) | −9.97 (−9.10) |

### 6.2 Variation of the band gap with composition

A band gap can be introduced in graphene by breaking its A–B sublattice symmetry. In h-BN, this symmetry is broken and a band gap opens at the Dirac point. The magnitude of the band gap in G–BN, and distribution of mid-gap states depend upon the number of B and N atoms doped in graphene.[65] As substitution of equal number of B (holes) and N (electrons) atoms preserves the electron–hole symmetry, the density of electronic states above and below the Fermi energy level is nearly symmetric. The electronic properties are affected by the domain structure. In BN-rich systems, the segregated carbon domains are responsible for narrow quantum-dot-type energy bands just above and below the Fermi level in the midgap of BN. In contrast, when B/N < 1 *i.e.* nitrogen rich G–BN, there exists a narrow half-filled band (disorder band) at the Fermi level. The width and height of this band are dependent upon the disorder associated with carbon atoms in the BN matrix. With decreasing disorder associated with carbon in BN, the width and height of the disorder band decrease.[62]

For the two types of domain structures (SZ and SD, described in Section 6.1), we have systematically determined the electronic gap as a function of concentration. Due to the zigzag type of the polar interface of BN with graphene in the SZ configuration, the electric field induced by the B and N edges results in the opening of a gap among the electronic states at the zigzag edges of graphene.[41] With increasing concentration of BN, the band gap increases, as expected. From the calculated band gap of the SZ-system with varying concentrations of carbon (see Fig. 25a), it is clear that the band gap decreases rapidly as the concentration of graphene ($x$) increases beyond 0.17. This is because (a) the electric field responsible for opening the gap decreases as the concentration of carbon increases and (b) states of graphene dominate the electronic structure near the Fermi level, and percolate.

For the SD configuration, the decrease of the band gap is gradual relative to that of the SZ configuration (Fig. 25b). At $x$ = 0.25, there is a sudden increase in band gap, which arises from the formation of a six membered hexagonal ring of carbon atoms. This structure is more stable[62] and exhibits a larger band gap[66] because a configuration with a domain is more favourable over the random configuration. Similarly, we have simulated a graphene lattice with domains of BN (*i.e.* $G_{(1-x)}BN_x$) in the SD configuration (Fig. 25c). As expected, the band gap vanishes at $x$ = 0, and increases with increasing concentration of BN, *i.e.* the size of the BN domain.

### 6.3 Stone–Wales defects and their effect on the electronic structure

A SW defect in the G–BN interface involves an in-plane 90° rotation of a carbon–carbon (C–C), boron–nitrogen (B–N), carbon–boron (C–B) or carbon–nitrogen (C–N) bond, resulting in the formation of a pair of pentagons and heptagons (pair of 5/-7 dislocations). Since a configuration with a single C atom doped in the BN region is energetically unfavourable compared to the segregated domains of BN in the graphene matrix, the SW defect is more likely to form at the G–BN interface. Hence, we examine SW defects at the armchair and zigzag interfaces of G–BN. Three configurations of the SW defect at the AC interface obtained by rotation of the C–C bond, C–B bond and C–N bond (Fig. 26a–c) have been investigated. For SC1 the formation energies of SW defects for these three configurations are 4.29 eV, 4.66 eV and 5.66 eV, respectively. For the ZZ interface, there exist two SW defect configurations at the B-terminated edge formed by C–B bond rotation and C–C bond rotation (Fig. 26d and e). The formation energies of the SW defect at the B-terminated edge in SC1 are 5.77 eV (C–B) and 4.38 eV (C–C), respectively. The two SW defect configurations at the N-terminated ZZ edge are formed by C–N bond rotation and C–C bond rotation (Fig. 26f and g). The formation energies of the SW defect at the N-terminated edge in SC1 are 4.48 eV (C–B) and 4.34 eV (C–C) respectively. The hierarchy in the energetic stability of the bonds (B–N > C–C > C–N > C–B > B–B > N–N)[64] dictates the formation energy of SW defects at the G–BN interface. The formation energy of the SW defect at the G–BN interface is dependent not only on the type of the interface or the type of the rotated bond, but also upon the size of graphene and BN stripes. Wider BN stripes lead to increased potential difference across the graphene stripe, hence increasing the formation energy of the SW defect at the interface. Since the BN stripe is wider in SC2 as compared to SC1, the formation energies of SW defects in SC2 are larger. The formation energies of SW defects in SC1 and SC2 are both in the range of 4–6.5 eV, hence showing that formation energies per SW defect at the G–BN interfaces are comparable to those of SW defects in graphene.[67]

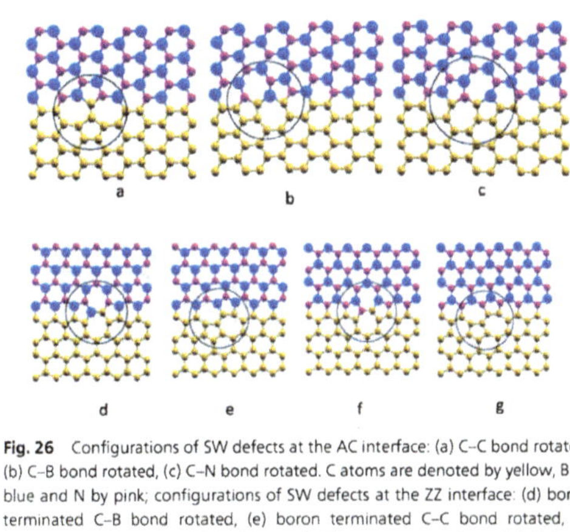


**Fig. 26** Configurations of SW defects at the AC interface: (a) C–C bond rotated, (b) C–B bond rotated, (c) C–N bond rotated. C atoms are denoted by yellow, B by blue and N by pink; configurations of SW defects at the ZZ interface: (d) boron terminated C–B bond rotated, (e) boron terminated C–C bond rotated, (f) nitrogen terminated C–N bond rotated, (g) nitrogen terminated C–C bond rotated. C atoms are denoted by yellow, B by blue and N by pink.

The effects of SW defects on the electronic structure of the G–BN interface are of interest. SC1 is non-magnetic with a direct band gap, however the magnitude of the gap reduces (never vanishes) on the inclusion of a SW defect. All the configurations considered here have equal number of B and N atoms, implying that electron-hole symmetry is preserved and the conduction and valence band states are symmetrically distributed around the Fermi level. Since all the configurations (AC and ZZ) corresponding to SC1 are non-metallic, we investigate their charge density distribution for the highest occupied molecular orbitals (HOMO) and lowest unoccupied molecular orbitals (LUMO) (Fig. 26). In the case of the AC interface, the HOMO and LUMO states get localized at the defect (Fig. 27a). In the case of ZZ interfaces without defects, the charge density plots of the HOMO and LUMO show states localized to the boron and nitrogen terminated interfaces, respectively. However with the inclusion of a SW defect (Fig. 27b), the HOMO state gets localized at the defect for the B-terminated ZZ edge and the LUMO state gets localized at the defect for the N-terminated ZZ edge. Since the HOMO and LUMO states get localized to the defect sites, SW defects should make good adsorption sites for ad-atoms and impurities.

For SC2, the AC configurations exhibit a finite band gap in the presence or absence of SW defects. However, the ZZ interface of

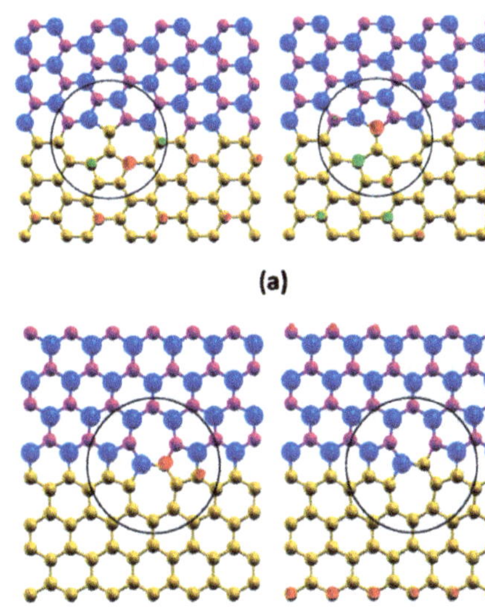


**Fig. 27** Charge density plots of HOMO (left panel) and LUMO (right panel) of the SW defect configurations. (a) Armchair configuration with the C–C bond rotated at the interface, and (b) zigzag configuration with the C–B bond rotated at the B-terminated edge. The HOMO and LUMO of all the armchair configurations are localized to the SW defect. The HOMO and LUMO of the B-terminated ZZ configurations are localized to the defect and nitrogen edge, respectively, whereas the LUMO and HOMO of the N-terminated ZZ configurations are localized to the defect and boron edge respectively.

SC2 without defects is a half-metallic antiferromagnet. Our calculations with the generalized gradient approximation[59] show that the AFM configuration is the lowest energy state for SC2 with a ZZ interface (in agreement with previous theoretical work).[68] To check for magnetic ground states in the presence of SW defects, we have initialized the spins of the C atoms parallel (FM) and anti-parallel (AFM) on the opposite edges for all defect configurations of SC2 with the ZZ interface. We observe that magnetic moments do not survive in the presence of a SW defect. Thus, SW defects annihilate half-metallicity and weaken the effect on magnetism in G–BN systems.

### 6.4 Gas adsorption characteristics

Graphene and other carbon-based materials with porous structures are of interest for gas adsorption, because of their low cost and ease of processing.[44,60] Dispersive or van der Waals (vdW) interactions influence details of the adsorption process such as the geometry and energetics of the adsorbing surface and adsorbed molecule,[69] and often determine chemical reaction pathways.[59,70] To study the interaction of $CO_2$ and $CH_4$ molecules with G–BN, a periodic supercell consisting of 24 atoms with an armchair interface separating BN and graphene domains is employed. The following sites of adsorption are considered: (a) above the atoms (B, C and N), (b) above the centre of a hexagonal ring (pore) and (c) bond centres. In addition, different orientations of molecules with the molecular axis (a) parallel and (b) perpendicular to the atomic plane, with molecular orientation along different in-plane crystallographic directions, are also considered.

We first examine the effect of inclusion of van der Waals interactions on the interaction of $CO_2$ and $CH_4$ molecules on a $C_1(BN)_1$ sheet with random distribution of B, C and N atoms (3 × 3 supercell of 18 atoms).[44,60] An adsorption energy of ~19 kJ mol$^{-1}$ is found for $CO_2$ (17 wt%), with the molecule remaining parallel to the G–BN sheet. The strongest adsorption sites are centres of hexagonal rings with the molecular axis along the C–B or C–N diagonal, in agreement with our earlier work.[60] On the other hand, $CH_4$ molecule (7 wt%) binds slightly weakly with a binding energy of ~13 kJ mol$^{-1}$, and the adsorption at the C–B bond is stronger than that at the atomic or pore sites. The shortest distance of $CO_2$ and $CH_4$ molecules from the G–BN plane is ~3.18 Å and 2.64 Å, respectively. These two gases adsorb on the G–BN interface of a $C_2(BN)_1$ sheet of 24 atoms with an energy of ~19 kJ mol$^{-1}$ for one $CO_2$ molecule (7 wt%), the strongest adsorption site being the one with its molecular axis parallel to the G–BN plane and the carbon atom above the center of a C–B bond. For the $CH_4$ molecule at a coverage of 3.7 wt% on $C_2(BN)_1$, the adsorption energy is 14 kJ mol$^{-1}$ with the $CH_4$ molecule centered on the hexagonal ring. The shortest distance of $CO_2$ and $CH_4$ molecule from the G–BN plane is ~3.20 Å and 2.79 Å, respectively. The adsorption energies of $CO_2$ and $CH_4$ at the G–BN interface are roughly similar for adsorption on randomly distributed C, B and N atoms in $C_1(BN)_1$ and domains of graphene and BN in $C_2(BN)_1$. The adsorption energies increase significantly by about 10–12 kJ mol$^{-1}$ or a factor of ~5 upon the inclusion of van der Waals interactions.

To study the adsorption of these two gases on two G–BN supercells (SC1) with Stone–Wales defects, two configurations, a rotated C–C bond ($AI_C$, Fig. 26a) and rotated C–N bond ($AI_N$, Fig. 26c), which have the minimum and maximum formation energies of SW defect for the armchair interface (AI) for the SC1 supercell among the configurations, can be considered. Simulated adsorption of $CO_2$ and $CH_4$ gases at various sites near the pentagonal ring of the SW defect shows that for one $CO_2$ molecule per cell, corresponding to ~3.6 wt%, adsorption energies are 83 kJ mol$^{-1}$ and 89 kJ mol$^{-1}$ for the $AI_C$ and $AI_N$ configurations, respectively (Fig. 28a and b), almost 5 times larger than those for adsorption on non-defective sites. The shortest distance of $CO_2$ to the surface is 3.25 and 3.14 Å, respectively, for $AI_C$ and $AI_N$ configurations, comparable to those in adsorption on non-defective sites. For the $CH_4$ molecule (1.35 wt%), the adsorption energies are 97 kJ mol$^{-1}$ and 84 kJ mol$^{-1}$ with distances between the molecule and G–BN sheet of ~2.73 Å and 2.75 Å, respectively, for $AI_C$ and $AI_N$ configurations (Fig. 28c and d). The G–BN sheet shows noticeable bending at the adsorption sites, and its bonds near the molecular adsorption site expand by 1.5 to 2%.

Although dispersion-corrected methods (DFT-D) overestimate the binding energies,[71] we note that the contribution of van der Waals interactions to the adsorption energy is ~20 to 25 kJ mol$^{-1}$. Also, the contribution of the bending of the G–BN sheet containing SW defects[72] to the adsorption energy is of the order of 63 kJ mol$^{-1}$. Hence, we attribute the strong binding of gas molecules on G–BN to the bending of the sheet on account of the SW defect. If the pristine G–BN sheet is allowed to buckle, the adsorption energy reduces to ≈19 kJ mol$^{-1}$. From the density of electronic states (DOS) of these structures, the interaction of molecules with the G–BN layer is found to result in the shifting of its states just above the Fermi energy towards

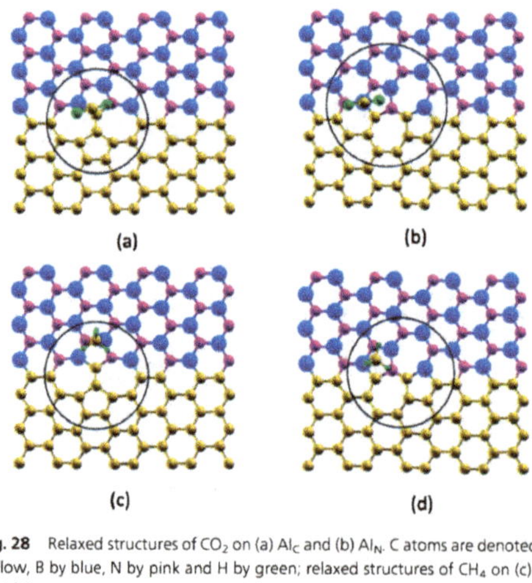


Fig. 28 Relaxed structures of $CO_2$ on (a) $AI_C$ and (b) $AI_N$. C atoms are denoted by yellow, B by blue, N by pink and H by green; relaxed structures of $CH_4$ on (c) $AI_C$ and (d) $AI_N$. C atoms are denoted by yellow, B by blue, N by pink and H by green.

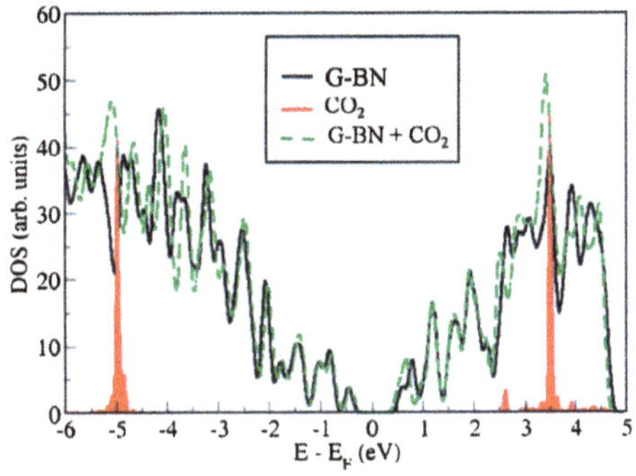


**Fig. 29** Electronic Density of States (DOS) plots for $CO_2$ adsorbed on the SW defect in the SC1 configuration. The Red filled curve is the projected DOS of $CO_2$ molecule on G-BN, the green curve is the DOS of G-BN without the molecule and the black curve is the DOS for Gs-BN + $CO_2$.

the Fermi level. Since all the configurations with $CO_2$ and $CH_4$ appear to show similar trends in the nature of the DOS, we discuss only the interaction of $CO_2$ with G-BN in the $Al_C$ configuration. From the DOS of G-BN with and without $CO_2$ (Fig. 29), the contribution of the $CO_2$ molecule to the total DOS is projected out (the HOMO states of G-BN + $CO_2$ and $CO_2$ were aligned with the HOMO of G-BN with zero as the Fermi level of G-BN). Since the $CO_2$ molecule is near the sites which contribute to the LUMO of $Al_C$ (Fig. 27a and 28a), the binding between $CO_2$ and the G-BN layer is strong. Comparing the DOS of G-BN and G-BN + $CO_2$ (Fig. 29), we find that there is a weak overlap of wave functions of $CO_2$ and G-BN at the SW defect leading to stronger binding and a decrease in the band gap of ~0.1 eV for $CO_2$ on $Al_C$. The origin of this lies in the fact that defect electronic states are localized in the plane of G-BN, but relatively more extended in the direction perpendicular to the plane.

## 7 Conclusions

Borocarbonitrides, $B_xC_yN_z$, containing boron, carbon and nitrogen in varying proportions can possess hexagonal BCN networks or be composites of graphene and BN. They can be prepared by different methods and exhibit novel features such as composition-dependent electronic properties and selective adsorption of $CO_2$ and $CH_4$. Potential supercapacitor applications of these materials are noteworthy. Semiconducting band gaps in borocarbonitrides are reported in the range 1.0–3.9 eV and the gap appears to be smaller in carbon-rich samples mainly based on photoluminescence measurements.[73–76] It would be most worthwhile to study the electronic properties of $B_xC_yN_z$ with a wide range of compositions. Optical properties would be equally of value, since some interesting nonlinear properties of BCN have been found.[77] The nature of these materials and the dependence of properties on the structure and composition are best revealed by theoretical calculations. Calculations show that the interfaces between domains of graphene and BN have important consequences on the properties showing composition-dependent electronic properties as found experimentally. Stone-Wales (SW) defects occur at the interfaces with formation energies ranging from 4 to 6 eV, comparable to that of SW defects in graphene. The variation and trends in the formation energies of SW defects at different interfaces can be understood in terms of the type of bonds involved and associated strain. The SW defects eliminate or weaken the half-metallic properties of the G-BN system. Adsorption of $CO_2$ or $CH_4$ does not depend much on the arrangement of B and N atoms in the graphene matrix, but the inclusion of van der Waals interactions affects the adsorption energies significantly. SW defects act as strong adsorption sites for $CO_2$ and $CH_4$. The adsorption energy of these gases is ~83 to 97 kJ $mol^{-1}$ (at coverages of 3.6 and 1.35 wt% of $CO_2$ and $CH_4$, respectively) for G-BN with SW defects, as compared to ~14 to 19 kJ $mol^{-1}$ (at coverages of 7 and 3.7 wt% of $CO_2$ and $CH_4$, respectively) for adsorption on G-BN without defects. It would be a good idea to introduce or engineer SW defects in G-BN to enhance the adsorption of these greenhouse gases.

## Acknowledgements

SNS is thankful to the Council of Scientific and Industrial Research, India, for Research Fellowship. AKM acknowledges the Department of Science and Technology, New Delhi for a fellowship under the Fast Track Scheme for Young Scientists. UVW acknowledges support from AOARD grants (no. FA2386-10-1-4062 and FA2386-10-1-4150).

## Notes and references

1 H. W. Kroto, J. R. Heath, S. C. O'Brien, R. F. Curl and R. E. Smalley, *Nature*, 1985, **318**, 162.
2 S. Iijima, *Nature*, 1991, **354**, 56.
3 C. N. R. Rao and A. Govindaraj, *Nanotubes and Nanowires*, RSC Publishing, 2nd edn, 2011.
4 K. S. Novoselov, A. K. Geim, S. V. Morozov, D. Jiang, M. I. Katsnelson, I. V. Grigorieva, S. V. Dubonos and A. A. Firsov, *Nature*, 2005, **438**, 197.
5 K. S. Novoselov, Z. Jiang, Y. Zhang, S. V. Morozov, H. L. Stormer, U. Zeitler, J. C. Maan, G. S. Boebinger, P. Kim and A. K. Geim, *Science*, 2007, **315**, 1379.
6 C. N. R. Rao, A. K. Sood, K. S. Subrahmanyam and A. Govindaraj, *Angew. Chem., Int. Ed.*, 2009, **48**, 7752.
7 C. N. R. Rao, H. S. S. R. Matte and K. S. Subrahmanyam, *Acc. Chem. Res.*, 2013, **46**, 149.
8 H. S. S. R. Matte, A. Gomathi, A. K. Manna, D. J. Late, R. Datta, S. K. Pati and C. N. R. Rao, *Angew. Chem., Int. Ed.*, 2010, **49**, 4059.
9 J. N. Coleman, M. Lotya, A. O'Neill, S. D. Bergin, P. J. King, U. Khan, K. Young, A. Gaucher, S. De, R. J. Smith, I. V. Shvets, S. K. Arora, G. Stanton, H.-Y. Kim, K. Lee, G. T. Kim, G. S. Duesberg, T. Hallam, J. J. Boland, J. J. Wang, J. F. Donegan, J. C. Grunlan, G. Moriarty, A. Shmeliov, R. J. Nicholls, J. M. Perkins, E. M. Grieveson,

 

K. Theuwissen, D. W. McComb, P. D. Nellist and V. Nicolosi, *Science*, 2011, **331**, 568.

10 N. G. Chopra, R. J. Luyken, K. Cherrey, V. H. Crespi, M. L. Cohen, S. G. Louie and A. Zettl, *Science*, 1995, **269**, 966.

11 A. Loiseau, F. Willaime, N. Demoncy, G. Hug and H. Pascard, *Phys. Rev. Lett.*, 1996, **76**, 4737.

12 R. Sen, B. C. Satishkumar, A. Govindaraj, K. R. Harikumar, G. Raina, J.-P. Zhang, A. K. Cheetham and C. N. R. Rao, *Chem. Phys. Lett.*, 1998, **287**, 671.

13 D. Golberg, Y. Bando, C. C. Tang and C. Y. Zhi, *Adv. Mater.*, 2007, **19**, 2413.

14 A. Nag, K. Raidongia, K. P. S. S. Hembram, R. Datta, U. V. Waghmare and C. N. R. Rao, *ACS Nano*, 2010, **4**, 1539.

15 D. Golberg, Y. Bando, Y. Huang, T. Terao, M. Mitome, C. Tang and C. Zhi, *ACS Nano*, 2010, **4**, 2979.

16 L. S. Panchakarla, K. S. Subrahmanyam, S. K. Saha, A. Govindaraj, H. R. Krishnamurthy, U. V. Waghmare and C. N. R. Rao, *Adv. Mater.*, 2009, **21**, 4726.

17 L. Song, Z. Liu, A. L. M. Reddy, N. T. Narayanan, J. Taha-Tijerina, J. Peng, G. Gao, J. Lou, R. Vajtai and P. M. Ajayan, *Adv. Mater.*, 2012, **24**, 4878.

18 Y. Chen, J. C. Barnard, R. E. Palmer, M. O. Watanabe and T. Sasaki, *Phys. Rev. Lett.*, 1999, **83**, 2406.

19 L. Liao, K. Liu, W. Wang, X. Bai, E. Wang, Y. Liu, J. Li and C. Liu, *J. Am. Chem. Soc.*, 2007, **129**, 9562.

20 K. Suenaga, C. Colliex, N. Demoncy, A. Loiseau, H. Pascard and F. Willaime, *Science*, 1997, **278**, 653.

21 P. Redlich, J. Loeffler, P. M. Ajayan, J. Bill, F. Aldinger and M. Rühle, *Chem. Phys. Lett.*, 1996, **260**, 465.

22 K. B. Shelimov and M. Moskovits, *Chem. Mater.*, 1999, **12**, 250.

23 J. D. Guo, C. Y. Zhi, X. D. Bai and E. G. Wang, *Appl. Phys. Lett.*, 2002, **80**, 124.

24 W. Q. Han, J. Cumings and A. Zettl, *Appl. Phys. Lett.*, 2001, **78**, 2769.

25 A. Vinu, M. Terrones, D. Golberg, S. Hishita, K. Ariga and T. Mori, *Chem. Mater.*, 2005, **17**, 5887.

26 D. Portehault, C. Giordano, C. Gervais, I. Senkovska, S. Kaskel, C. Sanchez and M. Antonietti, *Adv. Funct. Mater.*, 2010, **20**, 1827.

27 S. Y. Kim, J. Park, H. C. Choi, J. P. Ahn, J. Q. Hou and H. S. Kang, *J. Am. Chem. Soc.*, 2007, **129**, 1705.

28 M. Hubáček and T. Sato, *J. Solid State Chem.*, 1995, **114**, 258.

29 A. Gomathi and C. N. R. Rao, *Mater. Res. Bull.*, 2006, **41**, 941.

30 K. Raidongia, D. Jagadeesan, M. Upadhyay-Kahaly, U. V. Waghmare, S. K. Pati, M. Eswaramoorthy and C. N. R. Rao, *J. Mater. Chem.*, 2008, **18**, 83.

31 K. Raidongia, K. P. S. S. Hembram, U. V. Waghmare, M. Eswaramoorthy and C. N. R. Rao, *Z. Anorg. Allg. Chem.*, 2010, **636**, 30.

32 R. B. Kaner, J. Kouvetakis, C. E. Warble, M. L. Sattler and N. Bartlett, *Mater. Res. Bull.*, 1987, **22**, 399.

33 N. Kumar, K. Raidongia, A. K. Mishra, U. V. Waghmare, A. Sundaresan and C. N. R. Rao, *J. Solid State Chem.*, 2011, **184**, 2902.

34 M. P. Levendorf, C.-J. Kim, L. Brown, P. Y. Huang, R. W. Havener, D. A. Muller and J. Park, *Nature*, 2012, **488**, 627.

35 L. Ci, L. Song, C. Jin, D. Jariwala, D. Wu, Y. Li, A. Srivastava, Z. F. Wang, K. Storr, L. Balicas, F. Liu and P. M. Ajayan, *Nat. Mater.*, 2010, **9**, 430.

36 Z. Liu, L. Song, S. Zhao, J. Huang, L. Ma, J. Zhang, J. Lou and P. M. Ajayan, *Nano Lett.*, 2011, **11**, 2032.

37 C. R. Dean, A. F. Young, I. Meric, C. Lee, L. Wang, S. Sorgenfrei, K. Watanabe, T. Taniguchi, P. Kim, K. L. Shepard and J. Hone, *Nat. Nanotechnol.*, 2010, **5**, 722.

38 Z. Yan, Z. Peng, Z. Sun, J. Yao, Y. Zhu, Z. Liu, P. M. Ajayan and J. M. Tour, *ACS Nano*, 2011, **5**, 8187.

39 G. Gao, W. Gao, E. Cannuccia, J. Taha-Tijerina, L. Balicas, A. Mathkar, T. N. Narayanan, Z. Liu, B. K. Gupta, J. Peng, Y. Yin, A. Rubio and P. M. Ajayan, *Nano Lett.*, 2012, **12**, 3518.

40 L. Song, L. Balicas, D. J. Mowbray, R. B. Capaz, K. Storr, L. Ci, D. Jariwala, S. Kurth, S. G. Louie, A. Rubio and P. M. Ajayan, *Phys. Rev. B: Condens. Matter Mater. Phys.*, 2012, **86**, 075429.

41 J. M. Pruneda, *Phys. Rev. B: Condens. Matter Mater. Phys.*, 2010, **81**, 161409.

42 S. Bhowmick and U. V. Waghmare, *Phys. Rev. B: Condens. Matter Mater. Phys.*, 2010, **81**, 155416.

43 W. H. Brito, R. Kagimura and R. H. Miwa, *Phys. Rev. B: Condens. Matter Mater. Phys.*, 2012, **85**, 035404.

44 N. Kumar, K. S. Subrahmanyam, P. Chaturbedy, K. Raidongia, A. Govindaraj, K. P. S. S. Hembram, A. K. Mishra, U. V. Waghmare and C. N. R. Rao, *ChemSusChem*, 2011, **4**, 1662.

45 K. Jayaramulu, N. Kumar, A. Hazra, T. K. Maji and C. N. R. Rao, *Chem.–Eur. J.*, 2013, under publication.

46 Y. Goto, M. Sasaki, M. Hashizume and M. Suzuki, *J. Eur. Ceram. Soc.*, 1999, **19**, 2695.

47 D. Sauter, M. Weinmann, F. Berger, P. Lamparter, K. Müller and F. Aldinger, *Chem. Mater.*, 2002, **14**, 2859.

48 (*a*) T.-P. Fellinger, D. S. Su, M. Engenhorst, D. Gautam, R. Schlogl and M. Antonietti, *J. Mater. Chem.*, 2012, **22**, 23996; (*b*) K. Moses, S. Sen, A. Bhattacharyya and C. N. R. Rao, 2013, unpublished.

49 P. Chaturbedy, H. S. S. R. Matte, R. Voggu, A. Govindaraj and C. N. R. Rao, *J. Colloid Interface Sci.*, 2011, **360**, 249.

50 A. Sundaresan and C. N. R. Rao, *Nano Today*, 2009, **4**, 96.

51 C. N. R. Rao, H. S. S. R. Matte, K. S. Subrahmanyam and U. Maitra, *Chem. Sci.*, 2012, **3**, 45.

52 E. Iyyamperumal, S. Wang and L. Dai, *ACS Nano*, 2012, **6**, 5259.

53 K. Gopalakrishnan, K. Moses, A. Govindaraj and C. N. R. Rao, *Solid State Commun.*, 2013, under publication.

54 P. Giannozzi, S. Baroni, N. Bonini, M. Calandra, R. Car, C. Cavazzoni, D. Ceresoli, G. L. Chiarotti, M. Cococcioni, I. Dabo, A. Dal Corso, S. de Gironcoli, S. Fabris, G. Fratesi, R. Gebauer, U. Gerstmann, C. Gougoussis, A. Kokalj, M. Lazzeri, L. Martin-Samos, N. Marzari, F. Mauri, R. Mazzarello, S. Paolini, A. Pasquarello, L. Paulatto, C. Sbraccia, S. Scandolo, G. Sclauzero, A. P. Seitsonen, A. Smogunov, P. Umari and R. M. Wentzcovitch, *J. Phys.: Condens. Matter*, 2009, **21**, 395502.

55 D. M. Ceperley and B. J. Alder, *Phys. Rev. Lett.*, 1980, **45**, 566.

56 D. Vanderbilt, *Phys. Rev. B: Condens. Matter Mater. Phys.*, 1990, **41**, 7892.

57 H. J. Monkhorst and J. D. Pack, *Phys. Rev. B: Solid State*, 1976, **13**, 5188.
58 S. Grimme, *J. Comput. Chem.*, 2006, **27**, 1787.
59 J. P. Perdew, K. Burke and M. Ernzerhof, *Phys. Rev. Lett.*, 1996, **77**, 3865.
60 K. Raidongia, A. Nag, K. P. S. S. Hembram, U. V. Waghmare, R. Datta and C. N. R. Rao, *Chem.–Eur. J.*, 2010, **16**, 149.
61 K. Dileep, L. S. Panchakarla, K. Balasubramanian, U. V. Waghmare and R. Datta, *J. Appl. Phys.*, 2011, **109**, 063523.
62 J. d. R. Martins and H. Chacham, *Phys. Rev. B: Condens. Matter Mater. Phys.*, 2012, **86**, 075421.
63 V. V. Ivanovskaya, A. Zobelli, O. Stéphan, P. R. Briddon and C. Colliex, *J. Phys. Chem. C*, 2009, **113**, 16603.
64 H. Nozaki and S. Itoh, *J. Phys. Chem. Solids*, 1996, **57**, 41.
65 B. Xu, Y. H. Lu, Y. P. Feng and J. Y. Lin, *J. Appl. Phys.*, 2010, **108**, 073711.
66 M. S. C. Mazzoni, R. W. Nunes, S. Azevedo and H. Chacham, *Phys. Rev. B: Condens. Matter Mater. Phys.*, 2006, **73**, 073108.
67 L. Li, S. Reich and J. Robertson, *Phys. Rev. B: Condens. Matter Mater. Phys.*, 2005, **72**, 184109.
68 Y. Ding, Y. L. Wang and J. Ni, *Appl. Phys. Lett.*, 2009, **95**, 123105.
69 W. G. Schmidt, K. Seino, M. Preuss, A. Hermann, F. Ortmann and F. Bechstedt, *Appl. Phys. A: Mater. Sci. Process.*, 2006, **85**, 387.
70 E. Rauls, S. Blankenburg and W. G. Schmidt, *Phys. Rev. B: Condens. Matter Mater. Phys.*, 2010, **81**, 125401.
71 T. Kar, H. F. Bettinger, S. Scheiner and A. K. Roy, *J. Phys. Chem. C*, 2008, **112**, 20070.
72 S. N. Shirodkar and U. V. Waghmare, *Phys. Rev. B: Condens. Matter Mater. Phys.*, 2012, **86**, 165401.
73 M. O. Watanabe, S. Itoh, T. Sasaki and K. Mizushima, *Phys. Rev. Lett.*, 1996, **77**, 187.
74 J. Yu, J. Ahn, S. F. Yoon, Q. Zhang, Rusli, B. Gan, K. Chew, M. B. Yu, X. D. Bai and F. G. Wang, *Appl. Phys. Lett.*, 2000, **77**, 1949.
75 L.-W. Yin, Y. Bando, D. Golberg, A. Gloter, M.-S. Li, X. Yuan and T. Sekiguchi, *J. Am. Chem. Soc.*, 2005, **127**, 16354.
76 L. Qin, J. Yu, S. Kuang, C. Xiao and X. Bai, *Nanoscale*, 2012, **4**, 120.
77 S. Kumar, N. Kamaraju, K. S. Vasu, A. Nag, A. K. Sood and C. N. R. Rao, *Chem. Phys. Lett.*, 2010, **499**, 152.

# II. PHASE TRANSITIONS IN SOLIDS

*Commentary by C.N.R. Rao*

The study of phase transitions is an important aspect of modern physical science. Solids undergo a variety of phase transitions accompanied by significant changes in properties. Phase transitions in solids with implications for possible applications are of vital interest in materials chemistry.[1] Typical of phase transitions are those involving changes in magnetic, electrical and dielectric properties. Spin-state transitions occur both in molecular and ionic solids. Metal-insulator transitions in solids involving marked changes in electrical properties continue to be of interest. There are transitions such as the glass transition, which have taken a long time to be understood. The advent of Monte Carlo and molecular dynamics methods has made an impact on the study of phase transitions. I have carried out investigations of phase transitions in solids over the last 50 years. In this section, a few articles, dealing with phase transitions are included. These include transitions of $TiO_2$ and alkali halides as well as metal-insulator transitions. Phase transitions in fullerenes are described in Section IV of this book.

## References

1. C.N.R. Rao and K.J. Rao, *Phase Transitions in Solids — An Approach to the Study of Chemistry and Physics of Solids*, McGraw-Hill, New York, 1978.

Reprinted from Accounts of Chemical Research, 1984, *17*, 83-89


# Phase Transitions and the Chemistry of Solids†

C. N. R. Rao

*Solid State and Structural Chemistry Unit, Indian Institute of Science, Bangalore-560012, India*

*Received January 27, 1983 (Revised Manuscript Received September 1, 1983)*

A variety of solids exhibit transformations from one crystal structure to another (polymorphism) as the temperature or pressure is varied. Besides such phase transitions involving changes in atomic configuration, solids also undergo transformations where the electronic or the spin configuration changes. The subject of phase transitions has grown enormously in recent years, with new types of transitions as well as new approaches to explain the phenomena having been reported extensively in the literature.[1,2] Traditionally, metallurgists and physicists have evidenced keen interest in this subject, but it is equally of importance in solid-state chemistry. In this article, I shall discuss some interesting types of phase transitions of relevance to solid-state chemistry investigated by my co-workers and myself in order to illustrate the scope and vitality of the subject.

## General Features of Phase Transitions

During a phase transition, the free energy of the solid remains continuous, but thermodynamic quantities such as entropy, volume, and heat capacity exhibit discontinuous changes. Depending on which derivative of the Gibbs free energy, $G$, shows a discontinuous change at the transiton, phase transitions are generally classified as first order or second order. In a first-order transition where the $G(P,T)$ surfaces of the parent and product phases intersect sharply, the entropy and the volume show singular behavior. In second-order transitions, on the other hand, the heat capacity, compressibility, or thermal expansivity shows singular behavior.

We all know that when a liquid transforms to a crystal, there is a change in order; the crystal has greater order than the liquid. The symmetry also changes in such a transition; the liquid has more symmetry than a crystal since the liquid remains invariant under all rotations and translations. Landau introduced the concept of an order parameter, $\xi$, which is a measure of the order resulting from a phase transition. In a first-order transition (e.g., liquid–crystal), the change in $\xi$ is discontinuous, but in a second-order transition where the change of state is continuous, the change in $\xi$ is also continuous. Landau proposed that $G$ in a second-order (or structural) phase transition is not only a function of $P$ and $T$ but also of $\xi$ and expanded $G$ as a series in powers of $\xi$ around the transition point. The order parameter vanishes at the critical temperature, $T_c$, in such transitions. Landau also considered the symmetry changes across phase transitions. Thus, a transition from a phase of high symmetry to one of low symmetry is accompanied by an order parameter. In a second-order transition, certain elements of symmetry appear or disappear across the transition; for example, when the tetragonal, ferroelectric $BaTiO_3$ in which the dipoles are all ordered, transforms to the cubic, paraelectric phase where the dipoles are randomly oriented, there is an increase in symmetry (appearence of certain syymetry elements) but decrease in order. In a ferroelectric–paraelectric transition, electric polarization is the order parameter; in ferromagnetic–paramagnetic transition, magnetization is the order parameter.

Many physical properties diverge near $T_c$, i.e., show anomalously large values as $T_c$ is approached from either side. The divergences in different phase transitions are, however, strikingly similar. These divergences can be quantified in terms of critical exponents, $\lambda$:

$$\lambda = \lim_{\epsilon \to 0} \left| \frac{\ln f(\epsilon)}{\ln |\epsilon|} \right|$$

where $\epsilon = (T - T_c)/T_c$ and $\lambda$ is called an exponent since $f(\epsilon)$ is proportional to $\epsilon^{\lambda}$. The most important exponents are those associated with the specific heat ($\alpha$), the order parameter ($\beta$), the susceptibility ($\gamma$), and the range over which individual constituents like atoms and atomic moments are correlated ($\nu$). It so happens that the individual exponents for many different transitions are roughly similar (e.g., $\beta \approx 0.33$). More interesting is the fact that $\alpha + 2\beta + \gamma = 2$ in most transitions, independent of the detailed nature of the system. In other words, although individual values of exponents may vary from one transition to another, they all add up to 2. Such a universality in critical exponents is understood in the light of Kadanoff's concept[3] of scale invariance associated with the fluctuations near $T_c$. The exponents themselves can be calculated by employing the renormalization group method developed by Wilson.[4] Thanks to all these developments, we are now able to characterize all higher order phase transitions in terms of the physical dimensionality of the system, $d$, and the dimensionality of the order parameter, $n$. This is illustrated in the case of some observed phase transitions in Figure 1. It is noteworthy that there can be no phase transitions in one dimension if short-range forces alone operate.

C. N. R. Rao obtained his Master's degree in chemistry from the Banaras Hindu University, Doctor of Science degree from the University of Mysore, and Ph.D. (and later D.Sc. honoris causa) from Purdue University. He worked as a postdoctoral research associate with Professor K. S. Pitzer at the University of California at Berkeley. Before joining the Indian Institute of Science about 7 years ago to initiate a new department devoted to solid-state and structural chemistry, he was a Professor of Chemistry at the Indian Institute of Technology, Kanpur. Dr. Rao's interests are in solid-state chemistry, surface science, spectroscopy, and molecular structure. He is a Fellow of the Indian National Science Academy and of the Royal Society, London. He was selected a Centennial Foreign Fellow by the American Chemical Society in 1976. He is Vice President/President-Elect of the International Union of Pure and Applied Chemistry.

†Contribution No. 220 from the Solid State and Structural Chemistry Unit.

(1) H. K. Henisch, R. Roy, and L. E. Cross, Ed., "Phase Transitions", Pergamon Press, New York, 1973.
(2) C. N. R. Rao and K. J. Rao, "Phase Transitions in Solids", McGraw-Hill, New York, 1978.
(3) L. P. Kadanoff, *Physics*, **2**, 263 (1966).
(4) K. G. Wilson and J. Kogut, *Phys. Rep.*, **12C**, 77 (1974).

0001-4842/84/0117-0083$01.50/0 

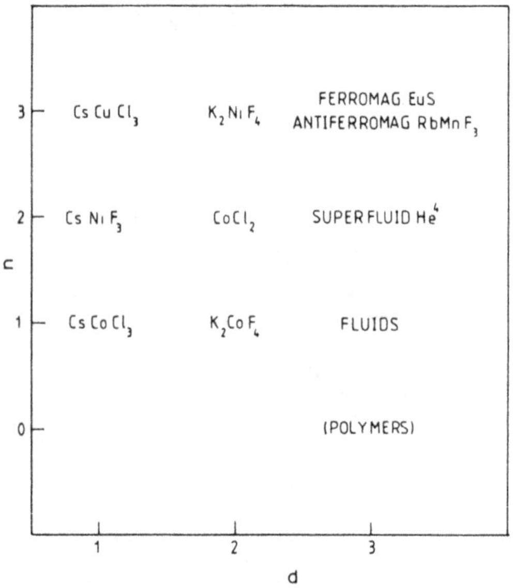


**Figure 1.** Some observed phase transitions in the d–n plane.

Another important aspect of phase transitions in solids is the presence of soft modes. Operationally, a soft mode is a collective excitation whose frequency decreases anomalously as the transition point is reached. In second-order transitions, the soft mode frequency goes to zero at $T_c$, but in first-order transitions, the phase change occurs before the mode frequency goes to zero. Soft modes have been found to accompany a variety of solid-state transitions, including those of superconductors and organic solids.[2,5] Occurrence of soft modes in phase transitions can be inferred from Landau's treatment wherein atomic displacements may themselves be considered to represent an order parameter.

It has been found convenient to classify phase transitions in solids on the basis of the mechanism.[2] Three important kinds of transitions of common occurrence are as follows: (i) nucleation and growth transitions, a typical example being the anatase-rutile transformation of $TiO_2$, (ii) positional and orientational order–disorder transitions, and (iii) martensitic transitions. A typical example of a positional order–disorder transition is that of AgI; orientational order–disorder transitions are exhibited by many solids such as ammonium halides, plastic (orientationally disordered) crystals, and salts of di- or polyatomic anions. A martensitic transition is a structural change caused by atomic displacements (and not by diffusion) corresponding to a homogeneous deformation wherein the parent and product phases are related by a substitutional lattice correspondence, an irrational habit plane and a pracise orientationl relationship. These transitions that occur with high velocities of the order of sound velocity were originally discovered in steel but are now known to occur in several inorganic solids such as $KTa_{0.65}Nb_{0.35}O_3$ and $ZrO_2$.

On the basis of our knowledge of crystal chemistry, we can predict the nature of structural changes in the

(5) J. F. Scott, *Rev. Mod. Phys.*, **46**, 83 (1974).

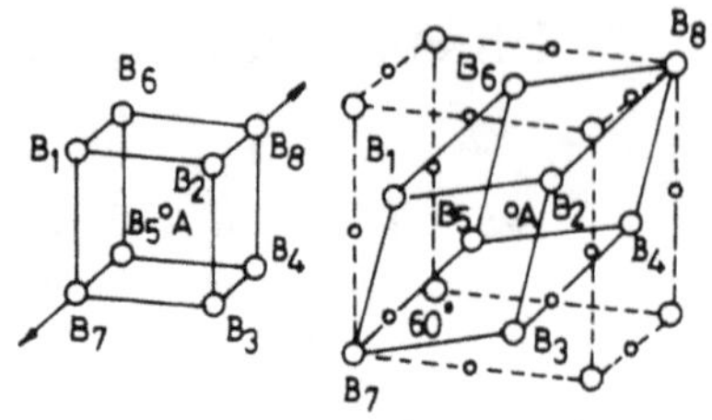


**Figure 2.** Dialatation transformation from CsCl structure to NaCl structure of an AB-type compound. Symmetry about the the unique axis of dilatation (3*m*) is preserved.

phase transitions of simple ionic solids. Thus, the Born model satisfactorily explains the relative stabilities of structures of simple ionic solids. On the basis of ionicity considerations, we can account for the structures of III–V, II–VI, and such binary compounds.[2] Some years ago, Buerger[6] classified phase transitions in solids on the basis of changes in the primary or higher coordination. Transformations involving primary coordination (e.g., CsCl structure–NaCl structure, aragonite–calcite) can be reconstructive or dilatational. Transformations involving second or higher coordination (e.g., $\alpha$–$\beta$ quartz, tetragonal-cubic transition of $BaTiO_3$) can be reconstructive or displacive. Buerger suggested that transformations involving changes in primary coordination such as in the CsCl–NaCl transition can occur more readily by a dilatational or deformational mechanism rather than by a drastic mechanism necessitating the breaking and making of bonds (Figure 2). It is interesting that the CsCl–NaCl transition is now considered to be martensitic with orientational relations between the two phases. It seems likely that phase transitions of many inorganic solids do not require diffusion and involve a deformational mechanism[2] wherein the parent and product phases have orientational relationships. Many such transitions may also exhibit soft mode behavior, an aspect that is worthy of investigation.

## Polytypism

Solids such as SiC, ZnS, $CdI_2$, $TaS_2$, mica, and perovskite oxides exhibit polytypism wherein the unit cells of the different polytypic forms differ from one another only in the *c* dimension; the *c* dimension of polytypes of certain substances vary anywhere between a few angstroms and a few thousand angstroms. This phenomenon arises because of differences in the sequence in which the atomic layers are stacked in different polytypes along the *c* axis. Although several theories have been put forward to explain the varied characteristics of polytypic substances,[7] it is difficult to understand some of their features. These include the existence of different types of unidimensional order (since one-dimensional systems should not normally show long-range ordering), which is an apparent violation of Gibbs' phase rule (owing to the coexistence of several phases under identical conditions), syntactic coalescence (growth of different polytypic forms in different parts of the same specimen), and the existence of varying extents of disorder.

(6) M. J. Buerger, *Forschr. Miner*, **39**, 9 (1961).

(7) A. R. Verma and G. C. Trigunayat in "Solid State Chemistry", C. N. R. Rao, Ed., Marcel Dekker, New York, 1974.

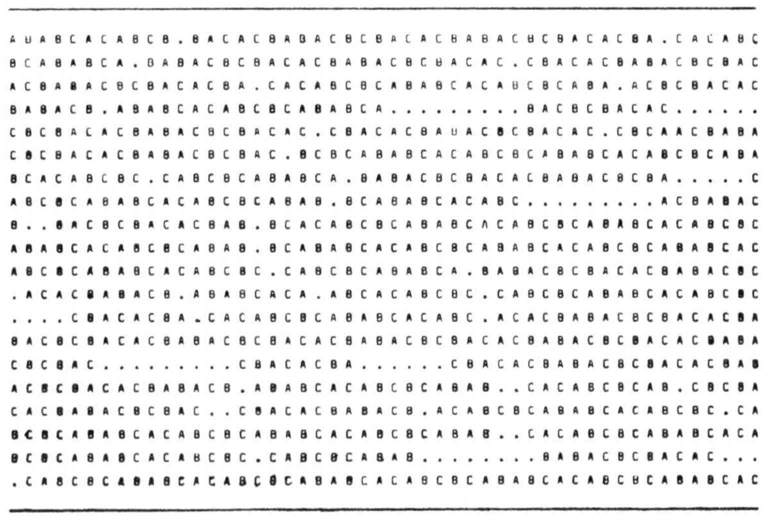

**Figure 3.** Display of long-range order as in 12R polytype (ABACBCBACACB) obtained by computer simulation (taken from ref 8b).

Since polytypism is essentially a one-dimensional phenomenon and the atomic layers exist in either cubic or hexagonal configurations, we can, in principle, treat polytypes as different ordered states of a spin-half Ising chain.[8] Since such an Ising chain should have long-range order at nonzero temperatures, it is essential to have an infinite-range interaction. At the same time, in order to obtain different ordered states of the chain, it is necessary to have a short-range interaction competing with the infinite-range interaction. Spin-half Ising chains with competing short-range and infinite-range interactions have been investigated by Theumann and Høye.[9] The Theumann–Høye Ising chain involves the nearest- and the next-nearest-neighbor antiferromagnetic interaction and an infinite-range ferromagnetic interaction of the Kac type. This chain exhibits different spin orderings at $T = 0$ for different interaction strengths and shows phase transitions as the interaction strengths are varied. In view of the close resemblance between the behavior of this Ising chain and that of polytypes, we have carried out[8] a Monte Carlo simulation of polytypes based on the competing interactions model. The basic Hamiltonian employed was that of Theumann and Høye.

The two states of the spin in an Ising chain can be taken to represent the two lowest energy configurations of a layer in a polytype, viz., the cubic (ABC) and the hexagonal (AB) configurations. Other layer configurations like AAB or ABB are of high energy and are not observed. We should, therefore, expect correspondence between the thermodynamic properties of a polytype and those of an appropriate Ising chain. In order to have an ordered spin arrangement in an Ising chain, it is necessary to introduce an infinite-range interaction among spins, the equivalent-neighbor type being most suitable for Monte Carlo simulations. The contribution to internal energy from this form of infinite-range interaction is given by $-J_{ir}M^2$. An analogous squared term has been suggested for the major part of the contribution to the elastic energy between atoms of different sizes.[10] It is, however, necessary to add a competing short-range interaction term to the equivalent neighbor interaction term in order that the spin orderings in the Ising chain describe polytypism.

We have found that a general double-layer mechanism (e.g., ABCACB → ABCBCB or ABABCB) connects different states of the polytype chain with about the same probability as the spin–flip mechanism in magnetic Ising chains.[8a] It has been possible to simulate various polytypes with periodicities extending up to 12 layers in this manner (see Figure 3). During the growth of a polytype (whether it be from melt or vapor), it is reasonable to assume that the atoms in the layers are jostling about, leading to fluctuations in interlayer interaction strength. A proper simulation of polytypes should take into account such fluctuations.[8b] We have therefore treated the interaction parameters in the simulation as random variables (varying with time) during any given simulation and carried out a computer simulation of polytype growth from vapor, employing both constant and fluctuating short-range interaction parameters. We have obtained short stretches of fairly long ordered polytypes such as 14H and 33R in such simulations. It is my feeling that the simulations discussed hitherto may be useful in understanding the occurrence of long period structures which include coherent intergrowth phases in systems such as hexagonal barium ferrites and intergrowth bronzes. It appears that long periodicities could originate from more than one cause, but it is likely to be thermodynamic in some systems. This indeed appears to be the case in alloys such as CuAu. Sato and co-workers[11] have shown that the operative factor in such a system is the lowering of the total electronic energy by decreasing the size of the Brillouin zone; the requisite periodicity for creation of

(8) (a) S. Ramasesha and C. N. R. Rao, *Philos. Mag.*, **36**, 827 (1977); (b) M. K. Uppal, S. Ramasesha, C. N. R. Rao, *Acta Crystallogr., Sect. A* **36**, 356 (1980).

(9) W. K. Theumann and J. S. Høye, *J. Chem. Phys.*, **55**, 4159 (1971).

(10) (a) P. W. Anderson and S. T. Chui, *Phys. Rev. B*, **9**, 3229 (1974); (b) J. Friedel in "Solid State Physics", F. Seitz, D. Turnbull, and H. Ehrenreich, Ed., Academic Press, New York, 1956, Vol. 3.

(11) H. Sato, R. S. Toth, and G. Honjo, *J. Phys. Chem. Solids*, **28**, 137 (1967), and the references cited therein.

energy gaps varies in some systems with the electron/atom ratio.

### Spin-State Transitions

Transitions of solids from magnetically ordered states to paramagnetic states are well-known. We have been interested in transitions between spin states of solids containing transition metal ions in the $d^4$–$d^8$ configuration. These transition metal ions can exist either in the low-spin or the high-spin ground state, depending upon the crystal field strength. In some of the transition-metal compounds, where the crystal field strength is close to the crossover point, interesting magnetic and structural behavior is observed. Spin-state transitions are known to occur in molecular systems of transition-metal complexes as well as in transition-metal oxides and other solids. Typical complexes belonging to the first category are $Fe(phen)_2(NCS)_2$ and $Fe(phen)_2(NCSe)_2$. The latter category consists of solid materials such as MnAs and rare-earth cobaltites, $LnCoO_3$ (Ln = La or rare earth). The low-spin ($t_{2g}^6$) to high-spin ($t_{2g}^4e_g^2$) transition in $LaCoO_3$ manifests itself as a plateau in the inverse susceptibility–temperature curve, the low- and high-spin ion ordering themselves on unique sites above a particular temperature.[12,13] The other rare-earth cobaltites, on the other hand, only show a maximum in the inverse susceptibility–temperature curve.[14] Spin-state transitions have been studied by several experimental techniques including measurements of heat capacity and magnetic susceptibility and Mössbauer spectroscopy. We have recently examined them by X-ray photoelectron spectroscopy[15] and NMR spectroscopy.[16]

Spin-state transitions have been found to occur in two-dimensional oxides of $K_2NiF_4$ structure. Thus, $La_4LiCoO_8$ shows a peak in the inverse susceptibility–temperature curve not unlike $NdCoO_3$ and other rare-earth cobaltites.[14] Trivalent cobalt in this system appears to transform from the low-spin state to the intermediate ($t_{2g}^5e_g^1$) as well as the high- ($t_{2g}^4e_g^2$) spin states.[17,18] Other oxides of cobalt of $K_2NiF_4$ structure[18] that seem to undergo low intermediate spin transitions are $Sr_4TaCoO_8$ and $Sr_4NbCoO_8$. in $LaMNiO_4$ where M is Sr or Ba, $Ni^{3+}$ is supposed to be in the low-spin state[19] when M is Sr and in the high-spin state[20] when M is Ba. We have examined the $LaSr_{1-x}Ba_xNiO_4$ system[18] and found that, with increase in $x$, the width of the $\sigma^*_{x^2-y^2}$ band (already present when $x = 0$) decreases, accompanying an increase in the unit cell volume; high-spin $Ni^{3+}$ ions are formed to a small extent with increasing $x$, but there appears to be no spin-state transition.

A successful model of spin-state transitions should be capable of explaining the following observations: (i) smooth as well as abrupt changes in the spin-state population ratio with temperature, (ii) occurrence of thermodynamically second-order transitions and first-order transitions in certain instances and a spin-state population ratio around unity near the transition, (iii) nonzero population of the high-spin state at low temperatures found in some systems, and (iv) a plateau or a peak in the plots of inverse magnetic susceptibility against temperature.

We have examined several models for spin-state transitions.[21] An earlier model described by Bari and Sivardiere[22] is static and can be solved exactly even when the dynamics of the lattice are included; the dynamic model does not, however, show any phase transition. We have investigated a dynamic model of spin-state transitions in which the high-spin and the low-spin states are mixed by a coupling to the lattice. The mode that can bring about such mixing is an ion-cage mode wherein the transition-metal ion moves off-center with respect to the octahedral cage in which it is placed. During such a vibration, the symmetry of the crystal field does not remain octahedral and hence the two spin states mix. This model predicts nonzero population of the high-spin state at low temperature but no spin-state transition. Susceptibility behavior of some $Fe^{2+}$ complexes can be explained by this model.

We have also examined a two-sublattice model, where the displacement on one sublattice is opposite to that on the other, but this model shows only second-order spin-state transitions. In order to explain the occurrence of both first- and second-order spin-state transitions, we have explored a two-sublattice model where the spin states are coupled to the cube of the breathing mode displacement. This model predicts first- or second-order transitions but only zero high-spin-state population at low temperatures. The most general model that predicts nonzero high-spin-state population at low temperatures, a first- or a second-order transition, and other features appears to be one where the coupling of the spin states to a breathing mode is linear and that to an ion-cage mode is quadratic. Nonetheless, spin-state transitions in extended solids need to be further explored to enable us to fully understand the mechanism of these transitions.

### Electronic Transitions

Among the phase transitions where electronic factors play a major role, the most well-known are the metal–insulator transitions exhibited by transition-metal oxides, sulfides, and so on. This subject has been discussed at length.[2,23,24] A recent observation[25] of some interest is that the metal–nonmetal transition occurs at a critical electron concentration as given by the particular form of the Mott criterion, $n_c^{1/3}a_H = 0.26 \pm 0.05$. The Verwey transition in $Fe_3O_4$ is associated with a marked jump in conductivity, but the material remains a semiconductor both above and below the transition temperature (123 K); below 123 K, there is

(12) P. M. Raccah and J. B. Goodenough, *Phys. Rev.*, **155**, 932 (1967).
(13) V. G. Bhide, D. S. Rajoria, G. Rama Rao, and C. N. R. Rao, *Phys. Rev. B*, **6**, 1021 (1972).
(14) W. H. Madhusudan, K. Jagannathan, P. Ganguly, and C. N. R. Rao, *J. Chem. Soc., Dalton Trans.*, 1397 (1980).
(15) S. Vasudevan, H. N. Vasan, and C. N. R. Rao, *Chem. Phys. Lett.*, **65**, 444 (1979).
(16) M. Bose, A. Ghoshray, A. Basu, and C. N. R. Rao, *Phys. Rev. B*, **26**, 4871 (1982).
(17) G. Demazeau, M. Pouchard, M. Thomas, J. F. Colombet, J. Grenier, L. Lournes, J. Souveyroux, and P. Hagenmuller, *Mater. Res. Bull.*, **15**, 451 (1980).
(18) R. Mohan Ram, K. K. Singh, W. H. Madhusudan, P. Ganguly, and C. N. R. Rao, *Mater. Res. Bull.*, **18**, 703 (1983).
(19) G. Demazeau, M. Pouchard, and P. Hagenmuller, *J. Solid State Chem.*, **18**, 159 (1976).
(20) G. Demazeau, J. L. Marty, B. Buffat, J. M. Dance, N. Pouchard, P. Dordor, and B. Chevalier, *Mater. Res. Bull.*, **17**, 37 (1982).
(21) S. Ramasesha, T. V. Ramakrishnan, and C. N. R. Rao, *J. Phys. C*, **12**, 1307 (1979).
(22) R. A. Bari and J. Sivardiere, *Phys. Rev. B*, **5**, 4466 (1972).
(23) N. F. Mott, "Metal–Insulator Transitions", Taylor and Francis, London, 1974.
(24) J. M. Honig, *J. Solid State Chem.*, **45**, 1 (1982), and the references cited therein.
(25) P. P. Edwards and M. J. Sienko, *Acc. Chem. Res.*, **15**, 87 (1982).

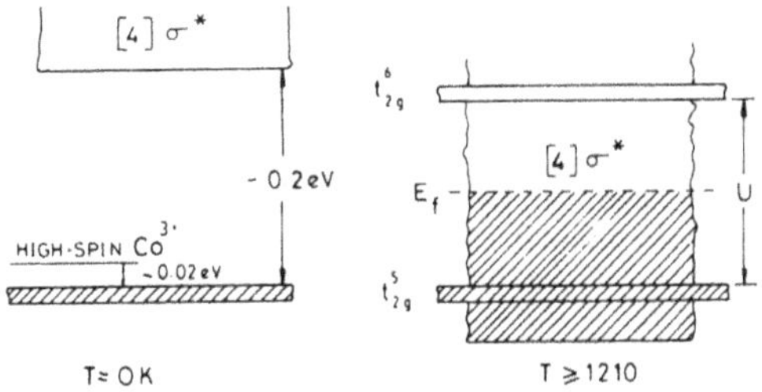


**Figure 4.** Energy band scheme for $LaCoO_3$ at 0 K and above the first-order electronic transition at 1200 K (taken from ref 13). At 0 K, all the 3d electrons are in the $t_{2g}$ level (low-spin state), and the high-spin $Co^{3+}$ level is empty. Above 1200 K, the $e_g$ electrons form a partially filled $\sigma^*$ band and the $t_{2g}$ electrons remain localized.

charge ordering of $Fe^{2+}$ and $Fe^{3+}$ ions. The Verwey transition has been a subject of intensive discussion,[24] but there is still some doubt with regard to the mechanism of the transition. In this section, I shall briefly present the features of a few other interesting electronic transitions investigated in this laboratory.

$LaCoO_3$ and other rare-earth cobaltites show first-order phase transitions around 1200 K, which seem to be essentially governed by the change in electronic entropy.[12,13,26] Temperature evolution of the electronic and spin configurations of cobalt in these cobaltites is interesting, and we have investigated this by employing Mössbauer spectroscopy[13,26] and X-ray photoelectron spectroscopy.[27] At low temperatures, cobalt ions are in the diamagnetic low-spin state ($t_{2g}^6$) and transform to the high-spin state ($t_{2g}^4e_g^2$) with increase in temperature, the two spin states being clearly distinguished in Mössbauer spectra. Electron hopping between the two spin states gives rise to charge-transfer states ($Co^{2+}$ + $Co^{4+}$) and associated increase in electrical conductivity. The magnitude of charge transfer depends on the acidity of the rare-earth ion. As the temperature is increased further, the $e_g$ electrons tend to form a $\sigma^*$ band; accordingly, the center shift in the Mössbauer spectra shows a decrease in this temperature region (~700–1000 K) due to progressive increase in the cation–anion orbital overlap. Mössbauer spectra show a single resonance with a chemical shift close to zero (corresponding to the band state of $e_g$ electrons), as we approach the first-order transition temperature (~1200 K). Above 1200 K, the cobaltites become metallic due to the change in the nature of d electrons from localized to itinerant behavior (Figure 4). Since no change in crystal symmetry was noticed at the transition, it was considered that the entire entropy change was electronic in origin. However, it seems likely that there is an increase in the symmetry of $LaCoO_3$ (change from rhombohedral to cubic structure) after the transition.[12,13] Even so, the large $\Delta S$ of the transition (over 4 J $K^{-1}$ $mol^{-1}$) can only arise because of a significant electronic contribution. The Lamb–Mössbauer factor (area under the resonance) decreases markedly before this transition, indicative of large ionic vibrations, and increases sharply above the transition, suggesting the establishment of long-range order.

Another interesting electronic transition that we have been interested in is that of $La_2NiO_4$, which crystallizes in the two-dimensional $K_2NiF_4$ structure. Unlike $K_2NiF_4$, $La_2NiO_4$ is not known to show any long-range antiferromagnetic order but only shows deviations from Curie–Weiss law due to short-range interactions.[28] What is more interesting is that $La_2NiO_4$ undergoes a gradual semiconductor-metal transition[29] around 550 K; no structural change accompanying the transition has been reported. The Ni–O–Ni distance in $La_2NiO_4$ (3.86 Å) is shorter than in NiO (4.18 Å). The transfer integral, $b$, in NiO is close to the critical transfer integral, $b_c$, at which the description for localized electron states breaks down. In $La_2NiO_4$, $b$ should be nearly equal to $b_c$, so that we obtain partially filled $\sigma^*(x^2 - y^2)$ bands of strongly correlated itinerant d electrons and hence the metallic behavior. The half-filled $d_{z^2}$ orbitals of $Ni^{2+}$, on the other hand, would be localized. A likely origin of the semiconductor-metal transition is the splitting of the $\sigma^*(x^2 - y^2)$ band on lowering the temperature because of strong electron correlations, which in turn may be accentuated by short-range antiferromagnetic ordering $d_{z^2}$ spins.[30]

In spite of the great interest in the 550 K electronic transition of this two-dimensional system, all the electrical measurements have hitherto been carried out on pellets of polycrystalline material. Preliminary measurements on single crystals[31] indicate the occurrence of a sharp transition (in the *ab* plane) with at least an order of magnitude jump in conductivity. Furthermore, $Ni^{3+}$ ions that are inevitably present (3–9%) in all preparations of $La_2NiO_4$ seem to influence the structure and properties quite significantly. Samples of $La_2NiO_4$ annealed in a $CO_2$ atmosphere at 1400 K (to removed $Ni^{3+}$) seem to show evidence for long-range antiferromagnetic ordering. In addition, electron diffraction studies show them to have a monoclinic distortion contrary to published crystallographic data. Careful investigations of the electronic and magnetic properties of well-characterized samples of $La_2NiO_4$ are therefore warranted.

The last type of electronic transition that I shall discuss is the one exhibited by silver chalcogenides, $Ag_{2+\delta}Ch$ (Ch = S, Se, or Te). These compounds transform to a symmetrical phase on heating wherein $Ag^+$ are randomly distributed, giving rise to superionic conductivity just as AgI. These materials are small-gap semiconductors at room temperature and exhibit interesting electronic behavior as a function of temperature as well as of composition.[32] Thus, in the high temperature phase ($T > 406$ K), $Ag_{2+\delta}Se$ shows metallic behavior of electronic conductivity for high values of X. With decrease in $\delta$, the electronic conductivity shows evidence for an interesting transition (Figure 5), the ionic conductivity of the high temperature phase being essentially independent of $\delta$. The magnitude of change in electronic conductivity at the phase transition is also

(26) (a) V. G. Jadhao, G. Rama Rao, D. Bahadur, R. M. Singru, and C. N. R. Rao, *J. Chem. Soc., Faraday Trans. 2*, 71, 1885 (1975); (b) V. G. Jadhao, R. M. Singru, G. Rama Rao, D. Bahadur, and C. N. R. Rao, *J. Phys. Chem. Solids*, **37**, 113 (1976).

(27) G. Thornton, A. F. Orchard, and C. N. R. Rao, *J. Phys. C*, **9**, 1991 (1976).

(28) P. Ganguly, S. Kollali, and C. N. R. Rao, *Magn. Lett.*, **1**, 107 (1978).

(29) P. Ganguly and C. N. R. Rao, *Mater. Res. Bull.*, **8**, 405 (1973).

(30) J. B. Goodenough and S. Ramasesha, *Mater. Res. Bull.*, **17**, 383 (1982).

(31) C. N. R. Rao, D. Buttrey, N. Otsuka, P. Ganguly, H. R. Harrison, C. J. Sandberg, and J. M. Honig, *J. Solid State Chem.*, in press.

(32) A. K. Shukla, H. N. Vasan, and C. N. R. Rao, *Proc. R. Soc. London, Ser. A*, **376**, 619 (1981).

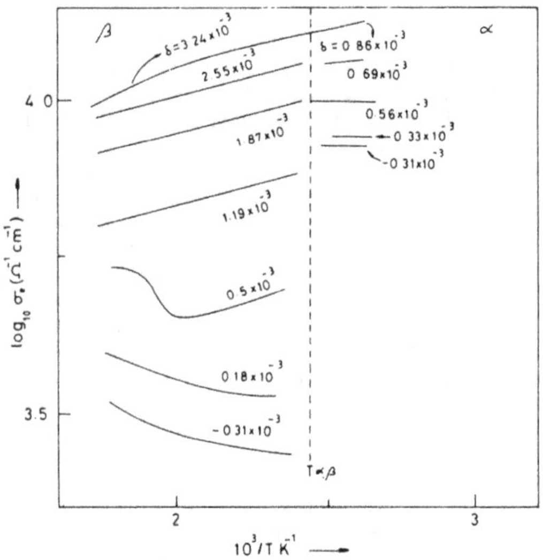


**Figure 5.** Plot of the logarithm of electronic conductivity against inverse of absolute temperature for various compositions of α- and β-silver selenide. The degree of nonstoichiometry, δ, was determined by solid-state coulometry (taken from ref 32).

determined by stoichiometry. In the low-temperature phase, the material conducts like a semimetal around 400 K and the conductivity decreases substantially at low temperatures, the magnitude of the decrease depending on the value of δ. This behavior as well as the δ-dependent electronic transition in the high-temperature phase is indeed interesting. We are now investigating $Ag_{2+\delta}Te$, which is a p-type material with a small gap (unlike $Ag_{2+\delta}Se$, which is a n-type material), at ordinary temperatures.

## Organic Solids and Related Systems

Although phase transitions in organic solids have been reported from time to time, extensive mechanistic studies have not been carried out as in the case of inorganic solids.[2] It was believed until recently by many workers that there are no structural relations between phases in the case of organic solids. Jones et al.,[33] however, have reported that the stress-induced phase transition of 1,8-dichloro-10-methylanthracene proceeds by a diffusionless displacive transition (somewhat similar to a martensitic transition) with definite orientational relationships. The irrational habit plane seems to be composed of close packed planes and the properties of the interface could be formulated in terms of slip dislocations. The reversible topotactic phase transition of 5-methyl-1-thia-5-azoniacyclooctane 1-oxide perchlorate has been explained by Parkinson et al.[34] in terms of recurrent glissile partial dislocations; it was earlier thought[35] that the transition involved a cooperative inversion and rotation of half of the molecular cations. The phase transition of paraterphenyl involving rotational disorder has been elucidated by Ramdas and Thomas[36] by evaluating pairwise interactions between nonbonded atoms.

In this laboratory, we have been studying phase transitions of a variety of organic solids. For example, we have investigated the α–γ–α–β transitions of *p*-dichlorobenzene by infrared spectroscopy.[37] The γ phase is characterized by unusually high intramolecular vibrations mode frequencies. The α–γ transition shows athermal nucleation behavior as in martensitic transitions; the α–β transition seems to be associated with some disorder. When the asymmetry of the molecules is not too high, organic crystals often exhibit order–disorder transitions (e.g., benzothiophene). We have investigated order–disorder transitions in benzothiophene and other compounds by making use of changes in infrared band intensities and half-widths.[38] Raman spectroscopy would be especially useful in investigating order–disorder transitions.

We have investigated the phase transitions of compounds of the type $(C_nH_{2n+1}NH_3)_2MCl_4$, M = Mn, Fe, Cd, or Cu, which provide interesting model systems to investigate magnetic phenomena in two dimensions. Earlier spectroscopic investigations[39] seemed to indicate that in $(CH_3NH_3)_2MCl_4$, the phase transitions are determined essentially by the motions of the methylammonium groups. We have examined the infrared spectra of several $(C_nH_{2n+1}NH_3)_2MCl_4$ systems through their phase transitions to investigate whether the intramolecular vibration modes show the expected changes.[40] In the high-temperature phase of these solids, the $(CH_3NH_3)^+$ ion has $C_{3v}$ symmetry, but the symmetry goes down to $C_1$ or $C_s$ in the ordered low-temperature phases. The spectra indeed show the expected site-groups as well as factor-group splittings in the low-temperature phases, the degenerate bending modes of $NH_3$ and $CH_3$ being particularly sensitive. The phase transitions of the tetrachlorometallates are similar to those of the corresponding alkylammonium chlorides, $C_nH_{2n+1}NH_3Cl$, thereby establishing that the phase transitions in the former are entirely controlled by the motions of the $(C_nH_{2n+1}NH_3)^+$ group. Accordingly, we find that the $(C_nH_{2n+1}NH_3)_2MBr_4$ system shows transitions similar to those of the chloro compounds.

We have been particularly interested in the study of the plastic states of organic compounds,[41] which are characterized by high values of ΔS of formation from the crystalline state, the ΔS of fusion (plastic–liquid transition) being much smaller. We find that the ΔH as well as the ΔS of the crystal–plastic transition generally decrease as the temperature range of stability of the plastic phase increases; the ΔH and the ΔS of the plastic–liquid transition, on the other hand, increase as the temperature range of stability of the plastic phase increases.[42] Neutron scattering, NMR spectroscopy, and several other techniques have been employed to study molecular reorientation in the plastic state.[41] We

(33) W. Jones, J. M. Thomas, and J. O. Williams, *Philos. Mag.*, **32**, 1 (1975).
(34) G. M. Parkinson, J. M. Thomas, J. O. Williams, M. J. Goringe, and L. W. Hobbs, *J. Chem. Soc., Perkin Trans. 2*, 836 (1976).
(35) I. C. Paul and K. T. Go, *J. Chem. Soc. B*, 33 (1969).
(36) S. Ramdas and J. M. Thomas, *J. Chem. Soc., Faraday Trans. 2*, **72**, 1251 (1976).

(37) S. Ganguly, J. R. Fernandes, G. Bahadur, and C. N. R. Rao, *J. Chem. Soc., Faraday Trans. 2*, **75**, 923 (1979).
(38) C. N. R. Rao, S. Ganguly, and H. R. Swamy, *Croat. Chem. Acta*, **55**, 207 (1982), and the references cited therein.
(39) (a) D. M. Adams and D. C. Stevens, *J. Phys. C*, **11**, 617 (1978); (b) R. Blinc, B. Zeks, and R. Kind, *Phys. Rev. B*, **17**, 3409 (1978).
(40) C. N. R. Rao, S. Ganguly, H. R. Swamy, and I. A. Oxton, *J. Chem. Soc., Faraday Trans. 2*, **77**, 1825 (1981).
(41) J. N. Sherwood, Ed., "The Plastically Crystalline State", Wiley, New York, 1979.
(42) S. Ganguly, J. R. Fernandes, and C. N. R. Rao, *Adv. Mol. Relaxation Interact. Processes*, **20**, 149 (1981).

have employed Raman band shape analysis to obtain rotational correlation function of plastic phases of a few organic compounds.[43] The correlation times are continuous through the plastic–liquid transition.

Molecular dynamics simulation of the plastic state of $CH_4$ has been carried out by Bounds et al.[44] who have evaluated the static and dynamic structure factors. The crystalline phase of $CH_4$ described by James and Keenan[45] has been investigated by several techniques. We have carried out a Monte Carlo simulation study[46] on the orientational glasses (or glassy crystalline phases) obtained by annealing or quenching the plastic phase. Different cooling rates lead to different states of the glassy crystalline phase. Temperature variation of the orientational parameter of molecules suggests the presence of a transition between the plastic and glassy crystalline phases.

Phase transitions of hydrogen-bonded solids such as ferroelectric hydrogen phosphates and Rochelle salt have been investigated widely in the literature.[2] We have been recently investigating phase transitions of alkanedioc acids employing vibrational spectroscopy.[38] The phase transition of malonic acid at 360 K is especially interesting. At ordinary temperatures, the unit cell of malonic acid contains two cyclic dimeric rings orthogonal to each other; above 360 K, the two hydrogen-bonded rings become similar as evidenced from IR and Raman spectra.[47,48] Hydrogen bonds in the high-temperature phase are on the average weaker than those in the low-temperature phase. The phase transition occurs at a higher temperature (366 K) in the fully deuterated acid, and the vibrational bands show a positive deuterium isotope effect. It appears that the transition is governed by librational and torsional modes of the hydrogen-bonded rings (around 90 and 50 $cm^{-1}$, respectively, below the transition temperature), which show a tendency to soften.

## Concluding Remarks

The above discussion should serve to indicate the variety of problems in the area of phase transitions that are of relevance to solid-state chemistry. Of course there are other important aspects of phase transitions, such as cooperative Jahn–Teller effect, commensurate–incommensurate structure transitions and transitions in ferroics, that I have not touched upon. Neither have I dealt with technological applications.[2] The main point that I have tried to make is that the study of structural and mechanistic aspects of phase transitions in the solid state constitutes an essential and interesting aspect of the chemistry of solids.[49]

A word about the so-called normal-superionic conducting state transitions would be in order. It seems to be wrong to refer to these transitions as superionic transitions by analogy with the electronic superconducting transitions. While AgI related materials show a marked jump in ionic conductivity at a transition temperature, there are many other superionic materials that do not.[50]

A transition that has eluded a proper understanding and has increasingly become an area of vital interest is the glass transition.[51] Although many models have been proposed, none of them is able to explain all the features of this transition. An interplay of kinetic and thermodynamic effects further complicates the situation. We have recently proposed a cluster model of the glass transition[52] based on the premise that glass is an ensemble of more ordered regions (clusters) embedded in a liquidlike, disordered tissue material. Toward the glass transition, clusters gradually melt and add on to the tissue material. The cluster model making use of the relative size of the cluster as an order parameter seems to explain several features of the transition. Another aspect of the glass transition that is worth noting is that it is not merely characteristic of the normal variety of (positionally disordered) glasses but also of orientationally disordered glasses and dipolar glasses. A satisfactory model would have to account for this feature as well. Some other aspects of phase transitions that deserve attention are transitions in organic solids, deformational mechanisms of phase transitions in complex solids, transitions of orientationally disordered crystals, and transitions involving ordering of defects. Computer simulation studies are bound to be increasingly useful in understanding the nature of many of the solid-state phase transitions.

(43) S. Ganguly, H. R. Swamy, and C. N. R. Rao, *J. Mol. Liq.*, **25**, 139 (1983).
(44) D. G. Bounds, M. L. Klein, and G. N. Patey, *J. Chem. Phys.*, **72**, 5348 (1980).
(45) H. M. James and T. A. Keenan, *J. Chem. Phys.*, **31**, 12 (1959).
(46) S. Yashonath and C. N. R. Rao, *Chem. Phys. Lett.*, **101**, 524 (1983).
(47) S. Ganguly, J. R. Fernandes, G. R. Desiraju, and C. N. R. Rao, *Chem. Phys. Lett.*, **69**, 227 (1980).
(48) J. de Villepin, M. H. Limage, A. Novak, M. LePostollec, H. Poulet, S. Ganguly, and C. N. R. Rao, *J. Raman Spectrosc.*, in press.
(49) C. N. R. Rao, *Chem. Scr.*, **19**, 124 (1982).
(50) R. A. Huggins and A. Rabenau, *Mater. Res. Bull.*, **13**, 1315 (1978).
(51) R. Parthasarathy, K. J. Rao, and C. N. R. Rao, *Chem. Soc. Rev.*, in press.
(52) K. J. Rao and C. N. R. Rao, *Mater. Res. Bull.*, **17**, 1337 (1982).

For a discussion of some aspects of the glass transition, see R. Parthasarathy, K.J. Rao and C.N.R. Rao, The Glass Transition: Salient Facts and Models, Chem. Soc. Revs., 12, 361-385 (1983)

Offprinted from the *Transactions of The Faraday Society*,
No. 476, Vol. 58, Part 8, August, 1962

# Mechanism of Crystal Structure Transformations

## Part 3.—Factors Affecting the Anatase-Rutile Transformation *

By S. R. Yoganarasimhan and C. N. R. Rao†

Dept. of Inorganic and Physical Chemistry, Indian Institute of Science, Bangalore 12, India

*Received* 20*th November*, 1961

The particle size and crystallite size of anatase increase markedly in the region of the crystal structure transformation. The unit cell of anatase seems to expand prior to the transformation to rutile. This expansion has been attributed to a displacive transformation of the type defined by Buerger. Smaller particle size and larger surface area seem to favour the transformation.

The kinetics of the transformation of anatase prepared by the hydrolysis of titanium sulphate have been studied at different temperatures and are found to be considerably different from the kinetics of the transformation of pure anatase. The transformation becomes immeasurably slow below $\sim 695 \pm 10$°C compared to $\sim 610$°C for pure anatase. An induction period is observed in the transformation of anatase obtained from sulphate hydrolysis and the duration decreases with increase in temperature. The activation energy is $\sim$120 kcal/mole, a value higher than that for the pure anatase-rutile transformation. The results have been interpreted in terms of the relative rates of nucleation and propagation processes. The activation energy for the nucleation process seems to be much larger than for the propagation process. The kinetics of the transformation of anatase samples doped with different amounts of sulphate ion impurity have also been studied and the transformation is found to be progressively decelerated with increase in the impurity concentration. The energy of activation for the transformation appears to increase progressively with increase in impurity concentration.

---

The transformation of pure anatase (tetragonal) to rutile (tetragonal) has been systematically investigated by Czanderna, Rao and Honig [1] and Rao,[2] and the transformation of brookite (orthorhombic) to rutile by Rao, Yoganarasimhan and Faeth.[3] These studies prompted us to determine the various parameters that affect crystal structure transformations and the nature and magnitude of their effects. The transformation of anatase to rutile is a reconstructive transformation [4] involving changes in secondary coordination. The rate of transformation of spectroscopically pure anatase [2] is exponential, with an energy of activation $\sim$90 kcal/mole. It has been shown that the ratio of the propagation rate constant to the nucleation rate constant is small and that the energy of activation is utilized mainly for the random production of nucleation sites.[5] We have now investigated the changes in particle size, crystallite size and lattice dimensions accompanying this crystal structure transformation and also the effect of particle size and surface area on the transformation. Such a study is considered important for understanding the features and mechanism of crystal-structure transformations. Anatase of the highest purity has been used in this investigation, since impurities are known to have marked effect on the transformation.[6, 7]

It is also known that the transformation characteristics depend on the method of preparation of anatase and the major cause for the wide variation in the transformation temperature reported in the literature (400-1000°C)[8-11] is probably due to impurity

* Contains material from the Ph.D. thesis of S. R. Yoganarasimhan to be submitted to Indian Institute of Science under the guidance of C. N. R. Rao. Part 1, ref. (2); part 2, ref. (3).

† To whom all the correspondence should be addressed.

effects. Thus the transformation of anatase prepared by the hydrolysis of titanium sulphate occurs at much higher temperature than that of pure anatase possibly due to the presence of sulphate ion impurity in the sample. The effect of impurity atoms is by no means specific to anatase-rutile transformation and is found in several systems such as aragonite + calcite, marcosite + pyrite, sphalerite + wurtzite, etc. Although the impurity effects on phase transformations have been noted qualitatively, no detailed investigations on the nature and magnitude of their effects have hitherto been reported. We have now investigated, in detail, the effect of sulphate ion impurity on the kinetics of the anatase-rutile transformation.

## EXPERIMENTAL

Spectroscopically pure peroxy titanium oxide was prepared by the method described in the literature,[12] using spectrographically standardized titanium powder supplied by Johnson and Matthey, London, England. Anatase was then obtained by the thermal decomposition of the peroxy compound. Differential thermal analysis of the peroxy compound was carried out using an apparatus described by Pask and Warner [13] (fig. 1*a*).

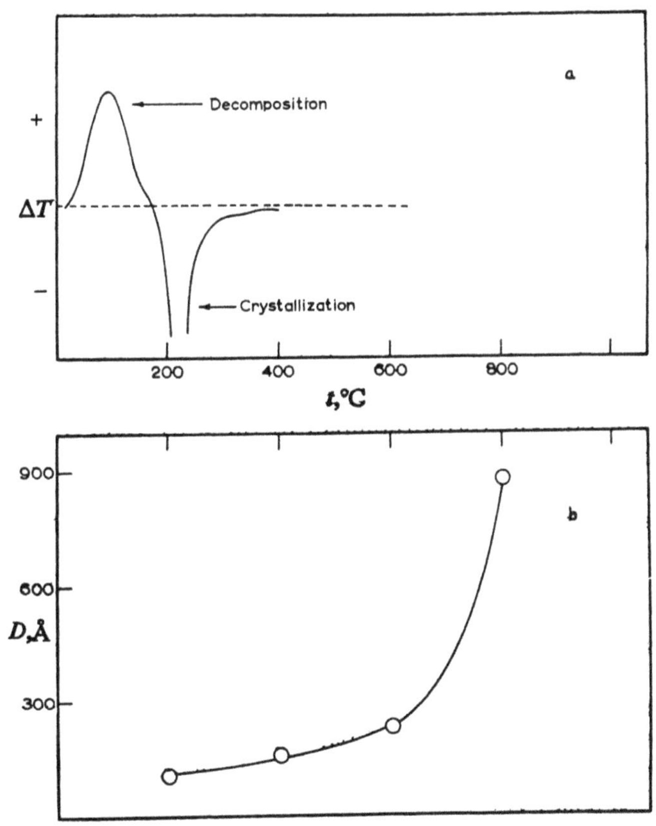


FIG. 1.

(*a*) Differential thermal analysis curve of peroxy titanium oxide.

(*b*) Variation of crystallite size with temperature.

Particle size was studied by means of photomicrographs taken with an Earnst Leitz Wetzlar unit with a magnification of 100. Surface areas of the samples were determined by employing a volumetric apparatus described by Srinivasan.[14] A North American Philips X-ray diffractometer equipped with a Geiger-counter recording device was employed for the

X-ray line-width measurements, while a Rich Seifert Debyeflex unit was used for the other powder patterns. A North American Philips back-reflection camera was employed for the precision determination of lattice constants. The uncertainty in the lattice constants varies between $\pm 0{\cdot}001$ and $\pm 0{\cdot}0025$ Å. Anatase samples were heated in quartz tubes placed in a vertical tubular furnace fitted with a thermo-regulator. The fraction of anatase in a mixture was determined by the equation,

$$f_A = (1 + 1{\cdot}26\, I_R/I_A)^{-1}, \quad (1)$$

where $I_R$ and $I_A$ are the intensities of rutile ($d = 3{\cdot}245$ Å) and anatase ($d = 3{\cdot}51$ Å) lines respectively. The approximate uncertainty in the energy of activation of the transformation is $\pm 10$ kcal/mole.

Titanium sulphate was prepared by digesting titanium hydroxide, obtained by hydrolysis of pure titanium tetrachloride, in sulphuric acid. A sample of anatase was prepared by hydrolyzing titanium sulphate solution with ammonia. The precipitate was washed thoroughly till the washings gave no test for sulphate ion. The precipitate was dried at $\sim 100$°C. Such precipitates were, however, found to contain some sulphate impurity.

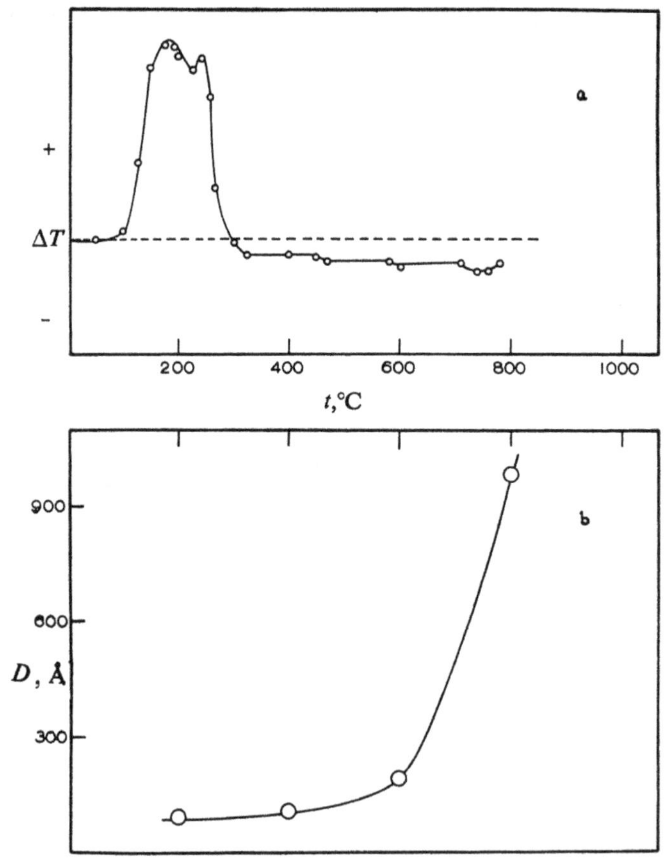


FIG. 2.

(*a*) Differential thermal analysis curve of titanium dioxide obtained from sulphate hydrolysis.
(*b*) Crystallite size variation of anatase obtained from sulphate hydrolysis.

Quantitative estimation of the sulphate content by the fusion method was found to be difficult because of the low percentage of the impurity. The anatase, thus prepared, was amorphous. The surface area of this anatase sample (B.E.T.) was $\sim 54$ m$^2$/g and the differential thermal analysis curve of the anatase sample is shown in fig. 2*a*. Although no exothermic peak due to crystallization was observed, the endothermic peak shows a definite splitting around

250°C which may be due to crystallization.[15] The sample, in fact, gives an X-ray pattern (though ill-defined) when heated at ~250°C. The crystallite size is very low (~100 Å) and increases markedly when heated to higher temperatures (fig. 2*b*) just as for pure anatase (fig. 1*b*). The kinetics of the transformation of this anatase sample were studied at different temperatures.

Anatase samples doped with 0·1, 1 and 5 atom % sulphate ion impurity were prepared by the method reported earlier.[6] Spectrographically standardized titanium metal powder supplied by Johnson and Matthey, London, was used for the purpose. The lattice constants of these anatase samples remained the same within experimental error ($a = 3{\cdot}771 \pm {\cdot}005$ Å; and $c = 9{\cdot}498 \pm {\cdot}013$ Å).

## RESULTS AND DISCUSSION

The amorphous peroxy titanium oxide, on heating, lost the peroxy oxygen and water around 100°C. The resulting oxide was poorly crystalline. The differential thermal analysis of peroxy titanium oxide (fig. 1*a*) showed an endothermic peak

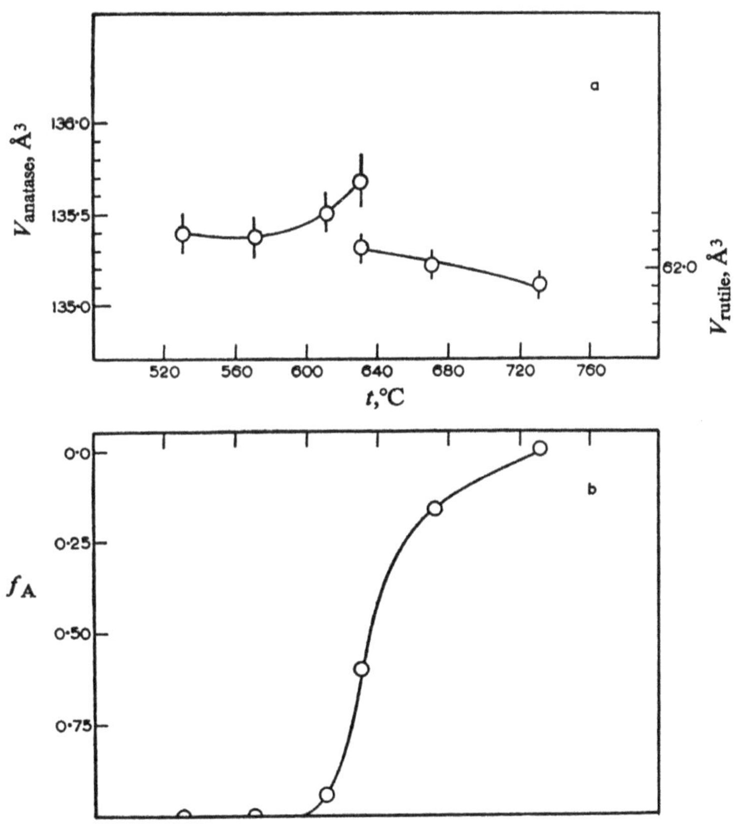


FIG. 3.

(*a*) Variation of unit cell volumes of anatase and rutile with temperature.

(*b*) Variation of fraction of anatase at different temperatures.

around 100°C due to the decomposition reaction and a very large exothermic peak around 250°C. While the oxide was very poorly crystalline before the exothermic peak, it gave a good X-ray diffraction pattern characteristic of crystalline anatase, just after passing the peak. The exothermic reaction (fig. 1*a*) is clearly associated with the crystallization reaction.[15]

In order to study the growth of particles with temperature, anatase powder (preheated to 150°C) was heated for a period of 3 h at 400, 600, 800 and 1000°C. Marked increase in particle size was noticed in the 600-1000°C region, as indicated by the photo-micrographs. The specific surface area (B.E.T.) of anatase heated at 400°C was ~55 $m^2/g$ and decreased markedly for samples heated to higher temperatures. The crystallite size normal to the (101) and (110) reflecting planes of anatase and rutile samples was calculated by measuring the X-ray diffraction line-widths of the samples heated at 200, 400, 600, 800 and 1000°C for 3 h. The Scherrer equation corrected for instrumental line-broadening by Warren's equation was employed for the calculation.[16] The line-width of the sample heated at 1000°C was taken as the reference. The crystallite size increases rapidly after 600°C (fig. 1*b*). The transformation of pure anatase also starts only above 600°C.

It was considered interesting to see if lattice dimensions change during the transformation of anatase to rutile. To study such an effect, lattice constants of anatase heated for 3 h at different temperatures in the transformation region were determined by taking Debye-Scherrer X-ray patterns at 25°C and subjecting the data to Cohen's extrapolation technique followed by least-square treatment. The results are presented in fig. 3*a*, where the unit cell, volumes $V$ of anatase and rutile are plotted against temperature. The composition at each temperature is indicated by the curve in fig. 3*b*. The results in fig. 3*a* seem to indicate that the anatase unit cell expands before transforming to rutile. An expanded rutile unit cell then contracts with increase in temperature. It is interesting to note that Schossberger [11] had noticed that the density (pyknometric) of anatase underwent a small but finite decrease during transformation to rutile. The expansion of the anatase lattice prior to transformation may be taken as an indirect indication that disordered lattices favour this crystal-structure transformation. The anatase-rutile transformation may be visualized as shown in scheme below.

*Scheme*

| | anatase I ; collapsed form (ordered lattice) |
|---|---|
| ~620°C | ↓ Displacive transformation (Buerger [4]). |
| | anatase II; open form (disordered lattice) |
| ~620-650°C | ↓ Reconstructive transformation (Buerger [4]). |
| | rutile II ; open form (disordered lattice) |
| above ~650°C | ↓ Displacive transformation (?) |
| | rutile I ; collapsed form (ordered lattice). |

## EFFECT OF PARTICLE SIZE ON THE TRANSFORMATION

Anatase with relatively large particle size was prepared by heating the freshly prepared oxide to 400°C and a sample of particle size $74 > P_1 > 53$ microns was obtained by employing Tyler standard sieves (−200+270 mesh). These particles were crushed to finer particles and sieved (mesh sizes −270+325 and −325) to provide anatase samples ($P_2$ and $P_3$) with particle sizes $53 > P_2 > 43$ and $P_3 < 43$ microns respectively. Since the thermal history of all the three samples is the same, it was assumed that any changes in transformation characteristics can be attributed to particle-size effects. The kinetics of transformation of the three samples were studied at 700°C and the transformation curves shown in fig. 4*a*, clearly indicate that smaller particle size of anatase favours the transformation. The values of the energy of activation of the transformation for $P_1$ and $P_3$, however, are found to be about the same within the limits of experimental error ($\sim$114 and $\sim$107 kcal/mole respectively for $P_1$ and $P_3$). Recently, larger particle size has been found not to favour quartz-cristobalite transformation.[17] The effect of particle size on the transformation is possibly due to dislocations, produced during the grinding operation, acting as nucleation sites.

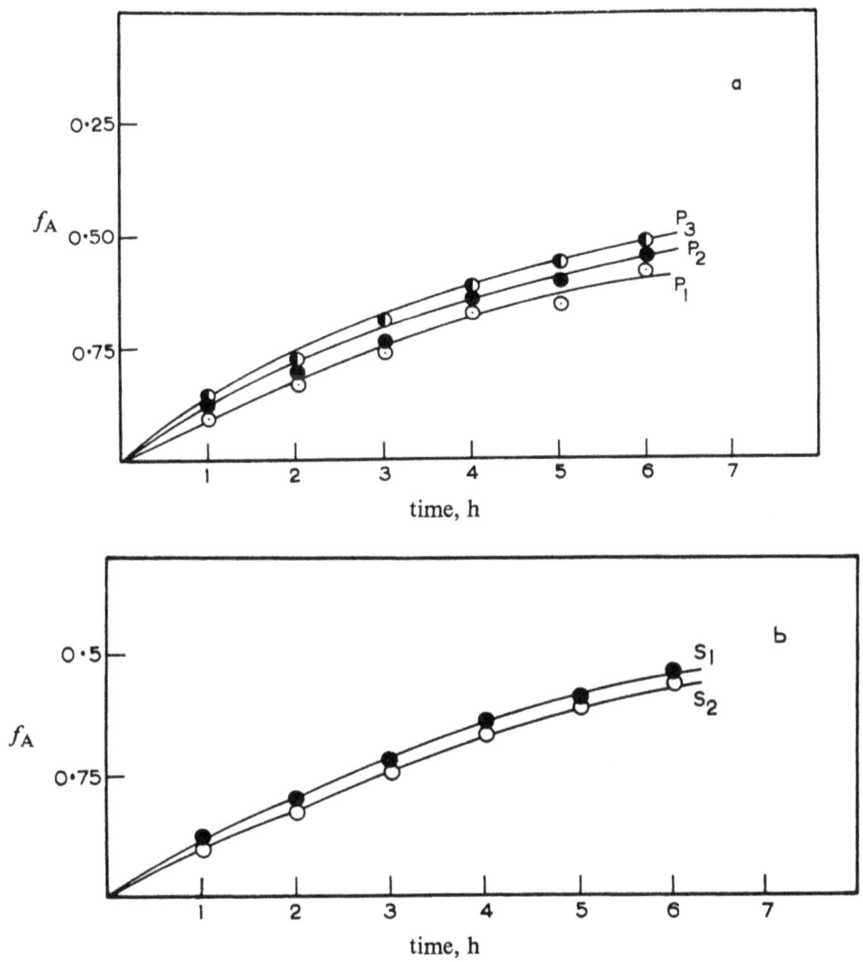


FIG. 4.

(*a*) Transformation curves for anatase samples of different particle sizes at 700°C: $P_1$ = 74 >$P_1$ >53 microns, $P_2$ = 53 >$P_2$ >43 microns, and $P_3$< 43 microns.

(*b*) Transformation curves of anatase samples of two different surface areas at 700°C: $S_1$, 55 $m^2/g$ and $S_2$, 25 $m^2/g$.

### EFFECT OF SURFACE AREA ON THE TRANSFORMATION

An anatase sample $P_2$ of specific surface area $\sim$55 $m^2/g$ was heated at 600°C for 3 h and the specific surface area was found to decrease to $\sim$25 $m^2/g$. The transformation characteristics of these two samples were studied (fig. 4*b*). Since the growth in the crystallite size is negligible in the 400-600°C region and the particle size is nearly the same for the two samples, the observed effect may be interpreted as due to the effect of surface area. It can be seen from fig. 4*b* that the effect of surface area is not very marked and, if any, larger surface area favours the transformation. While the energy of activation for the transformation of the sample of high surface area is ~110 kcal/mole, that for the lower surface area sample is ~102 kcal/mole.

## EFFECT OF IMPURITIES ON THE TRANSFORMATION

The kinetics and energetics of the transformation of anatase prepared by the hydrolysis of titanium sulphate are markedly different from those of pure anatase. The transformation of anatase prepared by sulphate-hydrolysis becomes immeasurably slow below ~695±10°C, compared to 610±10°C [2] for pure anatase (fig. 6). While the transformation of pure anatase follows the exponential rate law, that of anatase

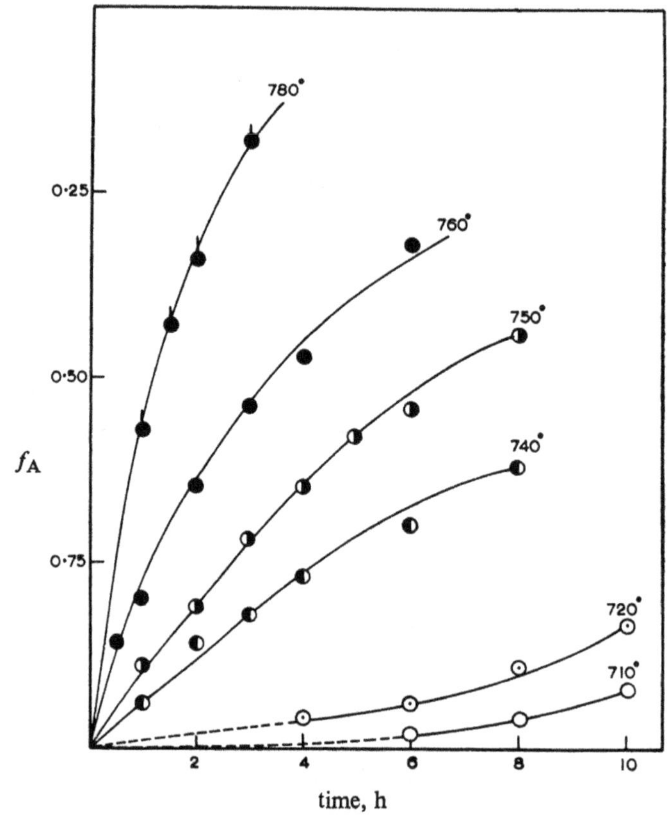


FIG. 5.—Plots of $f_A$ as a function of time at different temperatures (sulphate hydrolysis).

from sulphate hydrolysis shows a considerable induction time at lower temperatures (fig. 5). The induction period decreases with increase in temperature and at sufficiently high temperatures (>740° for this sample) the rate becomes exponential (fig. 5 and 7).

These observations can be represented as a special case of the general rate equation derived by the application of order-disorder theory to diffusionless transitions in solids.[3] According to this equation, the shape of the rate curve is determined by the relative numerical values of $zk_p/k_n$ and of $c$. The larger the factor $k_p/k_n$ is relative to $c$, the more sigmoidal the curves become. This is understandable since the propagation effect which is responsible for the autocatalytic character of the transformation becomes more noticeable when $k_p/k_n$ is large and $c$ small. Under these conditions some time elapses before a sufficient number of nucleation sites are formed; then the propagation becomes dominant. The duration of the induction period itself varies depending on the system. The observation of induction period in the rate curves for the transformation of anatase prepared by sulphate hydrolysis may thus be interpreted as due to a relatively large ratio of propagation rate constant to nucleation rate constant.

The decrease of induction time with increase in temperature is probably due to the greater sensitivity of the nucleation process to temperature changes. The nucleation process should, therefore, be associated with a high energy of activation $E_a$. This is indeed confirmed by the calculation of $E_a$ from the rate data presented in

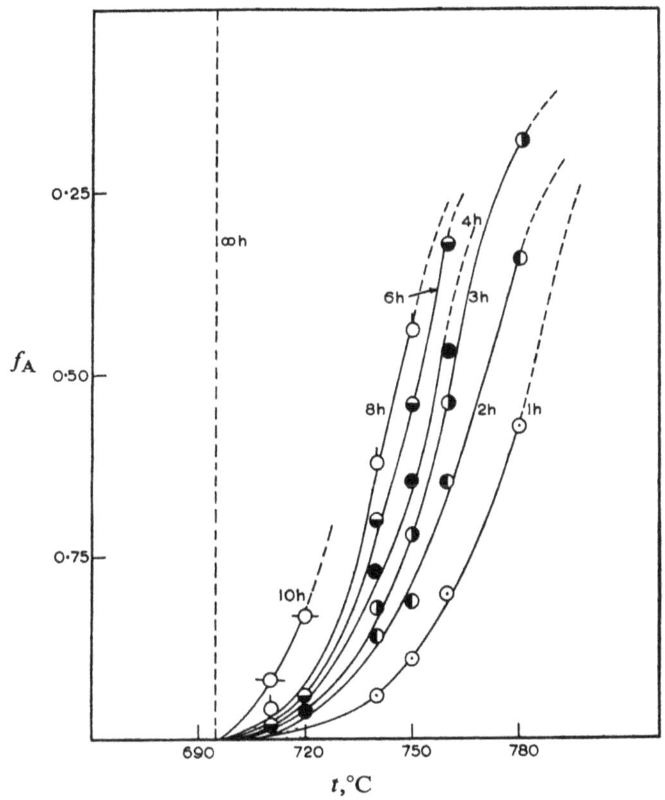


FIG. 6.—Transformation of anatase obtained from sulphate hydrolysis as a function of temperature at different times.

fig. 5. The $E_a$ of transformation for the rate data at high temperatures (where the rates are exponential) is $\sim 120 \pm 10$ kcal/mole. From the rate data at lower temperatures, two values of $E_a$ can be calculated; $\sim 215$ kcal/mole for the induction-time region (fig. 8) and $\sim 40$ kcal/mole after the induction-time region. The observation that the $E_a$ for the rate data in the induction-time region is much greater in magnitude than that for the data in the region where the rates are exponential seems to support the earlier postulate [2] that $E_a$ mainly represents the energy for establishing nucleation sites. The mean of these two values of $E_a$ is nearly equal to the observed energy of activation for the rate data at higher temperatures. These results are diagrammatically represented in fig. 8. It is, therefore, thought that the exponential rate curves observed in crystal structure transformations [2, 3] represent a combination of two processes, viz., the nucleation process associated with a very large $E_a$ and the propagation process associated with a relatively low $E_a$.

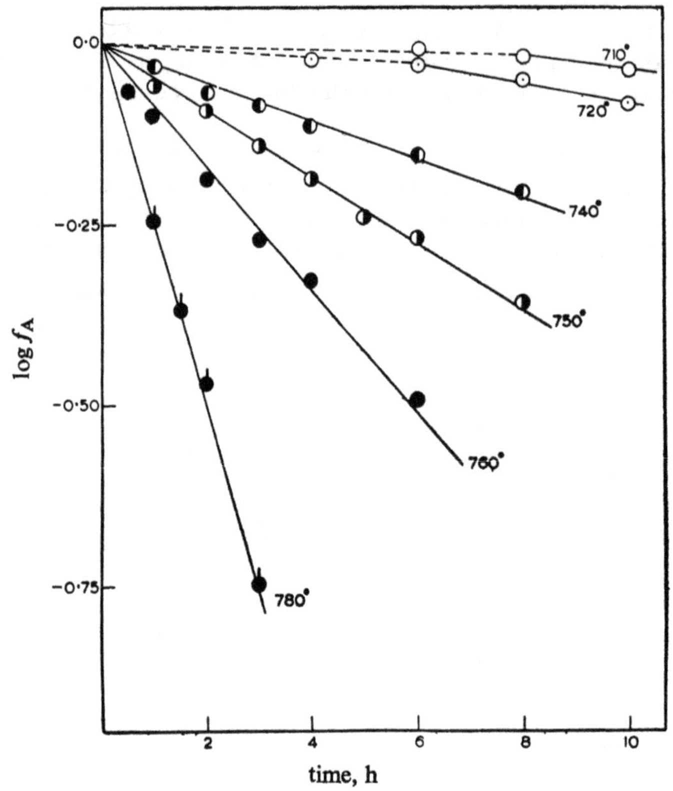


FIG. 7.—First-order plots of the data in fig. 5.

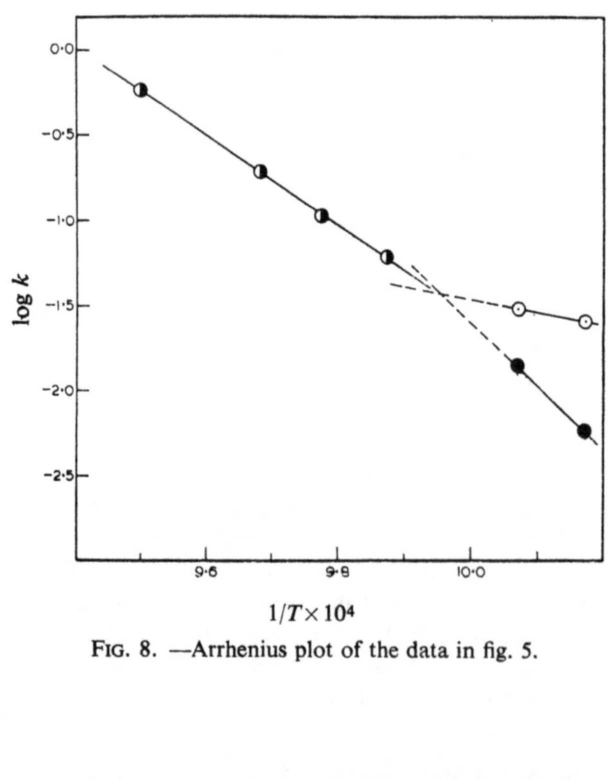


FIG. 8. —Arrhenius plot of the data in fig. 5.

The marked difference in the kinetics and energetics of the transformation of anatase prepared by sulphate hydrolysis as compared to that of pure anatase may be ascribed to the decelerating effect of sulphate impurity. It is interesting to note that $E_a$ of the transformation of anatase prepared by sulphate hydrolysis is considerably higher than that of pure anatase ($\sim 90 \pm 10$ kcal/mole). The widely different temperatures reported in the literature [8-11] for the transformation of anatase samples prepared by sulphate hydrolysis are considered to be due to the variation in the extent of washing of the precipitate (resulting in the variation in the concentration of sulphate impurity in the samples). In order to substantiate further that the observed effects are due to the presence of varying amounts of sulphate impurity, transformation of anatase samples doped with sulphate to different extents [6] (0·1, 1 and 5 atom %) has been studied (fig. 9). The sulphate ion progressively inhibits the transformation with

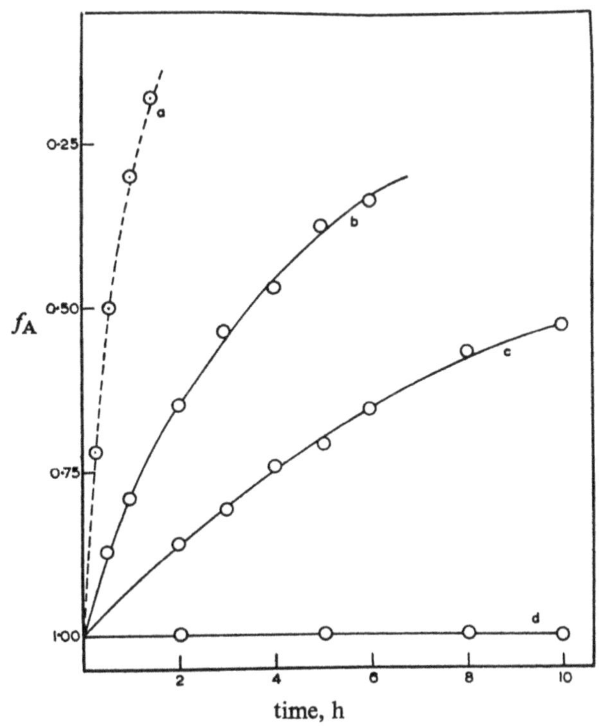


FIG. 9.—Influence of sulphate ion impurity on the anatase-rutile transformation.
(*a*) spectroscopically pure, (*b*) 0·1 % sulphate-doped,
(*c*) 1 % sulphate-doped, (*d*) 5 % sulphate-doped.

increasing concentration. The $E_a$ also increases with increase in the impurity concentration. Thus, the values of $E_a$ for the transformations of pure 0·1 % sulphate doped and 1 % sulphate doped samples of anatase were ~90, ~100, and ~120 kcal/mole respectively.

The results discussed show that crystal structure transformations are considerably dependent on the thermal history of the samples; to be more specific, the crystallite size, particle size and surface area have measurable effects on the transformation. It would, therefore, probably be difficult to reproduce strictly transformation data with different samples. The magnitudes of these effects are, however, not too great to result in the wide variability of temperatures of polymorphic transformations. The wide variations in transformation temperature can only be due to other factors such as impurity effects. Impurities not merely affect the transformation temperatures,[6, 7] but also profoundly affect the basic mechanism(s) underlying the transformation. The sulphate ion impurity decelerates the anatase-rutile transformation and markedly affects the rates of nucleation and propagation processes. However, the mechanism by which the sulphate ion impurity affects the transformation characteristics is not clear. In order to understand the mechanism of impurity effects, detailed investigations on the effects of various anionic and cationic impurities are presently in progress in these laboratories.

The authors are thankful to Prof. M. R. A. Rao for his interest and to Dr. P. A. Faeth for his assistance in some of the work. One of them (S. R. Yoganarasimhan) is indebted to the University Grants Commission of India for the award of a research Fellowship.

1 Czanderna, Rao and Honig, *Trans. Faraday Soc.*, 1958, **54**, 1069.
2 Rao, *Can. J. Chem.*, 1961, **39**, 498.
3 Rao, Yoganarasimhan and Faeth, *Trans. Faraday Soc.*, 1961, **57**, 504.
4 Buerger, *Phase Transformations in Solids*, ed. Smoluchowski (John Wiley, New York and Chapman and Hall, London, 1957), chap. 6.
5 Rao and Honig, *Symp. Rate Processes in Physico-chemical Reactions* (Bombay, India, Jan. 1960), to be published in the special volume of the *J. Indian Chem. Soc.*
6 Rao, Turner and Honig, *J. Physic. Chem. Solids*, 1959, **11**, 173.
7 Rao and Lewis, *Curr. Sci. (India)*, 1960, **29**, 52.
8 Barksdale, *Titanium: its Occurrence, Chemistry and Technology* (Ronald Press Co., New York, 1949).
9 Skinner, Johnston and Beckett, *Titanium and its Compounds* (Herrick L. Johnston Enterprises, 1954), pp. 22-23.
10 Gmelin, *Handbuch der anorganischen chemie* (Gmelin Institute, Verlag Chemie, Weinheim, Germany, 1951), **41**, pp. 233-34.
11 Schossberger, *Z. Krist.*, 1942, **104**, 358.
12 Czanderna, Clifford and Honig, *J. Amer. Chem. Soc.*, 1957, **79**, 5407.
13 Pask and Warner, *Bull. Amer. Ceram. Soc.*, 1954, **33**, 168.
14 Srinivasan, *Proc. Indian Acad. Sci. A*, 1957, **46**, 123.
15 Rao, Yoganarasimhan and Lewis, *Can. J. Chem.*, 1960, **38**, 2359.
16 Klug and Alexander, *X-Ray Diffraction Procedures* (John Wiley and Sons Inc., New York, 1954), chap. 9.
17 Roberts, *Kinetics of High Temperature Processes*, ed. Kingery (Technology Press and John Wiley, New York and Chapman and Hall, London, 1959), chap. 27.

For a study of cubic-monoclinic transformation of $ZrO_2$ (xerogels), see G.N. Subbanna and C.N.R. Rao, Eur. J. Solid State Chem. 26, 7 (1989)

PRINTED IN GREAT BRITAIN AT
THE UNIVERSITY PRESS
ABERDEEN

Offprinted from the *Transactions of The Faraday Society*
No. 574, Vol. 66, Part 10, October, 1970

# *Pm3m-Fm3m* Transformations of Alkali Halides

## Solid Solutions of CsCl with KCl, CsBr, $SrCl_2$

By M. Natarajan, K. J. Rao and C. N. R. Rao *

Dept. of Chemistry, Indian Institute of Technology, Kanpur, India

*Received* 31*st December*, 1968

*Pm3m-Fm3m* transformations of solid solutions of CsCl with KCl and CsBr exhibit different behaviours. With increasing percentages of KCl, the NaCl structure gets stabilized in the CsCl+KCl system. In the CsCl+CsBr system, the transformation temperature increases with % CsBr and $\Delta H$ essentially remains constant. Both these behaviours can be satisfactorily explained in terms of the Born treatment of ionic solids. The *Pm3m-Fm3m* transformation retains its first-order characteristics in the CsCl+KCl system, but higher-order components seem to be present in the CsCl+CsBr system. Incorporation of vacancies do not affect the transformation of CsCl markedly.

Caesium chloride transforms from the CsCl structure (*Pm3m*) to the NaCl structure (*Fm3m*) around 480°C.[1] The transformation is thermodynamically first order and is associated with considerable thermal hysteresis. The *Pm3m-Fm3m* transformations of CsCl and other alkali halides have been examined by employing the Born treatment of ionic solids.[2, 3] The Born-Mayer expression with four repulsive parameters explains the relative stabilities of the *Pm3m* and *Fm3m* structures satisfactorily. Further, this treatment also accounts for the stabilization of solid solutions of CsCl with RbCl in the *Fm3m* structure; the behaviour of CsCl+KCl solid solutions appears to be similar [2] to that of CsCl+RbCl solid solutions. On the other hand, in the transformations of CsCl+CsBr solid solutions, the transformation temperature increases with the percentage of CsBr.[4, 5]

We have studied the transformations of the CsCl+KCl and CsCl+CsBr solid solutions in order to find the limitations and applicability of the Born treatment in explaining the two entirely different behaviours of the solid solutions of these two systems. Such a study is of value since theoretical approaches to explain the relative stabilities of structures of ionic solids have not been quite successful, and it is important to explain the relative stabilities of at least the two simplest structure types in ionic solids, viz., the NaCl and CsCl structures. We also wished to find out whether the first order characteristics of *Pm3m-Fm3m* transitions are retained in the solid solutions. We have therefore examined the crystallography of the *Pm3m* and *Fm3m* phases of the solid solutions as functions of temperature; from these data, coefficients of expansion of the two structures have been calculated.

Menary, Ubbelohde and Woodward,[1] have reported that a large increase in vacancies occurs during the transformation of CsCl; and the formation energy of a Schottky pair in CsCl is considerably lower in the *Pm3m* phase (~1 eV) compared, to the *Fm3m* phase (~2 eV).[6, 7] The *Pm3m-Fm3m* transformations of two CsCl+$SrCl_2$ solid solutions have now been examined to find out whether the vacancies produced by the incorporation of $Sr^{+2}$ favour the transformation.

## EXPERIMENTAL

All the compounds were better than 99.99 % purity or of spectroscopic grade (J. & M. or Alfa). The preparation of solid solutions as well as the procedure for recording the

* to whom all the correspondence should be addressed.

differential thermal analysis (DTA) curves have been described.[2, 8] The procedures to obtain enthalpies of transformation $\Delta H_{tr}$ (from DTA peak areas) and activation energies $E_a$ are also described earlier.[2, 8]

The X-ray patterns were recorded (with Cu $K\alpha$ radiation) at different temperatures on a General Electric diffractometer fitted with a high-temperature camera and a temperature programmer. The $a_0$ values of the *Pm3m* and *Fm3m* phases of the CsCl+KCl solid solutions were determined at different temperatures, and that $a_0$ values of the CsCl+CsBr solid solutions were determined at 25 and 560°C. All the calculations employing the Born treatment were carried out on the IBM 1620 and IBM 7044 computers in this Institute.

## RESULTS AND DISCUSSION

### CsCl+KCl SOLID SOLUTIONS

The results of the transformation of various CsCl+KCl solid solutions are summarized in table 1 and typical differential thermograms are shown in fig. 1. The lattice dimensions (at 25°C) of the *Pm3m* and *Fm3m* forms of the solid solutions are also given in table 1. The $a_0$ values of both the forms decrease with increase in

TABLE 1.—*Pm3m-Fm3m* TRANSITIONS OF CsCl+KCl SOLID SOLUTIONS

| % KCl (mol) | $T_t$/°C forward (DTA peak) | $T/_t$°C reverse (DTA peak) | $\Delta H_{tr}$ cal mol$^{-1}$ | $E_a$ kcal mol$^{-1}$ | $a_0(Pm3m)$ 25°C (*a*) | $a_0(Fm3m)$ 570°C | $a_0(Fm3m)$ 25°C (*b*) (extra.) | $\Delta V$ (25°C) (*c*) | $\eta Pm3m$ deg.$^{-1}$ | $\eta Fm3m$ deg.$^{-1}$ (*e*) |
|---|---|---|---|---|---|---|---|---|---|---|
| 0 | 479 | 444 | 580 | 180 | 4.1210 | 7.0980 | 6.9050 | 7.34 | $1.25\times10^{-4}$ | $1.5\times10^{-4}$ |
| 2.5 | 470 | — | 520 | 114 | 4.1125 | 7.0750 | 6.8960 | 7.48 | $2.24\times10^{-4}$ (*d*) | $2.8\times10^{-4}$ |
| 5.0 | 470 | — | 465 | 100 | 4.1000 | 7.0630 | 6.8820 | 7.49 | $2.16\times10^{-4}$ (*d*) | $2.7\times10^{-4}$ |
| 10.0 | 470 | — | 420 | — | 4.0760 | 7.0250 | 6.8580 | 7.61 | $1.2\times10^{-4}$ | $1.2\times10^{-4}$ |
| 20.0 | 470 | — | 258 | — | 4.0400 | 7.0100 | 6.8100 | 7.83 | $2.6\times10^{-4}$ | $1.0\times10^{-4}$ |
| 25.0 | no transition | — | — | — | — | — | — | — | — | — |

(*a*) The $a_0(Pm3m)$ of pure KCl found by linear extrapolation was 3.65 Å (3.76 Å is the value calc. assuming $a_0(Pm3m)/a_0(Fm3m) = 0.598$).

(*b*) The $a_0(Fm3m)$ of pure KCl found by linear extrapolation was 6.30 Å (N.B.S. value = 6.294 Å).

(*c*) change in molar volume.

(*d*) The $\eta^{(Pm3m)}$ values in these solutions were calculated assuming $a_0^{(Pm3m)}/a_0^{(Fm3m)} = 0.598$; the $\eta^{Pm3m}/\eta^{Fm3m}$ values are in the range 0.8-2.6. the value increasing with % KCl in the solid solution. The $\eta^{Pm3m}/\eta^{Fm3m}$ values for pure CsCl and solutions with 10 % and 20% KCl are obtained from experimental $a_0$ values.

(*e*) calc. from $a_0$ values in the temperature range, 500-570°C.

% KCl. The calculated value of the change $\Delta V$ in molar volume accompanying the transformation increases slightly with the percentage of KCl suggesting that the transformation remains first order (thermodynamically) even in solid solutions.

The volume coefficient of expansion of the *Pm3m* and *Fm3m* phases of pure CsCl were $1.25\times10^{-4}$ and $1.5\times10^{-4}$ deg.$^{-1}$ respectively in the transition region. The $\eta^{Fm3m}$ values for the solid solutions determined experimentally are given in table 1, with the calculated values of the $\eta^{Pm3m}$. The ratio, $\eta^{Pm3m}/\eta^{Fm3m}$, is about 0.80 for pure CsCl, a value which satisfactorily accounts for the melting behaviour of CsCl.[9] The $\eta^{Pm3m}/\eta^{Fm3m}$ ratio for the CsCl+KCl solid solutions increases in the range 0.80-2.60 % KCl. The melting points of these solid solutions decrease slightly with % KCl.

The results from DTA and X-ray studies (table 1) indicate that the *Pm3m-Fm3m* transformation of CsCl+KCl solid solutions is reversible only up to 3 % KCl. With further increase in the proportion of KCl, the transformation becomes irreversible and at ~25 % KCl the *Fm3m* phase of CsCl is stable. At, or above, 25 % KCl, the solid solutions showed no indication of the transformation. Thus, the $\Delta H$ of the transformation decreases continuously with % KCl in the solid solutions up to 25 % KCl. While one may visualise how high percentages of KCl or RbCl stabilize the $Fm3m$ form of the solid solutions by causing an appreciable decrease in the interionic distance, it is difficult to understand why the reversibility of the $Pm3m$-$Fm3m$ transformation ceases above 3 % KCl (or 5 % RbCl). The energy of activation also decreases with % KCl indicating that the transformation is favoured by these additives.

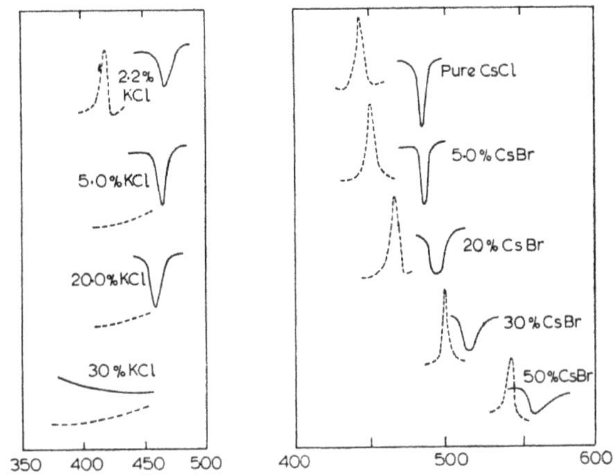


FIG. 1.—DTA curves of CsCl+KCl and CsCl+CsBr solid solutions.

The behaviour of the CsCl+KCl solutions may be understood in terms of the modified Born treatment [2, 3] as for the CsCl+RbCl system.[2] The lattice energy difference between the $Pm3m$ and $Fm3m$ structures of a solid solution is given by

$$\Delta U^{s.s} = U^{s.s}_{Pm3m} - U^{s.s}_{Fm3m} \tag{1}$$

where $U^{s.s}_{Pm3m}$ and $U^{s.s}_{Fm3m}$ are given by the weighted sum of the lattice energies of CsCl and KCl of the appropriate structures.[2] The lattice energy $U$ has the general form,

$$U = N_0\left[\frac{\alpha e^2}{r_0} + K\left\{\frac{C}{r_0^6} + \frac{D}{r_0^8}\right\} - B(r_0) - \phi_0\right] \tag{2}$$

where $B(r_0) = M_1 b_1 \exp(-r_0/\rho_1) + M_2 b_2 \exp(-ar_0/\rho_2)$. The terms in eqn. (2) have been described earlier.[2, 3] In the repulsive term $B(r_0)$, $a$ is the ratio of the distance between the second nearest neighbour to that between the nearest neighbours. Calculations employing eqn. (1) and (2) provide $\Delta U^{s.s}$ at 25°C since $r_0$ values are taken from X-ray data at this temperature. The experimental $\Delta H_{tr}$ values (table 1) are, however, at the transformation temperatures. Since no transformation was noticed when the fraction $f$ of KCl was 0.25, it may be taken that $\Delta U^{s.s}$ (25°C) approaches zero at this composition.

By choosing various values of $K$ and corresponding values of the repulsive parameters for CsCl and KCl,[2, 3] $\Delta U^{s.s}$ was calculated as a function of $f$ (fig. 2). $\Delta U^{s.s}$ becomes zero at $f \approx 0.25$, when $K = 2.8$ or 3.00. The values of the repulsive parameters when $K = 3.00$ are:

| | $b_1$ (ergs mol$^{-1}$) | $b_2$ (ergs mol$^{-1}$) | $\rho_1$ (cm) | $\rho_2$ (cm) |
|---|---|---|---|---|
| CsCl | $644\times10^{-12}$ | $795\times10^{-12}$ | $0.367\times10^{-8}$ | $0.424\times10^{-8}$ |
| KCl | $3879\times10^{-12}$ | $35.6\times10^{-12}$ | $0.345\times10^{-8}$ | $0.487\times10^{-8}$ |

These results indicate that the Born-Mayer expression with four repulsive parameters is satisfactory in predicting the general behaviour of the solid solution of CsCl with KCl or RbCl. Again, the hardness parameters $\rho_2$ and $\rho_1$ are related by the structure constant *a*.

Several workers [10, 11] have proposed revised van der Waals coefficients in alkali halides which eliminate the need for employing high van der Waals terms in the study

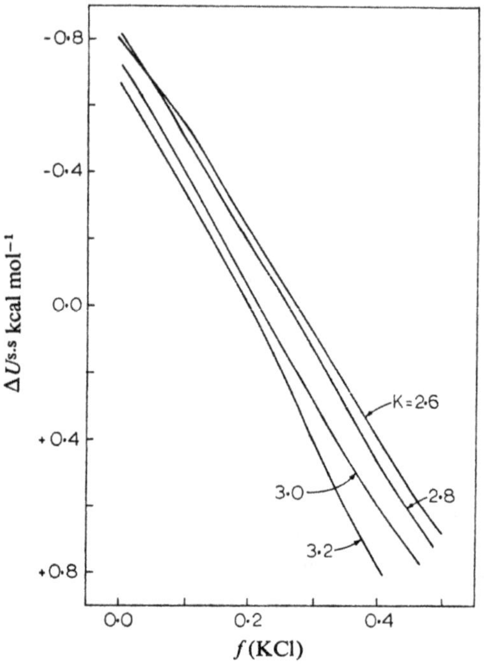


FIG. 2.—Variation of $\Delta U^{s.s}$ of CsCl+KCl solid solutions calculated for various values of $K$ for CsCl and KCl (the graphs correspond to $K_{CsCl} = K_{KCl} =$ the value indicated).

of phase transitions of alkali halides.[12] By employing these coefficients of Hajj [10] for CsCl (with $K = 1$), we find that $\Delta U^{s.s}$ becomes zero at $f \sim 0.40$; by employing the parameters of Lynch [11] for KCl along with Hajj's values for CsCl, $\Delta U^{s.s}$ becomes zero at $f \approx 0.25$.

TABLE 2.—*Pm3m-Fm3m* TRANSITIONS OF CsCl+CsBr SOLID SOLUTIONS

| % CsBr (mol) | $T_t$/°C forward (DTA peak) | $T_t$/°C reverse (DTA pek) | $\Delta H_{tr}$ cal mol⁻¹ | $a_0(Pm3m)$ 25°C | $a_0(Fm3m)$ 560°C | $a_0(Fm3m)$ 25°C (*a*) | $\Delta V$ (25°C) cm³ mol⁻¹ |
|---|---|---|---|---|---|---|---|
| 0 | 479 | 444 | 580 | 4.1210 | 7.0010 | 6.9050 | 7.34 |
| 5 | 480 | 450 | 600 | 4.1300 | 7.0130 | 6.9150 | 7.33 |
| 10 | 485 | 455 | 825 | 4.1350(*b*) | 7.0320(*b*) | 6.9250 | 7.31 |
| 20 | 495 | 462 | 950 | 4.1550 | 7.0600 | 6.9400 | 7.15 |
| 30 | 505 | 480 | 1025 | 4.1710 | 7.1000 | 6.9550 | 6.91 |
| 40 | 530 | 503 | 1100 | 4.1900 | 7.1300 | 6.9675 | 6.63 |
| 50 | 555 | 535 | 1050 | 4.2080 | 7.1515 | 6.9825 | 6.37 |
| 60 | 610(*c*) | — | — | — | — | — | — |

(*a*) calc. on the basis that $a_0$ $(Pm3m)/a_0(Fm3m) \approx 0.598$.

(*b*) The $a_0$ $(Pm3m)$ and $a_0$ $(Fm3m)$ values for this composition were determined at several temperatures; the $\eta^{Fm3m}$ and $\eta^{Pm3m}$ are $1.17 \times 10^{-4}$ and $0.68 \times 10^{-4}$ deg.⁻¹ repectively. The $\eta^{Pm3m}/\eta^{Fm3m}$ is thus 0.53

(*c*) very close to the m.p. (614°C).

## CsCl+CsBr SOLID SOLUTIONS

The results from DTA and X-ray analysis are shown in table 2. Typical differential thermograms are given in fig. 2. The transformation temperature increases markedly with % CsBr, the trend continuing up to the melting point of the solid solution (~65 % CsBr). The transformation temperatures reported here essentially agree with those of earlier workers.[4, 5] The $\Delta H$ of the transformation varies little in these solid solutions. The $\eta^{Pm3m}/\eta^{Fm3m}$ ratio in the solution with 10 % CsBr is ~0.5, a value much less than in pure CsCl.

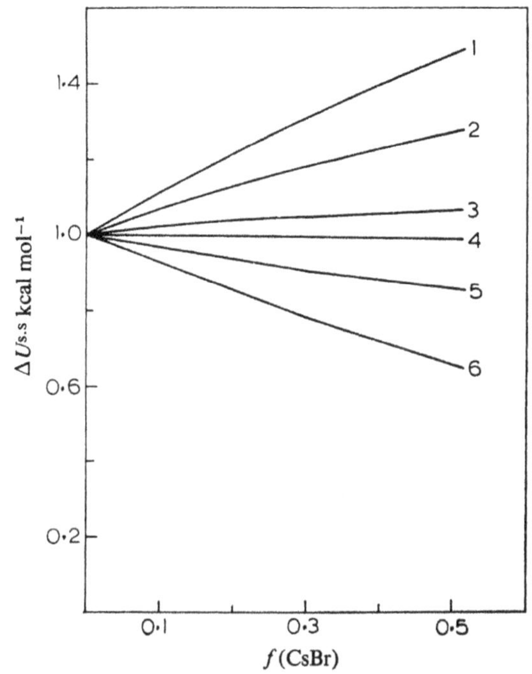


FIG. 3.—Variation of $\Delta H^{s.s}$ of CsCl+CsBr solid solutions of different $b_2$ values (in $10^{-12}$ erg $mol^{-1}$): 1, 0; 2, 200; 3, 500; 4, 468; 5, 600; 6, 800.

Since CsBr has the same *Pm*3*m* structure as CsCl, it does not stabilize the *Fm*3*m* structure of the olid solutions; accordingly, there is an increase in the $a_0$ values of the *Pm*3*m* and *Fm*3*m* structures with increasing % CsBr. On the other hand, in the solid solutions of CsCl with KCl (or RbCl), the $a_0$ values of both the structures decrease with the increases in the % KCl (or RbCl). The change $\Delta V$ in the molar volume accompanying the *Pm*3*m*-*Fm*3*m* transformation (table 2) decreases with increasing CsBr content and at 50 % CsBr the decreases in $\Delta V$ is about 14 %; $\Delta V$ is expected to decrease with % CsBr on the basis of ionic size considerations.[2] This may mean that the transformation has a component of higher or second order in CsCl+CsBr solid solutions. The DTA peaks at high % CsBr also become (less sharp) compared with those for pure CsCl. The magnitude $\Delta T$ of thermal hysteresis, decreases slightly with increasing % CsBr, thus indirectly substantiating that the transformation has components of higher order; $\Delta T$ would be expected to decrease with % CsBr since $\Delta T$ is proportional to the $\Delta V$ of the transformation.[8] Samples containing 30 % CsBr (or >30 %) showed evidence for the coexistence of the *Pm*3*m* and *Fm*3*m* phases in the transformation region (~500-550°C). The coexistence of phases and the formation of hybrid cystals have been noticed by Ubbelohde [13] in continuous or higher-order transformation.

In order to employ the Born treatment to explain the transformations of CsCl+CsBr solid solutions, we have used the revised van der Waals coefficients of CsBr proposed by Hajj.[10] The repulsive parameters of CsBr can be evaluated as follows (i) In the expression for the lattice energy (eqn. (3)) we have assumed that $\rho_2 = a\rho_1$ as proposed by Rao and Rao.[3] (ii) We calculate $(8b_1+6b_2)_{CsBr}$ by substituting the value of $\rho_1$ obtained from the compressibility relations.[2]

$$U^{Pm3m} = N_0[\alpha e^2/r_0 + C/r_0^6 + D/r_0^8 - (8b_1+6b_2)\exp(-r_0/\rho_1) - \phi_0]. \quad (3)$$

The value of $(8b_1+6b_2)$ thus calculated is $76867\times10^{-12}$ ergs mol$^{-1}$. The $\Delta U^{s.s}$ of solid solutions of CsCl and CsBr were evaluated (eqn. (1)) for different values of $b_2$ (in the range 0-900 $\times 10^{-12}$ erg mol$^{-1}$) using the $r_0$ values from table 2. The results are shown diagramatically in fig. 3, the values of $b_1$ and $b_2$ are $9257\times10^{-12}$ and $468\times10^{-12}$ erg mol$^{-1}$ respectively when $(6b_2/8b_1+6b_2)_{CsCl}\approx(6b_2/8b_1+6b_2)_{CsBr}$. The $\Delta U^{s.s}$ remains nearly constant with the % CsBr when $b_2$ is between 400 and $500\times10^{-12}$ erg mol$^{-1}$. Thus, curves 3 and 4 of fig. 3 closely represent the experimentally observed situation. Thus, the Born model of ionic solids describes the relative stabilities of structures of alkali halides in widely different situations as manifested here by the CsCl+KCl and CsCl+CsBr systems.

## $CsCl+SrCl_2$ SOLID SOLUTIONS

Although there is doubt with regard to the importance of vacancies in the mechanism of the *Pm3m-Fm3m* transformation in CsCl,[1, 14] experimental and theoretical studies on Schottky defects seem to indicate that vacancies could also facilitate the *Pm3m-Fm3m* transformation in CsCl, to some extent. The formation energy of Schottky defects in the *Pm3m* structure in CsCl is about half that in the *Fm3m* structure. The concentration of defects should therefore decrease after the *Pm3m-Fm3m* transformation. Studies on CsCl with different amounts of $Sr^{2-}$ show that the transformation temperature decreases slightly due to the increased vacancy concentrations. Thus, $T_t$ (forward) is 475 and 474°C with 2.5 and 5 % $Sr^{2-}$ respectively; $T_t$ (reverse) is unaffected. The results seem to indicate that *Pm3m-Fm3m* transformation of CsCl may essentially proceed by a dialatational mechanism.[15]

The authors thank the U.S. National Bureau of Standards for a research grant (G-51) under the Special International Programme, and the staff of the IIT Computer Center.


[1] J. W. Menary, A. R. Ubbelohde and I. Woodward, *Proc. Roy. Soc. A*, 1951, **208,** 158.
[2] K. J. Rao ,G. V. Subba Rao and C. N. R. Rao, *Trans. Faraday Soc.*, 1967, **63,** 1013.
[3] K. J. Rao and C. N. R. Rao, *Proc. Phys. Soc.*, 1967, **91,** 754.
[4] L. J. Wood, C. Seeney and M. T. Derbes, *J. Amer. Chem. Soc.*, 1959, **81,** 6148.
[5] L. J. Wood, W. Secunda and C. H. McBride, *J. Amer. Chem. Soc.*, 1958, **80,** 307.
[6] K. J. Rao and C. N. R. Rao, *Physica stat solidi*, 1968, **28,** 158.
[7] (*a*)M. Natarajan and C. N. R. Rao unpublished results.
[7] (*b*) Z. Morlin and J. Tremmel, *Acta Phys. Hungary*, 1966, **21,** 129.
[8] K. J. Rao and C. N. R. Rao, *J. Materials Sci.*, 1966, **1,** 238.
[9] K. J. Rao and C. N. R. Rao, *Chem. Phys. Letters*, 1967, **1,** 499.
[10] F. Hajj, *J. Chem. Phys.*, 1966, **44,** 4618
[11] D. W. Lynch, *J. Phys. Chem. Solids*, 1967, **28,** 1941.
[12] K. J. Rao, G. V. Subba Rao and C. N. R. Rao, *Proc. Phys. Soc.*, 1968, **1,** 134.
[13] A. R. Ubbelohde in *Reactivity in Solids* ed. by J. H. de Boer, (Elsevier Publishing Co., Amsterdam, 1961).
[14] I. M. Hoodless and J. A. Morrison, *J. Phys. Chem.*, 1962, **66,** 557.
[15] M. J. Buerger, in *Phase transformation in Solids*, ed., by R. Smoluchowski, (John Wiley and Sons. Inc., New York, 1951).


See M. Natarajan and C.N.R. Rao, J. Chem. Soc. A 3087 (1970) for a study of phase transitions of silver halides.

For a study of heats of formation of solid solutions of CsCl with other alkali halides, see A.K. Shukla, J.C. Ahluwalia and C.N.R. Rao, JCS Faraday 1 72, 1288 (1976)

JOURNAL OF SOLID STATE CHEMISTRY **68**, 193–213 (1987)

# Computer Simulation of Transformations in Solids*

C. N. R. RAO† AND S. YASHONATH

*Solid State and Structural Chemistry Unit, Indian Institute of Science, Bangalore 560 012, India*

Received August 18, 1986

Recent developments in molecular dynamics (MD) and Monte Carlo (MC) methods enable us to fruitfully investigate transformations in solids by employing appropriate potentials. The possibility of varying both the volume and the shape of the simulation cell in these simulation techniques is especially noteworthy. In this article we briefly describe some of the highlights of the recent MD and MC methods and show how they are useful in the study of transitions in monatomic solids, ionic solids, molecular solids (especially orientationally disordered solids), and glasses. The availability of reliable pair potentials will undoubtedly make these methods more and more useful for studying various aspects of condensed matter in the years to come. © 1987 Academic Press, Inc.

## 1. Introduction

Computer simulation is being used increasingly in diverse areas of science in the past few years. It has also emerged to become one of the powerful means for investigating condensed matter (*1*). The principal tools employed in computer simulation are the Monte Carlo and the molecular dynamics methods. In these methods, properties of a collection of particles, usually between 30 and 1000 in number, interacting via a potential $\phi(r)$ are obtained numerically. Reliable estimates of equilibrium and transport properties as well as microscopic properties can be obtained from such calculations.

In the Monte Carlo method, one performs stochastic averaging in the configuration space by means of the Metropolis importance sampling scheme (*2*). Monte Carlo calculations are generally performed in the canonical ensemble, the isothermal isobaric ensemble, or the grand canonical ensemble. In the canonical or the *NVT* ensemble calculations, the number of particles, *N*, the volume of the simulation cell, *V*, and the temperature, *T*, are constant during the course of the simulation. In isothermal isobaric or the *NPT* ensemble calculations, the number of particles, the pressure, *P*, and the temperature, *T*, are held constant. In the molecular dynamics method, Newton's equations of motion are solved numerically (*3*). The energy, *E*, of the system is conserved over the generated trajectory and the average of any property over this trajectory corresponds to the average in the microcanonical or the *NVE* ensemble.

There have been some significant advances in the methods of Monte Carlo and

* Contribution No. 356 from the Solid State and Structural Chemistry Unit.

† To whom all correspondence should be addressed.



0022-4596/87 $3.00

molecular dynamics in the last few years. The pioneering contribution of Andersen (*4*), provides a method for performing molecular dynamics calculations at constant pressure or constant temperature. The method enables the generation of trajectories corresponding to the isoenthalpic isobaric and isothermal isobaric ensembles. This is made possible by coupling the temperature and the pressure of the system to those of a bath. Another significant development is due to Parrinello and Rahman (*5*), who allowed for variation not only in the volume but also in the shape of the simulation cell in constant-pressure calculations. The modified molecular dynamics methods have been suitably employed by Nosé and Klein (*6*) for the study of polyatomic molecules. The Monte Carlo method has been modified by Yashonath and Rao (*7*) to include the variation in the shape of the simulation cell. These developments have rendered the molecular dynamics and the Monte Carlo methods most useful for the investigation of a variety of phenomena exhibited by condensed matter. The methods are important because they can be employed not only for the study of stable phases but also to examine systems undergoing transformations. The methods are therefore ideally suited to investigate microscopic as well as macroscopic changes accompanying phase transitions in solids.

In this article, we shall discuss studies of phase changes in solids carried out by the application of the generalized Monte Carlo and the molecular dynamics methods. This topic is of particular significance because of the recent modifications of the method to include both variation in size and shape of the simulation cell. What is especially gratifying is that we are now able to make meaningful predictions of phase transitions in real solids by employing reliable pair potentials. Besides phase transitions of molecular solids, we shall examine phase transitions in monatomic and ionic crystals. Some of the molecular systems discussed are $CCl_4$, $CF_4$, and adamantane. We shall present recent results on the glass transition obtained by the two methods, a noteworthy feature being the simulation of a real molecular glass formed by isopentane interacting via a realistic potential. The simulation study on isopentane glass has thrown some light on the structure of the glassy state. Andersen's cooling experiments on a Lennard-Jones fluid are also significant in this regard. We shall also indicate in the article the scope for further investigations of solids by computer simulation. Before discussing the various types of phase transitions of solids, we shall briefly review the recent developments in the Monte Carlo and molecular dynamics methods.

## 2. Recent Developments in Molecular Dynamics and Monte Carlo Methods

There are many excellent reviews on the standard molecular dynamics method dealing with calculations in the microcanonical ensemble as well as on the Monte Carlo method involving calculations in the canonical, isothermal isobaric, and grand canonical ensemble (*8*). In the present article, we shall limit ourselves exclusively to those developments that have taken place since the work of Andersen (*4*). In the molecular dynamics method, the developments are the constant-pressure, constant-temperature, constant-temperature–constant-pressure, variable shape simulation cell MD, and isostress calculations; in the Monte Carlo method, it is the variable shape simulation cell calculation.

### 2.1. Molecular Dynamics Methods

*Constant-pressure calculations.* In the traditional molecular dynamics calculations, the equations of motion for the coordinates $\mathbf{r}_1, \mathbf{r}_2, \ldots, \mathbf{r}_N$ of the $N$ particles confined to a cell of fixed volume $V$ are solved numerically. This corresponds to constant volume and constant energy or the microcanonical ensemble. In the constant-pressure calculations, the volume is considered to be a dynamical variable in addition to these (*4*). The well-defined lagrangian

$$\alpha = \frac{1}{2}\sum_i m_i \dot{\mathbf{s}}_i \cdot \dot{\mathbf{s}}_i - \sum_{i<j} \phi(r_{ij}) + \frac{1}{2} M\dot{U}^2 - PU, \quad (1)$$

first utilized by Haile and Graben (*9*), is employed for the purpose. Here, $m_i$ is the mass of the $i$th atom and the particle–particle interaction potential is denoted by $\phi(r)$. The scaled coordinate $\mathbf{s}_i$ of the $i$th particle and $\mathbf{r}_i$ are related by

$$\mathbf{s} = U^{1/3}\mathbf{r}. \quad (2)$$

If $U$ is taken to be the volume $V$ of the cell, the last two terms represent respectively the kinetic and the potential energies associated with motion of the simulation wall whose mass is $M$. The parameter $P$ is the uniform hydrostatic external pressure acting on the simulation cell wall.

The lagrangian equations of motion

$$\frac{d}{dt}\left(\frac{\partial \alpha}{\partial \dot{x}}\right) = \frac{\partial \alpha}{\partial x}, \quad (3)$$

where $x$ is a coordinate, are solved numerically to obtain the particle coordinate and momenta as a function of time $t$.

Andersen (*4*) has shown that the average of any function, $F$, over the trajectory thus generated is equal to the ensemble average for the isoenthalpic isobaric ensemble, i.e.,

$$\bar{f} = \bar{f}(N, P, H). \quad (4)$$

The enthalpy $H$ is conserved over this trajectory and the external pressure corresponds to the value of $P$.

The value for the quantity $M$ is chosen to approximately equal $U^{1/3}$ divided by the speed of sound in the system being simulated. Equilibrium properties of the system are independent of $M$ and dynamical properties are dependent on $M$ through the time scale of the volume fluctuations, the volume changes depending on the imbalance between the internal and the external pressure. The effect of $M$ on single-particle dynamical properties seems to be small as indicated by the work of Nosé and Klein (*6*).

*Constant-temperature calculations.* Andersen (*4*) also proposed a constant-temperature method in which the energy fluctuates during the simulation. The lagrangian in this method is the same as that in the standard MD method. However, it differs from the standard MD method in that the particles undergo random stochastic collisions. The collisions are considered to be instantaneous and bring about a change in the momentum of the atom. The number of collisions and outcome of the collisions is determined by $\nu$, the mean rate at which each particle suffers a collision and $T$ the required temperature at which the simulation is to be performed. Particles are chosen randomly and a new value of momentum is assigned while the momentum and coordinate of all the other particles are unaltered. The new value for the momentum of the particle undergoing collision is chosen at random from the Boltzmann distribution corresponding to the temperature $T$ of the simulation. The average of any property calculated from this trajectory equals the canonical ensemble average of that property for the specified values of $(T, V, N)$ provided the Markov chain generated is irreducible (*4*). At low temperature and high densities, the Markov chain may not satisfy the irreducibility condition as in the traditional MD and MC techniques.

Equilibrium properties are independent of the value of $\nu$, but the dynamical properties are not the same for two different values of $\nu$. It is necessary to choose the value of $\nu$ within a small range (*10*). If the value is too high, the diffusion coefficient decreases, and if it is too small, large fluctuations occur in the kinetic energy and hence the temperature of the system. An appropriate choice for $\nu$ would be that resulting in a time for the decay of energy fluctuations approximately equal to that expected for a small volume of liquid in a large volume. Andersen (*4*) has discussed the method to estimate the optimum value of $\nu$.

Nosé (*11*) has described another alternative formulation for performing calculations under constant-temperature conditions wherein an additional degree of freedom is introduced to permit energy fluctuations. The lagrangian for this system is given by

$$\alpha = \frac{1}{2}\sum_i m_i \dot{\mathbf{r}}^2 g^2 - \sum_{i<j} \phi(r_{ij}) + \frac{Q}{2}\dot{g}^2 - (f+1)kT_0 \ln g, \quad (5)$$

where the last two terms give the kinetic and potential energies associated with $g$, $T_0$ is the temperature at which simulation is carried out, and $f$ is the number of degrees of freedom associated with $N$ atoms. The velocities are scaled (*11*) as

$$\mathbf{v}_i = g\dot{\mathbf{r}}_i, \quad (6)$$

where $\mathbf{v}_i$ is the real velocity of particle $i$.

The physical system and the scaled system are related by Eq. (6). The scaling of velocities permits the exchange of heat between the simulated and the external heat reservoir. The equations of motion for the $N$ atoms and the scaling factor $g$ are solved numerically and averages calculated from the trajectory. The special choice $(f + 1)$ $kT_0 \ln g$ for the potential energy associated with $g$ guarantees that the averages of equilibrium quantities calculated from the MD trajectory are the same as those in the canonical ensemble.

While Andersen's method for the constant-temperature calculation introduces energy fluctuations in the simulated system stochastically, Nosé's method achieves the same goal by the introduction of an additional degree of freedom; the scaling of velocities by a factor $g$ can be interpreted as the scaling of time. This is similar to the scaling of coordinates in the constant-pressure method. The real time step $\Delta t'$ is given by

$$\Delta t' = \Delta t/g. \quad (7)$$

Thus, the real time steps, $\Delta t'$, are unequal in Nosé's formulation. As in constant-pressure calculations, the value of $Q$, in units of energy $(\text{time})^2$, determines the dynamical quantities, although the equilibrium quantities are independent of the value of $Q$. The procedure for the selection of the optimum value for $Q$ has been discussed by Nosé (*11*). It is believed that single-particle dynamics are less sensitive to the value of $Q$.

In the formulation of Nosé, the total hamiltonian, the total momentum, and total angular momentum are conserved. The conserved quantities help to monitor the precision of the calculation and also provide a powerful criterion to check for the correctness of the program.

Other methods for performing constant-temperature molecular dynamics calculations have been proposed recently. Evans (*12*) has introduced an external damping force in addition to the usual intermolecular force in order to keep the temperature constant in the simulation of a dissipative fluid flow. In another method, Haile and Gupta (*13*) have imposed the constraint of constant kinetic energy on the lagrangian equations of motion to perform calculations at constant temperature.

*Constant-temperature–constant-pressure calculations*. The trajectory for the atoms and volume are generated according to the solution of the equations of motion for the lagrangian (1) discussed earlier in connection with constant-pressure calculations. In addition, stochastic collisions are introduced to allow for fluctuations in enthalpy (*4*). The new momentum is sampled randomly from the Boltzmann distribution

$$\exp(-\boldsymbol{\pi}_i \cdot \boldsymbol{\pi}_i/2m_i V^{2/3}kT), \qquad (8)$$

where $\boldsymbol{\pi}_i$ is the momentum conjugate to $\mathbf{s}_i$. The average over this trajectory obtained thus equals the ensemble average of the property in the isothermal isobaric ensemble.

Another MD formulation for constant-temperature and -pressure calculations based on the constant temperature formulation of Nosé (*11*) has been described recently. The stochastic collisions which permit energetic fluctuations has been replaced by the dynamic method of scaling of velocities of the atoms, in addition to the scaling of velocities by $V^{1/3}$. The method, which is completely dynamical, requires one to choose appropriate values for $Q$ and $M$ which respectively determine the time scale of the temperature and volume fluctuations.

*Parrinello–Rahman method.* The constant pressure MD formulation of Andersen (*4*) provides a method for permitting the variation in the volume of the simulation cell. Parrinello and Rahman (*5, 14, 15*) generalized Andersen's method to include variations in the volume as well as the shape of the simulation cell. In the Parrinello–Rahman formulation, the simulation cell is represented by three vectors, **a**, **b**, and **c**. The lagrangian is given by

$$\alpha = \frac{1}{2}\sum_{i}^{N} m_i \dot{\mathbf{s}}_i G \mathbf{s}_i - \sum_{i<j} \phi(r_{ij}) + \frac{1}{2} M \,\mathrm{Tr}(\mathbf{h}'\mathbf{h}) - PV, \qquad (9)$$

where the simulation cell is represented by $\mathbf{h} = \{\bar{\mathbf{a}}, \bar{\mathbf{b}}, \bar{\mathbf{c}}\}$ and the scaled coordinates by $\mathbf{s}_i = \mathbf{h}^{-1}\mathbf{r}_i$. The metric tensor $G = \mathbf{h}'\mathbf{h}$.

The external hydrostatic pressure on the system is denoted by $P$. The equations of motion are

$$\ddot{\mathbf{s}}_i = m_i^{-1} \sum_{i \neq j}^{N} \frac{1}{r_{ij}} \frac{d\phi(r_{ij})}{dr_{ij}} (\mathbf{s}_i - \mathbf{s}_j) - G^{-1}\dot{G}\dot{\mathbf{s}}_i, \quad i, j = 1, 2 \ldots N$$

$$\ddot{\mathbf{h}} = W^{-1}(\boldsymbol{\pi} - P)\boldsymbol{\sigma}, \qquad (10)$$

where $\boldsymbol{\sigma}_{ij} = \partial V/\partial h_{ij}$. The internal stress tensor $\boldsymbol{\pi}$ is given by

$$V\boldsymbol{\pi} = \sum_i m_i \mathbf{v}_i \mathbf{v}_i + \sum_{i<j} \frac{1}{r_{ij}} \frac{d\phi(r_{ij})}{dr_{ij}} (\mathbf{r}_i - \mathbf{r}_j)(\mathbf{r}_i - \mathbf{r}_j), \qquad (11)$$

the vector $\mathbf{v}_i$ being $\mathbf{h}\dot{\mathbf{s}}_i$. The imbalance between the internal stress $\boldsymbol{\pi}$ and the external pressure $P$ acting on the cell wall given by $\boldsymbol{\sigma}$ determines the motion of the cell **h**. This method has been appropriately modified for calculations on molecular systems by Nosé and Klein (*6*).

Andersen's formulation is useful in the study of fluid–solid transitions where the change in volume is significant. The generalization of Parrinello and Rahman which allows variations in both the volume and the shape of the simulation cell is most suited for the study of phase transformations in solids. In the case where only the volume is allowed to change, the periodic boundary conditions do not permit the solid to undergo structural transformations involving a change in crystal symmetry. The variable shape MD method has been modified to include external stress rather than uniform hydrostatic pressure (*16*). A generalization to the isothermal isotension ensemble calculation giving a somewhat different interpretation of the additional degree of freedom introduced in the constant-temperature and -pressure formulation of Nosé has been discussed recently by Ray and Rahman (*17*).

### 2.2. *Modified Monte Carlo Method*

The traditional MC method in the isothermal isobaric ensemble has been modified to include variations in size as well as the shape of the simulation cell (*7*). The average of any quantity, $F$, in the usual isothermal isobaric calculation is given by

$$F(\mathbf{r}^N, V) = \frac{\int_0^\infty dV \int_V d\mathbf{r}^N f(\mathbf{r}^N, V) \exp[-\beta\Phi(\mathbf{r}^N)]}{\int_0^\infty dV \int_V d\mathbf{r}^N \exp[-\beta\Phi(\mathbf{r}^N)]}, \quad (12)$$

where $\mathbf{r}_1, \mathbf{r}_2, \ldots, \mathbf{r}_N$ are the coordinates of the $N$ atoms and are represented by $\mathbf{r}^N$, $\beta = 1/kT$, and $\Phi$ is the total potential energy. The simulation cell is a cube with edge length $L = V^{1/3}$. In order to allow both the size and shape of the cell to vary during the simulation, the cell is defined as in the Parrinello–Rahman MD method by the three vectors $\mathbf{a}$, $\mathbf{b}$, $\mathbf{c}$ (7). The matrix $\mathbf{h} = \{\bar{\mathbf{a}}, \bar{\mathbf{b}}, \bar{\mathbf{c}}\}$ represents the cell. For purposes of integration, the scaled coordinates $\boldsymbol{\alpha}_i = \mathbf{h}\mathbf{r}_i$ are employed rather than $\mathbf{r}_i$, where the components of $\boldsymbol{\alpha}_i$ vary between 0 and 1. The average of any quantity $f$ in the modified isothermal isobaric ensemble is given by

$$f([\mathbf{h}\boldsymbol{\alpha}]^N, \mathbf{h}) = \frac{\int_0^\infty d\mathbf{h} \int_{\mathbf{h}} d[\mathbf{h}\boldsymbol{\alpha}]^N f([\mathbf{h}\boldsymbol{\alpha}]^N, \mathbf{h}) \exp(-\beta\Phi([\mathbf{h}\boldsymbol{\alpha}]^N, \mathbf{h}))}{\int_0^\infty d\mathbf{h} \int_{\mathbf{h}} d[\mathbf{h}\boldsymbol{\alpha}]^N \exp(-\beta\Phi([\mathbf{h}\boldsymbol{\alpha}]^N, \mathbf{h}))}. \quad (13)$$

A trial move is made by displacing the particle and also the cell edge:

$$\begin{aligned} \boldsymbol{\alpha}_i &\rightarrow \boldsymbol{\alpha}_i \pm \varepsilon_P \overline{\mathbf{R}} \\ \mathbf{a}_i &\rightarrow \mathbf{a}_i \pm \varepsilon_c \overline{\mathbf{A}}, \end{aligned} \quad (14)$$

where $\varepsilon_P$ and $\varepsilon_c$ are random numbers between 0 and 1 and $\overline{\mathbf{R}}$ and $\overline{\mathbf{A}}$ specify respectively the maximum particle and cell displacements.

It is noteworthy that Monte Carlo simulations involve simpler coding than molecular dynamics simulations. The Monte Carlo method therefore provides a convenient alternative to the molecular dynamics method if one is not interested in the dynamics of the system. In recent years many workers have felt the need for employing more complicated intermolecular potentials in order to predict accurately the properties of matter, especially those in the solid state. In such situations, the MC method is to be preferred as it is easier to code programs which incorporate complicated intermolecular potentials. Another advantage of the MC method is the ease with which the number of degrees of freedom can be altered in any calculation.

The above method is easily extended to perform calculations in isostress isothermal ensemble employing an isotropic stress tensor in place of a uniform hydrostatic pressure (*18*). The extension of the MC method to the case of polyatomic systems is straightforward (*19*). In the case of molecular systems, however, there are instances where the simulation cell rotates as a whole in space. For example, in the simulation of the low-temperature, high-pressure phase of carbon tetrachloride carried out by us, the simulation cell rotated in space (*19*). The variation of the magnitudes of $\mathbf{a}$, $\mathbf{b}$, and $\mathbf{c}$ and angles $\alpha$, $\beta$, and $\gamma$ of the simulation cell, as well as the variation of the cell components, is shown in Fig. 1. It is apparent from the figure that while the cell components change continuously during the course of a run, the magnitudes and angles of the vectors fluctuate around their average values. Similar rotation has been observed in a molecular dynamics study of the high-pressure phase of nitrogen by Nosé and Klein (*6*). Only six degrees of freedom are required ($a$, $b$, $c$, $\alpha$, $\beta$, and $\gamma$) to define the cell. The rotation is a consequence of the three extra degrees of freedom associated with the simulation cell. The cell can be prevented from rotating by imposing a restriction on the cell so as to allow six degrees of freedom to be varied (*19*). Nosé and Klein (*6*) have proposed a method wherein the system is rotated to keep the

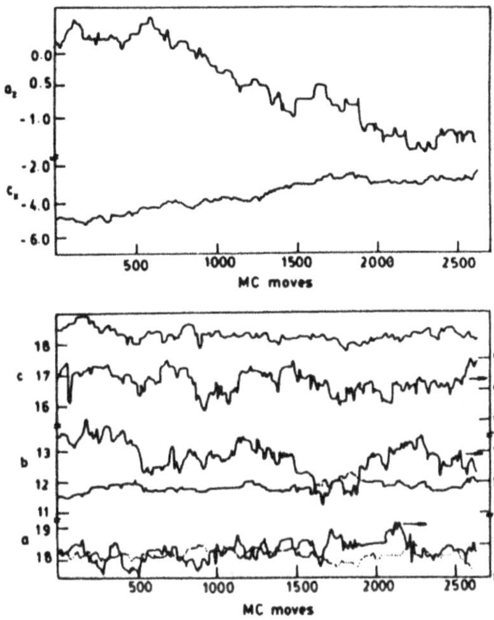


FIG. 1. Variation of the cell components $c_x$ and $a_z$, the cell edges $a$, $b$, and $c$, and angles $\alpha$, $\beta$, and $\gamma$ during the first 2500 MC moves in the simulation of carbon tetrachloride, indicating the rotation of the cell. Angles are in degrees and the lengths in Å. (From Yashonath and Rao (*19*).)

growth of the antisymmetric component to zero after each step. While both methods can be used in the simulation of polyatomic systems, the former has the advantage that it is simpler to implement and also requires significantly less computer time.

### *2.3. General Remarks Regarding the MD and MC Methods*

The MC and MD methods permitting the variation of the shape of the cell are best suited for the study of phase transitions in solids. These methods have been used to study phase transitions of a few solids in the last few years. Among these are monatomic solids such as rare gas solids, ionic solids, and molecular solids. There are, however, some inherent limitations in these methods. While certain transitions are readily investigated by these methods, others are more difficult. The b.c.c. to f.c.c. transformation of monatomic solids is an example of a transition that is readily observed (*5, 7*) (see Figs. 2 and 3). This transition has been observed as a function

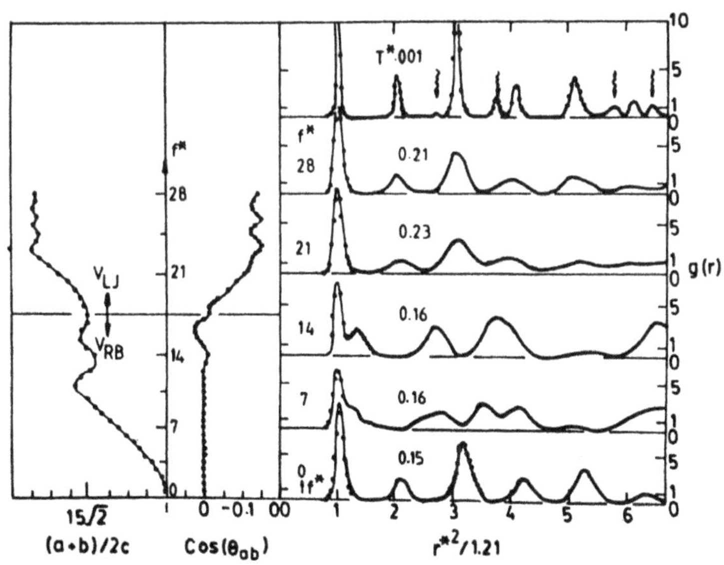


FIG. 2. Variation of the ratio $(a + b)/2c$ and $\cos\theta_{ab}$ (left), where $\theta_{ab}$ is the angle between the cell edges $a$ and $b$, and the r.d.f.'s at various times (right) from the MD study of Parrinello and Rahman (*5*), showing the transition from f.c.c. to b.c.c. and from b.c.c. to f.c.c. due to a change of interaction potential between the particles.

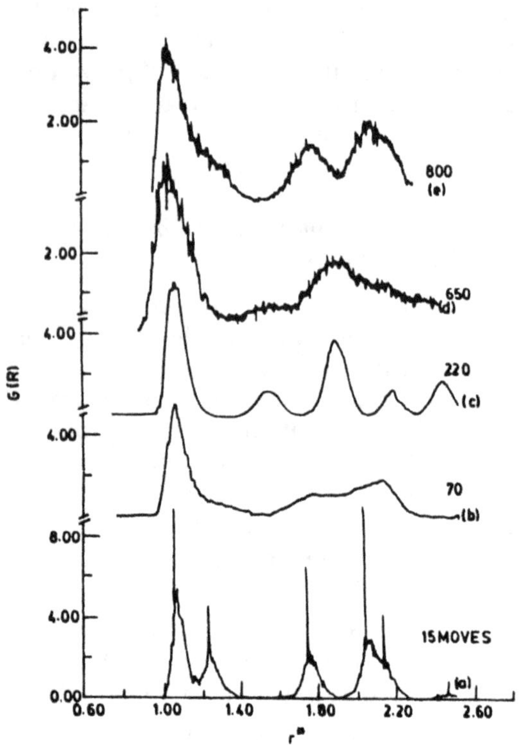


FIG. 3. Pair correlation functions as functions of reduced distance $r^*$. The b.c.c. arrangement transforms to f.c.c. at 220 MC moves when the Lennard-Jones potential is employed. If at this stage the interaction is changed to cesium potential, there is a change back to the b.c.c. at 450 MC moves. (From Yashonath and Rao (*7*).)

of pressure in helium and also on changing the interaction potential from that for cesium or rubidium to the Lennard-Jones potential. The transition from h.c.p. to f.c.c. is an example of a difficult transition. This transition requires a modification at the fourth shell of neighbors and both MD and MC calculations have failed to simulate the transformation (*20, 21*). There are also limitations imposed by the use of periodic boundary conditions and finite size effects (*22*).

In spite of the aforementioned limitations, the MD and MC methods can be most profitably employed in the investigations of a variety of systems. These methods provide fairly good approximations to average properties even when a relatively small number of particles (30 to 1000) are employed and are computationally quite economical. This is in contrast to the alternative method described in the literature (*22, 23*) which requires expensive hardware, considerable memory, and computer time, as it requires a much larger number of particles for the same accuracy. This method, however, has the advantage of being relatively free from effects due to finite size and periodic boundary conditions.

Calculations carried out by us (*7*) by the MC method where only the nine degrees of freedom were permitted and no particle displacements were allowed could successfully reproduce the f.c.c. to b.c.c. transition (Fig. 3). These and other considerations suggest that the MC method is to be preferred to the MD when only the equilibrium properties are of interest. For example, if one is interested in the determination of a phase diagram, the MC method would be more appropriate than the MD method. The MD method is preferable for obtaining dynamic properties even though it is not clear how one can obtain the exact dynamical behavior (*4, 6*). The technical details and aspects of programming for the MD and the MC methods are available (*6, 21*) and we shall not discuss them here.

## 3. Applications of MD and MC Methods to the Study of Transformations in Crystals

### *3.1. Monatomic Solids*

Parrinello and Rahman (*5*) in their MD study investigated the relation between structure and interaction potential by employing two different potentials. The (6–12) Lennard-Jones potential which predicts the f.c.c. arrangement as the stable structure and the rubidium potential which has the stable b.c.c. structure were employed for the purpose. Starting with a f.c.c. structure, they found that the solid transformed to a b.c.c. structure on changing the particle interaction from the Lennard-Jones to the rubidium potential. The radial distribution functions (r.d.f.'s) show distinct peaks corresponding to a b.c.c. structure within a few hundred MD steps of imposing the rubidium potential (see Fig. 2). The cubic cell edges transformed into a rectangular parallelopiped by a decrease in the cell edge, $\vec{c}$. The particle interaction was then changed back to the Lennard-Jones. It was found that the system transformed back to the f.c.c. arrangement within a few hundred MD steps. This calculation illustrates the usefulness of the Parrinello–Rahman method for studies involving the structure–potential relationships. We have carried out an isobaric isothermal calculation employing a variable shape MC cell (*7*) and the results are indeed similar (Fig. 3).

Najafbadi and Yip (*18*) have investigated the stress–strain relationship in iron under uniaxial loading by means of a MC simulation in the isostress isothermal ensemble. At finite temperatures, a reversible b.c.c. to f.c.c. transformation occurs with hysteresis. They found that the transformation takes place by the Bain mechanism and is accompanied by sudden and uniform changes in local strain. The critical values of stress required to transform from the b.c.c. to the f.c.c. structure or vice versa are lower than those obtained from static calculations. Parrinello and Rahman (*14*) investigated the behavior of a single crystal of Ni under uniform uniaxial compressive and tensile loads and found that for uniaxial tensile loads less than a critical value, the f.c.c. Ni crystal expanded along the axis of stress reversibly.

### *3.2. Silver Iodide*

Structural transformation in the superionic conductor silver iodide has been investigated by employing the modified molecular dynamics method in the constant-enthalpy–constant-pressure ensemble (*24*). A potential derived by considering the properties of both the $\beta$ and the superionic $\alpha$ phases was employed for the purpose. The calculated structure, diffusion coefficient, etc., for $\alpha$-AgI at 700 K were in agreement with the experiment (*25*). On cooling from 700 to 350 K, the b.c.c. lattice of iodine transformed to the close-packed h.c.p. structure accompanied by a marked decrease in $D_{Ag}$; the transformation was reversible on heating. Structural and dynamical properties as well as the transition temperature for the $\alpha$-$\beta$ transformation obtained from the MD study were in good agreement with experiment.

### *3.3. Molecular Crystals*

*Nitrogen*. Nosé and Klein (*26*) carried out a MD simulation of the various phases of nitrogen using a simple Lennard-Jones 6–12 potential with $\varepsilon$ = 37.3 K, $\delta$ = 3.31 Å. Starting with the high-pressure room-temperature structure (*Pm*3*n*), results have been reported for $T$ = 300 K and at lower temperatures and a presence of 70 kbar. The cell was allowed to vary in size as well as shape. Molecules in the $T_d$ sites perform free rotation while the remaining three-quarters in $D_{2d}$ sites rotate within a plane. Below 230 K, molecules in the $D_{2d}$ sites align themselves parallel to the direction of the unit cell vectors while those in the $T_d$ sites rotate almost freely. This phase has cubic symmetry (*R*3*c*) in which all the molecules are aligned along the same direction. The transitions were reversible.

*Carbon tetrafluoride*. Carbon tetrafluoride, which undergoes a transition to a plastically crystalline (orientationally disordered) phase, has been investigated by the Parrinello–Rahman molecular dynamics method under constant-pressure conditions (*6*). A simple intermolecular potential model of the Lennard-Jones form was derived by taking into account the experimen-

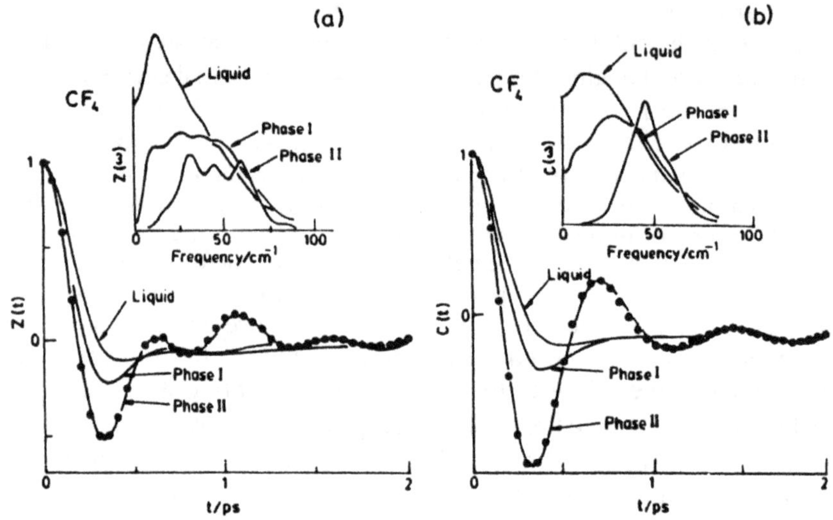


FIG. 4. (a) The velocity autocorrelation function and its power spectrum and (b) the angular velocity autocorrelation function and the power spectrum for different phases of $CF_4$. (From Nosé and Klein (*6*).)

tal molar volume and lattice energy. Calculations were carried out at 50.5 K and 1.2 kbar employing this potential with the starting arrangement of the ordered monoclinic phase II ($C2/c$) with $a = 8.43$, $b = 4.32$, $c = 8.48$ Å, $\beta = 120.7°$; calculations were also performed at higher temperatures and pressures. The calculated volume, lattice energy, and the r.d.f.'s of phase II were in agreement with experiment (*27*). On increasing the temperature, the system showed considerable increase in the molar volume around 75 K and the features in the radial distribution function became broad, suggesting increased molecular motion; the molecules in the lattice were orientationally disordered. The calculated change in enthalpy and molar volume were found to be in good accord with experiment (*28*). Properties of liquid carbon tetrafluoride have also been reported. Properties of solid carbon tetrafluoride have been studied by the MC method and found to be in agreement with the above results.

Time-dependent quantities have been examined (*6*) for the orientationally ordered and disordered phases as well as the liquid phases of $CF_4$. Power spectra for the translational and librational motions are reported (see Fig. 4). The peaks appear in the same region of frequency, suggesting strong translation–rotation coupling. The far-infrared spectrum in the ordered solid (*29*) has three peaks at 51, 57, and 66 $cm^{-1}$. On the basis of the significant intensity and temperature dependence of the 51-$cm^{-1}$ band, it has been assigned to the 43-$cm^{-1}$ band of the calculated $Z(\omega)$ and $C(\omega)$. This is a good illustration of the application of the Parrinello–Rahman MD method for the study of a crystal to plastic crystal transition. The study also shows how a simple potential can account for many of the properties exhibited by the solid. The dynamical properties seem to require a rather accurate description of intermolecular interactions. Constant-pressure and constant-volume calculations on carbon tetrafluoride (*6*) demonstrate that the single-particle dynamical properties are not very sensitive to the choice of the mass of the wall.

*Bicyclo*[2.2.2]*octane*. At low tempera-

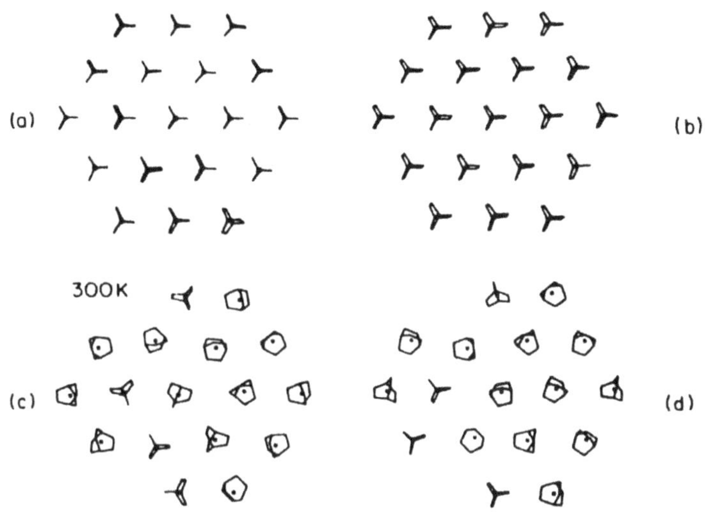


FIG. 5. A view of the bicyclo[2.2.2]octane down the trigonal axis (a) at 50 K for the 8-site model, (b) at 50 K for the 22-site model, (c) at room temperature for the 8-site model, and (d) at room temperature for the 22-site model. (From Neusy *et al.* (*32*).)

tures, bicyclo[2.2.2]octane exists as a trigonal phase with the individual molecular $C_3$ symmetry axes parallel to the crystal [111] directions. At 164.35 K, the ordered crystalline phase transforms to a cubic, orientationally disordered phase in which the quasi-spherical molecules exhibit increased molecular reorientation (*30, 31*). This order–disorder phase transition has been investigated by Neusy *et al.* (*32*) by means of constant-pressure MD employing two different potential models. In the first potential model of the Lennard-Jones form given by Jorgensen (*33*), each $CH_2$ or CH group is represented by a single interaction site and hence there are only eight interaction sites per molecule. In the model given by Williams (*34*), which is of the 6-exp form, each atom is represented by an interaction site that is placed on each atom.

Calculations carried out at 50 and 137.5 K on the ordered phase with an initial arrangement corresponding to the experimental structure ($a = 6.4$, $c = 15.1$ Å) resulted in a slightly elongated unique axis (*35*). The calculated $c/a$ ratio was 2.6 for both the 8-site and the 22-site model as compared to the experimental value of 2.3. The molar volume determined from the 22-site model was in better agreement with experiment although the calculated molar volume was about 5% higher than the experimental value for the 8-site model. On heating to room temperature, the molecules oriented themselves almost randomly with a preference for the [111] crystallographic direction (Fig. 5). The ordered trigonal crystal was found to transform to the disordered cubic phase at about 170 K, in good agreement with experimentally determined transition temperature (164.25 K).

Power spectra of the autocorrelation functions of the linear and angular velocities parallel and perpendicular to the $C_3$ symmetrical axes have also been examined by Neusy *et al.* (*32*). In the rotator phase, there is good agreement with the Raman data (*36*). The calculated characteristic time ($\tau_4$) for reorientation of the $C_3$ axes from one [111] direction to another and also the reorientation time ($\tau_3$) for rotation of molecules around the $C_3$ axes were similar

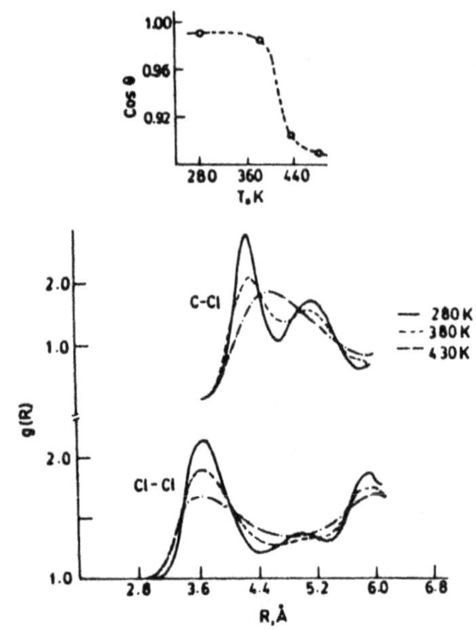


FIG. 6. Radial distribution functions C–C, C–Cl, and Cl–Cl in $CCl_4$ at 280, 380, and 430 K and 1.0 GPa pressure showing the changes from an ordered to an orientationally ordered phase. Inset shows a plot of cos $\theta$ against temperature. Here $\theta$ is the smallest angle between the $(\vec{a} + \vec{c})$ direction and the four Cl–Cl bonds. The dashed line is drawn to guide the eye. (From Yashonath and Rao (*19*).)

for both potential models. These results indicate that transferable intermolecular potential functions provide a reasonably good model for solids. Experimentally, from incoherent neutron scattering studies it is found that $\tau_4$ is much slower than $\tau_3$. It appears from these results that while simple potentials can provide a reasonable picture of the phase transitions, they cannot make detailed predictions especially with regard to the dynamics of the system.

*Carbon tetrachloride*. Carbon tetrachloride exists in at least five phases, Ia, Ib, II, III, and IV, of which two phases, Ia and Ib, are orientationally disordered (*37*); phases III and IV exist only at high pressures. We have carried out an investigation (*19*) of the transition from the high-pressure ordered phase III to an orientationally disordered phase by employing the generalized variable shape MC method in the isothermal isobaric ensemble. A simple Lennard-Jones potential used by McDonald *et al.* (*38*) in the simulation of liquid and solid phases with $\delta_{CC} = 4.6$ Å, $\varepsilon_{CC} = 51.2$ K, $\delta_{ClCl} = 3.5$ Å, and $\varepsilon_{ClCl} = 102.4$ K was employed in our calculations. Starting with the experimental structure (*39*) of phase III of $CCl_4$ with lattice parameters $a = 9.1$, $b = 5.8$, $c = 9.2$ Å and $\beta = 104.3°$, simulation was carried out at 280 K and 1.0 GPa. The calculated values of molar volume, configurational energy, lattice parameters, etc., were in agreement with the available experimental values (*39*). The r.d.f.'s for phase III are shown in Fig. 6. On increasing the temperature to 380 K, the thermodynamic properties were found to undergo small changes but, however, no significant changes were observed in the structure, the structure remaining ordered at this temperature (Fig. 7). Considerable changes in molar volume and intermolecular energy were observed on increasing the temperature to 430 K. The r.d.f.'s showed broad peaks similar to those of the liquid, suggesting increased molecular motion. The snapshot picture of the molecular arrangement in Fig. 7 shows that the crystal is orientationally disordered. This was con-

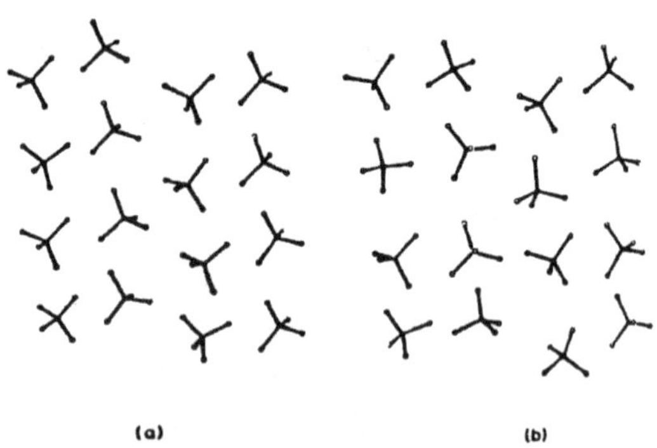


FIG. 7. A snapshot picture of the molecular arrangement viewed down the *b*-axis (a) at 380 K and (b) at 430 K in carbon tetrachloride. (From Yashonath and Rao (*19*).)

firmed by the drastic change in the minimum of the angle between the four $C_3$ axes of the molecule and the crystallographic [101] direction (see inset of Fig. 6). Calculations at an intermediate temperature of 405 K suggest that the transition temperature may be between 400 and 410 K, which compares well with the experimental value of 410 K (*37*).

*Adamantane.* Adamantane is a globular molecule in which the six methylene groups form an octahedron and the four methine groups projecting out form a tetrahedron. Adamantane exists as an ordered body-centered tetragonal solid at low temperatures (space group $P42_1c$; $a = 6.6$, $b = 8.8$ Å) (*40–42*). At 208.6 K, the solid undergoes a structural transition from the tetragonal to a cubic phase (*43*). Early X-ray studies of this phase by Nowacki (*40*) and Giocomello and Illuminati (*44*) suggested an ordered structure (space group $F43m$), but later work of Nordman and Schmitkons (*41*) showed the molecules to be orientationally disordered in the cubic phase ($Fm3m$). The nature of the high-temperature phase has, however, remained controversial (*45–47*).

We have carried out a MC simulation of adamantane in the isothermal isobaric ensemble permitting variation of the shape of the simulation cell (*48*). In these calculations, we have employed potential functions described by Jorgensen (*33*) in which there is a single interaction site for groups such as CH or $CH_2$. The potential parameters are, for $CH_2$–$CH_2$; $\varepsilon = 0.478$ kJ mole$^{-1}$, $\sigma = 3.983$ Å; for $CH_2$–CH: $\varepsilon = 0.312$ kJ mole$^{-1}$, $\sigma = 4.118$ Å; and for CH–CH: $\varepsilon = 0.203$ kJ mole$^{-1}$, $\sigma = 4.252$ Å. Calculations were carried out at 110, 213, and 298 K and atmospheric pressure. The calculated molar volume and the configurational energy are in good agreement with those from experiment (*41*). The molecules are completely ordered at 110 K and perform small librations. The $CH_2$–$CH_2$, CH–$CH_2$, and CH–CH r.d.f.'s (Fig. 8) show

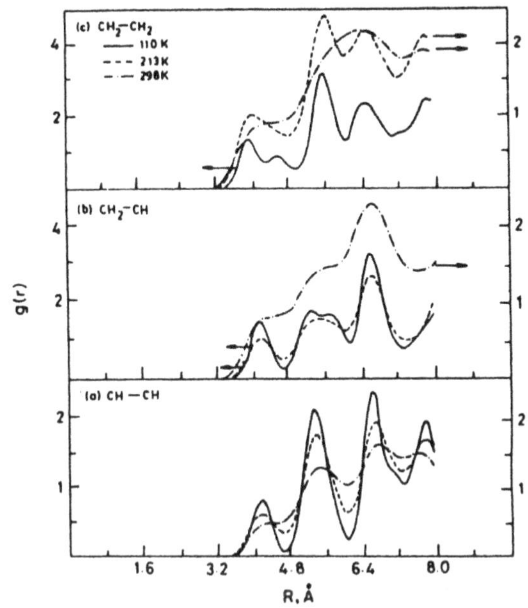


FIG. 8. Radial distribution functions for adamantane between (a) $CH_2$–$CH_2$, (b) $CH_2$–CH, and (c) $CH_2$–$CH_2$ groups at 110, 213, and 298 K and 1 atmosphere pressure. (From Yashonath and Rao (*48*).)

well-defined peaks at this temperature. At 213 K, the peaks become somewhat broad, but the arrangement at this temperature is still ordered. On increasing the temperature to 298 K, the r.d.f.'s show broad peaks unlike those of the ordered solid (Fig. 8). The molecules were found to perform large rotational jumps around the $C_4$ axis (Fig. 9). The calculated unit cell parameters are 9.31, 9.31, 9.25 Å with the angles close to 90°. These compare well with the experimental cubic cell parameter of 9.44 Å (*41*). The elongation of the $c$-axis on going from 213 to 298 K is also known experimentally. The calculated configurational energy (−58.0 kJ/mole) is in agreement with the heat of sublimation (−58.5 kJ/mole) (*49*). This study supports the view that the room-temperature phase of adamantane is orientationally disordered.

*Biphenyl.* In biphenyl, the angle between the normal to the two phenyl rings, $\theta$,

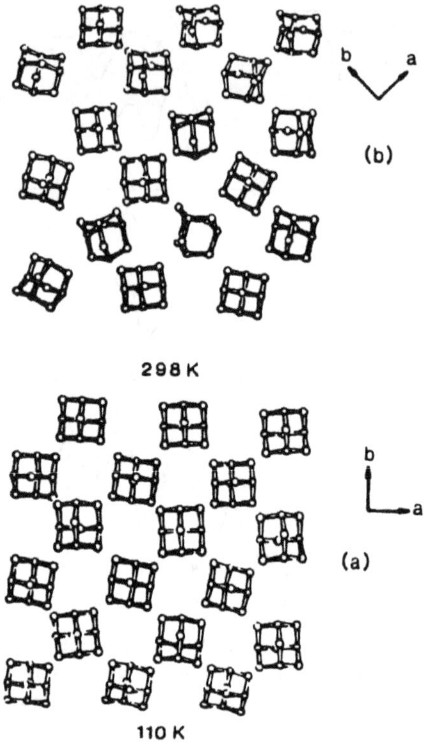


FIG. 9. A view of the instantaneous arrangement in adamantane looking down the *c*-axis: (a) tetragonal ordered phase at 110 K and (b) orientationally disordered cubic phase at 298 K. (From Yashonath and Rao (*48*).)

which is 42° in the vapor phase, decreases to about 32° in the melt (*50*). Upon cooling, the melt crystallizes in the space group $P2_1/a$, where the two phenyl rings are essentially planar. Further cooling results in a phase with the two rings deviating from this planar arrangement. Raman spectroscopic and other studies indicate that this solid–solid phase transformation is a second-order transition occurring over a range of temperatures (75 to 40 K) (*50, 52*). The conformations in the vapor and the liquid phases result from a balance between resonance stabilization and ortho–ortho hydrogen repulsion. In the solid, intermolecular interactions also play a very significant role and hence packing considerations lead to a planar conformation in the high-temperature solid phase (*53*). We have carried out some preliminary MC investigations on this interesting solid by employing a 6-exp atom–atom potential derived by Misskaya *et al.* (*54*) and used by Ramdas and Thomas (*55*) in their study of the structure of *p*-terphenyl. The same potential was used to calculate the intermolecular interactions as well as the intramolecular interactions arising out of ortho–ortho hydrogen repulsions. The potential parameters were

$$A_{CC} = 1761.9 \text{ kJ A}^6 \text{ mole}^{-1}; \quad B_{CC} = 299{,}646 \text{ kJ mole}^{-1}; \quad C_{CC} = 3.68 \text{ Å}^{-1};$$

$$A_{CH} = 493.8 \text{ kJ A}^6 \text{ mole}^{-1}; \quad B_{CH} = 77{,}841 \text{ kJ mole}^{-1}; \quad C_{CH} = 3.94 \text{ Å}^{-1};$$

$$A_{HH} = 121.4 \text{ kJ A}^6 \text{ mole}^{-1}; \quad B_{HH} = 20{,}506.5 \text{ kJ mole}^{-1}; \quad C_{HH} = 4.29 \text{ Å}^{-1}.$$

The initial arrangement was the experimental structure of the high-temperature phase with $a = 7.82$, $b = 5.58$, $c = 9.44$ Å and $\beta = 94.6°$ (*51*). The calculated intermolecular energy at 110 K was −68.2 kJ mole$^{-1}$ compared to the sublimation energy of −81.6 kJ mole$^{-1}$. The average angle between the two phenyl rings was 20° as against the experimentally observed planar conformation (*51*). The discrepancy in the observed angle could be due to the large repulsion in the region of 1.8–2.2 Å in the hydrogen–hydrogen interaction potential functions. Potential functions were derived by considering the lattice energy of aromatic hydrocarbons such as benzene, naphthalene, etc., where only intermolecular distances larger than 2.2 Å are present (*54*). It is, therefore, not surprising that the potential may not represent the interactions correctly at lower distances even though it may show excellent agreement at larger distances. A recent calculation on *n*-butane has shown that the conformational angle is much more sensitive to the potential than other properties (*56*). On cooling the solid to 40 K, we found that $\theta$ increased to 25.3° with significant decrease in libration around the long molecular axis. This trend is in agreement with experiment (*57*). We plan to carry out further calculations using a different potential model where the intermolecular interactions and intramolecular interactions are modelled by different potential functions. These calculations are expected to answer several questions, especially those concerning the distribution of the interplanar angle, the structure of the low-temperature solid, and the nature of the motion in the high-temperature phase.

### 4. Glasses

Computer simulation has contributed significantly to the understanding of equilibrium and dynamic properties of supercooled liquids and glasses. Hard-sphere, Lennard-Jones, and ionic glasses have been studied (*58–61*), providing some insights into structural and other aspects (*62, 63*). For example, it has been found that the split second peak, a characteristic feature of the r.d.f.'s of metallic glasses, is observed in computer-simulated Lennard-Jones glasses (*62*). The MD method was limited to the microcanonical ensemble in these studies until the constant-temperature and constant-pressure MD was introduced by Andersen. The constant-volume and energy conditions used in the earlier studies did not permit direct comparison with the laboratory glasses, where the conditions of preparation of glasses are very different (*64, 65*). Also, some of the typical characteristics of laboratory glasses were absent in computer-simulated glasses (*64*). The smeared nature of the transition at temperatures which are higher than the experimental transition temperature prompted Angell and co-workers (*64, 65*) to suggest that the transition observed in computer simulation studies bears little resemblance to laboratory glasses. More recent results obtained under isothermal- isobaric conditions by MC and MD methods indicate that there is, in fact, a close parallel between the laboratory and the computer-simulated glasses. The quenching rates employed in computer simulation are, however, orders of magnitude higher than those employed in the laboratory and, as a consequence, computer-simulated glasses differ in certain aspects from those obtained in the laboratory. Frenkel and McTague (*66*) have reviewed the earlier studies in this area. We shall limit the present discussion to certain significant findings based on some of the recent studies on monatomic, molecular, and orientational glasses.

*Lennard-Jones glass.* Extensive calculations on a monatomic Lennard-Jones argon-like system have been carried out by Fox and Andersen (*67*) under conditions of constant temperature and constant pressure with a variety of cooling rates. These workers have reported density, enthalpy, and self-diffusion coefficient at different pressures and temperatures. The glass transition temperature obtained from the density–temperature plot is essentially identical to that obtained from the enthalpy–temperature plots. The glass transition temperature is higher at higher pressures. Also the slower the rate of cooling, the higher was the density of the glass, indicating the existence of slow relaxation processes involving volume changes (see Fig. 10). The apparent self-diffusion coefficient could be fitted equally well to both Arrhenius and Doolittle equations. Interestingly, the pressure dependence of the structural relaxation was different from that of self-diffusion. Hysteresis in the transition was evident from a comparison of the heating and the cooling runs.

The above results indicate that even though the quenching rates employed in

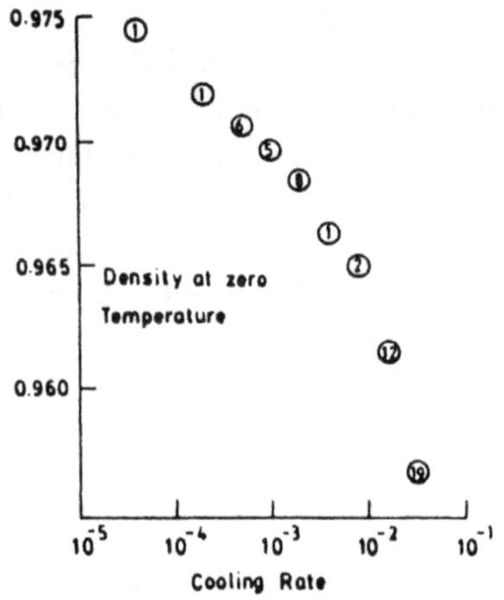


FIG. 10. The densities obtained at different cooling rates at $T = 0$ and $p = 1$ for the Lennard-Jones glass. The cooling rate is the value of the stochastic collision frequency per particle, $\nu$. The number of runs employed to obtain the reported average is shown inside each point. (From Fox and Andersen (*67*).)

computer simulation are very high, the general features of simulated glasses are quite similar to those of laboratory glasses. The demonstration of the history dependence of the properties and also the presence of hysteresis near the transition are significant. The radial distribution functions for glasses obtained by employing two different cooling rates shown in Fig. 11 are indeed similar; the r.d.f.'s are somewhat insensitive to the structural changes taking place in the glass. Our calculations on methane (*68*) in which only orientational degrees of freedom were permitted, however, show distinct changes in the r.d.f.'s of the glass before and after annealing (see Fig. 12). This is the only study that we are aware of that shows changes in r.d.f.'s even though it is for an orientationally disordered glass and not for a positionally disordered glass. This is probably because barriers for going from one state to another state are much smaller in the case of orientational glass. The history dependence of the properties of the glass has also been observed in a study on isopentane glass (*69*); the intermolecular energy of the glass is lowered by a small amount on annealing near the glass transition.

*Isopentane glass.* A recent MC simulation of isopentane glass (*70*) using realistic potentials has yielded some insight into glass structure and the nature of rearrangement occurring during the glass transition. Isothermal isobaric ensemble calculations have been performed on glasses obtained by instantaneous quenching of the liquid. There are four distinct groups in isopentane: CH, $CH_2$, the methyl group attached to CH, $CHCH_3^*$, and the methyl group attached to $CH_2$, $CH_2CH_3^*$. These give rise to ten r.d.f.'s. Four of these r.d.f.'s, between the somewhat inaccessible CH and other groups (referred to as the i-type r.d.f.'s), are similar in nature. They have a single peak within the 9-Å region at a somewhat larger distance of 5.3 to 6.2 Å. The remaining six r.d.f.'s (the p-type r.d.f.'s) in the liquid between the peripheral groups $CH_2$, $CH_2CH_3^*$, and $CHCH_3^*$ show a shoulder around 4 Å and a main peak around 6 Å. A

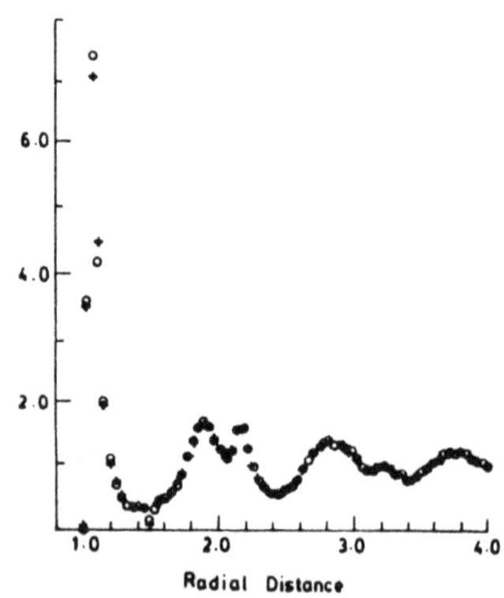


FIG. 11. Radial distribution functions for systems obtained by continuous cooling at $p = 1$. The stochastic collision frequencies were 3.2 for the + and 0.04 for the O points. (From Fox and Andersen (*67*).)

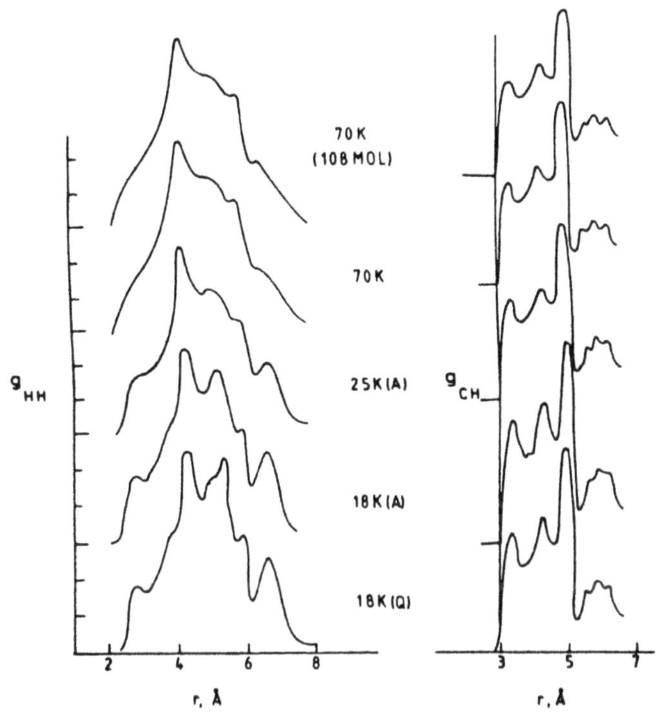


FIG. 12. Pair distribution functions $g_{HH}(r)$ and $g_{CH}(r)$ for the plastic and annealed (A) and quenched phases (Q) of methane. Only the orientational degrees of freedom were considered in the calculation. Note the distinct change in the r.d.f.'s in $g_{HH}(r)$ and to a smaller extent of $g_{CH}(r)$ on annealing. (From Yashonath and Rao (*68*).)

few typical r.d.f.'s are shown in Fig. 13. On cooling below the glass transition temperature, peaks generally become sharper and show more features. However, there are several interesting differences between the i-type and the p-type r.d.f.'s. While the

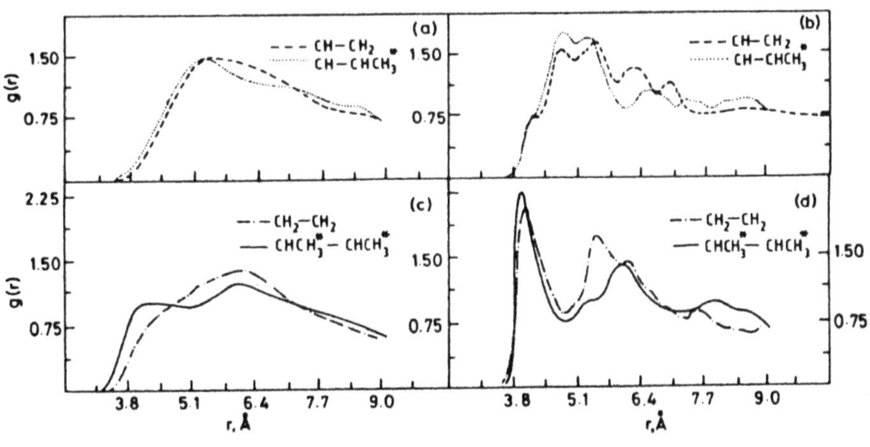


FIG. 13. Typical r.d.f.'s (i-type) between the inaccessible CH and other groups (a) for liquid isopentane at 301 K and (b) for the glass at 30 K. Typical r.d.f.'s (p-type) between the peripheral groups (c) for the liquid at 301 K and (d) for the glass at 30 K. (From Yashonath and Rao (*70*).)

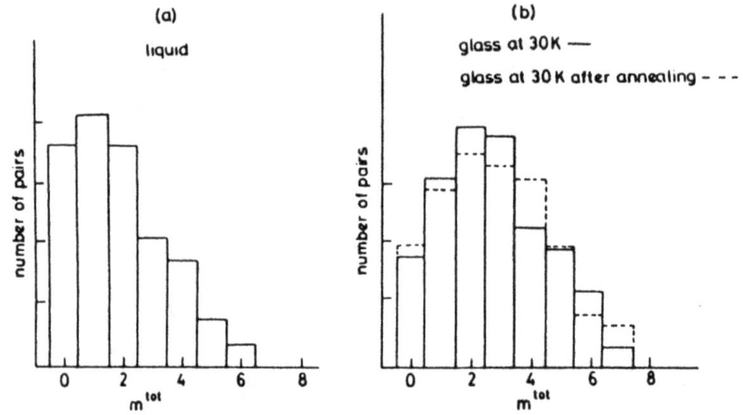


FIG. 14. Histograms for the distribution of the number of pairs of neighbors, in arbitrary units, for different values of $m_{tot}$ for (a) isopentane liquid at 301 K and (b) glass at 30 K, with and without annealing. (From Yashonath and Rao (*71*).)

For the role of molecular shape on orientational rearrangement during vitrification, see A. Chakrabarti, S. Yashonath and C.N.R. Rao, J. Phys. Chem., 96, 6762 (1992)

peak positions of the i-type r.d.f.'s are shifted towards lower distances, those of p-type r.d.f.'s do not show such shifts. This is surprising as one would have expected a more or less uniform shift of peaks in all the r.d.f.'s towards lower distances on vitrification. Furthermore, on glass formation, the shoulder in the p-type r.d.f.'s of the liquid around 4.2 Å gains significantly in intensity as compared to the main peak. These observations suggest the existence of a somewhat complex rearrangement of the neighbors. This rearrangement seems to involve the shift of the center of mass of the neighbors towards each other. In addition, the disproportionate increase in the first peak intensity on vitrification and the absence of shift towards lower distances for the p-type r.d.f.'s suggests the presence of a molecular reorientational mechanism.

The number of pairs of neighbors specified in arbitrary units is plotted against $m_{tot}$, where $m_{tot}$ is the sum of the number of groups in the region of the first peak (for the six p-type r.d.f.'s), in Fig. 14 for the liquid at 301 K and the glass at 30 K; the effect of annealing the glass is also shown in the figure. The shift of the histogram towards higher values of $m_{tot}$ indicates the overall increase in the number of neighbors in the region of the first peak. We have been able to estimate the reorientational contribution $E_r^g$ to the increase in energy from the r.d.f.'s and find the reorientational contribution to the increase in the intermolecular energy on vitrification (*71*) to be nearly 50%. In obtaining an estimate of $E_r^g$ we have neglected the reorientational contribution from the i-type r.d.f.'s and made the assumption that the p-type r.d.f.'s of the hypothetical glass are similar to those of the liquid but for the increased intensity due to the higher density. In other words,

$$g_h(r) = g_l(r) n_t^g / n_t^l, \tag{15}$$

where the $g_h(r)$ and $g_l(r)$ are the r.d.f.'s of the hypothetical glass and the liquid respectively and $n_t^g$ and $n_t^l$ are the total number of neighbors within a distance of 9 Å for the glass and the liquid.

The total intermolecular energy, $E_{inter}$, and the total reorientational contribution, $E_r$, are plotted against temperature in Fig. 15 for isopentane cooled from 301 K to 220, 120, 50, 40, and 30 K. The change in slope observed around 80 K is considerably more

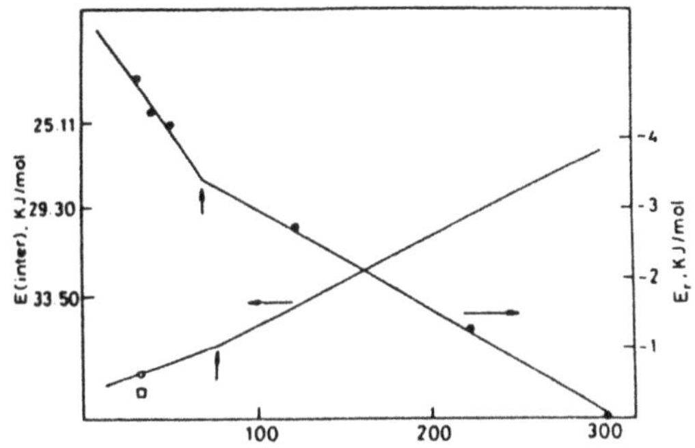


FIG. 15. Variation of the total intermolecular energy, $E_{inter}$, and the contribution from reorientation, $E_r$, for isopentane at different temperatures obtained by instantaneous cooling from 301 K. (From Yashonath and Rao (*70, 71*).)

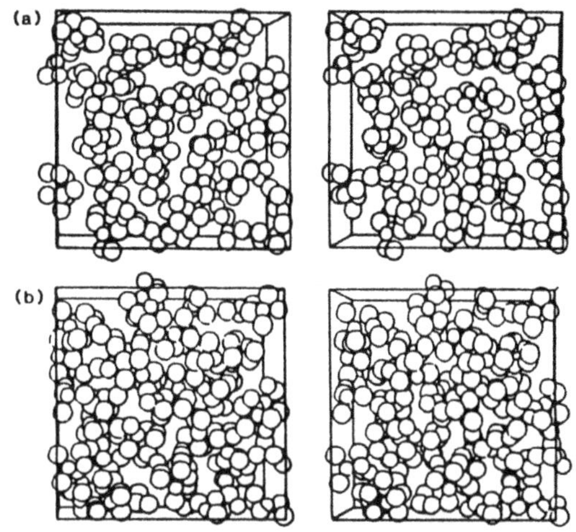


FIG. 16. Stereoplots showing the structure (a) in the liquid at 301 K and (b) in the glass at 30 K for isopentane. Note that the molecules at the surface interact with those at the opposite surface. (From Yashonath and Rao (*70*).)

marked in the case of $E_r$ as compared to $E_{inter}$. It appears, therefore, that the reorientational mechanism contributes significantly to the changes in properties near the glass transition. We have also found that significant structural rearrangements involving orientational degrees of freedom occur on annealing the glass. Stereoplots of the molecular arrangement in the liquid and the glass are shown in Fig. 16. The presence of considerable free volume even in the glass is evident from this figure.

Investigations into the properties of glassy water and methanol are presently in progress in our laboratory. Whether amorphous solid water exhibits a glass transition is an aspect of considerable interest (*72*). Preliminary studies (*73*) indicate that liquid water quenched to low temperatures does indeed show a glass transition, with the nature of species, intermolecular energy, heat capacity, etc., showing the expected changes at the transition temperature (Fig. 17).

## 5. Concluding Remarks

The generalized MD and MC methods can be used for studying different classes of materials and phenomena. The method introduced by Parrinello and Rahman (*5*) permits one to derive interparticle potentials taking into account more than one phase of

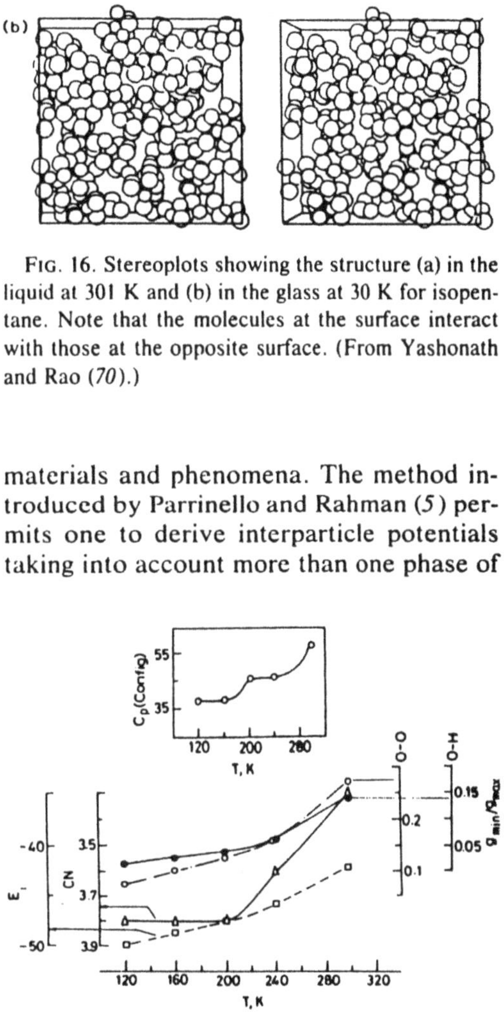


FIG. 17. Temperature variation of internal energy, $E_i$ (kJ mole$^{-1}$), coordination number (CN), and $g_{min}/g_{max}$ ratios of water showing the occurrence of the glass transition in the 200–240 K range. Volume also shows a similar change, but a slightly lower temperature. In the inset, variation of the configurational heat capacity, $C_p$ (J deg$^{-1}$ mole$^{-1}$), with temperature is shown. (From Chandrasekhar and Rao (*73*).)

the solid. These potentials are expected to be more accurate and to describe the solid over a wider range of temperatures and pressures. The new MD and MC methods provide a very useful means of studying a variety of solids. Study of different phases of elemental solids by these methods would be interesting. Such studies would, however, be limited by the availability of pair potentials. Recently, there has been an attempt to derive pair potentials for metals like sodium, magnesium, and aluminum (*74*). Phase transitions in molecular crystals such as benzene and *p*-dichlorobenzene should provide insight into the microscopic behavior in these solids and be useful in interpreting the large amount of experimental data available in the literature. The mechanism of transitions and the dependence of transition temperature and other properties on the intermolecular potential can be fruitfully investigated. In our opinion, systematic studies of orientationally disordered solids (see Refs. (*75–77*) for the very recent papers) as well as matrix isolated molecular systems such as $CH_4$ in Ar matrix should be worthwhile. The liquid crystalline state can be effectively investigated by MC and MD methods. Yet another interesting area would be to investigate the effect of defects on the phase transitions and other properties of solids. The above simulation methods could be very profitably employed in the area of "crystal engineering" and organic solid-state reactions as well as in the study of the liquid crystalline state. If the interaction potentials are known accurately, they could also be used in deriving the crystal structures of simple solids at finite temperatures.

## Acknowledgments

The authors thank the Department of Science and Technology and the University Grants Commission for support of this research.

## References


*1.* M. L. KLEIN, *Annu. Rev. Phys. Chem.* **36,** 525 (1985).

*2.* N. METROPOLIS, A. W. ROSENBLUTH, M. N. ROSENBLUTH, A. H. TELLER, AND E. TELLER, *J. Chem. Phys.* **21,** 1087 (1953).

*3.* B. J. ALDER AND T. W. WAINWRIGHT, *J. Chem. Phys.* **31,** 459 (1959).

*4.* H. C. ANDERSEN, *J. Chem. Phys.*.**72,** 2384 (1980).

*5.* M. PARRINELLO AND A. RAHMAN, *Phys. Rev. Lett.* **45,** 1196 (1980).

*6.* S. NOSÉ AND M. L. KLEIN, *Mol. Phys.* **50,** 1055 (1983); *J. Chem. Phys.* **78,** 6928 (1983).

*7.* S. YASHONATH AND C. N. R. RAO, *Mol. Phys.* **54,** 245 (1985).

*8.* (a) W. W. WOOD, *in* "Physics of Simple Liquids" (H. N. V. Temperley, J. S. Rowlinson, and G. S. Rushbrooke, Eds.), North-Holland, Amsterdam (1968); (b) J. P. VALLEAU AND S. G. WHITTINGTON, *in* "Statistical Mechanics, Part A: Equilibrium Techniques" (B. J. Berne, Ed.), p. 137, Plenum, New York (1977); (c) J. P. VALLEAU AND G. M. TORRIE, *ibid.*, p. 169; (d) J. J. ERPENBECK AND W. W. WOOD, *in* "Statistical Mechanics, Part B: Time-Dependent Processes" (B. J. Berne, Ed.), p. 1, Plenum, New York, (1977); (e) J. KUSHICK AND B. J. BERNE, *ibid.*, p. 41; (f) K. BINDER (Ed.), "Monte Carlo Methods," Methuen, London (1964); (g) K. BINDER (Ed.), "Monte Carlo Methods in Statistical Physics," Springer-Verlag, New York (1979).

*9.* J. M. HAILE AND H. W. GRABEN, *J. Chem. Phys.* **73,** 2412 (1980).

*10.* H. TANAKA, K. NAKANISHI, AND N. WATANABE, *J. Chem. Phys.* **78,** 2626 (1983).

*11.* S. NOSÉ, *Mol. Phys.* **52,** 255 (1984); *J. Chem. Phys.* **81,** 511 (1984).

*12.* D. J. EVANS, *J. Chem. Phys.* **78,** 3297 (1983).

*13.* J. M. HAILE AND S. GUPTA, *J. Chem. Phys.* **79,** 3067 (1983).

*14.* M. PARRINELLO AND A. RAHMAN, *J. Appl. Phys.* **52,** 7182 (1981).

*15.* M. PARRINELLO AND A. RAHMAN, *J. Chem. Phys.* **76,** 2622 (1982).

*16.* J. R. RAY AND A. RAHMAN, *J. Chem. Phys.* **80,** 4423 (1984).

*17.* J. R. RAY AND A. RAHMAN, *J. Chem. Phys.* **82,** 4243 (1985).

*18.* R. NAJAFBADI AND S. YIP, *Scr. Metall.* **17,** 1199 (1983).

*19.* S. YASHONATH AND C. N. R. RAO, *Chem. Phys. Lett.* **119,** 22 (1985).

*20.* M. PARRINELLO AND A. RAHMAN, *in* "Melting, Localization and Chaos" (R. K. Kalia and P. Vashishta, Eds.), Elsevier, Amsterdam/New York (1982).

21. S. YASHONATH, Ph.D. thesis, Indian Institute of Science, Bangalore, India (1985).
22. G. S. PAWLEY AND G. W. THOMAS, *Phys. Rev. Lett.* **48,** 410 (1982).
23. G. S. PAWLEY, *J. Mol. Struct.* **30,** 17 (1985).
24. M. PARRINELLO, A. RAHMAN, AND P. VASHISHTA, *Phys. Rev. Lett.* **50,** 1073 (1983).
25. W. ANDREONI AND J. C. PHILLIPS, *Phys. Rev. B* **23,** 6456 (1981).
26. S. NOSÉ AND M. L. KLEIN, *Phys. Rev. Lett.* **50,** 1207 (1983).
27. C. N. BOL'SHUTKIN, V. M. GASAN, A. I. PROKHAVTILOV, AND A. I. ERENBURG, *Acta Crystallogr., Sect. B* **28,** 3542 (1972).
28. J. W. STEWART AND R. I. LAROCK, *J. Chem. Phys.* **28,** 3542 (1972).
29. Y. A. SATATY, A. RON, AND F. H. HERBSTEIN, *J. Chem. Phys.* **62,** 1094 (1975).
30. J. N. SHERWOOD (Ed.), "The Plastically Crystalline State," Wiley, New York (1979); C. N. R. RAO, *Proc. Indian Acad. Sci. (Chem. Sci.)* **94,** 181 (1985).
31. E. F. WESTRUM, JR., W. K. WONG, AND E. MORAWETZ, *J. Phys. Chem.* **74,** 2542 (1970).
32. E. NEUSY, S. NOSÉ, AND M. L. KLEIN, *Mol. Phys.* **52,** 269 (1984).
33. W. L. JORGENSEN, *J. Amer. Chem. Soc.* **103,** 335 (1981).
34. D. E. WILLIAMS, *J. Chem. Phys.* **47,** 4680 (1967).
35. A. J. LEADBETTER, J. PIPER, R. M. RICHARDSON, AND P. G. WRIGHTON, *J. Phys. C* **15,** 5921 (1982).
36. J. L. SAUVAJOL, Thesis, Universite de Lille (1983).
37. R. G. ROSS AND P. ANDERSON, *Mol. Phys.* **36,** 39 (1978).
38. I. R. MCDONALD, D. G. BOUNDS, AND M. L. KLEIN, *Mol. Phys.* **45,** 521 (1982).
39. G. J. PIERMARINI AND A. B. BRAUN, *J. Chem. Phys.* **58,** 1974 (1973).
40. W. NOWACKI, *Helv. Chim.* **28,** 1283 (1945).
41. C. E. NORDMAN AND D. L. SCHMITKONS, *Acta Crystallogr.* **18,** 764 (1965).
42. J. DONOHUE AND S. H. GOODMAN, *Acta Crystallogr.* **22,** 352 (1967).
43. S. C. CHANG AND E. F. WESTRUM, *J. Phys. Chem.* **64,** 1547 (1960).
44. G. GIOCOMELLO AND G. ILLUMINATI, *Gazz. Chim. Ital.* **75,** 246 (1945).
45. P. A. REYNOLDS, *Acta Crystallogr., Sect. A* **34,** 362 (1978).
46. J. P. AMOUREUX, M. BEE, AND J-C. DAMIEN, *Acta Crystallogr., Sect. B* **36,** 2633 (1980).
47. P. A. REYNOLDS AND B. R. MARKEY, *Acta Crystallogr., Sect. A* **35,** 627 (1979).
48. S. YASHONATH AND C. N. R. RAO, *J. Phys. Chem.* **90,** 2552 (1986).
49. R. J. JOCHEMS, *Chem. Thermodyn.* **14,** 395 (1982).
50. A. M. PONTE GONCALVES, *Prog. Solid State Chem.* **13,** 1 (1980).
51. G. P. CHARBONNEAU AND Y. DELUGEARD, *Acta Crystallogr., Sect. B* **32,** 1420 (1976).
52. R. M. HOCHSTRASSER, G. W. SCOTT, A. H. ZEWAIL, AND H. F. FUESS, *Chem. Phys.* **11,** 273 (1975); C. A. HUTCHINSON, JR., AND V. H. MCCANN, *J. Chem. Phys.* **61,** 820 (1974).
53. G. CASALONE, C. MARIANI, A. MUGNOLI, AND M. SIMONETTA, *Mol. Phys.* **15,** 339 (1968).
54. K. V. MISSKAYA, I. E. KOZLOVA, AND V. F. BEREZNITSKAYA, *Phys. Status Solidi B* **62,** 291 (1974).
55. S. RAMDAS AND J. M. THOMAS, *J. Chem. Soc., Faraday Trans. 2* **72,** 1251 (1976).
56. A. BANBN, F. SERRANO ADAN, AND J. SANTAMERIA, *J. Chem. Phys.* **83,** 297 (1985).
57. A. S. CULLICK AND R. E. GERKIN, *Chem. Phys.* **23,** 217 (1977).
58. L. V. WOODCOCK, *Ann. N.Y. Acad. Sci.* **371,** 274 (1981).
59. J. H. R. CLARKE, *J. Chem. Soc., Faraday Trans. 2* **75,** 1371 (1979).
60. S. BREWER AND M. J. BISHOP, *J. Chem. Phys.* **75,** 3522 (1981).
61. L. V. WOODCOCK, C. A. ANGELL, AND P. CHEESMAN, *J. Chem. Phys.* **65,** 1656 (1976).
62. F. F. ABRAHAM, *J. Chem. Phys.* **72,** 359 (1980).
63. J. R. FOX AND H. C. ANDERSEN, *Ann. N.Y. Acad. Sci.* **371,** 123 (1981).
64. C. A. ANGELL, J-H-R. CLARKE, AND L. V. WOODCOCK, *Adv. Chem. Phys.* **48,** 397 (1981).
65. C. A. ANGELL, *Ann. N.Y. Acad. Sci.* **371,** 136 (1981).
66. D. FRENKEL AND J. P. MCTAGUE, *Annu. Rev. Phys. Chem.* **31,** 491 (1980).
67. J. R. FOX AND H. C. ANDERSEN, *J. Phys. Chem.* **88,** 4019 (1984).
68. S. YASHONATH AND C. N. R. RAO, *Chem. Phys. Lett.* **101,** 524 (1983).
69. S. YASHONATH, K. J. RAO, AND C. N. R. RAO, *Phys. Rev. B* **31,** 3196 (1985).
70. S. YASHONATH AND C. N. R. RAO, *Proc. R. Soc. London, Ser. A* **400,** 61 (1985).
71. S. YASHONATH AND C. N. R. RAO, *J. Phys. Chem.* **90,** 2581 (1986).
72. S. R. ELLIOTT, C. N. R. RAO, AND J. M. THOMAS, *Angew Chem. Int. Ed.* (English) **25,** 31 (1986).
73. J. CHANDRASEKHAR AND C. N. R. RAO, *Chem. Phys. Lett.*, in press.
74. D. G. PETTIFER AND M. A. WARD, *Solid State Commun.* **49,** 291 (1984).
75. R. D. MOUNTAIN AND A. C. BROWN, *J. Chem. Phys.* **82,** 4236 (1985).
76. M. FERRORIO, I. R. MCDONALD, AND M. L. KLEIN, *J. Chem. Phys.* **83,** 4726 (1985).
77. R. W. IMPEY, M. L. KLEIN, AND I. R. MCDONALD, *J. Chem. Phys.* **82,** 4690 (1985).

5228 Reprinted from The Journal of Physical Chemistry, 1995, 99.


# The Metal–Nonmetal Transition: A Global Perspective

**P. P. Edwards**

*The School of Chemistry, University of Birmingham, Birmingham, B 15 2TT, U.K.*

**T. V. Ramakrishnan**

*Department of Physics, Indian Institute of Science, Bangalore 560 012, India*

**C. N. R. Rao***

*CSIR Centre of Excellence in Chemistry, Indian Institute of Science, Bangalore 560 012, India, and Department of Chemistry, University of Wales, Cardiff CF1 3TB, U.K.*

*Received: October 24, 1994; In Final Form: January 26, 1995*[⊗]

A wide range of condensed matter systems traverse the metal–nonmetal transition. These include doped semiconductors, metal–ammonia solutions, metal clusters, metal alloys, transition metal oxides, and superconducting cuprates. Certain simple criteria, such as those due to Herzfeld and Mott, have been highly successful in explaining the metallicity of materials. In this article, we demonstrate the amazing effectiveness of these criteria and examine them in the light of recent experimental findings. We then discuss the limitations in our understanding of the phenomenon of the metal–nonmetal transition.

## 1. Introduction

Of all the physical properties of condensed materials, the electrical conductivity exhibits the widest range, anywhere from $10^{-22}$ ohm$^{-1}$ cm$^{-1}$ in the best nonmetals to around $10^{10}$ ohm$^{-1}$ cm$^{-1}$ in pure metals (not in the superconducting state). There are several situations where condensed phases transform from the metallic to the nonmetallic state on changing thermodynamic parameters such as temperature, pressure, and composition, with the electrical conductivity changing by factors of $10^3$–$10^{14}$ over a small range of the thermodynamic parameter.[1-3] In Figure 1, we show the temperature–composition plane with the electron density varying between $10^{12}$ and $10^{30}$, with the temperature going up to $10^{10}$ K. The normal experimental conditions where we find metals and semiconductors are indicated in the figure, as are also the conditions appropriate to stars. The elements H, Xe, Cs, and Hg shown by asterisks are at conditions close to the critical points attained by one of many routes such as shock waves, wire explosions, or MHD (magnetohydrodynamic implosion). The figure also shows the regime for a degenerate strongly coupled plasma, besides the values of the Wigner–Seitz radius, $r_s$, given by

$$r_s = \left(\frac{3}{4\pi n a_0^3}\right)^{1/3} \qquad (1)$$

where $n$ is the conduction electron density and $a_0$ is the Bohr radius. The largest value of $r_s$ in metals[5] is found for Cs. In all the systems in Figure 1, the possibility of a thermodynamically induced transition from the metallic to the nonmetallic regime exists. Clearly, the metal–nonmetal transition is a problem of vital interest, being concerned with a wide range of issues from the metallization of stars to the size-induced transition in small clusters of metals.[3,6,7] Between these two extrema is a myriad of examples in a variety of condensed matter systems.[1,2] The range of systems traversing the metal–nonmetal

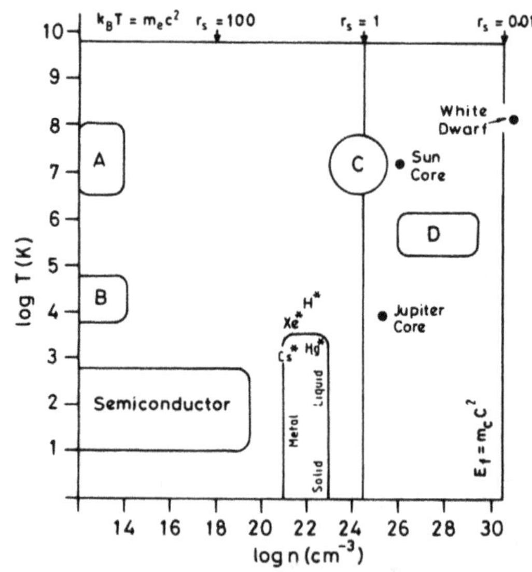


**Figure 1.** A nonrelativistic window of the temperature–composition plane, showing electron density ($n$) and temperature ($T$). Normal conditions (on earth) for semiconductors and elemental metals and conditions on the Sun, Jupiter, and the White Dwarf are shown. Experimental methods in A, B, C, and D are Tokamak, glow-discharge, laser fusion, and degenerate strongly coupled plasma, respectively. Wigner–Seitz radii, $r_s$, are also shown (adapted from Redmer[4]).

* To whom correspondence should be addressed at CSIR Centre of Excellence in Chemistry, Indian Institute of Science, Bangalore 560 012, India.

⊗ Abstract published in *Advance ACS Abstracts*, March 15, 1995.

0022-3654/95/2099-5228$09.00/0 

transition is continually increasing. Thus, oxides exhibiting high-temperature superconductivity are close to or at the metal–nonmetal boundary,[8,9] and some of them actually exhibit metal to nonmetal transitions with a change in composition.

In spite of the plethora of experimental findings, the status of our theoretical understanding of metal–nonmetal transitions is far from satisfactory. The difficulty is largely intrinsic to the phenomenon. Electronic states involved in charge transport(i.e., those near the Fermi energy) are spatially extended in the metal and are localized in the insulator. The localization may be due to static disorder (Anderson localization), to strong local electron–electron correlations which "freeze" the local electron number (Mott transition), or to strong electron–lattice coupling which traps the electron locally.[1,2,10,11] In all these cases, the natural modes of description of the electronic states in the different phases are diametrically opposite; it is difficult to find an approach which does both. Secondly, in many cases, more than one mechanism is operative, and one may reinforce the other. For example, in a disordered, strongly correlated oxide, Anderson localization due to disorder tends to increase the local correlation effect. In all electronic systems, the Coulomb interaction which is relatively weak and short ranged in a metal but strong and long ranged in an insulator (giving rise to bound electron–hole states) is present and can be important in promoting the insulating state. There is one mechanism for the metal–nonmetal (M–NM) transition, in a crystalline solid, that does not involve localized states. This is the transition of electrons from a fully filled band (insulator) to a partially filled band (metal) under pressure or structural change. This transition, however, appears to be uncommon. We discuss these mechanisms in some detail in section 6 (see also refs 1, 2, 10, and 11).

There are certain simple criteria for the occurrence of the metal–nonmetal transition, based on powerful physical concepts which turn out to be surprisingly successful. One such criterion is the idea due to Mott that in a low carrier density metal, the screened Coulomb attraction may be strong enough to bind an electron hole pair, thus destabilizing the metal.[2,12] Another useful criterion due to Mott[13] is the idea of a minimum conductivity, $\sigma_{min}$, that a metal can support, corresponding to the mean free path being equal to the de Broglie wavelength of the electron at the Fermi energy. Then, there is the Herzfeld metallization criterion of the dielectric catastrophe[14] which could occur for a dense collection of polarizable atoms. We discuss these criteria further in sections 3–5. We shall first present a brief overview of phenomena and systems associated with the M–NM transition.

## 2. Diverse Systems Exhibiting Metal–Nonmetal Transitions

As mentioned earlier, a large variety of systems exhibit M–NM transitions. The systems include[1-3] metal–ammonia (or amine) solutions, expanded metals, doped semiconductors, metal–noble gas films, metal–metal halide melts, alloys of gold with metals such as cesium, transition metal oxides and sulfides, and other inorganic and organic solids. These systems have been adequately reviewed, and we shall briefly examine only those findings that are new or directly relevant to the later discussion. Transition metal oxides[15] are especially noteworthy in that the M–NM transition in them can arise from one of many causes. Typical of the transitions found in oxide systems are the following: (i) pressure-induced transitions as in NiO, (ii) transitions as in $Fe_3O_4$ involving charge ordering, (iii) transitions as in $LaCoO_3$ that are initially induced because of the different spin configurations of the transition metal ion, (iv) transitions as in EuO arising from the disappearance of spin polarization band-splitting effects when the ferromagnetic Curie temperature is reached, (v) compositionally induced transitions, as in $La_{1-x}Sr_xCoO_3$ and $LaNi_{1-x}Mn_xO_3$, in which changes of band structure in the vicinity of the Fermi level are brought about by a change in composition or are due to disorder-induced localization, (vi) transitions as in $K_{0.3}MoO_3$ due to charge-density waves, and (vii) temperature-induced transitions in a large class of oxides such as $Ti_2O_3$, $VO_2$, and $V_2O_3$ involving more than one mechanism.

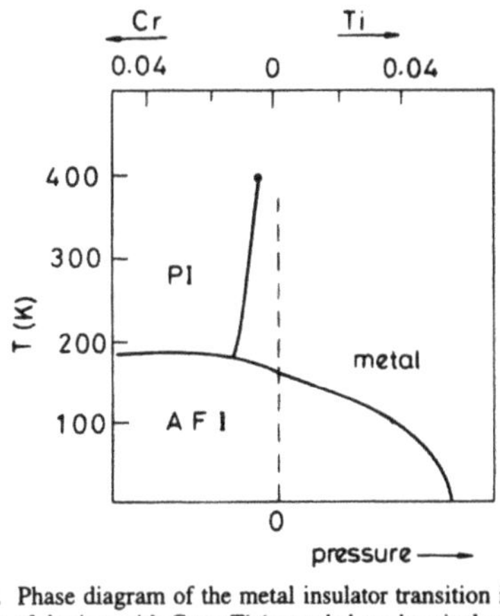


**Figure 2.** Phase diagram of the metal insulator transition in $V_2O_3$ as a function of doping with Cr or Ti (regarded as chemical pressure) as well as actual pressure, showing a critical point (after McWhan, D. B.; et al. *Phys. Rev. Lett.* **1971**, *27*, 941). The three phases are paramagnetic insulator (PI), metal, and antiferromagnetic insulator (AFI). The different kinds of resistivity behavior in Figure 3 correspond to different constant pressure (or Cr, Ti content) cuts in this phase diagram.

The last category of M–NM transitions has attracted considerable attention. In $Ti_2O_3$, a second-order transition occurs around 410 K, accompanied by a gradual change in the rhombohedral *c/a* ratio and a 100-fold jump in conductivity; the oxide remains paramagnetic throughout. A simple band-crossing mechanism occurring with the change in the *c/a* ratio can explain this transition. Accordingly, substitution of Ti by V up to 10% in $Ti_2O_3$ makes the system metallic; the *c/a* ratio of this metallic solid solution and the high-temperature phase of $Ti_2O_3$ are similar. In $VO_2$, a first-order transition occurs around 340 K, accompanied by a change in structure (monoclinic to tetragonal) and a $10^4$-fold jump in conductivity; the material remains paramagnetic throughout. A crystal distortion model wherein a gap opens up in the low-temperature low-symmetry structure adequately explains the transition. Substitution of trivalent ions such as $Cr^{3+}$ and $Al^{3+}$ for vanadium in $VO_2$ leads to a complex phase diagram with at least two insulating phases whose properties are significantly different from those of the insulating phase of pure $VO_2$. These phases are now fairly well understood.

The M–NM transition in $V_2O_3$ and its alloys has been the subject of a large number of publications.[1,2,15-19] Pure $V_2O_3$ undergoes a first-order transition (monoclinic–rhombohedral) at 150 K accompanied by a $10^7$-fold jump in conductivity and an antiferromagnetical–paramagnetic transition. The carrier effective mass and other properties also show large changes at this transition. Application of pressure makes $V_2O_3$ increasingly metallic, thus suggesting that it is near a critical point. Accordingly, doping with Ti or Cr has a marked effect on the transition; the former has a positive pressure effect and the latter a negative pressure effect (Figure 2). $V_2O_3$ also shows a second-

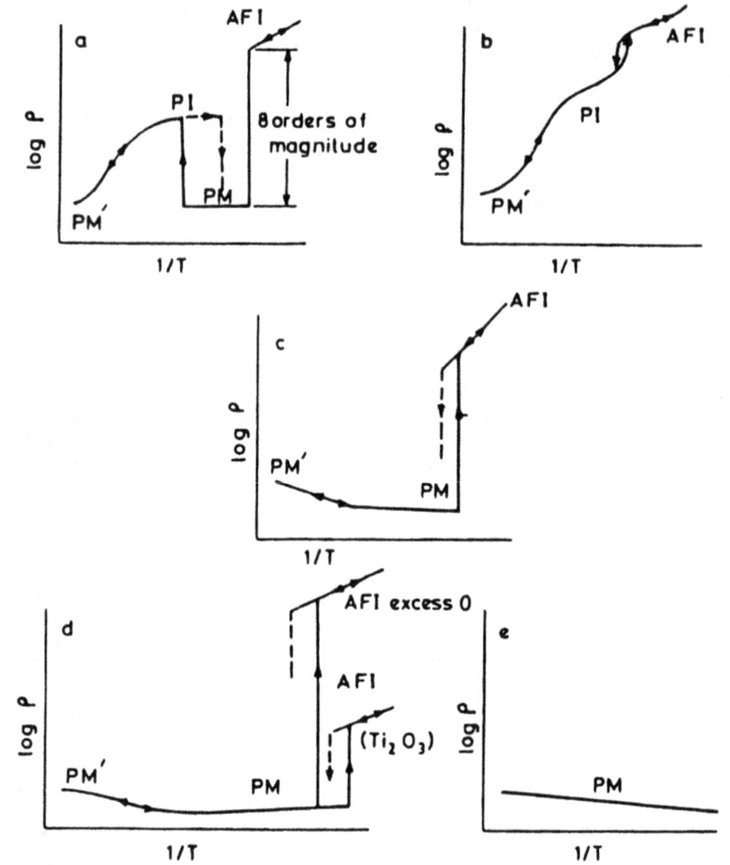


**Figure 3.** Schematic diagram depicting the changes of resistivity, $\varrho$, with temperature, $T$, in the $V_2O_3$ system: (a) $V_2O_3$ doped with 1.5% $Cr_2O_3$, (b) $V_2O_3$ doped with 3% $Cr_2O_3$, (c) pure $V_2O_3$, (d) $V_2O_3$ doped with 1% $Ti_2O_3$ and nonstoichiometric $V_2O_3$, and (e) 5.5% $Ti_2O_3$ doped $V_2O_3$. AFI, antiferromagnetic insulator; PM (PM′), paramagnetic metal; PI, paramagnetic insulator (after Honig and Spalek[16]).

order transition around 400 K with a small conductivity anomaly. Mere crystal distortion or magnetic ordering cannot explain the large connectivity jump at 150 K. The current status of the $V_2O_3$ transition can be represented in terms of Figure 3. This figure also serves to indicate the complexity of the metal−nonmetal transition in a relatively simple oxide system. There are many recent findings on $V_2O_3$ which are interesting. Thus, Carter et al.[17] have carefully measured the electrical and magnetic properties of single crystals of pure and doped $V_2O_3$ near the M−NM transition. Bao et al.[18] have examined the phase diagram of $V_{2-x}O_3$ in the $x$, $P$, $T$ space and identified an incommensurate spin density wave (SDW) in metallic $V_2O_3$ close to the transition. The optical conductivity of $V_2O_3$, $\alpha(\omega)$, has been investigated by Thomas et al.[19] In spite of extensive experimental and theoretical effort, a complete understanding of the transition in the $V_2O_3$ system is yet to emerge. We shall examine some of the factors responsible for this situation in section 6.

Compositionally controlled M−NM transitions in oxides are worthy of special mention. We shall examine two types of compositionally controlled transitions[20] as typified by $La_{1-x}A_xMO_3$ (A = Ca or Sr and M = V, Mn, or Co) and $LaNi_{1-x}M_xO_3$ (M = Mn or Fe). In $La_{1-x}A_xMO_3$, progressive substitution of trivalent La by divalent A brings about itinerant behavior of the d electrons, because every A ion creates an $M^{4+}$ ion and promotes electron hopping from $M^{3+}$ to $M^{4+}$ ions (impurity-band formation). This is to be contrasted with $LaNi_{1-x}M_xO_3$ where $LaNiO_3$ ($x = 0$), which is a correlated metal, transforms to an insulator on progressive substitution of Ni by M (somewhat like a deep-impurity situation). In Figure 4 we show typical electrical resistivity data in the two types of transitions. Unlike the above two systems, $A_{0.3}MoO_3$ (M = K or Rb) shows a metal−nonmetal transition associated with charge-density waves.[21]

Many of the high-temperature superconducting cuprates show compositionally controlled M−NM transitions.[8] Thus, in $TlCa_{1-x}Ln_xSr_2Cu_2O_y$ (Ln = Y or rare earth), the superconducting $T_c$ shows a maximum at an optimal value of $x$ (corresponding to the optimal value of the hole concentration). The system also shows a M−NM transition in the normal state as $x$ is varied (Figure 5). Thus, the cuprate is metallic when $x = 0.25$ and insulating when $x = 1.0$. Rather interesting behavior occurs at $x = 0.75$ when the superconducting transition occurs from a seemingly semiconducting state. A compositionally controlled M−NM transition is also exhibited by $Bi_2Ca_{1-x}Ln_xSr_2Cu_2O_{8+\delta}$ (Figure 5) which has a maximum $T_c$ at an optimal $x$ value of 0.25. $La_{2-x}Sr_xCuO_4$ shows a similar metal−nonmetal transition ($x = 0$ is an insulator and $x = 0.3$ is a metal) with change in $x$ and the maximum $T_c$ is at $x = 0.20$. $La_{2-x}Sr_2CuO_4$ and a few other systems traverse the insulator−superconductor−metallic regimes with change in composition (increase in $x$ from 0.0 to 0.3 in $La_{2-x}Sr_xCuO_4$). This suggests that the high-temperature

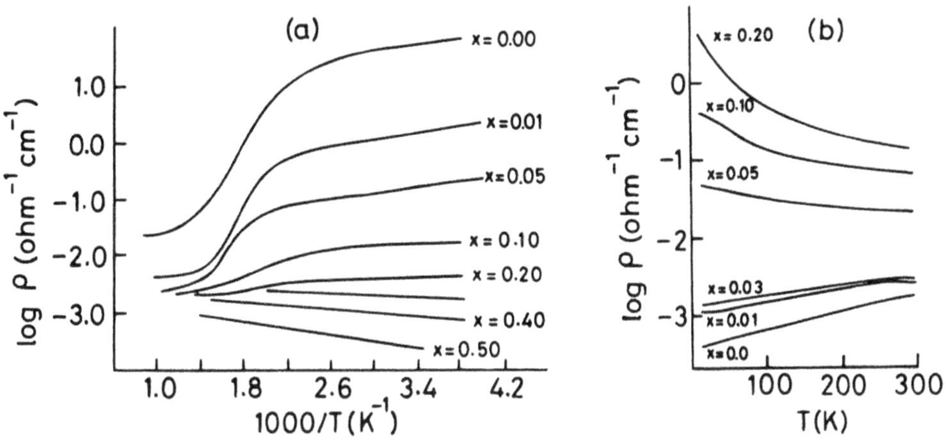


**Figure 4.** Compositionally controlled metal–nonmetal transitions in (a) $La_{1-x}Sr_xCoO_3$ and (b) $LaNi_{1-x}Mn_xO_3$ (from Rao and Ganguly[20]).

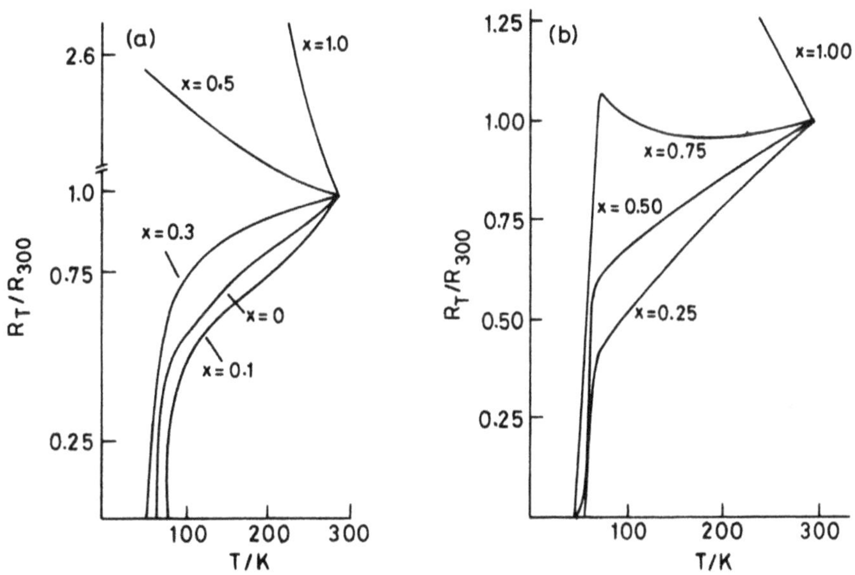


**Figure 5.** Compositionally controlled metal–nonmetal transition in superconducting cuprates: (a) $Bi_2Sr_2Ca_{1-x}Nd_xCu_2O_8$ and (b) $TlCa_{1-x}Nd_xSr_2CuO_4$ (after Rao[8]).

superconducting cuprates are at the boundary between metals and insulators.

One of the fundamental questions that has intrigued scientists is, "how many atoms maketh a metal?". The question is concerned with the possibility of a transition from the metallic to the nonmetallic state as the bulk metal is divided into finer particles. Recent studies of metal clusters[22,23] have attempted to answer the question. Careful investigations of gold clusters have shown that the binding energy of the core levels increases markedly (relative to the bulk metal value) when the cluster size decreases below 1 nm. That the effect is not merely due to final state effects but due to the occurrence of metal to nonmetal transitions as the cluster size decreases is reinforced by valence band and Bremstraahlung isochromat spectroscopic studies. Furthermore, tunneling conductance measurements show the existence of a gap in clusters smaller than 1 nm containing around 50 atoms or less; in Hg atom clusters, the 6s–6p atomic transition gives way to a collective metal-like plasmon absorption for a cluster size between 7 and 20 atoms.[23] The size-dependent transition from the metallic to the nonmetallic state does not occur abruptly. The possibility of a matrix-bound insulator–metal transition in alkali metal-doped zeolites at a critical stage of loading has been suggested by Edwards et al.[24] The dependence of the energetics and charge distribution of electron states on the cluster size and the dielectric constant has been examined by Rosenblit and Jortner,[7] to shed light on the cluster size-induced metal–nonmetal transition.

## 3. The Herzfeld Criterion

The earlier theoretical prediction of a M–NM transition is that derived from the work of Goldhammer[25] and Herzfeld.[14] These authors considered the effect of increasing density on the atomic polarizability and suggested that there would be a divergence in the polarizability or the dielectric constant causing the release of bound electrons. The Herzfeld criterion for dielectric catastrophe is given by,

5232 J. Phys. Chem., Vol. 99, No. 15, 1995 

$$\frac{4\pi N\alpha_0}{3V} = \frac{R}{V} = 1 \qquad (2)$$

where $\alpha_0$ is the low-density static polarizability of the atom, $N$ Avogadro's number, $V$ the molar volume, and $R$ the molar polarizability. As a result of cooperative polarization effects, valence electrons get delocalized from the lattice sites at very high (metallic) densities[26] and the Drude free electron model becomes applicable. In the metallic regime, $(R/V) > 1$. In Figure 6 we show how this criterion excellently delineates metals from nonmetals in the periodic table.[27] Rao and Ganguly[28] have pointed out that the latent heat of vaporization, $\Delta H_v$, of elements with metallic cohesion is larger than that of elements which are insulating because of relatively weak bonding (Figure 7). There are, however, exceptions such as strongly covalently bonded solids e.g., carbon.

A recent development is the realization of a link between the Herzfeld metallization view and the stress-induced transformations in solids, notably in semiconductors.[29] Under a diamond pressure indentor, ordinary (semiconducting) silicon transforms to the much denser $\beta$-tin (metallic) structure, the critical pressure being in the range 11–12 GPa. This is consistent with the experiments which reveal a large, reversible drop in the electrical resistivity in a thin layer surrounding a microindentation in Si, as would be anticipated because of the metallic characteristics of the $\beta$-tin structure. Good correlations are found between the experimental metallization pressure and the values calculated from the Herzfeld polarization catastrophe criterion. Experimental transition pressures also correlate with Vickers hardness numbers and activation energies for dislocation motion. We show in Figure 8, a comparison of the calculated critical pressure (Herzfeld model) with the measured Vickers hardness (expressed in kilobars) for the group IV elements and SiC. Equally good correlations are obtained for all tetrahedrally bounded semiconductors and alkali and alkaline earth oxides.

Fujii et al.[30] have reported experimental evidence for the molecular dissociation process in $Br_2$ near 80 GPa. This transition, which is coincident with the onset of pressure-induced metallization, was first discovered in molecular/metallic iodine.[31] A diatomic molecular crystal loses its molecular character in the limit when the intermolecular distance becomes equal to the intramolecular bond length. Fujii et al.[30] applied the Herzfeld criterion to $I_2$ and $Br_2$ and estimated that the molar refractivity reaches the atomic limit around 20 GPa in $I_2$ and 80 GPa in $Br_2$. In both cases, the computed pressure coincides with that for molecular dissociation accompanied by metallization.

## 4. The Mott Criterion

The Herzfeld criterion considers the M–NM transition as viewed from the nonmetallic side. Over 30 years ago, Mott[12] proposed a simple model of the M–NM transition which considers how electron localization occurs as the transition is approached from the metallic side. In Figure 9, we show a schematic representation of a lattice of one-electron hydrogenic centers (P donors in Si) in the two limiting electronic regimes of high and low donor densities. At a sufficiently high density (small interparticle distance), the system would be a metal;[32] in this state, the system has a finite electrical conductivity at the absolute zero of temperature, i.e., $\sigma(T = 0) \neq 0$. At large interparticle distances, the system must surely become nonmetallic, or insulating, having a conductivity of 0 at $T = 0$ K, viz., $\sigma(T = 0) \to 0$. Mott argued that at a critical interdonor distance ($d_c$) a first-order (discontinuous) transition from metal to nonmetal would occur (Mott transition). A discontinuous transition at the absolute zero of temperature will always remain a tantalizing theoretical prediction. Experimentally, however, even very close to $T = 0$ K (down to 0.03 K), the experimental situation is equivocal.[32–35] The physics of the problem is that, on the metallic side of the transition, the effective valence electron–cation potential in an atom is completely screened by the conduction electron gas and bound states are therefore nonexistent. However, if the conduction electron density, or equivalently the average separation between donor centers is changed, there comes a point at which bound levels (i.e., nonmetallic states) appear at some critical concentration of centers such that the following condition is satisfied.

$$n_c^{1/3} a_H^* \leq 0.25 \qquad (3)$$

Here, $a_H^*$ is the Bohr orbit radius of the isolated center and $n_c$ is the critical carrier density at the M–NM transition. Another way of viewing the transition is that of an electronic instability which ensues when the trapping of an electron into a localized level also removes one electron from the Fermi gas of electrons. This must clearly lead to a further reduction in the screening properties (which are themselves directly related to the conduction electron density) and a catastrophic situation then ensures the localization of electrons from the previously metallic electron gas.

There appears to be little doubt that the Mott criterion given by eq 3 is an effective indicator of the critical condition at the M–NM transition itself. At the least, this simple criterion provides a numerical prediction for the metal–nonmetal transition in many situations. Figure 10 summarizes some of the experimental data.[34,36] Interestingly, besides doped semiconductors, metal–ammonia and metal–noble gas systems and superconducting cuprates all follow the linear relation given by eq 3. This is truly remarkable.

## 5. Minimum Metallic Conductivity at the Metal–Nonmetal Transition

Mott[13] has argued that the M–NM transition in a perfect crystalline material at $T = 0$ K is discontinuous (Figure 8) and proposed that, at the transition, there exists a minimum conductivity, $\sigma_{min}$, for which the material could still be viewed as metallic, prior to the localization of electrons.[2] Mott's ideas were based on arguments developed earlier by Ioffe and Regel[37] for the breakdown of the theory of electronic conduction in semiconductors. The conventional Boltzmann transport theory becomes meaningless when the mean-free path, $l$, of the itinerant conduction electrons becomes comparable to, or less than, the interatomic spacing, $d$. The Ioffe–Regel mean free path, $l_{IR}$, at the minimum metallic conductivity is equal to $d$. Abrahams et al.[38] have, however, predicted a continuous M–NM transition on the basis of a scaling theory of noninteracting electrons in a disordered system,[39] and their results question the existence of $\sigma_{min}$ in both two and three dimensions.[32,39,40] The two possible scenarios of the transition are compared in Figure 11.

**5.1. The Situation in Doped Semiconductors.** There is an increasing belief amongst workers in the field that the M–NM transition is continuous, based on experimental measurements carried out at low temperatures down to 3 mK. In Figure 12, we show the experimental evidence in P-doped Si. Note that at a fixed (very low) temperature, the conductivity changes continuously with, for example, donor concentration. In addition, the extrapolated zero-temperature value of the conductivity ($\sigma(0)$) varies continuously with impurity concentration. An example showing the variation of the extrapolated "zero-temperature" conductivity[41] in the case of B-doped Si is

J. Phys. Chem., Vol. 99, No. 15, 1995 **5233**

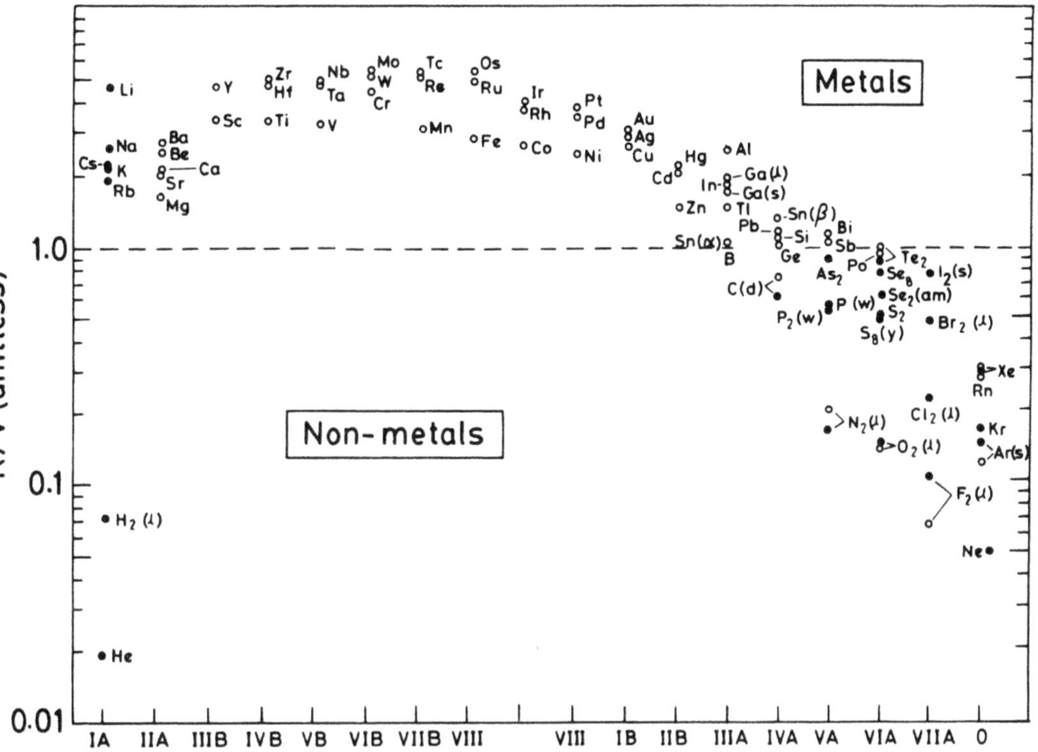


**Figure 6.** The occurrence of metallic vs nonmetallic character in the periodic table of the elements: the variation of $R/V$ for naturally occurring elements of the s, p, and d blocks (adapted from Edwards and Sienko[27]).

shown in Figure 13. This could be taken as strong experimental evidence for a continuous M–NM transition in doped semiconductors at $T = 0$ K. Möbius,[35,42] however, questions the reliability of such 0 K extrapolations and suggests that these findings do not disprove the existence of finite $\sigma_{min}$ at the transition. He argues that the data can be explained by a combination of $\sigma_{min}$ on the metallic side and a Coulomb interaction dominated $\sigma(T) \simeq \sigma_{min} \exp-\{(T_0/T)^{1/2}\}$ on the insulating side, with $T_0 \rightarrow 0$ as the disorder decreases to the critical value. As noticed by him,[35,42] not all the measurements are consistent with this suggestion; furthermore, the clear observation of characteristic weak localization or precursor effects in transport and magnetotransport weakens the argument for a $\sigma_{min}$. However, this analysis points out the fact that while the continuous conductivity transition prediction is for *noninteracting* disordered electrons, just on the insulating side the Coulomb interaction is by definition of long range and hence qualitatively important. It is possible that 3 mK may not be a low enough temperature since $k_BT$ may exceed the activation energy for conduction even at this temperature.

While the presence of a high degree of disorder may wipe out the discontinuous nature of the M–NM transition,[43,44] inclusion of strong correlation in scaling models could conversely change a continuous transition to a discontinuous one. In spite of such difficulties, however, $\sigma_{min}$ continues to be a useful experimental criterion[17,20,45] at least at the "high-temperature" limit. Earlier results providing experimental evidence for $\sigma_{min}$ have been reviewed by several authors. In Figure 14, we show some of the results. We note that $\sigma_{min}$ scales with $n_c$. As pointed out by Fritzsche,[46] $\sigma_{min}$ appears to satisfactorily represent the value of conductivity where the activation energy for conduction disappears. This aspect is specially borne out by investigations of transition metal oxide systems.

**5.2. The Situation in Transition Metal Oxides.** In many of the oxide systems,[15,20] especially those exhibiting compositionally controlled M–NM transitions (Figures 4 and 5), the temperature coefficient of the conductivity changes sign around $\sigma_{min}$ ($\sim 10^3$ ohm$^{-1}$ cm$^{-1}$). Most of these oxides, including the superconducting cuprates, follow the relation shown in Figure 14. The points corresponding to these oxides fall somewhere between those of fluid alkali metals and $La_{1-x}Sr_xVO_3$. The critical carrier concentrations in these materials from Figure 10 also give $\sigma_{min}$ values close to the observed values. Accordingly, $\sigma_{min}$ is often taken to represent the separation of localized and itinerant electron regimes.

Recent work of Raychaudhuri and co-workers[47,48] suggests the need to reevaluate the status of transition metal oxides with regard to $\sigma_{min}$. It appears that in these disordered oxide systems, genuine metallic states (with $\sigma(T = 0) \neq 0$) exist even when the conductivity is activated ($\sigma < \sigma_{min}$). This implies that the earlier values of the critical electron density at the transition in such oxides may be overestimates. In Figure 15, we show the behavior of $Na_xW_{1-y}Ta_yO_3$ where the temperature coefficient of the conductivity changes sign when $(x - y) \approx 0.20$, but the transition actually occurs at $(x - y) = 0.19$. The curves for $x = 0.35$ and 0.34 compositions both show "activated conductivity" at $T > 10$ K (i.e., a negative temperature coefficient of the resistivity), but the $\sigma(T)$ *saturates* at a fairly high residual value in the case of the $x = 0.35$ sample as expected of a metal ($\sigma(T = 0) \neq 0$). The $x = 0.34$ sample, however, shows a very much lower $\sigma$ value at low $T$, tending to infinity, and fits an activated

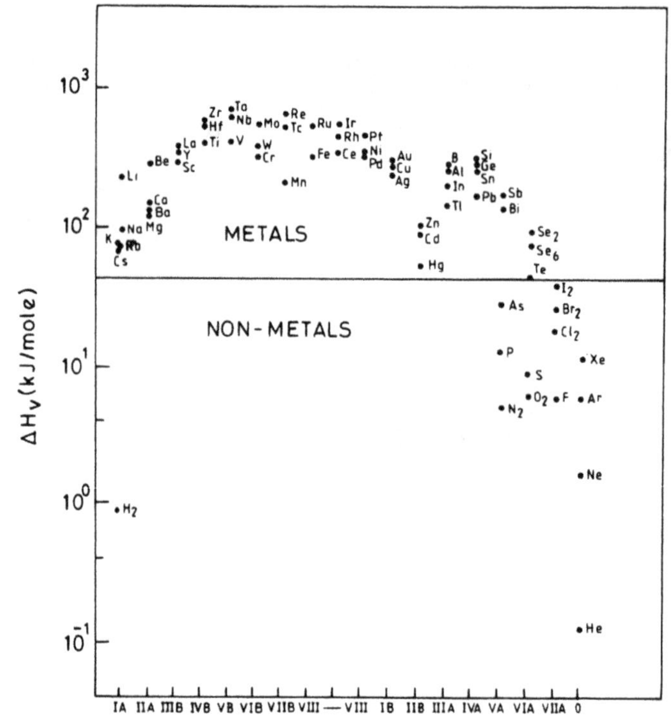


**Figure 7.** Plot of the latent heat of vaporization, $\Delta H_v$, for metals and nonmetals of the periodic table (from Rao and Ganguly[28]).

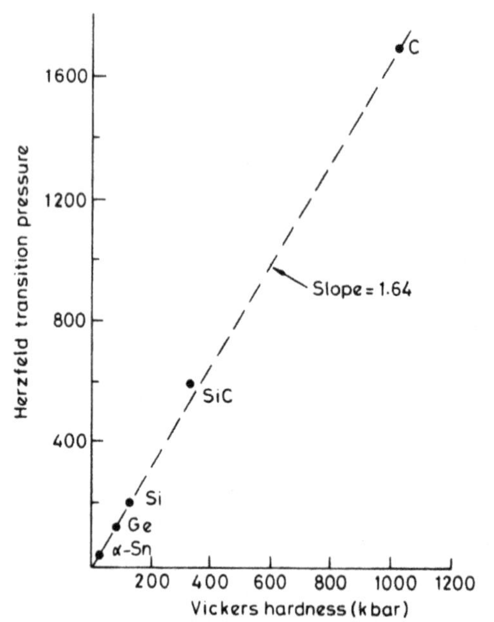


**Figure 8.** Correlation of the metallization pressure as calculated from the Herzfeld model and Vicker's indentation hardness number (from Gilman[29]).

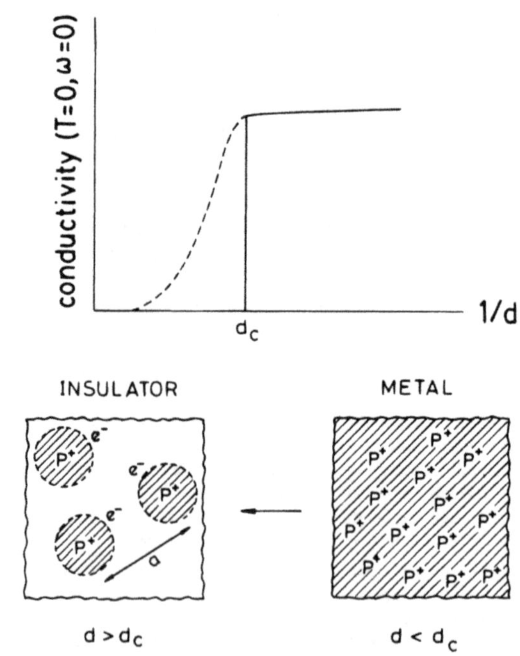


**Figure 9.** A schematic representation of the situation ($T = 0$ K) for P doped Si at both high and low donor densities. Also shown are two scenarios for the composition dependence of the electrical conductivity, showing the metal–nonmetal transition.

temperature dependence, with the behavior of a nonmetal ($\sigma(T = 0) = 0$). These results show that it is difficult to differentiate metallic *vs* nonmetallic behavior in oxides based on high-temperature measurements. What emerges, however, is that the "high-temperature" $\sigma_{min}$ value differentiates the (metallic *vs* nonmetallic) conductivity behavior in that temperature regime.

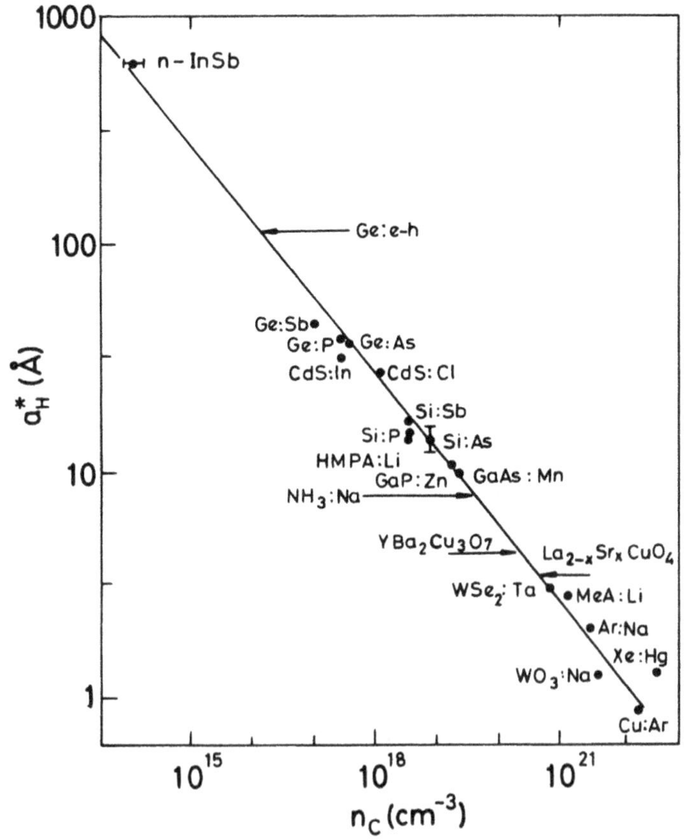


**Figure 10.** An experimental correlation of the critical density, $n_c$, with the Bohr radius, $a_H^*$. The solid line represents $n_c^{1/3}a_H^* = 0.26$ (from Edwards and Sienko[36] and Thomas[34]). Here, InSb refers to lightly doped n-InSb, Ge:e-h means germanium with photoexcited electrons and holes; Cu:Ar is a codeposited Cu and Ar mixture on a cold substrate.

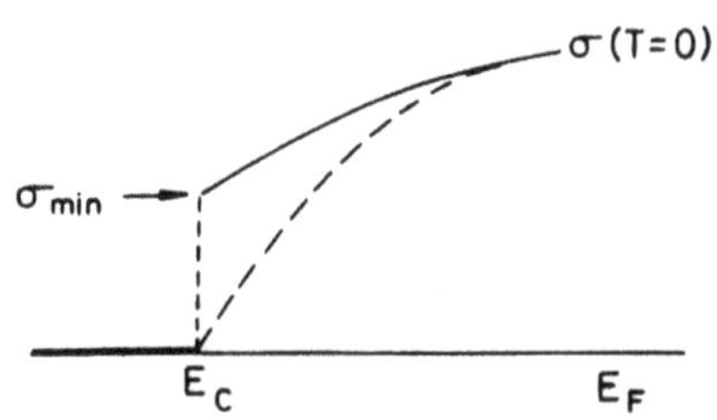


**Figure 11.** Schematic illustration of the two possibilities of a continuous or discontinuous metal–nonmetal transition at $T = 0$ K. The minimum metallic conductivity, $\sigma_{min}$, at the transition is also shown (from Lee and Ramakrishnan[39]). The conductivity at zero temperature (ordinate) and Fermi energy (abscissa) are shown. The discontinuous conductivity transition suggested by Mott is the full curve, with $\sigma_{min}$ occurring as $E_F$ crosses the mobility edge energy $E_c$. The dotted curve is the continuous conductivity transition predicted by the scaling theory.

## 6. Our Theoretical Understanding

The metal to nonmetal transition, a basic electronic change, has proven surprisingly difficult to understand in detail. The underlying reason is the diametrically opposite modes of description natural for the metal (extended electronic states) and for the insulator (localized states, often with local constraints on electron number). The observed diversity of systems and phenomena indicates that a number of causes may be at work, e.g., disorder, short-range electron correlation, long-range Coulomb interaction, and electron lattice coupling. The effects are often present simultaneously and interactively. Many reasonable scenarios as well as detailed theories of model systems do exist, but the distance between them and experiments is still large. In this section, we first look at the implications of the successful criteria for the M–NM transition. We then describe some theoretical models and approaches.

**6.1. Criteria for the Transition.** The three criteria discussed earlier are the dielectric catastrophe of Herzfeld, the Mott density criterion, and the idea (again due to Mott) that a metal is characterized by a minimum value of the electrical conductivity at the transition. The first two are related and consider how the Coulomb interaction and electron or atom density together can lead to the metal–insulator transition. The polarizability of an isolated atom depends on its size, being roughly proportional to its volume ($a_H^{*3}$). When such atoms are put together, the dielectric constant is enhanced because of local field effects, i.e., the large electric field at each atom due the collection of polarized atoms (or dipoles). At a density such that the polarizability proportional to $a_H^{*3}$ is comparable to the volume per atom, i.e., when $a_H^* n^{1/3} \simeq 1$, the dielectric constant diverges. It is difficult to develop a detailed theory along these lines, because at such high densities electron overlap effects are large. The electron energy level structure (band structure) and wave functions are very different from those at the atomic limit, so that the dielectric constant is not easily related to properties of isolated atoms. Further, disorder and lattice type can also have a large quantitative effect.

The Mott criterion, namely $a_H^* n^{1/3} \simeq 0.25$, is dimensionally identical to that of Herzfeld and is physically related. It considers the transition from the metallic side. If an electron–hole excitation is created in a metal, the attraction between them is screened, and they may not form a bound state unless the screened interaction is strong enough. This will happen at a low enough carrier density estimated by Mott to be such that $a_H^* n^{1/3} \lesssim 0.25$. The metal is then unstable against the formation of a bound electron–hole pair; a collection of these is an insulator or a nonmetal. Both the criteria thus point to the Coulomb interaction which is screened and effectively weak in the metal and is strong and long ranged in a nonmetal. A theory describing this crossover and its consequences does not exist. These approaches neglect electron spin and associated magnetic effects, *local* electron correlation, as well as disorder.

The minimum metallic conductivity criterion of Mott emphasizes the role of disorder. It was first realized by Anderson[49] that electronic states originally extended can become spatially localized if the disorder is strong enough. Mott argued that in a disordered metal, the minimum possible value of the mean free path is the interatomic spacing. For stronger disorder, the picture of electron waves scattered randomly is inappropriate; the electron state is localized, and one has a nonmetal. Thus, a minimum electrical conductivity, $\sigma_{min}$, characterizes the transition. The conductivity discontinuously jumps from this value to zero on the nonmetallic side. As mentioned earlier, this is a very good first approximation; very close to the critical disorder *and* at very low temperatures, $\sigma \to 0$ and values much less than $\sigma_{min}$ (but tending to a nonzero limit at $T = 0$) have been observed. A detailed theory for the disorder caused continuous M–NM transition exists.[38,39] The minimum metallic conductivity criterion emphasizes disorder and is silent about the effect of interactions. The Mott density criterion focuses on the Coulomb interaction and completely neglects disorder effects. The success of both these criteria, often in the same system, implies that both interactions and disorder are significant in real systems.

**6.2. Detailed Models.** *6.2.1. Anderson Localization.* It was pointed out by Anderson[49] that if the randomness in electronic state energies at different sites is large enough, electrons do not diffuse from one end of the system to another but are localized at potential fluctuations. Since for weak disorder electrons do diffuse, there must be a critical disorder at which an electron of a particular energy (say, the Fermi energy) does not diffuse, i.e., the dc electrical conductivity is zero. The energy separating the band of extended states from localized states is called the mobility edge. As disorder or electron density changes, the Fermi energy and the mobility edge coincide, and the system makes a transition from a metal to a nonmetal; this is the disorder-driven Anderson transition.

Predictions for physical properties near the transition were first made by Mott. He argued for a nonzero minimum conductivity of the metal at the transition point and for a divergence of localization length in the nonmetal with a consequent divergence in the dielectric constant.

A detailed theory of the transition, which identified singular backscattering as the mechanism of localization, was developed by Abrahams et al.[38] Quantum mechanically, an electron in a random medium can go from one point to another via infinitely many paths with appropriate amplitudes. Of these, self intersecting paths in which the loop is traversed in opposite directions have the same amplitude. Thus the probability (the square of the amplitude) for return to the origin is enhanced because of constructive interference. Nearby paths also contribute, depending on dimensionality. This process thus tends to localize electrons, depending on disorder and dimensionality. Detailed calculations show for example that the conductivity decreases from its value $\sigma$ at atomic length scales (of order the mean free path $l$) to smaller values at a larger length scale, $L$, on account of interference effects. In two dimensions, the decrease is logarithmic in $L$.

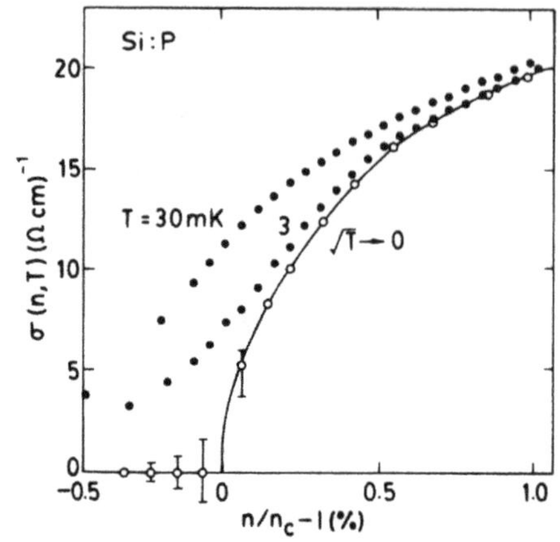


**Figure 12.** Electrical conductivity as a function of relative electron (donor) concentration in P doped Si to illustrate the metal–nonmetal transition at 0 K. The relative P density is varied by changing $n_c$ with uniaxial stress on a single sample. The open circles are extrapolated to $T = 0$ K (from Thomas[33]).

$$\sigma(L) = \sigma_0 - \left(\frac{e^2}{\hbar\pi^2}\right)\ln(L/l) \tag{4}$$

This means, from a scaling analysis, that for large enough systems, the conductivity (or conductance) is zero. There are no disordered *metal* films at absolute zero. In three dimensions, $\sigma(L)$ has the form

$$\sigma_{3d}(L) = \sigma_0 - \frac{e^2}{\hbar\pi^3}\left(\frac{1}{l} - \frac{1}{L}\right) \tag{5}$$

This implies for example that when $\sigma(l)$ has the Mott minimum value $(e^2/\hbar\pi^3 l)$, the dc or large length scale ($L \to \infty$) conductivity is zero. Thus, the conductivity goes continuously to zero at the Anderson transition, due to interference effects not envisaged in the Mott approximation. The exponent with which $\sigma$ vanishes is the same as that with which the localization length diverges.

The scaling picture, starting from weak localization due to interference, puts in perspective both the minimum metallic conductivity $\sigma_{min}$ and the possibility of a continuous conductivity transition. At nonzero temperatures, inelastic scattering limits the length scale $L$ to $L_{in} \sim T^{-\alpha}$ up to which interference effects take place. If $L_{in}$ is small, the dc conductivity even near critical disorder will be close to $\sigma_{min}$. As $T$ decreases, $L_{in}$ increases, the interference effects are stronger, and the conductivity decreases. This is considered "nonmetallic" behavior. However $\sigma(T = 0)$ is nonzero, so that one can naturally have strongly disordered metals with a negative temperature coefficient of the resistivity. The purely disorder driven transition is a continuous quantum ($T = 0$) transition. A number of predictions for quantities such as magnetoresistance, the Hall effect, and resistivity in the presence of magnetic and other scatterers have been made. These agree with experiment.[39,40] It is also realized that many of these phenomena are common to *all* waves in random media, e.g., light and sound waves.

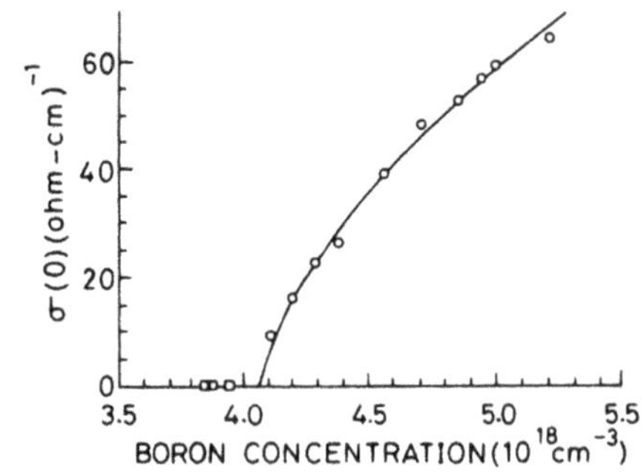


**Figure 13.** Zero-temperature conductivity $\sigma(T = 0)$ vs boron concentration in B doped Si. The circles represent metallic samples, while the three squares indicate samples which are nonmetallic (from Dai et al.[41]).

Real disordered electronic systems cannot be thought of as a collection of noninteracting particles moving independently in a random potential. With increasing disorder, electrons diffuse very slowly and so in effect interact strongly. The range of interaction increases until in the nonmetal it decreases only inversely with distance. It has proven very difficult to describe the passage of a disordered interacting electron gas to the insulating state. In the metallic regime, for effectively weak interaction, there are predictions for characteristic effects such as a dip in the density of states proportional to $(|E - E_F|)^{1/2}$ where $E$ is the energy and $E_F$ the Fermi energy. Scaling approaches to the transition starting from the perturbative regime show that there are several kinds of M–NM transitions, characterized by different critical behavior for density of states, conductivity, magnetic susceptibility, low-temperature specific heat, etc.[39,50] Unfortunately, the actual transition is often in the regime of strong coupling and so the perturbative scaling analysis is not reliable. Also, there is no theory which describes the Coulomb interaction realistically in the entire regime (weakly disordered metal–critical region–Coulomb glass). Experimentally, this transition has been recently explored in several systems.[47,48] The square root dip in the density of electronic states near the Fermi energy steepens, changes to a linear form, and becomes a soft quadratic gap just on the insulating side (i.e., the density of states goes as $(E - E_F)^2$). There are characteristic temperature and field dependences for the resistivity. These are not yet comprehensively understood.

*6.2.2. Correlation Effects.* The effects of electron correlation are strong in narrow band systems and can cause them to be insulating when they should be metallic according to band theory. Well-known examples are NiO and $La_2CuO_4$, which, with an odd number of electrons per site in the highest unfilled band, ought to be metallic. In the latter, for example, only the Cu–O configurations $d^9p^6$ and $d^{10}p^5$ are close in energy (the configuration $d^8p^6$ has very high energy). The former, containing one d hole per site, is more stable so that the system has a gap against transfer of one hole from Cu to O. This is a charge transfer insulator. An even simpler correlation driven one-band (d band) insulator requires the energy cost $U$ of local configurational changes involving sites $l$, $l + 1$, i.e., the change $(d_l^n - d_{l+1}^n$ to $d_l^{n-1} - d_{l+1}^{n+1})$ to be greater than the kinetic energy gain, KE, due to electron hopping. Thus, as $U$ decreases or increases (e.g., due to changes in physical conditions such as pressure) in relation to KE, the system can make a transition from insulator to metal. These ideas were extensively discussed by Mott. The one-band model of electrons on a lattice, with local correlations, is the well-known Hubbard model, and the correlation driven transition for one (or commensurate) electron per site, from a paramagnetic metal to a paramagnetic insulator is the Mott transition, mentioned earlier in section 4.

The Mott transition has proven difficult to describe in detail, though there are several plausible conclusions and results for spatial dimensionality $d = 1$ and $\infty$. On the metallic side close to the transition, for the case of one electron per site, the probability of two electrons being on the same site is small because of correlation. Using this as a variational parameter in a ground state wave function as was first proposed by Gutzwiller (see ref 40 for details) a number of conclusions have been drawn. As $U \rightarrow U_c$ (the critical correlation), electron hopping becomes infrequent so that the electron effective mass is large. Since each site has mostly one electron, local magnetic moments are very long lived, and the Pauli spin susceptibility is large, diverging as $U \rightarrow U_c$. On the insulating side, a gap (the Mott–Hubbard gap) opens up for charge excitations. This gap is different from that in a semiconductor. The former is

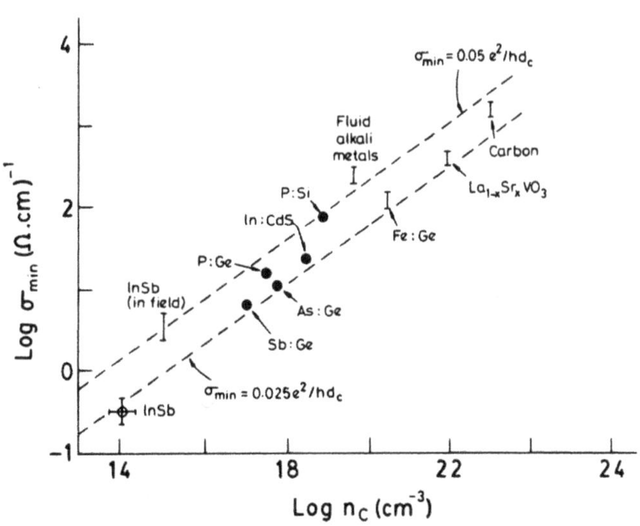


**Figure 14.** A plot of the minimum metallic conductivity, $\sigma_{min}$, against log $n_c$, the critical electron density at the metal–nonmetal transition. The two straight lines correspond to the values of the constant 0.05 and 0.026 in the equation relating $\sigma_{min}$ and $n_c^{1/3}$ (adapted from Mott[44] and Fritzche[46]).

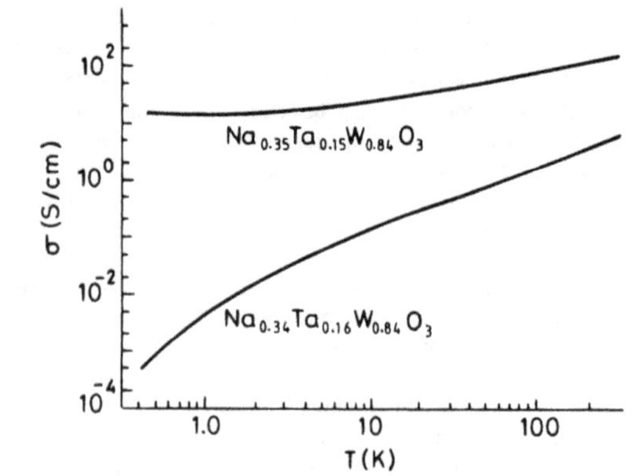


**Figure 15.** Electrical conductivity versus temperature for two samples of $Na_xTa_yW_{1-y}O_3$, one having metallic properties for $T \rightarrow 0$ (top curve) and one having nonmetallic status for $T \rightarrow 0$ (bottom curve). Notice the extremely small differences in concentration for the metallic and nonmetallic samples (from Raychaudhuri[47]).

due entirely to local correlations and the latter to periodicity. On the insulating side, local moments are well defined and interact antiferromagnetically. There is experimental evidence in metallic $V_2O_3$, near the M−NM transition, for mass enhancement, increased Pauli paramagnetism, a large $T^2$ term in the resistivity etc.[17−19,40]

Recently, there has been considerable progress in developing a theory of the Hubbard model in high spatial dimensions.[51] There are reasons to believe that the results can be usefully extrapolated to $d = 3$. The development of a Mott−Hubbard gap is due to the size and time dependence of the field on an electron at a given site. This field is calculated self-consistently. Some results are as follows: In the $(U,T)$ plane, there are three phases, metallic (PM), insulating paramagnetic (PI), and insulating antiferromagnetic (AFI). By suppressing the intermoment exchange coupling, the Mott transition (PM−PI) can be located and followed. It is discontinuous at $T \neq 0$ and is believed to become continuous as $T \rightarrow 0$. The physical properties of the heavy fermion metal for $U \lesssim U_c$ can also be calculated and agree with expectations. Electron−lattice coupling can drive the Mott transition discontinuous, with a $T \neq 0$ critical point.[52] Many of these results are similar to those found in $V_2O_3$.

The large $d$ approach developed so far concentrates exclusively on the single-site time-dependent electron potential. The other low-lying fluctuations (charge and spin) are integrated out so that we do not have a detailed picture of the quantum dynamics. There is little understanding of the effects of deviations from integer site occupation, i.e., of doping. This lacuna is serious because cuprate superconductors can be thought of as doped Mott insulators. The effect of disorder is also not known.

## 7. Concluding Remarks

The transition of a metal to an insulator is caused and accompanied by changes in the nature of chemical bonding as well as by changes in physical properties. The transition occurs in a wide variety of systems, as we have tried to indicate. In spite of this variety, simple criteria reflecting a basic competition between potential and kinetic energies have been remarkably successful in locating broad conditions under which such transitions may occur. A detailed understanding of transport, magnetic, optical, and other properties in the transition regime has proven difficult, since several causes seem to be at work simultaneously. However, a number of theoretical models valid for various limiting situations have given us valuable qualitative and quantitative insights.

The entire area has experienced a rebirth in the last few years, for several reasons. The most important perhaps is the realization that cuprate superconductors are not far from a largely correlation driven metal−insulator transition. Another is the appreciation that the transition is of wide occurrence, some recent examples being the superconductor−insulator, the quantum Hall fluid−insulator, and the stable quasicrystal metal−insulator transitions. Hopefully, the next few years should see considerable progress in the experimental and theoretical description of this basic electronic transition.

### References and Notes


(1) Edwards, P. P., Rao, C. N. R., Eds. *The Metallic and the Nonmetallic States of Matter*; Taylor and Francis: London, 1985.

(2) Mott, N. F. *Metal-Insulator Transitions*, 2nd ed.; Taylor and Francis: London, 1990.

(3) Rao, C. N. R.; Edwards, P. P. *Proc. Indian Acad. Sci. Chem. Sci.* **1986**, *96*, 473.

(4) Redmer, R. Thermodynamische und Transporteigenschafter dicter Alkali plasmen. Habilitationsschrift, Universitat Rostock, 1991.

(5) Ashcroft, N. W.; Mermin, N. D. *Solid State Physics*; Saunders College: Philadelphia, 1976.

(6) See articles of K. Kimura and P. P. Edwards in *Chemical Processes in Inorganic Materials: Metal and Semi-conductor Clusters and Colloids*; Persans, P. D., Bradley, J. S., Chianelli, R. R., Schmid, S., Eds.; *Mat. Res. Soc. Symp. Proc.* **1992**, *272*, 193, 311.

(7) Rosenblit, M.; Jortner, J. *J. Phys. Chem.* **1994**, *98*, 9365.

(8) Rao, C. N. R. *Philos. Trans. R. Soc. London* **1991**, *A336*, 595.

(9) Iye, Y. In *Physical Properties of High Temperature Superconductors*; Ginsberg, D. M., Ed.; World Scientific: Singapore, 1992; Vol. II.

(10) Logan, D. E.; Tusch, M. A. *J. Non-Cryst. Solids* **1993**, *156−158*, 639.

(11) Ramakrishnan, T. V. *J. Solid State Chem.* **1994**, *111*, 4.

(12) Mott, N. F. *Philos. Mag.* **1961**, *6*, 287.

(13) Mott, N. F. *Philos. Mag.* **1972**, *26*, 1015.

(14) Herzfeld, K. F. *Phys. Rev.* **1927**, *29*, 701.

(15) Rao, C. N. R. *Annu. Rev. Phys. Chem.* **1989**, *40*, 291.

(16) Honig, J. M.; Van Zandt, L. L. *Annu. Rev. Mater. Sci.* **1975**, *5*, 225. Honig, J. M.; Spalek, J. *Proc. Indian Natl. Sci. Acad.* **1986**, *A52*, 232.

(17) Carter, S. A.; Rosenbaum, T. F.; Honig, J. M.; Spalek, J. *Phys. Rev. Lett.* **1991**, *67*, 3440. Carter, S. A.; Rosenbaum, T. F.; Metcalf, P.; Honig, J. M.; Spalek, J. *Phys. Rev.* **1993**, *B48*, 16841.

(18) Bao, W.; Broholm, C.; Carter, S. A.; Rosenbaum, T. F.; Aeppli, G.; Trevino, S. F.; Metcalf, P.; Honig, J. M.; Spalek, J. *Phys. Rev. Lett.* **1993**, *71*, 766.

(19) Thomas, G. A.; et al. *Phys. Rev. Lett.* **1994**, *73*, 1529.

(20) Rao, C. N. R.; Ganguly, P. In *The Metallic and the Non-metallic States of Matter*; Edwards, P. P., Rao, C. N. R., Eds.; Taylor and Francis: London, 1985.

(21) Schlenker, C.; Dumas, J.; Escribe-Filippini, G. H.; Marcus, J.; Fourcandot, G. *Philos. Mag.* **1985**, *B52*, 643.

(22) Vijayakrishnan, V.; Chainani, A.; Sarma, D. D.; Rao, C. N. R. *J. Phys. Chem.* **1992**, *96*, 8679.

(23) Rao, C. N. R.; Vijayakrishnan, V.; Aiyer, H. N.; Kulkarni, G. U.; Subbanna, G. N. *J. Phys. Chem.* **1993**, *97*, 11157. Haberland, H.; von Issendorff, B.; Yufeng, Ji; Kolar, T. *Phys. Rev. Lett.* **1992**, *69*, 3212.

(24) Edwards, P. P.; Anderson, P. A.; Armstrong, A. R.; Slaski, M.; Woodall, L. J. *Chem. Soc. Rev.* **1993**, *22*, 305.

(25) Goldhammer, D. A. *Dispersion via Absorption des Lichtes*; Teubner: Leipzig, 1911.

(26) Sommerfeld, A. In *Gesammelten Schriften*; Vieweg: Braunschweig, 1968; Vol. 4, as cited by Ehrenreich, H. E. *Science* **1987**, *235*, 1029.

(27) Edwards, P. P.; Sienko, M. J. *J. Chem. Educ.* **1983**, *60*, 691.

(28) Rao, C. N. R.; Ganguly, P. *Solid State Commun.* **1986**, *57*, 5.

(29) Gilman, J. J. *J. Mater. Res.* **1992**, *7*, 535; see also *Philos. Mag.* **1993**, *67*, 207.

(30) Fujii, Y.; Hase, K.; Ohishi, Y.; Fujihisa, H.; Hamaya, N. H.; Takemura, K.; Shimomura, O.; Kikegawa, T.; Amemiya, Y.; Matsushita, T. *Phys. Rev. Lett.* **1989**, *63*, 536.

(31) Balchan, A. S.; Drickamer, H. G. *J. Chem. Phys.* **1961**, *34*, 1948.

(32) For a review see: Milligan, R. F.; Thomas, G. A. *Annu. Rev. Phys. Chem.* **1985**, *36*, 139.

(33) Thomas, G. A. *J. Phys. Chem.* **1983**, *88*, 3749.

(34) Thomas, G. A. In *Proceedings of the 39th Scottish Universities Summer School in Physics*; Tunstall, D. P., Barford, W., Eds.; Adam Hilger: Edinburgh, 1991.

(35) Möbius, A. *J. Phys. C: Solid State Phys.* **1985**, *18*, 4639.
(36) Edwards, P. P.; Sienko, M. J. *Phys. Rev.* **1978**, *B17*, 2575.
(37) Ioffe, A. F.; Regel, A. R. *Prog. Semicond.* **1960**, *4*, 237.
(38) Abrahams, E.; Anderson, P. W.; Liccardello, D. C.; Ramakrishnan, T. V. *Phys. Rev. Lett.* **1979**, *42*, 693.
(39) Lee, P. A.; Ramakrishnan, T. V. *Rev. Mod. Phys.* **1985**, *57*, 287.
(40) Ramakrishnan, T. V. In *The Metallic and Non-metallic States of Matter*; Edwards, P. P., Rao, C. N. R., Eds.; Taylor and Francis: London, 1985.
(41) Dai, P.; Zhang, Y.; Sarachik, M. P. *Phys. Rev. Lett.* **1991**, *66*, 1914.
(42) Möbius, A. *Phys. Rev.* **1989**, *B40*, 4194.
(43) Mott, N. F. *Philos. Mag.* **1978**, *B37*, 377.
(44) Mott, N. F. *Proc. R. Soc.* **1982**, *A382*, 1.
(45) Edwards, P. P.; Sienko, M. J. *Int. Rev. Phys. Chem.* **1983**, *3*, 83.
(46) Fritzche, H. In *The Metal-Nonmetal Transitions in Disordered Systems*; Friedman, L. R., Tunstall, D. P., Eds.; Scottish Universities Summer School in Physics, 1978.
(47) Raychaudhuri, A. K. *Phys. Rev.* **1991**, *B44*, 8572.
(48) Raychaudhuri, A. K.; Rajeev, K. P.; Srikanth, H.; Mahendiran, R. *Physica B* **1994**, *197*, 124.
(49) Anderson, P. W. *Phys. Rev.* **1958**, *109*, 1492.
(50) Belitz, D.; Kirkpactrick, T. R. *Rev. Mod. Phys.* **1994**, *66*, 261.
(51) Georges, A.; Krauth, W. *Phys. Rev. B* **1993**, *48*, 7167. Rozenberg, M. J.; Zhang, X. Y.; Kotliar, G. *Phys. Rev. B* **1994**, *49*, 10181.
(52) Majumdar, P.; Krishnamurthy, H. R. *Phys. Rev. Lett.* **1994**, *73*, 1525.

JP9428496

# III. TRANSITION METAL OXIDES

*Commentary by C.N.R. Rao*

Transition metal oxides constitute probably one of the most important categories of solids, exhibiting a variety of structures and properties.[1,2] The nature of metal-oxygen bonding can vary between nearly ionic to highly covalent or metallic. The unusual properties of transition metal oxides are clearly due to the unique nature of the outer *d*-electrons. The phenomenal range of electronic and magnetic properties exhibited by transition metal oxides is noteworthy. Thus, we find oxides with metallic properties (e.g. $RuO_2$, $ReO_3$, $LaNiO_3$) at one end and insulating oxides (e.g. MnO, $BaTiO_3$) at the other. There are also oxides that traverse both these regimes with change in temperature, pressure, or composition (e.g. $V_2O_3$, $La_{1-x}Sr_xCoO_3$). Interesting electronic properties also arise from charge density waves (e.g. $K_{0.3}MoO_3$), charge ordering (e.g. $Fe_3O_4$), and defect ordering (e.g. $Ca_2Fe_2O_5$). Oxides with diverse magnetic properties anywhere from ferromagnetism (e.g. $CrO_2$, $La_{0.5}Sr_{0.5}MnO_3$) to antiferromagnetism (e.g. NiO, $LaCrO_3$) are known. Many oxides possess switchable orientation states, as in ferroelectric (e.g. $BaTiO_3$, $KNbO_3$) and ferroelastic (e.g. $Gd_2(MoO_4)_3$) materials. No discovery in solid state science has created as much sensation as that of high-temperature superconductivity in cuprates. Although superconductivity in transition metal oxides has been known for some time, the highest $T_c$ reached was around 13K; we now have oxides with $T_c$'s approaching 160K. The discovery of high $T_c$ oxides has focused worldwide attention on the chemistry of metal oxides and at the same time revealed the inadequacy in our understanding of these fascinating materials. This section has a few representative articles on metal oxides, covering structure and properties, as well as high-temperature superconductivity.[3,4]

An important new addition to the fascinating phenomena exhibited by transition metal oxides is colossal magnetoresistance (CMR), discovered in 1993 in rare earth manganates of the formula $Ln_{1-x}A_xMnO_3$ (Ln = rare earth, A = Alkaline earth or other divalent ions).[5] These manganates containing Mn in the +3 and +4 states become ferromagnetic and metallic due to the double exchange mechanism of electron hopping. At the ferromagnetic $T_c$, they show a large decrease in resistance when applying magnetic fields. The $T_c$ and other properties of the manganates are sensitive to the average radius of the A-site cations, $\langle r_A \rangle$, as well as size disorder (mismatch). Besides CMR, the manganates undergo charge-ordering of the $Mn^{3+}$ and $Mn^{4+}$ ions depending on the $\langle r_A \rangle$, small values favoring charge-ordering.[5]

Charge-ordering competes with double exchange and gives rise to anti-ferromagnetic insulating states. Besides the ordering of charges and spins, the ordering of the $e_g$ orbitals of $Mn^{3+}$ ions in the rare earth manganates plays a significant role. In this section, articles dealing with CMR, charge-ordering and related aspects of rare earth manganates are presented. Other interesting aspects such as electron hole asymmetry and electronic phase separation are also examined in the papers. Electronic phase separation in solids is a phenomenon of importance being understood more clearly in the last few years.[6] It is likely that many transition metal oxides with correlated electrons will exhibit phase separation.

Multiferroics and magnetoelectrics constitute an important area of materials research.[7,8] Multiferroics are materials which are magnetic as well as ferroelectric. In principle, these two properties cannot occur together since one requires the presence of *d*-electrons and the other does not. Novel strategies to attain multiferroic and magnetoelectric properties in oxides are presented in one of the articles.

## References


1. C.N.R. Rao, *Ann. Rev. Phys. Chem.* **40**, 291 (1989).
2. C.N.R. Rao and B. Raveau, *Transition Metal Oxides*, Wiley-VCH, Second Edition, 1999.
3. T.V. Ramakrishnan and C.N.R. Rao, *Superconductivity Today*, Second Edition, University Press, Hyderabad, India, 1999.
4. C.N.R. Rao (Ed.), *Chemistry of High Temperature Superconductors*, World Scientific, Singapore, 1991.
5. C.N.R. Rao and B. Raveau (Eds.), *Colossal Magnetoresistance, Charge Ordering and Related Properties of Manganese Oxides*, World Scientific, Singapore, 1998.
6. V.B. Shenoy and C.N.R. Rao, *Phil. Trans. Royal. Soc. London*, **A366**, 63 (2008).
7. C.N.R. Rao and C.R. Serrao, *J. Mater. Chem.* **17**, 4931 (2008).
8. C.N.R. Rao, A. Sundaresan and R. Saha, *J. Phys. Chem. Lett.* **3**, 2237 (2012).

64 J. Chem. Soc. (A), 1970

# Thermodynamics and Kinetics of Oxidation of $Pr_2O_3$ and $Tb_2O_3$ to Non-stoicheiometric Oxides

By **S. Ramdas, K. C. Patil,** and **C. N. R. Rao,*** Department of Chemistry, Indian Institute of Technology, Kanpur, India

Isobaric gravimetric studies of $PrO_x$ and $TbO_x$ systems have shown the equilibrium existence of several non-stoicheiometric phases; the transformation temperatures found by differential thermal analysis agree well with the data from gravimetric studies. Oxidation reactions of $Pr_2O_3$ and $Tb_2O_3$ to non-stoicheiometric oxides generally show anomalously high entropy changes (*ca.* −25 e.u./mole of $O_2$) compared with the value of *ca.* −44 e.u./mole of $O_2$ generally found in oxidation of metals. The anomalous entropy changes are shown to arise from configurational factors. Cubic $Pr_2O_3$ with the defect structure is more readily oxidized than the hexagonal $Pr_2O_3$. Kinetic studies show that oxidation of $Pr_2O_3$ is a phase-boundary-controlled reaction.

THE phase equilibria and phase reactions in a few $LnO_x$-$O_2$ (Ln = Ce, Pr, and Tb) systems have revealed[1-7] a homologous series of oxides of the formula $LnO_{2n-2}$-($n \geqslant 4$). We have obtained some data[5-7] on the phase equilibria in the $PrO_x$ and $TbO_x$ systems from isobaric gravimetry and d.t.a. under different partial pressures of oxygen. Most of our results are in general agreement with those of others.[2-4] We are interested in the anomalous entropy of oxidation reported by Honig and his co-workers[2] for reaction (1). When $\nu = 0{\cdot}214$,

$$PrO_{1\cdot5} + \frac{\nu}{2}O_2 \longrightarrow PrO_{1\cdot5+\nu} \qquad (1)$$

$\Delta H - \Delta H^\circ$ was −43 kcal./mole of $O_2$ and $\Delta S - \Delta S^\circ$ was 28 e.u./mole of $O_2$. This value of $\Delta S - \Delta S^\circ$ is comparable with the entropy change of 25 e.u. found[8] for reaction (1) with $\nu = 0{\cdot}333$ based on e.m.f. results; thus, $-\Delta S^\circ$ ($\gtrsim \Delta S - \Delta S^\circ$) has nearly the same value for $\nu = 0{\cdot}214$ and $0{\cdot}333$. On the other hand, $\Delta S^\circ$ for removal of one mole of $O_2$ from the gas phase is known to be *ca.* −49 e.u.; indeed, in the oxidation of most metals, $\Delta S^\circ$ varies between −40 and −44 e.u./mole of $O_2$.[9] We have therefore examined $\Delta S - \Delta S^\circ$ for the oxidation of $Pr_2O_3$ and $Tb_2O_3$ to various non-stoicheiometric oxides employing the available isothermal and isobaric gravimetric data.

Sastry, Mehrotra, and Rao[6] reported that cubic (C type, $\phi$) $Pr_2O_3$ is oxidized at a lower temperature than the hexagonal (A type, $\theta$)$Pr_2O_3$. The $\theta$ phase is a close-packed structure without any anion vacancy while the $\phi$ phase has the defect fluorite structure of $PrO_2$ with a quarter of the anion positions vacant. We have compared the oxidation of the $\phi$ and $\theta$ phases of $Pr_2O_3$ employing thermogravimetry and d.t.a. Of special

[1] D. J. M. Bevan and J. Kordis, *J. Inorg. Nuclear Chem.*, 1964, **26**, 1509.
[2] J. M. Honig, A. F. Clifford, and P. A. Faeth, *Inorg. Chem.*, 1963, **2**, 791.
[3] B. G. Hyde, D. J. M. Bevan, and L. Eyring, *Phil. Trans.*, 1966, **259**, 583.
[4] B. G. Hyde and L. Eyring, in ' Rare Earth Research, vol. 3,' ed. by L. Eyring, Gordon and Breach, New York, 1965.
[5] R. L. N. Sastry, P. N. Mehrotra, and C. N. R. Rao, *J. Inorg. Nuclear Chem.*, 1966, **28**, 2167.
[6] R. L. N. Sastry, P. N. Mehrotra, and C. N. R. Rao, *Trans. Faraday Soc.*, 1966, **62**, 1632.
[7] G. V. Chandrashekhar, P. N. Mehrotra, G. V. Subba Rao, E. C. Subbarao, and C. N. R. Rao, *Trans. Faraday Soc.*, 1967, **63**, 1295.
[8] A. Kuznetsov and T. N. Rezukhina, *Zhur. fiz. Khim.*, 1962, **36**, 1364.
[9] A. Glassner, 'The Thermochemical Properties of the Oxides, Fluorides, and Chlorides to 2500°K,' ANL 5750, Argonne National Laboratory, Lamont, Ill.

interest were the kinetics and mechanism of oxidation of $\theta$ and $\phi$ phases to the $\beta$ phase ($PrO_{1\cdot833}$).

RESULTS AND DISCUSSION

*Phase Equilibria of* $PrO_x$ *and* $TbO_x$ *Systems.*—Our isobaric oxidation studies on the $PrO_x$ system starting with the $\theta$ phase showed clear evidence for the formation of the stable ordered equilibrium phases, $PrO_{1\cdot714}$ ($\iota$), $PrO_{1\cdot78}$ ($\zeta$), and $PrO_{1\cdot833}$ ($\beta$) as well as the disordered non-stoicheiometric phases $PrO_{1\cdot67}$ ($\sigma^m$) and $PrO_{1\cdot74}$ ($\alpha^m$). These results agree with those of Honig *et al.*[2] and Hyde *et al.*[3] and indicate the equilibrium existence of widely non-stoicheiometric phases.

From our isobaric gravimetric data on the $PrO_x$ system, the enthalpy ($\Delta H - \Delta H^\circ$) for the oxidation of $\theta$ to the $\iota$ phase was calculated to be $-48 \pm 5$ kcal./mole of $O_2$. This value is close to the value reported by Honig and his co-workers [2] based on their isothermal gravimetric data. This correspondence between the isothermal and isobaric data indicates that the measurements correspond to equilibrium conditions; this also confirms that the $PrO_{1\cdot5}$ used by Honig and his co-workers were of A-type.

We find that in the $PrO_x$ system, plots of $\Delta G$ [$= \boldsymbol{R}T \ln p(O_2)$] against composition show a much sharper break when $x = 1{\cdot}714$ than at $x = 1{\cdot}67$, $1{\cdot}78$, $1{\cdot}82$, or $1{\cdot}833$. This may imply that the transformation to the $\iota$ phase more closely corresponds to an equilibrium situation than the transformation to $\sigma^m$, $\alpha^m$, $\delta$, or $\beta$ phases. There is definite evidence for greater disorder in the $\sigma^m$ and $\alpha^m$ phases than in the $\iota$ phase. We suspect that in the $\delta$ and $\beta$ phases also there may be appreciable disorder compared with $\iota$, even though these phases show distinguishable $X$-ray diffraction patterns. Even the ordered intermediate phases show narrow homogeneity ranges indicating the presence of some disorder; [3] such disorder would be difficult to detect by $X$-ray diffraction. In all these oxides it is possible that there is substantial ordering of the vacancies in microdomains, the domains themselves being distributed randomly.[3] The $\sigma^m$ and $\alpha^m$ phases are probably highly disordered non-stoicheiometric phases while the $\iota$ phase is an ordered phase. The $\iota$ phase has a unique crystal structure (rhombohedral) where there is ordering of the anions as well as the cations.

In the oxidation of C-type ($\phi$) $Tb_2O_3$ we find evidence for the formation of $TbO_{1\cdot61}$ ($\sigma^m$), $TbO_{1\cdot67}$ ($\sigma^m$), $TbO_{1\cdot75}$ ($\alpha^m$), and $TbO_{1\cdot81}$ ($\delta$) phases; the $TbO_{1\cdot61}$ and $TbO_{1\cdot75}$ phases are clearly seen even in the gravimetric curve in dry air at atmospheric pressure [$p(O_2) = 138$ mm. Hg]. The composition $TbO_{1\cdot75}$ corresponds to the well-known $Tb_4O_7$ which is the normally stated composition of stable terbia. Hyde and Eyring [4] have noticed this phase to occur at relatively low oxygen pressures.

We have carried out the d.t.a. of $PrO_2$, $PrO_{1\cdot833}$, $PrO_{1\cdot71}$, $PrO_{1\cdot5}$, $TbO_2$, $TbO_{1\cdot81}$, and $TbO_{1\cdot5}$ from room temperature to *ca.* 1000° under different oxygen partial pressures. Contrary to earlier conclusions,[3] in all cases studied the d.t.a. temperatures for the oxidation or the decomposition agree well with the data from gravimetric studies [2-4] as seen in Table 1. In the d.t.a. curve of $PrO_{1\cdot833}$ we have found a peak around 450° due to disorder–order transformation besides the three prominent peaks arising from decomposition to the $\alpha^m$, $\iota$, and $\theta$ phases.

TABLE 1

D.t.a. temperatures

| $P(O_2)$ (mm.) | |
|---|---|
| 138 | $PrO_2$—360°→$\beta$—750°→$\zeta$—985→$\theta$ |
| 0·5 | $\phi$—410°→$\beta$—560°→$\alpha^m$—820°→$\zeta$—860°→$\theta$ |

*Entropy Changes for the Oxidation of* $Pr_2O_3$ *and* $Tb_2O_3$.—Honig and his co-workers [2] obtained $\Delta S - \Delta S^\circ$ for reaction (1) with $\nu = 0{\cdot}214$ using the thermodynamic relation (2), where $p$ is the equilibrium pressure of

$$(2{\cdot}303\boldsymbol{R}T \times \nu/2) \log p = T\,(\Delta S - \Delta S^\circ) + \log \frac{a_{1\cdot5+\nu}}{a_{1\cdot5}} - (\Delta H - \Delta H^\circ) \quad (2)$$

oxygen and $a$ is the activity of the phase. We have used Faeth's isothermal gravimetric data [10] for the other stable as well as the metastable phases of $PrO_x$ and obtained the value of $\Delta S - \Delta S^\circ$ from the plots of $\Delta G$ against temperature for each composition. The $\Delta S - \Delta S^\circ$ values thus calculated are given in Table 2. For

TABLE 2

Entropies of formation for the Oxidation

$$PrO_{1\cdot5} + \frac{\nu}{2}O_2 \longrightarrow PrO_{1\cdot5+\nu}$$

| $1{\cdot}5 + \nu$ | Final phase [a] | $(\Delta S - \Delta S^\circ)$ [b] (e.u. per mole of $O_2$) |
|---|---|---|
| 1·61 | $\sigma^m$ | 25 |
| 1·714 | $\iota$ | 21 [c] |
| 1·72 | $\alpha^m$ | 50 |
| 1·75 | $\alpha^m$ | 47 |
| 1·78 | $\alpha^m$ | 50 |
| 1·82 | $\delta$ | 34 |
| 1·833 | $\beta$ | 26 |

[a] We have used the nomenclature of Hyde, Bevan, and Eyring.[3] [b] $\Delta S^\circ$ will be negative since $-\Delta S^\circ \sim (\Delta S - \Delta S^\circ)$. [c] Honig and his co-workers [2] have reported this value to be 29 e.u./mole of $O_2$.

the stable phases $\Delta S - \Delta S^\circ$ is *ca.* 25 e.u./mole of $O_2$, a value much higher than the value of $\Delta S^\circ_{cxi}$ ($-44$ e.u./mole of $O_2$) generally found for the oxidation of metals.

The anomalously high $\Delta S^\circ$ ($\approx -[\Delta S - \Delta S^\circ]$) found in the oxidation of $Pr_2O_3$ to the different non-stoicheiometric oxides needs an explanation. This difference is much too large to be due to the $\phi$–$\theta$ phase transition of $Pr_2O_3$.[6] Further, our gravimetric studies clearly show that the $Pr_2O_3$ employed by both Honig and his co-workers [2] and Bevan, Hyde, and Eyring [3] was the $\theta$ phase.

We feel that the major contributions to the high $\Delta S^\circ$ value found in the oxidation of the sesquioxide to the non-stoicheiometric oxides may be due to configurational

[10] P. A. Faeth, Ph.D. Thesis, Purdue University, 1961.

factors. Since the incorporation of oxygen in the lattice of $Pr_2O_3$ alters the vacancy concentration and also causes drastic changes in the electrical properties,[7] we shall have entropy contributions from electronic disorder ($\Delta S^{el}$) as well as from vacancies ($\Delta S^{v}$). The variations of $\Delta S^{el}$ and $\Delta S^{v}$ for reaction (1) with the composition of $PrO_x$ are shown in Figure 1. $\Delta S^{el}$ is positive for all values of $x$ in $PrO_x$ and reaches the maximum for $x =$ 1·75. The value of $\Delta S^{v}$, however, depends on whether the initial reactant is the θ or the ϕ phase, since the θ phase has no vacancy in the structure. The initial

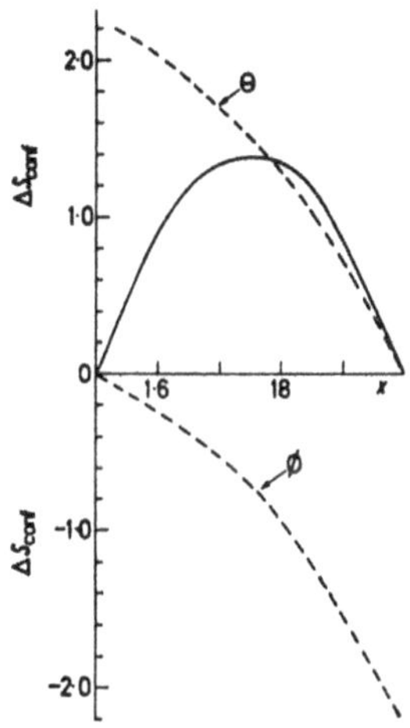


FIGURE 1 Configurational entropy changes for reaction (1) as a function of $x$ in $PrO_x$: Full line, $\Delta S_{conf}$ for electronic disorder $\Delta S^{el}$; dotted line, $\Delta S_{conf}$ for vacancies, $\Delta S^{v}$. If the initial reactant is θ-$Pr_2O_3$, the $\Delta S^{v}_{conf}$ will be positive for $x > 0$ but will become zero for $x = 0$. For the sake of convenience $\Delta S_{conf}$ values in the Figure are given in e.u./mole of Pr. To obtain $\Delta S_{conf}$ in units of e.u./mole of $O_2$ multiply the values in the Figure by $2/\nu_2$

reactant in all the studies of the phase equilibria of the $PrO_x$ system has been the θ phase and thus $\Delta S^{v}$ will be positive for all values of $x > 0$. Figure 1 assumes a random distribution of vacancies in the product phase; this is not completely correct because vacancy ordering in the product phases tends to decrease the magnitude of $\Delta S^{v}$. Further, electronic disorder and vacancy disorder may not be independent as assumed presently. The +3 and +4 ions probably show some site preference, thereby bringing down $\Delta S^{el}$. Thus, in the oxidation of θ to the highly ordered ι phase, the contribution to $\Delta S_{conf}$ may be essentially from $\Delta S^{el}$ alone; the observed $\Delta S°$ may thus be looked up as $\Delta S°_{oxi} + \Delta S^{el} \approx -44 + 13 = -31$ e.u./mole of $O_2$. The situation is similar for $PrO_{1\cdot833}$ (β), $PrO_{1\cdot82}$ (δ), and $PrO_{1\cdot61}$ ($\sigma^m$) phases; in the last case particularly, the agreement between the calculated $\Delta S°$ and the observed $\Delta S°$ is found only if we ignore the contribution from $\Delta S^{v}$. Apparently, there is appreciable vacancy ordering in all these phases. We therefore feel that at least for the oxidation of θ to the ι, β, δ, and $\sigma^m$ phases, the anomalous entropy changes can be understood in terms of electronic disorder in these phases compared with θ.

The situation is somewhat complex for the oxidation of θ to $PrO_{1\cdot75}$ ($\alpha^m$) and $PrO_{1\cdot78}$ ($\alpha^m$) phases, where the observed $\Delta S - \Delta S°$ values are close to the normal value of $\Delta S°_{oxi}$. If we were to consider these $\alpha^m$ phases to be formed through the ι intermediate, then $\Delta S°$ for the initial oxidation of ι to $\alpha^m$, involving only a small perturbation of the structure, should probably be close to the normal value of −44 e.u./mole of $O_2$. We would similarly expect to find normal $\Delta S°$ values for the initial oxidation of the ϕ phase as well. Our analysis of Faeth's data [10] in the region $PrO_{1\cdot5}$—$PrO_{1\cdot58}$ seems to bear evidence of this conclusion, but unfortunately the data were not too reliable in this region. The data on the initial oxidation of ϕ-$Tb_2O_3$, however, gives the normal value of $\Delta S°$ as will be seen later.

From the analysis of isobaric gravimetric data of Hyde and Eyring [4] on the $TbO_x$ system, we have calculated $\Delta S - \Delta S°$ for a few reactions as shown in Table 3.

TABLE 3

| Reaction | $\Delta S - \Delta S°$/mole of $O_2$ |
|---|---|
| Initial oxidation of $TbO_{1\cdot714}$(ι) | 54 |
| Initial reduction of $TbO_{1\cdot714}$(ι) | 31 |
| Oxidation or reduction of $TbO_{1\cdot719}$(ι′) | 37 |
| Initial oxidation of $TbO_{1\cdot5}$(ϕ) | 40 |

It can be seen that the reduction of ι is associated with a much higher $\Delta S°$ value than for the oxidation of ι. Further, the $\Delta S°$ for the initial oxidation of ϕ is near the normal $\Delta S°_{oxi}$. From the present studies, we feel that the anomalously high entropies of oxidation ($\Delta S°$) are generally associated with the formation of grossly non-stoicheiometric oxides from ordered initial phases. It may be noted that $\Delta S°$ for the oxidations involving wüstite [11] and $Nb_2O_{5-\nu}$ [12] are also large (*ca.* −30 e.u./mole of $O_2$).

*Comparative Study of the Oxidation of the* θ *and* ϕ *Phases of* $Pr_2O_3$.—Sastry, Mehrotra, and Rao [6] have studied the oxidation of the ϕ- and θ-$Pr_2O_3$ to $PrO_{1\cdot833}$ (β) at different partial pressures of oxygen and have reported d.t.a. evidence for the existence of two stages of oxidation of $Pr_2O_3$ at $p(O_2) =$ *ca.* 138 mm. Hg. We have repeated these experiments and have again found these two stages in the d.t.a. curve; however, we do not see these two stages in the isobaric gravimetric curve (Figure 2). When $p(O_2)$ was 0·1 mm. Hg., both the gravimetric and the d.t.a. curves showed only one stage in the oxidation to β. Unlike $Pr_2O_3$, $Tb_2O_3$ shows two stages of oxidation in the gravimetric curve at $p(O_2) = 138$ mm. Hg., but one broad peak in the d.t.a. curve (Figure 3). The observation of different stages of oxidation in the d.t.a. curves may be determined by kinetic factors; since d.t.a. is a dynamic technique kinetic factors would be important. The present results are in agreement with the

[11] E. G. King and K. K. Kelley, U.S. Bur. Mines, Rept. Invest., no. 1952, 4870.

[12] P. Kofstad and P. B. Anderson, *J. Phys. Chem. Solids*, 1961, **21**, 280.

conclusions reached earlier regarding the higher reactivity of the $\phi$ phase than the $\theta$ phase.

We have estimated the activation energies, $E_a$, for the oxidations from the gravimetric curves recorded at constant heating rates by employing the method of Dave and Chopra.[13] The $E_a$ value for the oxidation of $\phi$ to $\beta$ at $p(O_2)$ 138 mm. Hg is *ca.* 24 ± 8 kcal./mole in the temperature range 200—300°; for the oxidation of $\theta$,

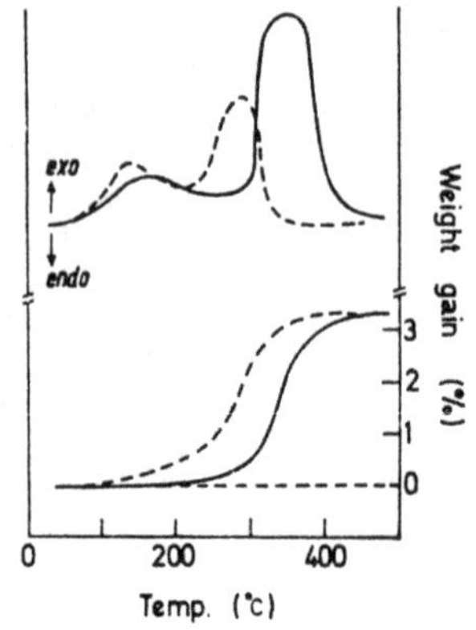


FIGURE 2 Gravimetric and d.t.a. curves for the oxidation of $Pr_2O_3$ to $PrO_{1\cdot833}$: broken line, $\phi$; full line, $\theta$; $p(O_2)$ = 138 mm. Hg. Rates of heating were 4° min.$^{-1}$

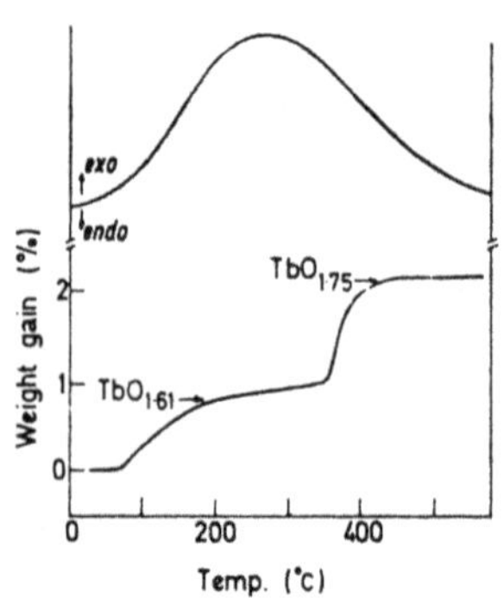


FIGURE 3 Gravimetric and d.t.a. curves for the oxidation of $\phi$-$Tb_2O_3$[$p(O_2)$ = 138 mm. Hg.]. Rates of heating were 4° min.$^{-1}$

the value is 43 ± 12 kcal./mole in the temperature range 275—375°. These $E_a$ values are in the right direction since it should be relatively easier to oxidize the defect $\phi$ phase than the close-packed $\theta$ phase. Anderson and Gallagher [14] have found similar differences in the chemisorption of $O_2$ on the $\phi$ and $\theta$ phases.

We have also obtained the $E_a$ values given in Table 4 for the oxidation of $\theta$-$Pr_2O_3$ to other stages from the gravimetric data. The $E_a$ values for the oxidation of $\phi$-$TbO_{1\cdot5}$ to $TbO_{1\cdot61}$ and $TbO_{1\cdot75}$ (Figure 3) are *ca.* 8 and 5 kcal./mole respectively at an oxygen partial pressure

TABLE 4

| Final composition | $p(O_2)$ (mm. Hg.) | Temp. range | $E_a$ (kcal./mole) |
|---|---|---|---|
| $PrO_{1\cdot67}$ | 0·2 | 450—550° | 27 ± 5 |
| $PrO_{1\cdot74}$ | 3 or 36 | 400—520 | 15 ± 5 |
| $PrO_{1\cdot78}$ | 71 or 89 | 370—470 | 36 ± 5 |

of 138 mm. Hg; when $p(O_2)$ is 23 or 70 mm. Hg, $E_a$ for the oxidation to $TbO_{1\cdot67}$ is *ca.* 11 kcal./mole. The $E_a$ values for the oxidation of $\phi$-$Tb_2O_3$ are generally much ower than those found for $\phi$-$Pr_2O_3$.

*Kinetics of Oxidation of the $\phi$ and $\theta$ Phases of* $Pr_2O_3$ *to the* $\beta$ *Phase.*—The $E_a$ values from the gravimetric data reported earlier do not provide any indication as to the mechanism of oxidation of the sesquioxides. We have therefore investigated the actual kinetics of oxidation of the $\phi$ and $\theta$ phases of $Pr_2O_3$ at atmospheric pressure [$p(O_2)$ = 138 mm. Hg] employing the procedure of Hyde, Garver, Kuntz, and Eyring.[15]

One of the commonly employed kinetic models for solid-state reactions involves a homogeneous collection of spherical particles with rapid surface nucleation accompanied by a constant linear growth rate.[16,17] The rate equation for such systems is given by equation (3)

$$(1-\alpha) = (1-Rt/r)^3 \qquad (3)$$

where $\alpha$ is the extent of the reaction, $R$ is the velocity of advancement of the reactant–product interphase, $r$ is the radius of the particle and $t$ is the time. Equation (3) can be rewritten in the form (4), where $x_t$, $x_0$, and $x_\infty$

$$(x_\infty - x_t)^{\frac{1}{3}} = kt + c \qquad (4)$$

denote the (average) composition of the sample at times $t$, 0, and $\infty$ respectively. If $(x_\infty - x_t)$ is temperature-

$$k = -V(x_\infty - x_t)^{\frac{1}{3}}/r \text{ and } C = (x_\infty - x_t)^{\frac{1}{3}} \qquad (5)$$

independent [(equation (5)], $k$ in equation (4) would be the rate constant. Since $x_\infty$ cannot be experimentally determined for an oxidation reaction of this kind, an optimum value is estimated. In our experiments, the best agreement was found when $x_\infty$ was 1·833. By making use of this value of $x_\infty$ we find that the agreement between the experimental data and equation (4) is satisfactory. Typical plots of $(x_\infty - x_t)^{\frac{1}{3}}$ against $t$ are shown in Figure 4. The tailing off of the reaction rate at large $t$ is probably because the phase-boundary controlled reaction gets throttled by the diffusion of oxygen in the interphase. A similar observation has been made by Hyde, Garver, Kuntz, and Eyring [15] for the reaction $PrO_{1\cdot833} \longrightarrow PrO_2$; these workers have discussed their kinetic results in terms of free-energy diagrams and Fick's law of diffusion.

Oxidation of $Pr_2O_3$ may involve at least four processes: (*a*) uniphasic oxidation of the reactant; (*b*) nucleation; (*c*) propagation of the product phase; and (*d*), uniphasic

[13] N. G. Dave and S. K. Chopra, *Z. phys. Chem.*, 1966, **48**, 257.
[14] J. S. Anderson and K. J. Gallagher, *J. Chem. Soc.*, 1963, 52.
[15] B. G. Hyde, E. E. Garver, U. E. Kuntz, and L. Eyring, *J. Phys. Chem.*, 1965, **69**, 1667.
[16] P. W. M. Jacobs and F. C. Tompkins, in 'Chemistry of the Solid State,' ed. by W. Garner, Butterworths, London, 1955.
[17] J. S. Anderson, L. E. J. Roberts, and E. A. Harper, *J. Chem. Soc.*, 1955, 3946.

oxidation of the product phase to the equilibrium composition. Unlike in the oxidation of the β phase to $PrO_2$,[15] we find no evidence for the fast initial stage controlled by diffusion (process *a*). Instead, we find a slow stage in the oxidation of $\phi$ at low temperatures (curve for 240° in Figure 4). These two kinetic stages in the oxidation of $\phi$ are likely to be the same as those observed by Anderson and Gallagher.[18]

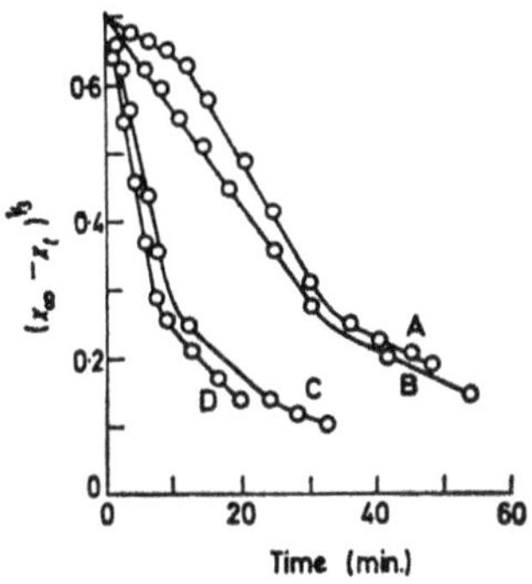


FIGURE 4 Kinetics of oxidation of $\phi$- and $\theta$-$Pr_2O_3$: $\phi$, curves A and C at 240° and 270° respectively; $\theta$, curves B and D at 300° and 320° respectively

The present results can be explained as follows. In the initial stages of the reaction the rate of nucleation is fairly fast and the propagation rate is nearly constant; the nucleation rate will however increase with temperature. At a critical value of α the phase boundary occurs and from then on the rate is given by the linear portions of the curves shown in Figure 4, typical of phase-boundary-controlled reactions. The critical value of α is apparently reached within a short time in the oxidation of $\phi$ and θ indicating that the mechanism involves nearly instantaneous diffusion and relatively rapid (and complete) nucleation followed by a rate-determining growth of the product phase. By taking the slopes of the linear curves (Figure 4) corresponding to the phase-boundary-controlled reaction, we have calculated the activation energies for the oxidation of $\phi$ and θ phases to be 25 and 50 kcal./g. atom of oxygen respectively. The higher value of $E_a$ for phase-boundary reaction of θ may be because there is a closer relationship between the $\phi$ and β phases rather than the θ and β phases.

## EXPERIMENTAL

Cubic and hexagonal forms of $Pr_2O_3$ were prepared by the reduction of 99·9% pure (Lindsay) $Pr_6O_{11}$ by procedures described earlier.[6] $Tb_2O_3$ was also prepared by hydrogen reduction of 99·9% terbia at 800°.[7] Isobaric gravimetric studies were carried out by employing an automatic recording semimicro-balance employing procedures similar to that of Hyde, Bevan, and Eyring.[3] D.t.a. curves were recorded by employing an automatic recording thermoanalyser.[19] The kinetics of oxidation of θ- and $\phi$-$Pr_2O_3$ were followed by a procedure similar to that of Hyde, Garver, Kuntz, and Eyring [15] by use of the automatic thermobalance.

We thank the U.S. National Bureau of Standards for a research grant.

[9/621 *Received, April 17th*, 1969]

[18] J. S. Anderson and K. J. Gallagher, Proc. Int. Symp. on Reactivity of Solids, Amsterdam, 1960, p. 223.
[19] K. J. Rao and C. N. R. Rao, *J. Materials Sci.*, 1966, **1**, 238.

JOURNAL OF SOLID STATE CHEMISTRY **2**, 377–384 (1970)

# Electrical Transport in Rare-Earth Oxides*

G. V. SUBBA RAO, S. RAMDAS, P. N. MEHROTRA, AND C. N. R. RAO†

*Department of Chemistry, Indian Institute of Technology Kanpur, Kanpur-16, India*

Received January 9, 1970

Sesquioxides and nonstoichiometric oxides of rare-earths (Ln) exhibit electrical conductivities in the range $10^{-9}$–$10^{-1}$ $\Omega^{-1}$ $cm^{-1}$. The sesquioxides exhibit mixed conduction with some contribution from ionic conductivity and major contribution from electronic conductivity. Seebeck coefficient data as well as the oxygen partial pressure dependence of conductivity indicate that $LnO_x$ compounds are mixed valence semiconductors where oxides with $1.50 \leqslant x \leqslant 1.75$ are *p*-type semiconductors and oxides with $1.75 < x \leqslant 2.00$ are *n*-type semiconductors. The conductivity of $LnO_x$ (Ln = Pr or Tb) goes through a maximum at $x \approx 1.75$; Seebeck coefficients are sensibly constant with temperature and approach zero value at $x \approx 1.75$. Employing the conductivities and Seebeck coefficients, transport parameters have been calculated. The mechanism of conduction in these oxides can be understood in terms of the hopping model and the small polaron theory. Fully ionised cation vacancies seem to be the predominant defects contributing to the defect structure in rare-earth sesquioxides.

## Introduction

Rare-earth oxides with filled or partially filled inner 4*f*-shells of the lanthanide ions are likely to be narrow band materials and the electrical conduction in these oxides would undoubtedly involve the 4*f*-electrons (*1*). In the literature there are a few reports on the electrical conductivity of rare-earth oxides. Noddack and Walch (*2*) measured the resistivities of a few rare-earth oxides and found the conduction to be predominantly electronic in nature. Eyring and Baenziger (*3*) have reported some preliminary measurements of the resistivities of a few praseodymium oxides, while Honig and coworkers (*4*) have studied both the electrical resistivities and Seebeck coefficients of a few oxides. Some evidence has been presented by Tare and Schmalzried (*5*) for the presence of ionic conductivity in rare-earth sesquioxides.

We have presently examined the semiconduction of several rare-earth sesquioxides and the nonstoichiometric oxides of praseodymium and terbium in some detail. We have measured the conductivities of a few oxides at different partial pressures of oxygen in order to establish the defect equilibria. Seebeck coefficient measurements have been carried out to find out the sign of the charge carriers. Employing the experimentally measured values of conductivity and Seebeck coefficients, we have explained the mechanism of conduction with the aid of the hopping model and the polaron theory.

* Taken in part from the Ph.D. Theses of G. V. Subba Rao, S. Ramdas and P. N. Mehrotra.

† To whom all the correspondence should be addressed.

## Experimental

$Eu_2O_3$, $Ho_2O_3$, $Y_2O_3$, and the other sesquioxides (all of better than 99.99% purity) obtained from American Potash Co. were ignited to constant weight and stored in a desiccator before use. Lindsay (American Potash Co.) 99.99% pure $Pr_6O_{11}$ and "$Tb_4O_7$," ignited to constant weight were used in the investigation. C–$Pr_2O_3$ and C–$Tb_2O_3$ were prepared by the reduction of $Pr_6O_{11}$ and $Tb_4O_7$ in a stream of dry hydrogen for 6 hr at 600 and 900°C, respectively (*6, 7*). $PrO_2$ was prepared by leaching $Pr_6O_{11}$ with 1.6*N* acetic acid for five days and $TbO_2$ by leaching $Tb_4O_7$ with 0.1*N* HCl for 24 hr (*8*). $Pr_7O_{12}$ was prepared by heating $Pr_6O_{11}$ in vacuum (~ $10^{-5}$ mm Hg) at 490°C for 4 hr (*8*). Lattice constants were calculated from the X-ray patterns recorded with a GE XRD diffractometer using CuK$\alpha$ radiation. The lattice constants agreed well with the literature data (*9*).

The electrical conductivities of polycrystalline pellets (pressed at 9000 psi and sintered at 950°C) of the various oxides were measured with a cell fabricated in this laboratory (*10, 11*). A GR-1608A impedance bridge fitted with a GR-1232A tuned amplifier and null detector was employed for the measurements. The pellets were coated with duPont silver paint to obtain good electrical contact. The conductivity values reported are reliable and reproducible. The conductivities of the pellets did not vary appreciably with the pressure employed in the preparation.

Seebeck coefficients, $\beta$, were measured with an apparatus fabricated in this laboratory (*12*) which is a slightly modified version of the cell employed by Czanderna (*13*) and Honig and coworkers (*4*). Corrections were made for the platinum wires attached to the samples employing procedures reported in the literature (*13*, *14*). In any case, these corrections do not significantly affect the present results except in the case of $TbO_{1.5}$; further, we have found high $\beta$ in $TbO_{1.52}$. The uncertainty in temperature determination was $\pm 0.5°$ and in the thermal emf, about $\pm 2\%$. Even though the uncertainty in the determination of the Seebeck coefficient is around $\pm 12\%$ (and possibly more in cases where the value is very small) the general reproducibility of the results was very good; the values quoted are the results of many repeated experiments.

Dielectric constants were determined by measuring the capacitance of the oxide pellets over a wide range of frequencies, employing a GR716C capacitance bridge and a Boonton Q-meter. The measurements were made at room temperature and the dielectric constants at zero ($\epsilon_0$) and high frequency ($\epsilon_\infty$) were obtained by an extrapolation procedure. The uncertainty in the determination is $\pm 5\%$.

The restrahlen frequencies were determined by taking the infrared absorption spectra of the samples in the form of nujol mulls employing a Perkin Elmer 521 and Beckman IR-11 spectrometers.

## Results and Discussion

### Rare-Earth Sesquioxides

The electrical conductivity data of a few rare-earth sesquioxides, $Ln_2O_3$, are shown as a function of temperature in Fig. 1. Conductivities, $\sigma$, at two temperatures along with the activation energies for conduction, $E_a$, are given in Table I. The conductivities are generally low, particularly so in the heavier rare-earth sesquioxides. The conductivity of $Y_2O_3$ is in between that of $Ho_2O_3$ and of $Yb_2O_3$. All the sesquioxides showed decrease in conductivity with decrease in oxygen partial pressure in the temperature range 400–900°C indicating the presence of *p*-type electronic conduction.

Most of the sesquioxides, especially those of the heavier rare-earths show breaks in the $\log\sigma$ vs $1/T$ plots around 550–600°C. $La_2O_3$, $Pr_2O_3$, and $Tb_2O_3$ show such breaks at still lower temperatures, $T_B$. These breaks in the conductivity curves do not correspond to any phase transformations. Phase transformations are noticed at much higher temperatures in some of the sesquioxides and the $\log\sigma - 1/T$ plots show distinct changes in slope at the transformation temperatures (*10*, *15*, *16*). The change in slope was seen at all values of $p_{O_2}$ studied (0.10–760 mm Hg) and hysteresis effects were negligible. The actual values of conductivity were somewhat more sensitive to $p_{O_2}$ below $T_B$.

From Table I we see that whereas the conductivity decreases down the rare-earth series, $E_a$ above the break temperature, $T_B$, increases. The variation of

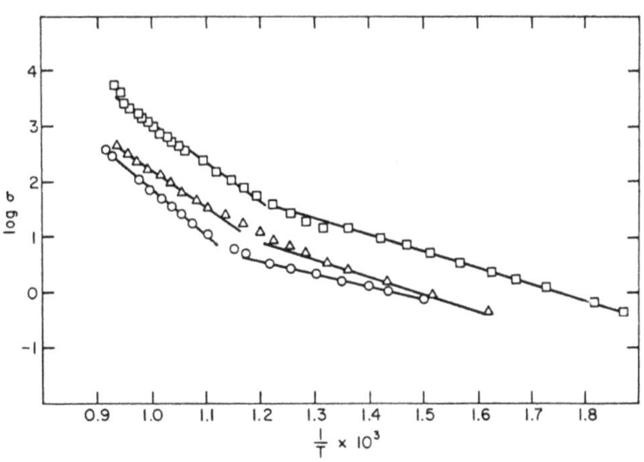


FIG. 1. Plots of electrical conductivity ($m\mu\Omega^{-1}\cdot cm^{-1}$) against the reciprocal of absolute temperature ($p_{O_2} = 150$ mm Hg; squares, $Sm_2O_3$; triangles, $Eu_2O_3$ and circles, $Yb_2O_3$.

TABLE I

ELECTRICAL CONDUCTIVITY DATA OF RARE-EARTH SESQUIOXIDES ($Ln_2O_3$)[a]

| Ln | $T_B$,°C | $E_1$,eV[b] ($T < T_B$) | $E_2$,eV ($T > T_B$) | $\sigma_{400°C}$ ($\times 10^{+9}\Omega^{-1}\cdot cm^{-1}$) | $\sigma_{650°C}$ ($\times 10^{+9}\Omega^{-1}\cdot cm^{-1}$) |
|---|---|---|---|---|---|
| La | 270 | 0.7 | 1.05 | 230 | 1700[c] |
| Pr[d] | 320 | 0.4 | 0.95 | 300 | 3450 |
| Nd | — | — | 1.15 | 25 | 1450 |
| Sm | 560 | 0.6 | 1.28 | 20 | 880 |
| Eu | 570 | 0.6 | 1.35 | 5 | 150 |
| Gd | 560 | 0.5 | 1.57 | 5 | 130 |
| Tb[d] | 280 | 0.4 | 0.95 | 3 | 200 |
| Ho | 575 | 0.7 | 1.61 | 5 | 160 |
| Yb | 605 | 0.5 | 1.61 | 3 | 50 |
| Y | — | — | 1.10 | — | 55 |

[a] DC conductivity measured at $p_{O_2} \approx 150$ mm Hg.
[b] Approximate values only since the conductivity is small at $T < T_B$.
[c] Values are for the hexagonal (A) form. In all the other cases, the oxides are in cubic (C) form (defect fluorite structure).
[d] Measured in dry hydrogen to prevent oxidation.

$\sigma$ and $E_a$ in the rare-earth series may be associated either with impurity effects or the variation in the ionic contribution to conductivity. We feel that ionic conductivity is more plausible in view of the following observations:

(i) The difference between $\sigma_{dc}$ and $\sigma_{ac}$ (at 10 KHz) is appreciable at $T > T_B$; $\sigma_{ionic}$ thus estimated (*17*) is about 30% at ~ 700°C in $Y_2O_3$ and $Nd_2O_3$ in agreement with other independent estimates. The percentage of $\sigma_{ionic}$ values are similar in other sesquioxides as well (*20*).

(ii) Continued electrolysis of sesquioxides at $T > T_B$ slightly reduces the conductivity.

(iii) Galvanic cell measurements of Tare and Schmalzried (*5*) clearly establish ionic conductivity to be appreciable in $Sm_2O_3$ at high temperatures.

These observations seem to suggest that the change in the slope of the conductivity-temperature plot at $T_B$ could arise from the onset of ionic conductivity. We therefore feel that rare-earth sesquioxides are most likely to be mixed conductors with some contribution from ionic conductivity and major contribution from electronic conductivity. We shall presently examine some aspects of electronic conduction in these sesquioxides.

In order to examine the electronic conduction and the associated defect chemistry in these sesquioxides, the conductivity of $Ho_2O_3$ was measured (*18*) at different oxygen partial pressures in the range 2–100mm Hg at a few temperatures (in the 400–700°C range). The data established that $\sigma$ was proportional to $p_{O_2}^{+1/5.3}$ in this pressure range. Schwab and Bohla (*19*) have recently found a $p_{O_2}^{+1/5.5}$ dependence of conductivity in $Eu_2O_3$ in the same pressure range. In $Y_2O_3$, Tallan and Vest (*20*) have observed a $p_{O_2}^{+1/5.33}$ dependence of conductivity. All these results show that the $p$-type electronic conduction in rare-earth sesquioxides arises from oxygen excess. The observed oxygen pressure dependence in the sesquioxides can be explained by a mechanism involving cation vacancies which would be consistent with the known crystal structures of these sesquioxides (*7*, *10*, *21*). Oxygen can enter the crystal structure of a sesquioxide creating metal ion vacancies and holes according to the equation,

$$3/4 O_2\,(g) \rightleftharpoons 1.5\, O_0^{2-} + \square_{Ln} + 3h^+, \quad (1)$$

where $O_0^{2-}$ is an oxygen ion at its lattice site, $\square_{Ln}$ is a cation vacancy and $h^+$ is a hole. Since $[h^+]/3 = [\square_{Ln}]$, we find that

$$\sigma \propto [h^+] \propto p_{O_2}^{+1/5.3} \quad (2)$$

as indeed observed experimentally in the case of $Ho_2O_3$ and $Y_2O_3$. Apparently, the $p_{O_2}$ dependence arises from fully ionized vacancies,; in $Eu_2O_3$ however, the $p_{O_2}^{+1/5.5}$ dependence of conductivity implies some contribution from trapped holes.

When oxygen was initially let into the evacuated conductivity cell containing the pellet of $Eu_2O_3$ or

$Ho_2O_3$, there was a marked decrease in the conductivity. It appears that at low pressures of oxygen, $Eu_2O_3$ and $Ho_2O_3$ may act as $n$-type semiconductors. Schwab and Bohla (*19*) and Tallan and Vest (*20*) have similarly found evidence for the $n$-type behavior of $Eu_2O_3$ and $Y_2O_3$ at low values of $p_{O_2}$. Tare and Schmalzried (*5*) report $n$-type conduction in $Sm_2O_3$ on the basis of emf measurements at low $p_{O_2}$. Thus, it appears that most of the rare-earth sesquioxides lose oxygen at low $p_{O_2}$ and high temperatures to produce anion vacancies ($\square_O$) and free electrons:

$$O_O^{2-} \rightleftharpoons 1/2O_{2(g)} + \square_O + 2e^-. \quad (3)$$

This would give rise to a $p_{O_2}^{-1/6}$ dependence of conductivity. The conductivity data of Schwab and Bohla (*19*) on $Eu_2O_3$ seem to conform to this behavior.

Except for $Pr_2O_3$, no Seebeck coefficient data are available for the rare-earth sesquioxides. We have measured the Seebeck coefficients of a few sesquioxides in the range 400–800°C at $p_{O_2} \approx 140$ mm Hg (Fig. 2). As expected, these oxides are $p$-type semiconductors in this temperature range. The $\beta$ values are relatively small at temperatures less than 600°C, but increase markedly above this temperature. Some of the rare-earth chalcogenides are known to exhibit a similar behavior (*22*). It is possible that ionic conductivity might be contributing significantly to $\beta$ at high temperatures (above $T_B$). Measurements on $Eu_2O_3$ in an atmosphere of nitrogen at ~ 800°C showed a decrease in $\beta$ (Fig. 2), but no change in sign. Studies at still lower values of $p_{O_2}$ may reveal a change in the sign of $\beta$.

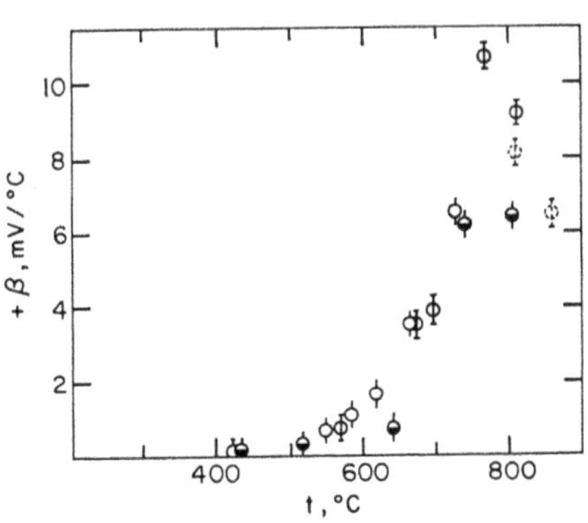


FIG. 2. Plots of Seebeck coefficient against temperature for various $Ln_2O_3$ measured in air ($p_{O_2} = 140$ mm Hg; ϕ: $Y_2O_3$; ϕ: $Ho_2O_3$; ϕ: $Eu_2O_3$; ϕ: $Eu_2O_3$ in nitrogen atmosphere.)

Nonstoichiometric Oxides of Praseodymium and Terbium

Phase equilibria in the praseodymium–oxygen (*23*, *24*) and terbium–oxygen (*6*, *25*) systems as well as the crystallography of the $Ln_nO_{2n-2}$ ($n \geqslant 4$) phases have been reported in the literature. Semiconduction in some of the equilibrium compositions of praseodymium and terbium oxides have been carefully studied in this laboratory. Among these oxides, the intermediate phase close to the composition $LnO_{1.75}$ (Ln = Pr or Tb) exhibits the highest electrical conductivity and the lowest activation energy for conduction. We shall not presently descuss the experimental data on $\sigma$ as a function $x$ since the essential features have already been presented earlier by Honig and coworkers (*4*). It is interesting that the conductivities of all the $LnO_x$ compositions fall on the $\sigma - x$ curves showing maxima at $x \approx 1.75$, immaterial of whether they are ordered (uniphasic) compounds with narrow homogeneity ranges or disordered (and biphasic) phases. Apparently, the main factor that governs the conductivity maximum at $x \approx 1.75$ is the stoichiometry ($Ln^{+3}$ and $Ln^{+4}$ concentrations) rather than the defect structure.

Conduction in $PrO_x$ and $TbO_x$ may be treated as a simple electron diffusion problem generally encountered in hoppers. Here, the mobility and diffusivity are related through a transition probability. At sufficiently high temperatures (*4*, *26*), we have

$$\begin{aligned} \sigma T &= C\tau^{-1} cs(1-s)\exp(-\Delta H/kT) \\ &= (n_e + n_p)e\mu(T), \end{aligned} \quad (4)$$

where $C$ is a collection of constants, $\tau_0^{-1}$ is the transition probability at infinite temperature, $c$ is the charge carrier density, $s$ is the concentration of available sites to which the charge carrier can jump and $\Delta H$ is the enthalpy change accompanying the transfer of a charged particle. Considering that the nonstoichiometric rare-earth oxides to be hoppers, the conductivity given by Eq. (4), should be proportional to the product $[Ln^{3+}][Ln^{4+}]$. Stated differently, $\sigma$ would be proportional to the product $(x-1.5)(2-x)$. Thus, in $LnO_x$, $\sigma$ should go through a maximum around $x = 1.75$ at a given temperature as found experimentally.

In praseodymium (*4*) and terbium oxides where $x$ is between 1.5 and 1.75, we would expect $p$-type conduction. For compositions near $LnO_2$ most of the cations will be present as $Ln^{4+}$ ions with occasional $Ln^{3+}$ ions; since the charge carriers in these oxides are electrons, we expect conduction to be of $n$-type. Midway between the extremes of $LnO_{1.5}$

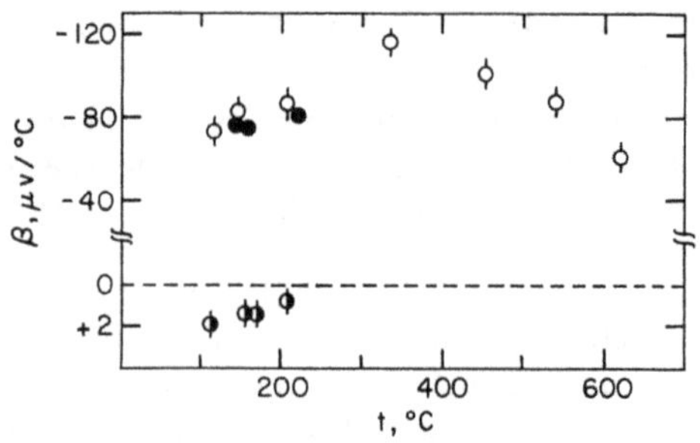


FIG. 3. Plot of Seebeck coefficient against temperature for $TbO_x$ in air ($p_{O_2} = 140$ mm Hg; open circles, $x = 1.81$; shaded circles, $x = 2.00$; half-shaded circles, $x = 1.50$).

and $LnO_2$ we would expect a change from $p$- to $n$-type behavior. This prediction is in agreement with the Seebeck coefficient data on $PrO_x$ reported by Honig and coworkers (*4*) and on $TbO_x$ obtained presently.

We have measured the Seebeck coefficients in four ordered narrow homogeneity phases, $TbO_{1.5}$, $TbO_{1.52}$, $TbO_{1.81}$ and $TbO_2$, over a range of temperatures. The results are presented in Fig. 3. We see that $\beta$ is positive for $TbO_{1.5}$ and $TbO_{1.52}$ typical of $p$-type semiconductors and negative in the case of the other two oxides indicating $n$-type behavior. In the case of $TbO_{1.5}$ (when the stoichiometry is nearly exact) $\beta$ is very small ($\sim 2\ \mu V/°C$ around 200°C), but when $x$ is slightly higher ($\sim 1.52$) $\beta$ increases markedly ($\sim +200\ \mu V/°C$ at $\sim 200°C$). A similar behavior has been noted by Honig and coworkers (*4*) in $PrO_{1.5}$. This may be because, at the stoichiometric composition $TbO_{1.5}$, we would expect a rapid change from a large negative to a large positive value of $\beta$; we note here that at $x < 1.5$ the oxide would be $n$-type. The expected high values of $\beta$ are indeed found in stoichiometric sesquioxides of other rare earths (Fig. 2). We also note from Fig. 3 that within experimental error $\beta$ is essentially independent of temperature for a given composition, a behavior typical of hoppers; $TbO_{1.81}$, however, shows a slight maximum around 350°C, the cause of which is not clear.

The variation of Seebeck coefficient with temperature in $LnO_x$ can be understood from relation (5) generally applicable to electron-transfer materials (*1, 27*):

$$\beta = -\frac{k}{e}\left[\ln\frac{c}{s} + \frac{S^*}{k}\right]. \tag{5}$$

Here $e$ is the electronic charge and $k$ is the Boltzmann constant. The entropy transported per particle, $S^*$, depends on the nature of the charge transfer process. Applying this to the case of $LnO_x$, we find that $\beta$ should vary with $x$ as,

$$\beta \approx -(k/e)\ln\{(x-1.5)/(2-x)\} \tag{6}$$

and $\beta$ will depend on $T$ only to the extent that $S^*$ does. To a first approximation, therefore, $\beta$ should be roughly independent of $T$. This theoretical prediction is in agreement with our experimental observations on $TbO_x$ (Fig. 3) and in a very limited temperature range for $Ln_2O_3$ (Fig. 2). We find that the value of $S^*/k$ is of the order of 0.2 in $TbO_x$; $S^*/k$ in $PrO_x$ has been reported to be 0.2 by Honig and coworkers (*4*).

The composition $LnO_{1.75}$ can be assumed to have a random distribution of $Ln^{3+}$ and $Ln^{4+}$ ions with a maximum of electronic disorder among the oxides of $LnO_x$ systems. The configurational entropy due to the electronic disorder would therefore be maximum at this composition (*28*). It is possible to conceive of a slightly different model according to which the conductivity maximum would be at the composition $x = 1.71$ instead of $x = 1.75$. The phase $LnO_{1.71}$ may be described as consisting of parallel 'strings' of irregular $LnO_6$ octahedra in the $\langle 111 \rangle$ direction surrounded by contiguous sheaths of seven-coordinated cations (*24, 25*). According to this model, the seven-coordinated cations with an average charge of +3.5 would be responsible for the conductivity. This would result in the conductivity maximum and zero value for $\beta$ at a composition of $LnO_{1.71}$ rather than at $LnO_{1.75}$. Although the second model appears to be reasonable, it is difficult to make such subtle distinctions on the basis of the available experimental data. This is because, (i) stable $LnO_{1.75}$ phase is nonexistent, (ii) experimental uncertainties in the determination of $\beta$ (particularly near zero) are large and most important of all, (iii) the cross-over point from $p$- to $n$-type behavior and the conductivity maximum are determined by the concentration of charge carriers, $n_i$, as well as their mobility, $\mu_i$. Thus, $\beta$ would be zero when $n_p\mu_p\beta_p = n_e\mu_e\beta_e$. Experimentally we find that the conductivity maximum is around the composition $LnO_{1.71}$ only at some temperatures while $\beta$ is zero at this composition in both $TbO_x$ and $PrO_x$. Oxygen partial pressure dependence of conductivity discussed next also throws some light on the cross-over from $p$- to $n$-type semiconduction in the $PrO_x$ system.

In order to examine possible defect equilibria in the $PrO_x$ system, electrical conductivities were measured at different partial pressures of oxygen. Preliminary measurements on $Pr_6O_{11}$ at two oxygen partial pressures clearly indicated it to be an $n$-type semiconductor as expected (Fig. 4). We see that the

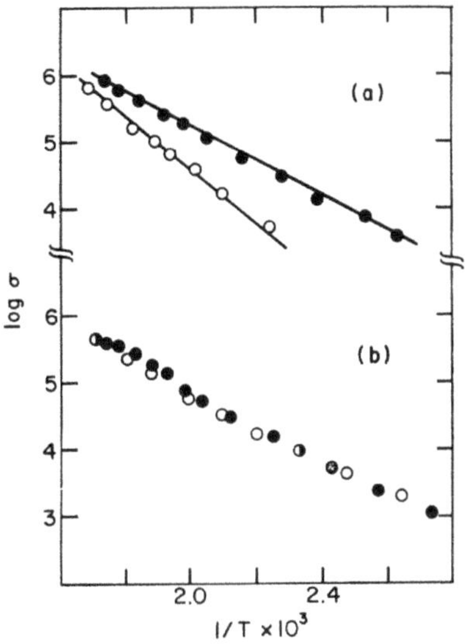


FIG. 4. Plot of electrical conductivity ($m\mu\Omega^{-1}\cdot cm^{-1}$) against the reciprocal of absolute temperature of $Pr_6O_{11}$ (a) and $Pr_7O_{12}$ (b). Open circles, $p_{O_2} = 140$ mm Hg; closed circles, $p_{O_2} = 0.2$ mm Hg.

dependence of $\sigma$ of $Pr_7O_{12}$ is not (where $Ln^{+3}$–$Ln^{+4}$ concentrations are close) at all appreciable (Fig. 4). More detailed investigations on $Pr_6O_{11}$ ($x = 1.833$) in the low pressure range ($p_{O_2} = 2$–$15$ mm Hg) indicated a $p_{O_2}^{-1/6}$ dependence of conductivity at $\sim 200°C$; this dependence can be interpreted in terms of anion vacancies (Eq. 3). The conductivity of the nonstoichiometric oxides in the composition range $1.68 < x < 1.72$, on the other hand, showed a dependence of $p_{O_2}^{+1/6}$; the $p_{O_2}^{+1/6}$ dependence is consistent with a cation vacancy mechanism. Although we can satisfactorily account for the oxygen pressure dependence of conductivity of $LnO_x$ in terms of defect equilibria, we are presently unable to eliminate the equally plausible interpretation based on the presence of mixed valence; these two mechanisms are, of course, interdependent.

### Small Polaron Model

Since band widths in rare-earth oxides are very small, we considered it interesting to examine the conduction in these compounds in terms of the polaron theory (*29–31*). The measure of the electron–phonon interaction in polaron materials is given by the dimensionless Fröhlich coupling constant (*29*), $\alpha_F^*$:

$$\alpha_F^* = e^2(m^*/2\hbar^3\,\omega_l)^{1/2}\,(\epsilon_\infty^{-1} - \epsilon_0^{-1}), \tag{7}$$

where $m^*$ is the effective mass of the charge carrier and $\omega_l$ is the longitudinal mode frequency. The dimension of the polaron, $\lambda_F^*$, defined by Feynman (*30*), is

$$\lambda_F^* = (\hbar/2m^*\,\omega_l)^{1/2}. \tag{8}$$

The effective mass of charge carriers in praseodymium and terbium oxides can be calculated from the experimental Seebeck coefficient data (*1*). Using the calculated $m^*$ values, the experimentally determined values of the dielectric constants and the longitudinal mode frequency, the values of the coupling constants ($\alpha_F^*$) were calculated for the oxides. The values of $\alpha_F^*$ in all the oxides is seen to be greater than unity (Tables II and III). The transition probability, $\tau^{-1}$, could be calculated making use of the conductivity data and the charge carrier density (*1*); the drift mobility, $\mu_D$, of the charge carriers was then calculated from $\tau^{-1}$.

An estimate of Szigeti's effective charge, $e^*$, of the charge carrier was obtained from the equation (*32*)

$$e^* = (3\nu_0/(\epsilon_\infty + 2))\,[(\epsilon_0 - \epsilon_\infty)\,\pi M V']^{1/2} \tag{9}$$

where $\nu_0$ is the observed restrahlen frequency, $M$ is

TABLE II

ELECTRICAL TRANSPORT PARAMETERS FOR $PrO_x$ AT 600°C

| Compound | $\epsilon_0$ | $\epsilon_\infty$ | $\nu_{obsd}$, $cm^{-1}$ | $E_a$, eV | $V$ in cc per ion ($\times 10^{-24}$) | $\sigma$, $\Omega^{-1}\cdot cm^{-1}$ | $\beta$, $\mu V/°C$ | $c$ | $m^*/m$ | $\alpha_F^*$ | $\tau^{-1}$, $sec^{-1}$ | $\mu_D$, $cm^2/V\,sec$ | $\lambda_F^*$, Å |
|---|---|---|---|---|---|---|---|---|---|---|---|---|---|
| $Pr_2O_3$ | 100 | 5.6 | 420 | 0.84 | 38.5 | $2.6 \times 10^{-6}$ | +10 | 0.42 | 25 | 6.6[a] (1.3) | $9.8 \times 10^5$ | $1.5 \times 10^{-8}$ | 0.83[a] |
| $PrO_{1.714}$ | 120 | 5.6 | 400 | 0.57 | 42.8 | $1.7 \times 10^{-2}$ | +9 | 0.43 | 23 | 6.3 (1.3) | $1.3 \times 10^{10}$ | $8.0 \times 10^{-4}$ | 0.85 (4.10) |
| $PrO_{1.833}$ | 60 | 15 | 640 | 0.81 | 41.0 | $8.4 \times 10^{-2}$ | −50 | 0.59 | 19 | 5.0 (1.0) | $5.4 \times 10^{10}$ | $2.5 \times 10^{-3}$ | 1.15 (4.89) |

[a] Value in parenthesis indicates for the free electron mass.

TABLE III

ELECTRICAL TRANSPORT PARAMETERS FOR $TbO_x$

| $T$,°C | $\sigma$, $\mu\Omega^{-1}\cdot cm^{-1}$ | $\beta$, $\mu V/°C$ | $c$ | $m^*/m$ | $\alpha_F^*$ | $\tau^{-1}$, $sec^{-1}$ | $\mu_D$, $cm^2/V\ sec$ | $\lambda_F^*$, Å |
|---|---|---|---|---|---|---|---|---|
| $TbO_{1.5}$ [a] | | | | | | | | |
| 200 | 0.3 | +1 | 0.42 | 45 | 8.4[b] (1.3) | $2.2 \times 10^5$ | $6.1 \times 10^{-8}$ | 0.96[b] (6.42) |
| $TbO_2$ [c] | | | | | | | | |
| 200 | 2.5 | −78 | 0.67 | 33 | 3.5 (0.6) | $9.9 \times 10^5$ | $6.6 \times 10^{-8}$ | 0.45 (2.61) |
| $TbO_{1.81}$ [d] | | | | | | | | |
| 200 | 6.3 | −92 | 0.70 | 29 | 3.0 (0.6) | $2.5 \times 10^6$ | $1.7 \times 10^{-7}$ | 0.62 (3.36) |
| 300 | 44.7 | −101 | 0.73 | 23 | 2.6 | $2.2 \times 10^7$ | $1.3 \times 10^{-6}$ | 0.70 |
| 400 | 177.8 | −102 | 0.73 | 20 | 2.5 | $1.0 \times 10^8$ | $5.0 \times 10^{-6}$ | 0.75 |
| 500 | 609.5 | −94 | 0.71 | 15 | 2.1 | $3.9 \times 10^8$ | $2.1 \times 10^{-5}$ | 0.87 |
| 600 | 1778.0 | −78 | 0.67 | 17 | 2.3 | $1.2 \times 10^9$ | $4.5 \times 10^{-5}$ | 0.82 |

[a] $\epsilon_0 = 8.4$, $\epsilon_\infty = 4.5$, $\nu_{obsd} = 545\ cm^{-1}$, $E_a = 0.95$ eV, $V = 38.6 \times 10^{-24}\ cm^3$.

[b] Value in parenthesis indicates for the free electron mass.

[c] $\epsilon_0 = 840.$ $\epsilon_\infty = 8$, $\nu_{obsd} = 440\ cm^{-1}$, $E_a = 0.5$ eV, $V = 35.5 \times 10^{-24}\ cm^3$.

[d] $\epsilon_0 = 285$, $\epsilon_\infty = 11.2$, $\nu_{obsd} = 540\ cm^{-1}$, $E_a = 0.66$ eV, $V = 37.5 \times 10^{-24}\ cm^3$.

the reduced mass of all the ions present in the crystal and $V'$ is the volume of the unit cell. The $\epsilon_0$ value is very high in many of the rare-earth oxides. Such large values have also been reported for other oxides like $TiO_2$ (*33*), $SrTiO_3$ (*34*), and $LaCoO_3$ (*35*).

The relevant experimental data and the calculated transport parameters for praseodymium and terbium oxides are presented in Tables II and III. For $TbO_{1.81}$ it has been possible to calculate all the transport parameters as a function of temperature which permits us to examine the applicability of the hopping model more closely. Hall data would have been valuable, but the small mobilities in these materials even at high temperatures would make such measurements less reliable.

The charge carrier density is higher in non-stoichiometric oxides than in stoichiometric oxides; in $TbO_{1.81}$ it is almost independent of temperature in the range studied (Table III). The effective mass of the charge carriers, $m^*$, is usually 20 to 40 times the free electron mass, indicating considerable interaction between the charge carriers and phonons; $m^*$ is slightly lower in non-stoichiometric oxides. Such high $m^*$ values have been encountered in the other oxides as well (*36*). For $TbO_{1.81}$, $m^*$ shows a decreasing trend with rise in temperature; Fröhlich coupling constant ($\alpha_F^*$) also shows a similar behavior.

Both $\tau^{-1}$ and $\mu_D$ are greater in the nonstoichiometric oxides compared to the stoichiometric oxides. In the oxides studied presently, $\mu_D$ is in the range $10^{-3}$–$10^{-8}$ $cm^2/V$ sec, which is in the region normally found in oxide semiconductors. In $TbO_{1.81}$, $\mu_D$ increases with temperature exponentially, the energy of activation being 0.5 eV. This monotonic increase in mobility can be taken to indicate hopping mechanism rather than 'polaron band' formation (*37*). The value of 0.5 eV noted above is comparable with the 0.6 eV obtained from the conductivity data. The polaron dimension is of the same order as the lattice parameter (small polarons) indicating that the charge carriers are localized at lattice sites. Szigeti's effective charge in the case of stoichiometric phases is not much different from the free electron charge, but considerably higher (~ 2 to 3$e$) in nonstoichiometric oxides.

## Acknowledgments

The authors are thankful to the U.S. National Bureau of Standards for a research grant (G-51) under their special international program and to the referee for many valuable suggestions.

## References

1. R. R. HEIKES, Rare Earth Res., **1**, 247, (1961) (Macmillan Co., New York).
2. W. NODDACK AND H. WALCH, *Z. Phys.* **211**, 194 (1959); *Z. Electrochem.* **63**, 269 (1959).

3. L. Eyring and N. C. Baenziger, *J. Appl. Phys.* **33**, 428 (1962).
4. J. M. Honig, A. A. Cella, and J. C. Cornwell, Rare Earth Res. **2**, 555 (1964) (Gordon and Breach, New York).
5. V. B. Tare and H. Schmalzried, *Z. Phys. Chem.* (NF) **43**, 30 (1964).
6. N. C. Baenziger, H. A. Eick, H. S. Schuldt, and L. Eyring, *J. Amer. Chem. Soc.* **83**, 2219 (1961).
7. R. L. N. Sastry, P. N. Mehrotra, and C. N. R. Rao, *Trans. Farad. Soc.* **62**, 1632 (1966).
8. R. L. N. Sastry, P. N. Mehrotra, and C. N. R. Rao, *J. Inorg. Nucl. Chem.* **28**, 2167 (1966).
9. L. Eyring and B. Holmberg, *Advan. Chem. Series* **39**, 46 (1963) (American Chemical Society, Washington).
10. P. N. Mehrotra, G. V. Chandrashekhar, E. C. Subbarao, and C. N. R. Rao, *Trans. Faraday Soc.* **62**, 3586 (1966).
11. G. V. Chandrashekhar, P. N. Mehrotra, G. V. Subba Rao, E. C. Subbarao, and C. N. R. Rao, *Trans. Faraday Soc.* **63**, 1295 (1967).
12. G. V. Subba Rao, Ph.D. Thesis, Indian Institute of Technology, Kanpur, 1969.
13. A. W. Czanderna, Ph.D. Thesis, Purdue University, Indiana, 1957.
14. J. E. Bauerle, P. H. Sutter, and R. W. Ure, Jr., *in* "Thermoelectricity-Science and Engineering," (R. R. Heikes and R. W. Ure, Jr., Eds.), Interscience, New York, 1961.
15. P. N. Mehrotra, Ph.D. Thesis, Indian Institute of Technology Kanpur, 1967
16. G. Brauer and R. Muller, *Z. Anorg. Allgem. Chem.* **321**, 234 (1963).
17. S. Iida, *J. Appl. Phys., Tokyo*, **6**, 77 (1967).
18. S. Ramdas, Ph.D. Thesis, Indian Institute of Technology, Kanpur, 1970.
19. G. M. Schwab and F. Bohla, *Z. Naturforsch. A* **23**, 1549 (1968).
20. N. M. Tallan and R. W. Vest, *J. Amer. Ceram. Soc.* **49**, 401 (1966); T. H. Etsell and S. N. Flengas, *Chem. Rev.* **70**, 339 (1970).
21. G. Brauer, *Progr. Sci. Technol. Rare Earths* **3**, 434 (1968).
22. D. J. Haase, H. Steinfink, and E. J. Weiss, Rare Earth Res. **3**, 535 (1965) (Gordon and Breach, New York).
23. J. M. Honig, A. F. Clifford, and P. A. Faeth, *Inorg. Chem.* **2**, 791 (1963).
24. B. G. Hyde, D. J. M. Bevan, and L. Eyring, *Phil. Trans. Roy. Soc. London* **259**, 583 (1966).
25. B. G. Hyde and L. Eyring, Rare Earth Res. **3**, 623 (1965) (Gordon and Breach, New York).
26. R. R. Heikes and W. D. Johnston, *J. Chem. Phys.* **26**, 582 (1957).
27. R. R. Heikes, *in* "Transition Metal Compounds," (E. R. Schatz, Ed.), Gordon and Breach, New York, 1964.
28. S. Ramdas, K. C. Patil, and C. N. R. Rao, *J. Chem. Soc. A* **64** (1970).
29. H. Fröhlich, *in* "Polarons and Excitons," (C. G. Kuper and G. C. Whitfield, Eds.), p. 12, Plenum Press, New York, 1962.
30. T. D. Schultz, *Phys. Rev.* **116**, 526 (1959).
31. T. Holstein, *Ann. Phys. New York* **8**, 343 (1959).
32. B. Szigeti, *Trans. Faraday Soc.* **45**, 155 (1949).
33. D. M. Eagles, *J. Phys. Chem. Solids* **25**, 1243 (1964).
34. A. H. Kahn and A. J. Leyendecker, *Phys. Rev. A* **135**, 1321 (1964).
35. H. G. Reik, E. Kauer, and P. Gerthsen, *Phys. Lett.* **8**, 29 (1964).
36. I. G. Austin, A. J. Springthorpe, B. A. Smith, and C. E. Turner, *Proc. Phys. Soc. London* **90**, 157 (1967).
37. J. Appel, *Solid State Phys.* **21**, 193 (1968).


For semiconduction of non-stoichiometric rare earth oxides, see G.V. Chandrashekhar, D.N. Mehrotra, G.V. Subba Rao and C.N.R. Rao, Trans. Faraday Soc., 63, 1295 (1967)

*Proc. R. Soc. Lond.* A. **355**, 301–312 (1977)
*Printed in Great Britain*

# Electron microscopy of ferroelectric bismuth oxides containing perovskite layers

BY J. L. HUTCHISON†, J. S. ANDERSON, F.R.S.† AND C. N. R. RAO‡
*Inorganic Chemistry Laboratory, University of Oxford*

(*Received* 12 *January* 1976 – *Revised* 23 *December* 1976)

[Plates 1–4]

Ferroelectric bismuth oxides of the general formula

$$(Bi_2O_2)^{2+}(A_{n-1}B_nO_{3n+1})^{2-}$$

(A = Ba, etc., B = Ti or other transition metal) have been examined by high-resolution lattice imaging electron microscopy. The lattice images show dark bands at the positions of the $Bi_2O_2$ layers, with $n-1$ lines between them due to the layers of the perovskite A cations or, in favourable circumstances, the fully resolved 0.4 nm square perovskite grid. Dislocations and domain boundaries have been imaged for the first time in ferroelectric crystals. The structure of the dislocations and domain walls is discussed in the light of the microstructural evidence.

## 1. INTRODUCTION

Lattice imaging methods of electron microscopy provide a powerful tool for investigating the local structure of crystals, in real space. They have been applied, to good effect, to a number of complex oxide systems and have yielded information, that would hardly be accessible to other methods, about stacking sequences, the regularity and ordering of layer structures, the existence of new intergrowth phases, superstructures and extended defects. Our study of the barium ferrite layer structures (McConnell, Hutchison & Anderson 1974) showed the value of examining structures in which a variety of large crystallographic units could be built up by the ordered repetition of a small number of distinct sub-units.

In this paper we describe some observations on a somewhat analogous family of layered structures, first described by Aurivillius (1949) as having the general chemical formula $Bi_2A_{m-2}B_{m-1}O_{3m-1}$, where A is a bulky cation ($Bi^{3+}$. $Ba^{2+}$, $Pb^{2+}$ etc.) and B is a transition metal cation normally found in 6-coordination with oxygen ($Ti^{4+}$, $Nb^{5+}$ etc.). The structural principle common to all the family is better expressed by the formulation $(Bi_2O_2)^{2+}(A_{n-1}B_nO_{3n+1})^{2-}$ (in which $n = m-1$); this emphasizes their construction from finite (and therefore anionically charged) slabs of

† Now at Edward Davies Chemical Laboratory, University College of Wales, Aberystwyth, SY23 1NE.

‡ Permanent address: Indian Institute of Science, Bangalore 560012, India.

perovskite structure, $n-1$ unit cells thick and thus having $n$ layers of B-site cations in corner-sharing octahedral coordination, enclosing $n-1$ layers of 12-coordinate A sites (Goodenough & Longo 1970). Between these slabs, and having their oxygen atoms in common with them, are double-sided $Bi_2O_2^{2+}$ sheets, of the kind frequently

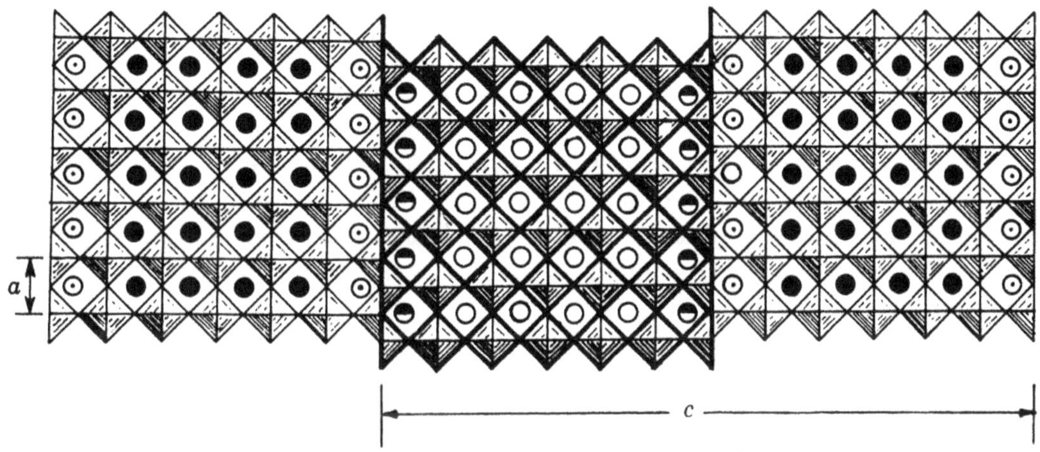


FIGURE 1. Structure of $Ba_2Bi_4Ti_5O_{18}$ (schematic), projected on (101) of the pseudotetragonal cell. Lightly drawn octahedra: B cations at $y = 0$; heavily drawn octahedra: B cations at $y = \frac{1}{2}$; ◓, Bi (in $Bi_2O_2$ layers) at $y = 0$; ⊙, Bi (in $Bi_2O_2$ layers) at $y = \frac{1}{2}$; ○, A cations at $y = 0$; ●, A cations at $y = \frac{1}{2}$.

found in bismuth oxy-compounds, made up of square pyramidal $[BiO_4]$ groups sharing their basal edges. Thus, in $Bi_4Ti_3O_{12}$ ($n = 3$) and $Bi_4BaTi_4O_{15}$ ($n = 4$) the perovskite units are slabs with the composition $BiTiO_3$ and $(Ba_{0.33}Bi_{0.67})TiO_3$ per perovskite cell respectively. An apparent variability of composition, as appeared from Aurivillius' papers, could arise if the perovskite slabs included between the $Bi_2O_2$ layers varied in thickness from layer to layer. All the compounds are tetragonal (or nearly so, with slight orthorhombic distortion); their a and $b$ axes lie along $[110]_c$ (where the suffix denotes the cubic perovskite subcell), so that

$$a \approx b \approx \sqrt{2}a_c \approx 0.54\,\text{nm}.$$

Their $c$ axes are inherently long: $\frac{1}{2}c_{ss} = (n-1)a_c +$ thickness of the $Bi_2O_2$ layer $\approx na_c$ (the suffix ss denoting dimensions, etc., of the large superstructure cell). For compounds with $n = 1$ to 5, $c_{ss}$ is in the range 1.54–5.04 nm.

Figure 1 shows a typical member of the series, $Ba_2Bi_4Ti_5O_{18}$ ($n = 5$), drawn as a polyhedron network structure. It may be noted that the bismuth atoms of the $Bi_2O_2$ sheets occupy positions close to virtual A sites, and also that the pyramidal $BiO_4$ groups (which could alternatively be regarded as half octahedra terminated by a Bi atom in place of an oxygen atom) share edges in such a way that successive slabs of perovskite structure are in antiphase, displaced by $\frac{1}{2}[110]_c$.

These layered perovskite compounds are ferroelectric, with fairly high Curie temperatures (Schmid 1975; Subbarao 1974). Some ambiguity still attaches to the interpretation of paraelectric/ferroelectric transitions, but it is clear that ionic polarization makes an important contribution to the permanent dipole in the ferroelectric state and that, in the perovskites and related structures, this ionic polarization involves a displacement of the highly charged B-site cations away from the centres of their coordination octahedra. Whether there is also a displacement of the oxide ions appears less certain. The displacements are small; in the case of $BaTiO_3$ the shift of the $Ti^{4+}$ ions has been estimated as being of the order of 0.006–0.010 nm, relative to the A sites. The structural consequence of this polarization is to distort the environment of the B cations from nearly regular octahedral to irregular octahedral coordination, and thereby to lower the crystal symmetry. The topology of the structure, however, as a network of vertex-linked octahedra, remains quite unaltered. It is permissible to extend these considerations from the perovskites proper to the Aurivillius phases considered here. It can be concluded that the only effects of ferroelectric polarization likely to show up in lattice images are distortions from orthogonality in the ferroelectric state; displacements of heavy atoms are small, as compared with the attainable resolution, and electron scattering (hence image contrast) will be dominated by the heavy atoms in the $Bi_2O_2$ layers and the A sites.

Regularity of microstructure is readily investigated, but it was of interest to see whether ferroelectric domains and boundaries could be revealed in lattice images. Evidence on this is given below. In addition, the materials afforded opportunity for observations of structural relaxation in the core regions of large dislocations.

## 2. Experimental

### *Materials*

Compounds with $n = 1$ to 5 were prepared from appropriate mixtures of the constituent oxides ($Bi_2O_3$, $WO_3$, $Nb_2O_5$, $TiO_2$; Johnson Matthey Specpure) and barium carbonate. The mixtures were ground thoroughly and precalcined at about 100 °C below the final firing temperature. They were then re-ground and heated at the temperatures shown in Table 1. The weight loss during the preparation was less than 0.5 %. Lattice parameters and ferroelectric Curie temperatures of the compounds thus prepared are listed in table 1. According to Ismailzade (1967) the $n = 8$ compound $Bi_9Ti_3Fe_5O_{27}$ is formed by heating an appropriate oxide mixture at 1300 K; a preparation following his procedure proved, however, to be structurally inhomogeneous (see below).

### *Electron microscopy*

Procedures for electron microscopy were those now customary in many-beam lattice imaging work. Crystals were finely crushed in an agate mortar and the finer fragments slurried on to carbon-coated grids, for examination in a Siemens Elmiskop 102 electron microscope, equipped with a high resolution, double tilt stage and a $z$-translation. Micrographs were recorded at 100 kV, at magnifications from 300000 to 700000, in through-focal series spanning the optimum defocus condition (about

70 nm under focus). As is well known, it is necessary to orient specimens with great accuracy, in order that corresponding structural units should be exactly lined up in projection. Suitably thin fragments were therefore tilted to bring the projection axis, usually $[100]_c$ or $[110]_c$, parallel to the electron beam, within 0.1°. A 40 μm objective aperture was used to combine the primary beam with all diffracted beams out to about 3.3 $nm^{-1}$; about 70 beams were thereby included for the $n = 5$ compound. Experience has shown that, in these conditions, the effective resolving power is about 0.35 nm.

TABLE 1. BISMUTH OXIDE LAYERED COMPOUNDS STUDIED

| n | compound | final heating temperature/K | lattice parameters/nm *a* | *b* | *c* | ferroelectric $T_c$/K |
|---|---|---|---|---|---|---|
| 1 | $Bi_2WO_6$ | 1040 | 0.5458 | 0.5438 | 1.5434 | 1223 |
| 2 | $BaBi_2Nb_2O_9$ | 1370 | 0.5533 | 0.5333 | 2.5550 | 473 |
| 3 | $Bi_4Ti_3O_{12}$ | 1280 | 0.5448 | 0.5411 | 3.283 | 498 |
| 4 | $BaBi_4Ti_4O_{15}$ | 1450 | 0.5461 | 0.5461 | 4.185 | 668 |
| 5 | $Ba_2Bi_4Ti_5O_{18}$ | 1450 | 0.5514 | 0.5526 | 5.0370 | 590 |
| 8 | $Bi_9Ti_3Fe_5O_{27}$ | 1450 | 0.5491 | 0.5502 | 7.62 | — |

Work on the calculation of $N$-beam lattice image contrast (Allpress & Sanders 1973; O'Keefe 1973; Lynch, Moodie & O'Keefe 1975) has shown that, for materials like the heavy metal oxides, in zone axis orientations, specimens less than 7–10 nm thick furnish images that approximate fairly closely to the projected potential, or the projected charge density. All the observations in this paper relate to crystal flakes or thin edges that fulfilled this condition. In terms of the projected charge density approximation, such images can therefore be interpreted in chemical, structural terms, bearing in mind that electron scattering is dominated by the heaviest atoms present (Bi atoms and A site cations). There is ample evidence (see, for example, the calculations and critical review of O'Keefe (1976)) that this approach is valid for regular structures, and empirical evidence from our own and other work, now being supplemented by calculations, that it can be extended to local modification of otherwise regular structures. We have therefore used the 'intuitive-structural' approach, bringing in what is already known of the crystal chemistry of these and related complex oxides, to interpret the micrographs. It would not be practicable to carry out a full $N$-beam multislice image contrast calculation for such a large image field as figure 3, but appropriate diffraction contrast calculations for crystals with dislocations would be useful. They would provide an assessment of the validity of inferences drawn from the intuitive approach.

## 3. Results and discussion

### *Lattice images*

The lattice images generally showed dark bands of contrast corresponding to the $Bi_2O_2$ layers. Between these were $n-1$ lines, arising from the layers of perovskite A-site cations (Bi and Ba), 0.38 nm apart. Under favourable conditions, a second set of fringes, perpendicular to the direction of the $Bi_2O_2$ bands and also (in $[010]_c$ projection) 0.38 nm apart fully delineated the square grid of the perovskite structure. These features are illustrated by the two-dimensional image in figure 2, plate 1, which can be understood in terms of the projected structure diagram, figure 1.

When a family of compounds can be generated by stacking of similar subunits, it is possible for a single crystal (in the macroscopic sense of a morphological entity) to vary in composition and structure at different points taken along the stacking direction. The lattice image (figure 2) shows that this is the case for the preparation which, replicating Russian work, should have yielded the $n = 8$ compound $Bi_9Ti_3Fe_5O_{27}$. No lattice images were found which showed the expected seven lines of A sites between $Bi_2O_2$ contrast bands. Instead, sequences corresponding to $n = 4$ and $n = 5$, ordered in short arrays of each kind, occurred randomly. The $n = 5$ layers predominated and the average multiplicity was probably around 4.7–4.8, but there were no long, homogeneous sequences. In X-ray diffraction, the superlattice lines would probably simulate an averaged multiplicity, but the electron diffraction pattern was of interest in view of the relatively small extent (15–20 nm) of any one uniform lamella. Although streaked, the diffraction spots corresponded to a superposition of the $n = 4$ and $n = 5$ patterns, not to an averaged multiplicity. It is also of interest – and relevant to the validity of the intuitive-structural interpretation of the lattice images – that although the sequence of $Bi_2O_2$ layers is faulted, the two-dimensional structure of the perovskite slabs is fully resolved.

### *Dislocations*

Several of the ferroelectric bismuth oxide perovskites showed the presence of dislocations in their lattice images: these were especially frequent in crystals of the $n = 2$ and $n = 5$ compounds. It may be relevant that these have the lowest Curie temperatures of the systems studied. It has been considered that dislocations play a role in the motion of domains in perovskite ferroelectrics and it is possible that, during microscopy, the electron beam raised the temperature of the samples close to $T_c$. On prolonged irradiation, electron beam damage gave rise to dislocation loops, probably as a result of chemical decomposition. Van Landuyt, Remaut and Amelinckx (1969) earlier reported observing dislocation ribbons in diffraction contrast micrographs of $Bi_4Ti_3O_{12}$ ($n = 3$); lattice imaging of dislocated regions in these and other polyhedron network structures affords, for the first time, the possibility of examining their intimate structure. In another paper (Anderson, Hutchison and Lincoln 1976) we have considered dislocations in the complex niobium oxide structures, which are also networks based on a simple cubic framework of vertex-linked octahedral groups around the cations. We there point out that (unlike metals and some simple binary compounds) the local chemistry imposed by dislocations cannot be neglected. In particular, strong near-neighbour atomic interactions are important in determining the local structure adopted as a result of atomic displacements: the maintenance of local charge balance around cations of high formal charge; atomic sizes, determining the packing around the cations; and the not negligible covalent anion–cation interactions. The effect of these constraints is that, except in the most highly distorted core region, the general topological characteristics of the structure must be conserved. This requirement restricts the most likely values for the displacement vectors of dislocations.

Figure 3, plate 1, shows a dislocation in orthorhombic $Ba_2Bi_4Ti_5O_{18}$. The clarity of the image, particularly in the core region, justifies an attempt to analyse the distorted region in some detail; for reasons discussed earlier, the lattice image will be treated as a projection of the charge density distribution, and the structure deduced accordingly. The immediate character of the dislocation and its strain field can be seen by comparison with a superimposed grid representing the $Bi_2O_2$ layers in an ideal, unperturbed structure. By using the 0.38 nm separation between rows of A sites as an internal calibration, the dimensional correctness of the grid is ensured, and the distortions in the defect itself, and in the surrounding good crystal, can be measured at every point.

In the upper half of the micrograph, and terminating in the less well resolved core, two extra slabs of perovskite, with their $Bi_2O_2$ layers, are present. Since these have the same thickness as in the bulk of the crystal, they represent the introduction of a half-slice of $n = 5$ structure, one complete unit cell in thickness. Strain due to this insertion appears to be accommodated in two ways. The first, clearly seen in regions a few nanometres away from the core, is by elastic deformation: a small change in the spacing between $Bi_2O_2$ layers (a small change in the local $c$ axis dimension). On the upper (compressed) side, this spacing is clearly diminished, to 2.40 nm as measured across five unit cells, a compression of 4–5 %. Below the dislocation the structure is slightly dilated, the average spacing of $Bi_2O_2$ layers, measured across several unit cells, being 2.54 nm; the stretching by about 1 % is just significant. These deformations imply a change in the spacing between the B-site cations of the framework, but not necessarily a corresponding change in cation–anion distances. The linear –Ti–O–Ti– chains must necessarily be stretched to produce the small dilation, but the shape of next-neighbour potential energy curves makes it unlikely that the larger compressional distortion involves a substantial decrease in Ti–O distances. It is more likely that the octahedral groups twist, to give puckered sheets of oxygen atoms. In terms of a local mode description of the lattice vibrations, the strain energy of dilation may be correlated with the restoring force of a local stretching mode, that of compression with a local bending mode of the $[TiO_6]$ groups. It is known that the former are usually the stiffer; in $BaTiO_3$ the frequency of the symmetric stretch is about 500 $cm^{-1}$ as compared with about 340 $cm^{-1}$ for the local bending modes. Thus a difference between the compressed and dilated sides of the

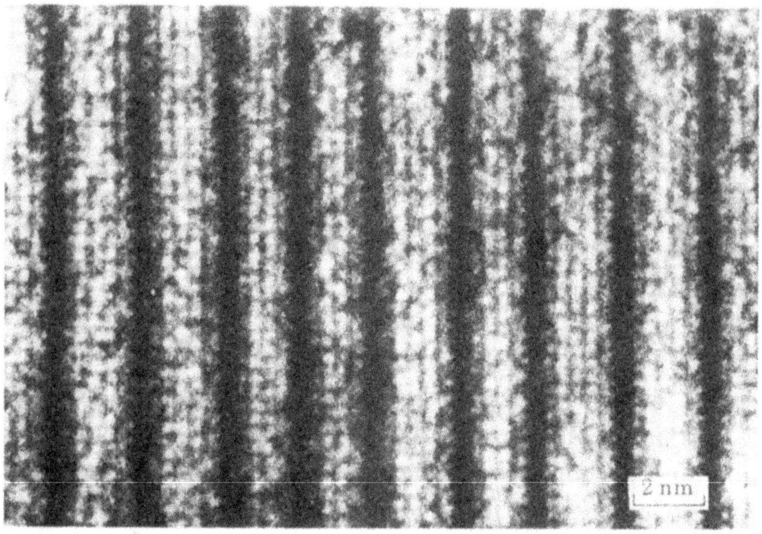


FIGURE 2. Lattice image showing faulted sequence in nominal $Bi_9 Ti_3Fe_5O_{27}$ ($n$ = 8), showing random occurrence of $n$ = 4 and $n$ = 5 sequences. The square grid of the perovskite structure can be seen.

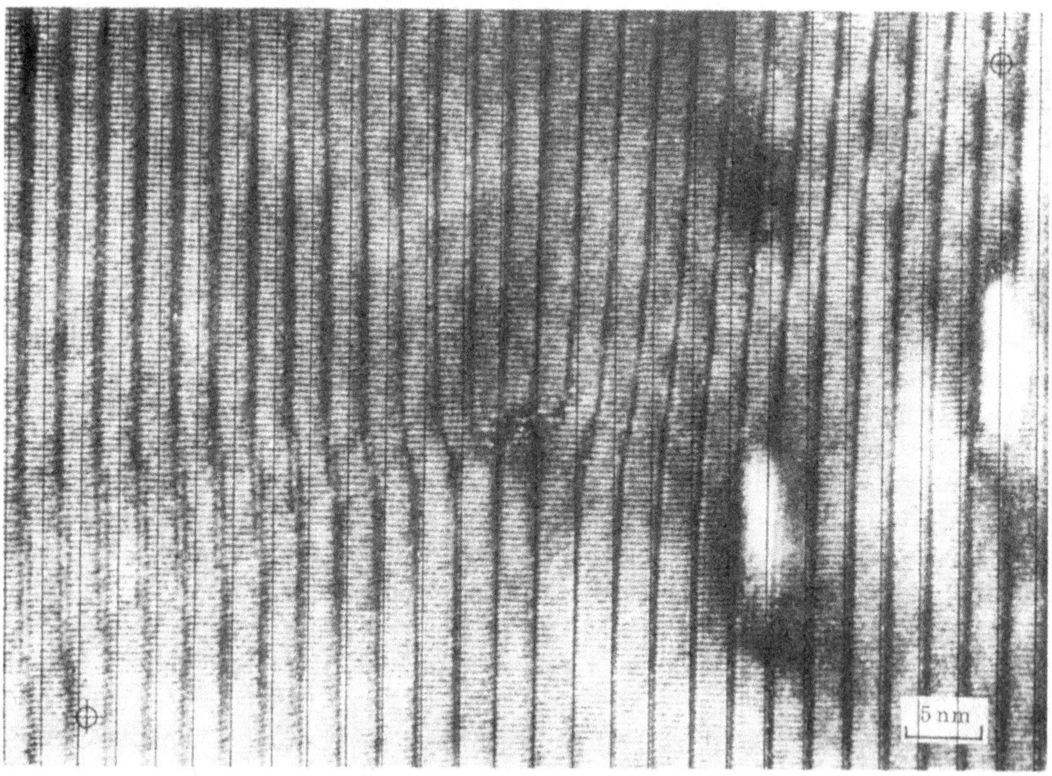


FIGURE 3. Lattice image of $Ba_2Bi_4Ti_5O_{18}$ ($n$ = 5) containing a dislocation. A grid showing positions of $Bi_2O_2$ layers in unperturbed structure is superimposed for comparison.

(*Facing p.* 306)

*Proc. R. Soc. Lond. A, volume* 355 *Hutchinson et al., plate* 2

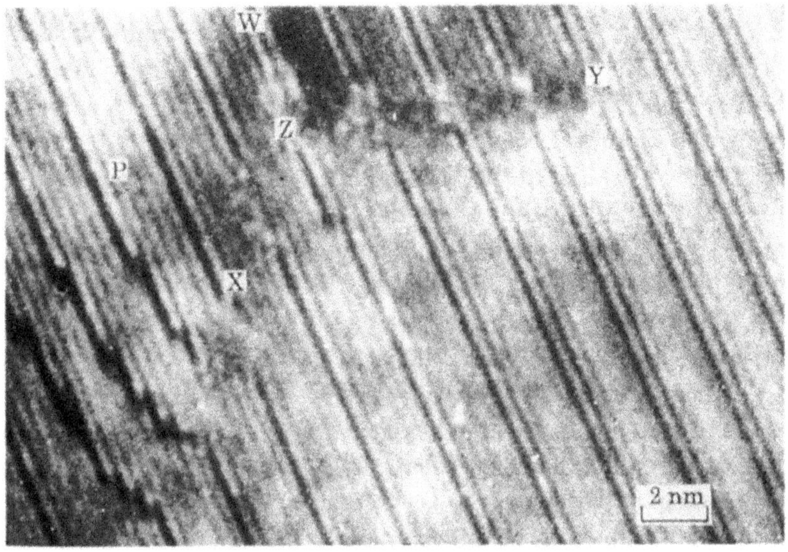


FIGURE 5. Fault in $Ba_2Bi_4Ti_5O_1$, with one extra half-slab of perovskite structure introduced, showing resultant displacement of $Bi_2O_2$ layers by several discrete steps.

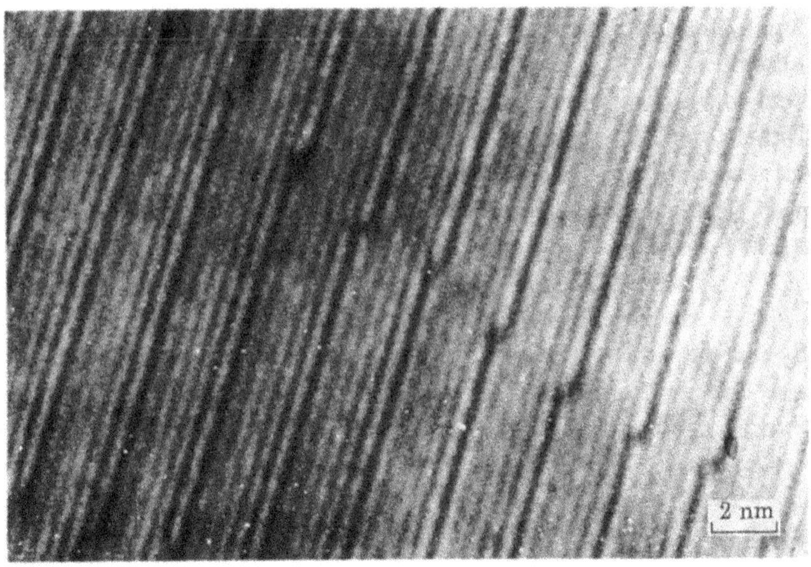


FIGURE 7. Stacking fault in $Ba_2Bi_4Ti_5O_{18}$, across which $Bi_2O_2$ layers are out of step by the width of one perovskite cell.

defect is explicable. The expansion of the structure produces a marked warping of the layers over considerable distances, as seen in the bending of the $Bi_2O_2$ bands over the whole field shown; this is particularly marked on the right-hand side.

There is a marked difference between the left-hand and right-hand sides of the dislocation. On the left, the $Bi_2O_2$ layers are much less severely bent in the $a_c$–$c_c$ plane, but the dark fringes showing their location are broadened, and this broadening increases with distance from the dislocation. It is possible that this broadening can be attributed to buckling in the $b_c$–$c_c$ plane: a general warping that destroys the exact alignment of bismuth atoms along the projection direction. Running left from the dislocation core is an extended fault of some kind, within which the $Bi_2O_2$ layers are seen as doubled fringes, each of the same width as in good crystal and separated by about 0.4 nm. The broadened contrast bands are offset by the same amount in crossing the fault. Such a doubling of the projected image would be observed if well aligned $Bi_2O_2$ layers were located at different positions, spaced one perovskite cell apart, near the upper surface and the lower surface of the crystal flake respectively. It is possible that buckling has been locally relieved by side-stepping of the $Bi_2O_2$ sheets, giving a strip in which two $Bi_2O_2$ layers are then seen in projection.

Distribution of strain about the dislocation is thus highly asymmetric. One contribution to this asymmetry, as between compressed and dilated sides, has been considered. There is little evidence on which to interpret the considerable differences between the strain on the right and left sides. Whether some ferroelectric distortion – bringing small changes in cell dimensions and a polar axis no longer collinear with the crystallographic *c* axis – could contribute to the asymmetry is speculative.

The extent to which a polyhedron network can be elongated or compressed is very limited. Dimensional changes in the core region are abrupt, and the image shows evidence of a second way in which they can be locally accommodated. In several places, most clearly on the left-hand side, the $Bi_2O_2$ layers side-step by about 0.4 nm, a unit step of one perovskite cell, corresponding to a change of $\pm 1$ in the number of $[TiO_6]$ octahedra between layers, as is shown schematically in figure 4. In effect, elastic strain in the most perturbed part of the transition region can be relieved by introducing elements of other members of the homologous series of compounds. Their presence implies a localized change of chemical composition.

Dislocations in these layered perovskite derivatives need structural readjustments very similar to those that we have analysed in the niobium oxide block structures. In each case, the basic structure is a simple cubic array of octahedrally coordinated B site cations; in order that the structure should join up as good crystal around the dislocation, all displacements must result in configurations that are compatible with that structure and with the superlattice, which is imposed, in the present case, by the $Bi_2O_2$ layers. The significant feature of this superlattice is the antiphase displacement of successive perovskite slabs by $\frac{1}{2}a_{ss}[100]_{ss}$ $(=\frac{1}{2}a_c[110]_c)$ across each $Bi_2O_2$ layer. Since dislocations such as that in figure 3 generate two extra perovskite slabs, with their bounding $Bi_2O_2$ layers, they interpolate one complete crystallographic unit cell; this preserves the matching of atomic positions at the

$Bi_2O_2$ layers. These are perfect dislocations of the superstructure, with the very large Burgers vector $\boldsymbol{b} = c_{ss}$, and they can have pure edge character.

Most of the dislocations found have been of this type. In principle, dislocations introducing only a single perovskite slab are possible. In these, the antiphase displacement across the $Bi_2O_2$ layers, i.e. $\frac{1}{2}a_{ss}$ must still be accommodated, and the smallest Burgers vector for such dislocations would appear to be $\frac{1}{2}[a\,0\,c]_{ss}$.

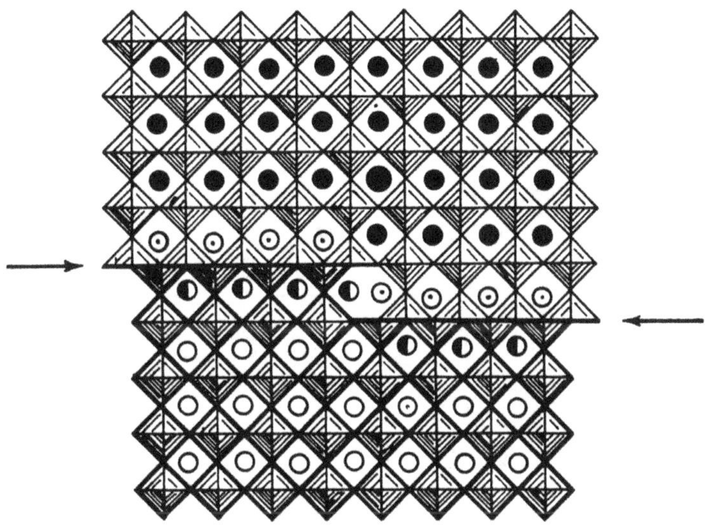

FIGURE 4. Idealized drawing of a unit shift of *ca.* 0.4 nm (one perovskite cell) in the $Bi_2O_2$ layer.

Dislocations of this type have not been identified with certainty. It can be shown that combinations of smaller faults enable an extra half slab of perovskite structure, with its $Bi_2O_2$ layer, to be accommodated with minimal strain. A fault that may be of this type is shown in figure 5, plate 2, with an extra half slab inserted at P into the upper half of the field shown. In this, unlike figure 3, the sheets of perovskite sites parallel to the $Bi_2O_2$ layers run through the structure without bending or interruption; the $Bi_2O_2$ layers on the left are laterally displaced by a succession of unit steps. The insertion of the extra half-slab is crossed, from a point below X–Y, by a narrow strip within which the resolution is poor, and at Z there is a strip of enhanced contrast, parallel to the $Bi_2O_2$ layers (i.e. to $a_c$) and with its lower edge almost parallel to $[101]_c$. With one extra $Bi_2O_2$ layer inserted, the topological requirements considered above could be fulfilled by invoking a single dislocation at $X$, with the displacement vector $\frac{1}{2}a_c[110]_c$. However, this would leave the other features of the micrograph unexplained. Consideration of the nature of antiphase boundaries (APBs), and the lattice images expected from them, suggests a possible interpretation. Permitted APBs in the perovskite structure lie on $\{110\}_c$; they have the structure shown in figure 6*a*. In a projection down $[010]_c$, APBs of the types $(110)_c$ and $(011)_c$ will be parallel to lines of perovskite sites and run into the crystal at 45° to the plane of projection. In a micrograph, their projection should be a band of changed contrast. There is no change in the spacing between lines of perovskite sites crossing the APB, although there is a spacing anomaly along each line of A sites; this would not be visible in the $a_c$ direction of figure 5, which affords only one-dimensional resolution. Within the band covered by the projection of the oblique APB, lines of perovskite sites parallel to it are spaced only 0.2 nm apart and would not be resolved.

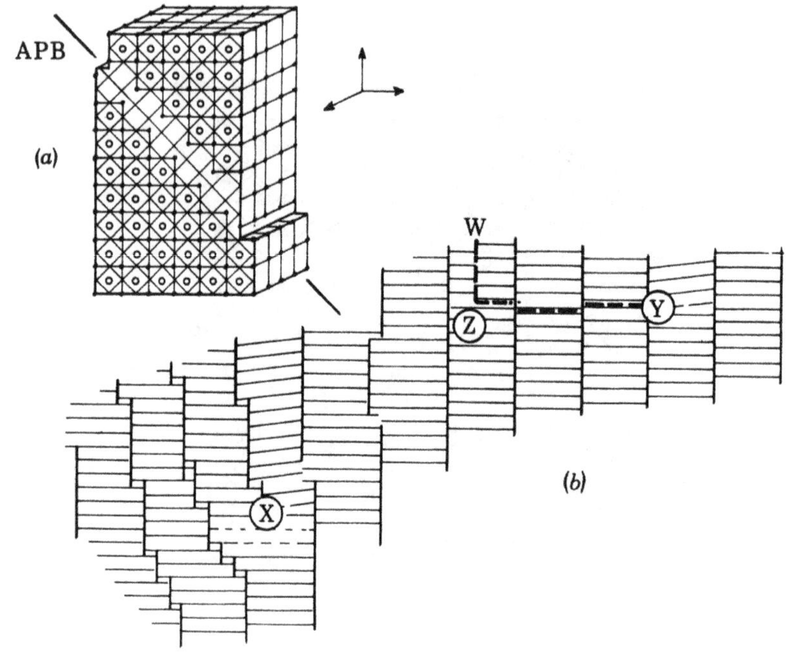


FIGURE 6. (*a*) Antiphase boundary in the perovskite structure. (*b*) Schematic representation of the possible structure of the faulted region in figure 5, with dislocations of $\frac{1}{2}$ $aa0_{ss}$ at $X$ and $Y$ and antiphase boundary $WZY$. Lines of perovskite sites perpendicular to the $Bi_2O_2$ layers (not resolved in the micrograph) are marked to show distortions imposed by the introduction of an odd number of $Bi_2O_2$ layers.

Hence, in figure 5, an APB on $(011)_c$ should be seen as a band of darker contrast. It is therefore possible that the feature YZ represents an APB on $(110)_c$, initiated by a dislocation of $\frac{1}{2}a_c[110]_c$ at Y, which at Z intersects another APB on $(011)_c$ (the dark band ZW). The structure of this faulted region, on that hypothesis, is schematically represented in figure 6*b*.

Ths dislocations considered above operate upon the superstructure; imperfect partials could introduce additional perovskite layers between $Bi_2O_2$ layers, which would thereby be displaced by about 0·4 nm. Stepped displacements of this kind, in the core region, have been referred to. Figure 7, plate 2, shows an area with a stacking fault, across which four of the five layers in each slab of $n = 5$ structure run without interruption, but every $Bi_2O_2$ layer is displaced by the width of one perovskite cell. This fault originates outside the field of the micrograph, but could be accounted for if a half-sheet of octahedra – a single layer of $BaTiO_3$ – had been interpolated at some point (to form an element of the $n = 6$ structure) or eliminated (an element of $n = 4$ structure) by a partial dislocation.

Tentative inferences about the direction of the dislocation lines can be drawn, on the basis that lattice imaging resolution is attained only when equivalent elements of structure are rather exactly aligned throughout the thickness of the crystal. Figure 3 is a projection down $[010]_c$. If the dislocation line were tilted in the $b_c$–$c_c$ plane, the termination of the added $Bi_2O_2$ layers would be offset, as between the top and bottom of the crystal. This should have the effect of changing the extent of distortion in the core region from level to level, and would smear out the projected image of the $Bi_2O_2$ layers laterally. A tilt in the $a_c$–$b_c$ plane would similarly spoil the alignment through the layers where they deviate from the original $a_c$ direction. Whilst ordering along the projection direction is imperfect near the core, there is little evidence for a progressive shift of the strain field through the thickness of the crystal. No firm assignment is possible, but it is likely that the dislocation line is nearly parallel to the projection direction.

The foregoing relates to the components of the Burgers vector in the $a_c$–$c_c$ plane. In principle, the lattice image would not detect a displacement along $b_c$. This would hold good in regions remote from the core, but any screw character would impose very severe distortions of the polyhedron network in the core itself and these would almost certainly impair the resolution of the image. Since resolution is preserved rather well into the core region, it is likely that dislocations such as that in figure 3 have a Burgers vector $c_{ss}$.

*Domain walls*

Some insight into the structure of domain walls may be expected if domain boundaries could be directly observed by lattice imaging. We have, for the first time, made such observations on these bismuth oxide ferroelectrics.

When a single crystal in the paraëlectric state transforms to a ferroelectric state of lower symmetry, stresses induced by dimensional changes can be relieved by domain twinning. For the tetragonal, pseudocubic $BaTiO_3$, there are two possible modes of twinning: 180° twins and 90° twins, with (100) and (101) respectively as composition planes. For the highly elongated structures considered in this paper, it may be expected that 180° twins could have $(100)_{ss}$, $(010)_{ss}$ or $(001)_{ss}$ as composition planes, but that 90° twins would be restricted to $(110)_{ss}$ as composition plane.

These compounds are generally pseudotetragonal or pseudoorthorhombic in the ferroelectric state. Cummins and Cross (1967) found a slight monoclinic distortion in $Bi_4Ti_3O_{12}$, where the monoclinic $a_m$–$c_m$ plane coincided with the orthorhombic (paraelectric) $b_o$–$c_o$ plane. The polar axis in these bismuth oxides has been considered to lie along, or close to, the crystallographic $c$ axis (Fang, Robbins & Aurivillius 1962; Ismailzade 1964); Cummins and Cross found that it was slightly tilted, by 7° or less, from the $b_o$–$c_o$ plane. By analogy with $BaTiO_3$ we would expect to find 180° domain walls parallel to the $a_o$–$c_o$ or $b_o$–$c_o$ planes and 90° domain walls with (110) contact planes. Whilst structural distortion along a 180° wall should be small, that along a 90° wall would be considerable.

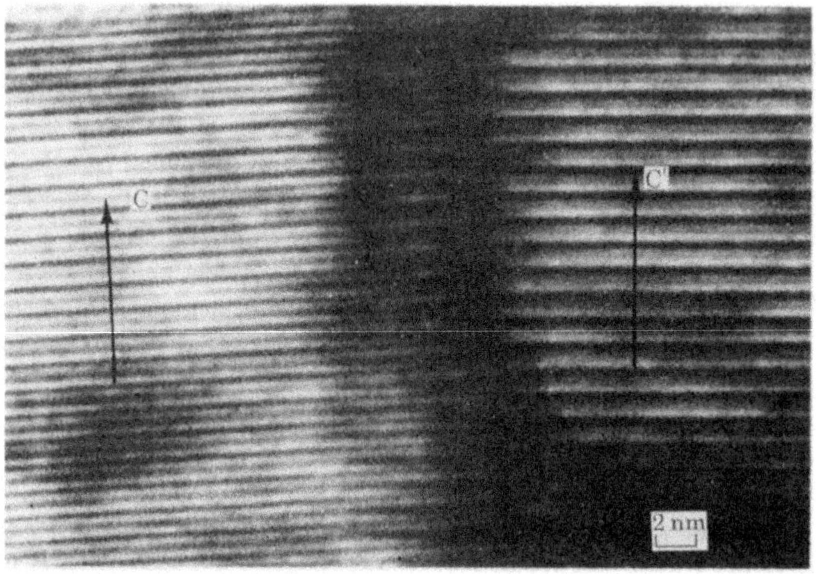


FIGURE 8. Lattice image of $BaBi_2Nb_2O_9$ ($n = 2$) showing a 180° domain wall.

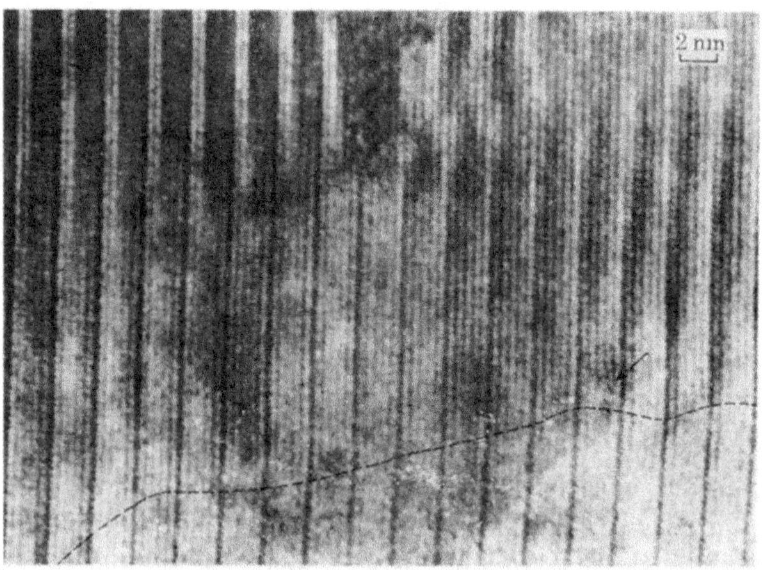


FIGURE 9. Lattice image of $Ba_2Bi_4Ti_5O_{18}$ showing a 180° domain wall.

(*Facing p.* 310)

*Proc. R. Soc. Lond. A*, *volume* 355 *Hutchinson et al.*, *plate* 4

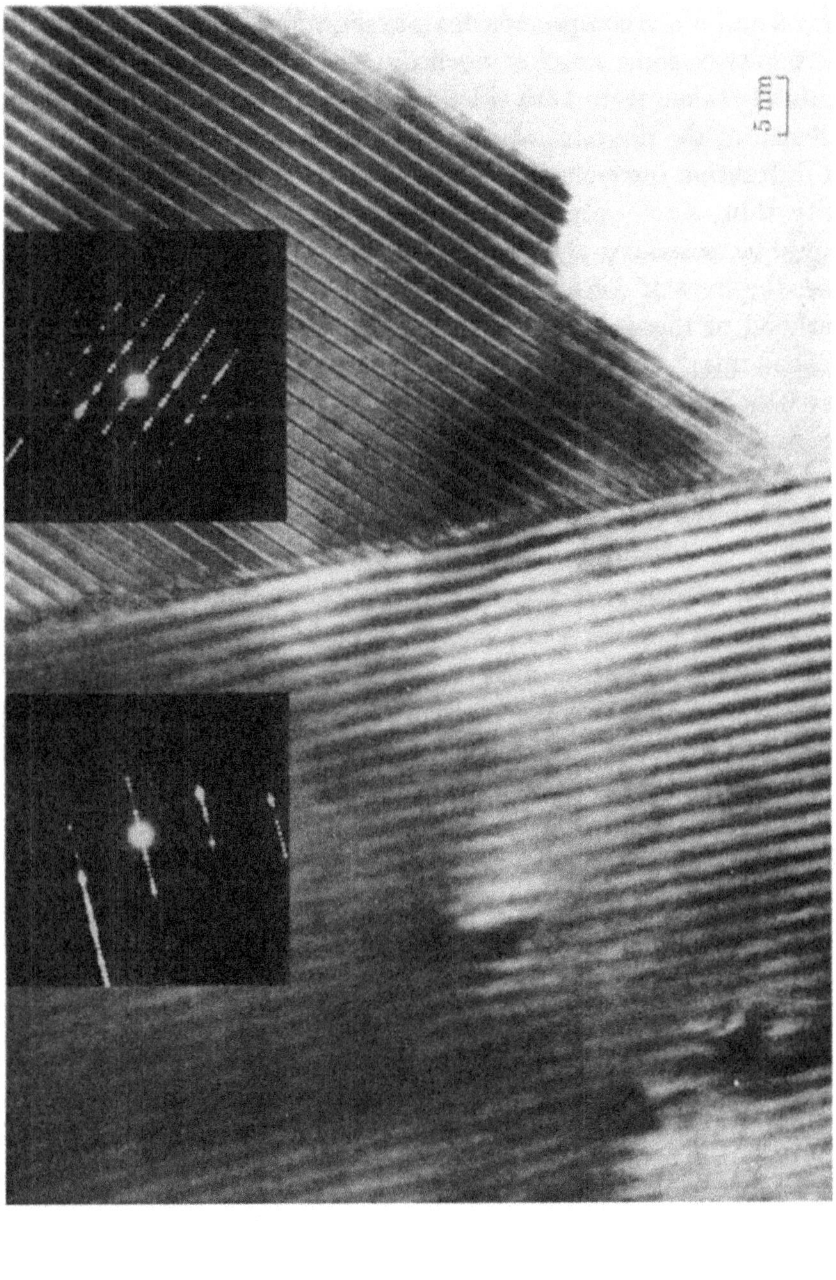


Figure 10. Lattice image of $BaBi_4Ti_4O_{15}$ ($n = 4$) showing an unusual domain wall, possibly of 90° type. Inserts: electron diffraction patterns from the two domains.

Figures 8 and 9, plate 3, are lattice images of two compounds ($n = 2$ and $n = 5$ respectively) showing evidence of 180° domain walls. The *c* axes, normal to the dark $Bi_2O_2$ sheets in the plane of the image, are nearly, but not exactly, parallel in the two domains; the two component *c* axes, C and C′, are inclined at about 3°. These observations suggest that the domain walls are of 180° type and that the polar axes of both the $n = 2$ and $n = 5$ compounds deviate slightly from the $c_t$ or $c_o$ axes. As in $Bi_4Ti_3O_{12}$, there may be some small monoclinic distortion. Selected area diffraction patterns, nominally taken from both sides of the domain boundary, were identical, but the small size of the domain, about 50 nm across, makes the evidence from selected area diffraction inconclusive. As expected for 180° domains, the domain walls are quite thin; since only a small adjustment or relaxation of the $[BO_6]$ octahedra would be necessary, the structure of the wall is undoubtedly simple. As a consequence, the lines of contrast due to the A site cations are clearly visible in the neighbourhood of the wall (figure 9), with only minor distortions around the boundary (dashed line).

A boundary which may represent a domain wall of a different type, in the tetragonal $n = 4$ compound, is shown in figure 10; plate 4; this presents some problems that cannot be resolved. In projection, the $Bi_2O_2$ layers are inclined at 45°; selected area diffraction patterns show that the two domains present different orientations. $[110]_{ss}$ of the tetragonal structure on the right hand side is exactly parallel to the electron beam, and the lines of A site cations between the $Bi_2O_2$ sheets are fully resolved. The left-hand domain is not thereby brought into an orientation in which its structure is resolved; the beam direction is approximately along $[210]_c$, but off-centre Laue zones show that it is tilted a few degrees away from this direction. The boundary between the domains, if it can be regarded as a domain wall, appears to be complex and thick; the interface may, indeed, be slightly inclined to the image plane, thereby giving rise to overlap contrast. Although this interface, which appears to be close to $(320)_c$, seems to be coherent, it is difficult to understand how that orientation could meet the prime requirement both for a physical composition plane and for a ferroelectric contact plane – continuity of structure – without severe distortions.

The authors thank the Science Research Council for support of this research and one of the authors (C.N.R.R.) is grateful to the Association of the Commonwealth Universities and the University of Oxford for the award of a Commonwealth Visiting Professorship during 1974–5.

## References

Allpress, J. G. & Sanders, J. V. 1973 *J. Applied Cryst.* **6**, 165.
Anderson, J. S., Hutchison, J. & Lincoln, F. J. 1976 *Proc. R. Soc. Lond.* A **352**, 303.
Aurivillius, B. 1949 *Arkiv Kemi* **1**, 463.
Cummins, G. E. & Cross, L. E. 1967 *Appl. Phys. Lett.* **10**, 14.
Fang, P., Robbins, C. R. & Aurivillius, B. 1962 *Phys. Rev.* **126**, 892.
Goodenough, J. G. & Longo, J. M. 1970 *Landolt-Bornstein, New Series Group III*, vol. 4*a*. New York: Springer.
Ismailzade, I. H. 1964 *Soviet Phys. Cryst.* **8**, 686.
Ismailzade, I. H. 1967 *Soviet Phys. Cryst.* **12**, 400.
Lynch, D. F., Moodie A. F. & O'Keefe, M. A. 1975 *Acta Cryst.* A **31**, 300.
McConnell, J. D. M., ,Hutchison, J. L. & Anderson, J. S. 1974 *Proc. R. Soc.* A **339**, 1.
O'Keefe, M. A. 1973 *Acta Cryst.* A **29**, 389.
O'Keefe, M. A. 1976 Thesis, University of Melbourne.
Schmid, H. 1975 *Magnetoelectric Interaction Phenomena in Crystals* (ed. H. J. Freeman & H. Schmid) New York: Gordon and Breach.
Subbarao, E. C. 1974 *Solid state chemistry* (ed. C. N. R. Rao). New York: Marcel Dekker.
Van Landuyt, J., Remaut, G. & Amelinckx, S. 1969 *Mat. Res. Bull.* **4**, 329.

*Proc. R. Soc. Lond.* A **395**, 127–139 (1984)
*Printed in Great Britain*

# Bismuth–tungsten oxide bronzes: a study of intergrowth phases and related aspects

By A. Ramanan[1], J. Gopalakrishnan[1], M. K. Uppal[2], D. A. Jefferson[2] and C. N. R. Rao, F.R.S.[1,2]

[1] *Solid State and Structural Chemistry Unit, Indian Institute of Science, Bangalore-560012, India*

[2] *Department of Physical Chemistry, University of Cambridge, Cambridge CB2 1EP, U.K.*

(*Received* 24 *January* 1984)

[Plates 1–10]

Reaction of bismuth metal with $WO_3$ in the absence of oxygen yields interesting bronze-like phases. From analytical electron microscopy and X-ray photoelectron spectroscopy, the product phases are found to have the general composition $Bi_xWO_3$ with bismuth in the 3+ state. Structural investigations made with high resolution electron micrscopy and cognate techniques reveal that when $x < 0.02$, a perovskite bronze is formed. When $x \geqslant 0.02$, however, intergrowth tungsten bronzes (i.t.b.) containing varying widths of the $WO_3$ slab are formed, the lattice periodicity being in the range 2.3–5.1 nm in a direction perpendicular to the $WO_3$ slabs. Image-matching studies indicate that the bismuth atoms are in the tunnels of the hexagonal tungsten bronze (h.t.b.) strips and the h.t.b. strips always remain one-tunnel wide. Annealed samples show a satellite structure around the superlattice spots in the electron diffraction patterns, possibly owing to ordering of the bismuth atoms in the tunnels. The i.t.b. phases show recurrent intergrowths extending up to 100 nm in several crystals. The periodicity varies considerably within the same crystal wherever there is disordered intergrowth, but unit cell dimensions can be assigned from X-ray and electron diffraction patterns. The maximum value of $x$ in the i.t.b. phases is *ca.* 0.07 and there is no evidence for the i.t.b. phase progressively giving way to the h.t.b. phase with increase in $x$. Hexagonal tungsten bronzes that contain bismuth with $x$ up to 0.02 can be formed by starting from hexagonal $WO_3$, but the h.t.b. phase seems to be metastable. Optical, magnetic and electron transport properties of the i.t.b. phases have been measured and it appears that the electrons become itinerant when $x > 0.05$.

## 1. Introduction

Bismuth forms two tungstates, $Bi_2WO_6$ and $Bi_2W_2O_9$. The third member of this homologus series, $Bi_2W_nO_{3n+3}$, has not been isolated as a single phase material, but $Bi_2W_3O_{12}$ is found occasionally as an intergrowth in $Bi_2W_2O_9$ (Jefferson *et al.* 1982). These homologous tungstates are similar to the well known Aurivillius phases (Aurivillius 1949; Hutchinson *et al.* 1977, $Bi_2A_{n-1}M_nO_{3n+3}$ (A = Ba, Bi; M = Ti, W, Nb), except that instead of the $A_{n-1}M_nO_{3n+1}$ perovskite layers, $W_nO_{3n+1}$ layers are present in between $Bi_2O_2$ layers in the tungstates. We considered it worthwhile to investigate the interaction between bismuth metal and $WO_3$ to see if reduced forms of the homologous tungstates of the type $Bi(WO_3)_n$ are formed. If such compounds are indeed formed, they can be solid solutions of the type $Bi_xW_{1-x}O_{3-y}$ or tungsten bronzes of the general formula $Bi_xWO_3$. The bronzes can be perovskite-like, tetragonal, hexagonal or, more interestingly, of the intergrowth variety. In spite of the extensive studies on tungsten bronzes reported in the literature, the Bi–$WO_3$ system has not been examined in any detail before and we have now made a detailed investigation of this system by using high resolution electron microscopy, X-ray diffraction and other techniques.

Various types of bronze structures have been reviewed by Ekstrom & Tilley (1980) and the structural principles of intergrowth bronzes have been discussed by Hussain & Kihlborg (1976). Intergrowth tungsten bronzes (i.t.b.) are characterized by the intergrowth of $WO_3$ slabs and strips of the hexagonal tungsten bronze (h.t.b.) structure. The i.t.b. phases described in the literature are rarely ordered, even on an ultramicrostructural scale, and the widths of the $WO_3$ slabs display a considerable variation, the h.t.b. regions ranging from one to three tunnels in width. The present study has shown that bismuth forms novel intergrowth bronze phases with wide-ranging periodicities, but with the h.t.b. region always limited to one tunnel in width. Bismuth atoms occupy the hexagonal tunnels and the maximum value of $x$ in $Bi_xWO_3$ appears to be *ca.* 0.10. We see no evidence for the i.t.b. phase giving way to the h.t.b. phase at large $x$. Recurrent coherent intergrowth in the i.t.b. phases extends up to 100 nm in some of the crystals and the largest cell parameter varies between 2 and 5 nm. More significantly, the present study on this i.t.b. system is probably the only one where ultrastructural studies have been made on samples that are well characterized, including *in situ* analysis. Results of the present study along with the observations on other coherent intergrowth systems in the recent literature raise questions about the traditional definitions of a phase and a unit cell. H.t.b. phases of bismuth ($x < 0.1$) prepared independently by starting from hexagonal $WO_3$ have a random distribution of bismuth atoms and seem to be metastable. Futhermore, h.t.b. phases with $x > 0.10$ cannot be made. In all the bronzes, bismuth is present in the 3+ state, each bismuth atom donating three electrons to $WO_3$. Accordingly, the electrical resistivity of i.t.b. $Bi_xWO_3$ phases decreases with increase in $x$, the d electrons becoming itinerant when $x > 0.05$.

## 2. Experimental

Samples of $Bi_xWO_3$ for nominal $x$ values of 0.01, 0.02, 0.05, 0.1 and 0.2 were prepared by heating $WO_3$ with metallic bismuth in the appropriate proportions in sealed and evacuated silica tubes at 1170 K for periods of six days. Initial characterization was by X-ray powder diffractometry employing a Philips diffractometer.

Samples were prepared for the electron microscope by liquid nitrogen embrittlement, followed by grinding in acetone suspension and deposition onto holey carbon films. Micrographs of regions of interest were recorded at magnifications in the range $\times 300000 - \times 600000$, astigmatism being corrected by observing the granularity of the holey carbon support film. Selected area electron diffraction (s.a.e.d) patterns were also recorded from the areas studied.

For the initial h.r.e.m. studies, samples were examined in a JEM-200CX electron microscope fitted with a side-entry goniometer and energy-dispersive X-ray detector. When operated at 200 kV, the objective lens characteristics of the instrument ($C_s = 2.8$ mm, $C_c = 3.4$ mm) were such that the interpretable point resolution just below the so-called Scherzer focus was of the order of 0.289 nm (figure 1*a*).

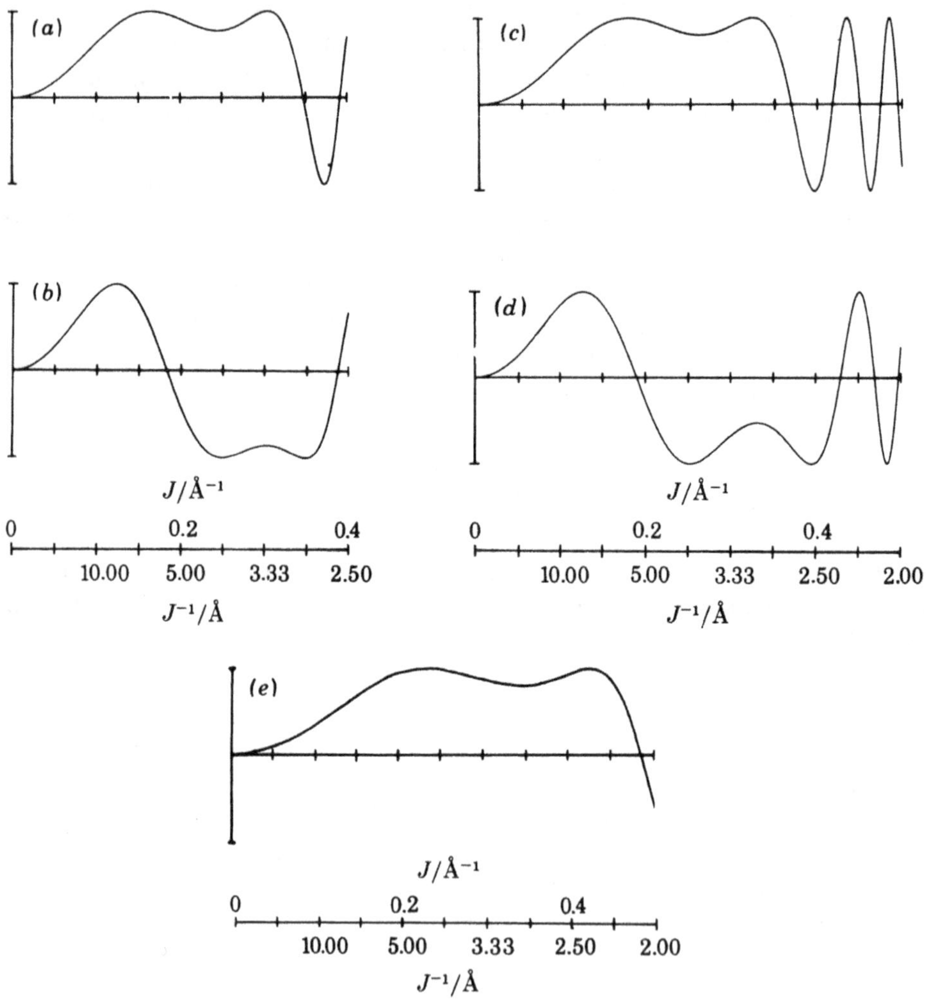


FIGURE 1. Phase contrast transfer functions for the electron-optical configurations used for image: (*a*) 200CX, initial configuration, $\Delta f = 85$ nm; (*b*) $\Delta f = 135$ nm; (*c*) 200CX, later configuration, $\Delta f = 72$ nm; (*d*) $\Delta f = 122$ nm; (*e*) Cambridge h.r.e.m., $\Delta f = 70$ nm. In all cases, the chromatic–beam divergence envelope is not shown because the conditions of validity of this approximation were not met in the actual images recorded.†

† $1\ \text{Å} = 10^{-10}\ \text{m} = 0.1$ nm.

Considering a simple damping envelope on the p.c.t.f. to indicate the effects of chromatic aberration and beam divergence (Fejes 1977), it could readily be shown that resolution up to *ca.* 0.25 nm was possible although this approximation was not directly applicable in the imaging conditions used. Consequently, most micrographs were recorded at a defocus well below the Scherzer position (figure 1*b*), to utilize the available resolution. Under such conditions, however, the atomic detail was imaged with reversed contrast and the p.c.d. approximation (Lynch *et al.* 1975) could not be used in image interpretation, image-matching studies using the multislice method (Goodman & Moodie 1974) being necessary. For a later part of the investigation the objective lens characteristics were improved to $C_s = 1.9$ mm and $C_c = 1.9$ mm. The corresponding 'optimum' and 'most useful' transfer function conditions for this later configuration are shown in figure 1*c*, *d*. To confirm the final model of the structure, images of the $Bi_{0.1}WO_3$ sample were recorded at 500 kV by using the Cambridge University high resolution electron microscope (h.r.e.m.) (Smith *et al.* 1982). In this case there was no advantage in going beyond the 'optimum' focus position (figure 1*e*), which was therefore used throughout. The interpretable point resolution of the images obtained was then of the order of 0.2 nm.

X-ray photoelectron spectra were recorded with a VG Escalab instrument using Mg K$\alpha$ radiation. The W (4f) signal was monitored with polycrystalline samples (Rao *et al.* 1979) and the relative intensities of $W^{5+}$ and $W^{6+}$ peaks were reproducible. The intensities were obtained by decomposing the 4f peaks into two Gaussians after subtracting the background. Reflectance spectra were measured with a Pye-Unicam SP8-100 spectrometer with MgO used as reference. Electrical resistivities were measured by the four-probe technique on pressed pellets of near-theoretical density. Seebeck coefficient measurements were made with a 'home-built' apparatus, while the magnetic measurements were made by the Faraday method with use of a Cahn electrobalance.

## 3. Results and discussion

### *Characterization*

Interaction of bismuth with $WO_3$ can give rise to $Bi_xW_{1-x}O_{3-y}$ or bronze-like $Bi_xWO_3$, depending on whether bismuth is in the 5+ or the 3+ state. In $Bi_xW_{1-x}O_{3-y}$, $Bi^{5+}$ would substitute for $W^{6+}$, while in $Bi_xWO_3$, $Bi^{3+}$ would occupy interstitial sites in $WO_3$. X-ray photoelectron spectra in the Bi(4f) region (figure 2) clearly show that bismuth is present in the 3+ state. Accordingly, the W(4f) spectra (figure 3) show the presence of both $W^{5+}$ and $W^{6+}$ states, the presence of each bismuth atom requiring three $W^{5+}$ ions. These observations suggest that the products of the reaction of Bi with $WO_3$ are most likely to be bronze-like. A scanning electron microscope examination showed that $Bi_xWO_3$ samples with $x \geqslant 0.05$ exhibit a plate-like habit, while samples with $x < 0.05$ show a three-dimensional habit similar to $WO_3$ (figure 4, plate 1). As expected of bronzes, the $Bi_xWO_3$ samples exhibit intense colours ranging from green to blue–black.

X-ray powder diffraction patterns of the $x = 0.01$ sample could be indexed on an orthorhombic unit cell with $a = 0.737(1)$, $b = 0.748(1)$ and $c = 0.3856(8)$ nm.

*Proc. R. Soc. Lond. A*, *volume* **395** *Ramanan et al.*, *plate* 1

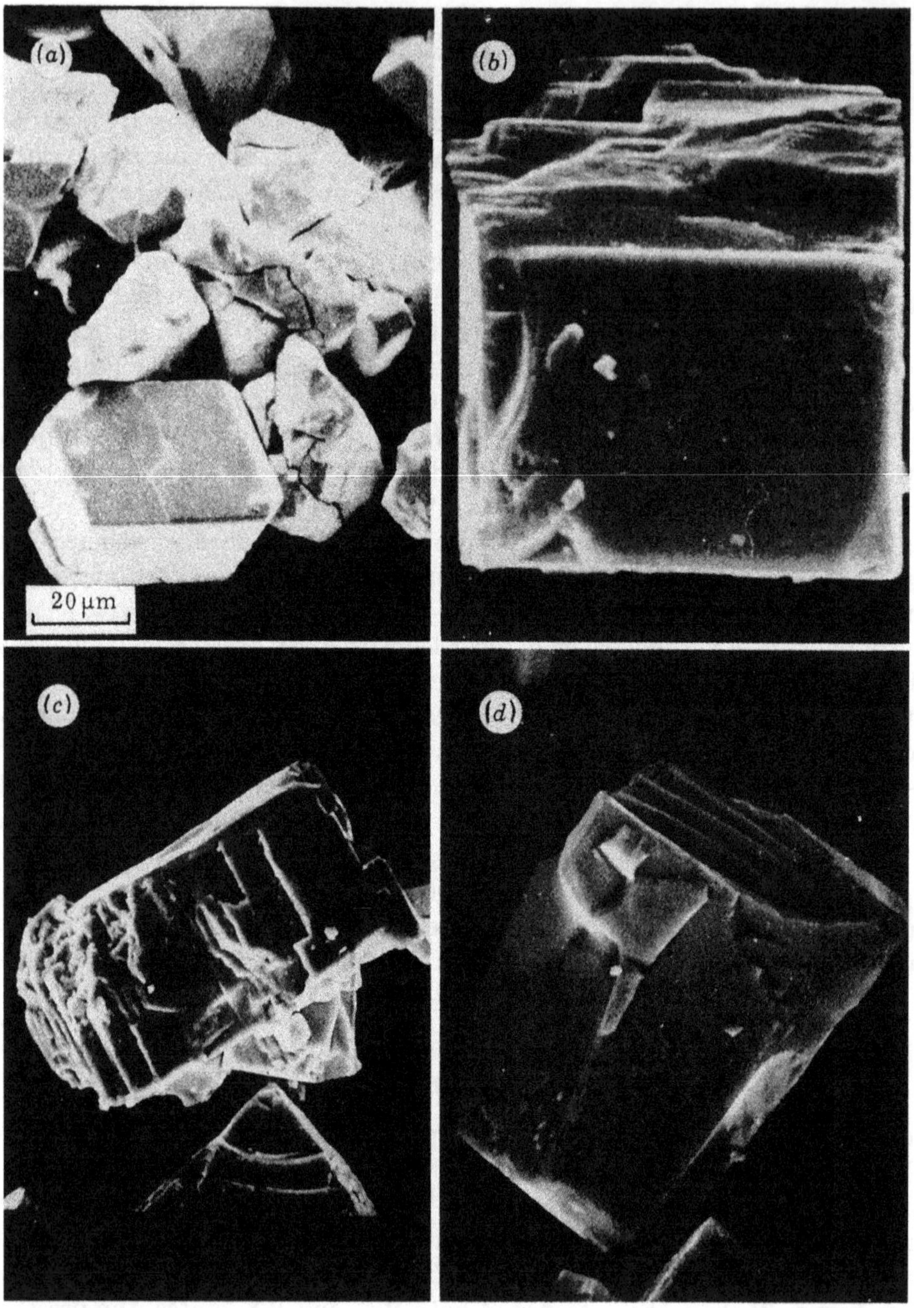


FIGURE 4. Scanning electron micrographs of $Bi_xWO_3$: (*a*) $x = 0.01$; (*b*) and (*c*) $x = 0.05$ and (*d*) $x = 0.10$. Magnifications are (*a*) $\times 2000$, (*b*) $\times 2000$, (*c*) $\times 1000$ and (*d*) $\times 5000$.

(*Facing p.* 130)

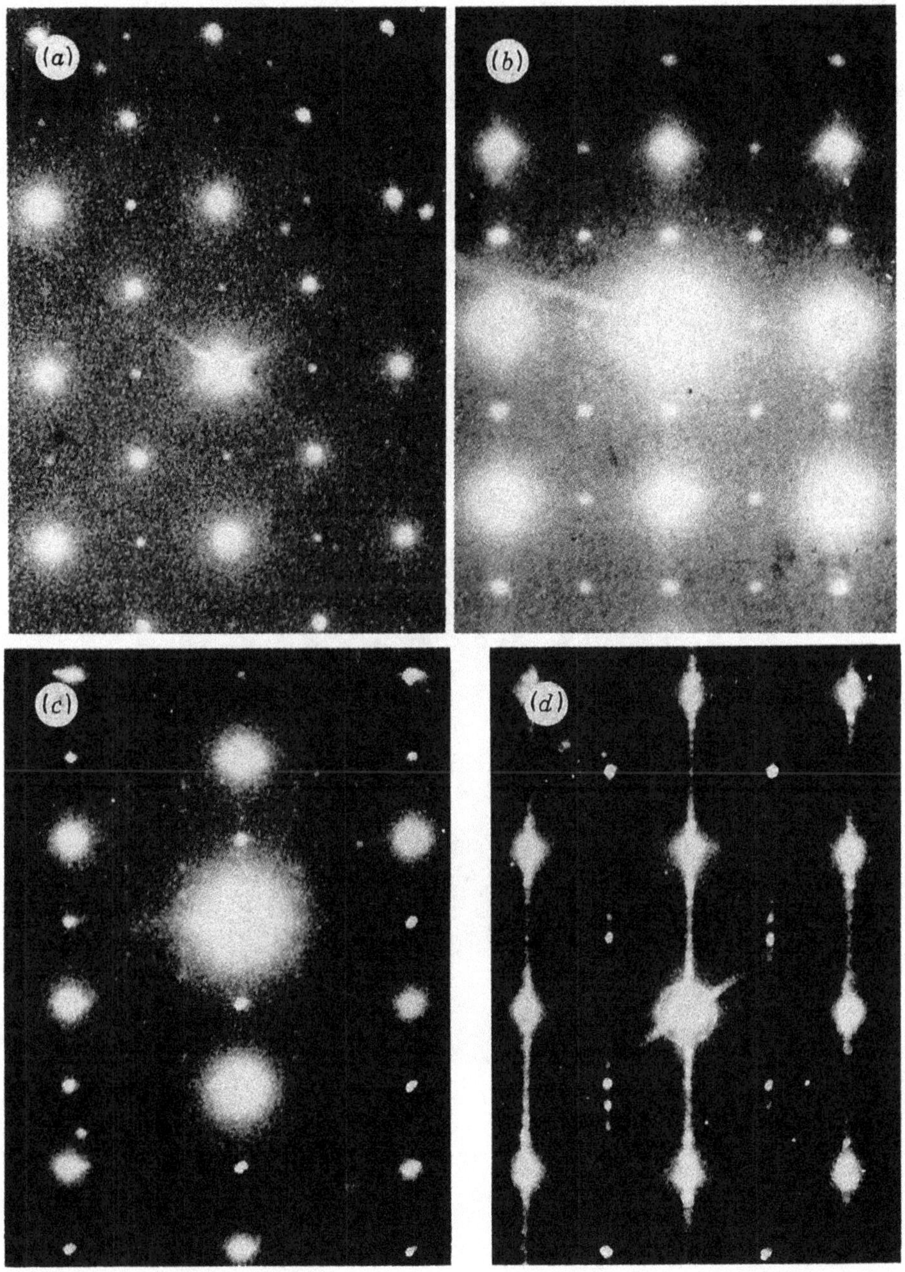


FIGURE 5. Electron diffraction patterns of $WO_3$-like crystals of $Bi_{0.01}WO_3$. Electron beam parallel to $[0\bar{1}0]$ in (*a*), [001] in (*b*) and $[01\bar{1}]$ in (*c*). The pattern of a minor phase showing streaking along $a^*$ is shown in (*d*).

*Proc. R. Soc. Lond. A, volume* 395 *Ramanan et al., plate* 3

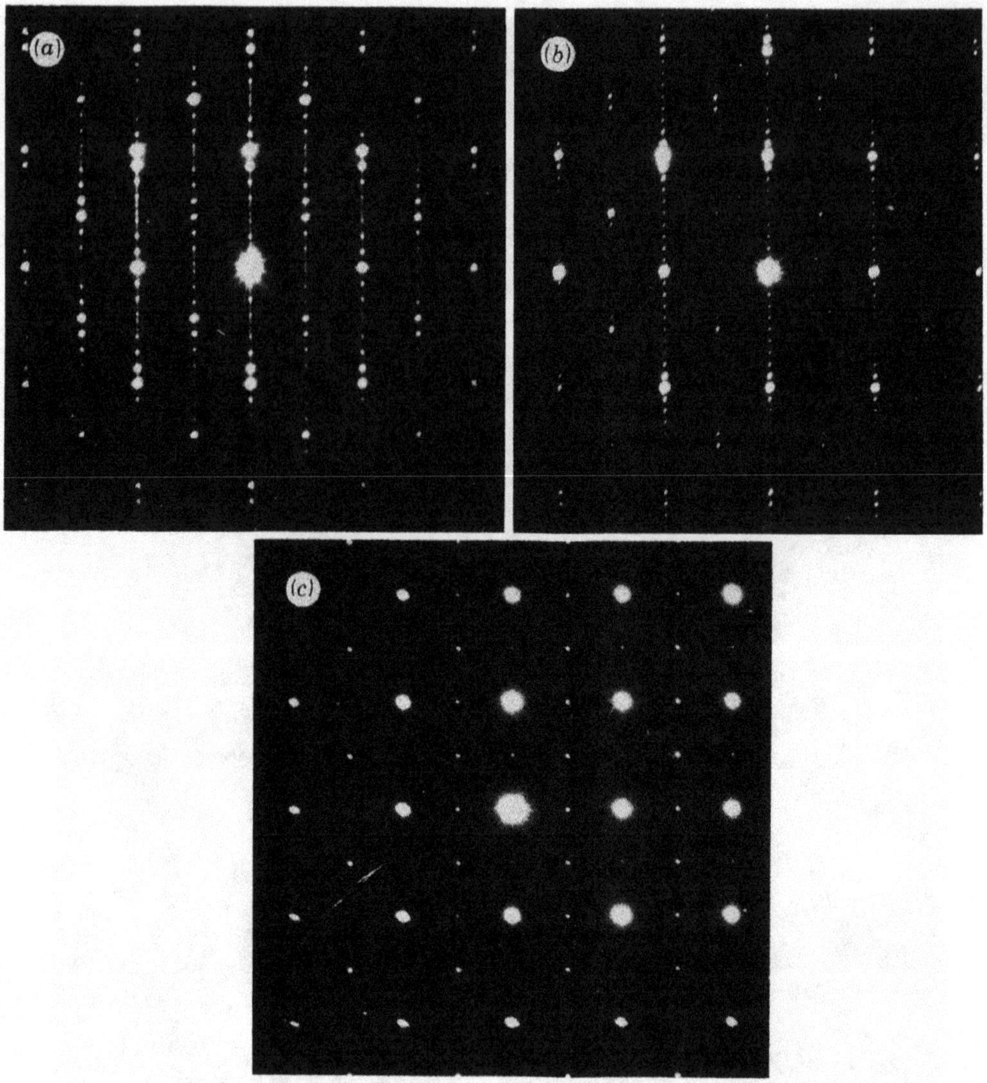

FIGURE 6. Typical s.a.e.d patterns of Bi–W-bronzes: (*a*) and (*b*) are principal sections showing the one-dimensional superlattice. These correspond to projection down [001] and [010] respectively, by the nomenclature of Hussain & Kihlborg (1976); (*c*) is the projection down the [100] axis.

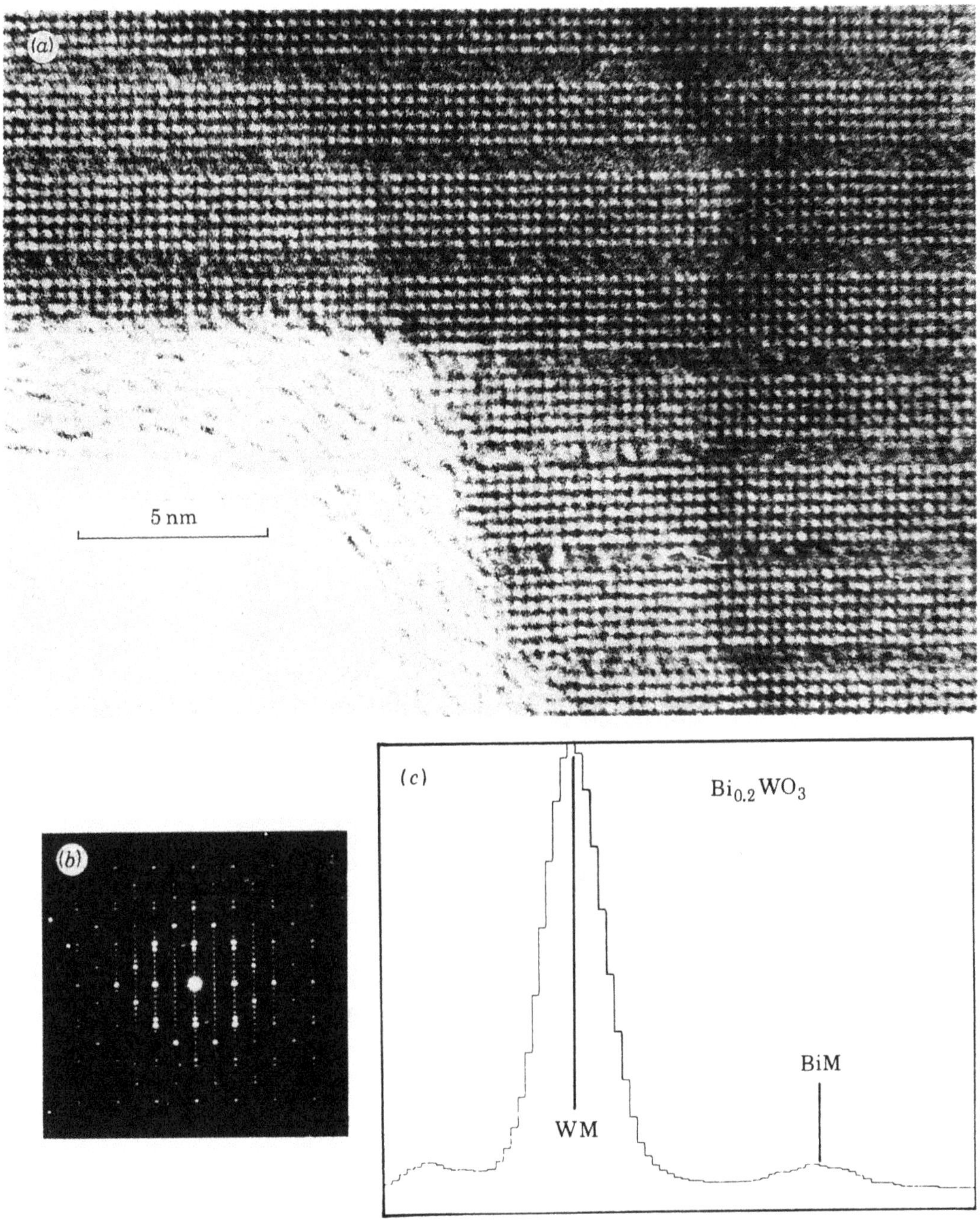


FIGURE 7. (*a*) h.r.e.m. (200 kV) image; (*b*) s.a.e.d. pattern and (*c*) part of the X-ray emission spectrum of a crystal of nominal composition of $Bi_{0.2}WO_3$. The image corresponds to the [001] projection.

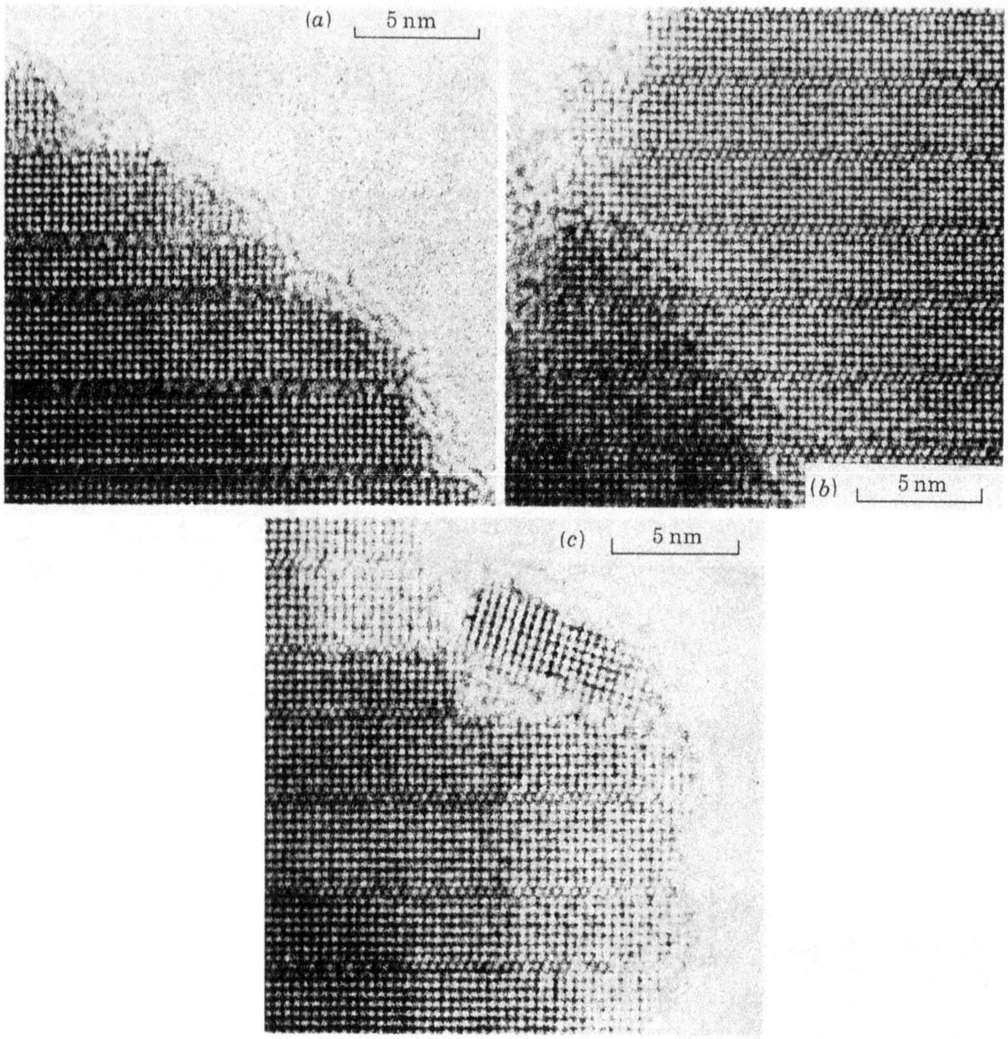


FIGURE 9. (*a*) H.r.e.m. image of a disordered intergrowth bronze of nominal composition $Bi_{0.1}WO_3$, $\Delta f \approx 150$ nm; (*b*) a more ordered crystal in the same sample, $\Delta f = 120$ nm; (*c*) crystal showing fragmentation at the edge, $\Delta f = 120$ nm. Axis of projection in each case is [001].

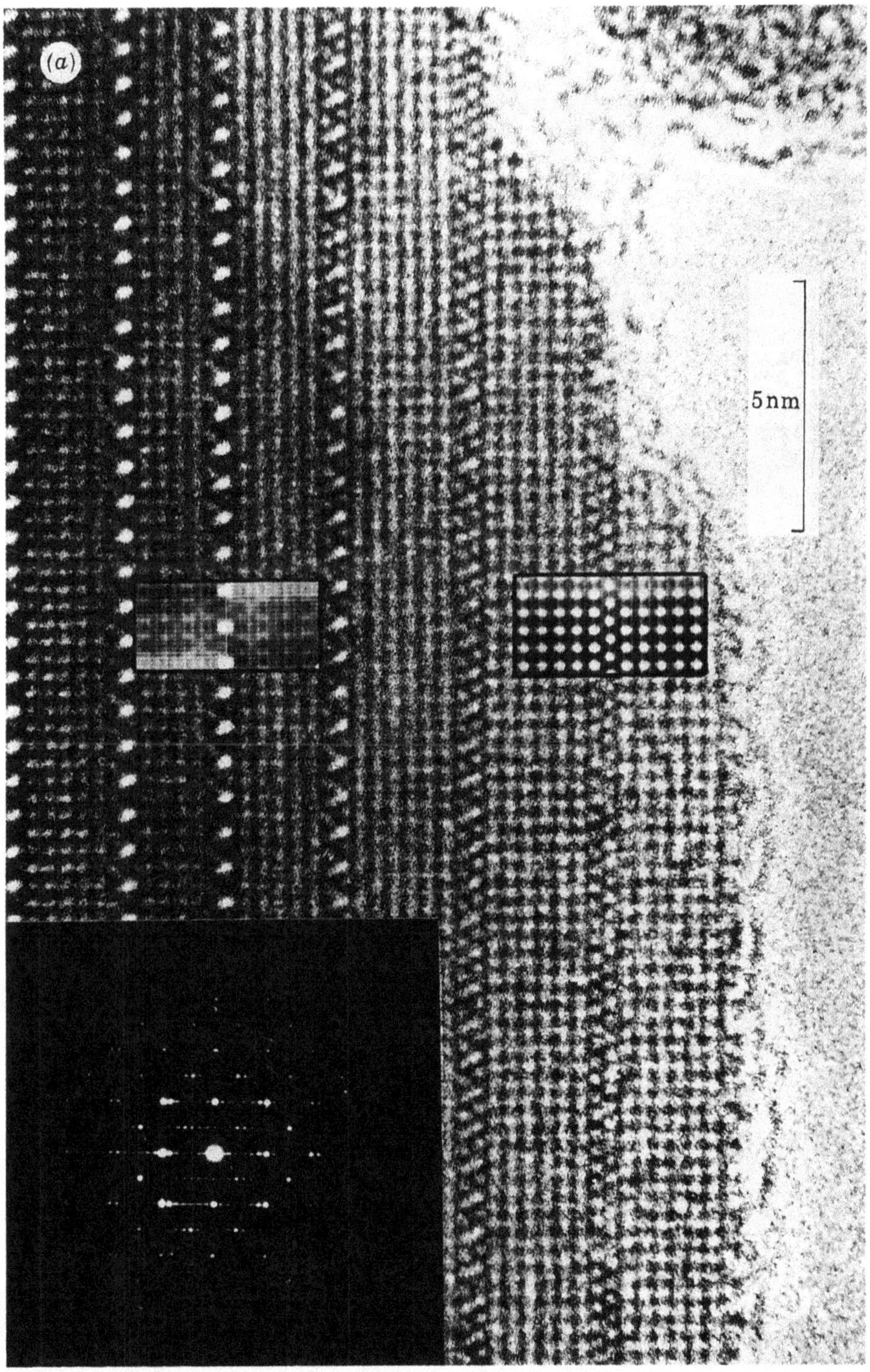


FIGURE 11. Image matching studies for the sample of nominal composition $Bi_{0.1}WO_3$. (*a*) [001] projection, simulations (inset) for crystal thickness of 3 nm (near edge) and 6 nm, $\Delta f = 120$ nm; (*b*) (opposite) [010] projection, simulation for crystal thickness of 4.5 nm and $\Delta f = 120$ nm. Completely filled tunnels in the simulations correspond to composition of $Bi_{0.067}WO_3$. The corresponding s.a.e.d. patterns are also shown.

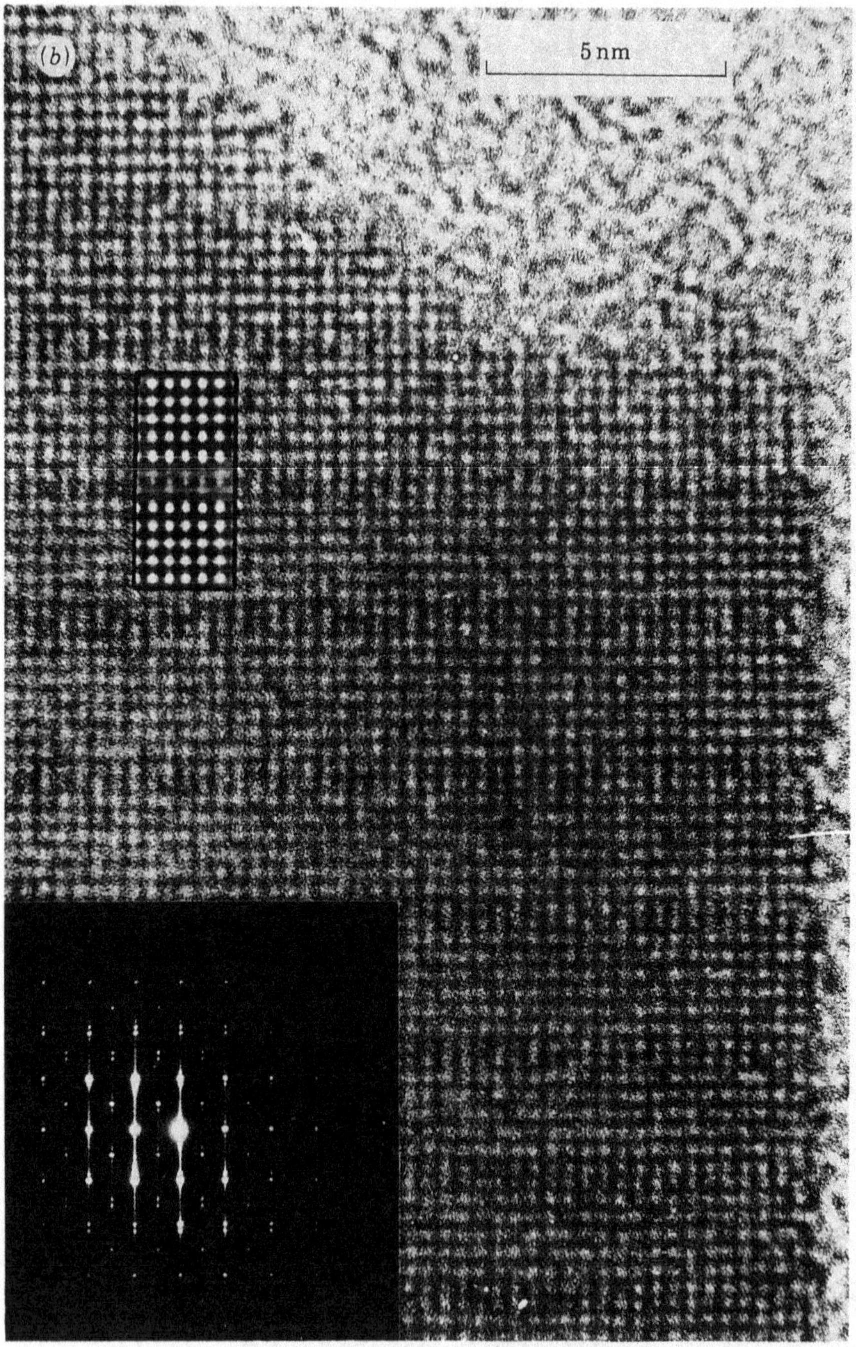


FIGURE 11(*b*). For full description see opposite.

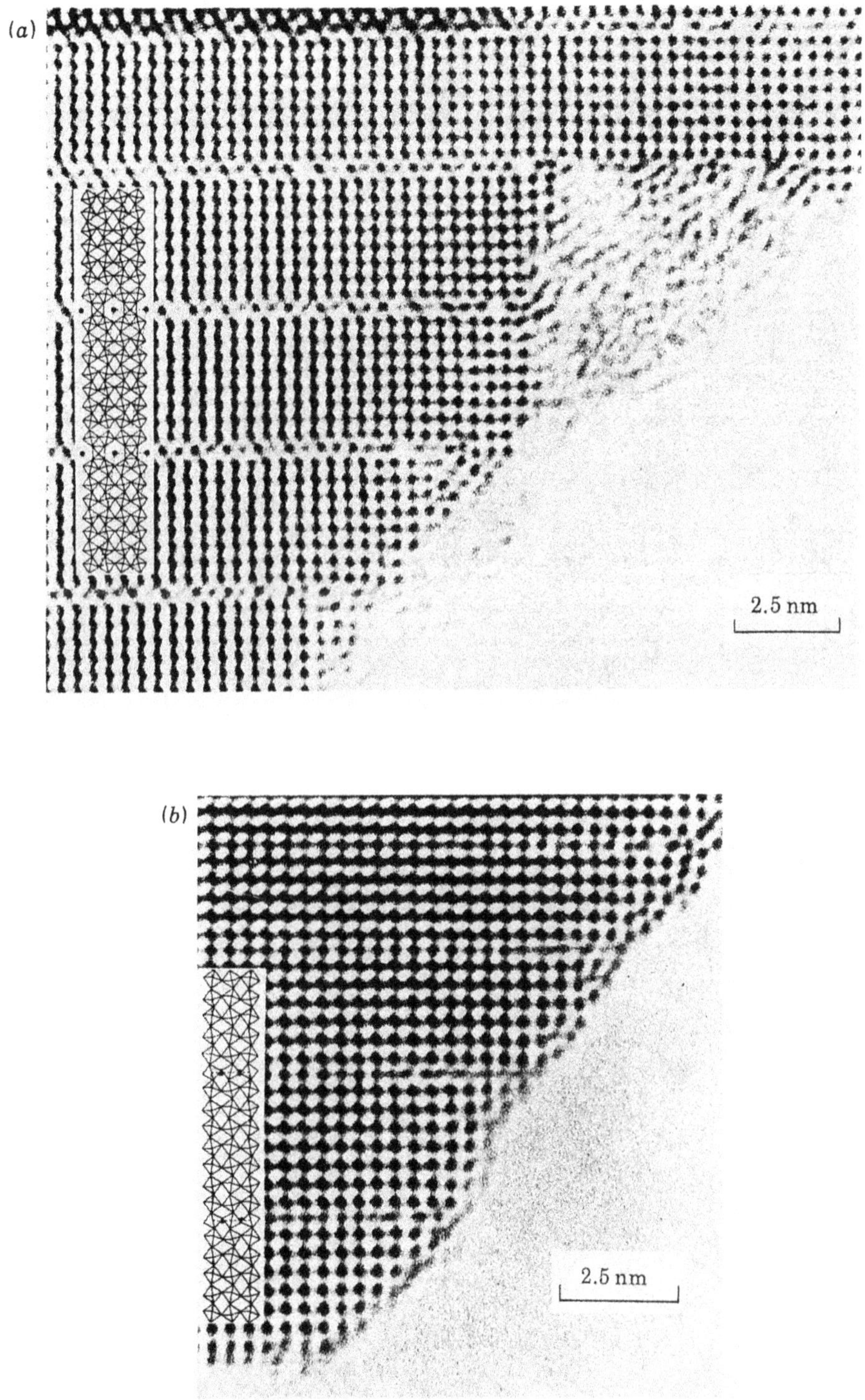


FIGURE 12. Five hundred kilovolt h.r.e.m. images of $Bi_{0.1}WO_3$ taken at $\Delta f \approx 70$ nm: (*a*) [001] projection showing apparently empty, or under occupied tunnels; (*b*) [010] projection. The corresponding structural models are also shown.

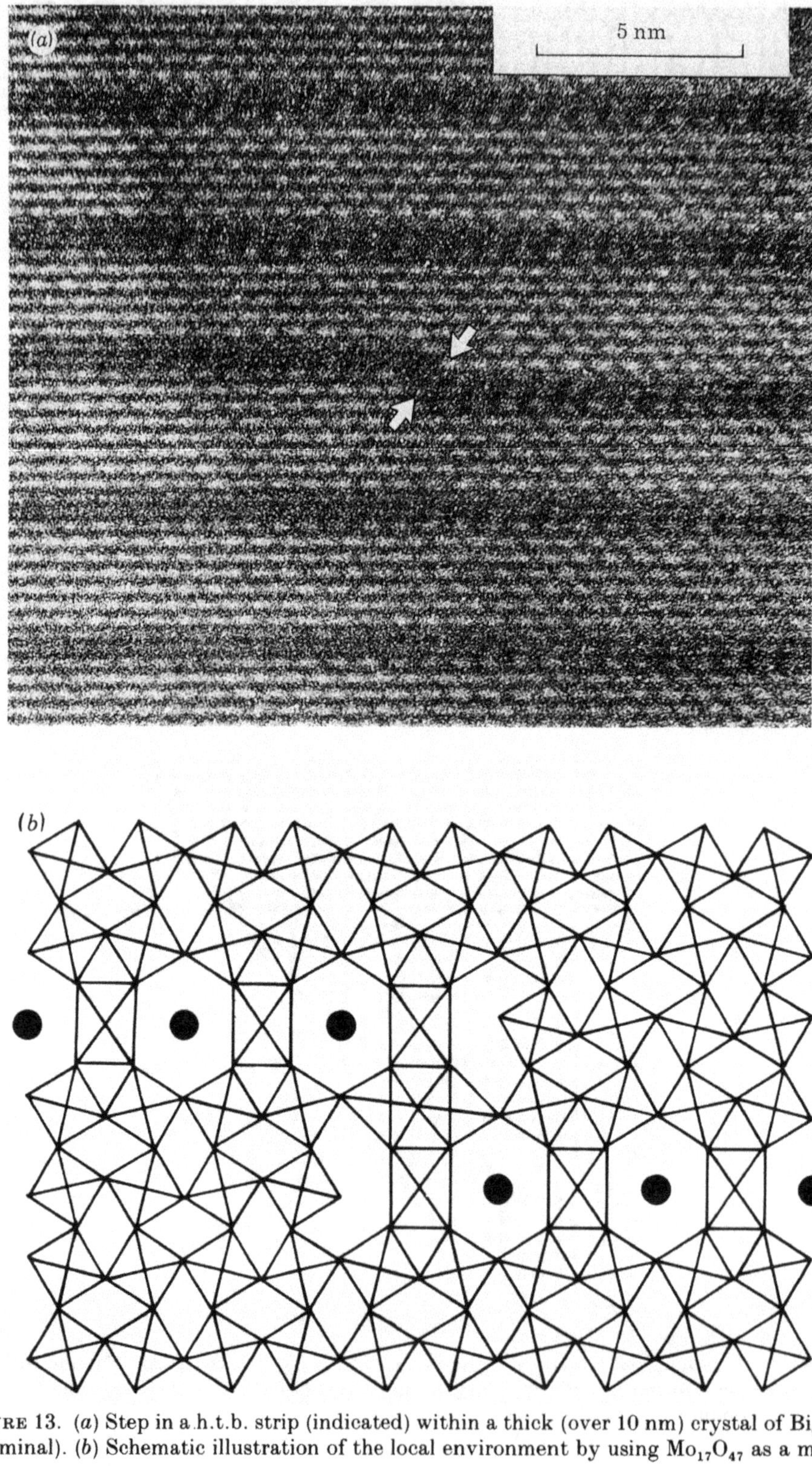


FIGURE 13. (*a*) Step in a h.t.b. strip (indicated) within a thick (over 10 nm) crystal of $Bi_{0.1}WO_3$ (nominal). (*b*) Schematic illustration of the local environment by using $Mo_{17}O_{47}$ as a model.

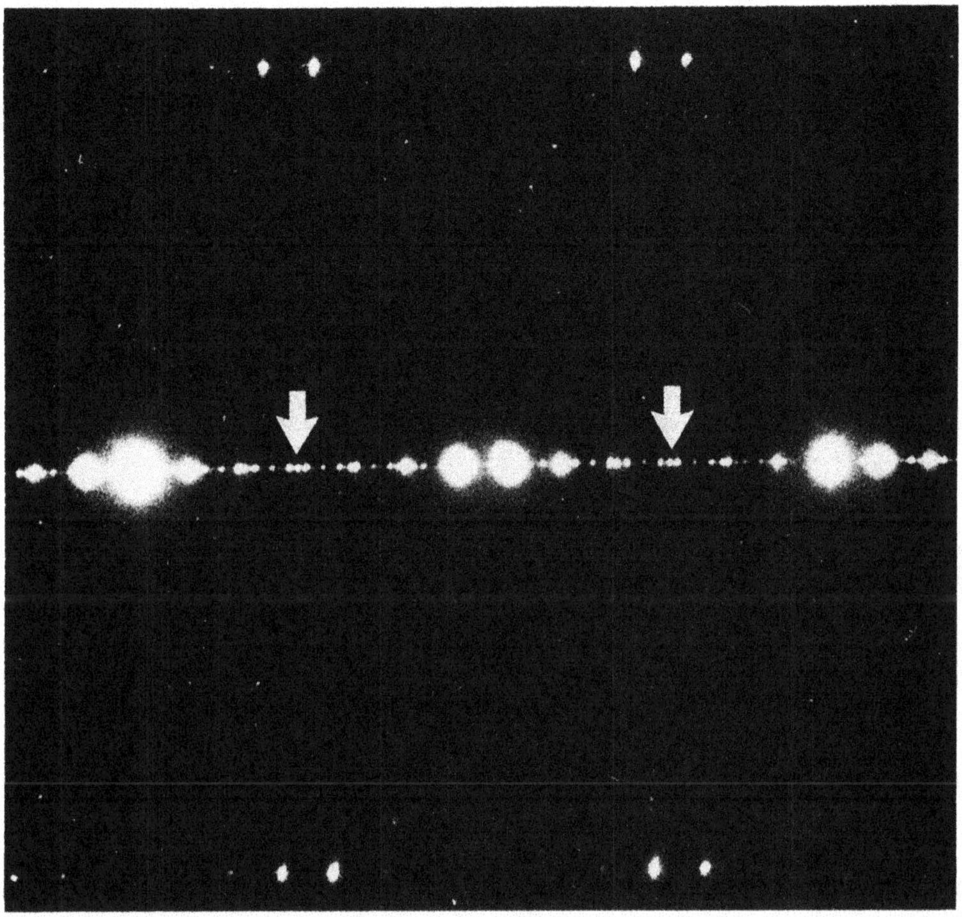

FIGURE 14. Satellites accompanying the superlattice spots in a crystal of $Bi_{0.1}WO_3$ (nominal). The overall periodicity of the satellite spot spacing corresponds to *ca.* 16 nm. The pattern corresponds to [011] projection.

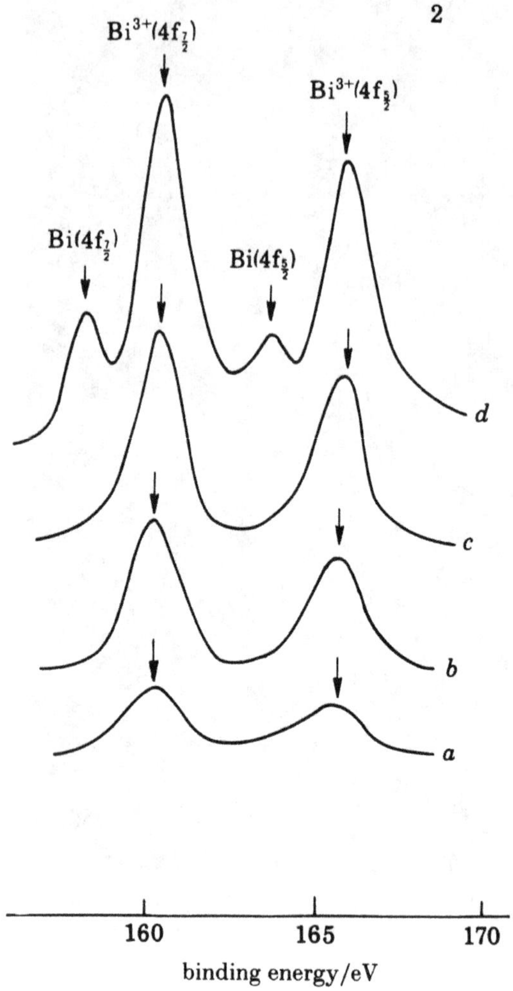


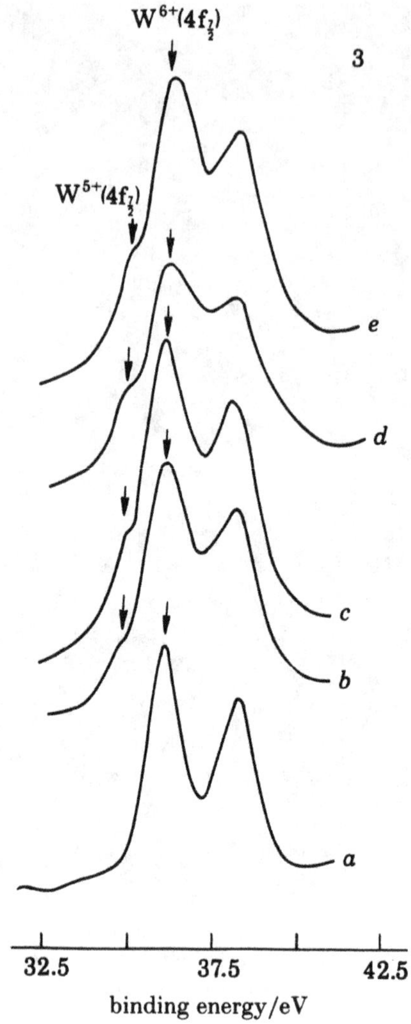


FIGURE 2. X-ray photoelectron spectra of $Bi_xWO_3$ in the Bi(4f) region: (*a*) $x = 0.01$; (*b*) $x = 0.05$; (*c*) $x = 0.10$ and (*d*) $x = 0.20$. Arrows indicate peak positions of $Bi^{3+}$ and $Bi^0$. Metallic $Bi^0$ is present in the samples since all the bismuth is not incorporated in the tunnels of $WO_3$. Progressive increase of $Bi^{3+}$ with increase in $x$ can be seen.

FIGURE 3. X-ray photoelectron spectra of $Bi_xWO_3$ in the W(4f) region: (*a*) $x = 0.00$; (*b*) $x = 0.01$; (*c*) $x = 0.05$; (*d*) $x = 0.10$ and (*e*) $x = 0.20$. Arrows indicate $4f_{\frac{7}{2}}$ peaks of $W^{5+}$ and $W^{6+}$.

Electron diffraction patterns of this specimen showed a superstructure of the basic $ReO_3$ structure, involving an approximate doubling of the pseudocubic $a_c$ axis in all the three directions (figure 5, plate 2). It is possible that this phase corresponds to the perovskite-like bronze structure reported by Parmentier & Gleitzer (1975). Some crystals of the $x = 0.01$ sample occasionally showed diffraction patterns (see figure 5*d*) similar to the patterns of the samples with $x \geqslant 0.02$. X-ray powder diffraction patterns of the samples with $x \geqslant 0.02$ were complex and markedly different from the pattern exhibited by the $x = 0.01$ sample. Their analysis could only be achieved through a study of the electron diffraction patterns.

Crystals from all samples with $x \geqslant 0.02$ regularly showed electron diffraction patterns characteristic of an i.t.b. phase. With the high-tilt side-entry stage, it was possible to establish that almost identical reciprocal lattice sections exhibiting a one-dimensional superlattice were obtained down two mutually perpendicular directions within the plane of the $WO_3$ slabs (figure 6*a*, *b*, plate 3). The average superlattice periodicity varied from *ca.* 2.3 nm in $Bi_{0.2}WO_3$ to approximately 4.5 nm in $Bi_{0.02}WO_3$, but wide variations were obtained from different crystals in each sample. When viewed down a third principal axis, perpendicular to the slabs, the reciprocal lattice sections of all samples were identical and resembled the [001] projection of $WO_3$ (figure 6*c*). Apart from fragments that were clearly metallic bismuth contaminants, the bismuth content of the samples $x = 0.02$, 0.05 and 0.1 was impossible to determine accurately by X-ray emission analysis. With the sample of nominal composition $Bi_{0.2}WO_3$, however, a bismuth signal could be detected and qualitatively estimated in the crystals used for h.r.e.m. studies (figure 7, plate 4). By using previously characterized samples of $Bi_2WO_6$ and $Bi_2W_2O_9$ as standards, both of which are beam-stable under normal conditions (Jefferson 1982), the average composition of the i.t.b.-like crystals in this specimen was assessed as $Bi_{0.07}WO_3$, with a mean variation of some 25%, over a sample of approximately twenty crystals.

Analysis of all the samples, however, became possible by means of X-ray photoelectron spectra (X.p.s.). The spectra (figures 2 and 3) established the presence of bismuth in all the samples ($x \geqslant 0.01$) and also enabled analysis of the composition based on the $W^{5+}/W^{6+}$ intensity ratios in the 4f spectra. The actual value of $x$ in $Bi_xWO_3$ so determined for the nominal compositions with $x = 0.01$, 0.02, 0.05, 0.10 and 0.20 was 0.01, 0.02, 0.03, 0.05 and 0.06 respectively. Some of the bismuth is obviously not used in the reaction, as indeed is shown by X.p.s. (figure 2). The uncertainties in $x$ values determined by X.p.s. would be around 15%, but it is gratifying that the actual value of $x$ in $Bi_{0.2}WO_3$ (nominal) by this method agrees so well with that obtained from X-ray emission analysis in the microscope. For convenience, we shall use the nominal compositions throughout this paper.

By making use of the electron diffraction data, especially the supercell spots, we have analysed the rather complex X-ray diffraction patterns (figure 8) of $Bi_xWO_3$ samples with $x \geqslant 0.02$. The results are presented in table 1, where we have also given the values of $x$ obtained from X-ray photoelectron spectroscopy. X-ray patterns of the samples with $x = 0.1$ and 0.2 appear to be due to mixtures of two predominant phases.

### *Ultrastructural studies of i.t.b. by h.r.e.m.*

Typical h.r.e.m. images recorded at 200 kV from a sample of $Bi_{0.1}WO_3$ nominal composition are shown in figure 9, plate 5. The electron beam direction in each case is that of figure 6*a*. Disordered intergrowths of the type shown in figure 4*a* were obtained most frequently, but the minimum overall periodicity observed was generally not less than 1.9 nm. Well ordered crystals were comparatively rare. Figure 9*b* shows a micrograph of one such crystal taken at the defocus corresponding to highest obtainable resolution. Crystals also commonly showed a tendency to fragment at the edges (figure 9*c*), but no *in situ* conversion to $WO_3$ under conditions

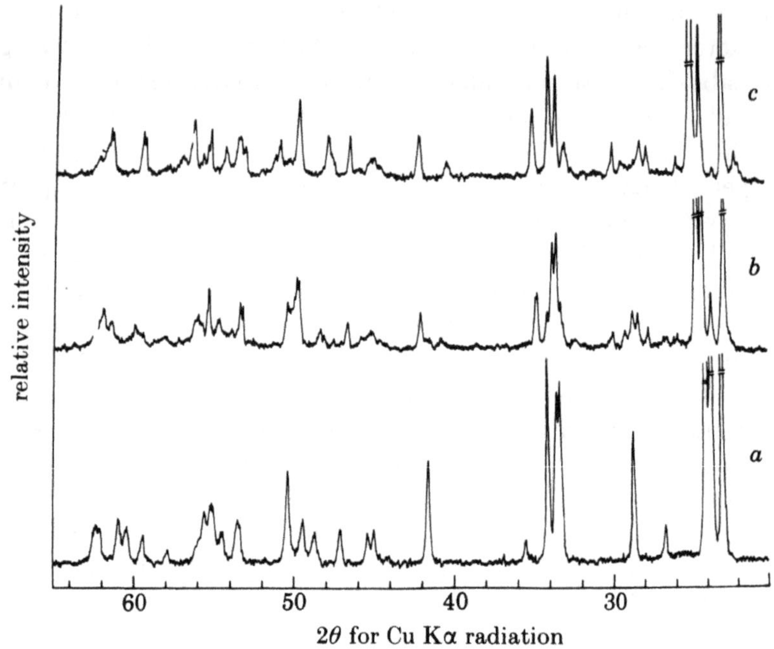


FIGURE 8. X-ray powder diffraction patterns of $Bi_xWO_3$: (*a*) $x = 0.01$; (*b*) $x = 0.05$ and (*c*) $x = 0.10$.

TABLE 1. UNIT CELL PARAMETERS OF $Bi_xWO_3$

| *x* | | | | | |
|---|---|---|---|---|---|
| nominal | actual[(a)] | $n^{(b)}$ | $a$/nm | $b$/nm | $c$/nm |
| 0.02 | 0.016 | 13 | 5.10 | 0.735 | 0.390 |
| 0.05 | 0.028 | 12 | 4.88 | 0.735 | 0.390 |
| 0.10 | 0.05 | 8 | 3.22 | 0.735 | 0.391 |
| | | 7 | 2.92 | 0.735 | 0.391 |
| 0.20 | 0.06 | 7 | 2.92 | 0.735 | 0.391 |
| | | 6 | 2.51 | 0.735 | 0.391 |

(*a*) From X-ray photoelectron spectroscopy. (*b*) Number of supercell spots between two subcell spots along the $a^*$ direction in the electron diffraction patterns.

of extremely high beam irradiation, as has been observed in h.r.e.m. images of $Bi_2W_2O_9$ (Bando *et al.* 1979), was noted.

H.r.e.m. evidence clearly indicated some form of i.t.b. structure. In particular, the regions of the crystals distinguished by their crossed 0.38 nm fringes of variable width were clearly slabs of $WO_3$, as expected. The structure of the intergrowth strip, however, was by no means clear. In particular, no well defined tunnels were observed, as for the known i.t.b. phases (Hussain & Kihlborg 1976; Ekstrom & Tilley 1980), the lack of clear detail in the strips being similar to that observed in $Pb_xWO_3$ at lower resolution (Ekstrom & Tilley 1978). As the existence of h.t.b. intergrowth strips was not conclusively established, image-matching studies were therefore made on three possible structural models, as shown in figure 10. The first of these (figure 10*a*) was essentially one involving intergrowth of perovskite

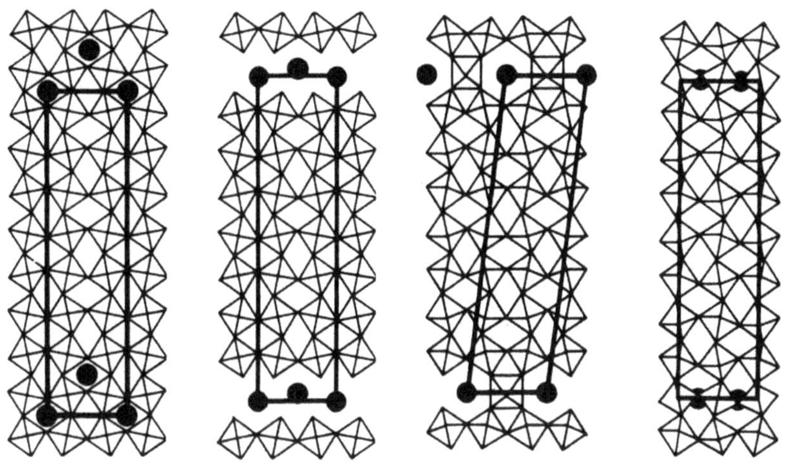

(a) (b) (c)

FIGURE 10. Possible models for the Bi–W-bronze structures: (a) 'perovskite strip' model; (b) 'isolated slab' model; (c) 'h.t.b.-strip' model, down both [001] and [010] projections.

strips, and did not correspond to observed images (at any values of crystal thickness) which require an excessive electron density along the intergrowth strips. The second model (figure 10*b*), involving isolated slabs of $WO_3$ separated by two-dimensional arrangements of bismuth atoms, gave reasonable agreement for thin crystals, but failed when the thickness exceeded 4.5 nm. Because of this partial agreement, and bearing in mind the closely similar electron scattering factors of tungsten and bismuth, a model was constructed by using h.t.b.-type intergrowth strips, but with completely filled channels (figure 10*c*). Although the projected charge density of this model was very similar to that of the previous one, the differences were sufficient to improve the fit at greater crystal thicknesses, and the overall stoichiometry of $Bi_{0.067}WO_3$ required was in good agreement with that observed experimentally for the nominal $Bi_{0.2}WO_3$ sample. The quality of the matching between observed and experimental images can be judged from figure 11*a*, plate 6, and 11*b*, plate 7, corresponding to the two directions of projection perpendicular to the superlattice axis. In particular, the way in which the model simulates apparently empty 'tunnels' at a crystal thickness of 6.0 nm in the projection of figure 11*a* is quite revealing, and suggests that in previous studies on other systems, lower than actual tunnel occupancies might have been assigned unless the crystals used for imaging were especially thin.

To confirm the structural model, 500 kV h.r.e.m. images were taken, trial simulations showing that at this level of resolution, it should be possible to distinguish between bismuth atoms within the tunnels and the tungsten atoms in the octahedra surrounding them. The results are indicated in figure 12, plate 8, for both projections, and with model structures inset. In the projection down the tunnel axis, a variation in the intensity of the dark dots within the intergrowth strips is noticeable. This phenomenon, which becomes more apparent upon prolonged exposure to the electron beam, appeared to be the beam-induced ejection of bismuth atoms from the tunnels in the very thin regions of the crystals.

It also provided a useful aid to characterizing conclusively the tunnel-like nature of the intergrowth strips. In the projection perpendicular to the tunnels, where the crystal fragments were extremely thin (figure 12*b*) a slight displacement of bismuth atoms within the tunnels might be inferred from the observed perturbations in the positions of the corresponding dark dots. In view of the demonstrated mobility of bismuth under 500 keV irradiation, however, such conclusions would have to be extremely speculative.

The structure derived for the $Bi_xWO_3$ phases fits in well with the data on other i.t.b. structures, and it would seem probable that the $Pb_xWO_3$ series also possess this structure. The retention of h.t.b.-type intergrowth strips for $0.01 < x < 0.05$ is, however, in conflict with previous data (Parmentier & Gleitzer 1975), which suggested a perovskite-like bronze within this composition range. The thermodynamic stability of regular bismuth bronze structures described above must, however, be in doubt, if only because of their apparent inability to form completely ordered structures, even upon prolonged re-annealing at 1170 K. The observed data indicates that bismuth will readily enter the $WO_3$ structure at this temperature to form isolated h.t.b. strips, which invariably extend completely across the crystals concerned. However, there is virtually no evidence for the mobility of the h.t.b. strips once formed. Very occasionally, steps can be observed in the intergrowth strips, as shown in figure 13*a*, plate 9. These invariably involve displacement of the strip by two $WO_3$ octahedra, and the break in the lines of tunnels can be reconciled with the grossly distorted hexagonal rings of $Mo_{17}O_{47}$ (Kihlborg 1960), as shown in figure 13*b*. However, such features appear to be relatively immobile, even under prolonged electron beam heating. Complete characterization was, however, impossible owing to the non-occurrence of these features in thin regions.

One other aspect of the $Bi_xWO_3$ structure that is not evident from the h.r.e.m. images is the actual positioning of bismuth atoms in the tunnels. In the model employed for the image simulations these are sited along the tunnel axes, at a level exactly between the surrounding layers of tungsten atoms, as in the rubidium and caesium bronzes (Kihlborg 1979). In the image simulation shown in figure 11*b*, the fact that the calculated contrast along the tunnel regions is less than that actually observed suggests that the height of the bismuth atoms in the tunnels may not be correct. This is further emphasized by the h.r.e.m. image of figure 12*b*, where, if the proposed model is adhered to absolutely, a continuous dark fringe along the tunnel axis might be expected, especially in the thin regions of the crystals. A further complication arises over the centring of bismuth atoms along the tunnel axes. By virtue of its lone pair of electrons, bismuth is a relatively large atom ($r = 0.102$ nm), but its electron density is unlikely to be spherically symmetric, and some displacement towards the tunnel sides is therefore likely. No evidence for this is found in h.r.e.m. images, nor would such small displacements be resolved, but if such displacements do exist, then the possibility of their ordering cannot be discounted. Some evidence for this can be seen in the selected area electron diffraction patterns of well annealed crystals, such as the one shown in figure 14, plate 10. This shows that the superlattice spots themselves possess a satellite structure, with an overall structural periodicity of *ca.* 16 nm. This is not merely due to an intergrowth of two bronzes with a different periodicity, which would affect all rows parallel to $X^*$ (to use the nomenclature of Hussain & Kihlborg (1976)). Instead, only rows with $l = 2n$ are affected, which suggests a more subtle explanation, possibly involving lateral displacements of the cations within the hexagonal tunnels. No perceptible contrast effects due to these satellites are seen in the h.r.e.m. images; consequently these features must await some form of single-crystal diffraction analysis, if sufficiently large crystals can be obtained.

Several features of the bismuth intergrowth tungsten bronzes are noteworthy. (i) The maximum value of $x$ in $Bi_xWO_3$ i.t.b. phases is *ca.* 0.10, just as in other i.t.bs. (ii) The h.t.b. strips are, however, always isolated, the width being limited to that of a single tunnel. This is in contrast to the behaviour of alkali metal i.t.b. phases where two-tunnel wide h.t.b. strips are most stable. (iii) Although i.t.b. phases with long-range order containing only one type of arangement of the h.t.b. strips with respect to the $WO_3$ slabs are not found, we have not infrequently observed recurrent ordered intergrowths extending up to 100 nm in several crystals. (iv) Presence of widely different periodicities in the same 'single' crystal is common, but we can assign unit cell dimensions based on diffraction patterns. Presence of a variety of periodicities (and hence compositions) in the same crystal has been noted earlier in other complex oxide phases. We could, in principle, consider such crystals as composed of solid solutions of different phases; since we can still assign unit cell dimensions, we can treat the single crystals as ones possessing considerable disorder. We cannot, however, stop wondering about the traditional definitions of terms such as 'a phase' and 'a unit cell'. Certainly, we have to make a distinction between ultramicrostructure as seen by h.r.e.m. and the average structure. Yet, it is not clear how extensive the repetitive order in the ultrastructure should be before it represents the average structure.

### *H.t.b. phase of bismuth*

Having failed to observe h.t.b. strips larger than a tunnel wide in the i.t.b. phases prepared by the reaction of bismuth with $WO_3$ at high temperatures, we attempted to make the pure h.t.b. phase by reaction of bismuth with hexagonal $WO_3$ (Gerand *et al.* 1979) below 500 °C. Two nominal compositions of $Bi_xWO_3$ ($x = 0.1$ and 0.2) so prepared were blue in colour and gave X-ray diffraction patterns characteristic of the h.t.b. phase. Based on the $W^{5+}/W^{6+}$ intensity ratio in the 4f bands, we find that the actual value of $x$ in the 0.1 and 0.2 samples work out to be 0.016 and 0.022 respectively. It appears that it is extremely difficult to incorporate bismuth in the hexagonal tunnel even by starting with hexagonal $WO_3$. This explains why we have not been able to obtain an i.t.b. phase to give way to a h.t.b. (or i.t.b.) phase at high concentrations of bismuth, by analogy with alkali metal i.t.b. systems.

X-ray lattice dimensions of the $x = 0.1$ and 0.2 h.t.b. phases were both close to $a = 0.74$ and $c = 0.76$ nm. Electron diffraction patterns showed no evidence for the ordering of bismuth atoms in the hexagonal tunnels of $WO_3$. This observation, along with the fact that the X-ray patterns showed the samples to be monophasic, suggest that bismuth atoms are randomly distributed in this h.t.b. The h.t.b. phase seems to be stable only up to 870 K, transforming to the i.t.b. phase at higher temperatures. The h.t.b. phase of bismuth is apparently metastable with respect to the i.t.b. phase.

*Properties of i.t.b. phases*

Optical absorption spectra of the i.t.b. phases (figure 15) show a progressive blue shift of the interband absorption edge around 450 nm with increasing $x$, just as in sodium–tungsten bronzes (Dickens *et al.* 1968). There is a broad band starting from 500 nm and extending to the near infrared in all these bronzes. These features are undoubtedly due to the creation of $W^{5+}$ by bismuth atoms. With increase in $x$, the W(5d) band gets progressively filled, shifting the interband transition edge around 450 nm to shorter wavelengths. Accordingly, magnetic susceptibility of $Bi_xWO_3$ shows the expected changes. Compositions with $x < 0.05$ are diamagnetic and those with $x > 0.05$ are paramagnetic; paramagnetism in the latter is temperature-independent, suggesting the presence of itinerant electrons.

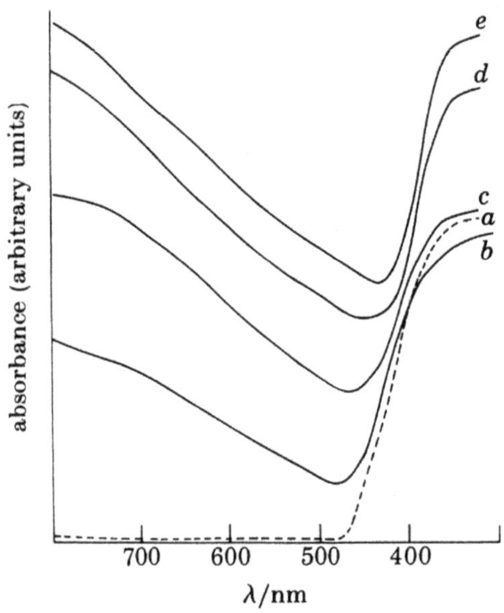


FIGURE 15. Reflectance spectra of $Bi_xWO_3$: (*a*) $x = 0.0$; (*b*) $x = 0.02$; (*c*) $x = 0.05$; (*d*) $x = 0.10$ and (*e*) $x = 0.20$.

Electron transport properties of the bronzes clearly show evidence for the itinerant nature of d electrons when $x > 0.05$. Electrical resistivity of $Bi_xWO_3$ decreases markedly with increase in $x$ (figure 16), and when $x = 0.2$ (nominal) the resistivity at 300 K is barely 0.20 Ω cm. Seebeck coefficient data also reflect the changes in the electronic structure of the bismuth i.t.b. phases with increase in $x$ (figure 17). The Seebeck coefficient (at 300 K) decreases sharply up to $x \approx 0.05$ (resistivity also drops markedly in this composition range). The absolute value of the Seebeck coefficient decreases with increase in temperature when $x < 0.05$, while the opposite holds when $x > 0.05$. It appears that the d electrons are localized when $x < 0.05$ and itinerant when $x > 0.05$. Based on the trends in transport properties reported here, we would expect $Bi_xWO_3$ to become metallic when $x$ is *ca.* 0.08. Interestingly, this value of $x$ is around one third the value in alkali metal bronzes, $M_xWO_3$, when they become metallic. Since bismuth can donate three electrons to $WO_3$, this is what we would expect.

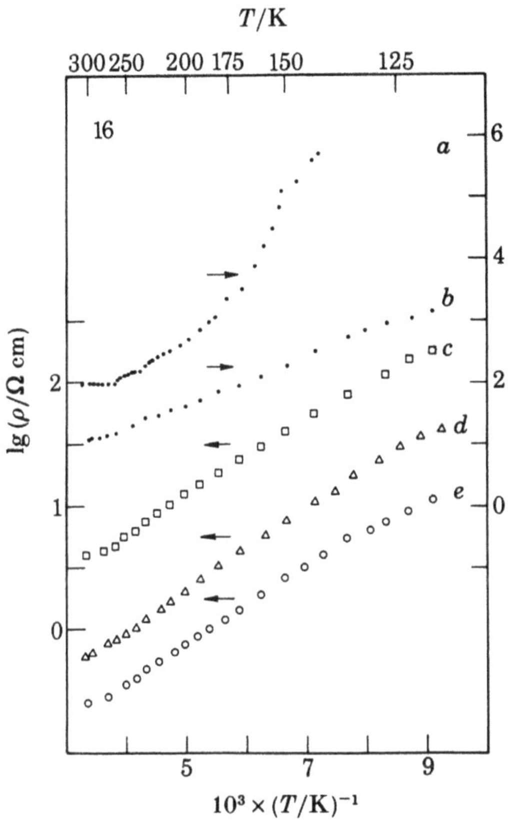


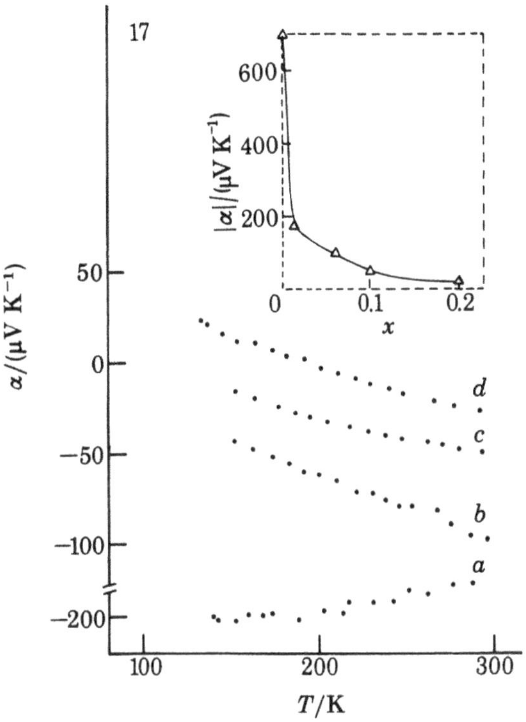


FIGURE 16. Variation of the electrical resistivity, $p$, of $Bi_xWO_3$ with temperature: (*a*) $x = 0.0$; (*b*) $x = 0.01$; (*c*) $x = 0.05$; (*d*) $x = 0.10$ and (*e*) $x = 0.20$. Resistivity data of $WO_3$ ($x = 0.0$) reflect the phase transitions exhibited by this oxide.

FIGURE 17. Variation of the Seebeck coefficient, $\alpha$, of $Bi_xWO_3$ with temperature: (*a*) $x = 0.01$; (*b*) $x = 0.05$; (*c*) $x = 0.10$ and (*d*) $x = 0.20$. In the inset $\alpha$ at room temperature is plotted against $x$.

The authors thank Dr D. J. Smith of the High Resolution Electron Microscope, University of Cambridge, for his assistance in obtaining images on the 500 kV microscope. C.N.R.R. thanks the University Grants Commission of India for support of this research and the University of Cambridge for the Nehru Visiting Chair.

## REFERENCES


Aurivillius, B. 1949 *Ark. Kemi* **1**, 463.

Bando, Y., Watanabe, A., Sekikawa, Y., Goto, M. & Horiuchi, S. 1979 *Acta Crystallogr.* A **35**, 142.

Dickens, P. G., Quillium, R. M. P. & Whittigham, M. S. 1968 *Mater. Res. Rull.* **3**, 941.

Ekstrom, T. & Tilley, R. J. D. 1978 *J. solid state Chem.* **24**, 209.

Ekstrom, T. & Tilley, R. J. D. 1980 *Chem. Scripta* **16**, 1.

Fejes, P. L. 1977 *Acta crystallogr.* A **33**, 109.

Gerand, B., Nowogrocki, G., Guenot, J. & Figlarz, M. 1979 *J. solid state Chem.* **29**, 429.

Goodman, P. & Moodie, A. F. 1974 *Acta crystallogr.* A **32**, 551.

Hussain, A. & Kihlborg, L. 1976 *Acta crystallogr.* A **32**, 551.

Hutchison, J. L., Anderson, J. S. & Rao, C. N. R. 1977 *Proc. R. Soc. Lond.* A **355**, 301.

Jefferson, D. A. 1982 *Phil. Trans. R. Soc. Lond.* A **305**, 535.
Jefferson, D. A., Gopalakrishnan, J. & Ramanan, A. 1982 *Mater. Res. Rull.* **17**, 269.
Kihlborg, L. 1960 *Acta chem. scand.* **14**, 1612.
Kihlborg, L. 1979 *Chem. Scripta* **14**, 187.
Lynch, D. F., Moodie, A. F. & O'Keefe, M. A. 1975 *Acta crystallogr.* A **31**, 300.
Parmentier, M. & Gleitzer, C. 1975 *C.r. hebd. Séanc. Acad. Sci. Paris* **281** C, 819.
Rao, C. N. R., Sarma, D. D., Vasudevan, S. & Hegde, M. S. 1979 *Proc. R. Soc. Lond.* A **367**, 239.
Smith, D. J., Camps, R. A., Cosslett, V. E., Freeman, L. A., Saxton, W. O., Nixon, W. C., Ahmed, H., Catto, C. J. D., Cleaver, J. R. A., Smith, K. C. A. & Timbs, A. E. 1982 *Ultramicroscopy* **9**, 203.

JOURNAL OF SOLID STATE CHEMISTRY **53**, 193–216 (1984)

# Crystal Chemistry and Magnetic Properties of Layered Metal Oxides Possessing the $K_2NiF_4$ or Related Structures*

P. GANGULY AND C. N. R. RAO†

*Solid State and Structural Chemistry Unit, Indian Institute of Science, Bangalore-560012, India*

Received June 30, 1983; in revised form January 20, 1984

There is increasing interest in recent years in the structural chemistry and properties of layered metal oxides possessing the $K_2NiF_4$ or related structures. Many new oxides of this structure exhibiting novel properties are being reported from time to time in the literature. The crystal chemistry of the oxides of the general formula $A_2BO_4$ with particular reference to the stability of the $K_2NiF_4$ structure and the relations between the different structures exhibited by this family of oxides is discussed. Non-stoichiometry in these oxides is another aspect of interest discussed in the article. While $K_2NiF_4$ itself is a well-known two-dimensional antiferromagnet, oxides of this structure with a variety of magnetic properties are examined in some detail. Besides the ternary $A_2BO_4$ oxides, the structure and magnetic properties of complex oxides, where the $A$ or/and the $B$ ions are partly substituted by other cations, is discussed. Some of the problems related to this family of oxides that are worth investigating are indicated. Much of the discussion in this article would have relevance in understanding the structure and properties of layered materials.

## 1. Introduction

$K_2NiF_4$ is a prototype, two-dimensional antiferromagnetic material (*1*). The tetragonal $K_2NiF_4$ structure consists of alternating layers of $KNiF_3$ perovskite layers and KF rock-salt layers. Antiferromagnetic interactions between the transition metal ions occur only in the planes containing the $KNiF_3$ perovskite layers. There is no Ni–F–Ni interaction in the direction parallel to the $c$ axis. A variety of oxides are known to crystallize in structures related to $K_2NiF_4$ (*2–14*). Besides the tetragonal $K_2NiF_4$ structure, $A_2BO_4$ type oxides with monoclinic, orthorhombic, and other tetragonal structures have been characterized in recent years (*6–9, 13–15*). The main objective of this article is to examine the stability of the $K_2NiF_4$ structure and the relations between the different structures of metal oxides in this family and to rationalize their properties in terms of the structures. We have discussed at length, the crystal chemistry of various oxides of the $A_2BO_4$ type in terms of the nature of $A$–O and $B$–O bonding and the consequences of the bonding on the structure and properties of the oxides. Non-stoichiometry in these oxides has also been discussed briefly.

Oxides with $K_2NiF_4$ or related structures exhibit interesting magnetic and electronic

* Contribution No. 213 from the Solid State and Structural Chemistry Unit.

† To whom all correspondence should be addressed at: Department of Physical Chemistry, University of Cambridge, Cambridge, CB2 1EP, United Kingdom.



0022-4596/84 $3.00

properties. It was reported a few years ago (*2*) that $La_2NiO_4$ shows an activated electrical conductivity behavior at low temperatures, but the resistivity has a positive temperature coefficient above 550 K. Such a semiconductor–metal transition is indeed of great interest. $La_2CuO_4$ is reported to exhibit a temperature-independent electrical resistivity ($\sim 10^{-1}$ ohm cm), but other rare earth derivatives of the general formula $Ln_2CuO_4$ are semiconductors (*2*). Electron transport properties of more complex oxides of the $K_2NiF_4$ family such as $LaSr_3Co_2O_8$ are quite different from those of the corresponding perovskite oxides.

Magnetic properties of oxides possessing $K_2NiF_4$ structure have been investigated more extensively than electron transport properties. Some of the oxides such as $Ca_2MnO_4$ show long-range antiferromagnetic ordering while others such as $La_2NiO_4$ do not. Ferromagnetism is known to occur in oxides such as $LaSr_3Co_2O_8$. The $K_2NiF_4$ structure appears to preferentially stabilize certain spin states of transitional metal ions and accordingly, some of the transition metal oxides of this family exhibit spin-state transitions. We shall discuss magnetic properties of a variety of oxides possessing $K_2NiF_4$ type structure in the light of their crystal chemistry.

## 2. Crystal Chemistry

Compounds of the general formula $A_2BX_4$ with the $K_2NiF_4$ structure (Fig. 1) may be considered to be built up of alternating layers of perovskite ($ABX_3$) and rock-salt ($AX$) structures. There are no close-packed $A_2X_4$ layers in $A_2BX_4$ similar to the close-packed $AX_3$ layers (Fig. 2a) present in perovskites. In a close-packed $A_2X_4$ layer, there would be considerable electrostatic repulsion between the two $A$ ions since they are forced to be adjacent to each other if the perovskite $AX_3$ layers are to be retained within the structure (see Fig. 2b).

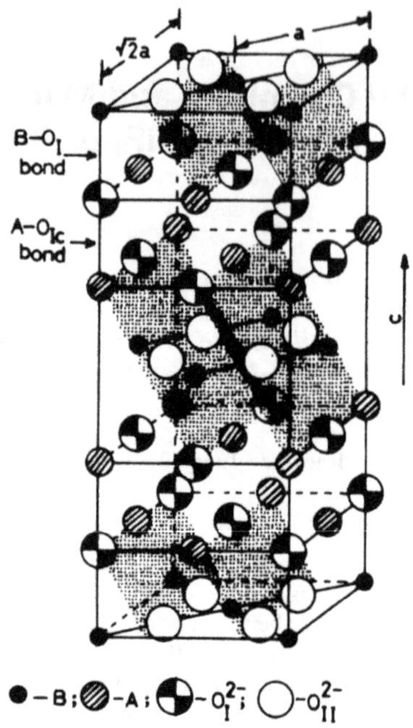


FIG. 1. The $K_2NiF_4$ structure of $A_2BO_4$ oxides showing $O_I$ and $O_{II}$ ions. $B$–$O_I$ and $A$–$O_{Ic}$ bonds are also shown. Shaded portion shows close-packed $AO_3$ layers.

The two $AX_3$ layers can, however, be displaced (Fig. 2c) to give alternating layers of rock-salt and perovskite structures. It is evident from Fig. 1 that along the {110} planes

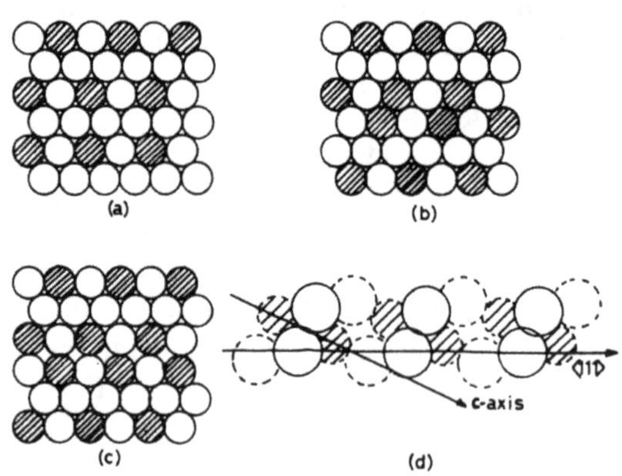


FIG. 2. (a) $AX_3$ close-packing: hatched circles are $A$ ions and open circles are $X$ ions; (b) hypothetical $A_2X_4$ close-packing built up from $AX_3$ layers; (c) displacement of $AX_3$ layers to give $A_2X_4$ packing as found in the ideal $K_2NiF_4$ structures; (d) corrugated packing of $A_2X_4$ layers as in $La_2NiO_4$.

of the $K_2NiF_4$ structure, the {110} planes of the perovskite layers are stacked alternately with the {100} planes of the rock-salt layer. In order to obtain the $A_2X_4$ packing shown in Fig. 2c, the {111} planes of the perovskite layer should be coplanar with the {100} planes of the rock-salt layer. This is roughly what is achieved along the {111} planes of the $K_2NiF_4$ structure (see Figs. 1 and 2d), the two planes being corrugated instead of being coplanar. This results in a better packing efficiency. The low $c/a$ ratio (3.25–3.30) found in several $A_2BO_4$ oxides (compared to the theoretical value of 3.414 ($c = (2 + \sqrt{2})a$) if the $BO_6$ octahedra were regular and the $A$–O distances were identical) is perhaps due to the corrugated nature of the packing in the {111} planes. The perovskite ($AX_3$) layers (Fig. 3a) are displaced in the $K_2NiF_4$ structure as represented in Fig. 3b. Structures of the related Ruddlesdon–Popper type compounds of the general formula $AX(ABX_3)_n$ with $n = 2$ and 3 are shown in Figs. 3c and d where 2 and 3 layers of perovskites are displaced, respectively. Insertion of an $AX$ layer in the $n = 2$ case (Fig. 3c) between the two perovskite bilayers and a reshuffling of the perovskite layers could give rise to the hypothetical compound $(AX)_2(ABX_3)_2$ with alternating bilayers of rock-salt and perovskite (Fig. 3e) instead of the monolayers found in the $K_2NiF_4$ structure. We shall discuss later why such compounds are not readily formed.

*2.1. Stability of the $K_2NiF_4$ structure in oxides*. Just as for perovskite oxides, a tolerance factor $t$ may be defined for $A_2BO_4$ oxides as

$$t = \frac{r(A\text{–O})}{\sqrt{2}r(B\text{–O})} \quad (1)$$

Here, $r(A$–O) and $r(B$–O) are distances obtained from ionic radii. Poix (*3*) has defined the tolerance factor for oxides as

$$t = \psi_A/\sqrt{2}\,\beta_B \quad (2)$$

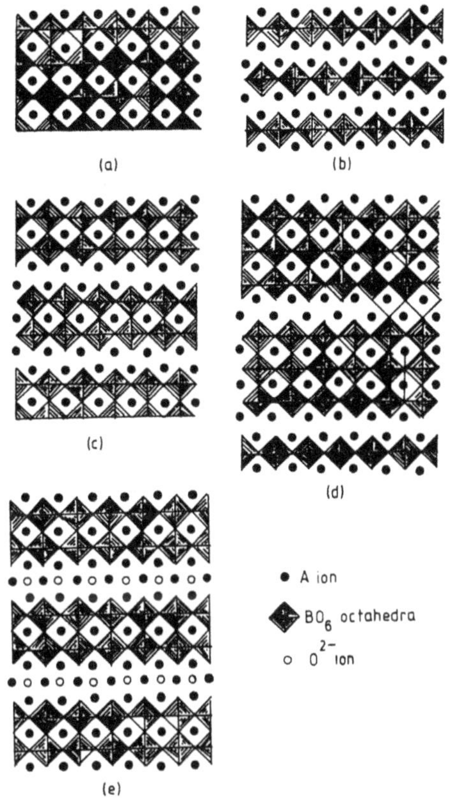


FIG. 3. Projection drawings of (a) $ABO_3$ perovskites and (b) $A_2BO_4$ oxides with $K_2NiF_4$ structure. Projection drawings of $n = 2$ and $n = 3$ members of the series $AO(ABO_3)_n$ are shown in (c) and (d). Projection drawing of hypothetical $(AO)_2(ABO_3)_2$ is shown in (e). After Tilley, Ref. (*45*).

where $\psi_A$ and $\beta_B$ are invariant values (*3, 4*) associated with $B$–O and $A$–O distances in six- and ninefold coordinations, respectively. The tetragonal $K_2NiF_4$ structure is assumed to be stable (*3*) within the limits $1.02 > t > 0.85$. This criterion is conceptually more appealing than Ganguli's criterion (*5*) according to which the tetragonal structure is stable when $r_A/r_B$ is between the limits 2.4 and 1.7 when one employs the ionic radii of Shannon (*16*). Such a criterion in terms of $A$–O and $B$–O distances is indeed equivalent to Poix's criterion. In this article, we shall adopt Poix's criterion while examining the stability of the tetragonal structure. We have calculated $t$ and $\psi_A$ for various oxides by using the following relationship proposed by Poix (*3*):

$$\beta_B + \sqrt{2}\,\psi_A = 0.996V^{1/3} \quad (3)$$

where $V$ is the volume of the unit cell and $\beta_B$ is available in the literature (*4*). In Table I, we have listed the values of $t$ thus obtained for several compounds. We note that $t$ values of $La_2NiO_4$, $La_2CuO_4$, and $La_2CoO_4$ are close to the lower limit. The orthorhombic structure (*6–8*) of $La_2CoO_4$ and $La_2CuO_4$ and the tetragronal structure (*6*) of $La_2NiO_4$ would seem to be consistent with $\beta_{Ni}$ being less than $\beta_{Cu}$ or $\beta_{Co}$. The monoclinic distortion in $Nd_2NiO_4$ and $Pr_2NiO_4$ (*9*) can be attributed to the low tolerance factor in these two oxides.

TABLE I

LATTICE PARAMETERS, $\psi_{Ln^{3+}}$ AND $t$ VALUES FOR SOME OXIDES WITH $K_2NiF_4$ STRUCTURE

| Compounds | $a$ (Å) | $c$ (Å) | $c/a$ | $\psi_{Ln^{3+}}$ | $t$ |
|---|---|---|---|---|---|
| $LaSrAlO_4$[a] | 3.761 | 12.649 | 3.363 | 2.586 | 0.978 |
| $LaSrVO_4$ | 3.87 | 12.65 | 3.270 | 2.563 | 0.914 |
| $LaSrCrO_4$ | 3.85 | 12.50 | 3.25 | 2.549 | 0.928 |
| $LaSrMnO_4$ | 3.804 | 13.10 | 3.44 | 2.537 | 0.901 |
| $LaSrFeO_4$[b] | 3.88 | 12.76 | 3.29 | 2.602 | 0.922 |
| $LaSrCoO_4$ | 3.806 | 12.503 | 3.285 | 2.626 | 0.989 |
| $LaSrNiO_4$ | 3.80 | 12.51 | 3.292 | 2.521 | 0.935 |
| $La_2CoO_4$ | 3.896[c] | 12.66 | 3.249 | 2.561 | 0.852 |
| $La_2NiO_4$ | 3.855 | 12.652 | 3.282 | 2.558 | 0.867 |
| $La_2CuO_4$ | 3.807[c] | 13.17 | 3.459 | 2.53 | 0.834 |
| $La_2Li_{0.5}Co_{0.5}O_4$[d] | 3.784 | 12.624 | 3.336 | 2.519 | 0.861 |
| $La_2Li_{0.5}Ni_{0.5}O_4$[d] | 3.756 | 12.87 | 3.426 | 2.54 | 0.877 |
| $PrSrAlO_4$ | 3.732 | 12.54 | 3.36 | 2.523 | 0.966 |
| $PrSrCrO_4$ | 3.836 | 12.377 | 3.226 | 2.485 | 0.909 |
| $PrSrFeO_4$ | 3.838 | 12.597 | 3.282 | 2.511 | 0.906 |
| $Pr_2NiO_4$ | 3.845[c] | 12.44 | 3.235 | 2.53 | 0.857 |
| $NdSrAlO_4$ | 3.726 | 12.49 | 3.352 | 2.504 | 0.963 |
| $NdSrCrO_4$ | 3.834 | 12.36 | 3.223 | 2.501 | 0.919 |
| $NdSrMnO_4$ | 3.768 | 12.98 | 3.445 | 2.461 | 0.888 |
| $NdSrFeO_4$ | 3.846 | 12.594 | 3.274 | 2.521 | 0.908 |
| $NdSrNiO_4$ | 3.786 | 12.26 | 3.238 | 2.448 | 0.922 |
| $Nd_2NiO_4$ | 3.810 | 12.31 | 3.231 | 2.491 | 0.844 |
| $GdSrAlO_4$[a] | 3.701 | 12.362 | 3.340 | 2.442 | 0.952 |
| $GdSrCrO_4$ | 3.823 | 12.263 | 3.207 | 2.442 | 0.902 |
| $GdSrMnO_4$ | 3.754 | 12.87 | 3.428 | 2.418 | 0.881 |
| $GdSrFeO_4$ | 3.853 | 12.554 | 3.258 | 2.523 | 0.908 |
| $GdSrNiO_4$ | 3.768 | 12.23 | 3.245 | 2.416 | 0.916 |
| $Gd_2CuO_4$ | 3.89 | 11.85 | 3.046 | 2.452 | 0.806 |

[a] Based on the present study.
[b] Lattice parameter data from Ref. (*54*).
[c] Pseudo-tetragonal lattice parameters.
[d] Lattice parameter data from Ref. (*25*). Rest of the lattice parameter data from Ref. (*35, 36*).

In compounds where $t$ is close to the lower limit of stability of the tetragonal $K_2NiF_4$ structure, it is possible to increase $t$ by several possible mechanisms. In compounds of the type $La_2BO_4$, $\beta_B$ can be reduced by the incorporation of a small proportion of $B^{3+}$ ions. Thus, $La_2NiO_4$ possessing the tetragonal structure almost always contains a finite proportion of $Ni^{3+}$ ions. Low $\beta_B$ can also arise if the $B$ ions are in the low-spin state instead of in the high-spin state. In $Sm_2CoO_4$, there is crystal structure evidence (*8*) for two different environments for the $Co^{2+}$ ions.

Investigations on polycrystalline as well as single crystal samples (*6, 10, 11*) of $La_2NiO_4$ have established the structure of this oxide to be tetragonal. Indeed, it was the first oxide to be shown to have the $K_2NiF_4$ structure. Recently (*12*), supercells corresponding to $\sqrt{2}\,a$ unit cells (with crystals grown from a skull melter) have been observed in the electron diffraction patterns. Subsequently, it has been found that electron diffraction patterns of $La_2NiO_4$ prepared by the ceramic method also show such superlattice reflections. These reflections are similar to those in the neutron diffraction pattern of three-dimensional, antiferromagnetically ordered $K_2NiF_4$. Such superlattice reflections are also observed in $La_2CuO_4$. Compounds such as $La_4LiBO_8$ which have similar tolerance factors show superlattice reflections similar to those observed in $La_2NiO_4$. In Fig. 4a we show electron diffraction patterns corresponding to the tetragonal [111] zone axis of $La_4LiCoO_8$ and to the tetragonal [110] zone axis of $La_2NiO_4$ (Fig. 4b) as well as of $La_4LiCoO_8$ (Fig. 4c). Such superlattice reflections have not been seen in the case of other $LaSrBO_4$ compounds including $La_2Sr_2NiTiO_8$; these oxides have higher tolerance factors. The superlattice reflections may be associated with two different envi-

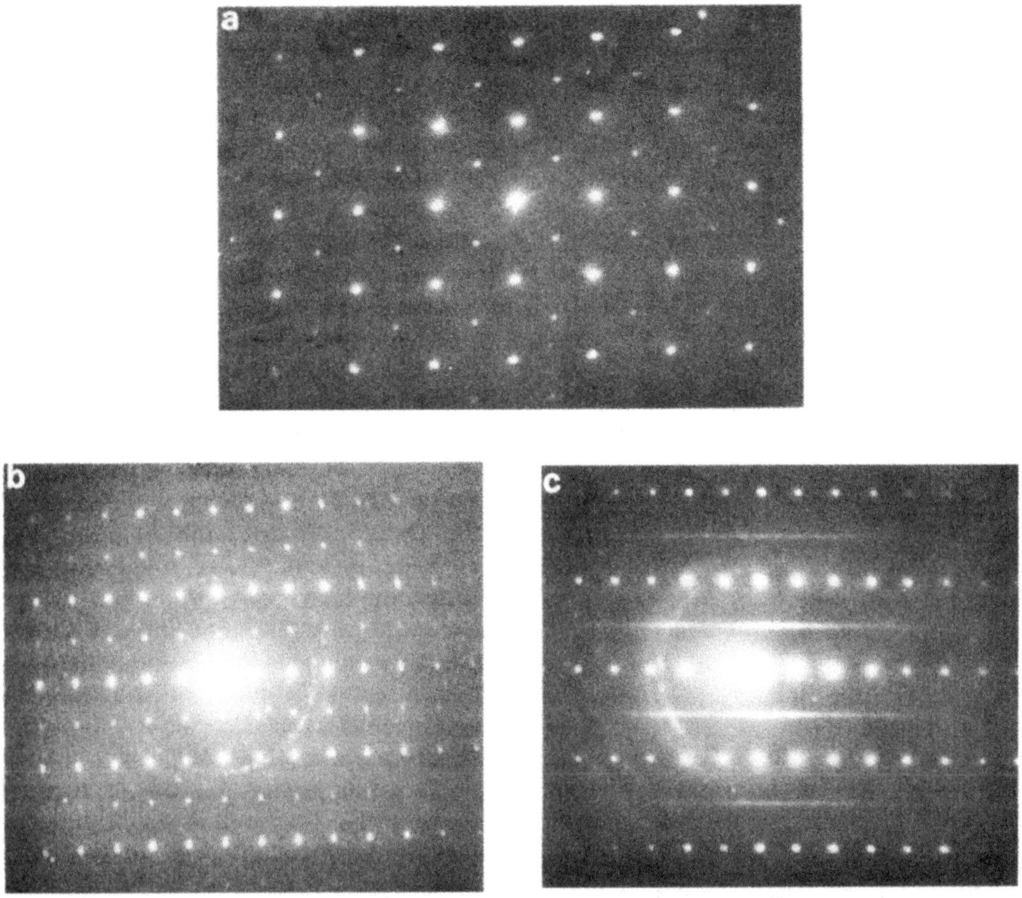

FIG. 4. Electron diffraction patterns corresponding to (a) the [111] zone axis of $La_4LiCoO_8$, (b) [110] zone axis of $La_2NiO_4$, and (c) [110] zone axis of $La_4LiCoO_8$.

ronments of the $B$ ions arising from the deviation of the $B$–$O_{II}$–$B$ angle from 180° or from changes in the $B$–O distance. The former can happen if $t \approx 0.85$ and the $BO_6$ octahedra get rotated about the $c$ axis (Fig. 5a) or if the octahedra are tilted in the ⟨100⟩ or the ⟨110⟩ directions (Figs. 5b and c). The latter is likely to happen in compounds such as $La_4LiBO_8$ where the nature of Li–O and $B$–O bonds is quite different. The distortions (and hence the superlattice reflections) in many of the $K_2NiF_4$ related compounds may be similar to those depicted in Fig. 5.

Several structures can result from a rearrangement of the oxygen ion positions in the $K_2NiF_4$ structure. Three of these arise from monoclinic distortions with $\gamma \neq 90°$. Of these, there are two structures with $a = b$ which can be indexed on the basis of a orthorhombic unit cell; two types of orthorhombic structures $O$ and $O'$ have been distinguished (*13*). For the $O$ structure, the conditions for the allowed reflections are $h + k = 2n$, $k + l = 2n$ and $h + l = 2n$ while for the $O'$ structure, the condition is $k + l = 2n$. In the $O'$ structure (*8, 13, 14*) found in $La_2CuO_4$, $La_2CoO_4$, and CaY $CrO_4$, the $BO_6$ octahedra are tilted as in Fig. 5c. The $O$ structure was established for

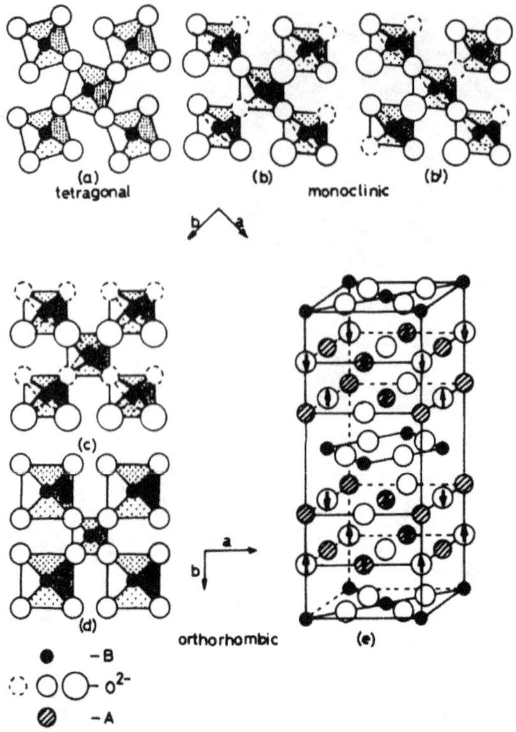


FIG. 5. (a) Rotation of $BO_6$ octahedra along the *c* axis; (b, b′) Tilting of the octahedra in the ⟨100⟩ direction by rotation along the *b* axis, the sense of rotation for adjacent *B* ions along the *b* axis being different in (b), and the same in (b′); (c) Tilting of octahedra in the ⟨110⟩ direction. Larger circles are above the plane of the paper and broken circles are below the plane. (d) $BO_6$ octahedra with different $B–O_{II}$ and possibly $B–O_I$ distances (e) Movement of $A–O_I–A$ linkages (Ref. (*13*)) to give rise to *O* or *O*′ structures.

$Sm_2CoO_4$ which has the *Fmm*2 symmetry with two $Sm–O_I$ and $Co–O_I$ distances (*8*); the $BO_6$ octahedra in this oxide may be arranged as in Fig. 5d. Although the orthorhombic nature of the unit cell is apparent in Fig. 5c (with the *a*′ parameter greater than the *b*′ parameter), the orthorhombic nature of the distortion in the *O* structure is not apparent from Fig. 5d. The *O* structure has been considered (*13*) to be due to a change in the $A–O_{Ib}–A$ angle from 180° along the *b* direction by a shift of the $A–O_{Ia}–A$ linkages along the *a* direction as shown in Fig. 5e ($a > b$).

In a study of the evolution of the tetragonal (*T*) $K_2NiF_4$ structure as a function of *x* in the series $La_{1-x}Y_xCaCrO_4$ and as a function of temperature in $CaYCrO_4$, it has been found that the sequence $O' \rightarrow O \rightarrow T$ occurs in both cases (*13*). Furthermore, as the *O* phase approaches the *O*′ phase, the *b* parameter decreases, but the *a* parameter remains constant. This is consistent with a continuous decrease in the $A–O_{Ib}–A$ angle. In the *O*′ phase, decreasing *t* or temperature, has the opposite effect (*b* remaining nearly constant and *a* decreasing rapidly). This can be understood by assuming that in the *O*′ phase, the $A–O_{Ib}–A$ angle resists further reduction and that the decrease in the tolerance factor imposes a strain on the $A–O_{Ia}–A$ linkage (*15*). Since the $A–O_{Ia}–A$ angle deviates from 180° in this phase, *a* approaches *b* with decreasing *t*. Accompanying these changes the $B–O_{II}–B$ angle also changes in the *O*′ phase and the $BO_6$ octahedra get tilted as shown in Fig. 5c. It would thus appear as though the *O* structure is superimposed on the *O*′ structure shown in Fig. 5c so that one may expect two different $B–O_I$ distances even in the *O*′ phase (*15*). This has, however, not been observed in the crystal structures of $La_2CoO_4$ or $La_2CuO_4$ (*8, 14*).

The truly monoclinic (*M*) distortion with $a \neq b \neq c$ and $\gamma \neq 90°$ has not been reported so far. $Pr_2NiO_4$ and $Nd_2NiO_4$ were reported to be monoclinic (*9*), but with $a = b$, so that the unit cell can be indexed on the basis of an orthorhombic cell. On rechecking the lattice parameters of $Pr_2NiO_4$ (*15*), it is found that the structure is indeed truly monoclinic with $a \neq b$. We have observed that $La_2NiO_{4+\delta}$ on heating for long periods in $CO_2$ at 1150°C gives rise to a monoclinic structure with $a \neq b$; in such a sample, δ would be zero or even slightly negative. As mentioned earlier, small amounts of $Ni^{3+}$ seems to stabilize the tetragonal structure of $La_2NiO_4$.

The tetragonal to monoclinic distortion

could be associated with changes in the tolerance factor. The monoclinic structure would be associated with the tilting of the $BO_6$ octahedra of the type shown in Fig. 5b; the sense of tilting of the octahedra between nearest neighbor $B$ ions may be assumed to be the same for a row of octahedra along the $b$ or $a$ axes (Fig. 5b′).

Another modification of the tetragonal $K_2NiF_4$ structure is the tetragonal $T'$ structure (Fig. 6) found in the copper oxides (*17, 18*) $Ln_2CuO_4$ ($Ln$ = Pr, Nd, Sm, Eu, Gd). The small $c/a$ ratio of the $Ln_2CuO_4$ compounds was initially associated with compressed $CuO_6$ octahedra (*7*) in the same manner as compressed $CuF_6$ octahedra were initially postulated in $K_2CuF_4$ (*19*). In the oxides, the $a$ parameter is ~3.97 Å and in order to obtain compressed octahedra, the Cu–$O_I$ distance should be less than 1.98 Å. The suggestion of Longo and Raccah (*7*) that $CuO_6$ octahedra are compressed is therefore subject to some doubt. The $T'$ structure is derived form the $T$ structure by a shift of the oxide ions from the *(0, 0, z)* positions in the $T$ structure to (0.5, 0.0, 0.25) position in the $T'$ structure. As a consequence, the oxide ion changes its coordination from 6 to 4 in the $T'$ structure. In such a structure, the rare earth and Cu ions have no intervening anion along the $c$ axis. The structure appears to be specific to $Cu^{2+}$ ions (*15*) as the $d_{z^2}^2$ electrons can provide sufficient screening to minimize repulsion between the $Cu^{2+}$ and $Ln^{3+}$ ions. The relationship between $K_2NiF_4$ related structures and the $T'$ structure has been examined by Singh *et al.* (*20*) in the case of the solid solutions, $La_{2-x}Ln_xCuO_4$ ($Ln$ = Pr, Nd). A first-order transition occurs between the two structures as a function of $x$ accompanied by a marked increase in the volume of the unit cell of the $T'$ phase across the critical value $x_c$; $x_c$ decreases with the decreasing size of the $Ln^{3+}$ ion. The $T'$ structure may be considered to be composed of alternating layers of $(CuO_2)^{2-}$ layers with the $Cu^{2+}$ ions in the square-planar, fourfold coordination and $(Ln_2O_2)^{2+}$ layers with the fluorite structure containing $Ln^{3+}$ ions in the eightfold coordination. The increase in volume in the $T'$ phase is attributed to the lower packing efficiency while the collapse in the $c/a$ ratio is attributed to the change from the rock-salt-like packing of the $(Ln_2O_2)^{2+}$ layers in the $K_2NiF_4$ structure to the fluorite type of packing in the $T'$ structure.

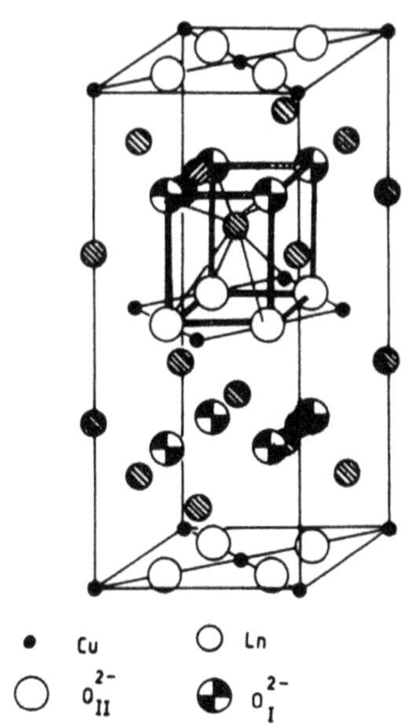


FIG. 6. The $T'$ structure of $Ln_2CuO_4$ compounds.

The driving force for the transition from the $K_2NiF_4$ related structure to the $T'$ structure in the $La_{2-x}Ln_xCuO_4$ compounds can perhaps be understood in terms of the competition between the $A$ and $B$ ions for covalency with the $O_I$ ion in the $A$–$O_I$–$B$ linkages. The higher acidity of the smaller $Ln^{3+}$ ion could further elongate the Cu–$O_I$ bond (compared to that in $La_2CuO_4$) and drive the $Cu^{2+}$ ions to a square-planar coordination.

It is interesting that the relationship between $Ln_2NiO_4$ and $Ln_2CuO_4$ ($Ln$ = Pr, Nd) with respect to the unit-cell volume and $c/a$ ratio is similar to that between $K_2NiF_4$ and

$K_2CuF_4$. This would suggest an antiferrodistortive ordering (Fig. 7a) of $CuO_6$ octahedra enhancing the $a$ parameter to be the driving force for the transition to the $T'$ structure in $Ln_2CuO_4$ compounds. An antiferrodistortive type of ordering of elongated $BO_6$ octahedra of the type found in $K_2CuF_4$ (*21, 22*) would be unstable in $A_2BO_4$ oxides. In the fluorides of $K_2NiF_4$ structures ($t \gtrsim 1$), a mechanism which enhances the $a$ parameter would be favored. In $A_2BO_4$ oxides, since $t < 1$, mechanisms that reduce the $a$ parameter should be operative and it is perhaps for this reason that $La_2CuO_4$ has a $c/a$ ratio suggestive of ferrodistortive ordering (*18, 22*) of elongated $CuO_6$ octahedra (Fig. 7b).

The relationship between the $O$ (or $O'$), $T$, $M$, and $T'$ structures may be obtained from a study of the solid solutions of $La_2NiO_4$ ($T$) and $La_2CuO_4$ ($O$) and that between $Pr_2NiO_4$ ($M$) and $Pr_2CuO_4$ ($T'$). In the series $La_2Ni_{1-x}Cu_xO_4$, the $O$ structure is obtained (*20*) from the $T$ structure for $x \geq 0.9$. Since $\beta_{Cu} > \beta_{Ni}$, the changes may be considered to be due to changes in the tolerance factor. However, in the $Pr_2Ni_{1-x}Cu_xO_4$ series (*15*), the $M$ structure which is stable in the range $0 < x \leq 0.1$ goes over to the $T$ structure for $0.25 \leq x \leq 0.50$. In the range $0.50 < x < 0.90$, a mixture of $T$ and $T'$ phases is obtained. For $0.9 \leq x \leq 1.0$, the $T'$ structure is obtained. The $M \rightarrow T$ transition in $Pr_2Ni_{1-x}Cu_xO_4$ is in the opposite direction to that expected from tolerance factor effects. Both the $O$ and $O'$ structures are found (*13*) in oxides of the formula $La_{1-x}Y_xCaCrO_4$ which have high tolerance factors. The fact that such a distortion has not been observed (*23*) in $CaYAlO_4$ probably implies that distortions in $La_{1-x}Y_xCaCrO_4$ are associated with the high energy required to distort the $CrO_6$ octahedra so that the Ca–$O_{Ic}$ and Y–$O_{Ic}$ distances are constrained to be different. This is in effect reduces the tolerance factor just as Jahn–Teller effects associated with $Cu^{2+}$ ions could enhance it. The immiscibility range in the $Pr_2Ni_{1-x}Cu_xO_4$ solid solutions is consistent with the requirement that two $d_{z^2}$ electrons are necessary to be stabilize the $T'$ structure.

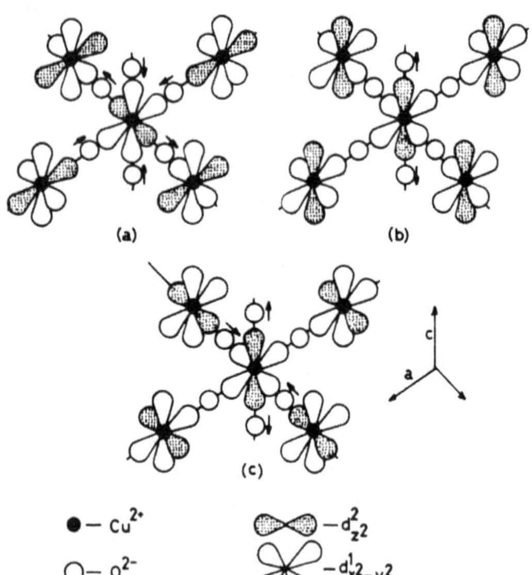


FIG. 7. (a) Antiferrodistortive ordering of elongated $CuO_6$ octahedra (arrows indicate direction of displacement of oxide ions). (b) Ferrodistortive ordering of elongated $CuO_6$ octahedra. (c) Ordering of elongated $CuO_6$ octahedra as in the *ac* or the *bc* planes of $KCuF_3$.

*2.2. Ratios c/a and the nature of A–O and B–O bonds in $A_2BO_4$ oxides*. When the tolerance factor $t = 1$, there is a perfect match of the $B$–O–$B$ and $A$–O–$A$ distances, in both perovskites and $A_2BO_4$ oxides. When $t < 1$, the situation in $A_2BO_4$ oxides is different from that in perovskites. In the latter, there is buckling of the three-dimensional corner-shared octahedral network tending to make the $B$–O–$B$ angle less than 180° so that the effective $B$–O–$B$ distance is reduced. In $A_2BO_4$ oxides, however, the intervening rock-salt layer imparts a rigidity to the two-dimensional octahedral network and prevents the buckling of the octahedra. Instead, there is a pressure effect on the $B$–$O_{II}$–$B$ bond tending to reduce the distance while the $A$–$O_I$–$A$ distance is stretched in order that the two distances match in the tetragonal $K_2NiF_4$ structure. In most oxides including $Sr_2TiO_4$, $t < 1$ and hence the $B$–$O_{II}$–$B$ distance would be expected to be smaller than that computed from the ionic radii. As a consequence, we would expect the $B$–$O_I$ bond to be lengthened and the $A$–$O_I$ bond to be shortened. Accordingly, fluoride ions in $Sr_2FeO_3F$ substitutes for the $O_I$ ions in keeping with the weaker $B$–$O_I$ bond strength (*24*).

Evidence for covalent $Ln$–$O_I$ bonding in $LnSrBO_4$ and $Ln_2BO_4$ compounds may also be obtained indirectly by the application of the method of invariants. Poix (*3*) has applied successfully the method of invariants (Eq. (3)) to oxides of the formula $Sr_2BO_4$ and to other oxides where the $B$ ions are in the 4+ state. We have evaluated the value of $\psi_{Ln}$ for a series of compounds of the type $Ln_2BO_4$ and $LnSrBO_4$ (Table I) and find that $\psi_{Ln}$ is not really an invariant. The reason for this may lie in the covalency of the $Ln$–$O_I$ bonds and the possible competition between $Ln$–$O_I$ and $B$–$O_I$ bonds.

Elongation of the $BO_6$ octahedra in $A_2BO_4$ type oxides can lead to the stabilization of unusual electronic configurations of the $B$ ions. Thus, intermediate-spin $Co^{3+}$ ions ($t_{2g}^5 d_{z^2}^1 d_{x^2-y^2}^0$) are found to exist at low temperatures in $La_4LiCoO_8$ and $LaSrCoO_4$ (*25–27*). At high temperatures, they are transformed to the high-spin configuration ($t_{2g}^4 e_g^2$). High-spin $Fe^{4+}$ ions are found in $La_3SrLiFeO_8$ (*28*). In $La_4LiNiO_8$, ESR and magnetic susceptibility studies (*29*) have shown that the $Ni^{3+}$ ions are in the low-spin ($t_{2g}^6 d_{z^2}^1 d_{x^2-y^2}^0$) configuration. In $Sr_2FeO_3F$, the $Fe^{3+}$ ions are in the low-spin configuration due to the apical positioning of the fluorines (*24*); $Sr_4FeTaO_8$ also seems to show $Fe^{3+}$ ions in the low-spin configuration (*16*).

Elongated $BO_6$ octahedra, short $B$–$O_{II}$–$B$ bonds as well as short $A$–$O_{Ic}$ bonds ($A$–$O_I$ bonds along the $c$ axis), are commonly found in many of the $A_2BO_4$ oxides. Thus, in $LaSrFeO_4$ and $LaSrCrO_4$, the values of the $a$ parameter (3.86 and 3.84 Å, respectively) are considerably smaller than the pseudo-cubic unit cell parameters of La $FeO_3$ and $LaCrO_3$ (3.931 and 3.883 Å, respectively); similarly, the $a$ parameter of $Sr_2TiO_4$ (3.88 Å) is smaller than that of Sr $TiO_3$ (3.90 Å). In general, pressure on the $B$–$O_{II}$–$B$ distance increases as $t$ decreases from unity or as the size of the $B$ ion increases (or the formal charge decreases). Accordingly, the $B$–$O_{II}$ distance in $La_2NiO_4$ is 1.93 Å (*6*) compared to the value of 2.09 Å in NiO (*30*). The A–$O_{Ic}$ distance would be expected to decrease as the charge of the $A$ ion increases. In $La_2NiO_4$, the La–$O_{Ic}$ distance (*2, 11*) is around 2.36 Å compared to the value of 2.616 Å computed from ionic radii (*16*).

Pressure effect on the $B$–$O_{II}$–$B$ bond can be viewed in another manner. Along the $c$ axis, there are $A$–$O_I$–$B$–$O_I$–A $\cdots$ $A$–$O_I$–$B$–$O_I$–$A$ linkages. There would be strong electrostatic repulsion between $A \cdots A$ ions with no intervening anions between them. The $O_{II}$ ions in the basal plane could come closer and screen the charge of the $A$ ion and thereby reduce the $a$ parameter. Electrostatic repulsion between the $A$ ions could push the $A$ ions closer toward the $O_{Ic}$ ions. Strong $A$–$O_I$ bonding would reduce the effective charge on the $A$ ion and hence the $A \cdots A$ electrostatic repulsion. If we consider the ionic potential of $A$ ions, we would expect the smaller ions to increase the $A \cdots A$ electrostatic repulsion. Furthermore, because of the competition between $A$ and $B$ ions for bonding with $O_I$ ions, the $A$–$O_{Ic}$ bond will get shorter (and the $B$–$O_I$ bond longer or $BO_6$ octahedra elongated) as the charge on the $A$ ion increases. Accordingly La–$O_{Ic}$ bond distances in $LaSrAlO_4$ and $La_2NiO_4$ are 2.53 and 2.40 Å, respectively (*6, 15*). A compression of the $BO_6$ octahedra may therefore be taken to indicate a weakening of the $A$–$O_{Ic}$ interaction.

It is interesting to examine the variation of the lattice parameters and the $c/a$ ratios in $LnSrBO_4$ ($Ln$ = rare earth) type compounds with the size of the rare earth ion. In general, the *a* parameter shows a linear dependence on the size of the rare earth ion. In Fig. 8, we have shown the variation of the *c*/*a* ratios in a series of compounds of the formula $LnSrBO_4$ ($B$ = Al, Cr, Fe, Ni) with the size of the rare earth ion. In $LnSrAlO_4$ (*15, 31*), there is a linear variation of the *a* parameter with the size of the *Ln* ion and a small change in the *c*/*a* ratio between Pr and Nd (Fig. 8a). Refinement of the positional parameters has shown (*15*) that the Al–$O_I$ distance in $LaSrAlO_4$ (2.01 Å) is larger than that in $GdSrAlO_4$ (1.95 Å); the Al–O distance computed from ionic radii (*16, 32*) is 1.935 Å. A compression of the $AlO_6$ octahedra implies that the Gd–$O_{Ic}$ interaction is weaker than the La–$O_{Ic}$ interaction. The average Al–O distance in $LaSrAlO_4$ is 1.92 Å while that in $GdSrAlO_4$ is 1.885 Å. The average ($Ln$,Sr)–$O_{Ic}$ distance is 2.53 Å in $LaSrAlO_4$ and 2.48 Å in $GdSrAlO_4$ compared to the computed values of 2.66 and 2.61 Å for the (La,Sr)–O and (Gd,Sr)–O distances and 2.62 and 2.51 Å for the La–O and Gd–O distances. While the Gd–$O_{Ic}$ distance in $GdSrAlO_4$ is close to that predicted by ionic radii, the La–$O_{Ic}$ distance in $LaSrAlO_4$ is considerably smaller. This would again imply that La–$O_{Ic}$ interaction is stronger than the Gd–$O_{Ic}$ interaction.

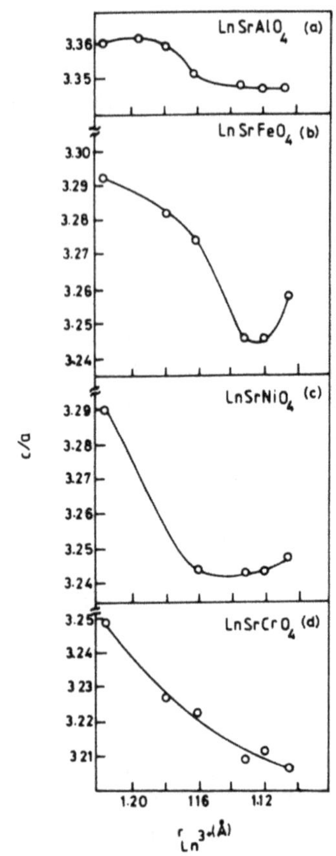


FIG. 8. Variation of the *c*/*a* ratio in $LnSrBO_4$ as function of the radius of the *Ln* ion.

In $LnSrNiO_4$ (*33*), the *a* parameter decreases linearly with the size of the *Ln* ion, but there is an abrupt change in the *c* parameter and in the *c*/*a* ratio (Fig. 8c). Since the size of low-spin $Ni^{3+}$ is comparable to that of $Al^{3+}$, the markedly different behavior of $LnSrNiO_4$ compounds cannot be due to ionic size effects; electronic factors seem to be important. A similar behavior is observed in $LnSrFeO_4$ and $LnSrCrO_4$ (*34*) as can be seen from Fig. 8.

When the *A* ion is kept constant and the *B* ion is varied, systematics in lattice parameters are not obvious. In Fig. 9, the lattice parameters and *c*/*a* ratios (*35, 36*) of some $Sr_2BO_4$ compounds are plotted against the ionic radius of the *B* ion. The *a* parameter varies linearly with the radius of the *B* ion provided that it has partially filled *d* orbitals. Thus, ions such as $Sn^{4+}$, $Hf^{4+}$, and $Zr^{4+}$ do not fall on this straight line. Poix (*3*) has, however, found a linear relationship using the $\beta_B$ parameters. What is important is that there is no linear relationship between the *c* parameters or the *c*/*a* ratios and the size of the *B* ion in these compounds. Furthermore, compounds containing *B* ions with partially filled *d* orbitals exhibit larger *c*/*a* ratios than those with filled or empty *d* orbitals. When the *B* ions have partially filled *d* orbitals, the *c*/*a* ratio

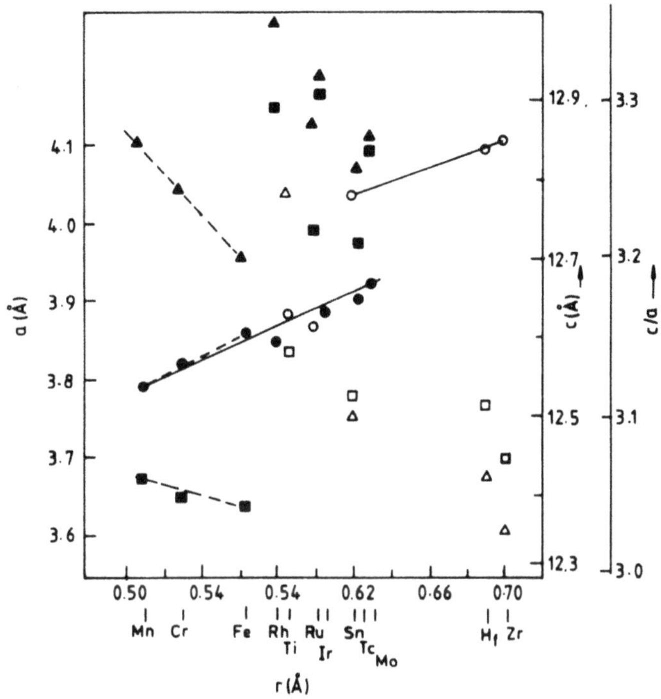


FIG. 9. Variation of the $a$ and $c$ parameters and the $c/a$ ratio of several $Sr_2BO_4$ compounds with the radius of the $B$ ion: circles, $a$ parameter; squares, $c$ parameter; triangles, $c/a$ ratio; closed symbols represent partly filled $d$ orbitals.

seems to increase with the decreasing size of the $B$ ion (of the same period). In the case of $LaSrBO_4$ compounds (*35, 36*) even the $a$ parameter does not show systematic changes with the radius of the $B$ ion. In $Sr_2BO_4$ compounds, the high charge of the $B^{4+}$ ion compared to that of $Sr^{2+}$ ensures that the $BO_6$ octahedra are regular since the $B$–$O_I$ bonds would be much stronger than the Sr–$O_{Ic}$ bonds. In $LaSrBO_4$ compounds, however, competition between La and $B$ ions for covalent bonding with the oxygen ions could complicate the situation. When the $B$ ion is a 3$d$ transition metal ion such as $Mn^{3+}$ or low-spin $Ni^{3+}$, further complications enter because of the possibility of static Jahn–Teller distortions.

In Fig. 10, we have plotted the $c/a$ ratios in some $LaSrBO_4$ compounds where the $B$ ion (Fe, Cr, V) is neither a Jahn–Teller ion nor expected to occur in the low-spin state, against the octahedral crystal field stabilization energy (*37*) and the optical electronegativity of the $B$ ion in $B_2O_3$ compounds (*38*). The linear relationships found here indicate that $B$ ions which do not form strong cova-

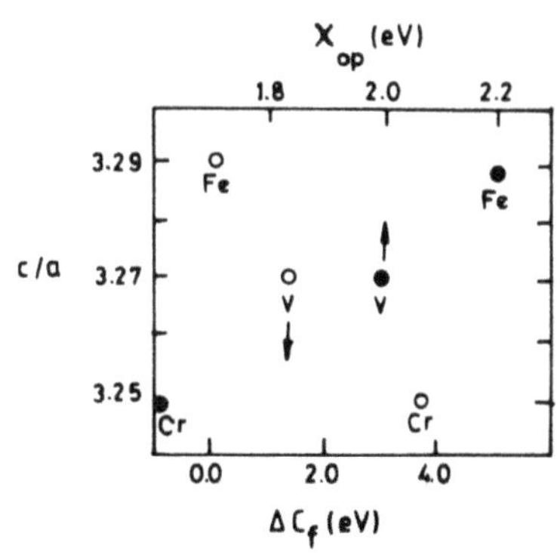


FIG. 10. Variation of the $c/a$ ratio of $LaSrBO_4$ ($B$ = Cr, V, Fe) with the octahedral crystal field stabilization energy, $\Delta C_f$, and the optical electronegativity of the $B$ ions in $B_2O_3$ compounds.

lent bonds (or which favor an octahedral environment) also favor a low *c/a* ratio. Thus, we may conclude that when $t < 1$, increasing covalency of the $B$–$O_{II}$ bond favors an elongation of the $BO_6$ octahedra. This elongation is a consequence of the pressure on the $B$–$O_{II}$ bond in the *ab* plane. In $Sr_2BO_4$ compounds, $t \approx 1$ and the pressure on the $B$–$O_{II}$ bond is considerably reduced. Nevertheless, the above considerations account for the high *c/a* ratios in $Sr_2BO_4$ compounds containing *B* ions with partially filled *d* orbitals. Elongation of the $BO_6$ octahedra associated with *B* ions which are Jahn–Teller ions, may favor a cooperative, ferrodistortive ordering of octahedra elongated parallel to the *c* axis. The unusually high *c/a* ratios (≈3.40) in $La_2CuO_4$ and $LaSrMnO_4$ (*7, 39*) could be indicative of such ordering.

Another feature of $A_2BO_4$ oxides with the $K_2NiF_4$ structure is that the average *B*–O distance is less than that computed from ionic radii especially when *t* is considerably less than unity. Thus, in $La_2NiO_4$ and $La_2CuO_4$, the average Ni–O and Cu–O distances are 2.03 and 2.07 Å, respectively (*6, 7*), while the values from ionic radii are 2.09 and 2.13 Å, respectively (*16, 32*). When the average *B*–O distance is very small, the question arises as to whether a disproportion of *B* ions can occur. Thus, $Cu^{2+}$ can disproportionate to $Cu^{1+}$ and $Cu^{3+}$. The average ($Cu^{1+}$, $Cu^{3+}$)–O distance is coincidentally 2.07 Å, which is the average Cu–O distance in $La_2CuO_4$ (*7*). Significantly again, the lattice parameters of $La_2CuO_4$ are close to that of $La_4Li^+Cu^{3+}O_8$ (*40*) and the radius of the $Li^+$ ion (0.76 Å) is only slightly smaller than that of $Cu^{2+}$ ion (0.77 Å). Although we do not propose that there is actually a static disproportionation of $Cu^{2+}$ ions in $La_2CuO_4$, the above arguments suggest the possible presence of charge-density waves. Transition metal ions with incompletely filled *d* orbitals have a mechanism to adjust their ionic radii toward a more favorable value of *t* by disproportionation or by forming charge-density waves. For example, disproportionation of $Fe^{4+}$ ions into $Fe^{3+}$ and $Fe^{5+}$ ions is known to occur in $CaFeO_3$ (*41*) but not in $SrFeO_3$ with a higher tolerance factor. Studies on $Ca_2FeO_4$ and $Sr_2FeO_4$ would be interesting to establish whether the disproportionation is associated with the tolerance factor.

It is clear from the above discussion that the *c/a* ratio in $A_2BO_4$ oxides is determined by several factors. Besides the covalency of the $B$–$O_{II}$-B bond and the competition between $A$–$O_I$ and $B$–$O_I$ linkages, other factors may also be important. Thus, $LaSrCuO_4$ in which the $Cu^{3+}$ ion is in the low-spin state has an unusually high *c/a* ratio (*40*); the preference for square-planar coordination of low-spin $Cu^{3+}$ ion is possibly an important factor here. $LaNaTiO_4$ and compounds of the formula $La_4LiBO_8$ ($B$ = Co, Cu, Ni) also have unusually high *c/a* ratios (*35, 36*). In the former, the Ti–O–Ti distance is unusually short (3.77 Å) compared to 3.88 Å in $Sr_2TiO_4$; in line with our earlier arguments, we expect the $TiO_6$ octahedra to be elongated. Since the $LiO_6$ octahedra would be elongated because of the short Li-$O_{II}$ distances we would expect increased *c/a* ratios in $La_4LiBO_8$ compounds.

*2.3. One-dimensional antiferromagnet model of $K_2NiF_4$ structure.* It is instructive to visualize the $K_2NiF_4$ structure as having a stacking sequence . . . *RPRPRPRPRPRP* . . . where *R* is a rock-salt layer and *P* is a perovskite layer. In so doing, we ignore the displacement of alternate perovskite layers mentioned earlier. Such a sequence is formally analogous to a one-dimensional antiferromagnetic Ising chain where *R* and *P* are treated as pseudospins with anisotropy. Long-range order is provided by an ordering field which can, in principle, arise from several mechanisms.

When *t* is less than unity, compression of the $B$–$O_{II}$–$B$ bond (and expansion of the $A$–$O_I$–$A$ bond) is necessitated. Long-range one-dimensional *RPRPRP* order would be favored because of the higher energy associated with interactions such as *RR* or *PP*.

In compounds with the Ruddlesdon–Popper structures (*42*) such as (SrO) (Sr $TiO_3)_n$, compounds with $n = 1$ have an $a$ parameter of 3.88 Å (compared to 3.906 Å in $SrTiO_3$) while the $a$ parameter of compounds with $n = 2$ and 3 is 3.90 Å, implying that in these compounds, the perovskite layer cannot be compressed. Furthermore, since the Ti–O–Ti distance in these oxides is larger than the $(1/\sqrt{2})$ $A$–O–$A$ distance, it would require considerable energy to stretch the Sr–O bond in a bilayer of SrO. Consequently, compounds of the type $(SrO)_m$ $(SrTiO_3)_n$ are not known including the type $m = n = 2$ shown in Fig. 3e. The $(AX)_m$ $(ABX_3)_n$ system may be possible, however, with larger $A$ ions and it would be interesting to investigate systems such as $LaMBO_4$ ($M$ = monovalent ion) and $Ba_2TiO_4$–$Sr_2TiO_4$ where $Ba_2TiO_4$ has the $K_2SO_4$ structure with Ti in the tetrahedral site (*43*). In compounds such as $La_2BO_4$, the $(LaO)^+$ layer is positively charged while the $(LaBO_3)^-$ layer is negatively charged. The stacking sequence in such oxides is therefore a favorable one for ordering along the $c$ axis, . . . $R^+$ $P^-$ $R^+$ $P^-$ $R^+$ $P^-$ $R^+$ . . . In oxygen excess $La_2BO_{4+\delta}$ compounds (e.g., $La_2NiO_4$), if the $\delta B^{3+}$ ions are randomly distributed, the $LaBO_3$ layers would be electricaly neutral ($P^0$). We can then have intergrowths of $P^-$ $P^0$ and $P^-$ $P^0$ $P^0$ type of layers; intergrowths of the composition $La_3Ni_2O_7$ and $La_4Ni_3O_{10}$ have been observed (*44*) in oxygen-excess $La_2NiO_4$. Tilley (*45*) has carried out an electron microscopic investigation of the $SrO$–$TiO_2$ system with special reference to $Sr_2TiO_4$. Besides the intergrowth of Ruddlesdon–Popper phases such as $Sr_3Ti_2O_7$ and $Sr_4$ $Ti_3O_{10}$, intergrowth of SrO-rich phases has been observed by Tilley. In the $La_2BO_4$ phases, intergrowth of two $(LaO)^+$ layers adjacent to each other would not be likely unless at least one of the LaO layers adjusts its oxygen and La content in such a manner as to make the layer neutral. Since this is unlikely, some $La_2O_3$ should be precipitated out in oxygen-excess $La_2BO_4$ compounds. Another possibility is that the excess oxygen is accommodated by $A$-site deficiency. In compounds such as $Sr_2BO_4$, $R$ and $P$ layers are both neutral and intergrowth of *RR* and *PP* sequence would be possible.

*2.4. Superlattice ordering in $A_2BO_4$ compounds*. Ordering of $B$ ions in compounds of the type $A_2BB'O_4$ due to charge difference potential between $B$ and $B'$ ions in different oxidation states would be analogous to antiferromagnetic ordering of spins in $K_2NiF_4$ due to an exchange potential. Since a three-dimensionally ordered antiferromagnetic structure of $K_2NiF_4$ is known (*1*), it seems reasonable to expect that $B$ and $B'$ ions may similarly get ordered. However, not all possible antiferromagnetic interactions can be satisfied in the $K_2NiF_4$ structure (*46*) and this frustration leads to a two-dimensional order.

In compounds such as $La_4LiBO_8$ ($B$ = Co, Ni, or Cu) or $Sr_4BB'O_8$ ($B$ = Co, Fe, Ni, etc.; $B'$ = Nb, Ta), X-ray diffraction studies do not reveal any evidence for an ordered superlattice (*47, 48*). Demazeau *et al.* (*25*) have found evidence for a $\sqrt{2}$ increase in the tetragonal $a$ parameter by employing neutron diffraction and X-ray diffraction (with monochromatized Cu $K_{\alpha 1}$ radiation) in $La_2LiCoO_8$. Our electron diffraction studies also reveal such ordering as mentioned earlier; the electron diffraction patterns of $La_4LiCoO_8$ show streaking parallel to the $c^*$ axis (Fig. 4c) similar to the ridges found in the neutron diffraction pattern of $K_2NiF_4$ in the temperature range where there is only two-dimensional antiferromagnetic ordering (*1*). We feel that the streaking in Fig. 4c may indeed be associated with two-dimensional ordering of $Li^+$ and $Co^{3+}$ ions.

Three-dimensional antiferromagnetic ordering in $K_2NiF_4$ is associated with an orthorhombic distortion. It is interesting that the distortion which gives rise to $M$, $O$, and $O'$ structures involves the movement of $O_I$ ions along the $c$ axis (Fig. 4) besides that of $O_{II}$ ions in the *ab* plane. By analogy with the three-dimensional antiferromagnetic ordering in $K_2NiF_4$, we would expect three-dimensional ordering of $B$ ions in $A_2BO_4$ only in distorted structures or when $t$ is close to 0.85.

In $A_2BO_4$ oxides, there are eight $A$ ions surrounding the $B$ ions as in the perovskite structure and two $A$ ions linked through $O_I$ ions with the $B$ ions along the $c$ axis. Random occupation of these sites by La and Sr ions in $LaSrBO_4$ (*33*) could give rise to a distribution of crystal fields. Of these, the most important are those involving the $O_I$ ions. La–$O_I$–B–$O_I$–La arrangement would give rise to the lowest crystal field while Sr–O–$B$–O–Sr arrangement would give rise to the highest crystal field. Evidence for a distribution of sites is seen in the $Fe^{3+}$ ESR spectra of $LaSrAl_{0.98}$ $Fe_{0.02}O_4$ (Fig. 11). There are two prominent lines at $g \approx 6$ and $g \approx 4.25$ which may be associated with $Fe^{3+}$ ions in axial and orthorhombic symmetry (*49–51*); the $g \approx 6$ line is also found in oxide glasses containing $Fe^{3+}$ (*51*). There is also a line at $g \approx 2.1$. The deviation from $g = 2$ may be attributed to spin-orbit coupling effects in $Fe^{3+}$ ions in distorted octahedra. Similarly, $LaSrAlO_4$ containing small amounts of $Ni^{3+}$ ions clearly shows evidence for the simultaneous existence of low-spin and high-spin $Ni^{3+}$ ions (*52*). Random ordering of $B$ ions in LaSr $(B,B')O_4$ has been used to explain percolation effects in electrical and magnetic properties of these compounds (*16*, *52*).

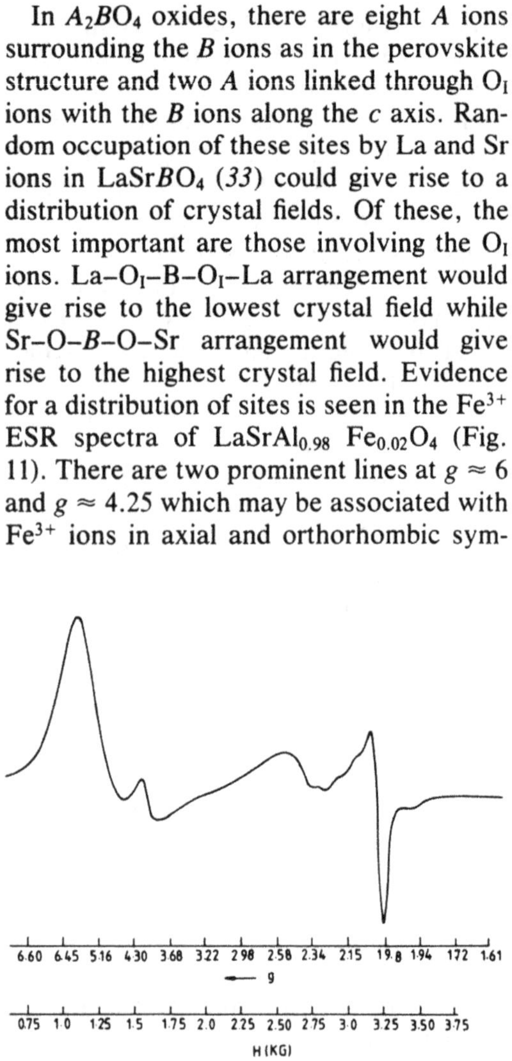


FIG. 11. ESR signal from $Fe^{3+}$ ions in $LaSrAl_{0.98}$ $Fe_{0.02}O_4$.

The only $AA'BO_4$ compound which shows ordering is $LaNaTiO_4$ (*53*). Ordering in this compound interestingly involves $(La_2O_2)^{2+}$ and $(Na_2O_2)^{2-}$ layers (*53*). There is considerable pressure on the Ti–$O_{II}$–Ti bonds in $LnNaTiO_4$ ($a \approx 3.78$ Å compared to 3.88 Å for $Sr_2TiO_4$), consistent with the low tolerance factor. As discussed earlier, a structure containing bilayers of rock-salt and perovskite (Fig. 3e) such as $(Ln_{0.5}Na_{0.5}O)_2$ $(Ln_{0.5}Na_{0.5}TiO_3)_2$ would also impose considerable pressure on the Ti–$O_{II}$–Ti bond. We have observed that X-ray diffraction line intensities calculated for such a structure by assuming a random distribution of $Ln$ and Na ions are comparable to those reported by Blasse (*53*).

*2.5. Non-stoichiometry.* Oxides with the $K_2NiF_4$ structure can accommodate considerable non-stoichiometry. This could lead to a significant variation in lattice parameters as indeed found in oxides such as LaSr $FeO_4$, $Pr_2NiO_4$, or $Nd_2NiO_4$, (*9*, *10*, *15*, *54*). In the $La_2NiO_{4+\delta}$ system, Drennan *et al.* (*44*) have shown that intergrowth of Ruddlesdon–Popper type phases such as $La_3Ni_2O_7$ and $La_4Ni_3O_{10}$ would account for anion-excess non-stoichiometry. Lewandowski *et al.* (*55*) have found $A$-site deficiency in lanthanum cobalt oxide; such a defect structure would be favored by $\delta B^{3+}$ ions.

Oxygen-deficient non-stoichiometry is more difficult to account for. Poeppelmeier *et al.* (*56*) have observed that $Ca_2MnO_{3.5}$ can be obtained topotactically from $Ca_2MnO_4$ by reduction, just as $CaMnO_{2.5}$ can be obtained from $CaMnO_3$. In $CaMnO_{2.5}$, $Mn^{3+}$ ions are in five-coordinated square-pyramidal coordination. It is assumed that the same situation may be present in $Ca_2MnO_{3.5}$. The idealized structure proposed for $Ca_2MnO_{3.5}$ is shown in Fig. 12 with the $O_{II}$ atoms being labile. The loss of $O_{II}$ atoms instead of $O_I$ atoms is consistent with $t < 1$; since the $B$–□–$B$ (□ = vacancy) distance would be smaller than the $B$—$O_{II}$—$B$ distance, it would favor the $K_2NiF_4$ structure. In $Sr_2CuO_3$, half the $O_{II}$ oxygen sites are vacant (*18*) and this is possible with $Cu^{2+}$ because of the presence of $d^2_{z^2}$ electrons. In this laboratory, a layered brownmillerite phase of the formula $Ca_2FeO_{3.5}$($a$ = 14.768, $b$ = 13.724, and $c$ = 12.20 Å) has been recently synthesized (*57*). This structure seems to have alternate columns of octahedra and tetrahedra in the *ab* plane with oxygen vacancies in both $O_I$ and $O_{II}$ positions. It is possible that in $Ca_2MnO_{3.5}$, fivefold coordination of Mn is achieved by the loss of $O_I$ oxygens. In any case, it is important to to note that anion-deficient non-stoichiometry can be achieved by the loss of $O_I$ or $O_{II}$ oxygens.

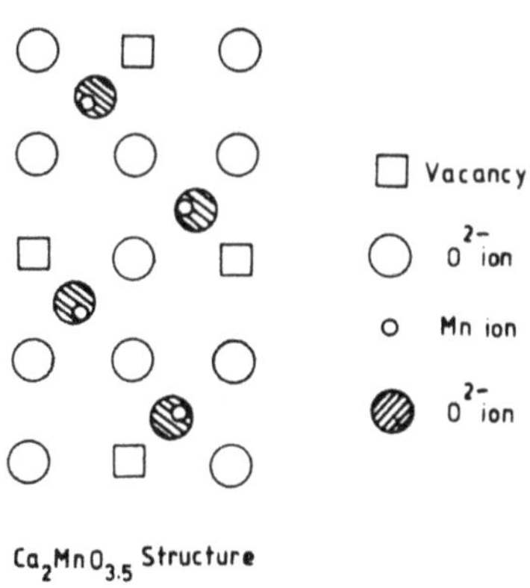


FIG. 12. Proposed ordering scheme in the *ab* plane of fivefold coordinated $Mn^{3+}$ ions in $Ca_2MnO_{3.5}$. Oxide ions above the plane of the paper and Ca ions are not shown (from Ref. (*56*)).

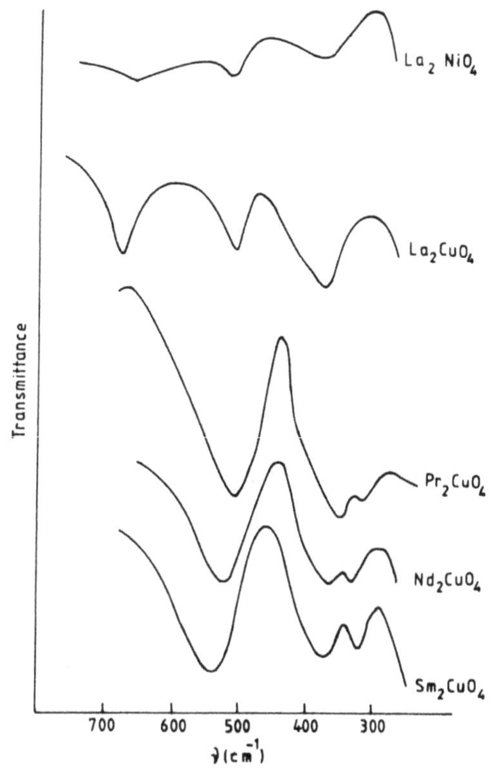


FIG. 13. Infrared spectra of some $Ln_2BO_4$ compounds.

## 3. Infrared Spectra

Infrared spectra of $A_2BO_4$ type oxides provide useful, diagnostic information on the nature of the $BO_n$ polyhedra. Thus, the spectra clearly distinguish $Ln_2CuO_4$ compounds of orthorhombic structure containing distorted $CuO_6$ octahedra from those of tetragonal structure containing square-planar $CuO_4$ polyhedra. We see from Fig. 13 that the Cu–O asymmetric stretching vibration band is split into a doublet (~690 and 520 $cm^{-1}$) in $La_2CuO_4$ (suggesting $D_{4h}$ symmetry of the isolated octahedron), while it appears as a single band in other $Ln_2CuO_4$ compounds with square-planar coordination. The spectrum of $La_2NiO_4$ possessing distorted $NiO_6$ octahedra is similar to that of $La_2CuO_4$. In the spectra of $La_{2-x}Pr_xCuO_4$, Cu–O asymmetric stretching is a doublet when $x < 0.75$ and a single band when $x \geq 0.75$ and Cu coordination is square-planar. In $LaSrBO_4$ ($B$ = Al or Fe), there seem to be three bands in the $B$–O stretching region suggesting a low symmetry of the $BO_6$ octahedra. The $B$–O stretching mode in $LaSrBO_4$ shifts to higher frequencies compared to that in the corresponding $La_2BO_4$. It is interesting that the $B$–O stretching frequency in $AA'BO_4$ oxides is higher compared to that in the corresponding $ABO_3$ perovskites. In $GdSrAlO_4$, for instance, the Al–$O_{II}$ stretching frequency increases by as much as 75 $cm^{-1}$ compared to that in $GdAlO_3$ and by about 50 $cm^{-1}$ compared to that in $LaSrAlO_4$. The Al–O stretching frequencies in $LaAlO_3$ and $GdAlO_3$ are similar, consistent with our earlier observation that in $ABO_3$ perovskites, lowering of $t$ leads to a buckling of the octahedra, the $B$–O distance remaining roughly the same. In $A_2BO_4$ oxides, the layered structure does not allow buckling and there is greater pressure on the $B$–$O_{II}$–$B$ bond with decreasing $t$.

## 4. Magnetic Properties

*4.1. Spin-state equilibria of transition metal ions.* Transitions between low-spin and high-spin states of transition metal ions have been found in perovskite oxides (*58, 59*) such as $LaCoO_3$. There is evidence for such transitions between spin states in oxides of $K_2NiF_4$ structure as well. The earliest evidence for such a transition was obtained with $La_4LiCoO_8$ by Blasse (*60*). Because of the elongated nature of the $BO_6$ octahedra in this oxide, the degeneracy of the $e_g$ orbital is expected to be lifted, shifting the $d_{x^2-y^2}$ orbital to higher energies. This orbital can remain unoccupied under certain conditions and this indeed appears to be the case with $Co^{3+}$ ions in $La_4LiCoO_8$. In Fig. 14, we have shown the inverse susceptibility–temperature curve of $La_4LiCoO_8$ and this is best interpreted in terms of a low-spin to intermediate-spin state transition. The intermediate-spin state with the configuration $t_{2g}^5 d_{z^2}^1 d_{x^2-y^2}^0$ seems to be stabilized (at intermediate temperatures) in compounds such as $La_4LiCoO_8$ (*25*) and also perhaps in $Sr_4CoNbO_8$ and $Sr_4CoTaO_8$ (*52*). In the last two compounds, there is some evidence of ordering of the two spin state (just as in $LaCoO_3$), the inverse susceptibility–temperature curve showing a plateau.

An interesting example of spin-state equilibrium between low- and high-spin $Ni^{3+}$ ions has been reported in $LaBaNiO_4$ on the basis of ESR evidence (*61*). The average Ni–O distance of 2.03 Å in this compound is consistent with the Ni–O distance expected from the ionic radius of high-spin $Ni^{3+}$. Magnetic susceptibility studies (*52*) on $LaSr_{1-x}Ba_xNiO_4$, however, reveal that the susceptibility can be entirely described on the basis of an equation of the form $\chi = [C/(T + \theta)] + \alpha$ where $\alpha$ is of the order of 6 $\times 10^{-4}$ emu and $C$ is of the order of 0.01–0.1 emu/K; $C$ increases with increasing $x$ while $\theta$ (in the range 10–20 K) decreases with increasing $x$. Magnetic susceptibility measurements show no evidence for an activated behavior. It would therefore seem that most of the $e_g$ electrons of the $Ni^{3+}$ ions are in extended states (as in $LaSrNiO_4$,

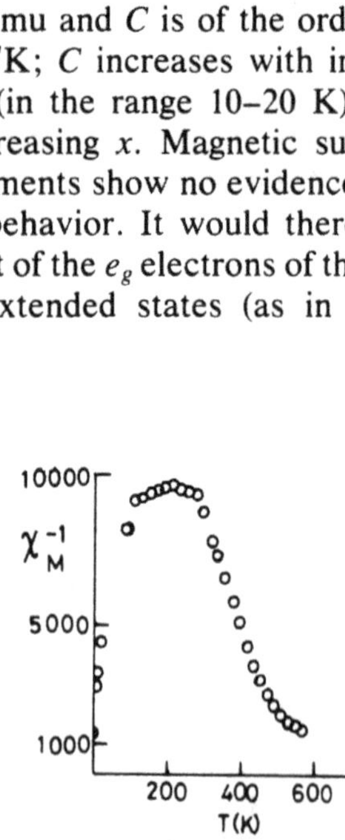


FIG. 14. $\chi_M^{-1}$ vs $T$ plot of $La_4LiCoO_8$ (after Ref. (*25*)).

with $x = 0$, where the $d_{x^2-y^2}$ electrons from a $\sigma^*_{x^2-y^2}$ band) with only a few of the electrons being localized. Localized electrons in $LaBaNiO_4$ may be associated with Ni sites with high crystal field, introduction of $Ba^{2+}$ in $LaSrNiO_4$ increasing the number of localized $Ni^{3+}$ states along with the unit-cell volume. It is possible that the ESR evidence of Demazeau *et al.* (*61*) is associated with such localized states.

An unusual configuration change of low-spin $Ni^{3+}$ has been observed (*52*) in the solid solution $LaSrAl_{1-x}Ni_xO_4$. For small values of $x$, the $Ni^{3+}$ ions are predominantly in the low-spin state showing an ESR signal similar to that observed in $La_4LiNiO_8$ (*29*) in which the $Ni^{3+}$ ions have the configuration $t^6_{2g}d^1_{z^2}d^0_{x^2-y^2}$. For $x \geq 0.75$, the ESR signal disappears and the magnetic susceptibility decreases sharply. The change in magnetic properties is accompanied by a significant decrease in the $c/a$ ratio. This points to a change to the configuration $t^6_{2g}$ $d^0_{z^2}d^1_{x^2-y^2}$ with the $d_{x^2-y^2}$ electrons forming $\sigma^*_{x^2-y^2}$ band-like states. It is significant that the magnitude as well as the temperature dependence of the susceptibility of the $x = 0.75$ sample (*52*) is close to that of pure $LaBaNiO_4$. It seems therefore extremely unlikely that the bulk of the $Ni^{3+}$ ions in the latter compound are involved in a low-spin to high-spin transition.

*4.2. Antiferromagnetic ordering.* When the *A* ion is nonmagnetic and the *B* ion is magnetic, $A_2BO_4$ oxides may be expected to behave as the fluorides with strong intralayer coupling and weak interlayer coupling. The earlier study (*62*) of such an oxide was on $Ca_2MnO_4$. In this oxide, $Mn^{4+}$ ions are coupled antiferromagnetically with the spins parallel to the $c$ axis. The magnetic structure determined by neutron diffraction (*62*) is shown in Fig. 15. Interplanar magnetic ordering leads to a doubling of the unit-cell parameter. Poeppelmeier *et al.* (*56*) have recently reported that the $c$ parameter of $Ca_2MnO_4$ is actually twice that of the value reported earlier. Considerable work has been carried out at Bordeaux on magnetic ordering in insulating $A_2BO_4$ oxides in which the spins couple antiferromagnetically. In Fig. 15, we show the magnetic structure of $\beta$-$Sr_2MnO_4$ (*63*) and $LaCaFeO_4$. In the latter, the magnetic moments are aligned along the $a$ axis.

Le Flem *et al.* (*64*) have pointed out that $\beta$-$Sr_2MnO_4$ (*63*), $LaSrCrO_4$ (*65*), and La-rich $La_{1-x}Y_xCaCrO_4$, show strong two-dimensional behavior as indicated by the critical exponents in the vicinity of the ordering

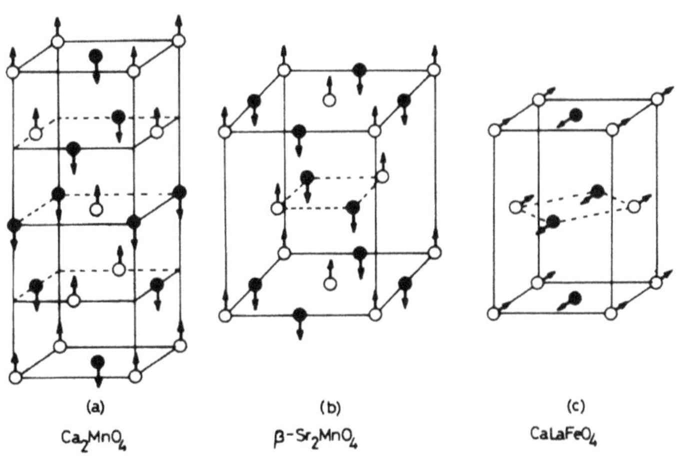


FIG. 15. Magnetically ordered structure of (a) $Ca_2MnO_4$, (b) $\beta$-$Sr_2MnO_4$, and (c) $LaCaFeO_4$ (from Ref. (*64*)).

temperatures as well as by neutron scattering investigations. Compounds such as $Ca_2MnO_4$ and $CaYCrO_4$ have essentially three-dimensional magnetic correlations. These authors also point out that the change from two-dimensional to three-dimensional correlation depends on the extent of covalency of the $A$–O bond. Making use of Goodenough's arguments in the case of perovskites (*66*), these authors suggest that there is competition between $t_{2g}$–$O_{II}$–$t_{2g}$ interaction (involving the $t_{2g}$ orbitals of the $B$ ions) and $\sigma_{A-O_{II}}$ bonding (involving the $A$ ions in the perovskite layer with a 90° $A$–$O_{II}$–$B$ linkage). The greater the covalency of the $A$–$O_{II}$ bonding, the weaker is the $B$–$O_{II}$–$B$ interaction and the stronger is the three-dimensional coupling. Thus, ions such as $Cr^{3+}$ and the isoelectronic $Mn^{4+}$ with only $t_{2g}$ electrons, have their magnetic interactions strongly determined by the nature of $A$–O interactions. The smaller the $A$ cation, the stronger is the $A$–$O_{II}$ linkage (and the three-dimensional correlations) and the lower is the antiferromagnetic ordering temperature. Applying the same arguments to oxides containing $Fe^{3+}$ ions, it is stated that since the magnetic interactions are strongly determined by the $e_g$ electrons, the influence of the $A$ ion is considerably diminished. Accordingly, there is not much difference in the ordering temperatures of $LaCaFeO_4$ (373 K) and $LaSrFeO_4$ (380 K). However, compounds such as $Ln$SrFeO$_4$ show a marked dependence of the magnetic ordering temperature on the size of the $Ln$ ion (*67*). Although the above arguments seem plausible, we should note that we have ignored the $A$–$O_I$–$B$ linkage which would be exected to play a role. The collapse in the $c/a$ ratio of $Ln$SrBO$_4$ compounds with the decreasing size of $Ln$ ion (Fig. 4) is consistent with a decreased $B$–$O_I$ distance. Since the $B$–$O_I$–$O_I$–$B$ interaction determines the interlayer coupling, we may expect the three-dimensional character to increase with decreasing $B$–$O_I$ bond length.

In $LaSrAl_{1-x}Fe_xO_4$ solid solutions (*15*), the Néel temperature drops to zero when $x > x_c$ where $x_c$ (0.59) is the critical percolation threshold for nearest-neighbor interactions in a square-planar array (*68*). The results are similar to those found in the $Rb_2Mg_{1-x}Mn_xF_4$ system (*69*). The surprising conclusion, therefore, is that even in oxides, long-range magnetic ordering is dominated by nearest-neighbor interactions.

*4.3. Unusual Behavior of $La_2NiO_4$, $La_2CuO_4$, and $La_2CoO_4$.* $La_2NiO_4$ shows a Curie–Weiss behavior at high temperatures with high $\theta$ ($\sim -500$ K) and $\mu_{eff}$ ($\sim$3.00 $\mu_B$) values (*70*). Below 200 K, there is deviation from the Curie–Weiss law, but neither neutron diffraction nor magnetic susceptibility studies down to the lowest temperature show any evidence for long-range antiferromagnetic ordering (*70, 71*). It has been found recently that below 100 K, the magnetic susceptibility again conforms to a Curie–Weiss behavior of (Fig. 16) with a $\mu_{eff}$ of

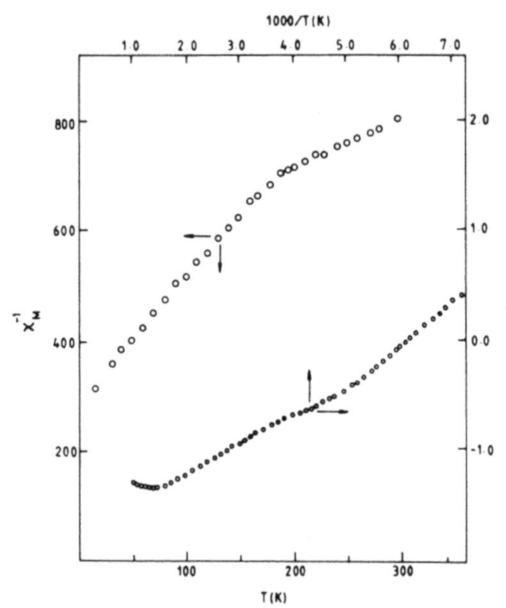


FIG. 16. Plots log $\rho$ vs $1/T$ and $\chi_M^{-1}$ vs $T$ of $La_2NiO_4$ (from Ref. (*2*) and unpublished results from this laboratry).

1.8 $\mu_B$ indicating the presence of one unpaired electron (*15*). The $\theta$ value obtained from the slope of the $\chi_M^{-1}$ vs $T$ plot is consistent with the expression $\theta = CW$ with a nearly temperature-independent $W$ and with a $C$ corresponding to two unpaired electrons above 200 K and one unpaired electron below 100 K. Such a behavior is also observed in the $La_2Ni_{1-x}Cu_xO_4$ system in the range $0.75 \geq x \geq 0.0$ below 100 K with a $\mu_{eff}$ value corresponding to one unpaired electron (*15*) per g. atom of Ni. Below 200 K, there is also an increase in the activation energy for electrical conduction (Fig. 16) suggesting that there is spin-pairing of the $d_{x^2-y^2}$ electron around 200 K. It is rather surprising that in $La_2NiO_4$, below 200 K, the $d_{x^2-y^2}$ electrons are coupled into nearly dimagnetic spin-paired states while the $d_{z^2}$ electrons are localized and coupled antiferromagnetically without the onset of long-range order. Since $La_2NiO_4$ prepared by ceramic techniques always has an excess of oxygen due to the presence of $Ni^{3+}$ ions, it is not clear whether long-range order is frustrated by the presence of such ions. Singh *et al.* (*15*) propose that a disproportionation of the $Ni^{2+}$ ions to $Ni^{1+}$ and low-spin $Ni^{3+}$ (which could result in the formation of charge-density waves) would stabilize the diamagnetic nature of the spin-paired states involving the $d_{x^2-y^2}$ electrons.

$La_2CuO_4$ was first reported to show nearly temperature independent magnetic susceptibility which was attributed to antiferromagnetism (*7*). Because of the low value the electrical resistivity, it was assumed that the oxide exhibited broad-band Pauli paramagnetism. It was subsequently shown (*72*) that there is an enhancement in the susceptibility at the lowest temperature with an anomaly around 200 K (Fig. 17). The anomaly in the susceptibility at low temperatures can be attributed to paramagnetic impurities and it has been shown by Saez-Puche *et al.* (*73*) that starting with high purity oxides, the paramagnetic behavior at low temperature is suppressed although there is a tendency toward an increase in the susceptibility below 100 K. Singh *et al.* (*15*) have since found that at low temperatures, there is a marked increase in the resistivity with $\rho(12\ K)/\rho(300\ K) \approx 10^2$. Careful magnetic susceptibility studies show that above 200 K, the susceptibility obeys a Curie law with a $\mu_{eff}$ corresponding to the contribution from about 16% of $Cu^{2+}$ ions. At low temperatures, point by point calculation of $\mu_{eff}$ (assuming a Curie law) shows a linear decrease in the concentration of paramagnetic $Cu^{2+}$ ions with decreasing temperature. Such a behavior seems to be consistent with the disproportionation of $Cu^{2+}$ to $Cu^{1+}$ and $Cu^{3+}$ as mentioned earlier (*15*) and the formation of charge-density waves. The two $B$–$O_I$ distances in the $O$ (or $O'$) structure of $La_2CuO_4$ would be consistent with orbital or-

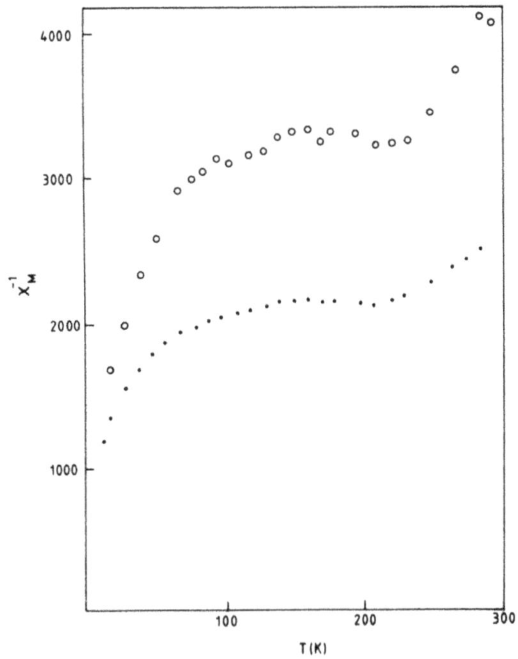


FIG. 17. $\chi_M^{-1}$ vs $T$ plot of $La_2CuO_4$ at 4000 G (dotted curve); after subtraction of the contribution from the ferromagnetic component (line of circles) (from Ref. (*15*)).

dering of the type shown in Fig. 7c and it is interesting to speculate on the consequences of such ordering.

There are several intriguing features associated with $Cu^{2+}$ ions in such oxides which are not readily understood. A comparison of the high-temperature susceptibilities of the oxides in the series $La_2Ni_{1-x}Cu_xO_4$ and $La_{2-2x}Sr_{2x}Ni_{1-x}Ti_xO_4$ shows that there is no contribution to the magnetic susceptibility at high temperatures from $Cu^{2+}$ ions (*15*). It has been observed that $Cu^{2+}$ ions do not contribute to the susceptibility of $Ln_2CuO_4$ compounds ($Ln$ = rare earth) at low temperatures and this has been taken to indicate antiferromagnetic ordering (*72–74*).

It is extremely difficult to prepare stoichiometric $La_2CoO_4$. Magnetic susceptibility of $La_2CoO_{4.04}$ measured in this laboratory (*75*) shows a broad maximum around 500 K and a sharp maximum around 400 K. DSC studies show a large number of the phase transitions in this temperature region. The results have been interpreted in terms of two-dimensional magnetic ordering with the onset of three-dimensional order at low temperatures. Further work on stoichiometric samples of $La_2CoO_4$ is necessary to establish the magnetic behavior of this oxide.

*4.4. Ferromagnetic oxides.* One of the earliest oxides to be studied was $LaSr_3Mn_2O_8$ and measurements on this oxide were reported in the same paper (*76*) dealing with the low-dimensional magnetic susceptibility behaviour of $K_2NiF_4$. The peculiarity of $LaSr_3Mn_2O_8$ is that although the high-temperature behavior is typical of ferromagnets (the susceptibility showing a Curie–Weiss plot with a high negative values of $\theta$), there is no spontaneous magnetization at low temperatures. This has been attributed (*77*) to the presence of small superparamagnetic clusters due to the absence of ordering of $Mn^{3+}$ and $Mn^{4+}$ ions (since only $Mn^{3+}$–O–$Mn^{4+}$ interactions are ferromagnetic). However, the corresponding cobalt analog $LaSr_3Co_2O_8$ is a true ferromagnet with a well-defined Curie temperature (*75*). In this oxide, $Co^{3+}$ ions could have the intermediate-spin configuration ($t_{2g}^5e_g^1$) while the $Co^{4+}$ ions are in the low-spin configuration ($t_{2g}^5$). $Co^{3+}$–O–$Co^{4+}$ would then be a Zener double exchange pair since the transfer of an electron from intermediate-spin $Co^{3+}$ to low-spin $Co^{4+}$ would have the initial and final state degenerate. Although this oxide is semiconducting, changes in the electron transport properties observed at the Curie temperature ($E_a$ decreases below $T_c$) is consistent with a double exchange mechanism. Such changes in transport properties at $T_c$ have not been observed in the corresponding three-dimensional perovskite compound $LaSrCo_2O_6$. The $\mu_{eff}$ value calculated from the $\chi_M^{-1}$ vs $T$ plot (Fig. 18) above the Curie temperature also supports the existence of intermediate-spin $Co^{3+}$ and low-spin $Co^{4+}$. At high temperatures (Fig. 18), the $\chi_M^{-1}$ vs $T$ plot shows a Curie behavior with the $\mu_{eff}$ corresponding to high-spin $Co^{3+}$ ions, and low-spin $Co^{4+}$ ions. This could be associated with a low-spin to high-spin transition. It should be mentioned that the nature of the $\chi_M^{-1}$ vs $T$ plot is similar to that predicted by Anderson and Hasegawa (*78*) for double exchange systems.

$LaSrMnO_4$ has been reported to be ferro-

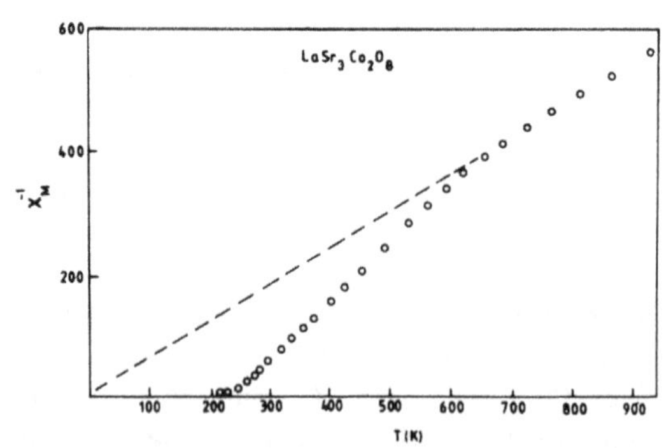


FIG. 18. $\chi_M^{-1}$ vs $T$ plot of $LaSr_3Co_2O_8$ above 300 K.

magnetic (*79*). The electronic configuration of $Mn^{3+}$ being the same as that of $Cr^{2+}$, we would expect ferromagnetism in $LaSrMnO_4$ by analogy with halides of the formula $A_2CrX_4$ which are transparent ferromagnets (*80*). Ferromagnetic $LaSrMnO_4$ is reported to have a tetragonal structure with a small $c/a$ ratio while the $A_2CrX_4$ compounds are orthorhombic with unequal Cr–$X$ distances in the basal plane indicative of antiferrodistortive ordering of $CrX_6$ octahedra. Recently, stoichiometric $LaSrMnO_4$ with a large $c/a$ ratio (value) has been reported (*39*); we would expect ferrodistortive ordering of the elongated $MnO_6$ octahedra in such an oxide with the $e_g$ electron in the $d_{z^2}$ orbital. Considerations based on the Goodenough–Kanamori rules (*81*) indicate that the interaction would be ferromagnetic if the transfer of the $d^1_{z^2}$ electron is to the empty $d^0_{x^2-y^2}$ orbital. Magnetic measurements on stoichiometric $LaSrMnO_4$ have not been reported. The small $c/a$ ratio of the earlier sample of $LaSrMnO_4$ suggests considerable oxygen excess non-stoichiometry and it is indeed known that the $c/a$ ratio of $La_{1-x}Sr_{1+x}MnO_4$ decreases with increasing $x$.

Two oxides that have been recently studied in this laboratory (*15*) which show evidence for ferromagnetic interactions are $La_2Sr_2MnNiO_8$ and $La_2Sr_2MnCoO_8$ which are the two-dimensional analogs of the three-dimensional ferromagnets $La_2Mn NiO_6$ and $La_2MnCoO_6$, respectively (*82*, *83*). $La_2Sr_2MnNiO_8$ shows a large negative value of the Weiss constant, but is like $LaSr_3Mn_2O_8$ in that it does not show spontaneous magnetization at low temperatures. There is no ion ordering in either of these oxides. The $c/a$ ratio of the former compound (3.27) is much less than that of stoichiometric $LaSrMnO_4$ (*39*), but similar to that of $LaSrNiO_4$ (*33*). This suggests that the electronic configurations of the Mn and Ni ions are $t^3_{2g}d^1_{x^2-y^2}$ and $t^6_{2g}d^1_{x^2-y^2}$, respectively, the $d_{z^2}$ orbitals being empty in both cases. $La_2Sr_2MnCoO_8$ shows the behavior of a true ferromagnet with a well-defined Curie temperature. The $c/a$ ratio in this compound is similar (3.28) to that of LaSr $CoO_4$ (*26*, *27*), but much less than that of $LaSrMnO_4$ (*39*). Electronic configurations of the $Mn^{3+}$ and $Co^{3+}$ ions in this oxide appear to be $t^3_{2g}d^1_{x^2-y^2}$ and $t^5_{2g}d^1_{z^2}$, respectively. The presence of a localized $d_{z^2}$ electron is significant. It is possible that the presence of a localized $d_{z^2}$ electron enhances the three-dimensional $B$–$O_I$–$O_I$–$B$ interlayer coupling. A mechanism that could be of relevance to compounds such as $LaSr_3Mn_2O_8$ and $La_2Sr_2MnNiO_8$ is one where the magnetic moments in the ferromagnetic state may be aligned perpendicular to the $ab$ plane. In the absence of intralayer coupling, demagnetization effects acting on spins aligned perpendicular to a layer would be considerable.

## 5. Concluding Remarks

It should be clear from the above discussion that oxides with the $K_2NiF_4$ structure offer considerable scope for research. The anisotropic bonding coupled with strong covalency effects in these oxides could give rise to unique properties which have not been adequately investigated. For example, we are yet to understand the nature of the semiconductor–metal transition in $La_2NiO_4$ and other rare earth nickelates. Although many models have been suggested (*84*) to explain the transition in $La_2NiO_4$ (*2*), all the measurements reported hitherto are on polycrystalline samples. Preliminary studies (*12*) on single crystals of $La_2NiO_4$ indicate that the transition occurs sharply (550 K) with about an order of magnitude jump in conductivity along the $ab$ plane. Anisotropic magnetic susceptibility of $La_2NiO_4$ needs to be investigated as also the effect of $Ni^{3+}$ ions on these properties. The structure of $La_2NiO_4$ itself seems to require a revision in the light of the superlattice spots found in the diffraction patterns. Electron transport properties of single crystals of $La_2CuO_4$ and related rare earth compounds are yet to be investigated. The possible occurrence of two-dimensional antiferromagnetic ordering in $La_2CoO_4$ is worth exploring.

An interesting feature of some of the $A_2BO_4$ oxides is that their electrical properties are considerably different from the corresponding perovskites, even though the magnetic properties are similar. For example, $LaSr_3Co_2O_8$ is a semiconducting ferromagnet while $LaSrCo_2O_6$ is an itinerant-electron ferromagnet. The corresponding manganese compounds also show a similar behavior. Another system showing such a behavior is that of vanadium, $La_{1-x}Sr_xVO_3$ and $La_{1-x}Sr_{1+x}VO_4$, where the latter is insulating while the former is metallic for $0.3 > x > 0.05$.

None of the $A_2BO_4$ oxides seems to exhibit a true metallic behavior down to the lowest temperatures. Most of these oxides show activated conduction and even those phases that have been considered to be metallic (e.g., $La_2CuO_4$) exhibit conductivities of the order of 10 $ohm^{-1}$ $cm^{-1}$. This is much less than the conductivities found in metallic oxides of perovskite structure (e.g., La $NiO_3$ or $LaSrCo_2O_6$ with $\sigma$ of $\sim 10^3$ $ohm^{-1}$ $cm^{-1}$). It is not clear whether the absence of true metallic conductivity in $A_2BO_4$ oxides has something to do with localization in two dimensions (*85, 86*).

In systems where the electrical properties are determined by the concentration of the component ions, the $A_2BO_4$ system shows an unusual concentration dependence of resistivity. For example, delocalization of $e_g$ electrons is found in systems like $LaSrAl_{1-x}Ni_xO_4$ when $x > 0.6$. In perovskite systems such as $La_{1-x}Sr_xCoO_3$ and $LaFe_{1-x}Ni_xO_3$, the oxides become metallic when $x = 0.25–0.30$. It is interesting to ponder whether such concentration limits are related to percolation limits in two-dimensional and cubic systems.

Some of the well-known ferroelectric materials are perovskite oxides. No ferroelectric oxide of $K_2NiF_4$ structure has been reported until now; similarly, other ferroic properties (*87*) are yet to be explored.

## Acknowledgments

The authors thank the Department of Science and Technology, Government of India and the University Grants Commission for support of this research.

## References

*1.* R. J. Birgeneau, H. J. Guggenheim, and G. Shirane, *Phys. Rev. B* **1,** 2211 (1970).
*2.* P. Ganguly and C. N. R. Rao, *Mater. Res. Bull.* **8,** 405 (1973).
*3.* P. Poix, *J. Solid State Chem.* **31,** 95 (1980).
*4.* P. Poix, *C.R. Acad. Sci. (Paris) C* **268,** 1139 (1969).
*5.* D. Ganguli, *J. Solid State Chem.* **30,** 353 (1979).
*6.* A. Rabeneau and P. Eckerlin, *Acta Crystallogr.* **11,** 304 (1958).
*7.* J. M. Longo and P. Raccah, *J. Solid State Chem.* **6,** 526 (1973).
*8.* V. U. Lehmann and H. R. Müller-Buschbaum, *Z. Anorg. Allg. Chem.* **470,** 59 (1980).
*9.* B. Willer and M. Dairre, *C.R. Acad. Sci. (Paris) C* **267,** 1482 (1968).
*10.* M. Föex, *Bull. Soc. Chim. Fr.* 109 (1961).
*11.* B. Grande and H. R. Müller-Buschbaum, *Z. Anorg. Allg. Chem.* **433,** 152 (1977).
*12.* C. N. R. Rao, D. Buttrey, N. Otsuka, P. Ganguly, H. R. Harrison, C. J. Sandberg, and J. M. Honig, *J. Solid State Chem.* **51,** 266 (1984).
*13.* R. Berjoan, J. P. Coutures, G. Le Flem, and S. Saux, *J. Solid State Chem.* **42,** 75 (1982), and references therein.
*14.* B. Grande, H. R. Müller-Buschbaum, and M. Schweizer, *Z. Anorg. Allg. Chem.* **428,** 120 (1977).
*15.* K. K. Singh, P. Ganguly, and J. B. Goodenough, *J. Solid State Chem.*, in press; also see K. K. Singh, Ph.D. thesis, Indian Institute of Science, Bangalore, India (1983).
*16.* R. D. Shannon, *Acta Crystallogr. Sect. A* **32,** 751 (1976).
*17.* H. R. Müller-Buschbaum and W. Wollschlager, *Z. Anorg. Allg. Chem.* **414,** 76 (1975).
*18.* H. R. Müller-Buschbaum, *Angew. Chem. (English transl.)* **16,** 674 (1977).
*19.* K. Knox, *J. Chem. Phys.* **30,** 991 (1959).

20. K. K. Singh, P. Ganguly, and C. N. R. Rao, *Mater. Res. Bull.* **17,** 493 (1982).
21. V. R. Haegele and D. Babel, *Z. Anorg. Allg. Chem.* **409,** 11 (1974).
22. D. Reinen and C. Friebel, *Struct. Bonding* **37,** 1 (1979).
23. J. P. Oudalov, A. Daoudi, J. C. Joubert, G. Le Flem, and P. Hagenmuller, *Bull. Soc. Chim. Fr.* 3408 (1970).
24. L. Fournes, N. Kinomura, and F. Menil, *C.R. Acad. Sci. (Paris) C* **291,** 235 (1980).
25. G. Demazeau, M. Pouchard, M. Thomas, J. F. Colombet, J. C. Grenier, L. Fournes, J. L. Soubeyroux, and P. Hagenmuller, *Mater. Res. Bull.* **15,** 451 (1980).
26. G. Demazeau, P. Courbin, G. Le Flem, M. Pouchard, P. Hagenmuller, J. L. Soubeyroux, I. G. Main, and G. A. Robins, *Nouv. J. Chim.* **3,** 171 (1979).
27. G. A. Robins, M. F. Thomas, J. D. Rush, G. Demazeau, and I. G. Main, *J. Phys. C* **15,** 233 (1982).
28. G. Demazeau, M. Pouchard, N. Chevreau, M. Thomas, F. Menil, and P. Hagenmuller, *Mater. Res. Bull.* **16,** 689 (1981).
29. G. Demazeau, J. L. Marty, M. Pouchard, T. Rojo, J. M. Dance, and P. Hagenmuller, *Mater. Res. Bull.* **16,** 47 (1981).
30. H. P. Rooksby, *Acta Crystallogr.* **1,** 266 (1948).
31. M. M. J. Fava, Y. Oudalov, J. M. Reau, G. Le Flem, and P. Hagenmuller, *C.R. Acad. Sci. (Paris) C* **274,** 183 (1972).
32. R. D. Shannon and C. T. Prewitt, *Acta Crystallogr. Sect. B* **25,** 925 (1969).
33. G. Demazeau, M. Pouchard, and P. Hagenmuller, *J. Solid State Chem.* **18,** 159 (1976).
34. J. C. Joubert, A. Collomb, D. Elmaleh, G. Le Flem, A. Daoudi, and G. Olliver, *J. Solid State Chem.* **2,** 343 (1970).
35. J. B. Goodenough and J. M. Longo, "Landolt-Bornstein" (K. H. Hellwege, Ed.), Group III/4a, p. 126, Springer-Verlag, Berlin (1970).
36. S. Nomura, "Landolt-Bornstein" (K. H. Hellwege and A. M. Hellwege, Eds.), Group III/12a, p. 425, Springer-Verlag, Berlin (1978).
37. D. S. McClure, *J. Phys. Chem. Solids* **3,** 311 (1957).
38. J. A. Duffy, *Struct. Bonding* **32,** 147 (1977).
39. A. Benabad, A. Daoudi, R. Salmon, and G. Le Flem, *J. Solid State Chem.* **22,** 121 (1977).
40. J. B. Goodenough, G. Demazeau, M. Pouchard, and P. Hagenmuller, *J. Solid State Chem.* **8,** 325 (1973).
41. M. Takano, N. Nakanishi, Y. Takada, S. Naka, and T. Takada, *Mater. Res. Bull.* **12,** 923 (1977).
42. S. N. Ruddlesdon and P. Popper, *Acta Crystallogr.* **10,** 538 (1957); **11,** 54 (1958).
43. J. A. Bland, *Acta Crystallogr.* **14,** 875 (1961).
44. J. Drennan, C. P. Tavares, and B. C. H. Steele, *Mater. Res. Bull.* **17,** 621 (1982).
45. R. J. D. Tilley, *J. Solid State Chem.* **21,** 293 (1977).
46. R. Plumier, *J. Appl. Phys.* **35,** 950 (1964).
47. G. Blasse, *J. Inorg. Nucl. Chem.* **27,** 2683 (1965).
48. J. F. Ackermann, *Mater. Res. Bull.* **14,** 487 (1979).
49. J. S. Griffiths, *Mol. Phys.* **8,** 213, 217 (1964).
50. M. S. Hadded, M. W. Lynch, W. D. Federer, and D. N. Hendrickson, *Inorg. Chem.* **20,** 123, 131 (1981).
51. J. T. Castner, G. S. Newell, W. C. Holton, and C. P. Slichter, *J. Chem. Phys.* **32,** 668 (1960).
52. R. Mohan Ram, K. K. Singh, W. H. Madhusudhan, P. Ganguly, and C. N. R. Rao, *Mater. Res. Bull.* **18,** 703 (1983).
53. G. Blasse, *J. Inorg. Nucl. Chem.* **30,** 656 (1968).
54. J. L. Soubeyroux, P. Courbin, L. Forunes, D. Fruchart, and G. Le Flem, *J. Solid State Chem.* **31,** 313 (1980).
55. J. T. Lewandowski, J. M. Longo, and R. A. McCauley, *Amer. Ceram. Soc. Bull.* **61,** 333 (1982).
56. K. Poeppelmeier, M. E. Leonwicz, and J. M. Longo, *J. Solid State Chem.* **44,** 89 (1982).
57. K. Vidyasagar, J. Gopalakrishnan, and C. N. R. Rao, *Inorg. Chem.*, in press.
58. W. H. Madhusudhan, K. Jagannathan, P. Ganguly, and C. N. R. Rao, *J. Chem. Soc. Dalton Trans.* 1397 (1980).
59. S. Ramasesha, T. V. Ramakrishnan, and C. N. R. Rao, *J. Phys. C* **12,** 1307 (1979).
60. G. Blasse, *J. Appl. Phys.* **36,** 879 (1965).
61. G. Demazeau, J. L. Marty, B. Buffat, J. M. Dance, N. Pouchard, P. Dordor, and B. Chevalier, *Mater. Res. Bull.* **17,** 37 (1982).
62. D. E. Cox, G. Shirane, R. J. Birgeneau, and J. B. MacChesney, *Phys. Rev.* **188,** 930 (1969).
63. J. C. Bouloux, J. L. Soubeyroux, M. Perrin, and G. Le Flem, *J. Solid State Chem.* **38,** 34 (1981).
64. G. Le Flem, G. Demazeau, and P. Hagenmuller, *J. Solid State Chem.* **44,** 82 (1982).
65. G. Olliver, These de Doctorates Sciences Physiques, Univ. Scientifique et Medicale, Grenoble (1973).
66. J. B. Goodenough, *Progr. Solid State Chem.* **5,** 145 (1972); "Solid State Chemistry," (C. N. R. Rao, Ed.), Dekker, New York (1973).
67. M. Shimada and M. Koizumi, *Mater. Res. Bull.* **11,** 1237 (1976).

68. B. K. Shante and S. Kirkpatrick, *Advan. Phys.* **20,** 325 (1971).
69. W. M. Walsh, R. J. Birgeneau, L. W. Rupp, and H. J. Guggenheim, *Phys. Rev. B* **20,** 4645 (1979).
70. G. A. Smolenskii, V. N. Yudin, and E. Sher, *Sov. Phys. Solid State* **4,** 2452 (1962).
71. G. A. Smolenskii, V. A. Bokov, S. A. Kizaev, E. I, Mal'teev, G. M. Nedhir, V. P. P. Plaknty, A. G. Tutov, and V. N. Yudin, "Proceedings International Conference on Magnetism, Nottingham, 1964," p. 354, Inst. Phys. and Phys. Soc. London (1965).
72. P. Ganaguly, S. Kollali, C. N. R. Rao, and S. Kern, *Magn. Lett.* **1,** 107 (1980).
73. R. Saez-Puche, M. Norton, and W. S. Glaunsinger, *Mater. Res. Bull.* **17,** 1523 (1982).
74. R. Saez-Puche, M. Norton, and W. S. Glaunsinger, *Mater. Res. Bull.* **17,** 1539 (1982).
75. P. Ganguly and S. Ramasesha, *Magn. Lett.* **1,** 131 (1980).
76. K. G. Srivatsava, *Phys. Lett.* **4,** 55 (1963).
77. J. C. Bouloux, J. L. Soubeyroux, A. Daoudi, and G. Le Flem, *Mater. Res. Bull.* **16,** 855 (1981).
78. P. W. Anderson and H. Hasegawa, *Phys. Rev.* **100,** 675 (1955).
79. J. B. MacChesney, J. F. Potter, and R. C. Sherwood, *J. Appl. Phys.* **40,** 1243 (1969).
80. P. Day, *Acc. Chem. Res.* **12,** 236 (1973), and references therein.
81. J. B. Goodenough, "Magnetism and the Chemical Bond," Wiley–Interscience, New York (1963).
82. J. B. Goodenough, A. Wold, R. J. Arnott, and N. Menyuk, *Phys. Rev.* **124,** 373 (1961).
83. N. Y. Vasanthacharya, K. K. Singh, and P. Ganguly, *Rev. Chim. Miner.* **18,** 333 (1981), and references therein.
84. J. B. Goodenough and S. Ramasesha, *Mater. Res. Bull.* **17,** 383 (1982).
85. E. Abraham, P. W. Anderson, D. C. Liccardello, and T. V. Ramakrishnan, *Phys. Rev. Lett.* **42,** 673 (1979).
86. E. Abraham and T. V. Ramakrishnan, *J. Non-Cryst. Solids* **35,** 15 (1980); *Philos. Mag.* [*Part*] *B* **42,** 827 (1980).
87. R. E. Newnham and L. E. Cross, *in* "Preparation and Characterization of Materials" (J. M. Honig and C. N. R. Rao, Eds.), Academic Press, New York (1981).

JOURNAL OF SOLID STATE CHEMISTRY **72**, 14–23 (1988)

# A Comparative Study of the Magnetic and Electrical Properties of Perovskite Oxides and the Corresponding Two-Dimensional Oxides of $K_2NiF_4$ Structure*

C. N. R. RAO,** P. GANGULY, K. K. SINGH, AND R. A. MOHAN RAM

*Solid State and Structural Chemistry Unit, Indian Institute of Science, Bangalore 560 012, India*

Received January 7, 1987

Electrical and magnetic properties of several oxide systems of $K_2NiF_4$ structure have been compared to those of the corresponding perovskites. Members of the $La_{1-x}Sr_{1+x}CoO_4$ system are all semiconductors with a high activation energy for conduction unlike $La_{1-x}Sr_xCoO_3$ ($x \geq 0.3$) which is metallic; the latter oxides are ferromagnetic. $La_{0.5}Sr_{1.5}CoO_4$ shows a magnetization of 0.5 $\mu_B$ at 0 K (compared to 1.5 $\mu_B$ of $La_{0.5}Sr_{0.5}CoO_3$), but the high-temperature susceptibilities of the two systems are comparable. In $SrO \cdot (La_{0.5}Sr_{0.5}MnO_3)_n$, both magnetization and electrical conductivity increase with the increase in $n$ approaching the value of the perovskite $La_{0.5}Sr_{0.5}MnO_3$. $LaSrMn_{0.5}Ni_{0.5}(Co_{0.5})O_4$ shows no evidence of long-range ferromagnetic ordering unlike the perovskite $LaMn_{0.5}Ni_{0.5}(Co_{0.5})O_3$; high-temperature susceptibility behavior of these two insulating systems is, however, similar. $LaSr_{1-x}Ba_xNiO_4$ exhibits high electrical resistivity with the resistivity increasing proportionately with the magnetic susceptibility (note that $LaNiO_3$ is a Pauli-paramagnetic metal). High-temperature susceptibility of $LaSrNiO_4$ and $LaNiO_3$ are comparable. Susceptibility measurements show no evidence for long-range ordering in $LaSrFe_{1-x}Ni_xO_4$ unlike in $LaFe_{1-x}Ni_xO_3$ ($x \leq 0.35$) and the electrical resistivity of the former is considerably higher. Electrical resistivity of $Sr_2RuO_4$ is more than an order of magnitude higher than that of $SrRuO_3$. Some generalizations of the properties of two- and three-dimensional oxide systems have emerged from these experimental observations. 

## Introduction

Electrical and magnetic properties of perovskite oxides, $ABO_3$ ($B$ = transition metal ion), have been investigated extensively (*1–6*). Quasi-two-dimensional oxides of the $K_2NiF_4$ structure, $A_2BO_4$, possessing $ABO_3$ layers in between rock-salt $AO$ layers wherein $B$–O–$B$ interaction occurs only in the *ab* plane have also been investigated in some detail (*7*). Electrical and magnetic properties of the $A_2BO_4$ oxides are considerably different from those of the $ABO_3$ perovskites. Thus, $LaNiO_3$ is a Pauli-paramagnetic metal while $La_2NiO_4$ exhibits two-dimensional canted antiferromagnetic ordering below 204 K and a semicon-

* Contribution No. 413 is dedicated to Professor John B. Goodenough on the occasion of his 65th birthday.

** To whom all correspondence should be addressed.

0022-4596/88 $3.00

ductor–metal transition above 650 K (*8, 9*). It is found that the evolution of metallic conductivity occurs as one goes from $La_2NiO_4$ to the higher members of the $LaO \cdot (LaNiO_3)_n$ or the $La_{n+1}Ni_nO_{3n+1}$ family (*10*). A strict comparison of the magnetic and electrical properties of the three-dimensional $ABO_3$ perovskites and two-dimensional oxides of the $A_2BO_4$ type would be possible only when the same oxidation state of the transition metal ($B$) ion is maintained. Such a comparison made with some of the available information (*7, 10, 11*) indicates fascinating differences between the two systems. For example, $LaSrNiO_4$ exhibits two orders of magnitude higher resistivity than $LaNiO_3$; the former also shows a higher susceptibility at low temperatures. $LaSr_3Co_2O_8$ is a semiconducting ferromagnet while $LaSrCo_2O_6$ is an itinerant-electron ferromagnet. We considered it most instructive to compare the magnetic and electrical properties of several two-dimensional oxide systems of $K_2NiF_4$ structure with those of the corresponding perovskites in order to arrive at possible generalizations. We have carried out investigations on several such systems: $La_{1-x}Sr_{1+x}CoO_4$ and $La_{1-x}Sr_xCoO_3$, $LaSrMn_{0.5}T_{0.5}O_4$ and $LaMn_{0.5}T_{0.5}O_3$ ($T$ = Ni or Co), $LaBNiO_4$ ($B$ = Sr or Ba) and $La$NiO$_3$, $SrO \cdot (La_{1-x}Sr_xMnO_3)_n$ and $La_{1-x}Sr_xMnO_3$, $LaSrFe_{1-x}Ni_xO_4$ and $LaFe_{1-x}Ni_xO_3$, and lastly $Sr_2RuO_4$ and $SrRuO_3$. In what follows, we briefly present the results of our studies on these systems and arrive at some important generalizations on the properties of these two- and three-dimensional oxides.

## Experimental

$La_{1-x}Sr_{1+x}CoO_4$ compounds were prepared by the decomposition of coprecipitated carbonates obtained by adding an aqueous solution of the nitrates of the corresponding metal ions, in the required stoichiometry, to a solution of $Na_2CO_3$ taken slightly in excess. The resulting precipitates were washed thoroughly, dried, and decomposed at 1150 K, after which they were pelletized and heated at 1200–1400 K for 24–72 hr.

Compounds of the $LaFe_{1-x}Ni_xO_3$ and $LaSrFe_{1-x}Ni_xO_4$ system were prepared by adding the required proportions of the estimated solutions of $La(NO_3)_3$, $Sr(NO_3)_2$, $Al(NO_3)_3$, $FeCl_3$, and $NiCl_2$ to a concentrated solution of $Na_2CO_3$ under constant stirring. The precipitate was washed thoroughly with hot water to obtain alkali-free precipitate. The precipitates were decomposed at 1000 and 1170 K, respectively, for 27 hr. The powder obtained was ground, pelletized, and heated at 1100 K (24 hr) and 1570 K (48 hr), respectively, with intermittent grinding and pelletizing. LaSr(Ba)Mn$TO_4$ ($T$ = Ni, Co) and $Sr_2RuO_4$ were prepared by a ceramic method. Required amounts of $La_2O_3$ (dried), $SrCO_3$, $NiC_2O_4 \cdot 2H_2O$, $CoC_2O_4 \cdot 2H_2O$, $MnC_2O_4 \cdot 2H_2O$, and $RuO_2$ were weighed, mixed well, ground, and decomposed at 1270 K for 24 hr. The resulting powders were ground, pelletized and heated at 1470 K for 24 hr with intermittent grinding and pelletizing. Oxygen stoichiometry of all the compounds was determined by iodimetric titrations and the titrable oxygen content corresponded to the formal oxidation states expected in most of the cases.

The monophasic nature of all the compounds was confirmed using JEOL X-ray diffractometer. Magnetic susceptibility measurements were carried out by the Faraday method using a Cahn RG electrobalance. Magnetization measurements were carried out with a PAR vibrating sample magnetometer in conjunction with a closed cycle refrigeration (AIR products Displex) unit. Electrical resistivity studies were carried out on pressed, sintered pellets using the four-probe dc method.

**Results and Discussion**

*$La_{1-x}Sr_{1+x}CoO_4$ and $La_{1-x}Sr_xCoO_3$*

The inverse magnetic susceptibility, $\chi_M^{-1}$–temperature ($T \geq 300$ K) plots of a few members of the $La_{1-x}Sr_{1+x}CoO_4$ ($0 \leq x \leq 0.5$) system are shown in Fig. 1. The $\mu_{eff}$ values calculated from the slopes of these plots in various temperature ranges are given in Table I along with the paramagnetic Curie temperatures. The $\chi_M^{-1}$–$T$ plot of the three-dimensional $La_{0.5}Sr_{0.5}CoO_3$ is well-documented in the literature (*12*). The important features of these results are as follows: (i) At high temperatures, there is an increase in $\mu_{eff}$, the value being close to that expected for low-spin $Co^{IV}$ ($t_{2g}^5$ $e_g^0$) and high-spin $Co^{3+}$ ($t_{2g}^4$ $e_g^2$). (ii) The $x = 0.5$ samples in both the two- and three-dimensional systems exhibit nearly the same $\mu_{eff}$, although the former is associated with a smaller value of $\theta$, consistent with a smaller number of nearest-neighbor interactions in lower dimensions. (iii) There is a decrease in $\mu_{eff}$ with increasing $x$ in the

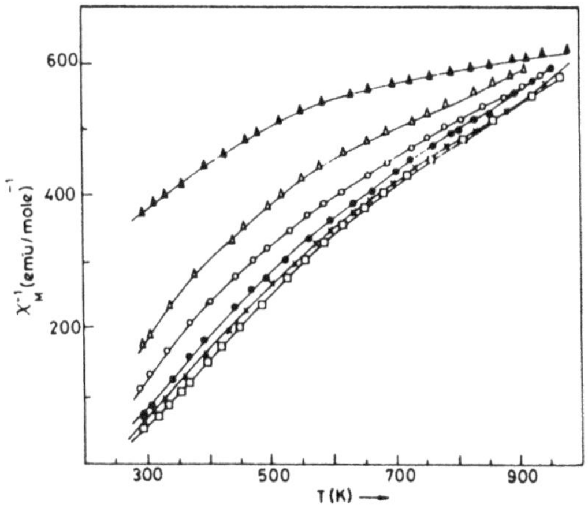


FIG. 1. $\chi_M^{-1}$ vs $T$ (K) plots of $La_{1-x}Sr_{1+x}CoO_4$; $x = 0.0$ (closed triangles), $x = 0.1$ (open triangles), $x = 0.2$ (open circles), $x = 0.3$ (closed circles), $x = 0.4$ (crosses), and $x = 0.5$ (boxes).

TABLE I

MAGNETIC PROPERTIES OF A FEW OXIDES OF $K_2NiF_4$ STRUCTURE

| Compound | Temperature range ($T$) | $\mu_{eff}$ ($\mu_B$) observed | $\mu_{eff}$ ($\mu_B$) calculated | $\theta$ (K) |
|---|---|---|---|---|
| $La_{1-x}Sr_{1+x}CoO_4$ | | | | |
| $x = 0.5$ | >600 | 3.51 | 3.67 | −15 |
| | <500 | 2.95 | | |
| $x = 0.4$ | >600 | 3.60 | 3.94 | 0 |
| | <500 | 2.95 | | |
| $x = 0.3$ | >600 | 3.72 | 4.20 | +100 |
| | <500 | 2.85 | | |
| $x = 0.2$ | >600 | 4.14 | 4.43 | +250 |
| $x = 0.1$ | >600 | 4.17 | 4.67 | +450 |
| $x = 0.0$ | >600 | 6.27 | 4.90 | +1400 |
| $LaSrMn_{0.5}Ni_{0.5}O_4$ | >600 | 5.37 | 5.19 | +190 |
| | <500 | 6.03 | (4.79) | |
| $LaBaMn_{0.5}Ni_{0.5}O_4$ | >600 | 5.246 | 5.19 | +140 |
| | <500 | 5.82 | (4.79) | |
| $LaSrMn_{0.5}Co_{0.5}O_4$ | >600 | 6.68 | 6.92 | +175 |
| | <500 | 7.41 | (5.47) | |
| $LaSrFe_{1-x}Ni_xO_4$ | | | | |
| $x = 0.2$ | >300 | 5.89 | 5.50 | +1350 |
| $x = 0.6$ | >300 | 4.82 | 4.71 | +1000 |
| $x = 0.8$ | >300 | 3.71 | 3.10 | +600 |

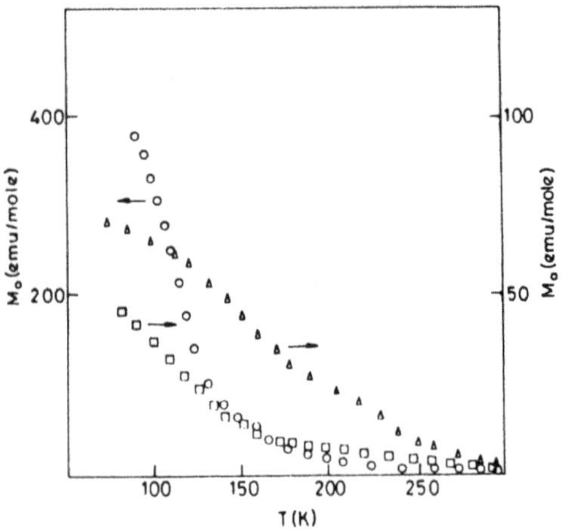


FIG. 2. Magnetization vs temperature of $La_2Sr_2MnCoO_8$ (circles), $La_2Sr_2MnNiO_8$ (boxes), and $La_{0.5}Sr_{1.5}CoO_4$ (triangles).

low-temperature region. Magnetization behavior of $La_{0.5}Sr_{1.5}CoO_4$ is shown in Fig. 2. The extrapolated value of magnetization at 0 K is only 0.5 $\mu_B$, which is much too small compared to 1.5 $\mu_B$ shown by $La_{0.5}Sr_{0.5}CoO_3$ (*12, 13*). The similarity of the high-temperature magnetic susceptibility behavior of the two- and three-dimensional oxides with a distinctly different low-temperature magnetization behavior has been noticed by us in other systems as well.

The electrical resistivity data of $La_{0.5}Sr_{1.5}CoO_4$ is compared with that of $La_{0.5}Sr_{0.5}CoO_3$ in Fig. 3. All the compositions of the two-dimensional $La_{1-x}Sr_{1+x}CoO_4$ system are semiconducting with a negative temperature coefficient of resistance (TCR) and a fairly high activation energy. The three-dimensional oxides, on the other hand, behave like metals with a positive TCR. The value of resistivity in the two-dimensional oxides is several orders of magnitude higher than that of the three-dimensional ones. The sign and the magnitude of the Seebeck coefficient at room temperature are similar in the two- and three-dimensional systems. However, in two dimensions, there is a clear anomaly in the Seebeck coefficient near the magnetic ordering temperature (see inset of Fig. 3). Such an anomaly is not seen in the three-dimensional system (*14*).

In $LaCoO_3$, the ground state of cobalt is the low-spin $Co^{III}$ ($t_{2g}^6\ e_g^0$) configuration (*3, 15*). In $LaSrCoO_4$, Demazeau *et al.* (*16*) argue that at low temperatures, there is an equal population of $Co^{III}$ and $Co^{3+}$ ($t_{2g}^4\ e_g^2$) states. In such a configuration the substitution of $La^{3+}$ by $Sr^{2+}$ should lead to the conversion of diamagnetic $Co^{III}$ to paramagnetic low-spin $Co^{IV}$ ($t_{2g}^5\ e_g^0$) causing an increase in $\mu_{eff}$. Since we do not observe this behavior, the spin configuration of Co ions in $LaSrCoO_4$ is likely to be the intermediate-spin $Co^{iii}$ ($t_{2g}^5\ e_g^1$). An examination of the Mössbauer, ferromagnetic resonance, and other physical properties of the $La_{1-x}Sr_xCoO_3$ system (*13, 14*) suggests that the properties are best explained by the existence of $Co^{iii}$ and $Co^{IV}$ ions. A ferromagnetic coupling of such spins would give a magnetization of 1.5 $\mu_B$ per cobalt as observed in $La_{0.5}Sr_{0.5}CoO_3$. Intermediate-spin $Co^{iii}$ may be favored in $LaSrCoO_4$ (instead of $Co^{III}$) by the presence of elongated octahedra.

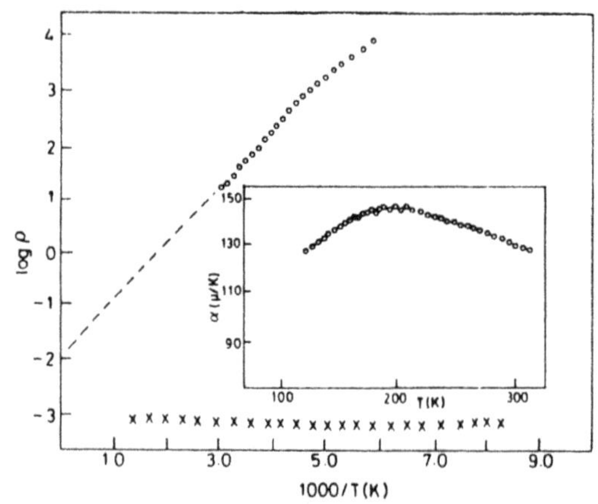


FIG. 3. log ρ vs 1000/*T* (K) plots of $La_{0.5}Sr_{1.5}CoO_4$ (circles) and $La_{0.5}Sr_{0.5}CoO_3$ (crosses). Inset shows the Seebeck coefficient vs temperature (K) plot of $La_{0.5}Sr_{1.5}CoO_4$.

### *SrO · (La$_{1-x}$Sr$_x$MnO$_3$)$_n$ and La$_{1-x}$Sr$_x$MnO$_3$*

Oxides of the $La_{1-x}Sr_xMnO_3$ ($0.25 \le x \le 0.5$) family are well-known itinerant-electron ferromagnets (*17*). In the $SrO \cdot (La_{1-x}Sr_xMnO_3)_n$ system, on the other hand, for $n = 1$, there is no spontaneous magnetization at low temperatures for any composition in the range $0.3 \le x \le 0.7$, although the paramagnetic Curie temperature is positive, indicative of ferromagnetic interactions. The paramagnetic Curie temperature is comparable to that of the corresponding three-dimensional oxides. When $n = 2$ or 3, there are two or three layers of adjacent perovskite layers so that with increasing $n$, the three-dimensional character increases. Magnetization and electrical resistivity data for the $x = 0.5$ composition are given in Figs. 4 and 5 for different values of $n$. We see that as the itinerant-electron behavior increases, the value of the magnetization increases. When $x = 0.3$, the low-temperature magnetization of the $n = 2$ composition matches that of the $n = \infty$ composition. Substitution of Sr by Ba or of La by Y does not bring about spontaneous long-range ferromagnetic ordering. Interestingly, Ba or Y substitution brings about an increase in the $c/a$ ratio of the oxide accompanied by a decrease in the paramagnetic Curie temperature.

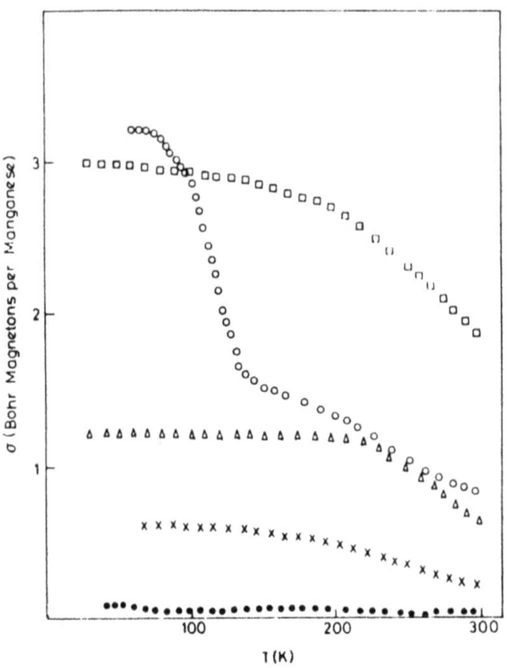


FIG. 4. Magnetization vs temperature plots of $La_{0.5}Sr_{1.5}Mn_3O_{10}$ (closed circles), $LaSr_2Mn_2O_7$ (crosses), $La_{1.5}Sr_{2.5}Mn_3O_{10}$ (triangles), $La_{0.5}Sr_{0.5}MnO_3$ (boxes), and $La_{1.4}Sr_{1.6}Mn_2O_7$ (open circles).

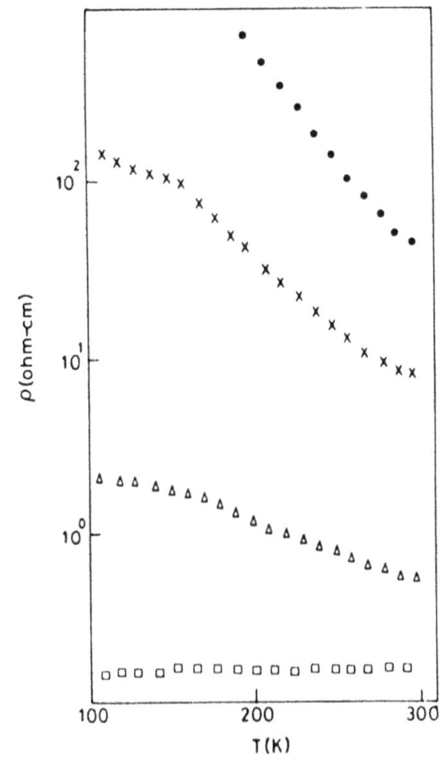


FIG. 5. $\rho$ vs $T$ (K) plots of $La_{0.5}Sr_{1.5}MnO_4$ (closed circles), $LaSr_2Mn_2O_7$ (crosses), $La_{1.5}Sr_{2.5}Mn_3O_{10}$ (triangles), and $La_{0.5}Sr_{0.5}MnO_3$ (boxes).

### *LaSrMn$_{0.5}$T$_{0.5}$O$_4$ and LaMn$_{0.5}$T$_{0.5}$O$_3$ (T = Ni or Co)*

The $LaMn_{0.5}T_{0.5}O_3$ ($T$ = Ni or Co) systems are the first examples of perovskites which are insulating and ferromagnetic (*18*). The magnetic susceptibility behavior

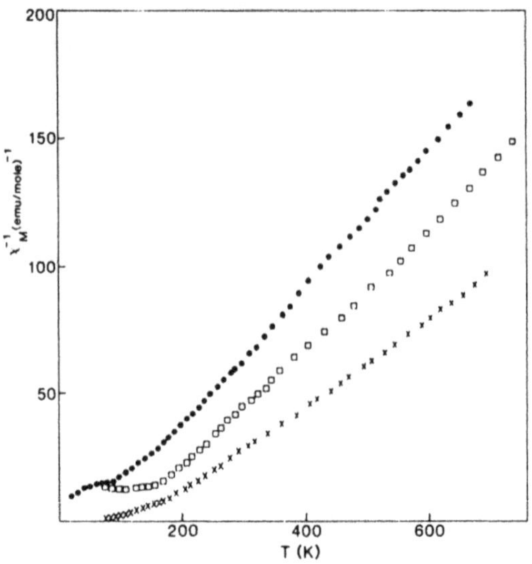


FIG. 6. $\chi_M^{-1}$ vs $T$(K) plots of $La_2Ba_2MnNiO_8$ (circles), $La_2Sr_2MnCoO_8$ (crosses), and $La_2Sr_2MnNiO_8$ (boxes).

of the corresponding two-dimensional compounds $LaSrMn_{0.5}T_{0.5}O_4$ are shown in Fig. 6. The slopes of the high-temperature $\chi_M^{-1}$–$T$ plots of these compounds are nearly identical to those of the corresponding three-dimensional compounds. The $\mu_{eff}$ values and the extrapolated paramagnetic Curie temperature are given in Table I. The decreases in $\theta$ in lower dimensions is consistent with the decrease in the number of nearest-neighbor magnetic ions. The magnetization behavior of these compounds as a function of temperature is given in Fig. 2. Despite a fairly high value of $\theta$, these compounds do not show any significant magnetization at low temperatures. Since both the three- and two-dimensional oxides are insulating, the absence of long-range ferromagnetic ordering in two dimensions does not seem to be related to the change in the nature of electrical conductivity (metallic vs nonmetallic). The magnetic susceptibility behavior of $LaSrMn_{0.5}Ni_{0.5}O_4$ at low temperatures seems to be consistent with a ferromagnetically coupled $Mn^{3+}Ni^{3+}$ ($S$ = 2) spin pair. In the case of $LaBaMn_{0.5}Ni_{0.5}O_4$, there appears to be a weak maximum around 80 K; *ac* susceptibility measurements are now being carried out on this compound.

*$LaNiO_3$ and $LaBNiO_4$ (B = Sr or Ba)*

The magnetic susceptibility of $LaNiO_3$ has been explained in terms of Pauli-paramagnetism because of the weak dependence of susceptibility at high temperatures (*19*). There is an enhancement in the susceptibility at low temperatures which can be accounted for in terms of the presence of a small amount of impurity. The metallic conductivity of $LaNiO_3$ is consistent with the itinerant-electron Pauli-paramagnetism picture. The electrical resistivity of LaSr (Ba)$NiO_4$ is much higher than that of La $NiO_3$. Magnetic susceptibilities of LaSr $NiO_4$ and $LaBaNiO_4$ at high temperatures are comparable to the susceptibility of La $NiO_3$ and there is no evidence for magnetic ordering in these oxides or in systems such as $LaSrAl_xNi_{1-x}O_4$. The slopes of the $\chi_M^{-1}$–$T$ plots of $LaBNiO_4$ at low temperatures increase as the electrical resistivity increases (*20*). In Fig. 7 we show the electrical resistivity behavior of the $LaSr_{1-x}Ba_xNiO_4$ system as a function of temperature. In the inset of Fig. 7, we show the log $\rho$ vs log $\chi_M$ plots at different temperatures to illustrate that the magnetic susceptibility and electrical resistivity behavior are correlated. That is, the low-temperature increase in the susceptibility in these oxides is an intrinsic property of the system and not due to extrinsic impurities.

Demazeau *et al.* (*21*) suggest, based on ESR spectroscopic studies, that there is a spin equilibrium between the low- and the high-spin $Ni^{3+}$ ions in $LaBaNiO_4$. We have not been able to obtain similar ESR results with any of our samples although the oxygen stoichiometry was good. Moreover, our susceptibility results do not show any evidence of a spin-state transition. In addition, the $\mu_{eff}$ values calculated from the

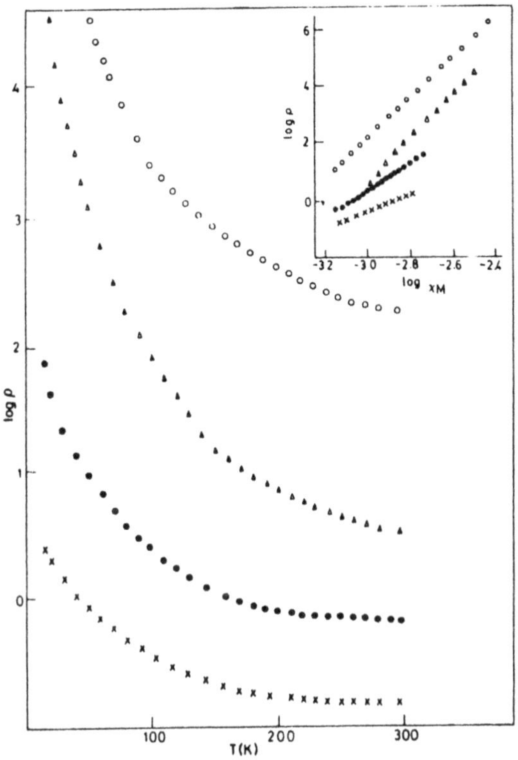


FIG. 7. log $\rho$ vs $T$ (K) plots of $LaSr_{1-x}Ba_xNiO_4$ system; $x = 1.0$ (circles), $x = 0.75$ (triangles), $x = 0.5$ (closed circles), and $x = 0.25$ (crosses). Inset shows the log $\rho$ vs log $\chi_M$ plots for the same compounds. The values of $\rho$ and $\chi_M$ were measured at the same temperature.

slopes of the $\chi_M^{-1}$ vs $T$ plots of $LaSrMn_{0.5}Ni_{0.5}O_4$ and $LaBaMn_{0.5}Ni_{0.5}O_4$ are nearly identical (see Table I), suggesting that the $Ni^{3+}$ ions remain in the low-spin state when Sr is substituted by Ba. We believe that the $Ni^{3+}$ ions in $LaBaNiO_4$ are in the low-spin state at all temperatures. The change in resistivity in the $LaSr_{1-x}Ba_xNiO_4$ system is therefore not likely to be due to changes in the spin state of $Ni^{3+}$ but rather because of an expansion of the lattice.

It has been noted by us that the high-temperature magnetic susceptibilities of $LaNiO_3$ and $LaSrNiO_4$ are similar. The electrical resistivity of the latter is, however, two orders of magnitude higher with a negative TCR. It seems that the increase in the resistivity in $LaSrNiO_4$ arises from increasing localization effects.

### $LaSrFe_{1-x}Ni_xO_4$ and $LaFe_{1-x}Ni_xO_3$

The three-dimensional perovskites $LaFeO_3$, $LaCrO_3$, $LaVO_3$ are antiferromagnetic (6) with their well-defined ordering temperatures (750, 290, and 140 K, respectively). The corresponding two-dimensional compounds such as $LaSrFeO_4$ and $LaSrCrO_4$ are also antiferromagnetic, although with reduced antiferromagnetic ordering temperatures. The solid solutions of metallic $LaNiO_3$ with insulating $LaTO_3$ ($T$ = Co, Mn, Fe, and Cr) have been studied

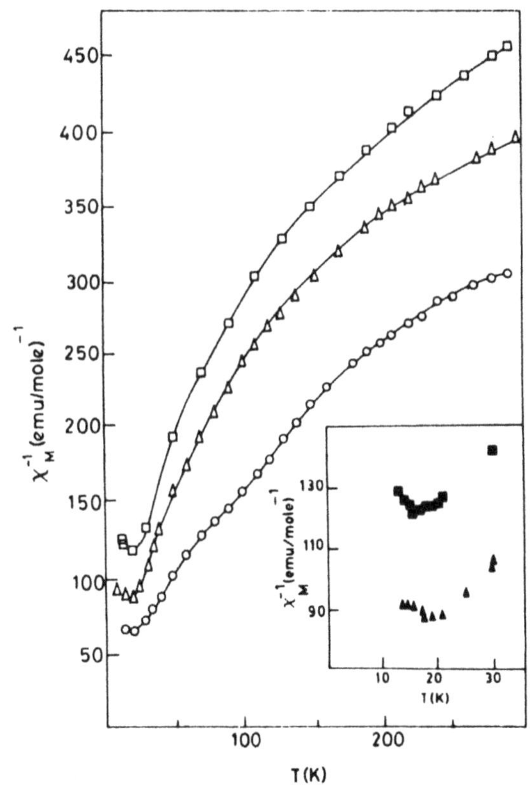


FIG. 8. $\chi_M^{-1}$ vs $T$ (K) ($T$ = 300 K) plots of $LaSrFe_{1-x}Ni_xO_4$ system; $x = 0.2$ (open circles), $x = 0.5$ (triangles), and $x = 0.8$ (boxes). Inset shows the $\chi_M^{-1}$ vs $T$ (K) plots in the temperature range of 10–30 K of $LaSrFe_{1-x}Ni_xO_4$ ($x = 0.5$ (triangles), $x = 0.8$ (boxes)).

by us with respect to compositionally controlled metal–insulator transitions as well as changes in magnetic ordering (*19, 22, 23*). The $LaFe_{1-x}Ni_xO_3$ system shows a well-defined magnetic ordering temperature up to $x \leq 0.35$; beyond this value of $x$, the susceptibility shows a broad maximum at lower temperatures. For $x \geq 0.65$, the system becomes metallic with a positive TCR (*23*). The magnetic susceptibility and electrical resistivity data of the $LaSrFe_{1-x}Ni_xO_4$ system are presented in Figs. 8 and 9, respectively. The high-temperature values of $\mu_{eff}$ and the paramagnetic Curie temperatures, $\theta$, are listed in Table I. These results show that the $Ni^{3+}$ ions in these oxides are in the low-spin state while the $Fe^{3+}$ ions are in the high-spin configuration; an increase in the $Ni^{3+}$ content decreases the value of $\theta$. There is no evidence for long-range antiferromagnetic ordering, but at low-temperatures there is a susceptibility maximum just as in the corresponding three-dimensional perovskites.

Electrical resistivities of all the members of the $LaSrFe_{1-x}Ni_xO_4$ family show negative TCR; the extrapolated high-temperature value of resistivity is generally $\sim 10^{-2}$ ohm cm. This value is about an order of magnitude less than that found in three dimensions (*23, 24*). Such behavior has been found in systems such as $La_{1-x}Sr_{1+x}CoO_4$ and $LaSrAl_{1-x}Ni_xO_4$ (*20*). The Seebeck coefficient of $LaSrFe_{1-x}Ni_xO_4$ compounds are negative for high values of $x$ and positive for low values of $x$. The room temperature Seebeck coefficient data for these oxides are given in the inset of Fig. 9 along with those for $LaSrAl_{1-x}Ni_xO_4$, $LaFe_{1-x}Ni_xO_3$, and $LaCr_{1-x}Ni_xO_3$. We see that the lowering of dimensionality changes the composition at which the sign of the Seebeck coefficient changes. There seems to be a correlation between the critical value of the site-percolation threshold (~0.3 for three-dimensional cubic structures and 0.59 for two-dimensional square-planar structures) and the value of $x$ at which the Seebeck coefficient changes sign. Evidence for such percolation effects has

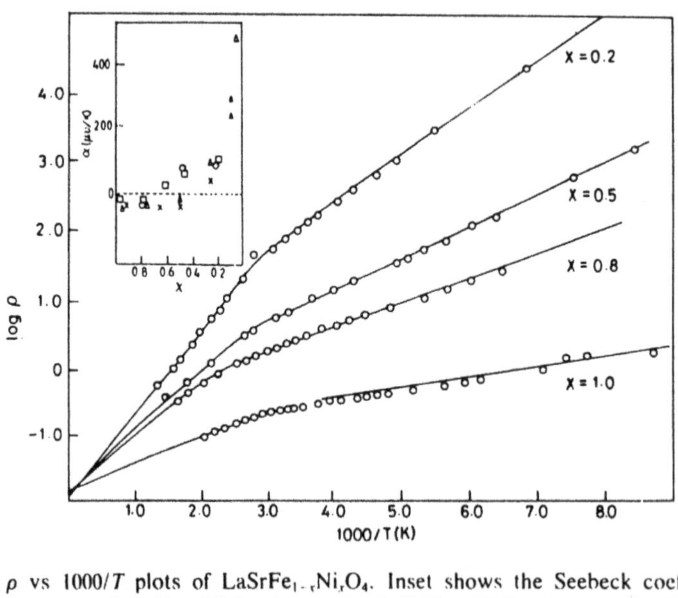


FIG. 9. log $\rho$ vs $1000/T$ plots of $LaSrFe_{1-x}Ni_xO_4$. Inset shows the Seebeck coefficient ($\alpha$) vs composition ($x$) plots of $LaFe_{1-x}Ni_xO_3$ (crosses), $LaSrFe_{1-x}Ni_xO_4$ (boxes), $LaCr_{1-x}Ni_xO_3$ (triangles), and $LaSrAl_{1-x}Ni_xO_4$ (open circles).

also been obtained by us in the magnetic susceptibility studies.

*$SrRuO_3$ and $Sr_2RuO_4$*

Magnetic susceptibility and electrical resistivity of $SrRuO_3$ and $Sr_2RuO_4$ are compared in Fig. 10. There is a rise in the suceptibility of $Sr_2RuO_4$ around 150 K due to the presence of ferromagnetic impurities of $SrRuO_3$; the steep rise persists after heating for long periods at high temperatures (even in the presence of excess Sr). The resistivity of $Sr_2RuO_4$ is more than an order of magnitude higher than that of $SrRuO_3$ and it has a negative TCR compared to the positive TCR of the latter. Since the magnetic and electrical properties of the $ARuO_3$ systems ($A$ = $Ca^{2+}$, $Sr^{2+}$, and $Ba^{2+}$) are sensitive to the nature of the $A$ ion, comparisons between the two- and three-dimensional systems would not be strictly valid. We note, however, that the increased resistivity and the absence of ferromagnetic ordering in $CaRuO_3$ is attributed to the more acidic nature of Ca ions. Since the $A$ ion in $Sr_2RuO_4$ and $SrRuO_3$ is the same, the effect of decreasing the lattice parameters would be to decrease the resistivity as observed in the $LaSr_{1-x}Ba_xNiO_4$ system. It is noteworthy that while $BaPbO_3$ is metallic, $Ba_2PbO_4$ is an insulator with a resistivity greater than $10^8$ at room temperature. It therefore seems to us that the higher resistivity of $Sr_2RuO_4$ is a real effect of dimensionality.

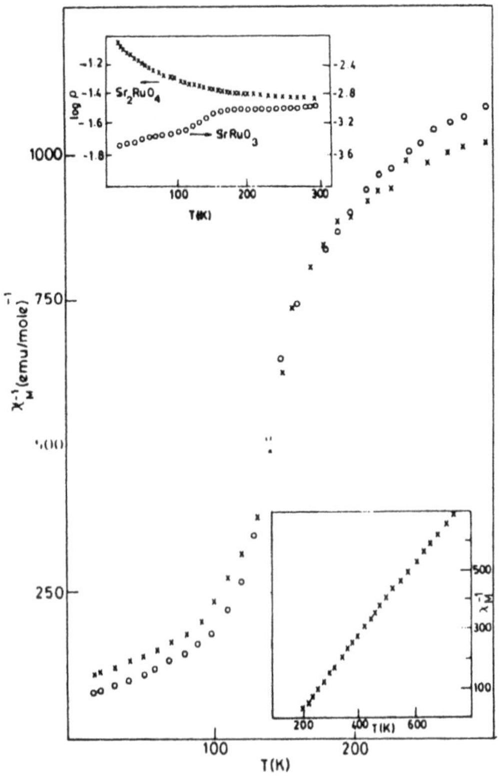

FIG. 10. $\chi_M^{-1}$ vs $T$ (K) plots of $Sr_2RuO_4$ after heating for 24 hr (crosses) and after 7 days of heating (circles). Top inset shows log $\rho$ vs $T$ (K) plots of $SrRuO_3$ (circles) and $Sr_2RuO_4$ (crosses). Lower inset shows the $\chi_M^{-1}$ vs $T$ (K) plot of $SrRuO_3$.

## Concluding Remarks

The following general conclusions emerge from the present studies: (i) Perovskite oxides which are ferromagnetic in three dimensions do not show long-range ferromagnetic ordering in the corresponding two-dimensional oxides. This behavior is observed in both itinerant and insulating oxide systems. (ii) The magnetic susceptibility behavior of these oxides in the high-temperature region is not affected by changing the dimensionality. The observed changes in the paramagnetic Curie temperature is consistent with the reduction of the number of nearest neighbors. (iii) Oxides which are metallic in three dimensions with a resistivity less than that corresponding to the "minimum metallic conductivity" ($2 \times 10^{-3}$ ohm cm) show a negative TCR in two dimensions with a resistivity always higher than $10^{-2}$ ohm cm. (iv) We have not encountered an oxide with the $K_2NiF_4$ structure which is metallic over a wide range of temperatures. (v) High temperature extrapolation of the electrical resistivity data indicates that $10^{-2}$ ohm cm may be the critical value for a change in the sign of TCR to occur in these oxides. So far, there is no solid known with the $K_2NiF_4$ structure which has a resistivity less than this value. What is somewhat curious in some of these oxide systems is that the magnetic properties reflect itinerant electron behavior while the electrical properties do not. (vi) In $K_2CuF_4$, $Rb_2CrCl_4$ (*25*), and such ferromagnetic halides of the $K_2NiF_4$ structure, the axis of elongation of the $MX_6$ octahedron and the easy axis of magnetization are in the *ab* plane. In the oxides of $K_2NiF_4$ structure, the easy axis of magnetization may be the axis of elongation of the $BO_6$ octahedra (*c* axis). If this indeed be the case, there would be no magnetization, a situation similar to that found in thin ferromagnetic disks with magnetization perpendicular to the plane of the disk. Elongated $BO_6$ octahedra could also give rise to strong polaronic effects leading to localization of charge carriers.

## Acknowledgments

The authors thank the University Grants Commission and the Department of Science and Technology for support of this research.

## References


*1*. J. B. Goodenough and J. M. Longo, "Landolt–Bornsteim Tabellen," Neve Serie III/4a, Springer-Verlag, Berlin (1970).
*2*. J. B. Goodenough, *Prog. Solid State Chem.* **5**, 145 (1972).
*3*. J. B. Goodenough, *in* "Solid State Chemistry" (C. N. R. Rao, Ed.) Dekker, New York (1974).
*4*. C. N. R. Rao and G. V. Subba Rao, *Phys. Status Solidi (A)* **1**, 597 (1970).
*5*. C. N. R. Rao, *J. Indian Chem. Soc.* **51**, 979 (1974).
*6*. P. Ganguly, Om Parkash, and C. N. R. Rao, *Phys. Status Solidi A* **36**, 339 (1976).
*7*. P. Ganguly and C. N. R. Rao, *J. Solid State Chem.* **53**, 193 (1984).
*8*. C. N. R. Rao, D. J. Buttrey, N. Otsuka, P. Ganguly, H. R. Harrison, C. J. Sandberg, and J. M. Honig, *J. Solid State Chem.* **52**, 266 (1984).
*9*. D. J. Buttrey, J. M. Homig, and C. N. R. Rao, *J. Solid State Chem.* **64**, 287 (1986).
*10*. R. A. Mohanram, L. Ganapathi, P. Ganguly, and C. N. R. Rao, *J. Solid State Chem.* **63**, 139 (1986).
*11*. R. A. Mohanram, P. Ganguly, and C. N. R. Rao, *J. Solid State Chem.* **70**, 82 (1987).
*12*. C. N. R. Rao, Om Parkash, D. Bahadur, P. Ganguly, and S. Nagabhushana, *J. Solid State Chem.* **22**, 353 (1977).
*13*. D. Bahadur, S. Kollali, C. N. R. Rao, M. J. Patini, and C. M. Srivastava, *J. Phys. Chem. Solids* **40**, 981 (1979).
*14*. V. G. Bhide, D. S. Rajoria, C. N. R. Rao, G. Ramarao, and V. G. Jadhao, *Phys. Rev. B* **12**, 2832 (1975).
*15*. V. G. Bhide, D. S. Rajoria, G. Ramarao, and C. N. R. Rao, *Phys. Rev. B* **6**, 1021 (1982).
*16*. G. Demazeau, P. Courbin, G. Le Flem, M. Pouchard, P. Hagenmuller, J. L. Soubeyroux, I. G. Main, and G. A. Robins, *Nouv. J. Chim.* **3**, 171 (1979).
*17*. G. H. Jonker and J. H. Van Santen, *Physica* **16**, 337, 599 (1950).
*18*. G. H. Jonker, *J. Appl. Phys.* **37**, 1424 (1967).
*19*. N. Y. Vasanthacharya, P. Ganguly, J. B. Goodenough, and C. N. R. Rao, *J. Phys. C* **17**, 2745 (1984).
*20*. R. A. Mohanram, K. K. Singh, W. H. Madhusudan, R. Ganguly, and C. N. R. Rao, *Master. Res. Bull.* **18**, 703 (1984).
*21*. G. Demazeau, J. L. Marty, M. Pouchard, T. Rojo, J. M. Dance, and P. Hagenmuller, *Mater. Res. Bull.* **16**, 47 (1981).
*22*. N. Y. Vasanthacharya, P. Ganguly, and C. N. R. Rao, *J. Solid State Chem.* **53**, 140 (1984).
*23*. P. Ganguly, N. Y. Vasanthacharya, C. N. R. Rao, and P. P. Edwards, *J. Solid State Chem.* **54**, 400 (1984).
*24*. C. N. R. Rao and P. Ganguly, *in* "The Metallic and the Non-metallic States of Matter" (P. P. Edwards and C. N. R. Rao, Eds.), Taylor & Francis, London, 1985.
*25*. P. Day, *Acc. Chem. Res.* **12**, 236 (1979) and references therein.

# The Blackett Memorial Lecture, 1991

# Chemical insights into high-temperature superconductors

By C. N. R. Rao

*Solid State and Structural Chemistry Unit and Jawaharlal Nehru Centre for Advanced Scientific Research, Indian Institute of Science, Bangalore 560 012, India*

## Contents



The high-temperature superconductors are complex oxides, generally containing two-dimensional $CuO_2$ sheets. Various families of the cuprate superconductors are described, paying special attention to aspects related to oxygen stoichiometry, phase stability, synthesis and chemical manipulation of charge carriers. Other aspects discussed are chemical applications of cuprates, possibly as gas sensors and copper-free oxide superconductors. All but the substituted Nd and Pr cuprates are hole-superconductors. Several families of cuprates show a nearly constant $n_h$ at maximum $T_c$. Besides this universality, the cuprates exhibit a number of striking common features. Based on Cu(2p) photoemission studies, it is found that the Cu–O charge-transfer energy, $\Delta$, and the Cu(3d)–O(2p) hybridization strength, $t_{pd}$, are key factors in the superconductivity of cuprates. The relative intensity of the satellite in the Cu(2p) core-level spectra, the polarizability of the $CuO_2$ sheets as well as the hole concentration are related to $\Delta/t_{pd}$. These chemical bonding factors have to be explicitly taken into account in any model for superconductivity of the cuprates.

## 1. Introduction

I am delighted that I have been asked to deliver the Blackett Memorial Lecture this year. Professor Blackett was a man of many parts and was keenly interested in India,

*Phil. Trans. R. Soc. Lond.* A (1991) **336**, 595–624
*Printed in Great Britain*  21-2

especially with regard to the policy for science and development. I have been myself involved in planning science and technology for development in the past few years and have greatly appreciated the concerns of Professor Blackett. Although I am not certain whether Professor Blackett would have predicted the science and technology scenario prevalent today, I feel that he would have been amused by a chemist from India talking about recent developments in warm superconductors. I have chosen this topic not only to demonstrate how chemists have much to do in this frontline area of condensed matter science, but also to show how certain classes of transition metal oxides – on which I have been working for many years (Rao & Subbarao 1970; Rao 1989) in a corner of the globe – have become so prominent because of high-temperature superconductivity. This area is no longer the one I pursued years ago out of curiosity, but one which has become so competitive that it is foolhardy to try to read all the literature. It has been really an exciting experience to witness this area develop explosively in so short a time.

Metal oxides themselves are not new to superconductivity. Superconducting transition temperatures of around 13 K were obtained some years ago in $Li_{1+x}Ti_{2-x}O_4$ (Johnston *et al.* 1973) and $BaBi_{1-x}PbO_3$ (Sleight *et al.* 1975). The discovery of superconductivity in the La–Ba–Cu–O system (Bednordz & Müller 1986) pushed the upper limit of the transition temperature from a stagnant value of 23 K to around 30 K and initiated an unprecedented pace of search for high-temperature superconductivity. The superconducting phase in the La–Ba–Cu–O system had the quasi two-dimensional $K_2NiF_4$ structure containing a perovskite layer, the parent compound being $La_2CuO_4$. I may recall here that my own interest in the structure and properties of this class of transition metal oxides goes back to several years (Ganguly & Rao 1973, 1984). When superconductivity above the liquid nitrogen temperature was reported in the Y–Ba–Cu–O system in early 1987 (Wu *et al.* 1987), we independently identified the phase responsible for superconductivity to be $YBa_2Cu_3O_7$ with a defect perovskite structure (Rao *et al.* 1987). We have since worked on several other families of superconducting cuprates, all containing perovskite layers, the highest $T_c$ till to date being 125 K. High-temperature superconductivity is not restricted to cuprates alone, but what is interesting is that all the high $T_c$ materials discovered up to now are metal oxides. It is not entirely unlikely that the ability of the metal–oxygen bond to traverse the entire range from the extreme ionic limit to the highly covalent limit is related to this feature. We cannot forget that transition metal oxides are versatile materials (Rao & Subbarao 1970; Goodenough 1971; Rao 1989) showing metallic behaviour at one end (e.g. $ReO_3$, $LaNiO_3$) and insulating behaviour at the other (e.g. $BaTiO_3$); then, we have oxides exhibiting metal–insulator transitions (e.g. $V_2O_3$, $LaNi_{1-x}Mn_xO_3$, $La_{1-x}Sr_xCoO_3$). Oxides can be ferromagnetic (e.g. $La_{0.5}Sr_{0.5}MnO_3$) or antiferromagnetic (e.g. $LaCrO_3$). Metal oxides exhibit interesting electronic properties arising from valence ordering (e.g. $Fe_3O_4$), charge-density-wave transitions (e.g. $Na_xWO_3$) and defect ordering (e.g. $Ca_2Fe_2O_5$). Properties of transition metal oxides depend on the dimensionality as well. Thus two-dimensional oxides do not exhibit ferromagnetism or real metallicity (Rao *et al.* 1988). The various cuprate families exhibiting superconductivity possess two-dimensional $CuO_2$ sheets just like $La_2CuO_4$.

I present here some of the highlights of the structure–property relations in the various families of high $T_c$ cuprates and illustrate how chemistry plays a major role in the development of such important materials and how solid state chemistry constitutes a fascinating and important branch of chemical science in its own right.

Chemical literature on high-temperature superconductors has become voluminous and I shall therefore cite only some of the very recent references, primarily those based on the work carried out in my laboratory. The earlier results have been adequately covered in many reviews and conference reports (Cava 1990; Goodenough & Manthiram 1990; Joshi *et al.* 1990; Kitazawa & Ishiguro 1989; Nelson *et al.* 1987; Ramakrishnan & Rao 1989; Rao 1988*a*, *b*; Rao & Raveau 1989; Sleight 1988). I shall attempt to point out the important commonalities among the various families of cuprate superconductors and examine such chemical factors as stoichiometry, oxygen disorder, oxidation states, phase stability and chemical manipulation of charge carriers as well as the crucial role of chemical bonding in understanding the phenomenon of high-temperature superconductivity. I shall also briefly touch upon certain synthetic aspects and possible chemical applications of these materials.

## 2. Cuprate families

The first family of high-temperature oxide superconductors are derived from $La_2CuO_4$ possessing the $K_2NiF_4$ structure (figure 1). Stoichiometric $La_2CuO_4$ is an antiferromagnetic insulator which when doped with holes (by formally creating trivalent Cu species through the substitution of $La^{3+}$ by divalent ions such as $Sr^{2+}$ or by incorporating excess oxygen) becomes superconducting. While $La_2CuO_4$ is orthorhombic at 300 K and becomes tetragonal at higher temperatures, superconducting $La_{2-x}M_xCuO_4$ ($x \approx 0.2$ when M = Sr) is tetragonal at 300 K and becomes orthorhombic around 180 K, well above the superconducting transition temperature (*ca.* 35 K). Accordingly, the Cu–O–Cu angle in the superconducting oxides of $K_2NiF_4$ structure are slightly bent (less than 180°) causing a buckling of the Cu–O sheets. In figure 2, the phase diagram of $La_{2-x}Sr_xCuO_4$ is shown to indicate the narrow range of the antiferromagnetic phase and the maximum in $T_c$ at a specific value of $x$ where the hole concentration is also a maximum (Torrance *et al.* 1988). Some doubt has been raised as to whether the maximum $T_c$ at $x \approx 0.2$ exhibited by $La_{2-x}Sr_xCuO_4$ is due to the presence of inhomogeneities in the compositions other than the one showing maximum $T_c$. This seems unlikely; as we will show later, the $T_c$ maximum occurs at an optimal hole concentration in all the families of cuprate superconductors (Rao *et al.* 1991*a*). In oxygen-excess $La_2CuO_{4+\delta}$, however, separation into superconducting and antiferromagnetic phases seems to occur (Chaillout *et al.* 1990; Jorgensen *et al.* 1988).

The $La^{3+}$ ion in $La_{2-x}M_xCuO_4$ can be substituted to some extent by $Pr^{3+}$ and other rare earth ions without losing superconductivity. It is to be noted that $Pr_2CuO_4$ and $Nd_2CuO_4$ themselves possess the so-called T′-structure with square-planar $CuO_4$ units unlike the T-structure of $La_2CuO_4$. The T and T′ cuprates form solid solutions over a reasonable range of compositions (Goodenough & Manthiram 1990; Singh *et al.* 1982).

The next homologue of the $La_{2-x}Sr_xCuO_4$ family containing two $CuO_2$ sheets had earlier been reported to be an insulator. However, recently Cava *et al.* (1990*a*) have synthesized $(La,Sr)_2CaCu_2O_6$ under a high oxygen pressure and found it to be superconducting with a $T_c$ of 60 K. This removes the discomfort one had related to the absence of superconductivity in the two-layer cuprate of this family and also underscores the importance of oxygen stoichiometry. Vijayaraghavan has recently synthesized analogous double-layer compounds of the general formula $LaSrLnCu_2O_6$ (Ln = Nd, Gd or Y) in my laboratory.

Figure 1

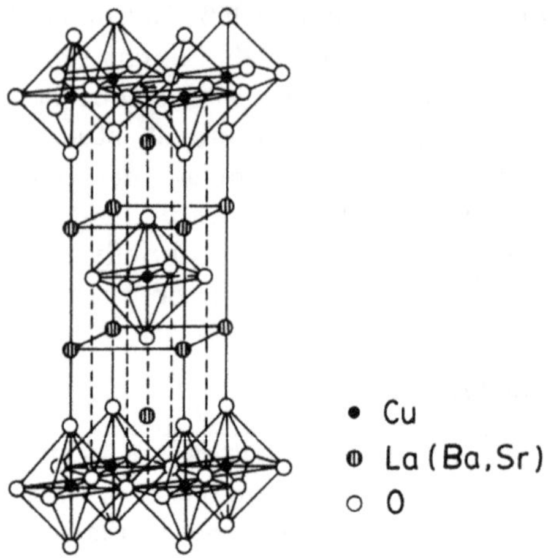


Figure 2

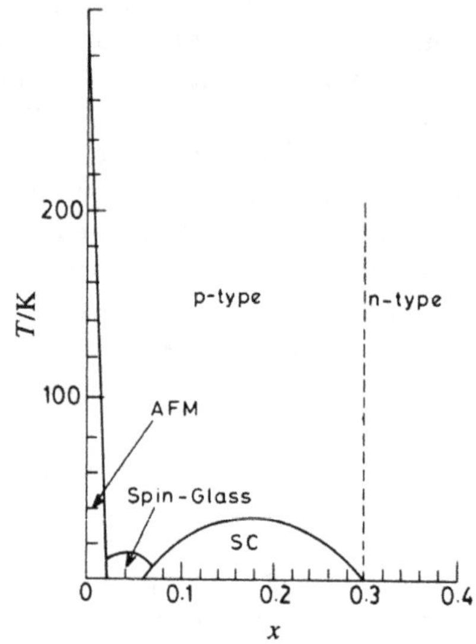


Figure 1. Structure of $La_{2-x}Sr_xCuO_4$.

Figure 2. Phase diagram of hole-doped $La_{2-x}Sr_xCuO_4$.

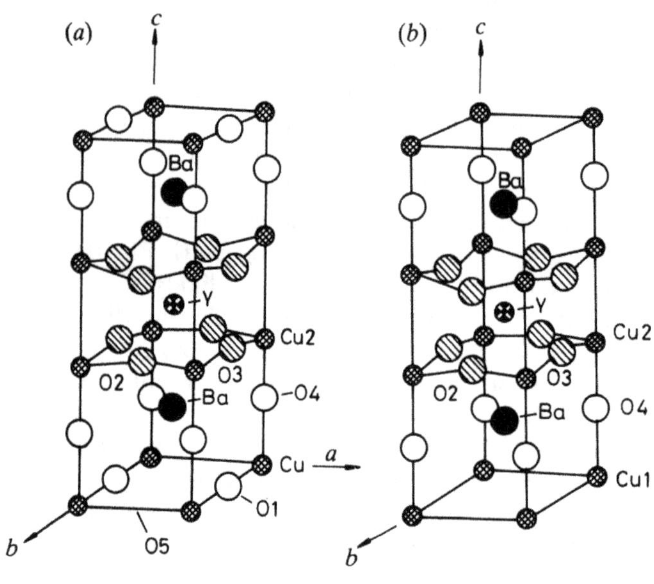


Figure 3. Structures of (*a*) $YBa_2Cu_3O_7$ and (*b*) $YBa_2Cu_3O_6$.

$YBa_2Cu_3O_{7-\delta}$ and the other 123 cuprates of the general formula $LnBa_2Cu_3O_7$ (Ln is a rare earth other than Ce, Pr and Tb) show superconductivity with a $T_c$ of *ca.* 90 K in the near stoichiometric compositions (not more than 0.2). These compounds have an orthorhombic structure. The structure as well as the superconductivity are sensitive to oxygen stoichiometry. Stoichiometric 123 cuprates contain Cu–O chains along the *b*-axis in addition to the $CuO_2$ sheets (figure 3). When the Cu–O chains are fully depleted of oxygen, we get the non-superconducting, tetragonal $YBa_2Cu_3O_6$.

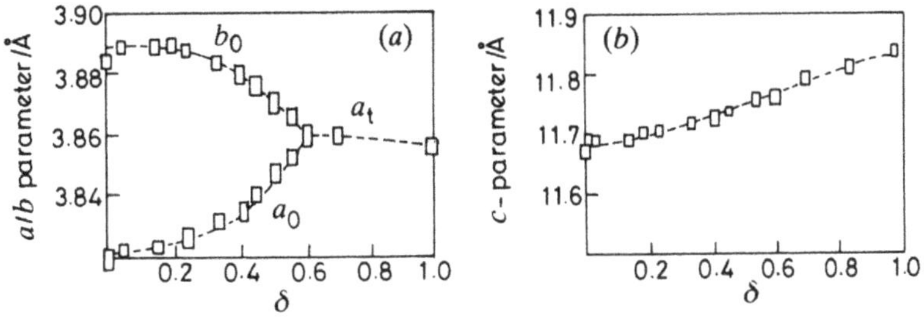


Figure 4. Variation of the lattice parameters of $YBa_2Cu_3O_{7-\delta}$ with $\delta$.

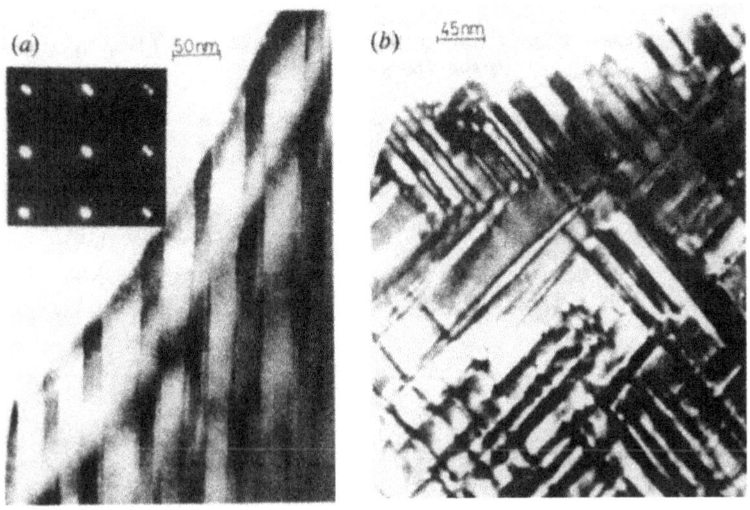


Figure 5. (*a*) Bright field electron microscopic images of twins in $YBa_2Cu_3O_7$. The splitting of the {110} Brugg spots is clearly seen in SAED pattern given in the inset. (*b*) 90° twins in $YBa_2Cu_3O_7$.

The variation of the unit cell parameters of $YBa_2Cu_3O_{7-\delta}$ with the oxygen stoichiometry is shown in figure 4 to illustrate how the orthorhombic structure gives way to the tetragonal structure around $\delta \approx 0.6$. All the $LnBa_2Cu_3O_7$ compounds undergo the orthorhombic tetragonal transition, the temperature of the transition depending on the Ln ion (Y = 970 K, La = 590 K). The orthorhombic phases show extensive twinning with a rotation of the *a* and *b* axes across the twin boundary. In figure 5 we show typical twins recorded by us in early March 1987. The twinning has no direct bearing on superconductivity; accordingly non-superconducting $PrBa_2Cu_3O_7$ shows twinning because of the orthorhombic structure.

The variation of $T_c$ with $\delta$ in $YBa_2Cu_3O_{7-\delta}$ is shown in figure 6. We see that $T_c$ is nearly constant (*ca.* 90 K) upto $\delta = 0.2$ and then drops sharply showing a sort of a plateau around 60 K for $\delta = 0.3$–$0.4$; the $T_c$ value reaches 45 K when $\delta = 0.5$. The formal valence of Cu in the sheets also shows a plateau in the 60 K region as shown in figure 6 (Cava *et al.* 1990*b*).

$YBa_2Cu_3O_{6.5}$ ($\delta = 0.5$) is a oxygen vacancy-ordered structure with fully oxidized Cu–O chains ($O_7$ units) alternating empty chain ($O_6$ units) as shown in figure 7. In $YBa_2Cu_3O_7$ and related 123 compounds, the orthorhombic *c* parameter is exactly equal to 3*b* (figure 8). $YBa_2Cu_3O_{6.75}$ also seems to have a vacancy-ordered structure (figure 7), but compositions with $\delta$ between 0.3 and 0.4 ($T_c \approx 60$ K) show no such

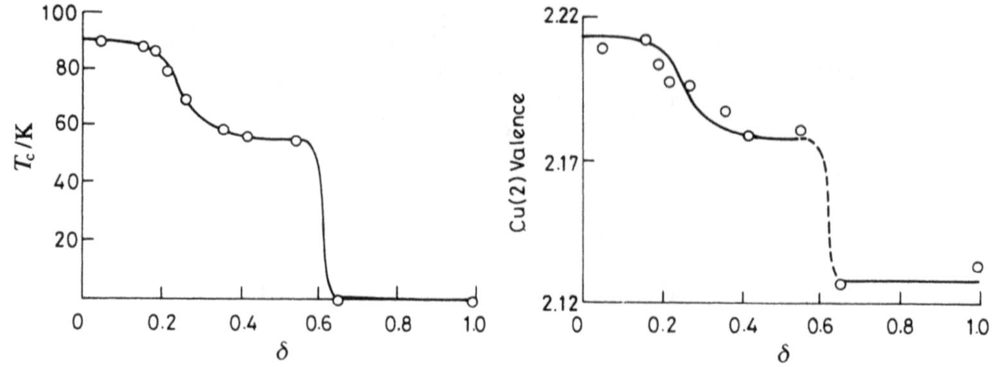


Figure 6. Variation of the $T_c$ and in-plane Cu(2) valence of $YBa_2Cu_3O_{7-\delta}$ with $\delta$ (From Cava *et al.* 1990*b*).

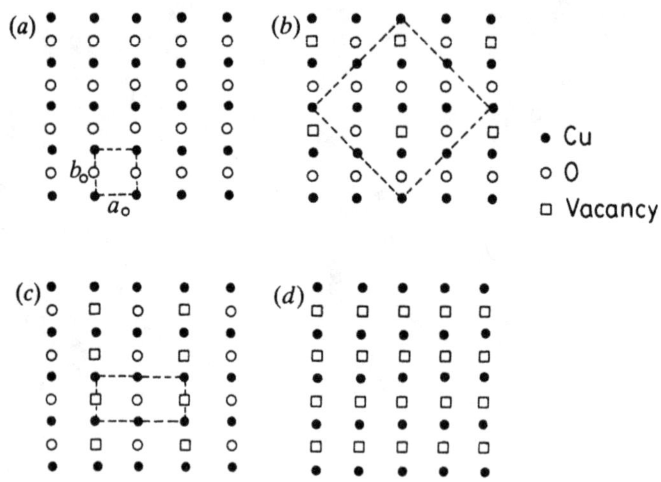


Figure 7. Structure of the basal (*ab*) plane of $YBa_2Cu_3O_{7-\delta}$: (*a*) $\delta = 0$, (*b*) $\delta = 0.25$, (*c*) $\delta = 0.5$, (*d*) $\delta = 1.0$. (From Rao *et al.* 1990*d*.)

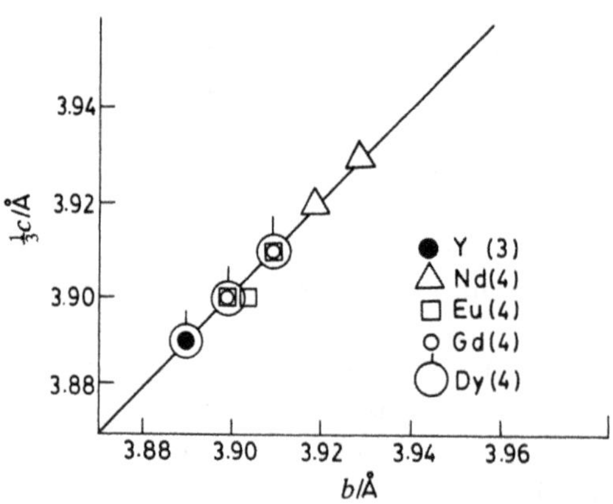


Figure 8. Relation between the *c* and *b* parameters in orthorhombic $LnBa_2Cu_3O_{7-\delta}$ (Ln = rare earth or Y, $\delta < 0.2$). ( From Rao *et al.* 1990*d*.)

ordering. It has indeed been shown that this composition range is metastable, disproportionating to orthorhombic and tetragonal phases on annealing at low temperatures of the order of 470 K (Rao *et al.* 1990*a*). Thus, on annealing $YBa_2Cu_3O_{6.7}$ at 470 K for a few hours, the X-ray diffraction pattern changes drastically while the electron diffraction pattern shows considerable streaking. Every oxygen added in this $\delta$ range oxidizes $Cu^+$ in the chains without adding extra holes to the $CuO_2$ sheets, thereby giving rise to the $T_c$ (as well as the Cu valence) plateau around 60 K (figure 6). It is noteworthy that tetragonal $YBa_2Cu_3O_{6.8}$ even though oxygen-rich and contains a fair proportion of $Cu^{III}$ species, is not superconducting since the different Cu layers get connected three dimensionally through the presence of orthorhombic microdomains (Caignaert *et al.* 1990). In $YBa_2Cu_3O_{6.4}$, local oxygen ordering seems to increase the $T_c$ from 0 to 20 K (Jorgensen *et al.* 1990).

The chemistry of $YBa_2Cu_3O_{7-\delta}$ is best understood starting with $YBa_2Cu_3O_6$ ($\delta = 1.0$) which is an antiferromagnetic insulator containing $Cu^I$ ions in the chain. As oxygen diffuses into tetragonal $YBa_2Cu_3O_6$, the linear O–$Cu^I$–O units are progressively converted into square-planar $CuO_4$ that share corners to form chains (figure 3); when *ca.* 40% of the Cu–O chains are thus oxidized, we get the orthorhombic structure. Fully oxidized $YBa_2Cu_3O_7$ itself can be formally considered to be $YBa_2Cu^{II}_2Cu^{III}O_7$. Substitution chemistry of $YBa_2Cu_3O_{7-\delta}$ has been examined by several workers who have partly substituted Cu by Fe, Ni or Zn; such substitutions are generally unfavourable to superconductivity just as in the case of $La_{2-x}Sr_xCuO_4$. By substitution of Ba by La or Y by Ca in $YBa_2Cu_3O_7$, one changes the oxygen stoichiometry and also brings about a decrease or an increase in the carrier concentration (Manthiram & Goodenough 1989; Tokura *et al.* 1988). We should note here that only the holes in the $CuO_2$ sheets are pertinent to superconductivity. It is therefore necessary to subtract out the holes in the chains from the total hole concentration as given by $n_h$ (plane) $= \frac{3}{2}$ ($n_h$ total $-n_h$ chain) (Shafer & Penney 1990).

As mentioned earlier, $PrBa_2Cu_3O_7$ is not superconducting (Ganguli *et al.* 1989*c*). This was first considered to be due to the presence of a small proportion of the Pr ions in the 4+ state. It now seems that this is not the case. It has been suggested that the presence of Pr (4f) levels in the vicinity of the Cu–O band could be responsible for the absence of superconductivity (Sarma *et al.* 1991). The recent observation (Norton *et al.* 1991) of superconductivity ($T_c \approx 43$ K) in films of $Pr_{0.5}Ca_{0.5}Ba_2Cu_3O_{7-\delta}$, however, suggests that absence of superconductivity in Pr cuprate $PrBa_2Cu_3O_7$ may be related to the small hole concentration. This aspect needs to be examined further.

Basically, $YBa_2Cu_3O_7$ is not thermodynamically stable. The more stable $YBa_2Cu_4O_8$ containing two Cu–O chains (figure 9) formed by $CuO_4$ units sharing edges has been prepared. This cuprate has a $T_c$ of 80 K. Other rare earth analogues of the general formula $LnBa_2Cu_4O_8$ are also known. Although the so-called 124 compounds were first prepared under high oxygen pressures (Karpinski *et al.* 1988), they can be prepared under ambient conditions (Cava *et al.* 1989; Liu *et al.* 1990; Rao *et al.* 1990*d*). Superconducting 247 compounds of the general formula $Ln_2Ba_4Cu_7O_{15}$ with a $T_c$ of *ca.* 90 K have been characterized. The structure of these compounds may be considered to be composed of an 1:1 ordered intergrowth of $LnBa_2Cu_3O_7$ and $LnBa_2Cu_4O_8$ (figure 9). Recall $LnBa_2Cu_3O_{6.5}$ is an 1:1 intergrowth of $LnBa_2Cu_3O_7$ and $LnBa_2Cu_3O_6$. The 247 as well as the 124 cuprates on thermal decomposition give 123 throwing out excess CuO. Effects of temperature and pressure on the structures of the 124 and 247 cuprates have been examined in detail

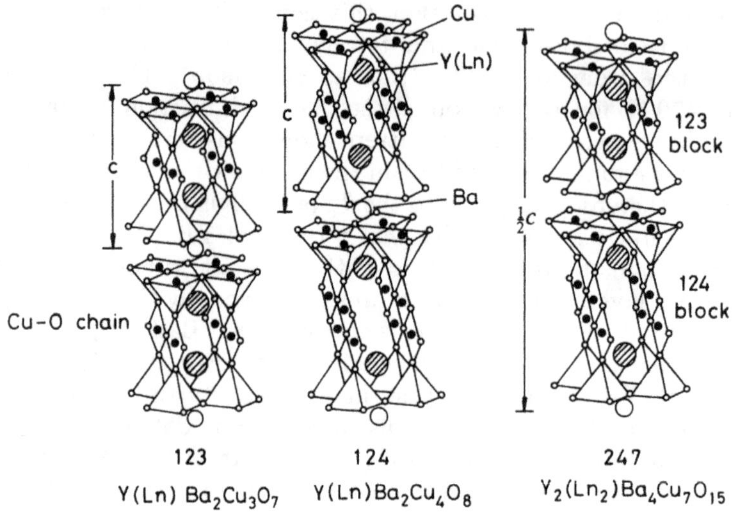


Figure 9. Comparison of the structures of 123, 124 and 247 cuprates. Note the presence of two chains in 124.

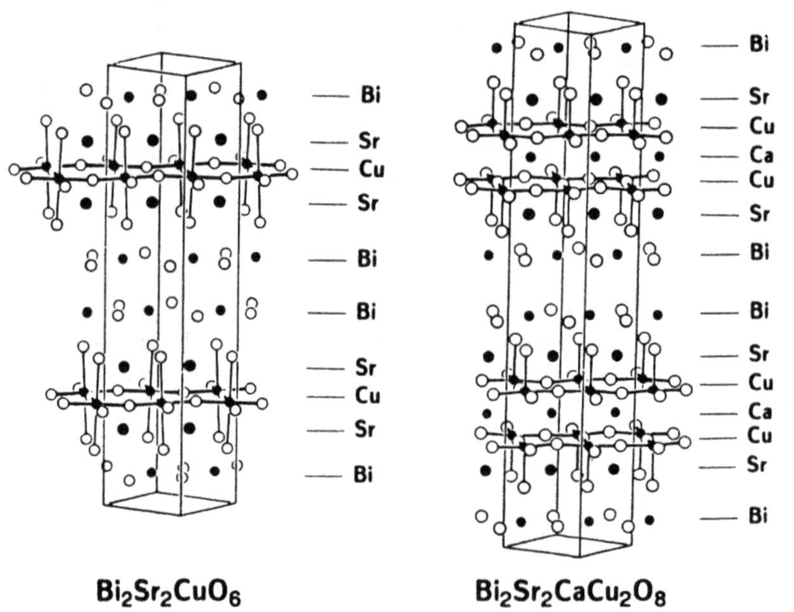


Figure 10. Structures of Bi cuprates. (From Torardi *et al.* 1989.)

by Hewat *et al.* (1990). Properties of the 124 cuprate have been reviewed by Miyatake *et al.* (1990).

Bismuth cuprates of the general formula $Bi_2(Ca,Sr)_{n+1}Cu_nO_{2n+4}$ possessing an orthorhombic structure and containing two rock-salt type layers of BiO constitute an important family of superconductors, with the $n = 2$ and the $n = 3$ members showing $T_c$s of 90 K and 110 K respectively (figure 10). The $n = 1$ member of the formula $Bi_{2+x}Sr_{2-x}CuO_6$ (without Ca) shows a maximum $T_c$ of around 20 K. The $n = 1$ member containing calcium has been reported, but it does not appear to be

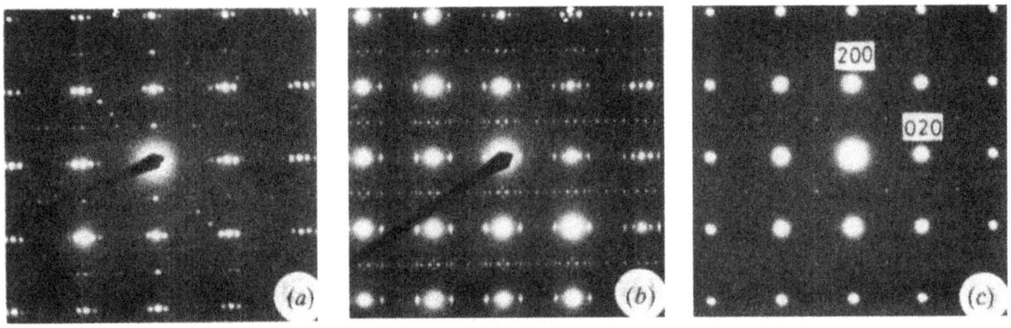


Figure 11. Electron diffraction patterns showing (*a*) superlattice modulation in superconducting $Bi_2CaSr_2Cu_2O_8$ recorded along the (001) direction, (*b*) superlattice modulation in non-superconducting $Bi_2YSr_2Cu_2O_8$, and (*c*) absence of superlattice modulation in superconducting $BiPbSr_2Y_{0.5}Ca_{0.5}Cu_2O_8$.

superconducting. One of the difficulties with the bismuth cuprates is that it is difficult to make them absolutely monophasic because of the presence of disordered intergrowths of different sequences. A unique feature of the bismuth cuprate superconductors is that they exhibit 4*b*-type superlattice modulation along the *b*-direction (figure 11) due to the insertion of extra oxygens in the BiO layers (LePage *et al.* 1989). It was first thought that superconductivity in these compounds had some relation to the superlattice modulation. However, it was possible to prepare compounds of the type $Bi_2Sr_2LnCu_2O_8$ (Ln = Y or rare earth) which are not superconducting but exhibited superlattice modulation (figure 11) of the 4*b* or the 8*b* type (Rao *et al.* 1990*b*). Similarly, non-superconducting modulation-free oxides of the type $BiPbSr_2MO_y$ (M = Mn, Fe or Co) have been prepared (Tarascon *et al.* 1990). Recently, we have been able to prepare two series of superconductors of the general formula $BiPbSr_{1+x}Ln_{1-x}CuO_6$ (maximum $T_c \approx 80$ K) both of which are modulation-free (Manivannan *et al.* 1991) as can be seen from figure 11. Since Pb in these cuprates is in the 2+ state, it helps to decrease the oxygen content in the BiO layers, thereby eliminating the modulation. In both these new series of superconductors, the hole concentration, $n_h$, varies with $x$ and the $T_c$ reaches a maximum value at an optimal value at $n_h$.

An interesting series of bismuth cuprates in terms of the variation of the $T_c$ as well as the hole concentration with composition is provided by $Bi_2Sr_2Ca_{1-x}Ln_xCu_2O_8$ where Ln = Y or rare earth (Rao *et al.* 1990*b*). The electrical resistivity data show a metal–insulator transition in the normal state with change in $x$ (figure 12). The $T_c$ as well as the $n_h$ show a maximum at a composition of $x = 0.25$ (figure 13). Note that when Ca is fully substituted by Ln, the material becomes a non-superconducting insulator. Hole concentration in these bismuth cuprates is readily determined by $Fe^{II}$–$Fe^{III}$ redox titrations.

Thallium cuprates of the general formula $Tl_2Ca_{n-1}Ba_2Cu_nO_{2n+4}$ with two Tl–O layers possessing a tetragonal structure (figure 14) show superconductivity with $T_c$s of 80, 110, and 125 K when $n = 1$, 2, 3 respectively. The corresponding $TlCa_{n-1}Ba_2Cu_nO_y$ series of cuprates with only a single Tl–O layer are also tetragonal (figure 14) with $T_c$s of 90 and 115 K respectively for $n = 2$ and 3. Higher members of the thallium cuprate families with $n > 3$ have been prepared, but the maximum $T_c$ is generally found for $n = 3$. The thallium cuprates also exhibit disordered intergrowths of different members in most of the preparations. Oxygen stoichiometry

Figure 13

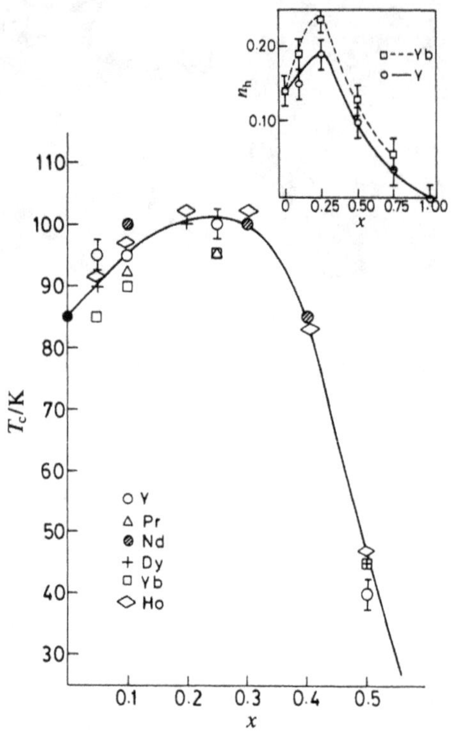


Figure 12

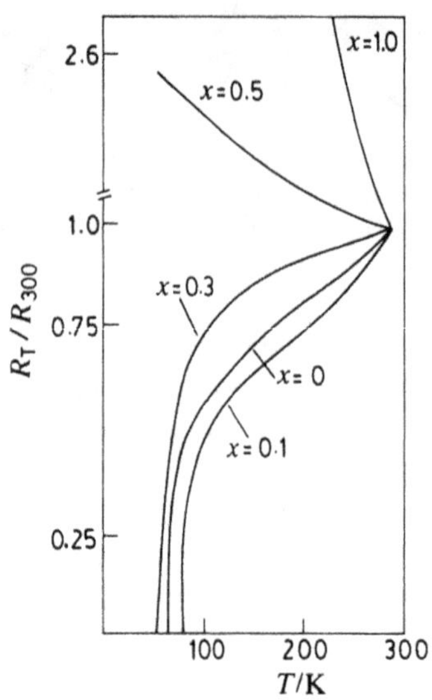


Figure 12. Resistivity data of superconducting $Bi_2Sr_2Ca_{1-x}Nd_xCu_2O_8$ showing the occurrence of a metal–insulator transition in the normal state. (From Rao *et al.* 1990*b*.)

Figure 13. Variation of $T_c$ with composition in $Bi_2Ca_{1-x}Ln_xSr_2Cu_2O_{8+\delta}$. Inset shows variation of hole concentration $n_h$ with DC. (From Rao *et al.* 1990*b*.)

plays an important role in the superconductivity of Tl cuprates as well. Thus hydrogen annealing of some of these materials increases the $T_c$ (Maignan *et al.* 1990). A serious problem with these Tl cuprates is the difficulty in exactly determining the hole concentration. A method has been suggested recently in the literature (Manthiram *et al.* 1990), but it does not appear to be entirely satisfactory. A more reliable method has been developed by Gopalakrishnan and others in this laboratory by making use of a reducing agent (HBr) which selectively reacts with $Cu^{III}$ giving $Br_2$ while thallium forms $TlBr_3$. The Tl cuprates are somewhat different from the corresponding bismuth cuprates although both Tl and Bi are essentially in the 3+ state. (Vijayakrishnan *et al.* 1990). The source of holes in the Tl cuprates is yet to be fully understood unlike in the bismuth cuprates (Goodenough & Manthiram 1990). It appears that one has to consider the presence of Tl vacancies as well as the possible overlap of the Tl(6s) band with the conduction band.

Although members of the Tl–Ca–Ba–Cu–O system of superconductors are readily prepared, the corresponding members of Tl–Ca–Sr–Cu–O cannot be prepared in pure form. It is, however, possible to prepare $Tl_{1-x}Pb_x(Ca,Sr)_{n+1}Cu_nO_y$ which are superconducting with the $n = 2$ member showing a $T_c$ of 90 K and the $n = 3$ member a $T_c$ of 120 K (Ganguli *et al.* 1988, 1989*b*; Subramanian *et al.* 1988). In these cuprates, Pb is in the 4+ state (Kulkarni *et al.* 1989), unlike in the Bi cuprates where it is in the 2+ state. A novel, analogous series of cuprates with the general formula,

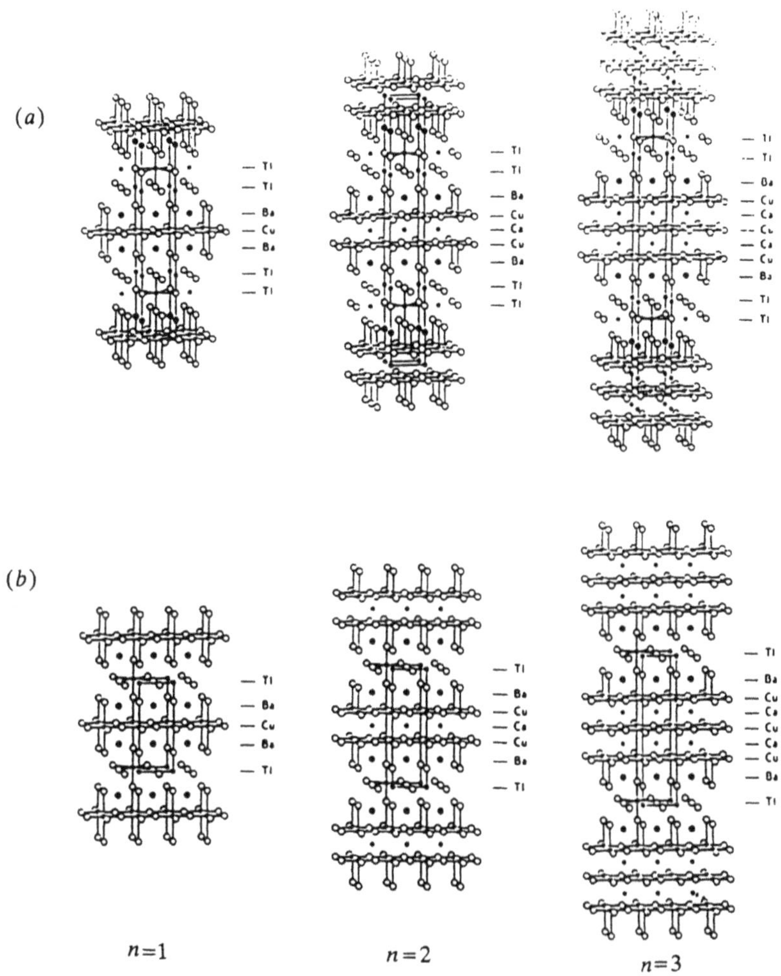


Figure 14. Structures of Tl cuprates of the type $Tl_mCa_{n-1}Ba_2Cu_nO_y$, where (*a*) $m = 1$ or (*b*) $m = 2$. (From Sleight *et al.* 1990.)

$TlCa_{1-x}Ln_xSr_2Cu_2O_y$ (Ln = Y or rare earth) has been prepared and characterized recently (Rao *et al.* 1989*a*, *c*). Here, $Ca^{2+}$ is replaced by $Ln^{3+}$ instead of $Tl^{3+}$ by $Pb^{4+}$ as in the earlier series. These cuprates also show a maximum $T_c$ of 90 K. In figure 15 we show the electrical resistivity behaviour of a series of these cuprates; note the occurrence of the metal–insulator transition in the normal state as $x$ is varied. The effect of substitution of Tl by Pb or of Ca by Ln is associated with the variation of the hole concentration. The interplay of chemical substitution with hole concentration can be nicely visualized in the $Tl_{1-y}Pb_yY_{1-x}Ca_xCu_2O_y$ system. This system may be considered as derived from the parent insulator $TlYSr_2Cu_2O_y$ wherein the substitution of Ca in place of Y increases the number of holes, while that of Pb in place of Tl does the opposite. Accordingly, $T_c$ becomes maximum at a higher value of $x$ (Ca concentration) as $y$ (Pb concentration) increases as shown in figure 16. We notice that both $T_c$ and the concentration of doped holes, $x$, reach a maximum at $y = 0.5$. The $T_c$ goes up to 105 K by the chemical manipulation of hole concentration.

All the cuprate superconductors discussed hitherto contain Ca, Ba or/and rare earth. Recently, superconducting thallium cuprates not containing Ca, Ba or even a

Figure 15

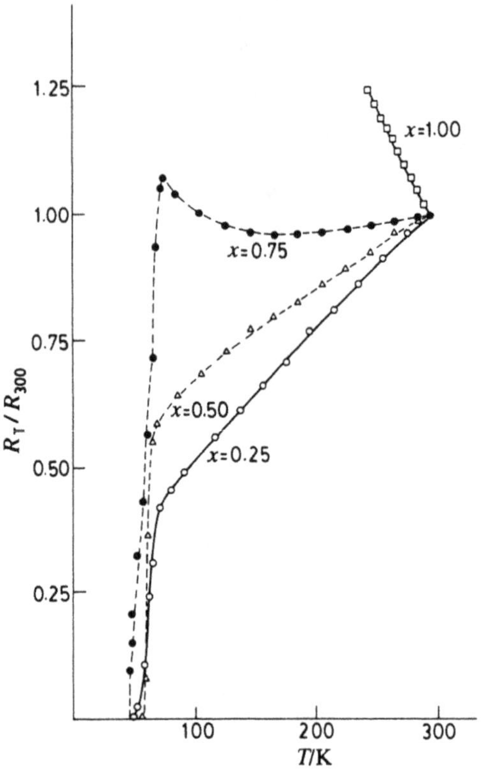


Figure 16

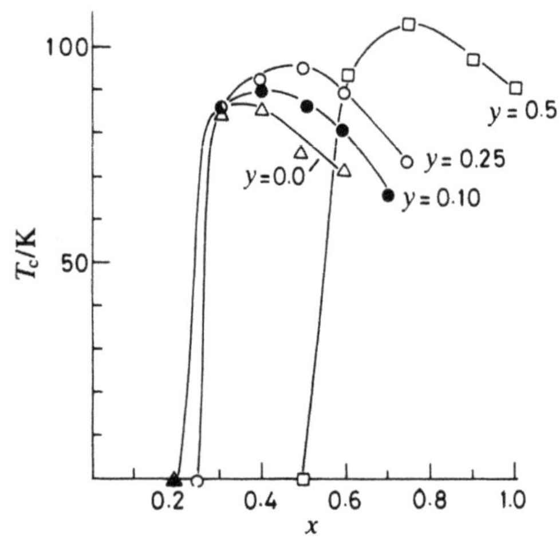


Figure 15. Resistivity data of superconducting $TlCa_{1-x}Nd_xSr_2Cu_2O_y$ showing composition-dependent metal–insulator transition in the normal state. (From Rao *et al.* 1989*c*.)

Figure 16. Variation of $T_c$ with composition in $Tl_{1-y}Pb_yY_{1-x}Ca_xSr_2Cu_2O_y$. (Unpublished results of Vijayaraghavan, Grantscharova and Rao.)

rare earth have been characterized. Thus, in the series, $TlSr_{n+1-x}Ln_xCu_nO_{2n+3+\delta}$, the $n = 1$ and 2 members with $T_c$s of 40 and 90 K have been prepared (Ganguli *et al.* 1989*a*). Similarly, the $n = 3$ member in the series $(Tl,Pb)Sr_{n+1-x}Ln_xCu_nO_{2n+3}$ with a $T_c$ of 60 K has been prepared though not in pure form (Manivannan *et al.* 1990). The $n = 1$ derivatives of these families may be considered to be derived from $TlSr_2CuO_5$. Substitution of $La^{3+}$ for $Sr^{2+}$ stabilizes the structure and reduces $Cu^{III}$, permitting superconductivity (Kovatcheva *et al.* 1991).

Lead cuprates of the general formula $Pb_2Sr_2(Ln,Ca)Cu_3O_8$ containing PbO layers and O–$Cu^{I}$–O sticks (figure 17) with a $T_c$ of about 60 K have been prepared (Cava *et al.* 1988*b*). In this system the average oxidation state of Cu is less than 2. The normal and superconducting state properties of $Pb_2Sr_2Y_{1-x}Ca_xCu_3O_{8+\delta}$ (note that the $x = 0.0$ composition is an insulator) have been examined in some detail (Koike *et al.* 1990). The superconducting lead cuprates have to be synthesized in an atmosphere deficient in oxygen ($N_2$ with 1% $O_2$) to prevent oxidation of Cu and Pb. Analogous to the single Tl–O layer compounds, lead cuprates of the formula $(Pb,Cu)Sr_2(Y,Ca)Cu_2O_y$ are found to exhibit a $T_c$ of 52 K (Maeda *et al.* 1990). Lower members of the Pb cuprate family of the type $(Pb,Cu)(SrLa)_2CuO_5$ ($T_c$ of 34 K) have also been synthesized. Recently, a lead cuprate of the formula

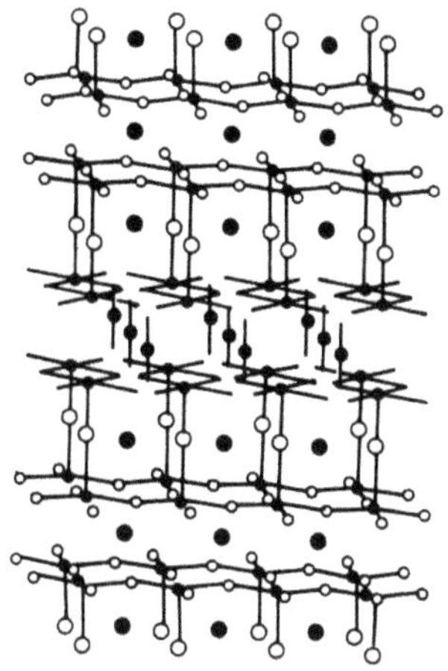


Figure 17. Structure of $PbSr_2(Ca,Y)Cu_3O_8$ (after Sleight 1988).

$(Pb,Cu)(Eu,Ce)_2(Sr,Eu)_2Cu_2O_9$ containing a fluorite layer has been prepared with a $T_c$ of 25 K (Maeda *et al.* 1990).

## 3. Nature of holes in the cuprate superconductors

All the cuprates described till now are hole superconductors. The nature of holes has been subject of considerable discussion (Chakraverty *et al.* 1988; Rao *et al.* 1989*b*; Sarma & Rao 1989). There has been no experimental evidence for the presence of $Cu^{III}$ type species in the doped cuprates. Instead, there is considerable evidence from electron and X-ray spectroscopies for the presence of hybridized oxygen holes which can be represented as $O^-$. The detailed description of the holes in terms of the d and p characters has been investigated (Bianconi 1990). Essentially, the mobile holes in the cuprates are present in the in-plane $\pi^*$ band which has O–2p character. The concentration of holes (in all but the Tl cuprates) are easily determined by iodometry or $Fe^{II}$–$Fe^{III}$ titrations (Rao *et al.* 1991*a*; Shafer & Penney 1990). Since the Hall coefficients are temperature dependent, the chemical titration method becomes invaluable.

In figure 18, we show the variation of $T_c$ with $n_h$ (obtained by chemical titrations) in a number of cuprate families. We see that in all these families, the $T_c$ goes through a maximum around the same hole concentration in the different series of cuprates containing the same number or $CuO_2$ sheets. Accordingly, $n_h \approx 0.2$ at the maximum $T_c$ in all the cuprates containing two $CuO_2$ sheets. This universality is noteworthy. In figure 19 we show the variation of $T_c$ in $Tl_{1-y}Pb_yY_{1-x}Ca_xSr_2Cu_2O_y$ against $(x-y)$ which is a direct measure of the hole concentration. We see that the maximum $T_c$ (105 K) is found for $y = 0.25$ and 0.5 at a $(x-y)$ value of *ca.* 0.22. This value is close to the $n_h$ value at maximum $T_c$ in the other cuprates; the maximum $T_c$ is lower when $(x-y)$ is larger as in the cases $y = 0.0$ and 0.1.

## 4. Electron-superconducting cuprates

Unlike the various cuprates discussed in §2 where the $CuO_2$ sheets could be doped with holes, $Pr_2CuO_4$ or $Nd_2CuO_4$ possessing the T′ tetragonal structure (figure 20) can be doped with electrons by partly substituting Nd by Ce or Th or oxygen by fluorine

Figure 18

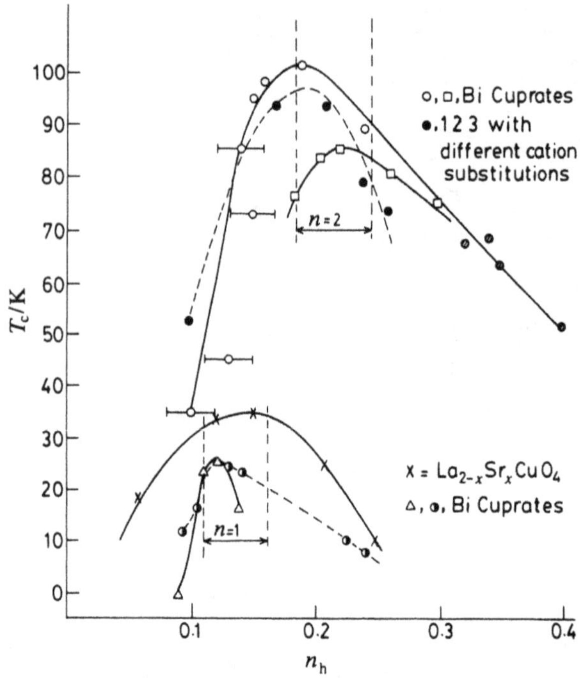


Figure 19

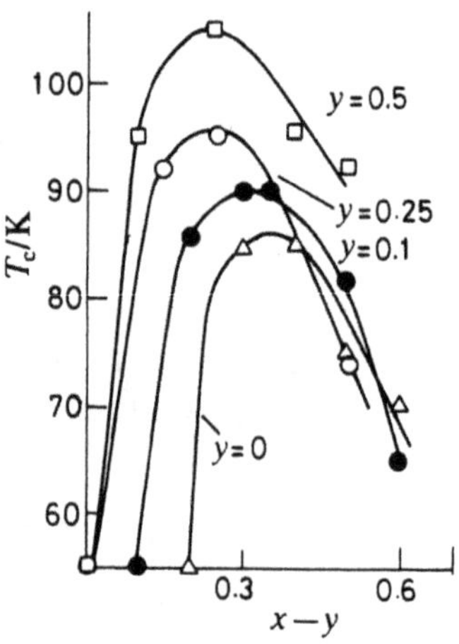


Figure 18. Variation of $T_c$ with hole concentration in cuprates (from Rao *et al.* 1991*a*).
Figure 19. Variation of $T_c$ with effective hole concentration, $(x-y)$, in $Tl_{1-y}Pb_yY_{1-x}Ca_xSr_2CuO_y$.

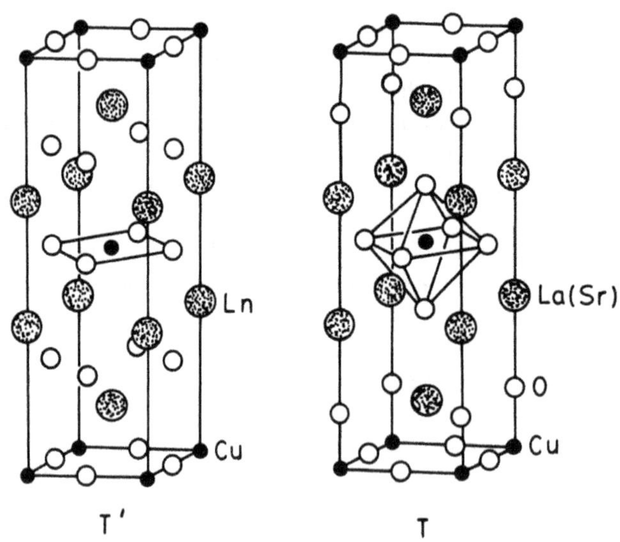


Figure 20. The T′ and T structure of cuprates (T, Ln = Nd, Pr, Ce).

(Maple 1990; Tokura *et al.* 1989*b*). The maximum $T_c$ that these compounds exhibit is around 25 K. Even though these cuprates are formally considered to be electron superconductors, the exact nature of the charge carriers is not fully certain. Some holes could also be present in such materials. In many ways, electron-superconducting cuprates are similar to the $La_2CuO_4$ family of hole superconductors. In $Nd_{2-x}Ce_xCuO_4$, it appears that the compositions are monophasic only for $x = 0.0$ and for the optimal $x$ value where $T_c$ is maximum (Lightfoot *et al.* 1990). Such inhomogeneity has been found in oxygen-excess $Ln_2CuO_4$ (Jorgensen *et al.* 1988). We do not have electron-superconducting oxides with $T_c$s comparable with those of the hole superconductors.

## 5. Synthetic aspects

In this section, I shall briefly present some preparative aspects of the cuprate superconductors based on the experience gained by us in the past four years. The cuprates are ordinarily made by the traditional ceramic method (mix, grind and heat), which involves thoroughly mixing the various oxides or/and carbonates (or any other salt) in the desired proportion and heating the mixture (preferably in pellet form) at a high temperature. The mixture is ground again after some time and reheated until the desired product is formed as indicated by X-ray diffraction. This method may not always yield the product with the desired structure purity or in oxygen stoichiometry. Variants of this method are often used. For example, decomposing a mixture of nitrates has been found to yield a better product in the case of the 123 compounds by some workers; some others prefer to use $BaO_2$ in place of $BaCO_3$ for the synthesis.

Coprecipitation and sol-gel methods are conveniently employed for the synthesis of 123 compounds and other cuprates. The sol-gel method provides a homogeneous dispersion of the various component metals when a solution containing the metal ions is transformed into a gel by adding an organic solvent such as a glycol or an alcohol often in the presence of other chemicals such as organic amines. The gel is then decomposed at relatively low temperatures to obtain the desired oxide, generally in fine particulate form. Materials prepared by such low-temperature methods may need to be annealed or heated under suitable conditions to obtain the desired oxygen stoichiometry as well as the characteristic high $T_c$. 124 cuprates, lead cuprates and bismuth cuprates have all been made by this method; the first two are particularly difficult to make by the ceramic method.

One of the problems with the bismuth cuprates is the difficulty in obtaining phasic purity (minimizing the intergrowth of the different layered phases). The glass or the melt route has been used to obtain better samples. The method involves preparing a glass by quenching the melt; the glass is then crystallized by heating it above the crystallization temperature. Thallium cuprates are best prepared in sealed tubes (gold or silver). Heating $Tl_2O_3$ with a matrix of the other oxides (already heated to 1100–1200 K) in a sealed tube is preferred by some workers. It is important that thallium cuprates are not prepared in open furnaces since $Tl_2O_3$ which readily sublimes is highly toxic. To obtain superconducting compositions corresponding to a particular copper content (number of $CuO_2$ sheets) by the ceramic method, one often has to start with various arbitrary compositions especially in the case of the Tl cuprates. The real composition of a bismuth or a thallium cuprate superconductor is not likely to be anywhere near the starting composition. The actual composition can be determined by analytical electron microscopy and other methods.

Heating oxidic materials under high oxygen pressures or in flowing oxygen often becomes necessary to attain the desired oxygen stoichiometry. Thus, $La_2CuO_4$ and $La_2Ca_{1-x}Sr_2Cu_2O_6$ heated under high oxygen pressure become superconducting with $T_c$s of 40 and 60 K respectively. In the case of the 123 compounds, one of the problems is that it loses oxygen easily. It therefore becomes necessary to heat the material in an oxygen atmosphere at an appropriate temperature below the orthorhombic–tetragonal transition temperature. Oxygen stoichiometry is, however, not a problem in the bismuth cuprates. The 124 superconductors were first prepared under high oxygen pressures. It was later found out that heating the oxide or nitrate mixture in the presence of $Na_2O_2$ in flowing oxygen is sufficient to obtain 124 compounds. Superconducting Pb cuprates, on the other hand, can only be prepared in presence of very little oxygen ($N_2$ with a small percentage of $O_2$). In the case of the electron superconductor, $Nd_{2-x}Ce_xCuO_4$, it is necessary to heat the material in an oxygen-deficient atmosphere; otherwise, the electron given by Ce will merely go into giving an oxygen excess material. It may be best to prepare $Nd_{2-x}Ce_xCuO_4$ by a suitable method (say decomposition of mixed oxalates or nitrates) and then reduce it with hydrogen.

## 6. Commonalities in the cuprates

There are many striking commonalities in the structure and properties of the high $T_c$ cuprates. All the cuprates can be considered to be a result of the intergrowth of defect perovskite layers of $ACuO_{3-x}$ with AO-type rock-salt layers leading to the general formula $[ACuO_{3-x}]_n$ $[AO]_{n'}$ as shown in figure 21. The 123 compounds, however, do not have rock-salt layers and may be considered as the $n' = 0$ member of this general family. The more important common features are the following:

(i) All the cuprates possess $CuO_2$ layers sandwiched between certain M–O layers (e.g. TlO, BiO) acting as charge reservoirs or spacers. The seat of superconductivity is in the $CuO_2$ layers. The $T_c$ in Bi and Tl cuprates increases up to $n = 3$ and then decreases, the cuprate with an infinite number of Cu–O layers being an antiferomagnetic insulator (figure 22). Interaction or spacing between the Cu–O layers is crucial. This is demonstrated by recent experiments where the introduction of a fluorite layer, $[Ln_{1-x}Ce_x]_2O_2$, between two $CuO_2$ sheets in Bi cuprates lowers the $T_c$ markedly (Tokura *et al.* 1989*a*). However, intercalation of iodine between the BiO layers in $Bi_2CaSr_2Cu_2O_8$ (causing a substantial increase in the *c*-parameter) does not affect the $T_c$ (Xiang *et al.* 1990).

(ii) The Cu–O bonds in the cuprates are highly covalent.

(iii) There is an interesting comparison between the Cu–O sheets in the hole and the electron superconductors. Cuprates with the T′-structure where Cu has a square-planar coordination can be doped with electrons while those with the T-structure as in $La_{2-x}Sr_xCuO_4$ can be doped with holes (figure 20). There is an interesting symmetry between these two situations as shown in figure 23. The Cu–O–Cu angle is less than 180° in the hole superconductors while it is close to 180° in the electron superconductors. The position of the apical oxygen in the Cu–O square-pyramids or octahedra in the hole superconductors seems to modulate the width of the conduction band.

(iv) The parent cuprates in all the superconductors are antiferromagnetic insulators. For example, $La_2CuO_4$, $YBa_2Cu_3O_6$, $Bi_2Sr_2LnCu_2O_8$ and $Pb_2Sr_2LnCu_3O_8$ are the antiferromagnetic insulators corresponding to the superconductors $La_{2-x}Sr_xCuO_4$, $YBa_2Cu_3O_7$, $Bi_2Ca_{1-x}Ln_xSr_2Cu_2O_8$ and $Pb_2Sr_2Ca_{1-x}Ln_xCu_3O_8$ respectively. In the case of the electron superconductor $Nd_{2-x}Ce_xCuO_4$, parent $Nd_2CuO_4$ is the antiferromagnetic insulator.

(v) All the cuprates nominally contain mixed valent copper which can disproportionate ($Cu^{II} \rightarrow Cu^{III} + Cu^{I}$). In other words, the phenomenon is associated

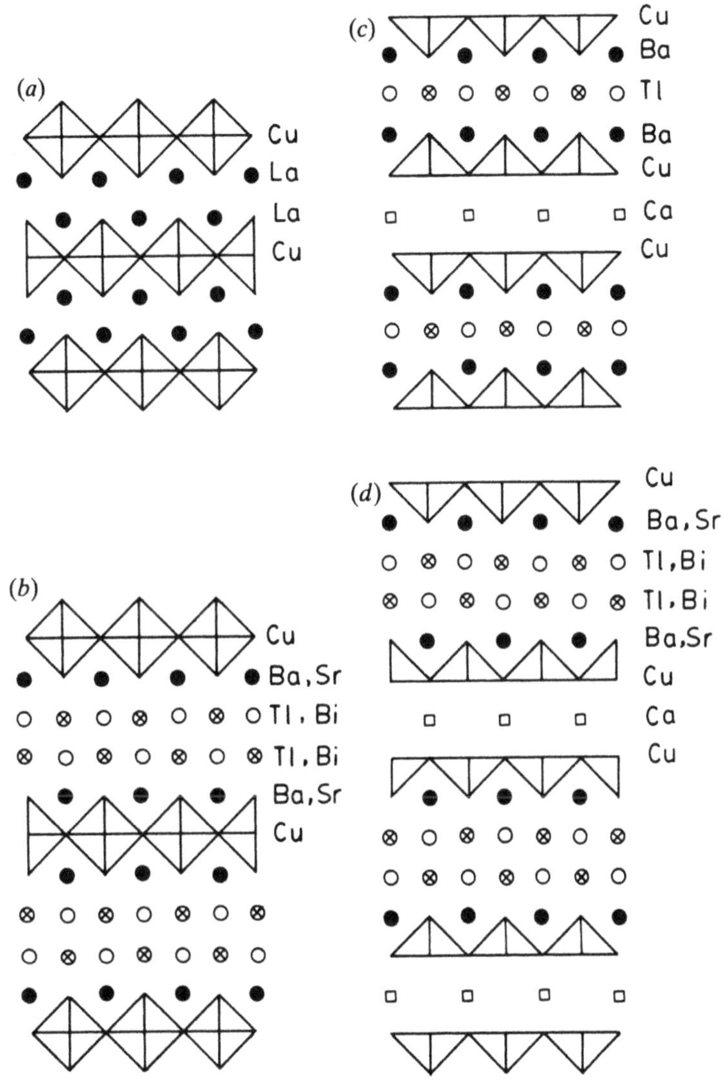


Figure 21. Schematic representation of the structures of (*a*) $La_2CuO_4$, (*b*) $Bi_2Sr_2CuO_6$ and $Tl_2Ba_2CuO_6$, (*c*) $TlCaBa_2Cu_2O_7$ and (*d*) $Bi_2CaSr_2Cu_2O_8$ and $Tl_2CaBa_2Cu_2O_8$. Oxygens are shown by open circles and Bi and Tl by circles with a cross. (From Rao & Raveau 1989.)

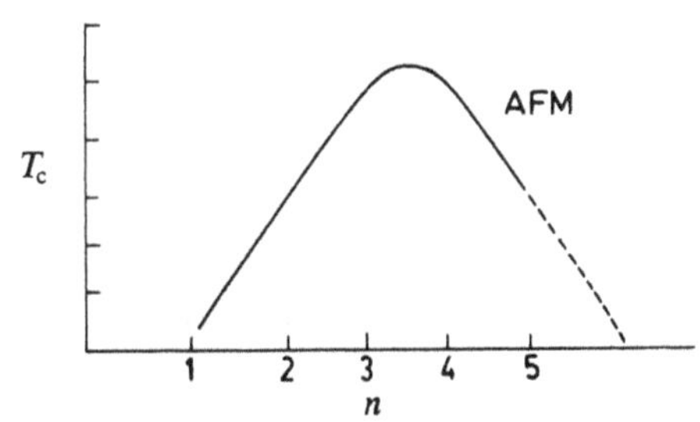


Figure 22. Variation of $T_c$ with the number of $CuO_2$ sheets, $n$ (schematic).

NOTE: Infinitely layered cuprates are since found to be superconducting

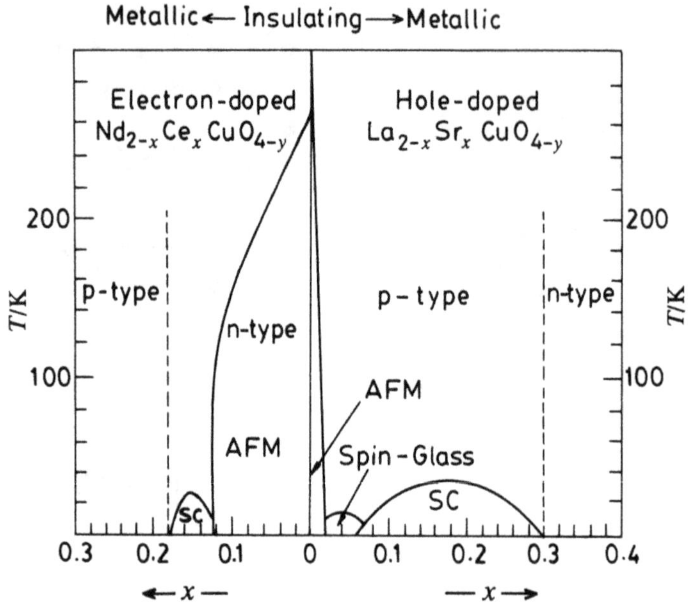


Figure 23. Symmetry in the phase diagrams of electron- and hole-superconductors, $Nd_{2-x}Ce_xCuO_4$ and $La_{2-x}Sr_xCuO_4$ (after Maple 1990.)

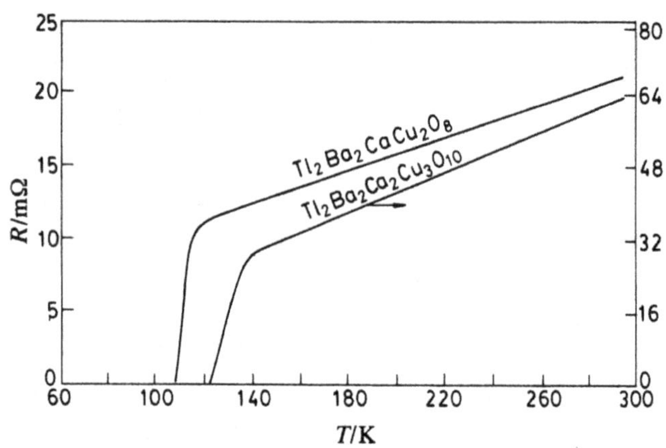


Figure 24. Linearity of the resistivity with temperature in the normal state of Tl cuprates (after Torardi *et al.* 1989).

with a band that gets filled with two electrons with stability associated with empty, half-filled and filled states.

(vi) Local charge distribution provides a basis to understand superconductivity in the cuprates. This is well demonstrated in $Pb_2Sr_2Ln_{1-x}Ca_xCu_3O_8$ where excess oxygen oxidizes $Pb^{2+}$ and $Cu^+$ without affecting the $CuO_2$ sheets. In this system, it is necessary to replace the yttrium by calcium in between the $CuO_2$ sheets to render it superconducting (Cava 1990).

(vii) Oxygen stoichiometry, homogeneity and disorder play an important role in the superconductivity of the cuprates (Hewat *et al.* 1989; Rao *et al.* 1990*d*; Raveau *et al.* 1990) as exemplified in the discussion of the various families.

(viii) All the superconducting cuprates are marginally metallic in the normal state, sitting on a metal–insulator boundary. We would, therefore, expect abnormal properties in the normal state. One of the striking abnormal normal-state properties of these materials is the linearity of resistivity over a wide range of temperatures as shown in figure 24 in the case of thallium cuprates.

## 7. Relation between the electronic structure and the superconductivity of cuprates

Photoemission spectroscopic studies (Allen & Olson 1990) of the high $T_c$ cuprates show superconducting gap formation and establish that the normal state is strongly correlated with atomic spectral features at high energy. There is strong Cu(3d)–O(2p) hybridization and the $E_F$ has nearly the same value for both hole and electron doping, lying in states filling the gap of the parent antiferromagnetic insulator. The gap filling states near $E_F$ obey the Luttinger counting theorem.

It is important to understand the relation between the electronic structure and superconductivity of the cuprates. The problem, however, is that it is not easy to exactly describe the electronic structure of such complex oxides or to develop meaningful models to describe the properties in the superconducting and normal states. Empirical relations and chemical intuition, however, continue to be useful. Accordingly, it has been shown that the difference in the Madelung site potential between a hole on copper and one on oxygen can control their site preference and the ability to delocalize (Torrance & Metzger 1989). A relation between the oxygen content and the average [Cu–O] charge of Cu valence shows the presence of a boundary between the insulators and superconductors. The [Cu–O] charge in the sheets has been shown by an large to determine $T_c$ (Tokura *et al.* 1988). The formal valence of Cu and O ions in the central $CuO_2$ planes has been calculated by de Leeuw *et al.* (1990) from experimental bond lengths following Zachariasen rules. The results show that in all the structures, the values of formal valence correlate well with the $T_c$; the $T_c$(max) increases when the holes prefer the oxygen sites over the copper sites in the $CuO_2$ sheets. This implies a higher value for $U-\Delta+\frac{1}{2}W$ where $U$ is the correlation energy, $\Delta$, the charge-transfer energy and $W$ the band width.

In the superconducting cuprates, the correlation energy, $U_{dd}$ within the Cu 3d manifold is considerably larger than the Cu–O charge-transfer energy, $\Delta$, or the Cu(3d)–O(2p) hybridization strength, $t_{pd}$. Hybridized oxygen holes in the $CuO_2$ sheets are the charge carriers responsible for the superconductivity. Although many of the models for superconductivity in the cuprates consider the parent compounds to be charge-transfer gap insulators and the $t_{pd}$ to be substantial, they do not explicitly take $\Delta$ and $t_{pd}$ as crucial parameters. Some of the phenomenological models consider the electronic polarizability, $\alpha$, to be important, but do not evaluate or relate $\alpha$ to $\Delta$, $t_{pd}$ or carrier concentration. Although it seems clear that chemical bonding factors such as $\Delta$ and $t_{pd}$ have to be given greater attention, there has been hitherto no experimental proof to show that these are indeed as important as they appear to be. Recent Cu core-level photoemission studies of the cuprates supported by theoretical calculations have, however, changed the picture (Rao *et al.* 1990*c*, 1991*b*; Rao & Sarma 1991; Santra *et al.* 1991).

The Cu $2p_{\frac{3}{2}}$ core-level spectra of cuprate superconductors show a main feature, M, around 933 eV due to the well-screened core-hole state of $2p^5\,3d^{10}$ configuration and a broad satellite, S, centred around 942 eV due to the poorly screened state of $2p^5\,3d^9$

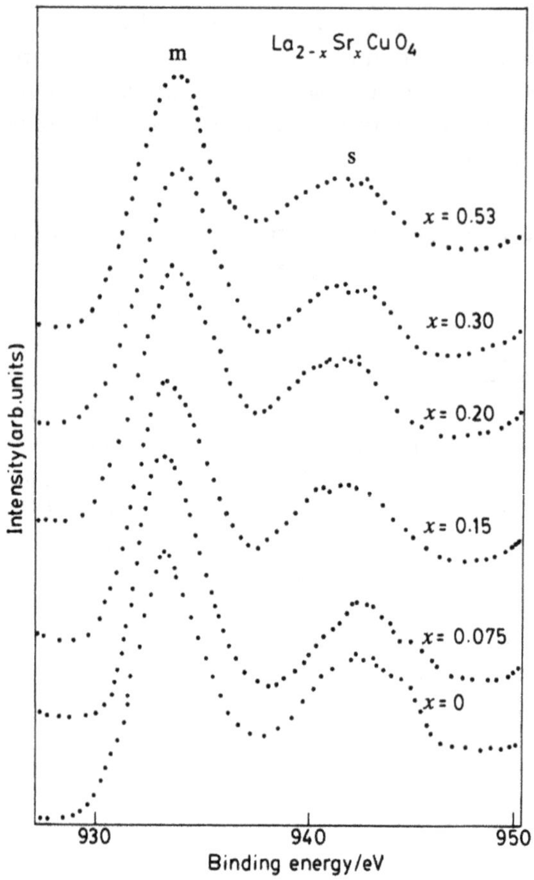


Figure 25. Cu 2p core-level spectra of $La_{2-x}Sr_xCuO_4$.

configuration. The relative intensity of the satellite with respect to that of the main feature, $I_s/I_m$, is determined by the charge-transfer energy, $\Delta$, and the Cu–O hybridization strength, $t_{pd}$. Making use of the $I_s/I_m$ ratio as an experimental handle, we have investigated the role of $\Delta$ and $t_{pd}$. For this purpose $I_s/I_m$ has been carefully measured in several series of cuprates with known hole concentration, $n_h$. To explain the nature of variation of $I_s/I_m$ with the composition of $n_h$, model calculations have been carried out on a $CuO_4$ cluster including configuration interaction. These calculations indeed reveal how $I_s/I_m$ depends sensitively on the $\Delta/t_{pd}$ ratio. We shall briefly examine the results of our Cu 2p core-level photoemission studies and theoretical calculations to appreciate how the Cu–O charge-transfer energy and the Cu(3d)–O(2p) hybridization strength play a crucial role in the superconductivity of the layered cuprates.

In figure 25 we show the X-ray photoemission spectra of $La_{2-x}Sr_xCuO_4$ for various values of $x$ in the Cu $2p_{3/2}$ region. The spectra exhibit the well-known two-peak structure with a peak at about 933 eV (the main peak) and the other at 941 eV (the satellite) binding energies. The ratio of the satellite to the main peak intensity, $I_s/I_m$, in figure 25 exhibits a systematic variation with $x$. We have quantitatively estimated $I_s/I_m$ as the ratio of the integrated areas under the main peak and the satellite after background subtraction of the spectra. The resulting $I_s/I_m$ is plotted as a function of $x$ in figure 26 for the $La_{2-x}Sr_xCuO_4$ series. In the same figure, we have also shown the

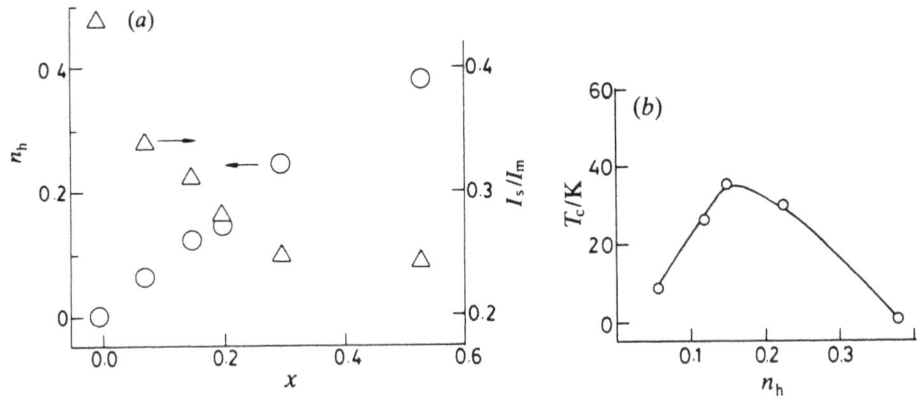


Figure 26. (*a*) Variation of $I_s/I_m$ and $n_h$ with $x$ in $La_{2-x}Sr_xCuO_4$. (*b*) Variation of $T_c$ with $n_h$.

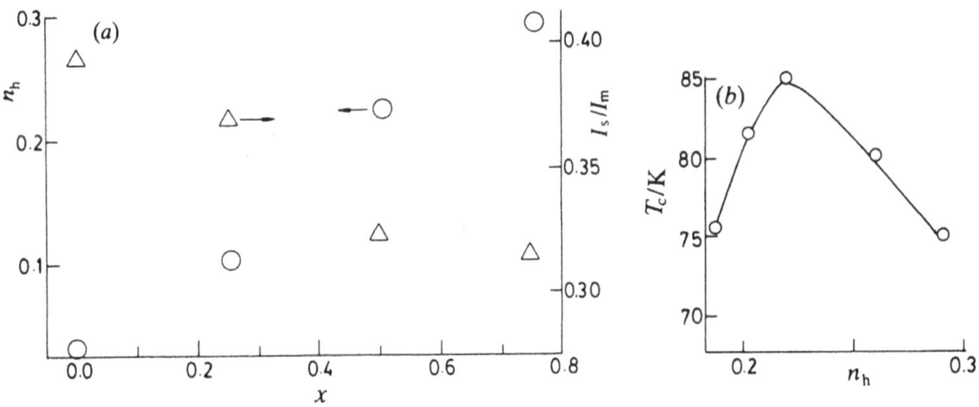


Figure 27. (*a*) Variation of $I_s/I_m$ and $n_h$ with $x$ in $BiPbSr_2Y_{1-x}Ca_xCu_2O_8$. (*b*) Variation of $T_c$ with $n_h$.

dependence of the experimentally obtained (by iodometric titrations) hole concentration, $n_h$, on $x$ in these compounds. The inset shows the variation of the superconducting transition temperature $T_c$ on $n_h$. This system traverses through the insulator–superconductor–metal régimes with increase in $x$. The $I_s/I_m$ decreases continuously though these régimes, while $n_h$ increases.

This behaviour seems to be common to all the series of superconducting cuprates that we have investigated. In figure 27 we plot the variation of $I_s/I_m$ and $n_h$ with $x$ in the $BiPbSr_2Y_{1-x}Ca_xCu_2O_8$ series. We also show the variation of $T_c$ with $n_h$ in this series in figure 27*b*. The $T_c$ reaches a maximum of about 85 K around $n_h = 0.22$; $n_h$ once again exhibits a linear dependence on $x$ with a slope of less than unity, while $I_s/I_m$ monotonically decreases with $x$.

We show the dependence of $n_h$ and $I_s/I_m$ on $x$ in the $Bi_2Ca_{1-x}Ln_xSr_2Cu_2O_8$ (Ln = Y or rare earth) series of compounds in figure 28. The inset shows the variation of $T_c$ with $n_h$ in this series. The $T_c$ appears to exhibit a broad maximum at about $n_h = 0.2$ in these compounds. Interestingly in this series, we obtain a non-monotonic dependence of $n_h$ on $x$. The $n_h$ increases with $x$ for small values of $x$ up to about 0.25 and then decreases continuously up to $x = 1.0$. This is in contrast to the dependence of $n_h$ on $x$ in the other two series shown in figures 26 and 27. It is significant that in this series, $I_s/I_m$ also exhibits a non-monotonic behaviour, showing a decrease up to

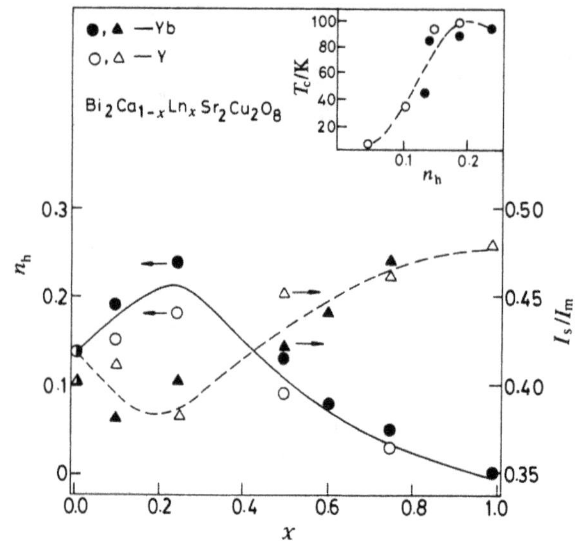


Figure 28. Variation of $I_s/I_m$ and $n_h$ with $x$ in $Bi_2Ca_{1-x}Ln_xSr_2Cu_2O_8$.

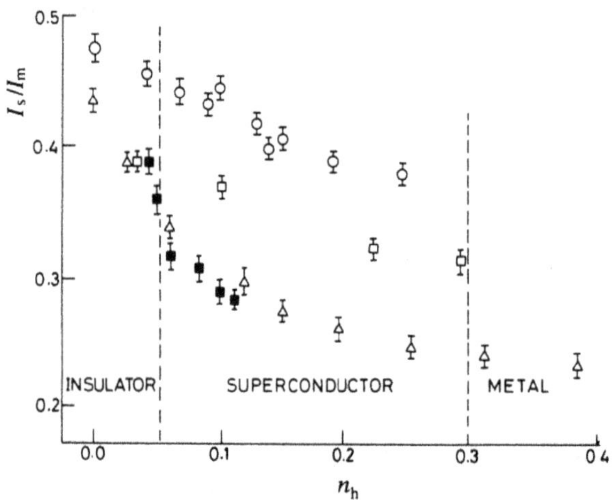


Figure 29. Variation of $I_s/I_m$ with $n_h$. ▪, $BiPbSr_{1+x}Pr_{1-x}CuO_6$; ○, $Bi_2Ca_{1-x}Ln_xSr_2Cu_2O_8$ (Ln, rare earth); △, $La_{2-x}Sr_xCuO_4$; □, $Bi_2PbSr_2Y_{1-x}Ca_xCu_2O_8$.

$x = 0.2$ and then a monotonic increase up to $x = 1.0$. This provides a crucial test of the fact that $I_s/I_m$ appears to have the complimentary dependence on $x$ compared with $n_h$. This implies that $I_s/I_m$ will monotonically decrease in all the series with increasing $n_h$.

There is a relation between the experimentally obtained $I_s/I_m$ and $n_h$ values as shown in figure 29 where we have plotted the $I_s/I_m$ ratios against the $n_h$ values for the three series, $La_{2-x}Sr_xCuO_4$, $BiPbSr_2Y_{1-x}Ca_xCu_2O_8$ and $Bi_2Ca_{1-x}Ln_xSr_2Cu_2O_8$ (Ln = Y or Yb) as well as for the series, $BiPbSr_{1+x}Pr_{1-x}CuO_6$. It becomes absolutely clear from this figure that in each of the three series, $I_s/I_m$ decreases monotonically with increased hole doping. The $La_{2-x}Sr_xCuO_4$ and the BiPb $Sr_{1+x}Pr_{1+x}CuO_6$ series exhibit the most pronounced dependence of $I_s/I_m$ on $n_h$ (over a narrow range of $n_h$), while the $Bi_2Ca_{1-x}Ln_xSr_2Cu_2O_8$ and $Bi_2PbSr_2Y_{1-x}Ca_xCu_2O_8$ series have a weaker dependence on $n_h$. In figure 29 we have also marked the

Figure 30

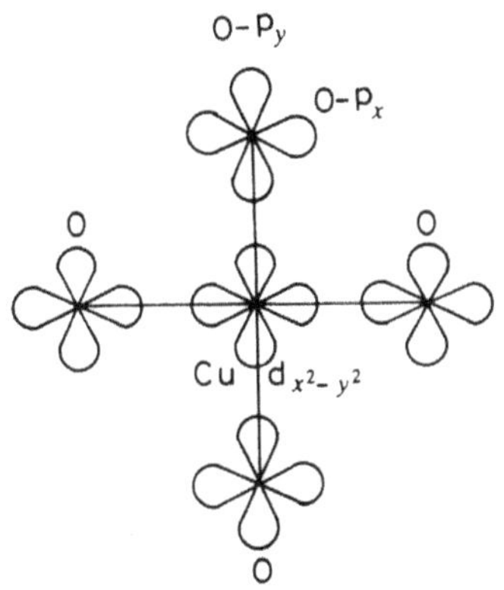


Figure 31

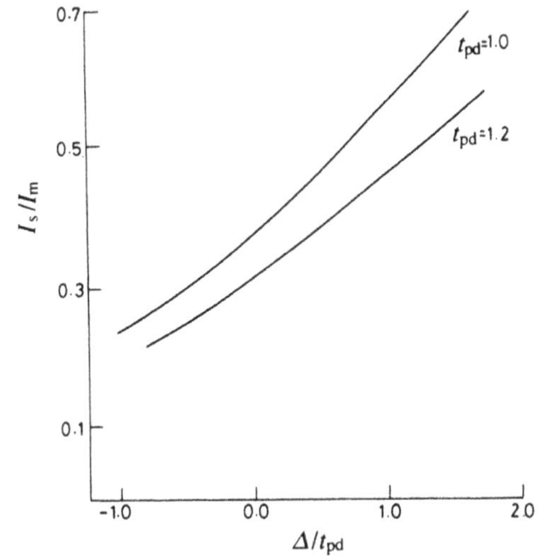


Figure 30. Bonding in a $CuO_4$ unit.
Figure 31. Variation of $I_s/I_m$ with $\Delta/t_{pd}$.

insulating, superconducting and metallic regions of the cuprates as $n_h$ is varied by chemical doping to demonstrate how $I_s/I_m$ varies continuously through these different régimes.

To understand the variation of $I_s/I_m$ in the cuprates, we have performed model calculations of the Cu $2p_{\frac{3}{2}}$ core level photoemission spectral features (Rao *et al.* 1991*b*; Rao & Sarma 1991; Santra *et al.*1991). For this purpose, we consider a $CuO_4$ cluster as shown in figure 30. We include the Cu $3d_{x^2-y^2}$ and the oxygen in-plane $2p_x$, $2p_y$ orbitals. In the $D_{4h}$ symmetry of the $CuO_4$ cluster, the Cu $3d_{x^2-y^2}$ orbital transforms according to the $b_{1g}$ irreducible representation. Out of the eight oxygen $2p_{x,y}$ orbitals shown in the figure, only one linear combination of these (namely $p_x^1 - p_y^2 - p_x^3 + p_y^4$) has the right symmetry ($b_{1g}$) to mix with the $3d_{x^2-y^2}$ orbital of Cu. Within this approximation the hybridization of the various oxygen orbitals do not play any important role as far as the spectral features are concerned, since these hybridization interactions will merely shift the $b_{1g}$ combination of the O 2p orbitals in energy and renormalize the charge-transfer excitation energy (Sarma & Ovchinnikov 1990). We consider only one hole per Cu, so that the Coulomb interaction strengths, $U_{dd}$, $U_{pp}$ and $U_{pd}$ do not play any role. Thus, the spectral features depend only on the charge-transfer energy, $\Delta$, between the Cu $3d_{x^2-y^2}$ and the oxygen-derived $b_{1g}$ orbitals, the hybridization interaction strength, $t$, between the 3d and ligand $b_{1g}$ level, and the 2p core-hole–3d-hole Coulomb repulsion, $U$. This problem can be exactly solved within the sudden approximation for the spectral features (Rao & Sarma 1991). In all our calculations we fix the value of the Coulomb repulsion, $U$, between the Cu 2p-core hole in the Cu 3d valence-hole at 8.0 eV.

In figure 31, we show the variation of the calculated $I_s/I_m$ with $\Delta/t_{pd}$, where $t_{pd} = \frac{1}{2}t$ is the hybridization strength between the Cu 3d and the O 2p orbitals for two different values of $t_{pd}$. We find from this figure that $I_s/I_m$ increases with increasing $\Delta/t_{pd}$. At large values of $t_{pd}$, the variation of $I_s/I_m$ with $\Delta/t_{pd}$ becomes less pronounced; the dependence of $I_s/I_m$ on $\Delta$ appears to be more pronounced when $t_{pd}$ is small. In this context, we note that the experimental $I_s/I_m$ values for $La_{2-x}Sr_xCuO_4$

Figure 32

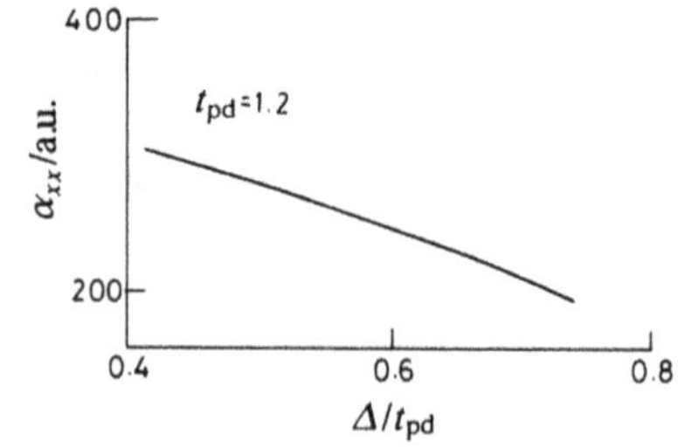


Figure 33

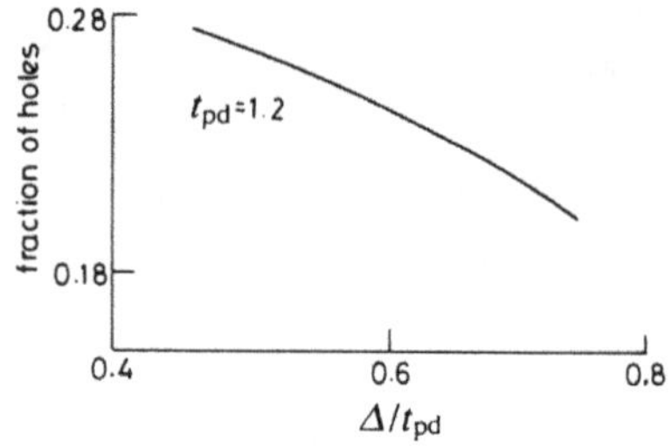


Figure 32. Variation of the polarizability of the $CuO_2$ sheets, $\alpha_{xx}$, with $\Delta/t_{pd}$ (from Rao *et al.* 1991).
Figure 33. Variation of the fraction of holes with $\Delta/t_{pd}$ (from Rao *et al.* 1991).

and $BiPbSr_{1-x}Pr_{1+x}CuO_6$ show a marked variation with $n_h$ and $x$, while $I_s/I_m$ for $Bi_2Ca_{1-x}Ln_xSr_2Cu_2O_8$ and $Bi_2PbSr_2Y_{1-x}Ca_xCu_2O_8$ exhibit a less pronounced variation. It, therefore, appears that the decrease in $I_s/I_m$ with $x$ or $n_h$ observed in these series of cuprates as well as the slopes of these variations are related to the magnitude of $\Delta/t_{pd}$.

Our calculations of $I_s/I_m$ in the cuprates indicate that the decrease in the charge-transfer energy is primarily responsible for the variation of the Cu 2p satellite intensity with increasing $n_h$. This is not an unreasonable expectation since increasing $n_h$ tends to renormalize the charge-transfer energy to smaller values in the presence of a finite interatomic Coulomb interaction strength ($U_{pd}$). Since we do not explicitly take into account $U_{dd}$ in our model, the renormalized $\Delta$ value should appear to be decreasing with increasing $n_h$. An earlier model (Sarma & Taraphder 1989) that includes $U_{pd}$ interactions explicitly has indeed shown that the $I_s/I_m$ is expected to decrease on hole-doping (increasing $n_h$) primarily due to the renormalization of $\Delta$ to a smaller value.

The $T_c$ in the various families of cuprates generally show a maximum at a certain $n_h$ value (figure 18). Accordingly, at a given $n_h$-value where the $T_c$ is maximum (say $n_h \approx 0.15$), increasing $I_s/I_m$ (going vertically in figure 29) is accompanied by an increase in $T_c$. Thus, $Bi_2Ca_{1-x}Ln_xSr_2Cu_2O_8$ exhibits the highest $T_c$ of around 100 K, while $La_{2-x}Sr_xCuO_4$ is associated with the lowest $T_c$; $BiPbSr_2Y_{1-x}Ca_xCu_2O_8$ falls in between, $YBa_2Cu_3O_7$ ($n_h = 0.2$ at maximum $T_c$ of *ca.* 90 K) shows an $I_s/I_m$ value (0.45) close to that of $Bi_2Ca_{1-x}Ln_xSr_2Cu_2O_8$. As we have already shown, an increase in the $I_s/I_m$ ratio is associated with an increase in $\Delta/t_{pd}$ (figure 31) which can arise from either a decrease in the $t_{pd}$ value or an increase in the $\Delta$ value. It therefore appears that $T_c$ in the different series of the cuprates are tuned by changing the value of $\Delta/t_{pd}$ *via* a change in the hole-doping level.

We show in figure 32 the static electronic polarizability, $\alpha_{xx}$, calculated (Rao *et al.* 1991*b*) for a $Cu_4O_8$ cluster as a function of $\Delta/t_{pd}$, while we show the fraction of the oxygen holes in the ground state wave function as a function of $\Delta/t_{pd}$ in figure 33. We find that $\alpha$ is rather large compared with the value for a single-band Hubbard model at $U = 4t$ of a similar size system at half-filling. The large static polarizability indicates a large dynamic polarizability as well, which would favour hole pairing in these systems. The polarizability increases with decreasing $\Delta/t_{pd}$, concomitant with the decrease of the $I_s/I_m$ ratio with decreasing $\Delta/t_{pd}$. At the same time, there is an increase in the weightage of the fraction of oxygen holes in the ground state wave function.

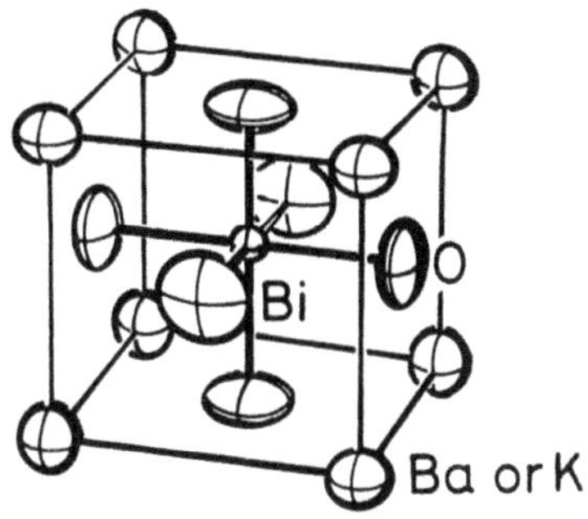


Figure 34. Structure of $Ba_{1-x}K_xBiO_3$.

Our studies of the Cu(2p) photoemission in cuprates combined with theoretical calculations enable us to come to the following conclusions.

(i) The intensity of the Cu 2p satellite, relative to the main feature, $I_s/I_m$, in the core-level spectra of the cuprates provides an experimental handle to investigate the role of important chemical bonding factors such as the Cu–O charge-transfer energy, $\Delta$, and the Cu(3d)–O(2p) hybridization strength, $t_{pd}$.

(ii) The $I_s/I_m$ ratio varies continuously with composition through the insulator–superconductor–metal régimes in the various series of cuprates. The experimental hole concentration $n_h$ and $I_s/I_m$ vary in opposite directions with composition suggesting that these two quantities are inversely related.

(iii) Theoretical calculations show that $I_s/I_m$ increases monotonically with increasing $\Delta/t_{pd}$. A relatively small $\Delta/t_{pd}$ gives rise to a large $n_h$ (and small $I_s/I_m$). The value of $n_h$ itself is nearly the same at maximum $T_c$ in all the cuprate superconductors containing the same number of $CuO_2$ sheets.

(iv) At small $\Delta/t_{pd}$ values, the electronic polarizability, $\alpha$, of the $CuO_2$ sheets will be large, favouring hole-pairing; $\alpha$ increases with increasing $n_h$ or decreasing $\Delta/t_{pd}$. Furthermore, Bose condensation of such pairs would be favoured by the two dimensionality of the $CuO_2$ sheets.

(v) Since all the cuprates containing the same number of $CuO_2$ sheets have roughly the same $n_h$ values at maximum $T_c$, it appears that they will also be associated with similar, low $\Delta/t_{pd}$ and high $\alpha$ values.

## 8. Copper-free oxide superconductors

Historically, the two oxide systems not containing copper which showed relatively high $T_c$s in the range of 13 K are $BaBi_{1-x}Pb_xO_3$ (Sleight *et al.* 1975) and $Li_{1+x}Ti_{2-x}O_4$ (Johnston *et al.* 1973). Both these oxides have mixed valent cations. In the bismuthate system, which has the perovskite structure, the nominal $Bi^{IV}$ disproportionates into $Bi^{III}$ and $Bi^{V}$ in the insulting phase ($x < 0.75$); in other words, there is a charge-density-wave (CDW) gap. There is a sharp insulator-metal transition at $x = 0.75$, when the CDW gap disappears and superconductivity manifests itself; $Bi^{IV}$ ions are delocalized in the metallic/superconducting phase (Kulkarni *et al.* 1990). In superconducting $Ba_{1-x}K_xBiO_3$ (figure 34) with a $T_c$ of *ca.* 30 K (Cava *et al.* 1988*a*), introduction of mixed valency by substitution of Ba by K competes with the CDW. These bismuthates are considered to be negative $U$ cases with $\Delta \gg U$. Although some of the features of bismuthates are similar to those of the cuprates, they show some properties which are quite different (Hinks 1990). Thus $Ba_{1-x}K_xBiO_3$ shows a large $^{18}O$ isotope effect (unlike the two-dimensional cuprates), but no static magnetic order. $Ba_{1-x}K_xBiO_3$ appears to be a superconductor in the weak to moderate coupling limit and the high $T_c$ is due to a large electron–phonon coupling constant. The mechanism of superconductivity in these materials seems to be different from that of cuprates. Other than $Ba_{1-x}K_xBiO_3$, there has not been much success in synthesizing three-dimensional oxides with high $T_c$s (see, for example, Nagarajan *et al.* 1991).

Among the other copper-free oxide superconductors, the lanthanum nickelates were suspected to show superconductivity because of the diamagnetic behaviour found in some samples of $La_{2-x}Sr_xNiO_4$ and related compounds (Nanjundaswamy *et al.* 1990). It has, however, not been possible to reproduce these findings universally and the origin of diamagnetism observed in some of the samples is not clear (Sreedhar & Rao 1990). Hopes were raised by a recent Japanese report of $T_c$s near 200 K in the Tl–Sr–V–O system, but we have found that these results are not reproducible. There was a Russian report that $LaCa_2Co_3O_y$ was superconducting with a $T_c$ of around 227 K, but we have not been able to reproduce this result as well. Clearly there must be other interesting oxides without copper which should exhibit high $T_c$s. Future investigations may bring into light such metal oxides.

## 9. Chemical applications

Potential applications of superconducting cuprates in electronics and other technologies are commonly known. These cuprates also exhibit significant catalytic activity. Thus, $YBa_2Cu_3O_{7-\delta}$ and related cuprates act as catalysts in oxidation or dehydrogenation reactions (Hansen *et al.* 1988; Halasz 1989; Mizuno *et al.* 1988). Carbon monoxide and alcohol are readily oxidized over the cuprates. $NH_3$ is oxidized to $N_2$ and $H_2O$ on these surfaces. Ammoxidation of toluene to benzonitrile has been found to occur on $YBa_2Cu_3O_7$ (Hansen *et al.* 1990).

The catalytic activity of the cuprate superconductors prompted us to examine their possible use as gas sensors. It should be noted that a good gas sensor would require not only the catalytic property of the oxide surface but also high sensitivity. For example, the electrical resistivity of the material should change sharply on contact with the gas or vapour in question. The superconducting compositions of the cuprates being metallic in the normal state, would therefore not be best suited for sensing while they may be good catalysts. With suitable compositional variation wherein the resistivity of the material is increased in order to have the right régime for sensing, it is possible to effectively use these materials. We have indeed found $La_{2-x}Sr_xCuO_4$ ($0 < x < 0.2$) and $Bi_2Ca_{1-x}Y_xSr_2Cu_2O_{8+\delta}$ ($x > 0.5$) are good sensors for alcohol and other vapours; the superconducting compositions are not. In figure 35, we show some of the recent results obtained by Grantscharova and Raju in this laboratory. We see that these materials are good sensors for alcohol and ether. Further studies on the gas sensor characteristics of superconducting materials would be worthwhile.

## 10. Concluding remarks

Structure–property relations and other aspects of the oxide superconductors that I have described so far should clearly indicate how chemistry becomes important in not only synthesizing novel materials of desired structures and properties, but also in understanding the phenomenon of high-temperature superconductivity. Our

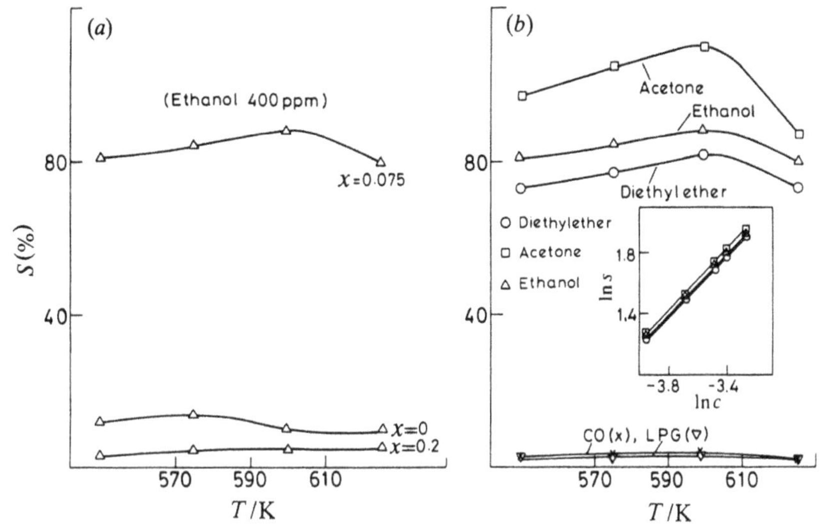


Figure 35. Gas sensing ability of $La_{2-x}Sr_xCuO_4$ ($S$ is sensitivity). Inset shows log sensitivity against log concentration plot (unpublished results of Grantscharova, Raju and Rao).

search for newer and better high $T_c$ materials is far from over. We are yet to investigate many other oxide and related systems, besides improving on the already known ones. Some of the important objectives of research in superconductivity today would be the following.

(i) To acquire better and more experimental data on the known high $T_c$ materials in order to improve our understanding of the properties.

(ii) To develop suitable models to understand the mechanism responsible for high $T_c$ superconductivity and carry out those experiments which would clearly delineate the various factors responsible for superconductivity.

(iii) To search for new materials exhibiting high $T_c$ and especially those not containing Cu (since most of present models require the d orbitals of Cu).

(iv) To improve our understanding of the chemistry of materials processing as well as the ceramic properties of the oxide superconductors.

(v) To prepare high quality films, tapes, wires, etc. with desired $J_c$ and other properties for applications.

(vi) To investigate catalytic, gas-sensing and other chemical applications, besides the well-known applications in electronics and other areas.

I thank the National Superconductivity Programme, the Department of Science and Technology, the University Grants Commission and the US National Science Foundation for support of this research. I would like to place on record my sincere appreciation of the dedicated effort made by my students and other coworkers in carrying out this research under difficult circumstances. This paper is contribution no. 771 from the Solid State and Structural Chemistry Unit.

## References

Allen, J. W. & Olson, C. G. 1990 *MRS Bull.* **15**, 34, and the references cited therein.

Bednorz, J. G. & Müller, K. A. 1986 *Z. Phys.* B **64**, 189

Bianconi, A. 1990 In *Superconductivity* (ed. S. K. Joshi, C. N. R. Rao & S. V. Subramanyam), p. 448. Singapore: World Scientific.

Caignaert, V., Hervieu, M., Wang, J., Desgardin, G., Raveau, B., Boterel, F. & Haussone, J. M. 1990 *Physica* C **170**, 139.

Cava, R. J. 1990 *Science, Wash.* **247**, 656.

Cava, R. J., Batlogg, B., Krajewski, J. J., Farrow, R., Rupp, L. W. Jr, White, A. E., Short, K., Peck, W. F. Jr & Kometani, T. 1988*a* *Nature, Lond.* **332**, 814.

Cava, R. J., Batlogg, B., Krajewski, J. J., Rupp, L. W., Schneemeyer, L. F., Siegrest, T., Van Dover, R. B., Marsh, P., Peok, W. F. Jr, Gallagher, P. K., Glarium, S. H., Marshall, J. H., Farrow, R. C., Waszezak, J. W., Hull, R. & Trevor, R. 1988*b* *Nature, Lond.* **336**, 221.

Cava, R. J., Batlogg, B., Van Dover, R. B., Krajewski, J. J., Waszezak, J. W., Fleming, R. M., Peck, W. F. Jr, Rupp, L. W. Jr, Marsh, P., James, A. C. W. P. & Schreemeyer, L. F. 1990*a* *Nature, Lond.* **345**, 602.

Cava, R. J., Hewat, A. W., Hewat, E. A., Batlogg, B., Marezio, M., Rabe, K. M., Krajewski, J. J., Peck, W. F. Jr & Rupp, L. W. Jr 1990*b* *Physica* C **165**, 419.

Cava, R. J., Krajewski, J. J., Peck, W. F. Jr, Batlogg, B., Rupp, L. W., Fleming, R. M., James, A. C. W. P. & Marsh, P. 1989 *Nature, Lond.* **338**, 328.

Chaillout, C., Chenavas, J., Cehong, S. W., Fisk, Z., Marezio, M., Morosin, B. & Schriber, J. E. 1990 *Physica* C **170**, 87.

Chakraverty, B., Sarma, D. D. & Rao, C. N. R. 1988 *Physica* C **156**, 413.

de Leeuw, D. M., Groen, W. A., Feiner, L. F. & Havinga, E. E. 1990 *Physica* C **166**, 133.

Ganguli, A. K., Nanjundaswamy, K. S. & Rao, C. N. R. 1988 *Physica* C **156**, 788.

Ganguli, A. K., Manivannan, V., Sood, A. K. & Rao, C. N. R. 1989*a* *Appl. Phys. Lett.* **55**, 2664.

Ganguli, A. K., Nanjundaswamy, K. S., Rao, C. N. R., Sequeira, A. & Rajagopal, H. 1989*b* *Mat. Res. Bull.* **24**, 883.

Ganguli, A. K., Rao, C. N. R., Sequeira, A. & Rajagopal, H. 1989*c* *Z. Phys.* B **74**, 215.

Ganguly, P. & Rao, C. N. R. 1973 *Mat. Res. Bull.* **17**, 493.

Ganguly, P. & Rao, C. N. R. 1984 *J. Solid State Chem.* **53**, 193, and the references cited therein.

Goodenough, J. B. 1971 *Prog. Solid State Chem.* **5**, 149.

Goodenough, J. B. & Manthiram, A. 1990 *J. Solid State Chem.* **88**, 115, and the references cited therein.

Halasz, I. 1989 *Appl. Catal.* **47**, L17.

Hansen, S., Otamiri, J. & Andersson, A. 1990 *Catal. Lett.* **6**, 33.

Hansen, S., Otamiri, J., Bovin, J. & Andersson, A. 1988 *Nature, Lond.* **334**, 143.

Hewat, A. W., Fischer, P., Kaldis, E., Karpinski, J., Physiecki, S. & Julek, E. 1990 *Physica* C **167**, 579.

Hewat, A. W., Hewat, E. A., Bordet, P., Capponi, J. J., Chaillout, C., Chenavas, J., Hodeau, J. L., Marezio, M., Strobel, P., Francois, M., Yoon, K., Fischer, P. & Tholence, J. L. 1989 *IBM J. Res. Dev.* **33**, 220.

Hinks, D. G. 1990 *MRS Bull.* **15**, 60, and the references cited therein.

Johnston, D. C., Prakash, H., Zachariasen, W. H. & Viswanathan, R. 1973 *Mat. Res. Bull.* **8**, 777.

Jorgensen, J. D., Dabrowski, B., Pei, S., Hinks, D. G., Soderholm, L., Morosin, B., Schriber, J. E., Venturini, E. L. & Ginley, D. S. 1988 *Phys. Rev.* B **42**, 6765.

Jorgensen, J. D., Pei, S., Lightfoot, P., Shi, H., Paulikas, A. P. & Veal, B. W. 1990 *Physica* C **167**, 571.

Joshi, S. K., Subramanyam, S. V. & Rao, C. N. R. 1990 *Superconductivity (ICSC)*. Singapore: World Scientific.

Koike, Y., Masuzawa, M., Noji, T., Sunagawa, H., Kawabe, H., Kobayashi, N. & Saito, Y. 1990 *Physica* C **170**, 130.

Karpinski, J., Kaldis, E., Jilek, E. & Bucher, B. 1988 *Nature, Lond.* **336**, 660.

Kitazawa, K. & Ishiguro, T. (eds) 1989 *Advances in superconductivity*. Tokyo: Springer-Verlag.

Kovatcheva, D., Hewat, A. W., Rangavittal, N., Manivannan, V., Guru Row, T. N. & Rao, C. N. R. 1991 *Physica* C **173**, 444.

Kulkarni, G. U., Sankar, G. & Rao, C. N. R. 1989 *Appl. Phys. Lett.* **55**, 388.

Kulkarni, G. U., Vijayakrishnan, V., Rao, G. R., Seshadri, R. & Rao, C. N. R. 1990 *Appl. Phys. Lett.* **57**, 1823, and the references cited therein.

Le Page, Y., McKinnon, W. R., Taraseon, J. M. & Barboux, P. 1989 *Phys. Rev.* B **40**, 4810.

Lightfoot, P., Richards, D. R., Dabroowski, B., Hinks, D. G., Pei, S., Marx, D. T., Mitchell, A. W., Zing, Y. & Jorgensen, J. D. 1990 *Physica* C **168**, 627.

Liu, R. S., Janes, R., Bennett, M. J. & Edwards, P. P. 1990 *Appl. Phys. Lett.* **57**, 920.

Maeda, T., Sakyuama, K., Koriyama, S. & Yamanchi, H. 1990 *ISTEC J.* **3**, 16, and the references cited therein.

Maignan, A., Martin, C., Huve, M., Provost, J., Hervieu, M., Michel, C. & Raveau, B. 1990 *Physica* C **170**, 350.

Manivannan, V., Ganguli, A. K., Subbanna, G. N. & Rao, C. N. R. 1990 *Solid State Commun.* **74**, 87.

Manivannan, V., Gopalakrishnan, J. & Rao, C. N. R. 1991 *Phys. Rev.* B **43**, 8686.

Manthiram, A. & Goodenough, J. B. 1989 *Physica* C **159**, 760.

Manthiram, A., Paranthaman, M. & Goodenough, J. B. 1990 *Physica* C **171**, 35.

Maple, M. B. 1990 *MRS Bull.* **15**, 60, and the references cited therein.

Miyatake, T., Itti, R. & Yamguchi, Y. 1990 *ISTEC J.* **3**, 23.

Mizuno, N., Yamato, M. & Misono, M. 1988 *J. chem. Soc. chem. Commun.* 887.

Nagarajan, R., Vasantacharya, N. Y., Gopalakrishnan, J. & Rao, C. N. R. 1991 *Solid State Commun.* **77**, 373.

Nanjundaswamy, K. S., Lewicki, A., Karol, Z., Gopalan, P., Metcalf, P., Honig, J. M., Rao, C. N. R. & Spalek, J. 1990 *Physica* C **166**, 361, and the references cited therein.

Nelson, D. L., Whittingham, M. S. & George, T. F. (eds) 1987 *Chemistry of high-temperature superconductors. ACS Symposium Series 351.* Washington, D.C.: American Chemical Society.

Norton, D. P., Lowndes, D. H., Sales, B. C., Budai, J. D., Chakoumakos, B. C. & Kerchner, H. R. 1991 *Phys. Rev. Lett.* **66**, 1537.

Ramakrishnan, T. V. & Rao, C. N. R. 1989 *J. Phys. Chem.* **93**, 4414, and the references cited therein.

Rao, C. N. R. 1988*a* *J. Solid State Chem.* **74**, 147, and the references cited therein.

Rao, C. N. R. (ed.) 1988*b* *Chemical and structural aspects of high-temperature superconductors.* Singapore: World Scientific.

Rao, C. N. R. 1989 *A. Rev. Phys. Chem.* **40**, 291.

Rao, C. N. R. & Raveau, B. 1989 *Acc Chem. Res.* **22**, 106, and the references cited therein.

Rao, C. N. R. & Sarma, D. D. 1991 In *Studies of high-temperature superconductors*, vol. 9. New York: Nova.

Rao, C. N. R. & Subbarao, G. V. 1970 *Physica Status Solidi* A **1**, 597.

Rao, C. N. R., Ganguli, A. K. & Vijayaraghavan R. 1989*a* *Phys. Rev.* B **40**, 2565.

Rao, C. N. R., Ganguly, P. & Mohan Ram, R. A. 1988 *J. Solid State Chem.* **72**, 14.

Rao, C. N. R., Sarma, D. D. & Rao, G. R. 1989*b* *Phase Transitions* **19**, 69, and the references cited therein.

Rao, C. N. R., Gopalakrishnan, J., Santra, A. K. & Manivannan, V. 1991*a* *Physica* C **174**, 11.

Rao, C. N. R., Ramasesha, S., Sarma, D. D. & Santra, A. K. 1991*b* *Solid State Commun.* **77**, 709.

Rao, C. N. R., Rao, G. R., Rajumon, M. K. & Sarma, D. D. 1990*c* *Phys. Rev.* B **42**, 1026.

Rao, C. N. R., Ganguly, P., Raychaudhuri, A. K., Mohan Ram, R. A. & Sreedhar, K. 1987 *Nature, Lond.* **326**, 856.

Rao, C. N. R., Nagarajan, R., Ganguli, A. K., Subbanna, G. N. & Bhat, S. V. 1990*a* *Phys. Rev.* B **42**, 6765.

Rao, C. N. R., Vijayaraghavan, R., Ganguli, A. K., Manivannan, V. & Vasanthacharya, N. Y. 1989*c* In *Studies of high-temperature superconductors*, vol. 4. New York: Nova.

Rao, C. N. R., Subbanna, G. N., Nagarajan, R., Ganguli, A. K., Ganapathi, L., Vijayaraghavan, R., Bhat, S. V. & Raju, A. R. 1990*d* *J. Solid State Chem.* **88**, 163.

Rao, C. N. R., Nagarajan, R., Vijayaraghavan, R., Vasanthacharya, N. Y., Kulkarni, G. U., Rao, G. R., Umarji, A. M., Somasundaram, P., Subbanna, G. N., Raju, A. R., Sood, A. K. & Chadrabhas, N. 1990*b* *Superconductor Sci. Technol.* **3**, 242.

Raveau, B., Michel, C. & Hervieu, B. 1990 *J. Solid State Chem.* **88**, 140.

Santra, A. K., Sarma, D. D. & Rao, C. N. R. 1991 *Phys. Rev.* B **43**, 5612.

Sarma, D. D. & Ovchinnikov, S. G. 1990 *Phys. Rev.* B **42**, 6817.

Sarma, D. D. & Rao, C. N. R. 1989 *Synthetic Metals* **33**, 131.

Sarma, D. D. & Taraphder, A. 1989 *Phys. Rev.* B **39**, 11570.

Sarma, D. D., Sen, P., Cimino, R., Carbone, C., Gudat, W., Sampathkumaran, E. V. & Das, I. 1991 *Solid State Commun.* **77**, 377.

Shafer, M. W. & Penney, T. 1990 *Eur. J. Solid State inorg. Chem.* **27**, 191.

Singh, K. K., Ganguly, P. & Rao, C. N. R. 1982 *Mat. Res. Bull.* **17**, 493.

Sleight, A. W. 1988 *Science, Wash.* **242**, 1519, and the references cited therein.

Sleight, A. W., Gillson, J. L. & Bierstedt, P. E. 1975 *Solid State Commun.* **17**, 27.

Sleight, A. W., Gopalakrishnan, J., Torardi, C. C. & Subramanian, M. A. 1989 *Phase Transitions* **19**, 149.

Sreedhar, K. & Rao, C. N. R. 1990 *Mat. Res. Bull.* **25**, 1235.

Subramanian, M. A., Torardi, C. C., Gopalakrishnan, J., Gai, P. L., Calabrese, J. C., Askew, T. R., Flippen, R. B. & Sleight, A. W. 1988 *Science, Wash.* **242**, 249.

Tarascon, J. M., Le Page, Y., McKinnon, W. R., Ramesh, R., Eibuschutz, M., Tselepis, E., Wang, E. & Hull, G. W. 1990 *Physica* C **167**, 20.

Tokura, Y., Arima, T., Takagi, H., Uchida, S., Ishigaki, T., Asamo, H., Beyers, R., Nazzal, A. I., Lacorre, P. & Torrance, J. B. 1989*a* *Nature, Lond.* **342**, 890.

Tokura, Y., Takagi, H. & Uchida, S. 1989*b* *Nature, Lond.* **337**, 345.

Tokura, Y., Torrance, J. B., Huang, T. C. & Wazzal, A. I. 1988 *Phys. Rev.* B **38**, 7156.

Torardi, C. C., Subramanian, M. A., Gopalakrishnan, J., McCarron, E., Calabrese, J. C., Morrissey, K. J., Askew, T. R., Flippen, R. B., Chowdhry, U., Sleight, A. W. & Cox, D. E. 1989 In *High-temperature superconductivity. Proc. Alabama Conf.* (ed. R. M. Metzger). New York: Gordon Breach..

Torrance, J. B. & Metzger, R. M. 1989 *Phys. Rev. Lett.* **63**, 1515.

Torrance, J. B., Tokura, Y., Nazzal, A. I., Bezinge, A., Huang, T. C. & Parkin, S. S. P. 1988 *Phys. Rev. Lett.* **61**, 1127.

Vijayakrishnan, V., Kulkarni, G. U. & Rao, C. N. R. 1990 *Mod. Phys. Lett.* B **4**, 451.

Wu, M. K., Ashburn, J. R., Torng, C. J., Hor, P. H., Meng, R. L., Gao, L. L., Huang, Z. J., Wang, Y. Q. & Chu, C. W. 1987 *Phys. Rev. Lett.* **58**, 908.

Xiang, X. D., McKernan, S., Vareka, W. A., Zettle, A., Corkill, J. L., Barbee, T. W. & Cohen, M. L. 1990 *Nature, Lond.* **348**, 145.

*Lecture delivered 25 March 1991; typescript received 15 April 1991*

Note

(a) For a review of synthesis of cuprate superconductors, see C.N.R. Rao et al, Supercond. Sci. Tech., 6, 1 (1993)

(b) For tunneling studies of $Bi_2Ca_{1-x}Y_xSr_2Cu_2O_{8+\delta}$ see: H. Srikant, A.K. Raychaudhuri, C.R.V. Rao, P. Ramasamy, H.N. Aiyer and C.N.R. Rao, Physica C, 200, 372 (1992)

(c) For analysis of thermopower data of superconducting Bi and Tl cuprates, see C.N.R. Rao, T.V. Ramakrishnan and N. Kumar, Physica C, 165, 183 (1990)

(d) For a discussion of polarizability of $CuO_2$ sheets, see S. Ramasesha and C.N.R. Rao, Phys. Rev. B, 44, 7046 (1991)

# Structure–Property Relationships in Superconducting Cuprates

**C. N. R. Rao***

*Department of Chemistry, University of Wales, Cardiff CF1 3TB, U.K., and Solid State and Structural Chemistry Unit, Indian Institute of Science, Bangalore 560 012, India*

**A. K. Ganguli**

*CSIR Centre of Excellence in Chemistry, Indian Institute of Science, Bangalore 560 012, India*

## 1 Introduction

The highest superconducting transition temperature known till 1986 was 23K and it seemed as though this barrier would not be broken.[1] The discovery of 30K superconductivity in an oxide of the La–Ba–Cu–O system by Bednorz and Müller[2] changed the picture. A variety of superconducting oxides, especially cuprates, have since been synthesized and characterized,[3—6] the highest transition temperature as of today being 155K in $HgBa_2Ca_2Cu_3O_{8+\delta}$. Studies of the various families of cuprates have shown many commonalities and unifying features.[6,7] Properties of the cuprates have been related to certain structural and electronic parameters. Although there is no simple relationship between the superconducting transition temperature and any specific structural feature of the cuprates, the various correlations help us to understand these materials better and to design newer ones. In this article, we shall briefly examine some of the significant structure–property relationships in superconducting cuprates along with the structural commonalities.

## 2 Common Structural Features

Many cuprate families have been discovered in the past seven years. The general features of the cuprates are shown schematically in Figure 1. The major families of cuprates are: (a) $La_{2-x}A_xCuO_4$ (A = alkaline earth) possessing the $K_2NiF_4$(T) structure; (b) $LnBa_2Cu_3O_{7-\delta}$ (Ln = Y or rare-earth other than Ce, Pr, and Tb) referred to as the 123 type (Figure 2) and the related $LnBa_2Cu_4O_8$ (124) and $Ln_2Ba_4Cu_7O_{15}$ (247) cuprates containing perovskite layers with $CuO_2$ sheets as well as Cu–O chains; (c) $Bi_2(Ca,Sr)_{n+1}Cu_nO_{2n+4}$ containing two BiO layers and perovskite layers with $CuO_2$ sheets (Figure 2); (d) $Tl_2A_{n+1}Cu_nO_{2n+4}$ and $TlA_{n+1}Cu_nO_{2n+3}$ (A = Ca, Ba, Sr *etc.*) containing Tl–O layers and perovskite layers with $CuO_2$ sheets (Figure 2); (e) lead-based superconducting cuprates such as $Pb_2Sr_2LnCu_3O_8$ containing $CuO_2$ sheets and CuI–O sticks; (f) Tl, Bi, and Pb cuprates containing fluorite layers and $CuO_2$ sheets; (g) mercury cuprates of the type $HgCa_{n-1}Ba_2Cu_nO_{2n+2+\delta}$; and (h) infinite layer cuprates such as $ACuO_2$ (A = Ca, Sr, Ba) and $Ln_{1-x}A_xCuO_2$ (Ln = Nd, Pr; A = Sr, Ba). All these cuprates have holes as charge carriers. The only well established superconducting cuprates with electrons as charge carriers are $Nd_{2-x}M_xCuO_4$ (M = Ce, Th) and related compounds with the T′ structure possessing square-planar $CuO_4$ units instead of the octahedra in the T structure (Figure 1).

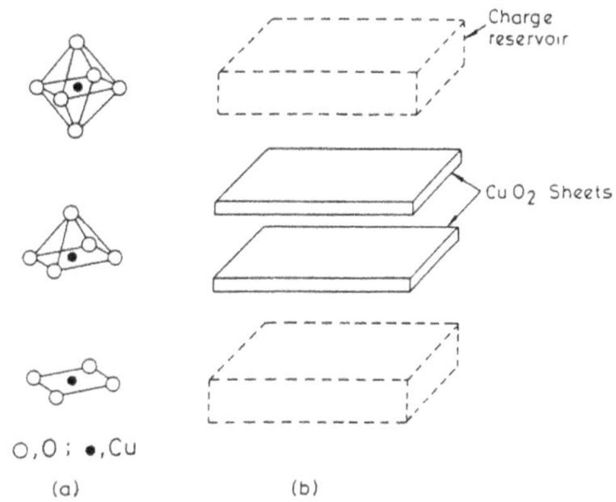


**Figure 1** (a) The three types of Cu–O polyhedra found in the superconducting cuprates (b) Schematic representation of cuprates.

* For correspondence

*A. K. Ganguli is a scientific officer at the Centre of Excellence in Chemistry at the Indian Institute of Science. He has an M.Sc. degree from the University of Delhi and a Ph.D. degree from the Indian Institute of Science and has carried out post-doctoral work at du Pont and Iowa State University.*

*C. N. R. Rao is a Professor of Chemical Science at the Indian Institute of Science, and President of the Jawaharalal Nehru Centre for Advanced Scientific Research, Bangalore, India. He is an Honorary Professor of Chemistry at the University of Wales, Cardiff. He was Commonwealth Visiting Professor at the University of Oxford and Nehru Professor at the University of Cambridge. He is a Fellow of the Royal Society, London, Foreign Associate of the U.S. National Academy of Sciences, and Foreign member of several other academies. He is a member of the Pontifical Academy of Sciences and an Honorary Fellow of the Royal Society of Chemistry. He is a recipient of the Marlow Medal of the Faraday Society and the RSC Medal for solid-state chemistry. His main research interests are in solid-state chemistry, spectroscopy, and molecular structure and surface science. He is the author of over 500 research papers and several books in solid state chemistry.*

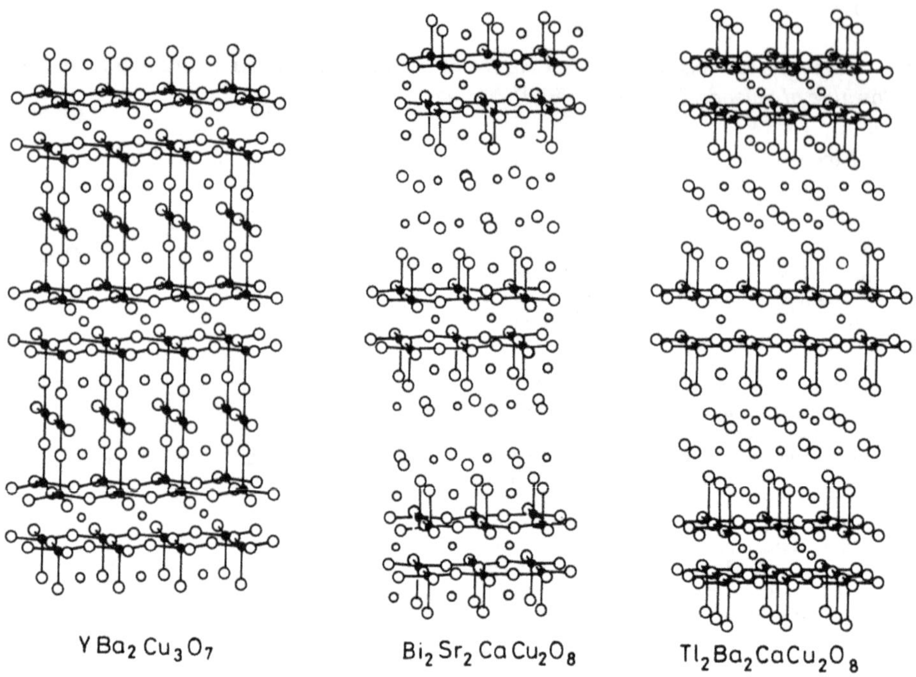


**Figure 2** Structures of $YBa_2Cu_3O_7$, $Bi_2CaSr_2Cu_2O_8$, and $Tl_2CaBa_2Cu_2O_8$. Notice the presence of $CuO_2$ sheets containing $CuO_5$ square-pyramids. Additionally, there are Cu–O chains in $YBa_2Cu_3O_7$.

Many of the cuprate families have an antiferromagnetic insulator member at one end of the composition (*e.g.* $La_2CuO_4$ in $La_{2-x}A_xCuO_4$, $YBa_2Cu_3O_6$ in $YBa_2Cu_3O_{7-\delta}$, and $TlYSr_2Cu_2O_7$ in $TlY_{1-x}Ca_xSr_2Cu_2O_{7-\delta}$. What is more interesting is that all the cuprates are at a metal–insulator boundary. Some of them undergo a metal–non-metal transition as a function of composition (*e.g* $Bi_2Ca_{1-x}Y_xSr_2Cu_2O_8$ and $TlY_{1-x}Ca_xSr_2Cu_2O_7$ with change in *x*).

The cuprates can be described on the basis of certain structural features common to many of them. For example, the structures of $La_{2-x}A_xCaO_4$, $Bi_2(Ca, Sr)_{n+1}Cu_nO_{2n+4}$ and the thallium cuprates can be considered to be intergrowths of oxygen-deficient perovskite layers, $ACuO_{3-x}$, with AO-type rock-salt layers.[7] The cuprates contain different types of Cu–O polyhedra with the hole superconductors necessarily having $CuO_5$ or $CuO_6$ units and the electron-superconducting $Nd_{2-x}M_xCuO_4$ containing only $CuO_4$ square-planar units (Figures 1 and 2). Thus, the essential feature of the cuprates is the presence of $CuO_2$ sheets with or without apical oxygens. The mobile charge carriers in the cuprates are in the $CuO_2$ sheets. All the cuprates have charge reservoirs as exemplified by the Cu–O chains in the 123 and 124 cuprates and the TlO, BiO, and HgO layers in the other cuprates. That the $CuO_2$ sheets are the seat of high-temperature superconductivity is demonstrated by the fact that intercalation of iodine between BiO layers in the bismuth cuprates does not affect the superconducting transition temperature while introduction of fluorite layers between the $CuO_2$ sheets adversely affects superconductivity. In the different series of cuprates with varying number of $CuO_2$ sheets studied hitherto, the $T_c$ reaches a maximum when $n = 3$ except in single thallium layer cuprates where the maximum is at $n = 4$. The infinite layered cuprates, where the $CuO_2$ sheets are separated by alkaline earth and other cations, show $T_c$'s in the 40—110K range.[8] Superconductivity in these materials appears to be due to the presence of Sr–O defect layers corresponding to the insertion of $Sr_3O_{2\pm x}$ blocks.[8]

Based on the interplanar Cu–Cu distances, one can classify cuprates into two categories.[9] In one category, *r*(Cu–Cu) lies between 3.0 and 3.6 Å with $T_c$'s varying between 50 and 133K and in another it is between ~6 and 12.5 Å encompassing superconductors with lower $T_c(<50K)$, except $Tl_2Ba_2CuO_6$ and $HgBa_2CuO_{4.065}$ with $T_c$'s of ~90K. In the first category with *r*(Cu–Cu) < 3.6 Å, the $T_c$ increases as the Cu–Cu distance decreases. In the 2222-type fluorite-based superconductors, there are three copper oxygen sheets, [$CuO_2$–$CuO_5$–$CuO_2$–fluorite–$CuO_2$–$CuO_5$–$CuO_2$], each block of three sheets separated by a $Ln_2O_2$ fluorite layer. The *r*(Cu–Cu) relevant to these compounds would be the distance between the $CuO_2$ sheets across the fluorite layer (~ 6.2 Å) and not the distance between two neighbouring sheets. Accordingly, these cuprates exhibit a low $T_c$(45—50K). It therefore appears that the distance between the $CuO_2$ sheets is a factor in determining the value of $T_c$, indicating that there is some interaction between the closely spaced $CuO_2$ sheets although the cuprates have quasi two-dimensional character.

Oxygen stoichiometry and ordering play a crucial role in determining the structure and properties of cuprates. The dependence of the structure and properties of $YBa_2Cu_3O_{7-\delta}$ on oxygen content has been studied in detail. Thus, $YBa_2Cu_3O_{7-\delta}$ which is orthorhombic ($c \simeq 3\,b$) with a $T_c$ of 90K when $0.0 \lesssim \delta \lesssim 0.25$, assumes another orthorhombic structure ($c \neq 3b$) when $0.3 \lesssim \delta \lesssim 0.4$ with a $T_c$ of 60K. When $\delta = 1.0$, all the oxygens in the CuO chains are depleted and the structure becomes tetragonal and the material is non-superconducting. When $\delta = 0.5$, there is an ordered arrangement of oxygen vacancies with the presence of fully oxidized ($O_7$) and fully reduced ($O_6$) chains alternately. The compositions showing 60K superconductivity are metastable and transform to a 124-type phase on heating at low temperatures.[10]

Oxygen-excess $La_2CuO_4$ is biphasic, consisting of the stoichiometric antiferromagnetic phase and an oxygen-excess superconducting phase.[11] In bismuth cuprates, excess oxygen in the BiO layers gives rise to incommensurate modulation. Modulation-free superconducting bismuth cuprates have been made[12] by replacing one $Bi^{3+}$ by $Pb^{2+}$. In $HgBa_2CuO_{4+\delta}$ oxygen excess in the Hg plane is necessary to render it superconducting.

## 3 The Relationship between $T_c$ and the Hole Concentration

As mentioned earlier, a majority of the cuprates have holes as charge carriers. These holes are created by the extra positive charge on copper (*e.g.* $Cu^{3+}$) or on oxygen (*e.g.* $O^{1-}$). The excess positive charge can be represented in terms of the formal valence of copper, which in the absence of holes will be + 2 in the $CuO_2$ sheets. In hole superconducting cuprates, it is generally around + 2.2. In electron superconductors, it would be less than + 2 as expected. The actual concentration of holes, $n_h$, in the $CuO_2$ sheets in $La_{2-x}A_xCuO_4$, $YBa_2Cu_3O_7$, and Bi cuprates is readily determined by redox titrations. In the 123 cuprates, the concentration of mobile holes in the $CuO_2$ sheets can be delineated from that in the Cu—O chains.[13] Determination of $n_h$ in thallium cuprates poses some problems, but in single Tl—O layer cuprates, chemical methods have been developed to obtain reasonable estimates.[14] Generally, $T_c$ in a given family of cuprates reaches a maximum value at an optimal value of $n_h$ as shown in Figure 3; the maximum is around $n_h \sim 0.2$ in most cuprates.[15] Notice that the points in the underdoped region in Figure 3 fall close to a straight line. Deviations occur in the overdoped region. Single layer thallium cuprates also show this behaviour. In $Tl_{1-y}Pb_yY_{1-x}Ca_xSr_2Cu_2O_7$ where the substitution of $Tl^{3+}$ by $Pb^{4+}$ has an effect opposite to that due to the substitution of $Y^{3+}$ by $Ca^{2+}$, the $T_c$ becomes a maximum at an optimal value of $(x - y)$, which is a measure of the hole concentration[16] (Figure 4). By suitably manipulating $x$ and $y$, the $T_c$ of this system can be increased from 85—90K up to 110K.

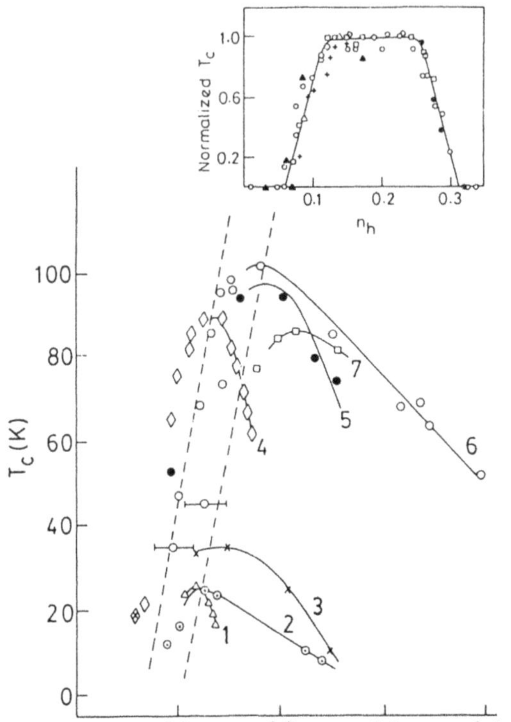


**Figure 3** Variation of $T_c$ with hole concentration $n_h$ in superconducting cuprate families (adapted from reference 15). Deviations in overdoped region are similar to those in the plots of muon spin relaxation rate ($n_h/m^*$) against $T_c$ of Uemura *et al.* (*Phys. Rev. Lett.*, 1991, **66**, 2665). 1, 2, 6, and 7, Bi cuprates; 5, 123 cuprates; 3, $La_{2-x}Sr_xCuO_4$, 4, $T_c$ cuprates. The variation of normalized $T_c$ with $n_h$ is shown in the inset (from reference 17).

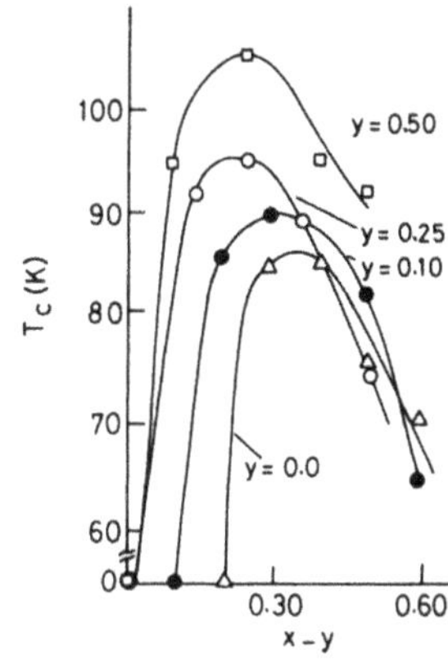


**Figure 4** Plots of $T_c$ against the effective hole concentration $(x - y)$ in $Tl_{1-y}Pb_yY_{1-x}Ca_xSr_2Cu_2O_7$ (from reference 16).

Another way of representing the variation[17] of $T_c$ with $n_h$ is to plot the reduced $T_c$ ($T_c$ observed/maximum value of $T_c$) against $n_h$ as shown in the inset of Figure 3. The ends of the plateau region in this curve correspond to insulating (possibly antiferromagnetic) and metallic regimes of the materials.

## 4 Relation between $T_c$ and the In-plane Cu—O Distance

The Cu—O bonds in the $CuO_2$ sheets involve an antibonding $\pi$ interaction and doping with holes reduces the bond distance. The in-plane Cu—O bond distance $r$(Cu—O), therefore reflects the hole concentration and a variation of $T_c$ with $r$(Cu—O) represents an alternative way of examining the $T_c$–$n_h$ relationship. In cuprates where $n_h$ cannot be determined, as for example in Tl cuprates, the $T_c$ *vs.* $r$(Cu—O) plots show maxima at an optimal distance. The value $r$(Cu—O) is around half that of the $a$-parameter in most cuprates. Whangbo *et al.*[18] find three distinct $T_c$–$r$(Cu—O) relationships depending on the cation located above and below the $CuO_2$ sheets, with each exhibiting a $T_c$ maximum at an optimal value of the distance (Figure 5). If we plot the reduced $T_c$ against $r$(Cu—O), we get the curves shown in Figure 6 where the highest $T_c$ values occur in the 1.89—1.94 Å

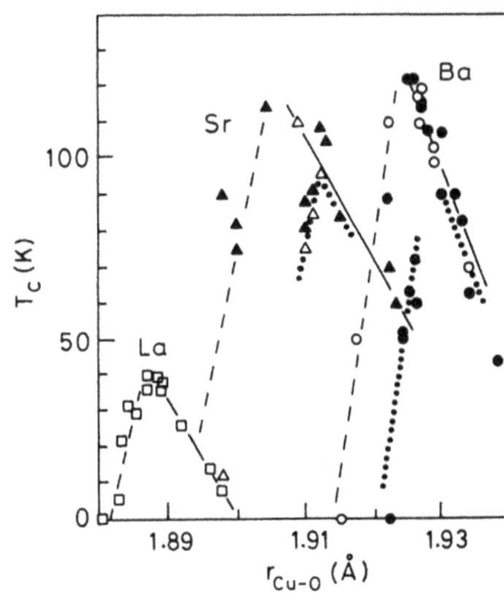


**Figure 5** Variation of $T_c$ with in-plane Cu—O distances in various families of cuprates (from reference 18).

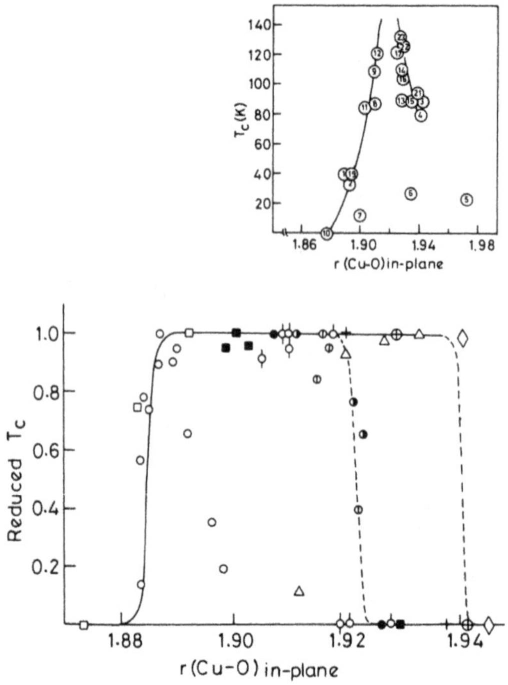


**Figure 6** Plot of reduced $T_c$ with in-plane $r$(Cu–O) distance: $La_{2-x}Sr_xCuO_4$ (open circles), $YBa_2Cu_3O_{7-\delta}$ (circles with cross), $Bi_2(Ca, Y, Sr)_3Cu_2O_8$ (half-shaded circles), $Bi_2Ca_{1-x}Y_xSr_2Cu_2O_8$ (circles with a line in the centre), modulation-free Bi-cuprates (filled squares), $TlSr_{2-x}La_xCuO_5$ (open squares), $Tl_2Ba_{2-x}Sr_xCuO_6$ (open triangles), $TlCa_{1-x}Y_xBa_2Cu_2O_7$ (crosses), $TlCa_{1-x}Ln_xSr_2Cu_2O_7$ (circles with one line on top), and $Tl_{0.5}Pb_{0.5}Ca_{1-x}Y_xSr_2Cu_2O_7$ (circles with lines above and below), $Pb_2Sr_2Y_{1-x}Ca_xCu_3O_{8+\delta}$ (filled circles), $HgBa_2CuO_{4+\delta}$ (open diamonds). Inset shows the variation of $T_c$ with the in-plane Cu–O distance where only the cuprate compositions showing the maximum $T_c$ in each family are taken into account. (1) $La_{1.85}Sr_{0.15}CuO_4$. (2) $La_{1.85}Ba_{0.15}CuO_4$. (3) $YBa_2Cu_3O_{6.91}$. (4) $YBa_2Cu_4O_8$. (5) $Nd_{1.85}Ce_{0.15}CuO_4$. (6) $Nd_{1.4}Ce_{0.2}Sr_{0.4}CuO_4$. (7) $Bi_2Sr_2CuO_6$. (8) $Bi_2CaSr_2Cu_2O_8$. (9) $Bi_2Ca_2Sr_2Cu_3O_{10}$. (10) $Tl_{0.5}Pb_{0.5}Sr_2CuO_5$. (11) $Tl_{0.5}Pb_{0.5}CaSr_2Cu_2O_7$. (12) $Tl_{0.5}Pb_{0.5}Ca_2Sr_2Cu_3O_9$. (13) $TlCaBa_2Cu_2O_7$. (14) $TlCa_2Ba_2Cu_3O_9$. (15) $Tl_2Ba_2CuO_6$. (16) $Tl_2CaBa_2Cu_2O_8$. (17) $Tl_2Ca_2Ba_2Cu_3O_{10}$. (18) $Tl_2Ca_3Ba_2Cu_4O_{12}$. (19) $TlSrLaCuO_5$. (2) $TlCa_{0.5}La_{0.5}Sr_2Cu_2O_7$. (21) $HgBa_2CuO_{4.065}$. (22) $HgCaBa_2Cu_2O_{6.22}$. (23) $HgCa_2Ba_2Cu_3O_{8.41}$.

range. When $r$(Cu–O) < 1.88 Å, the material is metallic; those with r(Cu–O) > 1.94 Å are certainly insulating, but there are different insulating boundaries for the different cation families, somewhat like in Figure 5. However, if we consider only the maximum $T_c$ value in each cuprate family and the corresponding $r$(Cu–O), we get the curve shown in the inset of Figure 6 which peaks at $r \approx 1.92$ Å. $Bi_2Sr_2CuO_6$ with a $T_c$ of 12K would not fall on the curve, but $Bi_2Sr_{2-x}La_xCuO_{6+\delta}$ with a $T_c$ of 30K would.

## 5 The Relationship between $T_c$ and the Apical Cu–O Distance

All the cuprates which are hole superconductors have apical oxygens which act as the link between the charge reservoirs and the $CuO_2$ sheets. (Note that electron superconductors such as $Nd_{2-x}Ce_xCuO_4$ contain only $CuO_4$ units without apical oxygens.) In $YBa_2Cu_3O_{7-\delta}$, the $T_c$–$r$(Cu2–O1) relationship (Figures 7a and b) mirrors the $T_c$–$\delta$ relationship.[19] In $YBa_2Cu_4O_8$, the $T_c$ increases with pressure from 80K to 90K, as the apical Cu–O distance decreases[20] (Figure 7c). We have sought to find relationships between $T_c$ and apical Cu–O distance in

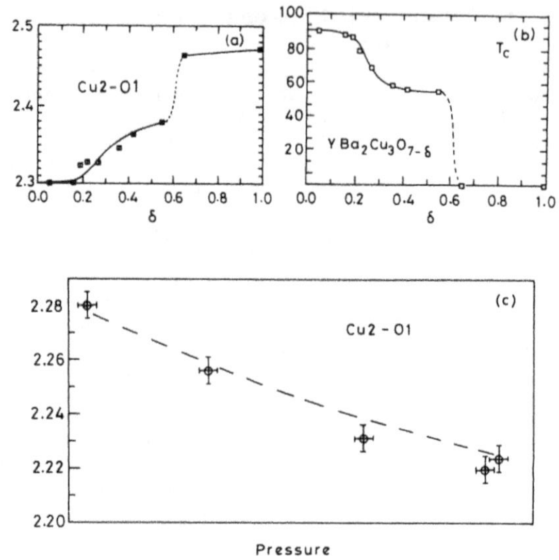


**Figure 7** Variation of (a) the apical Cu–O distance with $\delta$ in $YBa_2Cu_3O_{7-\delta}$ (b) $T_c$ with $\delta$ and (c) variation of the Cu–O apical distance with pressure in $YBa_2Cu_4O_8$ as shown by the points (the dashed line is obtained by the oxidation of $YBa_2Cu_3O_{7-\delta}$) (from references 19 and 20).

different cuprate families (Figure 8). Within a series of cuprates with varying number of $CuO_2$ sheets, (*e.g.*, $TlCa_{n-1}Ba_2Cu_nO_{2n+3}$), the apical (Cu–O) distance decreases with the increase in $n$ while $T_c$ increases linearly with the decrease in the apical distance. The slope of the $T_c$ *vs.* the apical Cu–O distance plot is nearly the same in $TlCa_{n-1}Ba_2Cu_nO_{2n+3}$, $Tl_{0.5}Pb_{0.5}Ca_{n-1}Sr_2Cu_nO_{2n+3}$, and $HgCa_{n-1}Ba_2Cu_nO_{2n+3}$, all of them having a single rock-salt layer (TlO, $Tl_{0.5}Pb_{0.5}O$, or $HgO_\delta$). The $Tl_2Ca_{n-1}Ba_2Cu_nO_{2n+4}$ family also shows increasing $T_c$ with the decrease in the apical Cu–O distance, but with a different slope. The mercury-based superconductors have larger apical distances compared to the other cuprates and they also show a large pressure dependence of $T_c$. Interestingly, if we consider the maximum $T_c$ points at the top of the plots for the different

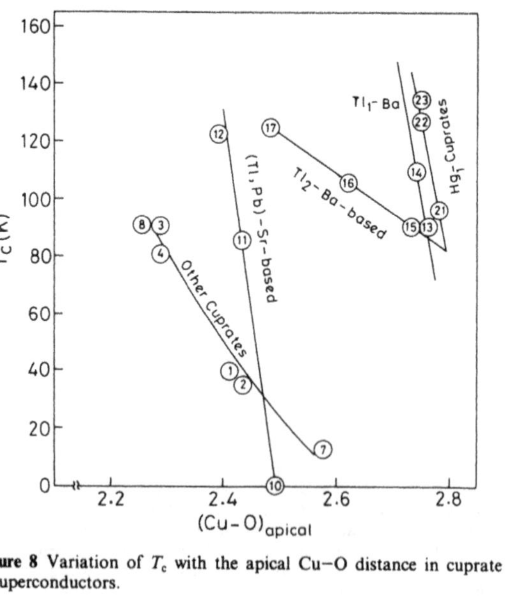


**Figure 8** Variation of $T_c$ with the apical Cu–O distance in cuprate superconductors.

groups of cuprates in Figure 8, we see that the $T_c$ increases with the increase in the apical Cu–O distance.

## 6 Covalency of the Charge Reservoir

All the superconducting cuprates have charge reservoirs and any damage to these reservoirs adversely affects superconductivity. The nature of the charge reservoir determines the carrier concentration and the ease of charge-transfer to the $CuO_2$ sheets. Covalent charge reservoirs can redistribute charge effectively through the apical oxygen of the $CuO_5$ square pyramids giving rise to high $T_c$'s. Ionic charge reservoirs, on the other hand, would be less flexible with regard to the charge states and do not favour high $T_c$'s. Structural mismatch as well as disorder in the reservoirs also adversely affect the superconducting properties. The covalency of the Hg–O bond could be related to the high $T_c$ of Hg cuprates.

The effect of covalency of the charge reservoir is clearly seen in $Tl_{0.5}Pb_{0.5}Sr_2Y_{1-x}Ca_xCu_2O_7$ which shows a $T_c$ of 110K at an optimal $x$ value, the material being an insulator when $x = 0.0$. The $a$ parameter decreases with the increase in $x$, because the population of the antibonding Cu $3d_{x^2-y^2}$ orbitals decreases with the increase in $x$, causing a strengthening of the Cu–O bond. The puckering of the $CuO_2$ sheets decreases with increasing hole concentration. The displacement of the apical oxygen is around 0.06 Å when $x = 1.0$ and 0.20 Å when $x = 0.0$. An increase in Y content (increased electron population), however, increases the puckering and pulls the apical oxygen away from the base of the pyramid.

## 7 The Relationship between $T_c$ and Madelung Potentials

The role of the Madelung site potential in the hole conductivity of the cuprate superconductors was first pointed out by Torrance and Metzger.[21] Two classes of cuprates can be delineated depending on the value of $\Delta V_M$ which is the difference in Madelung site potential for a hole on a Cu site and that on an oxygen site. Those with high $\Delta V_M$ ($\geq$ 47eV) are metallic and superconducting; those with lower $\Delta V_M$ are semiconducting with localized holes. It is possible to define a term $\Delta V_A$ which is the difference in the Madelung site potentials for a hole between the apex and the in-plane oxygen atoms and provides a measure of the position of the energy level of the $p_z$-orbital on the apical oxygens.[22] When the maximum values of $T_c$ of hole-doped superconductors are plotted against $\Delta V_A$ (Figure 9), one finds that nearly all the cuprates are located on a curve (with some width), the cuprates with large $\Delta V_A$ exhibiting high $T_c$'s. It appears that the energy level of the apical oxygen plays a significant role in the electronic states of the doped holes,

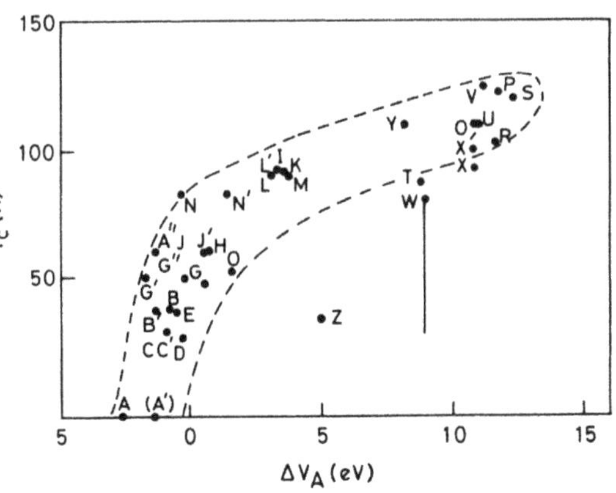


**Figure 9** Variation of $T_c$ with $\Delta V_A$. See reference 22 for the explanation of the letter symbols.

thereby affecting the $T_c$. The correlation probably owes its origin to the stability of local singlet states made up of two holes in the Cu $3d_{x^2-y^2}$ and O $2p$ orbitals in the $CuO_2$ sheet. The local singlet is well defined and stable when the energy level of the apical oxygen atom is sufficiently high. A comparison of the correlations of $T_c$ with $\Delta V_A$ and $\Delta V_M$ indicates that $\Delta V_A$ scales better with $T_c$.

It is instructive to correlate $T_c$ simultaneously with $\Delta V_M$ and the in-plane Cu–O bond length, $d_p$, in the $CuO_2$ planes. The maximum $T_c$ for each cuprate is shown in the $d_p$ vs. $\Delta V_M$ plot[23] in Figure 10. The data points are confined to a narrow strip running from the top left (high $d_p$ and low $\Delta V_M$) to the bottom right corner (low $d_p$ and high $\Delta V_M$). The $T_c$ value increases as we go from right to left (larger to smaller value of $\Delta V_M$) or from top to bottom (from higher $d_p$ to lower $d_p$). An important observation is that $T_c$ changes to a small extent as one goes from the top left to bottom right ($d_p$ decreasing and $\Delta V_M$ increasing) while $T_c$ increases drastically when traversing from the top right to bottom left of the map (both $d_p$ and $\Delta V_M$ decreasing). Clearly, the $T_c$ is governed by both $\Delta V_M$ and $d_p$. It appears that the difference in $T_c$ is a result of the difference in internal stress of the crystal (high $T_c$ when the $CuO_2$ planes are under compression and low $T_c$ when they are strained).

## 8 The Relationship between Bond Valence Sums and $T_c$

The bond-valence sum is a measure of the total charge on an atom in a structure. Its value changes with oxygen doping, cation substitution, or applied pressure indicating the occurrence of charge transfer within the structure. One defines, $V_- = 2 + V_{Cu2} - V_{O2} - V_{O3}$ where $V_-$ is the total excesss charge in the planes (O2 and O3 are the oxygens in the plane) and $V_+ = 6 - V_{Cu2} - V_{O2} - V_{O3}$. The $T_c$ vs. $V_-$ plot for $YBa_2Cu_3O_{7-\delta}$ is similar to the $T_c$ vs. $\delta$ plot; the plot of $T_c$ against $V_+$ is linear[24] (Figure 11).

## 9 The Importance of the Cu–O Charge-Transfer Energy

One of the unique features of the cuprates is the relatively small Cu–O charge-transfer energy. It is therefore of significance to relate this property with superconductivity. $X$-Ray photoemission spectra of cuprates in the Cu $2p_{3/2}$ region show a characteristic two-peak feature. The peak around 933 eV (main peak) is due to a final state with primarily $3d^{10}$ character, while the peak of weaker intensity at about 941 eV binding energy is mainly due to a $3d^9$ final state (satellite). Model calculations of the Cu $2p$ core-level photoemission within a $CuO_4$ square-planar cluster show that the $I_s/I_m$ ratio (relative intensity of the satellite to the main peak) is related to $\Delta/t_{pd}$ where $\Delta$ is the charge-transfer excitation energy and $t_{pd}$ is the hybridization strength between the Cu $3d$ and O $2p$ orbitals. Thus, any relationship between the experimentally observed $I_s/I_m$ ratio with $n_h$ would suggest a link between $n_h$ ($T_c$) and the $\Delta/t_{pd}$ ratio. It is indeed found that the $I_s/I_m$ ratio is related to the hole concentration $n_h$ in many of the superconducting cuprates.[25]

In Figure 12(a), we show the relative intensity of the satellite ($I_s/I_m$) as a function of $x$ in $La_{2-x}Sr_xCuO_4$. In the same plot, we also show the variation of the experimentally determined hole concentration with $x$ in this series. The inset of Figure 12(a) shows the dependence of $T_c$ on $n_h$ exhibiting a maximum around $n_h = 0.15$. We see that the value of $I_s/I_m$ decreases markedly with increasing $x$ until about 0.3, after which it changes slope. What is significant is that $I_s/I_m$ decreases as the hole concentration increases. In Figure 12(b) we show the variation of $n_h$ and $I_s/I_m$ with $x$ in $BiPbSr_2Y_{1-x}Cu_2O_8$. In the inset we show the variation of $T_c$ with $n_h$. Here also we see that the $I_s/I_m$ ratio decreases as the hole concentration increases. Figure 13 shows how in three series of high $T_c$ cuprates, the $I_s/I_m$ ratio decreases monotonically with the increase in magnitude of hole doping. The

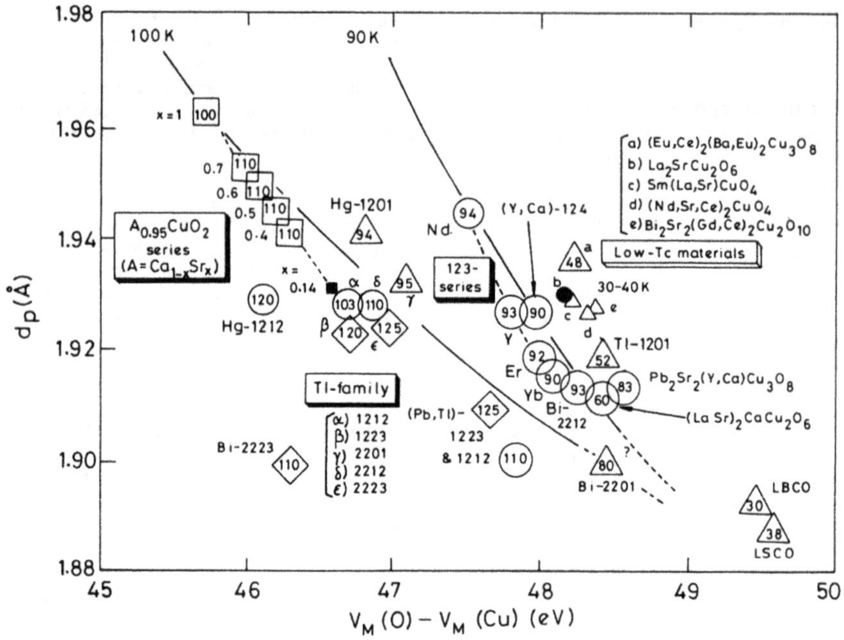


**Figure 10** Plot of $d_p$ *vs.* $\Delta V_M$ showing region of high $T_c$ and low $T_c$ superconductors (from reference 23).

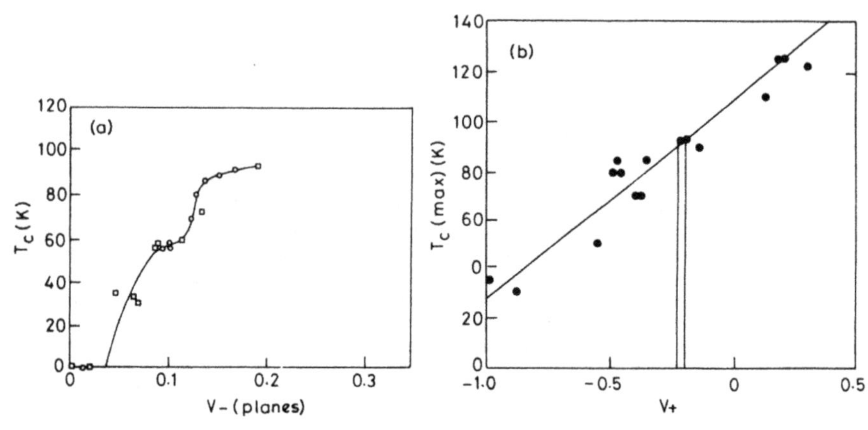


**Figure 11** Variation of (a) $T_c$ with $V_-$ (see text) and (b) $T_c$ with $V_+$ for $YBa_2Cu_3O_{7-\delta}$ (from reference 24).

variation is smooth across the insulator–superconductor–metal boundaries. Calculations show that the $I_s/I_m$ ratio increases with the $\Delta/t_{pd}$ ratio in the cuprates. Accordingly, at a given $n_h$ value where the $T_c$ is maximum, increasing $I_s/I_m$ is accompanied by an increase in $T_c$. Thus, $Bi_2Ca_{1-x}Ln_xSr_2Cu_2O_{8+\delta}$ exhibits the highest $T_c$ of around 100K while $La_{2-x}Sr_xCuO_4$ has the lowest $T_c$ value. The observation of a maximum $T_c$ around an optimal value of $n_h$ or Cu–O distance can therefore be traced to the optimal $\Delta/t_{pd}$ value or of the Cu–O charge-transfer energy.

## 10 Concluding Remarks

We have discussed several interesting correlations between the superconducting transition temperature and some of the crucial structural and electronic parameters. These correlations along with the structural commonalities in the cuprates show how superconductivity is intimately connected with the structural chemistry of these materials. They may also suggest many other relationships, some of which may be even more significant.

*Acknowledgement.* The authors thank the Science Office of the European Union, the Department of Science and Technology, and the University Grants Commission for support of this research.

## 7 References

1 C. N. R. Rao, K. J. Rao, and J. Gopalakrishnan, *Ann. Rep. Prog. Chem., Sect C*, 1985, **82**, 193.
2 J. Bednorz and K. A. Müller, *Z. Phys.*, 1986, **B64**, 189.
3 B. Raveau, C. Michel, M. Hervieu, and D. Groult, 'Crystal Chemistry of High $T_c$ Superconducting Copper Oxides', Springer-Verlag, Berlin, 1991.
4 'Chemistry of High-Temperature Superconductors', ed. C. N. R. Rao, World Scientific, Singapore, 1991.

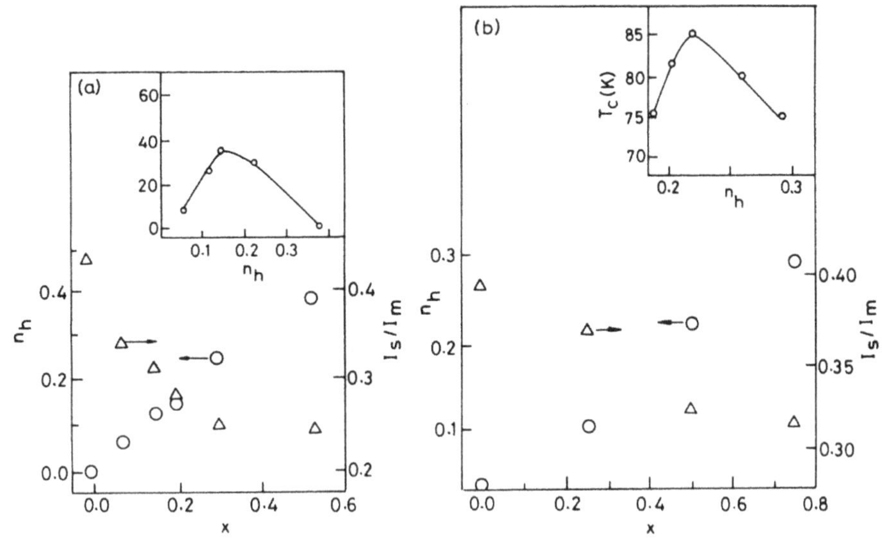


**Figure 12** (a) Variation of $I_S/I_M$ ratio with $x$ in $La_{2-x}Sr_xCuO_4$. (b) Also shown is the variation of $n_h$ with $x$ and $I_S/I_M$ and $n_h$ with $x$ in $BiPbSr_2Y_{1-x}Ca_xCu_2O_8$ (from reference 25).

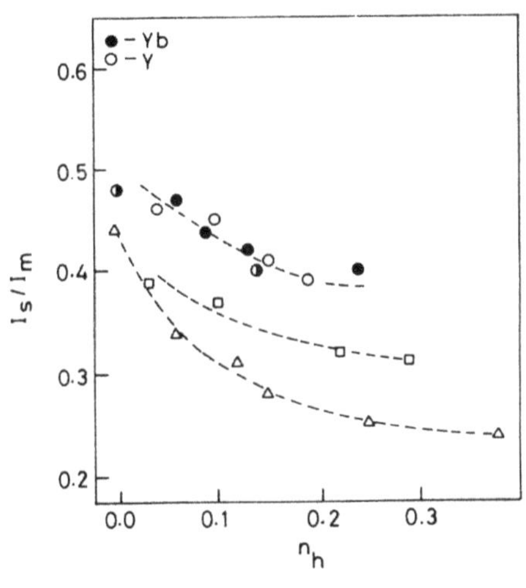


**Figure 13** Variation of the $I_S/I_M$ ratio with hole concentration ($n_h$): $La_{2-x}Sr_xCuO_4$ (triangles), $BiPbSr_2Y_{1-x}Ca_xCu_2O_8$ (squares), and $Bi_2Ca_{1-x}L_xSr_2CuO_2O_8$ (circles) (from reference 25).


5 M. Nunez-Regueiro, J. L. Tholence, E. V. Antipov, J. J. Capponi, and M. Marezio, *Science*, 1993, **262**, 976.

6 A. R. Armstrong and P. P. Edwards, *Ann. Rep. Prog. Chem., Sect. C*, 1991, **88**, 259.

7 C. N. R. Rao, *Philos. Trans. R. Soc. London*, 1991, **A336**, 595, 106.

8 H. Zhang, Y. Y. Wang, V. P. Dravid, L. D. Marks, P. D. Han, D. A. Payne, P. G. Radaelli, and J. D. Jorgensen, *Nature*, 1994, **370**, 352, and references therein.

9 H. Nobumasa, K. Shimizu, and T. Kawai, *Physica C*, 1990, **167**, 515.

10 R. Nagarajan and C. N. R. Rao, *J. Solid State Chem.*, 1993, **103**, 533.

11 J. D. Jorgensen, *Physics Today*, 1991, **44**, 34

12 V. Manivannan, J. Gopalakrishnan, and C. N. R. Rao, *Phys. Rev. B.* 1991, **43**, 8686.

13 R. Nagarajan and C. N. R. Rao, *J. Mater. Chem.*, 1993, **3**, 969.

14 C. N. R. Rao in 'Thallium-based High Temperature Superconductors , ed. A. M. Hermann and J. V. Yakhmi, Marcel Dekker, New York, 1994.

15 C. N. R. Rao, J. Gopalakrishnan, A. K. Santra, and V. Manivannan, *Physica C*, 1991, **174**, 11.

16 R. Vijayaraghavan, N. Rangavittal, G. U. Kulkarni, E. Grantscharova, T. N. Guru Row, and C. N. R. Rao, *Physica C*, 1991, **179**, 183.

17 H. Zhang and H. Sato, *Phys. Rev. Lett.*, 1993, **70**, 1697.

18 M. H. Whangbo, D. B. Kang, and C. C. Torardi, *Physica C*, 1989, **158**, 371.

19 R. J. Cava, A. W. Hewat, E. A. Hewat, B. Batlogg, M. Marezio, K. B. Rabe, J. J. Krajewski, W. F. Peck, and L. W. Rupp, *Physica C*, 1990, **165**, 419.

20 R. J. Nelmses, E. Loveday, E. Kaldis, and J. Karpinski, *Physica C*, 1992, **172**, 311.

21 J. B. Torrance and R. M. Metzger, *Phys. Rev. Lett.*, 1989, **63**, 1515.

22 Y. Ohta, T. Tohyama, and S. Maekawa, *Physica C*, 1990, **167**, 515.

23 M. Muroi, *Physica C*, 1994, **219**, 129.

24 J. L. Tallon and G. V. M. Williams, *J. Less Common Metals*, 1990, **164—165**, 60.

25 A. K. Santra, D. D. Sarma, and C. N. R. Rao, *Phys. Rev. B*, 1991, **43**, 5612.

MINIREVIEWS

# Stripes and Superconductivity in Cuprates – Is there a Connection?

N. Kumar[a] and C. N. R. Rao*[b]

KEYWORDS:

cuprates · scanning probe microscopy · soft X-ray scattering · stripes · superconductors

Charge stripes can be traced back to the 1990s or even earlier.[1] Stripes are a special case of the general phenomenon of electronic phase separation in strongly correlated transition metal oxide systems wherein regions with a high concentration of mobile carriers are separated from those with a low concentration due to Coulomb interactions. The carrier-rich regions are generally metallic and/or ferromagnetic, while the carrier-poor regions are antiferromagnetic and insulating. Such phase separation manifests itself in the form of metallic droplets in an antiferromagnetic insulating matrix or self-organization, for example, in one dimension as stripes.[1, 2] The phase-separated regions tend to be of nanometric dimensions.

Over the years, experimental evidence[1–12] has been steadily mounting in support of the theoretical[13–19] proposition that the cuprate superconductors may be intrinsically inhomogeneous in the distribution of their spins and charges in the $CuO_2$ sheets, and that this inhomogeneity may hold a clue to the mechanism of high-$T_c$ (high critical temperature) superconductivity in these materials. The growing realization that stripes matter has enlivened the debate on high temperature superconductivity in these ceramic materials, and emphasizes the incompleteness in our understanding of the phenomenon discovered in 1986 by Bednorz and Müller.[20]

Stripes have been reported in at least three families of hole-doped high-$T_c$ cuprates, all of which contain the $CuO_2$ sheets. These are the single-layered cation-doped 214 compound LSCO ($La_{2-x}Sr_xCuO_4$) or the oxygenated (anion-doped) LCO214 ($La_2CuO_{4+x}$), the 123 compound YBCO ($YBa_2Cu_3O_{7-x}$), which contains CuO chains in addition to the $CuO_2$ sheets, and the two-layered BSCCO-2212 ($Bi_2Sr_2CaCu_2O_{8+x}$) with two $CuO_2$ layers per unit cell. The strongest case for the stripes so far has been made for the model system NdLSCO ($La_{1.6-x}Nd_{0.4}Sr_xCuO_4$),[2] in which the stripes are structurally pinned, nearly static. The commonality of the all-important $CuO_2$ sheets, and the associated low-dimensional physics,[21–23] which is known to favor solitons and the solitonic doping, has been the compelling reason to believe that stripes may be generic to the layered cuprates—at least in the underdoped region of the phase diagram, which is shown schematically in Figure 1.

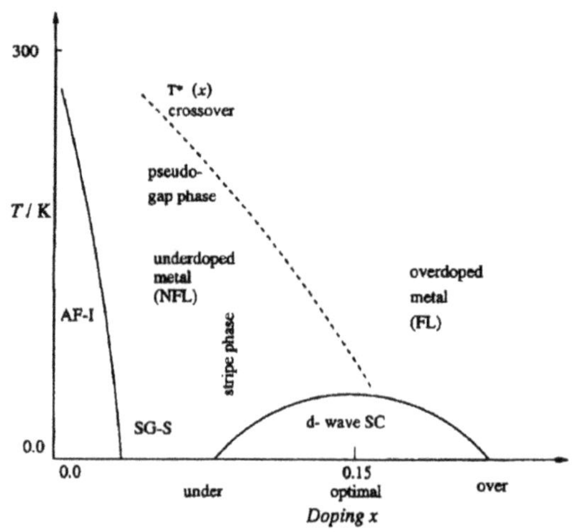


*Figure 1. Schematic temperature (T) – doping (x) phase diagram typical of hole-doped high-$T_c$ layered cuprate superconductors. AF-I denotes antiferromagnetically ordered insulator; SG-S, the spin-glass semiconductor; d-wave SC, the $d_{x^2-y^2}$ ordered superconductor; and T*(x), the crossover temperature to the pseudo-gapped, abnormal metallic underdoped phase. Stripe phase occupies an underdoped pseudogapped region, possibly extending into the superconducting phase. The overdoped region is a normal metal.*

Charge stripes have also been found in a few other transition metal oxides and were first reported in the isostructural hole-doped nickelates, $La_{2-x}Sr_xNiO_4$ and $La_2NiO_{4+x}$ which contain modulated $NiO_2$ sheets.[24] But there is a striking difference between the nickelates and the cuprates in that the stripe phase in the former is insulating, much less superconducting, unlike in

[a] Prof. Dr. N. Kumar
Raman Research Institute
Bangalore 560080 (India)

[b] Prof. Dr. C. N. R. Rao
CSIR Center of Excellence in Chemistry and Chemistry & Physics of Materials Unit
Jawaharlal Nehru Center for Advanced Scientific Research
Bangalore 560064 (India)
Fax: (+91) 80846 2760
E-mail: cnrrao@jncasr.ac.in

 DOI: 10.1002/cphc.200200601  

CHEMPHYSCHEM

the cuprates where it is conducting or superconducting, thus providing a case for a study in contrast between two related layered oxides systems with the spin-1/2 containing $Cu^{2+}$:3d$^9$ and the spin-1 containing $Ni^{2+}$:3d$^8$. Although cuprates are superconducting, while the nickelates are insulators, the characteristics of the stripes in the two materials are comparable in that for example, the stripes in both the systems are fluctuating, strongly correlated fluids. Interestingly, stripes have been reported in the rare earth manganates of the general formula, $Ln_{1-x}A_xMnO_3$ (Ln = rare earth metal, A = alkaline earth metal) as well.[25] Here, ferromagnetic metallic stripes are bounded by antiferromagnetic insulating stripes.

While the existence of charged stripes is not much in doubt, their relevance to high-temperature superconductivity (HTSC) is by no means established, particularly in the light of the recent anomalous X-ray scattering measurements on $La_2CuO_{4+\delta}$[26] (see, however, refs.[27, 28]). The presence of stripes raises many questions. Thus, we may ask whether they constitute a truly thermodynamic phase. Does the stripe order compete with superconductivity in the layered cuprates? Does it promote HTSC? Does it coexist with superconductivity—trivially, spatially separated from it, or nontrivially, overlapping with it? Or, as is likely, do the stripe order and the HTSC have a common underlying physics? How can the charged stripes, with 1D topology, subtend a 2D electrical conductivity, let alone superconductivity? Or, is it possible that the 1D stripes merely provide a solitonic route to hole doping into the 2D $CuO_2$ sheets at low levels of doping, as in its low-dimensional analogs in which point defects (zero-dimensional domain walls) provide solitonic doping, for example, polyacetylene.[21–23] Herein, we address some of these and related issues, including the provenance of stripes.

Figure 2 depicts schematically the geometry and the spin-charge order for an in-plane stripe as derived from low-energy neutron diffraction/scattering in a typical hole-doped cuprate superconductor,for example, $La_{2-x}Sr_xCuO_4$ (LSCO), or NdLSCO ($La_{2x-y}Nd_ySr_xCuO_4$) for $x = 1/8$ per $CuO_2$ unit.[2] The charge carriers (holes), which are introduced by doping, are segregated into linear (1D) domain walls—the charged stripes—that are barely a few atoms wide and separate the antiferromagnetically (AFM) ordered planar (2D) domains in all the $CuO_2$ sheets. Furthermore, the domain walls are antiphase, so that the AFM domain order is phase shifted by $\pi$ across the charged domain wall. This shift has the consequence of making the spatial period of the magnetic stripes twice that of the charge stripes. This factor of two follows from a general consideration of the ground state of the interacting spin–hole system. Such a real-space static stripe modulation gives the incommensurate elastic magnetic peaks, which are positioned in the reciprocal space at $(2\pi/a)$. The main Bragg diffraction peak for the parent AFM order is (0.5, 0.5). The incommensuration $\delta$ is a function of doping $x$ (at low doping, $x \approx \delta$) and is related to the real interstripe spacing ($= a/2\delta$). The tetragonal zone indexing $(h,k)$ is appropriate to the planar geometry of the $CuO_2$ sheets.

In simplistic terms, the doped holes in the cuprates are mostly on the oxygen in the $CuO_2$ sheets creating oxygen 2p-holes, while the spins reside on the copper (the $Cu^{2+}$:3d$^9$-spins). In Figure 2 we have only shown the copper atoms. The charge stripes are aligned parallel to the Cu–O bonds, but may be centered on the sites (on the Cu–O–Cu legs), on the bonds (between two Cu–O–Cu legs), or in between. The stripe phase is singly modulated, but the stripes in the adjacent $CuO_2$ planes are at 90° because of the symmetry-lowering lattice distortions. Hence, no incommensurate peaks occur at $(2\pi/a)$ $(1/2 + \delta, 1/2 + \delta)$, or $(2\pi/a)$ $(1/2 - \delta, 1/2 - \delta)$. Figure 2 assumes the stripes to be static; this is the case for low levels of doping (for the underdoped samples), where one observes narrow diffraction (elastic) peaks. In the optimally doped superconducting samples, on the other hand, the stripes fluctuate, giving relatively broad inelastic peaks, still positioned incommensurately about the AFM diffraction peaks. In the overdoped samples the stripes simply fade away. We may recall that the strongest evidence for the stripes comes from NdLSCO ($La_{2-x-y}Nd_ySr_xCuO_4$, with $x = 0.12$, $y = 0.4$),[2] in which the stripes are immobilized by commensuration pinning to the lattice distortion of the low-temperature tetragonal (LTT) phase. It is the modulation of the lattice distortion induced by the stripe charge order that is picked up in neutron scattering as an indirect signature of the charged stripes.

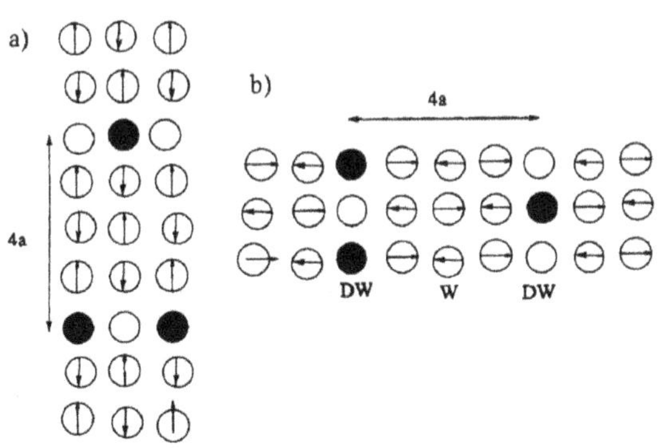


*Figure 2. A schematic diagram of the stripe showing spin-charge order in $CuO_2$ plane for hole doping $x = 1/8$. AFM ordered planar domains separated by charged linear domain walls (DW)—the charge stripes. The domain walls are antiphase: with AFM order phase shifted by $\pi$ across it, making the spacing of magnetic stripe (8a) twice that of the charged stripes (4a). The open circle with arrow denotes a spin 1/2 on copper, and the filled circle denotes a hole nominally on copper (actually on oxygen, not shown) with one hole per two copper atoms in the linear DW. The stripes in the adjacent planes (a) and (b) are orthogonal to each other.*

Let us briefly examine the nature of the stripes in cuprates. Several of the reciprocal-space bulk scattering probes and real-space surface-imaging tools have been used to study the stripes. The two methods are somewhat complementary. The scattering probes can detect fluctuating stripes as well, while surface imaging can only probe the static stripes at atomic resolution. The geometry of the stripe order in the bulk is best seen by neutron diffraction. If the stripes can be immobilized through pinning due to impurity atoms and structural commensuration as in NdLSCO, the applicability of elastic scattering allows a more thorough investigation. In order to probe spatially modulated dynamical correlations, low-energy ($\approx$few meV) inelastic neutron scattering is employed. The problem, however, is that neutrons couple directly to the electron spin and to the positions of the nuclei, whose forced displacement in response to any charge-ordering gives only an indirect evidence for charge order. Thus, the incommensurate, split superlattice peaks at $(2\pi/a)$ $(2\pm 2\delta)$ and their near absence at $(2\pi/a)$ $(2\pm 2\delta)$ constitute indirect evidence for the charged stripes (with a spacing half of that for the magnetic stripes). For direct evidence, one must turn to X-rays that directly couple to the charge. X-rays, however, couple nonspecifically to all the charges, including those of the core electrons that outnumber the mobile charge carriers (doped holes) by a factor of $\approx 500$ to 1, for example, in optimally doped LSCO. The normal X-ray diffraction cannot resolve this small proportion. This problem has been overcome recently by Abbamonte et al.[26] by the use of multiwavelength anomalous soft X-ray scattering, wherein the elastic (or the low-energy inelastic) scattering is resonantly and selectively enhanced many-fold (by over a factor of $\approx 10^3$) by operating close to an absorption pre-edge (pre-edge of the OK shell) because of the near-vanishing of the energy denominator that is associated with the virtual excitation of the O (1s) electron to the Fermi level vacated due to hole doping. Even this enhanced sensitivity and specificity have failed to detect incommensurate X-ray diffraction peaks corresponding to a charged-stripe modulation in $La_2CuO_{4+x}$ films of atomic perfection deposited on $SrTiO_3$ substrates. But, this new technique clearly needs to be tested out on samples proven to give the incommensurate magnetic peaks in neutron scattering. Also, Abbamonte et al.[26] studied the sample at a high temperature ($> T_c$), and the stripes may well appear below $T_c$. The issue is, therefore, not fully resolved.

Complementary to the soft X-ray scattering probe, which is a bulk probe of mesoscopic resolution, is the real-space surface imaging by scanning tunneling microscopy (STM) at atomic resolution. STM measures the differential tunneling conductance, which is in turn proportional to the local density of states (LDOS), $\alpha \sum_k |\varphi_k(r)|^2 \delta[E - e(k)]$, at the position $(r)$ of the STM tip and at the energy $E$ corresponding to the bias voltage for the charge-carrier injection. An earlier study of a single crystal of BSCCO 2212, cleaved along the BiO planes, reported charge-stripe modulations for optimal doping in the superconducting phase.[27] This result is subject to some doubt since STM studies find that the spatial modulation wavelength varies with the STM bias voltage, which suggests an extrinsic interpretation involving interference of the electron waves, which are scattered from defects/impurities—akin to Friedel oscillations or more accurately, von Laue-theoretic oscillations. Stripes should have given intrinsic voltage-independent spatial oscillations. It is, however, possible that the latter are masked by the extrinsic effect. It is important to note that, over and above the modulations due to any possible stripe order, there are microscopic inhomogeneities arising from local doping concentration (LDC) effects, because of the poor screening of the dopant potentials in these doped Mott insulators. The LDC effects have been isolated[29] by Fourier-filtering out of the stripe modulations, and remarkable correlations of, for example, the LDOS, the superconducting gap, and the doping, have been found; this is indeed intriguing.

The topological issue of one-dimensionality of the charge stripes, namely, that the mobile charge carriers must move along the linear tracks, defined laterally by the Mott-insulating AFM domains on the sides, is best probed through studies of charge motion in real space (for example, Hall effect measurements) and in the reciprocal (momentum) space (for example, through angle resolved photo electron spectroscopy; ARPES).

The 1D topology should block any off-diagonal Hall response making the Hall coefficient vanish, even for an arbitrary collection of nonpercolating stripes containing mobile holes. This is found to be the case in NdLSCO ($La_{1.4-x}Nd_{0.6}Sr_xCuO_4$).[6] At low doping ($x < 1/8$, the magic number at which the stripes are known to be structurally pinned in the LTT phase) the Hall coefficient decreases rapidly with decreasing temperature. For $x > 1/8$, however, where the stripes are known to be fluctuating, the Hall coefficient is large and featureless despite the charge-stripe order. The magic number $x = 1/8$ seems to mark a cross over from 1D to 2D-hole motion. The 1D electronic nature of the stripes with mobile carriers is strongly supported by the ARPES studies.[4]

ARPES measures the single-particle spectral function $A(\mathbf{k},w)$, within the photoionization weight factor $(w)$, and gives, on frequency integration the in-plane momentum $(\mathbf{k})$ distribution, function

$$n(\mathbf{k}) = \int_{-\infty}^{+\infty} A(\mathbf{k},w) f(w) dw/2\pi,$$

from which one constructs the Fermi surface that contains the high spectral weight region. Figure 3a shows such a Fermi surface, which is redrawn schematically from ARPES measurements on NdLSCO.[4] Figure 3b shows the Fermi surface expected of the orthogonal 1D charge stripes of the type shown in Figure 2, where the vertical dotted lines are along [100] and the horizontal lines (solid lines) are along [010]. (Recall that the

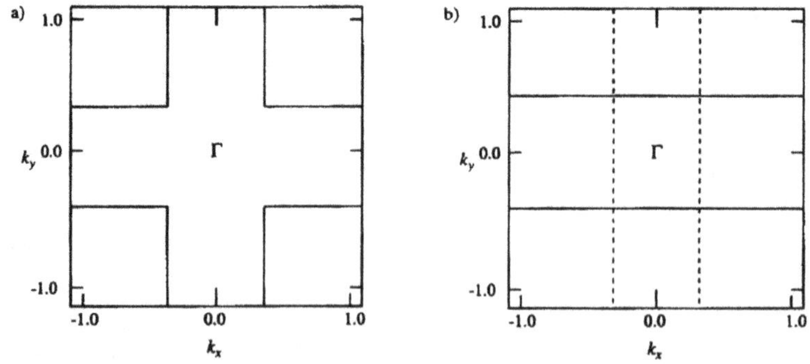


*Figure 3. a) A schematic of the ARPES results for the Fermi surface of NdLSCO for $x = 1/8$, taken and redrawn from ref.[4]. b) Fermi surfaces expected for 1/4-filled 1D bands, for comparison. Dashed lines (----) are for the horizontal stripes, and solid lines (——) for the vertical stripes.*

  

stripes in the adjacent singly modulated $CuO_2$ planes are orthogonal). The striking similarity suggests a 1/4-filled 1D band for a 1/2-filled charge stripe corresponding to one hole per two copper atoms in the linear stripe and satisfying the Luttinger theorem. The Fermi surfaces, or rather lines, are $(\pm\pi/4a, k_y)$ and $(k_x, \pm\pi/4a)$. A 2D homogeneous electronic structure clearly appears to be discredited, which thus suggests the striped nature of high-$T_c$ cuprates.

Whether there is any relation between stripes and cuprate superconductivity is the important question facing us today. Systematics of the dependence of the magnetic incommensuration $\delta$ and the superconductivity transition temperature $T_c$ on the hole concentration (the value of $x$) in LSCO and NdLSCO show striking correlations.[2, 3] First, superconductivity is anomalously suppressed ($T_c$ goes to almost zero) for $x = 1/8$, the magic number at which the stripes are structurally pinned, showing up in low-energy neutron scattering as incommensurate peaks of small momentum width. Second, away from this magic number, the dependence of $\delta$ and of $T_c$ on $x$ from underdoping to optimal doping closely track each other. Thus, the critical doping for the onset of superconductivity and for the appearance of the magnetic stripes are the same; signalled by the shifting of the neutron diffraction peaks away from the AFM position to the incommensurate split-peak positions. The incommensurate, low-energy inelastic magnetic peaks remain narrow in momentum in the superconducting phase, except in the weakly superconducting overdoped region, where they become broad and incoherent. Thus, fluctuating stripes are correlated with, and certainly can coexist with, superconductivity. Static stripes, on the other hand, appear to suppress superconductivity. $^{63}$Cu NQR (Nuclear Quadrupole Resonance) measurements that detect the charge dynamics in the proximate charge stripes through the electric field gradient they produce, however, indicate that static stripes and superconductivity can coexist.[30]

An intriguing aspect of the doping dependence of incommensuration $\delta$ is that the linear variation of $\delta$ with $x$ saturates at $x = 1/8$, while $T_c$ continues to increase up to optimal doping. While the linearity of $\delta$ with $x$ implies that the added holes are accommodated in the linear stripes, occurrence of saturation suggests an alternative mode of accommodation of the holes. This is reminiscent of the low-dimensional polyacetylene-like system,[22] where solitonic doping gives way to band-type doping beyond a critical doping. This would suggest that the stripes may well be a solitonic route to hole doping at low levels.

We now turn to microscopic considerations of the stripes and their implication for a theory of HTSC. Incommensurate modulations, structural or electronic (charge and spin), can result from competing length scales, for example, the length scale set by the nested Fermi surface and that of the lattice periodicity as in the Peierls or spin-Peierls instability that leading to spatial modulations (and gapping of the Fermi surface), well known in low-dimensional systems as the CDW (charge density wave) and the SDW (spin density wave) orders.[21, 23] In the case of cuprates, however, it is known that charge ordering occurs at a higher temperature than spin ordering, and so the stripe phase is charge driven[2]—as in $La_{2-x}Sr_xCuO_4$ as well as in the nickelates. This is not consistent with the usual Fermi surface instabilities, where the spins would order first. Incommensuration can also arise from competing interactions such as short-range attraction and the long-range repulsion, or short-range FM and the long-range AFM interactions in a spin-1 lattice gas that frustrate any macroscopic phase separation. Charge stripes, and the magnetic stripes driven by the charge order, ultimately result from the fact that the cuprates are essentially hole-doped Mott (or rather charge-transfer) insulators and the doped holes are energetically expelled from the AFM ordered domains, which they would otherwise disrupt through their motion. This basic physics is well captured by a short-range two-band 2D Hubbard model with strong on-site repulsion (the Hubbard correlation energy) on the copper atoms, and the $3d_{x^2-y^2}$-$2p_{x,y}$ hybridization between copper and oxygen, and the energy of charge transfer, $Cu^{2+}{:}3d^9 \rightarrow Cu^{1+}{:}3d^{10}$ + oxygen 2p-hole, in these strongly correlated insulators.[13] A mean-field treatment indeed gives charge-stripes, but they turn out to be insulating, as in the nickelates (unlike the conducting cuprate stripes). Antiphase spin-charge stripes of the form shown in Figure 2 result from a short-range one-band $t-J$ model.[19] Thus, a long-range repulsion is not a necessary condition for the occurrence of the stripes.

Next, we shall look at the possible connection between stripes and the pairing mechanism. Almost all the nonphonon pairing mechanisms proposed so far invoke interaction between spin fluctuations (magnetism) and hole motion (charge) in some form.[31] For a pairing mechanism involving spin–hole interaction, the electronic specific heat near the superconducting transition must track the temperature derivative of the average exchange energy of the spins. This follows from the thermodynamics of the superconducting phase transition, derived first in the context of the conventional electron–phonon pairing mechanism,[32] and generalized later to the spin-fluctuation mechanism.[33] This mechanism is really much the same as the temperature derivative of the spin-disorder resistivity that shows the critical behavior of the electronic specific heat in bad itinerant metals.[34] Recent inelastic neutron scattering measurements[35] of the spin-fluctuation spectrum of YBCO123, which are enhanced by the so-called $\pi$-resonance at the wavevector $(\pi/a, \pi/b, \pi/c)$, verify the above relation between the temperature dependence of the AFM exchange energy and the electronic specific heat.[36] While this does implicate magnetic fluctuations as a hole-pairing mechanism, it is too nonspecific to support any stripe-based theory.

A minimal model for the layered cuprates is the 2D short-range single-band Hubbard model that effectively describes the strongly correlated system.[37, 38] The undoped parent material, for example, $La_2CuO_4$, is a Mott insulator, as can be seen from the electron count, for example, $(La^{3+})_2(Cu^{2+})(O^{2-})_4$, that would give a half-filled band and is therefore a metal. It becomes conducting when doped away from stoichiometry as in, $La_{2-x}Sr_xCuO_4$. When the doping is not too small, the holes get unbound from the charged dopants as their potentials begin to overlap. The hole-doped Mott insulator is then well-described in the space of nondouble occupancy by the so-called $t-J$ Hamiltonian [Equation (1) and (2)][38]

$$H = -t\sum_{\langle ij\rangle,\sigma}(C^{+}_{i\sigma}C_{1\sigma} + Hc) + J\sum_{\langle ij\rangle}\left(S_iS_j - \frac{n_in_j}{4}\right) \quad (1)$$

$$S_i = \tfrac{1}{2}\sum_{\alpha,\beta} C_{i\beta}\sigma_{\beta\alpha}C_{i\alpha} \qquad (2)$$

that describes the motion of the holes in the active background of spins. This spin-hole Hamiltonian suggests a kinetic hole-pairing mechanism driven by the lowering of the kinetic energy due to delocalization, as described below.

A long-range or even a mesoscopic AFM order (Néel ordered) encumbers the hole motion. (As a hole moves to the nearest-neighbouring site, the spin at that site concomitantly moves to the site previously occupied by the hole). A succession of such moves shift-registers a line of spins, which become laterally misaligned with respect to the spins on the sides (Figure 4). The broken AFM bonds shown by the dashed lines create a linearly rising potential that impedes free hole motion. A compact

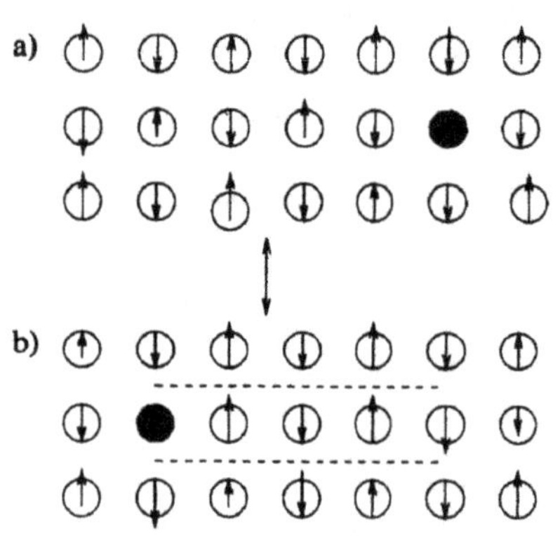

*Figure 4. a) A hole at the initial site in a Néel-ordered antiferromagnet; b) a hole at the final site after four hops. ---- show disrupted bonds between misaligned spins giving a linearly rising potential.*

pair of holes can, however, toboggan together, and thus delocalize quite freely. The trailing hole undoes what the leading hole does, thus healing the lateral misalignment of the spins. The lowering of the pair kinetic energy by the ease of pair delocalization then provides the kinetic energy mechanism for pairing.[39] (The quantum fluctuation of spins away from the Néel order must modify the above naive visualization, but one can still talk meaningfully about the hole-pair moving unencumbered by entanglement with the background spins). Clearly, for low doping the hole-hole repulsion will dominate over the kinetic mechanism that favors compact pairing. It is only for higher doping, when the mean hole separation matches the pair size that the kinetic pairing wins. For such a kinetic mechanism for pairing, stripe modulation in the zeroth approximation merely provides a modified spin background for the hole motion—not necessarily conducive to pairing. This basic idea of kinetic pairing seems to suggest a novel, highly elaborate quantum liquid-crystalline model for HTSC, in which the fluctuating stripes play a crucial role.[14–18]

The charge stripe is normally viewed as a 1D lane, along which the holes can freely move longitudinally, but are constrained in the transverse direction by the antiphase AFM-ordered magnetic domains. Indeed, the pinning at the magic number doping ($x = 1/8$) gives such a stripe which is static and gives a resolution-limited elastic neutron scattering. It is well known that an array of such static charge stripes will be insulating through a CDW instability. In the electronic quantum liquid-crystalline phase model of the doped Mott insulator the stripes show transverse quantum fluctuations that affect the motion of the charged particles which is guided by the fluctuating stripes. Thus, the charge stripes are viewed as a quantum mechanical deformable object with a topological integrity, and as having many degrees of freedom. The role of zero-point fluctuations was recognized quite early.[40] The mechanism for the mobile hole pairing here involves a magnetic analog of the superconducting proximity effect between the metallic stripe and the Mott insulators that bound it transversely. The mobile hole pairs virtually hop between the Mott-insulator domains and acquire a spin gap. This is in an underdoped system. At higher doping the hole pairs can tunnel between the stripes giving a global 3D phase stiffness to the local pairs. Several known features of HTSC such as the normal-state pseudogap in the underdoped samples, which have the same d-wave symmetry as the d-wave superconducting gap at lower temperatures, come out of this model. The electronic quantum liquid-crystalline model has many topological features of its underlying one-dimensionality that affect its charge and spin degrees of freedom. It certainly warrants a detailed examination—experimentally and theoretically, as something interesting in its own right.

Any discussion of the role of stripes in the high-$T_c$ layered cuprates is necessarily inconclusive at present. It is possible, however, that stripes may well be a solitonic route to hole doping. In any case, as of now the stripes cannot confidently signify superconductivity.

[1] For a critical discussion of stripes and electronic phase separation see a) J. Zaanen, *J. Phys. Chem. Solids* **1998**, *59*, 1769; b) J. M. Tranquada, *Physica B* **1998**, *241–243*, 745; c) J. M. Tranquada, *J. Phys. Chem. Solids* **1998**, *59*, 2150; d) *Proceedings of the workshop on Phase Separation in Cuprate Superconductors* (Eds.: K. A. Müller, G. Benedek), World Scientific, Singapore, **1993**; e) *Phase Separation in Cuprate Superconductors* (Eds.: E. Sigmund, K. A. Müller), Springer-Verlag, Heidelberg, **1994**; f) E. Dagotto, T. Hotta, A. Moreo, *Phys. Rep.* **2001**, *344*, 1; g) C. N. R. Rao, P. V. Vanitha, A. K. Cheetham, *Chem. Eur. J.* **2003**, *9*, 828..

[2] a) J. M. Tranquada, S. J. Sternlieb, J. D. Axe, Y. Nakamura, S. Uchida, *Nature* **1995**, *375*, 561; b) J. M. Tranquada, J. D. Axe, N. Ichikawa, Y. Nakamura, S. Uchida, B. Nachumi, *Phys. Rev. B* **1996**, *54*, 7489; c) J. M. Tranquada, J. D. Axe, N. Ichikawa, A. R. Moodenbaugh, Y. Nakamura, S. Uchida, *Phys. Rev. Lett.* **1997**, *78*, 338; d) *Stripes and Related Phenomena* (Eds.: A. Bianconi, N. L. Saini), Kluwer Academic Publishers, Dordrecht, **2000**.

[3] K. Yamada, C. H. Lee, K. Kurahashi, J. Wada, S. Wakimoto, S. Ueki, H. Kimura, X. Endoh, S. Hosaya, G. Shirane, R. J. Birgeneau, M. Greven, M. A. Kastner, Y. J. Kim, *Phys. Rev. B* **1998**, *57*, 6165.

[4] X. J. Zhou, P. Bogdanov, S. A. Kellar, T. Noda, H. Eisaki, S. Uchida, Z. Hussain, Z.-X. Shen, *Science* **1999**, *286*, 268.

[5] X. J. Zhou, T. Yoshida, S. A. Kellar, P. V. Bogdanov, E. D. Lu, A. Larzava, M. Nakamura, T. Noda, T. Takeshita, H. Eisaki, S. Uchida, A. Fujimori, Z. Hussain, Z.-X. Shen, *Phys. Rev. Lett.* **2001**, *86*, 5578.

[6] T. Noda, H. Eisaki, S.-I. Uchida, *Science* **1999**, *286*, 265.

[7] H. A. Mook, P. Dai, F. Dogan, *Phys. Rev. Lett.* **2002**, *88*, 097004.

[8] B. Lake, G. Aeppli, K. N. Clausen, D. F. McMorrow, K. Letmann, N. E. Hussey, N. Manykorntong, M. Nohava, H. Takagi, T. E. Mason, A. Schröder, *Science* **2001**, *291*, 1759.

  

[9] B. Lake, H. M. Ronnow, N. B. Christensen, G. Aeppli, K. Lefmann, D. F. McMorrow, P. Verderwisch, P. Smeibidl, N. Mangkorntong, T. Susagawa, M. Nohara, H. Takagi, T. E. Mason, *Nature* **2002**, *415*, 299.
[10] F. Mitrovic, E. E. Sigmund, M. Eschrig, H. N. Bachman, W. P. Halperin, A. P. Reyas, P. Kuhns, W. G. Moulton, *Nature* **2001**, *413*, 501.
[11] Y. S. Lee, R. J. Birgeneau, M. A. Kastner, Y. Endoh, S. Wakimoto, K. Yamada, R. W. Erwin, S. H. Lee, G. Shirane, *Phys. Rev. B* **1999**, *60*, 3643.
[12] C. Howald, P. Fournier, A. Kapitulnik, *Phys. Rev. B* **2001**, *64*, R100504.
[13] J. Zaanen, O. Gunnarsson, *Phys. Rev. B* **1989**, *40*, 7391.
[14] V. J. Emery, A. Kivelson, *Strongly Correlated Electronic Materials: The Los Alamos Symposium 1993* (Eds.: K. S. Bedell, Z. Wang, D. E. Meltzer, A. V. Balatsky, E. Abrahams), Addision-Wesley, Redwood City, **1994**, p. 619.
[15] A. Kivelson, E. Fradkin, V. J. Emery, *Nature* **1998**, *393*, 550.
[16] V. J. Emery, S. A. Kivelson, O. Zachar, *Phys. Rev. B* **1997**, *56*, 6120.
[17] a) V. J. Emery, A. Kivelson, *Phys. C* **1993**, *209*, 597; b) V. Löw, V. J. Emery, K. Fabricius, S. A. Kivelson, *Phys. Rev. Lett.* **1994**, *72*, 1918.
[18] a) M. Salkola, V. J. Emery, S. A. Kivelson, *Phys. Rev. Lett.* **1996**, *77*, 155; b) V. J. Emery, S. A. Kivelson, J. M. Tranquada, *Proc. Natl. Acad. Sci. USA* **1999**, *96*, 8814.
[19] a) S. R. White, D. J. Scalapino, *Phys. Rev. Lett.* **1998**, *81*, 3227; b) S. R. White, D. J. Scalapino, *Phys. Rev. Lett.* **1998**, *80*, 1272.
[20] J. G. Bednorz, K. A. Müller, *Z. Phys. B* **1986**, *64*, 189.
[21] W. P. Su, J. R. Schrieffer, A. J. Heeger, *Phys. Rev. B* **1980**, *22*, 2099.
[22] E. M. Conwell, *Phys. Today*, **1985**, 48 (June).
[23] V. L. Pokhrovsky, A. L. Talapov, *Theory of Incommensurate Crystals, Soviet Scientific Reviews Suppl. Sec. A, Vol. 1*, (Ed.: I. M. Khalatnikov), Harwood Academics, Switzerland, **1984**.
[24] a) J. M. Tranquada, J. E. Lorenzo, D. J. Buttrey, V. Sachan, *Phys. Rev. B* **1995**, *52*, 3581; b) S. W. Cheong, H. Y. Hwang, C. H. Chen, B. Batlogg, L. W. Rupp, Jr., S. A. Carter, *Phys. Rev. B* **1994**, *49*, 7088; c) H. Kuwahara, Y. Tomioka, Y. Moritomo, A. Asamitsu, M. Kasai, R. Kumai, Y. Tokura, *Science* **1996**, *272*, 80; d) S. H. Lee, J. M. Tranquada, K. Yamada, D. J. Buttrey, Q. Li, S. W. Cheong,24935 *Phys. Rev. Lett.* **2002**, *88*, 126401; e) M. Abu-Sheikah, O. Bakharev, H. B. Bram, J. Zaanen, *Phys. Rev. Lett.* **2001**, *87*, 237201.
[25] a) *Colossal magnetoresistance, Charge-ordering and Related Properties of Manganese Oxides* (Eds.: C. N. R. Rao, B. Raveau), World Scientific, Singapore, 1999; b) C. N. R. Rao, A. Arulraj, A. K. Cheetham, B. Raveau, *J. Phys. Condens. Matter* **2000**, *12*, R83; c) C. N. R. Rao, P. V. Vanitha, *Curr. Opin. Solid State Mater. Sci.* **2002**, *6*, 97.
[26] P. Abbamonte, L. Venema, A. Rusydi, G. A. Sawatzky, G. Logvenov, I. Bozovic, *Science* **2002**, *297*, 581.
[27] C. Howald, H. Eisaki, N. Kaneko, A. Kapitulnik, *cond-mat/0201546*.
[28] J. E. Hoffman, K. McElroy, D.-H. Lee, K. M. Lang, H. Eisaki, S. Uchida, J. C. Davis, *Science* **2002**, *297*, 1148.
[29] S. H. Pan, J. P. O'Neal, R. D. Badzey, C. Chamon, N. Ding, J. R. Engelbracht, Z. Wang, N. Eisaki, S. Uchida, A. K. Gupta, K.-W. Ng, E. W. Hudson, K. M. Lang, J. C. Davis, *Nature* **2001**, *413*, 282.
[30] P. M. Singer, A. W. Hunt, A. F. Cederstrom, T. Imai, *cond-mat/9906140*.
[31] J. R. Schrieffer, *J. Low Temp. Phys.* **1995**, *99*, 97.
[32] G. V. Chester, *Phys. Rev.* **1956**, *103*, 1693.
[33] D. J. Scalapino, S. R. White, *Phys. Rev. B* **1998**, *58*, 8222.
[34] L. Klein, J. S. Dodge, C. H. Ahn, G. J. Snyder, T. H. Geballe, M. R. Beasley, A. Kapitulnik, *Phys. Rev. Lett.* **1996**, *77*, 2774.
[35] a) P. Dai, H. A. Mook, S. M. Hayden, G. Aeppli, T. G. Perring, R. D. Hunt, F. Dogan, *Science* **1999**, *284*, 1344; b) E. Demler, S. C. Zhang, *Nature* **1998**, *396*, 733.
[36] J. W. Loram, K. A. Mirza, P. F. Freeman, *Phys. C* **1990**, *171*, 243.
[37] F. C. Zhang, T. M. Rice, *Phys. Rev. B* **1988**, *37*, 3759.
[38] E. Dagotto, *Rev. Mod. Phys.* **1994**, *66*, 763.
[39] a) J. E. Hirsch, *Phys. Rev. Lett.* **1987**, *59*, 228; b) M. M. Mohan, N. Kumar, *J. Phys. C* **1987**, *20*, L527; c) N. Kumar, *Phys. Rev. B* **1990**, *42*, 6138; d) Yu. A. Dimashko, *JETP Lett.* **1993**, *76*, 267.
[40] a) P. Prelovšek, X. Zotos, *Phys. Rev. B* **1993**, *47*, 5984; b) P. Prelovšek, I. Sega, *Phys. Rev. B* **1994**, *49*, 15241.

Received: November 28, 2002 [M601]

  

*J. Phys. Chem. B* **2000,** *104,* 5877–5889 **5877**

# FEATURE ARTICLE

## Charge, Spin, and Orbital Ordering in the Perovskite Manganates, $Ln_{1-x}A_xMnO_3$ (Ln = Rare Earth, A = Ca or Sr)

**C. N. R. Rao***

*Chemistry & Physics of Materials Unit and the CSIR Centre of Excellence in Chemistry, Jawaharlal Nehru Centre for Advanced Scientific Research, Jakkur P.O, Bangalore 560 064, India*

*Received: February 7, 2000; In Final Form: March 29, 2000*

Charge ordering occurs in some mixed-valent transition metal oxides. The perovskite manganates of the formula $Ln_{1-x}A_xMnO_3$ (Ln = rare earth; A = Ca, Sr) are especially interesting because long-range ordering of the $Mn^{3+}$ ($t_{2g}^3\, e_g^1$) and $Mn^{4+}$ ($t_{2g}^3\, e_g^0$) ions in these materials is linked to antiferromagnetic spin ordering, and also to the long-range ordering of the $Mn^{3+}$ ($e_g$) orbitals and the associated lattice distortions. Charge ordering occurs at a higher temperature than spin ordering in some of the manganates ($T_{CO} > T_N$), whereas in some others $T_{CO} = T_N$. Orbital ordering occurs without charge ordering in the A-type antiferromagnetic manganates, but in the manganates where charge ordering occurs, antiferromagnetism of CE-type is found along with orbital ordering. The subtle relations between charge, spin, and orbital ordering are discussed in the article, with special attention to the effects of cation size, chemical substitution, dimensionality, pressure, and magnetic and electric fields. Unusual features such as phase separation and electron–hole asymmetry are also examined.

### Introduction

Charge ordering is a phenomenon generally observed in mixed-valent transition metal oxides. When differently charged cations (i.e., 2+ and 3+) in an oxide order on specific lattice sites, the hopping of electrons between the cations is no longer favored. One therefore observes an increase in the electrical resistivity at the charge-ordering transition, often accompanied by a change in crystal symmetry. Because transition metal ions also carry spins, it is interesting to examine the magnetic (spin) ordering in the solids in relation to charge ordering. A well-known example of charge ordering is found in $Fe_3O_4$ (magnetite) where it occurs at a temperature well below spin ordering.[1,2] Charge and spin ordering in real space in metal oxides received renewed attention because of their role in cuprate superconductors.[3] In recent years, charge and spin ordering have been discovered in a few other transition metal oxides,[4] typical examples being $La_{1-x}Sr_xFeO_3$, $La_{2-x}Sr_xNiO_4$, and $LiMn_2O_4$. Charge ordering in the rare earth manganates of the perovskite structure, with the general composition $La_{1-x}A_xMnO_3$ (Ln = rare earth; A = alkaline earth), is considerably more interesting, because it is closely associated with spin and orbital ordering, giving rise to fascinating properties.[5–7]

The perovskite manganates originally became popular because of the discovery of colossal magnetoresistance (CMR).[5,6] CMR and related properties essentially arise from the double-exchange mechanism of electron hopping between the $Mn^{3+}$ ($t_{2g}^3\, e_g^1$) and $Mn^{4+}$ ($t_{2g}^3\, e_g^0$) ions.[8] In this mechanism, lining up of the spins (ferromagnetic alignment) of the incomplete $e_g$ orbitals of the adjacent Mn ions is directly related to the rate of hopping of the electrons, giving rise to an insulator–metal transition in the material at the ferromagnetic Curie temperature, $T_c$. In the ferromagnetic phase ($T < T_c$), the material becomes metallic, but is an insulator in the paramagnetic phase ($T > T_c$). In the insulating phase, a Jahn–Teller distortion associated with the $Mn^{3+}$ ions favors the localization of electrons. Charge ordering of the $Mn^{3+}$ and $Mn^{4+}$ ions competes with double-exchange and promotes insulating behavior and antiferromagnetism. It may be recalled that the $Mn^{3+}$–O–$Mn^{3+}$ and $Mn^{4+}$–O–$Mn^{4+}$ superexchange interaction, through the $e_g$ orbitals, is antiferromagnetic. Although charge ordering in the rare earth manganates was investigated by Jirak et al.[9] as early as1985, the subject has received renewed attention only in the past 5 years, for reasons described below.

Charge ordering occurs through a fairly wide range of compositions of $Ln_{1-x}A_xMnO_3$, provided the Ln and A ions are not too large. Large Ln and A ions (e.g., La, Sr) favor ferromagnetism and metallicity, whereas the smaller ones (e.g., La, Ca, or Pr, Ca) favor charge ordering. Charge ordering and spin (antiferromagnetic) ordering may or may not occur at the same temperature. Besides, the $Mn^{3+}$ ($d_{z^2}$) orbitals and the associated lattice distortions develop long-range order (as illustrated later in this article). Such orbital ordering may or may not occur with charge ordering in the manganates, but it generally accompanies antiferromagnetic (spin) ordering. In this article, we discuss the interplay of charge, spin, and orbital ordering in the rare earth manganates in some detail, and highlight some of the important recent results. Before discussing the manganates, we shall briefly review the properties of a few of the other transition metal oxides that exhibit charge ordering.

**Examples of Charge and Spin Ordering in Oxides.** $Fe_3O_4$ has a spinel structure, represented as $Fe^{3+}[Fe^{2+}, Fe^{3+}]O_4$ in which one third of the cations are tetrahedrally coordinated (A-sites) and two thirds of the cations are octahedrally coordinated (B-sites). It is ferrimagnetic below 858 K ($T_N$), with the spins

* For correspondence: e-mail: cnrrao@jncasr.ac.in.

10.1021/jp0004866 CCC: $19.00 
Published on Web 06/02/2000

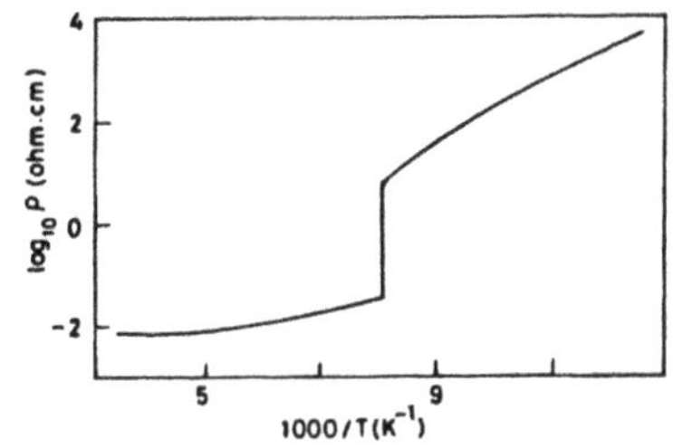


**Figure 1.** Verwey transition in $Fe_3O_4$ caused by charge ordering (from Honig[1b]).

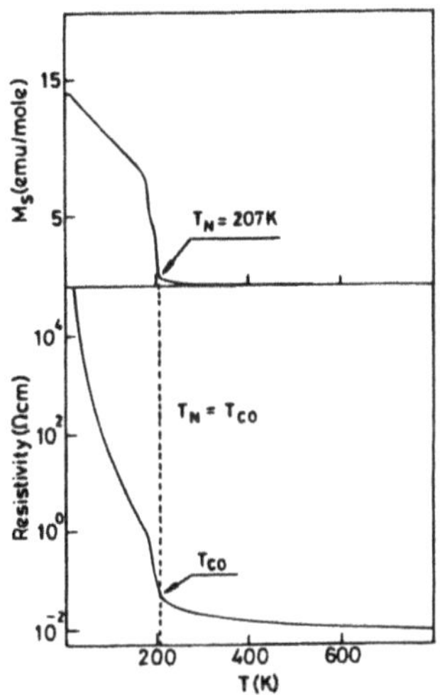


**Figure 2.** Temperature variation of electrical resistivity and magnetization of $La_{0.33}Sr_{0.67}FeO_3$ showing marked changes on charge ordering ($T_{CO} = T_N$) (from Park, Yamaguchi and Tokura, as quoted by Imada et al.[4]).

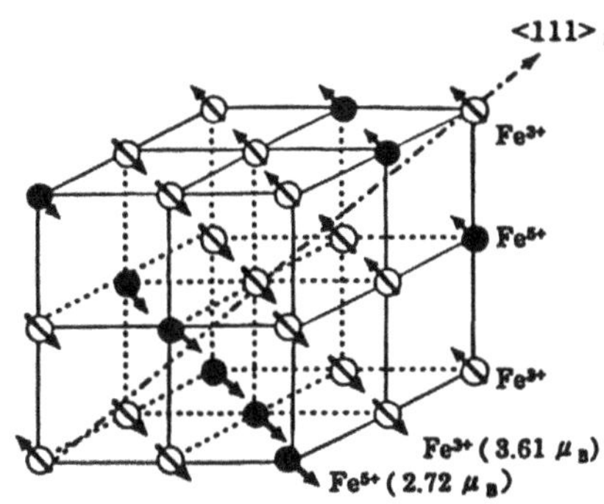


**Figure 3.** Structure showing charge and spin ordering in $La_{0.33}Sr_{0.67}FeO_3$ (from Battle et al.[10b]).

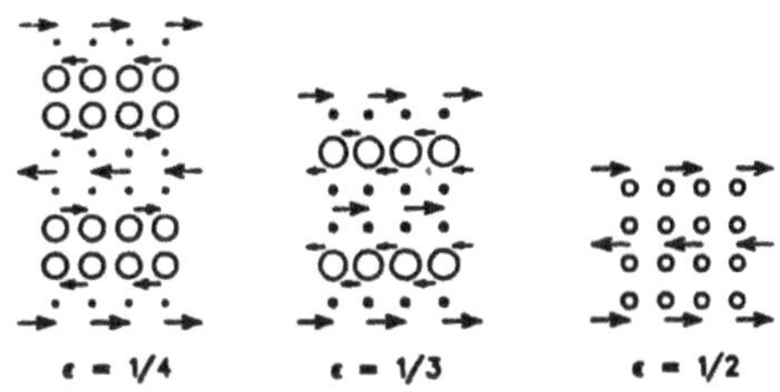


**Figure 4.** Spin and charge modulation in the $NiO_2$ planes of $La_{2-x}Sr_xNiO_{4+y}$ with increasing hole concentration. Here, the modulation wavenumber, $\epsilon$, is equivalent to the hole concentration, $p = x + 2y$ (from Tranquada et al.[11a]). Spin density on nickel ions is shown by arrows, and charge density on oxygen ions is shown by circles.

of the A- and B-sublattices being antiparallel. Around 120 K ($T_V$), $Fe_3O_4$ shows a sharp increase in resistivity, commonly referred to as the Verwey transition (Figure 1). Below $T_V$, the $Fe^{2+}$ and $Fe^{3+}$ ions are considered to be ordered, thereby giving rise to high electrical resistivity. Above $T_V$, conduction occurs on the B site sublattice. It actually turns out, however, that charge ordering in $Fe_3O_4$ is much more complicated, with some short-range ordering present even above $T_V$. Even today, there is argument about the symmetry of the low-temperature ordered phase. The Verwey transition in this oxide has been reviewed excellently by Honig[1] and Tsuda et al.[2]

$La_{1-x}Sr_xFeO_3$ ($x$ = 0.67) undergoes charge ordering at approximately 207 K ($T_{CO}$), at which temperature it also exhibits antiferromagnetic spin ordering. Thus, $T_{CO}$ is also the Néel temperature ($T_N$) in this oxide. The resistivity shows a marked increase at $T_{CO} = T_N$ (Figure 2). Although the formula requires the presence of $Fe^{3+}$ and $Fe^{4+}$ ions, the $Fe^{4+}$ ions disproportionate to give $Fe^{3+}$ and $Fe^{5+}$ ions.[10] The charge- and spin-ordered structure of $La_{0.33}Sr_{0.67}FeO_3$ is shown in Figure 3.

Quasi two-dimensional $La_{2-x}Sr_xNiO_{4+y}$ undergoes cooperative ordering of the Ni spins and of the charge carriers below a temperature ($T_{CO}$). Charge ordering occurs at nearly all doping levels ($p = x + 2y$), in the insulating regime of the material ($p$ < 0.7). Spin and charge modulations are observed in the $NiO_2$ plane and these vary with the hole concentration ($p$ value). This nickelate system is a typical instance of microscopic phase separation wherein the charge carriers (holes) localize in the domain walls in an antiferromagnetic system. Such phase separation causes stripe modulations. Here, the ordering may be viewed as alternating stripes of antiferromagnetic and hole-rich regions.[11] In Figure 4, we show the spin and charge modulations in the $NiO_2$ planes at two hole concentrations. The $\epsilon = 1/4$ case corresponds to $La_2NiO_{4.125}$ ($p = 0.25$), where the interstitial oxygens form a superlattice with a unit cell of $3a \times 5b \times 5c$. The compositional dependence of resistivity of $La_{2-x}Sr_xNiO_4$ shows peaks at $x = 0.33$ and 0.25, owing to charge ordering, which in the $x = 0.33$ composition occurs at 240 K. At 240 K, various properties show anomalies, as depicted in Figure 5. Superlattice peaks show up in the electron diffraction pattern at this temperature. Antiferromagnetic spin ordering in the nickelate occurs at a lower temperature (180 K). It appears that ordering is driven by charge. Charge ordering of holes, accompanied by spin ordering or the segregation of holes and spins in the stripe form, also occurs in $La_{2-x}A_xCuO_4$ (A = Sr, Ba, $x$ = 0.125), causing anomalous suppression of superconductivity.[3]

$LiMn_2O_4$ is a spinel with equal proportions of 3+ and 4+ Mn cations. It has been shown recently to undergo a Verwey-type transition with a resistivity anomaly at approximately 290 K where there is a structural (cubic-tetragonal) transition. Around 65 K, the material exhibits a complex long-range magnetic order.[13] Clearly, this charge-ordering transition requires further study.

**A Brief Description of the Different Types of Ordering in the Manganates.** $LaMnO_3$ has a layered antiferromagnetic structure, referred to as A-type antiferromagnetic ordering[14] (Figure 6a). Because there is a doubly degenerate $e_g$ orbital in each $Mn^{3+}$ ion ($t_{2g}^3\, e_g^1$), $LaMnO_3$ and the other analogous rare earth compounds such as $NdMnO_3$ show a fine interplay between spin and orbital ordering. The orbital ordering is coupled to the Jahn–Teller (JT) distortion. Figure 6b describes the JT distortion in $LaMnO_3$. The distortion disappears above

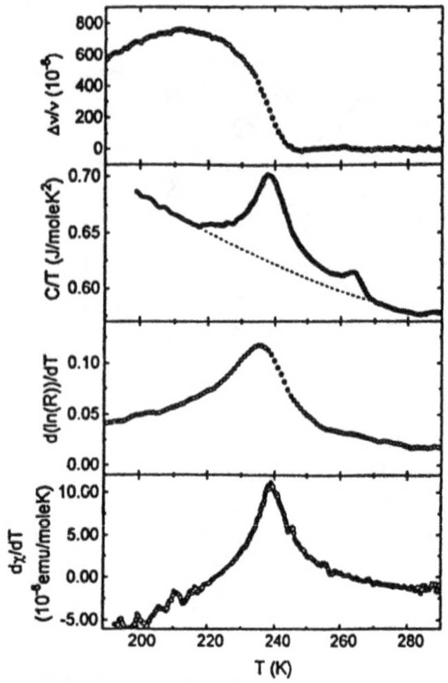


**Figure 5.** Normalized sound velocity ($\Delta v/v$), specific heat (c), as well as temperature derivatives of resistivity ($\rho$) and magnetic susceptibility ($\chi$) showing anomalies at the charge ordering transition (240 K) in $La_{1.67}Sr_{0.33}NiO_4$ (from Ramirez et al.[12]).

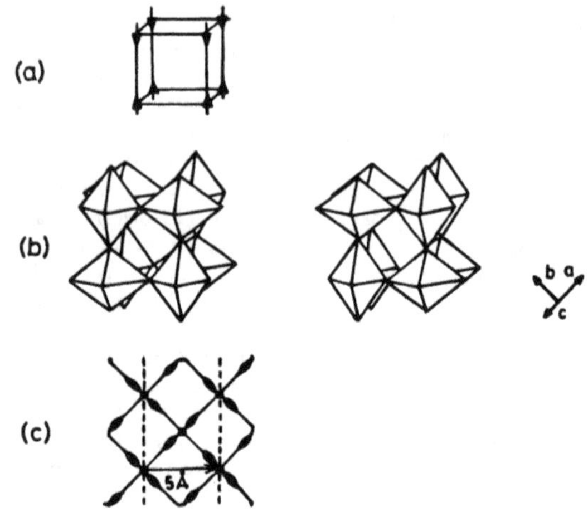


**Figure 6.** (a) A-type AFM ordering, (b) JT distortion, and (c) orbital ordering in $LaMnO_3$. The orientation of $e_g$ orbitals shown in panel c is consistent with the 5 Å lattice spacing.

750 K. Orbital ordering of $3x^2 - r^2$ or $3y^2 - r^2$ type accompanied by the JT distortion leads to a superexchange coupling in $LaMnO_3$, which is ferromagnetic (FM) in the planes and antiferromagnetic (AFM) between the planes (Figure 6a).[15] Orbital ordering in $LaMnO_3$ is shown in Figure 6c. Without the JT distortion, $LaMnO_3$ would have been a FM insulator; it is an A-type AFM insulator instead.

In $Ln_{1-x}A_xMnO_3$, besides orbital and spin ordering, we can have charge ordering because of the presence of $Mn^{3+}$ and $Mn^{4+}$ ions. Small Ln and A ions stabilize the charge-ordered (CO) state. Thus, $Pr_{0.7}Ca_{0.3}MnO_3$, charge-orders around 230 K in the paramagnetic state, becoming AFM at 170 K; it is an insulator and does not show ferromagnetism in the absence of a strong magnetic field. $La_{0.7}Ca_{0.3}MnO_3$, on the other hand, is a FM metal below the Curie temperature ($T_c \approx 230$ K) and a paramagnetic insulator above $T_c$. $La_{0.5}Sr_{0.5}MnO_3$ is metallic both in the FM and paramagnetic states, whereas $Nd_{0.5}Sr_{0.5}MnO_3$ shows a transition from a ferromagnetic metallic (FMM) state to an AFM charge-ordered state on cooling. The CO states in these manganates are associated with CE-type AFM ordering.[14] In the CE-type ordering, $Mn^{3+}$ and $Mn^{4+}$ ions are arranged as in a checker board and the $Mn^{3+}$ sites are JT distorted.[15] Along the *c*-axis, the in-plane arrangement mentioned above gets stacked and the neighboring planes are coupled antiferromagnetically. The exchange coupling between the $Mn^{3+}$ and $Mn^{4+}$ ions depends on the type of $e_g$ orbital at the $Mn^{3+}$ site, and hence the nature of orbital ordering becomes important. The CE-type AFM CO state in $Ln_{1-x}A_xMnO_3$ is associated with the ordering of $3x^2 - r^2$ or $3y^2 - r^2$ type orbitals at the $Mn^{3+}$ site. The JT distortion that follows such orbital ordering stabilizes the CE-type AFM state (relative to the FMM state). In Figure 7a, we show the spin, charge, and orbital ordering in the CE-type AFM state. The CO states in the manganates exhibit CE-type AFM ordering at the same temperature as the charge-ordering transition or at a lower temperature ($T_{CO} \geq T_N$). That is, spin ordering occurs concurrently or after charge ordering. Complete orbital ordering is achieved when there is both charge and spin ordering. Orbital and spin ordering occur without charge ordering in some of the manganates showing A-type antiferromagnetism.

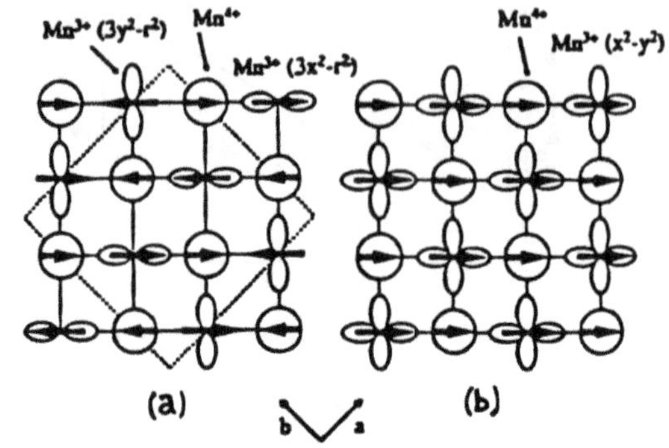


**Figure 7.** Charge, spin, and orbital ordering in (a) CE-type and (b) A-type AFM $Ln_{1-x}A_xMnO_3$. In part a, the broken line shows the unit cell for the CE-type AFM CO order. $Mn^{4+}$ ions are shown by circles.

The A-type AFM state described earlier in relation to $LaMnO_3$ (Figure 6a) is also encountered in the $Ln_{1-x}A_xMnO_3$ system. This state generally is not accompanied by charge ordering, because some electron transfer can occur between the Mn cations in the *ab* plane. Orbital ordering in A-type AFM ordering is depicted in Figure 7b. Here, the $x^2 - y^2$ type orbital is present at the $Mn^{3+}$ site. $Pr_{0.5}Sr_{0.5}MnO_3$ transforms from a FMM state to a A-type AFM state on cooling. $Nd_{0.45}Sr_{0.55}MnO_3$ is a A-type antiferromagnet, unlike $Nd_{0.5}Sr_{0.5}MnO_3$, which is a CE-type antiferromagnet exhibiting charge ordering at low temperatures; the former shows conductivity in the *ab* plane and is not charge-ordered.

Evidence for charge ordering can be observed in crystal structures at low temperatures. Thus, in $La_{0.5}Ca_{0.5}MnO_3$, the $Mn^{4+}$ environment is nearly isotropic, with all the Mn−O distances being nearly equal (~1.92 Å). In the $Mn^{3+}O_6$ octahedra, one of the Mn−O bonds is much longer (~2.07 Å) than the other bonds (~1.92 Å), consistent with orbital ordering. It must be noted that the Mn−O distances in the *ab* plane of the manganates are much longer than in the *c*-direction,

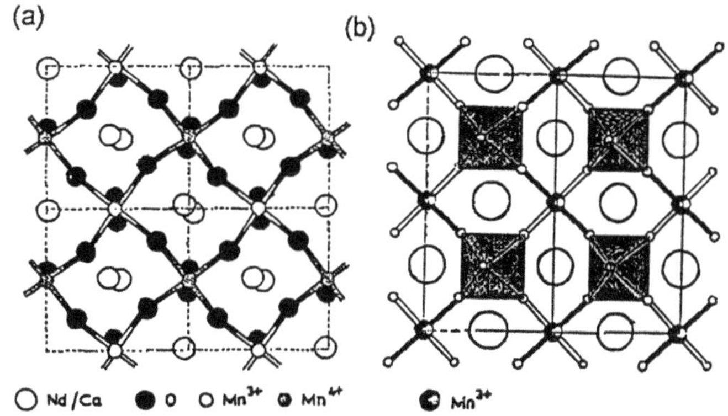


**Figure 8.** (a) Structure of charge-ordered $Nd_{0.5}Ca_{0.5}MnO_3$ in the *ab* plane at 10 K. $Mn^{4+}$ (black) located at (1/2, 0, 0) and $Mn^{3+}$ (grey) located (0, 1/2, 0) can be distinguished. The structure has zigzag chains with alternate long and short Mn−O distances. The distances are 1.921 and 2.021 Å ⟨110⟩ and 1.881 and 2.020 Å along ⟨−110⟩ (from Vogt et al.[16]). (b) Charge-ordered structure of $Nd_{0.5}Sr_{0.5}MnO_3$. The $Mn^{3+}$ ions are the filled circles, and Nd/Sr ions are large open circles. Oxygens are small open circles. The long $Mn^{3+}$−O bonds shown in black also represent the orientation of the $e_g$ orbitals. $Mn^{4+}$−O octahedra are shown in polyhedral representation (from Woodward et al. [50]).

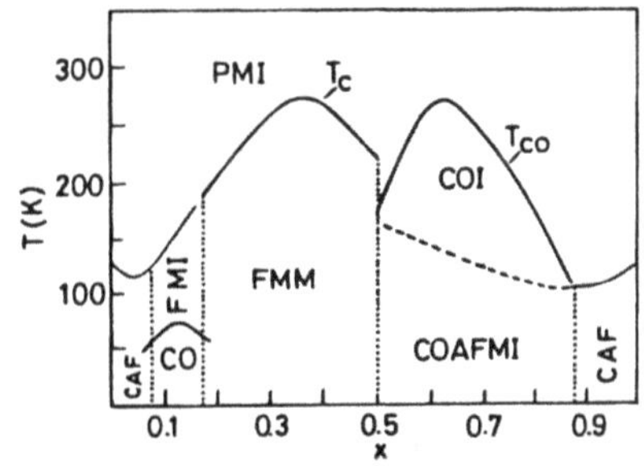


**Figure 9.** Phase diagram of $La_{1-x}Ca_xMnO_3$ (following Cheong). CAF, canted antiferromagnet; CO, charge-ordered phase; FMI, ferromagnetic insulator; PMI, paramagnetic insulator; FMM, ferromagnetic metallic state; CO AFMI, charge-ordered antiferromagnetic insulator; COI, charge-ordered insulator. Notice electron hole asymmetry by comparing the $x < 0.5$ and $x > 0.5$ regimes.

particularly in the AFM state. In Figure 8a, we show the projection of the structure of charge-ordered $Nd_{0.5}Ca_{0.5}MnO_3$ down the *c*-axis to illustrate the definitive identification of the unique sites occupied by $Mn^{3+}$ and $Mn^{4+}$ ions in the CO state. The structure of charge-ordered $Nd_{0.5}Sr_{0.5}MnO_3$, where the $Mn^{4+}$-centered octahedra are represented in polyhedral notation, is shown in Figure 8b.

**Representative Phase Diagrams of the Manganates.** In Figure 9 we show the phase diagram of $La_{1-x}Ca_xMnO_3$. In this system, charge ordering occurs in the $x \approx 0.5-0.8$ range. In Figure 10 we show the phase diagrams of $Pr_{1-x}Ca_xMnO_3$ and $Pr_{1-x}Sr_xMnO_3$. The latter system shows no charge ordering, but $Pr_{1-x}Ca_xMnO_3$ exhibits charge ordering over the $x \approx 0.3-0.8$ range. Note that $Pr_{0.7}Ca_{0.3}MnO_3$ exhibits charge ordering but $La_{0.7}Ca_{0.3}MnO_3$ does not. All such variations are essentially due to the effect of the size of the A-site cations, the smaller size favoring charge ordering. From Figures 9 and 10, we also see that the CO regime is prominent at large *x*. In fact, the $x > 0.5$ compositions in $Ln_{1-x}Ca_xMnO_3$ are almost entirely in the CO regime both when Ln = La and Pr. This regime can be considered as the electron-doped regime (substitution of trivalent rare earth in $CaMnO_3$), whereas the $x < 0.5$ compositions may be considered as the hole-doped regime (substitution of divalent Ca in $LnMnO_3$). Clearly, there is electron−hole asymmetry in these manganates.

It is surprising that ferromagnetism is not encountered in the electron-doped regime ($x > 0.5$). Effects of magnetic and electric

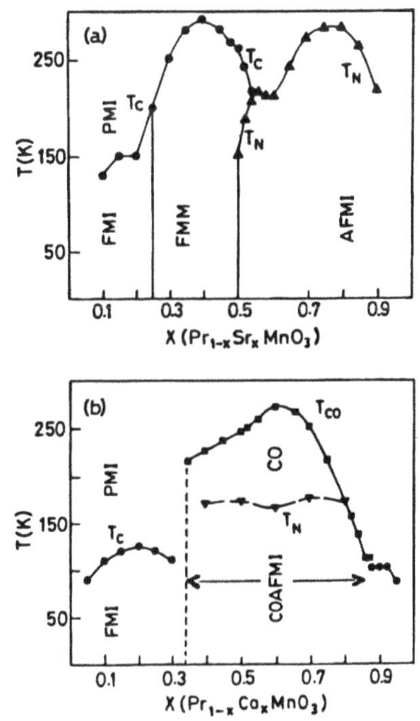


**Figure 10.** Phase diagrams of (a) $Pr_{1-x}Sr_xMnO_3$ and (b) $Pr_{1-x}Ca_xMnO_3$. Notice the wide charge ordering regime and electron−hole asymmetry in part b and the absence of charge ordering in part a. In part b, there is spin-glass or CAF behavior when $x \geq 0.8$.

fields on the hole- and electron-doped manganates (e.g., $Pr_{0.7}Ca_{0.3}MnO_3$ and $Pr_{0.3}Ca_{0.7}MnO_3$) are also different. There are some similarities between the hole- and electron-doped regimes. For example, in $Pr_{1-x}Ca_xMnO_3$, the charge-ordering transition temperature increases with hole concentration in the $0.3 < x \leq 0.5$ regime and with electron concentration in the $0.5 \leq x \leq 0.8$ regime.

**Case Studies.** To understand typical scenarios of charge-ordered manganates, it is instructive to examine the properties of two manganates with different sizes of the A-site cations. For this purpose, we choose $Nd_{0.5}Sr_{0.5}MnO_3$ with a weighted average radius of the A-site cations, $\langle r_A \rangle$, of 1.24 Å and $Pr_{0.6}Ca_{0.4}MnO_3$ with an $\langle r_A \rangle$ of 1.17 Å (Shannon radii are used here).

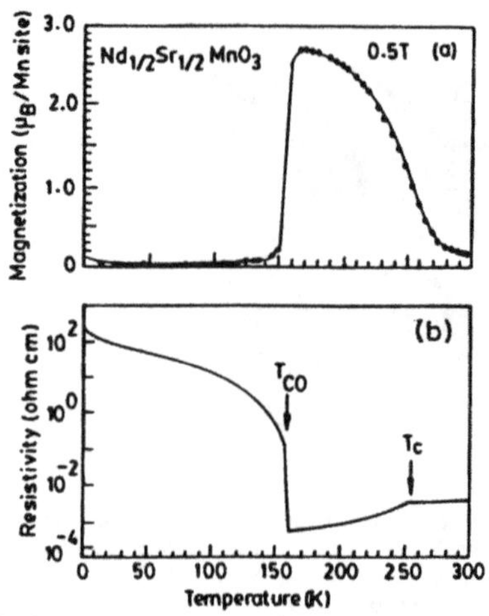


**Figure 11.** Temperature variation of (a) the magnetization and (b) the resistivity of $Nd_{0.5}Sr_{0.5}MnO_3$ (from Kuwahara et al.[17]).

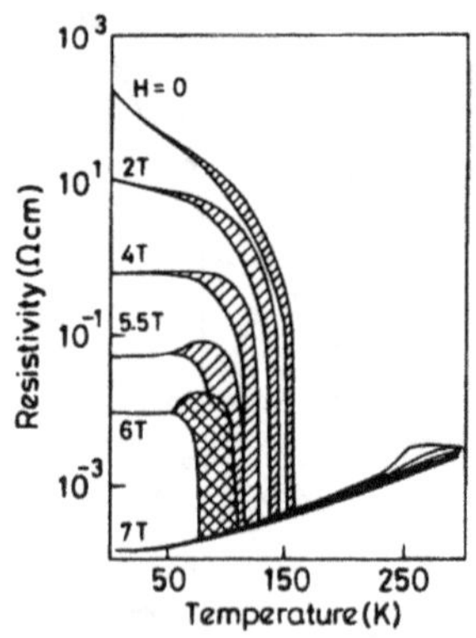


**Figure 12.** Effect of magnetic fields on the charge ordering transition of $Nd_{0.5}Sr_{0.5}MnO_3$ (from Kuwahara et al.[17]).

$Nd_{0.5}Sr_{0.5}MnO_3$ is a ferromagnetic metal with a $T_c$ of ~250 K and transforms to an insulating CO state at about 150 K (Figure 11). The CO transition is accompanied by spin ordering, and the CO insulator is a CE-type antiferromagnet.[17] The orbital ordering therefore involves $3x^2 - r^2/3y^2 - r^2$ orbitals as in Figure 7a. Application of a magnetic field of 7T destroys the CO state, and the material becomes metallic (Figure 12), the sharpness of the transition decreasing with increasing strength of the field. The transition is first-order, showing hysteresis, and is associated with changes in unit cell parameters. The unit cell volume of the CO state is considerably smaller than that of the FMM state. The properties of $Nd_{0.5}Sr_{0.5}MnO_3$ can be described in terms of the phase diagram shown in Figure 13. The *Imma* space group of this manganate renders the Mn−O−Mn angle in the *ab* plane closer to 180°, promoting the overlap of the Mn($e_g$) and O(2p) orbitals. Vacuum tunneling measurements[19] show that a gap of 250 meV opens up below $T_{CO}$ (Figure 14). The gap collapses on applying a magnetic field, suggesting that a gap in the density of states at $E_F$ is necessary for the stability of the CO state. Photoemission studies indicate a sudden change in electron states at the transition and give an estimate of 100 meV for the gap.[20] These estimates of the gap are considerably larger than the $T_{co}$ (12 meV); it is not clear how a magnetic field of 6T (1.2 meV) can destroy the CO state. $Nd_{0.5}Sr_{0.5}MnO_3$ shows anomalous magnetostriction behavior, with a large positive magnetovolume effect (Figure 15), owing

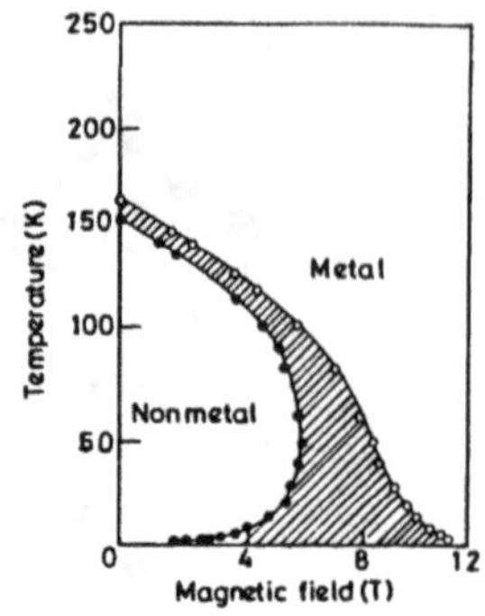


**Figure 13.** Temperature-magnetic field phase diagram for $Nd_{0.5}Sr_{0.5}MnO_3$ (from Tokura et al.[18]).

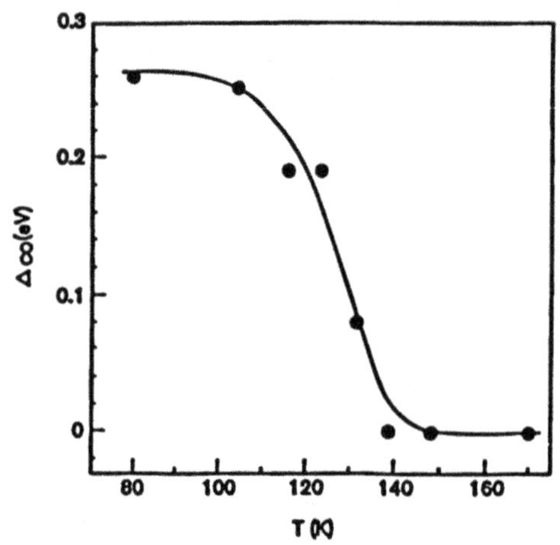


**Figure 14.** CO gap in $Nd_{0.5}Sr_{0.5}MnO_3$ as revealed by vacuum tunneling measurements (from Biswas et al.[19]).

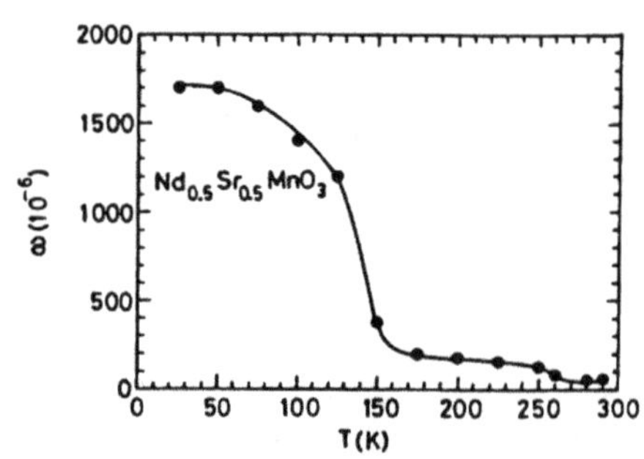


**Figure 15.** Temperature variation of the maximum volume magnetostriction in $Nd_{0.5}Sr_{0.5}MnO_3$ at 13.7T (from Mahendiran et al.[21]).

to the magnetic field-induced structural transition accompanying a change from the AFM CO state to the FMM state.[21]

$Pr_{0.6}Ca_{0.4}MnO_3$ is an insulator at all temperatures and becomes charge-ordered at about 230 K ($T_{CO}$). At this temperature, anomalies are found in the magnetic susceptibility, as well as in the resistivity, as shown in Figure 16. In the CO state, the $Mn^{3+}$ and $Mn^{4+}$ ions are regularly arranged in the *ab* plane with the associated ordering of the $3x^2 - r^2/3y^2 - r^2$ orbitals. On cooling, AFM ordering (CE-type) occurs at 170 K ($T_N$). At about 40 K, $Pr_{0.6}Ca_{0.4}MnO_3$ exhibits canted AFM ordering. Application of an external magnetic field transforms the CO state to a FMM state, as shown in Figure 16, but the field required is much larger than in $Nd_{0.5}Sr_{0.5}MnO_3$. The transition

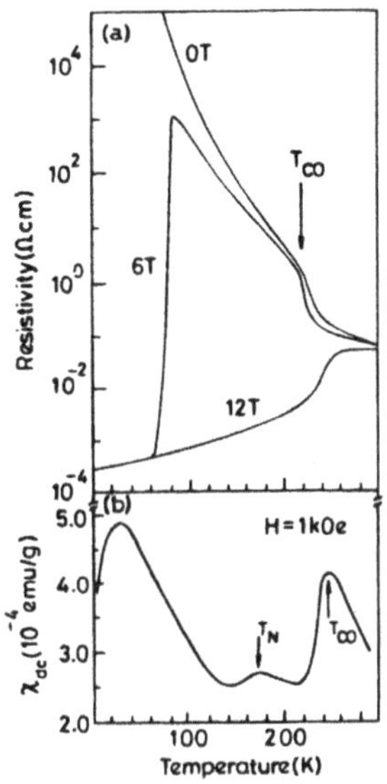


**Figure 16.** Temperature variation of (a) resistivity and (b) magnetic susceptibility of $Pr_{0.6}Ca_{0.4}MnO_3$ (from Tomioka et al.[22a] and Lees et al.[22b]).

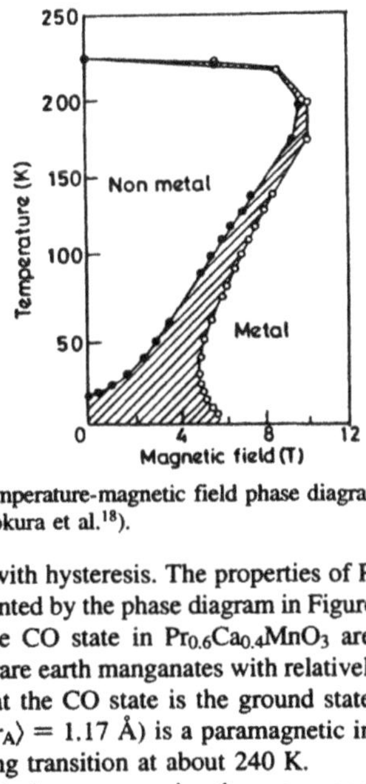


**Figure 17.** Temperature-magnetic field phase diagram for $Pr_{0.6}Ca_{0.4}MnO_3$ (from Tokura et al.[18]).

is associated with hysteresis. The properties of $Pr_{0.6}Ca_{0.4}MnO_3$ can be represented by the phase diagram in Figure 17. The basic features of the CO state in $Pr_{0.6}Ca_{0.4}MnO_3$ are exhibited by several other rare earth manganates with relatively small A-size cations, in that the CO state is the ground state. Thus, $Nd_{0.5}Ca_{0.5}MnO_3$ ($\langle r_A \rangle = 1.17$ Å) is a paramagnetic insulator with a charge-ordering transition at about 240 K.

Charge ordering occurs in the paramagnetic state in $Pr_{1-x}Ca_xMnO_3$ ($0.35 \leq x \leq 0.5$) with the $T_{CO}$ increasing with $x$. The paramagnetic state is characterized by FM spin fluctuations with a small energy scale.[23] At $T_{CO}$, these fluctuations decrease and disappear at $T_N$. Electron diffraction and dark-field transmission electron microscopy (TEM) images show the presence of incommensurate charge ordering in the paramagnetic insulating state (180−260 K) of the $x = 0.5$ composition.[24] At $T_N$, there is an incommensurate−commensurate CO transition. In the incommensurate CO structure, partial orbital ordering is likely to be present. Similar charge, orbital, and spin ordering has been found in the 0.3 composition as well.[25] Optical conductivity spectra of the $x = 0.4$ composition show evidence of spatial charge and orbital ordering at 10 K.[26] The CO state

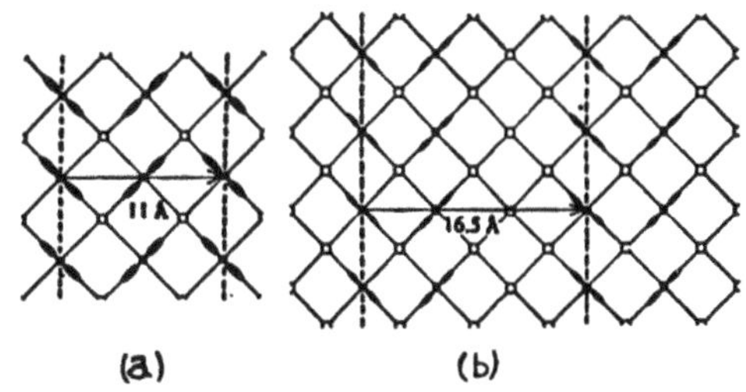


**Figure 18.** Charge and orbital ordering in $La_{1-x}Ca_xMnO_3$: (a) $x = 0.50$, (b) $x = 0.67$. Open circles represent $Mn^{4+}$ and lobes show $e_g$ orbitals of $Mn^{3+}$. Charge modulation wavelengths are ~11 and 16.5 Å for $x = 0.50$ and 0.67, respectively (following Cheong).

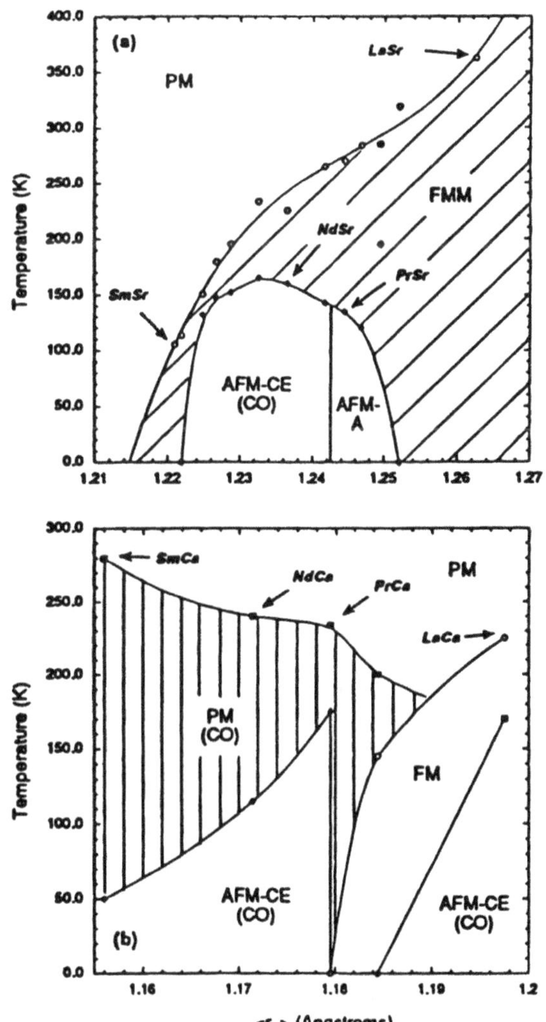


**Figure 19.** Temperature−$\langle r_A \rangle$ phase diagrams for (a) $Ln_{0.5}Sr_{0.5}MnO_3$ and (b) $Ln_{0.5}Ca_{0.5}MnO_3$ (from Woodward et al.[31]).

has a gap of ~0.2 eV, and the gap remains up to 4.5T. The gap value is the order parameter of the CO state and couples with spin ordering. $Pr_{0.67}Ca_{0.33}MnO_3$ shows thermal relaxation effects from the metastable FMM state (produced by the application of 10T magnetic field) to the CO state.[27] A metal−insulator transition is observed as an abrupt jump in resistivity at a well-

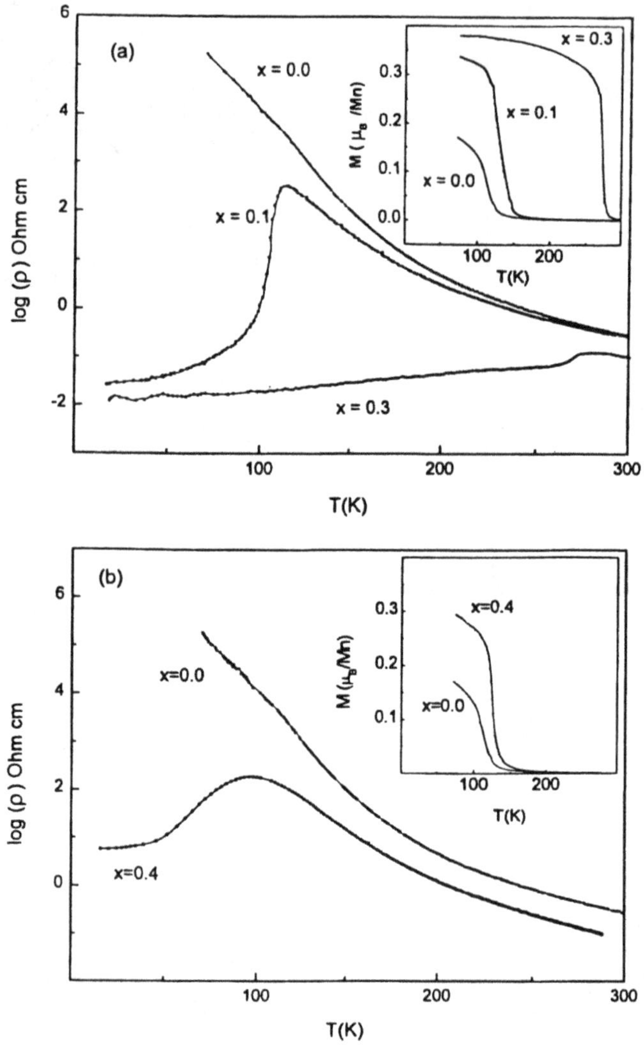


**Figure 20.** Effect of internal pressure on the properties of (a) $Pr_{0.7}Ca_{0.3-x}Sr_xMnO_3$ and (b) $Pr_{0.7-x}La_xCa_{0.3}MnO_3$. Note that substitution by a larger A-site cation renders the material FM (see insets). An insulator−metal transition is also observed (from Rao et al.[33]).

defined time, depending on the temperature. This observation seems to indicate a percolative nature of current transport.

Charge and orbital ordering in the manganates give rise to stripes.[28] In Figure 18 we show the modulation that can arise in $La_{1-x}Ca_xMnO_3$ with $x = 0.50$ and 0.67. At $x = 0.50$, the same number of $Mn^{3+}$ and $Mn^{4+}$ ions exist and diagonal stripes with a spacing of 11 Å are, therefore, to be expected (Figure 18a). When $x = 0.67$, there are twice as many $Mn^{4+}$ ions as $Mn^{3+}$ ions. Ordering of diagonal rows of $Mn^{3+}$ and $Mn^{4+}$ ions, besides the orientational ordering of the orbitals, would give rise to the striped pattern in Figure 18b, with a periodicity of 16.5 Å.

Paired JT distorted stripes or bistripes are believed to be present in $La_{0.33}Ca_{0.67}MnO_3$.[29] A supercell based on such ordering has been proposed, but a detailed structural investigation[30] has shown that the diffraction data can be explained satisfactorily without bistripes. Apparently, owing to orbital disordering, a mixture of paired and unpaired stripes seems to occur in $La_{0.5}Ca_{0.5}MnO_3$. Because such charge stripes generally are observed by electron microscopy, it is not clear whether these structures truly represent the bulk composition.

**Cation Size Effects.** In the manganates exhibiting CMR, the ferromagnetic $T_c$ increases with the average radius of the A-site cations, $\langle r_A \rangle$. Increasing $\langle r_A \rangle$ is equivalent to increasing the hydrostatic pressure and is therefore accompanied by an increase in the Mn − O − Mn angle and the $e_g$ bandwidth. If there is considerable mismatch in the radii of the different A-site cations, however, the $T_c$ does not increase with $\langle r_A \rangle$. Charge ordering is also highly sensitive to $\langle r_A \rangle$ and the $T_{CO}$ generally increases with decrease in $\langle r_A \rangle$. The sensitivity of $T_{CO}$ to $\langle r_A \rangle$ has been examined[7,31,32] and is generally attributed to an increased tilting of the $MnO_6$ octahedra as the $\langle r_A \rangle$ decreases. In Figure 19 we show that the phase diagrams[31] of $Ln_{0.5}Sr_{0.5}MnO_3$ and $Ln_{0.5}Ca_{0.5}MnO_3$ illustrate some of the important features. (Here, the $\langle r_A \rangle$ is varied by changing the Ln ion.) The Mn−O(eq)−Mn and Mn−O(ax)−Mn bonds are identical in $Ln_{0.5}Ca_{0.5}MnO_3$ (excect when Ln = La). For $La_{0.5}Ca_{0.5}MnO_3$ and all the $Ln_{0.5}Sr_{0.5}MnO_3$ compounds, the Mn−O(eq)−Mn angle is significantly larger (by 2−6°) than the Mn−O(ax)−Mn angle. Although the $Ln_{0.5}Ca_{0.5}MnO_3$ manganates crystallize in the *Pnma* symmetry, there is an evolution from *Pnma* to *I4/mcm* through *Imma* in the $Ln_{0.5}Sr_{0.5}MnO_3$ manganates, with increase in $\langle r_A \rangle$. The changes in the octahedral tilt system have consequences on the low-temperature magnetic structure. This is seen in $Nd_{0.5}Sr_{0.5}MnO_3$ where the charge ordering in the CE-type AFM state is associated with the *Imma* structure.

As pointed out earlier, the CO states of $Nd_{0.5}Sr_{0.5}MnO_3$ ($\langle r_A \rangle$ = 1.24 Å) and $Pr_{0.6}Ca_{0.4}MnO_4$ ($\langle r_A \rangle$ = 1.17 Å) are destroyed by magnetic fields. The field required to melt the CO state varies with $\langle r_A \rangle$ and the manganates with very small $\langle r_A \rangle$ remain charge-ordered even on the application of high magnetic fields.[32] Thus, $Y_{0.5}Ca_{0.5}MnO_3$ ($\langle r_A \rangle$ = 1.13 Å) has a robust CO state that is not affected by very high magnetic fields (>25T). We can thus distinguish three different categories of manganates with respect to their sensitivity to magnetic fields: (a) manganates that are FM and become charge-ordered at low temperatures (e.g., $Nd_{0.5}Sr_{0.5}MnO_3$ when $T_{CO} = T_N$), with the CO state transforming to a FMM state on the application of a moderate magnetic field; (b) manganates that are charge-ordered in the paramagnetic state ($T_N < T_{CO}$), and do not exhibit an FMM state, but transform to a FMM state under a magnetic field (e.g., $Pr_{1-x}Ca_xMnO_3$); and (c) those that are charge-ordered in the paramagnetic state ($T_N < T_{CO}$) as in b, but are not affected by magnetic fields up to 15T or greater (e.g., $Y_{0.5}Ca_{0.5}MnO_3$). Category c is encountered when $\langle r_A \rangle \leq 1.17$ Å. The apparent one-electron bandwidth estimated on the basis of the experimental Mn−O−Mn angle and the average Mn−O distance in $Ln_{0.5}A_{0.5}MnO_3$ does not vary significantly with $\langle r_A \rangle$, which suggests that other factors may be responsible for the sensitivity of the CO state to $\langle r_A \rangle$. One possibility is a competition between the A- and B-site cations for covalent mixing with the O(2p) orbitals.[32]

By increasing the size of the A-site cations or by the application of pressure, the CO state in the manganates can be transformed to the FMM state.[33,34] In Figure 20, we show the effect of internal pressure on the CO state of $Pr_{0.7}Ca_{0.3}MnO_3$ wherein Ca is substituted by the larger Sr or Pr is substituted by La. The $T_c$ in the $Pr_{0.5}Sr_{0.5-x}Ca_xMnO_3$ system, decreases with an increase in $x$ or a decrease in $\langle r_A \rangle$ up to $x = 0.25$; $T_{CO} = T_N$ from $x = 0.09$ to 0.25. When $\langle r_A \rangle$ is decreased further, $T_{CO}$ increases from 180 K for $x = 0.25$ to 250 K for $x = 0.30$; for $0.30 \leq x \leq 0.50$, $T_{CO} > T_N$.[35]

The effect of cation size disorder on the ferromagnetic $T_c$ of rare earth manganates exhibiting CMR has been investigated in detail. The disorder is quantified in terms of the variance in the distribution $\langle r_A \rangle$, as defined by Attfield.[36] The variance $\sigma^2$ is defined by, $\sigma^2 = \Sigma x_i r_i^2 - \langle r_A \rangle^2$, where $x_i$ is the fractional occupancy of A-site ions and $r_i$ is the ionic radius. The ferromagnetic $T_c$ decreases significantly with the variance, $\sigma^2$, based on the studies of rare earth manganates with fixed $\langle r_A \rangle$. A similar study of the variation of $T_{CO}$ with $\sigma^2$ in $Ln_{0.5}A_{0.5}MnO_3$, for fixed $\langle r_A \rangle$ values of 1.17 and 1.24 Å, has shown that $T_{CO}$ is not very sensitive to the size mismatch.[37] It appears that JT distortion and Coulomb interactions play a prominent role in determining the nature of the CO state in these materials.

Considering that the rare earth manganates with large $\langle r_A \rangle$, as exemplified by $Nd_{0.5}Sr_{0.5}MnO_3$ ($\langle r_A \rangle$ = 1.17 Å), exhibit entirely different characteristics of the CO state, and that $\langle r_A \rangle$ = 1.17 Å categorizes the manganates with respect to their insensitivity to magnetic fields, we would expect interesting and unusual properties in the intermediate range of $\langle r_A \rangle$ of 1.20 ± 0.20 Å. In this regime $T_{CO}$ approaches $T_c$, leading to a competition between charge ordering and ferromagnetism. Thus, $La_{0.5}Ca_{0.5}MnO_3$ ($\langle r_A \rangle$ = 1.20 Å) exhibits a region of coexistence of ferromagnetism and charge ordering ($T_c$ = 225 K, $T_{CO}$ = 135 K). At 135 K, the material also becomes AFM (CE-type) and the orbital ordering becomes commensurate.[38] In $(Nd_{1-x}Sm_x)_{0.5}Sr_{0.5}MnO_3$, $T_c$ decreases from 255 K to 115 K as $x$ increases from 0.0 to 0.875, as expected of a decrease in $\langle r_A \rangle$.[39] The $T_{CO}$ also decreases from 158 K to 0 K as $x$ changes from 0.0 to 0.875. This unusual behavior wherein $T_{CO}$ is suppressed as $T_c$ approaches $T_{CO}$ is interesting.

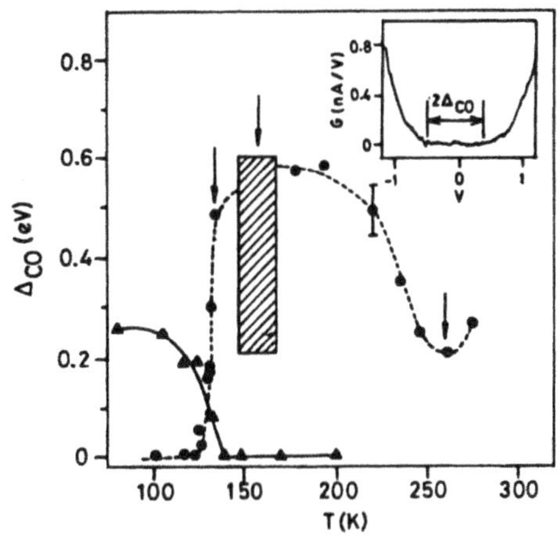


**Figure 21.** Temperature variation of the CO gap in $Nd_{0.25}La_{0.25}Ca_{0.5}MnO_3$ (closed circles) and $Nd_{0.5}Sr_{0.5}MnO_3$ (closed triangles). Inset shows a typical tunneling conductance curve (from Arulraj et al.[40]). Shaded region represents the coexistence regime.

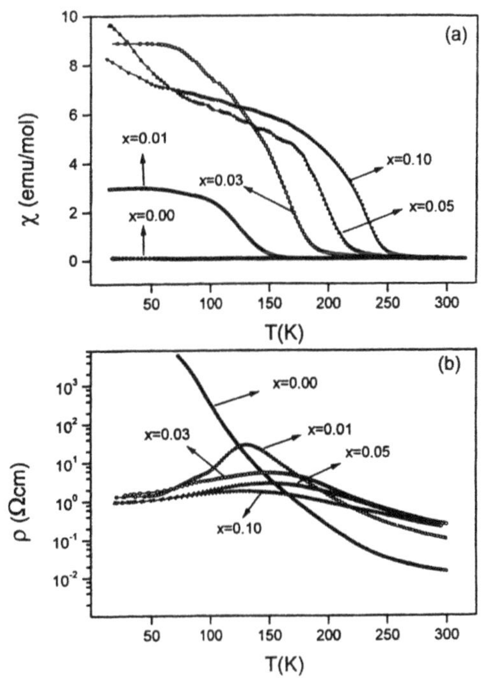


**Figure 22.** Temperature variation of the (a) magnetic susceptibility and (b) resistivity of $Nd_{0.5}Ca_{0.5}Mn_{1-x}Ru_xO_4$ (from Vanitha et al.[46b]).

$Nd_{0.25}La_{0.25}Ca_{0.5}MnO_3$ ($\langle r_A \rangle$ = 1.19 Å) reveals an intriguing sequence of phase transitions.[40] On cooling, this manganate develops an incipient CO state below 220 K. The formation of this state is accompanied by an increase in electrical resistivity and the opening up of a gap in the density of states near $E_F$. The orthorhombic distortion also increases, as a consequence of cooperative JT distortion of the lattice and short-range

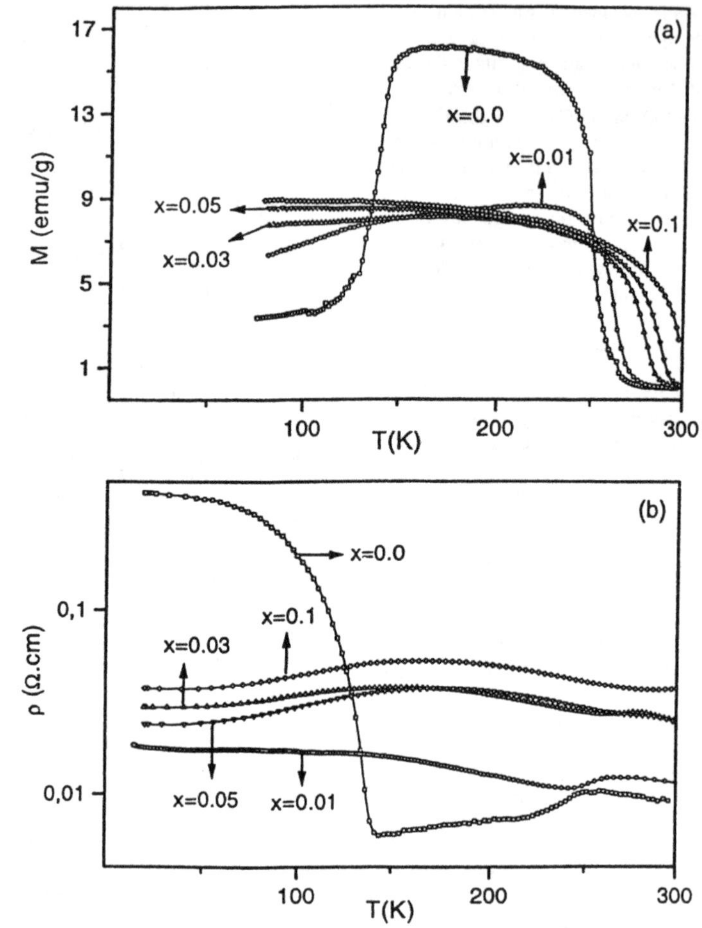


**Figure 23.** Temperature variation of the (a) magnetization and (b) resistivity of $Nd_{0.5}Sr_{0.5}Mn_{1-x}Ru_xO_4$ (from Vanitha et al.[46b]).

ordering of the $Mn^{3+}$ and $Mn^{4+}$ ions. At about 150 K, the incipient CO state becomes unstable and the material undergoes a reentrant transition to the FMM state. The transition is characterized by a sharp decrease in resistivity, collapse of the CO gap, development of FM ordering, and an abrupt decrease in the orthorhombic distortion. There is a two-phase coexistence region (150−220 K) around the CO−FMM transition. We show the fascinating reentrant transition in Figure 21 where the CO gap, obtained by vacuum tunneling measurements,[40] is plotted against temperature.

**Effects of Substitution in the Mn Site and by $^{18}O$.** Substitution of $^{16}O$ by $^{18}O$ (generally up to 85−90%) has a marked effect on the magnetic properties and CMR of the manganates, indicating the important role of electron−phonon coupling. Substituting $^{18}O$ for $^{16}O$ in $La_{0.175}Pr_{0.525}Ca_{0.3}MnO_3$ destroys the insulator−metal transition and renders the material insulating down to 4 K.[41] The CO transition in $La_{0.5}Ca_{0.5}MnO_3$ increases by 9 K upon replacing $^{16}O$ by $^{18}O$; the isotope shift increases with the magnetic field in $La_{0.5}Ca_{0.5}MnO_3$ and in $Nd_{0.5}Sr_{0.5}MnO_3$.[42] In $Pr_{0.67}Ca_{0.33}MnO_3$, the magnetic field−induced insulator−metal transition occurs at a higher field on $^{18}O$ substitution; the heavier isotope favors the insulating state.[43] The isotope effect on $T_{CO}$ is greater in $Nd_{0.5}Sr_{0.5}MnO_4$ than $Pr_{0.5}Ca_{0.5}MnO_3$, indicating the role of $\langle r_A \rangle$ as well.[44]

Substitution of Mn by cations such as $Al^{3+}$ and $Fe^{3+}$ in charge-ordered manganates destroys charge ordering at moderate doping ($x \geq 0.03$), but the materials remain insulating. However, substitution by $Cr^{3+}$ readily destroys charge ordering and renders the material FM and metallic.[45,46] In $Sm_{0.5}Ca_{0.5}Mn_{1-x}Cr_xO_3$, the $T_{CO}$ (275 K when $x = 0.0$) decreases with increasing x and charge ordering disappears at $x = 0.05$. The $x = 0.05$ composition shows an insulator−metal transition when the material becomes FM.[47] The effectiveness of $Cr^{3+}$ in destroying the CO state is considered to be due to its favorable electronic configuration ($t_{2g}^3$), which is the same as that of $Mn^{4+}$. However, $Cr^{3+}$ in the $Mn^{3+}$ site would be surrounded immediately by $Mn^{4+}$ ions, which would not allow for near-neighbor electron hopping. Hopping would be more favored if $Mn^{4+}$ were substituted by an appropriate quadrivalent cation such as $Ru^{4+}$ ($t_{2g}^4\ e_g^0$). Recent studies have shown that substitution of Mn by Ru in $Nd_{0.5}Ca_{0.5}MnO_3$ destroys charge ordering and renders the material FM, with the $T_c$ increasing with Ru content; the material also shows an insulator−metal transition[46] (Figure 22). The marked effect of Ru substitution is also seen in $Nd_{0.5}Sr_{0.5}MnO_3$ where the $T_c$ increases with the Ru content to well above 300 K, but charge ordering is destroyed (Figure 23).

**Phase Separation.** The formation of FM clusters in an AFM host matrix in the rare earth manganates has been noticed by many workers. Thus, spin glass behavior has been encountered in the $Ln_{1-x}A_xMnO_3$ system at either extreme, corresponding to large or small $x$. Electronic phase separation is also evidenced

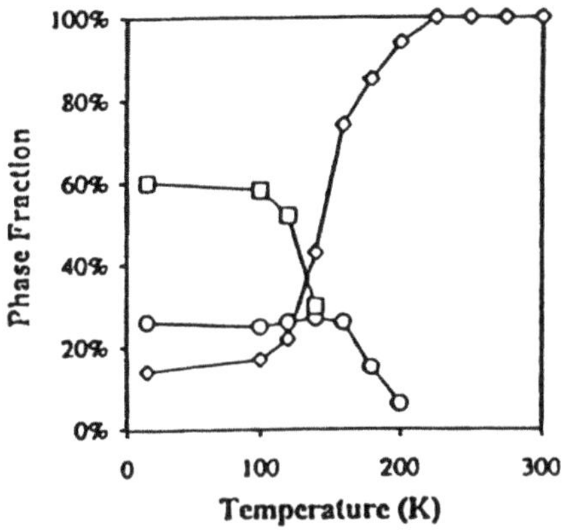


**Figure 24.** Variation in the percentage of the different phases of $Nd_{0.5}Sr_{0.5}MnO_3$ with temperature: FMM phase (diamonds); orbitally ordered A-type AFM phase (circles); charge-ordered CE-type AFM phase (squares) (from Woodward et al.[50]).

in the manganates. Thus, an electron microscopic study of $La_{1-x-y}Pr_yCa_xMnO_3$ ($x$ = 0.375) has shown electronic phase separation into a submicrometer-scale mixture of CO insulator regions and FMM domains. Perculative transport could occur between the two states.[48] The coexistence of FM and CO states has been observed in $La_{0.5}Ca_{0.5}MnO_3$ and $Nd_{0.25}La_{0.25}Ca_{0.5}MnO_3$.[38,40] Some phase separation is likely in many of the charge-ordered manganates. In Cr-doped $Nd_{0.5}Ca_{0.5}MnO_3$, submicrometer FMM domains are embedded in an AFM CO state, so the material shows a relaxor behavior.[49]

$Nd_{0.5}Sr_{0.5}MnO_3$, which shows evidence for separation into three macroscopic phases, is particularly interesting. These are the high-temperature FMM phase (*Imma*), the A-type AFM intermediate-temperature phase (*Imma*), and the CE-type AFM CO low-temperature phase ($P2_1/m$).[50] The A-type AFM phase starts manifesting itself around 220 K, whereas the CE-type CO phase first appears at 150 K. In Figure 24, we show the phase compositions at different temperatures. There are three phases at the so-called CO transition at 150 K. The presence of the high-temperature FMM phase, even at very low temperatures, is noteworthy. These results are of significance in interpreting many of the properties of this manganate. The fact that such phase separation is seen by X-ray/neutron diffraction implies that even the minority phases have large domains (≥100 nm). The FMM phase of $Nd_{0.5}Sr_{0.5}MnO_3$ has a larger volume than the average volume or the volume of the low-temperature CO phase. The phase-separation behavior of this system and the relative stabilities of the structures seem to depend crucially on the $Mn^{4+}/Mn^{3+}$ ratio. Thus, the regime $Mn^{4+}/Mn^{3+} > 1$ seems to stabilize the orbitally ordered AFM (A-type) phase. Unlike $Nd_{0.5}Sr_{0.5}MnO_3$, $Nd_{0.45}Sr_{0.55}MnO_3$ has an A-type AFM state below $T_N$ (220 K) and metallicity is confined to the ferromagnetic *ab* plane below $T_N$, whereas the material is insulating along the *c*-axis. The large anisotropy in this material implies that the carriers are confined to the FM layer by magnetic and orbital ordering.[51] An interesting experiment on phase-separated $Nd_{0.5}Sr_{0.5}MnO_3$ at low temperatures was performed recently wherein the phase composition was determined by neutron diffraction in the absence and presence of a magnetic field.[52] The results show how on applying a magnetic field, the CO state at low temperatures, by and large, transforms to the FMM state (Figure 25). For some reason, a small proportion of the intermediate-temperature A-type AFM phase persists.

**Effect of Electric Fields.** Charge-ordered manganates such as $Pr_{1-x}Ca_xMnO_3$ ($x \approx 0.3-0.4$) undergo a CO–FMM transition on application of moderate magnetic fields. However, manganates with small $\langle r_A \rangle$ such as $Y_{0.5}Ca_{0.5}MnO_3$ are, for all practical purposes, unaffected by magnetic fields. Laser irradiation has been reported to cause an insulator–metal transition in $Pr_{0.7}Ca_{0.3}MnO_3$, generating a localized conduction path, although the bulk of the sample is insulating.[53] It has been found recently that small d.c. currents induce insulator–metal transitions in thin films of several charge-ordered rare earth manganates, including $Y_{0.5}Ca_{0.5}MnO_3$, $Nd_{0.5}Ca_{0.5}MnO_3$, and $Pr_{0.7}Ca_{0.3}MnO_3$.[54] The current–voltage characteristics are nonohmic and show hysteresis. The I–M transition temperature decreases with increasing current (Figure 26). The hysteretic I–M transition in Figure 27 is specially noteworthy in that there is a reproducible memory effect in the cooling and heating cycles. The current-induced I–M transition occurs even in $Y_{0.5}Ca_{0.5}MnO_3$,

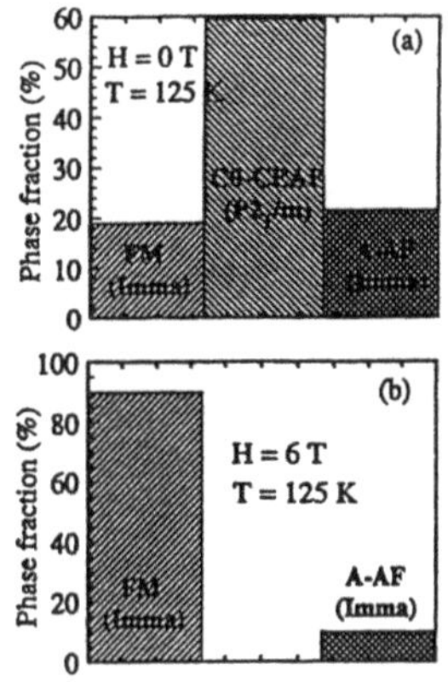


**Figure 25.** Effect of magnetic field (6T) on phase-separated $Nd_{0.5}Sr_{0.5}MnO_3$ at 125 K (from Ritter et al.[52]).

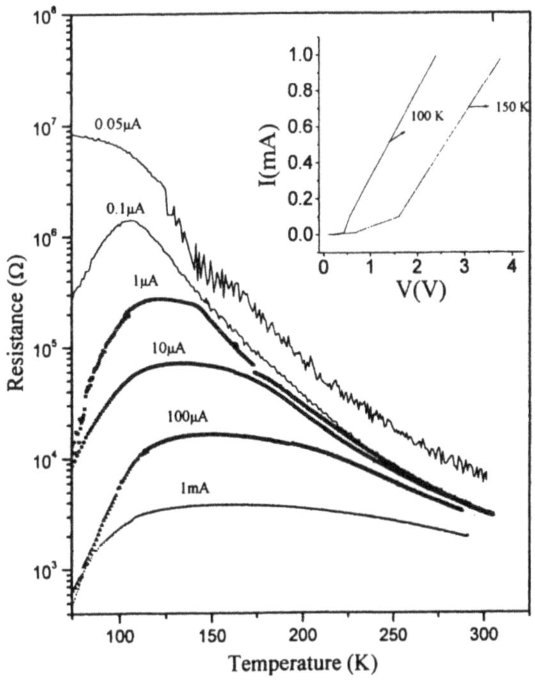


**Figure 26.** Electric current induced insulator–metal transition in $Nd_{0.5}Ca_{0.5}MnO_3$ films deposited on Si(100) at different values of the current. Inset show I–V curves at different temperatures (from Rao et al.[54]).

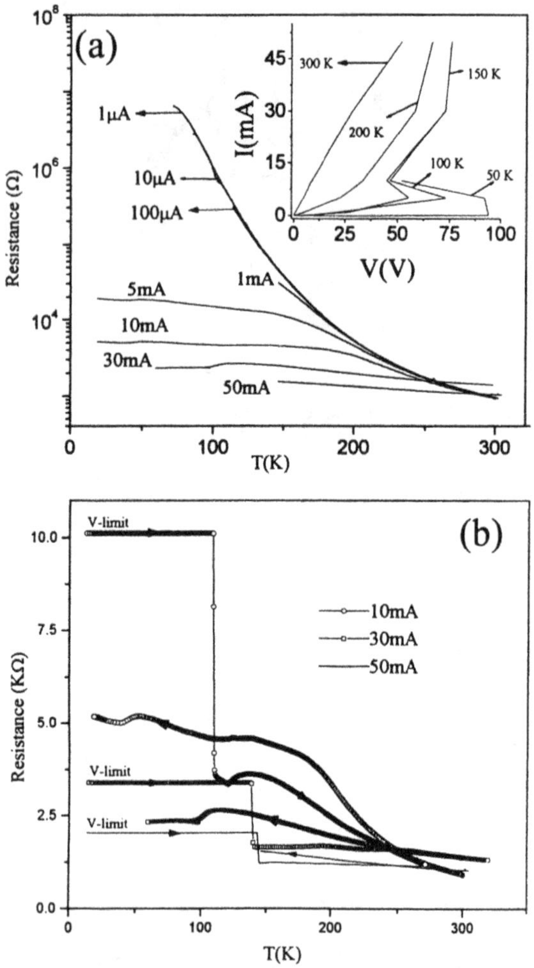


**Figure 27.** (a) Temperature variation of resistance of an oriented $Nd_{0.5}Ca_{0.5}MnO_3$ film deposited on $LaAlO_3(001)$ for different values of the current; (b) resistance−temperature plots for three current values recorded over cooling and heating cycles showing memory effect. Inset in part a shows I−V curves at different temperatures (from Rao et al.[54]).

which is not affected by large magnetic fields. Furthermore, there is no need for prior laser irradiation to observe the current-induced I−M transitions. It is proposed that electric fields cause depinning of the randomly pinned charge solid. There appears to be a threshold field in the CO regime beyond which nonlinear conduction sets in along with a large broad-band conductivity noise.[55] Threshold-dependent conduction disappears around $T_{CO}$, which suggests that the CO state gets depinned at the onset of nonlinear conduction. At small currents or low magnetic fields, resistance oscillations occur because of temporal fluctuations between resistive states.

**Layered Manganates.** The interplay of spin, orbital, and charge ordering in the rare earth manganates in determining their properties becomes even more prominent in the layered manganates. In the Ruddlesden−Popper series of manganates, $(Ln, A)_{n+1}Mn_nO_{3n+1}$, the $n = \infty$ phases are the three-dimensional perovskites. The $n = 2$ phases, such as $La_{2-2x}Sr_{1+2x}Mn_2O_7$, show a metallic ground state for $x \geq 0.17$ and high CMR; they also exhibit interesting properties arising from the layered nature of the materials. The $n = 1$ member, $Ln_{1-x}Sr_{1+x}MnO_4$, with the quasi two-dimensional $K_2NiF_4$ structure, becomes metallic only

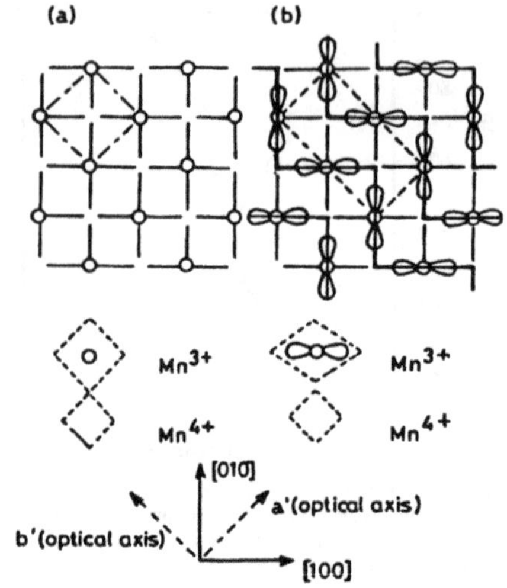


**Figure 28.** Charge ordering in $La_{0.5}Sr_{1.5}MnO_4$ showing oxide ion displacements. $Mn^{3+}$ ($e_g$) orbitals are shown (from Ishikawa et al.[56a]).

when $x = 0.7$. $La_{0.5}Sr_{1.5}MnO_4$ shows ordering of the $Mn^{3+}$ and $Mn^{4+}$ ions at about 220 K, accompanied by the ordering of the $e_g$ orbitals (Figure 28). The material becomes antiferromagnetic at 110 K and exhibits anisotropic properties arising from orbital ordering.[56] At the 220 K CO transition, a significant change occurs in the conductivity spectrum. The order parameter for orbital ordering increases at $T_{CO}$, grows further with decreasing temperature, and decreases on spin ordering at $T_N$. The order parameters for charge and orbital ordering seem to evolve together. Application of high magnetic fields gives rise to metamagnetic transitions below $T_{CO}$, and charge and orbital ordering are destroyed by the magnetic field. Charge ordering in $LaSr_2Mn_2O_7$ ($n = 2$) has been observed by electron diffraction.[57] There is a need to study charge ordering in this as well as in other rare earth analogues, including the calcium derivatives.

## Concluding Remarks

The rare earth manganates exhibit a variety of properties and phenomena, with an extraordinary sensitivity to various factors such as cation size, pressure, and magnetic and electric fields. In particular, the mutual relations between orbital ordering, charge ordering, and spin ordering in the rare earth manganates is truly fascinating. It is because of this interplay that charge ordering in these materials has turned out to be such an attractive area of research. In manganates such as $Pr_{0.6}Ca_{0.4}MnO_3$ and $Nd_{0.5}Ca_{0.5}MnO_3$ ($\langle r_A \rangle \approx 1.17$ Å), charge ordering occurs in the paramagnetic state, and commensurate orbital ordering accompanies a transition to the CE-type AFM state. In the paramagnetic CO state, there could be some orbital ordering that would be incommensurate, and FM correlations or clusters are generally present. In $Nd_{0.5}Sr_{0.5}MnO_3$ ($\langle r_A \rangle = 1.24$ Å), where charge ordering occurs on cooling a FMM state, however, it is possible to delineate transitions associated with orbital and charge ordering. In $La_{0.5}Ca_{0.5}MnO_3$ ($\langle r_A \rangle = 1.20$ Å), charge ordering and orbital ordering occur when the FM state transforms to an AFM state (CE type) on cooling. In the regime where the FM and CO states coexist, there is no long-range orbital ordering. As expected, the $La_{0.5}Ca_{0.5}MnO_3$ system gives rise to complex lattice images. In $La_{0.33}Ca_{0.67}MnO_3$, however, there is a clear-cut charge ordering transition (260 K) accompanied by orbital ordering, and followed by a transition to a CE-type AFM state, and paired stripes of $Mn^{3+}O_6$ octahedra have been observed in the lattice images. In manganates such as $Nd_{0.5}Ca_{0.5}MnO_3$ and $Y_{0.5}Ca_{0.5}MnO_3$ with a small $\langle r_A \rangle$, it is not clear whether ordered single or bistripes would be present. It is important to ensure that bistripes and other features found by electron miscroscopy truly represent the bulk structure. Low-temperature lattice imaging of these materials with high-resolution electron microscopy would be of great value.

Although we are able to distinguish spin, charge, and orbital ordering transitions, we do not understand some aspects. It is not entirely clear whether commensurate orbital ordering occurs only in the AFM (CE) state. The electron−hole asymmetry in the manganates, with respect to the FM and CO states, needs to be explained properly. Is it certain that we can never encounter ferromagnetism in the electron-doped materials? Also, what is the type of AFM ordering in the electron-doped CO manganates?

Phase separation in the manganates is especially interesting. Phase separation generally involves the formation of clusters of one phase in another (e.g., FM clusters in a CO or AFM phase). Macroscopic phase separation involving the presence of distinct domains of fairly large sizes (≥100 nm) with Bragg peaks in the diffraction patterns is, however, different from the formation of clusters or stripes. Such inhomogeneous distributions of CO or FM (charge and spin) phases in a material requires understanding. Coulomb forces would prevent accumulation of charge in a phase-separated regime in the absence of the means to compensate the charge. Electronic phase separation is known in $La_2NiO_{4+\delta}$ and $La_2CuO_{4+\delta}$.[3,11,28] In the rare earth manganates, there is increasing evidence for the coexistence of FM and AFM (CO) phases. Although it is possible that small changes in the $Mn^{3+}/Mn^{4+}$ ratio could affect phase separation, we do not understand the mechanism. Although one can consider the presence of large domains (e.g., CO and FM) of different phases of a given composition as a criterion for phase separation, the effects of phase separation, in contrast to those of chemical inhomogenieties or cluster formation, need to be investigated in greater detail.

The $\langle r_A \rangle$ regime of 1.20 ± 0.02 Å in $Ln_{0.5}A_{0.5}MnO_3$, which exhibits complex phenomena and properties, including reentrant transitions, deserves further study. It is useful to examine such systems with fixed $\langle r_A \rangle$ and variable cation size mismatch. Charge ordering in the layered manganates has to be investigated in detail. We do not yet have a full understanding of the extraordinary effect of electric fields on the CO state.

**Acknowledgment.** The author is thankful to the Department of Science and Technology and the Science Office of the European Union for support of this work.

### References and Notes


(1) (a) Honig, J. M. In *The Metallic and the Nonmetallic States of Matter*; Edwards, P. P., Rao, C. N. R., Eds.; Taylor and Francis: London, 1985. (b) Honig, J. M. *Proc. Indian Acad. Sci., Chem. Sci.* **1986**, *96*, 391.

(2) Tsuda, N.; Nasu, K.; Yanase, A.; Siratori, K. *Electronic Conduction in Oxides*; Springer: Berlin, 1991.

(3) (a) Kivelson, S. A.; Fradkin, E.; Emery, V. J. *Nature (London)* **1998**, *393*, 550. (b) Tranquada, J. M.; Sternleib, B. J.; Axe, J. D.; Nakamura, Y.; Uchida, S. *Nature (London)* **1995**, *375*, 561.

(4) Imada, M.; Fujimori, A.; Tokura, Y. *Rev. Mod. Phys.* **1998**, *70*, 1039.

(5) (a) von Helmolt, R.; Holzapfel, B.; Schultz, L.; Samwer, K. *Phys. Rev. Lett.* **1993**, *73*, 2331. (b) Chahara, K.; Ohno, T.; Kasai, M.; Kozono, Y. *Appl. Phys. Lett.* **1993**, *63*, 1990.

(6) (a) Rao, C. N. R.; Raveau, B., Eds. *Colossal Magnetoresistance, Charge Ordering and Related Properties of Manganese Oxides*; World Scientific: Singapore, 1998. (b) Ramirez, A. P. *J. Phys: Condens. Matter* **1997**, *9*, 8171.

(7) Rao, C. N. R.; Arulraj, A.; Santosh, P. N.; Cheetham, A. K. *Chem. Mater.* **1998**, *10*, 2714.

(8) Zener, C. *Phys. Rev.* **1951**, *82*, 403.

(9) Jirak, Z.; Krupicka, S.; Simsa, Z.; Dlouha, M.; Vratislav, S. *J. Magn. Magn. Mater.* **1985**, *53*, 153.

(10) (a) Takano, M.; Takeda, Y. *Bull. Inst. Chem. Res.* **1983**, *61*, 406. (b) Battle, P. D.; Gibb, T. C.; Lightfoot, P. *J. Solid State Chem.* **1990**, *84*, 271.

(11) (a) Tranquada J. M.; Loronzo, J. E.; Buttrey, D. J.; Sachan, V. *Phys. Rev.* **1995**, *B52*, 3581. (b) Tranquada, J. M.; Loronzo, J. E.; Buttrey, D. J.; Sachan, V. *Phys. Rev.* **1996**, *B54*, 12318.

(12) Ramirez, A. P.; Gammel, P. L.; Cheong, S. W.; Bishop, D. J.; Chandra, P. *Phys. Rev. Lett.* **1996**, *76*, 447.

(13) (a) Rodriguez-Carvajal, J.; Rousse, G.; Masquelier, C.; Hervieu, M. *Phys. Rev. Lett.* **1998**, *81*, 4660. (b) Willis, A. S.; Raju, N. P.; Greedan, J. E. *Chem. Mater.* **1999**, *11*, 1510.

(14) Wollan, E. O.; Koehler, W. C. *Phys. Rev.* **1955**, *100*, 545.

(15) (a) Goodenough, J. B. *Phys. Rev.* **1955**, *100*, 564. (b) Goodenough, J. B.; Wold, A.; Arnolt, R. J.; Menyuk, N. *Phys. Rev.* **1961**, *124*, 373.

(16) Vogt, T.; Cheetham, A. K.; Mahendiran, R.; Raychaudhuri, A. K.; Mahesh, R.; Rao, C. N. R. *Phys. Rev.* **1996**, *B54*, 15303.

(17) Kuwahara, H.; Tomioka, Y.; Asamitsu, A.; Moritomo, Y.; Tokura, Y. *Science* **1995**, *270*, 961.

(18) Tokura, Y.; Tomioka, Y.; Asamitsu, A.; Moritomo, Y.; Kasai, M. *J. Appl. Phys.* **1996**, *79*, 5288.

(19) Biswas, A.; Raychaudhuri, A. K.; Mahendiran, R.; Guha, A.; Mahesh, R.; Rao, C. N. R. *J. Phys. Condens. Matter* **1997**, *9*, L355.

(20) Sekiyama, A.; Suga, S.; Fujikawa; Imada, S.; Iwasaki, T.; Matsuda, K.; Matsushita, T.; Kaznacheyev, K. V.; Fujimori, A.; Kuwahara, H.; Tokura, Y. *Phys. Rev.* **1999**, *B59*, 15528.

(21) Mahendiran, R.; Ibarra M. R.; Maignan, A.; Millange, F.; Arulraj, A.; Mahesh, R.; Raveau, B.; Rao, C. N. R. *Phys. Rev. Lett.* **1999**, *82*, 2191.

(22) (a) Tomioka, Y.; Asamitsu, A.; Kuwahara, H.; Moritomo, Y.; Tokura, Y. *Phys. Rev.* 1996, *B53*, 1689. (b) Lees, M. R.; Barrat, J.; Balakrishnan, G.; Paul D, M. C. K.; Yethiraj, M. *Phys. Rev.* **1995**, *B52*, 14303.

(23) Kajimoto R.; Kakeshita T.; Oohara Y.; Yoshizawa H.; Tomioka Y.; Tokura, Y. *Phys. Rev.* **1998**, *B58*, R11837.

(24) Mori, S.; Katsufuji, T.; Yamamoto, N.; Chen, C. H.; and Cheong, S. W. *Phys. Rev.* **1999**, *B59*, 13573.

(25) Cox, D. E.; Radaelli, P. G., Marezio, M.; Cheong, S. W. *Phys. Rev.* **1998**, *B57*, 3305.

(26) Okimoto Y.; Tomioka, Y.; Onose, Y.; Otsuka, Y.; Tokura, Y. *Phys. Rev.* **1999**, *B59*, 7401.

(27) Anane, A.; Renard, J. P.; Reversat, L.; Dupas, C.; Veillet P.; Viret M.; Pinsard, L.; Revcolevsci, A. *Phys. Rev.* **1999**, *B59*, 77.

(28) Cheong, S. W.; Chen, C. H. In *Colossal Magnetoresistance, Charge Ordering and Related Properties of Manganese Oxides*; Rao, C. N. R., Raveau, B, Eds. World Scientific: Singapore, 1998.

(29) Mori, S.; Chen, C. H.; Cheong, S. W. *Nature (London)* **1998**, *392*, 473.

(30) Radaelli, P. G.; Cox, D. E.; Capogna, L.; Cheong, S. W.; Marezio, M. *Phys. Rev.* **1999**, *B59*, 14440.

(31) Woodward, P. M.; Vogt, T.; Cox, D. E.; Arulraj, A.; Rao, C. N. R.; Karen, P.; Cheetham, A. K. *Chem. Mater.* **1998**, *10*, 3652.

(32) Arulraj, A.; Santosh, P. N.; Gopalan, R. S.; Guha, A.; Raychaudhuri, A. K.; Kumar, N.; Rao, C. N. R. *J. Phys: Condens. Matter* **1998**, *10*, 8497.

(33) Rao, C. N. R.; Santosh, P. N.; Singh, R. S.; Arulraj, A. *J. Solid State Chem.* **1998**, *135*, 169.

(34) Moritomo, Y.; Kuwahara, H.; Tomioka, Y.; Tokura, Y. *Phys. Rev.* **1997**, *B55*, 7549.

(35) Damay, F.; Martin, C.; Maignan, A.; Hervieu, M.; Raveau, B.; Jirak, Z.; Andre, G.; Bouree, F. *Chem. Mater.* **1999**, *11*, 536.

(36) Attfield, J. P. *Chem. Mater.* **1998**, *10*, 3239.

(37) Vanitha, P. V.; Santosh, P. N.; Singh, R. S.; Rao, C. N. R.; Attfield, J. P. *Phys. Rev.* **1999**, *B59*, 13539.

(38) Radaelli, P. G.; Cox, D. E.; Marezio, M.; Cheong, S. W. *Phys. Rev.* **1997**, *B55*, 3015.

(39) Tokura, Y.; Kuwahara, H.; Moritomo, Y.; Tomioka, Y.; Asamitsu, A. *Phys. Rev. Lett.* **1996**, *76*, 3184.

(40) Arulraj, A.; Biswas, A.; Raychaudhuri, A. K.; Rao, C. N. R.; Woodward, P. M.; Vogt, T.; Cox, D. E.; Cheetham, A. K. *Phys. Rev.* **1998**, *B57*, R8115.

(41) Babushkina, N. A.; Belova, L. M.; Gorbenko, O. Yu; Kaul, A. R.; Bosak, A. A.; Ozhogin, V. I.; Kugel, KI. *Nature (London)* **1998**, *391*, 159.

(42) Zhao, G.; Ghosh, K.; Keller, H.; Greene, R. L. *Phys. Rev.* **1999**, *B59*, 81.

(43) Garcia-Landa, B.; Ibarra, M. R.; De Teresa, J. M.; Zhao, G.; Conder, K.; Keller, H. *Solid State Commun.* **1998**, *105*, 567.

(44) Mahesh, R.; Itoh, M. *J. Solid State Chem.* **1999**, *144*, 232.

(45) Raveau, B.; Maignan, A.; Martin, C.; Hervieu, M. *Chem. Mater.* **1998**, *10*, 2641.

(46) (a) Vanitha, P. V.; Singh, R. S.; Natarajan, S.; Rao, C. N. R. *Solid State Commun.* **1999**, *109*, 135. (b) Vanitha, P. V.; Arulraj, A.; Raju, A. R.; Rao, C. N. R. *C. R. Acad. Sci. Ser. II*, t.2 **1999**, 595.

(47) Barnabe, A.; Hervieu, M.; Martin, C.; Maignan, A.; Raveau, B. *J. Mater. Chem.* **1998**, *8*, 1405.

(48) Uehara M., Mori, S.; Chen, C. H.; Cheong, S. W. *Nature (London)* **1999**, *399*, 560.

(49) Kimura, T.; Tomioka, Y.; Kumai, R.; Okinoto, Y.; Tokura, Y. *Phys. Rev. Lett.* **1999**, *83*, 3940.

(50) Woodward, P. M.; Cox, D. E.; Vogt, T., Rao, C. N. R.; Cheetham, A. K. *Chem. Mater.* **1999**, *11*, 3528.

(51) Kuwahara, H.; Okuda, T.; Tomioka, Y.; Asamitsu, A.; Tokura, Y. *Phys. Rev. Lett.* **1999**, *82*, 4316.

(52) Ritter, C.; Mahendiran, R.; Ibarra, M. R.; Morellon, L.; Maignan, A.; Raveau, B.; Rao, C. N. R. *Phys. Rev.* **2000**, *B61*, R9229.

(53) Feibig, M.; Miyano, K.; Tomioka, Y.; Tokura, Y. *Science* **1998**, *280*, 1925.

(54) Rao, C. N. R.; Raju, A. R.; Ponnambalam. V.; Parashar, S.; Kumar, N. *Phys. Rev.* **2000**, *B61*, 591.

(55) Guha, A.; Ghosh, A.; Raychaudhuri, A. K.; Parashar, S.; Raju, A. R.; Rao, C. N. R. *Appl. Phys. Lett.* **1999**, *75*, 3381.

(56) (a) Ishikawa, T.; Ookura, K.; Tokura, Y. *Phys. Rev.* **1999**, *B59*, 8367. (b) Tokunaga, M.; Miura, N.; Moritomo, Y.; Tokura, Y. *Phys. Rev.* **1999**, *B59*, 11151.

(57) Li, J. Q.; Matsui, Y.; Kimura, T.; Tokura, Y. *Phys. Rev.* **1998**, *B57*, 3205.

PHILOSOPHICAL TRANSACTIONS OF THE ROYAL SOCIETY A

*Phil. Trans. R. Soc. A* (2008) **366**, 63–82
doi:10.1098/rsta.2007.2140
*Published online* 7 September 2007

# Electronic phase separation and other novel phenomena and properties exhibited by mixed-valent rare-earth manganites and related materials

BY VIJAY B. SHENOY[1] AND C. N. R. RAO[2,*]

[1]*Centre for Condensed Matter Theory, Department of Physics, Indian Institute of Science, Bangalore 560 012, India*
[2]*Chemistry and Physics of Materials Unit and CSIR Centre of Excellence in Chemistry, Jawaharlal Nehru Centre for Advanced Scientific Research, Jakkur Post, Bangalore 560 064, India*

Transition metal oxides, such as the mixed-valent rare-earth manganites $Ln_{(1-x)}A_xMnO_3$ (Ln, rare-earth ion, and A, alkaline-earth ion), show a variety of electronic orders with spatially correlated charge, spin and orbital arrangements, which in turn give rise to many fascinating phenomena and properties. These materials are also electronically inhomogeneous, i.e. they contain disjoint spatial regions with different electronic orders. Not only do we observe signatures of such electronic phase separation in a variety of properties, but we can also observe the different 'phases' visually through different types of imaging. We discuss various experiments pertaining to electronic orders and electronic inhomogeneities in the manganites and present a discussion of theoretical approaches to their understanding. It is noteworthy that the mixed-valent rare-earth cobaltates of the type $Ln_{(1-x)}A_xCoO_3$ also exhibit electronic inhomogeneities just as the manganites.

**Keywords: rare-earth cobaltates; rare-earth manganites; electronic phase separation; charge-ordering**

## 1. Introduction

Transition metal oxides such as rare-earth manganites, cuprates and cobaltates exhibit a plethora of properties and phenomena (Imada *et al.* 1998; Rao 2000) that have stimulated sustained investigations by materials chemistry and physics communities. Many of the fascinating properties of these oxides arise from the fact that the transition metal ions can exhibit mixed valence with strongly correlated electrons giving rise to many 'electronic orders'. By electronic orders, we mean spatially correlated arrangement of charge, spin, orbital and/or phase (as in a superconductor). In addition, there is overwhelming evidence that these oxides with high chemical homogeneity can show spatially inhomogeneous structures, i.e. regions showing different electronic orders (Mathur & Littlewood 2003;

* Author for correspondence (cnrrao@jncasr.ac.in).

One contribution of 15 to a Discussion Meeting Issue 'Mixed valency'.

 

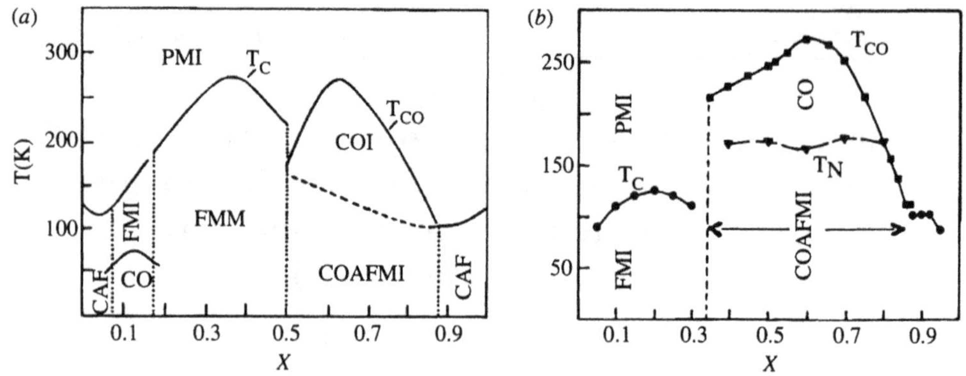


Figure 1. Phase diagram of (*a*) $La_{1-x}Ca_xMnO_3$ and (*b*) $Pr_{1-x}Ca_xMnO_3$ (adapted from Rao *et al.* (2004)).

Rao *et al.* 2004; Dagotto 2005; Shenoy *et al.* 2006). While many issues pertaining to these oxides have been understood, their science has many puzzles and challenges in addition to the promise of applications. The development of new oxide materials via the methods of materials chemistry with controlled electronic orders, and their theoretical understanding, therefore, continue to be at the forefront of research in condensed matter science.

In this article, we discuss spin, charge and orbital orders in transition metal oxides, in particular, in the rare-earth manganites which have become famous since 1993 owing to the phenomenon of colossal magnetoresistance (CMR) exhibited by them (Rao & Raveau 1998). We also touch upon the issue of electronic inhomogeneities; we discuss the signatures of the presence of such inhomogeneities and theoretical approaches to understand the same.

## 2. Electronic orders in rare-earth manganites

Rare-earth manganites of the type $Ln_{(1-x)}A_xMnO_3$ (Ln, rare-earth ion, and A, alkaline-earth ion) which crystallize in the perovskite structure were investigated several years ago by Wollan & Koehler (1955), but it was the discovery of CMR that stimulated their resurgence (Rao & Raveau 1998; Dagotto *et al.* 2001; Salamon & Jaime 2001; Tokura 2006). In the parent undoped compound $LaMnO_3$, the manganese ions exist in a $+3$ state, while the doped system such as $La_{(1-x)}Ca_xMnO_3$ contains (nominally) a fraction $(1-x)$ of $Mn^{3+}$ and $x$ of $Mn^{4+}$ ions. This mixed-valent character of Mn underlies the rich phase diagrams (figure 1) exhibited by the manganites. A study of the phase diagram reveals a large number of electronic orders present in these systems. In what follows, we discuss some of these electronic orders.

$LaMnO_3$ exhibits spin, orbital and lattice orders. The spin order is that of an A-type antiferromagnet (figure 2). This is accompanied by the orbital order, which couples strongly with the lattice distortion due to the Jahn–Teller (JT) effect, giving rise to the lattice distortion pattern shown in figure 2. Interestingly, the JT distortion pattern loses long-range order at approximately 750 K. This transition is also associated with a symmetry change of the crystal from an

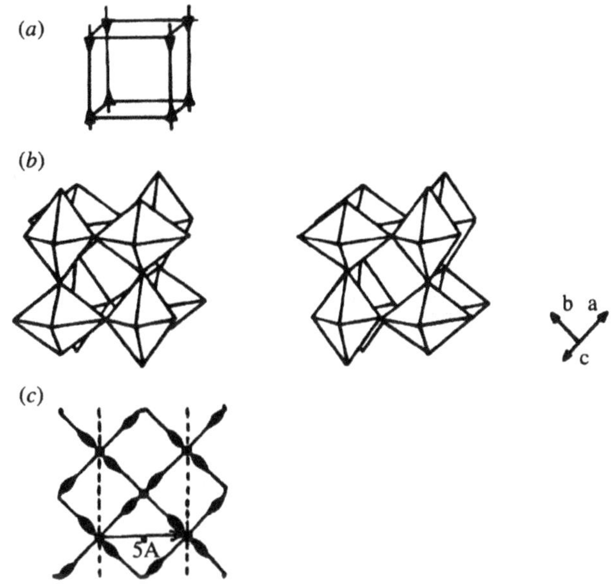


Figure 2. (*a*) A-type antiferromagnetic spin order, (*b*) Jahn–Teller distortion pattern and (*c*) orbital ordering in $LaMnO_3$ (after Rao (2000)).

orthorhombic structure at lower temperatures (due to the cooperative JT distortion) to a cubic structure at higher temperatures. At low temperatures, the combination of orbital ordering and JT distortion leads to an effective ferromagnetic (FM) exchange interaction between the spins on the plane, and a strong antiferromagnetic (AFM) exchange between spins between the planes, resulting in an A-type AFM arrangement of the spins. Other undoped manganites such as $NdMnO_3$ also show a similar electronic order.

Doping of manganites with alkaline-earth ions such as $Ca^{2+}$ and $Sr^{2+}$ results in a fraction of the manganese ions occurring in the +4 state, giving rise a new electronic order, charge order, wherein the +3 and +4 charges of the Mn ions are arranged in a periodic fashion (figure 3). Charge order is known to occur in other metal oxides as well, $Fe_3O_4$ being a well-known example (Rao 2000). The charge-ordered state is found predominantly in manganites with small Ln and A ions. Further, charge order disappears at high temperatures. Thus, in $Pr_{0.7}Ca_{0.3}MnO_3$, charge order appears only below 230 K. Interestingly, $Pr_{0.7}Ca_{0.3}MnO_3$ also shows AFM order below 170 K, where both antiferromagnetism and charge order coexist. At low temperatures, this manganite is electrically insulating. In contrast, $La_{0.7}Ca_{0.3}MnO_3$ does not show charge order, although it enters a FM metallic phase below a Curie temperature of 230 K, showing a PM insulating behaviour above this temperature. In the Sr-based $La_{0.7}Sr_{0.3}MnO_3$, the high-temperature PM phase is metallic. Note that the average radius of the A-site cations, $\langle r_A \rangle$, is larger in these two manganites than in $Pr_{0.7}Ca_{0.3}MnO_3$. CMR is observed at relatively moderate fields in $La_{0.7}Ca_{0.3}MnO_3$. Another important aspect of the doped manganites is that, when the size of the A-site cation (or the $e_g$ bandwidth) is large enough, electrons of the $Mn^{3+}$ hop to $Mn^{4+}$ by the double-exchange (DE) mechanism. The DE mechanism plays a crucial role in the occurrence of

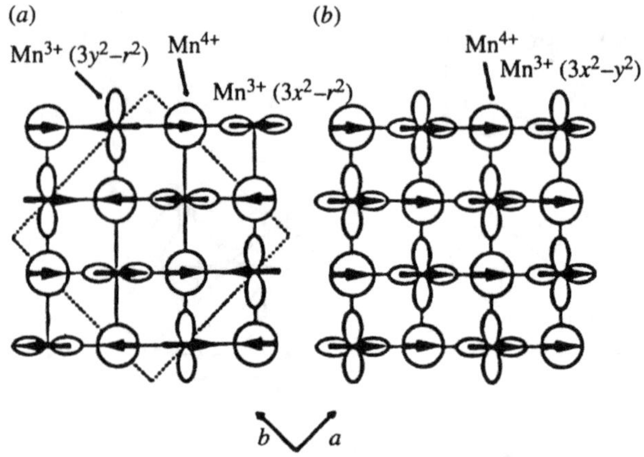


Figure 3. Ordering of charge, spin and orbital in doped manganites. (*a*) CE-type order, the dashed line shows the unit cell and (*b*) A-type order. In both figures, circles correspond to $Mn^{4+}$ (after Rao (2000)).

metallicity and ferromagnetism in these materials. The phenomenon of CMR is explained by invoking the DE mechanism along with certain competing interactions involving phonons. Charge ordering is exactly the opposite of DE, since it localizes electrons creating an AFM insulating state.

Half-doped manganites show many interesting electronic orders. For example, $Nd_{0.5}Sr_{0.5}MnO_3$, which is a FM metal at high temperatures, shows a transition to a charge-ordered AFM state at approximately 150 K. The charge-ordering temperature is usually somewhat higher than the magnetic-ordering temperature ($T_{CO} \gtrsim T_N$). Interestingly, the charge order can be 'melted' by a small magnetic field (Kuwahara *et al.* 1995). At low temperature, the charge-ordered state is associated with a CE-type AFM order (figure 3), orbital order and a regular pattern of JT distortions. The magnetic order consists of *ab* planes stacked along the *c*-axis (figure 3) with an AFM coupling between spins on neighbouring planes. This may be contrasted with A-type order (figure 2). Note that there is no charge ordering in an A-type antiferromagnet unlike in a CE-type antiferromagnet. Usually, orbital and spin orders that are not accompanied by charge order show A-type antiferromagnetism.

The presence of charge and orbital orders can be observed experimentally by a study of the structures of the manganites. For example, the charge-ordered structure of $Nd_{0.5}Sr_{0.5}MnO_3$ consists of distorted oxygen octahedra (figure 4) containing zigzag chains with alternate long and short Mn–O bonds. The bond lengths are 1.921 and 2.021 Å along $\langle 110\rangle$ and 1.881 and 2.020 Å along $\langle -110\rangle$ (figure 4). The emergence of a charge gap is also indicated by vacuum tunnelling measurements (Biswas *et al.* 1997). Intriguingly, the low-temperature gap (approx. 250 meV) is much larger than $T_{CO} \sim 12$ meV. Moreover, a magnetic field of 6 T (approx. 1.2 meV) can destroy the charge-ordered state. A consequence of this is the large magnetostrictive effect arising from the magnetic field-induced structural transition that accompanies the AFM charge-ordered insulator to FM-metal transition.

An interesting fundamental question that arises is, what controls the different electronic orders? At the level of the chemistry of the constituents, the mean A-site cation radius $\langle r_A\rangle$ is expected to play an important role. This arises from the

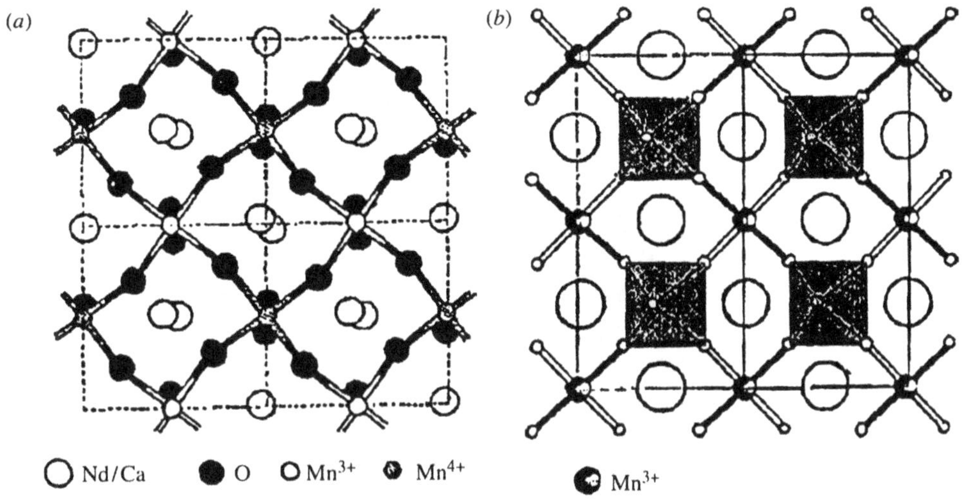


Figure 4. (*a*) Structure of charge-ordered $Nd_{0.5}Sr_{0.5}MnO_3$ in the *ab* plane at 10 K. $Mn^{4+}$ ions are located at (1/2, 0, 0) and $Mn^{3+}$ are located at (0, 1/2, 0). The structure contains zigzag chains with alternate long and short Mn–O bonds (see text). (*b*) The same structure as given in (*a*), shown in polyhedral representation ((*a*) after Vogt *et al.* (1996) and (*b*) after Woodward *et al.* (1999)).

fact that the bandwidth of the $e_g$ electrons of Mn ions is affected by the size of the A-site cation, i.e. increasing $\langle r_A \rangle$ is equivalent to increasing the hydrostatic pressure, which increases the Mn–O–Mn bond angle and consequently the bandwidth. Detailed studies of the effect of $\langle r_A \rangle$ on the electronic order have been reported by Woodward *et al.* (1998). Systems with very large $\langle r_A \rangle$ tend to be FM and metallic, with a Curie temperature $T_c$ that increases with $\langle r_A \rangle$. Charge ordering is also strongly affected by $\langle r_A \rangle$, with a smaller $\langle r_A \rangle$ stabilizing charge-ordered phase at low temperatures. It is evident that the tuning of $\langle r_A \rangle$ via chemical means provides an important means of controlling the phase of the manganite.

## 3. Electronic inhomogeneities in rare-earth manganites

As noted in §1, there is strong experimental evidence to indicate that many correlated oxides are electronically inhomogeneous (figure 5). Nowhere is it manifested better than the rare-earth manganites. Thus, these materials of high chemical homogeneity consist of different spatial regions with different electronic order (Mathur & Littlewood 2003; Rao *et al.* 2004; Dagotto 2005; Shenoy *et al.* 2006), a phenomenon that has come to be known as 'phase separation'. These regions can be static or dynamic and can be tuned by the application of external stimuli like a magnetic field. Moreover, the size scale of these inhomogeneities can vary from nanometres to as large as micrometres. The presence of electronic inhomogeneities raises many intriguing questions. What is the microscopic origin of these inhomogeneities? Why do they possess such a large range of length-scales? More importantly, are they responsible for the CMR and such responses exhibited by the manganites and related oxide materials? Indeed, there are suggestions in the literature that these features observed in correlated oxides indicate that they

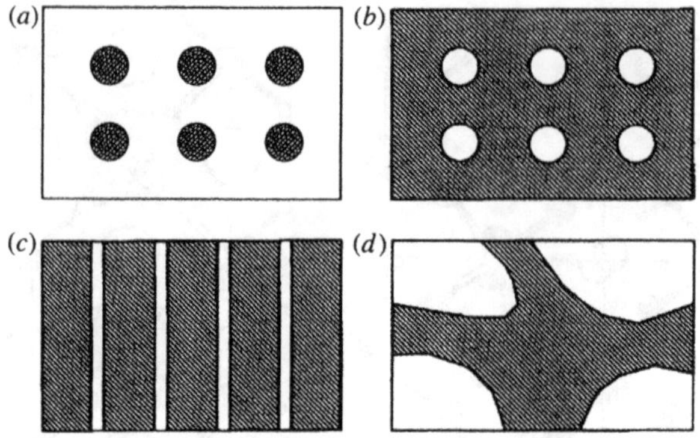


Figure 5. Schematic of electronic phase separation. Shaded portions indicate FM metallic regions; the unshaded portions correspond to AFM insulating regions. (*a*) FM metallic puddles in an insulating AFM background, (*b*) metallic regions with insulating droplets, (*c*) charged stripes and (*d*) phase separation on the mesoscopic scale.

are 'electronically soft' providing the possibility of spatial tuning of electronic properties, not unlike the liquid crystalline materials (Mathur & Littlewood 2003; Dagotto 2005; Milward *et al.* 2005). In this section, we discuss the phase separation phenomena especially in the rare-earth manganites. Following a description of the experimental signatures of electronic inhomogeneities, we discuss the theoretical ideas put forth to understand this phenomenon in §3*a*.

### (*a*) *Signatures of electronic inhomogeneities*

Investigations employing a variety of experimental probes indicate the presence of electronic inhomogeneities. When the length-scale of the inhomogeneities is large (a few hundred nanometres), structural and magnetic information from the X-ray and neutron diffraction patterns show clear signatures. Specialized local probes such as NMR and Mössbauer spectroscopies can be used to probe inhomogeneities at a smaller length-scale. Thermodynamic (e.g. magnetization) and transport measurements also indicate the presence of inhomogeneities. Direct observational evidence provided by images employing transmission electron microscopy, scanning probe microscopy and photoemission micro-spectroscopy makes for a convincing case for the presence of the electronic inhomogeneities.

The pioneering work on manganites by Wollan & Koehler (1955) revealed the presence of electronic inhomogeneities. They observed both FM and AFM peaks in the magnetic structure of $La_{1-x}Ca_xMnO_3$ obtained from neutron scattering. They concluded that there is the simultaneous presence of ferromagnetism and antiferromagnetism in this material. A more recent neutron diffraction study is that of Woodward *et al.* (1999) on $Nd_{0.5}Sr_{0.5}MnO_3$. This material first becomes FM at 250 K, partially transforming to an A-type AFM phase at approximately 220 K, followed by a transformation of a substantial fraction to a CE-type AFM phase at approximately 150 K. The CE-type AFM (abbreviated as CO-CEAF) phase has a simultaneous ordering of spins, orbitals and charge in a complex spatial arrangement as shown in figure 3. It is evident from the studies of Woodward *et al.* (1999) that the three phases coexist at low temperatures. The size scale of the inhomogeneities is at least in the mesoscopic range

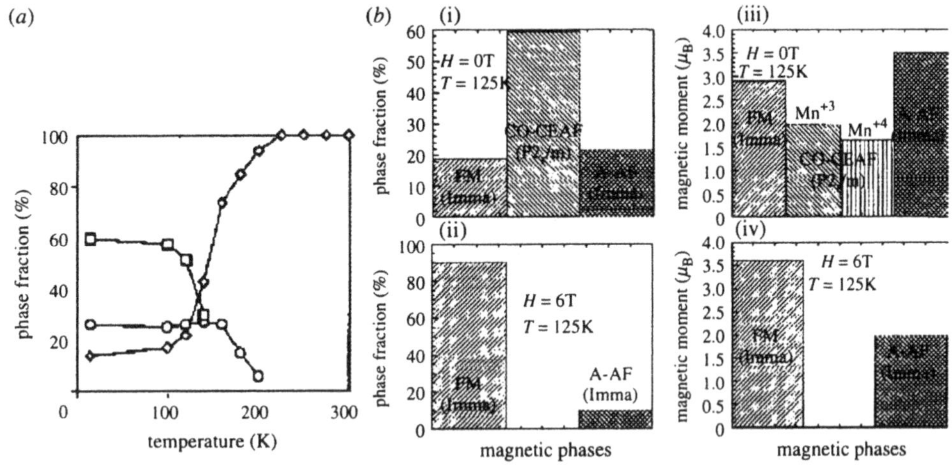


Figure 6. (*a*) Volume fraction of phases as a function of temperature in $Nd_{0.5}Sr_{0.5}MnO_3$. Open diamonds, open circles and open squares represent ferromagnet (FM), antiferromagnet (A-type, A-AF) and antiferromagnet (CE-type, CO-CEAF), respectively. (*b*) Phase fractions at 125 K at different magnetic fields (i) 0 T and (ii) 6 T; (iii) and (iv) show the magnetic moment in each of the phases ((*a*) after Woodward *et al.* (1999) and (*b*) after Ritter *et al.* (2000)).

(a few hundred nanometres or more), since they are large enough to produce well-defined reflections in neutron and X-ray diffraction patterns. The variation of the volume fraction of the three phases with temperature is shown in figure 6*a*. This phase diagram has an unusual feature with the three phases existing at the 150 K transition and below. The effect of the volume fractions of the electronic inhomogeneities due to an external magnetic field was studied by Ritter *et al.* (2000). The results shown in figure 6*b* demonstrate the evolution of the inhomogeneities where the FM fraction grows at the expense of the AFM regions in the presence of the applied magnetic field. Note the growth of one phase at the cost of another in figure 6. Clearly, the phase separation observed here is of a non-trivial type, in contrast to a situation where the sample is not phase-pure containing multiple phases that are both structurally and chemically distinct. Another interesting example (Radaelli *et al.* 2001) of phase separation is in $Pr_{0.7}Ca_{0.3}MnO_3$, which shows the presence of two distinct phases below the charge-ordering transition at 80 K. A charge-ordered AFM phase and a charge-delocalized phase have been observed by neutron diffraction.

Signatures of electronic inhomogeneities may be inferred from magnetic and electron transport measurements as well. Sudheendra & Rao (2003) studied these properties of $(La_{1-x}Ln_x)_{0.7}Ca_{0.3}MnO_3$, where Ln=Nd, Gd or Y. The lanthanide element was varied to control the average A-site cation radius $\langle r_A \rangle$. In the case of the La compound ($x$=0.0) with the largest value of $\langle r_A \rangle$, a clear FM transition with a saturation magnetization of $3\mu_B$ is observed (figure 7*a*) up to a critical Nd doping of $x \leq x_c \approx 0.5$. At higher values of Nd doping ($x > x_c$), there is a monotonic increase of the magnetization with decrease in temperature, although the magnetization never attains the maximum value of $3\mu_B$. Interestingly, reducing the value of $\langle r_A \rangle$ by Gd (intermediate value of $\langle r_A \rangle$) or Y (smallest value of $\langle r_A \rangle$) substitution decreases the $x_c$, i.e. the critical doping value $x$ above which the sharp FM transition vanishes,

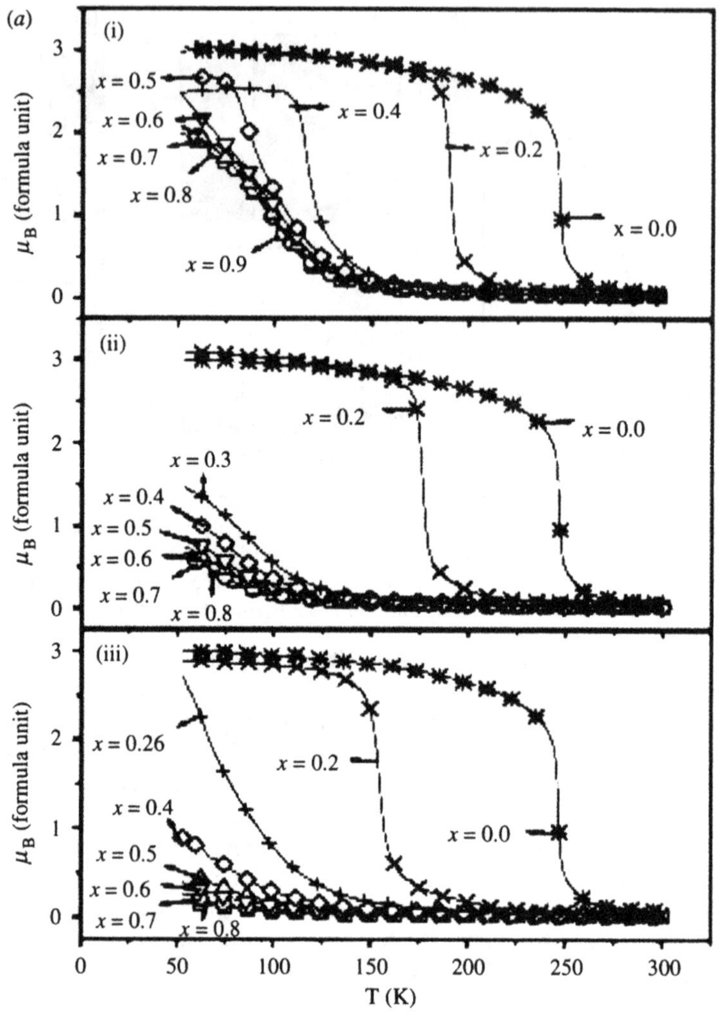


Figure 7. (*a*) Temperature dependence of the effective magnetic moment per formula unit in $(La_{1-x}Ln_x)_{0.7}Ca_{0.3}MnO_3$. (*b*) Temperature dependence of the resistivity in $(La_{1-x}Ln_x)_{0.7}Ca_{0.3}MnO_3$. The insets show the 'resistivity hysteresis' upon warming of the sample. In graphs in (*a*) and (*b*), (i), (ii) and (iii) correspond to Ln=Nd, Gd and Y, respectively. (*c*) Variation of (i) magnetization and (ii) resistivity as with temperature. The data shown are in $Ln_{0.7-x}Ln'_xA_{0.3-y}A'_y\,MnO_3$ with mean radius of the A-site cation are fixed at 1.216 Å. The different curves show results for different $\sigma^2$ ((*a*,*b*) after Sudheendra & Rao (2003) and (*c*) after Kundu *et al.* (2005*b*)).

reduces with the mean A-site cation radius $\langle r_A \rangle$. Moreover, the FM transition temperature $T_c$ for $x < x_c$ increases with $\langle r_A \rangle$ as expected. The precipitous reduction of the saturation magnetization at $x \gtrsim x_c$ is a signature of electronic and magnetic inhomogeneities induced by the A-site cation disorder. Transport measurements also show the signature of electronic inhomogeneities. In figure 7*b*, we show the variation of resistivity with temperature to demonstrate that all the three substitutions of manganites with Ln=Nd, Gd and Y exhibit a metal–insulator transition for $x < x_c$. The insets in figure 7*b* show the resistivity hysteresis across the

(*b*)

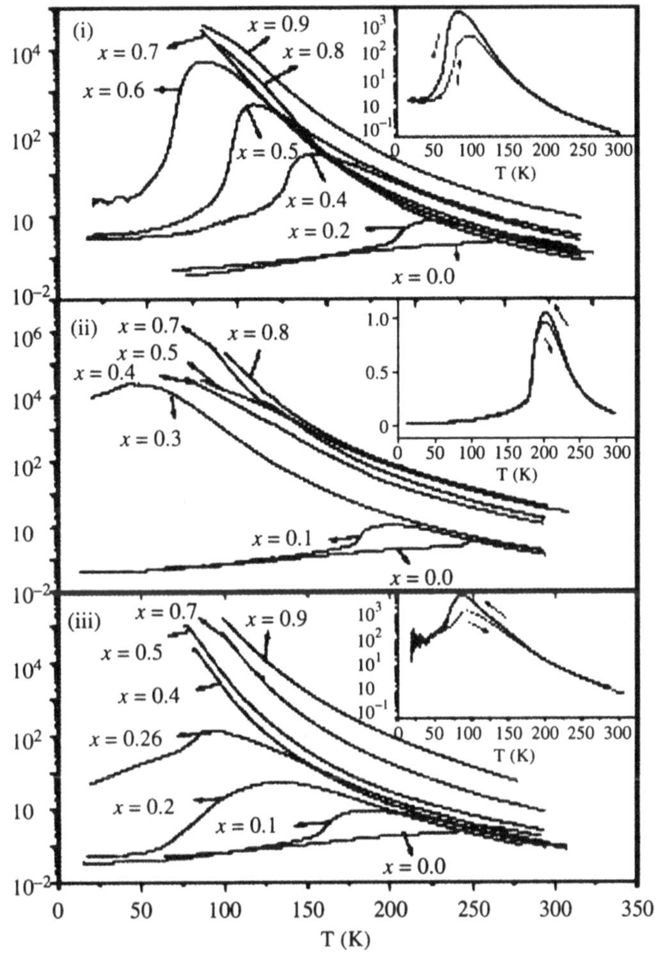


Figure 7. (*Continued.*)

transition. The insulator–metal transition temperature decreases linearly on doping and increases nearly linearly with $\langle r_A \rangle$. Electronic phase separation in such manganites generally occurs when $\langle r_A \rangle$ is relatively small ($<1.18$ Å) and is favoured by size disorder arising from the size mismatch between the A-site cations. This disorder is quantified (Rodríguez-Martínez & Attfield 2001) by the parameter $\sigma^2 = \sum_i x_i r_i^2 - \langle r_A \rangle^2$, where $x_i$ is the fractional A-site occupancy of species $i$ with radius $r_i$. A study of several series of manganites with fixed $\langle r_A \rangle$ and varying $\sigma^2$ shows that, as $\sigma^2$ is lowered, the phase-separated system transforms to a FM metallic state. It has been demonstrated recently (Kundu *et al.* 2005*b*) that in a series of manganites of the type $Ln_{0.7-x}Ln'_xA_{0.3-y}A'_yMnO_3$, where $\langle r_A \rangle$ always remains large (thereby avoiding effects due to bandwidth), a decrease in $\sigma^2$ transforms the insulating non-magnetic state to a FM metallic state (figure 7*c*).

The above examples, while being indicative of the existence of multiple electronic and magnetic phases within a single apparently structurally and chemically pure phase, do not provide direct evidence with regard to the length-scale of the

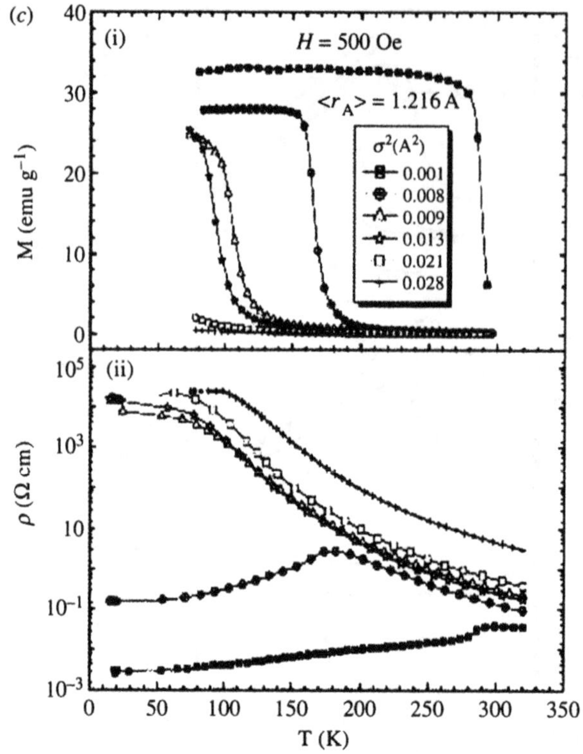


Figure 7. (*Continued.*)

inhomogeneities. While diffraction provides evidence of electronic phase separation on a mesoscale, NMR spectroscopy has revealed the possibility of phase separation at nanoscopic scales (Kuhns *et al.* 2003).

### (*b*) *Visual evidence for electronic phase separation*

We now turn to the experimental work that shows direct visual evidence of electronic inhomogeneities, thereby allowing one to probe the length-scale of the inhomogeneity. One of the earliest reports of the direct evidence of electronic patterns in manganites was in the hole-doped side of (La, Ca)$MnO_3$. Mori *et al.* (1998) observed charged stripes in thin films of $La_{1-x}Ca_xMnO_3$ with $0.5 \leq x \leq 0.75$ by high-resolution lattice images obtained by transmission electron microscopy. The spatial structure consisted of paired JT distorted oxygen octahedra surrounding the $Mn^{3+}$ ions, separated by stripes of $Mn^{4+}O_6$ octahedra. The spacing of the stripes, usually 5–10 lattice spacings, was found to be governed by the doping. The electronic inhomogeneity in this case was of a nanoscopic length-scale. Further evidence of nanoscopic electronic inhomogeneity in manganites was obtained by a scanning tunnelling microscopy study of $Bi_{1-x}Ca_xMnO_3$ ($x \approx 0.75$) by Renner *et al.* (2002). They found nanoscopic charge-ordered and metallic domains which correlated with the structural distortions.

The first direct evidence of mesoscale inhomogeneities in manganites was provided by Uehara *et al.* (1999), who observed electronic inhomogeneities in $La_{0.0625-y}Pr_yCa_{0.375}MnO_3$ by transmission electron microscopy. This work demonstrated the coexistence of charge-ordered (insulating) regions with interspersed FM metallic domains with a typical size of approximately 0.2 μm, bringing these features of inhomogeneity up to the mesoscopic scale. The authors suggested a possible mechanism for the colossal magnetoresistive response arising from such an electronic and magnetic inhomogeneity. The spin alignment of each of the FM metallic domains is random and the spin-polarized conduction electron of one domain cannot hop effectively via other domains due to the absence of available electronic states with the correct spin configuration on other randomly oriented magnetic domains. On application of a magnetic field, the spins of different magnetic domains align, enabling an unhindered hopping of electrons, thereby resulting in a drastic reduction of the resistance, known as the colossal negative magnetoresistance. This study suggested that such a mesoscopic phase separation is essential for colossal responses. Approximately at the same time, Fäth *et al.* (1999) used scanning tunnelling spectroscopy to study metal–insulator transitions in $La_{0.7}Ca_{0.3}MnO_3$ and found evidence of electronic inhomogeneities with a mesoscopic scale of approximately 0.2 μm, below the FM transition temperature, with FM metallic domains interspersed in insulating regions, just as in the work by Uehara *et al.* (1999). Most interestingly, the domains evolved as a function of the applied magnetic field, with the volume fraction of the magnetic (metallic) domains increasing at the cost of the non-magnetic (insulating) parts, closely accompanying the decrease in the resistivity with the magnetic field. This scenario contrasts with the spatially static phase separation scenario suggested by Uehara *et al.* (1999). A magnetic force microscope study of $La_{0.33}Pr_{0.34}Ca_{0.33}MnO_3$ by Zhang *et al.* (2002) has indicated that magnetic domains of mesoscopic scale evolved with temperature showing magnetic hysteresis coinciding with the resistivity hysteresis. This observation again suggested that mesoscale inhomogeneities are possibly crucial for colossal responses in magnetoresistance. Loudon *et al.* (2002) have studied $La_{0.5}Ca_{0.5}MnO_3$ using transmission electron microscopy and electron holography and found mesoscopic domains of FM regions interspersed in insulating regions. They also found evidence that some of the FM regions were charge ordered. A possible interpretation of the result is that the mesoscopic FM region is itself inhomogeneous at the nanoscale with coexisting metallic and charge-ordered regions.

A final example of mesoscale electronic inhomogeneities is from the experiments of Sarma *et al.* (2004), who used photoemission spectromicroscopy which has the advantage of spatially resolving the local metallic/non-metallic nature and simultaneously determining the local chemical composition with a resolution of approximately 0.5 μm. Using this technique, they studied $La_{0.25}Pr_{0.375}Ca_{0.375}MnO_3$. The key finding was that the material showed very large ($10\times5\ \mu m^2$) domains of insulating regions interspersed in a metallic background. These regions evolved on increasing temperature, with the metallic regions undergoing a metal-to-insulator transition at higher temperatures. On cooling the sample back to low temperatures, the insulating regions appeared essentially at the same locations, indicating, for the first time, a memory effect associated with the electronic inhomogeneities.

## 4. Electronic phase separation in rare-earth cobaltates

Electronic phase separation in rare-earth cobaltates of the type $Ln_{1-x}A_xCoO_3$ has come to the fore in the last 2 years. $La_{0.5}Sr_{0.5}CoO_3$ and other members of this family were once considered to be itinerant electron ferromagnets

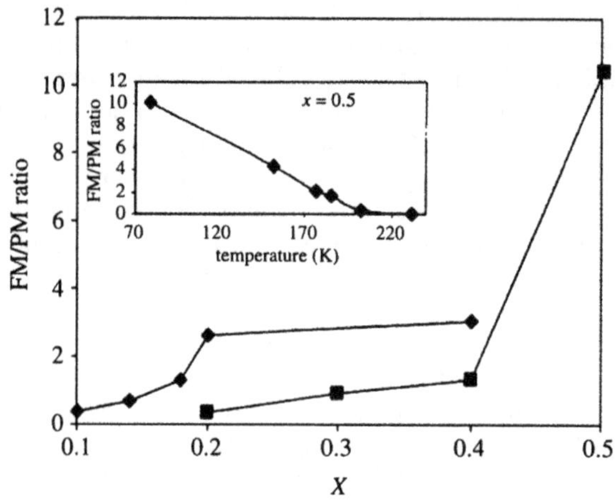


Figure 8. Variation of the ratio of the FM to PM species with composition in $La_{(1-x)}Sr_xCoO_3$ (filled squares, Mössbauer data at 78 K from Bhide *et al.* (1975); filled diamonds, NMR data at 1.9 K from Kuhns *et al.* (2003); inset shows the temperature variation of the FM/PM ratio in $La_{0.5}Sr_{0.5}CoO_3$ from the Mössbauer data).

showing metallic behaviour, but it was recognized later that the FM ordering was not long range (Rao *et al.* 1977; Señaris Rodríguez & Goodenough 1995). It is now established that $La_{0.5}Sr_{0.5}CoO_3$ exhibits a glassy magnetic behaviour (Wu *et al.* 2005). Recent studies of the magnetic and electrical properties of several members of the $Ln_{1-x}A_xCoO_3$ family have revealed that they are electronically phase-separated (Burley *et al.* 2004; Kundu *et al.* 2005*a*; Wu *et al.* 2005). The phase separation is favoured by small $\langle r_A \rangle$ as well as size disorder (Kundu *et al.* 2006). Measurements of magnetic relaxation and memory effects have shown that glassy behaviour is common among the cobaltates (Kundu *et al.* 2005*a*, 2006). Studies employing local probes such as NMR and Mössbauer spectroscopy have provided direct evidence for different distinct species involving large magnetic/metallic clusters and small non-magnetic clusters (Bhide *et al.* 1975; Kuhns *et al.* 2003). Thus, the NMR and Mössbauer studies of $La_{(1-x)}Sr_xCoO_3$ show distinct signals corresponding to FM and PM species. Cobalt Mössbauer spectra show a signal due to the CM species, but are always accompanied by a signal due to the PM species even at low temperatures ($T < T_c$). In figure 8, we show the variation of the FM to PM ratio with composition in $La_{(1-x)}Sr_xCoO_3$. In the inset, the variation of the FM to PM ratio of $La_{0.5}Sr_{0.5}CoO_3$ with temperature is shown (Kundu *et al.* 2007). Cobalt NMR studies of a cobaltate with small $\langle r_A \rangle$ has shown the presence of two distinct signals varying in intensity with temperature (Kundu *et al.* 2007).

## 5. Theoretical approaches

We now turn to a discussion of theoretical ideas that have been proposed to explain the origin of the electronic inhomogeneities, with specific focus on manganites. The simplest possible description of manganites at a microscopic level involves the Mn d-orbitals and the lattice distortion of the oxygen octahedron surrounding the Mn ion (Dagotto 2003). The octahedral crystal field splits the degeneracy of the five Mn d-orbitals into three degenerate $t_{2g}$ orbitals and two degenerate $e_g$ orbitals. In the doped manganites, both $Mn^{3+}$ and $Mn^{4+}$ configurations are present; both have three electrons in the $t_{2g}$ orbitals, and $Mn^{3+}$ has, in addition, a lone electron in one of the $e_g$ orbitals. The spins of the $t_{2g}$ electrons are aligned parallel due to strong Hund coupling, and it is only the resulting 'core spins' ($S=3/2$) of the $t_{2g}$ electrons that affect the low-energy physics of manganites. Thus, the relevant degrees of freedom at each manganese site are: an average of $(1-x)$ electrons per site populating the two $e_g$ orbitals, the $t_{2g}$ core spins and lattice (phonon) degrees of freedom corresponding to the distortion of the oxygen octahedra surrounding the manganese ion. The $e_g$ electrons hop from a Mn site to neighbouring Mn sites with an amplitude $t$ (approx. 0.2 eV). The spin of the $e_g$ electron has a strong FM Hund coupling $J_H$ (approx. 2.0 eV) with the local $t_{2g}$ core spin. Another important energy scale is the on-site Mott–Hubbard repulsion $U$ (approx. 5.0 eV) which forbids double occupancy of the local $e_g$ sector. Neighbouring $t_{2g}$ spins interact with each other via an AFM superexchange coupling $J_{SE}$ (approx. 0.02 eV). Finally, the energy gained by the JT distortion of the oxygen octahedron is given by $E_{JT}$ (approx. 0.5 eV). Competition between these different interactions, leading to a variety of states very close in energy, is responsible for the complex phase diagram of manganites with many electronic orders, and their extreme sensitivity to external perturbations such as temperature, magnetic field and strain (Sarma *et al.* 1995; Satpathy *et al.* 1996; Millis 1998; Ramakrishnan *et al.* 2003; Cepas *et al.* 2006).

Owing to the difficulties of dealing with all the degrees of freedom and competing interactions mentioned above, much of the early work aimed at understanding colossal responses and electronic inhomogeneities in manganites has been based on simplified models which neglect one or more of them. A prominent example is the work of Dagotto and coworkers (Dagotto 2003), who studied simple *magnetic Hamiltonians* with competing phases that are separated by a first-order transition. Based on these studies, they suggested that the system is prone to macroscopic phase separation, which is frustrated by disorder leading to the electronic inhomogeneities at various scales, the magnitude of the disorder determining the scale. From the real-space structure obtained from simulations, they constructed a random resistor network to explain the colossal responses by a mechanism similar to the one proposed by Fäth *et al.* (1999), again suggesting that the phase separation is key to colossal responses. Other simplified models have been studied, and alternate scenarios have been proposed. There are also suggestions that manganites close to half doping are near a multicritical point with competing phases affected by disorder (Motome *et al.* 2003; Tokura 2006). Ahn *et al.* (2004) consider a model Hamiltonian including electron–phonon interaction, and long-range elastic coupling between local lattice distortions. They present a scenario for mesoscopic/microscopic inhomogeneities and suggest that these are responsible for the colossal responses.

We now turn to a recent study of electronic inhomogeneities in manganites (Shenoy *et al.* 2007), which attempts to take into account all the degrees of freedom and their interactions based on the $\ell b$ model (Ramakrishnan *et al.* 2003, 2004; Krishnamurthy 2005).

### (*a*) *The ℓb model for manganites*

The ℓb model (Pai *et al.* 2003; Ramakrishnan *et al.* 2004) uses the idea that, under the conditions prevailing in doped manganites, the electrons populating the doubly degenerate $e_g$ states centred at the Mn sites spontaneously reorganize themselves into two types of coexisting electron fluids. One is obtained by populating essentially site-localized states labelled ℓ which are polaronic, with strong local JT distortions of the oxygen octahedra, an energy gain $E_{\mathrm{JT}}$ and exponentially reduced intersite hopping. The other, labelled *b*, is a fluid of broadband, non-polaronic electrons, with no associated lattice distortions and undiminished hopping amplitudes. There is a strong local repulsion between the two fluids, as double occupancy on a site costs an extra Coulomb of energy *U*. The spins of ℓb and *b* are enslaved to the Mn-$t_{2g}$ spins on site due to the large FM Hund's coupling $J_{\mathrm{H}}$. In the simplest picture, assuming all the $t_{2g}$ spins and the $e_g$ spins to be aligned parallel, the ℓ and *b* electrons can be regarded as spinless, leading to the Falicov–Kimball (Freericks & Zlatić 2003) like ℓb Hamiltonian

$$H_{\ell b} = -E_{\mathrm{JT}} \sum_i n_{\ell i} - t \sum_{\langle ij \rangle} (b_i^\dagger b_j + \mathrm{h.c.}) + U \sum_i n_{\ell i} n_{bi}. \tag{5.1}$$

Here, $\ell_i^\dagger$ and $b_i^\dagger$ create ℓ polarons and *b* electrons, respectively, at the sites *i* of a cubic Mn lattice, and $n_{\ell i} \equiv \ell_i^\dagger \ell_i$ and $n_{bi} \equiv b_i^\dagger b_i$ are the corresponding number operators.

### (*b*) *The extended ℓb model and the effects of long-range Coulomb interactions*

This model assumes that the Mn ions occupy the sites of a cubic lattice (taken to be of unit lattice parameters), while the dopant A ions occupy an *x* fraction of the 'body centre' sites of each unit cube formed by the Mn ions. Since the aim is to study the effect of long-range Coulomb interactions on phase separation, we make further simplifying assumptions. It is assumed that the $t_{2g}$ core spins are aligned ferromagnetically and that $J_{\mathrm{H}} \to \infty$; this effectively projects out ℓ or *b* electron spin opposite to that of the $t_{2g}$ core spins—we obtain an effectively spinless model. The above considerations lead us to the following *extended ℓb Hamiltonian*

$$H = H_{\ell b} + H_{\mathrm{C}} \quad \text{and} \quad H_{\mathrm{C}} = \sum_i \Phi_i q_i + \frac{V_0}{2} \sum_{i \neq j} \frac{q_i q_j}{r_{ij}}. \tag{5.2}$$

Here $\ell_i^\dagger$ and $b_i^\dagger$ create ℓ and *b* electrons, respectively, at site *i*, and $n_{\ell i} \equiv \ell_i^\dagger \ell_i$ and $n_{bi} \equiv b_i^\dagger b_i$ are the corresponding number operators. In terms of the hole operator ($h_i^\dagger \equiv \ell_i$ which removes an ℓ polaron at site *i*), the electron charge operator $q_i \equiv h_i^\dagger h_i - b_i^\dagger b_i$ and has the average value *x* per site owing to overall charge neutrality. The Coulomb term $H_{\mathrm{C}}$ has two parts; the charge at site *i* has energy $q_i \Phi_i$, where $\Phi_i$ is the electrostatic potential there due to $A^{2+}$ ions, and the interaction between the charges at sites *i* and *j* leads to an energy $V_0((q_i q_j)/r_{ij})$. In what follows, we take the short-range Coulomb correlation *U* to be large (∞).

We now discuss the results of full-scale numerical simulations of the Hamiltonian (5.2) on finite three-dimensional periodic lattices, allowing for a random distribution of the A ions following Shenoy *et al.* (2007). Here, all the energy scales are normalized by the bare intersite hopping amplitude *t*. Systems as large as 20×20×20 have been considered. The numerical determination of the groundstate of (5.2) requires further simplifying approximations. The most important approximation is the Hartree approximation, i.e. the charge operator $q_i$

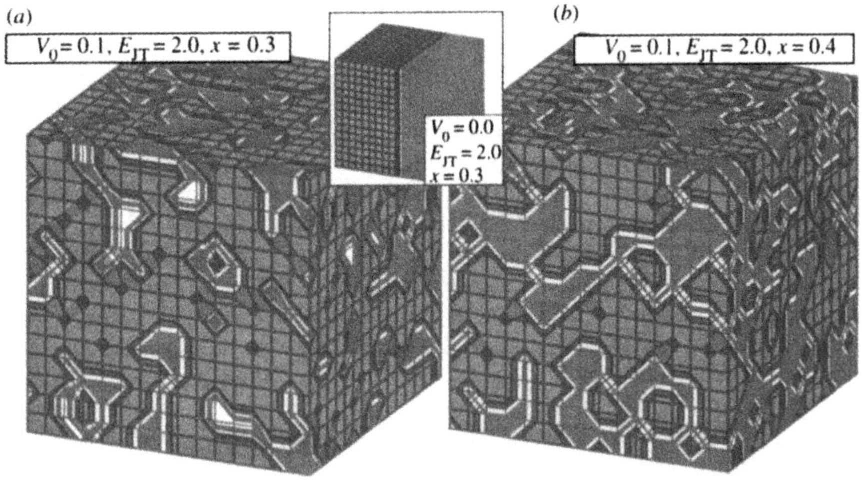


Figure 9. Real-space electronic distribution obtained from simulations on a $16^3$ cube. Magenta (darkest) denotes hole clumps with occupied $b$ electrons, white (lightest) denotes hole clumps with no $b$ electrons, cyan (2nd lightest) denote singleton holes, and light blue (2nd darkest) represents regions with $\ell$ polarons. (*a*) Isolated clumps with occupied $b$-electrons ($b$-electron puddles). (*b*) Larger doping; percolating clumps. Inset: 'macroscopic phase separation' *absence* of long-range Coulomb interaction ($V_0$=0.0) (after Shenoy *et al.* (2007)).

is replaced by its expectation value in the groundstate $\langle q_i \rangle = h_i^\dagger h_i - \langle b_i^\dagger b_i \rangle$. Since we have assumed that $U \to \infty$, the $b$ electrons do not hop to sites where an $\ell$ polaron is present. This leads to the segregation of the two types of electron into disjoint clusters. In a cluster of hole sites (which has at least two nearest neighbour hole sites), which we call a 'clump', the $b$ electron states are determined by solving the quantum kinetic energy Hamiltonian exactly. This is a new generalization of the common Coulomb glass simulation (Baranovskii *et al.* 1979; Davies *et al.* 1984; Vojta & Schreiber 2001), which includes the quantum mechanically obtained $b$ states within their clump or puddle. The electrostatic energy is calculated accurately using the Ewald technique and fast Fourier transform routines.

In the absence of the long-range Coulomb interaction ($V_0$=0), this procedure leads to the macroscopically phase-separated state (see inset of figure 9). The holes aggregate to one side of the simulation box and a fraction of the electrons occupying the $\ell$ states is promoted to the $b$ states in the one large hole clump—with their concentrations determined by the equality of the chemical potential in the two regions (i.e. the highest occupied $b$ level equals $-E_{\mathrm{JT}}$). This phase separation is due to the strong Coulomb repulsion between the two types of electron fluids and is in agreement with known results on the Falicov–Kimball model (Freericks *et al.* 2002; Freericks & Zlatić 2003).

Long-range Coulomb interaction 'frustrates' phase separation, i.e. macroscopic phase separation costs prohibitive energy in the presence of long-range Coulomb interactions. The precise nature of the resulting groundstate electronic configuration depends on the JT energy $E_{\mathrm{JT}}$ and the doping level $x$. There are two critical doping levels $x_{c1}$ and $x_{c2}$ for any given JT energy $E_{\mathrm{JT}}$. When the doping level is less than $x_{c1}$, i.e. $x < x_{c1}$, no $b$ states are occupied—the holes (and hence also the $\ell$ polarons) form a Coulomb glass (Efros & Shklovskii 1975). When the doping exceeds $x_{c1}$, some of the hole clumps are occupied and the groundstate consists of isolated *b*-electron puddles dispersed in a polaronic background (figure 9). There is a second critical doping level $x_{c2}$ (see Shenoy *et al.* 2007); when $x \gtrsim x_{c2}$ the occupied *b*-electron puddles percolate and the system attains metallicity (figure 9).

The size scale of the electronic inhomogeneity is nanometric. However, its dependence on $V_0$ differs fundamentally from the analytical results of the same model, which assumed a homogeneous distribution (jellium) of the dopant A ions. For the more realistic, random distribution of the A ions, the size of the electronic inhomogeneity is almost independent of the long-range Coulomb parameter $V_0$, in that even an extremely small $V_0$ produces clump sizes that are of the size scale of a few lattice spacings. The dependence of the average size of the electronic inhomogeneity is thus due to the long-range Coulomb interaction which *along with* the random distribution of A ions acts as a 'singular perturbation' that frustrates macroscopic phase separation. However, the sizes and the distribution of the clumps are determined by the random distribution of the A ions, and thus we conclude that doped manganites (and similarly, possibly many other correlated oxides) are necessarily and intrinsically electronically inhomogeneous, on a *nanometric scale.*

The results of the extended $\ell b$ model with realistic energy parameters presented here provide several new insights into the complex electronic inhomogeneities seen in correlated oxides, and in particular the low-bandwidth manganites with a large FM region in their phase diagram. These arguments may be sharply contrasted with earlier work. In particular, this work suggests that the nanoscale inhomogeneities in manganites arise not out of 'phase competition'-induced phase separation frustrated by disorder as argued from the studies of model spin Hamiltonians (Dagotto 2003), but from short-range Coulomb correlation-induced phase separation frustrated by long-range Coulomb interaction, similar to what has been suggested in cuprates (Emery *et al.* 1990; Emery & Kivelson 1993). Most importantly, nanoscale electronic inhomogeneities are present in *both* the insulating and metallic phases of doped manganites. As is evident from figure 9, each of these constitute a single thermodynamic phase that is homogeneous at mesoscales. One has a metal–insulator transition between two such nanoscopically inhomogeneous phases at $x_{c2}$ as a function of doping. These observations are in agreement with experiments. Indeed, the electron holography results of Loudon *et al.* (2002) show that even the ferro-metallic regions have interspersed in them charge-ordered insulating regions which can be interpreted as the cluster of $\ell$ states.

## 6. Concluding remarks

A number of competing energy scales are operative in mixed-valent correlated oxides giving rise to a large number of electronic orders. In the $Ln_{(1-x)}A_xMnO_3$ manganites, a variety of spin, charge and orbital (and associated lattice order) is found. The emergence of these orders can be understood based on a simple parametrization, i.e. the mean A-site cation radius which controls the bandwidth of the $e_g$ electrons. An intriguing aspect of these materials is that small perturbations can drive a transition from one type of electronic order to another, as is evident from the effect of a small magnetic field on the charge-ordered state.

In addition to the large number of electronic orders, it is found experimentally that there is coexistence of *different* electronic orders in disjoint spatial regions, i.e. these materials are electronically inhomogeneous showing signatures in thermodynamic and transport properties and can be seen directly in several experimental probes. These inhomogeneities can be of varied scales (nm to μm) and can be static or dynamic. Although the phenomenon has been called phase separation, it is clear that the regions of distinct electronic order may not necessarily be 'thermodynamic phases', particularly in the case where they are of nanometric scale. Indeed, the theoretical model discussed in the paper shows that a single 'phase' of the manganite itself is inhomogeneous on a nanometric scale.

Although, we have focused on manganites and cobaltates, electronic inhomogeneities are important in many other oxides. In the high $T_c$ cuprate $La_2CuO_{4-\delta}$, the electronic phase separation is caused by compositional variations with superconducting clusters of $La_2CuO_{3.06}$ and AFM clusters of $La_2CuO_{3.01}$ (Sigmund & Müller 1994). Even in the cuprates that possess chemical homogeneity on a mesoscale, it has been suggested that the unusual variation of zero-temperature superfluid density with the superconducting transition temperature is due to the possible coexistence of AFM and superconducting domains (Broun *et al.* 2005). Neutron-scattering measurements suggest the presence of charge stripes (Tranquada *et al.* 2004), while scanning probe microscopy indicates a spatially inhomogeneous superconducting gap (McElroy *et al.* 2005).

Theoretical understanding of the phenomenon of electronic inhomogeneities is still incomplete. Although a large body of previous work based on *model Hamiltonians* and phenomenological models exists, the microscopic mechanisms and the length-scales that arise in these systems remain poorly understood. The microscopically motivated theoretical model that we have discussed here clearly indicates that these materials are electronically inhomogeneous at nanoscales due to strong correlation physics. Naturally, the question of the origin of mesoscale inhomogeneities remains. There are several scenarios that could explain the origin of the mesoscale patterns. There is a possibility that, since many of the transitions are of first order, many of the inhomogeneities are due to an 'incomplete transition', i.e. they are metastable states. The second possibility is the presence of long-range elastic strain fields which are unscreened. Indeed, it is well known that pressure (strain) can drive transitions in the manganites (Soh *et al.* 2002; Postorino *et al.* 2003). The key question is whether strain effects are extrinsic (caused by defects, etc.) or intrinsic, i.e. the system reorganizes spontaneously into domains (much like a martensite).

The need for further exploration of electronic orders and inhomogeneities in oxides is evident. Experiments need to be designed so as to isolate long-range strain effects, with theoretical efforts towards building coarse-grained models to understand the mesoscale physics. This issue is particularly important for exploiting the electronic softness of these materials for applications.

V.B.S. gratefully acknowledges illuminating discussions and collaborative work with H. R. Krishnamurthy and T. V. Ramakrishnan.

## References

Ahn, K. H., Lookman, T. & Bishop, A. R. 2004 Strain-induced metal–insulator phase coexistence in perovskite manganites. *Nature* **428**, 401–404. (doi:10.1038/nature02364)

Baranovskii, S. D., Efros, A. L., Gelmont, B. L. & Shklovskii, B. I. 1979 Coulomb gap in disordered systems: computer simulation. *J. Phys. C: Solid State Phys.* **12**, 1023–1034. (doi:10.1088/0022-3719/12/6/015)

Bhide, V. G., Rajoria, D. S., Rao, C. N. R., Rama Rao, G. & Jadhao, V. G. 1975 Itinerant-electron ferromagnetism in $La_{1-x}Sr_xCoO_3$: mossbauer study. *Phys. Rev. B* **12**, 2832–2843. (doi:10.1103/PhysRevB.12.2832)

Biswas, A., Raychaudhuri, A. K., Mahendiran, R., Guha, A., Mahesh, R. & Rao, C. N. R. 1997 Direct measurement of the charge ordering gap in $Nd_{0.5}Sr_{0.5}MnO_3$ using vacuum tunnelling. *J. Phys. Condens. Matter* **9**, L355–L360. (doi:10.1088/0953-8984/9/24/002)

Broun, D. M., Turner, P. J., Huttema, W. A., Ozcan, S., Morgan, B., Liang, R., Hardy, W. N. & Bonn, D. A. 2005 In-plane superfluid density of highly underdoped $YBa_2Cu_3O_{6+x}$. arxiv.org:cond-mat/0509223.

Burley, J. C., Mitchell, J. F. & Short, S. 2004 Competing electronic ground states in $La_{1-x}Ca_xCoO_3$. *Phys. Rev. B* **69**, 054401. (doi:10.1103/PhysRevB.69.054401)

Cepas, O., Krishnamurthy, H. R. & Ramakrishnan, T. V. 2006 Instabilities and insulator–metal transitions in half-doped manganites induced by magnetic-field and doping. *Phys. Rev. B* **73**, 035218. (doi:10.1103/PhysRevB.73.035218)

Dagotto, E. 2003 *Nanoscale phase separation and colossal magnetoresistance.* Berlin, Germany: Springer.

Dagotto, E. 2005 Complexity in strongly correlated electronic systems. *Science* **309**, 257–362. (doi:10.1126/science.1107559)

Dagotto, E., Hotta, T. & Moreo, A. 2001 Colossal magnetoresistant materials: the key role of phase separation. *Phys. Rep.* **344**, 1–153. (doi:10.1016/S0370-1573(00)00121-6)

Davies, J. H., Lee, P. A. & Rice, T. M. 1984 Properties of the electron glass. *Phys. Rev. B* **29**, 4260–4271. (doi:10.1103/PhysRevB.29.4260)

Efros, A. L. & Shklovskii, B. I. 1975 Coulomb gap and low-temperature conductivity of disordered systems. *J. Phys. C: Solid State Phys.* **8**, L49–L51. (doi:10.1088/0022-3719/8/4/003)

Emery, V. J. & Kivelson, S. A. 1993 Frustrated electronic phase separation and high-temperature superconductors. *Physica C* **209**, 597–621. (doi:10.1016/0921-4534(93)90581-A)

Emery, V. J., Kivelson, S. A. & Lin, H. Q. 1990 Phase-separation in the t–J model. *Phys. Rev. Lett.* **64**, 475–478. (doi:10.1103/PhysRevLett.64.475)

Fäth, M., Freisem, S., Menovsky, A. A., Tomioka, Y., Aarts, J. & Mydosh, J. A. 1999 Spatially inhomogeneous metal–insulator transition in doped manganites. *Science* **285**, 1540–1542. (doi:10.1126/science.285.5433.1540)

Freericks, J. K. & Zlatić, V. 2003 Exact dynamical mean field theory of the Falicov–Kimball model. *Rev. Mod. Phys.* **75**, 1333–1382. (doi:10.1103/RevModPhys.75.1333)

Freericks, J. K., Leib, E. H. & Ueltschi, D. 2002 Phase separation due to quantum mechanical correlations. *Phys. Rev. Lett.* **88**, 106401. (doi:10.1103/PhysRevLett.88.106401)

Imada, M., Fujimori, A. & Tokura, Y. 1998 Metal–insulator transitions. *Rev. Mod. Phys.* **70**, 1039–1263. (doi:10.1103/RevModPhys.70.1039)

Krishnamurthy, H. R. 2005 A new theory of doped manganites exhibiting colossal magnetoresistance. *Pramana* **64**, 1063–1074. (Invited talk published in the *Proc. 22nd IUPAP Int. Conf. on Statistical Physics* (eds S. Dattagupta, H. R. Krishnmurthy, R. Pandit, T. V. Ramakrishanan & D. Sen).)

Kuhns, P. L., Hoch, M. J. R., Moulton, W. G., Reyes, A. P., Wu, J. & Leighton, C. 2003 Magnetic phase separation in $La_{1-x}Sr_xCoO_3$ by $^{59}Co$ nuclear magnetic resonance. *Phys. Rev. Lett.* **91**, 127202. (doi:10.1103/PhysRevLett.91.127202)

Kundu, A. K., Nordblad, P. & Rao, C. N. R. 2005*a* Nonequilibrium magnetic properties of single-crystalline $La_{0.7}Ca_{0.3}CoO_3$. *Phys. Rev. B* **72**, 144423. (doi:10.1103/PhysRevB.72.144423)

Kundu, A. K., Seikh, M. M., Ramesha, K. & Rao, C. N. R. 2005*b* Novel effects of size disorder on the electronic and magnetic properties of rare earth manganates of the type $La_{0.7-x}Ln_xBa_{0.3}Mn_{0.3}$ (Ln = Pr, Nd, Gd or Dy) with large average radius of the A-site cations. *J. Phys. Condens. Matter* **17**, 4171–4180. (doi:10.1088/0953-8984/17/26/015)

Kundu, A. K., Nordblad, P. & Rao, C. N. R. 2006 Spin-glass behavior in $Pr_{0.7}Ca_{0.3}CoO_3$ and $Nd_{0.7}Ca_{0.3}CoO_3$. *J. Solid State Chem.* **179**, 923–927. (doi:10.1016/j.jssc.2005.12.014)

Kundu, A., Sarkar, R., Pahari, B., Ghoshray, A. & Rao, C. N. R. 2007 A comparative study of the magnetic properties and phase separation behavior of the rare earth cobaltates, $Ln_{0.5}Sr_{0.5}CoO_3$ (*Ln*=rare earth). *J. Solid State Chem.* **180**, 1318–1324. (doi:10.1016/j.jssc.2007.02.003)

Kuwahara, H., Tomioka, Y., Asamitsu, A., Moritomo, Y. & Tokura, Y. 1995 A first-order phase transition induced by a magnetic field. *Science* **270**, 961–963. (doi:10.1126/science.270.5238.961)

Loudon, J. C., Mathur, N. D. & Midgley, P. A. 2002 Charge-ordered ferromagnetic phase in $La_{0.5}Ca_{0.5}MnO_3$. *Nature* **420**, 797–800. (doi:10.1038/nature01299)

Mathur, N. D. & Littlewood, P. B. 2003 Mesoscopic texture in manganites. *Phys. Today* **56**, 25–30. (doi:10.1063/1.1554133)

McElroy, K., Lee, J., Slezak, J. A., Lee, D.-H., Eisaki, H., Uchida, S. & Davis, J. 2005 Atomic-scale sources and mechanism of nanoscale electronic disorder in $Bi_2Sr_2CaCu_2O_{8+\delta}$. *Science* **309**, 1048–1052. (doi:10.1126/science.1113095)

Millis, A. J. 1998 Colossal magnetoresistance manganites: a laboratory for electron–phonon physics. *Phil. Trans. R. Soc. A* **356**, 1473–1480. (doi:10.1098/rsta.1998.0230)

Milward, G. C., Calderón, M. J. & Littlewood, P. B. 2005 Coexisting charge modulation and ferromagnetism produces long period phases in manganites: new example of electronic soft matter. *Nature* **433**, 607–610. (doi:10.1038/nature03300)

Mori, S., Chen, C. H. & Cheong, S.-W. 1998 Pairing of charge-ordered stripes in (La, Ca)$MnO_3$. *Nature* **392**, 473–476. (doi:10.1038/33105)

Motome, Y., Furukawa, N. & Nagaosa, N. 2003 Competing orders and disorder-induced insulator to metal transition in manganites. *Phys. Rev. Lett.* **91**, 167204. (doi:10.1103/PhysRevLett.91.167204)

Pai, G. V., Hassan, S. R., Krishnamurthy, H. R. & Ramakrishnan, T. V. 2003 Zero-temperature insulator–metal transition in doped manganites. *Europhys. Lett.* **64**, 696–702. (doi:10.1209/epl/i2003-00282-0)

Postorino, P., Congeduti, A., Dore, P., Sacchetti, A., Gorelli, F., Ulivi, L., Kumar, A. & Sarma, D. D. 2003 Pressure tuning of electron–phonon coupling: the insulator to metal transition in manganites. *Phys. Rev. Lett.* **91**, 175501. (doi:10.1103/PhysRevLett.91.175501)

Radaelli, P. G., Ibberson, R. M., Argyriou, D. N., Casalta, H., Andersen, K. H., Cheong, S.-W. & Mitchell, J. F. 2001 Mesoscopic and microscopic phase segregation in manganese perovskites. *Phys. Rev. B* **63**, 172419. (doi:10.1103/PhysRevB.63.172419)

Ramakrishnan, T. V., Krishnamurthy, H. R., Hassan, S. R. & Pai, G. V. 2003 Theory of manganites exhibiting colossal magnetoresistance. arxiv.org:cond-mat/0308396.

Ramakrishnan, T. V., Krishnamurthy, H. R., Hassan, S. R. & Pai, G. V. 2004 Theory of insulator metal transition and colossal magnetoresistance in doped manganites. *Phys. Rev. Lett.* **92**, 157203. (doi:10.1103/PhysRevLett.92.157203)

Rao, C. N. R. 2000 Charge, spin, and orbital ordering in the perovskite manganates, $Ln_{(1-x)}A_{(x)}MnO_{(3)}$ (Ln = rare earth, A = Ca or Sr). *J. Phys. Chem. B* **104**, 5877–5889. (doi:10.1021/jp0004866)

Rao, C. N. R. & Raveau, B. (eds) 1998 *Colossal magnetoresistance, charge ordering and related properties of manganese oxides*. Singapore: World Scientific.

Rao, C. N. R., Parkash, O., Bahadur, D., Ganguly, P. & Nagabhushana, S. 1977 Itinerant electron ferromagnetism in SR2+-doped rare-earth, CA2+-doped rare-earth, and BA2+-doped rare-earth orthocobaltites ($LN_{1-x(3+)}M_{x(2+)}CoO_3$). *J. Solid State Chem.* **22**, 353–360. (doi:10.1016/0022-4596(77)90011-1)

Rao, C. N. R., Kundu, A. K., Seikh, M. M. & Sudheendra, L. 2004 Electronic phase separation in transition metal oxides. *Dalton Trans.* 3003–3011.

Renner, C., Aeppli, G., Kim, B. G., Soh, Y.-A. & Cheong, S.-W. 2002 Atomic-scale images of charge ordering in a mixed-valence manganite. *Nature* **416**, 518–521. (doi:10.1038/416518a)

Ritter, C., Mahendiran, R., Ibarra, M. R., Morellon, L., Maignan, A., Raveau, B. & Rao, C. N. R. 2000 Direct evidence of phase segregation and magnetic-field-induced structural transition in $Nd_{0.5}Sr_{0.5}MnO_3$ by neutron diffraction. *Phys. Rev. B* **61**, R9229–R9232. (doi:10.1103/PhysRevB.61.R9229)

Rodríguez-Martínez, L. M. & Attfield, J. P. 2001 Disorder-induced orbital ordering in $L_{0.7}M_{0.3}MnO_3$ perovskites. *Phys. Rev. B* **63**, 024424. (doi:10.1103/PhysRevB.63.024424)

Salamon, M. B. & Jaime, M. 2001 The physics of manganites: structure and transport. *Rev. Mod. Phys.* **73**, 583–628. (doi:10.1103/RevModPhys.73.583)

Sarma, D. D., Shanthi, N., Barman, S. R., Hamada, N., Sawada, H. & Terakura, K. 1995 Band theory for ground-state properties and excitation spectra of perovskite $LaMO_3$ (M = Mn, Fe, Co, Ni). *Phys. Rev. Lett.* **75**, 1126–1129. (doi:10.1103/PhysRevLett.75.1126)

Sarma, D. D. *et al.* 2004 Direct observation of large electronic domains with memory effect in doped manganites. *Phys. Rev. Lett.* **93**, 097202. (doi:10.1103/PhysRevLett.93.097202)

Satpathy, S., Popović, Z. S. & Vukajlović, F. R. 1996 Electronic structure of the perovskite oxides: $La_{1-x}Ca_xMnO_3$. *Phys. Rev. Lett.* **76**, 960–963. (doi:10.1103/PhysRevLett.76.960)

Señaris Rodríguez, M. A. & Goodenough, J. B. 1995 Magnetic and transport properties of the system $La_{1-x}Sr_xCoO_{3-\delta}$ ($0 < x \leq 0.50$). *J. Solid State Chem.* **118**, 323–336. (doi:10.1006/jssc.1995.1351)

Shenoy, V. B., Sarma, D. D. & Rao, C. N. R. 2006 Electronic phase separation in correlated oxides: the phenomenon, its present status and future prospects. *Chemphyschem* **7**, 2053–2059. (doi:10.1002/cphc.200600188)

Shenoy, V. B., Gupta, T., Krishnamurthy, H. R. & Ramakrishnan, T. V. 2007 Coulomb interactions and nanoscale electronic inhomogeneities in manganites. *Phys. Rev. Lett.* **98**, 097201. (doi:10.1103/PhysRevLett.98.097201)

Sigmund, E. & Müller, K. A. (eds) 1994. *Phase separation in cuprate superconductors.* Heidelberg, Germany: Springer.

Soh, Y.-A., Evans, P., Cai, Z., Lai, B., Kim, C.-Y., Aeppli, G., Mathur, N., Blamire, M. & Isaacs, E. 2002 Local mapping of strain at grain boundaries in colossal magnetoresistive films using x-ray microdiffraction. *J. Appl. Phys.* **91**, 7742–7744. (doi:10.1063/1.1455609)

Sudheendra, L. & Rao, C. N. R. 2003 Electronic phase separation in the rare-earth manganates $(La_{1-x}Ln_x)_{0.7}Ca_{0.3}MnO_3$ (Ln = Nd, Gd and Y). *J. Phys. Condens. Matter* **15**, 3029–3040. (doi:10.1088/0953-8984/15/19/306)

Tokura, Y. 2006 Critical features of colossal magnetoresistive manganites. *Rep. Prog. Phys.* **69**, 797–851. (doi:10.1088/0034-4885/69/3/R06)

Tranquada, J. M., Woo, H., Perring, T. G., Goka, H., Gu, G. D., Xu, G., Fujita, M. & Yamada, K. 2004 Quantum magnetic excitations from stripes in copper oxide superconductors. *Nature* **429**, 534–538. (doi:10.1038/nature02574)

Uehara, M., Mori, S., Chen, C. H. & Cheong, S. W. 1999 Percolative phase separation underlies colossal magnetoresistance in mixed-valent manganites. *Nature* **399**, 560–563. (doi:10.1038/21142)

Vogt, T., Cheetham, A. K., Mahendiran, R., Raychaudhuri, A. K., Mahesh, R. & Rao, C. N. R. 1996 Structural changes and related effects due to charge ordering in $Nd_{0.5}Ca_{0.5}MnO_3$. *Phys. Rev. B* **54**, 15 303–15 306. (doi:10.1103/PhysRevB.54.15303)

Vojta, T. & Schreiber, M. 2001 Localization and conductance in the quantum Coulomb glass. *Philos. Mag. B* **81**, 1117–1129.

Wollan, E. O. & Koehler, W. C. 1955 Neutron diffraction study of the magnetic properties of the series of perovskite-type compounds $[La_{(1-x)}, Ca_x]MnO_3$. *Phys. Rev.* **100**, 545–563. (doi:10.1103/PhysRev.100.545)

Woodward, P. M., Vogt, T., Cox, D. E., Arulraj, A., Rao, C. N. R., Karen, P. & Cheetham, A. K. 1998 The influence of cation size on the structural features of $Ln_{1/2}A_{1/2}MnO_3$ perovskites at room temperature. *Chem. Mater.* **10**, 3652–3665. (doi:10.1021/cm980397u)

Woodward, P. M., Cox, D. E., Vogt, T., Rao, C. N. R. & Cheetham, A. K. 1999 Effect of compositional fluctuations on the phase transitions in $(Nd_{1/2}Sr_{1/2})MnO_3$. *Chem. Mater.* **11**, 3528–3538. (doi:10.1021/cm990281d)

Wu, J., Lynn, J. W., Glinka, C. J., Burley, J., Zheng, H., Mitchell, J. F. & Leighton, C. 2005 Intergranular giant magnetoresistance in a spontaneously phase separated perovskite oxide. *Phys. Rev. Lett.* **94**, 037201. (doi:10.1103/PhysRevLett.94.037201)

Zhang, L., Israel, C., Biswas, A., Greene, R. L. & de Lozanne, A. 2002 Direct observation of percolation in a manganite thin film. *Science* **298**, 805–807. (doi:10.1126/science.1077346)

THE JOURNAL OF PHYSICAL CHEMISTRY Letters

Perspective

pubs.acs.org/JPCL

# Multiferroic and Magnetoelectric Oxides: The Emerging Scenario

C. N. R. Rao,* A. Sundaresan, and Rana Saha

Chemistry and Physics of Materials Unit, International Centre for Materials Science and CSIR Unit of Excellence in Chemistry, Jawaharlal Nehru Centre for Advanced Scientific Research, Jakkur P.O., Bangalore 560064, India

**ABSTRACT:** Multiferroics were considered to be rare because magnetism and ferroelectricity require entirely different criteria for the materials. Several multiferroic oxides have, however, been discovered in the past few years by virtue of novel operating mechanisms, the most effective one being ferroelectricity driven by magnetism itself. Many such oxides where the magnetic and electric order parameters interact also exhibit magnetoelectric or magnetodielectric properties. In this Perspective, properties of manganites, ferrites, and other monophasic multiferroic oxides with spin-induced electric polarization are described. Multiferroic properties arising from charge ordering are examined. The present status of $BiMnO_3$, which is an unusual example of a ferromagnetic–ferroelectric, is presented. Recent findings suggest that it is likely that many more multiferroic and magnetoelectric oxide materials exhibiting magnetically induced ferroelectricity will be found in the future.

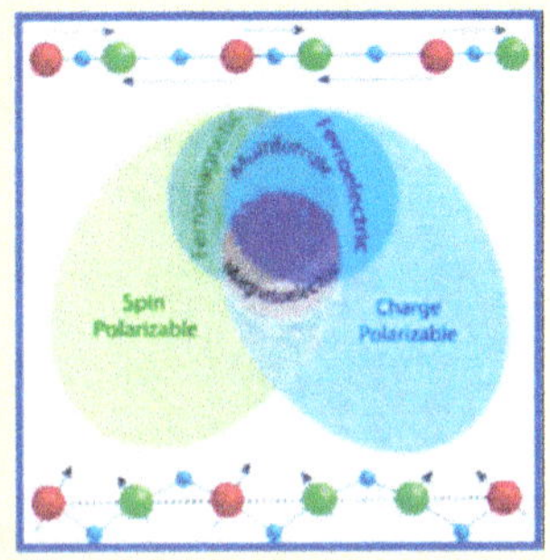


Multiferroics are defined as those materials that exhibit both ferroelectricity and ferromagnetism. Ferroelctricity and magnetism, however, have contradicting requirements. While magnetism requires d electrons, ferroelectric distortion is suppressed by d electrons. Ferroelectricity also requires that the material crystallize in a noncentrosymmetric space group. It is, therefore, not surprising that this led to an article by Nicola Hill[1] in *J. Phys. Chem.* (2000) entitled "Why Are There So Few Magnetic Ferroelectrics?". Despite the limitations mentioned above, several multiferroics have been discovered wherein ferroelectricity is induced by ingeneous routes.[2−5] Thus, in $YMnO_3$ and related manganites, ferroelectricity results from polyhedral tilting.[6] Lone pair (6s) electrons play a role in imparting ferroelectricity in bismuth-containing transition-metal oxides.[7] Frustrated magnetism and spiral magnetic ordering induces ferroelectricity in manganese oxides of the type $LnMn_2O_5$ (Ln = rare earth) and $Tb(Dy)MnO_3$.[3−8] Magnetically induced ferroelectricity in oxides has become an intense area of investigation in the past few years.

**It was believed that multiferroics can only be rare because magnetism cannot coexit with ferroelectricity in the same material.**

Magnetoelectric materials form a smaller subclass of multiferroics. In these materials, there is coupling between the electric and magnetic order parameters, and the application of a magnetic field induces electric polarization. In magnetodielectrics, the capacitance of a material is affected by an external magnetic field. A multiferroic may or may not be magnetoelectric but may exhibit magnetocapacitance. We must note that the magnetodielectric effect can also have a resistive origin arising from the Maxwell–Wagner effect and magnetoresistance.[9]

In the past few years, several oxides exhibiting multiferroic behavior due to novel types of magnetic interactions, cation charge ordering, and other causes have been reported. A good example of multiferroic oxides based on charge ordering are rare earth ferrites of the type $LnFe_2O_4$.[10,11] Especially noteworthy are the simple perovskite oxides that unexpectedly show spin-driven ferroelectricity. Some of the recent examples of magnetically induced ferroelctricity are $TbMnO_3$,[8] $GdFeO_3$,[12] $SmFeO_3$,[13] and $LnMn_2O_5$.[3,14]

In this Perspective, we briefly discuss the important features of monophasic multiferroic oxides exhibiting magnetically driven ferroelectricity. We then present the findings on charge-ordered oxides including rare earth manganites ($LnA_{1-x}Mn_xO_3$; A = alkaline earth).[15,16] We examine the curious case of $BiMnO_3$, which is the only ferromagnetic oxide reported to be ferroelecric.[17,18] Most of the multiferroic oxides reported are antiferromagnetic, with some of them exhibiting canted antiferromagnetism or ferrimagnetism. In this Perspective, we have attempted to present a fair picture of the recent developments in this exciting field. In doing so, we had to limit the number of references due to journal requirements, and we would like to be excused for any omission or error in judgement.

*Magnetically Induced Ferroelectricity in Manganites.* Magnetically induced ferroelectricity arises from the coupling between magnetization and electric polarization. Many interesting aspects of magnetism and the relation of magnetism to ferroelectric polarization in oxides have emerged from recent studies. Thus, the overlap of the electronic wave functions of two atoms with canted spins appears to cause spontaneous electronic polarization.[19] Unusual magnetic structures such as the spiral order in some of the rare earth manganites also give rise

Received: May 28, 2012
Accepted: July 26, 2012

   dx.doi.org/10.1021/jz300688b | J. Phys. Chem. Lett. 2012, 3, 2237−2246

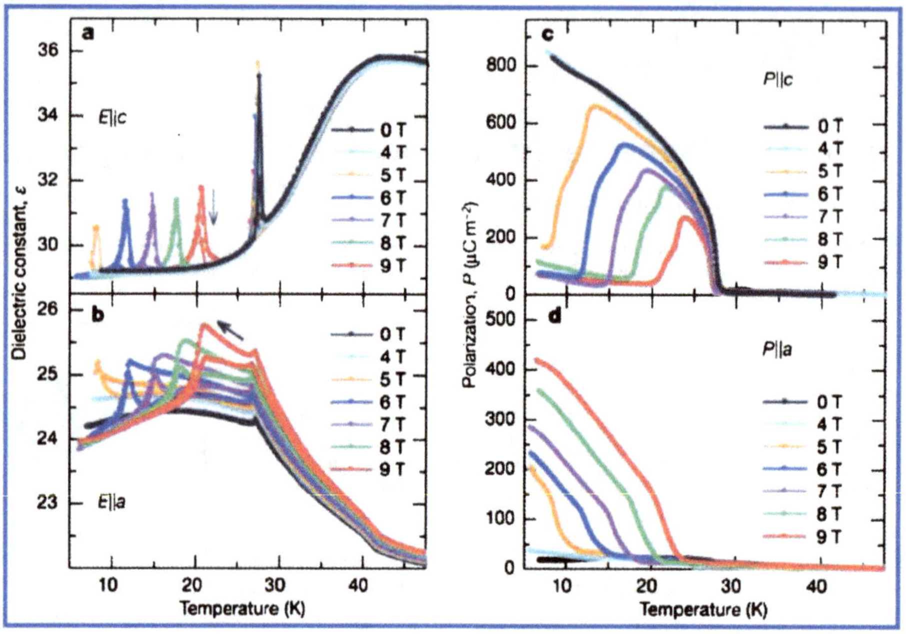


**Figure 1.** Magnetic-field-induced polarization switching in single crystals of $TbMnO_3$. Variation of the dielectric constant at 10 kHz (a,b) and the electric polarization along the *c* and *a* axes (c,d) with temperature, at various magnetic fields applied along the *b* axis. Reprinted with permission from ref 23, copyright 2003, by the Nature Publishing Group.

to ferroelectric polarization. Multiferroic manganites have been reviewed by a few workers.[20−22] We shall briefly present some of the highlights of the multiferroic properties of some of the manganites.

In the rare earth manganites of the type $LnMnO_3$ with Ln = Tb or Dy, competition between the nearest-neighbor (NN) ferromagnetic and next-nearest-neighbor (NNN) antiferromagnetic interactions gives rise to complex magnetic structures, with the long-wavelength antiferromagnetic order generating ferroelectricity through lattice modulations induced by magnetoelasticity.[23,24] Both ionic and electronic contributions, depending on the spiral plane and octahedral rotations, contribute to the ferroelectric polarization in these manganites.[25] Neutron diffraction studies reveal that $TbMnO_3$ undergoes successive magnetic phase transitions upon lowering the temperature.[26] The $Mn^{3+}$ moments exhibit incommensurate antiferromagnetic ordering with a collinear sinusoidal modulation along the *b* direction at $T_N$ (Mn, ~46 K). Upon further reducing the temperature, transition to a noncollinear spiral spin structure occurs at $T_{lock}$ (~28 K). NNN spin exchange, which is enhanced by the orthorhombic $GdFeO_3$-type distortion, causes spiral magnetic ordering, while the Dzyaloshinskii–Moriya interaction and single-ion anisotropy determine the spiral phase.[27] Anomalies in heat capacity and magnetization have been observed at around 46 and 28 K, with an additional anomaly at ~7 K due to sinusoidal ordering of the Tb spins. The spiral magnetic structure at the lock-in transition temperature (28 K) causes the spatial inversion symmetry to be broken and induces spontaneous electric polarization along the *c* direction of $TbMnO_3$. Upon application of a magnetic field of 4−8 T along the *a* or *b* axis, the polarization switches from $P||c$ to $P||a$, as shown in Figure 1. Such magnetic-field-induced ferroelctric polarization switching demonstrates the presence of strong coupling between magnetism and polarization. Application of a large magnetic field (≥8 T) along the *c* direction has a different effect than that along *a* or *b*. Application of a magnetic field results in the suppression of ferroelectricity at around 20 K, which is attributed to the magnetic transition from an incommensuarate antiferromagnetic structure to a canted A-type antiferromagnetic state of the Mn spins, which cannot support a lattice modulation with a nonzero wave vector.[24,28] At lower temperatures and higher magnetic fields, ferroelectricity again manifests itself due to the development of incommensurate ordering of the Tb and Mn spins.[28] Temperature variation of dielectric constant data shows a strong anomaly at the ferroelctric phase transition[23] (Figure 1).

**Of the mechanisms responsible for multiferroicity in oxide materials, magnetism-induced ferroelectricity is of great interest.**

$DyMnO_3$ also shows incommensurate to commensurate spin ordering below the Neel temperature ($T_N$(Mn) = 38 K), which induces ferroelctricity and the associated dielctric anomaly. It exhibits magnetic-field-induced polarization switching and a large value of $P_S \approx 0.2\ \mu C/cm^2$ with a magnetocapacitance ratio of ~500% at 18 K.[29] The high value of polarization above the ordering temperature of Dy (6.5 K) arises because of the increased amplitude of the Mn spin spiral.[30] Unlike $TbMnO_3$ and $DyMnO_3$, the ground state of $GdMnO_3$ is paraelectric in the absence of a magnetic field. However, application of a low magnetic field of 1 T or more along the *b* axis induces ferroelectricity with spontaneous polarization along the *a* axis. The absence of ferroelectricity is attributed to the disappearance of lattice modulation with zero wave vectors below $T_{lock} \approx 23$ K.

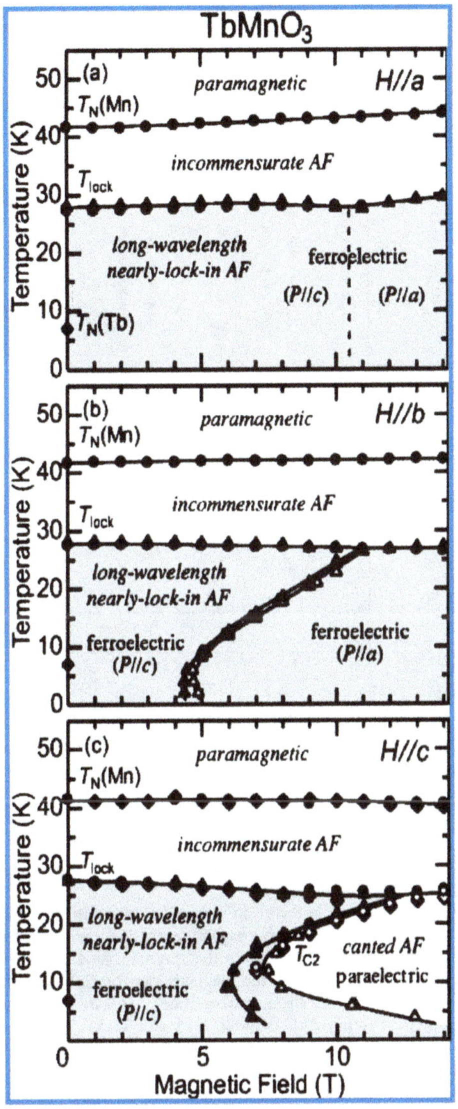


**Figure 2.** Magnetoelectric phase diagram of $TbMnO_3$ with the magnetic field along the *a*, *b*, and *c* axes. Circles, triangles, and diamonds indicate the dielectric constant, pyroelectric (or magnetoelectric) current, and magnetization data, respectively. Open and closed symbols indicate the data obtained with decreasing temperature (or increasing magnetic field) and increasing temperature (or decreasing magnetic field), respectively. Gray regions represent the ferroelectric states. Reprinted with permission from ref 24, copyright 2005, by the American Physical Society.

In Figure 2, we show a typical phase diagram,[24] representing the magnetic-field-induced changes in the dielectric and ferroelectric properties of $TbMnO_3$. Solid solutions of the type $Ln_{1-x}Y_xMnO_3$ (Ln = Eu, Gd) also exhibit remarkable multiferroic behavior accompanied by complex magnetic structures and ferroelctricity.[31]

Besides rare earth manganites of the type $LnMnO_3$, there are other manganites showing ferroelectric polarization induced by magnetic interactions. $LnMn_2O_5$ (Ln = Tb, Dy, Ho, Y) type materials exhibit ferroelectricity due to magnetostriction induced by collinear spins in the frustrated magnetic structures. These manganites undergo four sequential magnetic phase transitions with different transition temperatures indicating competing exchange interactions among the Mn spins and interaction between the Mn spins and the Ln moments. Long-range incommensurate sinusoidal antiferromagnetic ordering of the $Mn^{3+}/Mn^{4+}$ spins ($T_N$ = 42–45 K) followed by a ferroelectric transition at a lower temperature ($T_{N1}$ = 38–41 K) is observed due to the commensurate antiferromagnetic ordering that is almost collinear. One also observes a re-entrant transition into the incommensurate sinusoidal phase in the temperature range of 20–25 K and finally antiferromagnetic ordering of the rare earth ions below ~10 K. Very strong coupling between magnetism and electric polarization is exhibited by $TbMn_2O_5$, which shows highly reversible flipping of the spontaneous ferroelectric polarization at 3 K with the periodic variation of the external magnetic field between 0 and 2 T.[14] This magnetic-field-induced polarization switching phenomenon is also accompanied by a giant magnetocapacitance effect[14] (Figure 3). $DyMn_2O_5$ also shows successive

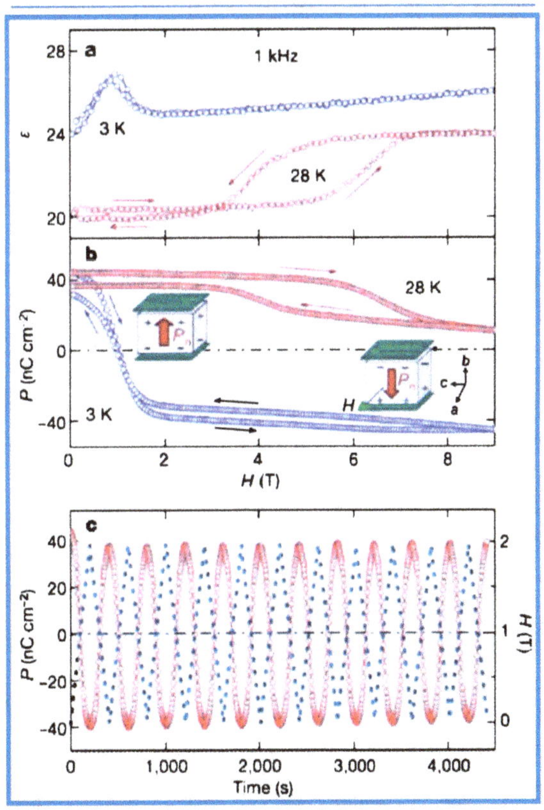


**Figure 3.** Variation of (a) dielectric constant and (b) electric polarization of $TbMn_2O_5$ as a function of magnetic field at 3 and 28 K; (c) highly reversible 180° polarization reversal at 3 K along the *b* axis induced by magnetic fields periodically varying between 0 and 2 T along the *a* axis. Reprinted with permission from ref 14, copyright 2004, by the Nature Publishing Group.

phase transitions associated with reordering of the spin state in the temperature range of 43–4 K, accompanied by three



dx.doi.org/10.1021/jz300688b | J. Phys. Chem. Lett. 2012, 3, 2237–2246 ferroelectric phases ($T_C$ = 40, 28, and 13 K) and the lowest-lying re-entrant paraelectric phase.[32] An interesting feature of this oxide is the induction of ferroelectric polarization in the paraelectric ground state upon application of a magnetic field of 1.5–2.5 T. This observation shows that the ferroelectric phases are strongly coupled to the antiferromagnetic Mn d electron spins with the large exchange striction affected by the f–d exchange interaction.[32] The magnetic-field-induced ferroelectric polarization is the cause of the large magnetodielectric effect exhibited by $DyMn_2O_5$. $DyMn_2O_5$ exhibits a very large change in the dielectric constant of ~109% (field-cooled, FC) and ~90% (zero-field-cooled, ZFC) at 3 K by the application of 7 T, compared to the Tb and Ho analogues.[33] The net $P$ in these manganites is composed of two collinear components,[34] one appearing below $T_C$ and another below $T_{N1}$, where these two components are parallel to each other for $HoMn_2O_5$ and antiparallel for $Tb(Dy)Mn_2O_5$.[34]

The case of geometric ferroelectrics as exemplified by the hexagonal manganites $YMnO_3$ and $LuMnO_3$, wherein octahedral tilting gives rise to ferroelectricity, was mentioned earlier. These materials exhibit large magnetoelectric coupling accompanying the large atomic displacements. Among the other manganese oxides exhibiting ferroelectricity, $CaMn_7O_{12}$ is interesting. In this material, improper ferroelectricity is induced by the incommensurate helical magnetic structure below $T_N$ (90 K). One observes large polarization due to ferroaxial coupling.[35]

*Magnetically Induced Ferroelectricity in Ferrites and Chromites.* In the quest for magnetically induced multiferroics, rare earth orthoferrites and orthochromites would be expected to be good candidates for exploration. Among the rare earth orthoferrites, $GdFeO_3$ is an example of a magnetic ferroelectric. The ferroelectric state is realized in the distorted orthorhombic perovskite, $GdFeO_3$, at a low temperature (~2.5 K) through a different mechanism.[12] The polar state in $GdFeO_3$ at the antiferromagnetic ordering temperature of Gd ions ($T_N$ = 2.5 K) is the result of spin-exchange striction between the Gd and Fe ions. The mutual controllability of the magnetization and polarization with the electric and magnetic fields shown in Figure 5 is explained in light of the composite structure of the domain wall, which is a combination of ferroelectric as well as ferromagnetic domains.[12] Exchange striction between adjacent

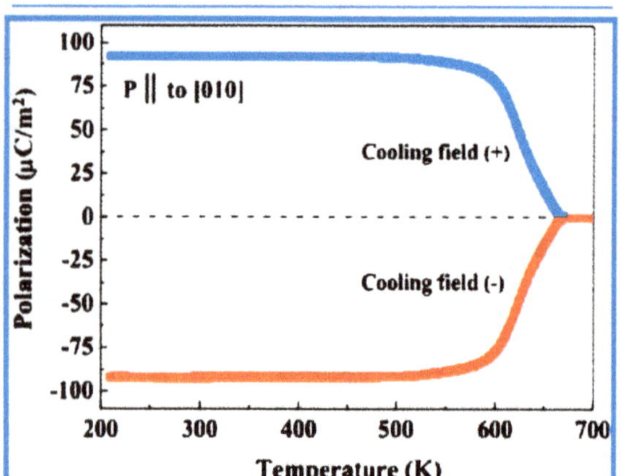


**Figure 4.** Variation of the ferroelectric polarization along [010] with temperature in $SmFeO_3$. Reprinted with permission from ref 13a, copyright 2011, by the American Physical Society.

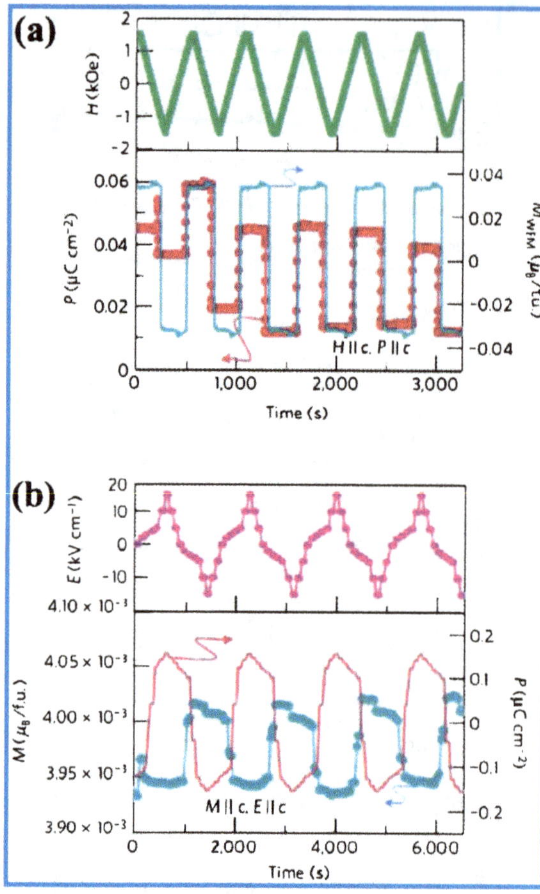


**Figure 5.** (a) Change in the electric polarization of $GdFeO_3$ with the magnetic field at 2.2 K; the ferromagnetic component of magnetization is repeatedly reversed by changing the magnetic field between −1.5 and 1.5 kOe. (b) Change in the magnetization of $GdFeO_3$ with the electric field at 2 K; the electric polarization is repeatedly reversed by changing the electric fields between −15 and 15 kV/cm. Reprinted with permission from ref 12, copyright 2009, by the Nature Publishing Group.

$Dy^{3+}$ and $Fe^{3+}$ layers with antiferromagnetic components gives rise to ferroelectric polarization in $DyFeO_3$.[36] $SmFeO_3$ is reported to be a multiferroic at room temperature. Although $SmFeO_3$ crystallizes in the centrosymmetric orthorhombic crystal structure, it exhibits improper ferroelctricity, with an onset temperature coinciding with $T_N$ (670 K).[13] The magnetic structure of $SmFeO_3$ consists of two nonequivalent Fe spin pairs, resulting in canted antiferromagnetism. Ferroelectricity in this ferrite was attributed to spin–orbit coupling driven inverse Dzyaloshinskii–Moriya-type antisymmetric mechanism, wherein the two nonequivalent canted AFM spin pairs are coupled, noncollinearly interacting with each other. It now appears that exchange striction is mainly responsible for the polarization in $SmFeO_3$ as well.[13] The appearance of the polar state shown in Figure 4 at the magnetic ordering temperature suggests that $SmFeO_3$ is a magnetically driven ferroelectric.

$AlFeO_3$, $GaFeO_3$, and their solid solutions crystallize in the noncentrosymmetric $Pna2_1$ structure, wherein there are four cationic sites Fe1, Fe2, Al(Ga)2, and Al(Ga)1, of which the first



dx.doi.org/10.1021/jz300688b | J. Phys. Chem. Lett. 2012, 3, 2237–2246 three are octahedral. They are collinear ferrimagnets with Néel temperatures in the 210–250 K range.[37] Cationic disorder is prominent in these oxides. They exhibit the magnetodielectric effect.[38] Studies based on pyroelectric current measurements indicate a maximum at around 100 K (below $T_N$) due to possible appearance of ferroelectricity, as shown in Figure 6.[39]

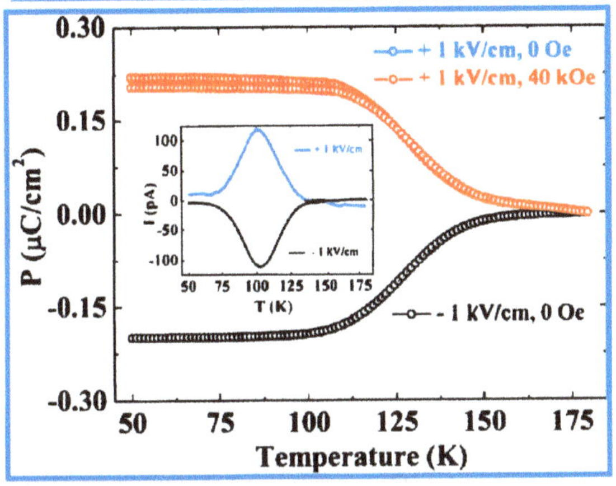


**Figure 6.** Temperature variation of the electric polarization ($P$) of $AlFeO_3$ with +1 and −1 kV/cm poling along with the effect of a 4 T magnetic field. (Inset) The variation of the pyroelectric current (corrected for leakage current) with temperature, showing a maximum at around 100 K. (From ref 39.)

The unit cell parameters as well as some of the Raman bands of $AlFeO_3$ do show anomalies around this temperature.[37,40] The origin of ferroelectricity could be the noncentrosymmetric magnetic ordering arising from the inherent magnetic frustration. There are, however, some questions about the ferroelectric nature (below $T_N$) of $AlFeO_3$ and $GaFeO_3$, suggested to occur on the basis of pyroelectric measurements, but then, we observe substantial polarization in addition to magnetoelectric and magnetodielectric properties. It is noteworthy that some of the hexaferrites have been recently found to exhibit the magnetoelectric effect because of magnetic frustration at room temperature and low magnetic fields.[41] A typical example of such a hexaferrite is $Sr_3Co_2Fe_{24}O_{41}$. The hexaferrites may find practical applications.

> Several manganites, ferrites, and chromites have been found to exhibit magnetically driven ferroelectricity.

Rare earth orthochromites, $LnCrO_3$, have been investigated for possible ferroelectricity.[42] These chromites with a centrosymmetric orthorhombic structure exhibit antiferromagnetic ordering with weak ferromagnetism in the 120–300 K range. Interestingly, the chromites with a magnetic rare earth Ln (Ln = Sm, Gd, Tb, Tm) become ferroelectric at the $T_N$(Cr) as shown in Figure 7. The cause of the ferroelectricity is considered to lie in the interaction between the magnetic Ln ion and the ferromagnetic moment of the Cr ion. The underlying mechanism could, however, be complex. On the basis of the results of $LnCrO_3$, we would expect all of the rare

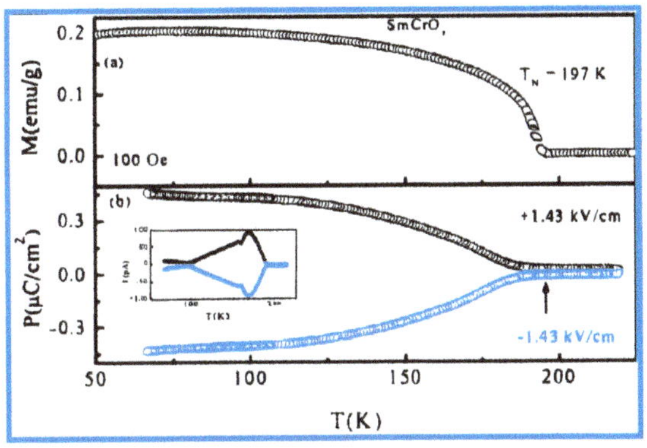


**Figure 7.** (a) Temperature dependence of the FC magnetization data of $SmCrO_3$ at 100 Oe. (b) Temperature dependence of the electric polarization ($P$) at +1.43 and −1.43 kV/cm. The inset in (b) shows the variation of the pyroelectric current with temperature at +ve and −ve poling fields. (From ref 42.)

earth ferrites, $LnFeO_3$, with magnetic Ln (as in $SmFeO_3$) to exhibit ferroelectricity at $T_N$.

Mandal et al.[43] have observed magnetoelectric and magnetodielectric effects in $YFe_{1-x}Mn_xO_3$ ($0.1 \leq x \leq 0.4$). These oxides exhibit a spin-reorientation transition at a temperature $T_{SR}$, which increases with $x$ while the $T_N$ decreases. The magnetodielectric effect is found at $T_{SR}$ as well as $T_N$ (Figure 8a), but ferroelectricity occurs at a lower temperature (Figure 8b). Thus, when $x$ = 0.4, $T_{SR}$ = 320 K, $T_N$ = 370 K, and the ferroelectric transition is at around 115 K. Recent studies of Sundaresan and Rajeswaran in this laboratory show that perovskite oxides of the type $LnA_{1-x}B_xO_3$ (Ln = Y, La, Lu), where A and B are transition-metal ions (Fe, Cr, Mn), become ferroelectric at $T_N$. The $T_N$ and hence the ferroelectric transition temperature can thus be varied depending on the A-site ion and the composition of the B-site. Ferroelectric polarization in these oxides is attributed to the nonequivalent disordered spins of these canted AFM oxides. It is noteworthy that adjacent atoms with canted antiferromagnetism coupled in a cycloidal spiral spin structure are expected to give rise to ferroelectric polarization[22] (Figure 9). This finding enlarges the scope of finding more multiferroic and magnetoelectric oxides.

*Other Transition-Metal Oxides.* Besides the above groups of oxides exhibiting spin-driven ferroelectricity, several other transition-metal oxides are found to exhibit multiferroic and magnetodielectric properties. Thus, CuO with a monoclinic $C2/c$ structure is reported to become ferroelectric at 230 K.[44] The delafossite $CuFeO_2$ becomes ferroelectric upon partial substitution of Fe by Rh, with the dielectric maximum coinciding with the Néel temperature.[45] In $YBaCuFeO_5$, a dielectric anomaly is observed near the incommensurate to commensurate antiferromagnetic transition due to magnetically induced electric polarization.[46]

$FeCr_2O_4$ and $CoCr_2O_4$ spinels show polarization anomalies at the ferrimagnetic transition.[47] There is a relation between the distortion due to Jahn–Teller $Fe^{2+}$ and the large polarization value at low temperatures in the case of $FeCr_2O_4$. The $CdV_2O_4$ spinel with a collinear magnetic structure is also multiferroic.[48] Magnetoelectric and magnetodielectric effects have been reported in $FeVO_4$.[49] A study of Ising chain magnetism and ferroelectricity in monoclinic $Ca_3CoMnO_6$ has shown the inequivalence of the Co–Mn distances to be responsible for

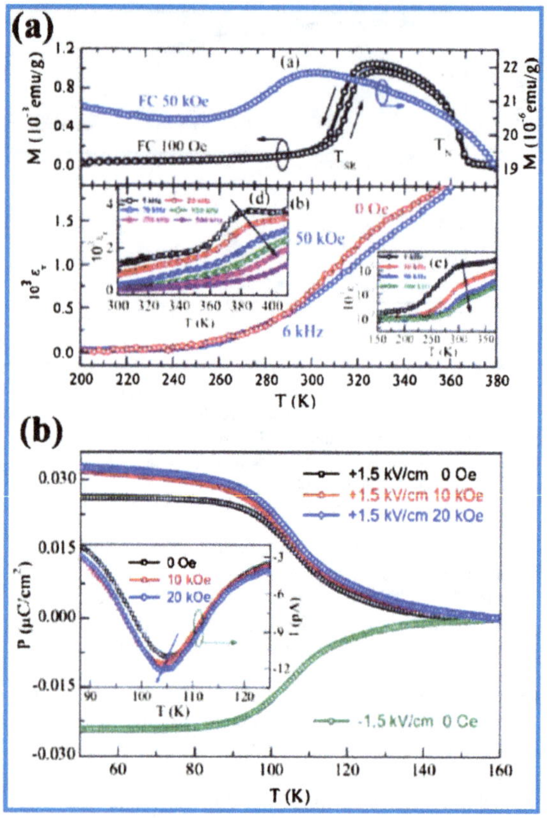


**Figure 8.** (a) Temperature dependence of (a) FC magnetization data of $YFe_{0.6}Mn_{0.4}O_3$ at 100 Oe and 50 kOe, exhibiting a change in the spin-reorientation temperature ($T_{SR}$) and (b) the dielectric constant at 0 Oe and 50 kOe, exhibiting a magnetocapacitance effect in $YFe_{0.6}Mn_{0.4}O_3$. The insets show dielectric constant anomalies near $T_{SR}$ and $T_N$. (b) Variation of ferroelectric polarization ($P$) in $YFe_{0.6}Mn_{0.4}O_3$ with temperature at positive and negative poling fields. (Inset) The effect of the magnetic field on the ferroelectric polarization. Reprinted with permission from ref 43, copyright 2011, by the American Physical Society.

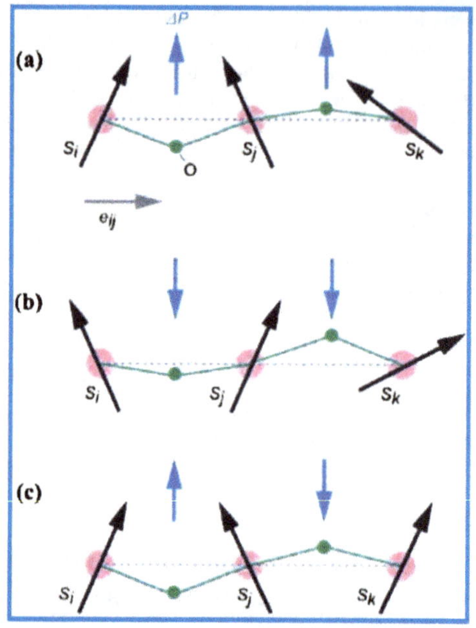


**Figure 9.** Schematic representation of spin-canting-induced changes in the local electric polarization in (a) counterclockwise (CCW) cycloidal spiral, (b) clockwise (CW) cycloidal spiral, and (c) normal canted antiferromagnetic[66] structures. The difference between the dark green and the light green dots represents the shift of oxygen ions due to the spin canting between the two neighboring spins. The resultant change of the local electric polarization is shown by blue arrows ($\Delta P$). Reprinted with permission from ref 22, copyright 2007 by Annual Reviews.

ferroelectricity.[50] A dielectric anomaly has been found at the magnetic ordering temperature of $CoV_2O_6$ as well.[51] In monoclinic $MnWO_4$ with a spiral magnetic order, the multiferroic state occurs at low temperatures.[52]

*Charge-Ordered Oxides.* One of the most well-known charge-ordered materials is magnetite, $Fe_3O_4$. This oxide itself is magnetoelectric at 4.2 K, as shown by Rado et al. several years ago.[53] There is some evidence for the occurrence of ferroelectricity in the low-temperature noncentrosymmetric triclinic phase of $Fe_3O_4$.[54] According to the model of Brink and Khomskii,[55] alternate arrangement of $Fe^{2+}$ and $Fe^{3+}$ ions along with alternating long and short Fe–Fe bonds along the *b* direction of the monoclinic cell results in polarization. A mechanism of ferroelectricity in the Verwey structure of magnetite in terms of three-site distortions, "trimerons", where the localized electrons are distributed over linear three-Fe-site units, has also been proposed.[56] The trimerons together with off-center atomic displacements in the charge-ordered phase give rise to electrical polarization, which couples with magnetization. Oxides derived from $Fe_3O_4$ such as $LnFe_2O_4$ (Ln = Y, Yb, Lu) show charge ordering and also electrical polarization at the ferrimagnetic transition (Figure 10a); these oxides also show a magnetodielectric effect[10,11] (Figure 10b).

An interesting charge-ordered oxide system that has been investigated is that of rare earth manganites. Efremov et al.[16] have proposed that charge-ordered manganites can be classified into bond-centered (BCO) and site-centered (SCO) systems (Figure 11). For certain compositions, for example, $Pr_{0.6}Ca_{0.4}MnO_3$, the ground state is ferroelectric. These manganites show dielectric anomalies around the charge-ordering temperature.[57] Magnetic fields markedly affect the dielctric properties of $Pr_{0.6}Ca_{0.4}MnO_3$, suggesting coupling between magnetization and polarization. In order to see whether magnetoresistance has any effect on the magnetocapacitance, investigations have been carried out on the highly insulating $Y_{1-x}Ca_xMnO_3$ ($x$ = 0.4, 0.45, 0.5), where the charge-ordered state is very robust and is not affected even by high magnetic fields (>10 T). These oxides also show dielectric constant maxima at around the charge-ordering temperature as well as magnetocapacitance (Figure 12).[58] Application of an electric field is known to melt the charge-ordered state of manganites into the metallic magnetic state.[5,15] Upon applying an electric field, charge-ordered manganites are found to emanate a magnetic flux. These manganites exhibit second-harmonic generation as well. Occurrence of ferroelectricity in charge-ordered half-doped manganites has been discussed on the basis of theoretical simulations.[59]

*Curious Case of $BiMnO_3$.* Among the bismuth compounds, $BiFeO_3$ is well-known as a multiferroic with a ferroelectric transition at 1100 K and antiferromagnetic transition at 643 K. There have been extensive studies on $BiFeO_3$ and compounds

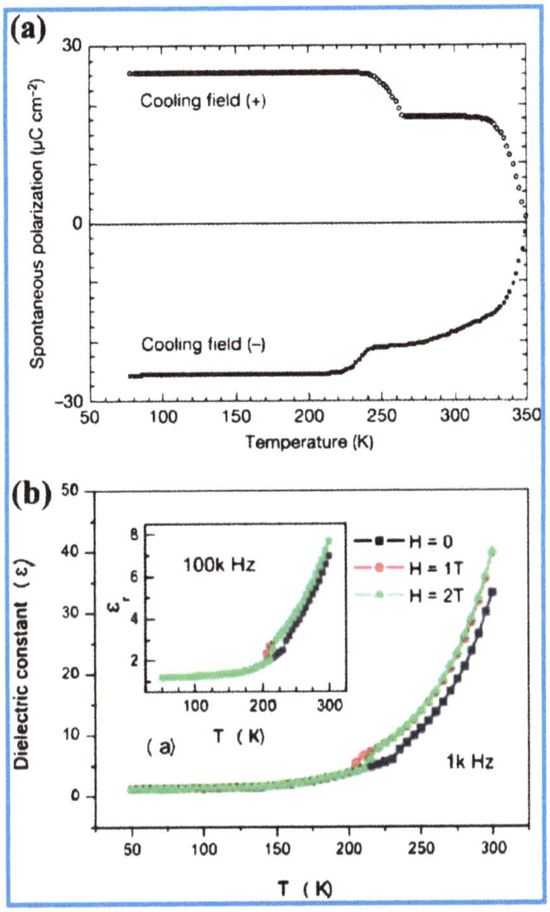


**Figure 10.** (a) Temperature dependence of the ferroelectric polarization of $LuFe_2O_4$. Reprinted with permission from ref 10, copyright 2005, by the Nature Publishing Group. (b) Temperature dependence of dielectric constant data of $YFe_2O_4$ measured at 0, 1, and 2 T at a frequency of 1 kHz. (Inset) The dielectric data measured at 100 kHz. Reprinted with permission from ref 11, copyright 2008, by the American Institute of Physics.

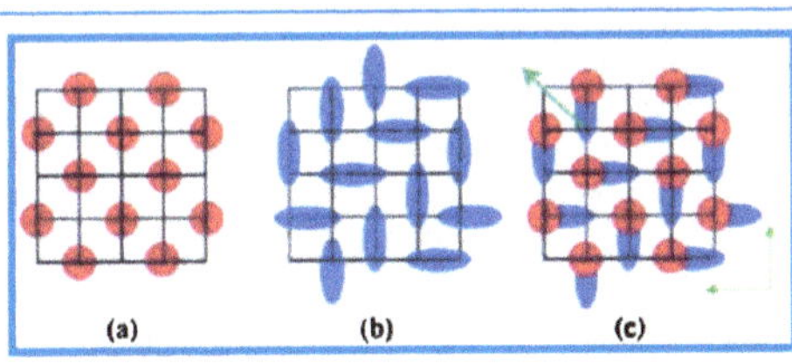


**Figure 11.** Different types of charge ordering: (a) site-centered charge order; (b) bond-centered charge order; and (c) a ferroelectric state. Green arrows indicate the dipole moments of horizontal and vertical dimers, and the diagonal arrow is the net spontaneous polarization. Reprinted with permission from ref 16, copyright 2004, by the Nature Publishing Group.

where Bi and Fe have been substituted partly by other cations. $BiFeO_3$ has an incommensurately modulated spin structure. Room-temperature electronic conductivity has been found at the ferroelectric domain walls of $BiFeO_3$.[60] $BiCrO_3$ becomes antiferromagnetic along with weak ferromagnetic order at 114 K and is antiferroelectric at room temperature.[61]

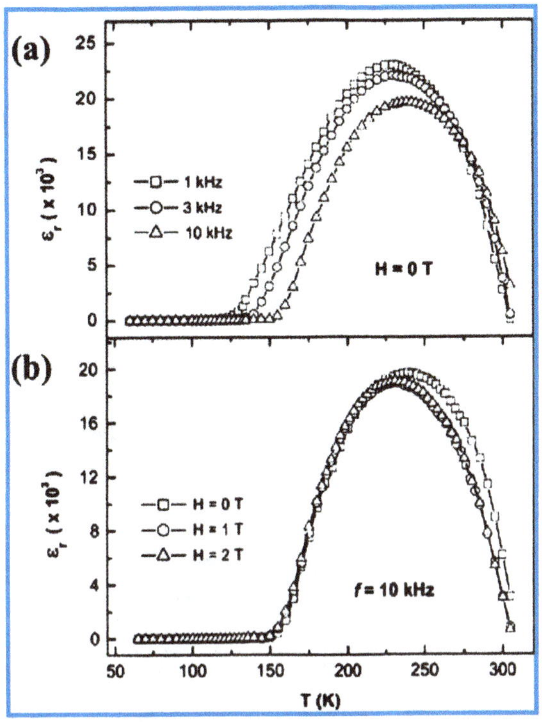


**Figure 12.** (a) Variation of the dielectric constant data of $Y_{0.5}Ca_{0.5}MnO_3$ with temperature at different frequencies. (b) Effect of magnetic fields on the temperature dependence of dielectric constant of $Y_{0.5}Ca_{0.5}MnO_3$ at 10 kHz. Reprinted with permission from ref 58, copyright 2009, by Elsevier.

Because of the influence of the stereochemically active 6s lone pair, $BiMnO_3$ was predicted to be ferroelectric.[7] $BiMnO_3$ would be one of the very few multiferroics that is ferromagnetic ($T_C$ = 450 K). $EuTiO_3$ is reported to become a ferromagnetic multiferroic at very low temperature upon application of stress. Ferroelectricity in $BiMnO_3$ was reported some time ago, but the observed polarization was small.[17] The magnetoelectric nature of $BiMnO_3$ has also been demonstrated.[18] Doubts about the ferroelectricity of $BiMnO_3$ arose because the experimentally observed and theoretically predicted space group is centrosymmetric, $C2/c$. Pair distribution function analysis, however, showed that $BiMnO_3$ contains domains corresponding to $P2_1$ and $P2$ space groups, whose atomic shifts break the symmetry of the C-center.[62] Because $BiMnO_3$ is prepared under high pressures, the sample generally contains some impurity phases. A careful synthesis paying attention to stoichiometry carried out by Sundaresan et al.[63] showed that the structure as well as magnetic and dielectric properties were dependent on the oxygen stoichiometry. Recently, Midgley and co-workers[64] have carried out a convergent beam electron diffraction (CBED) study of $BiMnO_3$ with differing oxygen stoichiometries. From the CBED patterns, they found that $BiMnO_{2.99}$ and $BiMnO_{2.94}$ both belong to the noncentrosymmetric space groups of $Pmn2_1$ and $Cmc2_1$, respectively, as shown in Figure 13. It appears that the $BiMnO_3$ compositions with oxygen deficiency crystallize in noncentrosymmetric space groups, consistent with the

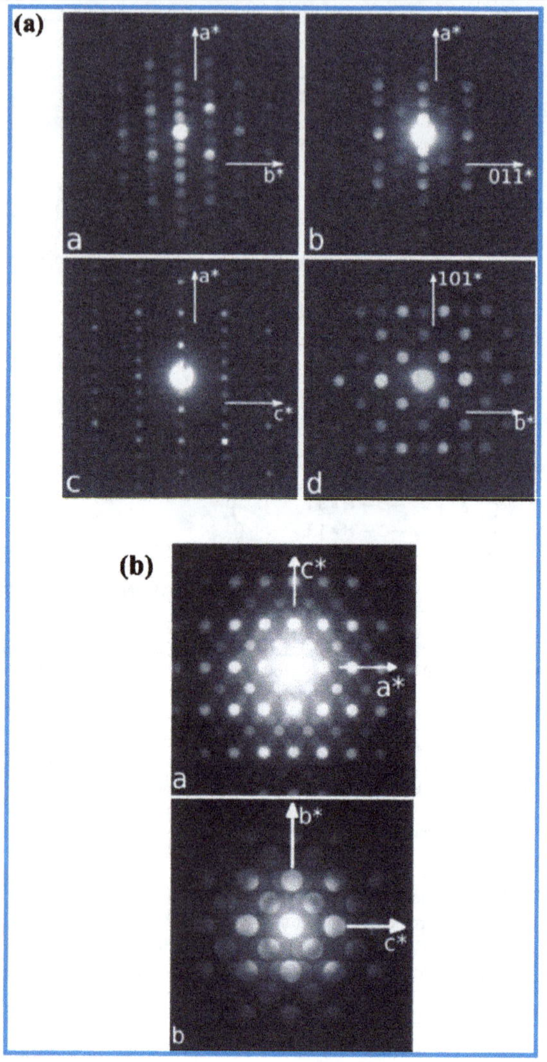


**Figure 13.** (a) Precession electron diffraction patterns from the orthorhombic phase of $BiMnO_{2.94}$, recorded parallel to a) [001], b) [01−1], c) [010], and d) [10−1] zone axes. (b) Convergent beam electron diffraction (CBED) patterns of $BiMnO_{2.99}$ recorded parallel to the a) [010] and b) [100] zone axes. Reprinted with permission from ref 64, copyright 2011, by the Royal Society of Chemistry.

occurrence of ferroelectricity. These phases are also ferromagnetic. $BiMn_7O_{12}$, which is antiferromagnetic, is found to show a magnetodielectric effect.[65]

> A good room-temperature ferromagnetic magnetoelectric would find many uses.

*Outlook.* The preceding discussion should clearly indicate that multiferroics are not as few and far between as expected a decade ago. The discovery of several monophasic multiferroic and magnetoelectric oxides in the past few years, exhibiting magnetically driven ferroelectricity, is quite impressive. It is indeed likely that many transition-metal oxides with fascinating polar and magnetic properties will be discovered in the future, as exemplified by the high-pressure polar corundum phase of $ScFeO_3$, which exhibits weak ferromagnetism.[66] Interestingly, hexagonal $LuFeO_3$ is a high-temperature ferroelectric just like $YMnO_3$.[67] It appears that there is yet need to unravel the origin of ferroelectricity in many of the oxides such as $LuFe_2O_4$.[68,69] Oxide materials such as the hexaferrites, which exhibit multiferroic properties at room temperature, would be of great value in terms of applications.[70] It would be worthwhile to look for ferromagnetic multiferroics, which are rare. Because the magnetic transition temperature and hence the ferroelectric transition temperature in some of the perovskite oxides can be varied by tuning the constituent rare earth or transition-metal ions, it may be possible to generate multiferroics operating over a wide range of temperatures. In terms of theory, it would be important to know whether the multiferroic properties of rare earth orthochromites, manganites, and ferrites have some commonality. Other mechanisms of inducing ferroelectricity such as charge ordering and the lone pair effect can also be exploited to design multiferroics. A noteworthy aspect relates to the multiferroic properties of nanoparticles of $BaTiO_3$.[71] Nanoparticles of $BaTiO_3$ exhibit room-temperature ferromagnetism just like all other inorganic nanostructures. It appears that magnetism in these nanoparticles couples with ferroelectricity. It would be worth exploring nanomultiferroics.

## ■ AUTHOR INFORMATION

### Corresponding Author

*E-mail: cnrrao@jncasr.ac.in. Fax: (+91) 80-2208-2760. Web-page: http://www.jncasr.ac.in/cnrrao/.

### Notes

The authors declare no competing financial interest.

### Biographies

**C. N. R. Rao** is National Research Professor and Linus Pauling Research Professor at Jawaharlal Nehru Centre for Advanced Scientific Research (JNCASR). His main interests are in materials chemistry. He is a member of several science academies including the U.S. National Academy of Sciences. He is the recipient of the Royal Medal of the Royal Society of London, the Dan David Prize for materials research, and the August Wilhelm von Hoffmann medal of the German Chemical Society. http://www.jncasr.ac.in/cnrrao/

**A. Sundaresan** obtained his Ph.D. degree from the Indian Institute of Technology Bombay and did postdoctoral work in Tsukuba, Caen, and Grenoble before joining JNCASR. He is an associate professor at JNCASR, where he works on oxide materials and magnetism. He is the recipient of medals of the Materials and Chemical Research Societies of India.

**Rana Saha** obtained his B.Sc. degree in chemistry from the University of Calcutta and his M.S. degree in materials science from JNCASR. He is carrying out his Ph.D. studies on multiferroics at JNCASR.

## ■ REFERENCES

(1) Hill, N. A. Why Are There So Few Magnetic Ferroelectrics? *J. Phys. Chem. B* **2000**, *104*, 6694−6709.

(2) Ramesh, R.; Spaldin, N. A. Multiferroics: Progress and Prospects in Thin Films. *Nat. Mater.* **2007**, *6*, 21−29.

(3) Cheong, S.-W.; Mostovoy, M. Multiferroics: A Magnetic Twist for Ferroelectricity. *Nat. Mater.* **2007**, *6*, 13−20.

(4) Eerenstein, W.; Mathur, N. D.; Scott, J. F. Multiferroic and Magnetoelectric Materials. *Nature* **2006**, *442*, 759−765.

(5) Rao, C. N. R.; Serrao, C. R. New Routes to Multiferroics. *J. Mater. Chem.* **2007**, *17*, 4931–4938.
(6) van Aken, B. B.; Palstra, T. T. M.; Filippetti, A.; Spaldin, N. A. The Origin of Ferroelectricity in Magnetoelectric $YMnO_3$. *Nat. Mater.* **2004**, *3*, 164–170.
(7) Seshadri, R.; Hill, N. A. Visualizing the Role of Bi 6s "Lone Pairs" in the Off-Center Distortion in Ferromagnetic $BiMnO_3$. *Chem. Mater.* **2001**, *13*, 2892–2899.
(8) Chapon, L. C.; Radaelli, P. G.; Blake, G. R.; Park, S.; Cheong, S. W. Ferroelectricity Induced by Acentric Spin-Density Waves in $YMn_2O_5$. *Phys. Rev. Lett.* **2006**, *96*, 097601/1–097601/4.
(9) Catalan, G. Magnetocapacitance without Magnetoelectric Coupling. *Appl. Phys. Lett.* **2006**, *88*, 1029021/–102901/3.
(10) Ikeda, N.; Ohsumi, H.; Ohwada, K.; Ishii, K.; Inami, T.; Kakurai, K.; Murakami, Y.; Yoshii, K.; Mori, S.; Horibe, Y.; et al. Ferroelectricity from Iron Valence Ordering in the Charge-Frustrated System $LuFe_2O_4$. *Nature* **2005**, *436*, 1136–1138.
(11) Serrao, C. R.; Sahu, J. R.; Ramesha, K.; Rao, C. N. R. Magnetoelectric Effect in Rare Earth Ferrites, $LnFe_2O_4$. *J. Appl. Phys.* **2008**, *104*, 016102/1–016102/3.
(12) Tokunaga, Y.; Furukawa, N.; Sakai, H.; Taguchi, Y.; Arima, T.-h.; Tokura, Y. Composite Domain Walls in a Multiferroic Perovskite Ferrite. *Nat. Mater.* **2009**, *8*, 558–562.
(13) (a) Lee, J.-H.; Jeong, Y. K.; Park, J. H.; Oak, M.-A.; Jang, H. M.; Son, J. Y.; Scott, J. F. Spin-Canting-Induced Improper Ferroelectricity and Spontaneous Magnetization Reversal in $SmFeO_3$. *Phys. Rev. Lett.* **2011**, *107*, 117201/1–117201/5. (b) Lee, J.-H.; Jeong, Y. K.; Park, J. H.; Oak, M.-A.; Jang, H. M.; Son, J. Y.; Scott, J. F. *Phys. Rev. Lett.* **2012**, *108*, 219702.
(14) Hur, N.; Park, S.; Sharma, P. A.; Ahn, J. S.; Guha, S.; Cheong, S. W. Electric Polarization Reversal and Memory in a Multiferroic Material Induced by Magnetic Fields. *Nature* **2004**, *429*, 392–395.
(15) Rao, C. N. R. Charge, Spin, and Orbital Ordering in the Perovskite Manganates, $Ln_{1-x}A_xMnO_3$ (Ln = Rare Earth, A = Ca or Sr). *J. Phys. Chem. B* **2000**, *104*, 5877–5889.
(16) Efremov, D. V.; van den Brink, J.; Khomskii, D. I. Bond- Versus Site-Centred Ordering and Possible Ferroelectricity in Manganites. *Nat. Mater.* **2004**, *3*, 853–856.
(17) Moreira dos Santos, A.; Parashar, S.; Raju, A. R.; Zhao, Y. S.; Cheetham, A. K.; Rao, C. N. R. Evidence for the Likely Occurrence of Magnetoferroelectricity in the Simple Perovskite, $BiMnO_3$. *Solid State Commun.* **2002**, *122*, 49–52.
(18) Kimura, T.; Kawamoto, S.; Yamada, I.; Azuma, M.; Takano, M.; Tokura, Y. Magnetocapacitance Effect in Multiferroic $BiMnO_3$. *Phys. Rev. B* **2003**, *67*, 180401/1–180401/4.
(19) Katsura, H.; Nagaosa, N.; Balatsky, A. V. Spin Current and Magnetoelectric Effect in Noncollinear Magnets. *Phys. Rev. Lett.* **2005**, *95*, 057205/1–057205/4.
(20) Tokura, Y.; Kida, N. Dynamical Magnetoelectric Effects in Multiferroic Oxides. *Philos. Trans. R. Soc. London, Ser. A* **2011**, *369*, 3679–3694.
(21) Dong, S.; Liu, J.-M. Recent Progress of Multiferroic Perovskite Manganites. *Condens. Matter* **2012**, arXiv:1204.0577v1.
(22) Kimura, T. Spiral Magnets as Magnetoelectrics. *Annu. Rev. Mater. Res.* **2007**, *37*, 387–413.
(23) Kimura, T.; Goto, T.; Shintani, H.; Ishizaka, K.; Arima, T.; Tokura, Y. Magnetic Control of Ferroelectric Polarization. *Nature* **2003**, *426*, 55–58.
(24) Kimura, T.; Lawes, G.; Goto, T.; Tokura, Y.; Ramirez, A. P. Magnetoelectric Phase Diagrams of Orthorhombic $RMnO_3$ (R=Gd, Tb, and Dy). *Phys. Rev. B* **2005**, *71*, 224425/1–224425/13.
(25) Malashevich, A.; Vanderbilt, D. Dependence of Electronic Polarization on Octahedral Rotations in $TbMnO_3$ from First Principles. *Phys. Rev. B* **2009**, *80*, 224407/1–224407/7.
(26) Kajimoto, R.; Yoshizawa, H.; Shintani, H.; Kimura, T.; Tokura, Y. Magnetic Structure of $TbMnO_3$ by Neutron Diffraction. *Phys. Rev. B* **2004**, *70*, 012401/1–012401/4.
(27) Mochizuki, M.; Furukawa, N.; Nagaosa, N. Theory of Spin-Phonon Coupling in Multiferroic Manganese Perovskites $Rmno_3$. *Phys. Rev. B* **2011**, *84*, 144409/1–144409/14.
(28) Argyriou, D. N.; Aliouane, N.; Strempfer, J.; Zegkinoglou, I.; Bohnenbuck, B.; Habicht, K.; Zimmermann, M. v. Melting of Incommensurate-Ferroelectric Phase with Magnetic Field in Multiferroic $TbMnO_3$. *Phys. Rev. B* **2007**, *75*, 020101/1–020101/4.
(29) Goto, T.; Kimura, T.; Lawes, G.; Ramirez, A. P.; Tokura, Y. Ferroelectricity and Giant Magnetocapacitance in Perovskite Rare-Earth Manganites. *Phys. Rev. Lett.* **2004**, *92*, 257201/1–257201/4.
(30) Prokhnenko, O.; Feyerherm, R.; Dudzik, E.; Landsgesell, S.; Aliouane, N.; Chapon, L. C.; Argyriou, D. N. Enhanced Ferroelectric Polarization by Induced Dy Spin Order in Multiferroic $DyMnO_3$. *Phys. Rev. Lett.* **2007**, *98*, 057206/1–057206/4.
(31) Noda, K.; Akaki, M.; Kikuchi, T.; Akahoshi, D.; Kuwahara, H. Magnetic-Field-Induced Switching between Ferroelectric Phases in Orthorhombic-Distortion-Controlled $RMnO_3$. *J. Appl. Phys.* **2006**, *99*, 08S905/1–08S905/3.
(32) Higashiyama, D.; Miyasaka, S.; Kida, N.; Arima, T.; Tokura, Y. Control of the Ferroelectric Properties of $DyMn_2O_5$ by Magnetic Fields. *Phys. Rev. B* **2004**, *70*, 174405/1–174405/7.
(33) Hur, N.; Park, S.; Sharma, P. A.; Guha, S.; Cheong, S. W. Colossal Magnetodielectric Effects in $DyMn_2O_5$. *Phys. Rev. Lett.* **2004**, *93*, 107207/1–107207/4.
(34) Inomata, A.; Kohn, K. Pyroelectric Effect and Possible Ferroelectric Transition of Helimagnet. *J. Phys.: Condens. Matter* **1996**, *8*, 2673–2678.
(35) Johnson, R. D.; Chapon, L. C.; Khalyavin, D. D.; Manuel, P.; Radaelli, P. G.; Martin, C. Giant Improper Ferroelectricity in the Ferroaxial Magnet $CaMn_7O_{12}$. *Phys. Rev. Lett.* **2012**, *108*, 067201/1–067201/4.
(36) Tokunaga, Y.; Iguchi, S.; Arima, T.; Tokura, Y. Magnetic-Field-Induced Ferroelectric State in $DyFeO_3$. *Phys. Rev. Lett.* **2008**, *101*, 097205/1–097205/4.
(37) Saha, R.; Shireen, A.; Bera, A. K.; Shirodkar, S. N.; Sundarayya, Y.; Kalarikkal, N.; Yusuf, S. M.; Waghmare, U. V.; Sundaresan, A.; Rao, C. N. R. Structure and Magnetic Properties of the $Al_{1-x}Ga_xFeO_3$ Family of Oxides: A Combined Experimental and Theoretical Study. *J. Solid State Chem.* **2011**, *184*, 494–501.
(38) Shireen, A.; Saha, R.; Mandal, P.; Sundaresan, A.; Rao, C. N. R. Multiferroic and Magnetodielectric Properties of the $Al_{1-x}Ga_xFeO_3$ Family of Oxides. *J. Mater. Chem.* **2011**, *21*, 57–59.
(39) Saha, R.; Shireen, A.; Shirodkar, S. N.; Waghmare, U. V.; Sundaresan, A.; Rao, C. N. R. On the Multiferroic and Magnetoelectric Nature of $GaFeO_3$, $AlFeO_3$ and Related Oxides. *Solid State Commun.* **2012**, doi:10.1016/j.ssc.2012.07.018.
(40) Kumar, P.; Bera, A.; Muthu, D. V. S.; Shirodkar, S. N.; Saha, R.; Shireen, A.; Sundaresan, A.; Waghmare, U. V.; Sood, A. K.; Rao, C. N. R. Coupled Phonons, Magnetic Excitations, and Ferroelectricity in $AlFeO_3$: Raman and First-Principles Studies. *Phys. Rev. B* **2012**, *85*, 134449/1–134449/6.
(41) Kimura, T. Magnetoelectric Hexaferrites. *Annu. Rev. Mater. Res.* **2012**, *3*, 93–110.
(42) Rajeswaran, B.; Khomskii, D. I.; Sundaresan, A.; Rao, C. N. R. Ferroelectricity at the Neel Temperature of Chromium in Rare-Earth Orthochromites: Magnetic Jahn-Teller Effect. *Condens. Matter* **2012**, arXiv:1201.0826v.
(43) Mandal, P.; Bhadram, V. S.; Sundarayya, Y.; Narayana, C.; Sundaresan, A.; Rao, C. N. R. Spin-Reorientation, Ferroelectricity, and Magnetodielectric Effect in $YFe_{1-x}Mn_xO_3$ ($0.1 \leq x \leq 0.4$). *Phys. Rev. Lett.* **2011**, *107*, 137202/1–137202/5.
(44) Kimura, T.; Sekio, Y.; Nakamura, H.; Siegrist, T.; Ramirez, A. P. Cupric Oxide As an Induced-Multiferroic with High-$T_c$. *Nat. Mater.* **2008**, *7*, 291–294.
(45) Pachoud, E.; Martin, C.; Kundys, B.; Simon, C.; Maignan, A. Spin-Driven Ferroelectricity in the Delafossite $CuFe_{1-x}Rh_xO_2$ ($0 \leq x \leq 0.15$). *J. Solid State Chem.* **2010**, *183*, 344–349.

(46) Kundys, B.; Maignan, A.; Simon, C. Multiferroicity with High-$T_c$ in Ceramics of the $YBaCuFeO_5$ Ordered Perovskite. *Appl. Phys. Lett.* **2009**, *94*, 072506/1–072506/3.
(47) Singh, K.; Maignan, A.; Simon, C.; Martin, C. $FeCr_2O_4$and $CoCr_2O_4$ Spinels: Multiferroicity in the Collinear Magnetic State? *Appl. Phys. Lett.* **2011**, *99*, 172903/1–172903/3.
(48) Giovannetti, G.; Stroppa, A.; Picozzi, S.; Baldomir, D.; Pardo, V.; Blanco-Canosa, S.; Rivadulla, F.; Jodlauk, S.; Niermann, D.; Rohrkamp, J.; et al. Dielectric Properties and Magnetostriction of the Collinear Multiferroic Spinel $CdV_2O_4$. *Phys. Rev. B* **2011**, *83*, 060402/1–060402/4.
(49) Dixit, A.; Lawes, G.; Harris, A. B. Magnetic Structure and Magnetoelectric Coupling in Bulk and Thin Film $FeVO_4$. *Phys. Rev. B* **2010**, *82*, 024430/1–024430/10.
(50) Wu, H.; Burnus, T.; Hu, Z.; Martin, C.; Maignan, A.; Cezar, J. C.; Tanaka, A.; Brookes, N. B.; Khomskii, D. I.; Tjeng, L. H. Ising Magnetism and Ferroelectricity in $Ca_3CoMnO_6$. *Phys. Rev. Lett.* **2009**, *102*, 026404/1–026404/4.
(51) Singh, K.; Maignan, A.; Pelloquin, D.; Perez, O.; Simon, C. Magnetodielectric Coupling and Magnetization Plateaus in $\alpha$-$CoV_2O_6$ Crystals. *J. Mater. Chem.* **2012**, *22*, 6436–6440.
(52) Heyer, O.; Hollmann, N.; Klassen, I.; Jodlauk, S.; Bohatý, L.; Becker, P.; Mydosh, J. A.; Lorenz, T.; Khomskii, D. A New Multiferroic Material: $MnWO_4$. *J. Phys.: Condens. Matter* **2006**, *18*, L471–L475.
(53) Rado, G. T.; Ferrari, J. M. Electric Field Dependence of the Magnetic Anisotropy Energy in Magnetite ($Fe_3O_4$). *Phys. Rev. B* **1975**, *12*, 5166–5174.
(54) Miyamoto, Y.; S, S.; Chikazumi, s. Crystal Symmetry of Magnetite in Low Temperature Phase Deduced from Magnetoelectric Measurements. *J. Phys. Soc. Jpn.* **1988**, *57*, 2040–2050.
(55) Brink, J. v. d.; Khomskii, D. I. Multiferroicity Due to Charge Ordering. *J. Phys.: Condens. Matter* **2008**, *20*, 434217/1–434217/12.
(56) Senn, M. S.; Wright, J. P.; Attfield, J. P. Charge Order and Three-Site Distortions in the Verwey Structure of Magnetite. *Nature* **2012**, *481*, 173–176.
(57) Serrao, C. R.; Sundaresan, A.; Rao, C. N. R. Multiferroic Nature of Charge-Ordered Rare Earth Manganites. *J. Phys.: Condens. Matter* **2007**, *19*, 496217/1–496217/5.
(58) Sahu, J. R.; Serrao, C. R.; Ghosh, A.; Sundaresan, A.; Rao, C. N. R. Charge-Order-Driven Multiferroic Properties of $YCa_{1-x}Mn_xO_3$. *Solid State Commun.* **2009**, *149*, 49–51.
(59) Yamauchi, K.; Picozzi, S. Charge Ordering and Ferroelectricity in Half-Doped Manganites. *Condens. Matter* **2012**, arXiv:1203.3359v1.
(60) Seidel, J.; Martin, L. W.; He, Q.; Zhan, Q.; Chu, Y. H.; Rother, A.; Hawkridge, M. E.; Maksymovych, P.; Yu, P.; Gajek, M.; et al. Conduction at Domain Walls in Oxide Multiferroics. *Nat. Mater.* **2009**, *8*, 229–234.
(61) Kim, D. H.; Lee, H. N.; Varela, M.; Christen, H. M. Antiferroelectricity in Multiferroic $BiCrO_3$ Epitaxial Films. *Appl. Phys. Lett.* **2006**, *89*, 162904/1–162904/3.
(62) Kodama, K.; Iikubo, S.; Shamoto, S.-i.; Belik, A. A.; Takayama-Muromachi, E. Local Crystal Structure of Multiferroic System $BiMnO_3$ by Atomic Pair Distribution Function Analysis. *Jpn J. Appl. Phys.* **2007**, *76*, 124605–124607.
(63) Sundaresan, A.; Mangalam, R. V. K.; Iyo, A.; Tanaka, Y.; Rao, C. N. R. Crucial Role of Oxygen Stoichiometry in Determining the Structure and Properties of $BiMnO_3$. *J. Mater. Chem.* **2008**, *18*, 2191–2193.
(64) Eggeman, A. S.; Sundaresan, A.; Rao, C. N. R.; Midgley, P. A. The Structure of Two New Non-Centrosymmetric Phases of Oxygen Deficient Bismuth Manganite. *J. Mater. Chem.* **2011**, *21*, 15417–15721.
(65) Imamura, N.; Singh, K.; Pelloquin, D.; Simon, C.; Sasagawa, T.; Karppinen, M.; Yamauchi, H.; Maignan, A. Magnetodielectric Response of Square-Coordinated $MnO_2$ Unit in Cubic $BiMn_7O_{12}$. *Appl. Phys. Lett.* **2011**, *98*, 072903/1–072903/3.
(66) Li, M.-R.; Adem, U.; McMitchell, S. R. C.; Xu, Z.; Thomas, C. I.; Warren, J. E.; Giap, D. V.; Niu, H.; Wan, X.; Palgrave, R. G.; et al. A Polar Corundum Oxide Displaying Weak Ferromagnetism at Room Temperature. *J. Am. Chem. Soc.* **2012**, *134*, 3737–3747.
(67) Jeong, Y. K.; Lee, J.-H.; Ahn, S.-J.; Jang, H. M. Epitaxially Constrained Hexagonal Ferroelectricity and Canted Triangular Spin Order in $LuFeO_3$ Thin Films. *Chem. Mater.* **2012**, *24*, 2426–2428.
(68) Groot, J. de.; Mueller, T.; Rosenberg, R. A.; Keavney, D. J.; Islam, Z.; Kim, J.-W.; Angst, M. Charge Order in $LuFe_2O_4$: An Unlikely Route to Ferroelectricity. *Phys. Rev. Lett.* **2012**, *108*, 187601/1–187601/5.
(69) Niermann, D.; Waschkowski, F.; Groot, J. de.; Angst, M.; Hemberger, J. Dielectric Properties of Charge-Ordered $LuFe_2O_4$ Revisited: The Apparent Influence of Contacts. *Phys. Rev. Lett.* **2012**, *109*, 016405/1–016405/4.
(70) Scott, J. F. Applications of Magnetoelectrics. *J. Mater. Chem.* **2012**, *22*, 4567–4574.
(71) Mangalam, R. V. K.; Ray, N.; Waghmare, U. V.; Sundaresan, A.; Rao, C. N. R. Multiferroic Properties of Nanocrystalline $BaTiO_3$. *Solid State Commun.* **2009**, *149*, 1–5.

# IV. MOLECULAR MATERIALS

*Commentary by C.N.R. Rao*

A few decades ago, research on molecular solids was limited to a few specialists, and properties of these solids were not considered that attractive. The situation changed when molecular solids with fascinating electronic and magnetic properties, including metallicity and superconductivity, began to be discovered. The discovery of fullerenes and the increasing attention being paid to supramolecular architectures have brought about a sea change in this area.[1,2]

The articles in this section cover some supramolecular hydrogen-bonded structures. Among the hydrogen bonded structures, the adduct between trithiocyanuric acid and bipyridyl to form shape-selective channels is noteworthy, just as the rosette structure between cyanuric acid and melamine. Phase transitions and other aspects of fullerenes are presented in two articles.

Experimental electronic charge densities of organic crystals obtained from X-ray crystallography are of value in understanding chemical bonds. This is specifically illustrated in the case of organic crystals showing nonlinear optical properties.

## References

1. C.N.R. Rao and J. Gopalakrishnan, *New Directions in Solid State Chemistry*, Cambridge University Press, Second Edition, 1997.
2. W. Jones and C.N.R. Rao (Eds.), *Supramolecular Organization and Materials Design*, Cambridge University Press, 2002.

# Phase Transitions, Superconductivity, and Ferromagnetism in Fullerene Systems

C.N.R. Rao and Ram Seshadri

## Introduction

By virtue of their unique structures, fullerenes exhibit novel chemical transformations. Particularly pertinent to this article are the interesting properties exhibited by fullerenes in the solid state. These molecules are spherical or near-spherical in shape. Molecules with high point-group symmetry, which are not bound strongly in the solid state, tend to crystallize into structures with long-range periodicity of the molecular centers of mass, but the molecular orientations are random or even dynamically disordered. When dynamically disordered, the molecules rotate about some preferred axis. $C_{60}$ and $C_{70}$ satisfy the criteria for such orientationally disordered solids and exhibit rich phase behavior in the solid state.[12] Since $C_{60}$ has high electron affinity, it forms anion salts with alkali and alkaline-earth metals as well as with strong organic donor molecules. With tetrakis dimethylaminoethylene (TDAE), which is a very powerful electron donor, $C_{60}$ forms a 1:1 solid that is ferromagnetic.[3] $C_{60}$-TDAE is the molecular organic ferromagnet with the highest $T_c$ (of 16 K) known to date. Some of the alkali and alkaline-earth fullerides, on the other hand, show superconductivity, with transition temperatures going up to 33K.[4] We shall briefly examine some of these solid-state properties.

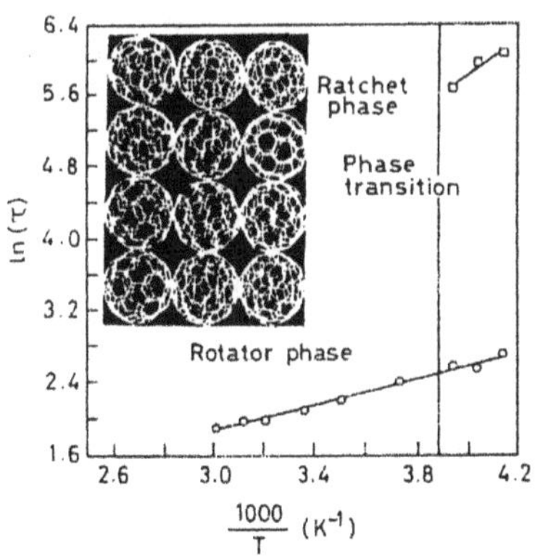


*Figure 1. Logarithm of the NMR orientational correlation time for solid $C_{60}$ (τ in picoseconds) plotted against 1/T.[5] Two regions of activated rotation are clearly discerned. The inset shows a snapshot of the high-temperature rotator phase from Monte Carlo studies.[7]*

## Phase Transitions

At ambient temperature, $C_{60}$, like solids of rare gas atoms such as argon or xenon, crystallizes in an fcc structure. The molecules in the solid rotate at frequencies almost comparable to those in the gas phase. Cooling the solid slows down and restricts the free rotation. The low-temperature phase of $C_{60}$ is simple cubic.[1] NMR spectroscopy permits one to probe the two regimes of rotational behavior as a function of temperature.[5] Figure 1 shows the activation plot for the orientational correlation time, obtained from NMR spectroscopy. Above 260 K, the barrier to rotation is small, giving rise to a free rotator phase. Below this temperature, the rotational activation barrier increases to 3000 K, and the molecules *ratchet* between preferred orientations. Computer simulation studies have thrown much light on the phase behavior. The inset in Figure 1 shows a snapshot of the high-temperature phase of $C_{60}$, obtained from Monte Carlo simulations.[7] One can see that the centers of the molecules define a cubic lattice, but that the orientations of individual molecules are random. Infrared[8] and Raman[9] spectroscopies have been used to follow the effect of the ordering transitions on the intramolecular vibrational modes. Raman spectra across the transition show unusual cycling effects attributed to laser-induced electronic excitation.[9]

The effect of pressure on $C_{60}$ is to increase the transition temperature and enrich the phase diagram. There is evidence that application of pressure results in the observation of two phase transitions instead of one.[10] Pressure also modulates the electronic properties of the solid. For example, with increasing pressure, the photoluminescence bandgap of solid $C_{60}$ decreases,[11] associated with a decrease in activation energy for conduction. There is also much recent interest in pressure-induced structural changes in $C_{60}$, including amorphization.[12] Under nonhydrostatic pressure, $C_{60}$ can be converted into diamond and diamondlike carbon.[14]

The lower symmetry of $C_{70}$ compared to $C_{60}$ gives rise to a phase diagram richer than that of $C_{60}$.[2] It has however taken longer for the structures of the different phases of $C_{70}$ to be ascertained, largely because the solid is highly disordered and there are many competing phases of nearly equal energies at ambient temperatures. Computer simulations suggest that the high-temperature free-rotator fcc phase, on cooling, should give rise to phases of lower symmetry, with concomi-

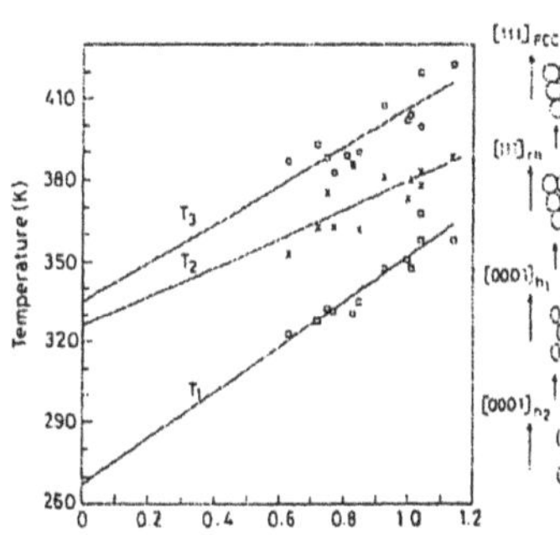


*Figure 2. Pressure dependence of the three-phase transition temperatures of $C_{70}$, obtained from resistance measurements. The best-fit straight lines are extrapolated to ambient pressure.[10] The possible structures of the various phases are shown in the schematic on the right.*

## Phase Transitions, Superconductivity, and Ferromagnetism in Fullerene Systems

tant freezing of the rotation.[15] At lower temperatures, rhombohedral, hcp, and monoclinic phases have been observed in experiments.[2] Application of pressure, as in $C_{60}$, clearly allows the delineation of as many as three transitions in $C_{70}$ (see Figure 2); extrapolation to ambient pressure gives 337, 330, and 280 K as the transition temperatures.[10] Calorimetric measurements[16] support these observations. The possible structures across the phase transitions are depicted in Figure 2. Infrared[17] and Raman[18] studies of $C_{70}$ across the phase transitions have proved to be useful probes of the structure and ordering. The frequencies of intramolecular vibrations as obtained from both Raman and infrared experiments, increase on traversing the phase transitions from high to low temperatures. The spectral linewidths and intensities also show signatures of the transitions.

At higher pressures than used in the preceding experiment, there is considerable structural change. Thus Raman studies[19] indicate that above a pressure of about 15 GPa, there is only a single broad absorption due to $sp^2$ carbon atoms, suggesting the formation of an amorphous phase, as shown in Figure 3a. Interestingly, the amorphization is reversible and decreasing the pressure results in the recovery of the original spectrum (Figure 3b).

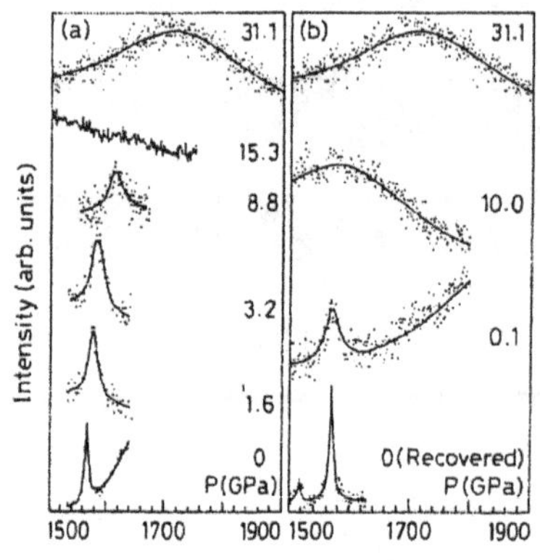


*Figure 3. (a) Raman spectra of a crystal of $C_{70}$ as the pressure is increased. Note the single broad feature at 31.1 GPa. (b) The spectra as the pressure is decreased, showing the perfectly reversible nature of the amorphization.[19]*

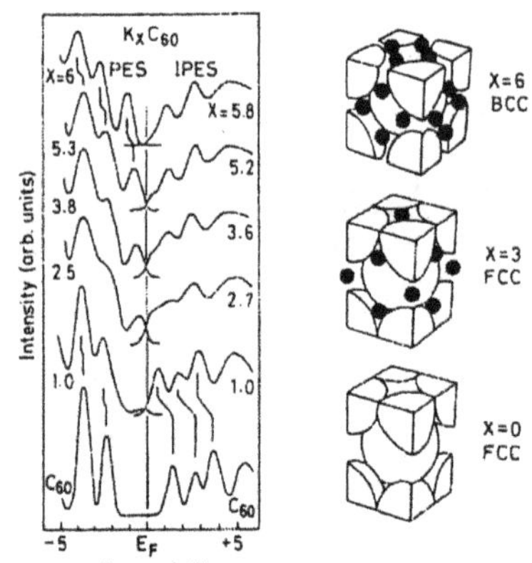


*Figure 4. Normal (PES) and inverse (IPES) photoemission density of states of $C_{60}$ as a function of exposure to potassium vapor. The gradual filling of the $C_{60}$ $t_{1u}$ LUMO is clearly seen. The spectral manifold shifts to lower energy with increasing exposure due to Fermi-level pinning.[20] On the right, body-centered tetragonal representations show structures of $C_{60}$ (x=0), $K_3C_{60}$ (x=3), and $Cs_6C_{60}$ (x=6).*

### Superconductivity in Alkali and Alkaline-Earth-Doped $C_{60}$

The ability of fullerenes to act as electron acceptors results in the formation of different types of salts. Alkali and alkaline-earth fullerides show interesting electronic properties.[4] Photoemission studies probing the occupied electronic states, as well as inverse photoemission measurements probing the unoccupied electronic states, have allowed direct monitoring of the nature of electron doping into the $C_{60}$ levels.[20] The valence band of solid $C_{60}$ is derived from a fivefold degenerate $h_u$ orbital. Since this orbital is well stabilized, removing electrons from $C_{60}$ is not easy. However, the threefold degenerate $t_{1u}$ and $t_{1g}$ levels are quite low-lying, making $C_{60}$ a good acceptor. Exposing $C_{60}$ to alkali metal vapor in the photoelectron spectrometer results in electron filling of the $t_{1u}$ level. The Fermi energy is pinned to the top of the filled level, so the spectral manifold is shifted to lower energy as filling increases (Figure 4). Since the level is threefold degenerate, half-filling corresponds to three electrons and $K_3C_{60}$ is a metal, which becomes superconducting below 19 K. Further filling produces $A_6C_{60}$ (e.g., $Cs_6C_{60}$) which, since it has a fully filled $t_{1u}$ LUMO, is insulating. The structures of $C_{60}$, $K_3C_{60}$, and $Cs_6C_{60}$, as determined by high-resolution powder x-ray diffraction and allied techniques,[4] are shown on the right in Figure 4. The structures are cubic, but have been represented here in tetragonal form. The alkali metal atoms sit in the tetragonal and octahedral voids of the fcc-$C_{60}$ structure. With alkaline-earth metals, the $t_{1g}$ orbital derived band is also partially filled. Thus $Ca_5C_{60}$, $Sr_6C_{60}$, and $Ba_6C_{60}$ are superconducting.[21]

It is popularly believed that the mechanism for high-temperature superconductivity in these doped fullerides is rather simple. The following factors, however, must be considered: First, the superconductors are isotropic with cubic lattices. High-symmetry systems resist the sort of low-temperature distortions that give rise to magnetic and/or insulating ground states. Second, the $C_{60}$ cage resists distortions that would lift the threefold degeneracy of the $t_{1u}$ and $t_{1g}$ levels, resulting in filled bands. Finally, the intramolecular vibrations, or phonons, are very strong, of the order of 1000 K. If these phonons could couple conduction electrons to give rise to Cooper pairs within the framework of traditional BCS theory, the high-transition temperatures would not be surprising. An important piece of evidence supporting the role of intramolecular phonons is that, in a whole range of $A_3C_{60}$ superconductors, the transition temperature is a function of only the separation between $C_{60}$ molecules and does not depend on the identity of the species in the A site. Separating the molecules results in a sharpening of the density of states and a concomitant increase in $T_c$. This is shown in Figure 5 as a plot of superconducting transition temperature against the $a$-lattice parameter of fcc-$A_3C_{60}$. Figure 5 combines ambient pressure data from alloying experiments in the A site as well as high-pressure data on $K_3C_{60}$ and $Rb_3C_{60}$. Despite being a more recent phenomenon, superconductivity in doped fullerides seems to be a more tractable problem than superconductivity in low-dimensional copper oxides, from a mechanistic viewpoint.

### Ferromagnetism in $C_{60}$-TDAE

Organic molecules that are effective electron donors have been known for some time, one of the more powerful of these being *tetrakis* dimethylaminoethylene. This molecule has an electron-rich $\pi$ bond because the substituent dimethylamino groups are good electron donors and because their bulk forces the molecule out of planarity, thereby destabilizing the bonding $\pi$ orbital. When mixed with $C_{60}$ in solution, a black 1:1 salt precipitates out.[3] What is surprising is that, on cooling to

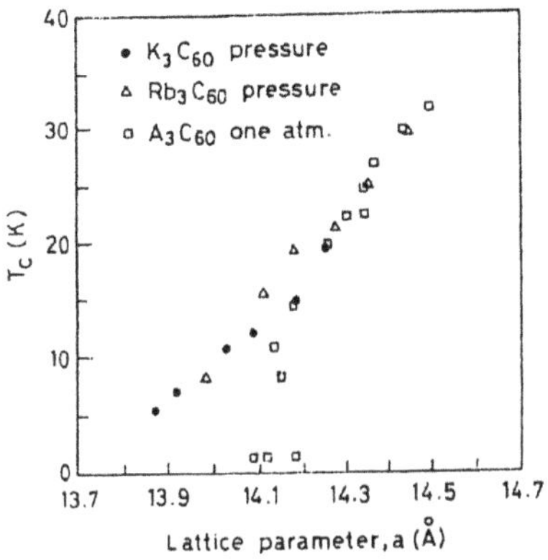


*Figure 5. Superconducting transition temperature plotted as a function of the a-lattice parameter for a variety of $A_3C_{60}$ phases. The data points represent ambient pressure alloying experiments as well as high-pressure experiments on $K_3C_{60}$ and $Rb_3C_{60}$.[4]*

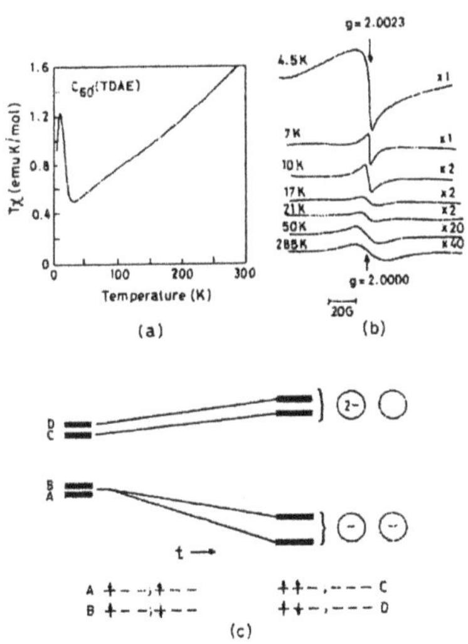


*Figure 6. (a) Product of the molar magnetic susceptibility, χ, and the temperature, T, plotted as a function of T for $C_{60}$-TDAE, showing a transition to a ferromagnetic state.[3] (b) ESR spectra of $C_{60}$-TDAE at different temperatures. Below the transition, the g value is seen to increase due to internal fields.[22] (c) Configuration interaction picture for the stabilization of a triplet ground state between two $C_{60}^-$ species. Hopping strength is given by t, and A,B,C,D represent the configurations of the $t_{1u}$ LUMO.*

below 16 K, the solid shows a transition to a ferromagnetic state (Figure 6a). $C_{60}$-TDAE is a soft ferromagnet; i.e., it shows no remanence in the magnetization curve. It holds the record for the highest Curie temperature of any purely organic molecular solid. Electron spin resonance (ESR) studies confirm that one electron is doped from TDAE to $C_{60}$.[22] As the temperature is lowered to below the transition at 16 K, the ESR $g$ value increases due to the development of an internal magnetic field (Figure 6b). Single-electron doping has been independently confirmed by comparing the shifts in the Raman phonon frequencies of $C_{60}$-TDAE with the shifts known for $A_xC_{60}$ systems.[23] A key property of $C_{60}$-TDAE is that the highly nonspherical structure of TDAE forces $C_{60}$-TDAE into adopting a low-symmetry monoclinic structure.[24] In the discussion on superconductivity, we mentioned that distortions favor insulating and/or magnetic rather than metallic ground states. The low-symmetry structure of $C_{60}$-TDAE ideally suits it for magnetic ordering. Another important aspect of $C_{60}$-TDAE is *single*-electron doping in the $C_{60}$-$t_{1u}$ orbital. With such a configuration, it is easy to visualize the stabilization of the ferromagnetic state. This is shown in Figure 6c as follows: Imagine two neighboring $C_{60}^-$ ions, each with one electron in the $t_{1u}$ orbital. The electrons may be either both spin-up (triplet state A) or spin-up on one $C_{60}$ and spin-down on the other (singlet state B). At this point, we cannot tell which is favored. To determine this, we admix some of the excited state by switching on a transfer integral, $t$. This serves to hop an electron from one $C_{60}$ to another, giving rise to the pair, $C_{60}^{2-}$-$C_{60}$. Hund's rule would favor the triplet state C rather than the singlet D. Since like configurations interact, the net result is the stabilization of the ground state triplet A. This demonstrates that the local interaction favors ferromagnetism. Since the structure is quasi-one-dimensional, this interaction propagates throughout the crystal.

## Conclusion

The above discussion illustrates the fascinating physical properties of $C_{60}$ and $C_{70}$. Undoubtedly, as chemists continue to prepare a wide variety of derivatives of these molecules, researchers will discover many more interesting solid-state properties of these systems. Solid-state properties of the higher fullerenes $C_{76}$, $C_{78}$, $C_{82}$, etc. are yet to be explored. At present, it is difficult to prepare them in large quantities, but this problem is also likely to be surmounted in the near future.

## References

1. P.A. Heiney, *J. Phys. Chem. Solids* 53 (1992) p. 1333.
2. P.A. Heiney and J.E. Fischer, *J. Phys. Chem. Solids* 54 (1993) p. 1725.
3. P.M. Allemand, K.C. Khemani, A. Koch, F. Wudl, K. Holczer, S. Donovan, G. Gruner, and J.D. Thompson, *Science* 253 (1991) p. 301.
4. D.W. Murphy, M.J. Rosseinsky, R.M. Fleming, R. Tycko, A.P. Ramirez, R.C. Haddon, T. Siegrist, G. Dabbagh, J.C. Tully, and R.E. Walstedt, *J. Phys. Chem. Solids* 53 (1992) p. 1321.
5. R.D. Johnson, D.S. Bethune, and C.S. Yannoni, *Acc. Chem. Res.* 25 (1992) p. 169.
6. A. Cheng, M. Klein, M. Parinello, and M. Sprik, *Philos. Trans. R. Soc. London* 341 (1992) p. 133.
7. A. Chakrabarti, S. Yashonath, and C.N.R. Rao, *Chem. Phys. Lett.* 215 (1993) p. 519.
8. L.R. Narasimhan, D.N. Stoneback, A.F. Hebbard, R.C. Haddon, and C.K.N. Patel, *Phys. Rev B* 46 (1992) p. 2591.
9. K. Akers, K. Fu, P. Zhang, and M. Moskovits, *Science* 259 (1993) p. 1152.
10. S. Ramasesha, A.K. Singh, R. Seshadri, A.K. Sood, and C.N.R. Rao, *Chem. Phys. Lett.* 220 (1994) p. 203.
11. A.K. Sood, N. Chandrabhas, D.V.S. Muthu, A. Jayaraman, N. Kumar, H.R. Krishnamurthy, T. Pradeep, and C.N.R. Rao, *Solid State Commun.* 81 (1992) p. 87.
12. D.W. Snoke, Y.S. Raptis, and K. Syassen, *Phys. Rev B* 45 (1992) p. 14419.
13. M. Nunez-Regueiro, P. Monceau, A. Rassat, P. Bernier, and A. Zahab, *Nature* 354 (1991) p. 289.
14. H. Hirai, K. Kondo, N. Yoshizawa, and M. Shiraishi, *Appl. Phys. Lett.* 64 (1994) p. 1797.
15. M. Sprik, A. Cheng, and M.L. Klein, *Phys. Rev Lett.* 69 (1992) p. 1660.
16. J. Sworakowski, K. Palewska, and M. Bertault, *Chem. Phys. Lett.* 220 (1994) p. 197.
17. V. Varma, R. Seshadri, A. Govindaraj, A.K. Sood, and C.N.R. Rao, *Chem. Phys. Lett.* 203 (1993) p. 545.
18. N. Chandrabhas, K. Jayaram, D.V.S. Muthu, A.K. Sood, R. Seshadri, and C.N.R. Rao, *Phys. Rev B* 47 (1993) p. 10963.
19. N. Chandrabhas, A.K. Sood, D.V.S. Muthu, C.S. Sundar, A. Bharathi, Y. Hariharan, and C.N.R. Rao (to be published); A.K. Sood, N. Chandrabhas, D.V.S. Muthu, Y. Hariharan, A. Bharathi, and C.S. Sundar, *Philos. Mag. A* (1994) in press.
20. J.W. Weaver, *J. Phys. Chem. Solids* 53 (1992) p. 1433.
21. A.R. Kortan, N. Kopylov, S. Glarum, E.M. Gyorgy, A.P. Ramirez, R.M. Fleming, F.A. Thiel, and R.C. Haddon, *Nature* 355 (1992) p. 529; 360 (1992) p. 566.
22. R. Seshadri, A. Rastogi, S.V. Bhat, S. Ramaseha, and C.N.R. Rao, *Solid State Commun.* 85 (1993) p. 971.
23. D.V.S. Muthu, M.N. Shashikala, A.K. Sood, R. Seshadri, and C.N.R. Rao, *Chem. Phys. Lett.* 217 (1994) p. 146.
24. P.W. Stephens, D. Cox, J.W. Lauher, L. Mihaly, J.B. Wiley, P.M. Allemand, A. Hirsch, K. Holczer, Q. Li, J.D. Thompson, and F. Wudl, *Nature* 355 (1992) p. 331. □

16814 *J. Phys. Chem.* **1995,** *99,* 16814–16816

# Polymerization and Pressure-Induced Amorphization of $C_{60}$ and $C_{70}$[†]

**C. N. R. Rao,* A. Govindaraj, Hemantkumar N. Aiyer, and Ram Seshadri**

*CSIR Centre of Excellence in Chemistry, Solid State & Structural Chemistry Unit and Materials Research Centre, Indian Institute of Science, Bangalore 560 012, India*

*Received: July 13, 1995; In Final Form: September 18, 1995*[⊗]

Scanning tunneling microscopy of solid films of $C_{60}$ and $C_{70}$ clearly demonstrate the occurrence of photochemical polymerization of these fullerenes in the solid state. X-ray diffraction studies show that such a polymerization is accompanied by contraction of the unit-cell volume in the case of $C_{60}$ and expansion in the case of $C_{70}$. This is also evidenced from the STM images. These observations help to understand the differences in the amorphization behavior of $C_{60}$ and $C_{70}$ under pressure. Amorphization of $C_{60}$ under pressure is irreversible because it is accompanied by polymerization associated with a contraction of the unit cell volume. Monte Carlo simulations show how pressure-induced polymerization is favored in $C_{60}$ because of proper orientation as well as the required proximity of the molecules. Amorphization of $C_{70}$, on the other hand, is reversible because $C_{70}$ is less compressible and polymerization is not favored under pressure.

It has been reported recently that $C_{60}$ undergoes polymerization in the solid state involving 2+2 cycloaddition induced photochemically[1] or at high pressures and temperatures.[2] $C_{60}$ is also known to undergo irreversible amorphization at high pressures, and the irreversibility has been attributed to the formation of cycloaddition products.[3] $C_{70}$ seems to undergo photopolymerization in the solid state involving 2+2 cycloaddition,[4] but unlike $C_{60}$, it undergoes reversible amorphization at high pressures.[5] We have been interested in understanding the polymerization of $C_{60}$ and $C_{70}$ in the solid state[6] and its relation to the nature of amorphization of these fullerenes. For this purpose, we have investigated photopolymerization of solid films of the two fullerenes by scanning tunneling microscopy (STM) and X-ray diffraction and carried out certain simulation studies on the effects of pressure on these solids. The study has indeed revealed the subtle differences between $C_{60}$ and $C_{70}$ with regard to features of their polymerization and amorphization.

Photopolymerization of solid films of $C_{60}$ and $C_{70}$ was studied by recording the STM images before and after UV irradiation. Typical STM images of $C_{60}$ and $C_{70}$ shown in Figure 1 bear clear evidence for the occurrence of photopolymerization. Photopolymerization of $C_{60}$ (Figure 1a) does not change the symmetry of the top surface. The internal features of the molecules become visible after photopolymerization because of the complete freezing of molecular motion.[7] Internal features become visible after photopolymerization of $C_{70}$ as well (Figure 1b), but the lattice symmetry of the top layer appears to change from fcc(111)-like to fcc(100)-like. STM results indicate that in $C_{60}$, a lattice contraction of $\approx 13\%$ occurs on polymerization while in $C_{70}$, an expansion of $\approx 10\%$ is observed. X-ray diffraction measurements on $C_{60}$ and $C_{70}$ films confirm similar changes in the unit-cell volume. Accordingly, we see from Figure 2 that the $2\theta$ values decrease on photopolymerization of $C_{70}$ while the opposite holds in the case of $C_{60}$. The changes in the unit cell volume can be understood by means of the model for the dimerization of $C_{70}$ and $C_{60}$ shown in Figure 2. The volume expansion in $C_{70}$ arises due to the ellipsoidal shape of the molecule as well as the high reactivity of the bonds radial to the capping pentagons toward 2+2 addition, i.e., the pentagons defining the 5-fold axis.[8] Since the molecules do not have to reorient for the polymerization of $C_{60}$ to occur, the only effect is due to intermolecular bonding resulting in the observed contraction. The observation of unit-cell expansion in the case of $C_{70}$ and contraction in the case of $C_{60}$ is indeed noteworthy and has direct bearing on the amorphization of these fullerenes.

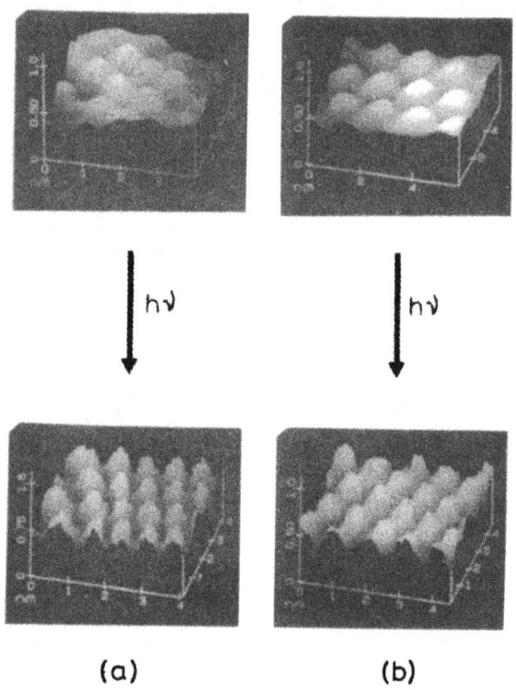


**Figure 1.** (a) STM images of $C_{60}$ films on highly oriented pyrolytic graphite (HOPG) before and after UV irradiation. (b) STM images of $C_{70}$ films on HOPG before and after UV irradiation. The conditions were 254 nm UV, 12 h, $10^{-5}$ Torr vacuum.

[†] Supported by the Jawaharlal Nehru Centre for Advanced Scientific Research.

* Correspondence to be addressed at the CSIR Centre of Excellence in Chemistry, Indian Institute of Science, Bangalore 560 012, India.

[⊗] Abstract published in *Advance ACS Abstracts,* October 15, 1995.

0022-3654/95/2099-16814$09.00/0 

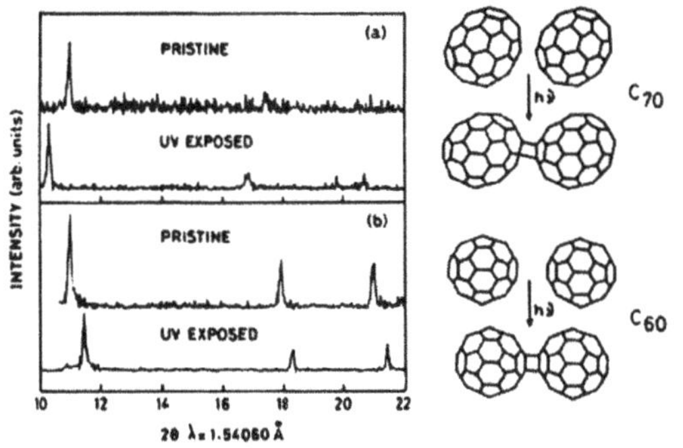


**Figure 2.** XRD patterns of (a) $C_{70}$ and (b) $C_{60}$ on glass substrates before and after UV irradiation. Whereas the $C_{60}$ lattice contracts on photopolymerization, for $C_{70}$, the lattice is seen to expand. The 2+2 addition of $C_{60}$ and $C_{70}$ (resulting in photopolymerized products) is shown schematically.

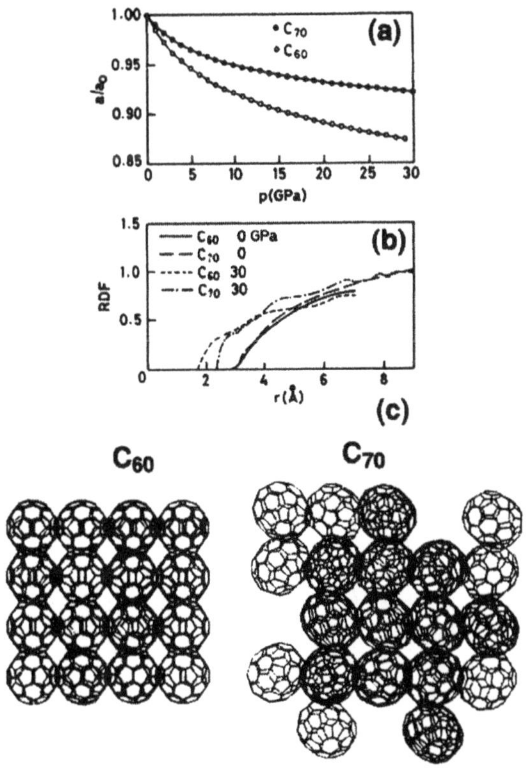


**Figure 3.** (a) Pressure dependence of the ratio of the pseudocubic lattice parameter $a$ of $C_{60}$ and $C_{70}$ to the parameter $a_0$ at ambient pressure as given by the Murnaghan equation of state. Experimental compressibilities were employed to produce these plots. It is assumed that there are no intervening phase changes and that $C_{60}$ remains cubic and $C_{70}$, rhombohedral throughout. (b) Atom−atom radial distribution functions of $C_{60}$ and $C_{70}$ molecules at 0 and 30 GPa and 300 K from NVT Monte Carlo studies. The molecules of $C_{60}$ approach closer at a given pressure than do $C_{70}$. (c) Snapshots of the $C_{60}$ lattice at 20 GPa and $C_{70}$ lattice at 25 GPa obtained from NVT-MC simulations, looking down the ⟨100⟩ direction.

What is also interesting about $C_{60}$ and $C_{70}$ is that their compressibilities are quite different.[9] The ratios of the pseudocubic lattice parameters of $C_{60}$ and $C_{70}$ with respect to their corresponding room-pressure values are shown as a function of pressure in Figure 3a. These plots were made using the Murnaghan equation of state and experimental compressibilities. We see that solid $C_{60}$ is more compressible than $C_{70}$. Our NVT Monte Carlo studies[10] also throw light on the nature of changes brought about on application of pressure. Among the various results obtained from such studies, the one most relevant to the context of polymerization under pressure is the following: On application of pressure, the $C_{60}$ molecules approach one another much more closely than $C_{70}$ molecules. The calculated closest approach between two $C_{60}$ molecules at 300 K and 30 GPa is ≈1.8 Å, while it is ≈2.2 Å in $C_{70}$ (Figure 3b). Although the $C_{60}$ molecules are randomly oriented in the solid up to moderate pressures, the molecules align themselves at higher pressures providing so as to be favorable for the 2+2 addition to occur. In the case of $C_{70}$ on the other hand, the alignment of the molecules under pressure occurs with their long axes along the face and body diagonals of the unit cell so as to render the reactive end caps of the molecule to be relatively farther apart. Snapshots of the simulated $C_{60}$ and $C_{70}$ lattices under pressure are shown in Figure 3c.

Clearly, polymerization under pressure is favored in the case of $C_{60}$ but not in $C_{70}$. Since the polymerization is also accompanied by volume contraction in the case of $C_{60}$, pressure would indeed favor such a process. This is not so in $C_{70}$, where polymerization is accompanied by volume expansion. On the basis of these observations, we are able to understand why the amorphization of $C_{60}$ under pressure is irreversible because of polymerization.

## References and Notes

(1) (a) Rao, A. M.; Zhou, P.; Wang, K. A.; Hager, G. T.; Holden, J. M.; Wang, Y.; Lee, W. T.; Bi, X. X.; Ecklund, P. C.; Cornett, D. S.; Duncan, M. A.; Amster, I. J. *Science* **1993**, *259*, 955. (b) Weaver, J. H.; Poirier, D. M. *Solid State Phys.* **1994**, *48*, 1. (c) Zhou, P.; Rao, A. M.; Wang, K. A.; Robertson, J. D.; Eloi, C.; Meier, M. S.; Rin, S. L.; Bi, X. X.; Ecklund, P. C.; Dresselhaus, M. S. *Appl. Phys. Lett.* **1992**, *60*, 2871. (d) Bacsa, W. S.; Lannin, J. S. *Phys. Rev.* **1994**, *B49* 14750. (e) Menon, M.; Subbaswamy, K. R. *Phys. Rev.* **1994**, *B49* 13966.

(2) (a) Nunez-Reguero, M.; Marques, L.; Hodeau, J. L.; Bethoux, O.; Perroux, M. *Phys. Rev. Lett.* **1995**, *74*, 278. (b) Xu, C. H.; Scuseria, G. E. *Phys. Rev. Lett* **1995**, *74*, 274.

(3) Rao, A. M.; Menon, M.; Wang, K. A.; Ecklund, P. C.; Subbaswamy, K. R.; Cornett, D. S.; Duncan, M. A.; Amster, I. J. *Chem. Phys. Lett.* **1994**, *224*, 106.

(4) (a) Yoo, C. S.; Nellis, W. J. *Chem. Phys. Lett.* **1992**, *198*, 379. (b) Moshary, F.; Chen, N. H.; Silvera, I. F.; Brown, C. A.; Dorn, H. C.; deVries, M. S.; Bethune, D. S. *Phys. Rev. Lett.* **1992**, *69*, 466. (c) Snoke, D. W.; Syassen, K.; Mittelbach, A. *Phys. Rev.* **1993**, *B47*, 4146.

(5) Chandrabhas, N.; Sood, A. K.; Muthu, D. V. S.; Sundar, C. S.; Bharathi, A.; Hariharan, Y.; Rao, C. N. R. *Phys. Rev. Lett.* **1994**, *73*, 3411.

(6) Since 2+2 cycloaddition is symmetry allowed under photochemical conditions but not under thermal conditions, under the application of pressure, it must take place by routes other than concerted, for example, via polar intermediates.

(7) Both $C_{60}$ and $C_{70}$ are orientationally disordered in the solid state at ordinary temperatures. Moreover, the disorder is dynamic. This prevents the probing of intramolecular features by techniques such as STM. However, when the fullerenes are deposited as films on surfaces such as single crystal metals, ordered structures can be obtained because the molecules are attached to the surface relatively strongly because of charge transfer from the metal to the fullerenes. When the molecules polymerize, they cannot rotate freely and the internal structure becomes visible. For details see: (a) Wang, X. D.; Hashizume, T.; Sakurai, T. *Mod. Phys. Lett.* **1994**, *8*, 1397. (b) Aiyer, H. N.; Govindaraj, A.; Rao, C. N. R. *Bull. Mater. Sci.* **1994**, *17*, 563. (c) Joachim, C.; Gimzewski, J.; Schittler, R.; Chavy, C. *Phys. Rev. Lett.* **1995**, *74*, 2102. (d) Aiyer, H. N.; Govindaraj, A.; Rao, C. N. R. *Philos. Mag. Lett.*, in print.

(8) (a) Henderson, C. C.; Cahill, P. A. *Science* **1994**, *263*, 397. (b) Karfunkel, H. R.; Hrisch, A. *Angew. Chem., Int. Ed. Engl.* **1992**, *31*, 1468. (c) Rathna, A.; Chandrasekhar, J. *Curr. Sci. (India)* **1993**, *65*, 768. (d) Govindaraj, A.; Rathna, A.; Chandrasekhar, J.; Rao, C. N. R. *Proc. Indian Acad. Sci. (Chem. Sci.)* **1993**, *105*, 303.

(9) (a) Duclos, S. J.; Brister, K.; Haddon, R. C.; Kortan, A. R.; Thiel, F. A. *Nature* **1991**, *351*, 380. (b) Christides, C.; Thomas, I. M.; Dennis, T. J. S.; Prassides, K. *Europhys. Lett.* **1993**, *22*, 545.

(10) (a) NVT Monte Carlo simulations were performed on cells containing 32 rigid $C_{60}$ or $C_{70}$ molecules. The intermolecular potential was of the Lennard-Jones form. The unit cells for the simulation were taken from experiments referred to in the text. Visualization was done using the InsightII software from BIOSYM Technologies, San Diego on IndigoII workstations. For details of the simulation as well as the potential parameters used, see: (a) Cheng, A.; Klein, M. L. *J. Phys. Chem.* **1991**, *95*, 6750. (b) Girifalco, L. A. *J. Phys. Chem.* **1992**, *98*, 858. (c) Chakrabarti, A.; Yashonath, S.; Rao, C. N. R. *Chem. Phys. Lett.* **1993**, *215*, 519.

JP951970P

1752 *J. Am. Chem. Soc.* **1999,** *121,* 1752–1753

# Hydrothermal Synthesis of Organic Channel Structures: 1:1 Hydrogen-Bonded Adducts of Melamine with Cyanuric and Trithiocyanuric Acids

Anupama Ranganathan, V. R. Pedireddi, and C. N. R. Rao*

*Chemistry & Physics of Materials Unit*
*Jawaharlal Nehru Centre for Advanced Scientific Research*
*Jakkur P.O. Box 6436, Bangalore 560 064, India*

*Received November 12, 1998*

Whitesides and co-workers[1] have extensively cited the 1:1 hydrogen-bonded adduct between cyanuric acid (**CA**) and melamine (**M**)-forming rosettes, as an outstanding case of noncovalent synthesis. The hydrogen-bonded **CA·M** lattice is the template on which these workers based the design of a family of self-assembled hydrogen-bonded aggregates, stable in solution.[2,3] The **CA·M** adduct is expected to form a hexagonal network through the formation of hydrogen bonds between the two molecules, but the structure has not been established by single-crystal X-ray crystallography.[3] The crystal structure of **CA·M·3HCl** reported by Wang et al.[4] reveals one-dimensional tapes rather than the rosette structure. The difficulty in growing crystals of the **CA·M** adduct is partly because **CA** and **M** are both highly hydrogen-bonded solids melting at high temperatures, with limited solubility in most organic solvents. Mixtures of **CA** and **M** in water or alcohol solution give precipitates containing the adduct, but do not yield crystals suitable for X-ray crystallography.[1,3] We felt that to obtain single crystals of the **CA·M** adduct, it may be necessary to employ more drastic conditions than normally employed. We have, therefore, made use of hydrothermal conditions,[5] commonly employed in the synthesis of quartz, zeolites, and inorganic open-frame structures, to obtain the crystals of the 1:1 adduct. We have indeed been able to obtain good crystals of the 1:1 **CA·M** adduct by this means. Crystals of the 1:1 adduct of trithiocyanuric acid (**TCA**) and **M** could also be

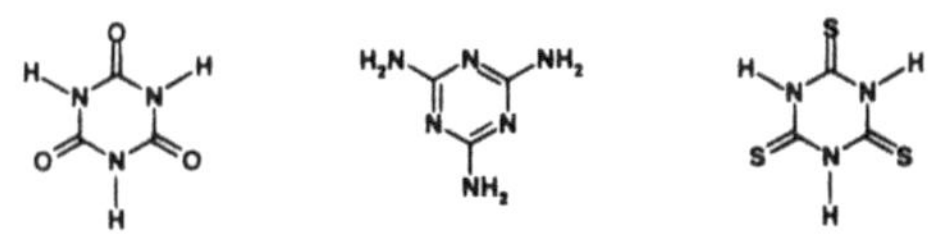

CA M TCA

prepared hydrothermally. We report the structures of both of these fascinating adducts of **M** in this paper.

The procedure for preparing the crystals of the **CA·M** adduct was as follows. Aqueous solutions of cyanuric acid (0.1 mmol) and melamine, (0.1 mmol) were mixed in a Teflon flask, and the mixture (15 mL) was kept in a stainless steel bomb. The bomb was sealed and maintained in a furnace at 180 °C. Rectangular platelike crystals of good quality separated from the solution, upon cooling the bomb to room temperature over a period of 4 h. These crystals were used for collecting intensity data on the single-crystal X-ray diffractometer. Crystals of the 1:1 **TCA·M** adduct were obtained by a similar procedure except that the temperature was maintained at 100 °C.[6]

The 1:1 **CA·M** adduct[7] has an asymmetric unit consisting of superimposed **CA** and **M** molecules.[8] Packing analysis shows that

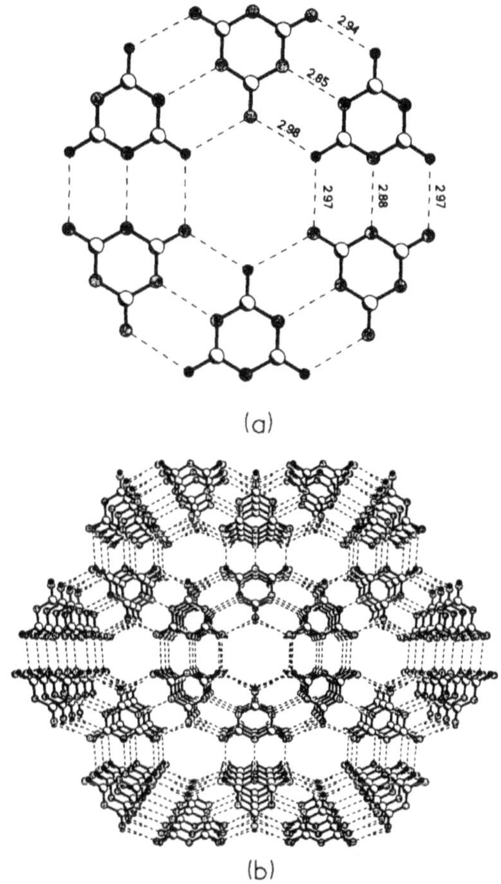


**Figure 1.** (a) A rosette formed between cyanuric acid (**CA**) and Melamine (**M**) in the adduct of **CA·M** with a cavity diameter of approximately 4 Å. Dashed lines represent intermolecular hydrogen bonds. (b) Three-dimensional arrangement of the **CA·M** adduct forming channels along the crystallographic *c*-axis.

(1) Whitesides, G. M.; Simanek, E. E.; Mathias, J. P.; Seto, C. T.; Chin, D. N.; Mammen, M.; Gordon, D. M. *Acc. Chem. Res.* **1995,** *28,* 37. Whitesides, G. M.; Mathias J. P.; Seto, C. T. *Science* **1991,** *254,* 1312.
(2) Seto C. T.; Whitesides, G. M. *J. Am. Chem. Soc.* **1993,** *115,* 905.
(3) Mathias, J. P.; Simanek, E. E.; Zerkowski, J. A.; Seto, C. T.; Whitesides, G. M. *J. Am. Chem. Soc.* **1994,** *116,* 4316. (b) Zerkowski, J. A.; Seto, C. T.; Wierda, D. A.; Whitesides, G. M. *J. Am. Chem. Soc.* **1990,** *112,* 9025.
(4) Wang, Y.; Wei, B.; Wang, Q. *J. Crystallogr. Spectrosc. Res.* **1990,** *20,* 79.
(5) Rao, C. N. R.*Chemical Approaches to the Synthesis of Inorganic Materials*; John Wiley: New York, 1993.

(6) **TCA** decomposed when the temperature was maintained at 180 °C as in the case of **CA**.

(7) Crystal data for the **CA·M** adduct: $(C_3H_3N_3O_3):(C_3H_6N_6)$, $M = 255.22$, crystal dimensions, 0.3 × 0.2 × 0.2 mm, monoclinic, space group, *C2/m*, $a = 14.853(2)$ Å, $b = 9.641(2)$ Å, $c = 3.581(1)$ Å, $\beta = 92.26(1)°$, $V = 512.4(2)$ Å$^3$, $Z = 2$, $\rho_{calcd} = 1.654$, $\mu$(Mo K$\alpha$) = 0.136 mm$^{-1}$, $F(000) = 264$, $\lambda = 0.710\,73$ Å, Smart CCD area detector, Siemens, $\omega - 2\theta$ scan, $2° < \theta < 24°$ ($-16 \leq h \leq 14$, $-8 \leq k \leq 10$, $-3 \leq l \leq 3$), 1095 total reflections, 395 independent reflections which were used for refinement. The structure was solved by direct methods (SHELXTL-PLUS) and refined by full-matrix least squares on $F^2$ (SHELX-93; G. M. Sheldrick, Gottingen, 1993) to R1 = 0.056 and wR2 = 0.141. Residual density, min/max −0.495/0.393 e·Å$^{-3}$. Hydrogen atoms were not included in the refinement, but including the hydrogen atoms in the calculated positions improves the *R*-factor by 1%.

(8) The best stacking arrangement was one with alternating **CA** and **M** molecules (with 50% occupation each), as established by the low *R*-factor and also by energy calculations using the MOPAC program. The least stable arrangement is the one with **CA** on top of another **CA**, as one would expect.

10.1021/ja983928o CCC: $18.00 
Published on Web 02/17/1999

*Communications to the Editor* *J. Am. Chem. Soc., Vol. 121, No. 8, 1999* 1753

**CA** and **M** are held together by N−H···O and N−H···N hydrogen bonds yielding the hexamer (rosette) shown in Figure 1a. The interatomic N···O and N···N distances corresponding to N−H···O and N−H···O bonds are in the range of 2.94−2.98 and 2.85−2.88 Å, respectively. The hexamers are arranged in two dimensions to form planar sheets, the structure being exactly as predicted by Whitesides.[1] A significant feature is that the planar sheets are stacked in three dimensions[8] to give channels with a diameter of 4 Å (Figure 1b). These channels are comparable to the cavities in cryptands.[9]

Crystal structure determination of the **TCA·M** adduct[10] reveals that it has features similar to those of the **CA·M** adduct, except that the N−H···O hydrogen bonds are replaced by the N−H···S hydrogen bonds as shown in Figure 2a. The N···S and N···N distances corresponding to N−H···S and N−H···N bonds are in the range of 2.96−2.99 and 2.86−2.88 Å, respectively. The sheets are stacked in a three-dimensional channel arrangement as depicted in Figure 2b. The diameter of the channel is approximately 4 Å here as well.

The present study illustrates the efficacy of the hydrothermal method to prepare crystals of organic materials in certain special situations, as in the present case where the compounds are highly hydrogen-bonded (with limited solubility in common solvents) and give precipitates when mixed in water.[11]

It is noteworthy that the adducts of **M** with **CA** and **TCA** constitute members of a rare class of organic solids containing channels.[12] Trimesic acid is a good example of a system forming hexagonal channels.[13] The other example of such a solid is the adduct of **TCA** with 4,4′-bipyridyl.[14] Experiments to explore whether the channels in the **CA·M** adduct can be filled by cations and other species are in progress.

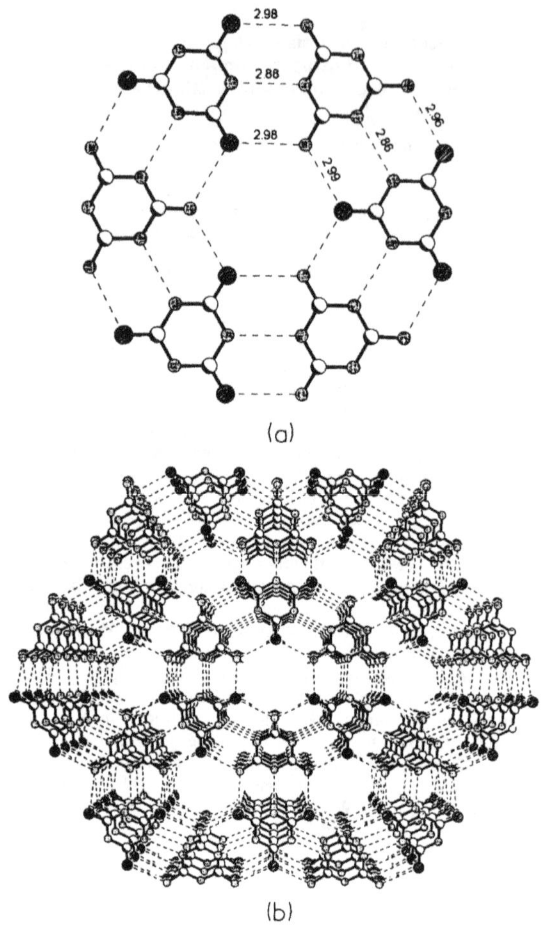


**Figure 2.** (a) Hexagonal network (rosette) formed between trithiocyanuric acid (**TCA**) and Melamine (**M**) in the adduct of **TCA·M** with a cavity diameter of approximately 4 Å. (b) Three-dimensional arrangement of the **TCA·M** adduct forming channels along crystallographic *c*-axis.

**Supporting Information Available:** Two figures (ORTEP drawings of the adducts **CA·M** and **TCA·M** with labeling scheme) and 6 tables containing general crystallographic information, fractional coordinates of non-hydrogen atoms, list of bond lengths and angles, and anisotropic displacement parameters (PDF). This material is available free of charge via the Internet at http://pubs.acs.org.

JA983928O

---

**CA** on top of **M** is more stable by 37 kcal mol$^{-1}$ relative to **CA** on **CA**. The packing coefficient of the **CA·M** adduct calculated by using Cerius is 67.5%.

(9) Vogtle, F. *Supramolecular Chemistry*; Wiley: Chichester, 1971.

(10) Crystal data for the **TCA·M** adduct: $(C_3H_3N_3S_3):(C_3H_6N_6)$ $M$ = 303.39, crystal dimensions, 0.3 × 0.25 × 0.35 mm, monoclinic, space group, *C*2/*m*, $a$ = 14.862(3) Å, $b$ = 9.650(2) Å, $c$ = 3.588(1) Å, $\beta$ = 92.32(1)°, $V$ = 514.2(2) Å$^3$, $Z$ = 2, $\rho_{calcd}$ = 1.901, $\mu$(Mo K$\alpha$) = 0.716 mm$^{-1}$, $F(000)$ = 294, $\lambda$ = 0.710 73 Å, Smart CCD area detector, Siemens, $\omega-2\theta$ scan, $2° < \theta < 24°$ ($-13 \le h \le 16$, $-10 \le k \le 10$, $-3 \le l \le 3$), 1091 total reflections, 397 independent reflections which were used for refinement. The structure was solved by direct methods (SHELXTL-PLUS) and refined by full-matrix least squares on $F^2$ (SHELX-93; G. M. Sheldrick, Gottingen, 1993) to R1 = 0.062 and wR2 = 0.134. Residual density, min/max −0.396/0.487e·Å$^{-3}$. Hydrogen atoms were not included in the refinement, but including the hydrogen atoms in the calculated positions improves the *R*-factor by 1%. The best stacking arrangement is **TCA** on top of **M** (see (8) for details). The packing coefficient of the adduct is 74.6% as calculated from Cerius.

(11) The powder X-ray diffraction pattern of the **CA·M** adduct simulated on the basis of structure found by us agrees with the experimental powder pattern of the precipitate or of the polycrystalline sample.

(12) Zimmerman, S. C. *Science* **1997**, *276*, 544.

(13) Kolotuchin, S. V.; Fenlon, E. E.; Wilson, S. R.; Loweth, C. J.; Zimmerman, S. C. *Angew. Chem., Int. Ed. Engl.* **1995**, *34*, 2654.

(14) Pedireddi, V. R.; Chatterjee, S.; Ranganathan, A.; Rao, C. N. R. *J. Am. Chem. Soc.* **1997**, *119*, 10867.

JOURNAL OF Materials CHEMISTRY

# An organic channel structure formed by the supramolecular assembly of trithiocyanuric acid and 4,4′-bipyridyl

**Anupama Ranganathan, V. R. Pedireddi, Swati Chatterjee and C. N. R. Rao***

*Chemistry & Physics of Materials Unit, Jawaharlal Nehru Centre for Advanced Scientific Research, Jakkur P.O., Bangalore 560 064, India*

***Received 18th February 1999, Accepted 5th July 1999***

Trithiocyanuric acid (TCA) and 4,4′-bipyridyl (BP) form hydrogen-bonded co-crystals with aromatic compounds such as benzene, toluene, *p*-xylene and anthracene. The TCA–BP co-crystal is composed of cavities formed by the N–H···N hydrogen bonds between the two molecules, and the three-dimensional structure contains channels of approximately 10 Å where aromatic molecules are accommodated. The molar ratios of TCA, BP and the aromatic compound in the co-crystals are 2 : 1 : 1 or 2 : 1 : 0.5. Benzene, toluene and *p*-xylene are removed from the channels around 190, 183 and 170 °C respectively, and these aromatic guests can be reintroduced into the empty channels of the apo-hosts. The apo-hosts with empty channels have reasonable thermal stability and exhibit shape selectivity in that the empty channels accommodate *p*-xylene but not *m*- or *o*-xylene or mesitylene.

## 1 Introduction

A variety of inorganic microporous solids have been synthesized in recent years because of their use in catalysis and separation technology.[1–3] These structures are generally prepared by using structure-directing template molecules such as organic amines. Organic amines are also employed extensively to synthesize open-framework metal phosphates and carboxylates.[3] A few organic porous structures containing transition metal ions have been synthesized.[4,5] The supramolecular assembly formed by trimesic acid (benzene-1,3,5-tricarboxylic acid) with cobalt nitrate and pyridine reported by Yaghi *et al.*[4] is a good example of such a metal–organic microporous material. The recently reported nanoporous material formed between trimesic acid and cupric nitrate reported by Chui *et al.*[6] highlights the role of metal–organic interactions in supramolecular assemblies. There is also considerable interest in designing organic microporous solids by making use of supramolecular assemblies. Thus, hexagonal channel structures formed by trimesic acid and its anion have been described.[7] There are also other organic channel structures comprising large pores in three-dimensional structures reported in the literature.[8,9] The nanoporous molecular sandwiches formed by guanidium ions and disulfonate ions serve as representative examples.[8] It has been shown recently that the hexagonal rosette structure formed through hydrogen bonding between cyanuric acid and melamine gives rise to channels in the three-dimensional structure.[10] Self-assembly of three-dimensional networks with large chambers has been reported by using a pyridone as the tecton.[11] Inclusion compounds of urea and thiourea as well as the quinol clathrates also belong to this class of organic solids, but they are not engineered supramolecular designs. We have been exploring several possible designs of organic porous solids for some time and have found that the hydrogen bonded co-crystal between trithiocyanuric acid (TCA) and 4,4′-bipyridyl (BP) has channels in the three-dimensional structure that can accommodate benzene.[12] We have carried out detailed investigations to explore whether other aromatic compounds can be introduced in these channels and if so, whether there is shape selectivity. The study has shown that the supramolecular hydrogen-bonded structure formed by TCA and BP accommodates several aromatic molecules, with some shape selectivity and thermal stability.

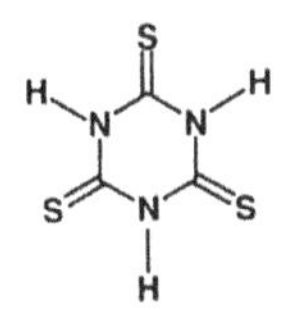

TCA

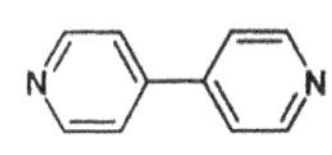

BP

## 2 Experimental

The 2 : 1 co-crystal of TCA with BP was first prepared by the co-crystallization of the two compounds in methanol solution, but the crystals were not very stable, because of the loss of methanol. After the loss of methanol, the crystallinity was not sufficiently good to carry out a single crystal study. In the presence of benzene, toluene, *o*-xylene, *m*-xylene, *p*-xylene or anthracene, however, the crystals of the 2 : 1 co-crystal of TCA and BP were stable. In a typical preparation, 106 mg of TCA and 47 mg of BP were taken up in 15 mL of methanol in the presence of an aromatic compound (5 mL). In the case of anthracene, the same quantities of TCA and BP as above were taken up in 15 mL of methanol along with 55 mg of anthracene. After slow evaporation, crystals containing the 2 : 1 co-crystal of TCA with BP, containing different proportions of the aromatic compounds were obtained. We designate the co-crystals containing benzene, toluene, *o*-xylene, *m*-xylene, *p*-xylene, and anthracene by **1**, **2**, **3**, **4**, **5** and **6** respectively. We could not prepare the 2 : 1 TCA–BP cocrystal with mesitylene. The crystals were generally needle-shaped and yellow in colour. The as-prepared crystals were dried at room temperature over a period of two days prior to the characterization studies.

Crystals of **1–6** prepared as above were employed for the determination of molecular structure by single crystal X-ray diffraction. The relevant details of the crystal structures of the co-crystals are listed in Table 1.† The intensity data were collected on a SMART system, Siemens, equipped with a CCD area detector,[13] using Mo-Kα radiation. The structures were solved and refined using SHELXTL[14] software. The refinements were uncomplicated and converged to good *R*-factors as

†CCDC reference number 1145/171. See http://www.rsc.org/suppdata/jm/1999/2407 for crystallographic files in .cif format.

**Table 1** Crystal data for the complexes, **1–6**, formed between trithiocyanuric acid and 4,4′-bipyridyl along with various guest molecules

| | **1** | **2** | **3** | **4** | **5** | **6** |
|---|---|---|---|---|---|---|
| Formulae | $2(C_3H_3N_3S_3)$: $C_{10}H_8N_2$:$C_6H_6$ | $2(C_3H_3N_3S_3)$: $C_{10}H_8N_2$:$C_7H_8$ | $2(C_3H_3N_3S_3)$: $C_{10}H_8N_2$:$C_8H_{10}$ | $2(C_3H_3N_3S_3)$: $C_{10}H_8N_2$:$C_8H_{10}$ | $2(C_3H_3N_3S_3)$: $(C_{10}H_8N_2)$: $0.5(C_8H_{10})$ | $2(C_3H_3N_3S_3)$: $C_{10}H_8N_2$: $0.5(C_{14}H_{10})$ |
| Mol. wt. | 588.82 | 602.85 | 616.87 | 616.87 | 872.23 | 599.82 |
| Crystal system | triclinic | triclinic | triclinic | triclinic | triclinic | triclinic |
| Space group | $P\bar{1}$ | $P\bar{1}$ | $P\bar{1}$ | $P\bar{1}$ | $P\bar{1}$ | $P\bar{1}$ |
| $a$/Å | 7.018(1) | 7.112(1) | 7.282(1) | 10.324(1) | 10.359(1) | 10.472(1) |
| $b$/Å | 10.344(1) | 10.489(1) | 10.621(1) | 11.615(1) | 11.467(1) | 11.419(1) |
| $c$/Å | 10.757(1) | 10.681(1) | 10.648(1) | 12.319(1) | 18.412(1) | 12.932(1) |
| $\alpha$/° | 63.75(1) | 65.40(1) | 66.30(1) | 102.16(1) | 83.65(1) | 100.46(1) |
| $\beta$/° | 75.78(1) | 76.58(1) | 76.53(1) | 93.46(1) | 75.00(1) | 107.69(2) |
| $\gamma$/° | 73.82(1) | 73.05(1) | 70.42(1) | 90.68(1) | 67.94(1) | 111.28(2) |
| Cell vol./Å$^3$ | 665.8(1) | 687.4(1) | 705.9(1) | 1441.0(2) | 1957.7(3) | 1296.5(2) |
| $Z$ | 1 | 1 | 1 | 2 | 1 | 2 |
| $\mu$/mm$^{-1}$ | 0.54 | 0.53 | 0.52 | 0.51 | 0.55 | 0.56 |
| $T$/K | 293 | 293 | 293 | 293 | 293 | 293 |
| Total reflectn. | 2581 | 2887 | 2781 | 6134 | 7548 | 4996 |
| Non-zero reflectn. | 1838 | 1928 | 1970 | 4109 | 5489 | 3556 |
| $R$ | 0.032 | 0.063 | 0.067 | 0.082 | 0.058 | 0.049 |
| $R_w$ | 0.077 | 0.183 | 0.184 | 0.219 | 0.132 | 0.098 |

shown in Table 1. Except for the atoms of the guest molecules (toluene, *o*-xylene and *m*-xylene) in the complexes **2**, **3** and **4**, all the other atoms are oriented in well resolved positions. Non-hydrogen atoms in all the structures except guest molecules in **2**, **3** and **4** were refined anisotropically while the hydrogen atoms were refined isotropically, except in the structures where the hydrogen atoms were placed in the calculated positions using AFIX routines of SHELXTL. The guest molecules in **2** and **5** were refined isotropically only owing to the disorder. The crystal structures reveal that the ratio of TCA, BP and the aromatic molecules in the co-crystals **1–4** was 2 : 1 : 1. The ratio was 2 : 1 : 0.5 in **5** and **6**. We see from Table 1 that the unit cells of **1–3** are similar and the unit cells of **4–6** form another set. As an illustrative example, we list the structural parameters of **2** in Table 2. These data typify the structures of the TCA–BP co-crystals examined here. We have not given the data for the other co-crystals for the purpose of brevity. Intermolecular interactions were calculated using the PLATON[15] programme.

Thermogravimetric analysis (TGA) of the co-crystals was carried out over the temperature range 25 °C to 200 °C by employing a Mettler Toledo instrument. These curves showed mass loss due to the removal of the aromatic guest molecules. The apo-hosts thus obtained did not dissolve in benzene and other aromatic solvents. The apo-hosts were immersed in the respective aromatic liquid for several hours. The crystals were then taken out, and the TGA repeated. This procedure was repeated more than once to find out whether the inclusion of the guest molecule was reversible and also whether there was any change in the temperature of decomposition or the proportion of the aromatic compound in the co-crystal, with such cycling.

**Table 2** Bond lengths and angles of **2**

| Distances/Å | | Angles/° | |
|---|---|---|---|
| S(23)–C(23) | 1.654(6) | C(21)–N(23)–C(23) | 123.3(5) |
| S(22)–C(22) | 1.651(6) | C(16)–C(11)–C(12) | 116.5(6) |
| S(21)–C(21) | 1.662(6) | N(14)–C(15)–C(16) | 122.4(7) |
| N(23)–C(21) | 1.357(8) | C(22)–N(22)–C(23) | 125.4(5) |
| N(23)–C(23) | 1.369(8) | C(13)–C(12)–C(11) | 119.6(6) |
| C(11)–C(16) | 1.387(9) | N(23)–C(23)–N(22) | 115.8(5) |
| C(11)–C(12) | 1.393(9) | N(23)–C(23)–S(23) | 123.4(4) |
| C(15)–N(14) | 1.339(8) | N(22)–C(23)–S(23) | 120.9(5) |
| C(15)–C(16) | 1.393(9) | C(11)–C(16)–C(15) | 120.5(6) |
| N(22)–C(22) | 1.363(7) | C(22)–N(21)–C(21) | 126.1(5) |
| N(22)–C(23) | 1.378(7) | N(21)–C(22)–N(22) | 113.4(5) |
| C(12)–C(13) | 1.392(9) | N(21)–C(22)–S(22) | 122.8(5) |
| N(21)–C(22) | 1.361(8) | N(22)–C(22)–S(22) | 123.7(5) |
| N(21)–C(21) | 1.369(7) | C(13)–N(14)–C(15) | 117.5(6) |
| N(14)–C(13) | 1.327(9) | N(23)–C(21)–N(21) | 115.9(6) |
| C(4)–C(7) | 1.26(3) | N(23)–C(21)–S(21) | 122.6(4) |
| C(4)–C(5) | 1.47(3) | N(21)–C(21)–S(21) | 121.5(5) |
| C(4)–C(6) | 1.78(3) | N(14)–C(13)–C(12) | 123.5(7) |
| C(3)–C(5) | 1.35(4) | C(7)–C(4)–C(5) | 71(2) |
| C(3)–C(6) | 1.53(4) | C(7)–C(4)–C(6) | 115(2) |
| C(3)–C(2) | 1.97(5) | C(5)–C(3)–C(6) | 50(2) |
| C(5)–C(6) | 1.23(3) | C(5)–C(3)–C(2) | 88(3) |
| C(5)–C(7) | 1.60(3) | C(6)–C(3)–C(2) | 38(2) |
| C(2)–C(6) | 1.20(3) | C(6)–C(5)–C(3) | 72(3) |
| | | C(6)–C(5)–C(4) | 82(2) |
| | | C(3)–C(5)–C(4) | 154(4) |
| | | C(6)–C(5)–C(7) | 130(3) |
| | | C(3)–C(5)–C(7) | 158(4) |
| | | C(6)–C(2)–C(3) | 51(2) |
| | | C(2)–C(6)–C(5) | 148(3) |
| | | C(2)–C(6)–C(3) | 91(3) |
| | | C(5)–C(6)–C(3) | 57(2) |
| | | C(2)–C(6)–C(4) | 157(3) |
| | | C(5)–C(6)–C(4) | 55(2) |
| | | C(3)–C(6)–C(4) | 112(3) |
| | | C(4)–C(7)–C(5) | 61(2) |

## 3 Results and discussion

Co-crystallization of TCA with BP from methanol solution gives a 2 : 1 hydrogen bonded co-crystal containing the solvent of crystallization. The crystal structure of the co-crystal reveals the presence of intermolecular N–H···N hydrogen bonds between TCA and BP as shown in Fig. 1. This hydrogen bonded structure has a cavity of 10 Å (calculated using CERIUS, version 3.0), occupied by methanol molecules. These crystals were unstable under ambient conditions because of the evaporation of methanol. However the 2 : 1 : 1 co-crystals of TCA with BP incorporating benzene, **1**, or toluene, **2**, were highly stable. These crystals gave the same two-dimensional hydrogen bonded structure with a cavity as in Fig. 1. Besides the strong N–H···N bonds (H···N, 1.82 Å) responsible for the formation of the cavity, there are weak C–H···S bonds (H···S, 2.91, 2.93 Å) between the TCA and BP molecules. In addition, there are N–H···S hydrogen bonds between TCA molecules (H···S, 2.5 Å). The three-dimensional structure of **2** is shown in Fig. 2. The structure clearly reveals the presence of channels formed by the stacking of the layers with cavities. The channels accommodate aromatic molecules as can be seen in Fig. 2(a). Fig. 2(b) shows another projection of **2** where we see layers of TCA·BP with toluene molecules residing in the regions of the

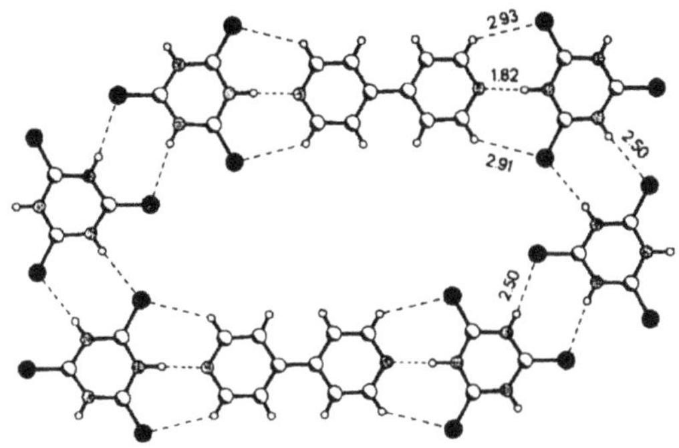


**Fig. 1** Self assembly of TCA and BP forming a layered network with a cavity.

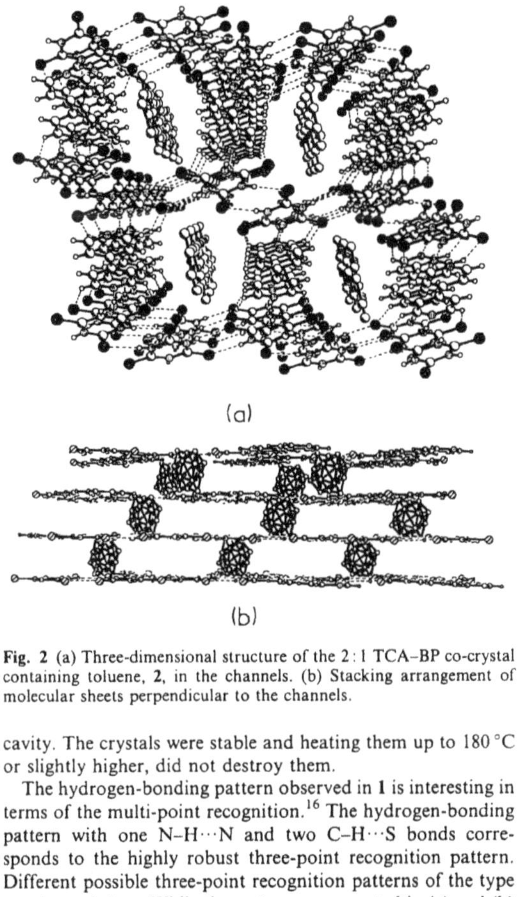


**Fig. 2** (a) Three-dimensional structure of the 2 : 1 TCA–BP co-crystal containing toluene, **2**, in the channels. (b) Stacking arrangement of molecular sheets perpendicular to the channels.

cavity. The crystals were stable and heating them up to 180 °C or slightly higher, did not destroy them.

The hydrogen-bonding pattern observed in **1** is interesting in terms of the multi-point recognition.[16] The hydrogen-bonding pattern with one N–H···N and two C–H···S bonds corresponds to the highly robust three-point recognition pattern. Different possible three-point recognition patterns of the type are shown below. While the patterns represented in (a) and (b) correspond to the motifs comprising strong hydrogen bonds, pattern (c) is a result of weak C–H···O hydrogen bonds.[17] A search of the Cambridge Structural Database[18] (CSD) suggests that the pattern present in **1** corresponding to (d) is indeed unique. No other such system appears to be reported in the literature.

Crystal structures of **3**–**6** show that the cavities in these co-crystals are similar to those in **2**, a minor exception being **4** containing *m*-xylene, where the size and shape of the cavity are somewhat different. In Fig. 3, we show the two-dimensional

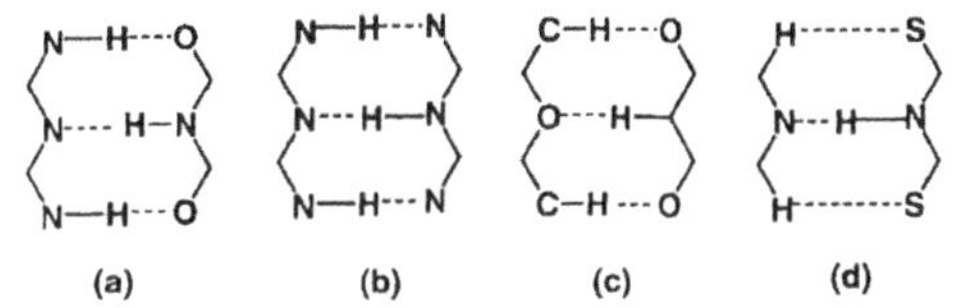


structure of the TCA–BP in **4** to illustrate the small differences between the cavities in this co-crystal and **1**. Here, the intermolecular N–H···N as well as the C–H···S bonds are slightly longer but the N–H···S bonds between the TCA molecules are similar to those in **2** (Fig. 1). A noticeable difference is the slight asymmetry in the cavity arising from the different intermolecular hydrogen bond distances on the two sides. In Fig. 4a, we show the three-dimensional structure of **6** where the guest molecule is anthracene. In Fig. 4b, we show another projection of **6** showing TCA·BP layers wherein anthracene molecules protrude through the cavity. Fig. 2 and 4 serve to demonstrate the similarity of the three-dimensional channels in the various co-crystals.

In Fig. 5, we show the TGA curve of the benzene co-crystal **1**, recorded at a heating rate of 2 °C $min^{-1}$. All the benzene is removed (mass loss, 13.9%) around 190 °C which is well above the boiling point of benzene (see curve 1). After the removal of benzene, the apo-host crystals were soaked in benzene for 36 hours and again subjected to TGA. (These crystals are insoluble in benzene as evidenced by the retention of the crystal morphology and absence of the relevant spots on the TLC plates.) The crystals showed the loss of benzene at a slightly lower temperature (158 °C), but the quantity of benzene in the channels is less by about 4% (see curve 2 in Fig. 5), compared to the as-prepared co-crystal. A repetition of this procedure showed that the removal of benzene continued to occur around 158 °C with the same mass loss (curve 3 in Fig. 5). It appears that after the benzene in the channels of the initial co-crystal is removed, there are some definitive changes in the channel structure, with the empty channels accommodating benzene to a smaller extent. The lower temperature at which benzene is removed also suggests that the guest molecules do not interact as strongly with the host channels. One difference that we notice is that the weak C–H···S interactions present in the as-formed co-crystal is absent after the removal of the guest molecules. The crystal structure of the TCA–BP co-crystal **1** heated to 200 °C (to remove all the benzene) gave a single crystal X-ray diffraction pattern, but the data were not sufficiently good to obtain the structure. We could not determine the structure after reintroducing benzene. Powder

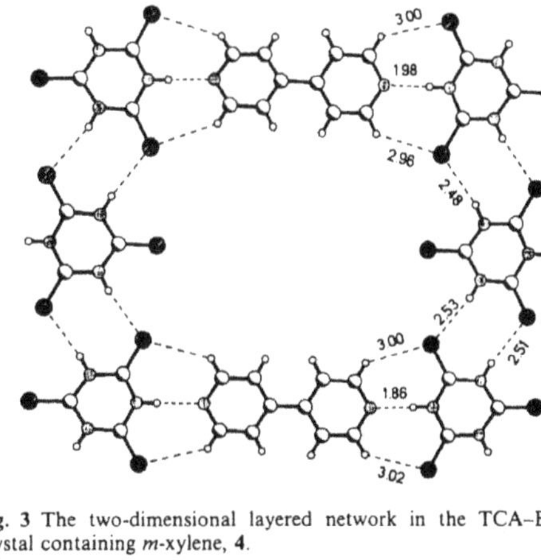


**Fig. 3** The two-dimensional layered network in the TCA–BP co-crystal containing *m*-xylene, **4**.

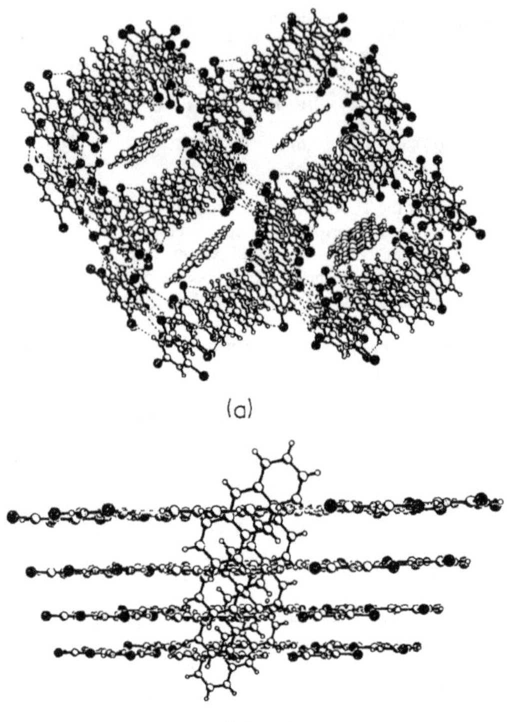

(a)

(b)

**Fig. 4** (a) Three-dimensional structure of the 2 : 1 TCA–BP co-crystal containing anthracene, **6**. (b) Stacking arrangement of planar molecular sheets perpendicular to the channels.

XRD patterns were identical after two cycles of guest removal, but the patterns of the apo-host were generally poorer in quality.

TGA of the toluene co-crystal **2** showed a mass loss of 14.8% at 183 °C. The host crystals with the empty channels were soaked in toluene for 36 hours and then subjected to TGA. These crystals showed a mass loss of 9.2% at 166 °C, a behaviour similar to that of the benzene co-crystal, **1**. TGA of the *p*-xylene cocrystal, **5**, showed a mass loss (12.2%) due to the removal of the aromatic guest around 167 °C. After reintroduction of *p*-xylene in the channels, the mass loss occurred around 139 °C, but the magnitude of mass loss was smaller

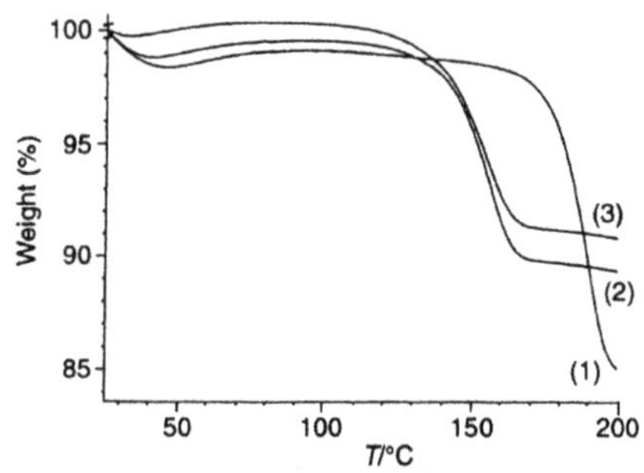


**Fig. 5** Thermogravimetic analysis (heating rate 2 °C min$^{-1}$) of the 2 : 1 TCA–BP co-crystal containing benzene. Curve 1 is that of the as-prepared sample. Curve 2 was obtained with a sample obtained by immersing the host crystal with empty channels in the aromatic liquid for several hours. Curve 3 was obtained after repeating this procedure for a second time.

(8%). The temperature of decomposition and the mass loss did not change appreciably after further cycling in these cases as well. The co-crystals of *o*-, *m*-xylene, **3** and **4**, however, behaved entirely differently. Once the guest molecules were removed, they could not be reintroduced in the channels even after soaking the host crystals with empty channels for long periods. This observation suggests that there is some selectivity in the channels of the TCA–BP co-crystal. This can be understood from the differences in the shapes of *p*-, *o*- and *m*-xylenes. In *p*-xylene, the two methyl groups are symmetrically positioned and favor the ready re-incorporation into the channels of the host crystal. This is not the case in the other two isomers. Accordingly, the empty channel of the TCA–BP co-crystal only takes in *p*-xylene from a mixture of the three xylene isomers. It should be recalled that the channel does not accommodate mesitylene as well. In the apo-host obtained by removing benzene from co-crystal **1**, toluene could be introduced readily, but not *p*-xylene. We have attempted to understand the selectivity of the apo-host channels by carrying out powder XRD measurements before and after the removal of the guest molecules. Although the apo-hosts are still crystalline, we have not yet been able to come to any meaningful conclusions from the XRD patterns since the data are not entirely sufficient.

## 4 Conclusions

The stable organic solid containing channels formed by the supramolecular hydrogen-bonded assembly of trithiocyanuric acid and 4,4′-bipyridyl can accommodate aromatic molecules such as benzene, toluene and *p*-xylene. The apo-host is thermally stable up to 200 °C and exhibits shape selectivity with respect to the xylene isomers. The channels do not accommodate mesitylene. It would be interesting to carry out chemical reactions in these channels.

## 5 References

1 (*a*) G. Ozin, *Adv. Mater.*, 1992, **4**, 612; (*b*) J. M. Thomas, *Angew. Chem., Int. Ed. Engl.*, 1994, **33**, 913.
2 (*a*) S. Zone and M. E. Davis, *Curr. Opin. Solid State Mater. Sci.*, 1996, **1**, 107; (*b*) A. K. Cheetham, *Science*, 1994, **264**, 794; (*c*) X. Bu, P. Feng and G. D. Stucky, *J. Chem. Soc., Chem. Commun.*, 1995, 1337.
3 (*a*) S. Ayyappan, A. K. Cheetham, S. Natarajan and C. N. R. Rao, *Chem. Mater.*, 1998, **10**, 3746 and references cited therein; (*b*) S. Natarajan, S. Ayyappan, A. K. Cheetham and C. N. R. Rao, *Chem. Mater.*, 1998, **10**, 1627 and references cited therein.
4 (*a*) O. M. Yaghi, G. Li and H. Li, *Nature*, 1995, **378**, 703; (*b*) O. M. Yaghi, H. Li and T. L. Groy, *J. Am. Chem. Soc.*, 1996, **118**, 9096; (*c*) O. M. Yaghi and H. Li, *J. Am. Chem. Soc.*, 1995, **117**, 10401; (*d*) H. Li, M. Eddaoudi, T. L Groy and O. M. Yaghi, *J. Am. Chem. Soc.*, 1998, **120**, 8571.
5 (*a*) D. Venkataraman, G. B. Gardner, S. Lee and J. S. Moore, *J. Am. Chem. Soc.*, 1995, **117**, 11600; (*b*) B. F. Abrahams, J. Coleiro, B. Hoskins and R. Robson, *Chem. Commun.*, 1996, 603.
6 S. S. Y. Chui, S. M. F. Lo, J. P. H. Charmant, A. G. Orpen and I. D. Williams, *Science*, 1999, **283**, 1148.
7 (*a*) S. V. Kolotuchin, E. E. Fenlon, S. R. Wilson, C. J. Loweth and S. C. Zimmerman, *Angew. Chem., Int. Ed. Engl.*, 1995, **34**, 2654; (*b*) R. Melendez, C. V. K. Sharma, M. J. Zaworotko, C. Baner and R. D. Rogers, *Angew. Chem., Int. Ed. Engl.*, 1996, **35**, 2213; (*c*) S. C. Zimmerman, *Science*, 1997, **276**, 543.
8 (*a*) V. A. Russel, C. C. Evans, W. Li and M. D. Ward, *Science*, 1997, **276**, 575; (*b*) V. A. Russel and M. D. Ward, *Chem. Mater.*, 1996, **8**, 1654.
9 J. S. Moore and S. Lee, *Nature*, 1995, **374**, 792.
10 A. Ranganathan, V. R. Pedireddi and C. N. R. Rao, *J. Am. Chem. Soc.*, 1999, **121**, 1752.
11 M. Simard, D. Su and J. D. Wuest, *J. Am. Chem. Soc.*, 1991, **113**, 4696.
12 V. R. Pedireddi, S. Chatterjee, A. Ranganathan and C. N. R. Rao, *J. Am. Chem. Soc.*, 1997, **119**, 10867.
13 Siemens, *SMART System*, Siemens Analytical X-ray Instruments Inc., Madison, WI, U.S.A., 1995.

14 G. M. Sheldrick, SHELXTL, *Users Manual*, Siemens Analytical X-ray Instruments Inc., Madison, WI, U.S.A., 1993.
15 A. L. Spek, PLATON, *Molecular Geometry Program*, University of Utrecht, The Netherlands, 1995.
16 (*a*) J. C. MacDonald and G. M. Whitesides, *Chem. Rev.*, 1994, **94**, 2383; (*b*) G. M. Whitesides, E. E. Simanek, J. P. Mathias, C. T. Seto, D. N. Chin, M. Mammen and D. M. Gordon, *Acc. Chem. Res.*, 1995, **28**, 37.
17 (*a*) G. R. Desiraju, *Angew. Chem., Int. Ed. Engl.*, 1995, **34**, 2311; (*b*) G. R. Desiraju, *Crystal Engineering: The Design of Organic Solids*, Elsevier, New York, 1989; (*c*) G. R. Desiraju, *Science*, 1997, **278**, 404; (*d*) G. R. Desiraju, *Curr. Opin. Solid State Mater. Sci.*, 1997, 451.
18 F. H. Allen and O. Kennard, *Chemical Design Automation News*, 1993, **8(1)**, 31.

*Paper 9/05344A*

CHEMPHYSCHEM

# An Experimental Charge Density Study of the Effect of the Noncentric Crystal Field on the Molecular Properties of Organic NLO Materials**

R. Srinivasa Gopalan,[a] Giridhar U. Kulkarni,[a] and C. N. R. Rao*[a]

*The structure, packing, and charge distribution in molecules of nonlinear optical materials have been analysed with reference to their counterparts in centrosymmetric structures based on low temperature X-ray measurements. The systems studied are the centric and noncentric polymorphs of 5-nitrouracil as well as the diamino, dithio, and thioamino derivatives of 1,1-ethylenedicarbonitrile; the latter possesses a noncentric structure. The molecular structure of 5-nitrouracil is invariant between the two forms, while the crystal packing is considerably different, leading to dimeric N–H···O rings in the centric polymorph and linear chains in noncentric one. There is an additional C–H···O contact in the centric form with a significant overlap of the electrostatic potentials between the alkenyl hydrogen atom and an oxygen atom of the nitro group. The dipole moment of 5-nitrouracil in the noncentric form is much higher ($\mu = 9$ D) than in the centric form ($\approx 6$ D). Among the 1,1-ethylenedicarbonitriles, there is an increased charge separation in the noncentric thioamino derivative, leading to an enhanced dipole of 15 D compared to the centric diamino (5 D) and dithio (6 D) derivatives. The effect of the crystal field is borne out by semiempirical AM1 calculations on the two systems. Dipole moments calculated for the molecules in the frozen geometries match closely with those obtained for centric crystals from the experimental charge densities. The calculated values of the dipole moment in the frozen or optimized geometries in the noncentric structures are, however, considerably lower than the observed value. Furthermore, the conformation of the S–$CH_3$ group in the noncentric crystal is anti with respect to the central C=C bond while the syn conformation is predicted for the free molecule in the optimized geometry.*

**KEYWORDS:**

charge density · dipole moment · materials science · nonlinear optics · semiempirical calculations

Several organic materials with nonlinear optical (NLO) properties have been described in the recent literature.[1–5] Nitroaniline and push–pull ethylenes are typical examples of NLO materials involving intramolecular charge transfer. Besides having a polarizable π-cloud with electron-donating and -accepting groups, an essential feature of such compounds is that they crystallize in noncentrosymmetric space groups. Espinosa et al.[6] have reported a charge density study of L-arginine phosphate. Puig-Molina et al.[7] have compared the topological properties of charge density of 2-amino-5-nitropyridinium dihydrogen phosphate from experiment and theory. Charge density studies of 2-methyl-4-nitroaniline[8] and urea[9] show that there is a considerable enhancement of the molecular dipole moment in the solid state. Based on a charge density study on *N*-(4-nitrophenyl)-L-prolinol, Fkyerat et al.[10, 11] extracted hyperpolarizability from the octapole moments. Such a possibility has been discussed in connection with inorganic NLO materials as well.[12, 13] While there is some speculation as to whether charge densities can provide hyperpolarizability,[14] it is generally accepted that reliable molecular dipole moments can be extracted from charge densities.[15] Thus, Hamazaoui et al.[16] obtained the dipole moment of 3-methyl-4-nitropyridine-*N*-oxide from its charge density, which agrees well with that from semiempirical calculations. Recently, Madsen et al.[17] have evaluated the dipole moment of phosphangulene from its charge density to be 42% higher than that measured in a chloroform solution. They also obtained the pyroelectric coefficient by combining the derived dipole moment with temperature-dependent measurements of the unit cell volume.

We have carried out an experimental charge density investigation of organic NLO systems in order to understand the manner in which the noncentric nature of the crystal field affects the molecular dipole moment and other properties in the solid. For this purpose, we have examined two systems—one where the center of symmetry is destroyed by crystallizing the molecule in a related polymorphic structure and the other where the noncentric nature is introduced by changing the pattern of substitution. For the first system, we have chosen the two polymorphic forms of 5-nitrouracil, **1**, which crystallizes in orthorhombic $P2_12_12_1$ and *Pbca* space groups, the former being noncentrosymmetric. The NLO activity of the noncentric form at 1.06 μm is ~160 times that of potassium dihydrogen phos-

[a] Prof. Dr. C. N. R. Rao, R. S. Gopalan, Dr. G. U. Kulkarni
Chemistry and Physics of Materials Unit
Jawaharlal Nehru Centre for Advanced Scientific Research
Jakkur P.O., Bangalore, 560064 (India)
E-mail: cnrrao@jncasr.ac.in

[**] The authors thank Dr. S. N. Ghanshyam Acharya for the synthesis of dicarbonitrile derivatives **2–4**.

phate.[18] In order to explore the effect of substitution, we have studied 1,1-ethylenedicarbonitriles (EDCN), in which the positions of the dimethylamino and methylthio substituents have been varied to give the symmetric and unsymmetric molecules bis(*N*,*N*-dimethylamino)-EDCN (**2**), 2,2-bis(methylthio)-EDCN (**3**), and 2-(*N*,*N*-dimethylamino)-2-methylthio-EDCN (**4**). The diamino and the dithio derivatives, **2** and **3**, crystallize in centric

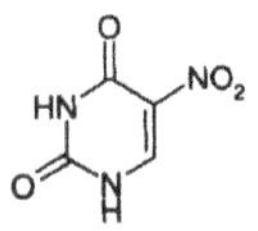

**1**

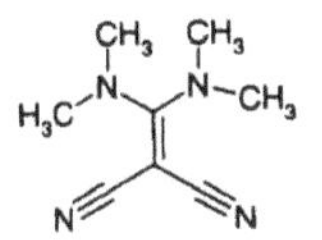

**2**

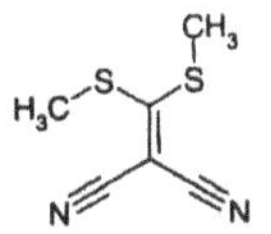

**3**

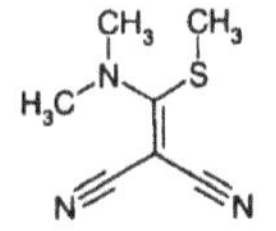

**4**

structures whereas the thioamino, **4**, occurs in a noncentrosymmetric structure exhibiting NLO properties. In both the systems, we have examined the intramolecular bonding using the Laplacian of the total electron density. Where necessary, we have analyzed the intermolecular interactions in terms of the electrostatic potential. We have computed the dipole moments from the experimental charge densities and compared the results with those from semiempirical AM1 calculations.

Editorial Advisory Board Member:[*] **C. N. R. Rao** received his M.Sc. in Chemistry from Banaras Hindu University (1953), Ph.D. in Chemistry from Purdue University (1958), and D.Sc. from the University of Mysore (1961). Following postdoctoral study in Berkeley, he worked as a faculty member at the Indian Institute of Technology, Kanpur, and the Indian Institute of Science, Bangalore. His main research interests are in the chemistry of materials and chemical spectroscopy. He is at present the Linus Pauling Research Professor at the JNCASR. He has been a Visiting Professor at the Universities of Cardiff, Cambridge, and California at Santa Barbara. He is a member of the Royal Society, US National Academy of Sciences, Pontifical Academy of Sciences, French Academy of Sciences, Japan Academy, and several others. He is a recipient of the Marlow Medal of the Faraday Society, Einstein Medal of UNESCO, Centenary Medal of the Royal Society of Chemistry, and the Hughes Medal of the Royal Society.

[*] *Members of the Editorial Advisory Board will be introduced to the readers with their first manuscript.*

*5-Nitrouracil*

The centric polymorph of **1** crystallizes with eight molecules per unit cell (*Pbca*; $a = 8.308(3)$, $b = 10.426(3)$, $c = 13.363(4)$ Å) while the unit cell of the noncentric form contains four molecules ($P2_12_12_1$; $a = 5.4342(1)$, $b = 9.8406(1)$, $c = 10.3659(1)$ Å). The densities of both forms are similar (Table 1).

The space groups *Pbca* and $P2_12_12_1$ contain the same set of symmetry elements but for the inversion center, which makes the problem interesting. In both polymorphs, the asymmetric unit comprises a whole molecule. The intramolecular geometry is essentially the same in both cases and the molecule is nearly planar (Table 2): The nitro group forms an angle of 1.5° with the plane of the uracil ring in the centric form and 2.0° in the noncentric polymorph.

In Figure 1, the packing diagram along with the atom labels are shown. In the centric form, the molecule is held by the dimeric N–H···O contacts of $R_2^2(8)$ type[19] with the two neighboring coplanar molecules (Figure 1 a). The N–H···O contacts in the noncentric form are of D(2) type arising from four different molecules (Figure 1 b). In both the cases, the N–H···O bonds are ~1.8 Å long, with ∢N–H···O angles of 155–174°. The centric polymorph, in addition, exhibits an intermolecular C–H···O interaction with the nitro group at a relatively short distance of 2.15 Å. The nitro group of the noncentric polymorph is in an unfavorable geometry to establish such a contact (Figure 1 b).

The results of the multipolar refinement (see Experimental Section) are described using the Laplacian of the total charge density. The topology of the Laplacian field allows one to obtain a chemical model of the bonded and nonbonded pairs and to characterize the local concentration ($\nabla^2\rho < 0$) and depletion ($\nabla^2\rho > 0$) of the molecular charge distribution.[20, 21] Figure 2 shows the contour maps of the negative Laplacian obtained for the two polymorphs. The lone electron pairs on the oxygen

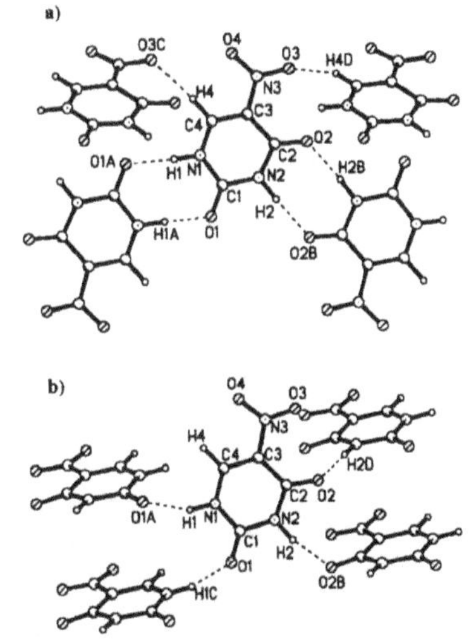


***Figure 1.** Molecular packing in a) centric and b) noncentric **1** showing the intermolecular hydrogen bonds. Atom labels are also shown.*

Table 1. Crystal data for 5-nitrouracil (1) and the 1,1-ethylenedicarbonitriles (2–4).[a]

| Crystal | 1 centric | 1 noncentric | 2 | 3 | 4 |
|---|---|---|---|---|---|
| Chemical formula (g mol$^{-1}$) | $C_4H_3N_3O_4$ (157.09) | | $C_8H_{12}N_4$ (164.22) | $C_6H_6N_2S_2$ (170.25) | $C_7H_9N_3S$ (167.23) |
| Crystal system | orthorhombic | orthorhombic | orthorhombic | monoclinic | orthorhombic |
| Space group | *Pbca* (No. 61) | $P2_12_12_1$ (No. 19) | *Pcab* (No. 6) | $P2_1/n$ (No. 14) | $Pna2_1$ (No. 33) |
| $a$ [Å] | 8.308(3) | 5.4342(1) | 7.6280(1) | 4.0261(1) | 7.904(2) |
| $b$ [Å] | 10.426(3) | 9.8406(1) | 14.455 | 13.227 | 8.630(2) |
| $c$ [Å] | 13.363(4) | 10.3659(1) | 16.3489(2) | 14.3906(1) | 12.856(3) |
| $\beta$ [°] | 90 | 90 | 90 | 95.445(1) | 90 |
| $V$ [cm$^3$] | 1157.5(6) | 554.32(1) | 1802.70(3) | 762.89(2) | 876.9(3) |
| $Z$ | 8 | 4 | 8 | 4 | 4 |
| $F_{000}$ | 640 | 320 | 704 | 352 | 352 |
| $\rho$ [g cm$^{-3}$] | 1.803 | 1.882 | 1.210 | 1.482 | 1.267 |
| No. of reflections for cell parameters | 45 | 37 | 60 | 60 | 60 |
| $\mu$ [mm$^{-1}$] | 0.16 | 0.17 | 0.079 | 0.617 | 0.309 |
| Crystal form | cuboidal | hexagonal | cuboidal | cuboidal | cuboidal |
| Crystal size [mm] | 0.3 × 0.2 × 0.2 | 0.15 × 0.15 × 0.1 | 0.2 × 0.1 × 0.1 | 0.2 × 0.15 × 0.1 | 0.2 × 0.1 × 0.1 |
| Crystal color | colorless | colorless | colorless | yellow | colorless |
| No. of measured reflections | 16262 | 16515 | 29153 | 12914 | 24387 |
| No. of independent reflections | 5026 | 8371 | 8589 | 6451 | 8576 |
| No. of observed reflections | 3879 | 3239 | 4894 | 4863 | 2874 |
| $R_{merge}$ | 0.0644 | 0.0434 | | | |
| $R_{int}$ | 0.0763 | 0.0454 | 0.0361 | 0.0301 | 0.0692 |
| $\theta_{min}$ [°] | 3.0 | 2.8 | 2.49 | 2.10 | 2.84 |
| $\theta_{max}$ [°] | 49.4 | 49.4 | 49.48 | 49.47 | 49.89 |
| Range of $h, k, l$ | $-12 \leq h \leq 16$ | $-10 \leq h \leq 11$ | $-15 \leq h \leq 15$ | $-8 \leq h \leq 7$ | $-16 \leq h \leq 15$ |
| | $-22 \leq k \leq 11$ | $-19 \leq k \leq 20$ | $-15 \leq k \leq 30$ | $-21 \leq k \leq 27$ | $-18 \leq k \leq 18$ |
| | $-28 \leq l \leq 27$ | $-21 \leq l \leq 21$ | $-34 \leq l \leq 34$ | $-29 \leq l \leq 28$ | $-27 \leq l \leq 27$ |
| After multipole refinement | | | | | |
| Weighting scheme | 0.02, 0.3 | 0.02, 0.1 | 0.05, 0.5 | 0.03, 0.3 | 0.04, 0.3 |
| $R1$ | 0.0345 | 0.0391 | 0.0501 | 0.0397 | 0.0332 |
| $wR2$ | 0.0416 | 0.0454 | 0.0775 | 0.0475 | 0.0513 |
| $S$ | 1.05 | 1.00 | 0.94 | 1.06 | 1.27 |
| No. of variables | 399 | 404 | 161 | 178 | 197 |
| $N_{ref}/N_v$ | 20.5 | 17.2 | 30.4 | 27.3 | 14.6 |
| Cambridge Crystallographic Database | 145990 | 145991 | 145987 | 145988 | 145989 |

[a] Experimental details: Siemens CCD diffractometer, crystal–detector distance 5.0 cm, $Mo_{K\alpha}$ radiation ($\lambda = 0.71073$ Å, 50 kV, 40 mA), 130 ± 1 K

atoms occur as (3, −3) critical points. There are some differences in the Laplacian maps of the two polymorphs. The atomic basins are generally linked in the centrosymmetric form but they appear to be disjoint in the noncentric case. The N3–O3 bond region in the centrosymmetric form contains nonoverlapping atomic lobes, which indicate a closed shell interaction. This is reflected in the properties of the bond critical points, listed in Table 3. This bond carries a much smaller Laplacian (−3.0 e Å$^{-5}$) compared to the other N–O bond of the molecule as well as the N–O bonds of the noncentric form, though the electron density itself is quite comparable. We also notice from Table 3 that the densities and the associated Laplacians of the non-hydrogen bonds are somewhat higher in the noncentric form while those of the N–H bonds are higher in the centric form. Moreover, the pseudo-atomic charges are generally higher in the noncentric polymorph. This implies that there is an increased charge separation in the noncentric case. The nitro group charge is −0.19 and −0.38 e respectively, in the centric and the noncentric forms, which compare well with −0.25 e obtained for 3-methyl-4-nitropyridine-*N*-oxide.[16] We also observe that the properties of the various bonds of the uracil moiety match closely with those reported in the case of 1-methyluracil.[22]

The nature of the intermolecular bonding is described in terms of the electrostatic potential. In Figure 3, contour maps of the electrostatic potential maps in the N–H···O and the C–H···O bond regions are shown. The contours originating from the N–H region overlap with those from the oxygen atom (Figures 3 a, b). The penetration is significant and indicates that these are strong contacts.[23] They are associated with (3, −1) critical points in $\rho$ as well as in electrostatic potential. The Laplacian at the CPs are small and positive (Table 4), as generally found in hydrogen bonds.[24] The C–H···O contact of the centric form also exhibits a noticeable overlap of the potentials ($\rho = 0.08(2)$ e Å$^{-3}$, $V = 0.28(1)$ e Å$^{-1}$), while in the noncentric form a zero potential surface passes between the proton and the oxygen atom to indicate no interaction between them (Figure 3 c). This is understandable, since in the presence of a carbonyl group, a nitro group acts as a secondary hydrogen bond acceptor.[25] The C–H···O contact of the centric form is special in that the oxygen O3, which participates in intermolecular bonding, tends to have nearly ionic interaction with the nitro group (Figure 2). Such bonding would be favored by the centrosymmetric packing of the molecules. For the same reason, this contact is absent in the noncentric form.

| Table 2. Bond lengths and angles of 1. | | |
|---|---|---|
| Bond length [Å] | centric **1** | noncentric **1** |
| O1-C1 | 1.2284(13) | 1.2310(13) |
| O2-C2 | 1.2252(13) | 1.2290(12) |
| O4-N3 | 1.2364(12) | 1.235(2) |
| O3-N3 | 1.2294(12) | 1.2306(14) |
| N2-C1 | 1.3667(13) | 1.3663(12) |
| N2-C2 | 1.3945(13) | 1.3900(14) |
| N2-H2 | 1.01 | 1.01 |
| N1-C4 | 1.3427(14) | 1.341(2) |
| N1-C1 | 1.3807(13) | 1.3786(14) |
| N1-H1 | 1.01 | 1.01 |
| N3-C3 | 1.4432(13) | 1.4412(14) |
| C2-C3 | 1.4568(14) | 1.454(2) |
| C3-C4 | 1.3575(14) | 1.3621(14) |
| C4-H4 | 1.08 | 1.08 |
| Bond angle [°] | centric **1** | noncentric **1** |
| C1-N2-C2 | 127.18(9) | 127.47(9) |
| C1-N2-H2 | 116.41(6) | 116.27(6) |
| C2-N2-H2 | 116.41(6) | 116.27(6) |
| C4-N1-C1 | 122.51(9) | 122.90(8) |
| C4-N1-H1 | 118.74(6) | 118.55(6) |
| C1-N1-H1 | 118.74(6) | 118.55(5) |
| O3-N3-O4 | 123.09(9) | 123.46(11) |
| O3-N3-C3 | 119.36(9) | 119.20(11) |
| O4-N3-C3 | 117.56(9) | 117.33(10) |
| O1-C1-N2 | 123.08(9) | 122.60(10) |
| O1-C1-N1 | 121.61(9) | 122.54(9) |
| N2-C1-N1 | 115.31(9) | 114.85(9) |
| O2-C2-N2 | 119.86(9) | 118.82(10) |
| O2-C2-C3 | 127.55(10) | 128.32(10) |
| N2-C2-C3 | 112.59(8) | 112.85(8) |
| C4-C3-N3 | 116.49(9) | 116.94(10) |
| C4-C3-C2 | 120.79(9) | 120.32(10) |
| N3-C3-C2 | 122.62(8) | 122.73(9) |
| N1-C4-C3 | 121.39(9) | 121.40(10) |
| N1-C4-H4 | 119.30(6) | 119.30(6) |
| C3-C4-H4 | 119.30(6) | 119.30(7) |

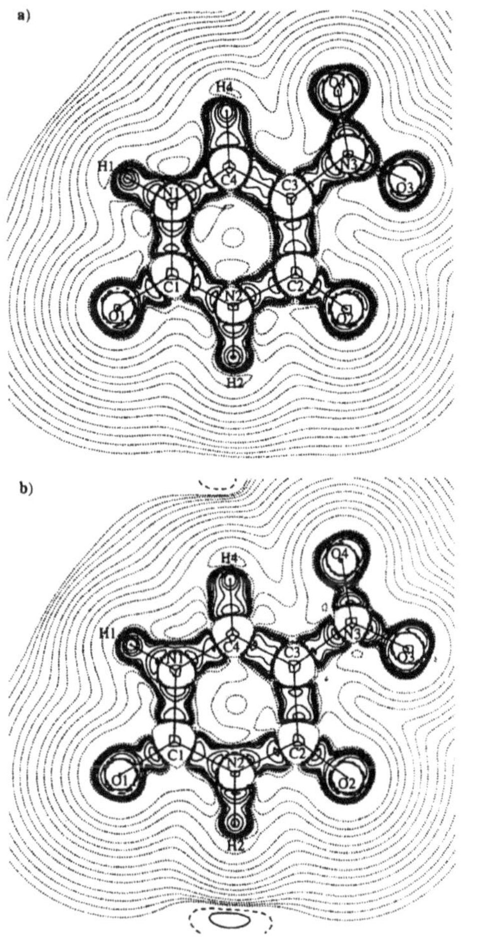


*Figure 2. Contour maps of the total density Laplacian in the C1-C2-C4 plane for a) centric and b) noncentric 1.*

In situ dipole moments of molecules in crystals can be obtained from charge densities by using Equation (1).

$$p_i = \sum_i z_i \mathbf{R}_i + \int \mathbf{r} \rho_i(\mathbf{r}_i) d\mathbf{r} \quad (1)$$

The XDPROP routine (the XD program) was used to calculate dipole moments and the dipole moment vectors are shown in Figure 4. The vectors lie close to the N(1)–H(1) bond direction in the plane of the molecule, with magnitudes of $\mu = 5.5(6)$ and 9(1) D in the centric and the noncentric polymorphs, respectively.[26] As the intramolecular geometry remains the same in the two cases (Table 2), the enhancement of the dipole moment in the noncentric structure is likely to arise mainly due to packing. The effect of crystal packing can be understood by calculating the dipole moment of the molecule ex situ in the frozen geometry. For this purpose, we carried out AM1, PRECISE calculations by freezing the molecular geometries obtained from X-ray diffraction. The calculated dipole moments are very similar in the two polymorphs ($\mu = 5.5$ and 5.6 D for centric and noncentric cases, respectively) and point in the same direction (Figure 4). Interestingly, the vectors obtained from the calculation lie close (in magnitude as well as in direction) to that from

| Table 3. Analysis of the bond critical points in 1. | | | | | | |
|---|---|---|---|---|---|---|
| | | centric | | | noncentric | |
| Bond | $\rho^{[a]}$ | $\nabla^2\rho^{[b]}$ | $\varepsilon^{[c]}$ | $\rho^{[a]}$ | $\nabla^2\rho^{[b]}$ | $\varepsilon^{[c]}$ |
| O1-C1 | 2.91(5) | −38.2(3) | 0.21 | 3.31(9) | −51.9(5) | 0.27 |
| O2-C2 | 2.77(6) | −26.8(4) | 0.18 | 3.21(97) | −51.3(4) | 0.17 |
| N1-C1 | 2.16(5) | −26.8(2) | 0.06 | 2.36(8) | −27.9(3) | 0.29 |
| N1-C4 | 2.29(5) | −27.3(2) | 0.25 | 2.50(9) | −31.5(4) | 0.28 |
| N1-H1 | 1.98(6) | −28.8(4) | 0.06 | 1.9(1) | −30.1(8) | 0.00 |
| N2-C1 | 2.12(4) | −23.9(2) | 0.21 | 2.34(8) | −30.1(3) | 0.38 |
| N2-C2 | 2.09(4) | −23.5(2) | 0.12 | 2.22(7) | −23.7(3) | 0.24 |
| N2-H2 | 1.92(6) | −29.8(4) | 0.04 | 1.8(1) | −36(1) | 0.04 |
| N3-C3 | 1.87(4) | −15.4(2) | 0.37 | 1.96(8) | −17.1(4) | 0.39 |
| C2-C3 | 1.93(3) | −16.0(1) | 0.27 | 1.98(7) | −7.0(2) | 0.46 |
| C3-C4 | 2.21(3) | −22.1(1) | 0.34 | 2.31(7) | −24.7(2) | 0.43 |
| C4-H4 | 1.79(5) | −20.1(2) | 0.05 | 1.9(1) | −29.5(7) | 0.07 |
| O3-N3 | 3.19(5) | −3.0(2) | 0.09 | 3.48(9) | −14.5(3) | 0.05 |
| O4-N3 | 3.28(5) | −11.9(2) | 0.08 | 3.61(9) | −18.6(4) | 0.08 |

[a] electron density, in e Å$^{-3}$. [b] Laplacian, in e Å$^{-5}$. [c] Ellipticity.

**Table 4.** Hydrogen bond critical points in 1.

| Polymorph | D-H···A | $\rho$ [e Å$^{-3}$] | $\nabla^2\rho$ [e Å$^{-5}$] | $V$ [e Å$^{-1}$] |
|---|---|---|---|---|
| centric | N(1)-H(1)···O(1)[a] | 0.18(3) | 3.32(2) | 0.55(2) |
| | N(2)-H(2)···O(2)[a] | 0.17(3) | 2.88(2) | 0.47(2) |
| | C(4)-H(4)···O(3)[b] | 0.08(2) | 1.39(1) | 0.28(1) |
| noncentric | N(1)-H(1)···O(1)[c] | 0.16(5) | 2.99(4) | 0.52(3) |
| | N(2)-H(2)···O(2)[d] | 0.15(6) | 3.05(5) | 0.61(4) |
| | O(1)···H(1)-N(1)[e] | 0.14(5) | 2.64(5) | 0.53(3) |
| | O(2)···H(2)-N(2)[f] | 0.13(5) | 2.69(5) | 0.60(3) |

Symmetries: [a] $-x, -y, 1-z$. [b] $^1/_2-x, ^1/_2+y, z$. [c] $^1/_2+x, ^1/_2-y, 1-z$. [d] $-^1/_2+x, ^1/_2-y, -z$; [e] $-^1/_2+x, ^1/_2-y, 1-z$. [f] $^1/_2+x, ^1/_2-y, -z$.

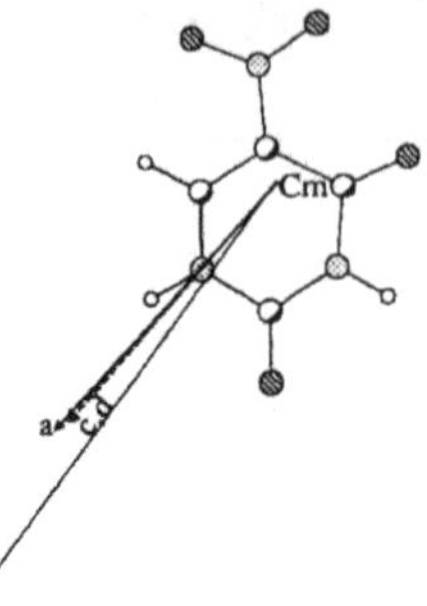


*Figure 4. Orientation of the molecular dipole moment of 1: a) centric, multipole; b) noncentric, multipole; c) centric, AM1; d) noncentric, AM1. Cm refers to the center of mass of the molecule.*

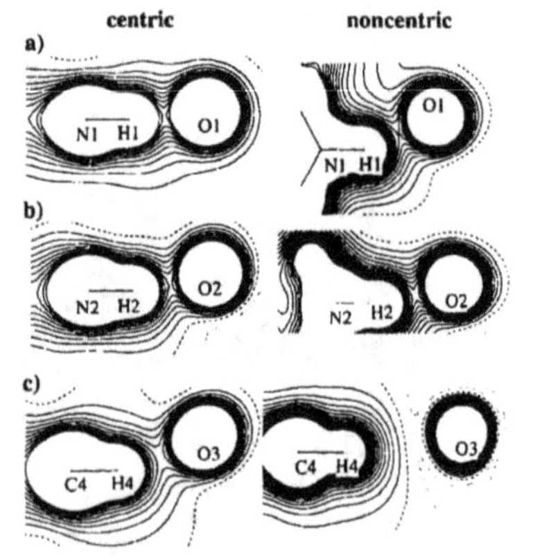


*Figure 3. Contour maps of the electrostatic potential in the intermolecular regions (contour level at 0.05 eÅ$^{-1}$): a) N1–H1···O1, b) N2–H2···O2, c) C4–H4···O3. The centric and noncentric forms are given on the left and right sides, respectively.*

experiment for the centric polymorph. This seems to suggest that the packing in the centric form of **1** has a minimal effect on the intramolecular charge distribution. This observation also demonstrates that an asymmetric crystal field in a noncentric structure can significantly enhance the dipole moment of a molecule. To our knowledge, this is the first instance of charge density study of a compound crystallizing in closely related space groups exhibiting differing molecular properties.

### *1,1-Ethylenedicarbonitriles*

The two nitriles, **2** and **3**, crystallize in the centrosymmetric *Pcab* and $P2_1/n$ space groups, respectively, whereas **4** crystallizes in the noncentric $Pna2_1$ space group (Table 1). We show the molecular diagrams along with the thermal ellipsoids in Figure 5, where the bond lengths and angles associated with the non-hydrogen atoms are also indicated. Diamine **2** possesses a pseudo two-fold symmetry along C3–C4 while the dithio **3** exhibits a *syn–anti* methylthio conformation. Thioamine **4** has methylthio group in the *anti* conformation. The 1,1-ethylenedicarbonitrile moiety, which contains a formal C3=C4 bond and two single C1–C3 and C2–C3 bonds, along with the cyano group, is essentially planar in all three molecules. The C≡N bond lengths are also similar (1.14(3) Å) and close to those reported in the literature. There are however, some differences between the formal C–C single and double bonds. In **2**, the single and the

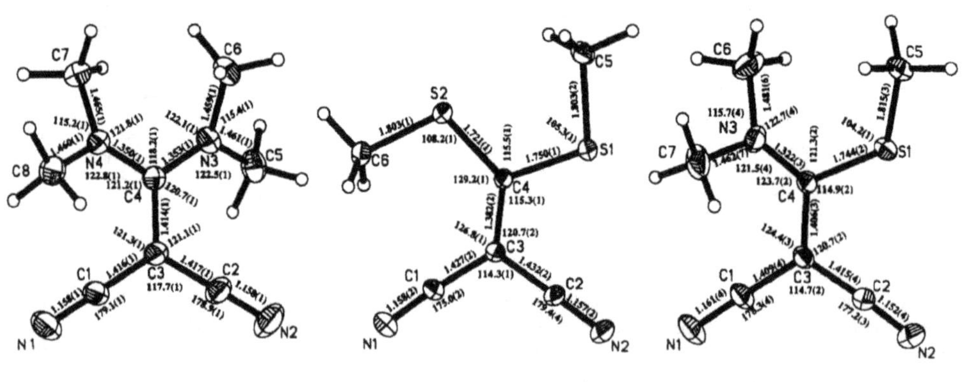


2 3 4

*Figure 5. Molecular diagrams of the diamino 2, dithio 3, and thioamino 4 EDCN derivatives, showing thermal ellipsoids. The bond lengths and angles of the non-hydrogen bonds are shown.*

double bonds are of equal length (~ 1.416 Å), while they differ in **3** (1.432(2) and 1.427(2), versus 1.382(2) Å). In **4**, the differences are moderate (see Figure 5). A survey of formally uncharged molecules containing the ethylenedicarbonitrile moiety (466 compounds with $r < 10\%$) listed in the Cambridge Crystallographic Database have mean C–C and C=C bond lengths of 1.43(2) and 1.37(3) Å, respectively. The shortening of the C–C single bonds in this moiety is noteworthy. The $C_{sp^2}$–N (or $C_{sp^2}$–S) bonds are shorter than $C_{sp^3}$–N (or $C_{sp^3}$–S). In **3**, there exist considerable differences between the bond lengths and the angles associated with the *syn* and *anti* conformations of the methylthio groups. The C–C≡N bond angle associated with the *syn*-methythio group is 175° instead of ~180°. An earlier structural report on this molecule suggested steric interaction to be responsible for the discrepancy.[27] However, 3,3-bis(methylthio)-2-nitro-2-propene-1-nitrile does not show significant differences in the bond lengths and angles between the *syn*- and the *anti*-configurations of the methylthio groups.[28]

Figure 6 shows the contour maps of the negative Laplacian of the total density in the plane defined by the N1-N2-C3 atoms. The lone pairs on the nitrogen and sulfur atoms exhibit (3, −3) as critical points. The lobes near sulfur are similar to those seen previously by Mallinson and co-workers in the case of 3,3,6,6-tetramethyl-*S*-tetrathiane.[29] In the diamine **2**, the contours associated with the amino groups are small, since this group is out of plane with the rest of the molecule. The C–S bonds of both **3** and **4** exhibit disjoint lobes.

Figure 7 depicts the critical points in the total density for the various bonds in the molecules. The values of total electron density, the Laplacian, and the ellipticity at the critical points are also shown. The C≡N group exhibits similar densities (~3.2 e Å$^{-3}$) in the three molecules although the Laplacian values are somewhat different, −36, −20, and −16 e Å$^{-5}$ in **1**, **2**, and **3**, respectively. Interestingly, the densities of the C–C single and double bonds are not significantly different in the three compounds. The Laplacian values are also similar, the differences are less than ~5 e Å$^{-5}$. This result is in contrast with what is expected of normal single and double bonds. The electron densities associated with the isolated single and double bonds are generally 1.71 and 2.5 e Å$^{-3}$ respectively.[30] In the present

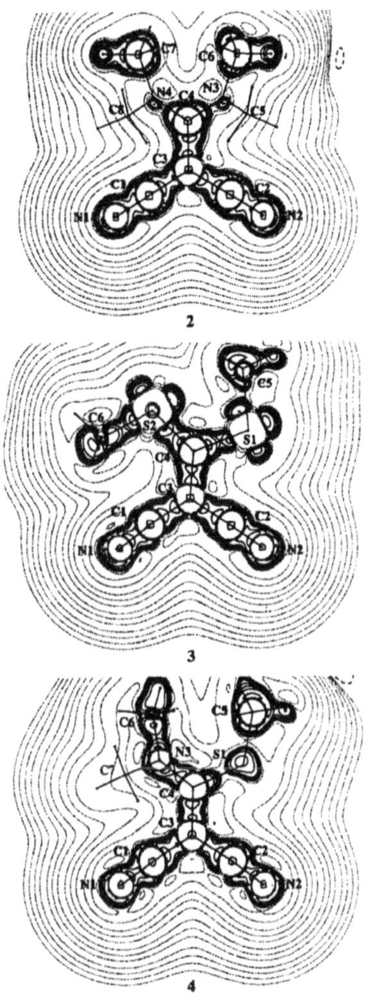

*Figure 6.* Contour maps of the Laplacian of the total electron density of the diamino **2**, dithio **3**, and thioamino **4** EDCN derivatives in the N1-C3-C4 plane.

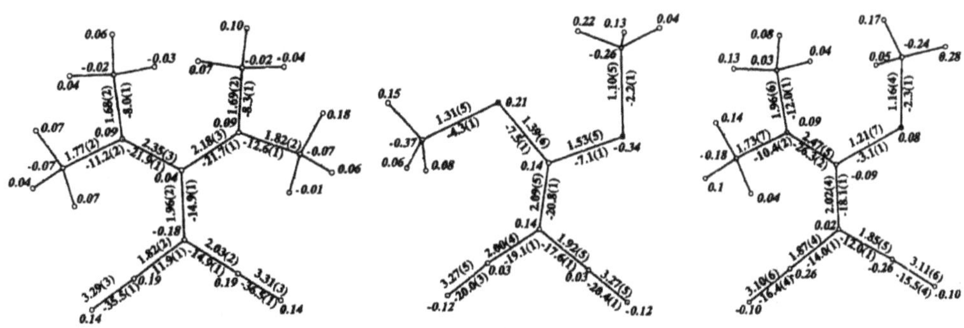

2 3 4

*Figure 7.* Stick diagrams showing the critical points for the various bonds of the diamino **2**, dithio **3**, and thioamino **4** EDCN derivatives. The electron density and the Laplacians associated with each bond are shown nearby. The pseudo-atomic charges are also indicated in italics. The atoms are labeled as in Figure 5.

case, there is probably some delocalization of electrons in the bonding region. A similar observation was made earlier by Espinosa et al.[31] in the case of the BTDMTTF-TCNQ complex. The low values of the densities and the Laplacians in the C–S bonds are characteristic of hypervalent species.[29] A recent study on tetrasulfurtetranitride[32] has reported low values of densities and Laplacians in the N–S bonds. The charge density analysis reflects the differences between the $C_{sp^2}$ and $C_{sp^3}$ bonding regions, as expected from the structural features discussed above. Accordingly, the $C_{sp^2}$ bonding regions carry higher densities and Laplacians. Based on the pseudoatomic charges shown in Figure 7, there appears to be an increased charge separation in the noncentric thioamine **4**.

The intermolecular contacts of the structures **2**–**4** are listed in Table 5. Both **2** and (**3**) exhibit five C–H···N contacts while the thioamine (**4**) makes only three such contacts. These are typical C–H···N contacts in terms of both the geometry and the charge density. In addition, there are intermolecular S···S and S···N contacts in the dithio compound and S···N contacts in the thioamine (Table 5). The distances as well as the charge densities associated with these contacts are similar to the contacts found in other molecules.[29, 32]

The in situ dipole moments of the molecules in the three crystals are projected in Figure 8. The dipole moment vectors lie close to the mean planes of the molecules. An important observation from Figure 8 is that the dipole moment of the thioamine **4**, in the noncentric structure, is much higher ($\mu$ = 15(2) D) than the values of the diamino (**2**) and the dithio (**3**) derivatives in the centric structures ($\mu$ = 5(2) and 6(2) D).[26] The enhancement of the dipole moment in the thioamine could arise from the asymmetric substitution of the electron donating groups or due to the noncentric crystal field. In order to shed more light, we have computed dipole moments of the three molecules in the free state in both the frozen and the optimized (AM1, PRECISE) geometries using MOPAC. In Figure 9, we depict the frozen and the optimized dipole moments projected along with the molecule. The calculated dipole moment remains essentially unchanged between the frozen and the optimized geometries in the case of **2** and **3**. Compared to the in-crystal dipole moment, these values are slightly higher in the diamino and are comparable in the case of the dithio compound (see Figures 8 and 9). The noncentric thioamine **4**, on the other hand, exhibits much smaller dipole moments outside the lattice. Both the optimized ($\mu$ = 5.9 D) and the frozen ($\mu$ = 8.1 D) moments are considerably lower than the in-crystal dipole moment of $\mu$ = 15 D. It is interesting that in the optimized geometry, the

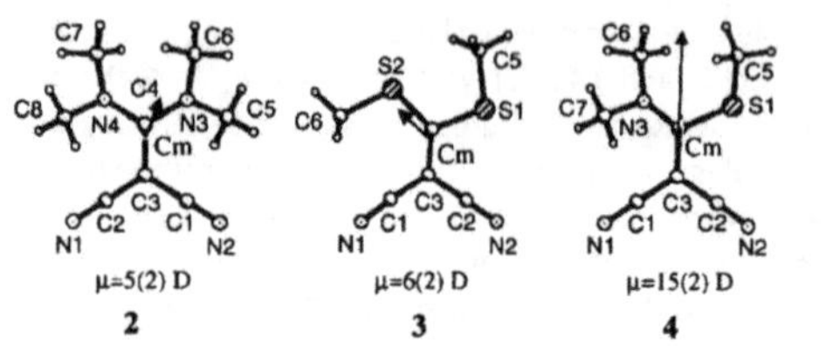


*Figure 8. Molecular diagrams showing the dipole moment vectors computed from charge density in 2–4. Cm refers to the center of mass of the molecule.*

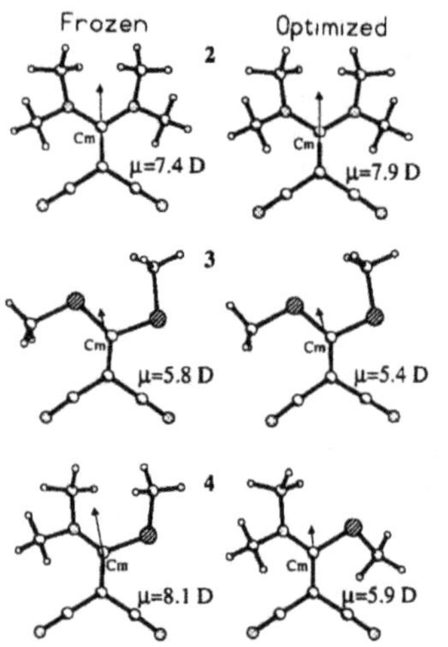


*Figure 9. Molecular dipole moments calculated using AM1 (MOPAC) in the frozen (left) and the optimized (right) geometries of the diamino 2, dithio 3, and thioamino 4 EDCN derivatives. Cm refers to the center of mass of the molecule. Atomic labeling is given in Figure 8.*

*Table 5. Intermolecular contacts in 2–4.*

| Hydrogen contacts between donor (D) and acceptor (A) | | | | | |
|---|---|---|---|---|---|
| D–H···A | H···A [Å] | D···A [Å] | ∡D-H···A [°] | $\rho$ [e Å$^{-3}$] | $\nabla^2\rho$ [e Å$^{-5}$] |
| **2** | | | | | |
| C5-H5A···N1[a] | 2.745 | 3.531 | 130.9 | 0.061(3) | 0.601(2) |
| C7-H7A···N1[b] | 2.510 | 3.467 | 149.6 | 0.103(5) | 0.967(2) |
| C8-H8C···N1[c] | 2.872 | 3.649 | 130.4 | 0.041(2) | 0.41(1) |
| C8-H8B···N2[d] | 2.720 | 3.353 | 137.2 | 0.059(3) | 0.616(1) |
| C7-H7B···N2[e] | 2.493 | 3.504 | 136.8 | 0.103(5) | 0.951(2) |
| **3** | | | | | |
| C6-H6C···N1[f] | 2.564 | 3.541 | 152.8 | 0.035(6) | 0.653(5) |
| C6-H6A···N1[g] | 2.840 | 3.500 | 120.6 | 0.042(5) | 0.523(3) |
| C5-H5B···N1[h] | 2.586 | 3.631 | 168.6 | 0.04(2) | 0.619(5) |
| C5-H5A···N2[i] | 2.560 | 3.603 | 167.8 | 0.040(2) | 0.522(2) |
| C6-H6A···N2[j] | 2.886 | 3.554 | 121.3 | 0.048(4) | 0.566(3) |
| **4** | | | | | |
| C5-H5B···N1[k] | 2.594 | 3.617 | 162.0 | 0.032(8) | 0.572(9) |
| C6-H6B···N1[k] | 2.780 | 3.633 | 137.5 | 0.025(8) | 0.420(4) |
| C5-H5A···N2[l] | 2.443 | 3.463 | 161.2 | 0.05(1) | 0.691(4) |
| Sulfur contacts | Distance [Å] | $\rho$ [e Å$^{-3}$] | $\nabla^2\rho$ [e Å$^{-5}$] | | |
| **3** | | | | | |
| S1···S1[m] | 3.515 | 0.067(1) | 0.654(2) | | |
| S1···S1[n] | 3.952 | 0.030(1) | 0.305(1) | | |
| N2···S2[o] | 3.307 | 0.055(1) | 0.603(1) | | |
| N2···S2[p] | 3.549 | 0.055(1) | 0.603(1) | | |
| **4** | | | | | |
| N1···S1[q] | 3.194 | 0.055(2) | 0.586(3) | | |

Symmetries: [a] $x, -1/2+y, 1/2+z$. [b] $3/2-x, -1/2+y, -z$. [c] $1/2+x, 3/2-y, z$. [d] $2-x, 3/2-y, -1/2+z$. [e] $3/2-x, y, -1/2+z$. [f] $1+x, y, z$. [g] $-x, 1-y, -z$. [h] $-1/2-x, 1/2+y, 1/2-z$. [i] $1-x, 1-y, 1-z$. [j] $1/2-x, 1/2+y, 1/2-z$. [k] $2-x, -y, 1/2+z$. [l] $2-x, 1-y, 1/2+z$. [m] $1-x, 1-y, 1-z$. [n] $-x, 1-y, 1-z$. [o] $1/2-x, -1/2+y, 1/2-z$. [p] $-1/2-x, -1/2+y, 1/2-z$. [q] $3/2-x, -1/2+y, -1/2+z$.

thioamine exhibits a *syn*-methylthio conformation with a smaller dipole moment compared to the *anti*-methylthio conformation found in the frozen geometry of the crystal. It appears that the noncentric structure induces significant distortions in the molecule and enhances the dipole moment.

In conclusion, the present charge density study on the centric and noncentric molecular systems has provided valuable insight into the effect of the crystal field on molecular properties. In the case of 5-nitrouracil, the dipole moment is 5.5(6) D in the centric structure and increases to 9(1) D in the noncentric lattice, while the molecular geometry essentially remains the same; the difference in the dipole moments clearly arises from the crystal packing. In the centric structure, there are N–H···O hydrogen bonded dimers, whereas in the noncentric the molecules form N–H···O linear chains. The centric crystal exhibits an additional contact, C–H···$O_{nitro}$, absent in the noncentric structure. The molecular dipole moment in the noncentric crystal is considerably greater than the value calculated for a free molecule of outside the crystal (frozen geometry). In centric polymorph, the dipole moment is close to that obtained from the free molecule.

In the ethylenedicarbonitriles, the crystal is centrosymmetric when the two substituents are the same, as in the diamino (**2**) and dithio (**3**) derivatives, and they have low dipole moments ($\mu \sim 6$ D). The dipole moment values are close to those obtained by AM1 calculations on the frozen molecular geometries. In contrast, the thioamine (**4**), with the *anti*-methylthio conformation of the S–$CH_3$ group, occurs in a noncentric structure where there is an increased charge separation within the molecule and, hence, a considerably higher dipole moment ($\mu = 15(2)$ D) than that calculated for the molecule of frozen geometry. When the molecule is subjected to optimization, it adopts a *syn*-methylthio conformation and the dipole moment falls to below $\mu = 5.9$ D. The enhancement of the dipole moment in the thioamino system is partly due to the *anti*-conformation forced on the molecule by the noncentric crystal field and partly due to the field itself.

## Experimental Section

On cooling an aqueous solution of 5-nitrouracil from 70 °C, crystals of the centrosymmetric polymorph were obtained. The noncentric crystals were obtained from an acetonitrile solution at room temperature. The dinitriles **2**–**4** were synthesized by employing reported procedures[33–35] and the crystals were grown from toluene, ethyl acetate, and benzene solutions, respectively. High quality crystals were chosen after examination under an optical microscope. X-ray diffraction intensities were measured by $\omega$ scans using a Siemens three-circle diffractometer.

The unit cell parameters and the orientation matrix of the crystal were initially determined using ~45 reflections from 25 frames collected over a small $\omega$ scan of 7.5° sliced at 0.3° intervals. A hemisphere of reciprocal space was then collected in two shells using the SMART software[36] with $2\theta$ at 28° and 70°. Data reduction was performed using the SAINT program[36] and the orientation matrix along with the detector and the cell parameters were refined for every 40 frames. Further experimental details are listed in Table 1. The crystal structures were first determined with the low-resolution data up to $(\sin\theta)/\lambda = 0.56$ Å$^{-1}$. The phase problem was solved by direct methods and the non-hydrogen atoms were refined anisotropically, by means of the full-matrix least-squares procedures using the SHELXTL program.[37] All the hydrogen atoms were located using the difference Fourier method and the temperature factors of these atoms were refined isotropically. The crystal structures agreed well with those reported previously.[27, 38, 39] The structure of the centrosymmetric 5-nitrouracil polymorph (*Pbca*) is reported for the first time and deposited in the Cambridge Crystallographic Database.

Charge density analysis was carried out based on multipole expansion of the electron density centered at the nucleus of the atom.[40] Accordingly, the aspherical atomic density can be described in terms of spherical harmonics [Eq. (2)]. Thus, for each atom, in which the origin at the atomic nucleus, [Eq. (3)] holds. The population coefficients, $P_{lmp}$, are to be refined along with the $\kappa$ and $\kappa'$ parameters which control the radial dependence of the valence shell density. The analysis was carried out in several steps.

$$\rho_{atom}(r) = \rho_{core}(r) + \rho_{valence}(r) + \rho_{def}(r) \tag{2}$$

$$\rho_{atom}(r) = \rho_{core}(r) + P_v \kappa^3 \rho_{valence}(\kappa r) + \sum_{l=0} \kappa'^3 R_l(\kappa' \zeta r) \sum_{m=0} \sum_{p=\pm l} P_{lmp} Y_{lmp}(\theta,\varphi) \tag{3}$$

The hydrogen atomic positions were found using the difference Fourier method and were adjusted to average neutron values,[41] as is usually done during the multipole refinement ($C_{sp^3}$–H 1.06, $C_{sp^2}$–H 1.083, N–H 1.01 Å). A high order refinement of the data was performed using reflections with $(\sin\theta)/\lambda \geq 0.6$ Å$^{-1}$ and $F_0 \geq 4\sigma$. All the hydrogen atoms were held constant throughout the refinement along with their isotropic temperature factors. Multipolar refinement for the charge density analysis was carried out using the XDLSM routine of the XD package;[42] details are given in Table 1. The atomic coordinates and the thermal parameters obtained from the high order refinement were used as input to XD refinement. Sulphur atoms were refined up to hexadecapole moments, carbon, nitrogen, and oxygen atoms up to octapole moments, while hydrogen atoms were restricted to dipole moments. During the refinement, the charge neutrality constraint was applied in all the cases. Kappa refinement was carried out on both the spherical and deformation valence shells of the non-hydrogen atoms, while those for the hydrogen atoms were restricted to the spherical valence. The multipolar refinement strategy was the following: a) scale factor, b) $P_v$, c) $P_{lm}$, d) repeat (b, c) until convergence, e) $\kappa$, f) repeat (b, c) until convergence, g) $\kappa'$, h) repeat (b, c) until convergence, i) positional and thermal parameters of all nonhydrogen atoms, and finally j) $P_v$ and $P_{lm}$ together. The quality of the refinement was monitored using the residual density; the magnitude of the highest residual peak in the final refinement was 0.19, 0.2, 0.2, 0.2, 0.15 eÅ$^{-3}$ for the centric **1**, noncentric **1**, **2**, **3**, and **4**, respectively. Difference mean-square displacement amplitudes for the various bonds obtained from XD were found to follow closely the Hirshfeld criterion.[43]

The XDPROP routine was used to calculate the total electron density $\rho(r)$, the Laplacian $\nabla^2\rho$, and the ellipticity $\varepsilon$, at the bond critical points (CPs). Electrostatic potential (*V*) and dipole moments were also obtained using this routine. The Laplacian maps have been plotted using the XDGRAPH routine. The MOPAC program[44] was used for the calculation of the dipole moments at the frozen and optimized molecular geometries. This calculation, although semiempirical, is known to provide reliable estimates of molecular dipole moments.[16]

[1] P. N. Prasad, J. D. Williams, *Introduction to Nonlinear Optical Effects in Molecules and Polymers*, Wiley, New York, **1990**.

[2] J. Zyss, *Molecular Nonlinear Optics: Materials, Physics, Devices*, Academic Press, Boston, **1994**.

[3] C. Bosshard, K. Sutter, P. Pretre, J. Hullinger, M. Florsheimer, P. Kaatz, P. Gunter, *Organic Nonlinear Optical Materials*, Gordon & Breach, Amsterdam, **1995**.
[4] D. S. Chemla, J. Zyss, *Nonlinear Optical Properties of Organic Molecules and Crystals*, Academic Press, Orlando, FL, **1987**.
[5] D. R. Kanis, M. A. Ratner, T. J. Marks, *Chem. Rev.* **1994**, *94*, 195.
[6] E. Espinosa, C. Lecomte, E. Molins, S. Veintemillas, A. Cousson, W. Paulus, *Acta Crystallogr. Sect. B* **1996**, *52*, 519.
[7] A. Puig-Molina, A. Alvarez-Larena, J. F. Piniella, S. T. Howard, F. Baert, *Struct. Chem.* **1998**, *9*, 395.
[8] S. T. Howard, M. B. Hursthouse, C. W. Lehmann, P. R. Mallinson, C. S. Frampton, *J. Chem. Phys.* **1992**, *97*, 5616.
[9] C. Gatti, V. R. Saunders, C. Roetti, *J. Chem. Phys.* **1994**, *101*, 10686.
[10] A. Fkyerat, A. Guelzim, F. Baert, W. Paulus, G. Heger, J. Zyss, A. Perigaud, *Acta Crystallogr. Sect. B* **1995**, *51*, 197.
[11] A. Fkyerat, A. Guelzim, F. Baert, J. Zyss, A. Perigaud, *Phys. Rev. B* **1996**, *53*, 16236.
[12] N. K. Hansen, J. Protas, G. Marnier, *C. R. Acad. Sci. Ser. B* **1988**, *307*, 475.
[13] N. K. Hansen, J. Protas, G. Marnier, *Acta Crystallogr. Sect. B*, **1991**, *47*, 660.
[14] M. A. Spackman, *Annu. Rep. Prog. Chem. Sect. C* **1998**, *94*, 177.
[15] P. Coppens in *X-ray Charge Densities and Chemical Bonding*, Oxford University Press, **1997**.
[16] F. Hamazaoui, F. Baert, J. Zyss, *J. Mater. Chem.* **1996**, *6*, 1123.
[17] G. K. H. Madsen, F. C. Krebs, B. Lebech, F. K. Larsen, *Chem. Eur. J.* **2000**, *6*, 1797.
[18] H. Youping, S. Genbo, W. Bochang, J. Rihong, *J. Cryst. Growth* **1992**, *119*, 393.
[19] M. C. Etter, *Acc. Chem. Res.* **1990**, *23*, 120.
[20] P. R. Mallinson, K. Wozniak, T. Garry, K. L. McCormak, *J. Am. Chem. Soc.* **1997**, *119*, 11502.
[21] P. R. Mallinson, K. Wozniak, C. C. Wilson, K. L. McCormak, D. M. Yufit, *J. Am. Chem. Soc.* **1999**, *121*, 4640.
[22] W. T. Klooster, S. Swaminathan, R. Nanni, B. M. Craven, *Acta Crystallogr. Sect. B* **1992**, *48*, 217.
[23] R. F. Stewart in *The Application of Charge Density Research to Chemistry and Drug Design* (Eds.: G. A. Jeffrey, J. F. Piniella), Plenum, New York, **1991**.
[24] R. F. W. Bader, *Atoms in Molecules—A Quantum Theory*, Clarendon Press, Oxford, **1990**.
[25] R. Srinivasa Gopalan, G. U. Kulkarni, S. Renganayaki, E. Subramanian, *J. Mol. Struct.* **2000**, *524*, 169.
[26] Following referee comments, $R_{int}$ for centric **1** and **4** were improved to $\sim$0.04 by eliminating reflections above 0.9 Å$^{-1}$. The dipole moments did not vary significantly.
[27] H. U. Hummel, H. Procher, *Acta Crystallogr. Sect. C* **1986**, *42*, 1602.
[28] N. U. Kamath, K. Venkatesan, *Acta Crystallogr. Sect. C* **1984**, *40*, 1211.
[29] K. L. McCormack, P. R. Mallinson, B. C. Webster, D. S. Yufit, *J. Chem. Soc. Faraday Trans.* **1996**, *92*, 1709.
[30] D. Cremer, E. Kraka, *Croat. Chem. Acta.* **1984**, *57*, 1259.
[31] E. Espinosa, E. Molins, C. Lecomte, *Phys. Rev. B* **1997**, *57*, 1820.
[32] W. Scherer, M. Spiegler, B. Pedersen, M. Tafipolsky, W. Hieringer, B. Reinhard, A. J. Downs, G. S. McGrady, *Chem. Commun.* **2000**, 635.
[33] E. Ericsson, J. Sandström, I. Wennerbeck, *Acta Chem. Scand.* **1970**, *24*, 3102.
[34] L. Dalgaard, H.-K. Andersen, S.-O. Lawesson, *Tetrahedron* **1973**, *29*, 2077.
[35] N. H. Nilsson, *Synthesis* **1974**, 433.
[36] Siemens Analytical X-ray Instruments, **1995**, Madison, WI, USA.
[37] SHELXTL (SGI version), Siemens Analytical X-ray Instruments, **1995**, Madison, WI, USA.
[38] J. G. Bergman, G. R. Crane, B. F. Levine, C. G. Bethea, *Appl. Phys. Lett.* **1972**, *20*, 21.
[39] D. Adhikesavulu, K. Venkatesan, *Acta Crystallogr. Sect. C* **1983**, *39*, 589.
[40] N. K. Hansen, P. Coppens, *Acta Crystallogr. Sect. A* **1978**, *34*, 909.
[41] F. H. Allen, O. Kennard, D. G. Watson, L. Brammer, A. G. Orpen, R. Taylor, *J. Chem. Soc. Perkin Trans. 2* **1987**, S1.
[42] T. Koritsansky, S. T. Howard, T. Richter, P. R. Mallinson, Z. Su, N. K. Hansen, *XD, A Computer Program Package for Multipole Refinement and Analysis of Charge Densities from Diffraction Data*, Cardiff University (UK), **1995**.
[43] H. L. Hirshfeld, *Acta Crystallogr. Sect. A* **1976**, *32*, 239.
[44] J. J. P. Stewart, *J. Comput. Aided Mol. Des.* **1990**, *4*, 1.

Received: June 26, 2000 [Z 57]

# V. METAL-ORGANIC FRAMEWORKS

*Commentary by C.N.R. Rao*

Zeolitic aluminosilicates and aluminophosphates are the most well-known examples of inorganic open-framework solids. In the last few years, a variety of metal phosphates with open architectures have been synthesized and characterized. Some of them exhibit interesting magnetic properties. These include one-, two- and three-dimensional structures. An important aspect of these materials concerns their mode of formation.[1] Are the complex 3D structures formed by making use of secondary structural building units?[2] Is there a progressive building up of structures from 1D to 3D? An attempt to answer such questions is presented in the papers of this section. Interestingly, even simple metal carboxylates crystallized under appropriate conditions exhibit novel structures.[3] Certain metal oxalates incorporate alkali halides in different forms and such hybrid structures are described.

Besides phosphates and silicates, oxyanions such as sulfate and selenite can be used to design open architectures.[4] Inorganic–organic hybrid materials[5] can have different dimensionalities of inorganic and organic connectivities. There is much to be done on hybrid-network materials and there is considerable scope for materials design. Organically templated compounds with the Kagome structure possessing unusual magnetic properties are part of this large family. A perspective on Kagome compounds is presented in this section.

## References

1. C.N.R. Rao, in *Frontiers of Solid State Chemistry*, Eds. S.H. Feng and J.S. Chen, World Scientific, Singapore, 2002.
2. R. Murugavel, M.G. Walawalkar, H.W. Roesky and C.N.R. Rao, *Acc. Chem. Res.* **37**, 763 (2004).
3. C.N.R. Rao, S. Natarajan and R. Vaidhyanathan, *Angew. Chem. Int. Ed.* **43**, 1466 (2004).
4. C.N.R. Rao, J.N. Behera and M. Dan, *Chem. Soc. Rev.* **35**, 58 (2007).
5. A.K. Cheetham and C.N.R. Rao, *Science* **318**, 58 (2007).

Acc. Chem. Res. **2001**, *34*, 80–87

# Aufbau Principle of Complex Open-Framework Structures of Metal Phosphates with Different Dimensionalities

C. N. R. RAO,*,† SRINIVASAN NATARAJAN, AMITAVA CHOUDHURY,† S. NEERAJ, AND A. A. AYI

*Chemistry and Physics of Materials Unit and CSIR Centre of Excellence in Chemistry, Jawaharlal Nehru Centre for Advanced Scientific Research, Jakkur P.O., Bangalore, 560 064, India*

Received July 25, 2000

**ABSTRACT**

Open-framework metal phosphates occur as one-dimensional (1D) chains or ladders, two-dimensional (2D) layers, and complex three-dimensional (3D) structures. Zero-dimensional monomers have also been isolated recently. These materials are traditionally prepared by hydrothermal means, in the presence of organic amines, but the reactions of amine phosphates with metal ions provide a facile route for the synthesis, and also throw some light on the mode of formation of these fascinating architectures. Careful studies of the transformations of monophasic zinc phosphates of well-characterized structures show that the 1D structures transform to 2D and 3D structures, while the 2D structures transform to 3D structures. The zero-dimensional monomers transform to 1D, 2D, and 3D structures. There is reason to believe that the 0D monomers, comprising four-membered rings, are the most basic structural units of the open-framework phosphates and that after an optimal precursor state, such as the ladder structure, is formed, further building may occur spontaneously. Evidence for the occurrence of self-assembly in the formation of complex structures is provided by the presence of the structural features of the one-dimensional starting material in the final products. These observations constitute the beginning of our understanding of the building-up principle of such complex structures.

## 1. Introduction

While supramolecular chemistry of organic compounds has developed to maturity in the past few years,[1] supramolecular inorganic chemistry is somewhat at a nascent stage. Supramolecular design provides a means to generate a variety of novel inorganic materials with complex, unusual structural features in areas such as host–guest chemistry, open-framework structures, and the like. In these compounds, one can visualize structures of different dimensionalities, the dimensionality varying from zero to three. Such complex structures would be expected to have subunits, which not only are structural motifs but also act as building blocks in the building up of the complex structures. Such building units can be considered to be synthons of complex structures. The synthons would be simple geometrical figures such as squares, cubes, or polyhedra, their corners acting as linkage points. The challenge that one faces in supramolecular inorganic materials chemistry is to establish whether such synthons exist in reality and, if so, whether one can demonstrate how they are involved in the formation of the larger structures. Müller et al.[2] have made use of the concepts of supramolecular inorganic chemistry to provide a beautiful description of large assemblies based on the chemistry of polyoxometalates. Recently, Férey[3] has described the concept of building units to understand inorganic solid-state structures, to visualize new topologies, and to conceive the formation of new solids with novel designs. On the basis of such building units, Férey also defines the notion of scale chemistry, which is concerned with the edification of the solids with building units and the consequences it has on the structure of the framework and the voids in them. Of particular interest is the fact that not only are the structures of open-framework compounds topologically interesting, but also the cavities present in them have potential applications. When one looks at the complexity as well as the beauty of the myriad of structures of both open-framework and host–guest compounds, one cannot escape the feeling that the building-up process cannot occur by conventional chemical means alone, involving simple making and breaking of chemical bonds. The formation of such organized structures would be expected to involve self-assembly or some such spontaneous process at a certain stage. In this Account, we demonstrate how the formation of complex open-framework metal phosphates with open architectures involves a building-up process from simple building units, possibly leading to an ultimate step where spontaneous self-assembly occurs.

Several classes of inorganic open-framework structures have been synthesized and characterized in the past several years. While zeolitic aluminosilicates constitute the best-known class of open-framework structures,[4] metal phosphates have been gaining considerable importance, and a variety of metal phosphates with open architectures have been reported in the past decade.[5] The open-framework phosphates are generally synthesized under hydrothermal conditions in the presence of organic

C. N. R. Rao obtained his Ph.D. degree from Purdue University and a D.Sc. degree from the Mysore University. He is Linus Pauling Research Professor at the Centre and Honorary Professor at the Indian Institute of Science. He is a member of several academies, including the Royal Society, London, the U.S. National Academy of Sciences, and the Pontifical Academy of Sciences.

Srinivasan Natarajan obtained his M.Sc. degree from the Madurai Kamaraj University and his Ph.D. from the Indian Institute of Technology, Madras. He did postdoctoral work at the Royal Institution London and at the University of California, Santa Barbara, and is a Faculty Fellow at the Centre.

Amitava Choudhury obtained his M.Sc. degree from North Bengal University and is now a Ph.D. scholar at the Indian Institute of Science.

S. Neeraj obtained his M.S. degree at the Centre and has just completed his Ph.D. work.

A. A. Ayi has an M.Sc. degree from the University of Calabar, Nigeria, where he is a Lecturer. He is on a visiting fellowship of the Third World Academy of Sciences.

* To whom correspondence should be addressed. E-mail: cnrrao@jncasr.ac.in. Fax: 91-80-8462766.

† Also at the Solid State & Structural Chemistry Unit, Indian Institute of Science, Bangalore 560012, India.



10.1021/ar000135+ CCC: $20.00 © 2001 American Chemical Society
Published on Web 11/15/2000

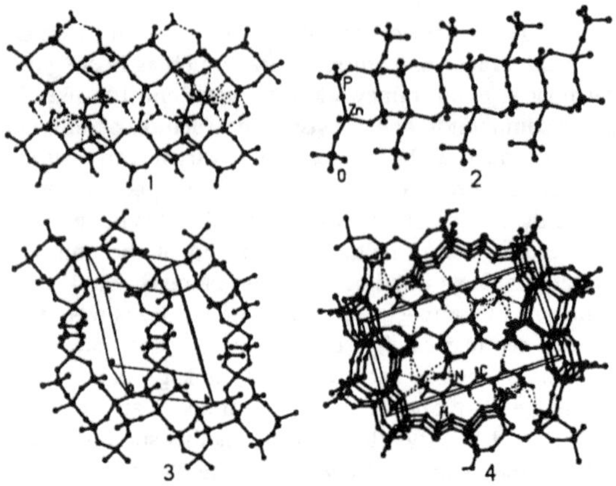

**FIGURE 1.** Examples of open-framework metal phosphates of different dimensionalities: **1**, one-dimensional (1D) linear chain phosphate, $[C_4N_2H_{10}][Zn(HPO_4)_2]\cdot H_2O$ with piperazine (PIP); **2**, one-dimensional (1D) ladder phosphate obtained with triethylenetetramine (TETA), $[C_6N_4H_{22}]_{0.5}[Zn(HPO_4)_2]$; **3**, two-dimensional (2D) layer phosphate $[C_6N_4H_{22}]_{0.5}[Zn_2(HPO_4)_3]$ with TETA; **4**, three-dimensional (3D) structure with 16-membered channels obtained with TETA. When counting the number of atoms forming a feature such as a ring or a channel, only metal and phosphorus atoms (T atoms) are taken into account.

amines. A noteworthy aspect of the metal phosphates is the occurrence of a hierarchy of structures with different dimensionalities. These include one-dimensional linear chains possessing corner-shared four-membered rings, one-dimensional ladders with edge-shared four-membered rings, two-dimensional layers, and three-dimensional structures with channels.[5,6] In Figure 1, we show typical examples of these structures (**1**−**4**) from the zinc phosphate family. Zero-dimensional metal phosphate monomers comprising four-membered rings have also been isolated recently.[7,8] Four-membered rings are generally the simplest units in the open-framework metal phosphates, and they seem to readily transform to six-membered, eight-membered, and higher membered rings.[9] Among the hierarchy of open-framework structures, the three-dimensional ones are most commonly observed, being associated with greater stability, and the lower dimensional architectures such as the linear chain and the ladder are somewhat rare. There has been considerable effort to understand the processes involved in the formation of open-framework structures.[10] Our knowledge of the mechanism(s) of formation of the beautiful architectures of metal phosphates with varying degrees of complexity, however, remains limited. It is difficult to readily obtain information related to the mechanism(s), partly because the materials are generally prepared under hydrothermal conditions. We know little about the nature of the species in solution or the exact role of the organic amine. The hydrothermal reaction vessel is the proverbial black box. The processes involved are kinetically controlled, and the energies associated with the different structures are likely to be comparable. Quite often, with a single organic amine, one obtains several open-framework phosphates with different structures. There have been some suggestions with regard to the role of the amine in the formation of these structures.[11] The amine could act as a structure-directing agent or merely fill the available voids and stabilize the structure through hydrogen-bonding and other interactions. Since the amine generally gets protonated in the reaction, it also helps in charge compensation with respect to the framework. Recently, it has been suggested that phosphates of the organic amines may act as intermediates in the formation of the metal phosphates.[12]

In situ synchrotron X-ray diffraction as well as NMR studies of gallium phosphates under hydrothermal conditions have revealed the spontaneous nature of the transformation of preformed precursor units into the open-framework structure, and also the initial formation of a four-membered ring phosphate.[13] It is possible to conceive of basic building units involved in the formation of open-framework metal phosphates, since a variety of complex three-dimensional structures with varying cell sizes are often produced by the arrangement of similar building units.[3] What is of vital importance, however, is to understand the relationship between the structures of different dimensionalities and complexity. For example, it is of value to explore whether it is possible to transform one-dimensional chain or ladder structures to two-dimensional layer or three-dimensional structures under certain conditions. If such transformations do occur, one can rationalize the formation of structures of different dimensionality in terms of a "building-up" (Aufbau) principle. In the case of aluminum phosphates, it has been proposed that a one-dimensional chain, on hydrolysis and rotation of bonds, may transform to higher dimensional structures.[6] In Figure 2, we show a few such transformations schematically, but there has not been sufficient experimental evidence in support of such a hypothesis.

We have been trying to understand the formation of the hierarchy of open-framework structures in metal phosphates by several methodologies. The amine phosphate route[12] has provided some insight into the mode of formation of these materials but does not adequately reveal the nature of interconvertibility of the structures. In this context, our recent finding that a zero-dimensional monomeric zinc phosphate transforms to a layered structure on heating[7] was encouraging and prompted us to explore the transformations of zero-, one-, and two-dimensional structures to higher dimensional ones under different conditions. In this Account, we shall briefly present some of the results obtained by the study of the reactions of organic amine phosphates with metal ions, to show how they throw some light on the formation of open-framework metal phosphates. We shall then discuss the results of our investigations of the transformations of well-characterized zinc phosphates of different dimensionalities and demonstrate how these studies provide valuable insight into the building-up (Aufbau) principle of these complex architectures. In particular, we discuss the transformation of the one-dimensional (1D) ladder structures to two-dimensional (2D) layer and three-dimensional (3D) structures with channels, and the trans-

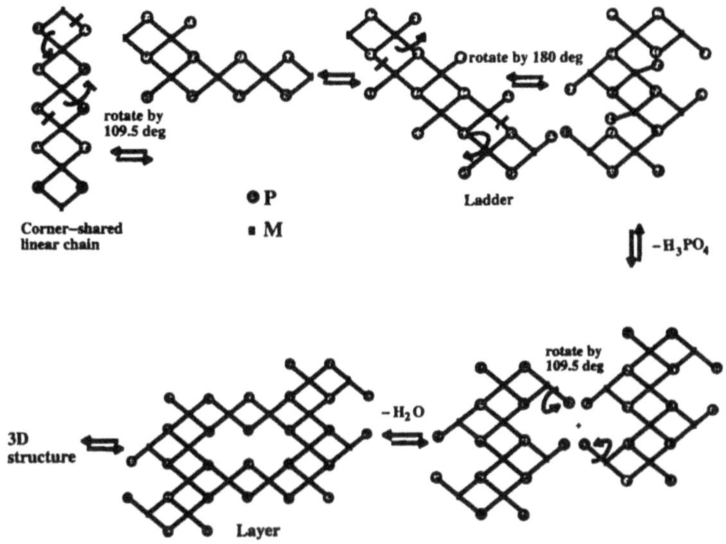


**FIGURE 2.** Schematic representation of possible types of transformations in open-framework phosphates.

formations of the 2D layer structures to 3D structures. We also examine the transformations of the zero-dimensional monomers to higher dimensional structures. We believe that these studies are of great significance not only to the area of open-framework structures, but also to our understanding of the nature of the processes involved in the building up of complex structures. Studies employing in situ methods and better facilities are necessary to make further progress. As pointed out by Müller et al.,[2] unraveling the Aufbau principle of complex inorganic systems and understanding such systems in terms of the topological classifications based on dimensionality constitute an important direction in the chemistry of materials.

## 2. Information Revealed by the Reactions of Organic Amine Phosphates

Phosphates of organic amines are often found as byproducts during the hydrothermal synthesis of open-framework metal phosphates. The role of the amine phosphates in the formation of metal phosphates was, however, not clear. Our recent studies of the reactions of well-characterized amine phosphates with metal ions have thrown some light in this regard.[12,14,15] Thus, amine phosphates react with $Zn^{2+}$ ions to yield open-framework metal phosphates of different dimensionalities. What is noteworthy is that many of these reactions can be carried out under ambient conditions, thereby avoiding the hydrothermal route. For example, piperazine phosphate (PIPP) on reaction with $Zn^{2+}$ ions gives the linear-chain phosphate $[C_4N_2H_{10}][Zn(HPO_4)_2]$ (**1**), formed by corner-shared four-membered rings (Figure 1) at temperatures as low as 85 °C. 1,3-Diaminopropane phosphate (DAPP) on reaction with $Zn^{2+}$ ions gives a ladder phosphate, $[C_3N_2H_{12}][Zn(HPO_4)_2]$ (**5**), comprising edge-shared four-membered rings (Figure 3) and on prolonged reaction yields a layered structure, $[C_3N_2H_{12}][Zn_2(HPO_4)_3]$ (**6**). These

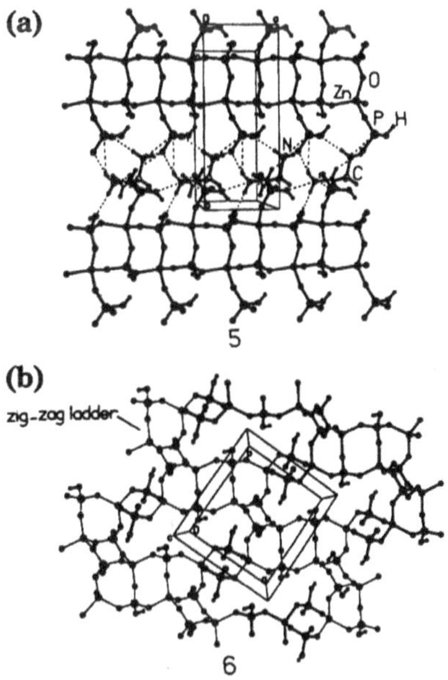


**FIGURE 3.** (a) One-dimensional ladder phosphate with 1,3-diaminopropane (DAP), $[C_3N_2H_{12}][Zn(HPO_4)_2]$ (**5**), and (b) the two-dimensional layer phosphate, $[C_3N_2H_{12}][Zn_2(HPO_4)_3]$ (**6**), obtained by the reaction of $Zn^{2+}$ ions with 1,3-diaminopropane phosphate (DAPP). We can see the features of a zigzag ladder in **6**.

reactions could be carried out in the 30–50 °C temperature range. The layers in **6** are formed from a zigzag chain of four-membered rings, constructed from two Zn and P atoms ($Zn_2P_2O_4$ units), that are connected to each other via two $PO_4$ units, creating a bifurcation within the layer.

Reaction of DAPP with $Zn^{2+}$ ions in aqueous solution at 30 °C for 24 h gives a product whose XRD pattern shows lines due to ladder structure **5** ($d_{011}$ = 9.76 Å), while the XRD pattern of the product obtained from the reaction

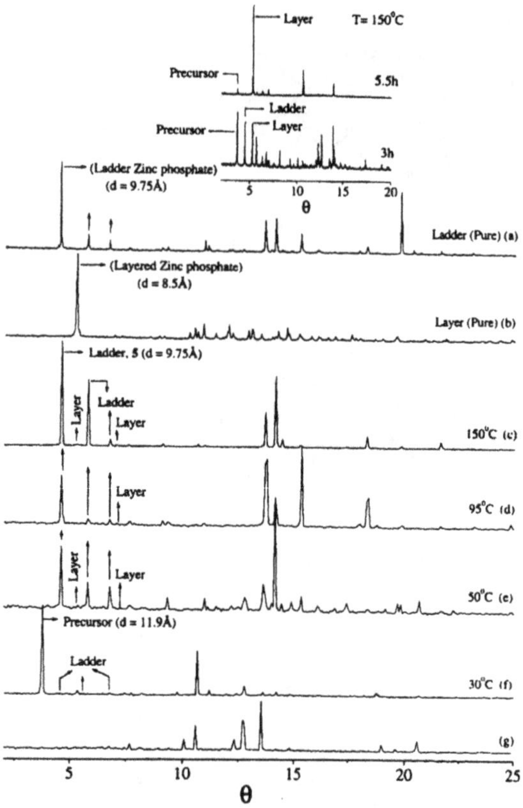


**FIGURE 4.** X-ray diffraction patterns of the products of the reaction of DAPP with $Zn^{2+}$ ions. (a) XRD pattern of the monophasic zinc phosphate with a ladder structure, and (b) XRD pattern of the monophasic layered zinc phosphate. The diffraction patterns (c)–(f) are those of the products obtained from the reaction of DAPP with $Zn^{2+}$ ions at different temperatures, as indicated (duration of reaction, ~24 h). Notice the presence of reflections due to the ladder and the layer structures in the patterns and the time evolution of phases. The XRD pattern (f), in addition, shows a unique reflection due to an unidentified precursor. The XRD pattern (g) is that of the amine phosphate (DAPP). The inset at the top of the figure shows the formation of the ladder and the layer structures and their time evolution at 150 °C.

at 50 °C (24 h) shows a reflection due to the ladder structure, **5**, as well as the layered structure, **6** ($d_{002}$ = 8.5 Å), as can be seen from Figure 4. Furthermore, immediately after mixing DAPP with $Zn^{2+}$ ions at room temperature, in addition to the weak line due to the ladder phase, one finds another reflection ($d$ = 11.9 Å) due to an uncharacterized precursor phase (Figure 4). This precursor phase subsequently transforms to the ladder, and finally to the layered structure (see Figure 4), suggesting that the transformation to the layer structure probably occurs through the ladder phase. These results are also supported by in situ $^{31}P$ NMR studies carried out at 85 °C, which showed the disappearance of the amine phosphate signal followed by the immediate appearance of a signal due to the precursor phase, before the ladder phase is formed. We have followed the time evolution of

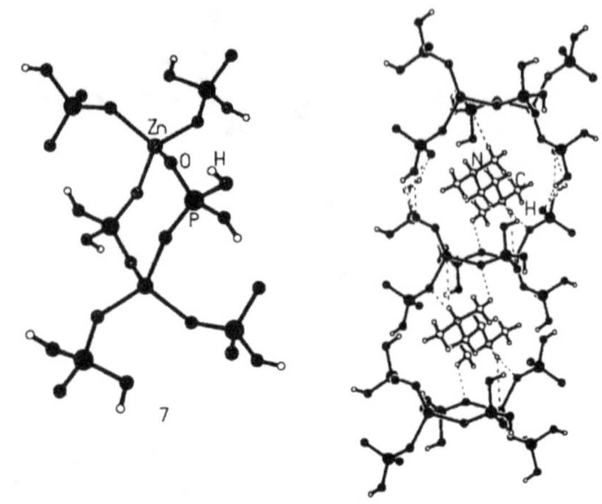


**FIGURE 5.** Monomeric zinc phosphate possessing a four-membered ring $[C_6N_2H_{18}][Zn(HPO_4)(H_2PO_4)_2]$ (**7**), formed by the reaction of *N,N,N',N'*-tetramethylethyelenediamine (TMED) with $Zn^{2+}$ ions in the presence of phosphoric acid.

the products of the reaction of DAPP with $Zn^{2+}$ ions by powder X-ray diffraction (XRD). The reaction at 150 °C initially gives a mixture of the ladder structure, **5**, and the layered structure, **6**. After a period of 3 h (see inset Figure 4), reflections due to the various phases can be seen. After 5.5 h, we mainly see reflections due to the layered phase, **6**, suggesting that the ladder, **5**, transforms to the layer, **6**.

We have been able to prepare a four-membered-ring zinc phosphate monomer, $[C_6N_2H_{18}][Zn(HPO_4)(H_2PO_4)_2]$ (**7**) by the reaction of *N,N,N',N'*-tetramethylethylenediamine phosphate (TMEDP) with $Zn^{2+}$ ions under ambient conditions (Figure 5). The structure consists of four-membered rings formed by $ZnO_4$ and $PO_2(OH)_2$ tetrahedra, with the $PO_3(OH)$ and $PO_2(OH)_2$ moieties hanging from the Zn center. Having discovered that the amine phosphates react with $Zn^{2+}$ ions in a facile manner, we have employed the amine phosphate route as an effective method for the synthesis of a variety of open-framework metal phosphates.[14–17] Thus, starting with PIPP, we have isolated a large number of open-framework structures with various metal ions.[14]

## 3. Transformations of One-Dimensional Zinc Phosphates

As mentioned earlier, one-dimensional metal phosphates can possess either linear chain or ladder structures. The linear chain zinc phosphates do not have pendant phosphate groups, and we have found them to be quite unreactive, unlike the one-dimensional ladder structures with pendant phosphate groups (see Figures 1 and 3). We have carried out transformations of the ladder structures **2** and **5** under different conditions.[18] On heating the ladder structure, $[C_6N_4H_{22}]_{0.5}[Zn(HPO_4)_2]$ (**2**), in water (**2**:$H_2O$ = 1:100) at 150 °C for 100 h, we obtained the 3D structure $[C_6N_4H_{22}]_{0.5}[Zn_3(PO_4)_2(HPO_4)]$ (**4**). The structure of **4** is built up of $ZnO_4$ and $PO_4$ tetrahedra sharing vertexes, forming a 3D structure (Figure 1), the connectivity between the $ZnO_4$ and $PO_4$ units giving rise to 16-membered 1D

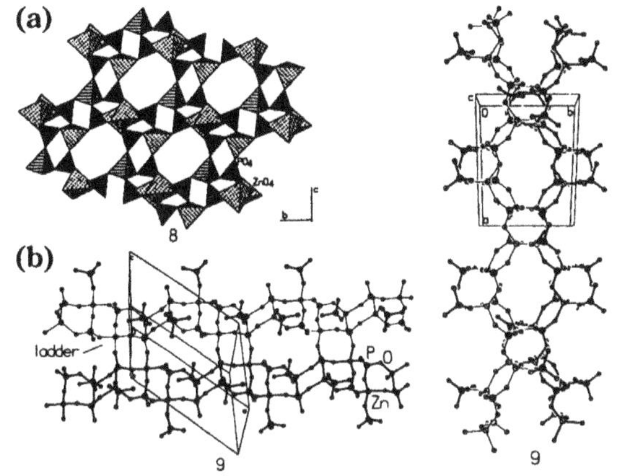


**FIGURE 6.** 3D zinc phosphates with eight-membered rings obtained from the transformation of the ladder compound **2**: (a) $[C_6N_4H_{22}]_{0.5}$-$[Zn_2(PO_4)_2]$ (**8**) and (b) $[C_2N_2H_{10}][Zn_2(PO_4)_2]$ (**9**). The features of the ladder structure can be clearly seen in the 3D structure of **9**. Such features are also present in **4** and **8**.

channels along the *bc* plane. However, by heating **2** in the presence of a base such as piperazine (PIP) or triethylenetetramine (TETA) at 150 °C in water (**2**:amine:$H_2O$ = 1:5:100) for relatively short periods (24 h), we obtained another 3D structure, $[C_6N_4H_{22}]_{0.5}[Zn_2(PO_4)_2]$ (**8**) (Figure 6a). The structure of **8** is also built up of $ZnO_4$ and $PO_4$ tetrahedra, but the connectivity between these units creates eight-membered channels along all the crystallographic directions. The same reaction carried out at a slightly higher temperature (165 °C) gave a new 3D structure, $[C_2N_2H_{10}]$ $[Zn_2(PO_4)_2]$ (**9**) (Figure 6b). The connectivity between $ZnO_4$ and $PO_4$ tetrahedra in **9** results in four-membered rings, which are connected to each other via oxygens. The connectivity between the four-membered rings is such that they form an edge-shared four-ring ladder. The ladders are then connected together within and out of the plane, forming eight-membered channels along the *a* axis. Along the *c* axis, the connectivity between the tetrahedra gives rise to another eight-membered channel. While forming **9**, the amine (TETA) decomposed to ethylenediamine. The 3D structure **4** with 16-membered rings appears to be the more stable phase, since we could transform **8** into **4** by heating it in water at 150 °C under acidic conditions. It is noteworthy that all the three 3D structures, **4**, **8**, and **9**, obtained by the transformation of **2** contain ladder-like features of the starting material.

By heating the ladder structure **2** in the presence of PIP at 165 °C, we were able to obtain a two-dimensional layer compound, $[C_4N_2H_{10}][Zn_2(PO_4)_2]$ (**10**), shown in Figure 7a. The ladder phosphate, **5**, obtained with DAP (Figure 3a) could be transformed to the layered structure, **6** (Figure 3b), by heating it in water (**5**:$H_2O$ = 1:100) at different temperatures (50–150 °C). We have followed this transformation by recording the X-ray powder diffraction patterns at different times. By heating **5** with zinc acetate in water at 85 °C (**5**:ZnAce:$H_2O$ = 1:5:100), we obtained the layered phosphate, $[C_3N_2H_{12}][Zn_4(PO_4)_2(HPO_4)_2]$ (**11**), shown in Figure 7b. Both of the layer structures, **10** and

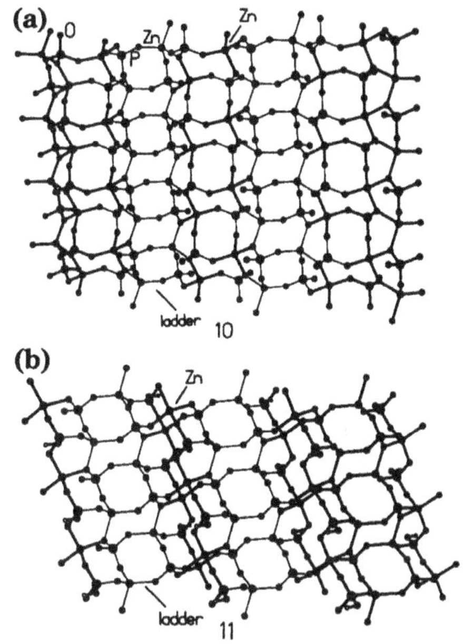


**FIGURE 7.** (a) 2D layer phosphate, $[C_4N_2H_{10}]$ $[Zn_2(PO_4)_2]$ (**10**), obtained from the transformation of the ladder compound **2**. (b) A 2D layer compound, $[C_3N_2H_{12}][Zn_4(PO_4)_2(HPO_4)_2]$ (**11**), obtained from the transformation of the ladder compound **5**. Both **10** and **11** exhibit features of the ladder structure, from which they are formed.

**11**, contain ladder-like features, as can be seen from Figure 7. The layers in **11** can be considered to be formed from the ladders in **2**, wherein the pendant groups connect the ladders, mediated by zinc ions, in an "*in*- and *out-of-plane*" fashion to result in chains with alternating three- and four-membered rings running in opposite directions. The connectivity between the chains and the parent ladder results in the formation of a tubular structure with six-membered rings capped by two or three three-membered rings on top and bottom and by two four-membered rings along the direction of the tubule. Thus, we have been able to transform the ladder structures **2** and **5** to 2D layers and 3D structures, all of which possess the structural features of the ladder, indicative of the likely occurrence of self-assembly in the formation of the complex structures.

We attempted to transform linear chain zinc phosphates of type **1** by carrying out reactions under different conditions but were generally unsuccessful. The only case of transformation of a linear chain phosphate was found with $[C_{10}N_4H_{26}][Zn(HPO_4)_2]$ (**12**), prepared with 1,4-bis(3-aminopropyl)piperazine, which on heating in the presence of phosphoric acid at 150 °C (**12**:$H_3PO_4$:$H_2O$ = 1:5:100) yielded $[C_{10}N_4H_{26}]$ $[Zn_3(PO_4)_2(HPO_4)]$ (**13**) with tubular layers (Figure 8). The tubules themselves are made of "strips" (formed by fusion of two linear chains via a three-coordinated oxygen) in which the three-membered rings get capped by Zn atoms alternately on top and bottom. We have also been able to transform the ladder phosphate **2** to the linear chain phosphate **1** (Figure 1a) by heating it with PIP in water. It appears that the transformation of

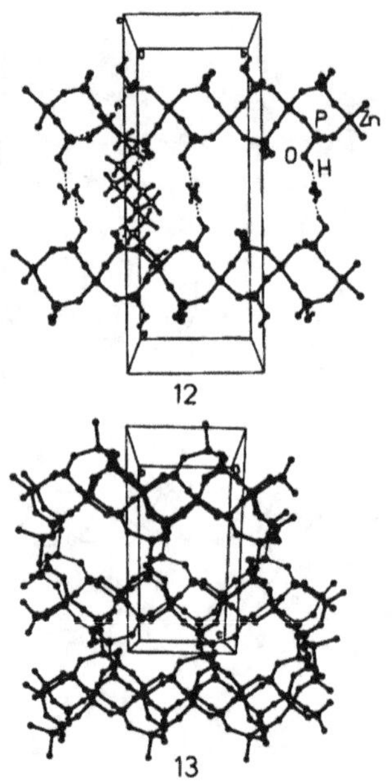


**FIGURE 8.** Linear chain phosphate, $[C_{10}N_4H_{26}][Zn(HPO_4)_2]$ (**12**), which transforms to a tubular layered phosphate, $[C_{10}N_4H_{26}]$ $[Zn_3(PO_4)_2(HPO_4)]$ (**13**). The features of the linear chain can be delineated in the tubular structure of **13**.

the ladder to other structures involves the deprotonation of the pendant $HPO_4$ groups when heated with an organic amine and/or water. In such a reaction, the ladder can go to a higher dimensional or to a one-dimensional linear chain structure where the pendant $HPO_4$ groups are absent.

## 4. Transformations of Two-Dimensional Zinc Phosphates

We have carried out the reactions of 2D layered zinc phosphates to see whether they transform to 3D structures. Thus, the layer structure $[C_6N_4H_{22}]_{0.5}[Zn_2(HPO_4)_3]$ (**3**), on heating in water at 150 °C (**3**:$H_2O$ = 1:200), gave the 3D structure **4** with 16-membered channels. It must be recalled that we could obtain this 3D structure from the ladder structure, **2**, as well. Heating the tubular layer phosphate obtained with TETA, $[C_6N_4H_{22}]_{0.5}[Zn_3(PO_4)_2(HPO_4)]$ (**14**), at 150 °C in water (**14**:$H_2O$ = 1:100), produced the 3D structure, **8**.

## 5. Transformations of Zero-Dimensional Monomers

The zero-dimensional monomeric zinc phosphate, **7**, in Figure 5, transforms to the layered compound, $[C_6N_2H_{18}][Zn_3(H_2O)_4(HPO_4)_4]$ (**15**), shown in Figure 9, on heating in water (**7**:$H_2O$ = 1:100) at 50 °C. On heating **7** with piperazine in water at 60 °C (**7**:PIP:$H_2O$ = 1:1:500), we obtained a 3D structure $[C_4N_2H_{10}][Zn(H_2O)Zn(HPO_4)(PO_4)]_2$, which has been described in the literature.[19] An in situ $^{31}P$ NMR study at 85 °C showed that the intensity of the signal due to the monomer decreases with time, accompanied by an

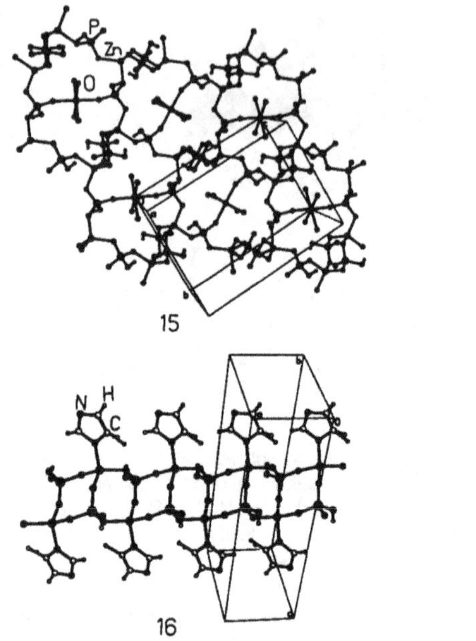


**FIGURE 9.** Layer phosphate, **15**, and ladder phosphate, **16**, obtained from the transformation of the zero-dimensional monomer, **7**.

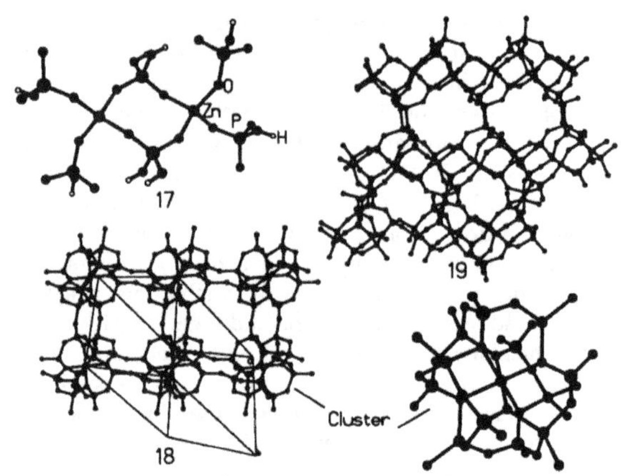


**FIGURE 10.** Monomeric zinc phosphate, **17**, which transforms to the 3D structure, **18**, and the tubular layer structure, **19**.

increase in the signal due to the ladder structure. The monomer structure disappears after ~4 h, and the intensity of the signal due to the ladder starts decreasing thereafter due to subsequent transformation. Heating the monomer **7** with imidazole in water (**7**:imidazole:$H_2O$ = 1:1:500) at 60 °C gave a ladder compound of the formula $[C_3N_2H_5][Zn(HPO_4)]$, **16**, where the pendant $HPO_4$ groups are replaced by imidazole molecules (Figure 9).

The monomeric zinc phosphate obtained with tris(2-aminoethyl)amine (TREN), $[C_6N_4H_{21}][Zn(HPO_4)_2(H_2PO_4)]$ (**17**), on heating in water (**17**:$H_2O$ = 1:100) at 180 °C, transformed to a 3D structure, $[C_6N_4H_{21}]_{1.33}[Zn_7(PO_4)_6]$ (**18**), shown in Figure 10. The structure of **18** is made of $ZnO_6$, $ZnO_4$, and $PO_4$ polyhedra, which connect in such a way as to give rise to a $Zn_7$ cluster. The clusters are arranged

**Scheme 1. Various Types of Transformations in Open-Framework Zinc Phosphates: (a) Transformations of One-Dimensional Ladder and 2D Layer Compounds and (b) Transformations of 0D Monomer[a]**

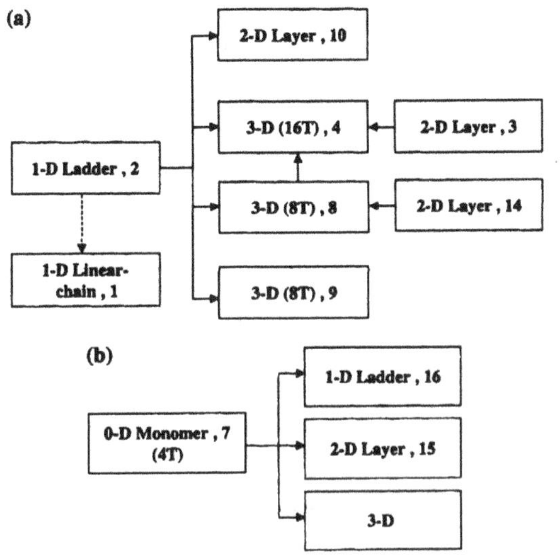


[a] T refers to tetrahedral framework atoms (Zn or P in the present case).

in such a manner that each cluster is displaced by half the length of *c*-axis from its neighbor, forming a honeycomb network. The next layer of clusters is identical to the first one but is displaced along the *a*-axis by half the unit cell, so that the honeycomb channels get capped. This type of three-dimensional AB-type stacking results in the formation of eight-membered channels along the *b*-axis between two such layers. On reaction of the monomer **17** with zinc acetate in water (**17**:ZnAce:$H_2O$ = 1:1:100), a tubular layer structure of the composition $[C_6N_4H_{21}][NH_4]$-$[Zn_6(PO_4)_4(HPO_4)_2]\cdot H_2O$ (**19**) was obtained (Figure 10).

## 6. Conclusions and Outlook

The transformations of lower dimensional structures to higher dimensional ones in the open-framework zinc phosphates are truly fascinating. In Scheme 1, we summarize the important observations, to highlight the changes in dimensionality and the interconvertibility of structures. Among the zinc phosphates investigated, the one-dimensional ladder structure is the most reactive, rather than the linear chain phosphate. The exact nature of the reactive low-dimensional structure may vary from one system to another, as exemplified by the linear chain gallium fluorophosphate described recently.[20] This one-dimensional gallium fluorophosphate, formed at room temperature, transforms to a 3D structure under hydrothermal conditions. This is comparable to the transformations of the ladder structures **2** and **5**, described in Figures 2 and 6 as well as Scheme 1. The transformations of the monomeric zero-dimensional zinc phosphates comprising four-membered rings to one-, two-, and three-dimensional structures are especially noteworthy, since the four-membered ring is the common structural unit in open-framework metal phosphates. It is possible that the four-membered ring is related to the building units described by Férey.[3] Thus, the hexameric unit of Férey has two four-membered rings.

Since the four-membered ring appears to be the first unit formed in the process of building of these complex open-framework structures,[13] it is possible that the four-membered rings initially form a one-dimensional chain or a ladder structure, which then transforms to 2D and 3D structures. In principle, we could consider the four-membered rings or/and the one-dimensional structures as synthons of the more complex structures. Self-assembly of ladder structures is suggested by the present study. The spontaneous assembly observed in in situ X-ray diffraction studies[13] could also correspond to the transformation of a 1D or 2D structure to a 3D structure by self-assembly. The formation of six-membered, eight-membered, and higher membered rings, commonly present in the open-framework phosphates, may follow that of the 0D/1D structures, the four-membered rings themselves transforming to the higher rings as suggested in the literature.[9] Although we are not yet in position to provide a definitive mechanism for the formation of complex 3D structures, it seems reasonable to state that the 3D and the 2D structures are likely to be formed through the 0D and 1D precursors. Clearly, there is a need for detailed in situ X-ray diffraction and NMR studies for fully characterizing the precursor states and the transformations. It would be of value to establish the occurrence of progressive 0D–1D–2D–3D transformations. It is therefore necessary to pursue the study of the transformations of well-characterized monophasic compounds of lower (zero,one, and two) dimensionalities.

In our studies of the open-framework zinc oxalates, we have recently isolated monomeric, dimeric, 1D linear chain, 2D layer, and 3D structures by the reaction of amine oxalates with $Zn^{2+}$ ions,[21] suggesting thereby that the presence of a hierarchy of structures is not unique to the phosphates alone. We believe that the evidence provided by our studies for the existence of an Aufbau principle of open-framework complex structures is of considerable significance. Many other complex inorganic structures are also likely to be formed by similar building-up processes, involving basic building units and self-assembly.

## References

(1) *Supramolecular Chemistry, Concepts and Perspectives* ; Lehn, J.-M., Ed.; VCH: Weinheim, 1995.
(2) Müller, A.; Reuter, H.; Dillinger, S. Supramolecular inorganic chemistry: small guests in small and large hosts. *Angew. Chem., Int. Ed. Engl.* **1995**, *34*, 2328.
(3) Férey, G. Building units, design and scale chemistry. *J. Solid State Chem.* **2000**, *152*, 37.
(4) *Hydrothermal Chemistry of Zeolites* ; Barrer, R. M., Ed.; Academic Press: London, 1982. *An Introduction to molecular sieves* ; Dyer, A., Ed.; Wiley: Chichester, 1988.
(5) Cheetham, A. K.; Ferey, G.; Loiseau, T. Open-framework inorganic materials. *Angew. Chem., Int. Ed.* **1999**, *38*, 3268.
(6) Oliver, S.; Kuperman, A.; Ozin, G. A. A new model for aluminophosphate formation: Transformation of a linear chain aluminophosphate to chain, layer and framework structures. *Angew. Chem., Int. Ed.* **1998**, *37*, 46.
(7) Neeraj, S.; Natarajan, S.; Rao, C. N. R. Isolation of a zinc phosphate primary building unit $[C_6N_2H_{18}]^{2+}[Zn(HPO_4)(H_2PO_4)_2]^{2-}$ and its transformation to open-framework phosphate $[C_6N_2H_{18}]^{2+}[Zn_3(H_2O)_4(HPO_4)_4]^{2-}$. *J. Solid State Chem.* **2000**, *150*, 417.

(8) Ayyappan, S.; Cheetham, A. K.; Natarajan. S.; Rao, C. N. R. A novel monomeric Tin(II) Phosphate: $[N(C_2H_5NH_3)_3]^{3+}[Sn(PO_4)(HPO_4)]^{3-}\cdot 4H_2O$, connected through hydrogen bonding. *J. Solid State Chem.* **1998**, *139*, 207.
(9) Chidambaram, D.; Neeraj, S.; Natarajan, S.; Rao, C. N. R. Open-framework zinc phosphates synthesized in the presence of structure-directing organic amines. *J. Solid State Chem.* **1999**, *147*, 154. Ayyappan, S.; Bu, X.; Cheetham, A. K.; Natarajan, S.; Rao, C. N. R. A simple ladder-tin phosphate and its layered relative. *Chem. Commun.* **1998**, 218.
(10) Thomas, J. M. New Microcrystalline Catalysts. *Philos. Trans. R. Soc. London A*. **1990**, *333*, 173. Davis, M. E. The quest for extra-large pore, crystalline molecular sieves *Chem. Eur. J.* **1997**, *3*, 1745. Francis, R. J.; O'Hare, D. The kinetics and mechanisms of the crystallization of microporous materials. *J. Chem. Soc., Dalton Trans.* **1998**, 3133.
(11) Davis, M. E.; Lobo, R. F. Zeolite and Molecular sieve synthesis. *Chem. Mater.* **1992**, *4*, 756. Feey, G. The new microporous compounds and their design. *C. R. Acad. Ser. Paris Ser. II* **1998**, 1.
(12) Neeraj, S.; Natarajan, S.; Rao, C. N. R. Amine phosphates as intermediates in the formation of open-framework structures. *Angew. Chem., Int. Ed.* **1999**, *38*, 3480.
(13) Francis, R. J.; O'Brien, S.; Fogg, A. M.; Halasyamani, P. S.; O'Hare, D.; Loiseau, T.; Férey, G. Time-resolved in-situ energy and angular dispersive X-ray diffraction studies of the formation of the microporous gallophosphates ULM-5 under hydrothermal conditions. *J. Am. Chem. Soc.* **1999**, *121*, 1002. Taulelle, F.; Haoqs, M.; Gerardin, C.; Estournes, C.; Loiseau, T.; Férey G. NMR of microporous compounds: From in-situ reactions to solid paving. *Colloids Interfaces* **1999**, *158*, 229.
(14) Rao, C. N. R.; Natarajan, S.; Neeraj, S. Exploration of a simple universal route to the myriad of open-framework metal phosphates. *J. Am. Chem. Soc.* **2000**, *122*, 2810.
(15) Rao, C. N. R.; Natarajan, S.; Neeraj, S. Building open-framework metal phosphates from amine phosphates and a monomeric four-membered ring phosphate. *J. Solid State Chem.* **2000**, *152*, 302.
(16) Natarajan, S.; Neeraj, S.; Choudhury, A.; Rao, C. N. R. Three-dimensional open-framework cobalt(II) phosphates by novel routes. *Inorg. Chem.* **2000**, *39*, 1426.
(17) Choudhury, A.; Natarajan, S.; Rao, C. N. R. A layered aluminum phosphate by the amine phosphate by the amine phosphate route. *Int. J. Inorg. Mater.* **2000**, *2*, 87. Natarajan, S.; Neeraj, S.; Rao, C. N. R. Three-dimensional open-framework Co $^{II}$ and Zn$^{II}$ phosphates synthesized via the amine phosphate route. *Solid State Sci.* **2000**, *2*, 87.
(18) The general procedure employed in these studies was to isolate a specific low-dimensional structure (monomer, ladder, or layer), solve its structure by single-crystal X-ray crystallography, and then study its transformation in an appropriate medium at a desired temperature. The medium could be just water or water with additives such as an amine, phosphoric acid, or zinc acetate. After the reaction, the product was analyzed by single-crystal X-ray crystallography.
(19) Feng, P.; Bu, X.; Stucky, G. D. Designed assemblies in open-framework materials synthesis: An interrupted sodalite and an expanded sodalite. *Angew. Chem., Int. Ed. Engl.* **1995**, *34*, 1745.
(20) Walton, R. I.; Millange, F.; Le Bail, A.; Loiseau, T.; Serre, C.; O'Hare, D.; Férey G. The room-temperature crystallization of a one-dimensional gallium fluorophosphate, a precursor to three-dimensional microporous gallium flurophosphates. *Chem. Commun*. **2000**, 203.
(21) Vaidhyanathan, R.; Natarajan, S.; Rao, C. N. R. Synthesis of a hierarchy of open-framework zinc oxalates from amine oxalates, communicated.

AR000135+

Acc. Chem. Res. **2004**, 37, 763–774

# Transformations of Molecules and Secondary Building Units to Materials: A Bottom-Up Approach

R. MURUGAVEL,*,† M. G. WALAWALKAR,‡ MEENAKSHI DAN,§ H. W. ROESKY,*,⊥ AND C. N. R. RAO*,§

*Department of Chemistry, IIT-Bombay, Powai, Mumbai-400 076, India, Applied Chemistry Division, Mumbai University Institute of Chemical Technology, Matunga, Mumbai-400 019, India, Chemistry and Physics of Materials Unit, Jawaharlal Nehru Center of Advanced Scientific Research, Jakkur, Bangalore-560 064, India, Solid State and Structural Chemistry Unit, Indian Institute of Science, Bangalore-560 012, India, and Institute of Inorganic Chemistry, University of Göttingen, D-37077 Göttingen, Germany*

Received March 18, 2004

**ABSTRACT**
A variety of complex inorganic solids with open-framework and other fascinating architectures, involving silicate, phosphate, and other anions, have been synthesized under hydrothermal conditions. The past few years have also seen the successful synthesis and characterization of several molecular compounds that can act as precursors to form open-framework and other materials, some of them resembling secondary building units (SBUs). Transformations of rationally synthesized molecular compounds to materials constitute an important new direction in both structural inorganic chemistry and materials chemistry and enable possible pathways for the rational design of materials. In this article, we indicate the potential of such a bottom-up approach, by briefly examining the transformations of molecular silicates and phosphates. We discuss stable organosilanols and silicate secondary building units, phosphorous acids and phosphate secondary building units, di- and triesters of phosphoric acids, and molecular phosphate clusters and polymers. We also examine the transformations of metal dialkyl phosphates and molecular metal phosphates.

## 1. Introduction

In 1756, Cronstedt discovered a class of minerals that on heating produced steam and called them *zeolites* from Greek, meaning *boiling stone*.[1,2] This was followed by the observations of Damour who noticed in 1840 that zeolites could be reversibly dehydrated and of Grandjean (1909) who reported that dehydrated zeolites adsorb small molecules. Weigel and Steinhoff found that the zeolite *chabazite* adsorbs methanol, ethanol, and formic acid but not acetone, ether, and benzene. MacBain interpreted this observation in 1926 in terms of molecular size differentiation at about 5 Å due to dehydration of the zeolites without the loss of the space lattice. Milton produced the first man-made zeolite in 1949.[1,2] Since then, zeolite science has emerged as one of the important areas attracting the attention of chemists, engineers, materials scientists, and physicists. Metal phosphates, on the other hand, exist in nature as minerals or as constituents of living animals and play an important role in energy conversion cycles. From a structural viewpoint, metal phosphates are similar to silicates and for this reason, studies on metal phosphate and related phosphonate materials have received considerable attention.

The search for new zeolite-like structures was initially extended to aluminophosphate-based molecular sieves,[3] and these explorations produced a variety of exotic compounds with open-framework structures, which include besides metal phosphates,[4] carboxylates,[5] sulfates,[6] selenites, and selenates.[7] Growth of this area has been rapid in the past decade necessitating new editions of the *Atlas of Zeolites*.[2] It is noteworthy that metal–organic framework (MOF) solids have been included as zeolite types in the latest edition of the *Atlas of Zeolites*.

While numerous framework structures incorporating a variety of metals and nonmetals have been assembled under laboratory conditions and structurally characterized and their properties have been explored, the present synthetic methodologies available for their preparation (viz., solvothermal and hydrothermal synthesis) do not offer much control over the final products. Typically, a metal source, a ligand, and an organic or inorganic structure-directing template are heated together in a sealed vessel for several days and cooled slowly to crystallize the framework solid. Although some generalizations can be made with regard to these reactions and to predicting the role played by the structure-directing

* Corresponding author. Fax: +91-080-23622760. E-mail: cnrrao@jncasr.ac.in (C. N. R. Rao).
† IIT-Bombay.
‡ Mumbai University Institute of Chemical Technology.
§ Jawaharlal Nehru Center of Advanced Scientific Research and Indian Institute of Science.
⊥ University of Göttingen.

Ramaswamy Murugavel is an Associate Professor of Chemistry at the Indian Institute of Technology-Bombay and has recently received the Swarnajayanti Fellowship from Government of India for his work on synthetic inorganic chemistry applied to materials science and catalysis.

Mrinalini Walawalkar is a Reader at the University Institute of Chemical Technology, Mumbai. Previously she was a Scientific Officer at IIT-Bombay and an Alexander-von-Humboldt Fellow at Bochum and Göttingen Universities. Her current research interests are in the area of inorganic materials.

Meenakshi Dan is a student of the Integrated Ph.D. program at the Indian Institute of Science, Bangalore. She has a M.S. degree in Chemistry from the Indian Institute of Science and works on open-framework materials.

Herbert Roesky is a Professor and Director at the Institute of Inorganic Chemistry at the University of Göttingen. He is a recipient of ACS Award for Creative Work in Fluorine Chemistry, ACS Award for Inorganic Chemistry, and Wilkinson Prize for Creativity in Inorganic Chemistry. He is a foreign member of the Indian Academy of Sciences and the French Academy.

C. N. R. Rao is Linus Pauling Research Professor at the Jawaharlal Nehru Centre for Advanced Scientific Research and Honorary Professor at the Indian Institute of Science. He is a member of several academies including the Royal Society, London, the U. S. National Academy of Sciences, the French Academy, the Japan Academy, and the Pontifical Academy of Sciences. He has received several awards including the Hughes medal from the Royal Society, Einstein Gold medal of UNESCO, and Somiya award of IUMRS.

10.1021/ar040083e CCC: $27.50 
Published on Web 08/06/2004

**Chart 1. Schematic Representation of the Formation of Zeolites from $TO_4$ Units**

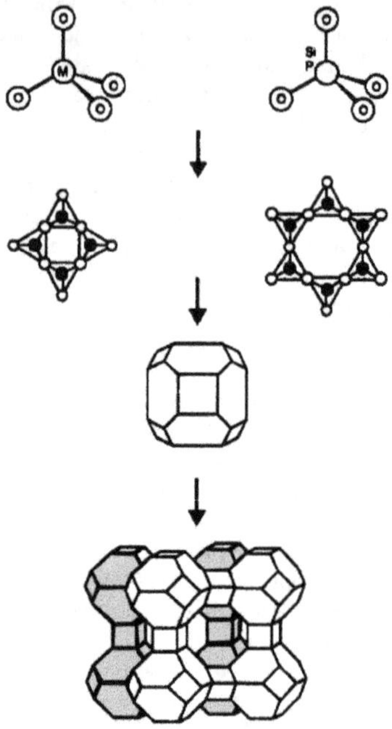

**Chart 2. Tetrahedral Primary Building Units**

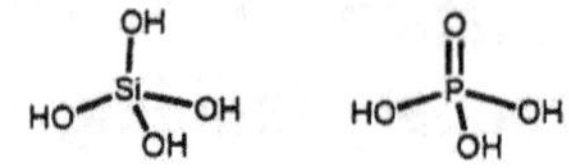

**Chart 3. Silanols and Phosphorous Acids Used as Building Blocks for the Synthesis of Model Compounds**

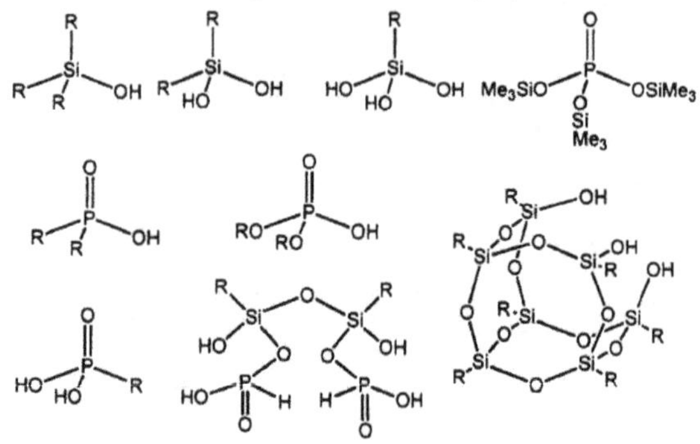

templates to some extent, it is not yet possible to choose a given set of reactants and reaction conditions and predict *before-hand* the major product that would form in preference to other possible products. In other words, although framework solids and synthetic zeolites are essentially formed by metal ions in association with anions such as $PO_4^{3-}$ and $SiO_4^{4-}$ (Chart 1), it is not yet possible to drive these reactions toward a particular framework solid.

The lack of suitable starting materials, combined with the fact that the reactions are generally carried out in a heterogeneous medium, explains the difficulties faced in designing framework solids. In light of recent developments in the molecular chemistry of siloxanes and phosphates, it is beginning to appear that routes to *designer framework solids* may indeed become possible soon. While there have been numerous literature reports of both hydrothermal synthesis of framework solids and laboratory preparation of model compounds often soluble in organic solvents,[3-8] there have been very few attempts to convert molecular compounds into framework structures through simple chemical transformations. The objective of this Account is to highlight some of the recent developments in the area of molecular compounds that resemble several of the secondary building units (SBUs) identified in extended structures[8] and demonstrate possible strategies and some successes wherein these small molecular units are transformed into framework solids or vice versa. In doing so, it is by no means possible to cover all aspects of this burgeoning area, and we shall therefore refer to pertinent reviews describing materials and structures and limit ourselves solely to the theme relating to molecule/SBU to materials transformations. We have restricted the discussion to silicates and phosphates due to the intense activity on these materials both at the molecular and extended framework levels. For the sake of completeness, a brief discussion on the conversion of the molecular compounds to condensed solids and inorganic–organic hybrid materials with potential use as solid-state catalysts is also included.

## 2. Stable Organosilanols and Silicate SBUs

While $Si(OH)_4$ and $P(O)(OH)_3$ could serve as ideal starting materials for the incorporation of $SiO_4^{4-}$ and $PO_4^{3-}$ anions (Chart 2), their use in synthesizing molecular siloxanes and phosphates is somewhat limited since their synthesis is conducted in a homogeneous organic medium. For example, orthosilicic acid is metastable and not available in a free state. Even if $Si(OH)_4$ could be made available, its insoluble nature precludes its chemistry from being explored in solution. Similar problems are encountered in using anhydrous $P(O)(OH)_3$. Synthetic chemists have sought to use alternative starting materials for the incorporation of Si–O and P–O linkages. This has led to the use of a variety of organosilanols,[9] esters of orthosilicic acid, phosphonic acids, and esters of orthophosphoric acid as starting materials (Chart 3).

Although the chemistry of metal-containing molecular silicates (*metallosiloxanes*) is more than a century old,[10] their use as model compounds for zeolites and as precursors for the preparation of metallosilicate materials has gained importance only in the past decade.[11] While the early work on metallosiloxanes provided insights into the nature of Si–O–M linkages, its direct application to zeolites was somewhat remote since monosilanols, $[R_3Si(OH)]$, produce metallosiloxanes, $[M(OSiR_3)_n]$, that barely resemble any of the SBUs of zeolites.[12] It is no wonder that these compounds have often been referred to as a special class of alkoxides rather than molecular silicates. During the 1970s and 1980s, the synthesis of cyclic metallosiloxanes was reported starting from silanediols of

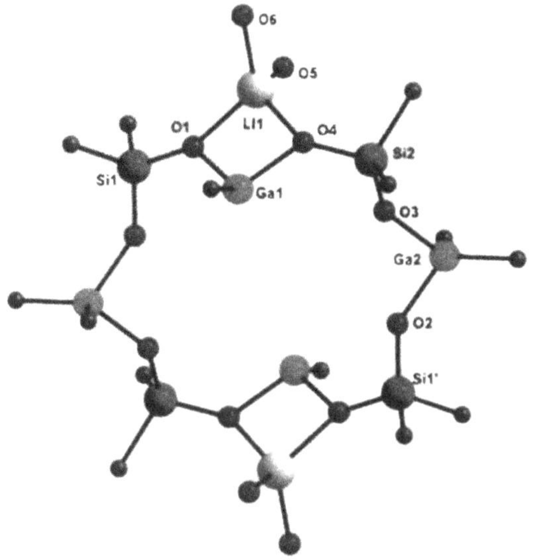


**FIGURE 1.** Molecular structure of the core of the lithium gallium siloxane showing the S8R.

the type $[R_2Si(OH)_2]$ and $[HOSiR_2{-}O{-}SiR_2OH]$.[13] We recently demonstrated that even a simple silanediol such as $Ph_2Si(OH)_2$ could be used for the preparation of the only known soluble compound with a single-8-ring (S8R) motif from its reaction with $LiGaMe_4$ (Figure 1). Interestingly, the cavity formed inside the 8-ring is large enough in this case to accommodate a $(LiOH)_2$ unit.[14]

Modern metallosiloxane chemistry *mainly* deals with three types of silanols: silanetriols,[11] incompletely condensed silasesquioxanes,[15] and tri(*tert*-butyl) ester of orthosilicic acid.[16] Research on these compounds has contributed to the development of model compounds for zeolites, new homogeneous catalysts, and low-temperature pathways to metal–silica materials.

**Silanetriols.** A major break-through in silanol research occurred in the 1990s with reports appearing on the high yield synthesis of silanetriols, $[RSi(OH)_3]$.[17] Silanetriols were traditionally considered to be highly unstable due to the well-known self-condensation reactions through which they form condensed siloxanes. However, the use of a sterically demanding and kinetically stabilizing substituent on silicon allowed the isolation of silanetriols in multigram quantities. For example, the silanetriol $[(2,6\text{-}{}^{i}Pr_2C_6H_3)(SiMe_3)NSi(OH)_3]$, is a unique material that is stable in air for at least 3 years![11] It is lipophilic and soluble in a variety of organic solvents. While the bulky 2,6-diisopropylphenyl group no doubt provides the kinetic impedance to Si–OH condensation, the role played by the $SiMe_3$ group in stabilizing this silanetriol is also crucial. This silanetriol has served as a versatile starting material for the synthesis of more than 100 metallosiloxanes, highlighting the use of these compounds as models for metal-embedded silica surfaces and single-source precursors for new catalytic materials. It is interesting that the reactions can often be fine-tuned by paying careful attention to the nature of the metal precursor, the reaction temperature, and the polarity of the medium. For example, the reactions of $RSi(OH)_3$ with organoaluminum reagents proceed at temperatures between −78 and 80 °C and produce cyclic and cage-soluble aluminosiloxanes, which model zeolites (Chart 4).[18] The structural varieties that can result from silanetriol and metal precursor reactions are shown in Chart 5.[19]

**Chart 4. Organic Soluble Molecular Silicates Derived from $RSi(OH)_3$**

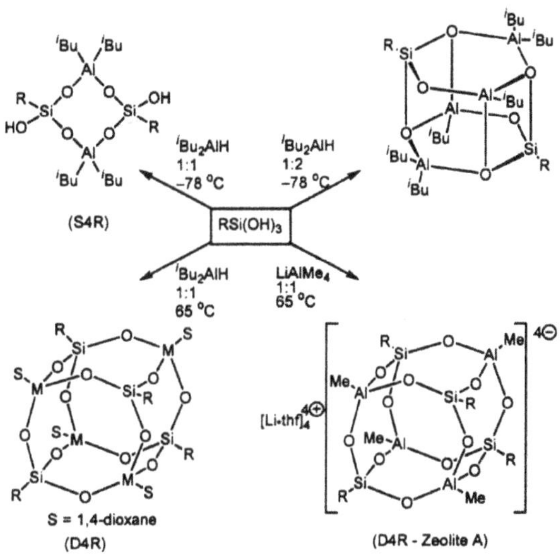


**Chart 5. Interesting Metallosiloxanes Derived from $RSi(OH)_3$**

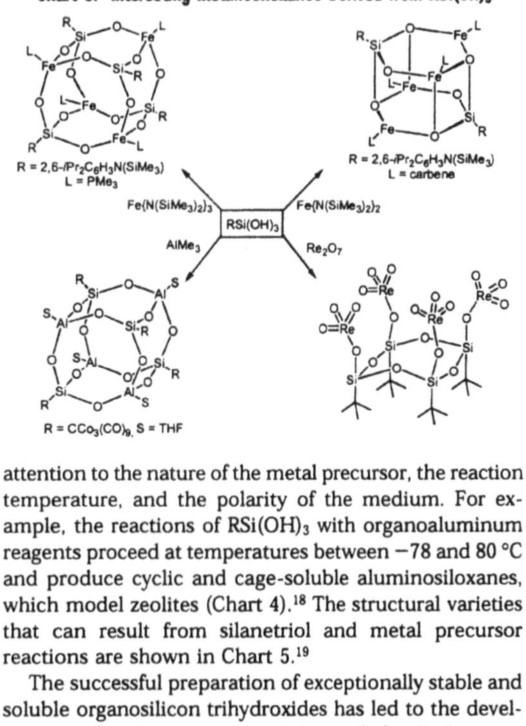


The successful preparation of exceptionally stable and soluble organosilicon trihydroxides has led to the development of exciting new chemistry, and clearly there are several opportunities for further work, especially in the synthesis of framework structures starting from the SBUs shown in Charts 4 and 5.

**Silsesquioxanes.** At the same time as silanetriols were stabilized, another significant observation was made in the hydrolysis of silyl trihalides. Incompletely condensed silsesquioxanes, $R_7Si_7O_9(OH)_3$ (Chart 6) are formed along with other products during the hydrolysis of $RSiCl_3$ (R =

**Chart 6. Silasesquioxanes Derived from $(cy)_7Si_7O_9(OH)_3$**

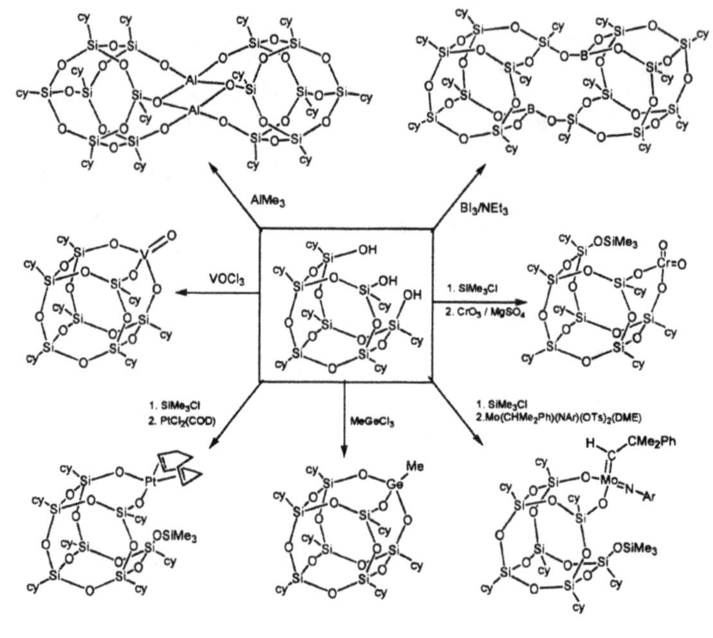


**Chart 7. Thermally Unstable Metal Siloxides Derived from Tri(*tert*-butoxy)silanol**

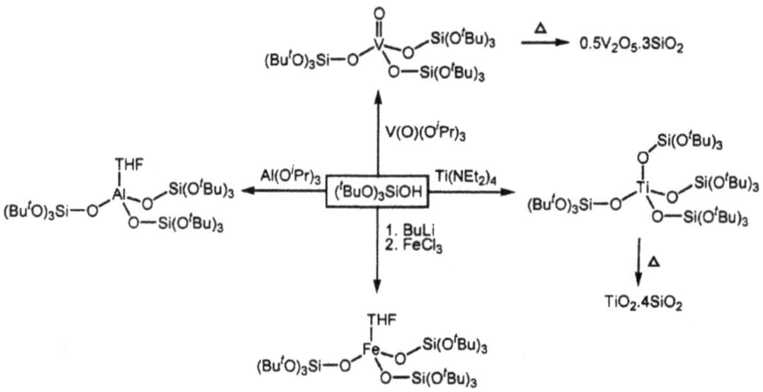


cyclohexyl) in commercial acetone.[20] Silsesquioxanes, $R_7Si_7O_9(OH)_3$, by the virtue of having three Si−OH hydroxyl groups pointing toward an open corner of a cube, are very important as model compounds for silica surfaces and also as starting materials for the preparation of metallasiloxanes through corner-capping reactions (Chart 6).[15] Some of these metallasilsesquioxanes are found to be useful as catalysts in organic transformations.[21] These aspects have been discussed elsewhere[15,21] and will not be elaborated here.

**Silicate and Hybrid Materials through Molecular Siloxanes.** Renewed interest in monosilanols and silanediols is primarily because compounds obtained from $[(tBuO)_3SiOH]$ and $[(tBuO)_2Si(OH)_2]$ serve as better single-source precursors for ceramic materials. A number of metallasiloxanes have been synthesized starting from $[(tBuO)_3SiOH]$, and a selection of these compounds is shown in Chart 7. Most of these metallosiloxanes serve as excellent precursors for the preparation of condensed-phase metallosilicate materials (or metal oxide/silica composites) with precise stoichiometries.[16,22,23] For example, direct thermolysis of the titanosiloxane $[Ti\{(OSi(OtBu)_3\}_4]$ results in the removal of organic material through $\beta$-hydride elimination and condensation of hydroxyl groups to result in $TiO_2$:$4SiO_2$. Using similar strategies, oxides of vanadium, chromium, molybdenum, tungsten, zinc, manganese, and several other metals have been stabilized in a silica matrix.[23] Aluminosilicates have been synthesized using this methodology starting from $[Al\{(OSi(OtBu)_3\}_3]$. Several of the metallosilicate materials prepared through this pathway are useful catalysts in organic transformations.[24]

An interesting variant of metal−tri(*tert*-butoxy)siloxide chemistry was described recently. Thus, the thermal

**Chart 8. Conversion of Zr[OSi(OtBu)$_3$]$_4$ into Ceramic and Inorganic/Organic Hybrid Materials**

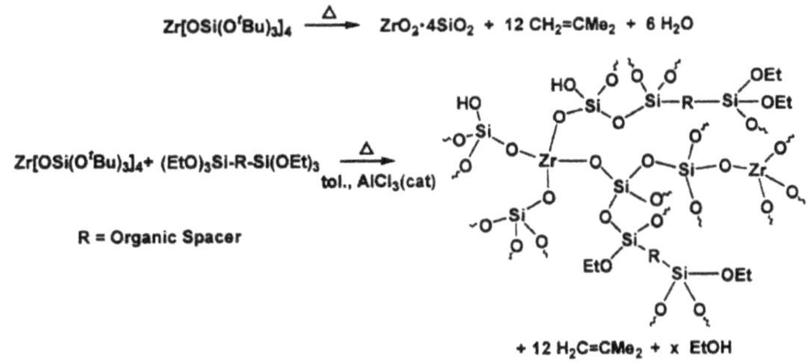


decomposition of Zr[OSi(O*t*Bu)$_3$]$_4$ occurs in toluene or in the solid state under mild conditions (150 °C) to give ($ZrO_2$·4$SiO_2$) with all of the component elements originating from a single molecular species (Chart 8). On the other hand, the cothermolytic synthesis of the hybrid network materials can be carried out in toluene at 155 °C in the presence of $AlCl_3$ catalyst.[25] It is postulated that the condensation of Zr[OSi(O*t*Bu)$_3$]$_4$ with (EtO)$_3$Si−R−Si(OEt)$_3$ proceeds through the initial thermal decompositon of Zr[OSi(O*t*Bu)$_3$]$_4$ to give Zr[OSi(OH)$_3$]$_4$. The reactive Si−OH groups then self-condense or condense with Si−OEt groups from (EtO)$_3$Si−R−Si(OEt)$_3$ to form an inorganic/organic network material via the concomitant elimination of ethanol. Gel formation occurs rapidly within ca. 10 min and affords high-surface-area xerogels upon drying. The as-synthesized materials are highly functionalized with surface OH groups up to 4.8 sites nm$^{-2}$. This method offers an opportunity to synthesize a wide variety of hybrid materials that combine transition metals with diverse organic spacers in an organic medium excluding water.

## 3. Phosphorus Acids and Phosphate SBUs

Unlike the silanols, alkyl phosphorus acids, such as phosphinic acids [$R_2P(O)(OH)$] and phosphonic acids [RP(O)(OH)$_2$] (Chart 2), are easier to prepare and handle under laboratory conditions.[26] Metal phosphates derived from these acids are similarly more stable than the corresponding metallasiloxanes. This has resulted in an outburst of activity in building molecular phosphinates and phosphonates that often resemble the smaller secondary building units of zeolites and other framework solids, apart from the synthesis of a large number of layered inorganic phosphonates. The chemistry of hundreds of layered metal phosphonates synthesized from phenylphosphonic acid, alkyl and aryl diphosphonic acids, and phosphonic acids containing additional carboxylic acid functional groups has been documented well in the literature in the last three decades.[26] Structures of these phosphonates do not resemble the SBUs of framework phosphates, nor are they soluble in organic solvents for further reaction chemistry. We will focus our attention here on soluble cyclic and cage-like organic phosphinates and phosphonate molecules, which resemble SBUs of framework solids.

**Chart 9. Silanetriolate Anion vs Phosphonate Anion**

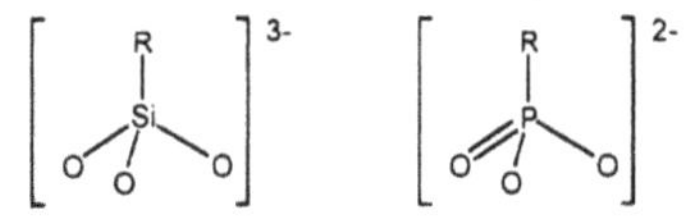

**Phosphonic Acids.** Phosphonic acids, RP(O)(OH)$_2$, when deprotonated afford $RPO_3^{2-}$ which, in principle, should behave similarly to the silanetriols, RSi(OH)$_3$ (Note that $SiO_2$ is isoelectronic to $AlPO_4$) (Chart 9).[26] For reasons of solubility, most of the efforts to generate SBUs have been carried out with *tert*-butylphosphonic acid, although reactions with phenyl- and methylphosphonic acid have also been examined. Thus, the reaction of *t*-BuP(O)(OH)$_2$ with any group 13 trialkyl leads to a three-dimensional cage with a double-4-ring (D4R) core (cf. Chart 4)[27] as in the case of reactions of silanetriols with triaalkylalumnium. While in compounds derived from the phosphonic acids one alkyl group is intact on each heteroatom, the fourth coordination site is a solvent molecule around the heteroelement in the analogous siloxane cubes.

There appears to be a steric control in determining whether a cubic cage (D4R) is always formed in these. A report by Mason suggests that in the reaction of bulky *t*-Bu$_3$Ga with phenylphosphonic acid, it is possible to obtain both an eight-membered ring compound (single-4-ring, S4R; Chart 10, structure **A**) and a cubic cage (D4R).[28] The reaction of Me$_3$Al with *tert*-butylphosphonic acid underscores such steric control.[29] In this reaction, apart from the cubic tetrameric aluminophosphonate **B** (D4R), a hexameric drum-like product with a double-6-ring (D6R) core (Chart 10, structure **C**) is obtained in considerable yields. While in the former case the bulkiness of the *tert*-butyl group on gallium is responsible for the formation of the S4R structure, in the latter the rather small size of the methyl group on aluminum is the main cause of the ready assembly of the D6R structure. The reactions of Cp*TiMe$_3$ with alkyl and aryl phosphonic acids offer easy access to novel oxygen bicapped distorted D4R structures (Chart 10, structure **D**).[30]

Ionic phosphonates, which contain loosely bound alkali metal ions, are of interest from the point of view of modeling zeolites containing dissolved metal ions displaying conductive hyperlattices. Reactions of *t*BuP(O)(OH)$_2$ with alkali metal salts of tetraalkylgallates provide a convenient route to ionic phosphates (Chart 11).[31] A

**Chart 10. Neutral Phosphonates Mimicking S4R, D4R, D6R, and Bicapped D4R SBUs**

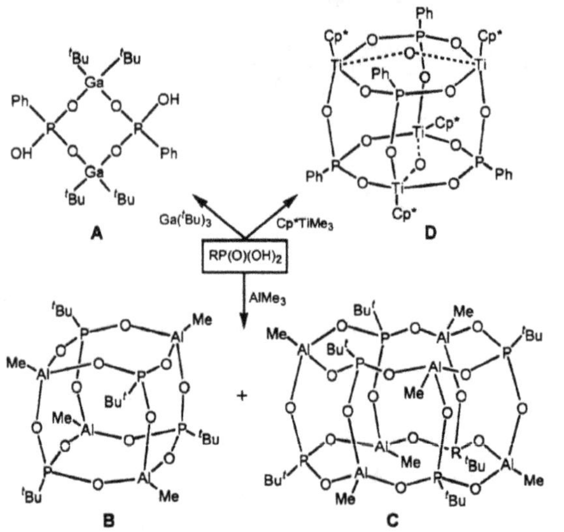

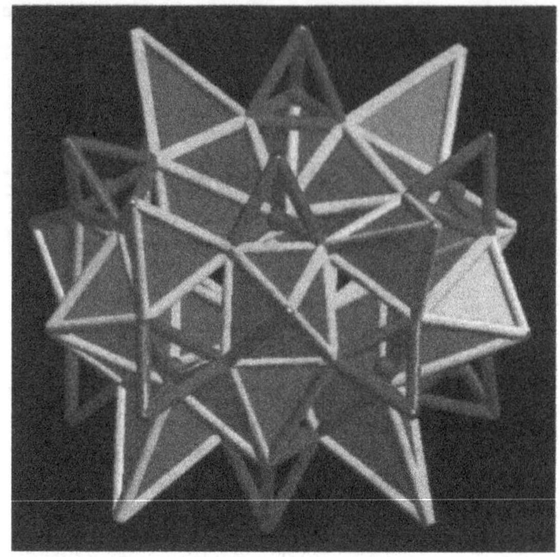

**FIGURE 2.** Polyhedral arrangement in $[(MeCd)_{10}((THF)Cd)_4Cd_6(\mu_4\text{-}O)_2(\mu_3\text{-}OH)_2(tBuPO_3)_{12}]$ showing cadmium (closed polyhedra with green interior and yellow edges) and phosphorus (open polyhedra in pink) atoms.

notable feature is that the size of the alkali metal ions regulates the final structure. Thus, the lithium compound is made of two 12-membered gallophosphonate bowls, which sandwich four lithium ions arranged in the form of a one-dimensional wire.[31a] The gallophosphonates derived from both $NaGaMe_4$ and $KGaMe_4$ are essentially made up of an eight-membered gallophosphonate bowl, which surrounds an ensemble of Na or K ions. Because of the larger size of Na and K ions, they can accommodate higher coordination numbers and form agglomerates rather than the linear wire as in the lithium case.[31b]

**Nanosized Metal Phosphonate Aggregates.** Several phosphonate clusters of a very large size have been synthesized in recent years. For example, the use of the organometallic precursor $Cs[Al\mathit{i}Bu_3F]$ has allowed the synthesis of a new supercluster $[Cs_3(THF)_3F(\mathit{i}BuAl)_3(tBuPO_3)_4]_2[(\mathit{i}BuAl)_2Al_2(\mu\text{-}F)_2(tBuPO_3)_4]$,[32] which contains several structural features closely related to the motifs present in layered and three-dimensional alumino- and gallophosphates (C6R or 6≡1 found in BPH and AFS types of layered aluminophosphates). The reaction of $Et_2Zn$ with $tBuP(O)(OH)_2$ leads to a rugby-ball shaped dodecanuclear Zn-cluster, $[Zn_4(\mu_4\text{-}O)(Zn{\cdot}THF)_2(ZnEt)_6(tBuPO_3)_8]$.[26] The reaction of $Me_2Cd$ with $tBuP(O)(OH)_2$ solution yields a still larger aggregate, $[(MeCd)_{10}((THF)Cd)_4Cd_6(\mu_4\text{-}O)_2(\mu_3\text{-}OH)_2(tBuPO_3)_{12}]$, which contains 68 heteroatoms arranged in the form of a hollow sphere (Figure 2).[33] This cluster contains a skeleton made up of 20 cadmium atoms and 12 phosphorus atoms held together by 36 surface oxygen atoms along with a pair of oxo and hydroxo groups in the interior of the sphere (Figure 2). The inorganic core is embedded in the organic sheath made up of 10 methyl groups, four THF, and 12 *tert*-butyl groups. Within the inorganic core, the diagonally opposite Cd–Cd and P–P distances are 10.9 and 9.8 Å, respectively. This giant molecule can encapsulate three $C_{60}$ molecules.

It is possible to use ancillary ligands in addition to phosphonic acids in building up nanosized cluster compounds of late transition metal ions. Thus, the reaction of $CuCl_2$ with *tert*-butylphosphonic acid in the presence of 3,5-dimethylpyrazole affords a dodecanuclear copper phosphonate with an interesting cage structure.[34] Similarly, large vanadium phosphonate clusters with up to 18 vanadium atoms have been assembled from phosphonic acids.[35]

## 4. Phosphoric Acid Esters as Primary Building Units

Much of the work on the use of phosphate esters for building up molecular metallaphosphates has emerged only in the past decade. While di(*tert*-butyl) phosphate (dtbp-H) has been the most widely used,[36,37] studies have also been carried out with the tris(trimethylsilyl) ester, $(Me_3SiO)_3P{=}O$.[38] Compared to phosphonic acid chemistry, the chemistry of phosphate esters assumes importance, because of the presence of the reactive/thermally labile P–OR linkages relative to the stable P–C linkages in the phosphonic acids. Although phosphonic acids provide model compounds that can act as SBUs of zeolite structures and nanoscopic clusters, expansion of these SBUs and clusters to framework solids would require further functionalization of the alkyl groups on phosphorus. Conversely, the generation of P–OH groups on the surface of the molecular cluster formed by phosphorus esters is a relatively easier task and is achieved either by $\beta$-hydride elimination of P–O*t*Bu groups or by the facile hydrolysis of $P{-}OSiMe_3$ groups in solution. Several metal phosphate molecules have been synthesized starting from these two

**Chart 11. Anionic Gallium Phosphonates**

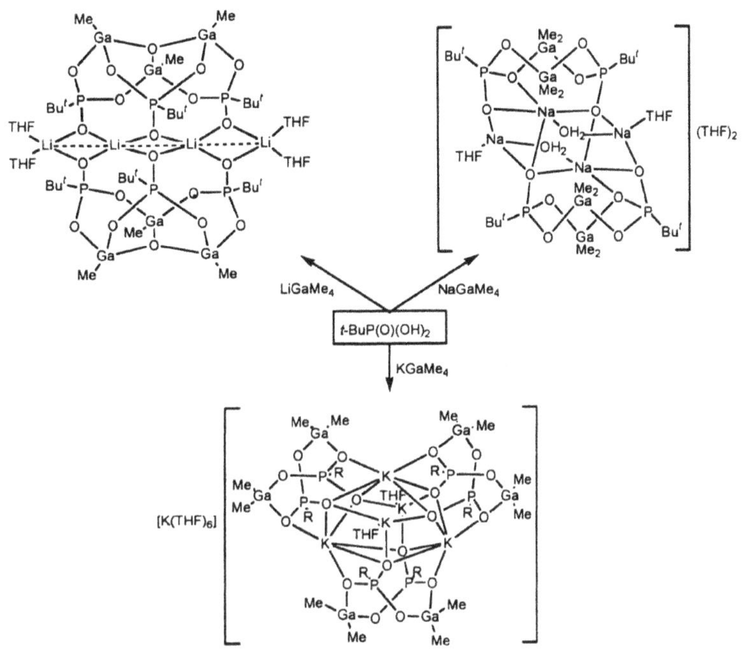


**Chart 12. Complexes of Di(*tert*-butyl) Phosphate with Aluminum**

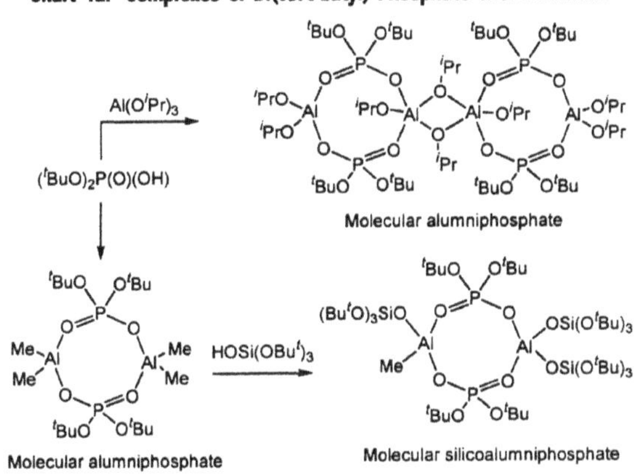


**Chart 13. Synthesis of Tetranuclear Phosphates**

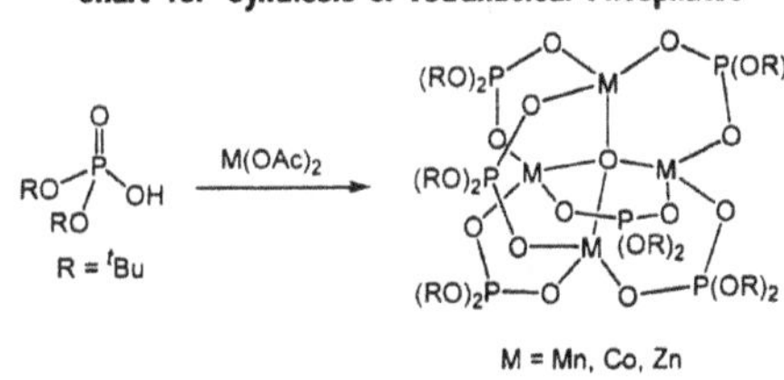


phosphate esters; we shall limit the discussion here to those that can act as precursors for ceramic phosphates or have the potential to act as starting materials for building up framework structures.

The reaction of 2 equiv of dtbp-H with $Al_2Me_6$ affords an eight-membered ring by the loss of 2 equiv of methane.[37c] On the other hand, the reaction with $[Al(O\textit{i}Pr)_3]_4$ yields a centrosymmetric tetramer with two eight-membered rings connected by bridging isopropoxides (Chart 12). The presence of reactive Al−Me, Al−O*i*Pr, and P−O*t*Bu linkages in these S4R products renders them useful as starting materials to build higher structures. For example, the cyclic aluminophosphate with Al−Me groups reacts readily with 3 equiv of $(\textit{t}BuO)_3SiOH$ to give a molecular silicoaluminophosphate (M-SAPO) (Chart 12), which on further thermal treatment or sol−gel processing is transformed to a SAPO-type material.[39]

## 5. Molecular Phosphate Clusters and Polymers

Tetranuclear clusters of divalent transition metal ions are generated from the direct reaction of a metal acetate with dtbp-H (Chart 13).[36a] Advantages of this reaction over the other reactions described in this article lie in the use of metal acetates instead of the more expensive and moisture-sensitive metal alkyls and the use of methanol as a solvent in aerobic conditions. It is to be noted, however, that organometallic routes can also be used to synthesize these tetrameric compounds.[37a] Small changes in the reaction conditions and the use of mild Lewis bases bring about changes in the structures. For example, polymeric phosphates are formed by the addition of THF or a mild base during the course of the reaction (Chart 14). These one-dimensional polymers are made up of M−O−P inorganic backbones covered with a hydrophobic sheath of *tert*-butyl groups.[36b] Thermal decomposition of the polymers in the 300−400 °C range produces the corresponding metal metaphosphate (Chart 15).[36b]

**Chart 14. Preparation of 1D Polymeric Phosphates**

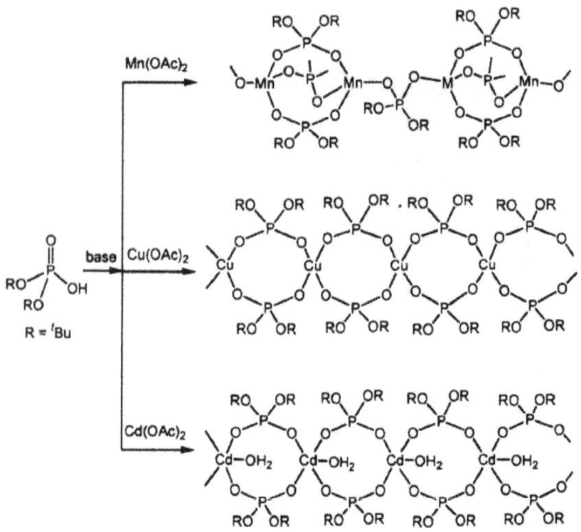


**Chart 15. Preparation of Phosphate Materials from Molecular Complexes**

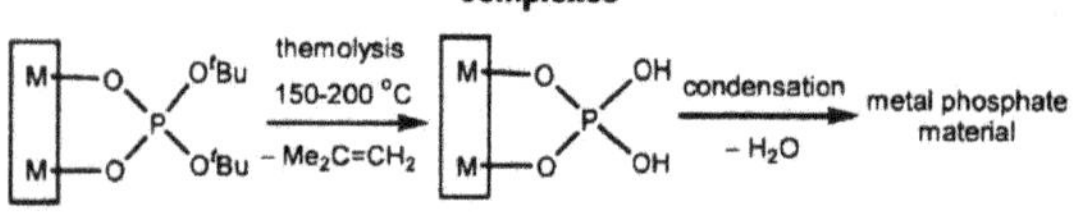


**Chart 16. Linear Polymer and Tetrameric Cage Phosphates from the Same Reaction**

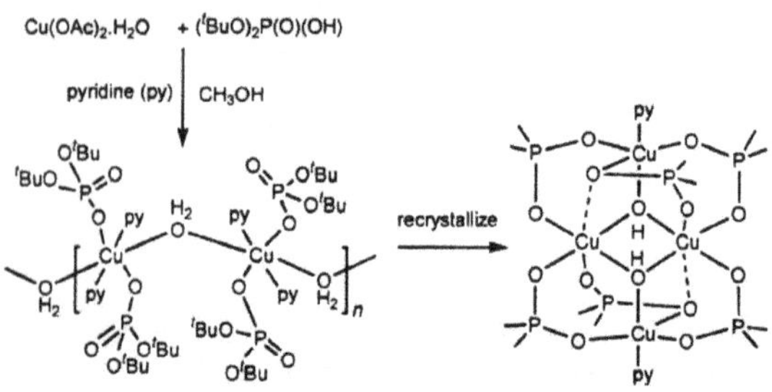


## 6. Structural Transformations in Metal Dialkyl Phosphates

If instead of a mild Lewis base such as THF a strong Lewis base such as pyridine is used in the reaction between copper acetate and dtbp-H, one obtains yet another type of one-dimensional polymeric phosphate (Chart 16).[36c] Recrystallization of this water-bridged polymer in a DMSO/THF/$CH_3OH$ mixture results in the transformation of the polymer to a more stable tetranuclear copper cluster $[Cu_4(\mu_3\text{-}OH)_2(dtbp)_6(py)_2]$ in about 60% yield (Figure 3). This observation demonstrates that even simple recrystallization can bring about an interesting structural transformation at room temperature in such materials.

Encouraged by the 1D polymer → tetrameric cage transformation, other possibilities of transforming metal dialkyl phosphates have been investigated. Interesting among the results obtained so far is the transformation of the 1D cobalt-dtbp polymer into a 2D polymeric grid structure by reaction with 2 equiv of 4,4-bipyridine in

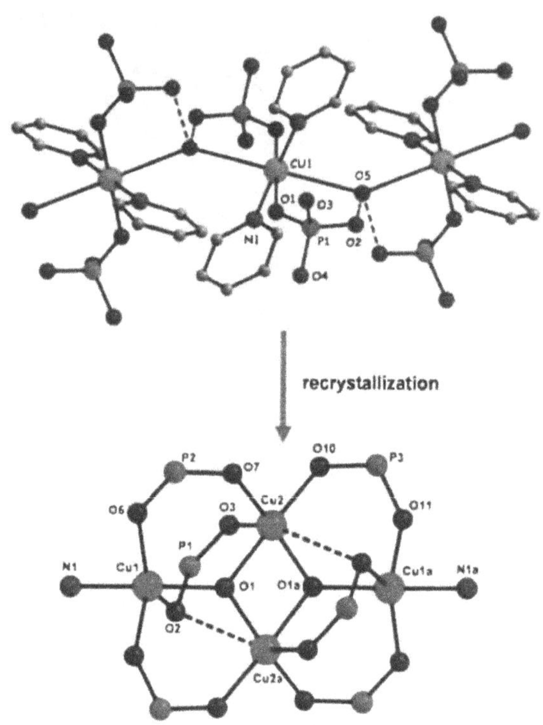


**FIGURE 3.** Conversion of a 1D polymer to a cage structure by simple recrystallization.

**Chart 17. Linear Polymer → 2D Grid Structure Conversion at Room Temperature**

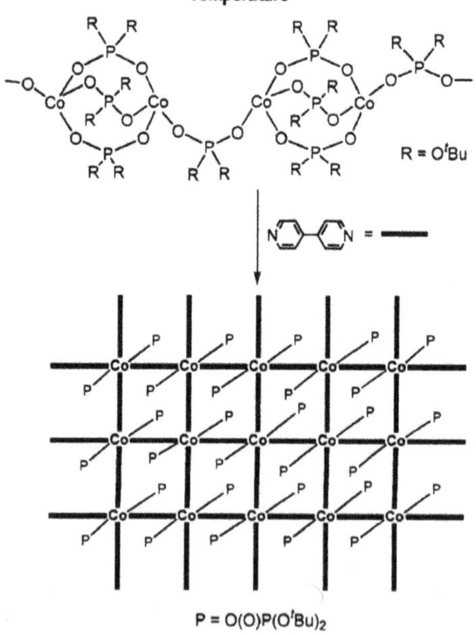


methanol at room temperature as shown in Chart 17.[40] This reaction is quantitative with no side products.

The above two examples are representative of how it is possible to either increase or decrease the dimensionality of a metal dialkyl phosphate, and clearly other possibilities exist in this area. Especially interesting would be the exploitation of the hydrolyzable P–O*t*Bu linkages in the presence of added amines (structure directors) to build up 3D framework phosphates.

## 7. Molecular Metal Phosphates from $P(O)(OH)_3$ and Their Conversion to Higher Dimensional Structures

While all the above molecular phosphates were prepared starting from phosphonic acids and phosphate esters, there are a few examples of molecular phosphates synthesized from phosphoric acid in aqueous medium. Although under hydrothermal conditions the reactions of phosphoric acid with metal ions generally result in extended open framework structures, it has been possible to isolate molecular zero-dimensional metal phosphates.[41,42]

A zero-dimensional tin phosphate comprising a four-membered ring was isolated a few years ago.[41] Similar molecular zinc phosphates were isolated by carrying out the reaction between zinc oxide, phosphoric acid, and an amine at 80 °C.[42a] Another four-membered ring zinc phosphate was obtained by the reaction between a zinc salt, phosphoric acid, and the amine at room temperature over an extended period of time.[42b] This four-membered ring phosphate could also be obtained by the reaction of the amine phosphate with the zinc salt. What is interesting about these four-membered ring metal phosphates is their close resemblance to the S4R secondary building unit. It has been shown recently that the four-membered ring zinc phosphate transforms to one-dimensional, two-dimensional, and three-dimensional structures under relatively mild conditions (Figure 4).[43,44] Interestingly, one-dimensional phosphates with linear chains or ladder structures also transform to two-dimensional and three-dimensional structures, while the layered two-dimensional structures transform to three-dimensional structures.[45,46] These studies show that the formation of complex three-dimensional structures may involve a progressive building up process, as indeed verified by synchrotron studies.[47] Acid degradation of three-dimensional zinc phosphates with channel structures have been found to yield low-dimensional structures, showing thereby the reversibility of the 1D ↔ 2D ↔ 3D transformation process.[48] What is especially noteworthy is the recent observation of the transformation of a four-membered ring zinc phosphate to a linear chain structure at room temperature on reaction with piperazine. The linear chain then transforms to a sodalite-related structure with a $4^6 8^8$ cage at 50 °C (Figure 5).[49] A transformation of metal squarates to sodalite structures has been observed recently,[50] showing thereby that the four-membered ring motif can generally act as a SBU.

In the case of aluminum phosphates, Ozin et al.[3b] have pointed out the possibility of the transformation of a chain structure to other chains or to layered and three-dimensional open-framework structures through a hydrolysis–condensation self-assembly pathway. We have recently carried out a few interesting molecule–material

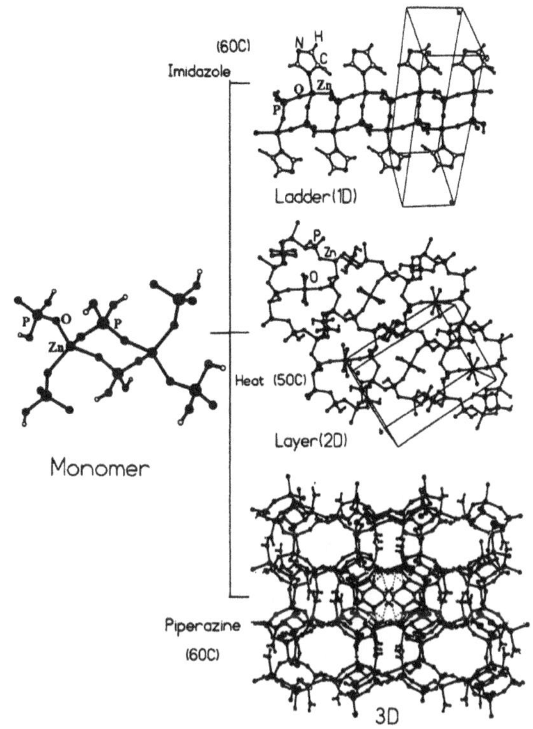


**FIGURE 4.** Transformations of a zero-dimensional monomeric zinc phosphate to 1D (ladder), 2D (layer), and 3D structures.

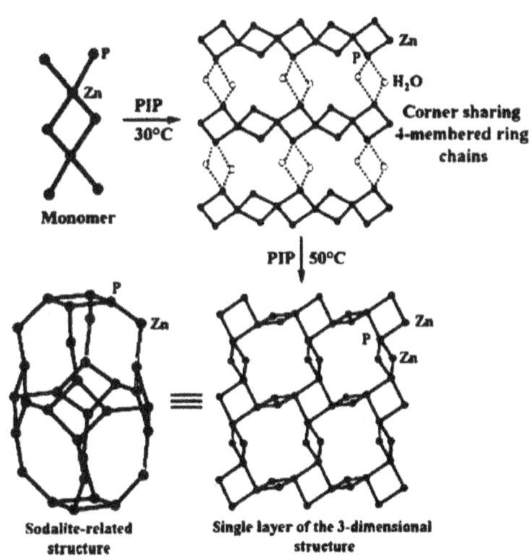


**FIGURE 5.** Transformation of a four-membered ring zinc phosphate (SBU) to a linear chain phosphate and the transformation of the latter to a 3D sodalite-related structure with a $4^6 8^8$ cage.

transformations in aluminophosphates. Thus, we found that the centrosymmetric, tetrameric molecular aluminophosphate (see Chart 12) transforms to a layered structure[51] of composition $[NH_3(CH_2)_2NH_3]^{2+}[Al_2(OH)_2(PO_4)_2(H_2O)]^{2-}\cdot H_2O$ on reaction with ethylenediamine in aqueous medium at 150 °C as shown in Figure 6. Interestingly, this structure can also be obtained as the transformation

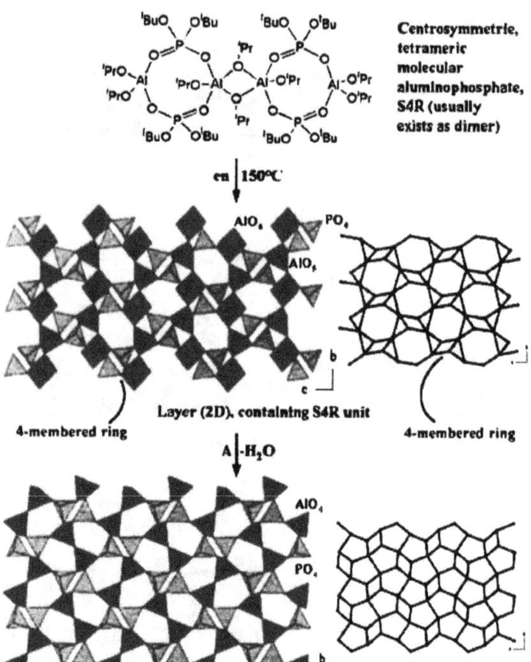


**FIGURE 6.** Transformation of a four-membered ring aluminophosphate (S4R) to a layered phosphate, which then undergoes transformation to another layered structure by the loss of water molecules. Note that the 2D component contains four-membered rings.

product of a one-dimensional ladder compound of composition $[NH_3(CH_2)_2NH_3][AlP_2O_8H]$.[52] This layered structure transforms to another layered structure by loss of water molecules, followed by reorganization of the aluminophosphate layer (Figure 6). The molecular aluminum phosphate compound gives rise to a three-dimensional structure with channels on reacting with piperazine in aqueous medium at 150 °C. We are exploring reactions of this molecular aluminum phosphate precursor and similar compounds in various media.

## 8. Conclusions and Outlook

The discussion in the previous sections should suffice to indicate the importance of the transformations of molecular compounds to materials. Studies of such transformations are only making a beginning, and there appears to be a great future for such investigations. Thus, the recent synthesis of a sodalite-related structure from a molecular zinc phosphate is noteworthy, besides the transformation of the centrosymmetric, tetrameric, molecular aluminophosphate to open-framework structures. Preliminary investigations[53] have shown that interesting transformations also occur in metal carboxylates. Thus, molecular zinc oxalate monomers and dimers are found to transform to chain or three-dimensional structures on heating with piperazine in an aqueous medium. The chain structure transforms to a layer structure. We believe that it would be rewarding to pursue investigations on the transformation of various molecular precursors including those involving aluminophosphate and silicate motifs.

## References


(1) Milton, R. M. Commercial Development of Molecular Sieve Technology. *Molecular sieves; papers read at the conference held at the School of Pharmacy, University of London, 4th–6th April, 1967*; Society of Chemical Industry: London, 1968; pp 199–203.

(2) (a) Barrer, R. M. *Hydrothermal Chemistry of Zeolites*; Academic Press: London, 1982. (b) Baerlocher, Ch.; Meier, W. M.; Olson, D. H. *Atlas of Zeolite Framework Types*, Elsevier: Amsterdam, 2001.

(3) (a) Wilson, S. T.; Lok, B. M.; Messina, C. A.; Cannan, T. R.; Flanigen, E. M. Aluminophosphate molecular sieves: A new class of microporous crystalline inorganic solids. *J. Am. Chem. Soc.* **1982**, *104*, 1146–1147. (b) Oliver, S.; Kuperman, A.; Ozin, G. A. A new model for aluminophosphate formation: Transformation of a linear chain aluminophosphate to chain, layer, and framework structures. *Angew. Chem., Int. Ed.* **1998**, *37*, 46–62.

(4) Cheetham, A. K.; Férey, G.; Loiseau, T. Open-framework inorganic materials. *Angew Chem., Int. Ed.* **1999**, *38*, 3268–3292.

(5) Rao, C. N. R.; Natarajan, S.; Vaidhyanathan, R. Metal Carboxylates with Open Architectures. *Angew. Chem., Int. Ed.* **2004**, *43*, 1466–1496.

(6) (a) Choudhury, A.; Krishnamoorthy, J.; Rao, C. N. R. An approach to the synthesis of organically templated open-framework metal sulfates by the amine-sulfate route. *Chem. Commun.* **2001**, 2610–2611. (b) Paul, G.; Choudhury, A.; Sampathkumaran, E. V.; Rao, C. N. R. Organically templated mixed-valent iron sulfates possessing Kagome and other types of layered networks. *Angew. Chem., Int. Ed.* **2002**, *41*, 4297–4300. (c) Paul, G.; Choudhury, A.; Rao, C. N. R. Organically templated linear and layered cadmium sulfates. *J. Chem. Soc., Dalton Trans.* **2002**, 3859–3867. (d) Doran, M.; Norquist, A. J.; O'Hare, D. $[NC_4H_{12}]_2[(UO_2)_6(H_2O)_2(SO_4)_7]$: the first organically templated actinide sulfate with a three-dimensional framework structure. *Chem. Commun.* **2002**, 2946–2947. (e) Behera, J. N.; Paul, G.; Choudhury, A.; Rao, C. N. R. An organically templated Co(II) sulfate with the kagome lattice. *Chem. Commun.* **2004**, 456–457.

(7) (a) Harrison, W. T. A.; Phillips, M. L. F.; Stanchfield, J.; Nenoff, T. M. $(CN_3H_6)_4[Zn_3(SeO_3)_5]$: The first organically templated selenite. *Angew Chem., Int. Ed.* **2000**, *39*, 3808–3809. (b) Choudhury, A.; Udayakumar, D.; Rao, C. N. R. Three-Dimensional Organically Templated Open-Framework Transition Metal Selenites. *Angew. Chem., Int. Ed.* **2002**, *41*, 158–161. (c) Udayakumar, D.; Rao, C. N. R. Organically templated three-dimensional open-framework metal selenites with a diamondoid network. *J. Mater. Chem.* **2003**, *13*, 1635–1638. (d) Behera, J. N.; Ayi, A. A.; Rao, C. N. R. The First Organically Templated Open-Framework Metal Selenate with a Three-Dimensional Architecture. *Chem. Commun.* **2004**, 456–457.

(8) (a) Férey, G. Building units design and scale chemistry. *J. Solid State Chem.* **2000**, *152*, 37–48. (b) Férey, G. Microporous solids: From organically templated inorganic skeletons to hybrid frameworks...ecumenism in chemistry. *Chem. Mater.* **2001**, *13*, 3084–3098.

(9) Lickiss, P. D. Polysilanols. In *The Chemistry of Organosilicon Compounds*; Rappoport, Z., Apeloig, Y. Eds.; John Wiley & Sons: 2001; Vol. 3, Chapter 12, pp 695–744.

(10) Murugavel, R.; Voigt, A.; Walawalkar, M. G.; Roesky, H. W. Hetero- and metallasiloxanes derived from silanediols, disilanols, silanetriols, and trisilanols. *Chem. Rev.* **1996**, *96*, 2205–2236.

(11) Murugavel, R.; Chandrasekhar, V.; Roesky, H. W. Discrete silanetriols: Building blocks for three-dimensional metallasiloxanes. *Acc. Chem. Res.* **1996**, *29*, 183–189.

(12) Schmidbaur, H. Advances in Heterosiloxane Chemistry. *Angew. Chem.* **1965**, *77*, 206–216.

(13) (a) Beckmann, J.; Jurkschat, K. Stannasiloxanes: from rings to polymers. *Coord. Chem. Rev.* **2001**, *215*, 267–300. (b) Lorenz, V.; Fischer, A.; Giessmann, S.; Gilje, J. W.; Gun'ko, Y.; Jacob, K.; Edelmann, F. T. Disiloxanediolates and polyhedral metallasilsesquioxanes of the early transition metals and f-elements. *Coord. Chem. Rev.* **2000**, *206–207*, 321–368. (c) King, L.; Sullivan, A. C. Main group and transition metal compounds with silanediolate $[R_2SiO_2]^{2-}$ and $\alpha,\omega$-siloxanediolate $[O(R_2SiO)_n]^{2-}$ ligands. *Coord. Chem. Rev.* **1999**, *189*, 19–57.

(14) Murugavel, R.; Walawalkar, M. G.; Prabusankar, G.; Davis, P. Synthesis and structure of a novel lithium gallosiloxane containing a $Ga_4Si_4O_8$ macrocycle analogous to the S8R building unit of zeolites. *Organometallics* **2001**, *20*, 2639–2642.

(15) (a) Feher, F. J.; Budzichowski, T. A. Silasesquioxanes as ligands in inorganic and organometallic chemistry. *Polyhedron*, **1995**, *14*, 3239–3253. (b) Duchateau, R. Incompletely condensed silsesquioxanes: Versatile tools in developing silica-supported olefin polymerization catalysts. *Chem. Rev.* **2002**, *102*, 3525–3542.

(16) Fujdala, K. L.; Tilley, T. D. New vanadium tris(*tert*-butoxy)siloxy complexes and their thermolytic conversions to vanadia-silica materials. *Chem. Mater.* **2002**, *14*, 1376–1384.

(17) Murugavel, R.; Chandrasekhar, V.; Voigt, A.; Roesky, H. W.; Schmidt, H.-G.; Noltemeyer, M. New lipophilic air-stable silanetriols: first example of an X-ray crystal structure of a silanetriol with Si–N bonds. *Organometallics* **1995**, *14*, 5298–5301 and references therein.

(18) (a) Montero, M. L.; Voigt, A.; Teichert, M.; Usón, I.; Roesky, H. W. Soluble aluminosilicates with frameworks of minerals. *Angew. Chem., Int. Ed. Engl.* **1995**, *34*, 2504–2506. (b) Montero, M. L.; Usón, I.; Roesky, H. W. Soluble organic derivatives of aluminosilicates with $Al_2Si_2O_4$ and $Al_4Si_2O_6$ frameworks. *Angew. Chem., Int. Ed. Engl.* **1994**, *33*, 2103–2104.

(19) Nehete, U. N.; Anantharaman, G.; Chandrasekhar, V.; Murugavel, R.; Walawalkar, M. G.; Roesky, H. W.; Vidovic, D.; Magull, J.; Samwer, K.; Sass, B. Polyhedral Ferrous and Ferric Siloxanes. *Angew. Chem., Int. Ed.* **2004**, *43*, 3832–3835.

(20) Feher, F. J.; Newman, D. A.; Walzer, J. F. Silsesquioxanes as models for silica surfaces. *J. Am. Chem. Soc.* **1989**, *111*, 1741–1748.

(21) Abbenhuis, H. C. L. Advances in homogeneous and heterogeneous catalysis with metal-containing silsesquioxanes. *Chem.—Eur. J.* **2000**, *6*, 25–32.

(22) Coles, M. P.; Lugmair, C. G.; Terry, K. W.; Tilley, T. D. Titania-silica materials from the molecular precursor $Ti[OSi((OBu^t)_3]_4$: Selective epoxidation catalysts. *Chem. Mater.* **2000**, *12*, 122–131.

(23) See also the cited references of Tilley et al. in ref 16.

(24) Fujdala, K. L.; Tilley, T. D. Design and synthesis of heterogeneous catalysts: the thermolytic molecular precursor approach. *J. Catal.* **2003**, *216*, 265–275.

(25) Brutchey, R. L.; Goldberger, J. E.; Koffas, T. S.; Tilley, T. D. Nonaqueous, molecular precursor route to hybrid inorganic/organic zirconia-silica materials containing covalently linked organic bridges. *Chem. Mater.* **2003**, *15*, 1040–1046.

(26) Walawalkar, M. G.; Roesky, H. W.; Murugavel, R. Molecular phosphonate cages: Model compounds and starting materials for phosphate materials. *Acc. Chem. Res.* **1999**, *32*, 117–126.

(27) Walawalkar, M. G.; Murugavel, R.; Roesky, H. W.; Schmidt, H.-G. The first molecular borophosphonate cage: Synthesis, spectroscopy, and single-crystal X-ray structure. *Organometallics* **1997**, *16*, 516–518.

(28) Mason, M. R. Molecular phosphates, phosphonates, phosphinates, and arsonates of the group 13 elements. *J. Cluster Sci.* **1998**, *9*, 1–23.

(29) Yang, Y.; Walawalkar, M. G.; Pinkas, J.; Roesky, H. W.; Schmidt, H.-G. Molecular aluminophosphonate: Model compound for the isoelectronic double-six-ring (D6R) secondary building unit of zeolites. *Angew. Chem., Int. Ed.* **1998**, *37*, 96–98.

(30) Walawalkar, M. G.; Horchler, S.; Dietrich, S.; Chakraborty, D.; Roesky, H. W.; Schaefer, M.; Schmidt, H.-G.; Sheldrick, G. M.; Murugavel, R. Novel organic-soluble molecular titanophosphonates with cage structures comparable to titanium-containing silicates. *Organometallics* **1998**, *17*, 2865–2868.

(31) (a) Walawalkar, M. G.; Murugavel, R.; Voigt, A.; Roesky, H. W.; Schmidt, H.-G. A novel molecular gallium phosphonate cage containing sandwiched lithium ions: Synthesis, structure, and reactivity. *J. Am. Chem. Soc.* **1997**, *119*, 4656–4661. (b) Walawalkar, M. G.; Murugavel, R.; Roesky, H. W.; Usón, I.; Kraetzner, R. Gallophosphonates containing alkali metal ions. 2. Synthesis and structure of gallophosphonates incorporating $Na^+$ and $K^+$ ions. *Inorg. Chem.* **1998**, *37*, 473–478.

(32) Yang, Y.; Pinkas, J.; Schaefer, M.; Roesky, H. W. Molecular model for aluminophosphates containing fluoride as a structure-directing and mineralizing agent. *Angew. Chem., Int. Ed.* **1998**, *37*, 2650–2653.

(33) Anantharaman, G.; Walawalkar, M. G.; Murugavel, R.; Gábor, B.; Herbst-Irmer, R.; Baldus, M.; Angerstein, B.; Roesky, H. W. A nanoscopic molecular cadmium phosphonate wrapped in a hydrocarbon sheath. *Angew. Chem., Int. Ed.* **2003**, *42*, 4482–4485.

(34) Chandrasekhar, V.; Kingsley, S. A dodecanuclear copper (II) cage containing phosphonate and pyrazole ligands *Angew. Chem., Int. Ed.* **2000**, *39*, 2320–2322.

(35) Salta, J.; Chen, Q.; Chang, Y. D.; Zubieta, J. The oxovanadium-organophosphonate system – Complex cluster structures $[(VO)_6(tBuPO_3)_8Cl]$, $[(VO)_4(PhPO_2OPO_2Ph)_4Cl]^-$, and $[V_{18}O_{25}(H_2O)_2(PhPO_3)_{20}Cl_4]^{4-}$ with encapsulated chloride anions prepared from simple precursors. *Angew. Chem., Int. Ed. Engl.* **1994**, *33*, 757–760.

(36) (a) Murugavel, R.; Sathiyendiran, M.; Walawalkar, M. G. Di-*tert*-butyl phosphate complexes of cobalt(II) and zinc(II) as precursors for ceramic $M(PO_3)2$ and $M_2P_2O_7$ materials: Synthesis, spectral characterization, structural studies, and role of auxiliary ligands. *Inorg. Chem.* **2001**, *40*, 427–434. (b) Sathiyendiran, M.; Murugavel, R. Di-*tert*-butyl phosphate as synthon for metal phosphate materials via single-source coordination polymers $[M(dtbp)_2]_n$ (M = Mn, Cu) and $[Cd(dtbp)_2(H_2O)]_n$ (dtbp-H = $(tBuO)_2P(O)OH$). *Inorg. Chem.* **2002**, *41*, 6404–6411. (c) Murugavel, R.; Sathiyendiran, M.; Pothiraja, R.; Walawalkar, M. G.; Mallah, T.; Riviére, E. Monomeric, tetrameric, and polymeric copper di-*tert*-butyl phosphate complexes containing pyridine ancillary ligands. *Inorg. Chem.* **2004**, *43*, 945–953. (d) Murugavel, R.; Sathiyendiran, M.; Pothiraja, R.; Butcher, R. J. O–H Bond elongation in coordinated water through intramolecular P=O···H–O bonding. 'Snapshots' in phosphate ester hydrolysis. *Chem. Commun.* **2003**, 2546–2547.

(37) (a) Lugmair, C. G.; Tilley, T. D.; Rheingold, A. L. Zinc di(*tert*-butyl)phosphate complexes as precursors to zinc phosphates. Manipulation of zincophosphate structures. *Chem. Mater.* **1997**, *9*, 339–348. (b) Lugmair, C. G.; Tilley, T. D. Di-*tert*-butyl phosphate complexes of titanium. *Inorg. Chem.* **1998**, *37*, 1821–1826. (c) Lugmair, C. G.; Tilley, T. D.; Rheingold, A. L. Di(*tert*-butyl)phosphate complexes of aluminum: Precursors to aluminum phosphate xerogels and thin films. *Chem. Mater.* **1999**, *11*, 1615–1620.

(38) Pinkas, J.; Chakraborty, D.; Yang, Y.; Murugavel, R.; Noltelmeyer, M.; Roesky, H. W. Reactions of trialkyl phosphates with trialkyls of aluminum and gallium: New route to alumino- and gallophosphate compounds via dealkylsilylation. *Organometallics* **1999**, *18*, 523–528.

(39) Fujdala, K. L.; Tilley, T. D. An efficient, single-source molecular precursor to silicoaluminophosphates. *J. Am. Chem. Soc.* **2001**, *123*, 10133–10134.

(40) Pothiraja, R.; Murugavel, R. Unpublished results.

(41) Ayyappan, S.; Cheethan, A. K.; Natarajan, S.; Rao, C. N. R. A Novel Monomeric Tin(II) Phosphate, $[N(C_2H_5NH_3)_3]^{3+}[Sn(PO_4)(HPO_4)]^{3-}\cdot 4H_2O$, Connected through hydrogen Bonding. *J. Solid State Chem.* **1998**, *139*, 207–210.

(42) (a) Harrison, W. T. A.; Hannooman, L. Two New Tetramethylammonium Zinc Phosphates: $N(CH_3)_4\cdot Zn(HPO_4)(H_2PO_4)$, an Open Framework Phase Built up from a Low-density 12-Ring topology, and $N(CH_3)_4\cdot Zn(H_2PO_4)_3$, a Molecular Cluster. *J. Solid State Chem.* **1997**, *131*, 363–369. (b) Neeraj, S.; Natarajan, S.; Rao, C. N. R. Isolation of a Zinc Phosphate Primary Building Unit, $[C_6N_2H_{18}]^{2+}[Zn(HPO_4)(H_2PO_4)_2]^{2-}$, and Its Transformation to an Open-Framework Phosphate, $[C_6N_2H_{18}]^{2+}[Zn_3(H_2O)_4(HPO_4)_4]^{2-}$. *J. Solid State Chem.* **2000**, *150*, 417–422.

(43) Rao, C. N. R.; Natarajan, S.; Choudhury, A.; Neeraj, S.; Ayi, A. A. Aufbau Principle of Complex Open-Framework Structures of Metal Phosphates with Different Dimensionalities. *Acc. Chem. Res.* **2001**, *34*, 80–87.

(44) Ayi, A. A.; Choudhury, A.; Natarajan, S.; Neeraj, S. Rao, C. N. R. Transformations of Low-Dimensional Zinc Phosphates to complex Open-Framework Structures. Part 1: Zero-dimensional to One-, Two- and Three-Dimensional Structures. *J. Mater. Chem.* **2001**, *11*, 1181–1191.

(45) (a) Choudhury, A.; Neeraj, S.; Natarajan, S.; Rao, C. N. R. Transformations of the Low-Dimensional Zinc Phosphates to Complex Open-Framework Structures. Part 2: One-Dimensional Ladder to Two- and Three-Dimensional Structures. *J. Mater. Chem.* **2001**, *11*, 1537–1546. (b) Choudhury, A.; Neeraj, S.; Natarajan, S.; Rao, C. N. R. Transformations of Two-Dimensional Layered Zinc Phosphates to Three-Dimensional and One-Dimensional Structures. *J. Mater. Chem.* **2002**, *12*, 1044–1052.

(46) Walton, R. I.; Millange, F.; Le Bail, A.; Loiseau, T.; Serre, C.; O'Hare, D.; Férey, G. The Room-Temperature Crystallization of a One-Dimensional Gallium Fluorophosphate, $Ga(HPO_4)_2F.\ H_3N(CH_2)_3NH_3\cdot 2H_2O$, a Precursor to Three-Dimensional Microporous Gallium Fluorophosphates. *Chem. Commun.* **2000**, 203–204.

(47) (a) Walton, R. I.; Norquist, A. J.; Neeraj, S.; Natarajan, S.; Rao, C. N. R.; O'Hare, D. Direct *in situ* Observation of Increasing Structural Dimensionality During the Hydrothermal Formation of Open-Framework Zinc Phosphates. *Chem. Commun.* **2001**, 1990–1991. (b) Francis, R. J.; O'Brein, S.; Fogg, A. M.; Halasyamani, P. S.; O'Hare, D.; Loiseau, T.; Ferey, G. Time-Resolved *In-Situ* Energy and Angular Dispersive X-ray Diffraction Studies of the Formation of the Microporous Gallophosphate ULM-5 under Hydrothermal Conditions. *J. Am. Chem. Soc.* **1999**, *121*, 1002–1015.

(48) Choudhury, A.; Rao, C. N. R. Understanding the Building-Up Process of Three-Dimensional Open-Framework Metal Phosphates: Acid Degradation of the 3D Structures to Lower Dimensional Structures. *Chem. Commun.* **2003**, 366–367.

(49) Dan, M.; Udayakumar, D.; Rao, C. N. R. Transformation of a 4-membered Ring Zinc Phosphate SBU to a Sodalite-Related 3-Dimensional Structure through a Linear Chain Structure. *Chem. Commun.* **2003**, 2212–2213.
(50) Neeraj, S.; Noy, M. L.; Rao, C. N. R.; Cheetham, A. K. Sodalite Networks Formed by Metal Squarates. *Solid State Sci.* **2002**, *4*, 1231–1236.
(51) Kongshaug, K. O.; Fjellvag, H.; Lillerud, K. P. Layered aluminophosphates II. Crystal Structure and Thermal Behaviour of the Layered Aluminophosphate UiO-15 and its High-Temperature Variants. *J. Mater. Chem.* **1999**, *9*, 1591–1598.
(52) Williams, I. D.; Yu, J.; Gao, Q.; Chen, J.; Xu, R. New Chain Architecture for a One-Dimensional Aluminophosphate, $[H_3NCH_2CH_2NH_3][AlP_2O_8H]$. *Chem. Commun.* **1997**, 1273–1274.
(53) Dan, M.; Ayi, A. A.; Rao, C. N. R. Unpublished results.

AR040083E

3524 *Chem. Mater.* **2001,** *13,* 3524–3533

# *Articles*

## Hybrid Inorganic–Organic Host–Guest Compounds: Open-Framework Cadmium Oxalates Incorporating Novel Extended Structures of Alkali Halides

R. Vaidhyanathan, Srinivasan Natarajan, and C. N. R. Rao*

*Chemistry and Physics of Materials Unit, Jawaharlal Nehru Centre for Advanced Scientific Research, Jakkur P.O., Bangalore 560 064, India*

*Received January 18, 2001*

Four unusual inorganic–organic nanocomposites containing extended alkali halide structures, present as layers or as three-dimensional units, have been synthesized by a metathetic reaction carried out under hydrothermal conditions. These compounds, with cadmium oxalate host lattices and the alkali halide structures as guests, have the compositions $[RbCl][Cd_6(C_2O_4)_6]\cdot 2H_2O$, **I;** $2[CsBr][Cd(C_2O_4)]\cdot H_2O$, **II;** $2[CsBr][Cd_2(C_2O_4)(Br)_2]\cdot 2H_2O$, **III;** and $3[RbCl][Cd_2(C_2O_4)(Cl_2)]\cdot H_2O$, **IV**. Crystal data for **I**, rhombohedral, space group = $R$-3 (no. 148), $a$ = 9.3859(3), $c$ = 23.9086(8) Å; $V$ = 1824.05(1) Å$^3$, $Z$ = 18, $M$ = 1359.51, $\rho_{calcd}$ = 3.702 g cm$^{-3}$, $\mu$(Mo K$\alpha$) = 7.375 mm$^{-1}$, $R_1$ = 0.04, $wR_2$ = 0.0904 [537 observed reflections with $I > 2\sigma(I)$]; for **II**, orthorhombic, space group = $Pbcm$ (no. 57), $a$ = 6.0854(6), $b$ = 11.0793(11), $c$ = 16.889(2) Å; $V$ = 1138.7(2) Å$^3$, $Z$ = 8, $M$ = 644.06, $\rho_{calcd}$ = 3.745 g cm$^{-3}$, $\mu$(Mo K$\alpha$) = 15.219 mm$^{-1}$, $R_1$ = 0.04, $wR_2$ = 0.08 [408 observed reflections with $I > 2\sigma(I)$]; for **III**, orthorhombic, space group = $Cmcm$ (no.63), $a$ = 23.6151(14), $b$ = 10.2528(6), $c$ = 7.8199(6) Å; $V$ = 1894.2(2) Å$^3$, $Z$ = 16, $M$ = 970.33, $\rho_{calcd}$ = 3.374 g cm$^{-3}$, $\mu$(Mo K$\alpha$) = 14.487 mm$^{-1}$, $R_1$ = 0.03, $wR_2$ = 0.0744 [651 observed reflections with $I > 2\sigma(I)$]; for **IV**, monoclinic, space group = $P2_1/c$ (no. 14), $a$ = 8.0648(2)(3), $b$ = 22.9026(4), $c$ = 9.3967(3) Å; $V$ = 1681.10(7) Å$^3$, $Z$ = 4, $M$ = 782.53, $\rho_{calcd}$ = 3.076 g cm$^{-3}$, $\mu$(Mo K$\alpha$) = 11.961 mm$^{-1}$, $R_1$ = 0.04, $wR_2$ = 0.093 [2071 observed reflections with $I > 2\sigma(I)$]. Compounds **I–IV** possess many unusual structural features. While **I** has interpenetrating lattices of the three-dimensional cadmium oxalate and RbCl with an expanded lattice, giving rise to a super rock-salt cell, **II** possesses alternating cadmium oxalate and CsBr layers, the latter comprising six-membered CsBr rings. Compounds **III** and **IV** have isolated cadmium oxalate units (zero-dimensional) linked and stabilized by the alkali halides.

### Introduction

A large variety of supramolecular organic and inorganic structures have been described in the recent literature.[1-4] These include several types of host–guest compounds,[2] as well as metal–organic and –inorganic open-framework structures.[4-7] In many of the host–guest compounds, ions, ion pairs, or molecules have been accommodated in the cages or cavities.[3,8] A novel departure from the traditional host–guest chemistry is the recent discovery of cadmium oxalates incorporating extended alkali halide structures of different dimensionalities.[9] These materials can be considered to be

* Corresponding author. E-mail: cnrrao@jncasr.ac.in.

(1) (a) Lehn, J.-M. Supramolecular Chemistry: Concepts and Perspectives; VCH: New York, 1995.
(2) (a) Dalton Discussion No. 3, Inorganic Crystal Engineering. *J. Chem. Soc., Dalton Trans.* **2000** (21). (b) Braga, D.; Grepioni, F. *Acc. Chem. Res.* **2000**, *33*, 601.
(3) Muller, A.; Reuterm, H.; Dillinger, S. *Angew. Chem., Int. Ed. Engl.* **1995**, *34*, 2328.
(4) (a) Cheetham, A. K.; Ferey, G.; Loiseau, T. *Angew. Chem., Int. Ed. Engl.* **1999**, *38*, 3268. (b) Serpaggi, F.; Ferey, G. *J. Mater. Chem.* **1998**, *8*, 2737. (c) Serpaggi, F.; Ferey, G. *Microporous and Mesoporous Mater.* **1999**, *32*, 311. (d) Riou, D.; Ferey, F. *J. Mater. Chem.* **1998**, *8*, 2733. (e) Livage, C.; Egger, C.; Nogues, M.; Ferey, G. *J. Mater. Chem.* **1998**, *8*, 2743. (f) Livage, C.; Egger, C.; Ferey, G. *Chem. Mater.* **1999**, *11*, 1546. (g) Kuhlman, R.; Schemek, G. L.; Kolis, J. W. *Inorg. Chem.* **1999**, *38*, 194.
(5) Chui, S. S.-Y.; Lo, S. M.-F.; Charmant, J. P. H.; Orpen, A. G.; Williams, I. A. *Science* **1999**, *283*, 1148.
(6) (a) Reineke, T. M.; Eddaoudi, M.; Fehr, M.; Kelly, D.; Yaghi, O. M. *J. Am. Chem. Soc.* **1999**, *121*, 1651. (b) Reineke, T. M.; Eddaoudi, M.; O'Keeffe, M.; Yaghi, O. M. *Angew. Chem., Int. Ed. Engl.* **1999**, *38*, 2590. (c) Li, H.; Eddaoudi, M.; O'Keeffe, M.; Yaghi, O. M. *Nature* **1999**, *402*, 276. (d) Reineke, T. M.; Eddaoudi, M.; Moler, D.; O_Keeffe, M.; Yaghi, O. M. *J. Am. Chem. Soc.* **2000**, *122*, 4843.
(7) (a) Ayyappan, S.; Cheetham, A. K.; Natarajan, S.; Rao, C. N. R. *Chem. Mater.* **1998**, *10*, 3746. (b) Natarajan, S.; Vaidhyanathan, R.; Rao, C. N. R.; Ayyappan, S.; Cheetham, A. K. *Chem. Mater.* **1999**, *11*, 1633. (c) Vaidhyanathan, R.; Natarajan, S.; Cheetham, A. K.; Rao, C. N. R. *Chem. Mater.* **1999**, *11*, 3636. (d) Vaidhyanathan, R.; Natarajan, S.; Rao, C. N. R. *Chem. Mater.* **2001**, *13*, 185.
(8) (a) Wragg, D. S.; Morris, R. E. *J. Am. Chem. Soc.*, **2000**, *122*, 11 246. (b) Taulelle, F.; Poblet, J.-M.; Ferey, G.; Benard, M. *J. Am. Chem. Soc.* **2001**, *123*, 111.
(9) Vaidhyanathan, R.; Neeraj, S.; Prasad, P. A.; Natarajan, S.; Rao, C. N. R. *Angew. Chem., Int. Ed. Engl.* **2000**, *39*, 3470.

10.1021/cm011011+ CCC: $20.00 
Published on Web 09/07/2001

**Table 1. Synthesis Conditions and Analysis for Compounds I–IV**

| reactant mole ratios | temp, °C | time, h | EDAX ratios | product |
|---|---|---|---|---|
| $CdCl_2$:0.5$Rb_2CO_3$:5$H_2C_2O_4$: 1.5$CH_3COOH$:88$n$-$C_4H_9OH$: 220$H_2O$ | 150 | 78 | Cd:Rb:Cl 6.5:1.0:1.1 | [RbCl][$Cd_6(C_2O_4)_6$]·2$H_2O$ (**I**) |
| $CdBr_2$:$Cs_2CO_3$:1.47$H_6C_4O_4$:70$n$-$C_4H_9OH$:150$H_2O$ | 150 | 108 | Cd:Cs:Br 1.1:1.5:1.5 | 2[CsBr][$Cd(C_2O_4)$] ·$H_2O$ (**II**) |
| $CdBr_2$:$(COOCs)_2$:1.95$CH_3COO$:40$n$-$C_4H_9OH$:62$H_2O$ | 150 | 80 | Cd:Cs:Br 1.1:1.1:1.8 | 2[CsBr][$Cd_2(C_2O_4)(Br)_2$]·4$H_2O$ (**III**) |
| $CdC_{l2}$:$Rb_2CO_3$:$H_2C_2O_4$:2$CH_3COOH$:33$n$-$C_4H_9OH$:33$H_2O$ | 150 | 80 | Cd:Rb:Cl 1.0:1.5:2.5 | 3[RbCl][$Cd_2(H_2O)C_2O_4)(Cl)_2$]·$H_2O$ (**IV**) |

organic–inorganic nanocomposites or hetero superlattices wherein the oxalate and alkali halide units are either intricately connected in three-dimensions or occur discretely in alternate layers. In view of the novelty of these hybrid structures, we considered it highly desirable to explore newer systems incorporating such alkali halide structures. In the present study, we describe a three-dimensional cadmium oxalate structure, [RbCl]-[$Cd_6(C_2O_4)_6$]·2$H_2O$, **I**, incorporating RbCl in the rock-salt (*Fm*3*m*) structure, with a unit-cell parameter double that of the normal RbCl. The structure of **I** is even more fascinating in that the oppositely charged cluster ions form a rock-salt supercell, by using the RbCl lattice as a template. We then have a layered cadmium oxalate, 2[CsBr][$Cd(C_2O_4)$]·$H_2O$, **II**, having CsBr in a graphitic layered architecture formed by six-membered rings, and two other cadmium oxalates, 2[CsBr][$Cd_2$-$(C_2O_4)(Br)_2$]·2$H_2O$, **III**, and 3[RbCl][$Cd_2(C_2O_4)(Cl_2)$]·$H_2O$, **IV**, where the alkali halide layers comprising different types of rings act as bridges between the oxalate units. The alkali halide units in **III** form a layered structure with eight-ring apertures. Compounds **I–IV** constitute members of a new class of hybrid host–guest materials and are likely to possess novel dielectric, optical, and other properties in view of the periodic arrangement of the organic and inorganic components.

**Table 2. X-ray Powder Data for I, [RbCl][$Cd_6(C_2O_4)_6$]·2$H_2O$**

| *H* | *K* | *L* | $2\theta_{obsd}$ | $\Delta(2\theta)^a$ | $d_{calcd}$ | $\Delta(d)^b$ | $I_{rel}{}^c$ |
|---|---|---|---|---|---|---|---|
| 0 | 0 | 3 | 11.102 | 0.000 | 7.970 | 0.000 | 11 |
| 1 | 0 | 1 | 11.498 | 0.000 | 7.696 | 0.000 | 9 |
| 1 | 0 | 2 | 13.163 | 0.008 | 6.722 | 0.004 | 52 |
| 1 | 0 | 4 | 18.485 | 0.060 | 4.815 | 0.015 | 8 |
| 1 | 1 | 0 | 19.007 | 0.097 | 4.693 | 0.024 | 20 |
| 1 | 1 | 3 | 21.996 | 0.017 | 4.044 | 0.003 | 14 |
| 0 | 0 | 6 | 22.369 | 0.059 | 3.985 | 0.010 | 11 |
| 2 | 0 | 2 | 23.179 | 0.065 | 3.848 | 0.011 | 8 |
| 2 | 0 | 4 | 26.560 | 0.040 | 3.361 | 0.005 | 100 |
| 1 | 0 | 7 | 28.480 | 0.137 | 3.149 | 0.015 | 11 |
| 2 | 1 | 1 | 29.384 | 0.075 | 3.047 | 0.008 | 17 |
| 1 | 1 | 6 | 29.444 | 0.040 | 3.038 | 0.004 | 13 |
| 2 | 1 | 2 | 30.135 | 0.105 | 2.976 | 0.010 | 9 |
| 2 | 1 | 4 | 32.804 | 0.030 | 2.732 | 0.002 | 7 |
| 0 | 0 | 9 | 33.739 | 0.000 | 2.657 | 0.000 | 15 |
| 2 | 0 | 7 | 34.395 | 0.101 | 2.615 | 0.007 | 6 |
| 2 | 1 | 5 | 34.751 | 0.046 | 2.585 | 0.003 | 18 |
| 3 | 0 | 3 | 35.045 | 0.068 | 2.565 | 0.005 | 7 |
| 2 | 0 | 8 | 37.439 | 0.091 | 2.408 | 0.006 | 11 |
| 1 | 1 | 9 | 39.086 | 0.128 | 2.312 | 0.007 | 6 |
| 3 | 1 | 1 | 40.211 | 0.034 | 2.244 | 0.002 | 48 |
| 3 | 0 | 6 | 40.248 | 0.001 | 2.241 | 0.000 | 27 |
| 3 | 1 | 2 | 40.703 | 0.024 | 2.215 | 0.001 | 11 |
| 3 | 1 | 5 | 44.424 | 0.002 | 2.039 | 0.000 | 4 |

[a] $2\theta_{obsd} - 2\theta_{calcd}$. [b] $d_{obsd} - d_{calcd}$. [c] $100I/I_{max}$. LSQ fitted lattice parameter (Cu K$\alpha$): $a$ = 9.3773(5), $b$ = 9.3773(5), $c$ = 23.8786(9) Å, $V$ = 1818.4(3) Å$^3$.

## Experimental Section

Compounds **I–IV** were prepared by metathetic reaction under hydrothermal conditions. Typically for the synthesis of **I**, 0.057 g of $Rb_2CO_3$ was dissolved in 4 mL of $n$-butanol and 2 mL of water mixture. Oxalic acid, 0.313 g, and $CdCl_2$, 0.1 g, were added, and the contents were stirred for 20 min. Finally 0.04 mL of glacial acetic acid was added and the mixture was stirred to homogeneity. The final mixture with the composition $CdCl_2$:0.5$Rb_2CO_3$:5$H_2C_2O_4$:1.5$CH_3COOH$:90$C_4H_9OH$:220$H_2O$ was sealed in a PTFE-lined stainless steel autoclave and heated at 150 °C for 72 h. The resulting product, a crop of truncated cubelike crystals suitable for single-crystal X-ray study, was filtered, washed, and dried at room temperature. A similar synthesis procedure was employed for the preparation of compounds **II–IV** and is presented in Table 1.

The initial characterization was carried out using powder X-ray diffraction (XRD), EDAX, and chemical analysis. The powder XRD patterns indicated that the products were new materials; the patterns were entirely consistent with the structures determined by using single-crystal X-ray diffraction. As a representative example, powder data for compound **I** is presented in Table 2.

Thermogravimetric analysis of compounds **I** and **II** were carried out in $O_2$ atmosphere (50 mL min$^{-1}$) in the range room temperature to 700 °C. While **I** showed two distinct mass losses, **II** had three distinct mass losses. For **I**, a gradual mass loss around 100 °C followed by a sharp mass loss in the range 350–400 °C (42.0%) was found due to the loss of the extra framework water and oxalate (calcd 41.5%). A mass loss above 500 °C was found due to the slow evaporation of CdO. For compound **II**, a sharp mass loss of 2.3% around 150 °C corresponds to the loss of extra framework water (calcd 2.8) and the second mass loss of 16% around 350 °C corresponds to the loss of oxalate units. Above 600 °C there was loss due to the evaporation of CdO. In both **I** and **II**, the powder XRD pattern of the product of decomposition was characteristic of CdO, mineral monteporite (JCPDS: 05-0640).

Infrared spectroscopic studies as a KBr pellet showed characteristic features of the bischelating oxalate.[10] The various bands for compounds **I–IV** are as follows: $\nu_{as}$(C=O) 1612–1672 cm$^{-1}$; $\nu_s$(O–C–O) 1372 and 1311 (s) cm$^{-1}$ (**I**), 1374 and 1310 (s) cm$^{-1}$ (**II**), 1372 (m) and 1310 (s) cm$^{-1}$ (**III**), 1373 and 1312 (s) cm$^{-1}$ (**IV**); $\delta$(O–C=O) 803 (m), 790 (w), and 774 (w) cm$^{-1}$ (**I**), 800 (w) and 773 (s) cm$^{-1}$ (**II**), 773 (s) and 731 (w) cm$^{-1}$ (**III**), 802 and 776 (m) cm$^{-1}$ (**IV**). The extraframework water molecules showed characteristic bands around 3550 (s) cm$^{-1}$ in all the cases. The M–O (M = Cd) stretching vibrations $\nu_s$(M–O) and $\nu_s$(C–C) are also observed at 512 and 441 (s) cm$^{-1}$ (**I**), 513 (m) cm$^{-1}$ (**II**), 514 (w) cm$^{-1}$ (**III**), 512 and 441 (s) cm$^{-1}$ (**IV**). The various infrared bands are consistent with the structure determined by single-crystal studies.

**Single-Crystal Structure Determination.** A suitable single crystal of each compound was carefully selected under a polarizing microscope and glued to a thin glass fiber with cyano-acrylate (super glue) adhesive. Single-crystal structure determination by X-ray diffraction was performed on a Siemens Smart-CCD diffractometer equipped with a normal focus, 2.4 kW sealed tube X-ray source (Mo K$\alpha$ radiation, $\lambda$ = 0.710 73 Å) operating at 50 kV and 40 mA. A hemisphere of intensity data was collected at room temperature in 1321 frames with $\omega$ scans (width of 0.30° and exposure time of 20s per frame) in the 2$\theta$ range 3–46.5°. Pertinent experimental details for the structure determinations are presented in Table 3.

(10) *Infrared and Raman Spectra of Inorganic and Coordination Compounds*, 5th ed.; Nakamoto, K., Ed.; John Wiley: New York, 1997.

**Table 3. Crystal Data and Structure Refinement Parameters for [RbCl][$Cd_6(C_2O_4)_6$]·$2H_2O$, I; 2[CsBr][$Cd(C_2O_4)$]·$H_2O$, II; 2[CsBr][$Cd_2(C_2O_4)(Br)_2$]·$4H_2O$, III; and 3[RbCl][$Cd_2(H_2O)(C_2O_4)(Cl)_2$]·$H_2O$, IV**

| parameters | I | II | III | IV |
|---|---|---|---|---|
| empirical formula | $H_4C_{12}Cd_6C_{11}O_{26}Rb_1$ | $H_2C_2Cd_1Br_2O_5Cs_2$ | $H_8C_2Cd_1Br_4O_5Cs_2$ | $H_4C_2Cd_2Cl_5O_6Rb_3$ |
| crystal system | rhombohedral | orthorhombic | orthorhombic | monoclinic |
| space group | *R*-3 (no. 148) | *Pbcm* (no. 57) | *Cmcm* (no. 63) | $P2_1/c$ (no. 14) |
| crystal size, mm | 0.09, 0.09, 0.09 | 0.08, 0.06, 0.07 | 0.08, 0.10, 0.12 | 0.08, 0.06, 0.09 |
| *a*, Å | 9.3859(3) | 6.0854(6) | 23.6251(14) | 8.0648(2) |
| *b*, Å | 9.3859(3) | 11.0793(11) | 10.2528(6) | 22.9026(4) |
| *c*, Å | 23.9086(8) | 16.889(2) | 7.8199 (58) | 9.3967(3) |
| α | 90° | 90° | 90° | 90° |
| β | 90° | 90° | 90° | 104.399(10)° |
| γ | 120° | 90° | 90° | 90° |
| vol, Å$^3$ | 1824.05(10) | 1138.7(2) | 1894.2(2) | 1681.10(7) |
| *Z* | 18 | 8 | 16 | 4 |
| formula mass | 1359.51 | 644.06 | 970.33 | 782.53 |
| $r_{calcd}$, g cm$^{-3}$ | 3.702 | 3.745 | 3.374 | 3.076 |
| λ (Mo Kα), Å | 0.71073 | 0.71073 | 0.71073 | 0.71073 |
| μ, mm$^{-1}$ | 7.375 | 15.219 | 14.487 | 11.961 |
| θ range | 2.56−23.26° | 2.41−23.27° | 1.72−23.24° | 1.78−23.25° |
| total data collected | 2598 | 4407 | 3894 | 6939 |
| index ranges | $-10 \le h \le 6, -7 \le k \le 10, -26 \le l \le 24$ | $-6 \le h \le 6, -12 \le k \le 11, -18 \le l \le 14$ | $-21 \le h \le 26, -11 \le k\ 11, -8 \le l \le 8$ | $-7 \le h \le 8, -21 \le k \le 25, -10 \le l \le 8$ |
| unique data | 589 | 855 | 770 | 2405 |
| obsd data ($I > 2\sigma(I)$) | 537 | 408 | 651 | 2071 |
| refinement method | full-matrix least-squares on $\|F^2\|$ | full-matrix least-squares on $\|F^2\|$ | full-matrix least-squares on $\|F^2\|$ | full-matrix least-squares on $\|F^2\|$ |
| $R_{int}$ | 0.0887 | 0.0647 | 0.0535 | 0.0950 |
| *R* indexes [$I > 2\sigma(I)$] | $R_1 = 0.0355,^a$ $wR_2 = 0.0904^b$ | $R_1 = 0.0377,^a$ $wR_2 = 0.0827^b$ | $R_1 = 0.0334,^a$ $wR_2 = 0.0744^b$ | $R_1 = 0.0409,^a$ $wR_2 = 0.0931^b$ |
| *R* (all data) | $R_1 = 0.0414$, $wR_2 = 0.0980$ | $R_1 = 0.0485$, $wR_2 = 0.0879$ | $R_1 = 0.0428$, $wR_2 = 0.0784$ | $R_1 = 0.0516$, $wR_2 = 0.1008$ |
| goodness of fit | 1.155 | 1.005 | 1.073 | 1.133 |
| no. of variables | 71 | 66 | 55 | 163 |
| largest difference map peak and hole eÅ$^{-3}$ | 0.806 and 1.595 | 1.129 and 1.625 | 1.338 and 0.618 | 0.775 and 1.652 |

$^a$ $R_1 = \Sigma||F_o| - |F_c||/\Sigma|F_o|$. $^b$ $wR_2 = \{\Sigma[w(F_o^2 - F_c^2)^2] / \Sigma[w(F_o^2)^2]\}^{1/2}$, $w = 1/[\sigma^2(F_o)^2 + (aP)^2 + bP]$, $P = [\max.(F_o^2, 0) + 2(F_c)^2]/3$; where $a = 0.0565$, $b = 9.1718$ for I; $a = 0.0493$, b = 0.0000 for **II**; $a = 0.0431$, $b = 14.3797$ for **III**; and $a = 0.0352$, $b = 10.3425$ for **IV**.

The structure was solved by direct methods using SHELXS-86[11] and difference Fourier syntheses. An absorption correction based on symmetry equivalent reflections was applied using SADABS program.[12]The hydrogen positions associated with the water molecules in the case of compounds **I**−**IV** could not be located from the difference Fourier map. The last cycles of refinement included atomic positions for all the atoms, anisotropic thermal parameters for all non-hydrogen atoms, and isotropic thermal parameters for all the hydrogen atoms. Full-matrix least-squares structure refinement against $|F^2|$ was carried out using SHELXTL-PLUS[13] package of programs. Details of the final refinements are given in Table 3. The final atomic coordinates, selected bond distances, and angles for compounds **I** are presented in Tables 4 and 5, for **II** in Tables 6 and 7, for **III** in Tables 8 and 9, and for **IV** in Tables 10 and 11.

**Table 4. Final Atomic Coordinates [x10$^4$] and Equivalent Isotropic Displacement Parameters [Å$^2$ x 10$^3$] for I, [RbCl][$Cd_6(C_2O_4)_6$]·$2H_2O$**

| atom | *x* | *y* | *z* | *U*(eq)$^a$ |
|---|---|---|---|---|
| Cd(1) | 2567(1) | 3862(1) | 96(1) | 15(1) |
| Rb(1) | 0 | 0 | 0 | 18(1) |
| Cl(1) | 3333 | 6667 | 1667 | 15(1) |
| O(1) | 1121(7) | 5001(7) | 587(2) | 19(1) |
| O(2) | 3151(8) | 3164(8) | 96(2) | 24(2) |
| O(3) | 2194(8) | 1212(8) | 1048(2) | 25(2) |
| O(4) | 4191(8) | 3795(7) | 1697(2) | 19(1) |
| C(1) | 342(11) | 4467(11) | 136(4) | 19(2) |
| C(2) | 2770(11) | 934(10) | 1481(4) | 15(2) |
| O(100) | 6667 | 3333 | 767(20) | 232(18) |

$^a$ *U*(eq) is defined as one-third of the trace of the orthogonalized $U_{ij}$ tensor.

## Results

**[RbCl][$Cd_6(C_2O_4)_6$]·$2H_2O$, I.** The asymmetric unit of **I** contains 10 non-hydrogen atoms, of which seven atoms make up cadmium oxalate host framework, and the remaining three are the guest molecules (Rb, Cl, and water). The Cd is six coordinated with respect to the oxygen atoms forming a triangular prism with Cd−O distances in the range 2.292(5)−2.384(6) Å (av 2.339 Å). The oxygens are linked to the carbon atoms forming the oxalate unit, with average C−O distances of 1.248 Å (Table 5). The O−Cd−O bond angles are in the range 69.8(2)−148.5(2)° (av 102.9°) (Table 5). The Rb ions are coordinated by 12 oxygen atoms at a distance of ~3 Å and the $Cl^-$ ions are surrounded by six cadmium atoms at ~2.9 Å (Table 5). These geometrical parameters are in good agreement with those reported for similar compounds.

The framework structure of **I** consists of a network of cadmium and oxalate units. The connectivity between these units is such that they form an unusual cadmium oxide cluster of the composition [$Cd_6O_{24}$]. The clusters are connected into three-dimensions by the oxalate units. The $Cl^-$ ions are located inside these clusters (Figure 1) with the charge compensating $Rb^+$ ions located in the cavities formed between these clusters (Figure 2). The $Rb^+$ and $Cl^-$ ions are perfectly ordered in three dimensions forming an independent rock-salt (*Fm*3*m*) structure with interpenetrating *fcc* lattices (Figure 3). The cubic unit cell parameter of RbCl is 13.452 Å, which is roughly double that in ordinary RbCl

(11) Sheldrick, G. M. *SHELXS-86 Program for Crystal Structure Determination*; University of Gottingen: Gottingen, Germany 1986; *Acta Crystallogr.* **1990**, *A35*, 467.
(12) Sheldrick, G. M. *SADABS Siemens Area Detector Absorption Correction Program*; University of Gottingen: Gottingen, Germany, 1994.
(13) Sheldrick, G. M. *SHELXS-93 Program for Crystal Structure Solution and Refinement*; University of Gottingen: Gottingen, Germany, 1993.

**Table 5. Selected Bond Distances and Bond Angles for I, $[RbCl][Cd_6(C_2O_4)_6]\cdot 2H_2O$**

| moiety[a] | distance, Å | moiety[a] | distance, Å |
|---|---|---|---|
| Cd(1)O(1) | 2.292(5) | Rb(1)O(3)#5 | 3.078(6) |
| Cd(1)O(2) | 2.327(6) | Rb(1)O(3) | 3.078(6) |
| Cd(1)O(3) | 2.341(6) | Cl(1)Cd(1)#7 | 2.8919(6) |
| Cd(1)O(4) | 2.342(6) | Cl(1)Cd(1)#1 | 2.8918(6) |
| Cd(1)O(4)#1 | 2.348(6) | Cl(1)Cd(1)#8 | 2.8918(6) |
| Cd(1)O(1)#2 | 2.384(6) | Cl(1)Cd(1)#7 | 2.8918(6) |
| Cd(1)Cl(1) | 2.8917(6) | Cl(1)Cd(1)#2 | 2.8918(6) |
| Rb(1)O(2) | 2.973(6) | O(1)C(1) | 1.259(11) |
| Rb(1)O(2)#3 | 2.973(6) | O(1)Cd(1)#7 | 2.384(6) |
| Rb(1)O(2)#4 | 2.973(6) | O(2)C(1)#6 | 1.237(11) |
| Rb(1)O(2)#5 | 2.973(6) | O(3)C(2) | 1.252(10) |
| Rb(1)O(2)#6 | 2.973(6) | O(4)C(2)#9 | 1.253(10) |
| Rb(1)O(2)#4 | 2.973(6) | O(4)Cd(1)#8 | 2.348(5) |
| Rb(1)O(3)#3 | 3.078(6) | C(1)O(2)#4 | 1.237(11) |
| Rb(1)O(3)#6 | 3.078(6) | C(1)C(1)#10 | 1.57(2) |
| Rb(1)O(3)#4 | 3.078(6) | C(2)O(4)#9 | 1.253(10) |
| Rb(1)O(3)#4 | 3.078(6) | C(2)C(2)#9 | 1.53(2) |

| moiety[a] | angle, deg | moiety[a] | angle, deg |
|---|---|---|---|
| O(1)Cd(1)–O(2) | 93.2(2) | O(2)Cd(1)–O(1)#2 | 69.8(2) |
| O(1)Cd(1)–O(3) | 136.1(2) | O(3)Cd(1)–O(1)#2 | 123.0(2) |
| O(2)Cd(1)–O(3) | 73.0(2) | O(4)Cd(1)–O(1)#2 | 87.7(2) |
| O(1)Cd(1)–O(4) | 148.5(2) | O(4)#1–Cd(1)O(1)#2 | 147.4(2) |
| O(2)Cd(1)–O(4) | 114.3(2) | O(2)#4–C(1)O(1) | 126.4(8) |
| O(3)Cd(1)–O(4) | 70.2(2) | O(2)#4–C(1)C(1)#10 | 117.3(10) |
| O(1)Cd(1)–O(4)#1 | 81.8(2) | O(1)C(1)–C(1)#10 | 116.2(10) |
| O(2)Cd(1)–O(4)#1 | 141.2(2) | O(3)C(2)–O(4)#9 | 124.8(8) |
| O(3)Cd(1)–O(4)#1 | 84.2(2) | O(3)C(2)–C(2)#9 | 118.0(9) |
| O(4)Cd(1)–O(4)#1 | 85.35(7) | O(4)#9–C(2)C(2)#9 | 117.3(9) |
| O(1)Cd(1)–O(1)#2 | 87.9(3) | | |

[a] Symmetry transformations used to generate equivalent atoms: #1, $y - 1/3, -x + y + 1/3, -z + 1/3$; #2, $-y + 1, x - y + 1, z$; #3, $-x, -y, -z$; #4, $x - y, x, -z$; #5, $-x + y, -x, z$; #6, $y, -x + y, -z$; #7, $-x + 2/3, -y + 4/3, -z + 1/3$; #8, $x - y + 2/3, x + 1/3, -z + 1/3$; #9, $-x + 2/3, -y + 1/3, -z + 1/3$; #10, $-x, -y + 1, -z$.

**Table 6. Final Atomic Coordinates [x10[4]] and Equivalent Isotropic Displacement Parameters [Å$^2$ x 10$^3$] for II, $2[CsBr][Cd(C_2O_4)]\cdot H_2O$**

| atom | *x* | *y* | *z* | *U*(eq)[a] |
|---|---|---|---|---|
| Cd(1) | 8496(2) | 871(1) | 2500 | 21(1) |
| Cs(1) | 6536(1) | 3910(1) | 3847(1) | 32(1) |
| Br(1) | 8483(2) | 809(1) | 896(1) | 40(1) |
| C(1) | 8168(20) | 1994(10) | 2500 | 15(3) |
| C(2) | 3271(21) | 1789(12) | 2500 | 23(3) |
| O(1) | 7119(15) | 1053(8) | 2500 | 28(2) |
| O(2) | 5319(13) | 1983(8) | 2500 | 30(2) |
| O(3) | 2291(15) | 816(7) | 2500 | 29(2) |
| O(4) | 9807(13) | 2882(8) | 2500 | 29(2) |
| O(100) | 2733(18) | 2500 | 5000 | 93(5) |

[a] *U*(eq) is defined as one-third of the trace of the orthogonalized $U_{ij}$ tensor.

(6.570 Å) as shown in Figure 4. The [100] RbCl planes run along the [111] direction of **I** (Figure 3). An alternate way to describe the structure of **I** would be to consider the $Cl^-$ ion as being at the center of a cluster of six Cd atoms arranged in such a manner that they form the six vertexes of an octahedron, with the next nearest oxygen neighbors forming the trigonal prisms with Cd atoms. Accordingly, **I** could be considered as a super rock-salt arrangement of two types of clusters, $[ClCd_6(C_2O_4)_3]^{5+}$ and $[Rb(H_2O)_2(C_2O_4)_3]^{5-}$. This type of arrangement is reminiscent of Chevrel phases.[14]

**$2[CsBr][Cd(C_2O_4)]\cdot H_2O$, II.** The asymmetric unit of **II** contains 10 non-hydrogen atoms, of which eight atoms make up cadmium bromo-oxalate host framework and the remaining two are the guest molecules (Rb and water). The Cd atom is in an octahedral coordination with four oxygen and two bromine atoms. The Cd–O distances are in the range 2.290(8)–2.367(9) Å (av 2.315 Å), and the Cd–Br distances have an average value of 2.710 Å. The oxygens, in turn, are linked with carbon

(14) (a) Subba Rao, G. V.; Balakrishnan, G. *Bull. Mater. Sci. (India)* **1984**, *6*, 283 and the references therein. (b) Subba Rao, G. V. *Proc. Ind. Natl. Sci. Acad.* **1986**, *52A*, 292.

**Table 7. Selected Bond Distances and Bond Angles for II, $2[CsBr][Cd(C_2O_4)]\cdot H_2O$**

| moiety[a] | distance, Å | moiety[a] | distance, Å |
|---|---|---|---|
| Cd(1)O(1) | 2.290(8) | Br(1)Cs(1)#10 | 3.6706(13) |
| Cd(1)O(2) | 2.293(8) | Br(1)Cs(1)#3 | 3.7153(12) |
| Cd(1)O(3)#1 | 2.310(9) | Br(1)Cs(1)#5 | 3.7336(13) |
| Cd(1)O(4) | 2.367(9) | C(1)O(1) | 1.223(13) |
| O(3)Cd(1)#11 | 2.310(9) | C(1)O(4)#3 | 1.240(13) |
| Cd(1)Br(1) | 2.7099(11) | C(1)C(2)#5 | 1.61(2) |
| Cd(1)Br(1)#2 | 2.7100(11) | C(2)O(3) | 1.233(14) |
| Cs(1)Br(1)#2 | 3.6603(13) | C(2)O(2) | 1.264(14) |
| Cs(1)Br(1)#8 | 3.6706(13) | C(2)C(1)#7 | 1.61(2) |
| Cs(1)Br(1)#9 | 3.7153(12) | O(3)Cs(1)#5 | 3.185(6) |
| Cs(1)Br(1)#7 | 3.7336(13) | O(3)Cs(1)#6 | 3.185(6) |
| Br(1)Cs(1)#2 | 3.6602(13) | O(4)C(1)#9 | 1.240(13) |

| moiety[a] | angle, deg | moiety[a] | angle, deg |
|---|---|---|---|
| O(1)Cd(1)–O(2) | 101.0(3) | O(3)#1–Cd(1)Br(1) | 90.14(3) |
| O(1)Cd(1)–O(3)#1 | 110.0(3) | O(4)Cd(1)–Br(1) | 91.43(3) |
| O(2)Cd(1)–O(3)#1 | 149.0(3) | O(1)C(1)–O(4)#3 | 127.8(11) |
| O(1)Cd(1)–O(4) | 178.2(3) | O(1)C(1)–C(2)#5 | 115.5(11) |
| O(2)Cd(1)–O(4) | 77.2(3) | O(4)#3–C(1)C(2)#5 | 116.7(10) |
| O(3)#1–Cd(1)O(4) | 71.8(3) | O(3)C(2)–O(2) | 128.7(12) |
| O(1)Cd(1)–Br(1) | 88.58(3) | O(3)C(2)–C(1)#7 | 118.0(11) |
| O(2)Cd(1)–Br(1) | 90.63(3) | O(2)C(2)–C(1)#7 | 113.3(11) |

[a] Symmetry transformations used to generate equivalent atoms: #1, $x + 1, y, z$; #2, $x, y, -z + 1/2$; #3, $-x + 2, y - 1/2, -z + 1/2$; #4, $-x + 2, y - 1/2, z$; #5, $-x + 1, y - 1/2, -z + 1/2$; #6, $-x + 1, y - 1/2, z$; #7, $-x + 1, y + 1/2, -z + 1/2$; #8, $x, -y + 1/2, z + 1/2$; #9, $-x + 2, y + 1/2, -z + 1/2$; #10, $x, -y + 1/2, z - 1/2$; #11, $x - 1, y, z$; #12, $x, -y + 1/2, -z + 1$.

**Table 8. Final Atomic Coordinates [x10[4]] and Equivalent Isotropic Displacement Parameters [Å$^2$ x 10$^3$] for III, $2[CsBr][Cd_2(C_2O_4)(Br)_2]\cdot 2H_2O$**

| atom | *x* | *y* | *z* | *U*(eq)[a] |
|---|---|---|---|---|
| Cd(1) | 1243(1) | 4005(1) | 2500 | 28(1) |
| Cs(1) | 3116(1) | 3170(1) | 2500 | 47(1) |
| Br(1) | 1692(1) | 1663(1) | 2500 | 52(1) |
| Br(2) | 3074(1) | 0 | 0 | 33(1) |
| C(1) | 0 | 3522(9) | 1499(13) | 22(2) |
| O(1) | 468(2) | 3522(5) | 772(6) | 32(1) |
| O(200) | 4173(3) | 1145(7) | 2500 | 33(2) |
| O(100) | 101(69) | 118(72) | 6151(186) | 739(227) |

[a] *U*(eq) is defined as one-third of the trace of the orthogonalized $U_{ij}$ tensor.

**Table 9. Selected Bond Distances and Bond Angles for III, $2[CsBr][Cd_2(C_2O_4)(Br)_2]\cdot 2H_2O$**

| moiety[a] | distance, Å | moiety[a] | distance, Å |
|---|---|---|---|
| Cd(1)O(1)#1 | 2.329(5) | Cs(1)Br(1) | 3.703(2) |
| Cd(1)O(1) | 2.329(5) | Br(1)Cs(1)#6 | 3.610(2) |
| Cd(1)O(200)#2 | 2.405(7) | Br(2)Cd(1)#6 | 2.7332(8) |
| Cd(1)Br(1) | 2.6248(14) | Br(2)Cd(1)#4 | 2.7332(8) |
| Cd(1)Br(2)#3 | 2.7332(8) | C(1)O(1) | 1.243(7) |
| Cd(1)Br(2)#4 | 2.7332(8) | C(1)O(1)#8 | 1.243(7) |
| Cs(1)Br(1)#2 | 3.610(2) | C(1)C(1)#1 | 1.57(2) |

| moiety[a] | angle, deg | moiety[a] | angle, deg |
|---|---|---|---|
| O(1)#1–Cd(1)O(1) | 70.9(2) | O(1)C(1)–C(1)#1 | 117.2(5) |
| O(1)#1–Cd(1)O(200)#2 | 82.7(2) | O(1)#8–C(1)C(1)#1 | 117.2(5) |
| O(1)Cd(1)–O(200)#2 | 82.7(2) | C(1)O(1)–Cd(1) | 115.7(5) |
| O(1)C(1)–O(1)#8 | 125.6(9) | Cd(1)#6–O(200)Cs(1) | 105.6(2) |

[a] Symmetry transformations used to generate equivalent atoms: #1, $x, y, -z + 1/2$; #2, $-x + 1/2, y + 1/2, -z + 1/2$; #3, $-x + 1/2, -y + 1/2, z + 1/2$; #4, $-x + 1/2, -y + 1/2, -z$; #5, $-x + 1/2, -y + 1/2, -z + 1$; #6, $-x + 1/2, y - 1/2, -z + 1/2$; #7, $x, -y, z - 1/2$; #8, $-x, y, z$; #9, $x, -y, -z + 1$.

**Table 10. Final Atomic Coordinates [x10⁴] and Equivalent Isotropic Displacement Parameters [Å² x 10³] for IV, $3[RbCl][Cd_2(C_2O_4)(Cl_2)]\cdot H_2O$**

| atom | *x* | *y* | *z* | *U*(eq)[a] |
|---|---|---|---|---|
| Cd(1) | 3176(1) | 4425(1) | 4647(1) | 24(1) |
| Cd(2) | 62(1) | 3703(1) | 9969(1) | 22(1) |
| Rb(1) | 393(1) | 3103(1) | 5741(1) | 33(1) |
| Rb(2) | 4892(1) | 2793(1) | 2861(1) | 36(1) |
| Rb(3) | 4799(1) | 4150(1) | 9398(1) | 36(1) |
| Cl(1) | 4605(3) | 3526(1) | 6111(3) | 30(1) |
| Cl(2) | 1497(3) | 3657(1) | 2765(3) | 35(1) |
| Cl(3) | 4832(3) | 5177(1) | 6845(3) | 33(1) |
| Cl(4) | 2435(3) | 3006(1) | 10294(3) | 32(1) |
| Cl(5) | 2212(3) | 3009(1) | 9392(3) | 28(1) |
| C(1) | 902(11) | 5002(4) | 10186(10) | 22(2) |
| C(2) | 47(12) | 5220(4) | 4394(10) | 23(2) |
| O(1) | 1282(8) | 5181(3) | 3804(7) | 30(2) |
| O(2) | 1511(8) | 4507(3) | 9666(7) | 28(2) |
| O(3) | 1605(8) | 4528(3) | 10268(7) | 29(2) |
| O(4) | 1083(8) | 5612(3) | 4100(7) | 29(2) |
| O(10) | 1496(8) | 3782(3) | 7500(7) | 28(2) |
| O(100) | 7572(10) | 3677(4) | 3508(9) | 51(2) |

[a] *U*(eq) is defined as one-third of the trace of the orthogonalized $U_{ij}$ tensor.

**Table 11. Selected Bond Distances and Bond Angles for IV, $3[RbCl][Cd_2(C_2O_4)(Cl_2)]\cdot H_2O$**

| moiety[a] | distance, Å | moiety[a] | distance, Å |
|---|---|---|---|
| Cd(1)O(4)#1 | 2.286(6) | Rb(2)Cl(4)#9 | 3.646(3) |
| Cd(1)O(1) | 2.315(6) | Rb(3)Cl(5) | 3.343(3) |
| Cd(1)Cl(3)#2 | 2.549(2) | Rb(3)Cl(3) | 3.365(3) |
| Cd(1)Cl(1) | 2.583(2) | Rb(3)Cl(1) | 3.371(3) |
| Cd(1)Cl(2) | 2.620(3) | Rb(3)Cl(4)#10 | 3.408(3) |
| Cd(1)Cl(3) | 2.765(3) | Rb(3)Cl(3)#11 | 3.796(3) |
| Cd(2)O(2) | 2.293(6) | Cl(1)Rb(2)#5 | 3.418(3) |
| Cd(2)O(3) | 2.316(6) | Cl(2)Cd(2)#8 | 2.618(2) |
| Cd(2)O(10) | 2.327(6) | Cl(3)Cd(1)#2 | 2.549(2) |
| Cd(2)Cl(4) | 2.569(2) | Cl(3)Rb(3)#11 | 3.796(3) |
| Cd(2)Cl(5) | 2.585(2) | Cl(4)Rb(2)#12 | 3.281(3) |
| Cd(2)Cl(2)#3 | 2.618(2) | Cl(4)Rb(1)#5 | 3.369(3) |
| Rb(1)Cl(5)#6 | 3.342(3) | Cl(4)Rb(3)#4 | 3.408(3) |
| Rb(1)Cl(4)#6 | 3.369(3) | Cl(4)Rb(2)#13 | 3.646(3) |
| Rb(1)Cl(2) | 3.386(3) | Cl(5)Rb(1)#5 | 3.342(3) |
| Rb(1)Cl(5) | 3.386(3) | Cl(5)Rb(2)#5 | 3.414(2) |
| Rb(1)Cl(1) | 3.465(3) | Cl(5)Rb(2)#3 | 3.470(3) |
| Rb(2)Cl(4)#7 | 3.281(3) | C(1)O(3) | 1.237(11) |
| Rb(2)Cl(2) | 3.362(3) | C(1)O(2)#14 | 1.249(11) |
| Rb(2)Cl(5)#6 | 3.414(2) | C(1)C(1)#14 | 1.58(2) |
| Rb(2)Cl(1)#6 | 3.418(3) | C(2)O(1) | 1.258(11) |
| Rb(2)Cl(5)#8 | 3.470(3) | C(2)O(4) | 1.259(11) |
| Rb(2)Cl(1) | 3.543(3) | C(2)C(2)#1 | 1.54(2) |

| moiety[a] | angle, deg | moiety[a] | angle, deg |
|---|---|---|---|
| O(4)#1−Cd(1)O(1) | 72.3(2) | O(3)C(1)−C(1)#14 | 118.1(9) |
| O(2)Cd(2)−O(3) | 72.0(2) | O(2)#14−C(1)C(1)#14 | 116.1(10) |
| O(2)Cd(2)−O(10) | 88.2(2) | O(1)C(2)−O(4) | 124.3(8) |
| O(3)Cd(2)−O(10) | 84.8(2) | O(1)C(2)−C(2)#1 | 118.0(10) |
| O(3)C(1)−O(2)#14 | 125.8(8) | O(4)C(2)−C(2)#1 | 117.7(10) |

[a] Symmetry transformations used to generate equivalent atoms: #1, −*x*, −*y* + 1, −*z* + 1; #2, −*x* + 1, −*y* + 1, −*z* + 1; #3, *x*, *y*, *z* + 1; #4, *x* − 1, *y*, *z*; #5, *x*, −*y* + 1/2, *z* + 1/2; #6, *x*, −*y* + 1/2, *z* − 1/2; #7, *x* + 1, −*y* + 1/2, *z* − 1/2; #8, *x*, *y*, *z* − 1; #9, *x* + 1, *y*, *z* − 1; #10, *x* + 1, *y*, *z*; #11, −*x* + 1, −*y* + 1, −*z* + 2; #12, *x* − 1, −*y* + 1/2, *z* + 1/2; #13, *x* − 1, *y*, *z* + 1; #14, −*x*, −*y* + 1, −*z* + 2.

forming the oxalate unit, with average C−O distances of 1.240 Å (Table 7). The various bond angles, viz., O−Cd−O, O−Cd−Br, and Br−Cd−Br have values in the range 71.8(3)−178.2(3)° (av 114.5°), 88.58(3)−91.43(3)° (av 90.2°), and 177.06(6)° (Table 7). The $Cs^+$ ions are eight-coordinated by four oxygen and bromine atoms at average distances of ~3.2 and 3.7 Å (Table 7). The various geometrical parameters found in **II** are in good agreement with those reported for similar compounds.

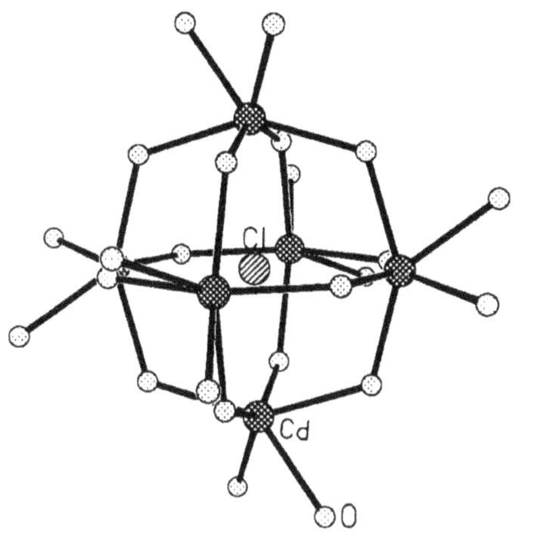


**Figure 1.** Figure showing a single $[Cd_6O_{24}]$ cluster with the $Cl^-$ ions at the center.

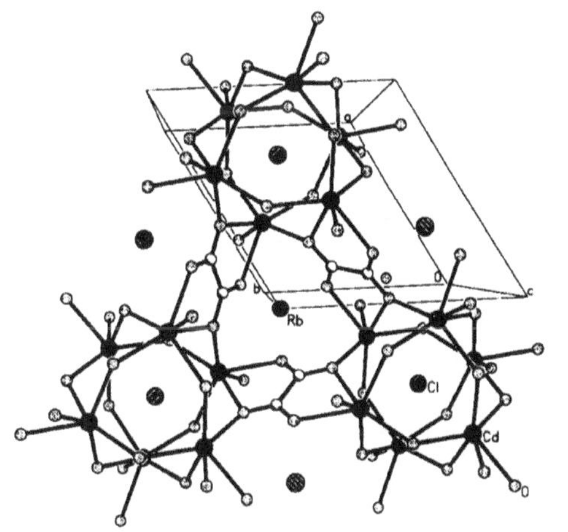


**Figure 2.** Structure of **I** in the *ab* plane. Note that the $Rb^+$ ions are positioned outside the cluster unit.

The framework structure of **II** consists of a network of cadmium and oxalate units forming layers (Figure 5). The $Br^-$ ions connected to the cadmium protrude into the interlamellar space. The $Cs^+$ ions along with a water molecule occupy the interlamellar space (Figure 5). The relative positions between the $Cs^+$ and $Br^-$ ions are such that they form a graphite-like layer (Figure 6). Such an arrangement creates two closely spaced hexagonal CsBr layers separated by 3.67 Å. The CsBr layers are made of neutral hexagonal units of $Cs_3Br_3$, with chair conformation as in cyclohexane. The unusual layered alkali halide structures in **II** are stabilized by the cadmium oxalate layers, present after every two CsBr layers, by providing additional coordination for the $Cs^+$ ions through the oxalate oxygens.

**$2[CsBr][Cd_2(C_2O_4)(Br)_2]\cdot 2H_2O$, III, and $3[RbCl][Cd_2(C_2O_4)(Cl_2)]\cdot H_2O$, IV.** The structures of **III** and **IV** have some common features and can be described together. The asymmetric units of **III** and **IV** contain 8 and 18 non-hydrogen atoms, respectively. The crystallographically distinct cadmium atom in **III** is octahedrally coordinated to oxygen and bromine atoms. There

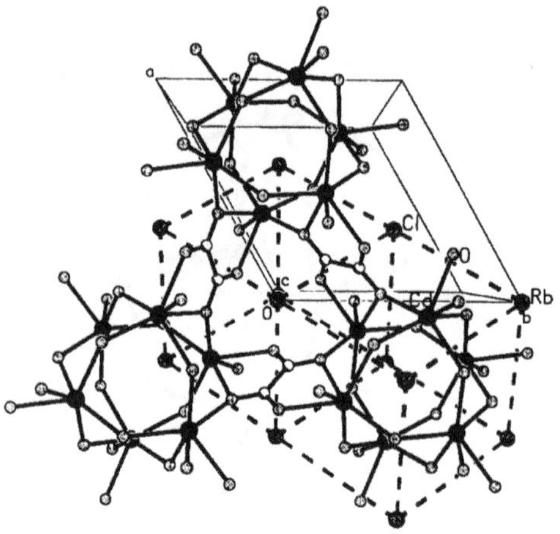


**Figure 3.** Structure showing the *Fm*3*m* lattice of RbCl within the parent cadmium oxalate structure.

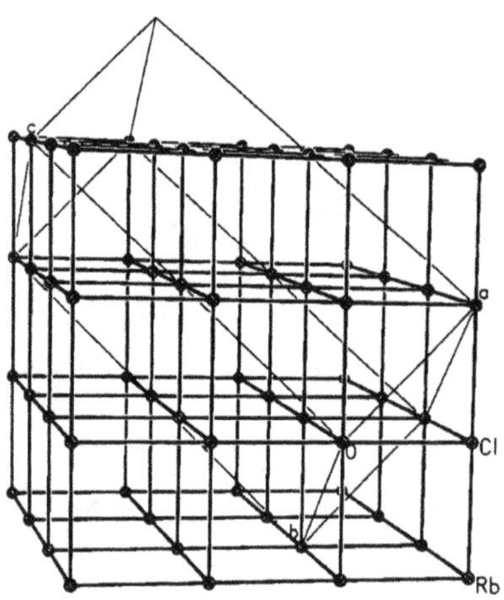


**Figure 4.** Structure of RbCl in **I**. Note that RbCl is along the [*111*] lattice.

are three Cd–O and three Cd–Br linkages with average distances of 2.354 and 2.697 Å. The longer Cd–O linkage with a distance of 2.405(7) Å belongs to the terminal water molecule. The oxygen atoms are connected to a carbon atom forming the oxalate unit, with average C–O distances of 1.243 Å (Table 9). The $Cs^+$ ions are eight coordinated by bromine atoms and possess an additional coordination through the terminal water molecule. The average Cs–Br distance is 3.824 Å, while the O–Cd–O, O–Cd–Br, and Br–Cd–Br bond angles have values in the ranges 70.9(2)–82.7(2)° (av 78.8°), 84.30(12)–179.7(2) (av 118.3°), and 91.33(4)–95.91(3)° (av 94.4°) (Table 9). The O–C–O bond angles and the other observed geometrical parameters are in satisfactory agreement with those reported for similar compounds (Table 9).

There are two crystallographically independent cadmium atoms in **IV**, both of which are octahedrally coordinated by oxygen and chlorine atoms. While Cd(1) possesses two Cd–O (av 2.3005 Å) and four Cd–Cl linkages (av 2.629 Å), Cd(2) has three Cd–O (av 2.312 Å) and Cd–Cl (av 2.591 Å) connections. The oxygen

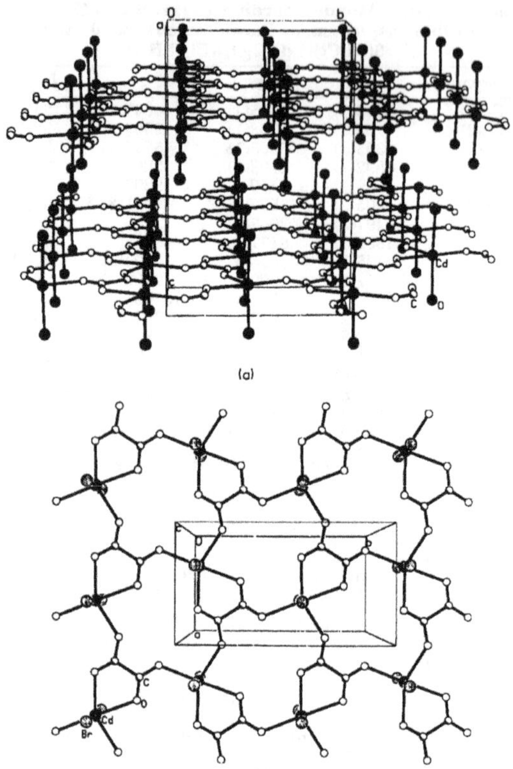


(a)

(b)

**Figure 5.** (a) Structure of **II** in the *bc* plane showing the layer arrangement. Note that the Br atoms point into the interlamellar space. Cs and water molecules are not shown for purposes of clarity. (b) Structure of **II** showing a single layer of the cadmium oxalate.

atoms are connected to carbon atoms and form the oxalate unit with average C–O distances of 1.251 Å (Table 11). There are three unique Rb atoms with Rb(1) and Rb(8), being eight coordinated, and Rb(3) is seven coordinated with chlorine and oxygen atoms. Thus, Rb(1) has five Rb–Cl (av 3.390 Å) and three Rb–O (av 2.978 Å) linkages, Rb(2) has seven Rb–Cl (av 3.448 Å) and one Rb–O (2.914(9) Å) bondings, and Rb(3) has five Rb–Cl (av 3.457 Å) and two Rb–O (av 2.893 Å) connectivities (Table 11). The average O–Cd–O, O–Cd–Cl, and Cl–Cd–Cl bond angles have values of 72.3, 107.24, and 107.5° for Cd(1) and 81.7, 117.1, and 95.18° for Cd(2). These are typical values and are in agreement with those observed earlier for similar compounds.

The framework structures of **III** and **IV** consist of linkages involving cadmium and oxalate units. Connectivity between the two moieties gives rise to interesting structural features. Thus, both **III** and **IV** possess isolated monomeric cadmium oxalate units connected together by the alkali halides, viz., CsBr and RbCl. In **III**, the isolated cadmium oxalate units are connected with the alkali halide in such a manner as to form a three-dimensionally extended channel structure containing water molecules (Figure 7). The alkali halide units are present as eight-membered rings with the chair conformation as in cyclooctane (Figure 8). The

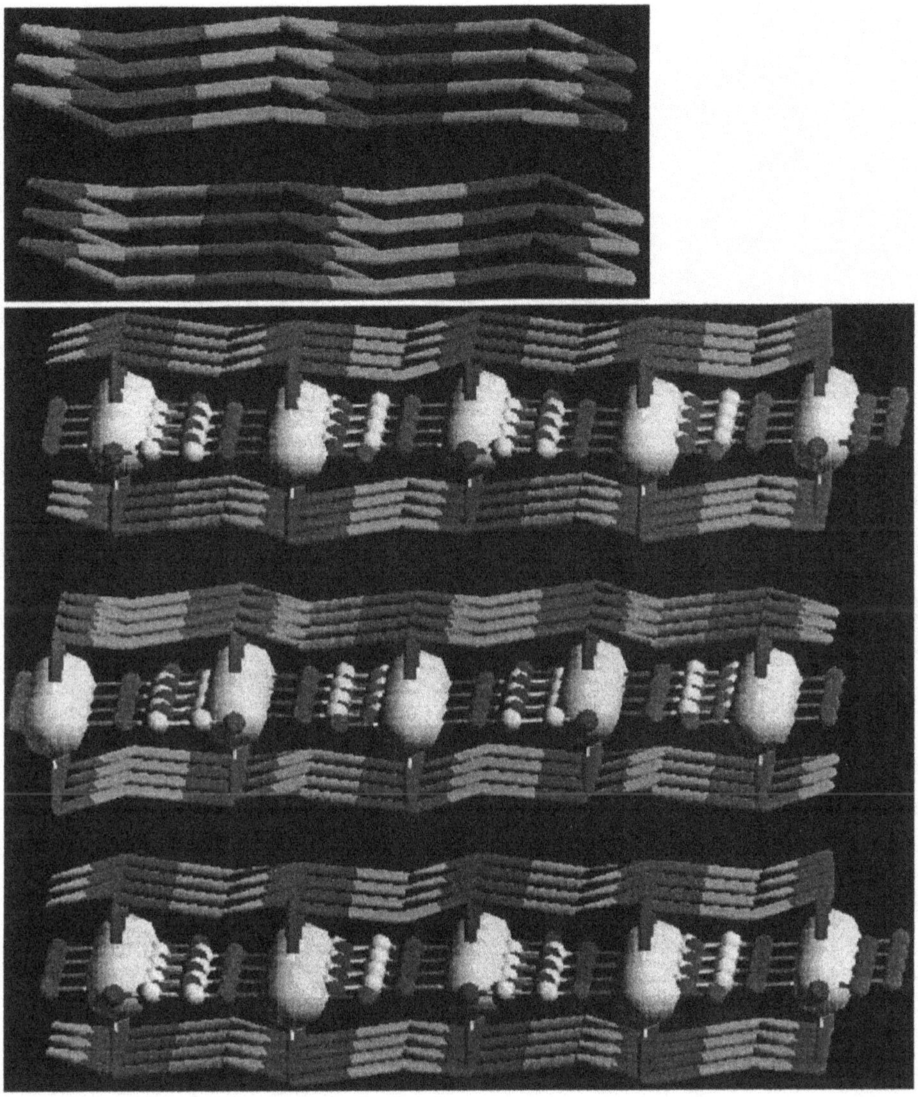

**Figure 6.** Structure of **II** showing the CsBr lattice within the layers. Inset shows the CsBr lattice alone. Note that alternate CsBr layers have larger separation. (Color scheme: yellow, Cd; white, C; red, O; purple, Br; green, Cs.)

connectivity between the isolated oxalate units along with the chloride ions in **IV** results in an interpenetrating layerlike arrangement (Figure 9). The layers are so positioned in a helical fashion, with the immediate next layer arranged in the opposite direction, thus canceling total helicity of the structure (Figure 10). Accordingly, the $Rb^+$ ions also reflect the helical nature of the layers and form a helical strand between the cadmium chlorooxalate layers (Figure 10). The alkali halide, however, does not form an independent lattice in **IV**.

## Discussion

Four novel cadmium oxalates incorporating extended alkali halide structures, $[RbCl][Cd_6(C_2O_4)_6]\cdot 2H_2O$, **I**, $2[CsBr][Cd(C_2O_4)]\cdot H_2O$, **II**, $2[CsBr][Cd_2(C_2O_4)(Br)_2]\cdot 2H_2O$, **III**, and $3[RbCl][Cd_2(C_2O_4)(Cl_2)]\cdot H_2O$, **IV**, have been prepared hydrothermally as good quality single crystals. The reaction itself involves simple metathetic exchange between a cadmium halide and an alkali oxalate. The cadmium oxalates **I** and **IV** are formed with RbCl; **II** and **III** are formed with CsBr. Though RbCl is part of the structure in both **I** and **IV**, the way the alkali halide units are arranged within the host cadmium oxalate is very different. In **II** and **III**, the CsBr has a layered structure with different apertures, but the host lattice possesses completely different structures. These fine differences in the structures are likely to result from the differences in experimental conditions such as the composition, time, and temperature (Table 1), but we are not yet in a position to relate the two.

The host oxalate structures have different architectures in **I**–**IV**, ranging from isolated cadmium oxalate

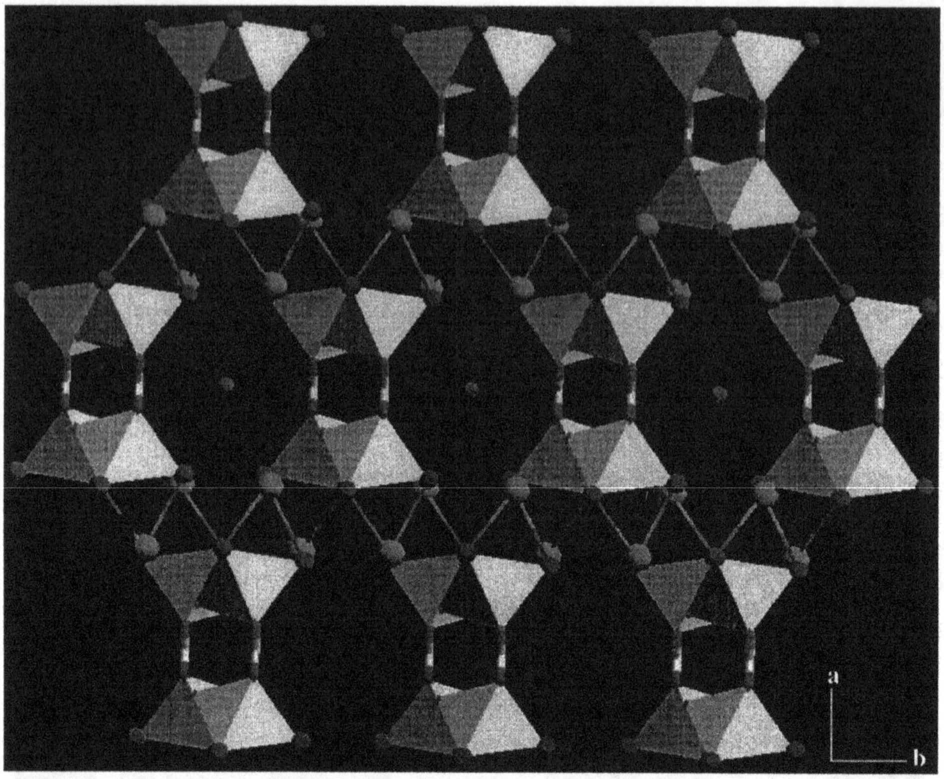


**Figure 7.** Structure of **III** showing the channels formed by the connectivity between the isolated cadmium oxalate and CsBr. Note that the water molecules occupy these channels. (Color scheme: yellow, Cd; white, C; red, O; purple, Br; green, Cs.)

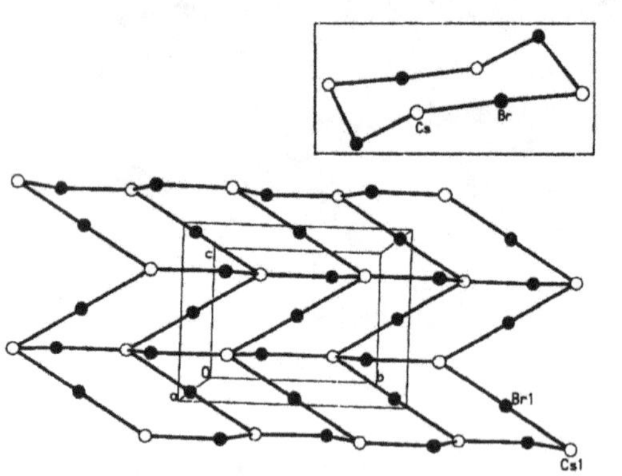


**Figure 8.** Layered structure of CsBr in **III**. Inset shows the CsBr in chair conformation.

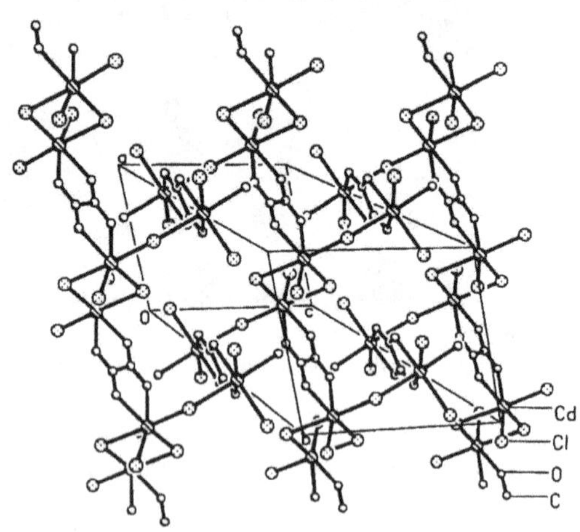


**Figure 9.** Structure of **IV** showing a single layer resulting from the connectivity between the isolated cadmium oxalate and chlorine atoms.

units to three-dimensionally extended architectures. While **I** possesses two interpenetrating lattices of three-dimensional cadmium oxalate and a $Fm3m$ RbCl lattice, **II** has a layered cadmium oxalate architecture (two-dimensional) with a layered CsBr structure. Compounds **III** and **IV** have isolated cadmium oxalate units (zero-dimensional) linked and stabilized by the alkali halides. The isolated cadmium oxalate units in **IV** are linked via a chlorine atom to form a sheet of cadmium chloro-oxalate that is arranged in a helical fashion. These are indeed unique architectures. In addition to the uniqueness of the host structures, the alkali halide guest species also possesses remarkable architectures.

In **I**, RbCl forms a face-centered lattice with unit-cell parameters which are approximately double that of the normal RbCl. In **II**, CsBr forms a layered architecture with the layers closely resembling that of graphite. The six-membered $Cs_3Br_3$ ring, in **II**, is in a chair conformation of the cyclohexane. The two closely spaced CsBr layers have a thickness of 3.67 Å. The thickness of the oxalate layers is 10.62 Å. Accordingly, **II** can be considered to be a genuine hetero-lattice or nanocomposite

**Figure 10.** Structure of **IV** along the *a* axis showing the cadmium chloro-oxalate layers and the Rb atoms. Note that both the cadmium chloro-oxalate and the Rb atoms form a helical structure. (Color scheme: yellow, Cd; white, C; red, O; purple, Cl; green, Rb)

with alternating organic and inorganic layers (Figure 6). In **III**, CsBr again forms a layered structure but is unusual in that it consists exclusively of eight-membered rings that resemble the chair conformation of cyclooctane (Figure 8). What is interesting in **III** is that the isolated cadmium oxalate units are situated between these CsBr layers and act as a pillar through Cd–Br connectivity. The thickness of the CsBr layers in **III** is ~3.7 Å, which is similar to the closely spaced CsBr layers in **II** (Figure 7). The oxalate layer has a thickness of 5.87 Å, which is roughly half that observed in **II**. In **IV**, the chlorine atoms connect the Cd ions of the isolated cadmium oxalate unit and form a two-dimensional layer with a right-handed helical strand with the immediate next layer having the opposite handedness, canceling the total helicity of the structure. The Rb atoms situated between these sheets reflect the helical nature of the cadmium chloro-oxalate layer (host), with the Rb atoms from the neighboring strand so positioned as to cancel the helicity of the structure. Thus, there is no independent alkali halide lattice in **IV** (Figure 10).

The distances between the alkali metal and the halogen ions in compounds **I**–**III** provide an interesting comparison. In **I**, the Rb–Cl distance is 6.726 Å, while in RbCl with rock-salt structure, it is 3.285 Å (Figure 4). In the case of compounds **II** and **III**, the average Cs–Br distance is ~3.7 Å, which is comparable to the Cs–Br distance (3.72 Å) in body-centered CsBr. It may be noted that in **I**, the RbCl forms a lattice identical to that of normal RbCl, but with double the unit-cell dimensions, whereas in **II** and **III** the CSBr forms unusual layered architectures despite having nearly similar distances between Cs and Br as in normal CsBr. These observations demonstrate the role of host–guest interactions in these structures.

Host–guest structures of the type observed in **I**–**IV** possessing expanded RbCl and layered CsBr are indeed most unusual. These structures are formed because the host structures provide the additional coordination necessary for the alkali halide ions. Thus, the expanded rock-salt structure of RbCl is made possible by the stability provided by the extra coordination due to the

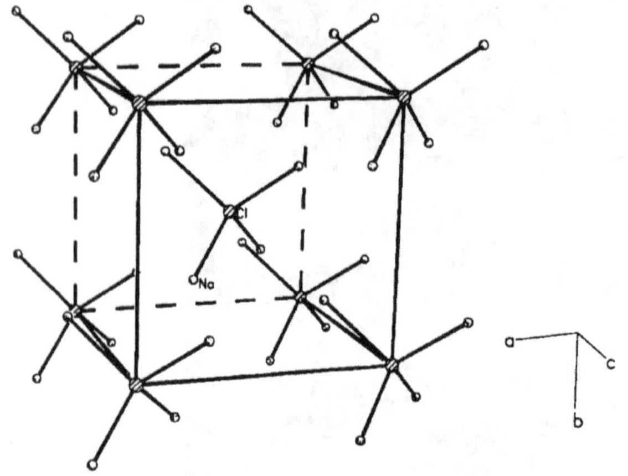


**Figure 11.** Structure of **V**, showing the body-centered cubic arrangement of the $[ClNa_4]$ units.

oxalate units. What is truly interesting, however, is that such beautiful structures are formed by a simple metathetic reaction carried out under hydrothermal conditions. Such a reaction under ambient conditions would clearly give a mixture of the Cd oxalate and the alkali halide. In this connection, it may be noted that the structures reported herein are different from those resulting from ion-exchange in layered solids, such as metal–anion arrays within the oxide hosts.[15] Thus, in the layered $RbLaNb_2O_7$, the $Rb^+$ ions, situated in the interlamellar region, are exchanged topotactically by copper halide, $CuX_2$ (X = Cl and Br), in evacuated and sealed quartz tubes at 325 °C.[15] This gives rise to a unique situation wherein the ability to prepare hosts at high temperatures, with varying interatomic spacings and topologies, is combined with the low-temperature ion-exchange methods to construct metal–anion arrays. The host–guest structures of **I**–**IV** are somewhat comparable to the KCl/CsCl structures reported in a phosphate framework.[16] Accordingly, in $A_2[M_3(X_2O_7)_2]$-[salt] (A = Rb, Cs; M = Mn, Cu; X = P, As), synthesized under molten salt conditions (>600 °C), the phosphate network is formed around the alkali metal chloride salts, the salt probably acting as a structure-directing template. The super rock-salt structure formed by the clusters in **I** can be considered to be a result of templating by the underlying RbCl lattice. This is reminiscent of organic molecules employed in conventional low-temperature hydrothermal methods for the synthesis of extended framework architectures.

It is likely that many more materials containing alkali and alkaline earth halides in unusual coordinations and structures may be formed by a suitable choice of the framework. In this direction, we have just isolated a new cadmium succinate, $[NaCl][Cd_5(H_4C_4O_4)_2Na_6]$, **V**, incorporating NaCl. In **V**, however, we have extra $Na^+$ ions in addition to the alkali halide, similar to the metal–anion arrays within the oxide hosts established in Dion-Jacobson structures.[15] The alkali halide does not form any independent lattice in **V**. The relative positions of $Na^+$ and $Cl^-$ ions in **V** are such that four $Na^+$ ions coordinate one $Cl^-$ ion in a tetrahedral fashion. The tetrahedral $[ClNa_4]$ units form a body-centered cube within the cadmium succinate structure (Figure 11).

The new types of hybrid host–guest compounds incorporating novel alkali halide structures are likely to possess novel properties. In particular, they would be expected to exhibit interesting optical and dielectric properties. The three-dimensional organic–inorganic periodic superlattice would also bestow certain unusual properties. It would be of interest to analyze the Infrarad and Raman spectra of the alkali halide guests in these compounds. Such studies will be pursued, although at this stage we primarily report the structures, which are unique in their own merit.

CM011011+

(15) Kodenkandath, T. A.; Lalena, J. N.; Zhou, W. I.; Carpenter, E. E.; Sangregorio, Falster, A. U.; Simmons, J. R.; O'Connor, C. J.; Wiley, J. B. *J. Am. Chem. Soc.* **1999**, *121*, 10 743.

(16) Huang, Q.; Ulutagy, M.; Michener, P. A.; Hwu, S.-J. *J. Am. Chem. Soc.* **1999**, *121*, 10323.

# Organically-templated metal sulfates, selenites and selenates

**C. N. R. Rao,†* J. N. Behera and Meenakshi Dan**

***Received 16th October 2005***
***First published as an Advance Article on the web 30th January 2006***
**DOI: 10.1039/b510396g**

The literature on inorganic open-framework materials abounds in the synthesis and characterization of metal silicates, phosphates and carboxylates. Most of these materials have an organic amine as the template. In the last few years, it has been shown that anions such as sulfate, selenite and selenate can also be employed to obtain organically templated open-framework materials. This tutorial review provides an up-to-date survey of organically templated metal sulfates, selenites and selenates, prepared under hydrothermal conditions. The discussion includes one-, two-, and three-dimensional structures of these materials, many of which possess open architectures. The article should be useful to practitioners of inorganic and materials chemistry, besides students and teachers. The article serves to demonstrate how most oxy-anions can be used to build complex structures with metal–oxygen polyhedra.

## Introduction

Inorganic open-framework compounds constitute an important class of materials that has attracted much attention as evidenced by the vast number of research papers published in the last few years. Although work in this area started with aluminium silicates[1,2] because of their important uses in sorption and catalysis, much of the recent work pertains to the structure and characterization of open-framework metal phosphates and carboxylates. Both these families of open-framework materials have been reviewed recently.[3–7] Since open-framework silicates and phosphates can essentially be considered as resulting from different arrangements of the metal–oxygen polyhedra and the anion tetrahedra, it seemed rational to examine whether open-framework compounds similar to metal phosphates are readily formed by an oxy-anions such as the sulfate. The main difference between the last two anions and the silicate and phosphate ions relates to the formal oxidation states or the charge on the central atom. In the bonds formed with metals, sulfur would have a higher charge compared to Si–O–M or P–O–M bonds. The sulfate bonds are also least covalent. Sulfates should, however, form stable structures with relatively low valent metal ions. Sulfates rarely form rings unlike silicates and the latter polymerize readily. In spite of such factors, it was not obvious why the sulfate ion should not form open-framework structures with metal–oxygen polyhedra. Recent efforts in this direction have indeed been successful in isolating several organically templated open-framework metal sulfates. A few amine-templated metal selenites and selenates have also been reported recently.

*Chemistry and Physics of Materials Unit and CSIR Centre of Excellence in Chemistry, Jawaharlal Nehru Centre for Advanced Scientific Research, Jakkur P.O., Bangalore 560064, India. E-mail: cnrrao@jncasr.ac.in; Fax: +91-80-2362-2760*
† Also at, Solid State and Structural Chemistry Unit, Indian Institute of Science, Bangalore 560012, India

**C. N. R. Rao**

*C. N. R. Rao obtained his PhD from Purdue University and the DSc degree from the University of Mysore. He has served as a faculty member at the Indian Institute of Technology (Kanpur) and the Indian Institute of Science (Bangalore) from 1959 to 1994 and has been associated since 1989 with the Jawaharlal Nehru Center for Advanced Scientific Research (Bangalore), where he is the Linus Pauling Research Professor. His research interests are mainly in solid-state and materials chemistry. He is a Fellow of the Royal Society, a Foreign Associate of the US National Academy of Sciences and French Academy of Sciences, and a Foreign Member of several academies including those of Russia and Japan. He is also a member of the Pontifical Academy of Sciences and an Honorary Fellow of the RSC. He is the recipient of several awards including the Dan-David prize from Israel, Hughes medal from the Royal Society, Einstein Gold Medal of UNESCO, Somiya award of IUMRS and the first India Science Award.*

**J. N. Behera**

*J. N. Behera obtained his MSc and BEd degrees from Utkal University and is now pursuing his PhD studies at the Indian Institute of Science, Bangalore. His work pertains to open-framework materials.*

 

Many of these sulfates, selenites and selenates have open-framework structures.

The open-framework metal sulfates, selenites and selenates are synthesized under hydro/solvothermal conditions in the presence of organic amines. The organic amines play different roles, as true templates, as structure directors and as space fillers.[8,9] In some instances the parent amine is taken as part of the reaction mixture while in some others the amine is taken in the form of an amine sulfate which serves as the source of both the amine and the sulfate.[10] The use of amine sulfates as reaction precursors and their reactions with metal ions is in its infancy, unlike the use of amine phosphates in open-framework phosphate chemistry.[11,12] Organically templated metal sulfates, selenites and selenates possessing 1D inorganic chain structures, 2D layer structures, as well as 3D structures with channels have been synthesized and characterized. The inorganic frameworks are generally connected by hydrogen bonds to the protonated amine molecules and the extra framework water molecules when present. In this article, we briefly review the present status of the organically templated metal sulfates, selenites and selenates of different dimensionalities.

## One-dimensional metal sulfates

**Single-stranded chain.** This simple topology is found in $[Zn(SO_4)(H_2O)_2(C_{10}N_2H_8)]$, **I**, which consists of chains formed by $[ZnO_4N_2]$ octahedra and $[SO_4]$ tetrahedra joined by common vertices.[13] The octahedral coordination around the zinc center is provided by two μ-O groups from two adjacent sulfate ligands in the chain and two nitrogens from the chelating 2,2′-bipyridine ligand. Each sulfate ligand uses two oxygen atoms to link two $ZnO_4N_2$ octahedra each of which, in turn, shares vertices with two $SO_4$ groups. This generates the –Zn–O–S–O–Zn– wire-like chain shown in Fig. 1a. Two adjacent chains interact through hydrogen bonds formed by the aqua ligand and terminal S=O bonds. A similar single stranded structure has been found in a vanadium sulfates.[14,15] The $[M(TO_4)\phi_4]$ chain backbone is the simplest possible 1D structure and is the basic motif in the mineral chalcanthite.[16]

**Double-stranded chain.** This topology is found in $[H_2N(CH_2)_2NH_2][FeF_3SO_4]$, **II**, comprising strictly alternating $FeF_3O_3$ octahedra and $SO_4$ tetrahedral units, to form a four-member ring ladder as shown in Fig. 1b.[17] The ladder runs along the *a*-axis with the $FeF_3O_3$ octahedra sharing vertices with triply-bridging $SO_4$ groups. The structure can be viewed as a $[M(TO_4)\varphi_4]$ chalcanthite chain, polymerized by the corner-sharing of polyhedra from the adjacent chains to form the (O–T) ladder. The structure of the mineral banattite, $[Cu^{2+}(SO_4)(H_2O)_3]$, is based on a skewed arrangement of chalcanthite chains where it forms six membered rings as in **II**. The formation of octahedral–tetrahedral (O–T) ladder in the metal sulfate family is noteworthy as there is only one other report of an O–T ladder in the literature,[18] unlike the relative large number of T–T ladders.[19–22] It is possible that higher dimensional structures can be obtained from these ladder structures as in the case of the zinc phosphates.[23]

**Kröhnkite-type chain.** In $[HN(CH_2)_6NH][(VO)_2(OH)_2(SO_4)_2].H_2O$, **III**, there are five-coordinated vanadium sites [V(1) and V(2)] with the square-pyramidal geometry with the

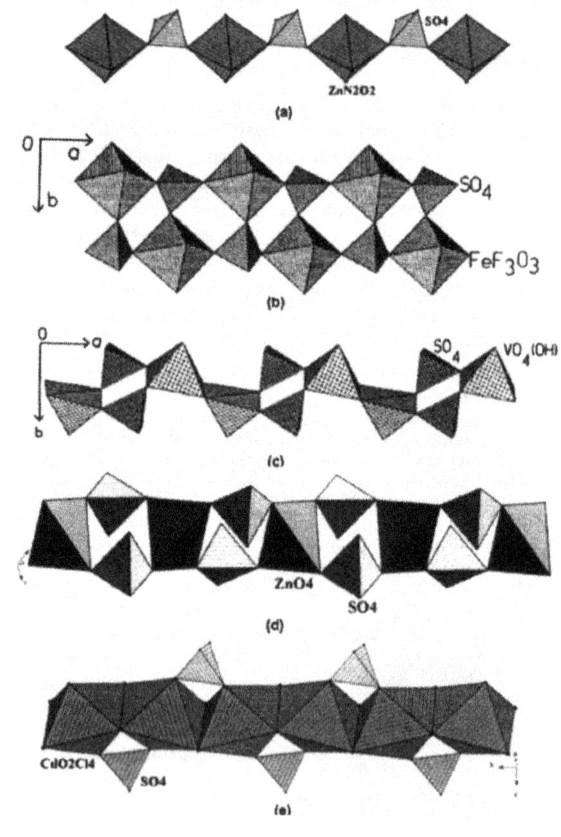


**Fig. 1** (a). Strands of $ZnN_2O_4$ octahedra and $SO_4$ tetrahedra forming wire like chains in $[Zn(SO_4)(H_2O)_2(2,2'\text{-bpy})]$, **I**, from ref. 13. (b) Double stranded chains of $[FeF_3(SO_4)]_n^{2n-}$ in **II**, from ref. 17. (c) Kröhnkite-type chain motif of $[(VO)_2(OH)_2(SO_4)_2]_n^{2n-}$ in **III**, from ref. 15. (d) View of the $[Zn(SO_4)_2]_n^{2n-}$ units along the *b*-axis showing the chains of corner sharing '4-ring' built from $ZnO_4$ and $SO_4$ tetrahedra in **IV**, from ref.13. (e) Linerite- type chain of $[CdCl_2(SO_4)]_n^{2n-}$ in **V**, redrawn from ref. 10.

**Meenakshi Dan**

*Meenakshi Dan is a student of the Integrated PhD program at the Indian Institute of Science, Bangalore. She has obtained the MS degree in chemistry from Indian Institute of Science and is working on open-framework materials for the PhD degree.*

basal plane defined by the oxygens of adjacent sulfate groups, two hydroxyl donors and a terminal oxide group.[15] Two square-pyramids, $V(1)O_3(OH)_2$ and $V(2)O_3(OH)_2$, share a common edge *via* two hydroxyl oxygens, with their apices on the opposite sides of the basal plane, to form $V_2O_8$ dimers. These dimers are capped by the sulfate tetrahedra by sharing the corners to form four-membered rings which propagate along the *a*-axis (Fig. 1c). This structure is related to that of the mineral kröhnkite $Na_2Cu^{II}(SO_4)_2{\cdot}2H_2O$.[16] Kröhnkite-type compounds contain infinite $[M(XO_4)_2(H_2O)_2]$ chains, where M is a divalent (Mn, Fe, Co, Ni, Cu, Zn or Cd) or a trivalent (Al, Fe or In) cation and X is pentavalent (P or As) or hexavalent (S, Se, Cr or Mo). In these chains, the $MO_6$ octahedra are corner-linked to bridging $XO_4$ tetrahedra to give the 1D-chains. The kröhnkite chain topology with tetrahedral atoms has been observed in open-framework metal phosphates[24–26] as well as in some metal sulfates and selenates.[27] The common occurrence of the kröhnkite-type chain is probably because the corner-link between the $MO_6$ and $SO_4$ groups is flexible, involving little steric constraint.

**Chain consisting of four-membered rings.** A typical case is that of $[Zn(SO_4)_2(H_2O)_2]^{2-}$, **IV**, consisting of infinite zinc sulfate chains containing four-membered rings obtained from strictly alternating $ZnO_4$ and $SO_4$ tetrahedral units, which are linked through their corners (Fig. 1d).[13] Two oxygen atoms of a $SO_4$ unit bridge adjacent $ZnO_4$ units while the other two form terminal S=O groups. Tetrahedra-based one-dimensional metal phosphates with a chain topology formed by $MO_4$ (M = Al, Ga and Zn) and $PO_4$ tetrahedra are known.[28–32] Such a corner-shared chain structure has been considered to be a precursor to 3D open-framework metal phosphates.[33,34]

**Linarite-type chain.** Four linear cadmium sulfates of linarite topology have been reported recently.[10,35] In the linarite chain, $[CdX_2SO_4]^{2-}$, **V**, $CdX_4O_2$ (X = Cl, Br) octahedra share edges in a *trans*-fashion with the sulfate tetrahedra grafted on to the chain as a symmetrical bridge. Two Cd–μ-X– Cd linkages between adjacent Cd atoms lead to infinite linear chains of *trans* edge-sharing $CdX_4O_2$ octahedra. Two of the sulfate oxygens bond to adjacent cadmium sites of the edge-shared $CdX_4O_2$ in the common symmetrical bridging mode, forming a synthetic analogue of the linarite chain[36] (Fig. 1e). In linarite, $Pb^{2+}[Cu^{2+}(OH)_2(SO_4)]$, octahedra share two *trans* edges to form a $[M\varphi_4]$ chain. This is decorated by flanking tetrahedra that adopt a staggered arrangement on either side of the chain. The structure of wherryite, $Pb_7^{2+}[Cu^{2+}(OH)(SO_4)(SiO_4)]_2$-$(SO_4)$, is based on linarite chains that are decorated by $(SiO_4)$ tetrahedra. The $[M(TO_4)_2\varphi]$ chains extend along the *c*-axis and pack along the vertices of a primitive orthogonal plane lattice. The additional $SO_4$ groups that do not link to the $(Cu^{2+}\varphi_6)$ octahedra are linked by [7]- and [8]- coordinated $Pb^{2+}$ cations. Infinite chains of $M(TO_4)\Phi_2$ stoichiometry in the linarite group of minerals are rare and occur in layered metal phosphates[37–39] such as tsumcorite and bermanite.

**Butlerite-type chain.** Open-framework metal sulfates form the $[M(TO_4)\varphi_3]$ chain which is the basic motif of butlerite,[40] wherein tetrahedra alternate along the chain and link to the *trans* vertices of the octahedra. The structure of $[H_2N(CH_2)_4NH_2][MF_3SO_4]$ (M = V, Fe), **VI**, consists of $MF_4O_2$ octahedra sharing vertexes with similar neighbors through the fluorine (Fig. 2a).[15] The sulfate tetrahedra are bridged onto the *trans* vertex of the metal octahedra along the chain. The *trans* orientation of the bridging F atom creates a zigzag {–F–M–F–M–} backbone to the linear chain of $MF_4O_2$ octahedra. In the sulfate tetrahedra, two oxygens bond to the adjacent M-sites of the vertex-shared $MF_4O_2$ octahedra in a symmetrical bridge, the remaining two forming terminal S=O bonds. The individual chains are held together by hydrogen bond interactions involving diprotonated amine molecules. A similar chain of *cis* corner-sharing octahedra occurs in fibroferrite, $[Fe^{3+}(OH)(H_2O)_2(SO_4)][H_2O]_3$, giving rise to a helical configuration. In butlerite $[Fe^{3+}(OH)(H_2O)_2(SO_4)]$, parabutlerite, $[Fe^{3+}(OH)(H_2O)(SO_4)]$, and fibroferite, the

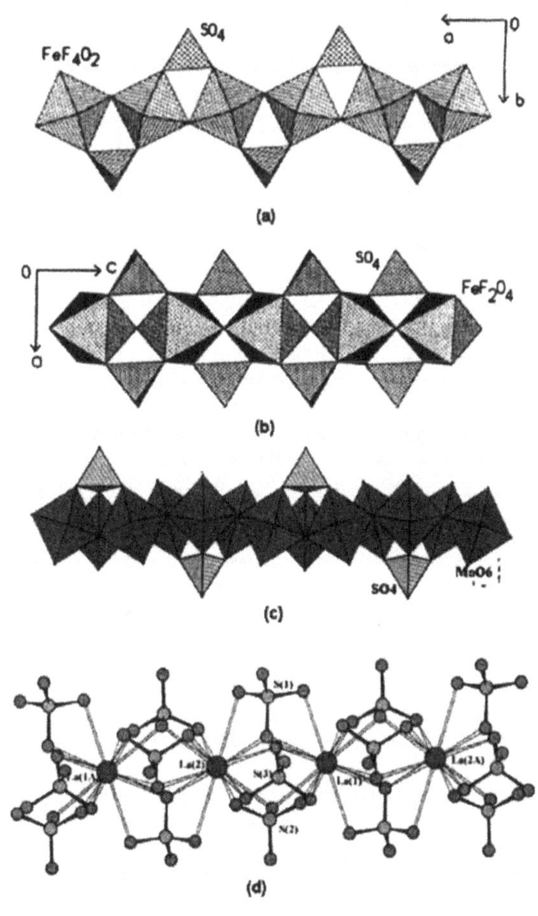


**Fig. 2** (a). Butlerite-type chains with alternating up-down bridging of sulfate tetrahedron in $[MF_3SO_4]^{2-}$, **VI**, from ref. 15. (b) Tancoite-type chains in metal sulfates in **VII**, from ref. 17. (c) α-Molybdena backbone in molybdenum sulfate, $[(MoO_3)_3(SO_4)]_n^{2n-}$, **VIII**, redrawn from ref. 46, copyright (2004) with permission from the American Chemical Society. (d) Infinite chains of $[La(SO_4)_3]_n^{3n-}$ in **IX** running along the *a*-axis, redrawn from ref. 48, copyright (2003) with permission from Elsevier.

chains are linked solely by hydrogen bonds as there are no interstitial cations whereas in organically templated compounds the chains are held together by the hydrogen bond assembly of the diprotonated amine molecules located in the inter-chain space to form the 3D assembly. The mineral uklonskovite consists of topologically identical $Mg(SO_4)(OH)(H_2O)_2$ chains, linked by 8-coordinated Na.

**Tancoite-type chain.** The tancoite chain is commonly found in open-framework materials. Tancoite is a phosphate mineral of the formula, $LiNa_2Hal(PO_4)_2(OH)$, possessing a one-dimensional chain structure.[41] The chain has the composition, $[M(TO_4)_2L]_n$ (M and T are cations of different coordination, usually octahedral and tetrahedral; L = anionic ligand, *e.g.* $O^{2-}$, $OH^-$ or $F^-$), and is found in vanadium and iron sulfates.[15,17] In $[FeF(SO_4)_2]^{3-}$, **VII**, the metal octahedra share vertices with similar neighbors to produce a linear zigzag chain, and the sulfate tetrahedra are grafted on to the chain in a symmetrical bridge. The *trans* orientation of the bridging OH/F group creates a {–M–(OH)/F–M–(OH)/F–M–} backbone in the linear chain and allows the sulfate moiety to bridge symmetrically to form a tancoite type topology as shown in Fig. 2b. Such a structure is also found in open-framework metal phosphates[42] as well as in several phosphate, sulfate and silicate minerals.[43] The 1-D tancoite type gallium phosphate chain has been shown to undergo facile transformation to 3-D open-framework structures under hydrothermal conditions.[44,45]

**α-Molybdena sulfate.** A sulfated α-molybdena, $[C_5H_{14}N_2][(MoO_3)_3(SO_4)].H_2O$, **VIII**, containing an organic amine has been reported.[46] It has three distinct $Mo^{6+}$ centers, each of which is coordinated to six oxygens in a distorted octahedral geometry. Three of the oxide ligands in each $MoO_6$ octahedron bridge between adjacent $Mo^{6+}$ centers in a $\mu^3$ fashion, exhibiting a structure analogous to that of α-molybdena (Fig. 2c). This chain topology has also been found in $Rb_2SMo_3O_{13}$.[47]

**Lanthanum and uranium sulfates.** Lanthanum forms a one-dimensional compound of the composition, $[C_4N_3H_{16}][La(SO_4)_3]\cdot H_2O$,[48] **IX**, where the edge linkage of the $LaO_{12}$ polyhedra and the $SO_4$ tetrahedra provides the building unit, the units being connected by the sharing of La atoms to give infinite chains of $[La(SO_4)_3]_n^{3-}$ along the *a*-axis. The $LaO_{12}$ polyhedra share faces *via* three-coordinated oxygens and the $SO_4$ tetrahedra are grafted on to the chains by sharing corners and edges (Fig. 2d). Several uranium sulfates are reported to occur in the presence of structurally related organic templating agents by O'Hare and coworkers.[49–56] One-dimensional chains are found in $[C_5N_2H_{14}][UO_2(H_2O)(SO_4)_2]$,[49] where the uranium ion is seven-coordinated in the pentagonal bipyramidal geometry. The two adjacent $[UO_7]$ pentagonal bipyramids are linked by the corner sharing sulfate tetrahedra. Adjacent $Ur\phi_5$ pentagonal bipyramids have their uranyl ions oriented approximately perpendicular to the chain length. The $Ur\phi_5$ polyhedra share corners with the sulfate tetrahedra in such a manner that each $Ur\phi_5$ pentaganal bipyramid is linked to four sulfate tetrahedra, the fifth equatorial ligand of the $Ur\phi_5$ pentaganal bipyramid being shared with an aqua ligand. This chain topology is analogous to the mineral $Mn[(UO_2)(SO_4)_2(H_2O)].4H_2O$ and $[(UO_2)(H_2PO_4)_2(H_2O)].(H_2O)_2$, and also occurs in arsenate and phosphate minerals such as brondtite, $Ca_2[Mn^{2+}(AsO_4)_2(H_2O)_2$, talmessite, $Ca_2[Mg(AsO_4)_2(H_2O)_2]$ and fairfieldite, $Ca_2[Mn^{2+}(PO_4)_2(H_2O)_2]$.

### Two-dimensional metal sulfates

Organically templated metal sulfates possessing two-dimensional structures are not that common. One example is that of $[H_3N(CH_2)_2NH_3)][Fe_2F_2(SO_4)_2(H_2O)_2]$, **X**, built up from $FeF_2O_4$, $FeF_2O_2(H_2O)_2$ octahedra and sulfate tetrahedra by sharing their vertices.[17] The $FeF_2O_4$ and $FeF_2O_2(H_2O)_2$ octahedra are alternately connected by their *trans*-fluorine vertexes to form an infinite {–F–Fe–F–Fe–} backbone analogous to the mineral butlerite[40] (Fig. 3a). The butlerite-type chains are fused together to form the layer architecture. This layer structure is similar to that of the iron fluorophosphates (ULM-10),[57] which consist of infinite chains of alternating corner-sharing $Fe^{3+}F_2O_4$ and $Fe^{2+}F_2O_2(H_2O)_2$ octahedra decorated by phosphate tetrahedra fused together to form the layered structure. ULM-11 and the mineral curetonite, $Ba_2[(Al, Ti)Al(PO_4)_2(OH, O_2)F]$, possess similar structures, being built up of the laucite motif.[58] Jacobson *et al.*[59] have reported two niobium phosphates and one titanium phosphate with similar structures.

In $[H_2N(CH_2)_4NH_2][Fe_2^{III}Fe_3^{II}F_{12}(SO_4)_2(H_2O)_2]$, **XI**, two iron octahedra are vertex-shared through *trans* –Fe–F–Fe– linkages to form an infinite chain along the *a*-axis.[60] These chains are connected by a trimer of edge-sharing octahedra along the <001> direction to form a layered network in the *ac*-plane. The triangular lattices formed by the trimers while connecting the chains are capped by the sulfate tetrahedra, thereby forming a 10-membered aperture within the layer (Fig. 3b). The layers are stacked one over the other along the *b*-axis in the AAA... fashion and are held together by hydrogen bonding interactions with the diprotonated amine molecules residing in the interlamellar space.

**Transition metal sulfates with the kagome lattice.** The most common iron sulfates with the kagome lattice are the jarosites which show magnetic frustration or low temperature anti-ferromagnetism.[61] The organically templated iron sulfate $[HN(CH_2)_6NH][Fe^{III}Fe_2^{II}F_6(SO_4)_2][H_3O]$, **XII**, is an unusual example of an iron compound with a perfect kagome structure.[60] The structure consists of anionic layers of vertex-sharing $Fe^{III}F_4O_2$ and $Fe^{II}F_4O_2$ octahedra and $SO_4$ tetrahedra units, which are fused together by $Fe^{III/II}$–F–$Fe^{III/II}$ and $Fe^{III/II}$–O–S moieties. Each $Fe^{III/II}F_4O_2$ unit shares four of its $Fe^{III/II}$–F vertices with similar neighbors with the $Fe^{III/II}$–O bonds roughly aligned in the *ab*-plane. The $Fe^{III/II}$–O bond is canted from the *ab*-plane and this $Fe^{III/II}$–O vertex effectively forces a three-ring trio of apical $Fe^{III/II}$–O bonds closer together to allow them to be capped by the $SO_4$ tetrahedra. The three- and six- rings of octahedra result from the in-plane connectivity and is typical of a kagome lattice with hexagonal tungsten bronze-type sheets (Fig. 3c). Simple calculations have been carried out to understand the properties of this mixed

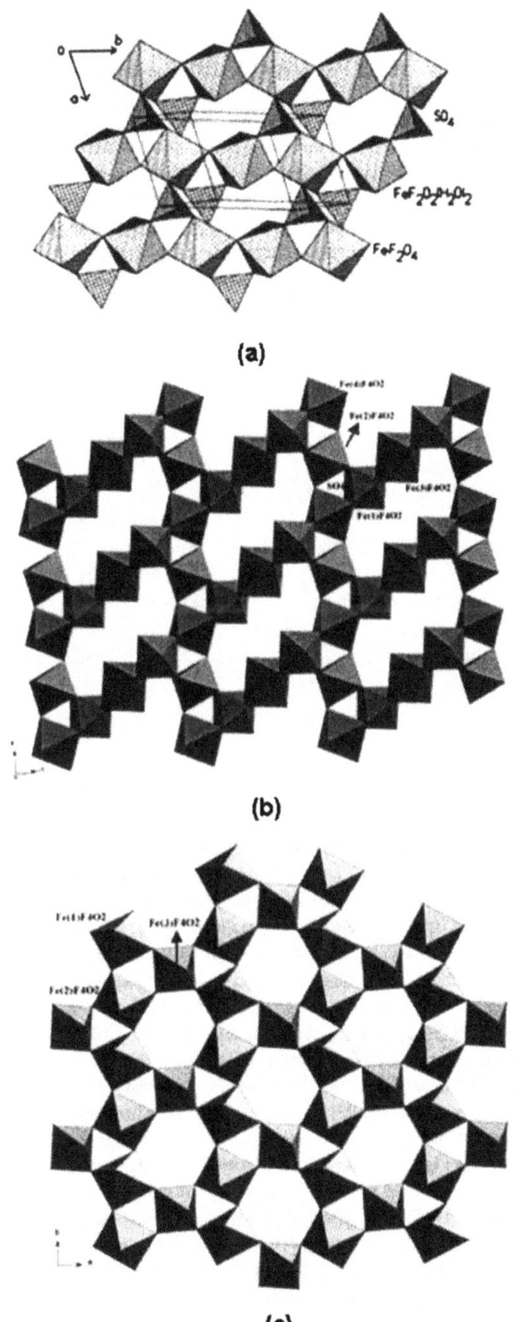

**Fig. 3** (a). Fusion of butlerite-type chains to form the layered structure in $[M_2F_2(SO_4)_2(H_2O)_2]_n^{2n-}$, **X**, (M = Fe, Mn) from ref. 17. (b) Layered in $[Fe_2^{III}Fe_3^{II}F_{12}(SO_4)_2(H_2O)_2]^{4-}$, **XI**, with symmetrical capping of the sulfate tetrahedra in the triangular lattice and the 10-membered aperture within it, redrawn from ref. 60. (c) Polyhedral view of the kagome layer in **XII**, from ref. 60.

valent kagome compound.[62] A layered iron sulfate with the kagome lattice with Fe in the +2 state has been reported.[63] An organically templated iron fluorosulfate $[H_3N(CH_2)_2NH_2(CH_2)_2NH_2(CH_2)_2NH_3][Fe_3^{II}F_6(SO_4)_2]$ with a slightly distorted kagome lattice arising from the presence of the three edge-sharing $Fe^{II}$ octahedra instead of corner-sharing octahedra in the layered network has also been characterized.[64] Magnetic properties of the Fe(II) and the mixed valent Fe kagome compounds are distinctly different from those of the Fe(III) sulfates possessing the kagome lattice. Unlike the Fe(III) kagome compounds which show magnetic frustration or antiferromagnetism at low temperatures, the Fe(II) compound as well as the mixed-valent Fe compounds with kagome structures show ferromagnetic interactions at low temperatures. Similar features are shown by a Fe(III) compound with a distorted kagome lattice. An amine-templated cobalt(II) sulfate with the kagome lattice, is, however, found to exhibit magnetic properties comparable to those of the Fe(III) compound.[65] It appears that only those kagome compounds containing transition metal ions with integral spins (*e.g.*, Fe(II), V(III)) exhibit ferri/ferromagnetic interactions at low temperatures.

**Nickel sulfate.** The two-dimensional Ni sulfate, $[C_4N_2H_{12}][Ni_3F_2(SO_4)_3(H_2O)_2]$,[66] **XIII**, is built up of $Ni(1)F_2O_4$ and $Ni(2)F_2O_4$ octahedra and $SO_4$ tetrahedra sharing edges as M–μ-F–M bonds and vertices as Ni–O–S bonds. The $Ni(1)F_2O_4$ octahedra occur as dimers by sharing edges with two fluorines to form the $Ni_2F_2O_8$ unit. The $Ni(2)F_2O_4$ octahedra link to the anion at one end of the shared edges to form the trimeric $Ni_3F_3O_{12}$ units. The sulfate tetrahedra cap the trimers in such a way that they share three corners with three octahedra giving rise to the formation of a hexameric unit. These units are connected to one another through corner-sharing Ni–O–S linkages to form sinusoidal chains. The chains are covalently bonded to each other by the $SO_4$ group forming the layer structure with an eight-membered aperture in the *ab*-plane (Fig. 4a). The inorganic layers are stacked along the *c*-axis in the ABAB fashion and the interlayer space is occupied by the diprotonated amine molecules, which ensure the stability of the structure through extensive hydrogen bonding. This structure is obtained with different amines such as piperazine, 1,4-diazobicyclo-[2.2.2]octane (DABCO) and 1,3-diamonoprane (DAP). The structure of **XIII** can be compared to that of the mineral sulfoborite,[67] which is made up of complex sheets of $MgO_6$ and $SO_4$ units. In the mineral amarantite, octahedral tetramers are polymerized to form octahedral chains parallel to [100] and linked by $SO_4$ tetrahedra, with the linkage between the sheets being provided by $[B(OH)_4]$ tetrahedra.

**Cadmium sulfate.** In the two-dimensional cadmium sulfate $[H_3N(CH_2)_3[Cd_2(H_2O)_2(SO_4)_3]$,[35] **XIV**, one cadmium is octahedrally coordinated to six oxygens, of which four are from the corner-sharing sulfate ions and two are from the terminal water molecules. The second cadmium atom is coordinated to six sulfate ions of which four are from the corner-sharing sulfate ions and two are from the edge sharing sulfate ions. The inorganic layer is built up of four-membered rings formed by $CdO_6$ octahedra and two $SO_4$ tetrahedra linked through the

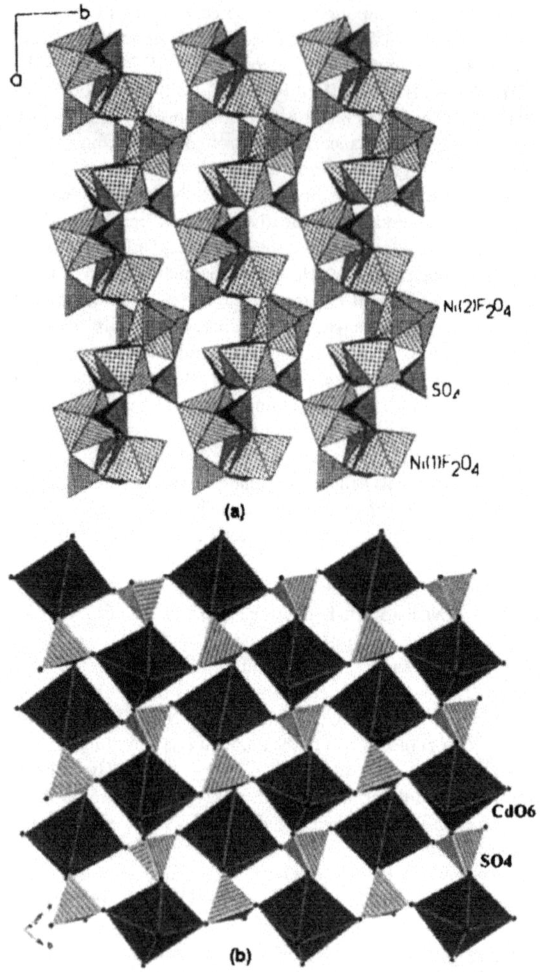


**Fig. 4** (a). Two-dimensional $[Ni_3F_2(SO_4)_3(H_2O)_2]^{2-}$, **XIII**, formed by the sinusoidal chains which, in turn, consist of hexameric units from ref. 66. (b) (O–T) two-dimensional cadmium sulfate of $[Cd_2(SO_4)_3(H_2O)_2]^{2-}$ in **XIV**, redrawn from ref. 35. Note the fusion of the four-membered ring ladders to form the layered structures.

vertices. The connectivity between the four membered rings leads to the formation of a four-membered ladder with strictly alternating $CdO_6$ octahedra and $SO_4$ tetrahedra. The ladder is then fused in the *ab*-plane to form the two-dimensional framework of the inorganic layer (Fig. 4b). The four-membered rings surround both the cadmium atoms with direct linkage to four $SO_4$ tetrahedra.

**Rare-earth sulfates.** The 2D lanthanum sulfates adopt a corrugated layer structure with 8-membered ring apertures, wherein the lanthanum polyhedra get connected by the sulfate tetrahedra in bidentate or monodentate fashion. The coordination around the metal centre is then completed by either coordinating to water, as observed in $[C_6N_2H_{14}]_2[C_2N_2H_8]$-$[Ln_2(H_2O)_4(SO_4)_4][SO_4]\cdot3H_2O$, **XV**,[68] or by coordinating to other terminal sulfate tetrahedra by sharing edges or corners.[69,70] The sulfate tetrahedra bonded to the Ln centres in the layers can get further bonded to other Ln atoms forming 4-membered ring ladders grafted onto the layers comprising 8-membered rings, as observed in $[C_2N_2H_{10}][La_2(H_2O)_4$-$(SO_4)_4]\cdot2H_2O$.[71] A layer with an eight-membered aperture formed by the joining of $NdO_9$ polyhedra and $SO_4$ tetrahedra is shown in Fig. 5a.The corrugated layer structure of the amine-templated lanthanum sulfates can be compared to that of β-$(NH_4)La(SO_4)_2$[72] or $(N_2H_5)Nd(H_2O)(SO_4)_2$,[73] except that in the latter hydrazinium groups are connected to the Nd atom. The interlamellar spacing depends on the size and orientation of the guest species, with the distance between the planes increasing with the increase in the size of the guest amine molecules.

The other structures of layered lanthanum sulfates are based on $LnO_3$ layers with (6, 3) nets as in $[C_3N_2H_{12}][La_2(SO_4)_4]$, **XVI**, (Fig. 5b). Here, the La centres are bridged by the sulfate tetrahedra to form 4-membered rings, which then join together by sharing of edges or corners. These chains or ladders of 4-membered rings fuse together through the three-coordinated oxygen atoms of the sulfate groups to form layers having infinite Ln–O–Ln linkages in two dimensions. The network formed by the linking of the Ln centres by the three-coordinated oxygens has the composition $LnO_3$ and can be described as (6, 3) nets with each Ln centre acting as a node.[68] An organically templated layered cerium phosphate sulfate, $[C_2N_2H_{10}][Ce^{III}(PO_4)(HSO_4)(H_2O)]$, has been reported.[74] Lanthanide compounds along with a few open-framework sulfates have been described in a review by Wickleder.[75]

**Uranium sulfates.** In contrast to the 2D-lanthanum sulfates, the uranium sulfates are made up of 4-membered uranium sulfate rings connected by the sulfate tetrahedra, giving rise to layers of different topologies. Depending on the mode of connectivity of the 4-membered rings, the layers comprise entirely of 4-membered rings as in $[C_3N_2H_5][(UO_2)_2(SO_4)_3]$, USO-8,[49] **XVII**, (Fig. 5c), contain 6-membered rings along with 4-membered rings as in $[C_3N_2H_{12}][(UO_2)_2(H_2O)(SO_4)_3]$, USO-2,[56] or 12-membered rings along with 4-membered rings as in $[C_6N_2H_{14}][UO_2(H_2O)(SO_4)_2]$, USO-5.[55] The inorganic layers in USO-2 and USO-5 contain bound water molecules which form additional hydrogen bonds. Two new layered actinide sulfates prepared recently by the amine-sulfate route, have structures similar to those of the lanthanum sulfates, wherein the uranium centres are bridged by the sulfate tetrahedra by the sharing of edges or corners to form layers with 8-membered ring apertures.[76] By employing hydrofluoric acid in the hydrothermal synthesis, six-layered uranium sulfates have been prepared.[51] The fluoride ligands join adjacent uranium centers through U–F–U linkages either by sharing an edge to form the dimer as in USFO-3 or corners to form a chain, which is then linked by the $SO_4$ tetrahedra to form layers as in USFO-5.

**Thorium sulfate.** In $[HN(CH_2)_6NH]_2[Th_2(SO_4)_6(H_2O)_2]\cdot$ $2H_2O$, **XVIII**,[13] two crystallographically distinct metal–oxygen polyhedra are capped by four sulfate ions in $Q_2$ connectivity to form the $Th_2S_4$ unit. The $Th(2)O_9$ polyhedra share edges with the $S(2)O_4$ tetrahedra to form a sinusoidal chain along the

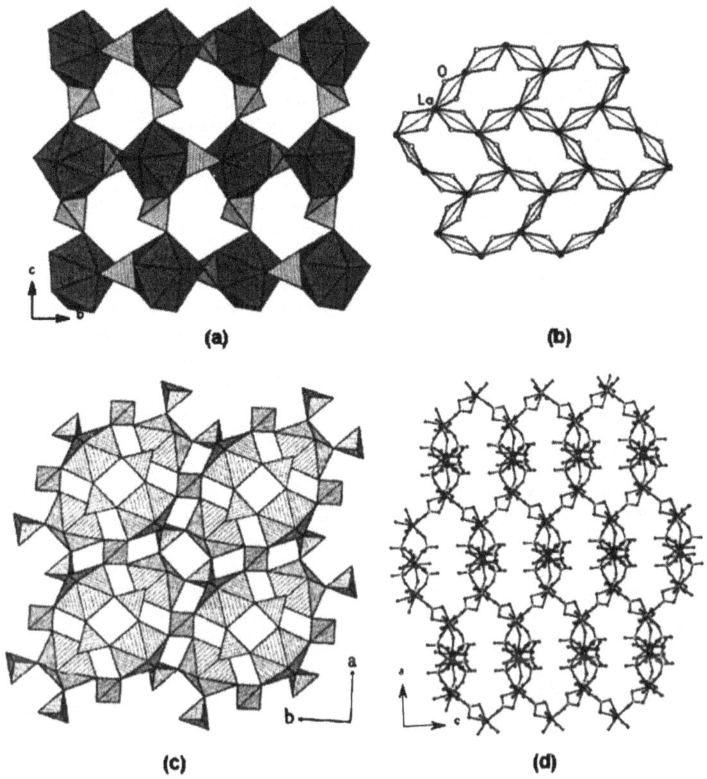


**Fig. 5** (a). Polyhedral view of the inorganic layer in $[Ln_2(SO_4)_4(H_2O)_4]^{4-}$ **XV** (Ln = La, Pr or Nd), formed by joining the $NdO_9$ polyhedra and $SO_4$ tetrahedra, redrawn from ref. 68. (b) The infinite two-dimensional La–O–La linkages in $[La_2(SO_4)_4]^{2-}$ in **XVI**, from ref. 68. (c) Two-dimensional uranium sulfate in $[UO_2(SO_4)_3]^{2-}$, **XVII**, formed by $[UO_7]$ and $[SO_4]$ units from ref. 49, copyright (2003) from the Royal Society of Chemistry. (d) Two-dimensional sheet of $[HN(CH_2)_6NH]_2[Th_2(SO_4)_6(H_2O)_2].2H_2O$, **XVIII** consist of cages along the *ac*-plane from ref. 13.

*c*-axis. Two such adjacent chains arranged in the ABAB... fashion connect the $Th(1)_2S_4$ unit at both their crests and troughs to form cages. The cages are extended in the *ac*-plane to generate a two-dimensional sheet as shown in Fig. 5d. The cages are occupied by amine and water molecules which form extensive hydrogen bonds with the framework oxygens.

### Three-dimensional metal sulfates

**Nickel sulfate.** The prominent feature of the three-dimensional nickel sulfate, $[C_4N_2H_{12}][Ni_2F_4(SO_4)\ H_2O]$,[66] **XIX**, is a dimer of edge-sharing octahedra. The Ni(1) and Ni(2) octahedral units share edges through fluorine (Ni(1)–F(1), F(2) and Ni(2)–F(3), F(4) ) to form the dimer units $Ni_2(1)F_6O_2(H_2O)_2$ and $Ni_2(2)F_6O_4$. The dimers are linked by sharing octahedral edges to form two distinct infinite sinusoidal chains along the [001] direction. These chains are alternately stacked one over the other along the [100] direction and are interlinked by sharing corners with the tetrahedral corners of the $SO_4$ groups to form a layer with an 11-membered aperture in the [101] plane. **XIX** seems to be the first example of an 11-membered aperture in open-framework compounds. In the layer, each sulfate shares three corners with three octahedra of the adjacent chains. It is connected to the Ni(1) octahedral chain by sharing one oxygen atom, whereas it shares two oxygen atoms with two neighboring octahedra of the Ni(2) chain. The layers are cross-linked by $SO_4$ tetrahedra along the *b*-axis through corner-sharing S–O–Ni linkages to form the three-dimensional structure. The cross inking of the layers gives rise to an elliptical channel along the *c*-axis as shown in Fig. 6a. The channels are filled by the amine molecules, which interact with the framework oxygen and fluorine atoms through F–H···O and C–H···O hydrogen bonds. This structure is somewhat related to the mineral phosphopherite,[77] which contains layers of sinusoidal chains made up of edge-shared octahedral trimers cross-linked by tetrahedral $PO_4$ units.

**Zinc sulfate.** $[(CN_3H_6)_2][Zn(SO_4)_2]$,[78] **XX**, is built up of alternating $ZnO_4$ and $SO_4$ tetrahedra, sharing vertices. The polyhedral connectivity of the $ZnO_4$ and $SO_4$ results in an infinite three-dimensional network. This gives a 12-ring system, built up of six $ZnO_4$ and six $SO_4$ units. The network possesses two-dimensional intersecting channels propagating along the <100> and <010> directions, with no channels apparent in the <001> direction. Well-ordered guanidinium cations occupy all the 12-ring windows (Fig. 6b) and interact with the zinc sulfate framework through N–H···O hydrogen

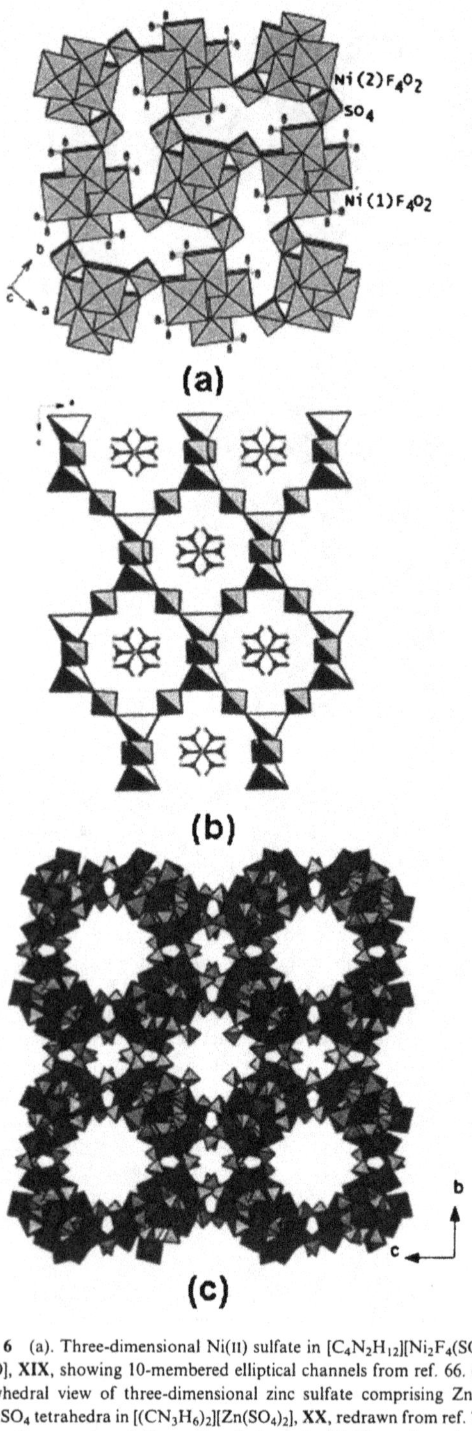


**Fig. 6** (a). Three-dimensional Ni(II) sulfate in $[C_4N_2H_{12}][Ni_2F_4(SO_4)H_2O]$, **XIX**, showing 10-membered elliptical channels from ref. 66. (b) Polyhedral view of three-dimensional zinc sulfate comprising $ZnO_4$ and $SO_4$ tetrahedra in $[(CN_3H_6)_2][Zn(SO_4)_2]$, **XX**, redrawn from ref. 78. Note the presence of 12-membered channels. (c) Three-dimensional structure of scandium with mixed sulfate and phosphate tetrahedra in **XXI**, redrawn from ref. 81, copyright (2002) from the Royal Society of Chemistry.

bonds. The structure of **XX** resembles that of diamond and is akin to that of a three-dimensional zinc phosphite[79] and of a guanidinium-templated zinc phosphate.[80]

**Scandium sulfate.** A three-dimensional scandium sulfate in which secondary building units of the formula $Sc_7(S,P)_{12}O_{48}$, **XXI**, templated with azamacrocycles has been synthesized hydrothermally.[81] The structure is made up of $ScO_6$ octahedra that share corners with the sulfate tetrahedra. The main feature of the framework is that supercages are formed at the corners and at the centre of the cubic unit cell. Each supercage is connected to six other supercages *via* smaller cages, in which protonated cyclen molecules are present. Each small cage links two supercages in such a way that in a unit cell there are two supercages and six smaller cages (Fig. 6c).

**Rare-earth sulfates.** Compared to the number of 2D rare-earth sulfates, very few 3D rare-earth sulfates are known. The three-dimensional lanthanum sulfates reported hitherto contain 8-membered ring units, which are connected by the sulfate tetrahedra or by the three-coordinating oxygens of the $SO_4$ unit to form the 3D framework. In $[C_4N_2H_{12}]_2[La_2(H_2O)_2(SO_4)_4]$, the 8-membered rings join together by sharing edges to form the 3D framework, giving rise to 8-ring apertures down the *b*-axis.[70] $[C_2N_2H_{10}]_2[Ln_2(H_2O)_2(SO_4)_5]$ possesses a α-Po structure wherein the 8-membered rings are bridged by the sulfate tetrahedra along the three directions, resulting in the formation of a network of interconnected 8-membered ring channels. Such a connectivity also results in the formation of 16-membered ring channels down the *a*-axis of the unit cell.[68] In $[C_4N_2H_{12}][Nd_2(H_2O)_2(SO_4)_4]$, (**XXII**), the 8-membered rings get connected by the sulfate tetrahedra in one direction and three-coordinated oxygen atoms in another direction, thereby resulting in the formation of channels with 12-ring apertures around the 8-ring channels[68] (Fig. 7a). In all the three 3D lanthanum sulfates, the protonated amine molecules reside in the channels and form hydrogen bonds with the framework oxygens.

Organically templated 3d–4f mixed metal sulfates of the formula $[C_2N_2H_{10}][La_2M(H_2O)_2(SO_4)_6]$ (M = Co, Ni), have been recently synthesized by Clearfield *et al.*[82] The three-dimensional anionic framework is formed by the bridging of La(III) and divalent transition metal ions by the sulfate anions. 8-Membered ring tunnels run along the *c*-axis, which are occupied by the protonated organic cations.

There is one report of a three-dimensional actinide sulfate, $[C_4NH_{12}]_2[(UO_2)_6(H_2O)_2(SO_4)_7]$, MUS-1 **XXIII**,[83] formed by the bridging of the uranium centres by the sulfate tetrahedra. Each sulfate tetrahedron binds to four different uranium atoms to form the three-dimensional network (Fig. 7b). MUS-1 is structurally similar to $Mg(UO_2)_6(MoO_4)_7(H_2O)_{15}$ and $Sr(UO_2)_6(MoO_4)_7(H_2O)_{19}$.[84] Channels run along the *c*-axis of the unit cell and are occupied by the tetramethylammonium cations.

## Open-framework metal selenites

Selenite-based frameworks are of interest due to the possible role of the lone pair of electrons as an invisible

 

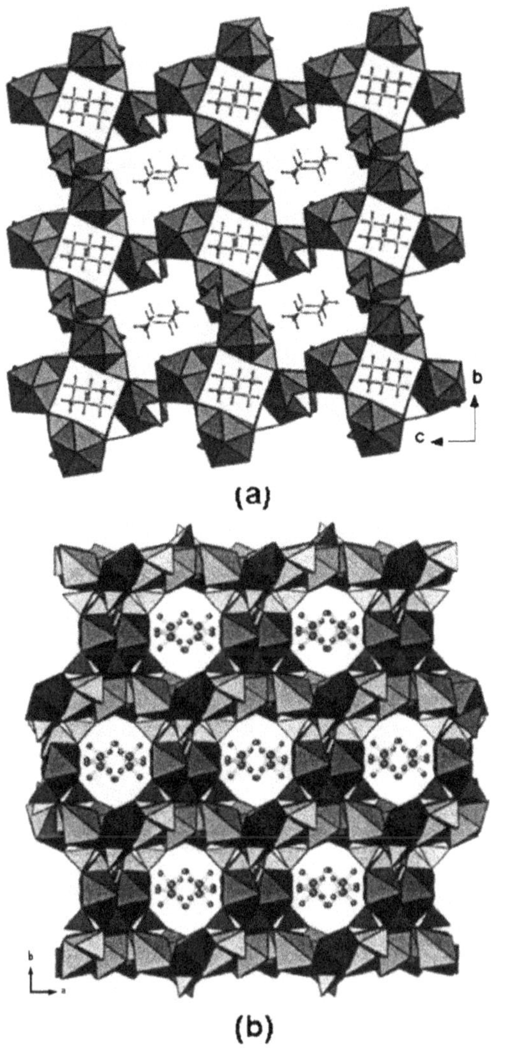


**Fig. 7** (a). The three-dimensional framework of neodymium sulfate with eight and 12-membered channels in $[C_4N_2H_{12}][Nd_2(H_2O)_2(SO_4)_4]$, **XXII**, redrawn from ref. 68. (b) Three-dimensional structure of MUS-1 in $[C_4NH_{12}]_2[(UO_2)_6(H_2O)_2(SO_4)_7]$, **XXIII**, as viewed along the [001] direction, redrawn from ref. 83, copyright (2002) from the Royal Society of Chemistry.

structure-directing agent. The stereochemically active lone pair of electrons in Se(IV) leads to a pyramidal coordination for the selenite species. Due to the presence of lone pair, metal selenites tend to crystallize in non-centrosymmetric structures giving rise to interesting physical properties, such as non-linear optical second harmonic generation (SHG).[85,86] The low reduction potential of the $Se^{IV}/Se^{0}$ couple, which under hydrothermal conditions may cause reduction to metallic Se, renders the synthesis of amine-templated open-framework metal selenites rather difficult. The first organically templated metal selenite with a layered structure was reported by Harrison *et al.*[87] Rao *et al.*[88,89] reported the first three-dimensional organically templated iron(III) and zinc selenites. Since then, there have been a few other reports of 2D and 3D metal selenites

**One-dimensional zinc selenite.** $[C_3N_2H_{12}]_4[Zn_4(SeO_3)_8]$ is formed by the connectivity of the strictly alternating $ZnO_4$ and $SeO_3$ units which results in the formation of corner sharing 4-membered ring chains, similar to that observed in metal phosphates and sulfates.[90]

**Two-dimensional selenites.** Depending on the mode of connectivity between the selenite units and the different metal atoms, the two-dimensional metal selenites show a rich variety of architectures. The connectivity of the metal centres by the selenite units wherein the selenite oxygens occupy the equatorial positions around the metal atom results in the formation of layers with 8-membered ring apertures (Fig. 8a). The coordination around the metal atom is completed by bonding to halo or oxo groups, as in $[C_2N_2H_{10}]$-$[CdCl_2(HSeO_3)_2]$ and $[C_2N_2H_{10}][(VO)(SeO_3)_2]$, **XXIV**.[91,92] Such a layered topology is found in a copper hydrogen selenite,[93] which is not an open-framework and in a few open-framework layered phosphates such as the layered tin(II) phosphate and phosphatoantimonate.[94,95] The metal centres can also be connected by the selenite groups to form

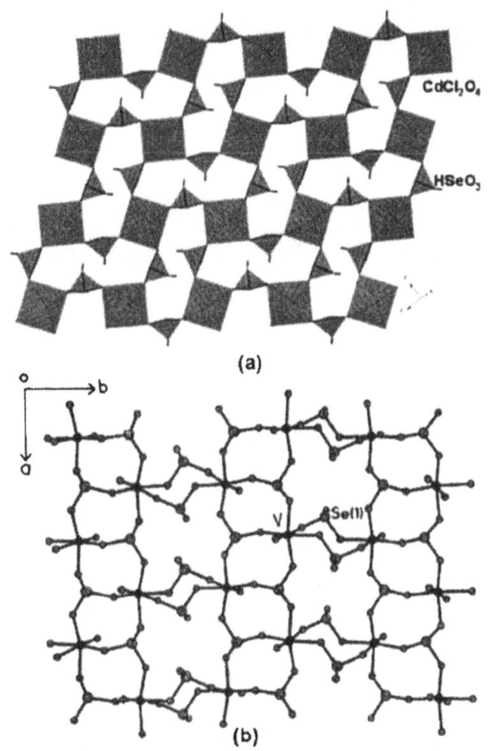


**Fig. 8** (a) The layer structure of $[C_2N_2H_{10}][CdCl_2(HSeO_3)_2]$, **XXIV**, with 8-membered ring aperture, redrawn from ref. 91. (b) Layer formed in $[C_6N_2H_{14}][(VO)_2(HSeO_3)_2(SeO_3)_2].2H_2O$, **XXV**, by the joining of the 4-membered ring ladders by $HSeO_3$ units from ref. 96.

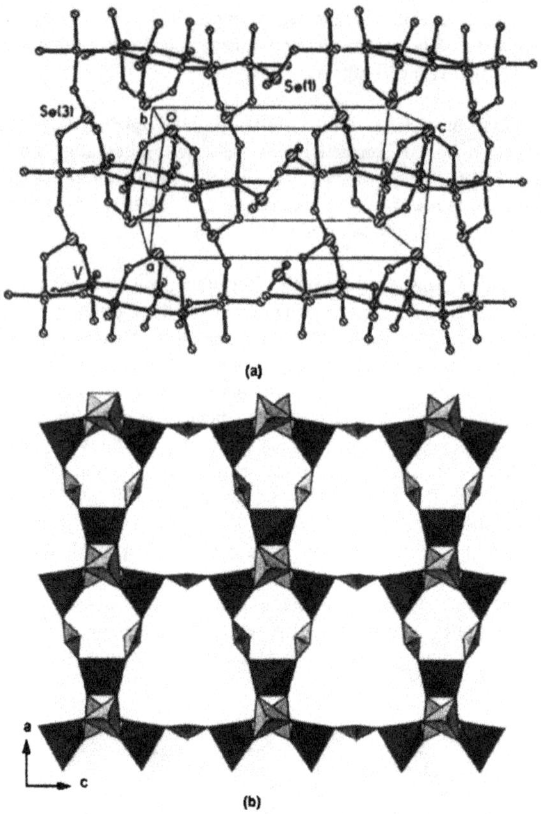


Fig. 9 (a) Layer structure formed by the joining of vanadium–oxygen metal clusters, which act as the secondary building units in $[C_2N_2H_{10}][(V^{IV}O)_2(V^{V}O)O_2(SeO_3)_3]\cdot 1.25H_2O$, **XXVI**, from ref. 96. (b) Polyhedral view of the sheet topology of $[CN_3H_6]_4[Zn_3(SeO_3)_5]$ in **XXVII**, redrawn from ref. 87, copyright (2000) with permission from Wiley-VCH.

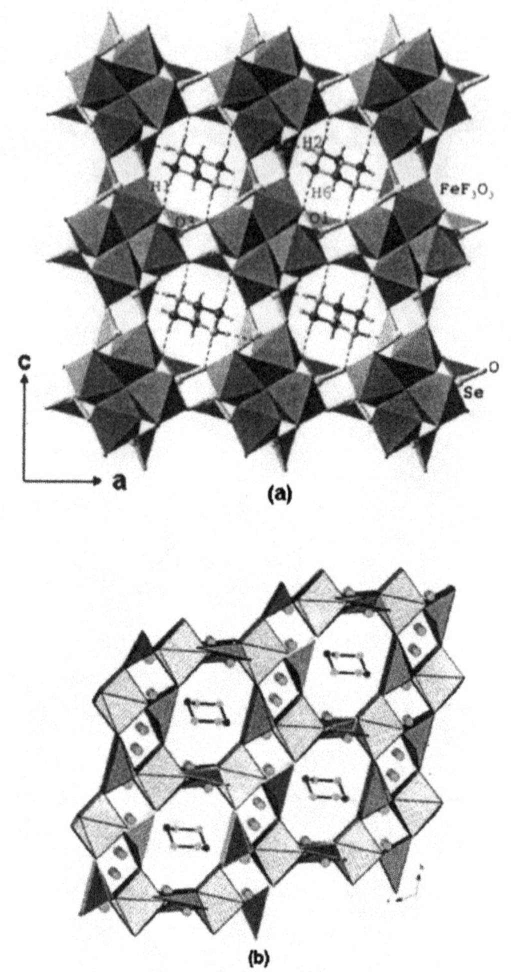


Fig. 10 (a) Polyhedral view of the inorganic framework of $[A][Fe_4F_6(SeO_3)_4]$, **XXVIII**, showing one-dimensional 8-ring channels from ref. 88. (b) Three-dimensional framework structure of $[C_4N_2H_{12}][M_2(HSeO_3)_2(Se_2O_5)_2]$, **XXIX**, showing 10-membered channels, redrawn from ref. 89.

4-membered ring units which get joined to form layers with 4- and 8-membered ring apertures (Fig. 8b) as in $[C_6N_2H_{14}][(VO)_2(HSeO_3)_2(SeO_3)_2]\cdot 2H_2O$, **(XXV)**,[96] Thus, the structure can be described as being formed by the joining of the four-membered ring ladders by the $HSeO_3$ groups. The importance of the ladder structure as a possible precursor of the 3D metal phosphates has been pointed out.[19] The layer structure in the metal selenite is comparable to that of the indium phosphate, $[C_5NH_6][In(HPO_4)_2]$.[97] The metal atoms and the selenite groups can join together to form clusters, which then get connected to form layers, as in the vanadium selenite, $[C_2N_2H_{10}][(V^{IV}O)_2(V^{V}O)O_2(SeO_3)_3]\cdot 1.25H_2O$, **XXVI**. In this compound, the clusters are joined by the selenite groups to form layers with 8-membered apertures[96] (Fig. 9a). Similarly, in $[CN_3H_6]_4[Zn_3(SeO_3)_5]$, **XXVII**, the layer is formed by the linking of $Zn_3Se_2O_{12}$ clusters by the selenite groups (Fig. 9b).[87]

**Three-dimensional selenites.** In the 3D selenites of the formula, $[A][Fe_4F_6(SeO_3)_4]$, **XXVIII**, obtained with different amines,[88] the Fe centres join together through bridging fluorine atoms to form tetrameric $Fe_4F_6O_{12}$ clusters, which then get connected by the selenite groups to form the three-dimensional architecture. Such a connectivity results in the formation of 3-, 4-, 5- and 6-membered rings in the structure. There is also a 8-ring 1D channel along the *b*-axis of the unit cell (Fig. 10a). The presence of the near-tetrahedral $Fe_4F_6O_{12}$ cluster makes it an interesting geometrically frustrated magnetic material.

A three-dimensional open-framework metal selenite of the formula $[C_4N_2H_{12}][M_2(HSeO_3)_2(Se_2O_5)_2]$, **XXIX**, (M = Zn, Co or Ni), containing both the selenite and the diselenite units has been reported.[89] The structure possesses a diamondoid network and has intersecting two-dimensional 10-membered channels. The three-dimensional framework is formed by the connectivity of the zinc-diselenite layers by the selenite groups acting as pillars. This connectivity results in the formation of

4- and 6-membered rings in the structure. In both the three-dimensional metal selenite frameworks, protonated amine molecules reside in the channels and forming extensive hydrogen bonds with the framework atoms (Fig. 10b).

Three-dimensional inorganic–organic hybrid metal selenites, wherein the metal selenite layers get crosslinked by the organic moiety to form the 3D architecture have been reported. In $[C_2N_2H_8][Zn_2(SeO_3)_2]$, the zinc selenite layers get connected by the pillaring ethylene diamine molecules in an end-to-end fashion resulting in the formation of channels along the *a*- and *c*-axes.[88] An allotropic form of $[C_2N_2H_8][Zn_2(SeO_3)_2]$, reported by Ferey *et al.*,[90] differs only in the topology of the zinc selenite layers. In $[M(C_{10}N_2H_{10})(H_2O)V_2Se_2O_{10}]$, M = Co, Ni, the Co/Ni–V–$SeO_3$ layers get crosslinked by the 4,4′-bipyridyl units to form the three-dimensional architecture.[98]

#### Open-framework metal selenates

It is only recently that it has been possible to prepare open-framework metal selenates. The difficulty in the synthesis of these compounds lies in the instability of Se(VI) in the basic medium. It, therefore, becomes necessary to prepare them in an acidic medium.

**One-dimensional selenates.** The framework structures in $[Cd(H_2O)_2(SeO_4)_2]^{2-}$ [99] and $[Zn(H_2O)_2(SeO_4)_2]^{2-}$, **XXX**,[100] has the kröhnkite-type topology. In this structure, a pair of selenate tetrahedra shares corners to form the one-dimensional chain structure bridging the neighboring metal octahedra as shown in Fig. 11a. Fleck *et al.*[101] have reviewed compounds with the kröhnkite-type topology. The structure of the kröhnkite-type infinite chain is composed of a distorted $MO_6$ octahedron corner-linked with the $XO_4$ tetrahedron, the remaining two opposite corners in the *trans* position being occupied by water molecules. Each tetrahedron connects the neighboring octahedra, forming infinite $[M(XO_4)_2(H_2O)_2]$ chains. The simplest chain of the formula $[Zn(SeO_4)(phen)(H_2O)]$, consisting of strictly alternating zinc octahedra and selenate tetrahedra similar to that of zinc sulfate $[Zn(SO_4)(H_2O)_2(2,2\text{-bpy})]$, **I**, has been reported.[100]

**Two-dimensional selenates.** The structure of $[C_4N_2H_{14}][La_2(SeO_4)_4].H_2O$, **XXXI**,[102] involves LaO sheets formed by La–μ-O–La linkages forming chains running parallel to the *b*-axis with the alternate La atoms of the chains being bridged by La–μ-O–La linkages along the perpendicular direction which forms a hexanuclear lanthanide ring. The hexanuclear lanthanide rings are surrounded by six similar rings resulting in the formation of a (6,3) net. Along the chain, two $LaO_9$ polyhedra are capped on either side by the $Se(2)O_4$ units which are connected to two La atoms in the neighboring chains through a three-coordinate oxygen (O8). In the perpendicular direction, the bridging La-atoms are capped by two $Se(1)O_4$ units which share an edge with a $LaO_9$ polyhedron of the neighboring chain. Accordingly, the $SeO_4$ units are involved in both capping and bridging the La atoms in the LaO sheets. Such a dual connectivity of the $SeO_4$ units produces corner-shared four-membered rings propagating along the *a*-direction as shown in Fig. 11b. $[C_2N_2H_{10}][Ln_2(SeO_4)_4(H_2O)_4]$, constructed by capping two edge-shared $LaO_9$ polyhedra by $SeO_4$ groups has also been described.[102]

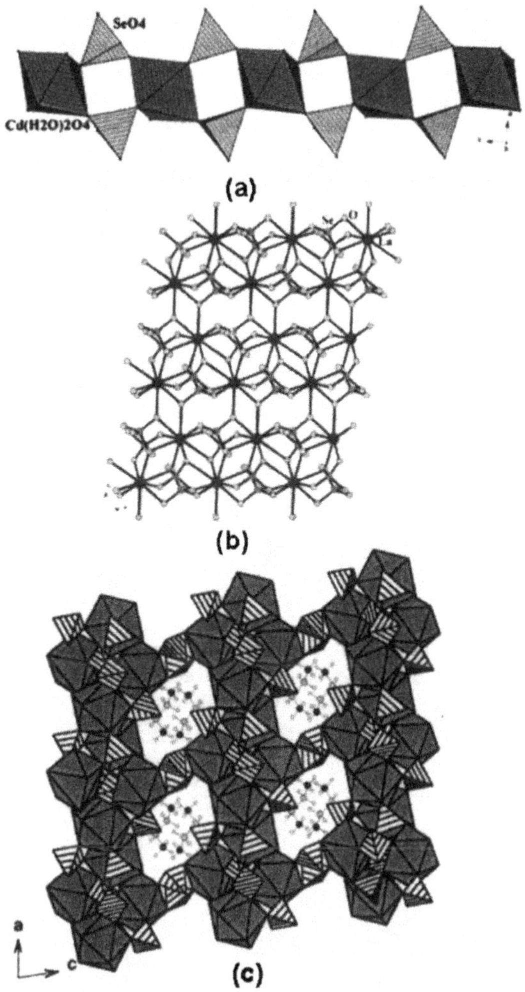


**Fig. 11** (a) Polyhedral structure of the $[Cd(SeO_4)_2(H_2O)_2]^{2-}$ chain in **XXX** along the *c*-axis from ref. 99. (b) The inorganic layer of $[La_2(SeO_4)_4]^{2-}$, in **XXXI** with the capping and bridging of the $LaO_9$ dimers by two selenate units, resulting in corner-shares four-membered rings along the *a*-axis from ref. 102. (c) Polyhedral view of three-dimensional lanthanum selenate comprising of $La_2Se_4$ building units in $[C_2N_2H_{10}][La_2(SeO_4)_4(H_2O)_3].H_2O$, **XXXII**, from ref. 103. Note the presence of 12-membered channels.

**Three-dimensional selenate.** A three-dimensional selenate of the formula, $[C_2N_2H_{10}][La_2(SeO_4)_4(H_2O)_3]\cdot H_2O$, **XXXII**, comprising $La_2Se_4$ building units and possessing 12-membered channels, has been prepared in an acidic medium under hydrothermal conditions.[103] The complex network of $LaO_8$, $LaO_9$ and $SeO_4$ moieties gives rise to a three-dimensional structure possessing channels along all the three axes. Connectivity between two building units with the $La_2Se_4$

stoichiometry through La–O–Se–O–La linkages gives rise to infinite chains along the <100> direction, which are connected through a bridging $Se(2)O_4$ to form a two-dimensional layer in the *ab*-plane, with 4- and 8-membered apertures. The layers are stacked one over the other along the *c*-axis and connected by the $Se(4)O_4$ tetrahedra which share corners with the La(1)–O polyhedra from the adjacent layers forming the 3-dimensional network. The 3D structure has 12-membered channels along the *a*-and *b*-axes, and 8-membered channels along the *c*-axis. Amine molecules reside in the 12-membered channels forming hydrogen bonds with the framework oxygens (Fig. 11c).

## Conclusions

The discussion in the earlier sections should suffice to demonstrate how oxyanions such as, sulfate, selenite and selenate could be used to design novel inorganic open-framework materials. The structures of many of the sulfates and selenates are close to those of the phosphates. Although several open-framework metal sulfates have been synthesized and characterized, there is need for further work to investigate the effects of various reaction parameters and solvents on the formation of these materials. Great effort is necessary to synthesize 3D metal selenates since we have one organically templated metal selenate to-date. To our knowledge, no organically-templated metal sulfite has been reported hitherto. It would be worth exploring reactions of organic amine sulfites with metal ions. It should be possible to make use combinations of these oxyanions with other anions such as the phosphate and the silicate to produce novel architectures. There is also need for efforts to obtain open-framework metal sulfates, selenites and selenates with empty channels, in order to study sorption and other properties. An important aspect that is worthy of study relates to the possible transformation of chain sulfates to 2D and 3D structures. It would be of interest to know if there is a secondary building unit for the sulfates similar to the 4-membered ring structures in the case of open-framework phosphates. The synthesis and transformation of such a 4-membered ring sulfate should be examined.

## Acknowledgements

The authors thank the DRDO (India) for support of this research.

## References


1 D. W. Breck, *Zeolite Molecular Sieves*, Wiley, New York, 1974.
2 W. M. Meier, D. H. Oslen and C. Baerlocher, *Atlas of Zeolite Structure Types*, Elsevier: London, 1996.
3 A. K. Cheetam, G. Ferey and T. Loiseau, *Angew. Chem., Int. Ed.*, 1999, **38**, 3268.
4 C. N. R. Rao, S. Natarajan, A. Choudhury, S. Neeraj and A. A. Ayi, *Acc. Chem. Res.*, 2001, **34**, 80.
5 C. N. R. Rao, S. Natarajan and R. Vaidyanathan, *Angew. Chem., Int. Ed.*, 2004, **43**, 1466.
6 O. M. Yaghi, M. O'Keefe, N. W. Ockwig, H. K. Chae, M. Eddaoudi and J. Kim, *Nature*, 2003, **423**, 705.
7 O. M. Yaghi, H. L. Li, C. Davis, D. Richardson and T. L. Groy, *Acc. Chem. Res.*, 1998, **31**, 474.
8 M. E. Davis and R. F. Lobo, *Chem. Mater.*, 1992, **4**, 756 and the references therein.
9 A. Choudhury, S. Natarajan and C. N. R. Rao, *Inorg. Chem.*, 2000, **39**, 4295.
10 A. Choudhury, J. Krishnamoorthy and C. N. R. Rao, *Chem. Commun.*, 2001, 2610.
11 S. Neeraj, S. Natarajan and C. N. R. Rao, *Angew. Chem., Int. Ed.*, 1999, **38**, 3480.
12 C. N. R. Rao, S. Neeraj and S. Natarajan, *J. Am. Chem. Soc.*, 2000, **122**, 2810.
13 J. N. Behera and C. N. R. Rao, *Z. Anorg. Allg. Chem.*, 2005, **631**, 3030.
14 M. I. Khan, S. Cevik and R. J. Doedens, *Chem. Commun.*, 2001, 1930.
15 G. Paul, A. Choudhury, R. Nagarajan and C. N. R. Rao, *Inorg. Chem.*, 2003, **42**, 2004.
16 F. C. Hawthrone, S. V. Krivovichev and P. C. Burns, *Rev. Mineral. Geochem.*, 2000, **40**, 1.
17 G. Paul, A. Choudhury and C. N. R. Rao, *Chem. Mater.*, 2003, **15**, 1174.
18 S. Chakrabarti and S. Natarajan, *Angew. Chem., Int. Ed.*, 2002, **41**, 1224.
19 A. Choudhury, S. Neeraj, S. Natarajan and C. N. R. Rao, *J. Mater. Chem.*, 2001, **11**, 1537.
20 X. Bu, T. E. Gier and G. D. Stutkey, *Chem. Commun.*, 1997, 2271.
21 W. T. A. Harrison, Z. Bircsak, L. Hannooman and Z. Zhang, *J. Solid State Chem.*, 1998, **136**, 93.
22 S. Neeraj, S. Natarajan and C. N. R. Rao, *Chem. Mater.*, 1999, **11**, 1390.
23 A. Choudhury, S. Neeraj, S. Natarajan and C. N. R. Rao, *J. Mater. Chem.*, 2001, **11**, 1537.
24 A. M. Chippindale and C. Turner, *J. Solid State Chem.*, 1997, **128**, 317.
25 T. Loiseau, F. Serpaggi and G. Férey, *Chem. Commun.*, 1997, 1093.
26 A. Choudhury, S. Natarajan and C. N. R. Rao, *J. Chem. Soc., Dalton Trans.*, 2000, 2595.
27 U. Kolitsch, *Acta Crystallogr., Sect. C: Cryst. Struct. Commun.*, 2004, **60**, i3–i6.
28 Q. Gao, J. Chen, S. Li, R. Xu, J. M. Thomas, M. Light and M. B. Hursthouse, *J. Solid State Chem.*, 1996, **127**, 145.
29 W. Tieli, Y. Long and P. Wenqin, *J. Solid State Chem.*, 1990, **89**, 392.
30 R. H. Jones, J. M. Thomas, R. Xu, Y. Xu, Q. Huo, A. K. Cheetham and J. Chen, *J. Chem. Soc., Chem. Commun.*, 1990, 1170.
31 A. M. Chippindale, A. D. Bond, A. D. Law and A. R. Cowley, *J. Solid State Chem.*, 1998, **136**, 227.
32 A. A. Ayi, S. Neeraj, A. Choudhury, S. Natarajan and C. N. R. Rao, *J. Phys. Chem. Solids*, 2001, **62**, 1481.
33 S. Oliver, A. Kuperman, A. Lough and G. A. Ozin, *Chem. Mater.*, 1996, **8**, 2391.
34 S. Oliver, A. Kuperman and G. A. Ozin, *Angew. Chem., Int. Ed.*, 1998, **37**, 46.
35 G. Paul, A. Choudhury and C. N. R. Rao, *J. Chem. Soc., Dalton Trans.*, 2002, 23859.
36 M. I. Khan, S. Cevik and R. J. Doedens, *Inorg. Chim. Acta*, 1999, **292**, 112.
37 M. Cavellec, D. Riou, J. M. Greneche and G. Ferey, *Inorg. Chem.*, 1997, **36**, 2187.
38 Z. A. D. Lethbridge, P. Lightfoot, R. E. Morris, D. S. Wragg, P. A. Wright, A. Kvick and G. Vaughan, *J. Solid State Chem.*, 1999, **142**, 455.
39 R. H. Jones, J. M. Thomas, H. Qisheng, M. B. Hursthouse and J. Chen, *J. Chem. Soc., Chem. Commun.*, 1991, 1520.
40 L. Fafani, A. Nunzi and P. F. Zanazzi, *Am. Mineral.*, 1971, **56**, 751.
41 R. A. Ramik, B. D. Sturman, P. J. Dunn and A. S. Poverennukh, *Can. Mineral.*, 1980, **18**, 185.
42 N. Simon, T. Loiseau and G. Ferey, *J. Mater. Chem.*, 1999, **9**, 585.
43 F. C. Hawthrone, *Acta Crystallogr., Sect. B: Struct. Sci.*, 1994, **50**, 481 and the references therein.
44 C. Livage, F. Millange, R. I. Walton, T. Loiseau, N. Simon, D. O'Hare and G. Ferey, *Chem. Commun.*, 2001, 994.
45 R. I. Walton, F. Millange, A. L. Bail, T. Loiseau, C. Serre, D. O'Hare and G. Férey, *Chem. Commun.*, 2000, 203.
46 J. R. Gutnick, E. A. Muller, A. N. Sarjeant and A. J. Norquist, *Inorg. Chem.*, 2004, **43**, 6528.

47 J. Fuch, H.-U. Kreusler and A. Z. Forster, *Z. Naturforsch., B*, 1979, **34**, 1683.
48 Y. Xing, Y. Liu, Z. Shi, H. Meng and W. Pang, *J. Solid State Chem.*, 2003, **174**, 381.
49 A. J. Norquist, M. B. Doran, P. M. Thomas and D. O'Hare, *Dalton Trans.*, 2003, 1168.
50 A. J. Norquist, P. M. Thomas, M. B. Doran and D. O'Hare, *Chem. Mater.*, 2002, **14**, 5179.
51 M. B. Doran, B. E. Cockbain and D. O'Hare, *Dalton Trans.*, 2005, 1774.
52 A. J. Norquist, M. B. Doran and D. O'Hare, *Solid State Sci.*, 2003, **5**, 1149.
53 A. J. Norquist, M. B. Doran, P. M. Thomas and D. O'Hare, *Inorg. Chem.*, 2003, **42**, 5949.
54 M. B. Doran, A. J. Norquist and D. O'Hare, *Inorg. Chem.*, 2003, **42**, 6989.
55 M. B. Doran, B. E. Cockbain and D. O'Hare, *Dalton Trans.*, 2004, 3810.
56 P. M. Thomas, A. J. Norquist, M. B. Doran and D. O'Hare, *J. Mater. Chem.*, 2003, **13**, 88.
57 M. Cavellec, D. Riou and G. Ferey, *J. Solid State Chem.*, 1994, **112**, 441.
58 M. Cavellec, D. Roiu and G. Ferey, *Eur. J. Solid State Inorg. Chem.*, 1995, **32**, 271.
59 X. Wang, L. Liu, H. Cheng, K. Ross and A. J. Jacobson, *J. Mater. Chem.*, 2000, **10**, 1203.
60 G. Paul, A. Choudhury, E. V. Sampathakumaran and C. N. R. Rao, *Angew. Chem., Int. Ed.*, 2002, **41**, 1224.
61 J. E. Dutrizac and S. Kaiman, *Can. Mineral.*, 1976, **14**, 151.
62 C. N. R. Rao, G. Paul, A. Choudhury, E. V. Sampathkumaran, A. K. Raychaudhuri, S. Ramasesha and I. Rudra, *Phys. Rev. B: Condens. Matter*, 2003, **67**, 134425.
63 C. N. R. Rao, E. V. Sampathkumaran, R. Nagarajan, G. Paul, J. N. Behera and A. Choudhury, *Chem. Mater.*, 2004, **16**, 1441.
64 G. Paul, A. Choudhury and C. N. R. Rao, *Chem. Commun.*, 2002, 1904.
65 J. N. Behera, G. Paul, A. Choudhury and C. N. R. Rao, *Chem. Commun.*, 2004, 456.
66 J. N. Behera, K. V. Gopalkrishnan and C. N. R. Rao, *Inorg. Chem.*, 2004, **43**, 2636.
67 F. C. Hawthrone, S. V. Krivovichev and P. C. Burns, *Rev. Mineral. Geochem.*, 2000, **40**, 55.
68 M. Dan, J. N. Behera and C. N. R. Rao, *J. Mater. Chem.*, 2004, **14**, 1257.
69 L. Liu, H. Meng, G. Li, Y. Cui, X. Wang and W. Pang, *J. Solid State Chem.*, 2005, **178**, 1003.
70 T. Bataille and D. Louer, *J. Mater. Chem.*, 2002, **12**, 3487.
71 Y. Xing, Z. Shi, G. Li and W. Pang, *Dalton Trans.*, 2003, 940.
72 P. Benard-Rocherulle, H. Tronel and D. Louer, *Powder Diffr.*, 2002, **17**, 290.
73 S. Govindarajan, K. C. Patil, H. Manohar and P. E. Werner, *J. Chem. Soc., Dalton Trans.*, 1986, 119.
74 D. Wang, R. Yu, Y. Xu, S. Feng, R. Xu, N. Kumada, N. Kinomura, Y. Matsumura and M. Takano, *Chem. Lett.*, 2002, 1120.
75 M. S. Wickleder, *Chem. Rev.*, 2002, **102**, 2011.
76 A. J. Norquist, M. B. Doran and D. O'Hare, *Inorg. Chem.*, 2005, **44**, 3837.
77 D. M. C. Huminicki and F. C. Hawthorne, *Rev. Miner. Geochem.*, 2002, **48**, 193.
78 C. N. Morimoto and E. C. Lingafelter, *Acta Crystallogr., Sect. B: Struct. Crystallogr. Cryst. Chem.*, 1970, **B26**, 341.
79 W.T.A. Harrison, M. L. F. Phillips and T. M. Nenoff, *J. Chem. Soc., Dalton Trans.*, 2001, 2459.
80 W.T.A. Harrison and M. L. F. Phillips, *Chem. Mater.*, 1997, **9**, 1837.
81 I. Bull, P. S. Wheatley, P. Lightfoot, R. E. Morris, E. Sastre and P. A. Wright, *Chem. Commun.*, 2002, 1180.
82 Y.-P. Yuan, R.-Y. Wang, D.-Y. Kong, J.-G. Mao and A. Clearfield, *J. Solid State Chem.*, 2005, **178**, 2030.
83 M. Doran, A. J. Norquist and D. O'Hare, *Chem. Commun.*, 2002, 2946.
84 V. V. Tabachenko, L. M. Kovba and V. N. Serezhkin, *Khoord Khim.*, 1984, **10**, 558.
85 P. S. Halasyamani and K. R. Poeppelmier, *Chem. Mater.*, 1998, **10**, 2753.
86 Y. Porter, N. S. P. Bhuvanesh and P. S. Halasyamani, *Inorg. Chem.*, 2001, **40**, 1172.
87 W. T. A. Harrison, M. L. F. Phillips, J. Stanchfield and T. M. Nenoff, *Angew. Chem., Int. Ed.*, 2000, **39**, 3808.
88 A. Choudhury, D. Udayakumar and C. N. R. Rao, *Angew. Chem., Int. Ed.*, 2002, **41**, 158.
89 D. Udayakumar and C. N. R. Rao, *J. Mater. Chem.*, 2003, **13**, 1635.
90 F. Millange, C. Serre, T. Cabourdin, J. Marrot and G. Ferey, *Solid State Sci.*, 2004, **6**, 229.
91 I. Pasha, A. Choudhury and C. N. R. Rao, *Solid State Sci.*, 2003, **5**, 257.
92 Z. Dai, Z. Shi, G. Li, X. Chen, X. Lu, Y. Xu and S. Feng, *J. Solid State Chem.*, 2003, **172**, 205.
93 H. Effenberger, *Z. Kristallogr.*, 1985, **173**, 265.
94 S. Natarajan and A. K. Cheetham, *J. Solid State Chem.*, 1998, **140**, 435.
95 Y. Piffard, V. Verbeare, S. Oyetola, S. Courant and M. Tournoux, *Eur. J. Solid State Inorg. Chem.*, 1989, **26**, 113.
96 I. Pasha and C. N. R. Rao, *Inorg. Chem.*, 2003, **42**, 409.
97 A. M. Chippindale and S. J. Brech, *Chem. Commun.*, 1996, 2781.
98 Z. Dai, X. Chen, Z. Shi, D. Zhang, G. Li and S. Feng, *Inorg. Chem.*, 2003, **42**, 908.
99 I. Pasha, A. Choudhury and C. N. R. Rao, *J. Solid State Chem.*, 2003, **174**, 386.
100 M.-L. Feng, J.-G. Mao and J.-L. Song, *J. Solid State Chem.*, 2004, **177**, 3529.
101 M. Fleck, U. Kolitsch and B. Hertweck, *Z. Kristallogr.*, 2002, **217**, 435.
102 D. Udayakumar, M. Dan and C. N. R. Rao, *Eur. J. Inorg. Chem.*, 2004, 1733.
103 J. N. Behera, A. A. Ayi and C. N. R. Rao, *Chem. Commun.*, 2004, 968.


 

FEATURE ARTICLE www.rsc.org/chemcomm | ChemComm

# Structural diversity and chemical trends in hybrid inorganic–organic framework materials

**Anthony K. Cheetham,*[a] C. N. R. Rao*[b] and Russell K. Feller[a]**

***Received (in Cambridge, UK) 18th July 2006, Accepted 20th October 2006***
***First published as an Advance Article on the web 7th November 2006***
**DOI: 10.1039/b610264f**

Hybrid framework compounds, including both metal–organic coordination polymers and systems that contain extended inorganic connectivity (extended inorganic hybrids), have recently developed into an important new class of solid-state materials. We examine the diversity of this complex class of materials, propose a simple but systematic classification, and explore the chemical and geometrical factors that influence their formation. We also discuss the growing evidence that many hybrid frameworks tend to form under thermodynamic rather than kinetic control when the synthesis is carried out under hydrothermal conditions. Finally, we explore the potential applications of hybrid frameworks in areas such as gas separations and storage, heterogeneous catalysis, and photoluminescence.

## 1 Introduction

The purpose of this feature article is to give an overview of developments in the field of hybrid inorganic–organic framework structures over recent years, especially during the last decade. We have not attempted to be comprehensive because of the huge amount of activity in the area, but instead we have focused on placing these developments in a broader context. We shall illustrate the enormous chemical and structural diversity of these materials and discuss some of the systematic trends that are starting to appear in synthetic routes for hybrids. We shall also examine some of the emerging application for materials in this exciting area.

There is an extensive class of purely inorganic framework materials based upon extended arrays such as chains, sheets or 3-D networks. The silicate and aluminosilicate minerals, which were classified by Pauling almost 70 years ago, constitute the most versatile group. Indeed, their dimensionalities can range from 0, as in simple silicates such as zircon that are based upon isolated orthosilicate $SiO_4^{4-}$ units, through 1-D silicate chains (*e.g.* pyroxenes), 2-D sheets (*e.g.* micas and clays) to 3-D arrays (*e.g.* quartz). Zeolites represent a particularly interesting sub-class of these aluminosilicate frameworks, since their architectures display nanoporosity that can be harnessed for applications in separations, catalysis and so on.[1] More recently it has been shown that a wide range of other inorganic families, especially phosphates, can form framework structures with varying dimensionalities. This is true, for example, of aluminium phosphates, tin(II) phosphates, zinc phosphates and so on.[2] Fig. 1 shows examples from the case of tin phosphates.

In the world of organic solids, by contrast, such structural diversity is less well represented. Molecular organics (*i.e.* 0-D) are ubiquitous, of course, but extended arrays are largely limited to 1-D chains, such as those found in polymer systems ranging from polyolefins to block copolymers and proteins. With the exception of covalent organic frameworks (COFs) that contain borate,[3] extended 2-D and 3-D organic arrays are essentially unknown, aside from cross-linked polymers and examples based upon molecular units that assemble into networks *via* hydrogen bonding rather than covalent bonding.[4] In the light of this basic distinction between inorganic and organic networks, it is interesting to examine the structural diversity of hybrid inorganic–organic frameworks.

We define hybrid inorganic–organic framework materials as compounds that contain both inorganic and organic moieties as integral parts of a network with infinite bonding connectivity in at least one dimension. This definition excludes systems that are molecular or oligomeric, such as the

[a]*Materials Research Laboratory, University of California, Santa Barbara, CA, 93106-5121, USA. E-mail: cheetham@mrl.ucsb.edu*
[b]*Jawaharlal Nehru Centre for Advanced Scientific Research, Jakkur P.O., Bangalore, 560 064, India. E-mail: cnrrao@jncasr.ac.in*

*Tony Cheetham was a member of the Chemistry faculty at Oxford, 1974–1991, and has been at the University of California at Santa Barbara since 1991. He is Professor in both the Materials and Chemistry Departments at UCSB, and since 2004 has been the Director of the new International Center for Materials Research (ICMR). Cheetham's research interests lie in the area of functional inorganic materials and currently include hybrid framework materials, phosphors for solid state lighting, and inorganic nanoparticles.*

*C. N. R. Rao is the National Research Professor of India, Linus Pauling Research Professor at the Jawaharlal Nehru Centre for Advanced Scientific Research, and Honorary Professor at the Indian Institute of Science. His research interests are in the chemistry of materials. He has authored nearly 1000 research papers and edited or written 30 books in materials chemistry.*

*Russell K. Feller graduated in Chemistry from the University of California at Los Angeles in 2003 and is currently a graduate student in the Chemistry Department at UCSB. His PhD research project focuses on hybrid framework materials of transition metals.*

 

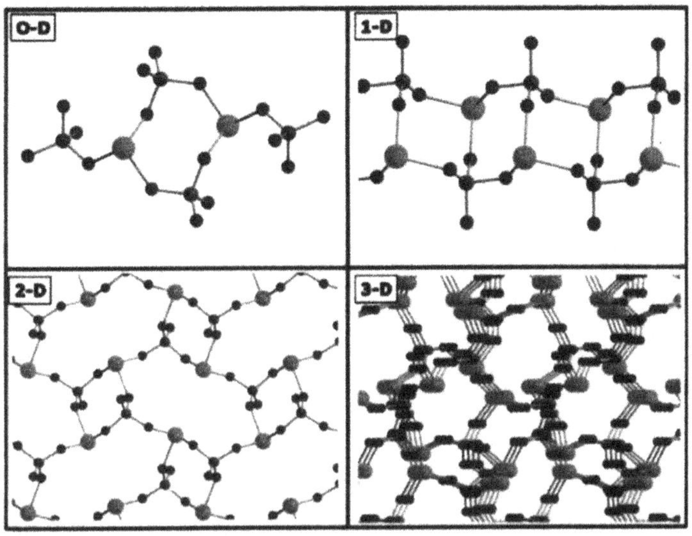


**Fig. 1** Four tin phosphates, each containing the same $Sn_2P_2O_4$ motif, with monomeric, 1-D, 2-D, and 3-D structures. Blue spheres denote tin, green phosphorus, and red oxygen.

supramolecular assemblies described by Lehn, Hosseini, Stang, Fujita and many others. It also excludes systems in which the organic is merely a guest inside an inorganic cavity, as is often observed in zeolites and mesoporous materials, and hybrid composites, in which the inorganic and organic components are present as separate phases. Most of the known hybrid frameworks may conveniently be divided into two categories.[5] The *coordination polymers*, or metal organic frameworks (MOFs) as they are also known, can be defined as extended arrays composed of isolated metal atoms or clusters that are linked by polyfunctional organic ligands, L; these are based upon M–L–M connectivity. Second, there are systems that contain extended arrays of inorganic connectivity, which we shall refer to for convenience as *extended inorganic hybrids*. At present, the vast majority of known materials in this area are based upon oxygen bridges. These *hybrid metal oxides*, which often contain infinite metal–oxygen–metal (M–O–M) arrays as a part of their structures, represent a sub-group of a larger class in which there is extended M–X–M bonding *via* other atoms such as Cl, N or S, or *via* inorganic groups such as phosphate.

Examples of 1-D coordination polymers are relatively common in the early literature, even though they were not seen at the time as part of a vast and remarkable family of materials. Examples include porphyrin coordination polymers (Fig. 2) with interesting magnetic properties that were first discovered by Basolo and co-workers in the 1970s and characterized by X-ray diffraction at a later date.[6] Early examples in the 3-D coordination polymer area can be found in the work of Gravereau, Garnier and Hardy in the late 1970s, in which zeolitic materials with ion-exchange properties were made by linking hexacyanoferrate units with tetrahedrally coordinated $Zn^{2+}$ cations.[7] There were also early examples of hybrid materials with extended inorganic connectivity, the most notable being the layered zirconium phosphonates such as the one shown in Fig. 3.[8]

Interest in the hybrid area began to accelerate in the 1990s, when several groups, particularly those of Robson, Hoffman and Yaghi, recognized that rigid, polyfunctional organic molecules could be used to bridge metal cations or clusters into extended arrays. Robson published a landmark paper in 1990,[9] laying the groundwork for an important part of the field of crystal engineering – the science of predicting basic

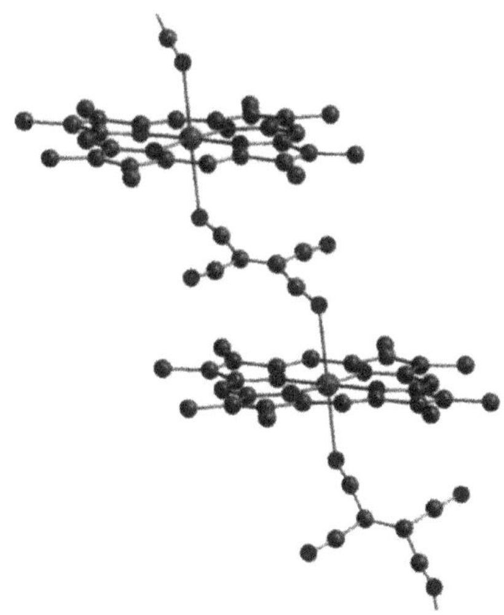

**Fig. 2** The 1-D coordination polymer octaethylporphyrinatomanganese(III) tetracyanoethenide (ethyl groups omitted for clarity).[6] The two tetracyanoethenide anions are not crystallographically equivalent, but slightly tilted with respect to each other. Pink spheres denote manganese, blue nitrogen, and gray carbon.

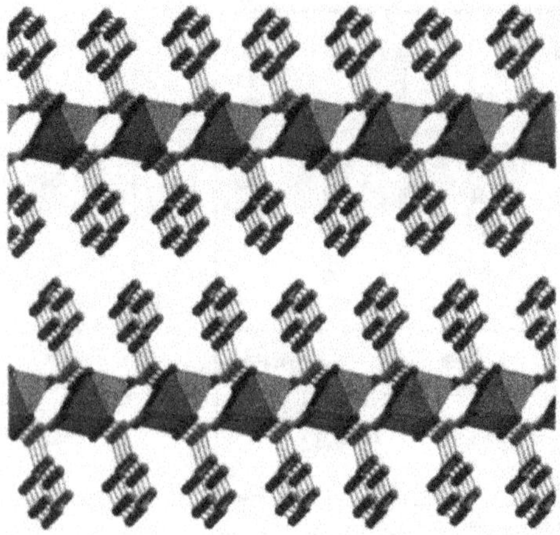

Fig. 3 A side view of the layered structure of zirconium bis(phenylphosphonate), with edge-to-edge interactions between phenyl rings on neighboring sheets.[8] Green spheres denote phosphorus, red oxygen, and gray carbon, with $ZrO_6$ octahedra in blue.

networks with potentially useful characteristics and then using appropriate molecular building blocks to synthesize them.[10] For the synthesis of *porous* materials, networks are often envisioned where rigid organic molecules and metal atoms or clusters replace bonds and atoms in classical inorganic structures.[11]

The purpose of our feature article is to discuss some of the emerging trends in this rapidly developing field rather than to provide a comprehensive review of hybrids. This is a very complex and diverse field, so we are suggesting a systematic classification of hybrid frameworks that places new and existing materials in a simple, rational context. We have also attempted to identify the structural and chemical trends that are beginning to emerge from the literature and to pinpoint some of the areas where there are important gaps and opportunities. Recent reviews of various aspects of hybrid materials include those of Clearfield, Rowsell and Yaghi, Rosseinsky, and Rao *et al.*[12]

## 2 Chemical and structural diversity

### 2.1 Chemical diversity

Hybrid frameworks are found for a wide range of metals and involve a diverse range of organic ligands. Most of the published work involves transition metals, including zinc, but there is a growing body of literature around rare-earth based systems, which are of interest for their optical properties. In addition, there has been a certain amount of effort with p-block elements, especially aluminium, gallium and tin, plus a recent growth of interest in magnesium, driven by the search for lightweight materials for hydrogen storage.

In terms of organic ligands, much of the recent focus has been on connectivity through oxygen atoms of carboxylic acid groups, and there has been a recent review of this field by Rao and colleagues.[12] Rigid dicarboxylic acids, such as benzene-1,4-dicarboxylic acid, have proved very versatile, as have the simple but more flexible aliphatic systems, such as succinic and glutaric acids. The simplest member of this family, oxalic acid, has been used extensively. As will be discussed later, monocarboxylic acids can also form hybrids, and there has been some recent effort with formic and acetic acids. Nor is the field limited to carboxylic acids, since phosphonic acids and phenolic acids can also form hybrid frameworks. Beyond network formation involving M–O linkages, there has been a reasonable amount of work with other types of ligands, such as pyridyls and imidazoles, as well as mixed ligands that offer the possibility of more than one type of connection, *e.g.* M–O plus M–N or M–S. Much remains to be explored in the area of these more complex linkages.

We shall aim to illustrate the diversity of chemical types in the choice of examples that will be given in subsequent sections. Clearly the structures are strongly influenced by the coordination preferences of the metals as well as by the variety of ways in which different ligands can coordinate to metals.

### 2.2 Coordination polymers

The term coordination polymer owes its origins to the analogy with coordination compounds, in which ligands, organic or otherwise, are coordinated to monomeric metal centers. Coordination compounds can be thought of loosely as the monomers of coordination polymers. Given the enormous volume of work on coordination compounds over the last century, starting with the pioneering work of Werner for which he was awarded the Nobel Prize in Chemistry in 1913, it is hardly surprising that the field of coordination polymers is turning out to be so rich and varied.

A vast range of coordination polymers or supramolecular architectures with different dimensionalities – 1-D, 2-D and 3-D – have been discovered in recent years. Fig. 4 illustrates an example of a 1-D chain system involving Ag–N bonding through a linear 4,4′-bipyridyl group.[13] Since the linkage through the silver is also linear, the chain, itself, is too. By

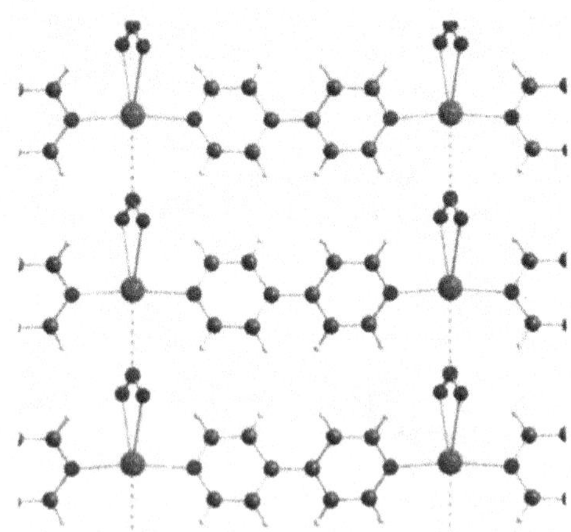

Fig. 4 A linear coordination polymer of silver with 4,4′-bipyridyl and bidentate nitrite anions.[13] There is a 3.0 Å contact between the silver and the nitrogen of the nitrite ion, shown as a dotted line. If this was a full covalent bond, the system would be two-dimensional. Gray spheres denote carbon, white hydrogen, dark blue nitrogen, red oxygen, and green silver.

 

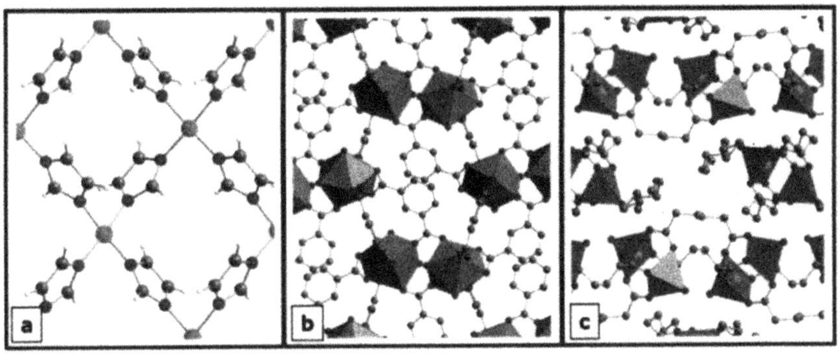


**Fig. 5** Some 2-D coordination polymers: (a) nickel bis(imidazolate), containing neutral sheets of square planar metal centers,[16] (b) a plan view of the europium isophthalate structure,[17] showing nine-coordinate $Eu^{3+}$, and (c) copper adipate,[18] which consists of both chains and sheets, viewed along the chain axis. Gray spheres denote carbon, white hydrogen, blue nitrogen, red oxygen, and green nickel, with $EuO_9$ polyhedra in orange and $CuO_5$ square pyramids in blue.

contrast, when tetrahedrally coordinated zinc ions are connected through 4,4′-bipyridyl, this gives rise to a zigzag chain.[14] The metal center does not have to be mononuclear. For example, the common copper acetate dimer, with its paddle wheel geometry, can form linear chains through simple bidentate ligands.[15] The scope of these 1-D coordination polymers is enormous, since they may be neutral or charged (in which case they require compensating cations or anions) and they often contain solvent molecules in voids or channels.

Turning to 2-D systems, Fig. 5(a) shows a very simple example in which nickel ions are connected *via* square planar coordination by imidazolate anions to form a very simple neutral layered structure.[16] A more complex case based upon a combination of rare-earth ions with isophthalic acid is shown in Fig. 5(b); the thiophene derivatized version of the same ligand combined with $Tb^{3+}$ yields a product with enhanced green luminescence.[17] As a final example to illustrate the versatility of this area, we show in Fig. 5(c) a 2-D coordination polymer based on the copper carboxylate dimer, linked through adipic acid, in which a 2-D layer alternates with 1-D chains that contain the same basic building blocks.[18] Here we note that the paddle wheels are linked *via* the dicarboxylic acid rather than by ligands in the axial position of the Jahn–Teller distorted copper coordination sphere.

Robson's early work yielded some elegant examples of crystal engineering in which 3-D networks of simple, known structure types, such as the diamondoid and $ReO_3$ structures, were built from suitable combinations of metal ions and rigid linkers. Fig. 6 shows a simple example based upon the PtS structure.[19] Others have exploited the same concept, including Carlucci *et al.*,[20] who reported some remarkable open-framework structures based upon silver in combination with ligands such as pyrazine (Fig. 7). Some of the most striking examples

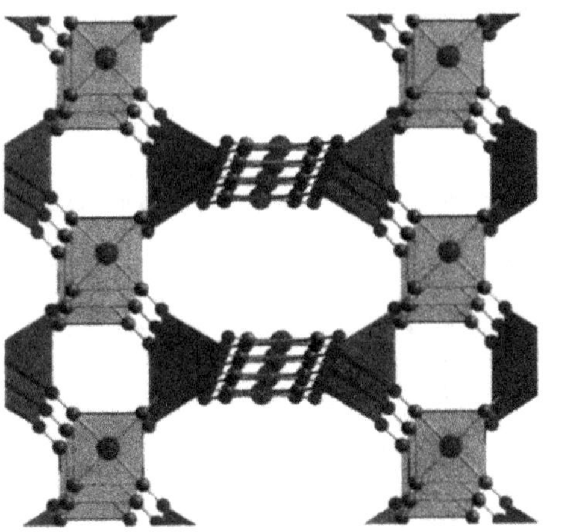

**Fig. 6** A mixed copper–platinum tetracyanide, structurally analogous to PtS, with sulphur replaced by a bridging copper tetracyano complex.[19] Pink spheres denote platinum, gray carbon, and blue nitrogen, with $PtC_4$ square planes in pink and $CuN_4$ tetrahedra in blue. Reproduced with permission. Copyright 1990, Royal Society of Chemistry.

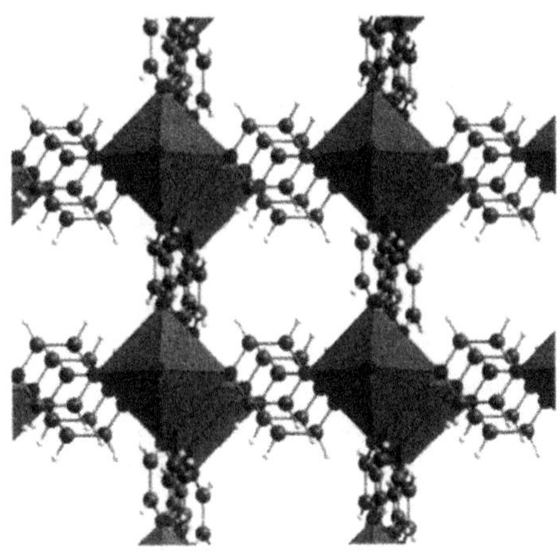

**Fig. 7** Silver pyrazine hexafluoroantimonate,[20] a porous framework with an $ReO_3$-like structure ($SbF_6^-$ groups omitted for clarity). Gray spheres denote carbon, blue nitrogen, and white hydrogen, with $AgN_6$ octahedra in green. Reproduced with permission. Copyright 1995, Wiley–VCH.

of porous 3-D coordination polymers can found in the work of Yaghi, O'Keeffe and co-workers, in which they have exploited bridging of simple $Zn_4O$ groups *via* rigid aromatic dicarboxylates such as benzene-1,4-dicarboxylic acid to build networks with remarkably low densities and high porosity, such as MOF-5 (Fig. 8).[21] They have shown that large families based upon the same architecture can be created by altering the length or other chemical details of the organic linker.[22] In the case of the *reticular* family based upon MOF-5, for example, they have made as many as 16 derivatives with the same basic architecture. It should also be noted that in addition to architectures based upon the topologies of simple inorganic structures, there has also been success in building porous hybrids based upon known zeolite structures. These include zinc, cadmium and indium coordination polymers that adopt the ABW, BCT, MTN, RHO and SOD topologies.[23] Imidazole-based ligands are particularly effective for this purpose since they can mimic the Si–O–Si angles that are found in typical zeolites.

One of the complications that can arise when the structures are very open is that the networks can interpenetrate or interweave, thereby reducing or eliminating the porosity. There are some remarkable examples of such behavior, including cases where multiple interpenetration is observed.[24] Paradoxically, Yaghi has shown that interpenetration can be attractive for certain gas storage applications, since it can increase the available surface area per unit volume (see section 5.1).

A very recent and exciting example of a 3-D coordination polymers can found in the work of Férey *et al.*[25] on the use of trinuclear chromium clusters in combination with benzene-1,4-dicarboxylic acid. This reaction yields porous structures with unit cells volumes of up to 700 000 $Å^3$, *i.e.* similar to that of a small protein! In the absence of single crystals for structure determination, the structures were solved by the ingenious use of Monte Carlo simulations with simulated annealing.

**Fig. 8** MOF-5,[21] a porous cubic zinc terephthalate which is topologically analogous to $ReO_3$. Gray spheres denote carbon, red oxygen, and white hydrogen, with $ZnO_4$ tetrahedra in blue. Reproduced with permission. Copyright 1999, Macmillan Publishers Ltd.

Finally in this brief section on the exploding field of coordination polymers, we should mention that there growing body of work on systems that contain more than one metal or more than one ligand type, or both. For example, some beautiful open architectures can form when transition metals are combined with rare-earths in the presence of glycolate and water as ligands.[26] Assemblies that contain more than one metal can often be facilitated by the use of ligands with multiple N and O donor atoms, taking advantage of the different ligand affinities of transition metals and rare earths. Equally, the use of more than one ligand with a single metal can yield interesting results, as in the work mixed oxalate–diphosphonates and oxalate–dicarboxylates.[27] A variation on this theme involves a single metal with a carboxylate ligand such as oxalate in combination with phosphate as a second anion.[28] Materials of this type were initially made serendipitously when oxalate salts were being used as precursors in the synthesis of metal phosphate frameworks. These examples illustrate the huge scope of this burgeoning area, with its unlimited permutations of metals and ligands.

### 2.3 Extended inorganic hybrids

The area of coordination polymers is only one sub-field of the broader domain of hybrid framework materials, since far more structural permutations become accessible if we allow for the possibility of extended inorganic connectivity. This idea is illustrated schematically in Fig. 9, which compares 1-D and 2-D coordination polymers with a system that has inorganic connectivity in two dimensions and is connected in the third dimension by organic linkers. Such extended inorganic hybrid materials not only open up a vast area of new chemical and structural permutations, but they also provide a basis for creating materials with properties that are traditionally found in metal oxides. Thus we have the tantalizing possibility of making hybrid materials that are metallic, superconducting, or high temperature ferromagnets. In this section of the paper, we describe some of the progress that has already been made in this fascinating area.

Early examples of 1-D hybrid metal halides include the famous Wolfram's red salt, which contains $\{Pt(EtNH_2)_4\}$ units linked into infinite chains by Pt–Cl–Pt bridges (Fig. 10(a)).[29] In this, and other more recent examples such as the zinc phosphonate chain of Stucky and co-workers (Fig. 10(b)),[30] the organic ligands simply decorate the inorganic chains, rather than cross-linking them. However, 1-D inorganic chains can also be cross-linked to make a layered structure rather than a 1-D network, as in the case of the metal succinate $Ni_7(OH)_6(H_2O)_3(C_4H_4O_4)_4 \cdot 7H_2O$ (Fig. 11).[31] A beautiful example of a cross-linked inorganic chain is found in transition-metal gallates, which comprises chains of *trans*-corner-sharing $MO_6$ octahedra cross-linked into a 3-D network by the gallate ions (Fig. 12).[32] Note that all the oxygen atoms of the $MO_6$ octahedra are supplied by the gallate ligands; the resulting topology of the inorganic network looks like the chain found in $Rb_2FeF_5$,[33] thereby revealing the striking resemblance between this area and classical solid-state chemistry. Note, too, that the array contains channels that accommodate zeolitic water molecules.

 

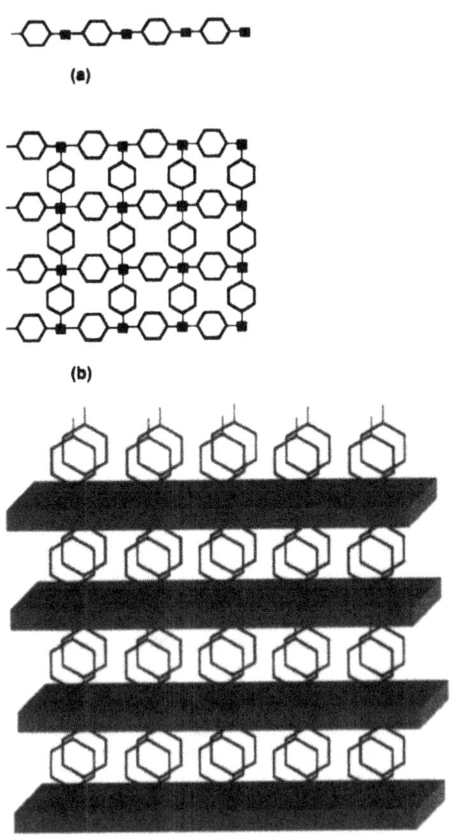


Fig. 9 Schematic representation of coordination polymers and extended inorganic hybrids: (a) and (b) show 1-D and 2-D coordination polymers, respectively, while (c) shows a system that has inorganic connectivity in two dimensions and is connected in the third dimension by organic linkers.

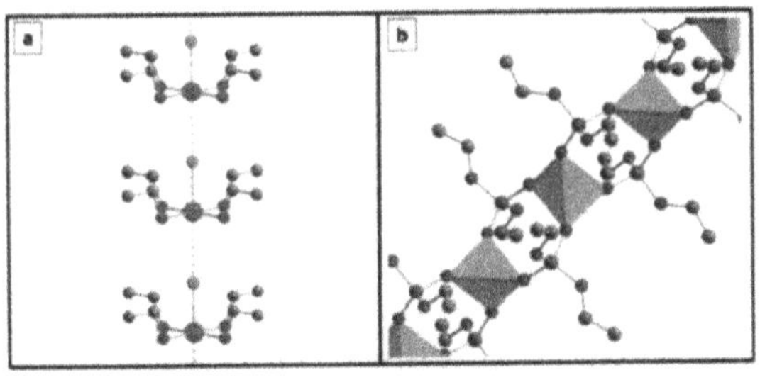


Fig. 10 Some 1-D extended inorganic hybrid materials: (a) Wolfram's Red salt, a 1-D platinum chloride polymer.[29] Controversy continues as to whether the coordination geometry is square pyramidal as shown here, or alternating octahedral and square planar; and (b) zinc diethylphosphate,[30] an inorganic chain decorated by organic groups. Light green spheres denote chlorine, gray carbon, dark blue nitrogen, pink platinum, red oxygen, and dark green phosphorus, with $ZnO_4$ tetrahedra in light blue.

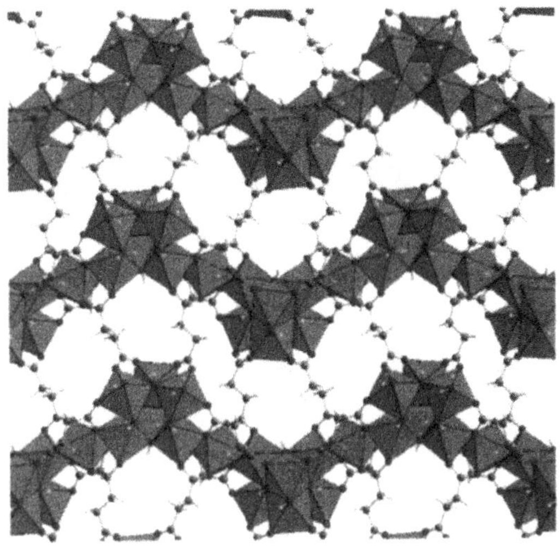

Fig. 11 A nickel succinate containing inorganic chains, bridged by organic groups along a second dimension to form sheets.[31] Gray spheres denote carbon, red oxygen, and white hydrogen, with $NiO_6$ octahedra in green. Reproduced with permission. Copyright 2003, Wiley-VCH.

Many nice examples of 2-D hybrid oxides can be derived from Clearfield's α-zirconium phosphate structure, which contains 2-D sheets of $ZrO_6$ octahedra sandwiched between phosphate layers.[34] Alberti *et al.*[8] was able to increase the interlayer spacing by taking advantage of monophosphonates, which acted as spacers between the layers, albeit with no bonding connection between them (Fig. 3). As with some of the examples in the previous paragraph, the organic groups are decorating the layers in the Alberti structures. In order to build three-dimensional frameworks, Dines *et al.* demonstrated that

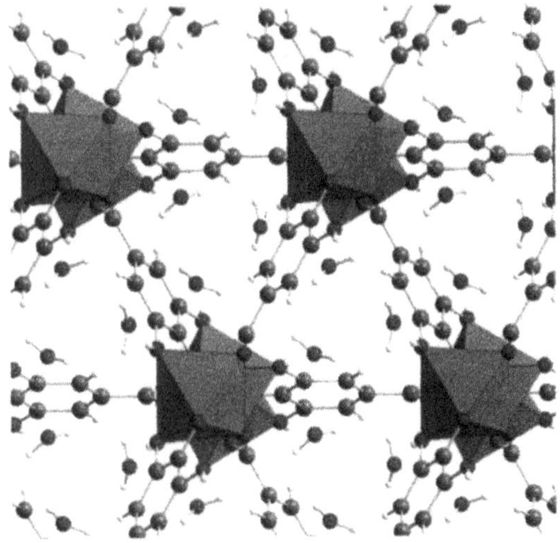

Fig. 12 Nickel gallate,[32] a framework containing inorganic chains bridged by organic groups into a 3-D network. Gray spheres denote carbon, red oxygen, and white hydrogen, with $NiO_6$ octahedra in green. Reproduced with permission. Copyright 2006, Elsevier Ltd.

 

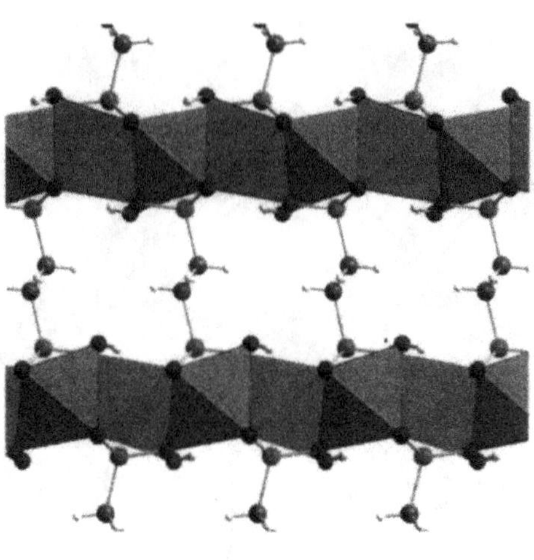

**Fig. 13** A pillared cobalt(II) ethanediphosphonate with inorganic sheets connected in the third dimension by organic groups.[35] Gray spheres denote carbon, red oxygen, white hydrogen, and green phosphorus, with $CoO_6$ octahedra in blue. Reproduced with permission. Copyright 2005, American Chemical Society.

is possible to create connections between the inorganic layers by using diphosphonates (Fig. 13).[36] These materials are not porous, but porosity can be created by using a combination of diphosphonate linkers and shorter monophosphonate groups.[37] There has also been some work to add functionality to the zirconium-based materials through careful choice of the diphosphonate used in the synthesis. For example, Vermuelen and Thompson used viologens, rigid organic molecules of the formula $[H_2O_3PCH_2CH_2\text{–bipyridinium–}CH_2CH_2PO_3H_2]^{2+}$, to bridge zirconium phosphonate layers.[38]

In view of the existence of 1-D and 2-D inorganic connectivity in hybrid materials, there has been considerable interest in the possible existence of 3-D systems. A striking example was found in 2002 with the discovery of a nickel succinate with a 3-D network of corner- and edge-sharing $NiO_6$ octahedra (Fig. 14).[39] Additional examples have since been found in other systems, including cadmium malonate[40] and nickel glutarate.[41] In all of these systems, the frameworks are open with the organic groups lining the pores, as in Fig. 14, and the structures show enhanced thermal stability on account of the inertness of the inorganic skeleton. Nickel succinate, for example, is stable to 400 °C in air.

### 2.3 Classification of hybrid framework structures

The examples of hybrid materials presented in the previous two sections underline the enormous structural diversity of this exciting class of materials. For example, while we have focused on the dimensionality with respect to either M–ligand–M or extended inorganic connectivity, it is clear that many systems exhibit both types of linkages. For example, the cobalt diphosphonate shown in Fig. 13[35] is 2-D with respect to inorganic connectivity and 1-D with respect to M–ligand–M connectivity. Overall, therefore, the dimensionality of the network is 3-D. Table 1 shows the full range of possibilities in terms of M–ligand–M or extended inorganic dimensionalities.

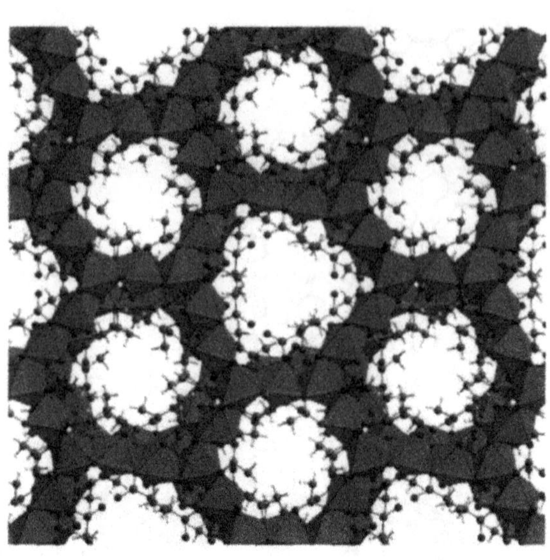

**Fig. 14** A nickel succinate containing a channelled 3D nickel–oxygen network decorated by organic groups.[38] Gray spheres denote carbon, red oxygen, and white hydrogen, with $NiO_6$ octahedra in green. Reproduced with permission. Copyright 2002, Wiley-VCH.

It is interesting to reflect that virtually the whole of organometallic chemistry and much of classical coordination chemistry is contained within a single box in this table (*i.e.* both M–L–M and inorganic connectivity = 0). For conciseness, we refer to this type as $I^0O^0$ (I = inorganic and O = organic); note that the sum of the exponents gives the overall dimensionality of the structure. The 3-D nickel succinate (Fig. 14) falls in the box with M–L–M = 0 and inorganic connectivity = 3 ($I^3O^0$) while the MOF-5 structure (Fig. 8) has M–L–M = 3 and inorganic connectivity = 0 ($I^0O^3$). There are known examples of all of these classes of hybrid material. Wolfram's red salt (Fig. 10(a)), for example, would be $I^1O^0$, since there is no bonding between the chains, whereas the 1-D coordination polymers of Basolo (Fig. 2) would be $I^0O^1$. However, some classes, such as $I^0O^3$, are quite common, while other, such as $I^1O^1$, are relatively rare.

The question arises as to whether there are materials that can be classified within the empty boxes in the bottom right part of the table. For example, can we have materials in which the connectivity is 2-D with respect to both M–L–M and M–X–M. We currently know of no such materials, but the possibility exists that these may be found in materials that have not yet been discovered.

### 2.4 Dense and open frameworks

In the previous three sections, we have not attempted to differentiate between dense and open hybrid framework structures. As with inorganic silicates and aluminosilicates, there is no fundamental chemical difference between the dense and open hybrid structures, though their properties and applications are often quite distinct. There are also differences in the synthetic strategies that are needed to synthesize them, and in section 4 we shall discuss some of the reaction

 

**Table 1** Proposed classification of hybrid materials, showing the dimensionality of different structures with respect to both organic connectivity between metal centers ($O^n$) and extended inorganic connectivity ($I^n$) (see text for explanation)

**Dimensionality of inorganic connectivity, $I^n$ ($n$ = 0–3)**

| Metal–organic–metal connectivity, $O^n$ ($n$ = 0–3) | 0 | 1 | 2 | 3 |
|---|---|---|---|---|
| 0 | Molecular complexes $I^0O^0$ | Hybrid inorg. chains $I^1O^0$ | Hybrid inorg. layers $I^2O^0$ | 3-D Inorg. hybrids $I^3O^0$ |
| 1 | Chain coordination polymers $I^0O^1$ | Mixed inorg.–organic layers $I^1O^1$ | Mixed inorg.–organic 3-D framework $I^2O^1$ | — |
| 2 | Layered coordination polymers $I^0O^2$ | Mixed inorg.–organic 3-D framework $I^1O^2$ | — | — |
| 3 | 3-D Coordination polymers $I^0O^3$ | — | — | — |

conditions that are likely to favor the formation of open structures. One strategy that should be mentioned here concerns the use of organic template molecules in the creation of open hybrid structures. Template molecules or structure-directing agents, especially quaternary amines, have been extensively used in zeolite synthesis, but the strategy has not been widely adopted in the hybrid area. Exceptions include the use of amines and other templates in the synthesis of transition-metal diphosphonates, aluminium monophosphonates, and cobalt squarates.[42] In the latter instance, Dan *et al.* prepared $[C_6N_2H_{14}]_2[Co_2(C_4O_4)_3(H_2O)_4]$ and $[C_3N_2H_5]_2$-$[Co_2(C_4O_4)_3(H_2O)_4]$ under hydrothermal conditions in the presence of quaternary amines. Both compounds contain chains formed by cobalt dimers linked by the squarate units, the chains being connected through hydrogen bonding interactions *via* the amines. These materials would be classified as $I^0O^1$ in the classification shown in Table 1, since we do not include organic connectivity through hydrogen bonding. Finally we should mention a rather unusual example of *inorganic* templating, in which Rao and co-workers prepared an open-framework cadmium oxalate that formed around an alkali halide assembly.[43]

## 3 Chemical trends

We now turn to the intriguing question of what chemical factors influence whether a particular system will form coordination polymers rather than frameworks with extended inorganic connectivity, or low dimensional rather than high dimensional networks. The findings so far in this area are relatively sparse, but a few systematic trends are beginning to emerge and will be discussed in the following sub-sections.

### 3.1 Effects of ligand geometry and flexibility on dimensionality

A growing number of materials have been made recently that involve the use of 1,2- 1,3- or 1,4-cyclohexanedicarboxylates (CHDCs) or cyclohexenedicarboxylates.[44] In the case of the CHDCs, the structural trends for the hybrids formed by the three different isomers have been examined systematically with cadmium- and manganese-containing systems;[45] each of the organics can be found as both a *cis*- and a *trans*-isomer. Two-dimensional layered structures of all three of the 1,2-, 1,3- and 1,4-cyclohexanedicarboxylates were made, but infinite metal-oxygen-metal linkages were observed only in the case of the 1,2-dicarboxylate (Fig. 15), the remaining phases being coordination polymers. Only with the close proximity of the carboxylate groups that is found in the 1,2 compound can the metals be sufficiently close to sustain infinite inorganic connectivity, while the 1,3 and 1,4 ligands all provide excellent linkages for coordination polymers with varying dimensionalities. This conclusion is further corroborated by work on cobalt and manganese 4-cyclohexene-1,2-dicarboxylates.[46] We note that the geometry of the 1,2 compound is similar to that of succinic acid, which readily forms extended inorganic connectivity (Fig. 14).

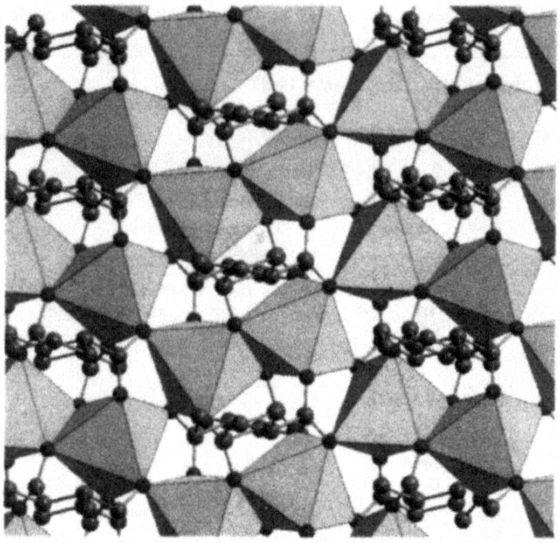

Fig. 15 A plan view of the inorganic sheet structure of cadmium cyclohexane-1,2-dicarboxylate.[45] Gray spheres denote carbon and red oxygen, with $CdO_6$ octahedra in yellow. Reproduced with permission. Copyright 2006, Royal Society of Chemistry.

Chirality offers another facet of ligand geometry that is very interesting. Many organic ligands, such as tartaric acid, can be obtained in an enantiomerically pure form (many occur naturally, of course), and there is much interest in the formation of chiral open frameworks that might be used as enantiomerically-selective catalysts (see section 5.2). One concern has been whether the chiral ligands are sufficiently robust to survive the reaction conditions that are required for hybrid framework formation without racemization, but this does not appear to be a general problem.[47] A considerable amount of work has been done with naturally occurring amino acids.[48] A nice example of the subtlety that can arise in this area is illustrated by a recent study with nickel aspartate.[49] The structure obtained with the L-aspartate contains a chiral helix based upon a chain of edge-sharing $NiO_6$ octahedra (space group $P2_12_12_1$), whereas the structure of the racemic analogue contains both left-handed and right-handed forms of the same chain (space group $I4_1/a$). In the more general case, it seems possible that the topologies of chiral structures might in some cases be quite different from those of their racemic analogues.

In addition to the geometry of the ligand, the degree of flexibility is also important in determining the type of structure that can be formed. In particular, ligands with greater flexibility are more likely to be able to adapt to the geometries found in extended inorganic hybrids, such as metal–oxygen–metal linkages. This has been seen in transition-metal diglycolates and iminodiacetates, where a range of hybrid metal oxides has been reported.[50] Furthermore, the use of monocarboxylates, which are clearly not ideal as linkers for coordination polymers, entirely eliminates the need for ligand flexibility, thereby facilitating the formation of frameworks with infinite inorganic connectivity. This is found, for example, in the case of transition-metal cyclopropane monocarboxylates[51] as well as in formates.[101] Lack of ligand flexibility is not only a disadvantage for forming extended inorganic hybrids, but there is also evidence that it may restrict the range of structures that may form in any particular system. Our recent work on transition-metal gallates illustrates this point, because in spite of many attempts to synthesize alternative hybrid frameworks, we have never succeeded in making anything other than the one shown in Fig. 12.[32] Other systems with more flexible ligands, such as the cobalt succinates, form large numbers of different structures (see section 4.1). We ascribe this difference to the limitations on the bonding options, which are very severe for the gallate ion compared with flexible ligands such as succinate.

### 3.3 Influence of metal ion properties

Turning to the role played by the metal ions in determining the types of structures that can form, a number of important factors are apparent. Most obviously, the preferred coordination number and geometry of the metal ion is a key issue, just as it is in classical coordination chemistry. Divalent and trivalent first-row transition metals, *e.g.* $Mn^{2+}$, $Fe^{2+}$, $Co^{2+}$, $Ni^{2+}$, $Mn^{3+}$, $Fe^{3+}$ and so on, all have well known coordination preferences that often depend upon the identity of the ligand environment. $Mn^{2+}$ is typically octahedrally coordinated by oxygen, while $Co^{2+}$ is more versatile and ranges from tetrahedral through pentacoordinated to octahedral. The optical properties change accordingly as the geometry influences the ligand field splitting. $Cr^{3+}$ is always octahedral, while $Ni^{2+}$ is usually octahedral when surrounded by oxygen, but may be square planar in nitrogen environments (this is apparent in the imidazole network shown in Fig. 5(a)). $Zn^{2+}$, which has been widely used in studies on hybrid frameworks, is very versatile and behaves somewhat like $Co^{2+}$. Isolated $Cu^{2+}$, by contrast, is relatively inflexible and is constrained by its need to accommodate the Jahn–Teller distortion that is characteristic of $d^9$ ions. The rare-earths ions, however, are entirely different from the transition metals, preferring coordination numbers greater than 6 and often 7, 8 or 9 with a wide variety of geometries (see Fig. 16 for a typical example[52]).

Another factor that is frequently apparent is that certain metal ions form well-defined and robust clusters that recur in many hybrid materials. $Cu^{2+}$, for example, is well known for forming a large number of molecular carboxylate clusters that contain the characteristic paddle wheel dimer shown for the case of copper acetate in Fig. 17(a), so it is not a surprise that this unit is ubiquitous in copper coordination polymers such as the highly porous 3-D Cu trimesate system described by Williams and co-workers[53] and the mixed 1-D and 2-D adipate illustrated in Fig. 5(c). Similarly, Yaghi and O'Keeffe have utilized the $Zn_4O$ cluster (Fig. 17(b)) that is a found in basic zinc acetate, $Zn_4O(OCOCH_3)_6$, using this unit as the primary inorganic node in a huge range of zinc dicarboxylate coordination polymers,[21,22] of which MOF-5 is a prototypic example (Fig. 8). A third example of such a cluster is the trimeric $Cr^{3+}$ cluster, $Cr_3O$ (Fig. 17(c)), which was used in combination with terephthalic acid linkers to create the huge unit cells that are found in MIL-101 and related materials.[25] Silica clusters based upon silsesquioxanes have also been

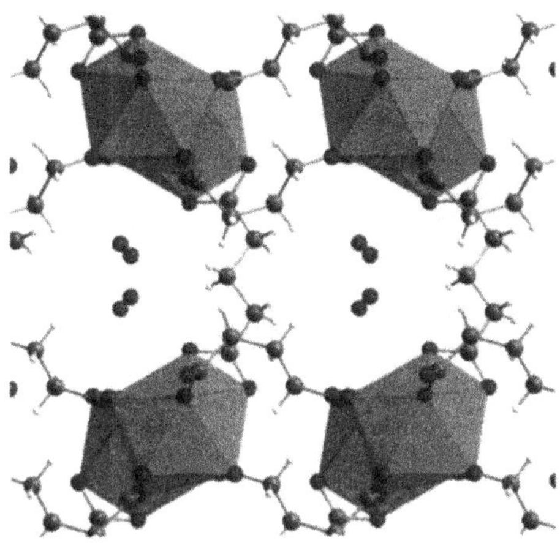

Fig. 16 Neodymium glutarate, a coordination polymer containing chains of edge-sharing $NdO_9$ polyhedra.[52] Gray spheres denote carbon, red oxygen, and white hydrogen, with $NdO_9$ polyhedra in blue. Reproduced with permission. Copyright 1998, Royal Society of Chemistry.

exploited, though many of the resulting hybrid frameworks are non-crystalline.[54]

A third point connected with the identity of the metal ions concerns their kinetic stability, or lack thereof. Most $M^{2+}$ and $M^{3+}$ aquo-coordination complexes show rapid ligand exchange with rate constants in the range $10^3$ to $10^8$ $s^{-1}$, but a small number of very inert ions have extremely slow ligand exchange rates. $Cr^{3+}$ is the most common and striking example, with a rate constant of $\sim 10^{-5}$ $s^{-1}$, *i.e.* about eight orders of magnitude slower than $Fe^{3+}$! The origin of this huge difference lies in the large ligand field stabilization energy of the $d^3$ $Cr^{3+}$ ion that is also responsible for its strong octahedral coordination preference. Clearly the $Cr^{3+}$ ion provides an excellent basis for creating very stable hybrid frameworks, as has been demonstrated by the work on Férey on the MIL-101, though it can also pose synthetic challenges due to the inertness of the starting materials.

A final point concerning the nature of the metal ion relates to those cations, mainly $d^{10}$, that readily form linear complexes, *e.g.* $Ag^+$. As with certain types of ligands, these ions are only able to support the formation of 1-D coordination polymers when they have linear coordination (Fig. 4). This behavior was exploited in some of the very early work by Robson,[9] and continues to be a useful approach for creating low-dimensional structures, as the recent work of Abu-Youssef *et al.* on silver quinoxalines shows.[55]

## 4 Synthetic trends

### 4.1 Effect of reaction temperature and pH

It is reasonable to ask how over 100 years of effort in the field of coordination chemistry failed to uncover the existence of the whole world of hybrid framework materials until very recently. With the knowledge of hindsight, we can see that this extraordinary omission arose primarily because the classical coordination chemists did not explore the use of temperature as a variable during synthesis. The first clear insight into the influence of reaction temperature on hybrid formation arose from a series of experiments in which cobalt(II) hydroxide was reacted with succinic acid in a 1 : 1 molar ratio at five temperatures between 60 and 250 °C.[56] Remarkably, and perhaps serendipitously, this yielded a series of five different phases with clear trends in the structures and compositions. In particular, the phases became more dense and less hydrated with increasing temperature, transitioning from a hydrated 1-D coordination polymer at the lowest temperature to an anhydrous, 2-D hybrid metal oxide at the highest (Fig. 18). This showed for the first time that hybrid framework formation is strongly influenced by classical thermodynamic factors, such as condensation due to entropy-driven dehydration reactions at higher temperatures.

In a more comprehensive study of the cobalt succinate system by high throughput experimentation, the trends as a function of pH and time were also examined.[57] It was found that extended inorganic hybrid structures are also favored at high pH, where the formation of M-O-M linkages arise due to the elimination of water or hydroxide groups by condensation reactions. The evolution of reaction products as a function of time, however, showed relatively few changes, further supporting the idea that thermodynamic factors can be very important in hybrid synthesis. The work also lead to the discovery of new

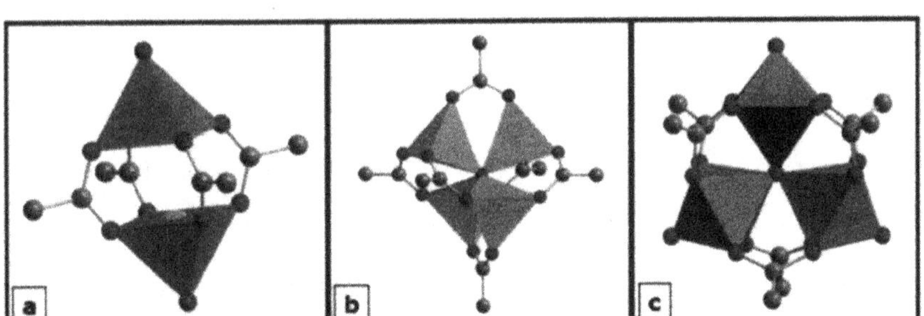


Fig. 17 Commonly recurring structural motifs: (a) the "paddlewheel" dimer of $CuO_5$ square pyramids, (b) the tetrahedron of $ZnO_4$ tetrahedra sharing a central oxygen, and (c) the trigonal planar trimer of $CrO_6$ octahedra sharing a central oxygen. Gray spheres denote carbon and red oxygen; $CuO_5$, $ZnO_4$ and $CrO_6$ polyhedra are shown in dark blue, pale blue and green, respectively.

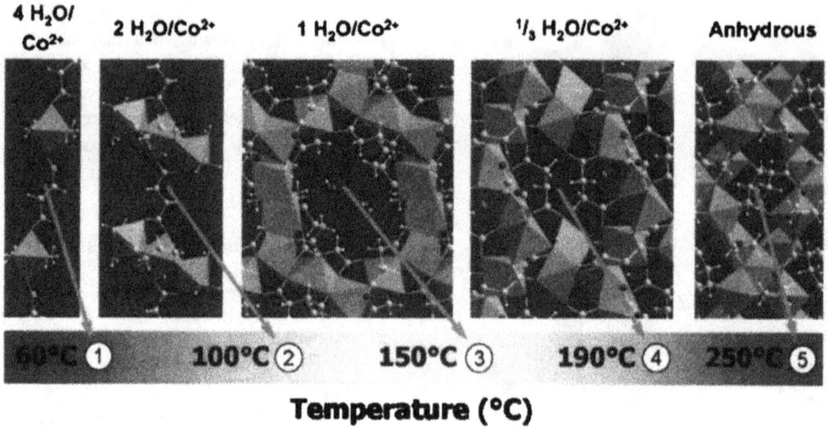


**Fig. 18** Formation temperatures of five cobalt succinates, showing the trend toward greater inorganic connectivity and less hydration at higher temperatures.[56] Gray spheres denote carbon, white hydrogen, and red oxygen, with $CoO_6$ octahedra in pink.

phases,[58] underlining the power of high throughput methods in materials discovery.

Another interesting example of the influence of pH is apparent in the work of Stock and Bein on manganese with the phosphonocarboxylic acid, *p*-$H_2O_3PCH_2C_6H_4CO_2H$.[59] At low pH values, only one oxygen of the phosphonate group is deprotonated and the system is limited to forming a 1-D coordination polymer (Fig. 19). As the pH is raised, the second proton of the phosphonate group is released and the dimensionality increases to form a 2-D coordination polymer, while at the highest pH, the carboxylate becomes deprotonated and the dimensionality can increase to 3-D.

It is now becoming clear that the trends observed in the cobalt succinate and zinc phosphonocarboxylate systems are quite typical of hybrid materials synthesized under hydrothermal conditions. For example, looking back at work on the nickel(II) diphosphonates, $Ni_4(O_3PCH_2PO_3)_2{\cdot}nH_2O$ ($n$ = 3, 2, 0), we see both an increase of dimensionality and multiple coordination changes during a temperature-driven quasi-topotactic dehydration reaction *in the solid state*.[60] With the knowledge of hindsight, we can also ascribe the increase in dimensionality with temperature in the system cobalt pyridine-3,4-dicarboxylate to the same effect.[61] However, we do not wish to imply that the synthesis of hybrid frameworks always proceeds under thermodynamic control; we shall return to this point in section 4.3.

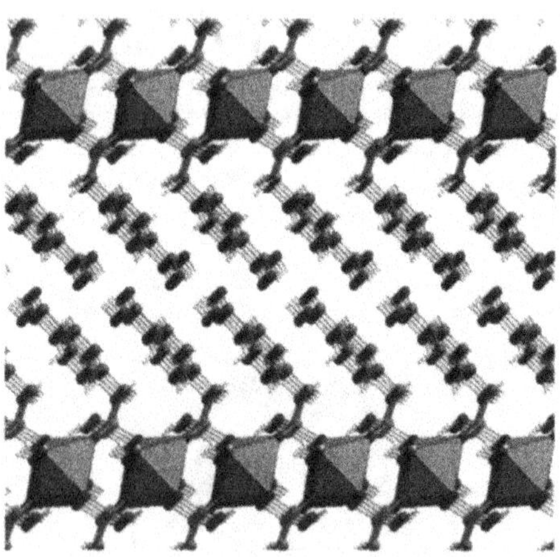

**Fig. 19** The low-pH chain structure of manganese *p*-(phosphonomethyl)benzoate, with all carboxylate oxygens and half the phosphonate oxygens protonated.[59] Gray spheres denote carbon, red oxygen, white hydrogen, and green phosphorus, with $MnO_6$ octahedra in blue. Reproduced with permission. Copyright 2006, Royal Society of Chemistry.

### 4.2 Influence of solvent

The nature of the solvent is an important parameter in hybrid synthesis, especially as it can sometimes be problematic to identify a solvent that is suitable for both the inorganic and the organic reactants. Obviously, the use of non-aqueous or mixed solvents has been widely adopted, just as it has in the case of purely inorganic frameworks, and the use of immiscible biphasic solvents,[18] whereby the products form at the solvent interface, is an interesting strategy. Ionic liquids, which have been shown to be effective in the synthesis of inorganic framework materials,[62] are just starting to be applied to hybrid frameworks.[63]

### 4.3 Kinetic *vs.* thermodynamic factors

The emergence of clear trends in the synthesis of hybrid framework materials, as discussed in section 4.1, points to the likelihood that thermodynamic factors are more dominant than is found to be the case in zeolites and other inorganic framework materials. In the synthesis of aluminosilicate zeolites, it is well-known that hydrothermal crystallization often proceeds under kinetic control, with successive crystallization of increasingly stable phases as a function of time, according to the Ostwald step rule.[64] The high throughput work on cobalt succinates[57] reveals very little change as a function of time, aside from reactions that start with a large percentage of solid cobalt(II) hydroxide, in which case the kinetics of dissolution appear to be important. Other recent work also points towards the importance of thermodynamic control. A combined experimental and computational study of layered aluminium alkyldiphosphonates has shown that the stacking sequence is controlled by packing considerations that depend upon the number of carbon atoms in the alkyl chain; with even numbers the inorganic layers are more stable if they stack in an AAAA sequence, and with odd numbers the ABAB packing is more stable.[65] In a very recent study of zinc 4-cyclohexene-*cis*-1,2-dicarboxylates,[66] a similar combination of experimental and computational methods has shown that the temperature dependent behavior is controlled by thermodynamic considerations, with a hydrated phase being formed at temperatures below 100 °C and an anhydrous phase at higher temperatures. However, time-dependent effects are seen due to competition between the rate of product crystallization and the rate of isomerization of the *cis* ligand to the more stable *trans* form (Fig. 20).

The likelihood that thermodynamics play a strong role in hybrid framework formation is not very surprising with the knowledge of hindsight. In the case of aluminosilicate zeolites, crystallizations and transformations involve the making and breaking of very strong Si–O or Al–O bonds, whereas transition metal-ligand bonds, *e.g.* M–O, M–N *etc.* in hybrids are relatively weak. For example, M–$OH_2$ bonds for most divalent and trivalent transition metals give rates of water exchange that are in the range $10^3$–$10^8$ $s^{-1}$, as mentioned in section 3.3, though a small number of very inert ions such as $Cr^{3+}$ have extremely slow ligand exchange rates. We might therefore expect to see stronger kinetic control in systems containing these kinetically inert ions. Furthermore, it seems probable that very open frameworks such as MOF-5 and MIL-101, which are synthesized at relatively low temperatures, form under kinetic control, as is found with the zeolitic aluminosilicates. Further research is needed to clarify this question.

The use of microwave radiation in hybrid synthesis is a closely related issue that is beginning to attract attention, and in some cases the enhancement of the reaction rates is two–three orders of magnitude.[67] The reasons for this strong enhancement are not yet clear, and nor is it apparent whether the use of microwaves affects the overall outcome of the reaction in terms of which products are formed. Further work in this area should be very illuminating.

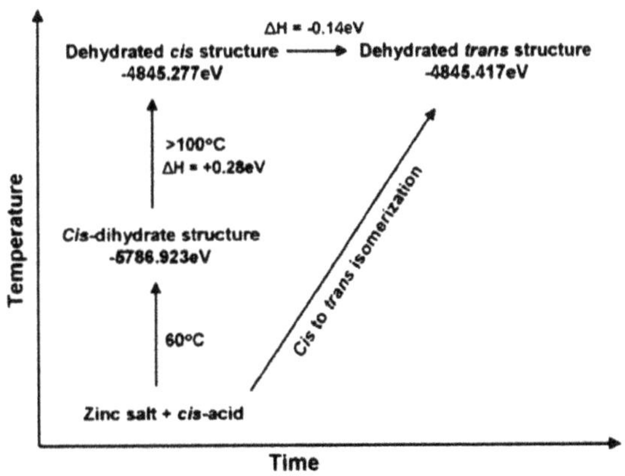


**Fig. 20** The role of thermodynamics and kinetics in the zinc cyclohexene-1,2-dicarboxylate system as a function of temperature and time.[66]

### 4.4 Reaction pathways and building-up processes

Although a wide variety of hybrid framework materials with different dimensionalities have been synthesized and characterized in recent years, we do not yet know the mode of formation of these materials. In the case of inorganic open-framework structures, several workers have identified discrete secondary building units (SBUs), such as SBU-4 (formed by two metal oxygen polyhedra and two anionic polyhedra), which are believed to be involved in the building up process.[68–70] In the case of metal phosphates, for example, zero-dimensional units comprising four-membered rings can be transformed to chains, sheets and three-dimensional structures under relatively mild conditions. A recent study of zinc oxalates has shown that zero-dimensional dimeric units undergo transformations of this type as a function of temperature and time (Fig. 21).[71] Further evidence supporting such a mechanism has been found in the tin phosphonates, where the dimensionality can be controlled by blocking certain reaction pathways by means of using unreactive substituent

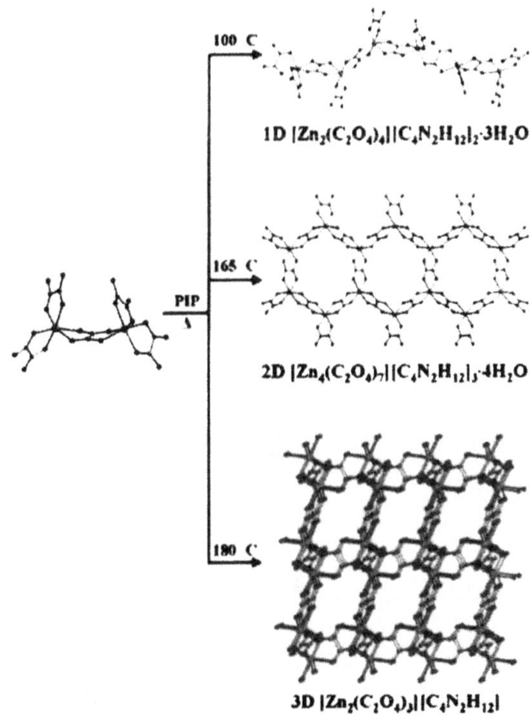


**Fig. 21** Zero-dimensional dimeric zinc oxalate species transform progressively to chain, layered and 3-D structures with increasing temperature.[71] Reproduced with permission. Copyright 2005, Wiley-VCH.

groups.[72] Specially noteworthy is the transformation found in metal squarates, where it has been possible to form a sodalite structure from the assembly of six squarate units with divalent transition metals.[73] There is currently no *in situ* evidence for a building up process of the type that has been found in the purely inorganic systems, but a recent EXAFS study indicates that a trimeric iron oxide SBU remains intact during the crystallization of MIL-89.[74]

## 5 Properties and potential applications

### 5.1 Adsorption and separation processes

In comparison to the enormous number of commercial applications of the aluminosilicate zeolites and related inorganic materials in the fields of catalysis, separations, ion-exchange, and so on, the potential uses of hybrid frameworks are only gradually beginning to emerge. The most obvious possibilities concern highly porous hybrids, such as the MOF-5 and MIL-101 types of structures. In the case of MOF-5, which is a very low-density material, there is a good deal of data on gas storage capacities, including hydrogen and methane.[75] The locations of the hydrogen adsorption sites in MOF-5 have been determined at 4 K by single crystal neutron diffraction and inelastic neutron scattering.[76] The total hydrogen adsorption capacity of the MOF materials is not as high as was originally expected, no doubt because the pores are too large and molecules prefer to adsorb at surfaces. However, this problem can be alleviated and the capacity increased by using interpenetrating networks[77] or by using lightweight hybrid materials with smaller cavities, such as magnesium formate.[78] Another attractive strategy is to use hybrids that contain coordinatively unsaturated metal sites that are able to bind dihydrogen through a weak chemisorptive interaction.[79] Systems with smaller cavities are also of interest for adsorbing other molecules; for example, $N_2$ and $CO_2$ have been studied both experimentally and computationally in aluminium methylphosphonates.[80]

The system MIL-53, based upon trivalent ions, *e.g.* Cr(III), Fe(III), Sc(III) *etc*, in combination with benzene-1,4-dicarboxylate, is another highly porous materials with quite different properties.[81] The structure comprises parallel 1-D chains of corner-sharing $MO_6$ octahedra that are cross-linked to form a 3-D network by the benzene-1,4-dicarboxylate groups (Fig. 22). In terms of the classification in section 2.3, MIL-53 would be a ($I^1O^2$) system. The surface area of the dehydrated form of the first member of the MIL-53 family, the Cr(III) phase, is $\sim$1500 $m^2\ g^{-1}$ and it is stable to 500 °C. What is remarkable about the MIL-53 architecture is that it is sufficiently flexible that it can adapt its structure to accommodate sorbates of different sizes by means of a so-called "breathing effect".[81] The dehydrated structure is very open and actually contracts when water is adsorbed due to hydrogen bonding between the water molecules and the oxygen atoms of the benzene-1,4-dicarboxylate groups (Fig. 22). The case of MIL-88, which is an iron(III) fumarate structure, is similar in the sense that it can contract and expand with a considerable change in volume.[82] However, unlike MIL-53, it expands with the addition of solvent due to the unusual flexibility of the framework. For example, the cell volume of the anhydrous form is 1135 Å$^3$, while that of the fully hydrated form is 2110 Å$^3$.

As discussed in section 3.2, hybrid frameworks provide a unique opportunity to create interesting enantiomerically pure (homochiral), porous networks. One of the motivations for so doing is the possible applications of such networks in the area of chiral separations, which was first demonstrated in 2000 by Rosseinsky and co-workers.[83] In more recent work, a homochiral network based upon nickel benzene-1,3,5-tricarboxylate showed a modest enantiomeric excess (ee) of $\sim$8% for the adsorption of a simple naphthol derivative.[84] In general, it is found that the enantiomeric discrimination depends upon the relative sizes of the cavities and the sorbate molecules, with better selectivity being found when the size match is close. It also appears that ee values are higher for catalytic applications than chiral separations, as described in the following section.

### 5.2 Catalytic applications

It might seem likely that porous hybrid frameworks would be generally inferior to conventional zeolitic materials for applications in heterogeneous catalysis, given their relative instability, their lack of strong acidity, and their relatively costly synthesis. However, hybrids offer certain advantages

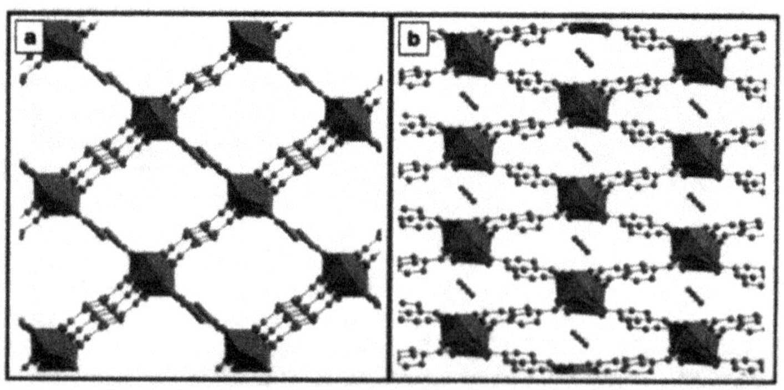


**Fig. 22** Expanded (anhydrous) and contracted (hydrated) forms of the chromium terephthalate MIL-53, drawn to the same scale.[81] Gray spheres denote carbon and red oxygen, with $CrO_6$ octahedra in green. Reproduced with permission from Serre *et al.* Copyright 2002, American Chemical Society.

 

because of (i) the relative ease with which they can be functionalized, *post-synthesis*, (ii) the simplicity with which the pore size can be tuned over a wide range of sizes, and (iii) the manner in which *enatiomerically-pure* chiral frameworks can be created. The examples shown below will illustrate that many of these advantages are now beginning to be realized; further details are given in a recent review of the area.[85]

The first study of catalysis by a nanoporous coordination polymer used a simple lamellar coordination polymer of Cd and 4,4′-bipyridine in which several aldehydes were tested for cyanosilation with cyanotrimethylsilane.[86] Good yields were found for smaller molecules, with progressively poorer yields for larger ones, while control experiments using $Cd(NO_3)_2$ and 4,4′-bipyridine as catalysts resulted in no reaction, confirming the role of the coordination polymer's porosity in catalyzing the reaction.

Catalysis with organically pillared zirconium phosphate/phosphonate catalysts has recently been reviewed by Clearfield and Wang.[87] In one of the earlier demonstrations, a highly acidic catalyst was created by post-synthesis sulfonation of aromatic rings in zirconium phosphate-based systems,[88] though the acid groups are only stable up to ~130 °C, thereby limiting the utility of these systems. However, a sample saturated by acetone at room temperature produced the polymerization products mesityl oxide, isophorone, and 1,3,5-trimethoxybenzene as verified by *in situ* NMR,[89] confirming that the samples are indeed catalytic. Similarly, Pt- or Pd-containing viologen-bridged phosphonates are active catalysts for the photochemical production of $H_2$ gas from water[90] and for the production of $H_2O_2$ from streams of $H_2$ and $O_2$,[91] though there is uncertainty as to whether the reactions take place in the micropores of the catalysts.

One of the major limitations of aluminosilicate zeolites is their current inability to impart shape selectivity based on molecular handedness, primarily because it is extremely difficult to make enantiomerically pure materials. Because of the relative ease of synthesizing accessible chiral channels in hybrid materials, a significant amount of attention is now being devoted to developing materials with chiral pores and studying their catalytic activity. The area has recently been reviewed by Lin.[92] One potential problem is that chiral organics may not survive hydrothermal reaction conditions enantiomerically intact, but Williams and co-workers[47] have shown that this is not a general problem.

In the first demonstration of enantiomerically selective catalysis with hybrids, Kim and coworkers[93] demonstrated that a tartaric acid derivative of a well known oxo-bridged trinuclear zinc carboxylate cluster forms a chiral, layered phase, POST-1, which shows good activity toward transesterification. Tests on a racemic mixture of reactants resulted in a modest 8% enantiomeric excess of either *S* or *R* enantiomers depending on the chirality of the framework. Although a chiral catalyst is not always sufficient to promote a preference for a particular handedness in products, some of the recent results from Lin and co-workers are very impressive. For example, a cadmium-containing hybrid with large channels and over 50% void space gave 93% ee for the addition of diethyl zinc to 1-naphthaldehyde, rivaling the results obtained from homogeneous analogs.[94]

In an unexpected finding, it has been shown that the size of products is not necessarily limited by the size of molecules that are able to escape from the pores (in stark contract to the shape-selectivity found in zeolites). Li and coworkers generated a 3D porous structure from a known 1D cobalt biphenyldicarboxylate chain by replacing ligand water molecules with 4,4′-bipyridine under solvothermal conditions.[95] This 3D structure is very stable and porous, but decomposes back to the 1D hydrated chain when immersed in water. Treatment of the 1D chains with amine solvothermally regenerates the 3D structure. The ability to break and rebuild this phase reversibly leads to the name RPM-1, for Rutger's Recyclable Porous Material #1. Recognizing that the recyclable nature of the material might allow for the formation of larger molecules than the cavities would normally permit, Li and coworkers tested the material for photochemical catalysis of dibenzyl ketone. The reaction resulted in a significantly different mixture of products than is seen in zeolite hosts.

### 5.3 Other applications

Many applications beyond gas storage and catalysis have been proposed for hybrid systems. For example, several groups, particularly those of Ward[96] and Lin,[97] have developed strategies to engineer non-centrosymmetric frameworks for use as nonlinear optical materials. There has also been considerable interest in using hybrid systems to create porous magnets.[98] Although no hybrids currently display long-range ordering temperatures above 100 K,[99] several promising strategies have been proposed, such as using radical organic ligands to couple metal centers.[100] Other interesting observations include the modulation of the magnetic ordering temperature by guest molecules in porous manganese formate,[101] ferrimagnetic-ferromagnetic transformations in nickel cyclohexane-1,4-dicarboxylates,[102] and complex exchange interactions in trimetallic 4f–3d coordination polymers based upon Cu–Gd–Fe.[103]

There is also considerable interest in the optical properties of hybrid frameworks on account of their tunability and their capacity to incorporate a wide range of metal ion and organic ligand chromophors. There are a number of examples of photoluminescent behavior in rare-earth containing hybrids that could be of interest for applications as phosphors or fluorescent probes.[17,104] In the case of porous materials, the photoluminescent lifetimes of $Eu^{3+}$-doped gadolinium glutarates have been shown to depend on the degree of dehydration because the coordinating water molecules act as relaxation agents.[105] Other systems show fluorescent emissions due to charge transfer between metal ions and ligands or $\pi$ to $\pi^*$ transitions within the ligands.[106] For commercial applications, of course, it will be necessary for the materials to have sufficient chemical and optical stability.

Other interesting avenues are also being explored. For example, soluble 1-D coordination polymers based upon dendrimers in combination with palladium have been made;[107] such soluble, low-dimensional polymers are of interest for liquid crystalline behavior and use in nanocomposites. One- and two-dimensional coordination polymers have also been used as templates for the formation of zinc oxide nanorods and radial nanoneedles,[108] while silver-containing coordination polymers are being explored for their antimicrobial activity.[55] Finally, we would like to note that thin films of layered metal diphosphonates have been used as intercalation sensors for small molecules.[109]

## 6 Future prospects

The purpose of this brief overview has been to illustrate the progress that has been made in many aspects of the hybrid frameworks area in the last decade. The diversity of chemical and structural types is enormous and grows by the day, and a better understanding of the factors that influence hybrid formation is beginning to emerge. As this understanding improves, our ability to design new materials for specific uses will also improve, and this will be reflected in a greater range of applications than we see at present. This is most certainly an extraordinarily rich area that will be seen in the future as one of the most important developments in the history of materials chemistry. Many avenues still remain to be explored, including some that are mentioned in the above discussion, and we encourage the community to put further effort into this exciting field.

## Acknowledgements

This work was supported by the National Science Foundation under Award No. DMR05-20415 to the MRSEC center at UCSB and Award No. DMR04-09848 to the International Center of Materials Research at UCSB.

## References

1 A. Corma, *J. Catal.*, 2003, **216**, 298; M. E. Davis, *Nature*, 2002, **417**, 813; J. M. Thomas, *Sci. Am.*, 1992, **266**, 112.
2 A. K. Cheetham, G. Férey and T. Loiseau, *Angew. Chem., Int. Ed.*, 1999, **38**, 3268.
3 A. P. Côté, A. I. Benin, N. W. Ockwig, M. O'Keeffe, A. J. Matzger and O. M. Yaghi, *Science*, 2005, **310**, 1166.
4 G. R. Desiraju, *Angew. Chem., Int. Ed. Engl.*, 1995, **34**, 2311.
5 A. K. Cheetham and P. M. Forster, in *Chemistry of Nanomaterials*, ed. C. N. R. Rao, A. Müller and A. K. Cheetham, Wiley–VCH, Weiheim, 2003, pp. 589–619.
6 D. A. Summerville, T. W. Cape, E. D. Johnson and F. Basolo, *Inorg. Chem.*, 1978, **17**, 3297; J. S. Miller, C. Vazquez, N. L. Jones, R. S. McLean and A. J. Epstein, *J. Mater. Chem.*, 1995, **5**, 707.
7 P. Gravereau, E. Garnier and A. Hardy, *Acta Crystallogr., Sect. B: Struct. Crystallogr. Cryst. Chem.*, 1979, **35**, 2843.
8 G. Alberti, U. Constantino, S. Allulli and N. Tomassini, *Inorg. Nucl. Chem.*, 1978, **40**, 1113; M. D. Poojary, H.-L. Hu, F. Campbell, III and A. Clearfield, *Acta Crystallogr., Sect. B: Struct. Sci.*, 1993, **49**, 996.
9 B. F. Hoskins and R. Robson, *J. Am. Chem. Soc.*, 1990, **112**, 1546.
10 R. Robson, *J. Chem. Soc., Dalton Trans.*, 2000, 3735; B. Moulton and M. J. Zaworotko, *Chem. Rev.*, 2001, **101**, 1629; B. Moulton and M. J. Zaworotko, *Curr. Opin. Solid State Mater. Sci.*, 2002, **6**, 117.
11 M. O'Keeffe, M. Eddaoudi, H. Li, T. Reineke and O. M. Yaghi, *J. Solid State Chem.*, 2000, **152**, 3.
12 A. Clearfield, *Curr. Opin. Solid State Mater. Sci.*, 2002, **6**, 495; J. L. C. Rowsell and O. M. Yaghi, *Microporous Mesoporous Mater.*, 2004, **73**, 3; M. J. Rosseinsky, *Microporous Mesoporous Mater.*, 2004, **73**, 15; C. N. R. Rao, S. Natarajan and R. Vaidhayanathan, *Angew. Chem., Int. Ed.*, 2004, **43**, 1466.
13 A. J. Blake, N. R. Champness, M. Crew and S. Parsons, *New J. Chem.*, 1999, 13.
14 J. T. Sampanthar and J. J. Vittal, *J. Chem. Soc., Dalton Trans.*, 1999, 1993.
15 W. J. Belcher, C. A. Longstaff, M. R. Neckenig and J. W. Steed, *Chem. Commun.*, 2002, 1602.
16 N. Masciocchi, F. Castelli, P. M. Forster, M. M. Tafoya and A. K. Cheetham, *Inorg. Chem.*, 2003, **42**, 6147.
17 A. de Bettencourt-Dias, *Inorg. Chem.*, 2005, **44**, 2734.
18 P. M. Forster, P. Thomas and A. K. Cheetham, *Chem. Mater.*, 2002, **14**, 17.
19 R. W. Gable, B. F. Hoskins and R. Robson, *J. Chem. Soc., Chem. Commun.*, 1990, 762.
20 L. Carlucci, G. Ciani, D. M. Proserpio and A. Sironi, *Angew. Chem., Int. Ed. Engl.*, 1995, **34**, 1895.
21 H. Li, M. Eddaoudi, M. O'Keeffe and O. M. Yaghi, *Nature*, 1999, **402**, 276.
22 N. L. Rosi, M. Eddaoudi, J. Kim, M. O'Keeffe and O. M. Yaghi, *CrystEngComm*, 2002, **4**, 401.
23 Q. R. Fang, G. S. Zhu, J. Y. Sun, Y. Wei, S. L. Qiu and R. R. Xu, *Angew. Chem., Int. Ed.*, 2005, **44**, 3845; Q. R. Fang, G. S. Zhu, M. Xue, J. Y. Sun and S. L. Qiu, *Dalton Trans.*, 2006, 2399; Y. L. Liu, V. C. Kravtsov, R. Larsen and M. Eddaoudi, *Chem. Commun.*, 2006, 1488; K. S. Park, Z. Ni, A. P. Cote, J. Y. Choi, R. D. Huang, F. J. Uribe-Romo, H. K. Chae, M. O'Keeffe and O. M. Yaghi, *Proc. Natl. Acad. Sci. U. S. A.*, 2006, **103**, 10186.
24 L. Carlucci, G. Ciani, P. Macchi and D. M. Proserpio, *Chem. Commun.*, 1998, 1837.
25 G. Férey, C. Mellot-Draznieks, C. Serre, F. Millange, J. Dutour, S. Surblé and I. Margiolake, *Science*, 2005, **309**, 2040; C. Mellot-Draznieks, J. Dutour and G. Férey, *Angew. Chem., Int. Ed.*, 2004, **43**, 6290.
26 A. C. Rizzi, R. Calvo, R. Baggio, M. T. Garland, O. Pena and M. Perec, *Inorg. Chem.*, 2002, **41**, 5609.
27 B. Adair, S. Natarajan and A. K. Cheetham, *J. Mater. Chem.*, 1998, **8**, 1477; N. Stock, G. D. Stucky and A. K. Cheetham, *Chem. Commun.*, 2000, 2277; M. Dan, G. Cotterau and C. N. R. Rao, *Solid State Sci.*, 2005, **7**, 437.
28 Y. F. Huang and K. H. Lii, *J. Chem. Soc., Dalton Trans.*, 1998, 4085; Z. A. D. Lethbridge and P. Lightfoot, *J. Solid State Chem.*, 1999, **143**, 58; A. Chowdhury, S. Natarajan and C. N. R. Rao, *Chem. Mater.*, 1999, **11**, 2316; J. Do, R. P. Bontchev and A. J. Jacobson, *Inorg. Chem.*, 2000, **39**, 3230.
29 B. M. Craven and D. Hall, *Acta Crystallogr.*, 1961, **14**, 475; S. Sato and K. Kobayashi, *Acta Crystallogr., Sect. C: Cryst. Struct. Commun.*, 1987, **43**, 1863.
30 W. T. A. Harrison, T. M. Nenoff, T. E. Gier and G. D. Stucky, *Inorg. Chem.*, 1992, **31**, 5395.
31 N. Guillou, C. Livage, W. van Beek, M. Nogues and G. Férey, *Angew. Chem., Int. Ed.*, 2003, **42**, 644.
32 R. K. Feller and A. K. Cheetham, *Solid State Sci.*, 2006, **8**, 1121; C. H. Wunderlich, R. Weber and G. Bergerhoff, *Z. Anorg. Allg. Chem.*, 1991, **598**, 371.
33 A. Tressaud, J. L. Soubeyroux, J. M. Dance, R. Sabatier and P. Hagenmuller, *Solid State Commun.*, 1981, **37**, 479.
34 A. Clearfield and J. A. Stynes, *J. Inorg. Nucl. Chem.*, 1964, **26**, 117.
35 C. A. Merrill and A. K. Cheetham, *Inorg. Chem.*, 2005, **44**, 5273.
36 M. B. Dines, P. M. DiGiacomo, K. P. Callahan, P. C. Griffith, R. H. Lane and R. E. Cooksey, in *Chemically Modified Surfaces in Catalysis and Electrocatalysis*, ed. J. S. Miller, ACS Symposium Series 192, American Chemical Society, Washington, D. C., 1982, p. 223; G. Alberti, R. Viviani and S. Murcia Mascarós, *J. Mol. Struct.*, 1998, **470**, 81.
37 A. Clearfield, *Chem. Mater.*, 1998, **10**, 2801.
38 L. A. Vermeulen and M. E. Thompson, *Nature*, 1992, **358**, 656.
39 P. M. Forster and A. K. Cheetham, *Angew. Chem., Int. Ed.*, 2002, **41**, 457.
40 R. Vaidhayanathan, S. Natarajan and C. N. R. Rao, *Dalton Trans.*, 2003, 1459.
41 N. Guillou, C. Livage, M. Drillon and G. Férey, *Angew. Chem., Int. Ed.*, 2003, **42**, 5314.
42 V. Soghomonian, Q. Chen, R. C. Haushalter and J. Zubieta, *Angew. Chem., Int. Ed. Engl.*, 1995, **34**, 223; V. J. Carter, P. A. Wright, J. D. Gale, R. E. Morris, E. Sastre and

 

J. Perez Pariente, *J. Mater. Chem.*, 1997, **7**, 2287; L. M. Zheng, H. H. Song, C. H. Lin, S. L. Wang, Z. Hu, Z. Yu and X. Q. Xin, *Inorg. Chem.*, 1999, **38**, 4618; L. M. Zheng, H. H. Song, C. Y. Duan and X. Q. Xin, *Inorg. Chem.*, 1999, **38**, 5061; H. H. Song, L. M. Zheng, Z. M. Wang, C. H. Yan and X. Q. Xin, *Inorg. Chem.*, 2001, **40**, 5025; M. Dan, K. Sivashanker, A. K. Cheetham and C. N. R. Rao, *J. Solid State Chem.*, 2003, **174**, 60.
43 R. Vaidhyanathan, S. Natarajan and C. N. R. Rao, *Chem. Mater.*, 2001, **13**, 3524.
44 Y. Kim and D. Y. Jung, *Chem. Commun.*, 2002, 908; M. Kurmoo, H. Kumagai, M. Akita-Tanaka, K. Inoue and S. Tagaki, *Inorg. Chem.*, 2006, **45**, 1627; J. Chen, M. Ohba, D. Zhao, W. Kaneko and S. Kitagawa, *Cryst. Growth Des.*, 2006, **6**, 664.
45 A. Thirumurugan, M. B. Avinash and C. N. R. Rao, *Dalton Trans.*, 2006, 221.
46 D. S. Kim, P. M. Forster, G. Diaz de Delgardo, S.-E. Park and A. K. Cheetham, *Dalton Trans.*, 2004, 3365.
47 S. Thushari, J. A. K. Cha, H. H.-Y. Sung, S. S.-Y. Chui, A. L.-F. Leung, Y.-F. Yen and I. D. Williams, *Chem. Commun.*, 2005, 5515.
48 J. B. Weng, M. C. Hong, Q. Shi, R. Cao and A. C. S. Chan, *Eur. J. Inorg. Chem.*, 2002, 2553.
49 E. V. Anohkhina and A. J. Jacobson, *J. Am. Chem. Soc.*, 2004, **126**, 3044.
50 P. M. Forster and A. K. Cheetham, *Microporous Mesoporous Mater.*, 2004, **73**, 57.
51 P. M. Forster and A. K. Cheetham, *Solid State Sci.*, 2003, **5**, 635.
52 F. Serpaggi and G. Férey, *J. Mater. Chem.*, 1998, **8**, 2737.
53 S. S. Y. Chui, S. M. F. Lo, J. P. H. Y. Charmant, A. G. Orpen and I. D. Williams, *Science*, 1999, **283**, 1148.
54 J. J. Morrison, C. J. Love, B. W. Manson, I. J. Shannon and R. E. Morris, *J. Mater. Chem.*, 2002, **12**, 3208; R. E. Morris, *J. Mater. Chem.*, 2005, **15**, 931.
55 M. A. M. Abu-Youssef, V. Langer and L. Öhrström, *Dalton Trans.*, 2006, 2542.
56 P. M. Forster, A. R. Burbank, C. Livage, G. Férey and A. K. Cheetham, *Chem. Commun.*, 2004, 368.
57 P. M. Forster, N. Stock and A. K. Cheetham, *Angew. Chem., Int. Ed.*, 2005, **44**, 7608.
58 P. M. Forster, A. R. Burbank, M. C. O'Sullivan, N. Guillou, C. Livage, G. Férey, N. Stock and A. K. Cheetham, *Solid State Sci.*, 2005, **7**, 1549.
59 N. Stock and T. Bein, *J. Mater. Chem.*, 2005, **15**, 1384.
60 Q. M. Gao, N. Guillou, M. Nogues, G. Férey and A. K. Cheetham, *Chem. Mater.*, 1999, **11**, 2937.
61 M.-L. Tong, S. Kitagawa, H.-C. Chang and M. Ohba, *Chem. Commun.*, 2004, 418.
62 E. R. Cooper, C. D. Andrews, P. S. Wheatley, P. B. Webb, P. Wormald and R. E. Morris, *Nature*, 2004, **430**, 1012.
63 J.-H. Liao, P.-C. Wu and Y.-H. Bai, *Inorg. Chem. Commun.*, 2005, **8**, 390; C.-Y. Sheu, S.-F. Lee and K.-H. Lii, *Inorg. Chem.*, 2006, **45**, 1891; Z. J. Lin, D. S. Wragg and R. E. Morris, *Chem. Commun.*, 2006, 2021.
64 R. A. van Santen, *J. Phys. Chem.*, 1984, **88**, 5768.
65 H. G. Harvey, B. Slater and M. P. Attfield, *Chem.–Eur. J.*, 2004, **10**, 3270.
66 C. Lee, C. Mellot-Draznieks, B. Slater, G. Wu, W. T. A. Harrison, C. N. R. Rao and A. K. Cheetham, *Chem. Commun.*, 2006, 2687.
67 S. H. Jhung, J.-H. Lee, P. M. Forster, G. Férey, A. K. Cheetham and J.-S. Chang, *Chem.–Eur. J.*, 2006, **12**, 7699.
68 G. Férey, *J. Solid State Chem.*, 2000, **152**, 37.
69 S. Oliver, A. Kuperman and G. A. Ozin, *Angew. Chem., Int. Ed.*, 1998, **37**, 46; C. N. R. Rao, S. Natarajan, A. Choudhury, S. Neeraj and A. A. Ayi, *Acc. Chem. Res.*, 2003, **34**, 80; A. J. Norquist and D. O'Hare, *J. Am. Chem. Soc.*, 2004, **126**, 6673.
70 R. Murugavel, M. G. Walawalkar, M. Dan, H. W. Roesky and C. N. R. Rao, *Acc. Chem. Res.*, 2004, **37**, 763.
71 M. Dan and C. N. R. Rao, *Angew. Chem., Int. Ed.*, 2005, **45**, 281.
72 B. A. Adair, S. Neeraj and A. K. Cheetham, *Chem. Mater.*, 2003, **15**, 1518.
73 S. Neeraj, M. L. Noy, C. N. R. Rao and A. K. Cheetham, *Solid State Sci.*, 2002, **4**, 1231.
74 S. Surble, F. Millange, C. Serre, G. Férey and R. I. Walton, *Chem. Commun.*, 2006, 1518.
75 J. L. C. Rowsell, A. R. Millward, K. S. Park and O. M. Yaghi, *J. Am. Chem. Soc.*, 2004, **126**, 5666; T. Duren, L. Sarkisov, O. M. Yaghi and R. Q. Snurr, *Langmuir*, 2004, **20**, 2683.
76 E. C. Spencer, J. A. K. Howard, G. J. McIntyre, J. L. C. Rowsell and O. M. Yaghi, *Chem. Commun.*, 2006, 278; J. L. C. Rowsell, E. C. Spencer, J. Eckert, J. A. K. Howard and O. M. Yaghi, *Science*, 2005, **309**, 1350.
77 J. L. C. Rowsell and O. M. Yaghi, *Angew. Chem., Int. Ed.*, 2005, **44**, 4670.
78 J. A. Rood, B. C. Noll and K. W. Henderson, *Inorg. Chem.*, 2006, **45**, 5521.
79 P. M. Forster, J. Eckert, B. Heiken, J. B. Parise, J. W. Yoon, S. H. Jhung, J.-S. Chang and A. K. Cheetham, *J. Am. Chem. Soc.*, in press.
80 C. Schumacher, J. Gonzalez, P. A. Wright and N. A. Seaton, *Phys. Chem. Chem. Phys.*, 2005, **7**, 2351.
81 C. Serre, F. Millange, C. Thouvenot, M. Nogues, G. Marsolier, D. Louer and G. Férey, *J. Am. Chem. Soc.*, 2002, **124**, 13519; T. Loiseau, C. Serre, C. Huguenard, G. Fink, F. Taulelle, M. Henry, T. Bataille and G. Férey, *Chem.–Eur. J.*, 2004, **10**, 1373; T. R. Whitfield, X. Q. Wang, L. M. Liu and A. J. Jacobsen, *Solid State Sci.*, 2005, **7**, 1096.
82 C. Mellot-Draznieks, C. Serre, S. Surble, N. Auderbrand and G. Férey, *J. Am. Chem. Soc.*, 2005, **127**, 16273.
83 C. J. Kepert, T. J. Prior and M. J. Rosseinsky, *J. Am. Chem. Soc.*, 2000, **122**, 5158.
84 D. Bradshaw, T. J. Prior, E. J. Cussen, J. B. Claridge and M. J. Rosseinsky, *J. Am. Chem. Soc.*, 2004, **126**, 6106.
85 P. M. Forster and A. K. Cheetham, *Top. Catal.*, 2003, **24**, 79.
86 M. Fujita, Y. J. Kwon, S. Washizu and K. Ogura, *J. Am. Chem. Soc.*, 1994, **116**, 1151.
87 A. Clearfield and Z. K. Wang, *J. Chem. Soc., Dalton Trans.*, 2002, 2937.
88 C.-Y. Yang and A. Clearfield, *React. Polym.*, 1987, **5**, 13.
89 A. Clearfield, Z. Wang and P. Bellinghausen, *J. Solid State Chem.*, 2002, **167**, 376.
90 H. Byrd, A. Clearfield, D. Poojary, K. P. Reis and M. E. Thompson, *Chem. Mater.*, 1996, **8**, 2239.
91 A. Dokoutchaev, V. V. Krishnan, M. E. Thompson and M. Balasubramanian, *J. Mol. Struct.*, 1998, **470**, 191; K. P. Reis, V. K. Joshi and M. E. Thompson, *J. Catal.*, 1996, **161**, 62.
92 W. Lin, *J. Solid State Chem.*, 2005, **178**, 2486.
93 J. Soo Seo, D. Whang, H. Lee, S. I. Jun, J. Oh, Y. J. Jeon and K. Kim, *Nature*, 2000, **404**, 982.
94 C.-D. Wu, A. Hu, L. Zhang and W. Lin, *J. Am. Chem. Soc.*, 2005, **127**, 8940.
95 L. Pan, N. Ching, X.-Y. Huang and J. Li, *Inorg. Chem.*, 2000, **39**, 5333.
96 K. T. Holman, A. M. Pivovar, J. A. Swift and M. D. Ward, *Acc. Chem. Res.*, 2001, **34**, 107.
97 O. R. Evans and W. B. Lin, *Acc. Chem. Res.*, 2002, **35**, 511.
98 G. Férey, *Chem. Mater.*, 2001, **13**, 3084.
99 K. Barthelet, J. Marrot, D. Riou and G. Férey, *Angew. Chem., Int. Ed.*, 2002, **41**, 281.
100 D. Maspoch, D. Ruiz-Molina, K. Wurst, N. Domingo, M. Cavallini, F. Biscarini, J. Tejada, C. Rovira and J. Veciana, *Nat. Mater.*, 2003, **2**, 190.
101 Z. M. Wang, B. Zhang, H. Fujiwara, H. Kobayashi and M. Kurmoo, *Chem. Commun.*, 2004, 416.
102 M. Kurmoo, H. Kumagai, M. Akita-Tanake, K. Inoue and S. Takagi, *Inorg. Chem.*, 2006, **45**, 1627.
103 H. Z. Kou, B. C. Zhou and R. J. Wang, *Inorg. Chem.*, 2003, **42**, 7658.
104 X. D. Gou, G. S. Zhu, Q. R. Fang, M. Xue, G. Tian, J. Y. Sun, X. T. Li and S. L. Qiu, *Inorg. Chem.*, 2005, **44**, 3850.
105 F. Serpaggi, T. Luxbacher, G. Férey and A. K. Cheetham, *J. Solid State Chem.*, 1999, **145**, 580.
106 Q. R. Fang, G. S. Zhu, M. Xue, J. Y. Sun, F. X. Sun and S. L. Qiu, *Inorg. Chem.*, 2006, **45**, 3582; J. Lu, K. Zhao, Q. R. Fang, J. Q. Xu, J. H. Yu, X. Zhang, H. Y. Bie and T. G. Wang, *Cryst. Growth Des.*, 2005, **5**, 1091.
107 H. Tokuhisa and M. Kanesato, *Langmuir*, 2005, **21**, 9728.
108 Z. Li, Y. Xiong and Y. Xie, *Nanotechnology*, 2005, **16**, 2303.
109 T. E. Mallouk and J. A. Gavin, *Acc. Chem. Res.*, 1998, **31**, 209.

FEATURE ARTICLE www.rsc.org/chemcomm | ChemComm

# Kagome network compounds and their novel magnetic properties

**Swapan K. Pati*[a] and C. N. R. Rao*[b]**

*Received (in Cambridge, UK) 29th April 2008, Accepted 19th June 2008*
*First published as an Advance Article on the web 27th August 2008*
**DOI: 10.1039/b807207h**

Compounds possessing the Kagome network are truly interesting because of their unusual low-energy properties. They exhibit magnetic frustration because of the triangular lattice inherent to the hexagonal bronze structure they possess, as indeed demonstrated by some of the $Fe^{3+}$ jarosites, but this is not the general case. Kagome compounds formed by transition metal ions with varying spins exhibit novel magnetic properties, some even showing evidence for magnetic order and absence of frustration. We describe the structure and magnetic properties of this interesting class of materials and attempt to provide an explanation for the variety of properties on the basis of theoretical considerations.

## 1. Introduction

Kagome is a Japanese word meaning *hole in the basket* and metal compounds possessing the hexagonal bronze structure with a triangular lattice are referred to as Kagome network compounds. Many transition metal compounds with Kagome networks have been synthesized and characterized in recent years. One of the most important characteristics of the Kagome network compounds, as typified by the $Fe^{3+}$ jarosites, is magnetic frustration.

Frustration is an old topic that can be traced back to the work of Linus Pauling and others on the frozen disorder of crystalline ice.[1–3] Frustration arises when the geometry of a system allows for a set of degenerate ground states. In the case of ice, this corresponds to the different ways of orienting the water molecules without changing the energy of the system. Such a ground state entropy in ice was observed experimentally.[2,3] The same approach can be extended to antiferromagnetically coupled tetrahedra, or systems with triangular symmetry which show a rich manifestation of magnetic frustration.[4–9]

Frustration has a geometrical origin wherein a set of degrees of freedom becomes incompatible with the extended symmetry of the system. A good example is the impossibility of close-packing pentagons in two-dimensional space. The under-constraint of a lattice geometry to the number of independent constraints of local geometry gives rise to geometrical frustration. In magnetic systems, this results in certain unique low energy characteristics in the magnetic properties and thus has generated tremendous interest.[10–15] Examples of geometrically frustrated magnetic systems include one-dimensional zigzag, ladder, and sawtooth lattice structures, two-dimensional layered, triangular and Kagome lattices and three dimensional pyrochlore lattice systems.[16–26] In all these systems, geometrical frustration is manifested due to the competition of the antiferromagnetic interactions between the neighboring spins.

The propensity to adopt many unusual magnetic structures is mostly found in two-dimensional frustrated lattice systems, the Kagome and the triangular lattices. While the triangular lattice point has six nearest neighbors and the adjacent triangles share one side or two lattice points, the Kagome lattice point has four nearest neighbors with the adjacent triangles on the lattice sharing only one lattice point. If the interactions are

[a] *Theoretical Sciences Unit, Jawaharlal Nehru Center for Advanced Scientific Research, Bangalore 560064, India. E-mail: pati@jncasr.ac.in; Fax: +91 80 22082767; Tel: +91 80 22082839; http://www.jncasr.ac.in/pati/*
[b] *Chemistry and Physics of Materials Unit, Jawaharlal Nehru Center for Advanced Scientific Research, Bangalore 560064, India. E-mail: cnrrao@jncasr.ac.in; Fax: +91 80 22082760; Tel: +91 80 22082761*

**Swapan K. Pati**

*Swapan K. Pati obtained his PhD from the Indian Institute of Science, Bangalore, followed by postdoctoral work at University of California and Northwestern University. He is currently an Associate Professor at the Jawaharlal Nehru Centre for Advanced Scientific Research. His research interests include quantum magnetism and developing new theoretical tools for a holistic understanding of structure–property correlations.*

**C. N. R. Rao**

*C. N. R. Rao obtained his PhD from Purdue University and DSc degree from the University of Mysore. He is the National Research Professor of India, Linus Pauling Research Professor at the Jawaharlal Nehru Centre for Advanced Scientific Research and Honorary Professor at the Indian Institute of Science and a member of several academies. His research interests are in the chemistry of materials.*

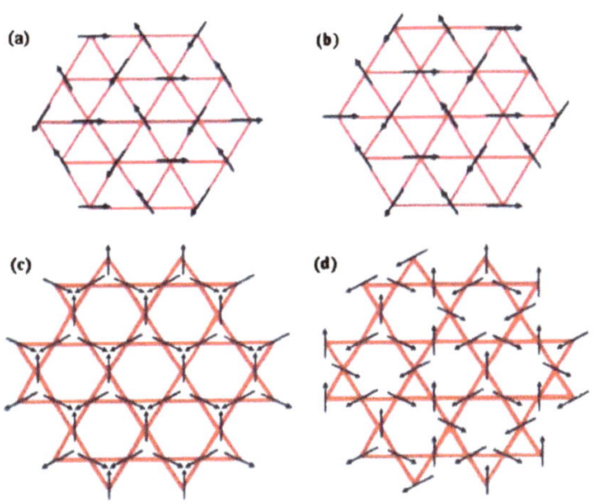


**Fig. 1** Two degenerate ground state structures of an antiferromagnetically coupled triangular lattice (a and b). Kagome lattice with two spin orientations: $Q = 0$ (c) and $Q = \sqrt{3} \times \sqrt{3}$ (d).

considered to be nearest neighbor antiferromagnetic within the Heisenberg model, for a lattice with triangular symmetry frustration would simply lead to a 120° array of spin configurations, as shown in Fig. 1a. Such a state will be doubly degenerate (Fig. 1b) associated with the handedness or chirality of the vector array, in addition to the trivial degeneracy associated with the global rotation of spins. Since this configuration may be extended infinitely in the plane of the paper in a unique fashion to produce a long-range ordered spin array, the ground state and excitations are believed to be rather conventional.[27]

When the triangular plaquettes are connected through corner sharing vertices to form the Kagome lattice, the order within an isolated plaquette cannot be propagated in an unambiguous fashion. Fig. 1c and d display two of the many manifolds of possible spin states which satisfy the requirement that all spins are at 120° to each other, the so-called $Q = 0$ state (Fig. 1c) and the $Q = \sqrt{3} \times \sqrt{3}$ state (Fig. 1d), where $Q$ corresponds to the wave vector. If thermal or quantum fluctuations are considered, it is possible that a subset of the manifold of ground states would be selected. For classical (large) spins, small quantum or thermal fluctuations would select spin configurations in which all the spins are coplanar and possibly at $T = 0$ it adopts the structure shown in Fig. 1d. Interestingly, there is also a strong link between Kagome systems with frustration and hole-doped high $T_c$ cuprate materials. In both cases, the number of nearest neighbors is the same (4) and the hole doping becomes equivalent to ground state degeneracy due to magnetic frustration.[26] The research efforts in this area are, therefore, highly interdisciplinary as one would expect.

With such a variety of magnetic ground state structures, the Kagome system offers new, interesting and exotic possibilities of low-energy and low-temperature properties. Thus, the Kagome lattice has emerged to be a classical case of a Heisenberg antiferromagnet which could represent a quantum spin liquid phase and offers the possibility of exploring both classical and quantum spin ground and excited states together with the correlated electron problems.

There has been a large number of theoretical and experimental studies recently on the various aspects of magnetic frustration in a large class of Kagome systems.[20–24] Although the ideal Kagome lattice compounds synthesized so far are only a few, there have been many efforts to synthesize and characterize different types of Kagome structures in the past few years. For example, there exist Kagome systems varying from $Rb_2M_3S_4$[28–31] (M is a magnetic ion such as Ni, Co, Mn) to $NaFe_3(SeO_4)_2(OH)_6$,[32] magnetoplumbite $SrCr_xGa_{12-x}O_{19}$,[33] staircase $M_3V_2O_8$ (M is a magnetic ion),[34] spherical Keplerates[35] and jarosites with formula $AM_3(OH)_6(SO_4)_2$.[6,24,25] Besides these, the spin lattices in $La_4Cu_3MoO_{12}$[36] and $RCuO_{2.66}$[37] are reported to be identical with the Kagome lattice. Two-dimensional solid $^3He$ adsorbed on graphite has also been studied as a Kagome antiferromagnetic system.[38] Although all these materials belong to the family of Kagome systems, it will be beyond the scope of this *feature article* to discuss all the structures. The Kagome staircase structure, however, is different in that the layers are not two-dimensional but are buckled in and out of the plane. We shall concentrate more on the Kagome network compounds of different ions possessing different site spins.

Theoretically, the ideal $S = \frac{1}{2}$ Kagome lattice antiferromagnet has been studied extensively. Studies based on the exact diagonalization of finite systems and series expansions[39–41] suggest that the system has a quantum disordered ground state. Finite size studies show a gap to spin excitations, but the most fascinating aspect of these numerical studies is the existence of a large number of singlet excitations below the lowest triplet state. Their number appears to grow exponentially with the size of the system.[42,43] Although there exist volumes of theoretical work on Kagome lattices, most of these are restricted to $S = \frac{1}{2}$ Kagome systems. We have recently attempted to explain the magnetic properties of Kagome systems on the basis of the magnitude of spins.[44]

In this *feature article*, we discuss the structures and magnetic properties of a number of Kagome systems synthesized with several transition metal ions such as $Fe^{3+}$, $Cu^{2+}$, $Mn^{2+}$, $V^{3+}$, $Co^{2+}$, $Fe^{2+}$, $Ni^{2+}$ and a $Fe^{2+}/Fe^{3+}$ mixed valent complex. We then try to rationalize the observation on magnetic properties in the jarosite class of materials based on theoretical analysis. We conclude with a summary and future outlook.

## 2. Experimental scenario

There has been much progress in the synthesis of the jarosite class of compounds, with the general formula $AM_3(OH)_6(XO_4)_2$. The A site is usually occupied by a monovalent, divalent or a trivalent cationic species which has a coordination number greater or equal to nine. The M site is the metal site in an octahedral coordination and the X site is occupied by a $P^{5+}$, $As^{5+}$ or $S^{6+}$ in a tetrahedral oxygen environment. The layered Kagome hexagonal sheets are separated by the A cations.[6] In general, the corner-shared tetrahedral lattice, which occurs in both spinel and pyrochlore structure materials, is called the pyrochlore lattice. For planar lattices, the mode of stacking in three dimensions of edge-shared triangular sheets with ···ABCABC··· stacking of Kagome

 

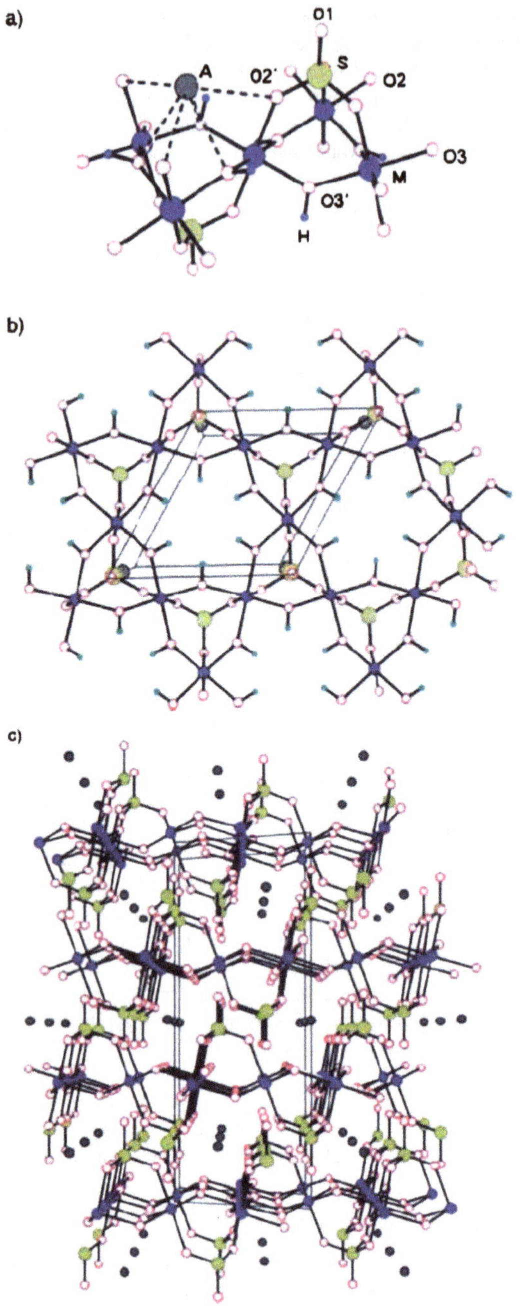


**Fig. 2** The X-ray crystal structure of (a) the magnetic subunit of $Fe_3(\mu\text{-}OH)_3$ triangles; (b) the in-plane segment of the Kagome layer; and (c) a side-view of stacked Kagome layers. In (a) the sphere coding of the atoms is presented.[25]

layers yields the pyrochlore lattice, as shown Fig. 2. The elementary triangles in the two-dimensional sheet are capped by the $XO_4$ tetrahedra which are positioned above and below the sheet in alternate fashion (see Fig. 2). The magnetic ion at each of the vertices of the triangular array, with four neighbors, confers unusual correlated magnetic properties.[45]

Jarosite-type materials are difficult to obtain in pure single crystal form, but highly stoichiometric and pure samples are essential to obtain consistent and meaningful results. Intense research in the past decade has sought to develop a few new synthetic routes for the preparation of pure and highly crystalline samples without many defects or with almost full coverage. While redox-based techniques have been employed by Nocera's group to prepare single crystalline Kagome samples with almost full coverage of the magnetic ions, many organically templated Kagome compounds have been synthesized by hydrothermal techniques by us. The employment of new synthetic routes has resulted in a large class of new materials. For example, a class of Kagome materials with the general formula $MCu_3(OH)_6Cl_2$ with various metal (M) ions has recently been prepared.[46,47] It has also been possible to prepare Kagome compounds with two different metals or the same metal with two different oxidation states.[48] Depending on the charge imbalance of the metal ions and the preparatory conditions, the resultant compounds exhibit different crystal symmetries and chemical formulae. The magnetic properties of these materials have been investigated after establishing the structures by X-ray crystallography together with compositional analysis and spectroscopic data.[49–55]

### 2.1 $S = \frac{5}{2}$ Kagome compounds

The alunite superfamily subgroup, with composition $KFe_3(SO_4)_2(OH)_6$, show layered Kagome network features. The crystal structure consists of $Fe_3^{III}(\mu\text{-}OH)_3$ triangles (Fig. 2a), capped by sulfate anions positioned alternately up and down about a hexagonal network forming the Kagome layers (Fig. 2). To date, there has been much progress with new redox based methods, and synthesis of Fe jarosites with general formula $AFe_3(OX)_6(TO_4)_2$ featuring different cations and anionic groups ($A = Na^+, K^+, Rb^+, NH_4^+, \frac{1}{2}Pb^{2+}, Ag^+, Tl^+$ and $TO_4 = SO_4^{2-}$ and $(OX)_6 = (OH)_6$ and $(OD)_6$; $A = K^+$ and $Rb^+$ and $TO_4 = SeO_4^{2-}$ and $(OX)_6 = (OH)_6$) with almost complete coverage has been possible.[45,56–62] In the crystal, while the $FeO_6$ octahedra are clearly puckered, the Fe ions form a perfectly planar Kagome net. The layer stacking sequence is ···ABCABC··· along the *c*-axis, shown in Fig. 2. The magnetic properties of these materials are very interesting (see Fig. 3). Except for $(D_3O)Fe_3(SO_4)_2(OD)_6$, which shows no long range order[63,64] down to 1.5 K (although there exists an apparent transition at 13.8 K), all the other Fe jarosites show long range magnetic ordering with a very large negative Curie–Weiss constant ($\theta$) value, even with 90% coverage of the Kagome net.[45,61,62] In Table 1, the Curie–Weiss constant and ordering temperature are listed for a number of these $Fe^{3+}$ jarosite compounds.

There have been attempts to synthesize Kagome compounds with $S = \frac{5}{2}$ with a transition metal ion other than $Fe^{3+}$. One such system synthesized is an amine templated manganese(II) sulfate of the composition $[C_4N_2H_{12}][NH_4]_2[Mn_3^{II}F_6(SO_4)_2]$, with the Kagome structure, under solvothermal conditions.[65] The structure consists of vertex-sharing $Mn^{II}F_4O_2$ octahedra

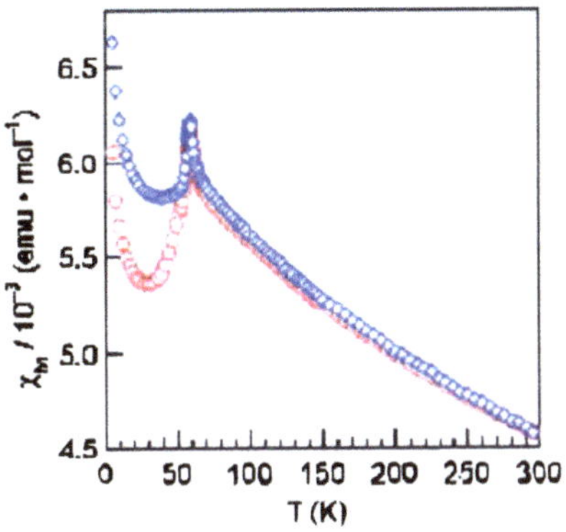


**Fig. 3** Field-cooled (FC, ◇) and zero-field-cooled (ZFC, ○) susceptibilities for $AgFe_3(OH)_6(SO_4)_2$ at a 100 Oe measuring field.[45]

and $SO_4$ tetrahedral units, which are fused together by Mn–F–Mn and Mn–O–S bonds. The distorted octahedron around $Mn^{2+}$ has the oxygen atoms from the sulfate along the $z$-axis and the F atoms along the $xy$-plane (Fig. 4). The magnetic properties of this compound differ from those of $Fe^{3+}$ jarosites. Down to very low temperature, the magnetic susceptibility does not show any magnetic ordering while the $\theta$ value is found to be negative (−30 K).

### 2.2 $S = \frac{1}{2}$ Kagome compounds

Nocera *et al.* have hydrothermally synthesized a series of compounds of general formula $Zn_xCu_{4-x}(OH)_6Cl_2$ with $0 < x < 1$, with variable Cu/Zn composition.[46,47] The crystal symmetry depends crucially on the $x$ value; for $x < 0.33$, the crystal symmetry is monoclinic with distorted Kagome lattice and for $x > 0.33$, the crystal symmetry increases to rhombohedral, with equilateral Cu triangular plaquettes. The magnetic measurements for the series with variance of $x$ show that for $x = 0$, there exists a ferromagnetic transition at $T_c = 6.5$ K and as $x$ increases from 0, the ferromagnetic transition is maintained, but the corresponding $T_c$ decreases (Fig. 5). For $x = 1$, *i.e.*, at full $Zn^{II}$ occupancy, no magnetic ordering has been observed. The Curie–Weiss constant is found to be large and negative and its magnitude decreases with an increase in $x$, as expected.

**Table 1** Curie–Weiss constants ($\theta_c$ in K) and antiferromagnetic ordering temperatures ($T_0$ in K) for various $Fe^{3+}$ jarosites

| Compound | $\theta_c$/K | $T_0$/K |
|---|---|---|
| $(H_3O)Fe(SO_4)_2(OH)_6$ | −820 | 65 |
| $(D_3O)Fe_3(SO_4)_2(OD)_6$ | −700 | — |
| $(ND_4)Fe_3(SO_4)_2(OD)_6$ | −640 | 46 |
| $NaFe_3(SO_4)_2(OD)_6$ | −667 | 42 |
| $RbFe_3(SO_4)_2(OD)_6$ | −688 | 47 |
| $KFe_3(SO_4)_2(OD)_6$ | −600 | 46 |
| $AgFe_3(SO_4)_2(OD)_6$ | −677 | 51 |
| $NaFe_3(SO_4)_2(OH)_6$ | −825 | 61.5 |
| $RbFe_3(SO_4)_2(OH)_6$ | −829 | 64.5 |
| $KFe_3(SO_4)_2(OH)_6$ | −828 | 65.5 |
| $AgFe_3(SO_4)_2(OH)_6$ | −798 | 59.5 |
| $TlFe_3(SO_4)_2(OH)_6$ | −838 | 63.5 |
| $Pb_{0.5}Fe_3(SO_4)_2(OH)_6$ | −818 | 56.5 |

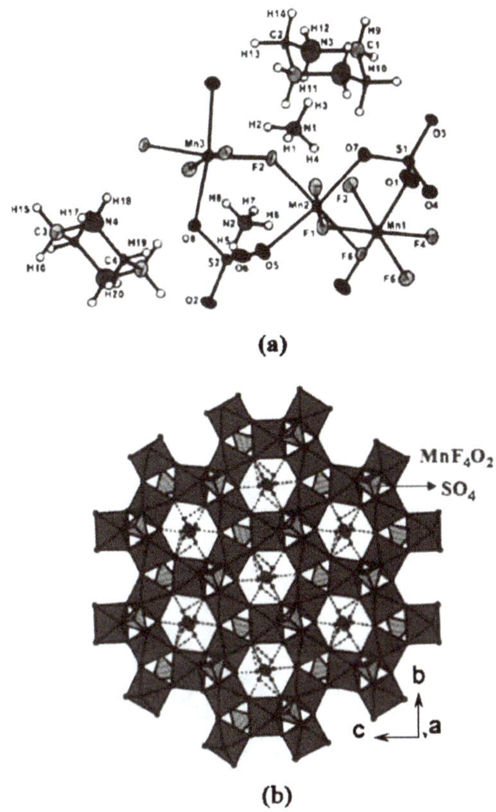


**Fig. 4** (a) Labeled asymmetric unit of $[C_4N_2H_{12}][NH_4]_2[Mn_3^{II}F_6(SO_4)_2]$. Thermal ellipsoids are given at 50% probability. (b) Polyhedral view of the hexagonal Kagome layer. Note the presence of the ammonium ion in the hexagonal channel.[65]

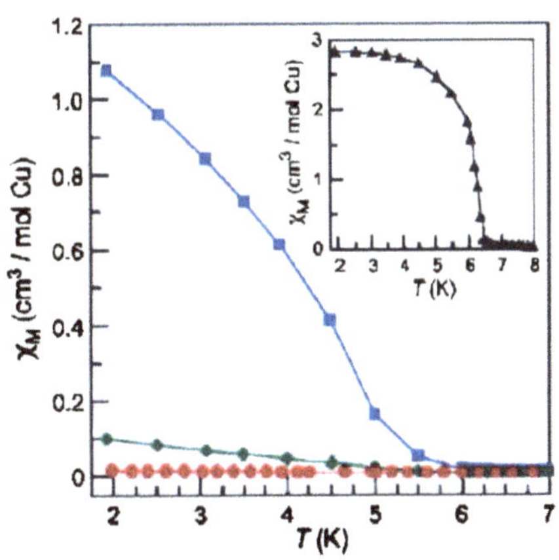


**Fig. 5** (a) Low temperature dependence of $\chi_M$ for compounds with formula $Zn_xCu_{4-x}(OH)_6Cl_2$ for $x = 0$ (▲), 0.50 (■), 0.66 (◆) and 1.00 (●) as measured under ZFC conditions at 100 Oe ($\chi_M$ in the inset is for $x = 0$).[46]

## 2.3 $S = 1$ Kagome compounds

Recently, a phase-pure compound of the formula $[C_6N_2H_8][NH_4]_2[Ni_3F_6(SO_4)_2]$, was prepared under solvothermal conditions, where the magnetic ion Ni(II) is in the $S = 1$ state.[66]

The structure of this compound (Fig. 6a) consists of anionic layers of vertex-sharing $Ni^{II}F_4O_2$ octahedra and $SO_4$ tetrahedra fused together by Ni–F–Ni and Ni–O–S bonds. Each $NiF_4O_2$ unit shares four of its Ni–F vertices with similar neighbors, with the Ni–F–Ni bonds roughly aligned in the *ab*-plane. The Ni–O bond is canted from the *ab*-plane, which forces a three ring trio of apical Ni–O bonds closer together to allow them to be capped by the $SO_4$ tetrahedra. The sulfate groups are positioned alternately up and down about the hexagonal network. The three- and six-membered rings formed by the octahedra give rise to the in-plane connectivity of the hexagonal tungsten bronze sheets, characteristic of the Kagome lattice, as shown in Fig. 6b. The $[Ni_3F_6(SO_4)_2]^{4-}$ framework is balanced by the presence of protonated amine and $NH_4^+$ ions in the interlayer space and hexagonal channels, respectively. Magnetic measurements show that the

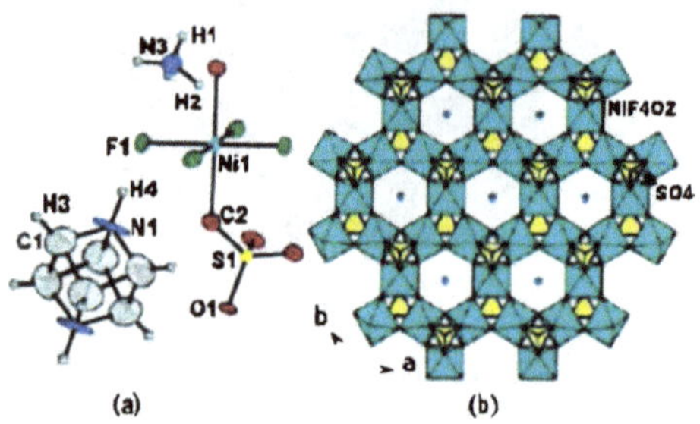


**Fig. 6** (a) Labeled asymmetric unit of $[C_6N_2H_8][NH_4]_2[Ni_3F_6(SO_4)_2]$. Thermal ellipsoids are given at 50% probability. (b) Polyhedral view of the hexagonal Kagome layer. Note the presence of the ammonium ion in the hexagonal channel.[66]

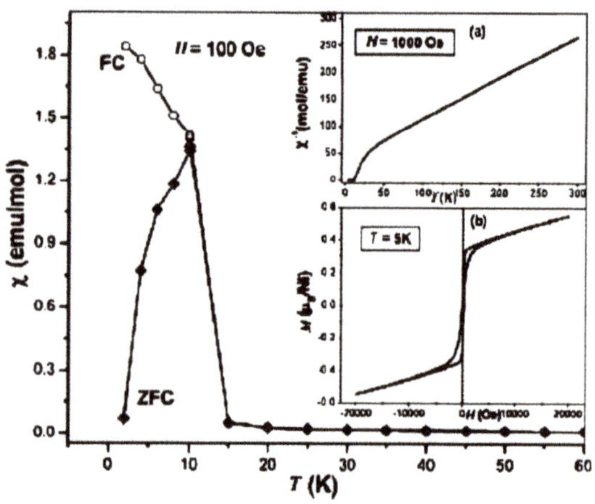


**Fig. 7** Temperature dependence of the magnetic susceptibility of the Ni Kagome compound (100 Oe) under FC and ZFC conditions. Inset (a) shows the temperature variation of the inverse susceptibility at 1000 Oe. Inset (b) shows magnetic hysteresis at 5 K.[66]

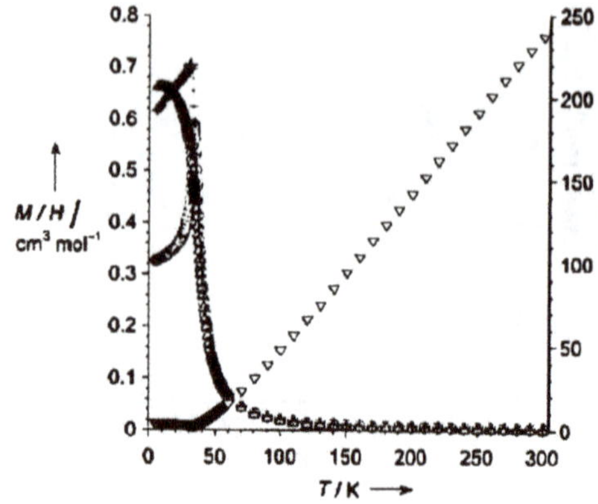


**Fig. 8** Temperature dependence of $M/H$ per mole of vanadium atoms for $NaV_3(OH)_6(SO_4)_2$ at applied magnetic field strengths of 0.5 (●), 5.0 (+), and 10.0 (△) kOe. The $H/M$ *versus* $T$ plot (▽) was measured at an applied field of 0.5 kOe.[68]

Curie–Weiss $\theta$ value is antiferromagnetic (−60 K), and the low-temperature magnetization data show hysteresis, possibly due to canted antiferromagnetic ordering (Fig. 7).[67]

There also exists an earlier example of a spin-1 Kagome lattice system with formula $NaV_3(OH)_6(SO_4)_2$, which was synthesized hydrothermally by Nocera *et al.*[68,69] The compound crystallizes in the rhombohedral space group $R\bar{3}m$, in which the vanadium ion is in the 3+ state and resides in a distorted octahedron, composed of four equatorial oxygen atoms from the bridging hydroxy groups and two axial oxygen atoms from the capping sulfate groups. Such a vanadium pseudo-octahedron is tetragonally elongated along its $z$-axis. The magnetic properties are found to depend crucially on the interlayer couplings (Fig. 8). The Curie–Weiss constant ($\theta$) is found to be +53 K, which indicates that the intralayer coupling between vanadium centers is strongly ferromagnetic, with $T_c = 33$ K. From the magnetic field dependent susceptibility data, it has been concluded that the interlayer coupling is weakly antiferromagnetic.

## 2.4 $S = \frac{3}{2}$ Kagome compounds

An organically templated cobalt(II) sulfate of the composition, $[H_2N(CH_2)_4NH_2][NH_4]^{2-}[Co_3^{II}F_6(SO_4)_2]$, with the Kagome lattice was isolated by Rao *et al.*[70] The structure consists of anionic layers of vertex-sharing $Co^{II}F_4O_2$ octahedra and tetrahedral $SO_4$ units, which are fused together by Co–F–Co and Co–O–S bonds. Each $CoF_4O_2$ unit shares four of its Co–F vertices with similar neighbors, with the Co–F–Co bonds roughly aligned in the *bc*-plane. The Co–O bond is canted from the *bc*-plane and the Co–O vertex forces a three-ring trio of apical Co–O bonds closer together to allow them to be capped by the $SO_4$ tetrahedra. The three- and six-membered rings of octahedra from the in-plane connectivity are shown in Fig. 9. The magnetic measurements give a $\theta$ value of −35.5 K and an effective magnetic moment of 5.2 $\mu_B$ per Co from the Curie–Weiss fit, which is a little higher than the spin-only value due to orbital contributions as found in many other $Co^{2+}$ compounds.[71]

There also exists another type of $S = \frac{3}{2}$ compound, the layered garnets $SrCr_xGa_{12-x}O_{19}$ (SCGO), which have been

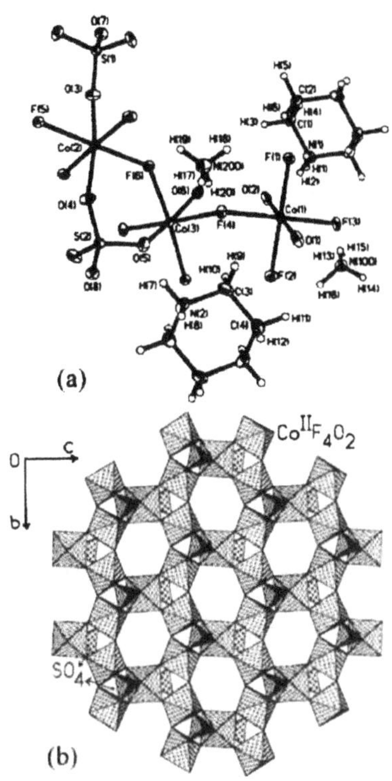

Fig. 9 (a) Asymmetric unit of $[H_2N(CH_2)_4NH_2][NH_4]^{2-}[Co_3^{II}F_6(SO_4)_2]$ (b) Polyhedral view of the Kagome layer.[70]

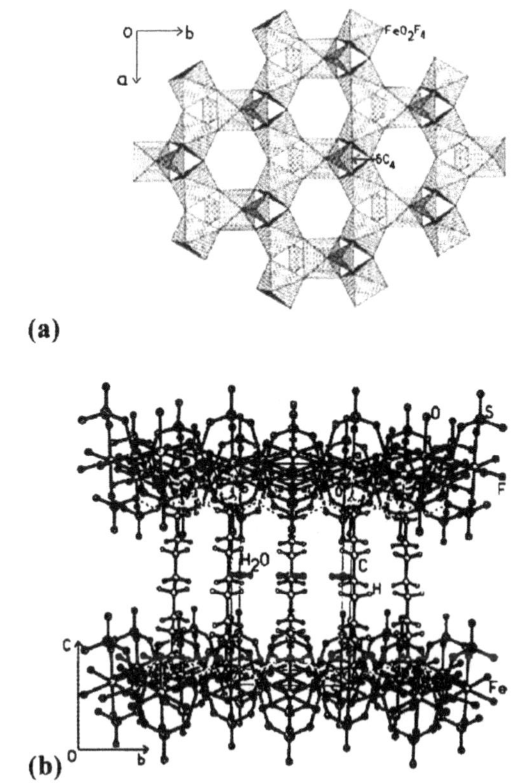

Fig. 10 The polyhedral view of $[H_3N(CH_2)_6NH_3][Fe_{1.5}^{II}F_3(SO_4)]\cdot 0.5H_2O$. The three-membered apertures are capped by sulfate moieties. (b) View along the $a$-axis showing the stacking of the layers pillared by the amine. The dotted lines indicate H-bond interactions.[77]

investigated quite extensively.[72–76] However, these materials contain an additional edge-sharing triangular (non-Kagome) lattice interposed between Kagome layers.

### 2.5 $S = 2$ Kagome compounds

A compound with the formula $[H_3N(CH_2)_6NH_3][Fe_{1.5}^{II}F_3(SO_4)]\cdot 0.5H_2O$ has been synthesized hydrothermally, where the magnetic ion is in the +2 state with integer spin.[77] The structure consists of anionic layers of vertex-sharing $Fe^{II}F_4O_2$ octahedra and tetrahedral $SO_4$ units, which are fused together by Fe–F–Fe and Fe–O–S bonds. Each $FeF_4O_2$ unit shares four of its Fe–F vertexes with similar neighbors, with the Fe–F–Fe bonds nearly aligned in the $ab$-plane. The Fe–O bond is canted from the $ab$-plane and the Fe–O vertex effectively forces a three-ring trio of apical Fe–O bonds closer together to allow them to be capped by the $SO_4$ tetrahedra. The three- and six-membered rings of octahedra, resulting from the in-plane connectivity, are shown in Fig. 10. Interestingly, the magnetic measurements show a magnetic polarization at low temperature reminiscent of ferrimagnetic ordering. The Curie–Weiss constant is found to be −70 K.

However, with a different set of initial conditions, a monophasic pale yellowish needle shaped crystal of an iron compound with the formula $[H_3N(CH_2)_2NH_2(CH_2)_2NH_2(CH_2)_2NH_3][Fe_3^{II}F_6(SO_4)_2]$, has been isolated.[78] In this system, there are five crystallographically distinct Fe atoms and two S atoms with all the Fe atoms having octahedral geometry. The Fe atoms have fluorine and oxygen neighbors to form $FeF_4O_2$

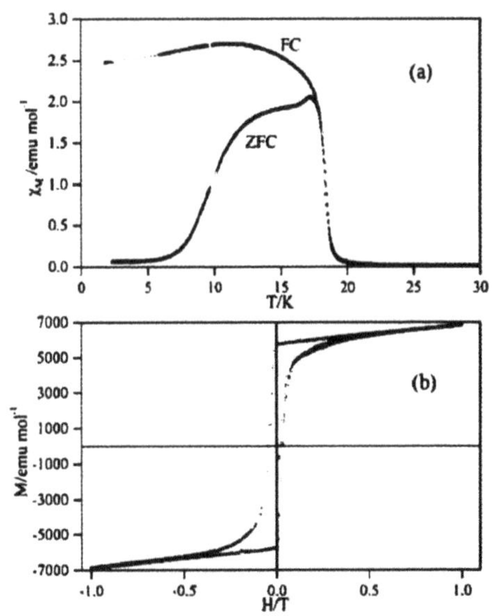

Fig. 11 (a) DC susceptibility data showing divergence between the ZFC and FC (at 100 Oe) behaviours. (b) Magnetic hysteresis at 10 K.[78]

 

octahedra. The structure consists of anionic framework layers of $[Fe_3F_6(SO_4)_2]$. Due to the presence of 50% edge-sharing octahedra, two types of triangular lattices are created. One type of three-ring trio is formed by the vertex-sharing of the $FeF_4O_2$ octahedra, creating a perfect three-membered ring, which is capped by the sulfate tetrahedron, as in a jarosite. On the other hand, the three-ring trio formed by two edge-sharing and one corner-sharing octahedra, creates a distorted triangular lattice which is bridged by the sulfate tetrahedron. This compound shows magnetic hysteresis (Fig. 11) below a critical temperature (18 K) due to the distortion caused by the presence of the edge sharing octahedra.

### 2.6 Mixed-valent iron Kagome compounds

In the process of finding new Kagome materials, a mixed-valent compound, $[HN(CH_2)_6NH][Fe^{III}Fe^{II}_2F_6(SO_4)_2]\cdot[H_3O]$, was synthesised by solvothermal methods.[48] The structure consists of anionic layers of corner sharing $Fe^{III}F_4O_2$ and $Fe^{II}F_4O_2$ octahedra and $SO_4$ tetrahedra. In Fig. 12, we show the structure of the compound to demonstrate how the Kagome lattice is derived from three and six-membered rings of octahedra, giving rise to the hexagonal tungsten-bronze-type sheets. There is 100% occupation, rendered possible by the presence of fluoride instead of the OH groups normally found in jarosites. The presence of the $Fe^{II}$ and $Fe^{III}$ states in the ratio 2 : 1 is consistent with the formula derived from crystallography and chemical analysis. Interestingly, the magnetic measurements show hysteresis at 10 K, however, the hysteresis data show a step-like behavior and a metamagnetic transition, the field required for the transition depending on the temperature.

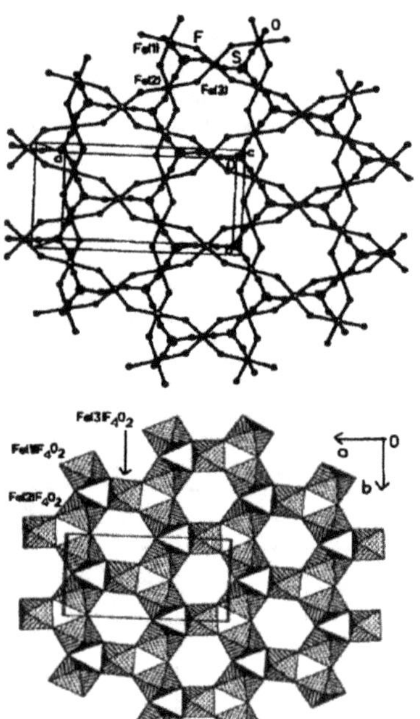

**Fig. 12** (a) Ball and stick representation of a section of $FeFe_2F_6(SO_4)_2$, showing the presence of three- and six-membered rings, where the three rings are capped by sulfate tetrahedra. (b) Polyhedral view of the 2D network of corner sharing iron octahedra in the mixed valent compound in the *ab*-plane. Here Fe(1) and Fe(2) are in the +2 state and Fe(3) is in the +3 state.[48]

## 3. Approaches to understand frustration and related magnetic properties

As can be seen, experimental magnetic behavior varies considerably depending on the magnetic ions present and on the percentage of coverage area. We summarize here the magnetic properties of various Kagome materials with a variety of magnetic metal ions in Table 2. As can be seen, while the $Fe^{3+}$ Kagome systems exhibit characteristics of frustrated low-temperature antiferromagnetism and occasional long-range antiferromagnetic order in the ground state,[6,45,61,62] vanadium ($V^{3+}$) jarosites are shown to exhibit ferromagnetism in the Kagome layers ($T_c \sim 33$ K).[68] On the other hand, chromium ($Cr^{3+}$) jarosite shows low-temperature antiferromagnetic behavior[72–76] while a Kagome compound of $Co^{2+}$[70] show strongly anisotropic high-spin Co(II) with appreciable spin–orbit coupling. A $Cu^{2+}$ Kagome compound shows spin frustration,[46] while a $Ni^{2+}$ sulfate with the Kagome structure[66] exhibits canted antiferromagnetism with magnetic polarization at low temperatures. An amine-templated mixed valent iron sulfate with the Kagome structure[48] becomes ferrimagnetic at low temperatures, although it shows some evidence for magnetic frustration. Interestingly, an organically templated $Fe^{2+}$ sulfate with the Kagome structure also shows ferrimagnetic properties at low temperatures.[24,78] The question arises as to the cause for the marked differences in the magnetic properties of the $V^{3+}$ Kagome compounds with the $Fe^{3+}$ or $Cr^{3+}$ and those with $Fe^{2+}$, mixed-valent Fe, $Co^{2+}$, $Cu^{2+}$ and $Ni^{2+}$, although, all of them possess more or less the common Kagome network structure.

Since the Kagome network structure shares one spin between two adjacent triangles, as the basic unit, let us consider a two triangle structure with five spins, where one spin is common to both the triangles, as shown in Fig. 13. For a

**Table 2** Magnetic properties of Kagome compounds with transition metal ions possessing different spins

| Metal ion compound | $S$ | $\theta_c$/K | $T_0$/K | Magnetic properties |
|---|---|---|---|---|
| $Cu^{2+}$ | 0.5 | −314 | 2 | AFM interactions, no ordering[46] |
| $Ni^{2+}$ | 1 | −60 | 9 | AFM interactions, low-T hysteresis[66] |
| $V^{3+}$ | 1 | +53 | 33 | FM layers coupled antiferromagnetically[68] |
| $Co^{2+}$ | 1.5 | −35 | 4 | AFM interactions[70] |
| $Cr^{3+}$ | 1.5 | −70 | 4 | AFM interactions[72–76] |
| $Fe^{2+}$ (perfect) | 2 | −75 | 19 | AFM interactions, low-T hysteresis[24] |
| $Fe^{2+}$ (distorted) | 2 | | 18 | AFM interactions, low-T hysteresis[78] |
| $Fe^{3+}$ | 2.5 | −820 | 65 | AFM interactions, 3D ordering[45] |
| $Mn^{2+}$ | 2.5 | −30 | — | AFM interactions, no ordering[65] |
| $Fe^{3+}/Fe^{2+}$ | 2.5/2 | −180 | 13 | Ferrimagnet[48] |

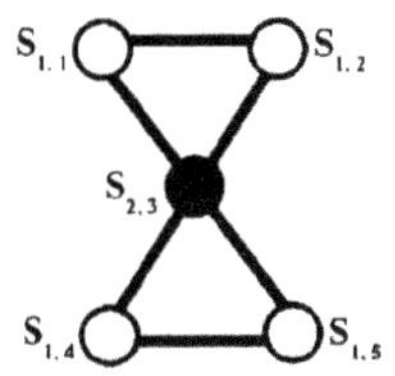


**Fig. 13** Simplest corner-sharing two triangle structure. $S_{i,j}$ corresponds to spin $s_i$ at site $j$.

general understanding, we consider spins $s1$ at the edges (numbered $S_{1,1}$, $S_{1,2}$, $S_{1,4}$ and $S_{1,5}$ in Fig. 13) of the two triangles and spin $s2$ as the corner shared (numbered $S_{2,3}$), so that there are four $s1$ spins and one $s2$ spin in this five-spin system. It is easy to solve this system exactly if we assume that the site spins interact *via* a nearest neighbor Heisenberg exchange Hamiltonian, which can be explicitly written as, $H = J[s_{1,1}\cdot s_{1,2} + s_{1,4}\cdot s_{1,5} + s_{2,3}\cdot(s_{1,1} + s_{1,2} + s_{1,4} + s_{1,5})]$, where each term is a dot product between two spin operators. We will consider the cases with $s1 = s2$ and $s1 \neq s2$ separately within this simple superexchange Hamiltonian. For $s1 = s2 = \frac{1}{2}$, the ground state is sixfold degenerate, with each doublet configuration of the shared spin giving rise to threefold degenerate states for each triangle. The ground-state magnetization is finite in each lattice point due to strong frustration. For $s1 = s2 = 1$, the ground state is threefold degenerate and is a singlet. There are three ways one can produce a singlet structure, and in each case, apart from the total magnetization, the magnetization in each of the edges as well as in the corner spin vanishes separately. Next we consider the cases with $s1 \neq s2$. There are two possibilities; each can be either an integer or a half-integer spin. For $s1 = \frac{1}{2}$ and $s2 = 1$, the ground state is a singlet and is threefold degenerate. The states are similar to the $s1 = s2 = 1$ case, although in this case, antiferromagnetic coupling in each edge is between the two spin-$\frac{1}{2}$ sites. For $s1 = \frac{1}{2}$ and $s2 = \frac{3}{2}$, the ground state is a doublet, with double degeneracy. Both the edges together form a triplet, which couples with the $Sz = \frac{1}{2}$ component of the spin $\frac{3}{2}$, and thereby removes the frustration completely in the corner-shared structure. By increasing the value of $s2$ further and keeping $s1 = \frac{1}{2}$ fixed, we get the same ferrimagnetic state with ground-state spin $S_G = s2-4s1$.

For a detailed theoretical study, we have considered 12- and 18-site Kagome lattices. They are shown in Fig. 14. Since experimentally there are examples of mixed-spin jarosite systems, we have considered a number of mixed spin together with unique site spin Kagome systems. In both cases, the integer as well as the half-integer spins are included.[79–82] The Hamiltonians correspond to antiferromagnetic exchanges between the spins connected by the solid lines (see Fig. 14). While the filled circles correspond to the $s1$ site spin, the empty circles have site spins $s2$. The site spins, $s1$ and $s2$, can be the same or different with integer (I) or half-integer (HI) values. Specifically, we consider the following four cases based on the above results for two-triangle systems: (a) $s1 = s2 = \frac{1}{2}$, (b) $s2 = \frac{1}{2}$ but $s1 = 1$, (c) $s1 = s2 = 1$, and (d) $s2 = 1$ and $s1 = \frac{1}{2}$. The periodic boundary conditions of the finite lattice systems ensure that these are small clusters of infinite two-dimensional Kagome structure. Note that the following results are with the

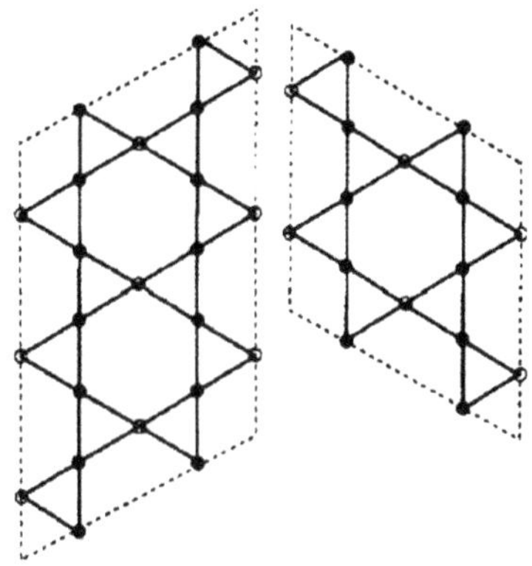

**Fig. 14** Schematics of 12- and 18-site Kagome lattices. The solid lines represent the antiferromagnetic exchange interactions, $J$. The dashed line defines the periodic boundary conditions. The filled and empty circles represent the site spins $s1$ and $s2$, respectively.

Hamiltonian corresponding to an ideal Kagome lattice with localized site spins interacting *via* isotropic nearest-neighbor superexchange pathways. Later, we consider additional interaction terms corresponding to other energy scales due to lack of symmetry in the real systems. However, since most of these systems consist of 3d transition-metal narrow band ions, often the Heisenberg superexchange processes between nearest-neighbor localized moments set the dominant energy scale.

We have diagonalized the Hamiltonian matrix corresponding to the nearest-neighbor Heisenberg model for the above four systems with varying site spin $s1$, $s2$ values for both lattice sizes. Low-energy spectra in terms of total spin quantum number, $S$, can be obtained with the Hamiltonian matrix in a smallest $S^z_{tot}$ sector, since every total spin has a projection in the smallest possible $S^z_{tot}$ value. In Fig. 15, we compare and contrast the energy-level diagram as a function of total spin, $S$, for these four cases for an 18-site system. Although, we present results only for a fixed system size $N = 18$, for smaller spin values, we have verified that the low-energy spectrum remains qualitatively similar even for system sizes up to $N = 21$. For example, for $(\frac{1}{2}, \frac{1}{2})$ Kagome, the ground-state energies per site for ($N$ sites) are −0.454 (12), −0.440 (15), −0.447 (18),

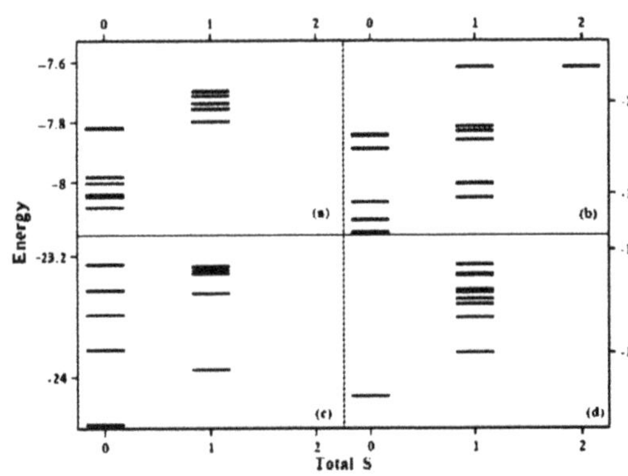


**Fig. 15** Energy levels as a function of total spin $S$, for the ($s1$, $s2$) systems, for an 18-site cluster: (a) $(\frac{1}{2}, \frac{1}{2})$, (b) $(1, \frac{1}{2})$, (c) (1, 1) and (d) $(\frac{1}{2}, 1)$.

and −0.437 (21). Similarly, for (1, 1) and ($\frac{1}{2}$, 1), they are −1.468 (12) and −0.824 (12), −1.62 (15) and −0.81 (15), −1.34 (18) and −0.801 (18), and −1.39 (21) and −0.79 (21). Interestingly, the ground-state energy per site extrapolated to $N \to \infty$ for a spin-$\frac{1}{2}$ Kagome lattice was reported to be −0.434 by Zeng and Elser.[83] This shows that, for such small system sizes, the low-lying energy states are quite clearly defined, even for the smallest ($s1$, $s2$) values.

For cases (a) and (b) in Fig. 15, we find that the ground state is a non-degenerate singlet. The lowest excitations are to the singlet states for both cases although there are more numbers of excited singlet states very close to the ground state for case (a) than for case (b). Moreover, there exists a finite gap between the ground and the lowest magnetic states in both cases. In case (c), with $s1 = s2 = 1$, the ground state is a non-degenerate singlet and the lowest excited state is a magnetic triplet state. The low energy excitation spectrum consists of alternate singlet and triplet states and the system behaves more like a regular one-dimensional antiferromagnetic chain.[81,82] In fact, the ground state consists of a resonating valence bond (RVB) state with nearest-neighbor spins forming alternating singlet combinations and the lowest excitation involves energy cost (therefore the finite gap) due to disappearance of one of the singlets.[84] The situation becomes completely different for case (d). Although the ground state in this case is a singlet, the low-energy excitations are to the triplet states. Further, the number of low-energy triplet states increases rapidly with increase in system size. This is because in this case, the couplings between $s2$ and $s1$ spins become strongly antiferromagnetic forcing the $s1$–$s1$ couplings to be weakly ferromagnetic, since the $s1$ spins have smaller magnitude.[44] The in-plane and $z$-component spin–spin correlation functions are shown in Fig. 16.

Thus far, we have discussed the magnetic energy spectrum of various Kagome systems with only nearest neighbor

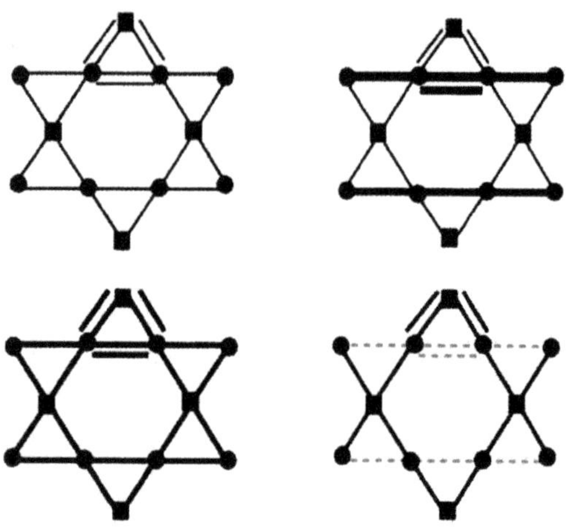

**Fig. 16** $z$-Component spin–spin correlation functions (the additional lines in the upper triangle for each case are the in-plane correlation), where the four systems are in the following order: upper row from left ($\frac{1}{2}$, $\frac{1}{2}$) and (1, $\frac{1}{2}$); lower row from left (1, 1) and ($\frac{1}{2}$, 1). The thickness indicates the correlation strength. The solid line is for antiferromagnetic correlation and the dashed lines are for ferromagnetic correlations.

Heisenberg interactions. However, in many of these systems, the magnetic ions are connected by anionic octahedral species through shared vertices, which remain most often tilted resulting in lowering of the overall hexagonal symmetry. For such geometry, another interaction term, known as the Dzyaloshinskii–Moriya (DM) Hamiltonian becomes important, in addition to the Heisenberg interaction. The DM Hamiltonian is of the type:

$$H_{\mathrm{DM}} = \sum_{ij} D_{ij}(S_i \times S_j)$$

Since this Hamiltonian is a vector, not only the spin-components become anisotropic, according to Moriya's rules,[85] but both the in-plane and the out-of-plane Dzyaloshinskii–Moriya (DM) vectors also become nonzero.[67,85–87] However, the out-of-plane coupling would not result in breaking the global rotational symmetry. The in-plane components ($XY$) of DM interactions, on the other hand, can induce canted spin moments in the out-of-plane directions ($Z$). Note that, inclusion of this term lowers the overall symmetry of the spin interactions in the system.

Since many of these systems show lowering of the crystal symmetry, we have analyzed the 12 site spin systems in terms of a Hamiltonian including $XY$ components of DM interactions to the nearest-neighbor Heisenberg term discussed earlier. The Hamiltonian is:

$$H_{\mathrm{total}} = \sum_{\langle ij \rangle} J S_i \cdot S_j + \mathrm{d}_z(S_i^x S_j^y - S_j^x S_i^y)$$

where the DM part is written in terms of the spin components. By diagonalizing the Hamiltonian matrix, we obtain the energy spectrum and various ground state correlation functions. We find that with inclusion of the DM-term, the low lying excitations behave differently depending on the strength of out-of-plane ($\mathrm{d}_z$) DM interactions. For our finite size ($\frac{1}{2}$, $\frac{1}{2}$) system, the number of singlets lying within the spin gap is finite, however, with an increase in the strength of DM interactions, this number reduces. In general, from the energy spectrum, we conclude that in cases where nonmagnetic states enter the spin-gap region, their number reduces with DM interactions, while for cases where a triplet is the lowest excitation, the DM interactions stabilize the triplet. This is because, with an increase in DM interaction strength, the nearest neighbour correlation between the $z$-components reduces, while the correlation between the in-plane component spins increases. This suggests that the spins become more and more tilted which enforces a finite spin moment for every magnetic ion triangle with the spins lying above and below the Kagome plane, resulting in magnetic polarizations. In fact, the magnitude of such a polarization depends crucially on the angle between the spins and the Kagome plane in the actual geometry and the resulting DM vector directions.

Although our model Hamiltonian calculations are for general Kagome systems, without consideration of details of each of the above experimental systems, the prediction of the model, however, confirms that within the nearest-neighbor superexchange processes and DM interactions, the variation of the magnitude of magnetic moments of ions plays a dominant role in governing the nature of the low-energy spectrum. Note that partially polarized magnetization or long-range magnetic order in the ground state may arise due to DM interactions determined by the magnetic anisotropy induced canted spin moments within a non-planar geometry. However, as has been mentioned, the DM-type interactions are generally present in systems without local inversion centers, which is reported to be the case for distorted Fe(III) Kagome systems. On the other hand, a V(III) Kagome layer is magnetically polarized due to geometrically favored ferromagnetic exchange couplings.[68] However, the half-integer spin systems without distortion, such as $Fe^{3+}$ ($S = \frac{5}{2}$),[45] $Cr^{3+}$ ($S = \frac{3}{2}$)[72–76] and $Co^{2+}$ ($S = \frac{3}{2}$)[70] compounds, show low temperature antiferromagnetism, as our model predicts. A $Cu^{2+}$ compound also has been reported to show strong manifestation of magnetic frustration, although at low temperature, due to small ferromagnetic interlayer coupling, the system develops some magnetic polarization.[46] Interestingly, two of the integer spin Kagome compounds reported so far, $Fe^{2+}$ ($S = 2$)[77,78] and $Ni^{2+}$ ($S = 1$),[66] show low-temperature magnetic polarization. Note that, removal of frustrations is crucial for the realization of a fully polarized state within a Kagome layer. The DM interactions, being weak in magnitude, can at best give small polarization at low temperature. In fact, the interplay between geometric factors, magnetic anisotropy, lattice disorder, and interplanar coupling and restrictive exchange processes in such a geometry governs the detailed nature of the low-energy and thereby the low-temperature properties of the Kagome network systems.

## 4. Summary and outlook

In conclusion, we have attempted to present the current state of knowledge in the area of frustrated Kagome network systems, emphasizing the structure and the magnetic properties of various Kagome systems with different magnetic ions. These systems are shown to exhibit magnetic phenomena ranging from long range order to spin glass and spin liquid behaviors. The experimental magnetic behaviors have been analyzed in terms of a generalized magnetic Hamiltonian with Heisenberg and inclusion of the Dzyaloshinskii–Moriya interactions arising due to local non-centrosymmetry in many of these systems. Our analysis shows that the role of the spin magnitudes of the magnetic ions is profound, giving rise to a variety of ground and excited state magnetic configurations and thereby varying the nature of the low-temperature thermodynamic properties, although the theoretical results described here are applicable for specialized isotropic Kagome lattices, with or without local non-centrosymmetry.

As indicated in the introduction, there also exist systems, called Keplerates, which are topologically similar to Kagome networks but occur in spherical shape with icosidodecahedral or cuboctahedral symmetry. Muller *et al.* have demonstrated that it is possible to build the spherically symmetrical structures with triangular or pentagonal units as building blocks.[35,88,89] Magnetic clusters with a $Mo_{72}M_{30}$ formula unit, where M = Fe, V, Cr, have been synthesized so far. Since these share similar topological features with Kagome networks, such a constrained geometry would be ideal to provide an alternate and rigorous understanding of the basic aspects of spin frustration-related magnetic properties and their relationship to the intrinsic magnetic moments of the magnetic ions.[90]

Another class of frustration-related systems are pyrochlore magnetic materials which are expected to show a range of phenomena including spin ice, spin-Peierls, heavy-fermion, high-$T_c$ *etc.*[91–94] Particularly interesting are those systems where strong correlations might perturb a section of the system thereby giving the possibility of multiferroic behavior with or without ordering of spin transition. Interestingly, the change in ground state entropy during a transformation from a non-degenerate magnetically polarized state to a highly degenerate ground state upon application of field may lead to a large magnetocaloric effect. Such an adiabatic demagnetization may thus be useful in prospective refrigerant materials at low temperature which are widely considered for space applications.[95]

There are issues of quantum fluctuations, long-range exchange, chiral vectors and microscopic disorder which need to be understood in a more general sense. We may hope that the current experimental and theoretical studies of multidimensional frustrated systems will lead to a deeper understanding of frustration-related effects in rare-earth and transition metal magnetic systems in quasi-two and three dimensions. Such an understanding might be relevant to as yet to be discovered systems where various competing interactions may be useful for a range of applications in thermal, mechanical, optical, electrical and magnetic phenomena. We expect that discrete and collective behavior of these and related materials will continue to capture the attention of solid-state and materials chemists and physicists alike.

## Notes and references


1 L. Pauling, *J. Am. Chem. Soc.*, 1935, **57**, 2680.
2 W. F. Giauque and M. Ashley, *Phys. Rev.*, 1933, **43**, 81.
3 W. F. Giauque and J. W. Stout, *J. Am. Chem. Soc.*, 1936, **58**, 1144.
4 J. B. Goodenough, *Magnetism and the Chemical Bond*, Wiley-Interscience, New York, 1963.
5 A. P. Ramirez, *Annu. Rev. Mater. Sci.*, 1994, **24**, 453.
6 J. E. Greedan, *J. Mater. Chem.*, 2001, **11**, 37.
7 A. P. Ramirez, in *Handbook on Magnetic Materials*, ed. K. J. H. Busch, Elsevier Science, Amsterdam, 2001, vol. 13, pp. 423.
8 K. Binder and A. P. Young, *Rev. Mod. Phys.*, 1986, **58**, 801.
9 R. de Pape and G. Ferey, *Mater. Res. Bull.*, 1986, **21**, 971.
10 A. B. Harris, C. Kallin and A. J. Berlinsky, *Phys. Rev. B: Condens. Matter Mater. Phys.*, 1992, **45**, 2899.
11 O. Kahn, *Molecular Magnetism*, Wiley-VCH, NY, 1993.
12 J. Renaudin, G. Ferey, M. Lahlou-Mimi, J. M. Greneche, Y. Mary and A. Kozac, *J. Magn. Magn. Mater.*, 1991, **92**, 381.
13 C. N. R. Rao and B. Raveau, *Transition Metal Oxides*, VCH Publishers, NY, 1995.
14 S. K. Pati, S. Ramasesha and D. Sen, in *Magnetism: Molecules to Materials IV*, ed. J. S. Miller and M. Drillon, Wiley-VCH, Weinheim, 2002, pp. 119–171,.
15 P. Day, *Molecules into Materials: Case Studies in Materials Chemistry – Mixed Valency, Magnetism and Superconductivity*, World Scientific, Singapore, 2007.
16 S. K. Pati, *Phys. Rev. B: Condens. Matter Mater. Phys.*, 2003, **67**, 184411.
17 D. Parihari and S. K. Pati, *Phys. Rev. B: Condens. Matter Mater. Phys.*, 2004, **70**, 180403(R).
18 S. Soulette, M. Drillon, P. Rabu and S. K. Pati, *Phys. Rev. B: Condens. Matter Mater. Phys.*, 2004, **70**, 054410.
19 L. Capogna, M. Mayr, P. Horsch, M. Raichle, R. K. Kremer, M. Sofin, A. Maljuk, M. Jansen and B. Keimer, *Phys. Rev. B: Condens. Matter Mater. Phys.*, 2005, **71**, 140402.
20 A. C. Mclaughlin, J. A. McAllister, L. D. Stout and J. P. Attfield, *Phys. Rev. B: Condens. Matter Mater. Phys.*, 2002, **65**, 172506.
21 A. Koga and N. Kawakami, *Phys. Rev. B: Condens. Matter Mater. Phys.*, 2001, **63**, 144432.
22 L. Capriotti, A. E. Trumper and S. Sorellla, *Phys. Rev. Lett.*, 1999, **82**, 3899.

23 S. Nakatsuji, Y. Nambu, H. Tonomura, O. Sakai, S. Jonas, C. Broholm, H. Tsunetsugu, Y. Qiu and Y. Maeno, *Science*, 2005, **309**, 1697.
24 C. N. R. Rao, E. V. Sampathkumaran, R. Nagarajan, G. Paul, J. N. Behera and A. Choudhury, *Chem. Mater.*, 2004, **16**, 1441.
25 D. G. Nocera, B. M. Barlett, D. Grohol, D. Papoutsakis and M. P. Shores, *Chem.–Eur. J.*, 2004, **10**, 3850.
26 S. T. Bramwell and M. J. P. Gingras, *Science*, 2001, **294**, 1495.
27 R. R. P. Singh and D. A. Huse, *Phys. Rev. Lett.*, 1992, **68**, 1766.
28 F. Nori, E. Gagliano and S. Bacci, *Phys. Rev. Lett.*, 1992, **68**, 240.
29 A. S. Wills, V. Dupuis, E. Vincent, J. Hammann and R. Calemczuk, *Phys. Rev. B: Condens. Matter Mater. Phys.*, 2000, **62**, R9264.
30 T. Kato, K. Hondou and K. Iio, *J. Magn. Magn. Mater.*, 1998, **177**, 591.
31 T. Hasegawa, M. Inui, K. Hondou, Y. Fujiwara, T. Kato and K. Iio, *J. Alloys Compd.*, 2004, **364**, 199.
32 N. Oba, C. Michioka, M. Kato, K. Yoshimura and Ko Mibu, *J. Phys. Chem. Solids*, 2005, **66**, 1438.
33 X. Obradors, A. Labarta, A. Isalgué, J. Tejada, J. Rodriguez and M. Pernet, *Solid State Commun.*, 1988, **65**, 189; F. Ladieu, F. Bert, V. Dupuis, E. Vincent and J. Hammann, *J. Phys.: Condens. Matter*, 2004, **16**, 5735.
34 N. Rogado, G. Lawes, D. A. Huse, A. P. Ramirez and R. J. Cava, *Solid State Commun.*, 2002, **124**, 229; N. Rogado, M. K. Haas, G. Lawes, D. A. Huse, A. P. Ramirez and R. J. Cava, *J. Phys.: Condens. Matter*, 2003, **15**, 907.
35 A. M. Todea, A. Merca, H. Bogge, J. Slageren, M. Dressel, L. Engelhardt, M. Luban, T. Glaser, M. Henry and A. Muller, *Angew. Chem., Int. Ed.*, 2007, **119**, 6218.
36 D. A. V. Griend, S. Boudin, V. Caignaert, K. Poeppelmeier, Y. Wang, V. P. Dravid, M. Azuma, M. Takano, Z. Ho and J. D. Jorgensen, *J. Am. Chem. Soc.*, 1999, **121**, 4787.
37 M. D. Nunez-Regueiro, C. Lacroix and B. Canals, *Phys. Rev. B: Condens. Matter Mater. Phys.*, 1996, **54**, 736.
38 Y. R. Wang, *Phys. Rev. B: Condens. Matter Mater. Phys.*, 1992, **45**, 12608, and references cited therein.
39 R. R. P. Singh and D. A. Huse, *Phys. Rev. Lett.*, 1992, **68**, 1766.
40 V. O. Cheranovski, T. G. Schmalz and D. J. Klein, *J. Chem. Phys.*, 1994, **101**, 5841.
41 S. K. Pati and R. R. P. Singh, *Phys. Rev. B: Condens. Matter Mater. Phys.*, 1999, **60**, 7695.
42 F. Mila, *Phys. Rev. Lett.*, 1998, **81**, 2356.
43 C. Waldtmann, H. U. Everts, B. Bernu, P. Sindzingre, C. Lhuillier, P. Lecheminant and L. Pierre, *Eur. Phys. J. B*, 1998, **2**, 501.
44 S. K. Pati and C. N. R. Rao, *J. Chem. Phys.*, 2005, **123**, 234703.
45 B. M. Bartlett and D. G. Nocera, *J. Am. Chem. Soc.*, 2005, **127**, 8985.
46 M. P. Shores, E. A. Nytko, B. M. Bartlett and D. G. Nocera, *J. Am. Chem. Soc.*, 2005, **127**, 13462.
47 E. A. Nytko, J. S. Helton, P. Muller and D. G. Nocera, *J. Am. Chem. Soc.*, 2008, **130**, 2922.
48 C. N. R. Rao, G. Paul, A. Choudhury, E. V. Sampathkumaran, A. K. Raychaudhuri, S. Ramasesha and I. Rudra, *Phys. Rev. B: Condens. Matter Mater. Phys.*, 2003, **67**, 134425.
49 I. D. Brown and D. Altermatt, *Acta Crystallogr., Sect. B: Struct. Sci.*, 1985, **41**, 244.
50 (*a*) J. Selbin, L. H. Holmes and S. P. McGlynn, *J. Inorg. Nucl. Chem.*, 1963, **25**, 1359; (*b*) K. Nakamoto, *Infrared and Raman Spectra of Inorganic and Coordination Compounds*, Wiley, New York, 1978.
51 (*a*) J. E. Dutrizac and S. Kaiman, *Can. Mineral.*, 1976, **14**, 151; (*b*) F. C. Hawthrone, S. V. Krivovichev and P. C. Burns, *Rev. Mineral. Geochem.*, 2000, **40**, 1.
52 N. E. Brese and M. O'Keeffe, *Acta Crystallogr., Sect. B: Struct. Sci.*, 1991, **47**, 192.
53 N. N. Greenwood and T. C. Gibb, *Mössbauer Spectroscopy*, Chapman and Hall, London, 1971.
54 G. M. Sheldrick, *SHELXS-86, Computer Program For Crystal Structure Determination*, University of Gottingen, Gottingen, Germany, 1986.
55 G. M. Sheldrick, *SHELXS-PLUS, Computer Program For Crystal Structure Solution And Refinement*, University of Gottingen, Gottingen, Germany, 1993.
56 A. S. Wills and A. Harrison, *J. Chem. Soc., Faraday Trans.*, 1996, **92**, 2161.
57 A. S. Wills, A. Harrison, C. Ritter and R. I. Smith, *Phys. Rev. B: Condens. Matter Mater. Phys.*, 2000, **61**, 6156.
58 A. Harrison, A. S. Wills and C. Ritter, *Phys. B (Amsterdam, Neth.)*, 1998, **241–243**, 6156.
59 A. S. Wills, A. Harrison, S. A. M. Mentink, T. E. Mason and Z. Tun, *Europhys. Lett.*, 1998, **42**, 325.
60 G. S. Oakley, D. Visser, J. Frunzke, K. H. Andersen, A. S. Wills and A. Harrison, *Phys. B (Amsterdam, Neth.)*, 1999, **267–268**, 142.
61 D. Grohol and D. G. Nocera, *J. Am. Chem. Soc.*, 2002, **124**, 2640.
62 D. Grohol, D. G. Nocera and D. Papoutsakis, *Phys. Rev. B: Condens. Matter Mater. Phys.*, 2003, **67**, 064401.
63 T. Inami, S. Maegawa and M. Takano, *J. Magn. Magn. Mater.*, 1998, **177–181**, 752.
64 T. Inami, M. Nishiyama, S. Maegawa and Y. Oka, *Phys. Rev. B: Condens. Matter Mater. Phys.*, 2000, **61**, 12181.
65 J. N. Behera and C. N. R. Rao, *Dalton Trans.*, 2007, 669.
66 J. N. Behera and C. N. R. Rao, *J. Am. Chem. Soc.*, 2006, **128**, 9334.
67 J. N. Behera, A. Sundaresan, S. K. Pati and C. N. R. Rao, *ChemPhysChem*, 2007, **8**, 217.
68 D. Grohol, D. Papoutsakis and D. G. Nocera, *Angew. Chem., Int. Ed.*, 2001, **40**, 1519.
69 D. Papoutsakis, D. Grohol and D. G. Nocera, *J. Am. Chem. Soc.*, 2002, **124**, 2647.
70 J. N. Behera, G. Paul, A. Choudhury and C. N. R. Rao, *Chem. Commun.*, 2004, 456.
71 C. Livage, C. Egger and G. Ferey, *Chem. Mater.*, 1999, **11**, 1546.
72 C. Broholm, G. Aeppli, G. P. Espinosa and A. S. Cooper, *Phys. Rev. Lett.*, 1990, **65**, 3173.
73 S. H. Lee, C. Broholm, G. Aeppli, T. G. Perring, B. Hessen and A. Taylor, *Phys. Rev. Lett.*, 1996, **76**, 4424.
74 A. Keren, P. Mendels, M. Horvatic, F. Ferrer, Y. J. Uemura, M. Mekata and T. Asano, *Phys. Rev. B: Condens. Matter Mater. Phys.*, 1998, **57**, 10745.
75 A. Keren, Y. J. Uemura, G. Luke, P. Mendels, M. Mekata and T. Asano, *Phys. Rev. Lett.*, 2000, **84**, 3450.
76 A. P. Ramirez, G. P. Espinosa and A. S. Cooper, *Phys. Rev. Lett.*, 1990, **64**, 2070.
77 C. N. R. Rao, E. V. Sampathkumaran, R. Nagarajan, G. Paul, J. N. Behera and A. Choudhury, *Chem. Mater.*, 2004, **16**, 1441.
78 G. Paul, A. Choudhury and C. N. R. Rao, *Chem. Commun.*, 2002, 1904.
79 R. Chitra, S. K. Pati, D. Sen, H. R. Krishnamurthy and S. Ramasesha, *Phys. Rev. B: Condens. Matter Mater. Phys.*, 1995, **52**, 6581.
80 S. Mohakud, S. K. Pati and S. Miyashita, *Phys. Rev. B: Condens. Matter Mater. Phys.*, 2007, **76**, 014435.
81 S. K. Pati, R. Chitra, D. Sen, H. R. Krishnamurthy and S. Ramasesha, *Europhys. Lett.*, 1996, **33**, 707.
82 S. K. Pati, R. Chitra, D. Sen, S. Ramasesha and H. R. Krishnamurthy, *J. Phys.: Condens. Matter*, 1997, **9**, 219.
83 C. Zeng and V. Elser, *Phys. Rev. B: Condens. Matter Mater. Phys.*, 1990, **42**, 8436.
84 K. Hida, *J. Phys. Soc. Jpn.*, 2000, **69**, 4003.
85 T. Moriya, *Phys. Rev.*, 1960, **120**, 91.
86 M. Elhajal, B. Canals and C. Lacroix, *Phys. Rev. B: Condens. Matter Mater. Phys.*, 2002, **66**, 014422.
87 S. Mohakud, K. Hijii, S. Miyashita and S. K. Pati, unpublished.
88 A. Muller, M. Luban, C. Shroder, R. Modler, P. Kogerler, M. Axenovich, J. Schnack, P. Canfield, S. Bud'ko and N. Harrison, *ChemPhysChem*, 2001, **2**, 517.
89 A. Muller, A. M. Todea, J. Slageren, M. Dressel, H. Bogge, M. Schmidtmann, M. Luban, L. Engelhardt and M. Rusu, *Angew. Chem., Int. Ed.*, 2005, **44**, 3857.
90 I. Rousochatzakis, A. M. Lauchli and F. Mila, *Phys. Rev. B: Condens. Matter Mater. Phys.*, 2008, **77**, 094420.
91 M. A. Subramanian, G. Aravamudan and G. V. Subba Rao, *Prog. Solid State Chem.*, 1983, **15**, 55.
92 M. J. Harris and M. P. Zinkin, *Mod. Phys. Lett. B*, 1996, **10** 417.
93 M. Enjalran, M. J. P. Gingras, Y.-J. Kao, A. D. Maestro and H. R. Molavian, *J. Phys.: Condens. Matter*, 2004, **16**, S673.
94 R. Moessner and A. R. Ramirez, *Phys. Today*, 2006, **59**, 24.
95 S. Maekawa, T. Tohyama, S. E. Barnes, S. Ishihara, W. Koshibae and G. Khaliullin, *Physics of Transition Metal Oxides*, Springer-Verlag, Berlin, 2004.

# VI. NANOMATERIALS

*Commentary by C.N.R. Rao*

Chemistry plays a major role in the study of nanomaterials.[1,2] Nanoparticles which constitute an important class of nanomaterials are not new to chemistry.[3,4] One of the earliest experiments on nanoparticles was carried out by Michael Faraday who prepared sols of gold and other noble metals. Nanoparticles of a variety of metals, semiconductors and ceramic materials have been prepared and characterized in recent years.[4,5] An aspect of great interest concerns the size-dependent properties of nanoparticles.[3,4] Quantum confinement in metal and semiconductor nanoparticles have been investigated widely. Thus, very small metal nanoparticles cease to be metallic and show different reactivities compared to the bulk. Semiconductor nanoparticles exhibit size-dependent energy gaps and the resulting shifts in electronic absorption and emission spectra. Metal quantum dots exhibit Coulomb blockade. The articles in this section discuss the synthesis, physical properties, electronic structure, assembly and chemical reactivity of nanoparticles. Certain unusual properties such as ferromagnetism exhibited universally by nanoparticles of the otherwise nonmagnetic oxides are presented.

Soon after the laboratory synthesis of fullerenes in 1990, carbon nanotubes were described by Iijima. The advent of carbon nanotubes gave great impetus to investigate various one-dimensional materials.[6] We initiated investigations of carbon nanotubes (CNTs) in the early 1990s that included synthesis, opening, filling and functionalization of nanotubes. An important discovery was the precursor synthesis of carbon nanotubes by employing organometallics. This method also yielded aligned CNT bundles and junction nanotubes. Doping CNTs with nitrogen was accomplished for the first time in 1997, followed by boron doping. Properties of CNT's including those of single-walled and double-walled nanotubes as well as Y-junction nanotubes have been investigated. In this section, representative articles covering some of the above investigations of CNTs are included.

In the last few years, nanotubes of several inorganic layered materials such as $MoS_2$, $NbS_2$ and BN have been prepared and characterized.[6,7] The area of inorganic nanotubes has been reviewed in one of the articles and nanotubes of $BC_4N$ have been described in another.

Nanowires constitute a major family of one-dimensional materials. Besides confinement, they exhibit several interesting properties. Inorganic nanowires can be prepared by several methods.[6] The growth mechanism of nanorods has been examined in one of the articles.

Functionalization and solubilization of nanomaterials are important components of nanomaterials chemistry. Equally, the use of the liquid–liquid interface to generate nanofilms of inorganic materials has been a useful discovery. This method is described in detail in one of the articles.

Graphene is a current hot topic in condensed matter science, along with its inorganic analogues. A few of the articles in this volume relate to graphene and the inorganic analogues.[8]

## References

1. C.N.R. Rao, A. Müller and A.K. Cheetham (Eds.), *Chemistry of Nanomaterials*, Vols. 1 & 2, Wiley-VCH, 2004.
2. C.N.R. Rao, A. Müller and A.K. Cheetham (Eds.), *Nanomaterials Chemistry, Recent Developments and New Directions*, Wiley-VCH, 2007.
3. J. Jortner and C.N.R. Rao, *Pure Appl. Chem.* **74**, 1491 (2002).
4. C.N.R. Rao, P.J. Thomas and G.U. Kulkarni, *Nanocrystals: Synthesis, Properties and Applications*, Springer-Verlag, 2007.
5. C.N.R. Rao, H.S.S.R. Matte, R. Voggu and A. Govindaraj, *Dalton Trans.* **41**, 5089 (2012).
6. C.N.R. Rao and A. Govindaraj, *Nanotubes and Nanowires*, Royal Society of Chemistry, London, Second Edition, 2012.
7. C.N.R. Rao and A. Govindaraj, *Adv. Mater.* **21**, 4208 (2009).
8. C.N.R. Rao and A.K. Sood (Eds.), *Graphene*, Wiley-VCH, Weinheim, 2013.

CONCEPTS

# Size-Dependent Chemistry: Properties of Nanocrystals

C. N. R. Rao,*[a] G. U. Kulkarni,[a] P. John Thomas,[a] and Peter P. Edwards*[b]

**Abstract:** Properties of materials determined by their size are indeed fascinating and form the basis of the emerging area of nanoscience. In this article, we examine the size dependent electronic structure and properties of nanocrystals of semiconductors and metals to illustrate this aspect. We then discuss the chemical reactivity of metal nanocrystals which is strongly dependent on the size not only because of the large surface area but also a result of the significantly different electronic structure of the small nanocrystals. Nanoscale catalysis of gold exemplifies this feature. Size also plays a role in the assembly of nanocrystals into crystalline arrays. While we owe the beginnings of size-dependent chemistry to the early studies of colloids, recent findings have added a new dimension to the subject.

**Keywords:** colloids · nanostructures · self-assembly · semiconductors

## Introduction

Steric effects—arising from bulky groups—are well known as key factors in determining the reactivity of organic molecules. Otherwise, we do not ordinarily concern ourselves with the physical dimension of a system as a factor in determining its intrinsic properties except in intercalation chemistry or some such situation where the pore or cavity in a host lattice molecule can accommodate guest species of a particular size. Size, however, becomes *the* sole controlling factor while dealing with the science and application of the so-called nanoparticles, that cover a size range between 1–100 nm. Nanoparticles within this size domain are thus intermediate between the atomic and molecular size regimes on one hand, and the macroscopic, bulk on the other. In this regime, size-dependent properties manifest themselves when the size of an individual particle is sufficiently small. A key aspect of the rapidly developing area of nanoscience and nanotechnology concerns itself with the size dependence of intrinsic properties of materials. It is, therefore, instructive to look at typical properties of metals and semiconductors in the size-dependent regime of nanoparticles. We shall discuss three important aspect of the nanocrystals of these materials, electronic structure and properties, chemical reactivity and self-assembly. Size dependent structural and thermodynamic properties of nanoparticles such as bond lengths, melting point and specific heat have already been reviewed[1] in the literature and will not be discussed in any detail in the present article.

Two centuries ago the study of nanoscale solid particles, dispersed within a liquid host, played a pivotal role in establishing colloid science. During the final decades of the last century, colloid science was, perhaps, something of an intellectual backwater—with one or two notable exceptions. However, significant advances in both experimental and theoretical aspects of the subject, and, of course, the emergence of, and explosion of interest in, the broad area of nanoscience and nanotechnology, have now set the scene for a renaissance in colloid science. Interestingly, it does appear that many scientists in the modern field of nanoscience may not even recognize its colloidal source. This is unfortunate, since there is a wealth of information and expertise in that old and venerable science.

An important example of the parentage of the modern subject of metal nanoparticles derives from the work of Faraday in the 1850s. During that period, Faraday carried out groundbreaking studies of nanoscale gold particles in aqueous solution. He established the first quantitative basis for the area, noting that these colloidal metal sols ("pseudosolutions") are thermodynamically unstable, and that the individual gold nanoparticles must be stabilized kinetically against aggregation. Once the nanoparticles coagulate, the process cannot be reversed. Remarkably, Faraday also identified the very essence of the nature of colloidal, nanoscale particles of metals; specifically, for the case of gold, he concluded (in

[a] Prof. Dr. C. N. R. Rao, Prof. Dr. G. U. Kulkarni, P. J. Thomas
Chemistry and Physics of Materials Unit
Jawaharlal Nehru Center for Advanced Scientific Research
Jakkur P.O., Bangalore, 560 064 (India)
E-mail: cnrrao@jncasr.ac.in

[b] Prof. Dr. P. P. Edwards
School of Chemistry
University of Birmingham, Edgabaston,
Birmingham, B15 2TT (UK)
E-mail: p.p.edwards@bham.ac.uk

  0947-6539/02/0801-0029 $ 17.50+.50/0 

1857!) ... "*the gold is reduced in exceedingly fine particles which becoming diffused, produce a beautiful fluid ... the various preparations of gold whether ruby, green, violet or blue ... consist of that substance in a metallic divided state*".

During that century, colloidal phenomena played a pivotal role in the genesis of physical chemistry by establishing a connection between descriptive chemistry and theoretical physics. For example, Einstein provided the relationship between Brownian motion and diffusion coefficient of colloidal particles.

Today, the overwhelming importance now associated with the nanoscale in both science and technology means that the scene is once again set for this key subject to impact upon the development of not only chemistry, but also physics and materials science. In the following sections we attempt to highlight a few of the key issues relating to nanparticles where size determines their properties.

**Electronic structure and properties**: The electronic structure of a nanocrystal critically depends on its very size. For small particles, the electronic energy levels are not continuous as in bulk materials, but discrete, due to the confinement of the electron wavefunction because of the physical dimensions of the particles (see Figure 1). The average electronic energy level spacing of successive quantum levels, $\delta$, known as the so-called Kubo gap, is given by, $\delta = 4E_f/3n$, where $E_f$ is the Fermi energy of the bulk material and $n$ is total number of valence electrons in the nanocrystal. Thus, for an individual silver nanoparticle of 3 nm diameter containing approximately one thousand silver atoms, the value of $\delta$ would be 5–10 meV. Since the thermal energy at room temperature, $kT \cong 25$ meV, a 3 nm particle would be metallic ($kT > \delta$). At low temperatures, however, the level spacings specially in small particles, may become comparable to $kT$, rendering them nonmetallic.[2] Because of the presence of the Kubo gap in individual nanoparticles, properties such as electrical conductivity and magnetic susceptibility exhibit quantum size effects.[3] The resultant discreteness of energy levels also brings about fundamental changes in the characteristic spectral features of the nanoparticles, especially those related to the valence band.

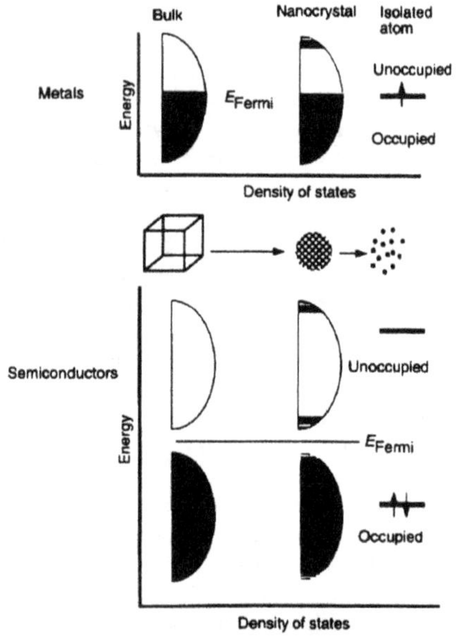


Figure 1. Density of states for metal (a) and semiconductor (b) nanocrystals. In each case, the density of states is discrete at the band edges. The Fermi level is in the center of a band in a metal, and so $kT$ may exceed the electronic energy level spacing even at room temperatures and small sizes. In contrast, in semiconductors, the Fermi level lies between two bands, so that the relevant level spacing remains large even at small sizes. The HOMO–LUMO gap increases in semiconductor nanocrystals of smaller sizes.

Extensive investigations of metal nanocrystals of various sizes obtained, for example by the deposition of metals on amorphised graphite and other substrates, by X-ray photoelectron spectroscopy and related techniques[4, 5] have yielded valuable information on their electronic structure. An important result from these experiments is that as the metal particle size decreases, the core-level binding energy of metals such as Au, Ag, Pd, Ni and Cu increases sharply. This is shown in the case of Pd in Figure 2, where the binding energy

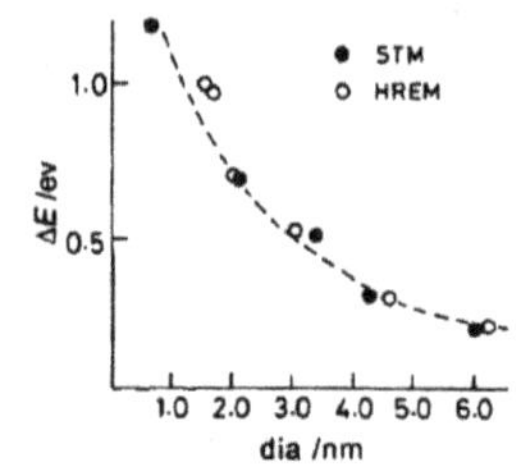


Figure 2. Variation of the shift, $\Delta E$, in the core-level binding energy (relative to the bulk metal value) of Pd with the nanoparticle diameter. The diameters were obtained from HREM and STM images (reproduced with permission from ref. [3]).

increases by over 1 eV at small size. The variation in the binding energy is negligible at large coverages or particle size, since the binding energies are close to those of the bulk, macroscopic metals. The increase in the core-level binding energy in small particles occurs due to the poor screening of the core-hole and is a manifestation of the *size-induced metal-nonmetal transition* in nanocrystals. Further evidence for the occurrence of such a metal-nonmetal transition driven by the size of the individual particle is provided by other electron spectroscopic techniques such as UPS, BIS.[5-7] All these measurements indicate that an electronic gap manifests itself for a nanoparticle having diameters of 1–2 nm possessing $300 \pm 100$ atoms.

Photoelectron spectroscopic measurements[6] on mass-selected $Hg_n$ nanoparticles ($n = 3$ to 250) in the gas phase reveal that the characteristic HOMO–LUMO (s–p) energy gap decreases gradually from ~3.5 eV for $n = 3$ to ~0.2 eV for $n = 250$, as shown in Figure 3. The band gap closure is predicted at $n \sim 400$. The metal–nonmetal transition in gaseous Hg nanoparticles was examined by Rademann and co-workers[7] by measuring the ionization energies (IE). For $n < 13$, the dependence of IE on $n$ suggested a different type of bonding. A small Hg particle with atoms in the $6s^26p^0$ configuration

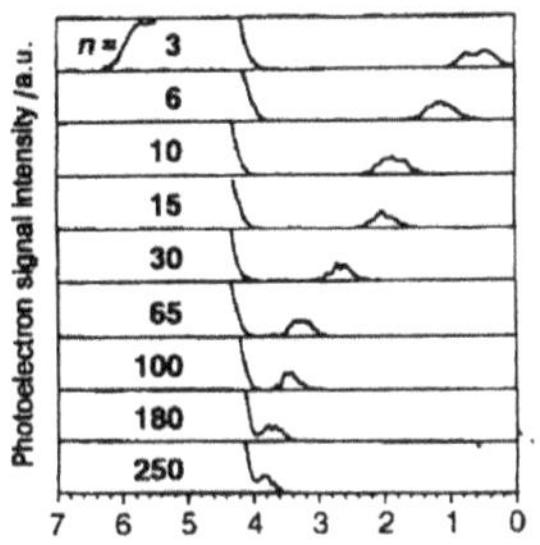


Figure 3. Photoelectron spectra of Hg clusters of varying nuclearity. The 6p feature moves gradually towards the Fermi level, emphasizing that the band gap shrinks with increase in cluster size (reproduced with permission from ref. [6]).

held together by relatively weak van der Waals forces, is essentially nonmetallic. As the nanoparticle grows in size, the atomic 6s and 6p levels broaden into bands and an insulator–metal transition appears to occur—driven by the physical dimensions of the individual particle. Note that this is the same element, Hg, behaving as either a metal or a nonmetal, depending upon its physical size!

The electronic absorption spectrum of metal nanocrystals in the visible region is dominated by the plasmon band. This absorption is due to the collective excitation of the itinerant electron gas on the particle surface and is characteristic of a metal nanocrystal of the given size. In colloids, surface plasmon excitations impart characteristic colours to the metal sols, the beautiful wine-red colour of gold sols being well-known. Gold nanocrystals of varying diameters between 2 and 4 nm exhibit distinct bands around ~525 nm, the intensity of which increases with size.[9, 10] The intensity of this feature becomes rather small in the case of 1 nm diameter particles basically due to a reduced less number of "itinerant" electrons in the electron cloud. With a change in temperature, the intensity of the plasmon band decreases as seen in the case of Au nanocrystals.[10] In contrast to the situation for metals, exciton peaks dominate the absorption of semiconductor nanocrystals in the visible region.[11, 12a] Thus CdS, a yellow solid, exhibits an exciton peak around 600 nm, which gradually shifts into the UV region as the nanocrystal diameters are varied below 10 nm (see Figure 4).[12b] Similar effects have been observed in the case of PbS and ZnO.[12b]

Direct information on the gap states in nanocrystals of metals and semiconductors is obtained by scanning tunneling spectroscopy (STS). This technique provides the desired sensitivity and spatial resolution making it possible to carry out tunneling spectroscopic measurements on individual particles. A systematic STS study of Pd, Ag, Cd and Au nanoparticles of varying sizes deposited on a graphite substrate has been carried out under ultrahigh vacuum conditions, after having characterized the nanoparticles by XPS and STM.[13] The I-V spectra of bigger particles were featureless while those of the small particles (< 1 nm) showed well-defined peaks on either side of zero-bias due to the presence of a gap. Ignoring gap values below 25 meV (~$kT$), it is seen that small particles of ≤ 1 nm diameter are in fact

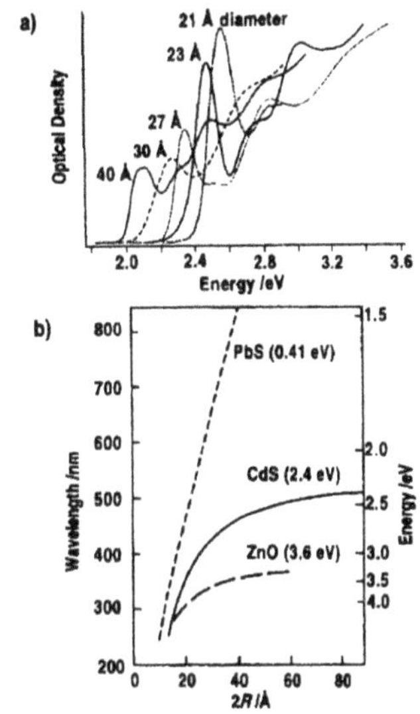


Figure 4. a) Absorption spectra of CdSe nanocrystals (at 10 K) of various diameters (reproduced with permission from ref. [12a]). b) Wavelength of the absorption threshold and band gap as a function of the particle diameter for various semiconductors. The corresponding energy gap in the bulk state is given in parenthesis (reproduced with permission from ref. [12b]).

nonmetallic! (Figure 5a) From the various studies discussed hitherto, it appears that the size-induced metal–insulator transition in metal nanocrystals (Figure 1) occurs in the range of 1–2 nm diameter or 300 ± 100 atoms. The band gap of CdS nanocrystals estimated by the above method yielded a value of 2.9 eV for a 3.1 nm diameter nanocrystal[14] and the gap increases with the decrease in size (Figure 5b).

Theoretical calculations of the electronic structure of metal and semiconductor nanocrystals throw light on the size-

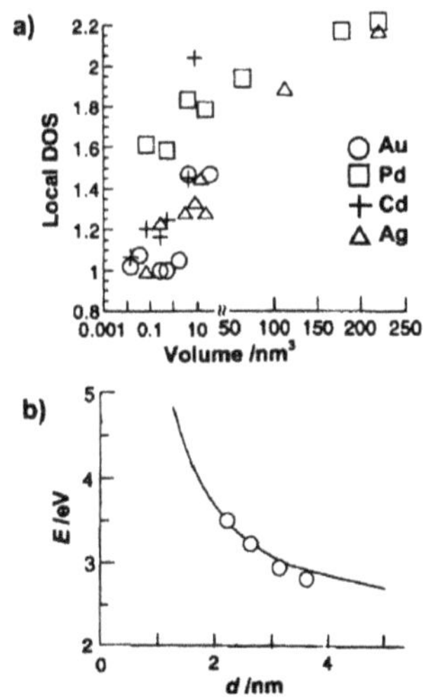


Figure 5. a) Variation of the nonmetallic band gap with nanocrystal size, b) in CdS nanocrystals (reproduced with permission from refs. [13, 14]).

  0947-6539/02/0801-0031 $ 17.50+.50/0 

induced changes in the electronic structure. Rosenblit and Jortner[15] calculated the electronic structure of a model metal cluster and predicted electron localization to occur in a cluster of diameter ~0.6 nm. A molecular orbital calculation on $Au_{13}$ cluster[16] in icosahedral and cuboctahedral structures shows that the icosahedral structure undergoes Jahn–Teller distortion while the cuboctahedral structure does not distort. The onset of the metallic state is barely discernible in the $Au_{13}$ cluster. Relativistic density functional calculations of gold clusters,[17] with $n = 6$ to 147 show that the average interatomic distance increases with the nuclearity of the cluster. The HOMO–LUMO electronic gap decreases with particle size from 1.8 eV for $Au_6$ (~0.5 nm diameter) to 0.3 eV for $Au_{147}$ (~2 nm diameter). Ab initio molecular dynamics simulations of aluminum clusters,[18] with $n = 2-6$, 12, 13, 55 and 147 reveal that the minimum energy structures of $Al_{13}$ and $Al_{55}$ to be distorted icosahedra whereas $Al_{147}$ is a near cuboctahedron. The HOMO–LUMO gap increases from ~0.5 eV for $Al_2$ to ~2 eV for $Al_{13}$; the gap is around 0.25 eV for $Au_{55}$ and decreases to ~0.1 eV for $Au_{147}$. In the case of semiconductor nanocrystals, it is shown using tight binding approximation that the band gaps nearly reach the bulk values at sizes of around 5 nm.[19] The convergence of the cluster properties towards those of the corresponding bulk materials with increase in size is noteworthy.

In a bulk metal, the energy required to add or remove an electron is its work function. In a molecule, the corresponding energies, electron affinity and ionization potential, respectively, are, however, nonequivalent. Nanocrystals being intermediary, the two energies differ only to a small extent,[20] the difference being the charging energy, $U$. This is a Coulombic energy and is different from electronic energy gap. Further, Coulombic states can be similar for both semiconductor and metallic nanocrystals unlike the electronic states. A manifestation of single electron charging is the Coulomb staircase behaviour observed in the tunneling spectra,[21] when a nanocrystal, covered with an insulating ligand shell is held between two tunnel junctions. A typical staircase along with its theoretical fit is shown in Figure 6a. Such measurements have also been carried out on Pd and Au nanocrystals in the size range, 1.5–6.5 nm.[22] The charging energies follow a scaling law[23] of the form, $U = A + B/d$, where A and B are constants, characteristic of the metal and $d$ is the particle diameter(see Figure 6b).

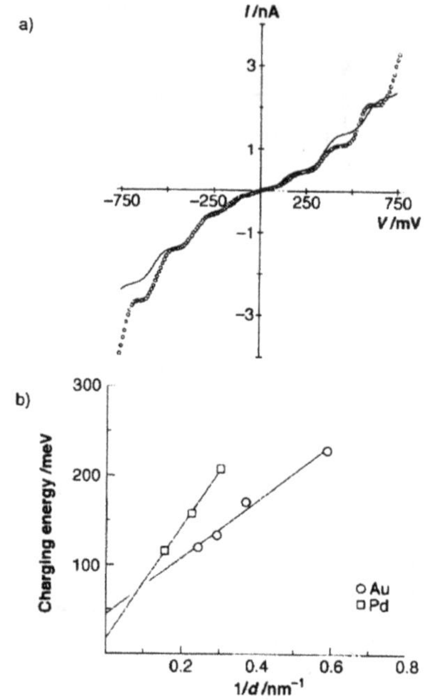


Figure 6. a) I-V characteristics of an isolated 3.3 nm Pd nanocrystal (dotted line) and the theoretical fit (solid line) obtained at 300 K using a semiclassical model according to which, the observed capacitance ($C$) may be resolved into two components $C1$ and $C2$ and the resistance ($R$) into $R1$ and $R2$, such that $C = C1 + C2$, $R = R1 + R2$. For $C1 \ll C2$ and $R1 \ll R2$, the model predicts steps in the measured current to occur at critical voltages, $V_c = n_c e/(C + (1/C)q_0 + e/2)$, where $q_0$ is the residual charge. b) Variation of the charging energies of Pd and Au nanocrystals with inverse diameters ($d$) (reproduced with permission from ref. [22]).

Magnetic properties of nanoparticles of transition metals such as Co, Ni show marked variations with size. It is well known that in the nanometric domain, the coercivity of the particles tends to zero.[23] Thus, the nanocrystals behave as superparamagnets with no associated coercivity or retentivity. The blocking temperature which marks the onset of this superparamagnetism also increases with the nanocrystal size. Further, the magnetic moment per atom is seen to increase as the size of a particle decreases[25] (see Figure 7).

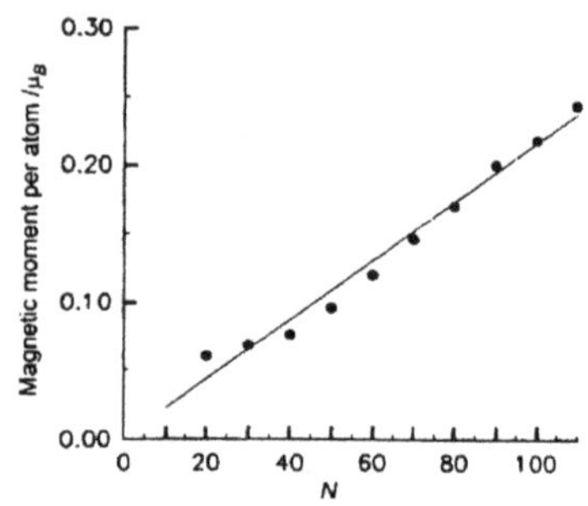


Figure 7. Variation of the magnetic moment per atom as a function of the number of atoms in Co clusters at $T = 170$ K and $H = 0.53$ T. The continuous curve is the theoretical value and the solid circles are the experimental results (reproduced with permission from ref. [25b]).

**Chemical reactivity**: The surface area of nanocrystals increases markedly with the decrease in size. Thus, a small metal nanocrystal of 1 nm diameter will have ~100% of its atoms on the surface. A nanocrystal of 10 nm diameter on the other hand, would have only 15% of its atoms on the surface. A small nanocrystal with a higher surface area would be expected to be more reactive. Furthermore, the qualitative change in the electronic structure arising due to quantum confinement in small nanocrystals will also bestow unusual catalytic properties on these particles, totally different from those of the bulk metal. We illustrate these important aspects with a few examples from the recent literature.

A low temperature study[26] of the interaction of elemental $O_2$ with Ag nanocrystals of various sizes (Figure 8) has



0947-6539/02/0801-0032 $ 17.50+.50/0

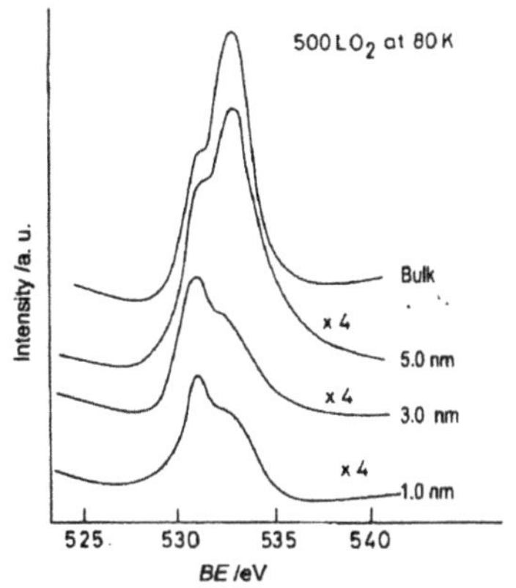


Figure 8. Change in the O(1s) spectra of Ag clusters exposed to 500 L $O_2$ at 80 K. The diameters of the clusters have been estimated from metal coverage. The lower binding energy peak at 531 eV corresponds to $O^-$ while that at 533 eV arises due to molecular oxygen (reproduced with permission from ref. [26]).

revealed the capability of smaller nanocrystals to dissociate dioxygen to atomic oxygen species. On bulk Ag, the adsorbed oxygen species at 80 K is predominantly $O_2^-$. This interaction of $O_2$ with Ag—*dependent on its particle size*—is remarkable. Another important example is the reaction of $H_2S$ with Ni nanocrystals giving rise to $S^{2-}$ species, with nanocrystals of different sizes exhibiting different temperature profiles (see Figure 9). Unlike bulk nickel, small nanocrystals show less dependence in their catalytic activity on ambient temperature.

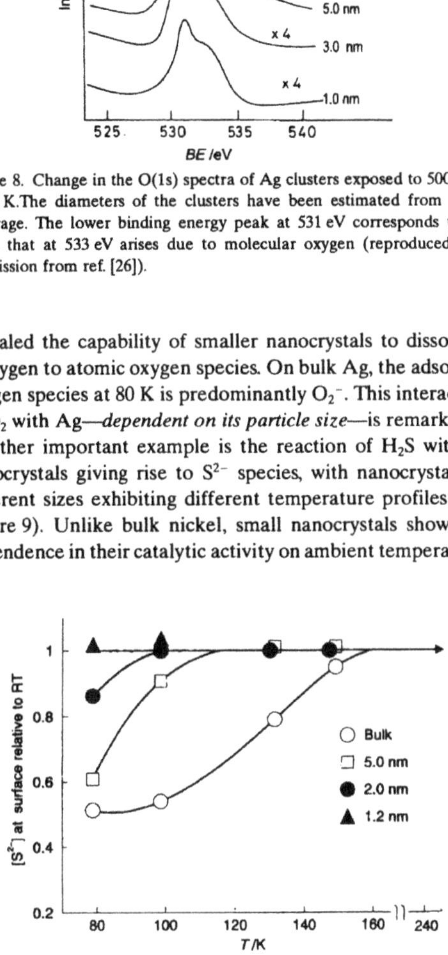


Figure 9. Variation of the normalized areas of the signal for the core-level transitions of $S^{2-}$ with temperature for different sizes of Ni clusters deposited on graphite (reproduced with permission from ref. [26]).

The ability of Cu, Pd and Ni nanoparticles to absorb CO has been thoroughly investigated. Carbon monoxide from a bulk Cu surface desorbs above 250 K. Small Cu particles, however, retain CO up to much higher temperatures (Figure 10).[27] A similar observation has been made in case of Pd particles.[28] The results obtained with Ni particles are more even interesting. In addition to showing a trend similar to the above, small Ni particles are also capable of dissociating CO to form carbidic species on the particle surface (Figure 11).[29]

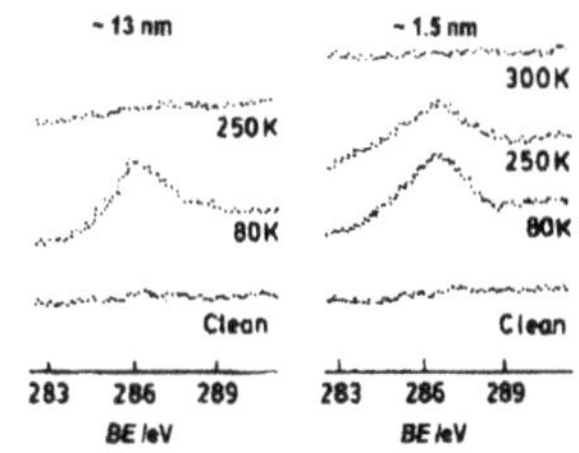


Figure 10. Change in the C(1s) spectra of CO adsorbed on Cu. The feature at 286 eV corresponds to molecularly adsorbed CO (reproduced with permission from ref. [27]).

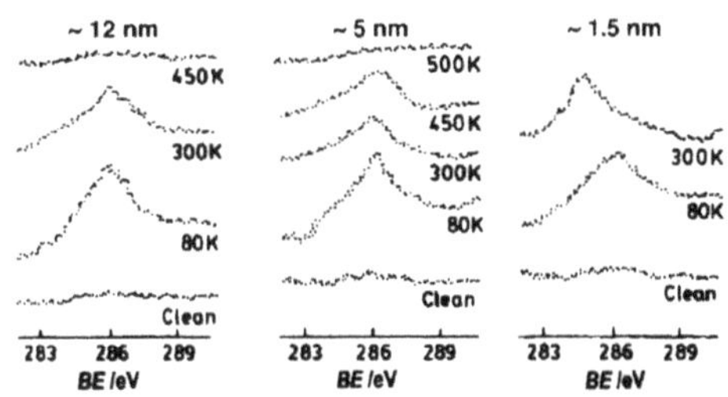


Figure 11. Change in the C(1s) spectra of CO adsorbed on Ni clusters with temperature. The feature at 286 eV corresponds to molecularly adsorbed CO while that at 284 eV arises due to the formation of carbidic species (reproduced with permission from ref. [27]).

This could be due to the Ni(3d) level in small clusters coming close to the anti-bonding energy level of CO($2\pi^*$).

Bulk Au is a noble metal. Goodman and co-workers,[30] however, found that Au nanocrystals supported on a titania surface show a marked size-effect in their catalytic ability for CO oxidation reaction, with Au nanoparticles in the range of 3.5 nm exhibiting the maximum chemical reactivity (Figure 12a). A metal to non-metal transition as observed in the I-V spectra (Figure 12b), as the cluster size is decreased below 3.5 $nm^3$ (consisting of ca. 300 atoms). This result is quite similar to that obtained with Pd particles supported on oxide substrate.[31] In another study of Au particles supported on a zinc oxide surface, smaller particles (<5 nm) exhibited a marked tendency to adsorb CO while those with diameters above 10 nm did not significantly adsorb CO (Figure 13).[32] The increased activity of these metal particles is attributed to the charge transfer between the oxide support and the particle surface.

**Self-assembly of nanocrystals**: Just as individual atoms aggregate to form crystals, nanocrystals themselves act as building units to form particle superlattices. Thus, monodispersed nanocrystals suitably covered by ligands such as alkane thiols, when transferred to a flat substrate, spontaneously assemble into two-dimensional lattices.[33–37] In Figure 14, we show typical arrays of 2.5 and 3.2 nm Pd nanocrystals coated with octane thiol. The diameter of the nanocrystal, *d*, and the length of the protecting ligand, *l*, play an important role in determining the very nature of the assembly.[38, 39] A study of the two-dimensional arrays formed by Pd nanocrystals of varying diameters covered with alkane thiols of different chain lengths has enabled to obtain an experimental stability

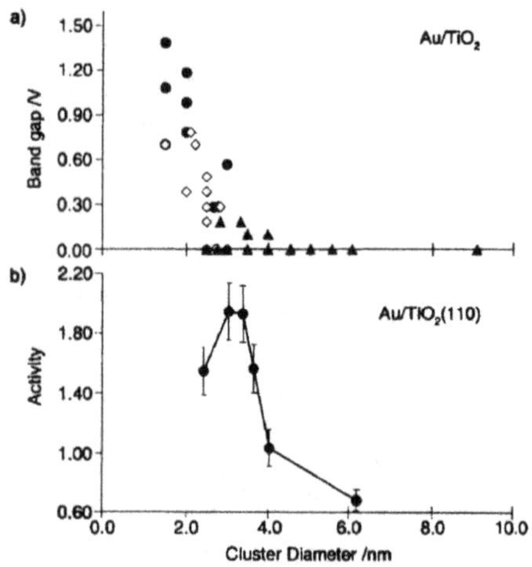


Figure 12. a) Cluster band gap measured by STS as a function of Au cluster size. The band gaps were obtained while the corresponding topographic scan was acquired on various Au coverages ranging from 0.2 to 4.0 ML. two-dimensional clusters ○; three-dimensional clusters, two atom layers in height □; three-dimensional clusters with three atom layers or greater in height △. b) The activity (*CO atoms/total Au atoms*) for CO oxidation at 350 K as a function of the Au cluster diameter supported on a $TiO_2$ surface. The $CO/O_2$ mixture was 1:5 at a total pressure of 40 Torr (reproduced with permission from ref. [30]).

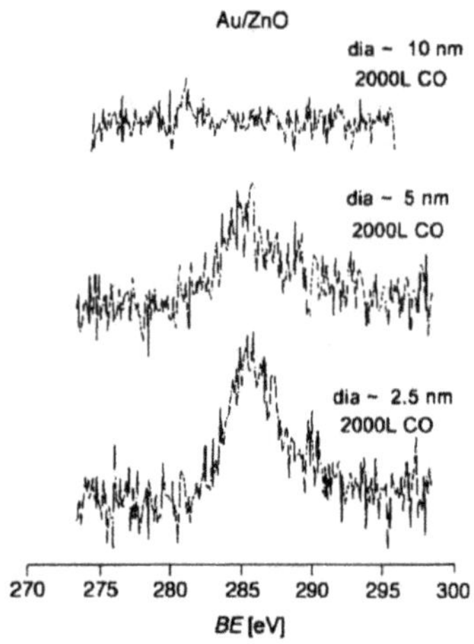


Figure 13. C(1s) core-level spectra of CO adsorbed on Au particles supported on a ZnO substrate. The feature at ~285 eV corresponds to molecularly absorbed CO. The diameters have been obtained from the metal coverages.

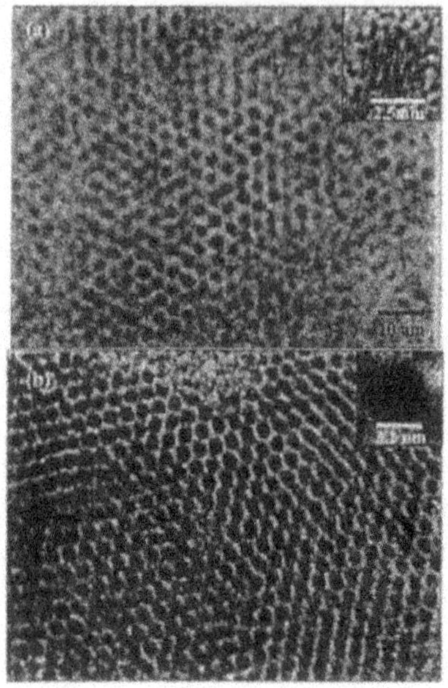

Figure 14. TEM micrograph showing hexagonal arrays of thiolized Pd nanocrystals: a) 2.5 nm, octane thiol, b) 3.2 nm, octane thiol (reproduced with permission from ref. [40]).

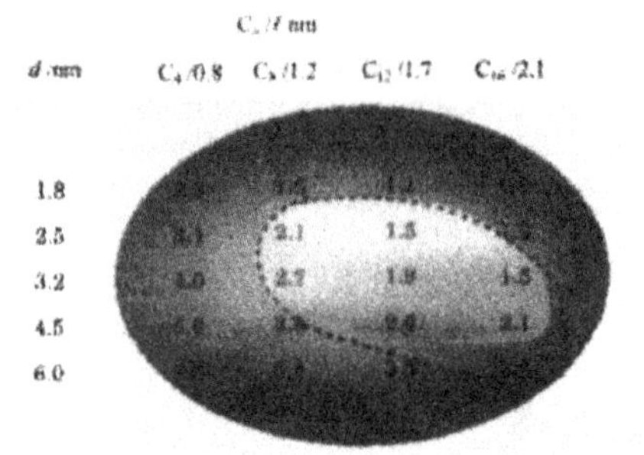

Figure 15. The $d-l$ phase diagram for Pd nanocrystals thiolized with different alkane thiols. The mean diameter, $d$, was obtained from the TEM measurements on as-prepared sols. The length of the thiol, $l$, is estimated by assuming an all-*trans* conformation of the alkane chain. The thiol is indicated by the number of carbon atoms, $C_n$. The bright area in the middle encompasses systems which form close-paced organizations of nanocrystals. The surrounding darker area includes disordered or low-order arrangements of nanocrystals. The area enclosed by the dashed line is derived from calculations from the soft sphere model (reproduced with permission from ref. [40]).

diagram (Figure 15) of the superlattices in terms of $d$ and $l$.[40] In this Figure, the bright area in the middle is the most favourable $d/l$ regime, corresponding to extended close packed organization, such as those illustrated in Figure 14. The $d/l$ values for the most favorable regions are in the range 1.5–3.8. The area shaded dark in Figure 15 includes $d/l$ regime giving rise to various short-range organizations; these are formed when the particles are small and the chain length is large or vice versa. The experimental results have been compared with empirical calculations based on a soft sphere model,[41] involving an attractive van der Waals term and a repulsive steric term (see Figure 15).

The ability to synthesize lattices of nanocrystals have led to explorations of their collective physical properties. Thus, it is observed in the case of Co nanocrystals (5.8 nm) that, accompanying lattice formation, the blocking temperature increases.[42] FePt alloy nanocrystals yield ferromagnetic assemblies for which the coercivity is tunable by controlling the parameters such as Fe:Pt ratio and the particle size.[43] The evolution of collective electronic states in CdSe nanocrystals have been followed by optical spectroscopic methods. Compared with isolated nanocrystals, those in the lattice exhibited

  0947-6539/02/0801-0034 $ 17.50+.50/0 

broader bands.[44] Various investigations have been carried out on the electrical transport properties of nanocrystalline lattices.[45, 46] Heath and co-workers have successfully demonstrated a reversible Mott–Hubbard metal–nonmetal transition with Ag nanocrystals (3 nm) capped with octane thiol.[47] Since then, this has been a subject of intense theoretical study.[48]

## Conclusion

We have hitherto discussed in the earlier sections, electronic structure and properties, chemical reactivity and self-assembly of nanocrystals, particularly those of metals. The discussion should suffice to illustrate how size if a crucial factor in deciding the chemistry in the nano regime. These size dependent properties form the basis of nanoscience, where the properties are exploited for possible applications.


[1] a) J. S. Vermaak, L. W. Mays, D. Kuhlmann-Wilsdorf, *Surf. Sci.* **1968**, *12*, 128; b) G. Apai, J. F. Hamilton, J. Stohor, A. Thompson, *Phys. Rev. Lett.* **1979**, *40*, 165; c) A. A. Montano, G. K. Shenoy, E. E. Apl, W. Schulze, J. Urban, *Phys. Rev. Lett.* **1986**, *56*, 2076; d) H. J. Wasswemann, J. S. Vermaak, *Surf. Sci.* **1970**, *22*, 164; e) P. A. Buffat, J. P. Borel, *Phys. Rev. A* **1976**, *13*, 2287; f) W. P. Halperin, *Rev. Mod. Phys.* **1986**, *58*, 533.

[2] a) P. P.Edwards, R. L. Johnston, C. N. R. Rao in *Metal Clusters in Chemistry* (Eds.: P. Braunstein, G. Oro, P. R. Raithby), Wiley-VCH, **1998**; b) A. I. Kirkland, D. A. Jefferson, D. G. Duff, *Annual Reports C*, Royal Society of Chemistry, **1993**, p. 247.

[3] H. N. Aiyer, V. Vijayakrishnan, G. N. Subanna, C. N. R. Rao, *Surf. Sci.* **1994**, *313*, 392.

[4] a) D. C. Johnson, R. E. Benfield, P. P. Edwards, W. J. H. Nelson, M. D. Vargas, *Nature* **1985**, *314*, 231; b) Y. Volokitin, J. Sinzig, L. J. de Jongh, G. Schmid, M. N. Vargaftik, I. I. Moiseev, *Nature* **1996**, *384*, 624.

[5] V. Vijayakrishnan, A. Chainani, D. D. Sarma, C. N. R. Rao, *J. Phys. Chem.* **1992**, *96*, 8679.

[6] R. Busani, M. Folker, O. Chesnovsky, *Phys. Rev. Lett.* **1998**, *81*, 3836.

[7] K. Rademann, O. D. Rademann, M. Schlauf, V. Even, F. Hensel, *Phys. Rev. Lett.* **1992**, *69*, 3208.

[8] H. Haberland, B. von Issendrof, Y. Yufeng, J. Kolar, G. Thanner, Z. *Phys. D: At. Mol. Clusters* **1993**, *26*, 8.

[9] K. V. Sarathy, G. Raina, R. T. Yadav, G. U. Kulkarni, C. N. R. Rao, *J. Phys. Chem. B* **1997**, *101*, 9876.

[10] S. Link, M. A. El-Sayed, *J. Phys. Chem. B* **1999**, *103*, 4212.

[11] A. P. Alivisatos, *J. Phys. Chem.* **1996**, *100*, 13226.

[12] a) D. M. Mittleman, R. W. Schoenlein, J. J. Shiang, V. L. Colvin, A. P. Alivisatos, C. V. Shank, *Phys. Rev. B* **1994**, *49*, 14435; b) A. Henglein, *Ber. Bunsenges. Phys. Chem.* **1995**, *99*, 903.

[13] C. P. Vinod, G. U. Kulkarni, C. N. R. Rao, *Chem. Phys. Lett.* **1998**, *289*, 329.

[14] M. Miyake, T. Torimoto, T. Sakata, H. Mori, S. Kuwabata, H. Yoneyama, *Langmuir* **1997**, *13*, 742.

[15] M. Rosenblit, J. Jortner, *J. Phys. Chem.* **1994**, *98*, 9365.

[16] R. A. Perez, A. F. Ramos, G. L. Malli, *Phys. Rev. B* **1989**, *39*, 3005.

[17] O. D. Haberlen, S. C. Chung, M. Stener, N. J. Rosch, *J. Chem. Phys.* **1997**, *106*, 5189.

[18] S. H. Yang, D. A. Drabold, J. B. Adams, A. Sachdev, *Phys. Rev. B* **1993**, *47*, 1567.

[19] P. E. Lippens, M. Lannoo, *Phys. Rev. B* **1989**, *39*, 10935.

[20] C. P. Collier, T. Vossmeyer, J. R. Heath, *Annu. Rev. Phys. Chem.* **1998**, *49*, 371.

[21] Single Charge Tunneling, Coulomb Blockade Phenomena in Nanostructures (Eds.: H. Grabert, M. H. Devoret), *NATO-ASI Ser. B* **1992**, *294*.

[22] P. J. Thomas, G. U. Kulkarni, C. N. R. Rao, *Chem. Phys. Lett.* **2000**, *321*, 163.

[23] J. Jortner, *Z. Phys. D: At. Mol. Clusters* **1992**, *24*, 247.

[24] C. P. Bean, J. D. Livingston, *J. Appl. Phys.* **1959**, *30*, 1208.

[25] a) Van de Heer, P. Milani, A. Chatelain, *Z. Phys. D: At. Mol. Clusters* **1991**, *19*, 241; b) S. N. Khanna, S. Linderoth, *Phys. Rev. Lett.* **1991**, *67*, 742.

[26] C. N. R. Rao, V. Vijayakrishnan, A. K. Santra, M. W. J. Prins, *Angew. Chem.* **1992**, *104*, 1110; *Angew. Chem. Int. Ed. Engl.* **1992**, *31*, 1062.

[27] A. K. Santra, S. Ghosh, C. N. R. Rao, *Langmuir* **1994**, *10*, 3937.

[28] E. Gillet, S. Channakhone, V. Matolin, M. Gillet, *Surf. Sci.* **1986**, *152/153*, 603.

[29] a) D. L. Doering, J. T. Dickinson, H. Poppa, *J. Catal.* **1982**, *73*, 91; b) D. L. Doering, H. Poppa, J. T. Dickinson, *J. Catal.* **1982**, *73*, 104.

[30] M. Valden, X. Lai, D. W.Goodman, *Science* **1998**, *281*, 1647.

[31] C. Xu, X. Lai, G. W. Zajac, D. W. Goodman, *Phys. Rev. B* **1997**, *56*, 13464.

[32] Unpublished results from our laboratory.

[33] C. N. R. Rao, G. U. Kulkarni, P. J. Thomas, P. P. Edwards, *Chem. Soc. Rev.* **2000**, *29*, 27.

[34] A. N. Shipway, E. Katz, I. Willner, *Chem. Phys. Chem.* **2000**, *1*, 18.

[35] M. P. Pileni, *New J. Chem.* **1998**, 693.

[36] G. Schmid, L. F. Chi, *Adv. Mater.* **1998**, *10*, 515.

[37] W. P. Wuelfing, F. P. Zamborini, A. C. Templeton, X. Wen, H. Yoon, R. W. Murray, *Chem. Mater.* **2001**, *13*, 87.

[38] R. L. Whetten, M. M. Shafigullin, J. T. Khoury, T. G. Schaaf, I. Vezmar, M. M. Alvarez, A. Wilkinson, *Acc. Chem. Res.* **1999**, *32*, 397.

[39] P. C. Ohara, D. V. Leff, J. R. Heath, W. M. Gelbart, *Phys. Rev. Lett.* **1995**, *75*, 3466.

[40] P. J. Thomas, G. U. Kulkarni, C. N. R. Rao, *J. Phys. Chem. B* **2000**, *104*, 8138.

[41] B. A. Korgel, S. Gullam, S. Conolly, D. Fitzmaurice, *J. Phys. Chem. B* **1998**, *102*, 8379.

[42] a) V. Russier, C. Petit, J. Legrand, M. P. Pileni, *Phys. Rev. B* **2000**, *62*, 3910; b) C. Petit, M. P. Pileni, *Appl. Surf. Sci.* **2000**, *162–163*, 519; c) C. Petit, A. Taleb, M. P. Pileni, *Adv. Mater.* **1998**, *10*, 259.

[43] S. Sun, C. B. Murray, D. Weller, L. Folks, A. Maser, *Science* **2000**, *287*, 1989.

[44] a) M. V. Artemyev, A. I. Bibik, L. I. Gurinovich, S. V. Gopenko, U. Woggon, *Phys. Rev. B* **1999**, *60*, 1504; b) C. R. Kagan, C. B. Murray, M. G. Bawendi, *Phys. Rev. B* **1996**, *54*, 8633; c) M. V. Artemyev, U. Woggon, H. Jaschinski, L. I. Gurinovich, S. V. Gopenko, *J. Phys. Chem. B* **2000**, *104*, 11617.

[45] M. Brust, D. Bethell, D. J. Schiffrin, C. J. Kiely, *Adv. Mater.* **1995**, *7*, 795.

[46] a) D. B. Janes, V. R. Kolagunta, R. G. Osifchin, J. D. Bielefeld, R. P. Andres, J. I. Henderson, C. P. Kubiak, *Superlattices Microstruct.* **1995**, *18*, 275; b) A. W. Snow, H. Wohltjen, *Chem. Mater.* **1998**, *10*, 947; c) Y. Liu, Y. Wang, R. O. Claus, *Chem. Phys. Lett.* **1998**, *298*, 315; d) H. M. Jaeger, personal communication, **2001**.

[47] a) G. Markovich, C. P. Collier, S. E. Hendricks, F. Remacle, R. D. Levine, J. R. Heath, *Acc. Chem. Res.* **1999**, *32*, 415; b) C. Mederios-Riberio, D. A. A. Phlberg, R. S. Williams, J. R. Heath, *Phys. Rev. B* **1999**, *59*, 1633.

[48] F. Remacle, R. D. Levine, *Chem. Phys. Chem.* **2001**, *2*, 20.

PERSPECTIVE www.rsc.org/dalton | Dalton Transactions

# Synthesis of inorganic nanomaterials

C. N. R. Rao,*[a,b] S. R. C. Vivekchand,[a] Kanishka Biswas[a,b] and A. Govindaraj[a,b]

*Received 1st June 2007, Accepted 9th July 2007*
*First published as an Advance Article on the web 6th August 2007*
**DOI: 10.1039/b708342d**

Synthesis forms a vital aspect of the science of nanomaterials. In this context, chemical methods have proved to be more effective and versatile than physical methods and have therefore, been employed widely to synthesize a variety of nanomaterials, including zero-dimensional nanocrystals, one-dimensional nanowires and nanotubes as well as two-dimensional nanofilms and nanowalls. Chemical synthesis of inorganic nanomaterials has been pursued vigorously in the last few years and in this article we provide a perspective on the present status of the subject. The article includes a discussion of nanocrystals and nanowires of metals, oxides, chalcogenides and pnictides. In addition, inorganic nanotubes and nanowalls have been reviewed. Some aspects of core–shell particles, oriented attachment and the use of liquid–liquid interfaces are also presented.

## 1. Introduction

Nanoscience involves a study of materials where at least one of the dimensions is in the 1–100 nm range. Properties of such materials are strongly dependant on their size and shape. Nanomaterials include zero-dimensional nanocrystals, one-dimensional nanowires and nanotubes and two-dimensional nanofilms and nanowalls. Synthesis forms an essential component of nanoscience and nanotechnology. While nanomaterials have been generated by physical methods such as laser ablation, arc-discharge and evaporation, chemical methods have proved to be more effective, as they provide better control as well as enable different sizes, shapes and functionalization. Chemical synthesis of nanomaterials has been reviewed by a few authors,[1–6] but innumerable improvements and better methods are being reported continually in the last few years. In accomplishing the synthesis and manipulation of the nanomaterials, a variety of reagents and strategies have been employed besides a wide spectrum of reaction conditions. In view of the intense research activity related to nanomaterials synthesis, we have prepared this perspective to present recent developments and new directions in this area. In doing so, we have dealt with all classes of inorganic nanomaterials. In writing such an article, it has been difficult to do justice to the vast number of valuable contributions which have appeared in the literature in last the two to three years. We had to be necessarily succinct and restrict ourselves mainly to highlighting recent results.

## 2. Nanocrystals

Nanocrystals are zero-dimensional particles and can be prepared by several chemical methods, typical of them being reduction of salts, solvothermal synthesis and the decomposition of molecular precursors, of which the first is the most common method used in the case of metal nanocrystals. Metal oxide nanocrystals are generally prepared by the decomposition of precursor compounds such as metal acetates, acetylacetonates and cupferronates in appropriate solvents, often under solvothermal conditions. Metal chalcogenide or pnictide nanocrystals are obtained by the reaction of metal salts with a chalcogen or pnicogen source or the decomposition of single source precursors under solvothermal or thermolysis conditions. Addition of suitable capping agents such as long-chain alkane thiols, alkyl amines and trioctylphosphine oxide (TOPO) during the synthesis of nanocrystals enables the control of size and shape. Monodisperse nanocrystals are obtained by post-synthesis size-selective precipitation.

### 2.1 Metals

Reduction of metal salts in the presence of suitable capping agents such as polyvinylpyrrolidone (PVP) is the common method to generate metal nanocrystals. Solvothermal and other reaction conditions are employed for the synthesis, to exercise control over their size and shape of the nanocrystals.[1,2,7] Furthermore, the sealed reaction conditions and presence of organic reagents reduce the possibility of atmospheric oxidation of the nanocrystals. The popular citrate route to colloidal Au, first described by Hauser and Lynn,[8] involves the addition of chloroauric acid to a boiling solution of sodium citrate.[9] The average diameter of the nanocrystals can be varied over the 10–100 nm range by varying the concentration of reactants. Au nanocrystals with diameters between 1 and 2 nm are obtained by the reduction of $HAuCl_4$ with tetrakis(hydroxymethyl)phosphonium chloride (THPC) which also acts as a capping agent.[10] Following the early work of Brust and co-workers,[11] the general practice employed to obtain organic-capped metal nanocrystals is to use a bi-phasic mixture of an organic solvent and the aqueous solution of the metal salt in the presence of a phase-transfer reagent. The metal ion is transferred across the organic–water interface by the phase transfer reagent and subsequently reduced to yield sols of metal nanocrystals. Metal nanocrystals in the aqueous phase can also be transferred to a nonaqueous medium by using alkane thiols to obtain organosols.[12,13] This method has been used to thiolize

[a]*Chemistry and Physics of Materials Unit, DST unit on nanoscience and CSIR Centre of Excellence in Chemistry, Jawaharlal Nehru Centre for Advanced Scientific Research, Jakkur P. O., Bangalore, 560064, India. E-mail: cnrrao@jncasr.ac.in; Fax: +91 80 22082760*
[b]*Solid State and Structural Chemistry Unit, Indian Institute of Science, Bangalore, 560012, India*

 

*C. N. R. Rao obtained his PhD degree from Purdue University and DSc degree from the University of Mysore. He is the Linus Pauling Research Professor at the Jawaharlal Nehru Centre for Advanced Scientific Research and Honorary Professor at the Indian Institute of Science (both at Bangalore). His research interests are in the chemistry of materials. He has authored nearly 1000 research papers and edited or written 30 books in materials chemistry. A member of several academies including the Royal Society and the US National Academy of Sciences, he is the recipient of the Einstein Gold Medal of UNESCO, Hughes Medal of the Royal Society, and the Somiya Award of the International Union of Materials Research Societies (IUMRS). In 2005, he received the Dan David Prize for materials research from Israel and the first India Science Prize.*

*S. R. C. Vivekchand received his BSc degree from The American College, Madurai in 2001. He is a student of the integrated PhD programme of Jawaharlal Nehru Centre for Advanced Scientific Research, Bangalore and received his MS degree in 2004. He has worked primarily on material chemistry aspects of one-dimensional nanomaterials.*

*Kanishka Biswas received his BSc degree from Jadavpur University, Kolkata in 2003. He is a student of the integrated PhD programme of Indian Institute of Science, Bangalore and received his MS degree in 2006. He has worked primarily on the synthesis and characterization of inorganic nanomaterials.*

*A. Govindaraj obtained his PhD degree from University of Mysore and is a Senior Scientific Officer at the Indian Institute of Science, and Honorary Faculty Fellow at the Jawaharlal Nehru Centre for Advanced Scientific Research. He works on different types of nanomaterials. He has authored more than 100 research papers and co-authored a book on nanotubes and nanowires.*

**C. N. R. Rao**

**S. R. C. Vivekchand**

**Kanishka Biswas**

**A. Govindaraj**

Pd nanocrystals with magic numbers of atoms.[14,15] A method to produce gold nanocrystals free from surfactants would be to reduce $HAuCl_4$ by sodium napthalenide in diglyme.[16]

Liz-Marźan and co-workers[17] have prepared nanoscale Ag nanocrystals by using dimethylformamide as both a stabilizing agent and a capping agent. By using tetrabutylammonium borohydride or its mixture with hydrazine, Jana and Peng[18] obtained monodisperse nanocrystals of Au, Cu, Ag, and Pt. In this method, $AuCl_3$, $Ag(CH_3COO)$, $Cu(CH_3COO)_2$, and $PtCl_4$ were dispersed in toluene with the aid of long-chain quaternary ammonium salts and reduced with tetrabutylammonium borohydride which is toluene-soluble. The reaction can be scaled up to produce gram quantities of nanocrystals. Mirkin and co-workers[19–21] have devised two synthetic routes for nanoprisms of Ag. In the first method, Ag nanoprisms (Fig. 1a) are produced by irradiating a mixture of sodium citrate and bis(*p*-sulfonatophenyl) phenylphosphine dihydrate dipotassium capped Ag nanocrystals with a fluorescent lamp. In the second method, $AgNO_3$ is reduced with a mixture of borohydride and hydrogen peroxide.[21] The latter method has been extended to synthesize branched nanocrystals of Au of the type shown in Fig. 1b and 1c.[22,23]

The shape and colour of Au nanoparticles can be altered by NAD(P)H-mediated growth in the presence of ascorbic acid.[24] The method yields dipods, tripods and tetrapods (nanocrystals with 2, 3 and 4 arms respectively, the one with 4 arms generally being tetrahedral). Icosahedral Au nanocrystals are obtained by the reaction of $HAuCl_4$ with PVP in aqueous media.[25] Right-bipyramid (75–150 nm in edge length) nanocrystals of Ag (Fig. 2) have been prepared by the addition of NaBr during the polyol reduction of $AgNO_3$ in the presence of PVP.[26] Cu nanoparticles of pyramidal shape have been made by an electrochemical procedure.[27]

Nanoparticles of Rh and Ir have been prepared by the reduction of the appropriate compounds in the ionic liquid, 1-*n*-butyl-3-methylimidazoliumhexafluorophosphate, in the presence of hydrogen.[28] Synthesis and functionalization of gold nanoparticles in ionic liquids is also reported, wherein the colour of the gold nanoparticles can be tuned by changing the anion of ionic liquid.[29] Ru nanoparticles, stabilized by oligoethyleneoxythiol, are found to be soluble in both aqueous and organic media.[30] While Rh multipods are obtained through the seeded-growth mechanism on reducing $RhCl_3$ in ethylene glycol in the presence of PVP,[31] Ir nanocrystals have been prepared by the reduction of an organometallic precursor in the presence of hexadecanediol and different capping agents.[32] Ru, Rh and Ir nanocrystals and other nanostructures are prepared by carrying out the decomposition of

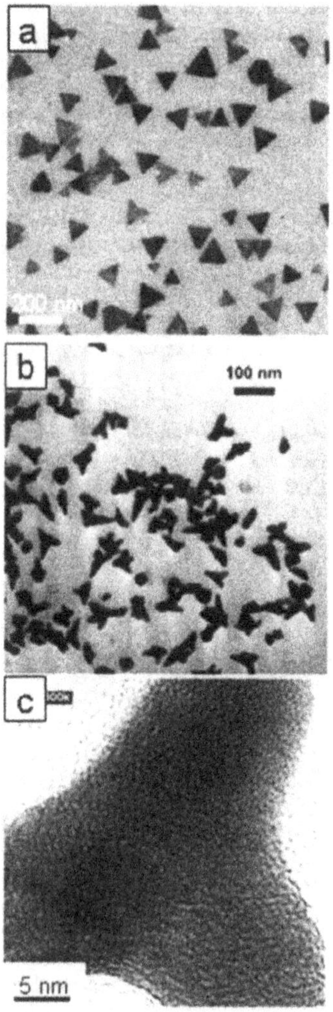


Fig. 1 (a) Ag nanoprisms obtained by controlled irradiation of bis(*p*-sulfonatophenyl) phenylphosphine dihydrate dipotassium capped Ag nanocrystals. (b) Low and (c) high-magnification (scale bar = 5 nm) TEM images of branched Au nanocrystals. Fig. 1a reprinted by permission from Macmillan Publishers Ltd.: *Nature*, 2003, **425**, 487, © 2003. Fig. 1b reprinted with permission from E. Hao, R. C. Bailey and G. C. Schatz, *Nano Lett.*, 2004, **4**, 327. © 2004 American Chemical Society.

the respective metal acetylacetonates in a hydrocarbon (decalin or toluene) or an amine (*n*-octylamine or oleylamine) around 573 K.[33] Cobalt nanoparticles of ~3 nm diameter have been synthesized by the reaction of di-isobutyl aluminium hydride with Co-($\eta^3C_8H_{13}$)($\eta^4C_8H_{12}$) or $Co[N(SiMe_3)_2]_2$.[34] Monodisperse Pt nanocrystals with cubic, cuboctahedral and octahedral shapes with diameters of ~9 nm have been obtained by the polyol process.[35] The polyol process has also been employed to obtain PtBi nanoparticles.[36] AuPt nanoparticles have been successfully incorporated in $SiO_2$ films.[37] FePt nanocubes with ~7 nm diameter (Fig. 3) have been synthesized by the reaction of the oleic acid and $Fe(CO)_5$ with benzyl ether/octadecene solution of $Pt(acac)_2$.[38]

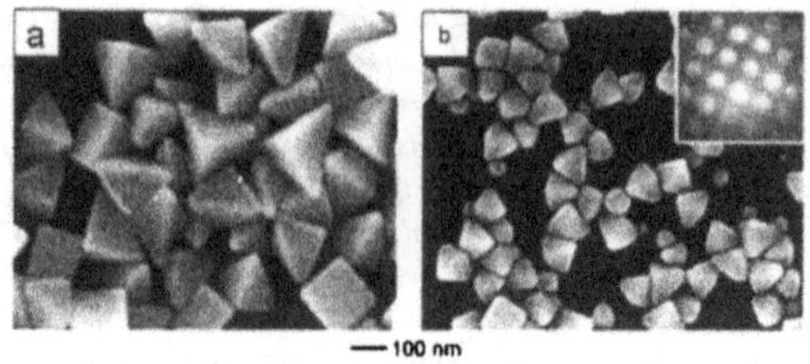


Fig. 2 SEM images of right-bipyramids approximately (a) 150 nm and (b) 75 nm in edge length. The inset in (b) shows the electron diffraction pattern obtained from a single right-bipyramid, indicating that it is bounded by (100) facets. Reprinted with permission from B. J. Wiley, Y. Xiong, Z.-Y. Li, Y. Yin and Younan Xia, *Nano Lett.*, 2006, **6**, 765. © 2006 American Chemical Society.

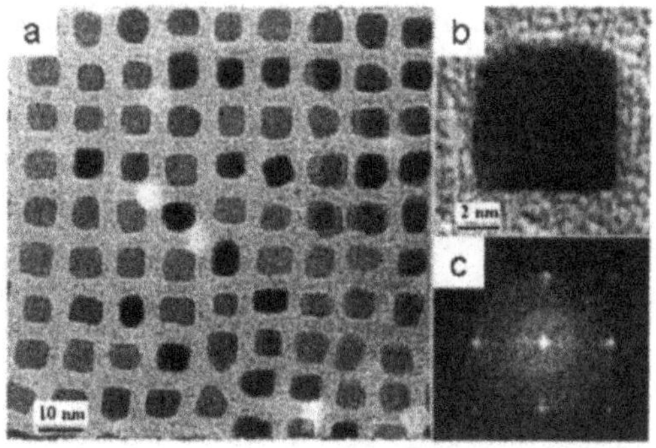


Fig. 3 TEM bright field images of (a) 6.9 nm $Fe_{50}Pt_{50}$ nanocubes; (b) HREM image of a single FePt nanocube; (c) Fast-Fourier transform (FFT) of the HREM in (b). Reprinted with permission from M. Chen, J. Kim, J. P. Liu, H. Fan and S. Sun, *J. Am. Chem. Soc.*, 2006, **128**, 7132. © 2006 American Chemical Society.

### 2.2 Metal oxides

Metal oxide nanocrystals are mainly prepared by the solvothermal decomposition of organometallic precursors. Solvothermal conditions afford high autogenous pressures inside the sealed autoclave that enable low-boiling solvents to be heated to temperatures well above their boiling points. Thus, reactions can be carried out at elevated temperatures and the products obtained are generally crystalline compared to those from other solution-based reactions.

Rockenberger *et al.*[39] described the use of cupferron complexes as precursors to prepare $\gamma$-$Fe_2O_3$, $Cu_2O$ and $Mn_3O_4$ nanocrystals. CoO nanocrystals with diameters in 4.5–18 nm range have been prepared by the decomposition of cobalt cupferronate in decalin at 543 K under solvothermal conditions.[40] Magnetic measurements indicate the presence of ferromagnetic interaction in the small CoO nanocrystals. Nanocrystals of MnO and NiO are obtained from cupferronate precursors under solvothermal conditions.[41] The nanocrystals exhibit superparamagnetism accompanied by magnetic hysteresis below a blocking temperature. Nanocrystals of CdO and CuO are prepared by the solvothermal decomposition of metal-cupferronate in presence of trioctylphosphine oxide (TOPO) in toluene.[42] ZnO nanocrystals have been synthesized from the cupferron complex by a solvothermal route in toluene solution.[43] $\gamma$-$Fe_2O_3$ and $CoFe_2O_4$ nanocrystals can also be produced by the decomposition of the cupferron complexes.[44]

Metallic $ReO_3$ nanocrystals with diameters in the 8.5–32.5 nm range are obtained by the solvothermal decomposition of the $Re_2O_7$-dioxane complex under solvothermal conditions.[45] Fig. 4a shows a TEM image of $ReO_3$ nanocrystals of 17 nm average diameter with the size distribution histogram as an upper inset. The lower inset shows a HREM image of 8.5 nm nanocrystal. The nanocrystals exhibit a surface plasmon band around 520 nm which undergoes blue-shifts with decrease in size (Fig. 4b). Such blue-shifts in the $\lambda_{max}$ with decreasing particle size is well-known in the case of metal nanocrystals.[46] Inset in Fig. 4b shows the photograph of four different sizes of $ReO_3$ nanocrystals soluble in $CCl_4$. Surface-enhanced Raman scattering of pyridine, pyrazine and pyrimidine adsorbed on $ReO_3$ nanocrystals has been observed.[47] Magnetic hysteresis is observed at low temperatures in the case of the 8.5 nm particles suggesting a superparamagnetic behaviour.

Apart from solvothermal methods, thermolysis of precursors in high boiling solvents, the sol–gel method, hydrolysis and use of micelles have been employed to synthesize the metal oxide nanocrystals. Thus, Park *et al.*[48] have used metal-oleates as precursors for the preparation of monodisperse $Fe_3O_4$, MnO and CoO nanocrystals. 1-Octadecene, octyl ether and trioctylamine have been used as solvents. Hexagonal and cubic CoO nanocrystals can be prepared by the decomposition of cobalt acetylacetonate in oleylamine under kinetic and thermodynamic conditions respectively.[49] Hexagonal pyramid-shaped ZnO nanocrystals have been obtained by the thermolysis of the Zn-oleate complex.[50] ZnO nanocrystals have been prepared from zinc acetate in 2-propanol by the reaction with water.[51] ZnO nanocrystals with cone (Fig. 5), hexagonal cone and rod shapes have been obtained by the non-hydrolytic ester elimination sol-gel reactions.[52] In this reaction, ZnO nanocrystals with various shapes were obtained by the reaction of zinc acetate with 1,12-dodecanediol in the presence of different surfactants. In this laboratory, it has been found that reactions of alcohols such as ethanol and *t*-butanol with Zn powder readily yield ZnO nanocrystals.[53]

Nanocrystals of $BaTiO_3$ are obtained by the thermal decomposition of MOCVD reagents (alkoxides such as $BaTi(O_2CC_7H_{15})[OCH(CH_3)_2]_5$) in diphenyl ether containing oleic acid, followed by the oxidation of the product with $H_2O_2$.[54] Thermal decomposition of uranyl acetylacetonate in a mixture solution of oleic acid, oleylamine, and octadecene at 423 K gives uranium oxide nanocrystals.[55] Treatment of metal acetylacetonates under solvothermal conditions produces nanocrystals of metal oxides such as $Ga_2O_3$, ZnO and cubic $In_2O_3$.[56] Nearly monodisperse $In_2O_3$ nanocrystals have been obtained starting with indium acetate, oleylamine and oleic acid.[57] $TiO_2$ nanocrystals can be prepared

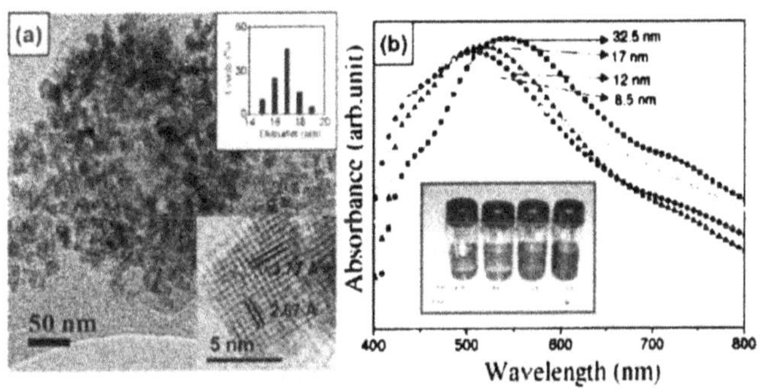


**Fig. 4** (a) TEM image of $ReO_3$ nanocrystals of average diameter 17 nm. Upper inset shows the size distribution histogram. Lower inset shows the HREM image of a single 8.5 nm nanocrystal. (b) Optical absorption spectra of $ReO_3$ nanocrystals with average diameters of 8.5, 12, 17 and 32.5 nm. Inset in (b) shows a picture of four different sizes of $ReO_3$ nanocrystals dissolved in $CCl_4$. Reprinted with permission from K. Biswas and C. N. R. Rao, *J. Phys. Chem. B*, 2006, **110**, 842. © 2006 American Chemical Society.

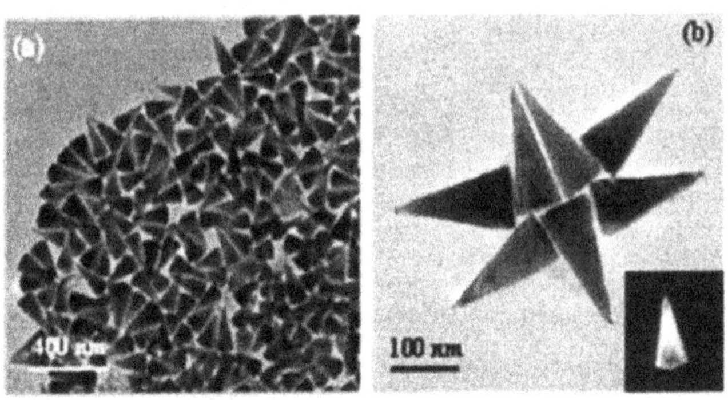


**Fig. 5** (a) and (b) TEM images of cone-shaped ZnO nanocrystal. Inset in (b) shows a dark field image of a single cone-shaped nanocrystal. Reproduced with permission from J. Joo, S. G. Kwon, J. H. Yu and T. Hyeon, *Adv. Mater.*, 2005, 17, 1873. © 2005 Wiley-VCH Verlag GmbH & Co. KGaA.

by the low-temperature reaction of low-valent organometallic precursors.[58] Pure anatase $TiO_2$ nanocrystals have been prepared by the hydrolysis of $TiCl_4$ with ethanol at 273 K followed by calcination at 360 K for 3 days.[59] The growth kinetics and the surface hydration chemistry have also been investigated. Pileni and co-workers[60,61] have pioneered the use of oil in water micelles to prepare particles of $CoFe_2O_4$, $\gamma$-$Fe_2O_3$, and $Fe_3O_4$. The basic reaction involving hydrolysis is now templated by a micellar droplet. The reactants are introduced in the form of a salt of a surfactant such as sodium dodecylsulfate (SDS). Thus, by adding $CH_3NH_3OH$ to a micelle made of calculated quantities of $Fe(SDS)_2$ and $Co(SDS)_2$, nanocrystals of $CoFe_2O_4$ are obtained.

### 2.3 Metal chalcogenides

Nearly monodisperse Cd-chalcogenide nanocrystals (CdE; E = S, Se, Te) have been synthesized by the injection of organometallic reagents such as alkylcadmium into a hot coordinating solvent in the presence of silylchalcogenides/phosphinechalcogenides.[62] Alivisatos and coworkers[63] have produced Cd-chalcogenide nanocrystals by employing tri-butylphosphine at higher temperatures. Nanocrystals of metal chalcogenides are generally prepared by the reaction of metal salts with an appropriate sulfiding or seleniding agent under solvothermal or thermolysis conditions. Thus, toluene-soluble CdSe nanocrystals with a diameter of 3 nm have been prepared solvothermally by reacting cadmium stearate with elemental Se in toluene in the presence of tetralin.[64] The key step in the reaction scheme is the aromatization of tetralin to naphthalene in the presence of Se, producing $H_2Se$. Organic-soluble CdS nanocrystals are similarly prepared by the reaction of a cadmium salt with S in toluene in the presence of tetralin.[65] Fig. 6 shows the TEM images and electron diffraction patterns of TOPO-capped CdS nanocrystals prepared by this method. PbS and PbSe crystallites and nanorods can also be prepared by this method.[66] Nanocrystals of the transition metal dichalcogenides

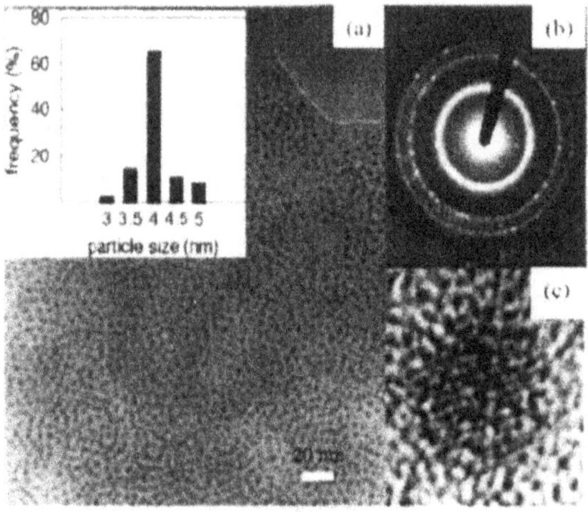


**Fig. 6** (a) TEM image of TOPO-capped CdS particles. Inset shows the size distribution of the nanocrystals. The electron diffraction pattern of the nanocrystals is shown in (b) and a HREM image of a nanocrystal is shown in (c). Reprinted from U. K. Gautam, R. Seshadri and C. N. R. Rao, *Chem. Phys. Lett.*, 2003, **375**, 560, © 2003 with permission from Elsevier. http://www.sciencedirect.com/science/journal/00092614

($ME_2$; M = Fe, Co, Ni, Mo; E = S or Se) with diameters in the range 4–200 nm have been prepared by a hydrothermal route.[67]

Peng *et al.*[68–71] have proposed the use of greener Cd sources such as cadmium oxide, carbonate and acetate instead of the dimethylcadmium. These workers suggest that the size distribution of the nanocrystals is improved by the use of hexadecylamine, a long-chain phosphonic acid or a carboxylic acid. The method can be extended to prepare CdS nanoparticles by the use of tri-*n*-octylphosphine sulfide (TOP-S) and hexyl or tetradecyl phosphonic acid in mixture with TOPO–TOP. Hyeon and co-workers[72] have prepared nanocrystals of several metal sulfides such as CdS, ZnS, PbS, and MnS with different shapes and sizes by the thermolysis of metal–oleylamine complexes in the presence of S and oleylamine (Fig. 7).

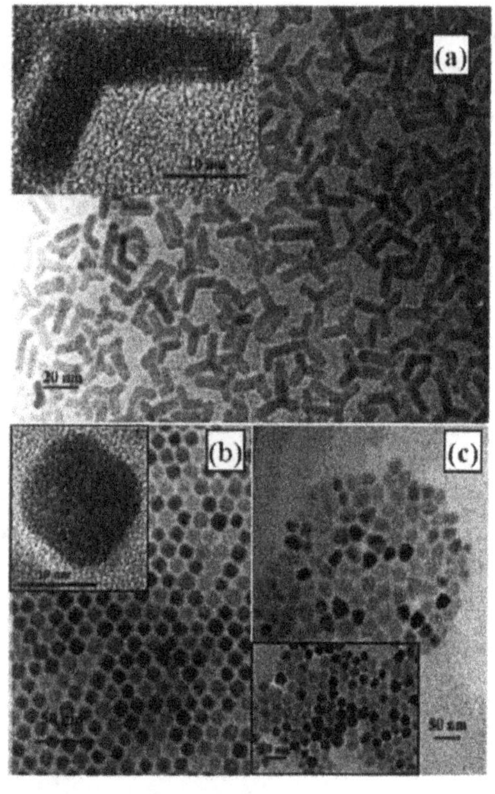


**Fig. 7** (a) TEM images of a mixture of rods, bipods, and tripods of CdS nanocrystals with an average size of 5.4 nm (thickness) 20 nm (length). Inset is a HREM image of a single CdS bipod-shaped nanocrystal. (b) Low-magnification TEM image of 13 nm PbS nanocrystals. Inset shows a HREM image of a single 13 nm PbS nanocrystal. (c) Short bullet-shaped MnS nanocrystals. Inset shows hexagon-shaped MnS nanocrystals. Reprinted with permission from J. Joo, H. B. Na, T. Yu, J. H. Yu, Y. W. Kim, F. Wu, J. Z. Zhang and T. Hyeon, *J. Am. Chem. Soc.*, 2003, **125**, 11100. © 2003 American Chemical Society.

CdSe and CdTe nanocrystals can be prepared without precursor injection.[73] The method involves refluxing the cadmium precursor with Se or Te in octadecene. CdSe nanocrystals have also been synthesized using elemental selenium dispersed in octadecene without the use of trioctylphosphine.[74] ZnSe nanocrystals have been prepared in a hot mixture of a long-chain alkylamine and alkylphosphines.[75] Highly monodisperse cubic-shaped PbTe nanocrystals (Fig. 8) have been prepared with size distributions less than 7% by a rapid injection technique.[76] PbS nanocrystals are obtained by reacting the $PbCl_2$–oleylamine complex with the S–oleylamine complex without the use of any solvent.[77] Homogeneously alloyed $CdS_xSe_{1-x}$ ($x$ = 0–1.) nanocrystals are prepared by the thermolysis of metal–oleylamine complexes in the presence of S and Se.[78] The band gap of these nanocrystals can be tuned by varying the composition. Thermolysis of a mixture composed of Cu and In oleates in alkanethiol yields copper indium sulfide nanocrystals with acorn, bottle, and larva shapes.[79] Nanocrystals of $Ni_3S_4$ and $Cu_{1-x}S$ have been prepared by adding elemental sulfur to metal precursors dissolved in dichlorobenzene or oleylamine at relatively high temperatures.[80]

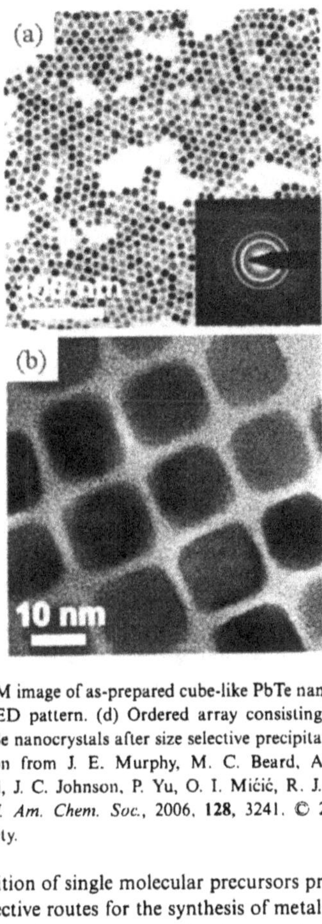


**Fig. 8** (a) TEM image of as-prepared cube-like PbTe nanocrystals. Inset shows the SAED pattern. (d) Ordered array consisting of 15 nm cubic-shaped PbSe nanocrystals after size selective precipitation. Reprinted with permission from J. E. Murphy, M. C. Beard, A. G. Norman, S. P. Ahrenkiel, J. C. Johnson, P. Yu, O. I. Mićić, R. J. Ellingson and A. J. Nozik, *J. Am. Chem. Soc.*, 2006, **128**, 3241. © 2006 American Chemical Society.

Decomposition of single molecular precursors provides convenient and effective routes for the synthesis of metal chalcogenide nanocrystals. In this method, a molecular complex consisting of both the metal and the chalcogen is thermally decomposed in a coordinating solvent. For example dithiocarbamates and diselenocarbamates have been found to be good air stable precursors for sulfides and selenides of Cd, Zn and Pb.[81] Nanocrystals of Cd, Hg, Mn, Pb, Cu, and Zn sulfides have been obtained by thermal decomposition of metal hexadecylxanthates in hexadecylamine and other solvents at relatively low temperatures (323–423 K) under ambient conditions.[82]

Hexagonal CdS nanocrystals have been obtained by the reaction of cadmium acetate dihydrate with thioacetamide in imidazolium[BMIM]-based ionic liquids.[83] Fig. 9(a) shows the TEM image along with HREM image as a top inset of CdS nanocrystals prepared in [BMIM][$MeSO_4$]. Particle size of the CdS nanoparticles varies between 3 and 13 nm with the anion of imidazolium based ionic liquid under the same reaction conditions. Addition of TOPO to the reaction mixture causes greater monodispersity as well as smaller particle size. Hexagonal ZnS and cubic PbS nanoparticles with average diameters of 3 and 10 nm respectively have been prepared by the reaction of the metal acetates with thioacetamide in [BMIM][$BF_4$]. Hexagonal CdSe nanocrystals with an average diameter of 12 nm were obtained by the reaction of cadmium acetate dihydrate with dimethylselenourea in [BMIM][$BF_4$]. CdSe nanocrystals have also been prepared using the phosphonium ionic liquid trihexyl(tetradecyl)phosphoniumbis(2,4,4 trimethylpentylphosphinate) as a solvent and capping agent.[84]

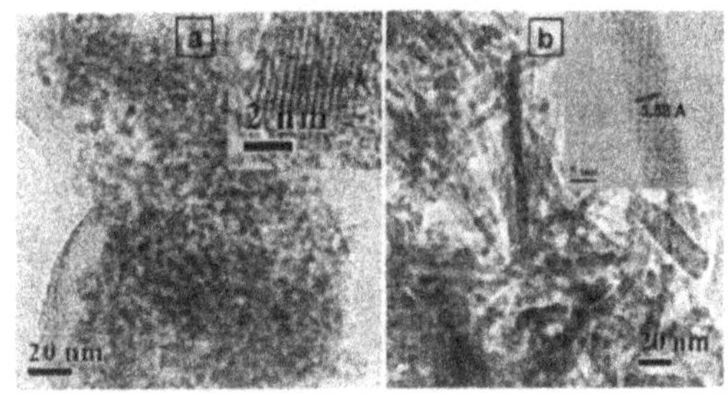


**Fig. 9** (a) TEM images of 4 nm CdS nanoparticles prepared in [BMIM][$MeSO_4$]. Insets show a HREM image of the 4 nm CdS nanoparticle. (b) TEM image of CdS nanorods prepared in a [BMIM][$BF_4$] and ethylenediamine mixture. Inset shows a HREM image of a nanorod. Reproduced with permission from K. Biswas and C. N. R. Rao, *Chem.-Eur. J.*, 2007, **13**, 6123. © 2007 Wiley-VCH Verlag GmbH & Co. KGaA.

### 2.4 Metal pnictides

Large GaN nanocrystals (32 nm) were prepared by Xie *et al.*,[85] by treating $GaCl_3$ with $Li_3N$ in benzene under solvothermal conditions. GaN nanocrystals of various sizes have been prepared under solvothermal conditions, by employing gallium cupferronate or chloride as the gallium source and 1,1,1,3,3,3-hexamethyldisilazane (HMDS) as the nitriding agent and toluene as solvent.[86] By employing surfactants such as cetyltrimethylammonium bromide (CTAB), the size of the nanocrystals could be controlled (Fig. 10a). In the case of cupferronate, the formation of the nanocrystals is likely to involve nitridation of the nascent gallium oxide nanoparticles formed by the decomposition of the cupferronate.

$GaO_{1.5} + (CH_3)_3SiNHSi(CH_3)_3 \rightarrow GaN + (CH_3)_3SiOSi(CH_3)_3 + 1/2H_2O$

With $GaCl_3$ as precursor, the reaction is,

$GaCl_3 + (CH_3)_3SiNHSi(CH_3)_3 \rightarrow GaN + Si(CH_3)_3Cl + HCl$

This method has been applied for the synthesis of AlN and InN nanocrystals (Fig. 10b).[87] The procedure yields nanocrystals with an average diameter of 10 nm for AlN, 15 nm for InN and as low

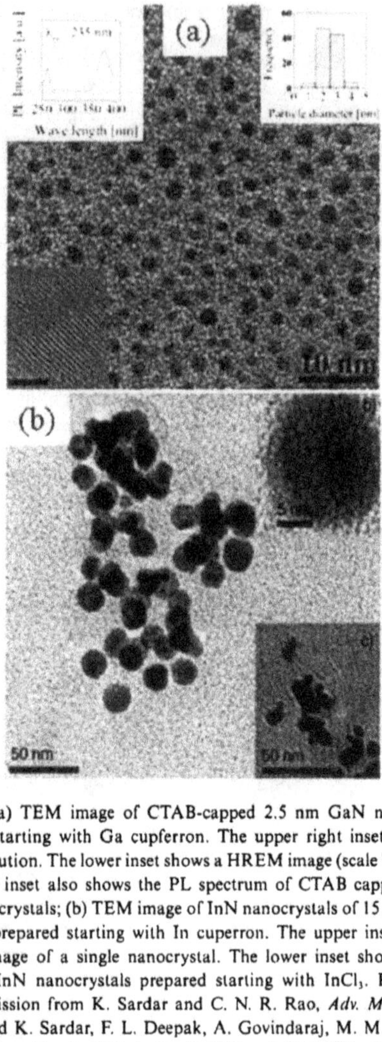


Fig. 10 (a) TEM image of CTAB-capped 2.5 nm GaN nanocrystals prepared starting with Ga cupferron. The upper right inset shows the size distribution. The lower inset shows a HREM image (scale bar is 2 nm) and upper inset also shows the PL spectrum of CTAB capped 2.5 nm GaN nanocrystals; (b) TEM image of InN nanocrystals of 15 nm average diameter prepared starting with In cuperron. The upper inset shows a HREM image of a single nanocrystal. The lower inset shows a TEM image of InN nanocrystals prepared starting with $InCl_3$. Reproduced with permission from K. Sardar and C. N. R. Rao, *Adv. Mater.*, 2004, **16**, 425 and K. Sardar, F. L. Deepak, A. Govindaraj, M. M. Seikh and C. N. R. Rao, *Small*, 2005, **1**, 91. © 2004 and 2005 Wiley-VCH Verlag GmbH & Co. KGaA.

as 4 nm for GaN. Indium-doped GaN nanocrystals with 5% and 10% In as well as 3% and 5% Mn-doped GaN nanocrystals have been prepared by this method.[88,89]

GaN nanocrystals have also been obtained by the thermal decomposition of precursor compounds such as $(C_2H_5)N{\cdot}Ga(N_3)_3$ and polymeric $[Ga(NH)_{3/2}]_n$.[90–92] Thus, by the thermolysis of azido compounds, Manz *et al.*[93] have obtained nanocrystals of hexagonal GaN of varying diameters. Group 13 metal nitrides (GaN, AlN, InN) have been prepared by a single source precursor route.[94] The precursors are the adducts of the metal chlorides and urea. Hexagonal nanocrystals of GaN, AlN, and InN were obtained by refluxing the precursors in tri-*n*-octylamine. Fig. 11 shows the TEM images and the PL spectrum of GaN nanocrystals prepared by the urea route. This method has also been extended for the synthesis of BN, TiN and NbN nanoparticles.[95] Solvothermal synthesis involving the reaction of $GaCl_3$ and $NaN_3$ yields a

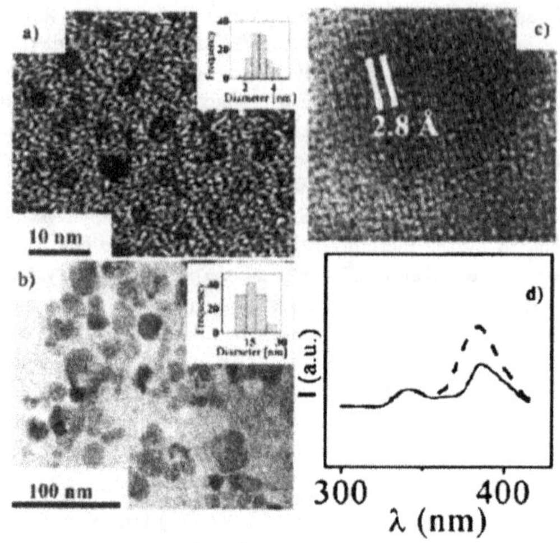


Fig. 11 (a), (b) TEM images of 3 and 15 nm GaN nanocrystals respectively prepared by the urea route. Insets at the top show particle size distributions (c) HREM image of a nanocrystals; (d) PL spectrum of 3 nm size particles at two different excitation wavelengths (solid curve 260 nm and broken curve 250 nm) (From Ref. [94]).

poorly crystalline initial product that crystallizes as relatively large nanocrystals (~50 nm in diameter) on annealing.[96]

Nanocrystalline InN powders are obtained by the metathesis reaction of $InBr_3$ and $NaN_3$ in superheated toluene and refluxing hexadecane solvents near 553 K.[97] This method has been extended to prepare $Ga_{1-x}In_xN$ ($x = 0.5$ and 0.75) nanocrystals. A benzene-thermal route has been developed to prepare nanocrystalline InN at 453–473K by choosing $NaNH_2$ and $In_2S_3$ as novel nitrogen and indium sources.[98] This route has been extended to synthesize other group III nitrides. AlN nanocrystals have been prepared by the benzene-thermal reaction between $AlCl_3$ and $Li_3N$.[99]

An early procedure for the preparation of phosphides and arsenides of gallium, indium and aluminium involved the de-hydrosilylation reaction. Alivisatos and co-workers[100] adapted this method to prepare GaAs nanoparticles using $GaCl_3$ and $As(SiMe_3)_3$ in quinoline. Using a similar scheme, GeSb, InSb, InAs, and InP nanocrystals were obtained.[101] This method has been modified to prepare InP, InAs, GaP, and $GaInP_2$ nanocrystals as well.[102–104] In a typical reaction, $InCl_3$ is complexed with TOPO/TOP and is reacted with a silylated pnictide such as $E(SiMe_3)_3$ (E = As, P) at 536 K, followed by growth at elevated temperatures for several days. Phase-pure FeP nanocrystals have been synthesized by the reaction of iron(III) acetylacetonate with tris(trimethylsilyl)phosphine at temperatures of 513–593 K using trioctylphosphine oxide as a solvent and dodecylamine, myristic acid, or hexylphosphonic acid as additional capping groups (ligands).[105] The treatment of $Mn_2(CO)_{10}$ with $P(SiMe_3)_3$ in (TOPO)/myristic acid at elevated temperatures produced MnP as discrete nanocrystals.[106] In the presence of a surfactant, potassium stearate, quantum-confined InP nanocrystals were hydrothermally synthesized in aqueous ammonia.[107] High quality InP nanocrystals are obtained by the reaction of $(Me)_3In$ and $P(SiMe_3)_3$ in a coordinating ester solvent.[108] O'Brien and co-workers developed a single molecular precursor route to synthesize InP and GaP nanocrystals using diorganophosphides-$M(PBut_2)_3$ (M=Ga, In). This method has also been adapted to synthesize $Cd_3P_2$ using $[MeCdP(But)_2]_3$. The dimer $[t\text{-}Bu_2AsInEt_2]_2$ has been synthesized and used as a single-source organometallic precursor to grow InAs nanocrystals.[109] Reduction of transition metal pnictates yields metal pnictides. Using this method nanoparticle of FeP, $Fe_2P$[110] and NiAs[111] has been prepared. Monodisperse iron, cobalt and nickel monoarsenide nanocrystals were obtained under high-intensity ultrasonic irradiation from the reaction of transition metal chlorides, arsenic and zinc in ethanol.[112]

## 3. Core@shell nanoparticles

Core@shell particles involving metal, semiconductor or oxide nanocrystals in the core, with shells composed of different materials have been investigated widely.[113] The method of Murray *et al.*[62] involving the decomposition of dimethyl cadmium has been adapted to synthesize nanocrystals of the type CdSe@ZnS, CdSe@ZnSe, and CdSe@CdS.[114] Core@shell growth is achieved by injecting the precursors forming the shell materials into a dispersion containing the core nanocrystals. The injection is carried out at a slightly lower temperature to force shell growth, avoiding independent nucleation. Thus, a mixture of diethylzinc and bis(trimethylsilyl)sulfide is injected into a hot solution containing the core CdSe nanocrystals to encase them with a ZnS layer.[115] O'Brien and co-workers[116] have used single-source methods to prepare core@shell nanocrystals. By successive thermolysis of unsymmetrical diseleno and dithiocarbamates, core@shell nanocrystals of the type CdSe@ZnS and CdSe@ZnSe have been prepared.[116] CdSe@CdS core@shell nanoparticles with a core diameter of ~1.5 nm have been prepared at the liquid–liquid interface starting with cadmium myristate and oleic acid in toluene and selenourea/thiourea in the aqueous medium.[117] Luminescent multi-shell nanocrystals of the composition CdSe-core CdS/$Zn_{0.5}Cd_{0.5}S$/ZnS-shell have been prepared by successive ion layer adhesion and reaction technique,[118] Where in the growth of the shell is carried out one monolayer at a time, by alternately injecting cationic and anionic precursors into the reaction mixture with core nanocrystals. Water-soluble CdSe@CdS core@shell nanocrystals with dendron carbohydrate anchoring groups have been prepared.[119] Cao and Banin[120] have successfully coated InAs nanocrystals with shells of InP, GaAs, CdSe, ZnSe, and ZnS. In Fig. 12, we show the TEM and HREM images of InAs@InP and InAs@CdSe core@shell nanocrystals. Using the shell layers, the bandgap of InAs can be tuned in the near-IR region.

Metal on metal core@shell structures provide a means for generating metal nanocrystals with varied optical properties. Morriss and Collins[121] prepared Au@Ag nanocrystals by reducing Au with P by the Faraday's method and Ag with hydroxylamine hydrochloride. They observed a progressive blue shift of the Au plasmon band with incorporation of Ag, accompanied by a slight broadening. For sufficiently thick shells, the plasmon band resembled that of pure Ag particles. Large Au nanoparticles prepared by the citrate method have been used as seeds for the reduction of Ag nanocrystals using ascorbic acid, with CTAB as the capping agent.[122] Au@Ag as well as Ag@Au nanocrystals were prepared by the sequential reduction using sodium citrate.[123] Au@Ag as well as Ag@Au nanoparticles are also prepared by a UV-photoactivation technique.[124] Mirkin and co-workers[125] have coated Ag nanocrystals with a thin Au shell to provide stability against precipitation under physiological conditions. A thin shell has little effect on the optical properties. Au@Pd nanoparticles with controllable size from 35 to 100 nm were prepared by the chemical deposition of Pd over pre-formed 12 nm Au seeds.[126] Reactive magnetic nanocrystals are rendered passive and made easy to handle by coating them with a layer of noble metals. For example, a layer of Ag was grown *in situ* on Fe and Co nanocrystals synthesized using reverse micelles.[127] A similar procedure has been used to coat Au as well.[128] Fe@Au nanocrystals are also prepared by sequential citrate reduction followed by magnetic separation.[129] Fe@Au nanocrystals were synthesized by a wet chemical procedure involving laser irradiation of Fe nanoparticles and Au powder in a liquid medium.[130] The nanoparticles were superparamagnetic with a blocking temperature of 170 K. $Fe_3O_4$ of selected size were used as seeding materials for the reduction of Au precursors to produce monodisperse $Fe_3O_4$@Au nanoparticles.[131]

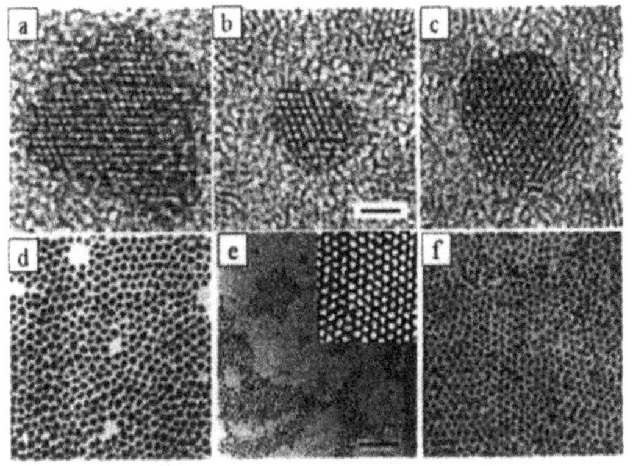


**Fig. 12** HREM images of InAs@InP core@shell (frame (a), core radius 1.7 nm, shell thickness 2.5 nm), InAs core ((b), core radius 1.7 nm), and InAs@CdSe core@shell ((c), core radius 1.7 nm, shell thickness 1.5 nm). The scale bar is 2 nm. (d), (e) and (f) are low magnification TEM images of InAs@InP, InAs core and InAs@CdSe core/shell nanoparticles respectively. The scale bar is 50 nm. The inset in (e) (70 × 70 nm), displays a portion of a superlattice structure formed from the InAs cores. Reprinted with permission from Y.-W. Cao and U. Banin, *J. Am. Chem. Soc.*, 2000, **122**, 9692. © 2000 American Chemical Society.

A dielectric oxide layer (*e.g.* silica) is useful as a shell material because of the stability it lends to the core and its optical transparency. The classic method of Stober for solution deposition of silica are adaptable for coating of nanocrystals with silica shells.[132] This method relies on the pH and the concentration of the solution to control the rate of deposition. The natural affinity of silica to oxidic layers has been exploited to obtain silica coating on a family of iron oxide nanoparticles including hematite and magnetite.[133] Such a deposition process is not readily extendable to grow shell layers on metals. The most successful method for silica encapsulation of metal nanoparticles is that due to Mulvaney and co-workers.[134] In this method, the surface of the nanoparticles is functionalized with aminopropyltrimethylsilane, a bifunctional molecule with a pendant silane group which is available for condensation of silica. The next step involves the slow deposition of silica in water followed by the fast deposition of silica in ethanol. Fig. 13 shows the TEM images

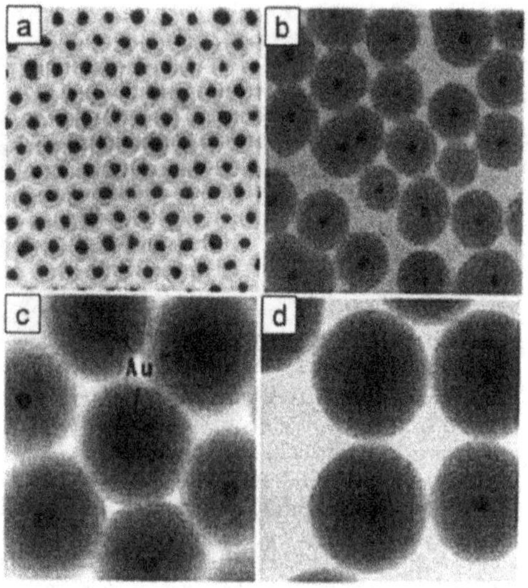


Fig. 13 TEM images of $Au@SiO_2$ particles produced during the growth of the silica shell around 15 nm Au particles. The shell thicknesses are (a) 10 nm, (b) 23 nm, (c) 58 nm, and (d) 83 nm. Reprinted with permission from L. M. Liz-Marzán, M. Giersig and P. Mulvaney, *Langmuir*, 1996, **12**, 4329. © 1996 American Chemical Society.

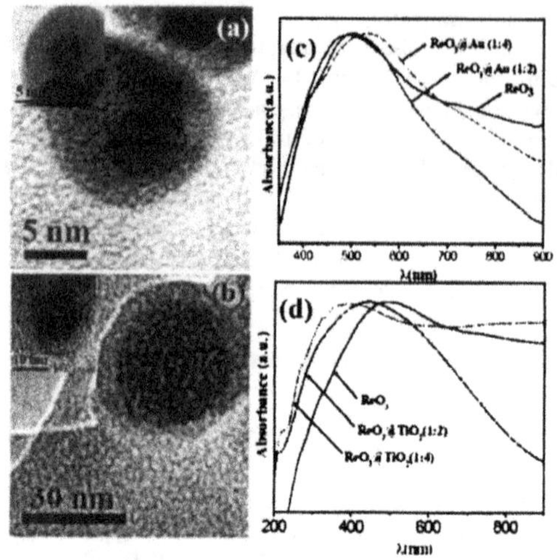


Fig. 14 TEM images of core–shell nanoparticles of (a) $ReO_3@Au$ formed with a 5 nm $ReO_3$ particle. Inset shows $ReO_3@Au$ formed over an 8 nm $ReO_3$ particle. (b) $ReO_3@TiO_2$ core–shell nanoparticle formed over a 32 nm $ReO_3$ particle with the inset showing a core-shell nanoparticle formed over a 12 nm $ReO_3$ nanoparticle. UV-visible absorption spectra of (c) $ReO_3@Au$ core-shell nanoparticles (1 : 2 and 1 : 4). and (d) $ReO_3@TiO_2$ core-shell nanoparticles (1 : 2 and 1 : 4) with a 12 nm $ReO_3$ particle (From Ref. [140]).

of $Au@SiO_2$ nanocrystals with various shell thickness reported by Mulvaney's group.[135] $Au@TiO_2$ core@shell nanocrystals have been prepared by complexation of a negatively charged titanium precursor, titanium(IV) bis(ammonium lactato)dihydroxide, with poly(dimethyldiallylammonium chloride).[136] Silver nanoparticles coated with a uniform, thin shell of titanium dioxide have been synthesized by a one-pot route, where the reduction of $Ag^+$ to $Ag^0$ and the controlled polymerization of $TiO_2$ on the surface of silver crystallites occurs simultaneously.[137] Pradeep and co-workers[138,139] have coated Au and Ag nanoparticles with $TiO_2$ and $ZrO_2$ in a single-step process.

$ReO_3@Au$ (Fig. 14a ) and $ReO_3@Ag$ were prepared by the reduction of metal salts over $ReO_3$ nanoparticle seeds.[140] $ReO_3@SiO_2$ and $ReO_3@TiO_2$ (Fig. 14b ) core–shell nanocrystals were prepared by hydrolysis of the organometallic precursors over the $ReO_3$ nanoparticles. $ReO_3@Au$ and $ReO_3@Ag$ core–shell nanoparticles show composite plasmon absorption bands comprising contributions from both $ReO_3$ and Au (Ag) (Fig. 14c ) whereas $ReO_3@SiO_2$ and $ReO_3@TiO_2$ show shifts in the plasmon bands depending on the refractive index of the shell material (Fig. 14d). $Co@SiO_2$ nanocrystals were prepared combining the sodium borohydride reduction in aqueous solution, the Stober method, and/or the layer-by-layer self-assembly technique.[141] Nanocrystals with the ferrimagnetic $CoFe_2O_4$ core and the antiferromagnetic MnO shell have been obtained by a high-temperature decomposition route with seed-mediated growth.[142]

## 4. Nanowires

There has been considerable interest in the synthesis, characterization and properties of nanowires of various inorganic materials.[4,143,144] Nanowires have been prepared using vapour phase methods such as vapour–liquid–solid (VLS) growth, vapour-solid (*VS*) growth, oxide-assisted growth and the use of carbothermal reactions. A variety of solution methods such as seed-assisted growth, polyol method, and oriented attachment have also been developed for the synthesis of one-dimensional nanostructures.

### 4.1 Metals

Metal nanowires are commonly prepared using templates such as anodic alumina or polycarbonate membranes, carbon nanotubes and mesoporous carbon.[145–148] The nanoscale channels are first impregnated with metal salts and the nanowires obtained by reduction, followed by the dissolution of the template. Nanowires of metals and semiconductors have also been grown electrochemically. This method has been employed to prepare linear Au–Ag nanoparticle chains.[149] Here, sacrificial Ni segments are placed between segments of noble metals (Au, Ag). The template pore diameter fixes the nanowire width, and the length of each metal segment is independently controlled by the amount of current passed before switching to the next plating solution for deposition of the subsequent segments. Nanowires are released by the dissolution of the template, and subsequently coated with the $SiO_2$.

Au nanorods and nanowires have been alternatively prepared by a simple solution based reduction method making use of nanoparticle seeds.[150] Au nanoparticles with ~4 nm diameter react with the metal salt along with the weak reducing agent such as ascorbic acid in the presence of a directing surfactant yielding Au nanorods. This method was extended to prepare dog-bone like nanostructures.[151] The reaction is carried out in two-steps, wherein the first step involves the addition of an insufficient amount of ascorbic acid to the growth solution, leaving some unreacted metal salt after the reaction, which is later deposited on the Au nanorods by the second addition of ascorbic acid. Addition of nitric acid enhances the proportion of Au nanorods with high aspect ratios (~20) in seed-mediated synthesis.[152] The growth of Au nanorods by the seed-assisted method does not appear to follow any reaction-limited or diffusion limited growth mechanism.[153]

A layer-by-layer deposition approach has been employed to produce polyelectrolyte-coated gold nanorods.[154] Au-nanoparticle-modified enzymes act as biocatalytic inks for growing Au or Ag nanowires on Si surfaces by using a patterning technique such as dip-pen-nanolithography.[155] Single-crystalline Au nanorods shortened selectively by mild oxidation using 1 M HCl at 343 K.[156] Aligned Au nanorods can be grown on a silicon substrate by employing a simple chemical amidation reaction on $NH_2$-functionalized Si substrates.[157] A seed-mediated surfactant method using a cationic surfactant has been developed to obtain pentagonal silver nanorods.[158]

A popular method for the synthesis of metal nanowires is the use of the polyol process,[159,160] wherein the metal salt is reduced in the presence of PVP to yield nanowires of the desired metal. For example, Ag nanowires have been rapidly synthesized using a microwave-assisted polyol method.[161] CoNi nanowires are obtained by heterogeneous nucleation in liquid polyol.[162] While Bi nanowires have been prepared employing $NaBiO_3$ as the bismuth source.[163] Pd nanobars are synthesized by varying the type and concentration of reducing agent as well as reaction temperature.[164]

Metal nanowires are obtained in good yields by the nebulized spray pyrolysis of a methanolic solution of metal acetates.[165] This method has been employed for the synthesis of single-crystalline nanowires of zinc, cadmium and lead (see Fig. 15). The nanowires seem to grow by the vapour–solid mechanism. ZnO nanotubes shown in Fig. 15d can be obtained by the oxidation of zinc nanowires in air at 723 K.

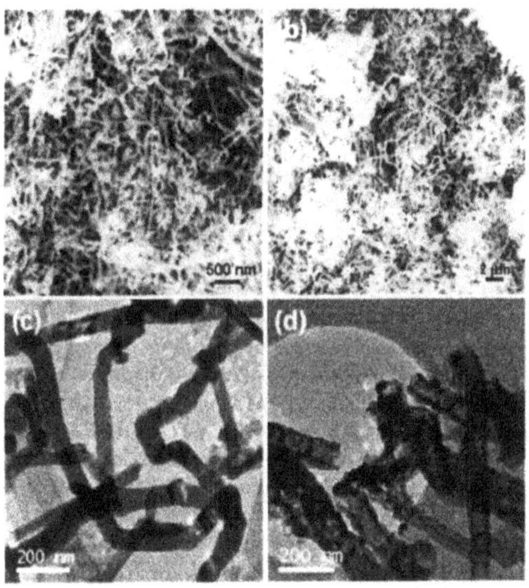


**Fig. 15** (a) and (b) SEM images of zinc and cadmium nanowires obtained by the pyrolysis of the corresponding metal acetates at 1173 K. (c) TEM image of zinc nanowires and (d) TEM image of ZnO nanotubes obtained by the oxidation of Zn nanowires at 723 K. Reproduced with permission from S. R. C. Vivekchand, G. Gundiah, A. Govindaraj and C. N. R. Rao, *Adv. Mater.*, 2004, **16**, 1842. © 2004 Wiley-VCH Verlag GmbH & Co. KGaA

### 4.2 Elemental semiconductors

Silicon nanowires with diameters in the 5–20 nm range have been prepared along with nanoparticles of 4 nm diameter by arc-discharge in water.[166] The Si nanowires shown in Fig. 16 were prepared in solution by using Au nanocrystals as seeds and silanes as precursors by the VLS mechanism.[167] Aligned Si nanowires are obtained by chemical vapour deposition (CVD) of $SiCl_4$ on a gold colloid deposited Si(111) substrate.[168] Gold colloids have been used for nanowire synthesis by the VLS growth mechanism. Using anodic alumina membranes as templates, Si nanowires have been synthesized on Si substrates.[169] In this method, porous anodic alumina is grown on the Si substrate followed by the electrodeposition of the gold catalyst. Epitaxial Si nanowires are then obtained subsequently by VLS growth. Presence of oxygen is important for the growth of long untappered Si nanowires by the VLS mechanism.[170]

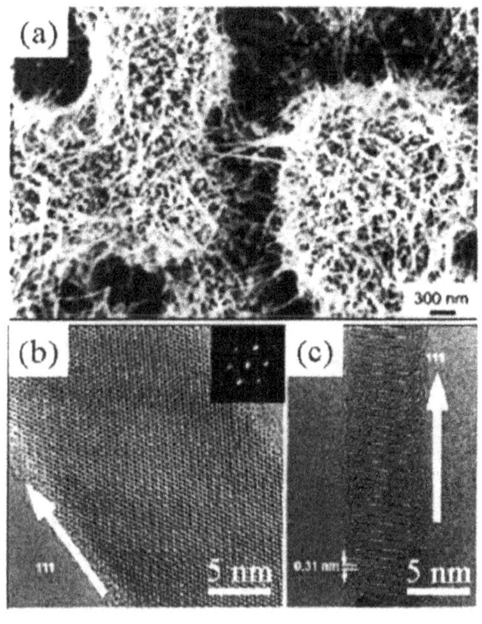


**Fig. 16** (a) SEM image of Si nanowires produced from Au nanocrystals and diphenylsilane at 723 K. (b), (c) HREM images of Si showing predominantly <111> orientation. Inset in (b) shows a FFT of the image. Reproduced with permission from D. C. Lee, T. Hanrath and B. A. Korgel, *Angew. Chem., Int. Ed.*, 2005, **44**, 3573. © 2005 Wiley-VCH Verlag GmbH & Co. KGaA

Solution–liquid–solid synthesis of germanium nanowires gives high yields.[171] In this work, Bi nanocrystals were used as seeds for promoting nanowire growth in TOP, by the decomposition of $GeI_2$ at 623 K. A solid-phase seeded growth with nickel nanocrystals yields Ge nanowires by the thermal decomposition of diphenylgermane in supercritical toluene.[172] A patterned growth of freestanding single-crystalline Ge nanowires with uniform distribution and vertical projection has been accomplished.[173]

High yields of Ge nanowires and nanowire arrays have been obtained by low-temperature CVD by using patterned gold nanoseeds.[174] Ge nanowires have been synthesized starting from the alkoxide, by using a solution procedure involving the injection of a germanium 2,6-dibutylphenoxide solution in oleylamine into a 1-octadecene solution heated to 573 K under an argon atmosphere.[175]

A simple solution-based procedure has been discovered for the synthesis of nanowires of *t*-Se.[176] In this method, selenium powder is first reacted with $NaBH_4$ in water to yield NaHSe which being unstable decomposes to give amorphous selenium. The nascent selenium imparts a wine red colour to the aqueous solution. On standing for a few hours, the solution transforms into amorphous Se in colloidal form. A small portion of the dissolved selenium precipitates as *t*-Se nanoparticles which act as nuclei to form the one-dimensional nanowires as seen in Fig. 17. The same reaction carried out under hydrothermal conditions yields nanotubes as shown in Fig. 17e. Extending the above strategy, Te nanorods, nanowires, nanobelts and junction nanostructures have been obtained.[177] Micellar solutions of nonionic surfactants can be employed to prepare nanowires and nanobelts of *t*-Se.[178] Se nanowires have been prepared at room temperature by using ascorbic acid as a reducing agent in the presence of β-cyclodextrin[179] and also generated at liquid–liquid interface between water and *n*-butyl alcohol.[180]

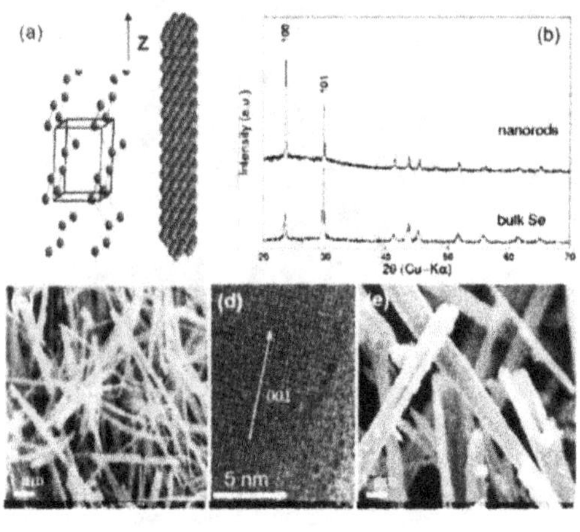


**Fig. 17** (a) Crystal structure of *t*-Se showing a unit cell with helical chains of covalently bonded Se atoms extended along the *c*-axis. The growth direction of the 1D nanostructures is shown along with an atomic model of a rod. (b) XRD patterns of the *t*-Se nanorods and bulk selenium powder used as the starting reagent. (c) SEM image of the Se nanorods obtained after 4 days by reacting 0.025 g of Se with 0.03 g of $NaBH_4$ in 20 ml water. (d) HREM image of a nanorods (arrow indicates the growth direction of the nanorods). (e) SEM image of *t*-Se scrolls obtained under hydrothermal conditions (From Ref. [176]).

### 4.3 Metal oxides

A seed-assisted chemical reaction at 368 K is found to yield uniform, straight, thin single-crystalline ZnO nanorods on a hectogram scale.[181] Zinc oxide nanowires have been synthesized in large quantities using plasma synthesis.[182] Variable-aspect-ratio, single crystalline, 1-D nanostructures (nanowires and nanotubes) can be prepared in alcohol/water solutions by reacting $Zn^{2+}$ precursor with an organic base, tetraammonium hydroxide.[183] It has been found recently that reaction of water with zinc metal powder or foils at room temperature gives ZnO nanowires.[184] A multi-component precursor has been used to produce nanoparticle nanoribbons of ZnO.[185] Porous ZnO nanoribbons are produced by the self-assembly of textured ZnO nanoparticles. Nanobelts of ZnO can be converted into superlattice-structured nanohelices by a rigid lattice rotation or twisting as seen in Fig. 18.[186] Well-aligned crystalline ZnO nanorods along with nanotubes can be grown from aqueous solutions on Si wafers, poly(ethylene terephthlate) and sapphire.[187] Atomic layer deposition was first used to grow a uniform ZnO film on the substrate of choice and to serve as a templating seed layer for the subsequent growth of nanorods and nanotubes. On this ZnO layer, highly oriented two-dimensional (2-D) ZnO nanorod arrays were obtained by solution growth using $Zn(NO_3)_2$ and hexamethylenetetramine in aqueous solution. Controlled growth of aligned ZnO nanorod arrays has been accomplished by an aqueous ammonia solution method.[188] In this method, an aqueous ammonia solution of $Zn(NO_3)_2$ is allowed to react with a zinc-coated silicon substrate at a growth temperature of 333–363 K. 3-D interconnected networks of ZnO nanowires

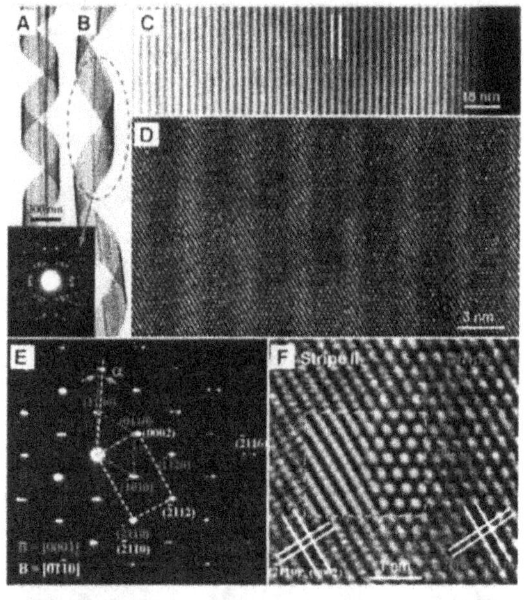


**Fig. 18** (A) Typical low-magnification TEM image of a ZnO nanohelix, showing its structural uniformity. (B) Low-magnification TEM image of a ZnO nanohelix with a larger pitch to diameter ratio. The selected-area ED pattern (SAED, inset) is from a full turn of the helix. (C) Dark-field TEM image from a segment of a nanohelix. The edge at the right and side is the edge of the nanobelt. (D and E) High-magnification TEM image and the corresponding SAED pattern of a ZnO nanohelix with the incident beam perpendicular to the surface of the nanobelt, respectively, showing the lattice structure of the two alternating stripes. (F) Enlarged high-resolution TEM image showing the interface between the two adjacent stripes. From P. X. Gao, Y. Ding, W. Mai, W. L. Hughes, C. Lao and Z. L. Wang, *Science*, 2005, **309**, 1700. Reprinted with permission from AAAS. http://www.sciencemag.org

and nanorods have been synthesized by a high temperature solid–vapour deposition process.[189] Templated electrosynthesis of ZnO nanorods wherein, electroreduction of hydrogen peroxide or nitrate ions is carried out to alter the local pH within the pores of the membrane, with the subsequent precipitation of the metal oxide within the pores.[190]

1-D ZnO nanostructures have been synthesized by oxygen assisted thermal evaporation of zinc on a quartz surface over a large area.[191] Pattern- and feature-designed growth of ZnO nanowire arrays for vertical devices has been accomplished by following a pre-designed pattern and feature with controlled site, shape, distribution and orientation.[192]

The ionic liquid 1-*n*-butyl-3-methylimidazolium tetrafluoroborate has been used to synthesize nanoneedles and nanorods of manganese dioxide ($MnO_2$).[193] Crystalline silica nanowires were prepared by Deepak *et al.*[194] by a carbothermal procedure. Crystalline $SiO_x$ nanowires have also been prepared by a low-temperature iron assisted hydrothermal procedure.[195]

$IrO_2$ nanorods can be grown by metal–organic chemical vapour deposition on sapphire substrates consisting of patterned $SiO_2$ as the non-growth surface.[196] By employing the hydrothermal route, uniform single-crystalline $KNbO_3$ nanowires have been obtained.[197]

MgO nanowires and related nanostructures have been produced by carbothermal synthesis, starting with polycrystalline MgO or Mg with or without the use of metal catalysts.[198] This study has been carried out with different sources of carbon, all of them yielding interesting nanostructures such as nanosheets, nanobelts, nanotrees and aligned nanowires. Orthogonally branched single-crystalline MgO nanostructures have been obtained through a simple chemical vapour transport and condensation process in a flowing $Ar/O_2$ atmosphere.[199]

$Ga_2O_3$ powder reacts with activated charcoal, carbon nanotubes or activated carbon around 1373 K in flowing Ar to give nanowires, nanorods and other novel nanostructures of $Ga_2O_3$ such as nanobelts and nanosheets.[200] Catalyst-assisted VLS growth of single-crystal $Ga_2O_3$ nanobelts has been accomplished by graphite-assisted thermal reduction of a mixture of $Ga_2O_3$ and $SnO_2$ powders under controlled conditions.[201] Zig-zag and helical one-dimensional nanostructures of $\alpha$-$Ga_2O_3$ have been produced by the thermal evaporation of $Ga_2O_3$ in the presence of GaN.[202] Large scale synthesis of $TiO_2$ nanorods has been achieved by the non-hydrolytic sol–gel ester elimination reaction, wherein the reaction is carried out between titanium(IV) isopropoxide and oleic acid.[203] Single-crystalline and well facetted $VO_2$ nanowires with rectangular cross sections have been prepared by the vapour transport method, starting with bulk $VO_2$ powder.[204] Copious quantities of single-crystalline and optically transparent Sn-doped $In_2O_3$ (ITO) nanowires have been grown on gold-sputtered Si substrates by carbon-assisted synthesis, starting with a powdered mixture of the metal nitrates or with a citric acid gel formed by the metal nitrates.[205] Vertically aligned and branched ITO nanowire arrays which are single-crystalline have been grown on yttrium-stabilized zirconia substrates containing thin gold films of 10 nm thickness.[206]

Bicrystalline nanowires of hematite ($\alpha$-$Fe_2O_3$) have been synthesized by the oxidation of pure Fe.[207] Single-crystalline hexagonal $\alpha$-$Fe_2O_3$ nanorods and nanobelts can be prepared by a simple iron–water reaction at 673 K.[208] Mesoporous quasi-single crystalline nanowire arrays of $Co_3O_4$ have been grown by immersing Si or fluorine doped $SnO_2$ substrates in a solution of $Co(NO_3)_2$ and concentrated aqueous ammonia.[209] Networks of $WO_{3-x}$ nanowires are obtained by the thermal evaporation of W powder in the presence of oxygen.[210] The growth mechanism involves ordered oxygen vacancies (100) and (001) planes which are parallel to the (010) growth direction. A general and highly effective one-pot synthetic protocol for producing one-dimensional nanostructures of transition metal oxides such as $W_{18}O_{49}$, $TiO_2$, $Mn_3O_4$ and $V_2O_5$ through a thermally induced crystal growth process starting from mixtures of metal chlorides and surfactants, has been described.[211] Self-coiling nanobelts of $Ag_2V_4O_{11}$ have been obtained by the hydrothermal reaction between $AgNO_3$ and $V_2O_5$.[212] Polymer-assisted hydrothermal synthesis of single crystalline tetragonal perovskite PZT ($PbZr_{0.52}Ti_{0.48}O_3$) nanowires has been carried out.[213] Nanowires of the type II superconductor $YBa_2Cu_4O_8$ have been synthesized by a biomimetic procedure.[214] The nanowires produced by the calcination of a gel containing the biopolymer chitosan and Y, Ba and Cu salts have mean diameters of 50 ± 5 nm with lengths up to 1 μm. Nanorods of $V_2O_5$ prepared by the polyol process self-assemble into microspheres.[215]

### 4.4 Metal chalcogenides

ZnS nanowires and nanoribbons with wurtzite structure can be prepared by the thermal evaporation of ZnS powder onto silicon substrates, sputter-coated with a thin (~25 Å) layer of Au film.[216] Thermal evaporation of a mixture of ZnSe and activated carbon powders in the presence of a tin-oxide based catalyst yields tetrapod-branched ZnSe nanorod architectures.[217] 1-D nanostructures of CdS have been formed on Si substrates by a thermal evaporation route.[218] The shapes of the 1-D CdS nanoforms were controlled by varying the experimental parameters such as temperature and position of the substrates. Nanorods of luminescent cubic CdS are obtained by injecting solutions of anhydrous cadmium acetate and sulfur in octylamine into hexadecylamine.[219] CdSe nanowires have been produced by the cation-exchange route.[220] By employing the cation-exchange reaction between $Ag^{2+}$ and $Cd^{2+}$, $Ag_2Se$ nanowires are transformed into single-crystal CdSe nanowires. A single-source molecular precursor has been used to obtain blue-emitting, cubic CdSe nanorods (~2.5 nm diameter and 12 nm length) at low temperatures.[221] Thin aligned nanorods and nanowires of ZnS, ZnSe, CdS and CdSe can be produced by using microwave-assisted methodology by starting from appropriate precursors.[222] An organometallic preparation of CdTe nanowires with high aspect ratios in the wurtzite structure has been described.[223] Thermal decomposition of copper-diethyldithiocarbamate ($CuS_2CNEt_2$) in a mixed binary surfactant solvent of dodecanethiol and oleic acid at 433 K gives rise to single-crystalline high aspect ratio ultrathin nanowires of hexagonal $Cu_2S$.[224] CdX(X = S, Se) nanorods have been synthesized in ionic liquids in the presence of ethylenediamine (Fig. 9b ).[87] ZnSe nanorods can also be prepared by the decomposition of zinc acetate in the presence of dimethylselenourea in [BMIM][$MeSO_4$] ionic liquid.

Atmospheric pressure CVD can be employed to prepare arrays and networks of PbS nanowires.[225] Monodisperse PbTe nanorods of sub-10 nm diameter are obtained by sonoelectrochemical means by starting with a lead salt and $TeO_2$ along with nitrilotriacetic acid.[226] Taking bismuth citrate and thiourea in DMF, well-segregated, crystalline $Bi_2S_3$ nanorods have been synthesized by a reflux process.[227] Single-crystalline $Bi_2S_3$ nanowires have also been obtained by using lysozyme which controls the morphology and directs the formation of the 1D inorganic material.[228] In this method, $Bi(NO_3)_3 \cdot 5H_2O$, thiourea and lysozyme are reacted together at 433 K under hydrothermal conditions. A solvent-less synthesis of orthorhombic $Bi_2S_3$ nanorods and nanowires with high aspect ratios (> 100) has been accomplished by the thermal decomposition of bismuth alkylthiolates in air around 500 K in the presence of the capping agent, octanoate.[229] Single-crystalline $Bi_2Te_3$ nanorods have been synthesized by a template free method at 373 K by the addition of thioglycolic acid or L-cysteine to a bismuth chloride solution.[230] $GeSe_2$ nanowires have been obtained by the decomposition of organoammonium selenide (See Fig. 19).[231] GeTe nanowires are obtained by a VLS process starting with GeTe powder using a Au nanoparticle catalyst.[232] Nanowires of copper indium diselenide have been prepared by the reaction of Se powder with $In_2Se_3$ and anhydrous $CuCl_2$ under solvothermal conditions.[233]

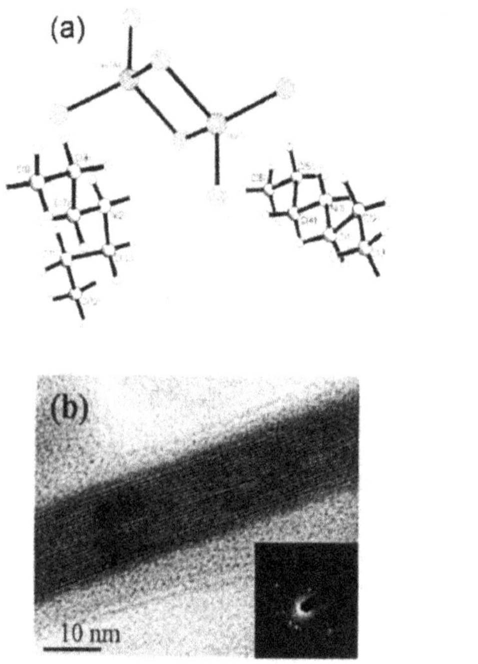


**Fig. 19** (a) Crystal structure of the organoammonium germanium selenide precursor and (b) HREM image of a $GeSe_2$ nanowire. The inset shows the typical SAED pattern. (From Ref. [231]).

### 4.5 Metal pnictides and other nanowires

Single crystalline AlN, GaN and InN nanowires illustrated in Fig. 20 can be deposited on Si substrates covered with Au islands by using urea complexes formed with the trichlorides of Al, Ga and In as the precursors.[94] Single crystalline GaN nanowires are also obtained by the thermal evaporation/decomposition of $Ga_2O_3$ powders with ammonia at 1423 K directly onto a Si substrate coated with a Au film.[234] Direction-dependent homoepitaxial

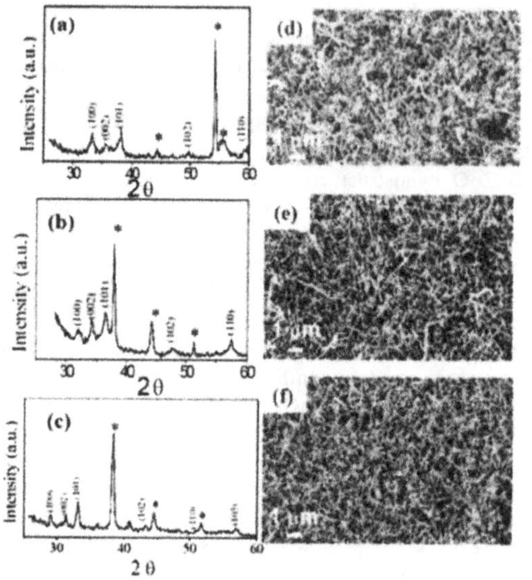


**Fig. 20** XRD patterns of (a) AlN, (b) GaN and (c) InN nanowires respectively (* indicates peaks arising due to substrate or gold). SEM images of (d) AlN, (e) GaN, (f) InN nanowires. (From Ref. [94]).

growth of GaN nanowires has been achieved by controlling the Ga flux during direct nitridation in dissociated ammonia (Fig. 21).[235] The nitridation of Ga droplets at a high flux leads

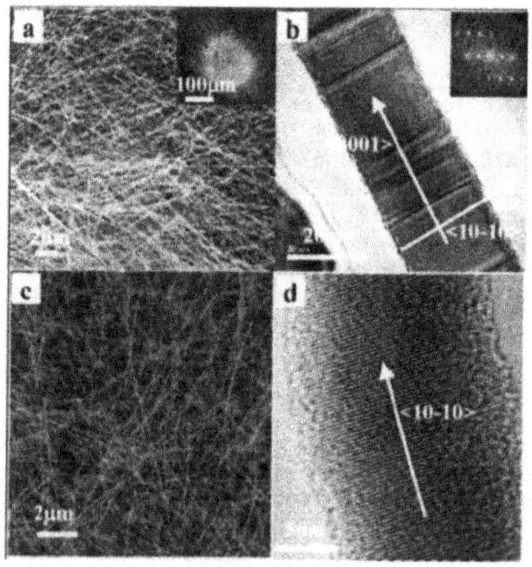


**Fig. 21** (a) SEM images of GaN nanowires. The inset shows the spontaneous nucleation and growth of multiple nanowires directly from a larger Ga droplet. (b) HREM image of a GaN nanowire. The inset shows FFT of the HREM image, (c) SEM image showing GaN nanowires with diameters less than 30 nm resulting from the reactive-vapour transport of a controlled Ga flux in a $NH_3$ atmosphere. (d) HREM image of a GaN nanowire from the sample shown in (c) indicating that the growth direction is <10-10>. Reproduced with permission from H. Li, A. H. Chin and M. K. Sunkara, *Adv. Mater.*, 2006, **18**, 216. © 2006 Wiley-VCH Verlag GmbH & Co. KGaA

to GaN nanowire growth in the $c$-direction (<1000>) as seen in Fig. 21b, while the nitridation with a low Ga flux leads to growth in the $a$-direction (<10-10>) (Fig. 21d). InN nanowires with uniform diameters have been obtained in large quantities by the reaction of $In_2O_3$ powders in ammonia.[236] A general method for the synthesis of Mn-doped nanowires of CdS, ZnS and GaN based on metal nanocluster-catalyzed chemical vapour deposition has been described.[237] Vertically aligned, catalyst-free InP nanowires have been grown on InP(111)B substrates by CVD of trimethylindium and phosphine at 623–723 K.[238]

Nanowires of $InAs_{1-x}P_x$ and $InAs_{1-x}P_x$ heterostructure segments in InAs nanowires, with the P concentration varying from 22% to 100%, have been grown by VLS method.[239] Single-crystalline nanowires of $LaB_6$, $CeB_6$ and $GdB_6$ have been deposited on a Si substrate by the reaction of the rare-earth chlorides with $BCl_3$ in the presence of hydrogen.[240] Starting from $BiI_3$ and $FeI_2$, $Fe_3B$ nanowires were synthesized on Pt and Pd (Pt/Pd) coated sapphire substrates by CVD at 1073 K.[241] The morphology of the $Fe_3B$ nanowires can be controlled by manipulating the Pt/Pd film thickness and growth time, the typical diameter is in the 5–50 nm range and the length in the 2–30 μm range. Nanowires and nanoribbons of $NbSe_3$ have been obtained by the direct reaction of Nb and Se powders.[242] A one-pot metal–organic synthesis of single-crystalline CoP nanowires with uniform diameters has been reported.[243] The method involves the thermal decomposition of cobalt(II) acetylacetonate and tetradecylphosphonic acid in a mixture of TOPO and hexadecylamine.

### 4.6 Oriented attachment

Oriented attachment of nanocrystals is employed to fabricate one-dimensional as well as complex nanostructures. Thus, nanotubes and nanowires of II–VI semiconductors have been synthesized using surfactants.[244] In Fig. 22, we show TEM image of CdS nanowires and CdSe nanotubes obtained using Triton 100-X as surfactant. Oriented attachment-like growth has been observed in CdS, ZnS and CuS nanorods prepared by using hydrogels as templates.[245] The nanorods or nanotubes of CdS and other materials produced in this manner actually consists of nanocrystals. The synthesis of $SnO_2$ nanowires from nanoparticles has been investigated.[246] CdSe nanorods can be formed by redox-assisted asymmetric Ostwald ripening of CdSe dots to rods.[247]

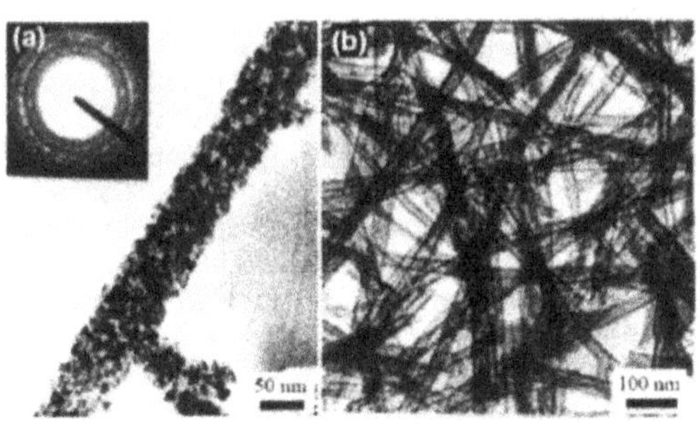


**Fig. 22** TEM images of (a) CdS nanowires (b) CdSe nanotubes obtained by using Triton 100-X as the surfactant. Inset in (a) shows the SAED pattern of the CdS nanowires. Reprinted with permission from C. N. R. Rao, A. Govindaraj, F. L. Deepak, N. A. Gunari and M. Nath, *Appl. Phys. Lett.*, 2001, **78**, 1853. © 2001, American Institute of Physics.

PbSe and cubic ZnS nanowires as well as complex one-dimensional nanostructures can be obtained in solution through oriented attachment of nanocrystals.[248,249] In Fig. 23, star-shaped PbSe nanocrystals and branched nanowires are shown.

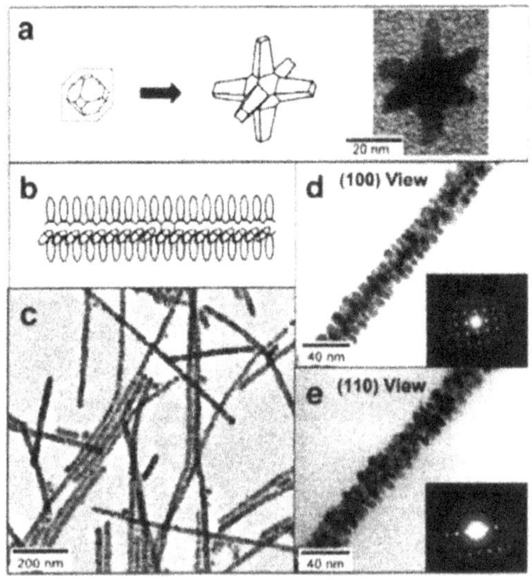


**Fig. 23** (a) Star-shape PbSe nanocrystals and (b–e) radially branched nanowires. (d) TEM image of the (100) view of the branched nanowire and the corresponding selected area electron diffraction pattern. (e) TEM image of the (110) view of the branched nanowire and the corresponding selected area electron diffraction pattern. Reprinted with permission from K.-S. Cho, D. V. Talapin, W. Gaschler and C. B. Murray, *J. Am. Chem. Soc.*, 2005, **127**, 7140. © 2005 American Chemical Society.

### 4.7 Coaxial cables and other hybrid nanostructures

A general procedure has been proposed for producing chemically bonded ceramic oxide coatings on carbon nanotubes and inorganic nanowires wherein reactive metal chlorides are reacted with acid-treated carbon nanotubes or metal oxide nanowires, followed by hydrolysis with water.[250] On repeating the above process several times followed by calcination, oxide coatings of the desired thickness are obtained. Core-sheath CdS and polyaniline (PANI) coaxial nanocables with enhanced photoluminescence have been fabricated by an electrochemical method using a porous anodic alumina membrane as the template.[251] SiC nanowires can be coated with Ni and Pt nanoparticles (~3 nm) by plasma-enhanced CVD.[252] Single and double-shelled coaxial core-shell nanocables of GaP with $SiO_x$ and carbon (GaP/$SiO_x$, GaP/C, GaP/$SiO_x$/C), with selective morphology and structure, have been synthesized by thermal CVD.[253] Silica-sheathed 3C-$Fe_7S_8$ has been prepared on silicon substrates with $FeCl_2$ and sulfur precursors at 873–1073 K.[254]

Nanowires containing multiple GaP–GaAs junctions are grown by the use of metal–organic vapour phase epitaxy on $SiO_2$.[255] Silica-coated PbS nanowires have been deposited by CVD using $PbCl_2$ and S on silicon substrates at temperatures between 650 and 973 K.[256] A novel silica-coating procedure has been devised for CTAB-stabilized gold nanorods and for the hydrophobation of the silica shell with octadecyltrimethoxysilane (OTMS).[257] A combination of the polyelectrolyte layer-by-layer technique and the hydrolysis followed by condensation of tetraethoxylorthosilicate in a 2-propanol–water mixture leads to homogeneous coatings with control on shell thickness. On the other hand, the strong binding of CTAB molecules to the gold surface makes surface hydrophobation difficult but the functionalization with OTMS, which contains a long hydrophobic hydrocarbon chain, allows the particles to be transferred into nonpolar organic solvents such as chloroform.

Fabrication of InP/InAs/InP core-multishell heterostructure nanowire arrays shown in Fig. 24 has been achieved by selective area metal–organic vapour phase epitaxy.[258] These core–multishell nanowires were designed to accommodate a strained InAs quantum well layer in a higher band gap InP nanowire. Precise control over the nanowire growth direction and the heterojunction formation enabled the successful fabrication of the nanostructure in which all the three layers were epitaxially grown without the assistance of a catalyst.

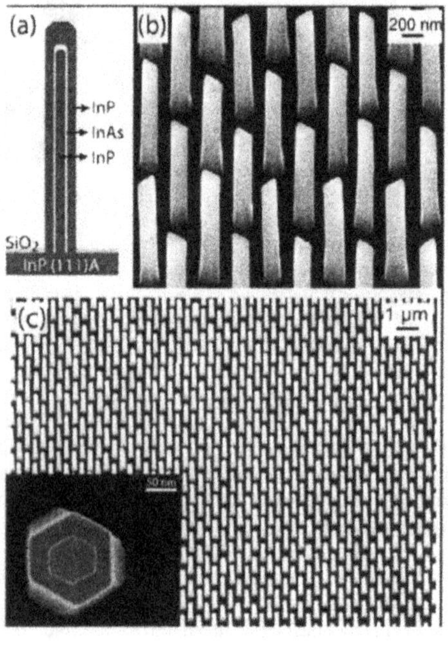


**Fig. 24** (a) Schematic cross-sectional image of InP/InAs/InP core-multishell nanowire. (b) SEM image of periodically aligned InP/InAs/InP core-multishell nanowire array. (c) SEM image showing highly dense ordered arrays of core-multishell nanowires. Schematic illustration and high resolution SEM cross-sectional image of a typical core–multishell nanowire observed after anisotropic dry etching and stain etching. Inset shows the top view of a single nanowire. Reprinted with permission from P. Mohan, J. Motohisa and T. Fukui, *Appl. Phys. Lett.*, 2006, **88**, 133105. © 2006, American Institute of Physics.

## 5. Inorganic nanotubes

There has been considerable interest in the synthesis of fullerenes and nanotubes of inorganic materials.[4,259] These efforts have primarily been focused on layered inorganic compounds such as metal dichalcogenides (sulfides, selenides, and tellurides), halides (chlorides, bromides, and iodides), oxides and boron nitride, which possess structures comparable to the structure of graphite. Tenne and coworkers[260–261] first recognized that nanosheets of Mo and W dichalcogenides are unstable against folding and closure and that they can form fullerene-like nanoparticles and nanotubes. Nanotubes and fullerene-like nanoparticles of dichalcogenides such as $MoS_2$, $MoSe_2$ and $WS_2$ have been prepared by processes such as arc-discharge and laser ablation.[262–264] Chemical routes for the synthesis of fullerenes and nanotubes of metal chalcogenides are more versatile and popular. Nanotubes of $MoS_2$ and $WS_2$ can be obtained by heating $MoO_3/WO_3$ in a reducing atmosphere and then reacted with $H_2S$.[265] In the case of metal selenide nanotubes, $H_2Se$ is used instead of $H_2S$.[266] Recognizing that $MoS_3$ and $WS_3$ are likely intermediates in the formation of the disulfides, the trisulfides have been directly decomposed in a $H_2$ atmosphere to obtain the disulfide nanotubes.[267] Similarly, diselenide nanotubes have been obtained by the decomposition of metal triselenides.[268] The trisulfide route provides a general route for the synthesis of the nanotubes of many metal disulfides such as $NbS_2$ and $HfS_2$.[269,270] The decomposition of precursor ammonium salts $(NH_4)_2MX_4$ (X = S, Se; M = Mo, W) also yields nanotubes.[267] The trichalcogenides are intermediates in the decomposition of the ammonium salts.

Bando and co-workers[271] have prepared BN nanotubes by the reaction of MgO, FeO and B in the presence of $NH_3$ at 1400 °C. Reaction of boric acid or $B_2O_3$ with $N_2$ or $NH_3$ at high temperatures in the presence of carbon or catalytic metal particles has been employed in the preparation of BN nanotubes.[272] Boron nitride nanotubes can be grown directly on substrates at 873 K by a plasma-enhanced laser-deposition technique.[273] Recently, GaN nanotube brushes have been prepared using amorphous carbon nanotubes templates obtained using AAO membranes.[274]

Large-scale synthesis of Se nanotubes has been carried out in the presence of CTAB.[275] Nanotubes and onions of GaS and GaSe have been generated through laser and thermally induced exfoliation of the bulk powders (see Fig. 25).[276] Single-wall nanotubes of $SbPS_{4-x}Se_x$ ($0 < x < 3$) with tuneable bandgap have been synthesized.[277] GaP nanotubes with zinc blende structure have been obtained by the VLS growth.[278] Open-ended gold nanotube arrays have been obtained by the electrochemical deposition of Au on to an array of nickel nanorod templates followed by selective removal of the templates.[279] Free-standing, electro-conductive nanotubular sheets of indium tin oxide with different In/Sn ratios have been fabricated by using cellulose as the template.[280] A low-temperature route for synthesizing highly oriented ZnO nanotubes/nanorod arrays has been reported.[281] In this work, a radio frequency magnetron-sputtering technique was used to prepare ZnO-film-coated substrates for subsequent growth of the oriented nanostructures. Controllable syntheses of $SiO_2$ nanotubes with dome shaped interiors have been prepared by pyrolysis of silanes over Au catalysts.[282] High aspect-ratio, self-organized nanotubes of $TiO_2$ are obtained by anodization of titanium.[283] These self-organized porous structures consist of pore arrays with a uniform pore diameter of ~100 nm and an average spacing of 150 nm. The pore mouths are open on the top of the layer while on the bottom of the structure the tubes are closed by the presence of an about 50-nm thick barrier of $TiO_2$. Electrochemical etching of titanium under potentiostatic

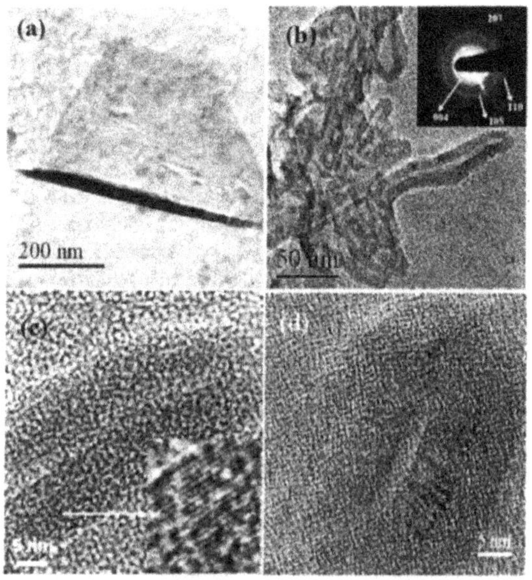


**Fig. 25** TEM images revealing (a) rolling of an exfoliated GaS layer and (b) GaS nanotubes obtained by solvent irradiated laser irradiation. (c) HREM image of a nanotube ($d$ = 3.15 Å). Insets shows GaS layer (d) TEM image of a GaSe onion obtained by laser irradiation in *n*-octylamine. Reprinted with permission from U. K. Gautam, S. R. C. Vivekchand, A. Govindaraj, G. U. Kulkarni, N. R. Selvi and C. N. R. Rao, *J. Am. Chem. Soc.*, 2005, **127**, 3658. © 2005, American Chemical Society.

conditions in fluorinated dimethyl sulfoxide and ethanol (1 : 1) under a range of anodizing conditions gives rise to ordered $TiO_2$ nanotube arrays.[284] $TiO_2$–B nanotubes can be prepared by the hydrothermal method.[285] Lithium is readily intercalated into the $TiO_2$–B nanotubes up to a composition of $Li_{0.98}TiO_2$ compared with $Li_{0.91}TiO_2$ for the corresponding nanowires. Intercalation of alkali metals into titanate nanotubes has also been investigated.[286] Highly crystalline $TiO_2$ nanotubes have been synthesized by hydrogen peroxide treatment of low crystalline $TiO_2$ nanotubes prepared by hydrothermal methods.[287] $TiO_2$ nanotubes with rutile structure have been prepared by using carbon nanotubes as templates.[288] Anatase nanotubes can be nitrogen doped by ion-beam implantation.[289]

$RuO_2$ nanotubes have been synthesized by the thermal decomposition of $Ru_3(CO)_{12}$ inside anodic alumina membranes.[290] Transition metal oxide nanotubes have been prepared in water using iced lipid nanotubes as the template.[291] Self-assembled cholesterol derivatives act as a template as well as a catalyst for the sol–gel polymerization of inorganic precursors to give rise to double-walled tubular structures of transition metal oxides.[292] Hydrothermal synthesis of single-crystalline γ-$Fe_2O_3$ nanotubes has been accomplished.[293] Nanotubes of single crystalline $Fe_3O_4$ have been prepared by wet-etching of the MgO inner cores of MgO/$Fe_3O_4$ core-shell nanowires.[294] Cerium oxide nanotubes are prepared by the controlled annealing of the as-formed $Ce(OH)_3$ nanotubes.[295]

Long hollow inorganic nanoparticle nanotubes with a nanoscale brick wall structure of clay mineral platelets have been synthesized by templating of block copolymer electrospun fibers with clay mineral platelets followed by interlinking of the platelets using condensation reactions.[296] Similarly, the construction of hollow inorganic nanospheres and nanotubes (inorganic nanostructures) with tunable wall thicknesses (with hollow interiors) is demonstrated by coating on self-assembled polymeric templates (nano-objects) with a thin $Al_2O_3$ layer by ALD, followed by removal of the polymer template upon heating.[297] The morphology of the nano-object (*i.e.*, spherical or cylindrical) is controlled by the block lengths of the copolymer. The thickness of the $Al_2O_3$ wall is controlled by the number of ALD cycles. Formation of ultra-long single-crystalline $ZnAl_2O_4$ spinel nanotubes, through a spinel-forming interfacial solid-state reaction of core-shell ZnO–$Al_{2O3}$ nanowires involving the Kirkendall effect, has been reported.[298] Polycrystalline lead titanate nano- and microtubes with diameters ranging from a few tens of nanometers up to one micron were fabricated by wetting ordered porous alumina and macroporous silicon with precursor oligomers coupled with templated thermolysis.[299] WC nanotubes can be synthesized by the thermal decomposition of $W(CO)_6$ in the presence of Mg powder at 1173 K under the autogenic pressure of the precursors in a closed Swagelok reactor.[300]

## 6. Nanocrystaline films generated at the liquid–liquid interface

The liquid–liquid interface has been exploited recently for obtaining ultra-thin nanocrystalline films of a variety of materials. The method involves dissolving an organic precursor of the relevant metal in the organic layer and the appropriate reagent in the aqueous layer. The product formed by the reaction at the interface contains ultra-thin nanocrystalline films of the relevant material formed by closely packed nanocrystals. This simple technique has been shown to yield nanocrystals of metals such as Au, Ag, Pd and Cu, chalcogenides such as CdS, CdSe, ZnS, CoS, NiS, CuS and PbS and oxides such as γ-$Fe_2O_3$, ZnO and CuO.[301–310] In a typical preparation of Au nanocrystalline films, a solution of $Au(PPh_3)Cl$ in toluene was allowed to stand in contact with aqueous alkali in a beaker at 300 K. Once the two liquid layers were stable, tetrakis(hydroxymethyl)phosphonium chloride (THPC) was injected into the aqueous layer using a syringe with minimal disturbance to the toluene layer.[304] The interface first appears pink, finally growing a robust Au film at the interface. This film could be converted either to a gold organosol or a hydrosol by using appropriate capping agents in the organic and aqueous layers. The thickness and the particle size of the nanocrystals (shown in Fig. 26) is found to be dependent on the reaction conditions employed, such as reactant concentrations, reaction time and temperature.[305] The mean diameters of the nanocrystals formed at 30, 45, 60 and 75 °C are 7, 10, 12 and 15 nm respectively. The liquid–liquid interface has also been employed to prepare nanocrystalline films of binary alloys of Au–Ag and Au–Cu, and also ternary Au–Ag–Cu alloys,[306] by starting with an appropriate mixture of the corresponding metal precursors. Nanocrystalline films of metal chalcogenides such as CdS have been prepared at the toluene–water interface. In a typical preparation of a CdS nanocrystals, $Na_2S$ was dissolved in water in a beaker and cadmium cupferronate [$Cd(cup)_2$], was dissolved in toluene by

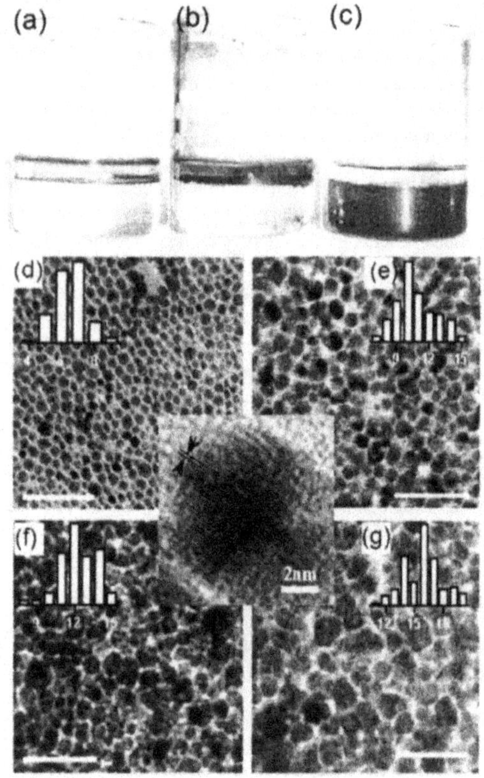

**Fig. 26** (a) Nanocrystalline film of Au formed at the toluene–water interface, (b) When dodecanethiol is added to the toluene layer, the film breaks up, forming an organosol of Au, (c) Au hydrosol obtained when mercaptoundecanoic acid is added to water. TEM images of the ultra-thin nanocrystalline Au films obtained at the liquid–liquid interface after 24 h: (d) 30 °C, (e) 45 °C, (f) 60 °C and (g) 75 °C. The histograms of particle size distribution are shown. The scale bars correspond to 50 nm. A high-resolution image of an individual particle is shown at the center. Reprinted with permission from V. V. Agrawal, G. U. Kulkarni and C. N. R. Rao, *J. Phys. Chem. B*, 2005, **109**, 7300. © 2005, American Chemical Society.

ultrasonication.[307] A few drops of *n*-octylamine were added to the $Cd(cup)_2$ solution in order to make it completely soluble. The toluene solution was slowly added to a beaker containing the aqueous $Na_2S$ solution. The interface started appearing yellow within a few minutes and a distinct film was formed after 10 h. $\gamma$-$Fe_2O_3$ nanocrystals have also been prepared at the toluene–water interface.

While many metals and metal sulfides yield films of nanocrystal assemblies at the liquid–aqueous interface, extended ultra-thin single-crystalline films are obtained in the case of CuS.[308] Single-crystalline films of CuSe, ZnS, PbS, $Cu(OH)_2$ and ZnO have also been prepared similarly. To prepare a CuS film, 75 mL of 0.12 mM $Cu(cup)_2$ solution in toluene was slowly added to an aqueous solution of 0.5 mM $Na_2S$ (75 ml) taken in a crystallization dish (10 cm diameter). An excess of $Na_2S$ was required in order to prevent the formation of $Cu_2S$. The interface gradually turns green, and the CuS film formed at the interface after 12 h, while the two liquid phases remained colourless. The film grows slowly with time, first appearing as green-islands at the interface (in the initial 1–2 h) and slowly covering the entire interface. The film is fairly continuous and extends over a wide area as can be seen in Fig. 27. Small fragments on the edges of the film are also seen as the film breaks while lifting from the interface (Fig. 27a ). The single-crystalline and essentially defect-free nature of the film can be inferred from the HREM image and the SAED pattern (Fig. 27b). The lattice spacing of 2.7 Å in the HREM image corresponds to the separation between (006) planes of the hexagonal CuS phase. The diffraction spots could be indexed on the basis of the hexagonal structure ($P63/mmc$, $a$ = 3.792 Å and $c$ = 16.34 Å). The thickness of the film was estimated to be ~50 nm from AFM and ellipsometric studies. Thicker films could be formed using higher concentration of reactants. The films prepared at higher temperatures are, however, less continuous and form flakes and rods. Nanorods and nanocrystals of various sizes and shapes as seen in Fig. 27d) and Fig. 27e were obtained upon sonication of the CuS films.

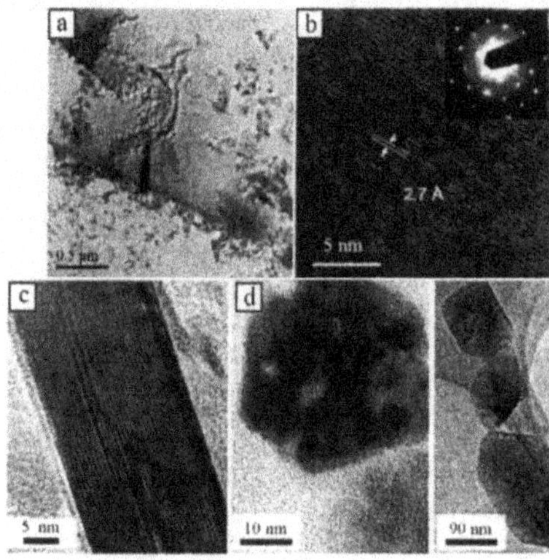

**Fig. 27** (a) TEM images of an ultra-thin 1 µm of CuS obtained at the toluene–water interface. (b) HREM image of the film. Inset shows the corresponding SAED pattern. (c) HREM image of a rod like fragment obtained at 343 K. (d) Hexagonal CuS nanocrystals obtained by sonication of the film obtained with 2 mg of $Cu(cup)_2$ and (e) the nanocrystals obtained by using 1 mg $Cu(cup)_2$. Reprinted with permission from U. K. Gautam, M. Ghosh and C. N. R. Rao, *Langmuir*, 2004, **20**, 10775. © 2004, American Chemical Society.

Extended single-crystalline films of ZnS are obtained by this method starting with $Zn(cup)_2$ and $Na_2S$.[309] The UV-visible absorption spectrum of a ZnS film obtained by reacting 5 mg of $Zn(cup)_2$ in 50 ml of toluene at ~30 °C for 12 h shows an absorption band at ~320 nm, close to that of bulk ZnS. The PL spectrum of the ZnS film displays a broad emission band centered at ~425 nm due to the presence of sulfur vacancies in the ZnS lattice. A blue-shift of the absorption band to 285 nm was observed when the reaction time was restricted to 1 h, suggesting a thin, incompletely formed film. Unlike chemical methods such as the LB technique, where non-single-crystalline films are obtained by assembling nanocrystals or CVD and related techniques where stringent conditions as well as substrates are required, the interface method is simple and can be extended to a variety of materials. Using a similar approach CdSe nanofilms have also been prepared.[310]

## 7. Nanowalls

Thermal exfoliation of GaS and GaSe gives rise to nanowalls (Fig. 28). Nanowalls of carbon, which are actually interconnected 2D nanosheets of carbon vertically standing on a substrate has been described by Wu and co-workers.[311] There are reports of ZnO nanowalls and ZnO nanorods grow from the nodes of nanowalls.[312] As mentioned in the previous section, thermal exfoliation of GaSe gave rise to solid deposits in the cooler end of the sealed tube.[313] Deposits with similar morphology were obtained in the case of GaS as well. The deposits contained wall structures with smooth curved surfaces as revealed by the SEM images. The XRD pattern of the sample could be indexed on the hexagonal phase of GaSe. The EDAX spectrum recorded at various locations of the sample confirmed the Ga : Se ratio to be 1 : 1. TEM images revealed that the walls are transparent, especially at the edges, indicating a thickness of around a few nanometers. The nanowalls are single-crystalline, as established by the HREM images as well as the SAED patterns. The lattice spacing observed in the HREM image of 3.229 Å corresponds to the separation between the [100] planes of GaSe in the space group $P6_3/mmc$. In the initial stage, the exfoliated sheets deposited in the cooler end of the tube melt forming droplets. Flower-like nanostructures form around these droplets and grow with time forming extended network structures. The $Ga_2O_3$ nanowalls were obtained by heating GaS and GaSe nanowalls in air at 823 K. On the other hand, heating the GaS and GaSe nanowalls in $NH_3$, GaN nanowalls were obtained.

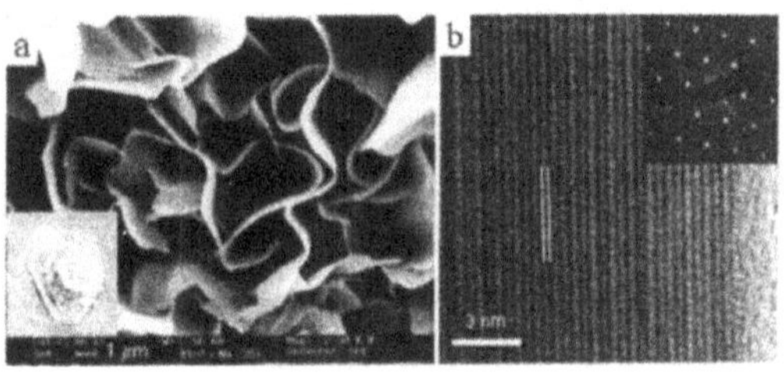


**Fig. 28** (a) SEM image of GaSe nanowalls deposited at 673 K. (b) HREM images of a nanowall. The separation between the lattice planes (3.229 Å ≈ $d_{100}$) is shown in the image. Inset shows the SAED pattern. (From Ref. [313]).

## 8. Conclusions

The preceding presentation reflects the tremendous contribution of chemists to the synthesis of inorganic nanomaterials. Beside the use of known methods, the synthesis of nanomaterials has required many newer strategies geared to prepare objects of different dimensionalities. Clearly, there will be continued efforts by a large body of chemists in preparing inorganic nanomaterials, both known and new, by employing novel and improved methods. In addition to synthesis, functionalization, solubilization and assembly of nanomaterials are aspects of great relevance where again chemistry plays a crucial role. While we have not touched on these aspects in this article, it is necessary that we recognise their importance in the manipulation and use of nanomaterials.

## References


1 G. Schmid, Clusters and Colloids, From *Theory to Applications*, VCH, Weinheim, 1994.
2 *The Chemistry of Nanomaterials*, Ed. C. N. R. Rao, A. Muller and A. K. Cheetham, 2004, Wiley-VCH Verlag, Weinheim, vol. 1 & 2.
3 *Nanomaterials Chemistry: Recent Developments*, Ed. C. N. R. Rao, A. Muller and A. K. Cheetham, 2007, Wiley-VCH-Verlag, Weinheim.
4 C. N. R. Rao and A. Govindaraj, *Nanotubes and Nanowires*, RSC series on Nanoscience, London, 2005.
5 C. N. R. Rao, A. Govindaraj and S. R. C. Vivekchand, *Ann. Rep. Prog. Chem.*, Royal Society of Chemistry: London, 2006, **102**, 20.
6 C. N. R. Rao, P. J. Thomas and G. U. Kulkarni, *Nanocrystals: Synthesis, Properties and Applications*, Springer series on material science: 95, 2007.
7 (*a*) C. Burda, X. Chen, R. Narayanan and M. A. El-Sayed, *Chem. Rev.*, 2005, **105**, 1025; (*b*) M. Rajamathi and R. Seshadri, *Curr. Opin. Solid State Mater. Sci.*, 2002, **6**, 337.
8 E. A. Hauser and J. E. Lynn, *Experiments in Colloid Chemistry*, p. 18 (McGraw-Hill: New York 1940).
9 (*a*) J. Turkevich, P. C. Stevenson and J. Hillier, *Spec. Discuss. Faraday Soc.*, 1951, **11**, 55; (*b*) J. Turkevich, *Gold Bull.*, 1985, **18**, 86.
10 D. G. Duff, A. Baiker and P. P. Edwards, *Langmuir*, 1993, **9**, 2301.
11 M. Brust, M. Walker, D. Bethell, D. J. Schiffrin and R. Whyman, *J. Chem. Soc., Chem. Commun.*, 1994, 801.
12 K. V. Sarathy, G. Raina, R. T. Yadav, G. U. Kulkarni and C. N. R. Rao, *J. Phys. Chem. B*, 1997, **101**, 9876.
13 K. V. Sarathy, G. U. Kulkarni and C. N. R. Rao, *Chem. Commun.*, 1997, 537.
14 P. J. Thomas, G. U. Kulkarni and C. N. R. Rao, *J. Phys. Chem. B*, 2000, **104**, 8138.
15 T. Teranishi and M. Miyake, *Chem. Mater.*, 1998, **10**, 594.
16 M. Schulz-Dobricks, K. V. Sarathy and M. Jansen, *J. Am. Chem. Soc.*, 2005, **127**, 12816.
17 (*a*) I. Pastoriza-Santos and L. M. Liz-Marzan, *Langmuir*, 2002, **18**, 2888; (*b*) I. Pastoriza-Santos and L. M. Liz-Marzan, *Pure Appl. Chem.*, 2000, **72**, 83.
18 N. R. Jana and X. Peng, *J. Am. Chem. Soc.*, 2003, **125**, 14280.
19 R. Jin, Y. Cao and C. A. Mirkin, *Science*, 2001, **294**, 1901.
20 R. Jin, Y. Cao, E. Hao, G. S. Métraux, G. C. Schatz and C. A. Mirkin, *Nature*, 2003, **425**, 487.
21 G. S. Metraux and C. A. Mirkin, *Adv. Mater.*, 2005, **17**, 412.
22 E. Hao, R. C. Bailey and G. C. Schatz, *Nano Lett.*, 2004, **4**, 327.
23 C. H. Kuo and M. H. Huang, *Langmuir*, 2005, **21**, 2012.
24 Y. Xiao, B. Shylahovsky, I. Popov, V. Pavlov and I. Wilner, *Langmuir*, 2005, **21**, 5659.
25 M. Zhou, S. Chen and S. Zhao, *J. Phys. Chem. B*, 2006, **110**, 4510.
26 B. J. Wiley, Y. Xiong, Z.-Y. Li, Y. Yin and Younan Xia, *Nano Lett.*, 2006, **6**, 765.
27 T. Tsukatani and H. Fujihara, *Langmuir*, 2005, **21**, 12093.
28 G. S. Fonseca, A. P. Umpierre, P. F. P. Fichtner, S. R. Teixeira and J. Dupont, *Chem.–Eur. J.*, 2003, **9**, 3263.
29 H. Itoh, K. Naka and Y. Chujo, *J. Am. Chem. Soc.*, 2004, **126**, 3026.
30 O. Margeat, C. Amiens, B. Chaudret, P. Lecante and R. E. Banfield, *Chem. Mater.*, 2005, **17**, 107.
31 J. D. Hoefelmeyer, K. Niesz, G. A. Somarjai and T. D. Tilley, *Nano Lett.*, 2005, **5**, 435.
32 C. A. Stowell and B. A. Korgel, *Nano Lett.*, 2005, **5**, 1203.
33 S. Ghosh, M. Ghosh and C. N. R. Rao, *J. Cluster Sci.*, 2007, **18**, 97.
34 O. Margeat, C. Amiens, B. Chaudret, P. Lecante and R. E. Banfield, *Chem. Mater.*, 2005, **17**, 107.
35 H. Song, F. Kim, S. Connor, G. A. Samorjai and P. Yang, *J. Phys. Chem. B*, 2005, **109**, 188.
36 C. Roychowdhury, F. Matsumoto, P. F. Mutolo, H. A. Abruna and F. J. DiSalvo, *Chem. Mater.*, 2005, **17**, 5871.
37 G. De and C. N. R. Rao, *J. Mater. Chem.*, 2005, **15**, 891.
38 M. Chen, J. Kim, J. P. Liu, H. Fan and S. Sun, *J. Am. Chem. Soc.*, 2006, **128**, 7132.

39 J. Rockenberger, E. C. Scher and A. P. Alivisatos, *J. Am. Chem. Soc.*, 1999, **121**, 11595.
40 M. Ghosh, E. V. Sampathkumaran and C. N. R. Rao, *Chem. Mater.*, 2005, **17**, 2348.
41 M. Ghosh, K. Biswas, A. Sundaresan and C. N. R. Rao, *J. Mater. Chem.*, 2006, **16**, 106.
42 M. Ghosh and C. N. R. Rao, *Chem. Phys. Lett.*, 2004, **393**, 493.
43 M. Ghosh, R. Seshadri and C. N. R. Rao, *J. Nanosci. Nanotechnol.*, 2004, **4**, 136.
44 S. Thimmaiah, M. Rajamathi, N. Singh, P. Bera, F. C. Meldrum, N. Chandrasekhar and R. Seshadri, *J. Mater. Chem.*, 2001, **11**, 3215.
45 K. Biswas and C. N. R. Rao, *J. Phys. Chem. B*, 2006, **110**, 842.
46 S. Link and M. A. El-Sayed, *Int. Rev. Phys. Chem.*, 2000, **19**, 409.
47 K. Biswas, S. V. Bhat and C. N. R. Rao, *J. Phys. Chem. C*, 2007, **111**, 5689.
48 J. Park, K. An, Y. Hwang, J. G. Park, H. J. Noh, J. Y. Kim, J. H. Park, N. M. Hwang and T. Hyeon, *Nat. Mater.*, 2004, **3**, 891.
49 W. S. Seo, J. H. Shim, S. J. Oh, E. K. Lee, N. H. Hur and J. T. Park, *J. Am. Chem. Soc.*, 2005, **127**, 6188.
50 S-H Choi, E-G Kim, J. Park, K. An, N. Lee, A. C. Kim and T. Hyeon, *J. Phys. Chem. B*, 2005, **109**, 14792.
51 Z. Hu, D. J. E. Ramirez, B. E. H. Cervera, G. Oskam and P. C. Searson, *J. Phys. Chem. B*, 2005, **109**, 11209.
52 J. Joo, S. G. Kwon, J. H. Yu and T. Hyeon, *Adv. Mater.*, 2005, **17**, 1873.
53 L. S. Panchakarla, A. Govindaraj and C. N. R. Rao, *J. Cluster Sci.*, 2007, (in print).
54 S. O'Brien, L. Brus and C. B. Murray, *J. Am. Chem. Soc.*, 2001, **123**, 12085.
55 H. Wu, Y. Yang and Y. C. Cao, *J. Am. Chem. Soc.*, 2006, **128**, 16522.
56 N. Pinna, G. Garweitner, M. Antonietti and M. Niederberger, *J. Am. Chem. Soc.*, 2005, **127**, 5608.
57 Q. Liu, W. Lu, A. Ma, J. Tang, J. Lin and J. Fang, *J. Am. Chem. Soc.*, 2005, **127**, 5267.
58 J. Tang, F. Redl, Y. Zhu, T. Siegrist, L. E. Brus and M. L. Steigerwald, *Nano Lett.*, 2005, **5**, 543.
59 G. Li, L. Li, J. B. Goated and B. F. Woodfield, *J. Am. Chem. Soc.*, 2005, **127**, 8659.
60 N. Moumen and M.-P. Pileni, *J. Phys. Chem.*, 1996, **100**, 1867.
61 N. Moumen and M.-P. Pileni, *Chem. Mater.*, 1996, **8**, 1128.
62 C. B. Murray, D. J. Norris and M. G. Bawendi, *J. Am. Chem. Soc.*, 1993, **115**, 8706–8715.
63 J. E. Bowen-Katari, V. L. Colvin and A. P. Alivisatos, *J. Phys. Chem.*, 1994, **98**, 4109.
64 U. K. Gautam, M. Rajamathi, F. Meldrum, P. Morgan and R. Seshadri, *Chem. Commun.*, 2001, 629.
65 U. K. Gautam, R. Seshadri and C. N. R. Rao, *Chem. Phys. Lett.*, 2003, **375**, 560.
66 U. K. Gautam and R. Seshadri, *Mater. Res. Bull.*, 2004, **39**, 669.
67 X. Chen and R. Fan, *Chem. Mater.*, 2001, **13**, 802.
68 L. Qu and X. Peng, *J. Am. Chem. Soc.*, 2002, **124**, 2049.
69 Z. A. Peng and X. Peng, *J. Am. Chem. Soc.*, 2001, **123**, 183.
70 Z. A. Peng and X. Peng, *J. Am. Chem. Soc.*, 2002, **124**, 3343.
71 L. Qu, Z. A. Peng and X. Peng, *Nano Lett.*, 2001, **1**, 333.
72 J. Joo, H. B. Na, T. Yu, J. H. Yu, Y. W. Kim, F. Wu, J. Z. Zhang and T. Hyeon, *J. Am. Chem. Soc.*, 2003, **125**, 11100.
73 Y. A. Yang, H. Wu, K. R. Williams and Y. J. Cao, *Angew. Chem., Int. Ed.*, 2005, **44**, 6712.
74 J. Jasieniak, C. Bullen, J. V. Embden and P. Mulvaney, *J. Phys. Chem. B*, 2005, **109**, 20665.
75 P. D. Cozzoli, L. Manna, M. L. Curri, S. Kudera, C. Giannini, M. Striccoli and A. Agostiano, *Chem. Mater.*, 2005, **17**, 1296.
76 J. E. Murphy, M. C. Beard, A. G. Norman, S. P. Ahrenkiel, J. C. Johnson, P. Yu, O. I. Mićić, R. J. Ellingson and A. J. Nozik, *J. Am. Chem. Soc.*, 2006, **128**, 3241.
77 L. Cademartiri, J. Bertolotti, R. Sopienza, D. S. Wiersma, G. Freymann and G. A. Ozin, *J. Phys. Chem. B*, 2006, **110**, 671.
78 L. A. Swafford, L. A. Weigand, M. J. Bowers II, J. R. McBride, J. L. Rapaport, T. L. Watt, S. K. Dixit, L. C. Feldman and S. J. Rosenthal, *J. Am. Chem. Soc.*, 2006, **128**, 12299.
79 S.-H. Choi, E.-G. Kim and T. Hyeon, *J. Am. Chem. Soc.*, 2006, **128**, 2520.
80 A. Ghezelbash and B. A. Korgel, *Langmuir*, 2005, **21**, 9451.
81 See for example: B. Ludolph, M. A. Malik, P. O'Brien and N. Revaprasadu, *Chem. Commun.*, 1998, 1849.
82 N. Pradhan and S. Efrima, *J. Am. Chem. Soc.*, 2003, **125**, 2050.
83 K. Biswas and C. N. R. Rao, *Chem.–Eur. J.*, 2007, **13**, 6123.
84 M. Green, P. Rahman and D. S. Boyle, *Chem. Commun.*, 2007, 574.
85 Y. Xie, Y. Qian, W. Wang, S. Zhang and Y. Zhang, *Science*, 1996, **272**, 1926.
86 K. Sardar and C. N. R. Rao, *Adv. Mater.*, 2004, **16**, 425.
87 (*a*) U. K. Gautam, K. Sardar, F. L. Deepak and C. N. R. Rao, *Pramana*, 2005, **65**, 549; (*b*) K. Sardar and C. N. R. Rao, *Solid State Sci.*, 2005, **7**, 217; (*c*) K. Sardar, F. L. Deepak, A. Govindaraj, M. M. Seikh and C. N. R. Rao, *Small*, 2005, **1**, 91.
88 S. V. Bhat, K. Biswas and C. N. R. Rao, *Solid State Commun.*, 2007, **14**, 325.
89 K. Biswas, K. Sardar and C. N. R. Rao, *Appl. Phys. Lett.*, 2006, **89**, 132503.
90 O. I. Mícic, S. P. Ahrenkiel, D. Bertram and A. J. Nozik, *Appl. Phys. Lett.*, 1999, **75**, 478.
91 J. L. Coffer, M. A. Johnson, L. Zhang, R. L. Wells and J. F. Janik, *Chem. Mater.*, 1997, **9**, 2671.
92 A. C. Frank, F. Stowasser, H. Sussek, H. Pritzkow, C. R. Miskys, O. Ambacher, M. Giersig and R. A. Fischer, *J. Am. Chem. Soc.*, 1998, **120**, 3512.
93 A. Manz, A. Brikner, M. Kolbe and R. A. Fischer, *Adv. Mater.*, 2000, **12**, 569.
94 K. Sardar, M. Dan, B. Schwenzer and C. N. R. Rao, *J. Mater. Chem.*, 2005, **15**, 2175.
95 A. Gomathi and C. N. R. Rao, *Mater. Res. Bull.*, 2006, **41**, 941.
96 (*a*) L. Grocholl, J. Wang and E. G. Gillan, *Chem. Mater.*, 2001, **13**, 4290; (*b*) L. Grocholl, J. Wang and E. G. Gillan, *Nano Lett.*, 2002, **2**, 899.
97 J. Choi and E. G. Gillan, *J. Mater. Chem.*, 2006, **16**, 3774.
98 J. Xiao, Y. Xie, R. Tang and W. Luo, *Inorg. Chem.*, 2003, **42**, 107.
99 M. Yu, X. Hao, D. Cui, Q. Wang, X. Xu and M. Jiang, *Nanotechnology*, 2003, **14**, 29.
100 M. A. Olshavsky, A. B. Goldstein and A. P. Alivisatos, *J. Am. Chem. Soc.*, 1990, **112**, 9438.
101 R. L. Wells, S. R. Aubuchon, S. S. Kher and M. S. Lube, *Chem. Mater.*, 1998, **7**, 793.
102 O. I. Mícic, C. J. Curtis, K. M. Jones, J. R. Sprague and A. J. Nozik, *J. Phys. Chem.*, 1994, **98**, 4966.
103 O. I. Mícic, J. R. Sprague, C. J. Curtis, K. M. Jones, J. L Machol, A. J. Nozik, H. Giessen, B. Fluegel, G. Mohs and N. Peyghambarian, *J. Phys. Chem.*, 1995, **99**, 7754.
104 A. A. Guzelian, J. E. B. Katari, A. V. Kadavanich, U. Banin, K. Hamad, E. Juban, A. P. Alivisatos, R. H. Wolters, C. C. Arnold and J. R. Heath, *J. Phys. Chem.*, 1996, **100**, 7212.
105 S. C. Perera, P. S. Fodor, G. M. Tsoi, L. E. Wenger and S. L. Brock, *Chem. Mater.*, 2003, **15**, 4034.
106 S. C. Perera, G. Tsoi, L. E. Wenger and S. L. Brock, *J. Am. Chem. Soc.*, 2003, **125**, 13960.
107 S. Wei, J. Lu, W. Yu and Y. Qian, *J. App. Phys.*, 2004, **95**, 3683.
108 S. Xu, S. Kumar and T. Nann, *J. Am. Chem. Soc.*, 2006, **128**, 1054.
109 M. A. Malik, P. O'Brien and M. Helliwell, *J. Mater. Chem.*, 2005, **15**, 1463.
110 K. L. Stamm, J. C. Garno, G. Liu and S. L. Brock, *J. Am. Chem. Soc.*, 2003, **125**, 4038.
111 P. Arumugam, S. S. Shinozaki, R. Wang, G. Maob and S. L. Brock, *Chem. Commun.*, 2006, 1121.
112 J. Lu, Y. Xie, X. Jiang, W. He and G. Du, *J. Mater. Chem.*, 2001, **11**, 3281.
113 W. Schartl, *Adv. Mater.*, 2000, **12**, 1899.
114 See for example: X. Peng, M. C. Schlamp, A. V. Kadavanich and A. P. Alivisatos, *J. Am. Chem. Soc.*, 1997, **119**, 7019.
115 M. A. Hines and P. Guyot-Sionnest, *J. Phys. Chem.*, 1996, **100**, 468.
116 M. A. Malik, P. O'Brien and N. Revaprasadu, *Chem. Mater.*, 2002, **14**, 2004.
117 D. Pan, Q. Wang, S. Jiang, X. Ji and L. An, *Adv. Mater.*, 2005, **17**, 176.
118 R. Xie, U. Kolb, J. Li, T. Basche and A. Mews, *J. Am. Chem. Soc.*, 2005, **127**, 7480.
119 Y. Liu, M. Kim, Y. Wang, Y. A. Wang and X. Peng, *Langmuir*, 2006, **22**, 6341.
120 Y.-W. Cao and U. Banin, *J. Am. Chem. Soc.*, 2000, **122**, 9692.
121 R. H. Morriss and L. F. Collins, *J. Chem. Phys.*, 1964, **41**, 3357.

122 L. Lu, H. Wang, Y. Zhou, S. Xi, H. Zhang, J. Hub and B. Zhaob, *Chem. Commun.*, 2002, 144.
123 L. Rivas, S. Sanchez-Cortes, J. V. Garcia-Ramos and G. Morcillo, *Langmuir*, 2000, **16**, 9722.
124 K. Mallik, M. Mandal, N. Pradhan and T. Pal, *Nano Lett.*, 2001, **1**, 319.
125 Y. W. Cao, R. Jin and C. A. Mirkin, *J. Am. Chem. Soc.*, 2001, **123**, 7961.
126 J.-W. Hu, Y. Zhang, J.-F. Li, Z. Liu, B. Ren, S.-G. Sun, Z.-Q. Tian and T. Lian, *Chem. Phys. Lett.*, 2005, **408**, 354.
127 J. Rivas, R. D. Sanchez, A. Fondado, A. J. Garcia-Bastida, J. Garcia-Otero, J. Mira, D. Baldomir, A. Gonzhlez, I. Lado, M. A. L. Quintela and S. B. Oseroff, *J. Appl. Phys.*, 1994, **76**, 6564.
128 C. T. Seip and C. J. O'Connor, *Nanostruct. Mater.*, 1999, **12**, 183.
129 J. Lin, W. Zhow, A. Kumbhar, J. Wiemann, J. Fang, E. E. Carpenter and C. J. O'Connor, *J. Solid State Chem.*, 2001, **159**, 26.
130 J. Zhang, M. Post, T. Veres, Z. J. Jakubek, J. Guan, D. Wang, F. Normandin, Y. Deslandes and B. Simard, *J. Phys. Chem. B*, 2006, **110**, 7122.
131 L. Wang, J. Luo, Q. Fan, M. Suzuki, I. S. Suzuki, M. H. Engelhard, Y. Lin, N. Kim, J. Q. Wang and C.-J. Zhong, *J. Phys. Chem. B*, 2005, **109**, 21593.
132 W. Stober, A. Fink and E. Bohn, *J. Colloid Interface Sci.*, 1968, **26**, 62.
133 See for example: M. A. Correa-Duarte, M. Giersig, N. A. Kotov and L. M. Liz-Marzán, *Langmuir*, 1998, **14**, 6430.
134 See for example: T. Ung, L. M. Liz-Marzán and P. Mulvaney, *J. Phys. Chem. B*, 1999, **103**, 6770.
135 L. M. Liz-Marzán, M. Giersig and P. Mulvaney, *Langmuir*, 1996, **12**, 4329.
136 K. S. Mayya, D. I. Gittins and F. Caruso, *Chem. Mater.*, 2001, **13**, 3833.
137 I. Pastoriza-Santos, D. S. Koktysh, A. A. Mamedov, M. Giersig, N. A. Kotov and L. M. Liz-Marzán, *Langmuir*, 2000, **16**, 2731.
138 V. Eswaranand and T. Pradeep, *J. Mater. Chem.*, 2002, **12**, 2421.
139 R. T. Tom, A. S. Nair, N. Singh, M. Aslam, C. L. Nagendra, R. Philip, K. Vijayamohanan and T. Pradeep, *Langmuir*, 2003, **19**, 3439.
140 S. Ghosh, K. Biswas and C. N. R. Rao, *J. Mater. Chem.*, 2007, **17**, 2412.
141 V. Salgueiriño-Maceira and M. A. C. Duarte, *J. Mater. Chem.*, 2006, **16**, 3539.
142 O. Masala and R. Seshadri, *J. Am. Chem. Soc.*, 2005, **127**, 9354.
143 C. N. R. Rao, F. L. Deepak, G. Gundiah and A. Govindaraj, *Prog. Solid State Chem.*, 2003, **31**, 5.
144 Y. Xia, P. Yang, Y. Sun, Y. Wu, B. Mayers, B. Gates, Y. Yin, F. Kim and H. Han, *Adv. Mater.*, 2003, **15**, 353.
145 C. R. Martin, *Science*, 1994, **266**, 1961.
146 D Almawlawi, C. Z. Liu and M. Moskovits, *J. Mater. Res.*, 1994, **9**, 1014.
147 A. Govindaraj, B. C. Satishkumar, M. Nath and C. N. R. Rao, *Chem. Mater.*, 2000, **12**, 202.
148 M. Zheng, L. Zhang, X. Zhang, J. Zhang and G. Li, *Chem. Phys. Lett.*, 2001, **334**, 298.
149 J. A. Sioss and C. D. Keating, *Nano Lett.*, 2005, **5**, 1779.
150 B. D. Busbee, S. O. Obare and C. J. Murphy, *Adv. Mater.*, 2003, **15**, 414.
151 L. Gou and C. J. Murphy, *Chem. Mater.*, 2005, **17**, 3668.
152 H.-Y. Wu, H.-C. Chu, T.-J. Kuo, C.-L. Kuo and M. L. H. Huang, *Chem. Mater.*, 2005, **17**, 6447.
153 A. Gulati, H. Liao and J. H. Hafner, *J. Phys. Chem. B*, 2006, **110**, 22323.
154 A. Gole and C. J. Murphy, *Chem. Mater.*, 2005, **17**, 1325–1330.
155 B. Basnar, Y. Weizmann, Z. Cheglakov and I. Willner, *Adv. Mater.*, 2006, **18**, 713–718.
156 C.-K. Tsung, X. Kou, Q. Shi, J. Zhang, M. H. Yeung, J. Wang and G. D. Stucky, *J. Am. Chem. Soc.*, 2006, **128**, 5352.
157 A. J. Mieszawska, G. W. Slawinski and F. P. Zamborini, *J. Am. Chem. Soc.*, 2006, **128**, 5622.
158 C. Ni, P. A. Hassan and E. W. Kaler, *Langmuir*, 2005, **21**, 3334.
159 Y. Sun, B. Gates, B. Mayers and Y. Xia, *Nano lett.*, 2002, **2**, 165.
160 Y. Chen, B. J. Wiley and Y. Xia, *Langmuir*, 2007, **23**, 4120.
161 L. Gou, M. Chipara and J. M. Zaleski, *Chem. Mater.*, 2007, **19**, 1755.
162 D. Ung, G. Viau, C. Ricolleau, F. Warmont, P. Gredin and F. Fievet, *Adv. Mater.*, 2005, **17**, 338.
163 W. Z. Wong, B. Poudel, Y. Ma and Z. F. Ren, *J. Phys. Chem. B*, 2006, **110**, 25702.
164 Y. Xiong, H. Cai, B. J. Wiley, J. Wang, M. J. Kim and Y. Xia, *J. Am. Chem. Soc.*, 2007, **129**, 3665.
165 S. R. C. Vivekchand, G. Gundiah, A. Govindaraj and C. N. R. Rao, *Adv. Mater.*, 2004, **16**, 1842.
166 S.-M. Liu, M. Kobayashi, S. Sato and K. Kimura, *Chem. Commun.*, 2005, 4690.
167 D. C. Lee, T. Hanrath and B. A. Korgel, *Angew. Chem., Int. Ed.*, 2005, **44**, 3573.
168 A. I. Hochbaum, R. Fan, R. He and P. Yang, *Nano Lett.*, 2005, **5**, 457.
169 T. Shimizu, T. Xie, J. Nishikawa, S. Shingubara, S. Senz and U. Gosele, *Adv. Mater.*, 2007, **19**, 917.
170 S. Kodambaka, J. B. Hannon, R. M. Tromp and F. M. Ross, *Nano Lett.*, 2006, **6**, 1296.
171 X. Lu, D. D. Fanfair, K. P. Johnston and B. A. Korgel, *J. Am. Chem. Soc.*, 2005, **127**, 15718.
172 H-Y Tuan, D. C. Lee, T. Hanrath and B. A. Korgel, *Chem. Mater.*, 2005, **17**, 5705.
173 P. Nguyen, H. T. Ng and M. Meyyappan, *Adv. Mater.*, 2005, **17**, 549.
174 D. Wang, R. Tu, L. Zhang and H. Dai, *Angew. Chem., Int. Ed.*, 2005, **44**, 2925.
175 H. Gerung, T. J. Boyle, L. J. Tribby, S. D. Bunge, C. J. Brinker and S. M. Han, *J. Am. Chem. Soc.*, 2006, **128**, 5244.
176 U. K. Gautam, M. Nath and C. N. R. Rao, *J. Mater. Chem.*, 2003, **13**, 2845.
177 U. K. Gautam and C. N. R. Rao, *J. Mater. Chem.*, 2004, **14**, 2530.
178 Y. Ma, L. Qi, W. Shen and J. Ma, *Langmuir*, 2005, **21**, 6161.
179 Q. Li and V. W.-W. Yam, *Chem. Commun.*, 2006, 1006.
180 J. M. Song, J. H. Zhu and S. H. Yu, *J. Phys. Chem. B*, 2006, **110**, 23790.
181 H. Zhang, D. Yang, X. Ma, N. Du, J. Wu and D. Gue, *J. Phys. Chem. B*, 2006, **110**, 827.
182 B. Cheng, W. Shi, J. M. R-Tanner, L. Zhang and E. T. Samulski, *Inorg. Chem.*, 2006, **45**, 1208.
183 H. Peng, Y. Fangli, B. Liuyang, L. Jinlin and C. Yunfa, *J. Phys. Chem. C*, 2007, **111**, 194.
184 L. S. Panchakarla, M. A. Shah, A. Govindaraj and C. N. R. Rao, *unpublished results*.
185 Z- Gui, J. Liu, Z. Wang, L. Song, Y. Hu, W. Fan and D. Chen, *J. Phys. Chem. B*, 2005, **109**, 1113.
186 P. X. Gao, Y. Ding, W. Mai, W. L. Hughes, C. Lao and Z. L. Wang, *Science*, 2005, **309**, 1700.
187 Q. Li, V. Kumar, Y. Li, H. Zhang, T. J. Marks and R. P. H. Chang, *Chem. Mater.*, 2005, **17**, 1001.
188 Y. Tak and K. Yong, *J. Phys. Chem. B*, 2005, **109**, 19263.
189 P. X. Gao, C. S. Lao, W. L. Hughes and Z. L. Wang, *Chem. Phys. Lett.*, 2005, **408**, 174.
190 M. Lai and D. J. Riley, *Chem. Mater.*, 2006, **18**, 2233.
191 S. Kar, B. N. Pal, S. Chaudhuri and D. Chakravorty, *J. Phys. Chem. B*, 2006, **110**, 4605.
192 J. H. He, J. H. Hsu, C. W. Wang, H. N. Lin, L. J. Chen and Z. L. Wang, *J. Phys. Chem. B*, 2006, **110**, 50.
193 L.-X. Yang, Y.-J. Zhu, W.-W. Wang, H. Tong and M.-L. Ruan, *J. Phys. Chem. B*, 2006, **110**, 6609.
194 F. L. Deepak, G. Gundiah, Md. M. Shiekh, A. Govindaraj and C. N. R. Rao, *J. Mater. Res.*, 2004, **19**, 2216.
195 P. Chen, S. Xie, N. Ren, Y. Zhang, A. Dong, Y. Chen and Y. Tang, *J. Am. Chem. Soc.*, 2006, **128**, 1470.
196 G. Wang, D.-S. Tsai, Y.-S. Huang, A. Korotcov, W.-C. Yeh and D. Susanti, *J. Mater. Chem.*, 2006, **16**, 780.
197 A. Magrez, E. Vasco, J. W. Seo, C. Dieker, N. Setter and L. Forro, *J. Phys. Chem. B*, 2006, **110**, 58.
198 K. P. Kalyanikutty, F. L. Deepak, C. Edem, A. Govindaraj and C. N. R. Rao, *Mater. Res. Bull.*, 2005, **40**, 831.
199 Y. Hao, G. Meng, C. Ye, X. Zhang and L. Zhang, *J. Phys. Chem. B*, 2005, **109**, 11204.
200 G. Gundiah, A. Govindaraj and C. N. R. Rao, *Chem. Phys. Lett.*, 2002, **351**, 189.
201 J. Zhang, F. Jiang, Y. Yang and J. Li, *J. Phys. Chem. B*, 2005, **109**, 13143.
202 J. Zhan, Y. Bando, J. Hu, F. Xu and D. Goldberg, *Small*, 2005, **1**, 883.
203 J. Joo, S. G. Kwon, T. Yu, M. Cho, J. Lee, J. Yoon and T. Hyeon, *J. Phys. Chem. B*, 2005, **109**, 15297.
204 B. S. Guiton, Q. Gu, A. L. Prieto, M. S. Gudiksen and H. Park, *J. Am. Chem. Soc.*, 2005, **127**, 498.

205 K. P. Kalyanikutty, G. Gundiah, C. Edem, A. Govindaraj and C. N. R. Rao, *Chem. Phys. Lett.*, 2005, **408**, 389.
206 Q. Wan, M. Wei, D. Zhi, J. L. MacManus-Driscoll and M. G. Blamire, *Adv. Mater.*, 2006, **18**, 234.
207 R. Wang, Y. Chen, Y. Fu, H. Zhang and C. Kisielowski, *J. Phys. Chem. B*, 2005, **109**, 12245.
208 Y. M. Zhao, Y.-H. Li, R. Z. Ma, M. J. Roe, D. G. McCartney and Y. Q. Zhu, *Small*, 2006, **2**, 422.
209 Y. Li, B. Tan and Y. Wu, *J. Am. Chem. Soc.*, 2006, **128**, 14258.
210 J. Zhou, Y. Ding, S. Z. Deng, L. Gong, N. S. Xu and Z. L. Wang, *Adv. Mater.*, 2005, **17**, 2107.
211 J.-W. Seo, Y.-W. Jun, S. J. Ko and J. Cheon, *J. Phys. Chem. B*, 2005, **109**, 5389.
212 G. Shen and D. Chen, *J. Am. Chem. Soc.*, 2006, **128**, 11762.
213 G. Xu, Z. Ren, P. Du, W. Weng, G. Shen and G. Han, *Adv. Mater.*, 2005, **17**, 907.
214 S. R. Hall, *Adv. Mater.*, 2006, **18**, 487.
215 A.-M. Cao, J.-S. Hu, H.-P. Liang and L.-J. Wan, *Angew. Chem., Int. Ed.*, 2005, **44**, 4391.
216 S. Kar and S. Chaudhuri, *J. Phys. Chem. B*, 2005, **109**, 3298.
217 J. Hu, Y. Bando and D. Goldberg, *Small*, 2005, **1**, 95.
218 S. Kar and S. Chaudhuri, *J. Phys. Chem. B*, 2006, **110**, 4542.
219 P. Christian and P. O'Brien, *Chem. Commun.*, 2005, 2817.
220 Y. Jeong, Y. Xia and Y. Yin, *Chem. Phys. Lett.*, 2005, **416**, 246.
221 S. G. Thoma, A. Sanchez, P. Provencio, B. L. Abrams and J. P. Wilcoxon, *J. Am. Chem. Soc.*, 2005, **127**, 7611.
222 A. B. Panda, G. Glaspell and M. S. El-Shall, *J. Am. Chem. Soc.*, 2006, **128**, 2790.
223 S. Kumar, M. Ade and T. Nann, *Chem.–Eur. J.*, 2005, **11**, 2220.
224 Z. Liu, D. Xu, J. Liang, J. Shen, S. Zhang and Y. Qian, *J. Phys. Chem. B*, 2005, **109**, 10699.
225 J.-P. Ge, J. Wang, H-X Zhang, X. Wang, Q. Peng and Y. Li, *Chem.–Eur. J.*, 2005, **11**, 1889.
226 X. Giu, Y. Lou, A. C. S. Samia, A. Devadoss, J. D. Burgess, S. Dayal and C. Burda, *Angew. Chem., Int. Ed.*, 2005, **44**, 5855.
227 R. Chen, M. H. So, C.-M. Che and H. Sun, *J. Mater. Chem.*, 2005, **15**, 4540.
228 F. Gao, Q. Lu and S. Komarneni, *Chem. Commun.*, 2005, 531.
229 M. B. Sigman and B. A. Korgel, *Chem. Mater.*, 2005, **17**, 1655.
230 A. Purkayastha, F. Lupo, S. Kim, T. Borca-Tasciuc and G. Ramanath, *Adv. Mater.*, 2006, **18**, 496.
231 M. Nath, A. Choudhury and C. N. R. Rao, *Chem. Commun.*, 2004, 2698.
232 D. Yu, J. Wu, Q. Gu and H. Park, *J. Am. Chem. Soc.*, 2006, **128**, 8148.
233 Y. H. Yang and Y. T. Chen, *J. Phys. Chem. B*, 2006, **110**, 17370.
234 B. Liu, Y. Bando, C. Tang, F. Xu, J. Hu and D. Goldberg, *J. Phys. Chem. B*, 2005, **109**, 17082.
235 H. Li, A. H. Chin and M. K. Sunkara, *Adv. Mater.*, 2006, **18**, 216.
236 S. Luo, W. Zhou, Z. Zhang, L. Liu, X. Dou, J. Wang, X. Zhao, D. Liu, Y. Gao, L. Song, Y. Xiang, J. Zhou and S. Xie, *Small*, 2005, **1**, 1004.
237 P. V. Radonanvic, C. J. Barrelet, S. Gradecak, F. Qian and C. M. Lieber, *Nano Lett.*, 2005, **5**, 1407.
238 C. J. Novotny and P. K. L. Yu, *Appl. Phys. Lett.*, 2005, **87**, 203111.
239 A. I. Persson, M. T. Björk, S. Jeppesen, J. B. Wagner, L. R. Wallenberg and L. Samuelson, *Nano Lett.*, 2006, **6**, 403.
240 H. Zhang, Q. Zhang, G. Zhao, J. Tang, O. Zhou and L.-C. Qin, *J. Am. Chem. Soc.*, 2005, **127**, 13120.
241 Y. Li, E. Tevaarwerk and R. P. H. Chang, *Chem. Mater.*, 2006, **18**, 2552.
242 Y. S. Hor, Z. L. Xiao, U. Welp, Y. Ito, J. F. Mitchell, R. E. Cook, W. K. Kwok and G. W. Crabtree, *Nano Lett.*, 2005, **5**, 397.
243 Y. Li, M. A. Malik and P. O'Brien, *J. Am. Chem. Soc.*, 2005, **127**, 16020.
244 C. N. R. Rao, A. Govindaraj, F. L. Deepak, N. A. Gunari and M. Nath, *Appl. Phys. Lett.*, 2001, **78**, 1853.
245 K. P. Kalyanikutty, M. Nikhila, U. Maitra and C. N. R. Rao, *Chem. Phys. Lett.*, 2006, **432**, 190.
246 E. J. H. Lee, C. Ribeiro, E. Longo and E. R. Leite, *J. Phys. Chem. B*, 2005, **109**, 20842.
247 R. Li, Z. Luo and F. Papadimitrakopoulos, *J. Am. Chem. Soc.*, 2006, **128**, 6280–6281.
248 K.-S. Cho, D. V. Talapin, W. Gaschler and C. B. Murray, *J. Am. Chem. Soc.*, 2005, **127**, 7140.
249 J. H. Yu, J. Joo, H. M. Park, S-II. Baik, Y. W. Kim, S. C. Kim and T. Hyeon, *J. Am. Chem. Soc.*, 2005, **127**, 5662.
250 A. Gomathi, S. R. C. Vivekchand, A. Govindaraj and C. N. R. Rao, *Adv. Mater.*, 2005, **17**, 2757.
251 Y. Xi, J. Zhou, H. Guo, C. Cai and Z. Lin, *Chem. Phys. Lett.*, 2005, **412**, 60.
252 A. D. LaLonde, M. G. Norton, D. N. McIlroy, D. Zhang, R. Padmanabhan, A. Alkhateeb, H. Man, N. Lane and Z. Holman, *J. Mater. Res.*, 2005, **20**, 549.
253 S. Y. Bae, H. W. Seo, H. C. Choi, D. S. Han and J. Park, *J. Phys. Chem. B*, 2005, **109**, 8496.
254 H.-X. Zhang, J.-P. Ge, J. Wang, Z. Wang, D.-P. Yu and Y.-D. Li, *J. Phys. Chem. B*, 2005, **109**, 11585.
255 M. A. Verheijen, G. Immink, T. de. Smet, M. T. Borgström and E. P. A. M. Bakkers, *J. Am. Chem. Soc.*, 2006, **128**, 1353.
256 M. Afzaal and P. O'Brien, *J. Mater. Chem.*, 2006, **16**, 1113.
257 I. Pastoriza-Santos, J. Pérez-Juste and L. M. Liz-Marźan, *Chem. Mater.*, 2006, **18**, 2465.
258 P. Mohan, J. Motohisa and T. Fukui, *Appl. Phys. Lett.*, 2006, **88**, 133105.
259 C. N. R. Rao and M. Nath, *Dalton Trans.*, 2003, 1.
260 R. Tenne, L. Margulis, M. Genut and G. Hodes, *Nature*, 1992, **360**, 444.
261 Y. Feldman, E. Wasserman, D. J. Srolovitch and R. Tenne, *Science*, 1995, **267**, 222.
262 M. Chhowalla and G. A. J. Amaratunga, *Nature*, 2000, **407**, 164.
263 P. A. Parilla, A. C. Dillon, K. M. Jones, G. Riker, D. L. Schulz, D. S. Ginley and M. J. Heben, *Nature*, 1999, **397**, 114.
264 P. A. Parilla, A. C. Dillon, B. A. Parkinson, K. M. Jones, J. Alleman, G. Riker, D. S. Ginley and M. J. Heben, *J. Phys. Chem. B*, 2004, **108**, 6197.
265 R. Tenne, *Chem.–Eur. J.*, 2002, **8**, 5296.
266 T. Tsirlina, Y. Feldman, M. Homyonfer, J. Sloan, J. L. Hutchison and R. Tenne, *Fullerene Sci. Technol.*, 1998, **6**, 157.
267 M. Nath, A. Govindaraj and C. N. R. Rao, *Adv. Mater.*, 2001, **13**, 283.
268 M. Nath and C. N. R. Rao, *Chem. Commun.*, 2001, 2336.
269 M. Nath and C. N. R. Rao, *J. Am. Chem. Soc.*, 2001, **123**, 4841.
270 M. Nath and C. N. R. Rao, *Angew. Chem., Int. Ed.*, 2002, **41**, 3451.
271 C. Y. Zhi, Y. Bando, C. Tang and D. Goldberg, *Solid State Commun.*, 2005, **135**, 67.
272 F. L. Deepak, C. P. Vinod, K. Mukhopadhyay, A. Govindaraj and C. N. R. Rao, *Chem. Phys. Lett.*, 2002, **353**, 345.
273 J. Wang, V. K. Kayastha, Y. K. Yap, Z. Fan, J. G. Lu, Z. Pan, I. N. Ivanov, A. A. Puretzky and D. B. Geohegan, *Nano Lett.*, 2005, **5**, 2528.
274 J. Dinesh, M. Eswaramoorthy and C. N. R. Rao, *J. Phys. Chem. C*, 2007, **111**, 510.
275 S.-Y. Zhang, Y. Li, X. Ma and H.-Y. Chen, *J. Phys. Chem. B*, 2006, **110**, 9041.
276 U. K. Gautam, S. R. C. Vivekchand, A. Govindaraj, G. U. Kulkarni, N. R. Selvi and C. N. R. Rao, *J. Am. Chem. Soc.*, 2005, **127**, 3658.
277 C. D. Malliakas and M. G. Kantzidis, *J. Am. Chem. Soc.*, 2006, **128**, 6538.
278 Q. Wu, Z. Hu, C. Liu, X. Wang, Y. Chen and Y. Lu, *J. Phys. Chem. B*, 2005, **109**, 19719.
279 M. S. Sander and H. Gao, *J. Am. Chem. Soc.*, 2005, **127**, 12158.
280 Y. Aoki, J. Huang and T. Kunitake, *J. Mater. Chem.*, 2006, **16**, 292.
281 H. Yu, Z. Zhang, M. Han, X. Hao and F. Zhu, *J. Am. Chem. Soc.*, 2005, **127**, 2378.
282 C. Li, Z. Liu, C. Gu, X. Xu and Y. Yang, *Adv. Mater.*, 2006, **18**, 228.
283 J. M. Macak, H. Tsuchiya and P. Schumuki, *Angew. Chem., Int. Ed.*, 2005, **44**, 2100.
284 C. Ruan, M. Paulose, O. K. Varghese, G. K. Mor and C. A. Grimes, *J. Phys. Chem. B*, 2005, **109**, 15754.
285 G. Armstrong, A. R. Armstrong, J. Canales and P. G. Bruce, *Chem. Commun.*, 2005, 2454.
286 R. Ma, T. Sasaki and Y. Bando, *Chem. Commun.*, 2005, 948.
287 M. A. Khan, H.-T. Jung and O.-B. Yang, *J. Phys. Chem. B*, 2006, **110**, 626.
288 D. Eder, I. A. Kinloch and A. H. Windle, *Chem. Commun.*, 2006, 1448.
289 A. Ghicov, J. M. Macak, H. Tsuchiya, J. Kunze, V. Haeublein, L. Frey and P. Schmuki, *Nano Lett.*, 2006, **6**, 1080.
290 H. Tan, E. Ye and W. Y. Fan, *Adv. Mater.*, 2006, **18**, 619.
291 Q. Ji and T. Shimizu, *Chem. Commun.*, 2005, 4411.

 

292 J. H. Jung, T. Shimizu and S. Shinkai, *J. Mater. Chem.*, 2005, **15**, 3979.

293 C.-J. Jia, L.-D. Sun, Z.-G. Yan, L.-P. You, F. Luo, X.-D. Han, Y.-C. Pang, Z. Zhang and C. H. Yan, *Angew. Chem., Int. Ed.*, 2005, **44**, 4328.

294 Z. Liu, D. Zhang, S. Han, C. Li, B. Lei, W. Lu, J. Fang and C. Zhou, *J. Am. Chem. Soc.*, 2005, **127**, 6.

295 C. Tang, Y. Bando, B. Liu and D. Goldberg, *Adv. Mater.*, 2005, **17**, 3005.

296 R. H. A. Ras, T. Ruotsalainen, K Laurikainen, M. B. Linder and O. Ikkala, *Chem. Commun.*, 2007, **13**, 1366.

297 R. H. A. Ras, M. Kemell, J. de Wit, M. Ritala, G. ten Brinke, M. Leskela and O. Ikkala, *Adv. Mater.*, 2007, **19**, 102.

298 H. J. Fan, M. Knez, R. Scholz, K. Nielsch, E. Pippel, D. Hesse, M. Zacharias and U. Gosele, *Nat. Mater.*, 2006, **5**, 627.

299 L. Zhao, M. Steinhart, J. Yu and U. Goesele, *J. Mater. Res.*, 2006, **21**, 685.

300 S. V. Pol, V. G. Pol and A. Gedanken, *Adv. Mater.*, 2006, **18**, 2023.

301 C. N. R. Rao, G. U. Kulkarni, V. V. Agrawal, U. K. Gautam, M. Ghosh and U. Tumkurkar, *J. Colloid Interface Sci.*, 2005, **289**, 305.

302 C. N. R. Rao, G. U. Kulkarni, P. J. Thomas, V. V. Agarwal and P. Saravanan, *Curr. Sci.*, 2003, **85**, 1041.

303 Y. Lin, H. Skaff, T. Emrick, A. D. Dinsmore and T. P. Russell, *Science*, 2003, **299**, 226.

304 C. N. R. Rao, G. U. Kulkarni, P. J. Thomas, V. V. Agarwal and P. Saravanan, *J. Phys. Chem. B*, 2003, **107**, 7391.

305 V. V. Agrawal, G. U. Kulkarni and C. N. R. Rao, *J. Phys. Chem. B*, 2005, **109**, 7300.

306 V. V. Agrawal, P. Mahalakshmi, G. U. Kulkarni and C. N. R. Rao, *Langmuir*, 2006, **22**, 1846.

307 U. K. Gautam, M. Ghosh and C. N. R. Rao, *Chem. Phys. Lett.*, 2003, **381**, 1.

308 U. K. Gautam, M. Ghosh and C. N. R. Rao, *Langmuir*, 2004, **20**, 10775.

309 K. P. Kalyanikutty, U. K. Gautam and C. N. R. Rao, *Solid State Sci.*, 2006, **8**, 296.

310 K. P. Kalyanikutty, U. K. Gautam and C. N. R. Rao, *J. Nanosci. Nanotechnol.*, 2007, **7**, 1916.

311 Y. Wu, B. Yang, B. Zong, H. Sun, Z. Shen and Y. Feng, *J. Mater. Chem.*, 2004, **14**, 469.

312 H. T. Nan, J. Li, M. K. Smith, P. Nguyen, A. Cassell, J. Han and M. Meyyappan, *Science*, 2003, **300**, 1249.

313 U. K. Gautam, S. R. C. Vivekchand, A. Govindaraj and C. N. R. Rao, *Chem. Commun.*, 2005, 3995.

Dalton Transactions

Cite this: *Dalton Trans.*, 2012, **41**, 5089

www.rsc.org/dalton

**PERSPECTIVE**

# Recent progress in the synthesis of inorganic nanoparticles

C. N. R. Rao,*[a,b] H. S. S. Ramakrishna Matte,[a] Rakesh Voggu[a] and A. Govindaraj[a,b]

***Received 25th November 2011, Accepted 27th January 2012***
**DOI: 10.1039/c2dt12266a**

Nanoparticles probably constitute the largest class of nanomaterials. Nanoparticles of several inorganic materials have been prepared by employing a variety of synthetic strategies. Besides synthesizing nanoparticles, there has been considerable effort to selectively prepare nanoparticles of different shapes. In view of the great interest in inorganic nanoparticles evinced in the last few years, we have prepared this perspective on the present status of the synthesis of inorganic nanoparticles. This article includes a brief discussion of methods followed by reports on the synthesis of nanoparticles of various classes of inorganic materials such as metals, alloys, oxides chalcogenides and pnictides. A brief section on core–shell nanoparticles is also included.

## 1. Introduction

Synthesis of nanomaterials constitutes a crucial aspect of nanoscience and technology. Several strategies have been employed for the synthesis of nanomaterials of different dimensionalities. In particular, zero-dimensional nanoparticles (NPs) have received greater attention because of the wide-spectrum of uses. Chemical methods have been very useful in the generation of nanoparticles of various inorganic materials such as metals, oxides, chalcogenides and pnictides and some of the reviews cover various aspects of these nanocrystals.[1,2] A perspective article that appeared a few years ago in this journal[3] covered the synthesis of inorganic NPs, nanotubes and nanowires. Since then, there has been an explosion in publications dealing with the synthesis of nanomaterials, specially of NPs. We, therefore, considered restricting ourselves to the synthesis of inorganic NPs in this article. This perspective provides an up-to-date report on the synthesis of inorganic NPs covering the period 2007–2011. We have briefly introduced some of the important methods employed for the synthesis of NPs in Section 2, citing typical

[a]*Chemistry and Physics of Materials Unit, CSIR Centre of Excellence and International Centre for Materials Science, Jawaharlal Nehru Centre for Advanced Scientific Research, Jakkur P. O., Bangalore 560 064, India. E-mail: cnrrao@jncasr.ac.in; Fax: (+91) 80-2208 2760*
[b]*Solid State and Structural Chemistry Unit, Indian Institute of Science, Bangalore 560 012, India*

**C. N. R. Rao**

*C. N. R. Rao obtained his PhD degree from Purdue University (1958) and DSc degree from the University of Mysore (1961). He is the National Research Professor and Linus Pauling Research Professor at the Jawaharlal Nehru Centre for Advanced Scientific Research and Honorary Professor at the Indian Institute of Science (both at Bangalore). His research interests are mainly in the chemistry of materials. He is a member of many of the science academies and an honorary fellow of RSC. He is the recipient of the Einstein Gold Medal of UNESCO, the Hughes and Royal Medals of the Royal Society, the August Wilhelm von Hofmann medal of the German Chemical Society, the Dan David Prize for materials research and the first India Science Prize.*

**H. S. S. Ramakrishna Matte**

*H. S. S. Ramakrishna Matte obtained his BSc degree from Government College (Autonomous), Rajahmundry in 2006 and MS degree in chemical science in 2009 from the Jawaharlal Nehru Centre for Advanced Scientific Research. He is completing his PhD work related to synthesis, characterization and properties of graphene and its inorganic analogues.*

examples from the literature. We then discuss the synthesis of NPs of metals, alloys, metal oxides, metal chalcogenides and metal pnictides in separate sections. We have included a brief section on the synthesis of core–shell NPs. We trust that practitioners as well as beginners will find this article useful in their efforts to prepare inorganic NPs. We apologize for any omissions and errors in judgment in citing articles from the literature.

## 2. Some synthetic methods

### 2.1. Use of microemulsions

Microemulsions consist of a ternary mixture of water, a surfactant or a mixture of surface-active agents and oil. Micelles are aggregates of surfactant molecules formed above a critical micellar concentration (CMC).[4,5] Micelles are responsible for the enhancement of the solubilization of organic compounds in water (oil-in-water (o/w) emulsion) or hydrophilic compounds in the oil phase (water-in-oil (w/o) emulsion). This process has been used for the synthesis of various NPs by mixing two microemulsions containing appropriate reactants (see schematic representation in Scheme 1).

Au NPs have been prepared by the reduction of $HAuCl_4$ with $NaCN–BH_3$ in sodium 1,4-bis(2-ethylhexoxy)-1,4-dioxobutane-2-sulfonate (AOT) water-in-hexane microemulsion.[6] Ag NPs are prepared by mixing reverse micellar solutions of 30% Ag(AOT), 70% Na(AOT) and $NaBH_4$ respectively.[7] Cu NPs are obtained by the reduction of a micellar solution of $Cu(DS)_2$(DS, dodecyl sulfate) with $NaBH_4$. Below the CMC, $CuO_2$ NPs occur along with the metal NPs where as above the CMC, only metallic nanostructures are obtained.[8] Pd NPs are obtained by the inverse micelle formed by AOT in isooctane, by the addition of capping agents to the suspension.[9] Bimetallic Au–Ag NPs are synthesized in a water–AOT–isooctane microemulsion by the co-reduction of $HAuCl_4$ and $AgNO_3$ with hydrazine at room temperature.[10] $SnO_2$ NPs are prepared by mixing microemulsions formed by cetyltrimethylammonium bromide (CTAB), 1-butanol, isooctane, $SnCl_4{\cdot}5H_2O$ and 0.1 M $NH_3$ solution, followed by heating the precipitate formed at 500 °C.[11] Addition of aqueous methylamine, to mixed micelles of $Co(DS)_2$ and Fe $(DS)_2$ results in the formation of $CoFe_2O_4$ NPs.[12] $Cu_2S$ NPs are prepared by the reaction of the copper ammonia complex with an equimolar thiourea solution in Triton-X 100–cyclohexane water-in-oil microemulsion.[13]

### 2.2. Thermal decomposition

Thermal decomposition of organometallic precursor compounds and metal–surfactant complexes in hot surfactant solutions has been used to synthesize inorganic NPs. Metal NPs are generated by the thermal decomposition of molecules containing zero-valent metals, such as metal carbonyls.[14] Thus, Pd NPs with particle sizes in the 3.5–7 nm range are prepared by the thermal decomposition of Pd–trioctylphosphine (TOP) complex at 300 °C for 30 min.[15] Monodisperse Co NPs can be synthesized *via* a high-temperature thermal decomposition of $Co_2(CO)_8$ in the presence of oleic acid (OA) and triphenylphosphine at 220 °C.[16] Bimetallic FePt and CoPt NPs are obtained by reacting $Pt(acac)_2$ with 1,2-hexadecanediol (HDD) and trioctylphosphine oxide (TOPO) at 100 °C followed by the addition of $Fe(CO)_5$ or $Co_2(CO)_8$ at 286 °C for 30 min respectively.[17] Thermal decomposition of metal–oleate precursors in high boiling solvents produces monodisperse NPs (Fig. 1).[18] Thus, the iron–oleate complex in octadecene (ODE) is slowly heated to 320 °C and aged at that temperature for 30 min to generate iron oxide NPs. Thermal decomposition of copper cupferronate (Cu $(cupf)_2$), on a silicon substrate at 250 °C for 10 min produced an assembled film of Cu NPs which on oxidation transformed to $Cu_2O$ with retention of the assembly.[19] Cadmium complexes of *N*,*N*-diethyl-*N*$^1$-benzoylthiourea and *N*,*N*-diethyl-*N*$^1$-benzoylselenourea are used as the precursors for the synthesis of CdS and CdSe NPs respectively in TOP and hexadecylamine at 200 °C for 60 min.[20] Thermal decomposition of $Cl_2GaP(SiMe_3)_2$ prepared from $GaCl_3$ with $P(SiMe_3)_3$ produces GaP NPs at 300 °C for 3 hours.[21] Green and O'Brien[22] report the synthesis of InP and GaP NPs by the thermal decomposition of $M(PtBu_2)_3$ (M = Ga, In). The decomposition of $Fe(CO)_4[PPh_2CH_2CH_2–Si(OMe)_3]$ in a silica xerogel matrix results in the formation of $Fe_2P$ nanoclusters.[23] Thermal decomposition of long-chain iron carboxylates such as Fe(III) oleate, palmitate, and myristate influences the morphology of iron oxide NPs varying from spheres to nanostars.[24] Synthesis of metal oxide nanoparticles by combustion synthesis is well documented.[25] $CoFe_2O_4$ NPs are synthesized by the decomposition of a mixture of $Fe(acac)_3$ and Co

**Rakesh Voggu**

*Rakesh Voggu received his MSc (Chemistry) degree from University of Hyderabad, India in 2005. He obtained his PhD degree in 2010 from Jawaharlal Nehru Centre for Advanced Scientific Research (JNCASR). He is presently working as a Research Associate at JNCASR and his work pertains to various aspects of nanocrystals and carbon nanostructures.*

**A. Govindaraj**

*A. Govindaraj obtained his PhD degree from University of Mysore and is a Principal Scientific Officer at the Indian Institute of Science, and Honorary Associate Professor at the Jawaharlal Nehru Centre for Advanced Scientific Research. He works on different types of nanomaterials. He has authored more than 100 research papers and co-authored a book on nanotubes and nanowires with Prof. Rao.*

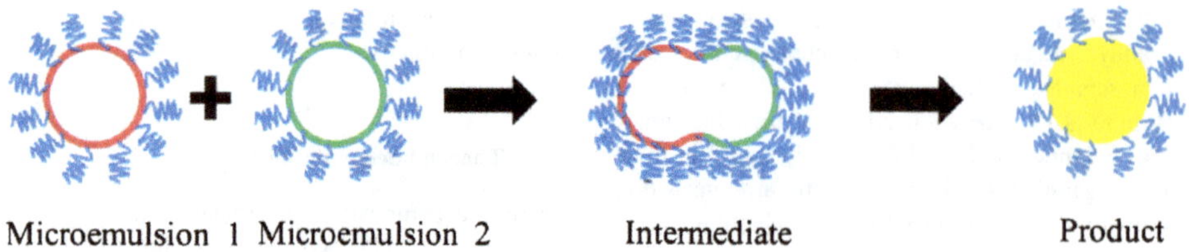


**Scheme 1** Proposed mechanism for the formation of nanoparticles by microemulsion approach.

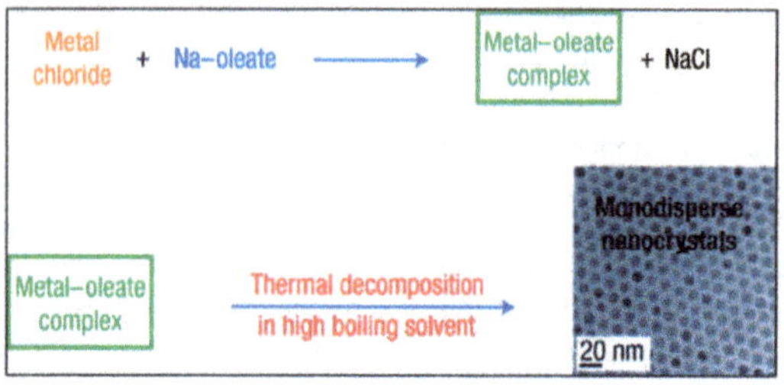


**Fig. 1** The overall scheme for the ultra-large-scale synthesis of monodisperse NPs and TEM of magnetite. Reprinted with permission from Park *et al.*, *Nat. Mater.*, 2004, **3**, 891. ©2004 Nature Publishing Group.

$(acac)_3$ in a mixture of ODE, OA and oleyl amine (OAm) at different temperatures ranging from 150–300 °C.[26] GaN, AlN and InN NPs are readily prepared by decomposing the urea complexes of the corresponding trichlorides in trioctyl amine under refluxing conditions.[27–30]

### 2.3. Hydrothermal and solvothermal synthesis

Hydrothermal–solvothermal synthesis is a common method to synthesize inorganic NPs. Properties of the reactants, such as their solubility and reactivity change in the solvent or water at elevated temperatures and pressures. Taking advantage of the high reactivity of metal salts and complexes at elevated temperatures and pressures, inorganic nanomaterials are conveniently prepared at temperatures considerably lower than in solid-state reactions. Reaction parameters such as time, temperature, pressure, reactant concentration, pH, reaction cell fill volume can be tuned to attain satisfactory nucleation rates and particle size distribution. NPs of metals, oxides, chalcogenides, pnictides and various materials have been prepared by the hydrothermal or solvothermal synthesis. Thus, an aqueous solution containing $HAuCl_4$, trisodium citrate, and CTAB is heated at 110 °C for 6, 12, 24, 48, and 72 h to obtain Au octahedra with average sizes of 30, 60, 90, 120, and 150 nm respectively.[31] Nanoanatase $TiO_2$ is obtained by using tetrabutyl titanate as the titanium precursor, acetic acid as an inhibitor, and diethyl ether anhydrous as a solvent at a temperature 100 °C.[32] Monodispersed ZnO nanorods with an average diameter less than 50 nm are prepared solvothermally by using zinc nitrate, NaOH, ethanol and ethylenediamine at 180 °C for 20 h.[33] $ReO_3$ NPs (diameter range of 8.5–32.5 nm) can be prepared by the decomposition of the $Re_2O_7$–dioxane complex at 200 °C for 4 h under solvothermal conditions.[34] To

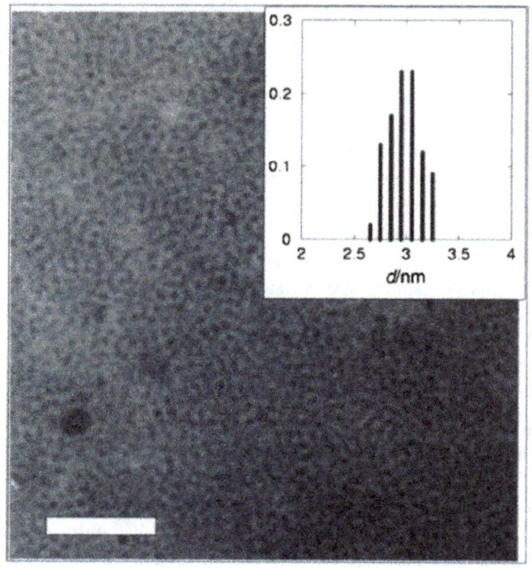


**Fig. 2** TEM image of a relatively dense arrangement of CdSe NPs showing a tendency to close packing in the plane (bar = 50 nm). The inset shows a histogram of particle sizes. Reprinted with permission from U. K. Gautam, M. Rajamathi, F. Meldrum, P. Morgan and R. Seshadri, *Chem. Commun.*, 2001, 629. ©2001 Royal Society of Chemistry.

synthesize PbS and CdS NPs, the molecular precursors, 2,2′-bipyridyl$(Pb(SC(O)(C_6H_5)_2)$ and 2,2′-bipyridyl$(Cd(SC(O)(C_6H_5)_2)$ respectively are heated in aqueous media at 100 °C for 30 min.[35] Monodisperse 3 nm CdSe particles are prepared solvothermally using $Cd(sterate)_2$, Se, tetralin, dodecanethiol and toluene in a stainless steel autoclave heated to 250 °C for 5 h (Fig. 2).[36] $\gamma$-$Fe_2O_3$ (10 nm dia.) and $CoFe_2O_4$ (7 nm dia.) are prepared by hydrothermal heating of the reaction mixtures containing $Fe(cupf)_3$–$Co(cupf)_2$, n-octylamine and toluene respectively at 220 °C for 1 hour.[37]

### 2.4. Sol–gel method

The sol–gel method is generally employed for the synthesis of metal oxide NPs as well as oxide nanocomposites. The sol–gel process involves the hydrolysis and condensation of metal precursors.[38,39] Further condensation and polymerization lead to three-dimensional metal oxide networks forming the gel. The sol–gel process can be either in aqueous or non-aqueous medium. In the aqueous sol–gel process, oxygen for the formation of the oxide is supplied by water molecules. In the non-aqueous process, oxygen is provided by a solvent (ethers, alcohols, ketones, or aldehydes) or by a organic constituent of the precursor (alkoxides or acetylacetonates).[40]

The aqueous sol–gel process can be represented as follows:

$$M(OR)_x + mH_2O \rightarrow M(OR)_{x-m}(OH)_m + mROH \qquad (OH)$$

If $x = m$, the reaction is said to be of total hydrolysis. This can happen in two stages, either by water or alcohol condensation.[4]

$$2M(OR)_{x-m}(OH)_m \rightarrow (OH)_{m-1}(OR)_{x-m}\text{–M–O–}M(OR)_{x-m}(OH)_{m-1} + H_2O \qquad (OH)$$

$$2M(OR)_{x-m}(OH)_m \rightarrow (OH)_m - 1(OR)_{x-m}\text{–M–O–}M(OR)_{x-m-1} + ROH \qquad (OR)$$

The total process is expressed as follows

$$M(OR)_x + x/2H_2O \rightarrow MO_{x/2} + xHOR \qquad (OR)$$

The condensation process in non-aqueous sol–gel synthesis involves metal–oxygen–metal bonds formed either by alkyl halide elimination, ether elimination, ester elimination, or by aldol-like condensation.

del Monte *et al.*[41] obtained γ-$Fe_2O_3$ NPs in a size range between 6 and 15 nm by direct heat treatment of gels at 400 °C. Ceria NPs are prepared by a reverse micelle sol–gel technique by using cerium isopropoxide in cyclohexane.[42] A nonhydrolytic sol–gel method to prepare $TiO_2$ NPs, involving condensation of titanium halide with titanium alkoxide has been reported by Trentler *et al.*[43] Various metal oxides such as ZnO,[44] $ZrO_2$,[45] iron oxide,[46] $CeO_2$[47] and ferrites[48] have been synthesized by a nonhydrolytic sol–gel method. Niederberger *et al.*[18] have developed a general nonaqueous sol–gel route to prepare NPs of binary metal oxide such as γ-$Ga_2O_3$, ZnO and cubic $In_2O_3$ by the reaction between the corresponding metal acetylacetonates and benzylamine.

### 2.5. Phase-transfer method

Phase-transfer mediated synthesis is widely used for the preparation of various NPs. This method involves the transfer of reactants from a polar medium to a non-polar medium, followed by further processing in a non-polar medium. NPs prepared in a nonpolar medium can also be phase-transferred to a polar medium. One of the earliest reports of this method is by Brust *et al.*[49] for the synthesis of Au NPs. The method involves the phase-transfer of chloroaurate ions into toluene using a phase-transfer reagent such as tetraalkylammonium bromide, followed by the reduction of gold ions using sodium borohydride, giving rise to Au NPs capped with alkanethiol molecules. The gold ions from the aqueous solution are transferred to toluene due to electrostatic interaction with tetraoctylammonium bromide. The movement of aqueous gold NPs into non-polar organic solvents requires hydrophobization of the NPs. Yang *et al.*[50] report a general protocol for the phase transfer of various metal ions from an aqueous to an organic medium for the synthesis of NPs. Here, phase-transfer of metal ions from water to an organic medium is carried out by dodecylamine by using ethanol as solvent. Using this method, it was possible to synthesize noble metal NPs, semiconductor–noble-metal composite NPs. Heterogeneous deposition of noble metals on semiconductor NPs, and homogeneous growth of semiconductors on noble metal NPs could also be accomplished. Fig. 3 shows TEM images of various NPs prepared by the transfer of metal ions from water to organic media. Thiol-derivatized NPs of Au, Pt, and Ag have been prepared as organosols by the acid-facilitated transfer of the NPs from a hydrosol to the toluene layer containing an alkanethiol.[51] Phase-transfer of nanostructures from aqueous medium to fluorous solvent has been carried out.[52] By using this approach, it has been possible to phase-transfer Au NPs from water, and semiconductor NPs such as CdSe and CdS from toluene, to perfluorohexane using perfluoroalkanethiol as capping agent. A one-step method for synthesis and phase-transfer of NPs from hydrocarbon medium such as toluene to a fluorous medium such as perfluorohexane has been reported recently.[53] This method is based on the fact that fluorous and organic solvents become miscible at elevated temperatures (Fig. 4). In this manner, NPs of metal chalcogenides such as CdSe, CdS, PbSe, ZnSe and metal oxides such as γ-$Fe_2O_3$ and ZnO and bimetallic FePt have been transferred to fluorous media.

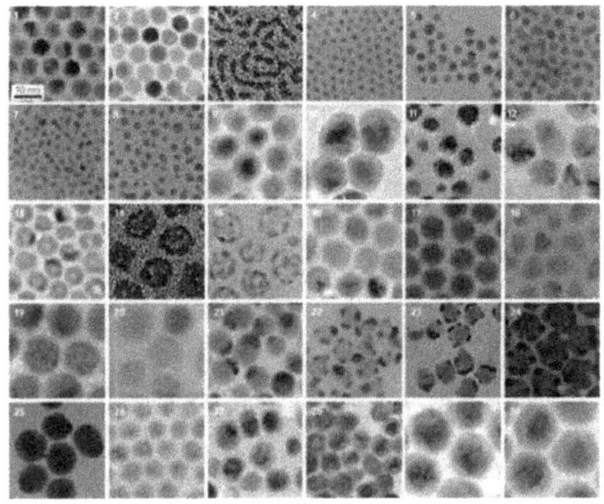

**Fig. 3** (1) Ag derived with HDD, (2) Au, (3) worm-like Pd and (4) Pt from Pt(IV), derived with TBAB alloy NPs of (5) Ag–Au, (6) Pd–Pt, (7) Pt–Rh and (8) Pt–Ru, synthesized by co-reduction of the metal precursors with tetrabutylammonium borohydride. Core–shell NPs of (9) 7.4 nm Au@Ag, (10) 12.7 nm Au@Ag, (11) 3.9 nm Pt@Ag and (12) 9.2 nm Pt@Ag, synthesized by seed-mediated growth. Core–shell NPs of (13) Ag@Au and (14) Ag@Pt, synthesized by the replacement reaction. (15) Pt hollow spheres synthesized by bis(*p*-sulphonatophenyl)phenylphosphane dihydrate dipotassium salt treatment of Ag@Pt NPs. (16) Ag–Pd alloy synthesized by the replacement reaction. Semiconductor NPs of (17) $Ag_2S$, (18) CdS, (19) HgS and (20) PbS. Hybrid NPs of (21) $Ag_2S$–Au, (22) CdS–Au, (23) CuS–Au, (24) PbS–Au, (25) $Ag_2S$–Ag, (26) CdS–Ag, (27) CuS–Ag and (28) PbS–Ag. Core–shell NPs of Au@$Ag_2S$ synthesized with Au–$Ag_2S$ precursor molar ratios of (29) 1 : 1 and (30) 1 : 3. The scale of each image is identical. Reprinted with permission from J. Yang, E. Sargent, S. Kelley and J. Y. Ying, *Nat. Mater.*, 2009, **8**, 683. ©2009 Nature Publishing Group.

### 2.6. Microwave synthesis

Microwave synthesis is commonly employed for the synthesis of inorganic NPs since it is fast, simple and energy-efficient. Microwaves are generally generated using a magnetron. Microwave energy is absorbed by the reactants, which is then converted to heat thereby causing the rapid decomposition of the precursors and creating highly supersaturated solutions, where nucleation and growth occur to produce the desired nanocrystalline products.[54] Au NPs (spherical, hexagons, truncated prisms) have been prepared in a microwave using $HAuCl_4$ in HCl as the gold precursor and OAm and OA as capping agents (Fig. 5).[55] $CdSO_4$, $Pb(Ac)_2$, and $CuSO_4$ are reacted with $Na_2SeSO_3$ in water in the presence of potassium nitrilotriacetate to synthesize of CdSe and PbSe NPs, or using triethanolamine to obtain $Cu_{2-x}Se$ NPs, in 10 minutes.[56] Core–shell Au–Pd NPs are prepared by the simultaneous reduction of palladium(II) chloride and hydrogen tetrachloroaurate(III) in ethylene glycol (EG) for 1 hour.[57] FePt NPs are obtained from microwave process using $Fe(acac)_3$ and $Pt(acac)_2$ as metal precursors and tetraethyleneglycol (TEG) as solvent as well as reducing agent at 250 °C.[58] Amorphous $Fe_2O_3$ NPs have been synthesized by microwave heating of an aqueous solution containing ferric chloride, polyethyleneglycol (PEG) and urea.[59]

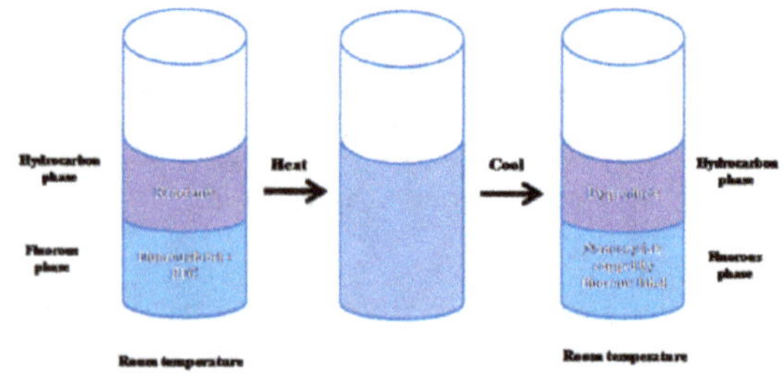


**Fig. 4** Schematic showing the thermomorphic nature of fluorous and hydrocarbon solvents. Reproduced with permission from R. Voggu, A. Shireen and C. N. R. Rao, *Dalton Trans.*, 2010, **39**, 6021. ©2010 Royal Society of Chemistry.

### 2.7. Use of the liquid–liquid interface

The liquid–liquid interface is conveniently used for the preparation of assemblies or thin films of inorganic NPs. The interface is a nonhomogeneous region with a thickness of a few nanometers. NPs are highly mobile at the interface and rapidly achieve an equilibrium assembly due to reduced interfacial energy. The parameters that influence assembly process at the liquid–liquid interface are: (i) nature of the interface, (ii) surface modification of NPs at the interface, and (iii) the effective radius of NPs, also smaller NPs generally attached more weakly to interface than larger ones.[60] To synthesize inorganic nanomaterials, an organic precursor of relevant metal is taken in the organic layer and an appropriate reagent in the aqueous layer. The product formed by the reaction at the interface contains ultra-thin nanocrystalline films of the nanomaterial.[61] This technique has been used to prepare NPs of metals such as Au, Ag, and Pd, of metal chalcogenides such as CdS and NiS, as well as extended ultrathin single crystalline films of CuO, ZnO, CuS, PbS, and ZnS.[62] In a typical preparation of Au–Cu–Ag nanocrystalline film, a solution of $Au(PPh_3)Cl$–$Cu(PPh_3)Cl$ or $Ag_2(PPh_3)_4Cl_2$ in toluene is allowed to stand in contact with aqueous alkali in a beaker at 300 K. Once the two liquid layers are stable, tetrakishydroxymethylphosphonium chloride (THPC) is injected into the aqueous layer using a syringe with minimal disturbance to the toluene layer.[63] By taking mixtures of the corresponding metal precursors in organic layer, nanocrystalline films of binary Au–Ag and Au–Cu alloys and ternary Au–Ag–Cu alloys are prepared.[64] Polycrystalline thin films of CdSe–CuSe have been prepared at the organic–water interface by reacting $Cd(cupf)_2$–$Cu(cupf)_2$ in the toluene layer with dimethylselenourea in the aqueous layer.[65] By reacting $Cu(cupf)_2$ in the organic layer with an aqueous NaOH–$Na_2S$ solution, a single-crystalline film of CuO–CuS is formed at the organic–aqueous interface.[66] Fig. 6 shows electron microscopy images of Au, Ag, CdS and CuS NPs formed at the liquid–liquid interface.

### 2.8. Use of ionic liquids

There have been several attempts during last five to six years to use ionic liquids as media for the synthesis of inorganic

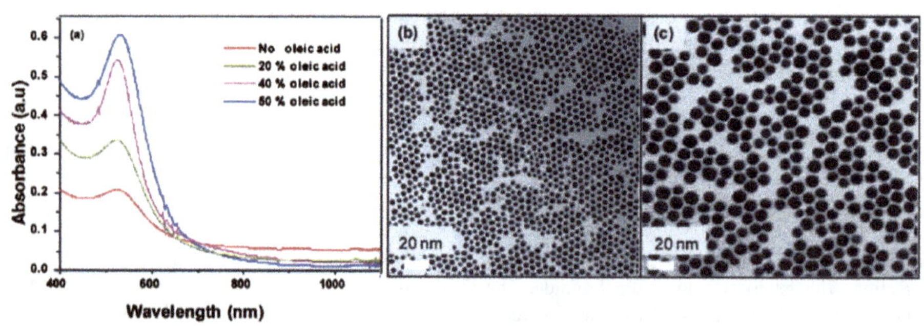


**Fig. 5** (a) UV-vis spectra for Au NPs prepared using different ratios of OA–OAm varied from 0 to 50% after 1 min. microwave time. TEM images of gold NPs prepared using pure amine (b) and in a 1 : 1 ratio of OA–OAm (c). The scale bar is 20 nm. Reprinted with permission from M. B. Mohamed, K. M. AbouZeid, V. Abdelsayed, A. A. Aljarash and M. S. El-Shall, *ACS Nano*, 2010, **4**, 2766. ©2010 American Chemical Society.

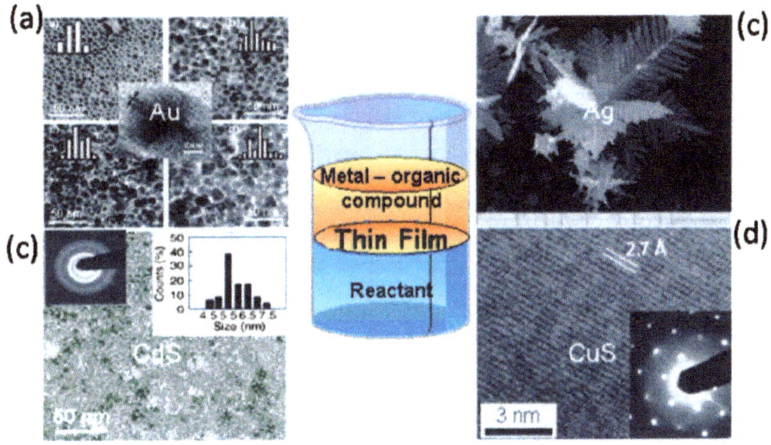


**Fig. 6** TEM images of (a) gold (b) Ag (c) CdS and (d) CuS NPs formed at the liquid–liquid interface. Reprinted with permission from C. N. R. Rao and K. P. Kalyanikutty, *Acc. Chem. Res.*, 2008, **41**, 489–499. ©2008 American Chemical Society.

nanomaterials.[67,68] Ionic liquids are organic salts with low melting points and no measurable vapour pressure, and do not evaporate. They remain as liquids over a wide range of temperatures, in some cases in excess of 400 °C. The low interfacial tension and the associated high nucleation rate as well as high thermal stability make ionic liquids attractive to carry out the synthesis of nanostructures. Ionic liquids have received much attention due to their potential use as green recyclable alternatives to traditional organic solvents. Polar reactions can be carried out in these liquids in the absence of water, and crystalline NPs can be synthesized conveniently at relatively low temperatures. The pronounced self-organization of a solvent is made use of in synthesis of self-assembled, highly organized hybrid nanostructures. Synthesis and functionalization of gold NPs in ionic liquids has been reported, wherein the color of the gold NPs can be tuned by changing the anion of ionic liquid.[69] NPs of Rh and Ir have been prepared by the reduction of appropriate compounds in 1-n-butyl-3-methylimidazolium-hexafluorophosphate, ([BMIM][$PF_6$]), in the presence of hydrogen.[70] $RuO_2$ NPs of 2 to 3 nm in diameter have been obtained by $NaBH_4$ reduction of $RuCl_3$ dissolved in the ionic liquid, 1-n-butyl-3-methylimidazolium-hexaflurophosphate.[71] Biswas and Rao[72] successfully synthesized nanostructures of CdS, CdSe, ZnS, ZnSe and PbS in ionic liquid media. They have examined effects of changing the anion of the ionic liquid as well as addition of surfactants and capping agents on the size and shape of nanostructures.

### 2.9. Other methods

There are many other methods employed for the synthesis of inorganic NPs.[1–3] Some of these are: photochemical syntheses as typified by radiation-induced decomposition and reduction of gold compounds, electrochemical synthesis of metal NPs, arrested precipitation, the sonochemical method and solvated metal atom dispersion involving trapping of metal atoms in a cold matrix of a liquid followed by growth on warming the matrix. Biological media and reagents have been used to prepare NPs of metals and other inorganic materials.

## 3. Metal nanoparticles

The first scientific study of metal NPs was by Michael Faraday[73] in 1857. Enüstün and Turkevich[74] in 1963 prepared spherical NPs of Au and Ag of tunable sizes by the reduction of gold or silver salts by citrate in aqueous solutions. During the past three decades research on NPs has greatly intensified due to the development of improved synthetic procedures which enable precise control over the size and shape.[1,2,75] Metal NPs can now be made in large quantities at low cost, there by expanding the possible range of applications. Besides the methods described earlier, metal NPs have been prepared by employing many variants and by using a variety of reducing agents ranging from alcohols to fungi.[76]

### 3.1. Gold

The popular citrate route to synthesize Au NPs was first reported by Hauser and Lynn.[77] The procedure involves the addition of chloroauric acid to a boiling solution of sodium citrate,[74,78] and enables one to obtain NPs with an average diameter in the 10–100 nm range by varying the concentration of the reactants.[64,69] Au NPs with diameters between 1 and 2 nm are obtained by the reduction of $HAuCl_4$ with alkaline tetrakis (hydroxymethyl)phosphonium chloride (THPC) which also acts as a capping agent.[79] Brust and co-workers[49] developed a general method to obtain organic-capped metal NPs in biphasic mixture of an organic solvent and aqueous solution of a metal salt in the presence of a phase-transfer reagent. Ji *et al.*[80] have studied the growth kinetics and temporal size/shape evolution of the Au NPs by citrate reduction in boiling water. Kanehara *et al.*[81] report the size control and evolution of decanethiol-protected Au NPs at room-temperature using a quaternary ammonium salt and proton acid (Fig. 7). Ohyama *et al.*[82] have used a one-phase method for the reduction of $HAuCl_4$ in a DMF solution containing a porphyrin-cored tetradentate passivant giving horizontal porphyrin monolayer-coated gold NPs.

Redel *et al.*[83] obtained Au NPs by thermal, photolytic and microwave-assisted decomposition–reduction under argon from

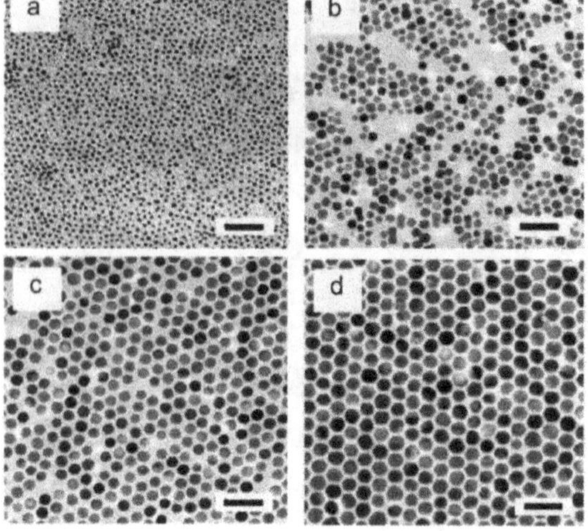

**Fig. 7** TEM images of (a) decanethiol(DT)–Au seed NPs acid treated DT–Au NPs with trifluoroacetic acid and tetraoctylammonium bromide for (b) 5 h at room temperature (rt) (c) 75 h at rt, and (d) 21 h at 55 °C. The scale bar length is 20 nm. Reprinted with permission from M. Kanehara, J.-I. Sakurai, H. Sugimura and T. Teranishi, *J. Am. Chem. Soc.*, 2009, **131**, 1630. ©2009 American Chemical Society.

Au(CO)Cl or $KAuCl_4$ in the presence of n-butylimidazol dispersed in ionic liquids. Han *et al.*[84] have developed a reactive template method to synthesize Au (or Ag) NPs with controllable size and morphology supported on both inner and outer surfaces of poly(*o*-phenylenediamine) hollow microspheres. The size of the Au NPs could be tuned from 3 to 15 nm by changing the concentration of the Au source. Goulent and Lennox[85] provide new insight to the Brust–Schiffrin method for the synthesis of metal NPs. It was believed that M(I) thiolate polymers are the precursors in both one- and two-phase reactions in this method. It was proposed that tetraammonium metal complexes are precursors in two-phase reactions, whereas M(I) thiolates are precursors in one-phase reactions.

Au NPs are prepared by using Irgacure-2959 (1-[4-(2-hydroxyethoxy) phenyl]-2-hydroxy-2-methyl-1-propane-1-one) and $HAuCl_4$ under illumination of 350 nm wavelength light.[86] Christian and Bromfield[87] report a method to prepare colloids of Ag, Au, and Cu NPs in a variety of solvents from water to toluene where the reduction of Ag, Au or copper ions is carried out in the presence of a PEG-based stabilizer. Biswas *et al.*[88] have investigated growth of gold NPs by reduction of $HAuCl_4$ using THPC, by combined use of small-angle X-ray scattering (SAXS) and isothermal titration calorimetry and transmission electron microscopy techniques. They show that the growth of the NPs does not follow the diffusion-limited Ostwald ripening but instead follows a sigmoidal rate curve. Fig. 8 shows TEM images of gold NPs grown at different time intervals and their diameter distribution as insets.

Egusa *et al.*[89] observed the spontaneous formation of hexagonal and square arrays of sub-10 nm Au NPs by asymmetric heating of alkanethiol-passivated atomic Au cluster solutions on amorphous carbon films. Personick *et al.*[90] synthesized Au nanostructures with different shapes (Fig. 9) (such as octahedra, rhombic dodecahedra, truncated ditetragonal prisms, and concave cubes) using a seed-mediated growth method by varying the $Ag^+$ concentration in reaction medium. Niu *et al.*[91] employed a seed-mediated growth method for selectively synthesizing single-crystalline rhombic dodecahedral, octahedral and cubic Au NPs by using cetylpyridinium chloride capped 41.3 nm Au NPs as seeds. Icosahedral Au NPs exhibiting 5-fold symmetry are formed on a glass surface using an aerosol spray process at 600 °C.[92] Yavuz *et al.*[93] describe a protocol to generate 18 nm Au icosahedra by mixing aqueous solutions of $HAuCl_4$ and PVP. Polyhedral Au NPs with decahedral, icosahedral, and truncated tetrahedral shapes are synthesized by the polyol process by reduction of $HAuCl_4$ in the presence of PVP by using either diethylene or tetraethylene glycols.[94] Ma *et al.*[95] have reported use of AuCl–OAm as a precursor to produce Au seeds to form truncated Au decahedrons in tetrahydrofuran (THF) at room temperature. Well-defined Au NPs of icosahedral shape are obtained in high yields by the rapid reduction of Au precursors with EG in the presence of PVP under hydrothermal conditions for 1 h.[96] Li *et al.*[97] used a modified polyol process in which a small amount of $NaBH_4$ was added to a PEG 600 solution of PVP to a hot aqueous gold(III) chloride solution to obtain octahedral NPs. Trisoctahedral Au NPs have been prepared by the reduction of an aqueous solution of $HAuCl_4$ with ascorbic acid in the presence of cetyltrimethylammonium chloride (CTAC) at room temperature.[98]

Huang *et al.*[99] describe a surfactant-promoted reductive route for the shape-controlled synthesis of Au nanostructures by hydrothermal treatment of chloroauric acid in presence of the surfactant CTAB. A simple low-cost electrochemical route to synthesize hierarchical flower-like Au microstructures has been reported using Au nanoplates or nanopricks as building blocks, without introducing any template or surfactant.[100] Monodisperse star-shaped Au NPs are obtained by the reduction of $HAuCl_4$ by L-ascorbic acid at room temperature in the deep eutectic solvents (DES) and various shapes and surface structures of Au are produced by adjusting the content of water in the DES.[101] Santos and Liz Marzán[102] used DMF as a solvent and reducing agent for preparing Au and Ag metal NPs, with tailored morphologies. A procedure for shape and size selection of Au NPs through formation of reversible flocculates by surfactant micelle-induced depletion interaction has been described by Park *et al.*[103] Polygonal Au NPs can be produced from an aqueous $HAuCl_4$ solution by a modified citrate reduction technique using ferric ammonium citrate instead of sodium citrate as the reducing agent, in absence of a surfactant or polymeric template.[104] Table 1 lists some of the methods of synthesis of gold NPs.

### 3.2. Silver

Chen *et al.*[110] have synthesized a series of nano/submicrometer Ag particles using ascorbic acid to reduce AgCl in an aqueous solution at ambient conditions. Ag NPs are prepared by reducing $AgNO_3$ by EG in the presence of PVP, followed by the addition of the so-obtained Ag seeds to a DMF solution containing $Ag^+$ ions for obtaining overgrowth of Ag crystals with {111} facets.[111] Richards *et al.*[112] have carried out a mechanistic study of Ag-NPs grown by reaction of $[(PPh_3)_2Ag(O_2CC_{13}H_{27})]$ and

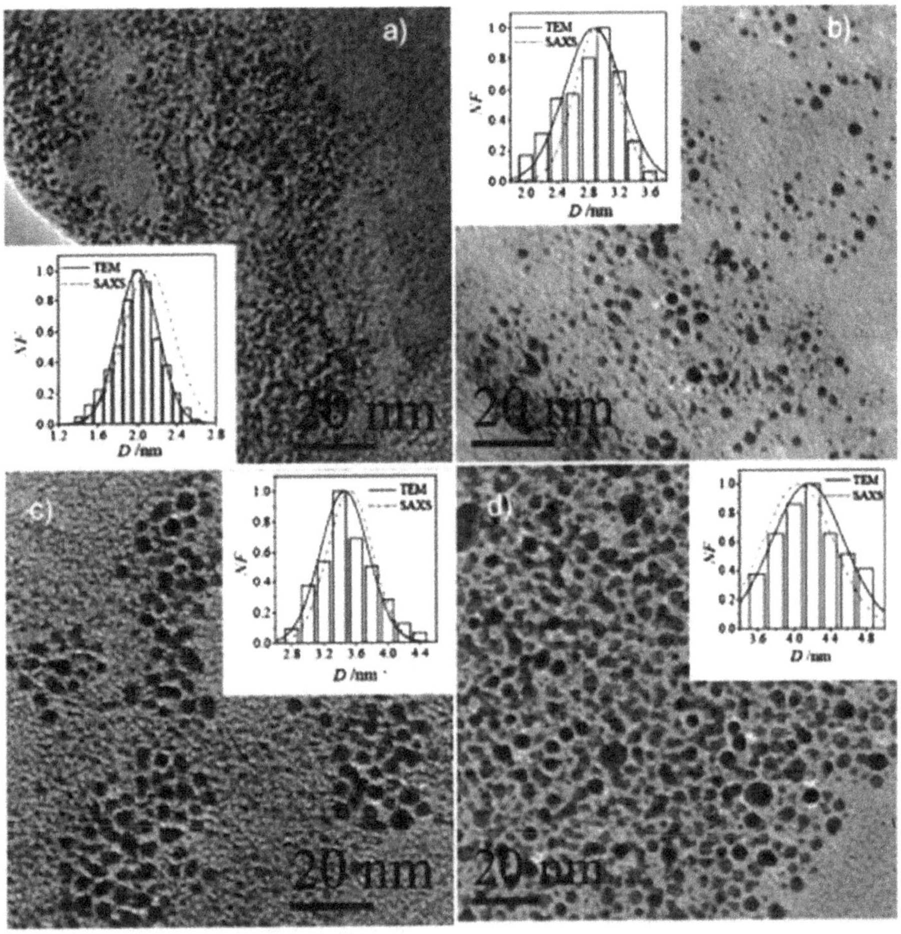


**Fig. 8** TEM images of gold NPs prepared from 125 mL (0.006 mmol) of 50 mM $HAuCl_4$, obtained after (a) 15, (b) 65, (c) 185, and (d) 630 min of growth. Insets compare the diameter distributions obtained from TEM and SAXS for the same time of growth. NF: normalized frequency. Reprinted with permission from K. Biswas, N. Varghese and C. N. R. Rao, *Small*, 2008, **4**, 649–655. ©2008 Wiley-VCH Verlag GmbH & Co. K GaA.

azoisobutyronitrile at 130 °C under reflux conditions to demonstrate the occurrence of an aggregative nucleation and growth process. Dong *et al.*[113] have examined the growth of Ag NPs by the stepwise citrate reduction of silver nitrate in the pH range 5.7–11.1 and have found that the reduction of the silver precursor ($Ag^+$) is promoted with an increase in pH. Kundu *et al.*[114] have prepared Ag NPs and nanochains in large quantities in the presence of alkaline 2,7-dihydroxy naphthalene as a reducing agent under microwave heating in a nonionic surfactant (Triton X-100) in just 60 s. Liang *et al.*[115] describe the synthesis of monodisperse Ag NPs in high yields by using PVP–$AgNO_3$ with a molar ratio of 8 at 260 °C, in which the PEG is used both as a solvent and reducing reagent and PVP acts as capping agent. A crude mixture of red- and blue-green-emitting clusters $Ag_8(H_2MSA)_8$ and $Ag_7(H_2MSA)_7$ ($H_2MSA$: mercaptosuccinic acid), are obtained in gram quantities by an interfacial etching reaction conducted at an aqueous–organic interface starting from the $H_2MSA$-protected Ag NPs (Ag@$H_2MSA$) as precursor.[116]

Kind *et al.*[117] prepared nanoscale Ag hollow spheres *via* a microemulsion approach using $[Ag(PPh_3)_4]NO_3$ as the starting material. Stable, fluorescent, water-soluble Ag nanoclusters have been obtained by a sonochemical approach, using polymethylacrylic acid (PMAA), as the capping agent and $AgNO_3$ as the silver source.[118] Li *et al.*[119] describe a polyol method that generates dimers of Ag nanospheres in one-step in which EG serves as the solvent. Ag nanocubes with controllable edge lengths in the range of 30–200 nm have been prepared by using $AgNO_3$ as the precursor.[120] Xue *et al.*[121] have developed a method in which they use pseudo-spherical Au NPs as plasmonic cores and experimental labels to determine role of plasmon excitation in the photochemical conversion of Ag NPs into nanoprisms.

Ag cubes and tetrahedra have been prepared by Wiley *et al.*[122] reducing silver nitrate with ethylene glycol heated to 148 °C in the presence of PVP and a trace amount of sodium chloride. An oxidative environment (chloride and oxygen from air) leads to preferential etching and dissolution of multiply twinned particles

 

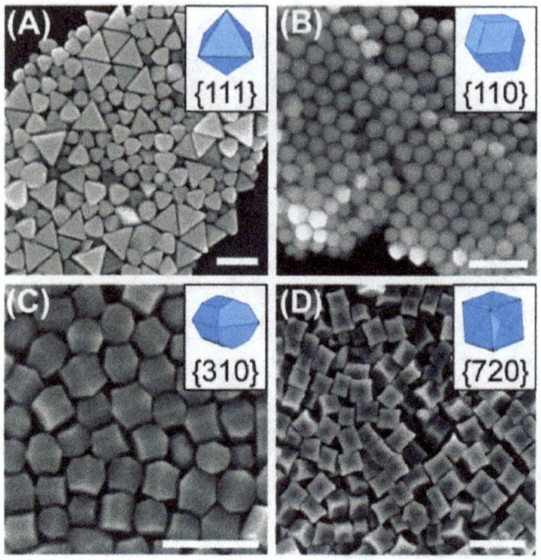


**Fig. 9** SEM images of (A) octahedra, (B) rhombic dodecahedra, truncated ditetragonal prisms, and (D) concave cubes synthesized from reaction solutions containing different ratios of $Ag^{+}$–$Au^{3+}$. Scale bars: 200 nm. Reprinted with permission from M. L. Personick, M. R. Langille, J. Zhang and C. A. Mirkin, *Nano Lett.*, 2011, **11**, 3394. ©2011 American Chemical Society.

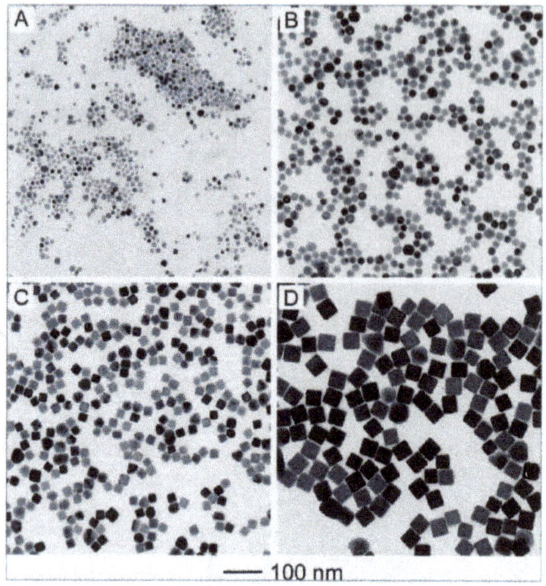


**Fig. 10** TEM images of four aliquots taken from the same synthesis (under argon) after a $AgNO_3$ solution had been added for different periods of time (min): (A) 5; (B) 10; (C) 15; (D) 20. Reprinted with permission from Q. Zhang, C. Cobley, L. Au, M. McKiernan, A. Schwartz, L.-P. Wen, J. Chen and Y. Xia, *ACS Appl. Mater. Interfaces*, 2009, **1**, 2044. ©2009 American Chemical Society.

**Table 1** Typical synthetic routes to prepare gold nanoparticles

| Method | Reactants and reaction conditions | Reference |
|---|---|---|
| Reduction | $HAuCl_4$ gold precursor, $Na_2[Mo_3(\mu_3\text{-}S)(\mu\text{-}S)_3(Hnta)_3]$ serves both as a reducing and a capping agent, at room temperature | 105 |
| | $HAuCl_4$ as gold precursor, *o*-diaminobenzene as reducing agent, PVP as capping agent, at room temperature | 106 |
| | $HAuCl_4$ as gold precursor, 2-[4-(2-hydroxyethyl)-1-piperazinyl]ethanesulfonic acid, serving as both reducing and shape-directing agents, at room temperature | 107 |
| | $HAuCl_4$ as gold precursor, cellulose as reducing agent, 1-butyl-3-methyl imidazolium chloride as solvent at 110 to 200 °C | 108 |
| | $HAuCl_4$ as gold precursor, hydroquinone as the reducing agent | 109 |
| Hydrothermal | $HAuCl_4$ as gold precursor, trisodium citrate as reducing agent and CTAB as capping agent, 110 °C for 6–72 h | 31 |

leaving only the single crystalline ones to grow into nanoscale cubes and tetrahedra. Later the same group replaced the chloride ion by bromide ion, which resulted in etching the seeds with multiple twin defects leading to the formation of bipyramids of silver.[123] Changing the concentration of Fe(II) or Fe(III) ions in the polyol reduction of silver nitrate allows production of either nanocubes or nanowires. A high concentration of iron ions facilitates the growth of multiply twinned seeds to form nanowires by preventing oxidative etching whereas at low concentrations oxidative etching dominates resulting in single-crystal nanocubes.[124] Ag nanocubes with sharp corners and edges could be obtained by using PVP as the capping agent in mixture with citric acid in aqueous solution under ambient conditions. As the reaction proceeds, the nanocubes transform into spheres.[125] Both PVP and Br-ions can selectively bind to the {100} facets of Ag resulting in the formation of Ag nanocubes and nanobars.[126] Ag octahedra and cubes were selectively obtained by introducing sodium citrate and PVP respectively as the capping agent.[127]

Courty *et al.*[128] report formation of triangular single crystals of Ag with a fcc structure by mild annealing of self-organized, ordered mono- and multi-layers of 5 nm Ag NPs, under atmospheric pressure at 50 °C. Monodisperse Ag right-triangular bipyramids are obtained by photoinduced or plasmon-mediated particle growth and the edge length of bipyramids is controlled by the wavelength of exciting radiation.[129] It has been shown that heating EG in air results in its oxidation to glycolaldehyde which is a dominant reductant in polyol synthesis and is capable of reducing Ag ions to Ag nanostructures.[130] Ag nanorice with uniform shape has been prepared in high yield by a polyol route.[131] Au nanocages are obtained from Ag nanocubes by a NaHS-mediated polyol synthesis with argon protection (Fig. 10).[132]

Ag NPs (truncated cubes, cuboctahedra, and octahedral) have been synthesized using the polyol method in the presence of PVP. The growth of polyhedral NPs seems to be the result of kinetically limited equilibrium conditions.[133] The hydroxyl groups of PVP appears to reduce the metal salts of Au, Ag, Pt and Pd at a sufficiently slow rate so that the growth becomes kinetically controlled, leading to the formation of thin plates.[134] Liu *et al.*[135] report the use of a polyol process to produce Ag nanohexapods with six branches. Monodisperse decahedral Ag NPs with excellent shape selectivity are obtained by the photochemical transformation of aqueous Ag NP precursors.[136] Giuffrida *et al.*[137] have prepared water-soluble, uncapped Ag NPs in a single step, by spontaneous decomposition of Ag(acac), in the absence of an external reducing agent and conventional stabilizing ligands. Uniform hexagonal Ag nanoplates can be prepared from triangular Ag nanoplates which are in turn obtained by the seed-mediated growth of Ag particles in the presence of trisodium citrate.[138] Branched flower like Ag nanostructured particles have been prepared using rutin, which acts as a reducing and stabilizing agent.[139]

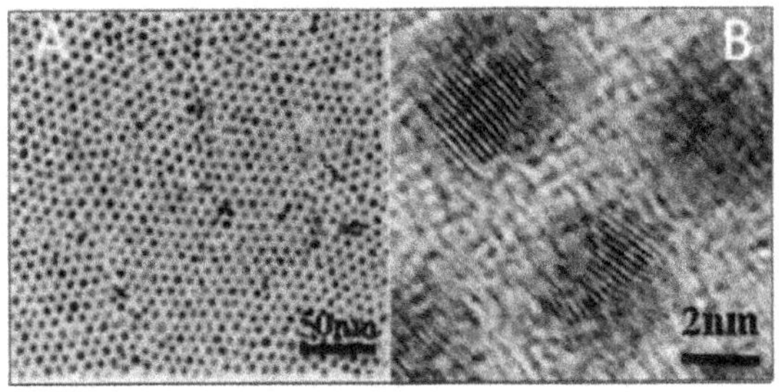


**Fig. 11** (A) TEM and (B) HRTEM images of the 4.5 nm Pd NPs prepared by the reduction of $Pd(acac)_2$ in OAm and the boranetributylamine complex. Reprinted with permission from V. Mazumder and S. Sun, *J. Am. Chem. Soc.*, 2009, **131**, 4588. ©2009 American Chemical Society.

### 3.3. Platinum and palladium

Synthesis of NPs of Pt and Pd is of importance because of their use in fuel cells, sensors, petroleum and in various industrial applications. Teranishi and Miyake[140] synthesized monodisperse Pd NPs by refluxing $H_2PdCl_4$ in a mixture of water and alcohol, in the presence of PVP. Platinum and palladium NPs have been prepared by various chemical methods.[141–144] In general, the synthesis involves the reduction of Pt or Pd precursors in the presence of a polymeric capping agent, using reducing agents such as alcohol, sodium borohydride, or hydrogen gas. Ahmadi *et al.*[145] reported the formation of well-faceted platinum NPs by using hydrogen gas as the reducing agent and sodium polyacrylate as the capping agent. Pt NPs were prepared by $NaBH_4$ reduction of $K_2PtCl_6$ in an aqueous medium. By using HCl, Pt nanoparticles were transferred to a toluene solution containing dodecanethiol, forming a toluene organosol of Pt nanoparticles protected by dodecanethiol.[146,147] Nagy and co-workers[148] prepared Pt colloids in the range of 4 to 6.6 nm, by varying the concentration of $K_2PtCl_4$ from 0.12 to 0.55 mM. Kim *et al.*[149] synthesized monodisperse Pd NPs with particle sizes of 3.5, 5, and 7 nm from the thermal decomposition of a Pd–surfactant complex where the particle size of Pd NPs was controlled by varying the concentration of stabilizing surfactant. Xia and co-workers[150] reported the preparation of palladium nanocubes between 8 to 50 nm by controlling the number of seeds formed at the nucleation stage. Various sizes and shapes of palladium NPs have been obtained by means of the aqueous-phase reaction between $Pd^{2+}$ ions and triblock copolymers.[151]

Pd NPs are obtained by reducing $Pd(acac)_2$ with OAm and the boranetributylamine complex in the temperature range of 60–90 °C (Fig. 11).[152] Sun *et al.*[153] synthesized highly crystalline rectangular Pd NPs by the reduction of $K_2PdCl_4$ using ascorbic acid in presence of CTAB at room temperature. By changing injection sequence of reductant and surfactant, cubic, multi-armed, and dendritic Pd NPs with efficient electrocatalytic activity towards formic acid oxidation could be prepared in high yields. Nearly monodisperse single-crystalline fcc Pd nanocubes and nanodendrites have been successfully prepared in aqueous solution at room temperature by utilizing the equilibrium between dissolution and precipitation of the Pd–CTAB complexes, where the morphology of the Pd NPs can be tuned by the addition of halide ions.[154] Bisson *et al.*[155] have studied the mechanism of formation of the Pd NPs of different shapes (cubes, penta-twinned nanorods, pyramids, and penta-twinned decahedra) by the seed-mediated reduction of a palladium salt in aqueous medium, in the presence of CTAB surfactant, through an oriented-attachment-directed aggregation. Coronado *et al.*[156] prepared water soluble, monodisperse Pd NPs with a narrow particle size distribution by the controlled reduction of $[PdCl_4]^{2-}$, by an aqueous solution of sodium borohydride with tetraoctylammonium bromide acting as the capping agent. The size of NPs could be reduced from the nanometer to the subnanometer size range.

Cubooctahedral NPs of Pd were prepared by a modified polyol process, in which $[PdCl_4]^{2-}$ was reduced by EG at 110 °C in the presence of PVP. It was found that oxidative etching of Pd NPs by air might lead to the removal of twinned particles and the formation of cubooctahedra.[157] Pd icosahedra were formed when citrate was used as the capping agent because of its strong affinity to the {111} facets of Pd.[158] When PVP was used, formation of nanocubes with truncated corners was favored.[150]

Chen *et al.*[159] have synthesized 5 nm Pt NPs by the reduction of $H_2PtCl_6$ in EG in the presence of PVP at 110 °C, while Tian *et al.*[160] report a novel electrochemical approach to prepare tetrahexahedral (THH) Pt NPs with high-index facets on glass carbon substrates (Fig. 12). Faceted Pt NPs are obtained on reducing $H_2PtCl_6$ with PVP in aqueous solution with a trace amount of $FeCl_3$.[161] Wang *et al.*[162] produced Pt nanodendrites by sonicating an aqueous solution containing $K_2PtCl_4$ precursor and formic acid in the presence of nonionic organic molecules. Pt NPs are obtained by the decomposition of $Pt_2(dba)_3$ (dba = dibenzylidene acetone) by using HDA as the stabilizer.[163]

Pt nanopetals can be electrodeposited on Si nanocone arrays in an aqueous solution of 1 M $H_2PtCl_6$ and 1 M $H_2SO_4$ at room temperature by potentiostatic pulse plating in a three-electrode cell system.[164] Pt NPs supported on carbon black are prepared from $Cs_2PtCl_6$ by applying an electrochemical square-wave potential method.[165] Dendritic Pt NPs are synthesized by the reduction of a Pt complex by reducing agent ascorbic acid in aqueous solution, mediated by block copolymer.[166] Pt nanocubes with porous morphology are prepared by using EG, HCl, PVP and $H_2PtCl_6$ under refluxing conditions.[167] Pt spheres with sizes ranging from 10 to 100 nm are prepared from an aqueous solution containing $H_2PtCl_6$, sodium citrate and sodium borohydride.[168]

Soft template synthesis of Pt nanoballs with hexagonal cells formed by connected nanorods has been accomplished by using

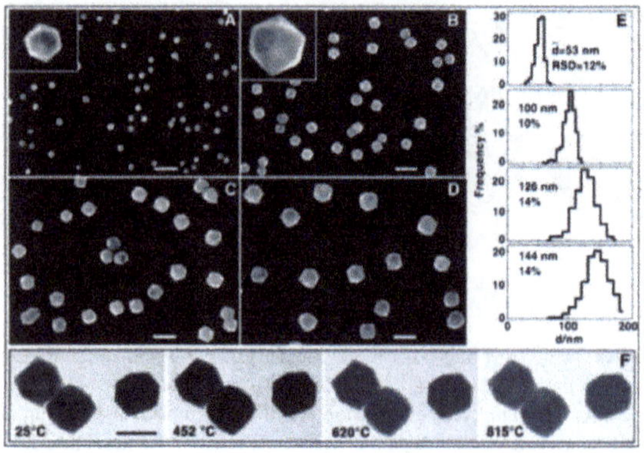


**Fig. 12** Size control of THH Pt NPs and their thermal stability. SEM images of THH Pt NPs grown at (A) 10, (B) 30, (C) 40, and (D) 50 min. The insets in (A) and (B) are the high-magnification SEM images that confirm the shape of THH. Scale bars, 200 nm. (E) Size distributions of THH Pt NPs in (A), (B), (C), and (D), respectively, after counting more than 500 particles for each sample. (F) *In situ* TEM observation on the thermal stability of THH Pt NPs. The images were recorded at various temperatures in TEM at a heating rate of 7 °C $min^{-1}$. The NP preserves its shape to ~815 °C and even higher with a slight truncation at the corners and apexes, as seen in the TEM image. Reprinted with permission from N. Tian, Z.-Y. Zhou, S.-G. Sun, Y. Ding and Z. L. Wang, *Science*, 2007, **4**, 732. ©2007 AAAS. http://www.sciencemag.org.

a mesophase with CTAB as the surfactant and $Pt(NH_3)_4Cl_2$ as the platinum source.[169] Demortière *et al.*[170] employ liquid–liquid phase transfer to synthesize Pt NPs with a cubic morphology, wherein a quaternary ammonium salt serves as the phase-transfer agent for the metallic platinum complex, sodium borohydride is the reducing agent and an alkylamine is a capping agent. Zheng *et al.*[171] have studied the growth of Pt NPs by *in situ* transmission electron microscopy and have proposed that the growth is either by monomer attachment from solution or by particle coalescence through the combination of these two processes. An initial broad size distribution spontaneously narrows into a nearly monodisperse distribution, suggesting that colloidal NPs take different pathways of growth based on their size- and morphology-dependent internal energies. Nanoflowers of Pt, Pd and Au, are prepared by mixing metal precursors and sodium *N*-(4-n-dodecyloxybenzoyl)-L-isoleucinate (SDBIL) at pH 5.0, using L-ascorbic acid as a reducing agent.[172] Cheong *et al.*[173] have reviewed different approaches in solution phase synthesis adopted for achieving shape controlled Pd and Pt NPs with specific crystallographic facets and catalytic properties depends on the shape of the NPs.

Varghese and Rao[174] have investigated platinum NPs prepared by two different methods by employing SAXS and TEM. One of the methods of synthesis involved the reduction of chloroplatinic acid by sodium citrate while in the other method reduction was carried out in the presence of PVP as the capping agent. The growth of platinum NPs prepared by citrate reduction in the absence of any capping agent follows Ostwald ripening growth with a $D^3$ dependence ($D$ = diameter) while in the presence of PVP, the growth of platinum NPs does not follow the Ostwald ripening model, making it necessary to include a surface reaction term in growth equation. The growth of platinum NPs in the presence of PVP has contributions both from diffusion and surface reaction, exhibiting a $D^3 + D^2$ type dependence.

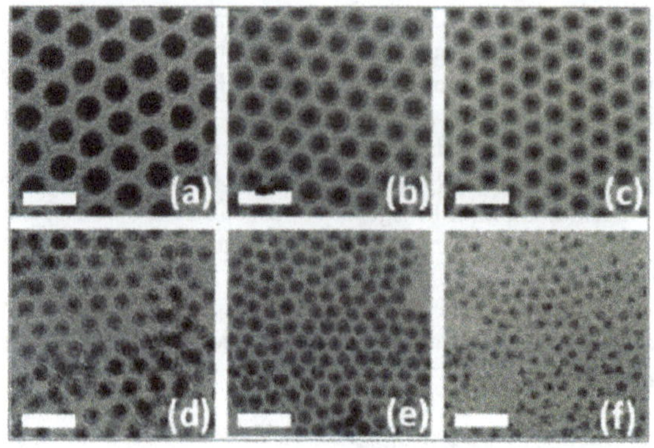


**Fig. 13** TEM images of Fe NPs synthesized from $Fe(CO)_x$–OAm at various reaction temperatures and times: (a) 30 °C for 1 min, (b) 30 °C for 60 min, (c) 30 °C for 180 min, (d) 70 °C for 60 min, (e) 100 °C for 60 min, and (f) 130 °C for 60 min. All scale bars represent 20 nm. Reprinted with permission from H. Kura, M. Takahashi and T. Ogawa, *J. Phys. Chem. C*, 2010, **114**, 5835. ©2010 American Chemical Society.

### 3.4. Iron, cobalt, nickel, copper and other transition metals

Kura *et al.*[175] synthesized monodisperse Fe NPs with high saturation magnetization by the thermal decomposition of the Fe $(CO)_x$–OAm precursor, where CO ligands in $Fe(CO)_5$ are partially replaced by OAm. This procedure provides a control over the size of the NPs over the range 2.3 to 10 nm by changing the reaction temperature (Fig. 13). Ling *et al.*[176] deposited fcc Fe NPs with a variety of morphologies on a carbon film coated copper TEM grids by drop casting of PEG–$Fe(NO_3)_3$ solution and annealing at 400 °C for 1 h in vacuum. Ling *et al.*[177] have also synthesized icosahedral fcc Fe NPs using PVP and ferric nitrate. By applying a voltage of 20 kV, a PVP–$Fe^{3+}$ fluid jet is sprayed on an amorphous carbon or silicon monoxide film coated Cu grid and the grid heated at 400 °C for 1 h in vacuum to yield icosahedral fcc-Fe NPs. Hollow Fe nanoframes are obtained by thermal decomposition of the Fe(II)–stearate complex in the presence of sodium oleate and OA.[178]

Scariot *et al.*[179] report the synthesis of ligand-free Co NPs with either cubic or spherical shapes by *in situ* decomposition of $Co_2(CO)_8$ dispersed in 1-n-decyl-3-methylimidazolium trifluorotris-(pentafluoroethane) phosphate at 150 °C. De Silva *et al.*[180] have investigated the influence of surfactants such as OA, TOPO, ODA and TOP on the reaction pathway, crystal structure, particle size, size distribution and magnetic properties of the Co NPs obtained from acetylene-bridged dicobalthexacarbonyl. Monodisperse water-soluble Co NPs have been prepared by the reduction method in aqueous media in the presence of alkylthioether end-functionalized poly(methacrylic acid).[181] Co NPs (diameter, <5 nm) are obtained by pulsed laser irradiation of $Co_2(CO)_8$ in 1,2-dichlorobenzene in the presence of stabilizers OA and TOPO.[182] Single-crystalline Co hollow nanoparallelepipeds have been prepared by the thermolysis of CoO solid nanoparallelepipeds in OAm at 290 °C for 2 h and 270 °C for 1 h.[183] Wang *et al.*[184] report the synthesis of cubic Co nanoskeletons from a dispersion of cobalt acetate tetrahydrate and sodium fluoride in absolute ethanol at 180 °C for 48–60 h.

Metin *et al.*[185] synthesized monodisperse Ni NPs from the reduction of $Ni(acac)_2$ with borane tributylamine in OAm and OA. Size-tunable monodispersed hierarchical metallic Ni NPs (58–190 nm in diameter) are obtained by the reduction of $Ni^{2+}$ with HDA under atmospheric pressure.[186] Plouet *et al.*[187] report the formation of metastable hcp and air-stable fcc Ni NPs from organically functionalized layered phyllosilicates of nickel(II) at temperatures less than 573 K in an inert atmosphere. Carenco *et al.*[188] have investigated the preparation of size-tunable monodisperse Ni NPs, by the reaction of $Ni(acac)_2$ with OAm and TOP. Monodisperse Ni NPs with tunable phase (fcc, hcp) and size are prepared by the thermal-decomposition of nickel acetate in the presence of hexadecylamine which is used as the solvent, reducing agent as well as stabilizing agent.[189] Ni NPs in the size range of 4–16 nm are obtained by high-temperature reduction of $Ni(CH_3COO)_2$ or $Ni(acac)_2$ followed by annealing.[190]

Nanoparticles of Cu, Ag, and Au have been prepared by mechanical stirring of solid mixtures of the metal salt and ammonia borane at 60 °C.[191] Engels *et al.*[192] have reported the synthesis of Cu NPs by using modified polyol-based reduction method in EG that incorporates PVP as polymeric anti-agglomerant and a reducing additive ($N_2H_4{\cdot}H_2O$, $NaBH_4$, $NaH_2PO_2{\cdot}H_2O$). Chang *et al.*[193] have reported the growth of arrays of tetragonal Cu nanopillars on Au–glass by the electrochemical reduction of an aqueous solution of $CuCl_2$ by Al metal in dodecyltrimethylammonium chloride. Haas *et al.*[194] prepared Cu dendrite nanostructures by a sonoelectrochemical process from an aqueous solution of $Cu^{2+}$ in the presence of PVA. Cu NPs with controllable size and shape are obtained by the reduction of copper(II) in an aqueous solution of hydrazine under an air atmosphere in the presence of polyacrylic acid (PAA) as a capping agent.[195] Huang *et al.*[196] obtained nanosize particles of Cu, Ag, Pd and Ni as well as CuPd and CuAg alloys encapsulated in inert carbon shells by the carbonization of metal–cyclodextrin complexes.

Bondi *et al.*[197] report a solution chemistry method for the synthesis of air-stable NPs of α-Mn, in which n-butyllithium is used as the reducing agent in diphenyl ether medium and OA is used to bind to the Mn NPs to render them air-stable. Redel *et al.*[198] report the synthesis of stable Cr, Mo and W NPs by thermal or photolytic decomposition of mononuclear metal carbonyl precursors under argon in ionic liquid media. Ruthenium(0) nanoclusters are prepared by the borohydride reduction of a colloidal solution of ruthenium(III) chloride at room temperature and stabilized by a nanozeolite framework. Rh multipods can be synthesized by taking $Na_3RhCl_6$ as the precursor of Rh. The branched morphology is retained only under inert conditions, while a spherical shape occurs in an oxidation environment.[199] NPs, nanorods and other nanostructures of nickel, ruthenium, rhodium and iridium have been prepared by a simple solvothermal procedure involving the decomposition of corresponding metal acetate or acetylacetonate in the presence of n-octyl amine.[200] Zhang *et al.*[201] report the synthesis of five-fold twinned, starfish-like Rh NPs by a facile polyol approach by replacing $RhCl_3$ or $Na_3RhCl_6$ with $[\{Rh(CF_3COO)_2\}_2]$ to eliminate oxidative etching. Ru(0) NPs have been prepared in imidazolium ionic liquids from $[Ru(COD)(2\text{-methylallyl})_2]$ in hydrogen atmosphere at 50 °C.[202] Vollmer *et al.*[203] have carried out microwave irradiation of metal–carbonyls for the facile synthesis of transition-metal NPs in ionic liquids. Yuan *et al.*[204] report tunable rhodium nanostructures with highly selective morphologies such as cubes, horned particles, dendrites and network-shaped wires from $RhCl_3$, PVP, SDS and sodium salts of different halides in water by heating at 220 °C for the desired time. Zhang *et al.*[205] report the synthesis of nafion-stabilized Ir NPs by taking nafion, $H_2IrCl_6$, NaOH, water and ethanol as reactants and heating at 125 °C for 5 hours in an autoclave.

### 3.5. Other metals

Metallic Mg NPs of 4 nm size can be prepared in polar aprotic solvents such as THF and dibutyldiglyme by the sonoelectrochemical method using a Grignard reagent. Large scale synthesis of hexagonal prismatic sub-micro sized Mg crystals can be grown *via* thermal decomposition of $MgB_2$ at 800 °C and for 1 h, followed by further heating to 1400 °C for 3 h.[206] Mg NPs are prepared by room temperature digestive ripening in the presence of HDA resulting in highly monodisperse colloidal Mg-HDA NPs of size 2.8 nm.[207] Meziani *et al.*[208] have investigated the wet-chemical synthesis of Al NPs by using dimethylethylamine alane and 1-methylpyrrolidine. Decomposition of an organometallic precursor of indium such as cyclopentadienyl indium ($InC_5H_5$) in methanol and PVP at room temperature to prepare small indium NPs.[209] Indium NPs are also prepared by reduction of indium chloride using lithium borohydride as the reducing agent in primary amine media.[210] Indium NPs are synthesized by the drop-wise addition of a solution of $NaBH_4$ in TEG to a solution of $InCl_3$ and PVP in isopropyl alcohol.[211] Ge NPs (1 to 15 nm) are obtained by the thermal decomposition trichlorogermane, tetraethylgermane or tetrabutylgermane in high boiling organic solvents such as trioctylamine, squalene, and octacosane.[65] Ge NPs stabilized by PVP are synthesized by borohydride reduction of $GeCl_4$ in EG and triglime.[212] Prabakar *et al.*[213] report size-controlled synthesis of Ge NPs by a hydride reduction of germanium tetrachloride in inverse micelles.

Adams *et al.*[214] prepared Cd NPs by the reduction of cadmium acetate around 200 °C in three different media (ODE, diphenyl ether, and dioctyl ether) resulting in nanostructures ranging from monodisperse spheres to hexagons. Resa *et al.*[215] have reported a method to produce a large quantity of monodisperse superconducting Pb NPs by the alcohol reduction of a mixture of lead oleate and lead acetate in the presence of TOP in ODE at 80 °C. Hu *et al.*[216] performed electron-beam irradiation in a TEM to fabricate NPs of low-melting-point metals such as Zn, Ga, Sn, and Cd and IV–VI semiconducting lead chalcogenides PbE (E = S, Se, and Te). Bi spheres are prepared *via* the thermolysis of a bismuth–thiolate precursor in a mild solventless condition with the help of organic capping reagents such as 1-dodecanethiol and PVP.[217]

## 4. Metal alloy nanoparticles

Alloy NPs consisting of two or more metals show additive as well as new properties of interest.[10,218,219] For example,

**Fig. 14** TEM images of the intermetallic CuAu with different sizes: (a) (4.22 ± 0.2) nm, (b) (6.0 ± 0.2) nm, (c) (7.4 ± 0.3) nm; (d) TEM image of a multilayer hcp superlattice consisting of (6.00.2) nm CuAu NPs; inset shows the corresponding fast Fourier transform (FFT) pattern. Reprinted with permission from W. Chen, R. Yu, L. Li, A. Wang, Q. Peng and Y. Li, *Angew. Chem., Int. Ed.*, 2010, **49**, 2917–2921. ©2009 Wiley-VCH Verlag GmbH & Co. K GaA.

magnetic alloy NPs are preferred to single metal NPs because of their relative stability under ambient conditions. Hollow triangular AuAg nanoboxes have been prepared from Ag nanoprisms by employing galvanic replacement reactions using ascorbic acid and $HAuCl_4$.[220] Silver in these nanoboxes was removed with an aqueous etchant, $Fe(NO_3)_3$ or $NH_4OH$.[221] AuAg bimetallic NPs have been prepared by the sonochemical co-reduction of Au(III) and Ag(I) ions in aqueous solutions.[222] Co-reduction of Au(III) and Ag(I) produces bimetallic NPs with a Au core–Ag shell morphology. AuAg alloy NPs can be prepared in a cellulose matrix by the co-reduction of corresponding metal ions using $NaBH_4$.[223] One-pot synthesis of OAm-coated AuAg alloy NPs are obtained by the simultaneous reduction of $HAuCl_4$ and $AgNO_3$ in the presence of octadecane and OAm at 120 °C for 0.5 h in a $N_2$ atmosphere.[224]

AuCu NPs are obtained by heating a mixture of Cu $(CH_3COO)_2$ and presynthesized Au NPs in OA and TOA at an appropriate temperature for 50–100 min.[225] AuCu and $AuCu_3$ NPs have been prepared by Chen *et al.*[226] by a seed-based diffusion route using Au NPs and $Cu(CH_3COO)_2$ as precursors (Fig. 14).

AuPd hybrid nanostructures are obtained from the galvanic replacement reaction between single-crystal Pd nanorods and $AuCl_4^-$ ions.[227] Colloidal dispersions of the PdAu bimetallic clusters are obtained by successive alcoholic reduction of their constituent metal ions refluxed in EG at 160 °C in the presence of PVP have a three layer core–shell morphology composed of an evenly alloyed inner core, an Au-rich intermediate layer, and a Pd-rich outer shell.[228] $Cu_{60}Pd_{40}$ NPs are prepared by the co-reduction of $Cu(acac)_2$ and $Pd(acac)_2$ using the superhydride $LiBEt_3H$ in THF.[229]

$Au_{21}Fe_{10}$, $Au_{22}Fe_{12}$, $Au_{28}Fe_{14}$, and $Au_{34}Fe_{14}$ clusters have been synthesized by the oxidation of $[Fe_3(CO)_{11}]^{2-}$ with $[AuCl_4]^-$ in acetone in an inert atmosphere.[230] Intermetallic compounds of $Au_3Fe$, $Au_3Co$, and $Au_3Ni$ with the cubic ($Cu_3Au$) structure are formed by the reaction of $HAuCl_4{\cdot}3H_2O$, OAm, and an appropriate quantity of $Fe(acac)_3$, $Co(acac)_2$, or Ni $(acac)_2$ in octyl ether and n-butyllithium at 80 °C under Ar.[231]

Synthesis of bimetallic AuPt NPs is carried out by using $HAuCl_4$ and $H_2PtCl_6$ as precursors and OAm as the solvent, stabilizer, and reducing reagent.[232] Heterostructured $PtAu_n$ NPs are obtained through a seed-mediated process in the presence of Pt NPs as seeds, wherein $HAuCl_4$ is reduced by OAm in ODE.[233] AuPt nanodendrites can be prepared by the hydrothermal co-reduction of $H_2PtCl_6$ and $HAuCl_4$ in ammonium formate along with etched Ti plates in a Teflon lined autoclave at 180 °C for 8 h.[234] AuPt alloy NPs have been prepared by the reduction of DMF-coordinated $Au^{3+}$ and $Pt^{4+}$.[235] In this process, DMF serves as the solvent, ligand, and reductant.

PtPd alloy NPs with well-defined twinned structures are synthesized by simple co-reduction of $Na_2PdCl_4$ and $K_2PtCl_4$ with PVP in aqueous solution.[236] PtPd bimetallic nanodendrites consisting of a dense array of Pt branches on a Pd core have been prepared by Lim *et al.*[237] by reducing $K_2PtCl_4$ with L-ascorbic acid in the presence of Pd NP seeds in aqueous solution (Fig. 15). A wet-chemical approach has been used for the synthesis of $Pt_3Ni$ nanooctahedra and nanocubes, wherein the reaction is carried out between $Ni(acac)_2$ and $Pt(acac)_2$ in a mixture of OA and OAm at ~200 °C in the presence of $W(CO)_6$.[238]

Intermetallic PtPb nanodendrites can be obtained by hydrothermal-assisted co-reduction of Pt and Pb inorganic precursors by formic acid in aqueous media in the absence of a surfactant or polymer.[239] Yang *et al.*[240] synthesized PtPb nanostructures by the simultaneous reduction of $Pt(acac)_2$ and $Pb(acac)_2$ by the *tert*-butylamine–borane complex in the presence of adamantanecarboxylic acid, hexadecanethiol and HDA in diphenylether at a temperature between 180 and 200 °C. Liu *et al.*[241] describe the synthesis of PtSn NPs by the co-reduction of $Pt(acac)_2$ and $SnCl_4$ in ODE by using $NaBEt_3H$ and with OAm and OA as capping agents to control the particle size and shape. $Pt_3Sn$ NPs are obtained starting with $Pt(dba)_2$ and $(n\text{-}Bu)_3SnH$ and carrying out the reaction under $H_2$ at room temperature.[242] RuPt bimetallic NPs have been deposited on the surface of alumina by the atomic layer deposition technique using 2,4-(dimethylpentadienyl) (ethylcyclopentadienyl)-ruthenium and trimethyl(methylcyclopentadienyl) platinum as the precursors.[243]

Sun and coworkers[244] have reported a general method to synthesize monodisperse FePt alloy NPs by the simultaneous reduction of $Pt(acac)_2$ and the thermal decomposition of Fe $(CO)_5$ in the presence of OA and OAm. A similar method has been employed for the synthesis of monodisperse, crystalline $CoPt_3$ NPs by the reduction of $Pt(acac)_2$ and thermal decomposition of $Co_2(CO)_8$ in the presence of 1-adamantanecarboxylic acid.[245] Thermal decomposition of anionic bimetallic carbonyl clusters $[FeCo_3(CO)_{12}]^-$, $[Fe_3Pt_3(CO)_{15}]^{2-}$, $[FeNi_5(CO)_{13}]^{2-}$ and $[Fe_4Pt(CO)_{16}]^2$ gives alloy NPs of $FeCo_3$, FePt, $FeNi_4$, and $Fe_4Pt$ respectively.[246] Wellons *et al.*[247] demonstrate a one-step synthesis of ferromagnetic Fe–Pt NPs by the reductive decomposition of the single-source precursor, FePt $(CO)_4$dppmBr, on a water-soluble support ($Na_2CO_3$) at 600 °C. $FePt_3$ NPs are also prepared by the thermal decomposition of the bimetallic precursor $[(CO)_3Fe(\mu\text{-dppm})(\mu\text{-CO})PtCl_2]$ in the

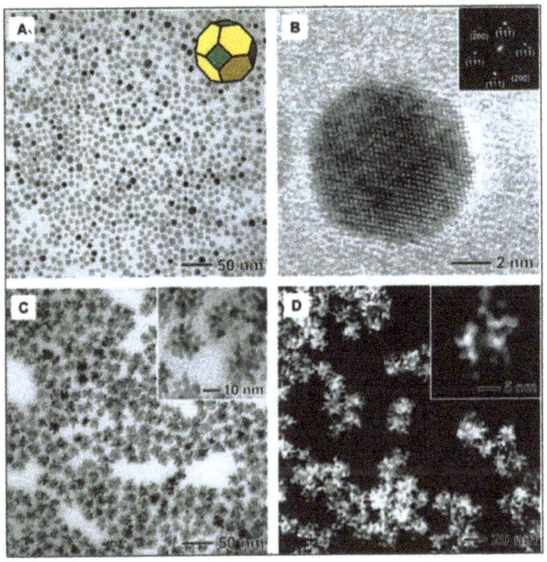

**Fig. 15** (A) TEM image of truncated octahedral Pd NPs synthesized by reducing $Na_2PdCl_4$ with L-ascorbic acid in an aqueous solution. The inset shows a geometrical model of the truncated octahedron, where the green and yellow colors denote the {100} and {111} facets, respectively. (B) HRTEM image of a single truncated octahedron of Pd recorded along the [011] zone axis and the corresponding FT pattern (inset). (C) TEM image of Pd–Pt nanodendrites synthesized by reducing $K_2PtCl_4$ with L-ascorbic acid in the presence of truncated octahedral Pd NP seeds in an aqueous solution. (D) HAADF-STEM image of Pd–Pt nanodendrites. Reprinted with permission from B. Lim, M. Jiang, P. H. C. Camargo, E. C. Cho, J. Tao, X. Lu, Y. Zhu and Y. Xia, *Science*, 2009, **5**, 324. ©2009 AAAS. http://www.sciencemag.org.

presence of OAE and OA in nonadecane.[248] Fe–Pt NPs can be prepared at 350 °C using $Na_2Fe(CO)_4$ and $Pt(acac)_3$ as reagents and nonadecane and tetracosane as solvents.[249] Hydrophobic to superhydrophobic surfaces are obtained by having fluorinated ligands on the surface of FePt.[250] FePt NPs capped with perfluorinated ligands (Fig. 16) are prepared by simultaneous thermal decomposition of $Fe(CO)_5$ and reduction of $Pt(acac)_2$.[53]

Hollow CoPt NPs are prepared wherein Co NPs are first reduced by $NaBH_4$ which act as *in situ* template, followed by the spontaneous galvanic displacement reaction with $PtCl_6^{2-}$ with Co NPs in the presence of a citrate stabilizer.[251] $Ni_xPt_{1-x}$ NPs are synthesized *via* reduction of $Pt(acac)_2$ and $Ni(acac)_2$ in the presence of OA, OAm and HDD.[252] By dissolving $Pt(acac)_3$ and Ni $(acac)_2$ in diphenyl ether, followed by the addition of a mixture of the borane–*tert*-butylamine complex and HDD along with long-alkane-chain amines, truncated-octahedral $Pt_3Ni$ NPs are obtained.[253]

CoFe NPs are synthesized *via* an organometallic route using $Fe(CO)_5$ and $[Co(h_3\text{-}C_8H_{13})(h_4\text{-}C_8H_{12})]$ in toluene in the presence of stabilizing agents such as a mixture of OA and HDA at 150 °C under $H_2$ for 48 h.[254] Nanocubes of FeCo alloys are formed on a large scale by the reduction of aqueous $Fe^{2+}$ and $Co^{2+}$ with hydrazine in the presence of PEG and cyclohexane at 80 °C.[235] CoFe NPs embedded in a MgO crystal matrix have been prepared starting from $MgCo^{II}Fe^{III}$ layered double hydroxides (LDHs) by hydrogen gas phase reduction at 450 °C.[255] Decomposition of $Fe(acac)_3$ and $Co(acac)_2$ in a mixture of surfactants and HDD in Ar + $H_2$ atmosphere at 300 °C yields FeCo NPs.[256]

Cable and Schaak[257] have reported a general strategy for the solution-mediated synthesis of nanocrystalline zinc-based intermetallic compounds, M–Zn (M = Au, Cu, Pd), involving the low-temperature chemical conversion of preformed metal NPs into intermetallics *via* reaction with zero-valent organozinc reagents such as $Et_2Zn$ and $Ph_2Zn$. NPs of the bismuth–antimony–telluride alloy, $Bi_{0.5}Sb_{1.5}Te_3$, are obtained by a direct chemical solution route, in which the bismuth and antimony precursors dissolved in a phenyl ether and coordinated with dodecanethiol, are reacted with tellurium in TOP.[258]

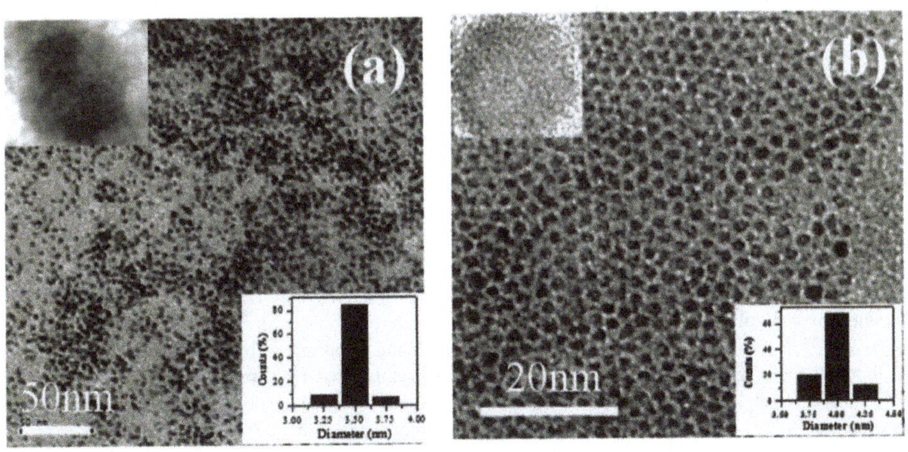


**Fig. 16** TEM images of fluorous ligand capped (a) 4 nm $\gamma\text{-}Fe_2O_3$ and (b) 3.5 nm FePt NPs with size distribution histograms as inset. Reproduced with permission from R. Voggu, A. Shireen and C. N. R. Rao, *Dalton Trans.*, 2010, **39**, 6021. ©Royal Society of Chemistry.

 

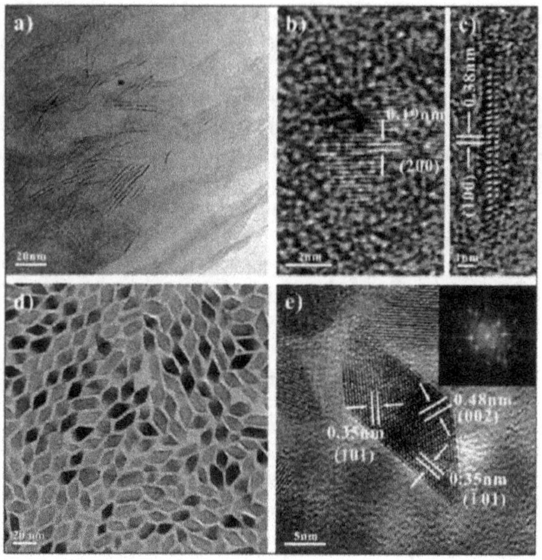


**Fig. 17** Representative TEM images of NS-$TiO_2$ and NP-$TiO_2$. (a) Low magnification image of as-made NS-$TiO_2$. (b and c) High-magnification images of nanosheets obtained by treating NS-$TiO_2$ with tetrabutylammonium hydroxide. (d and e) Low-magnification and high-resolution TEM images of NP-$TiO_2$. The inset in (e) is the Fourier transfer image corresponding to HRTEM of NP $TiO_2$ (NS – nanosheets; NP-nanoparticles). Reproduced with permission from Wu, C. Guo, N. Zheng, Z. Xie and G. D. Stucky, *J. Am. Chem. Soc.*, 2008, **130**, 17563. ©2009 American Chemical Society.

Iron silicide NPs are prepared *via* solution phase synthesis by the reaction of silicon tetrachloride with ironpentacarbonyl in the presence of HDD at 180 °C.[259] Single-crystal intermetallic M–Sn (M = Fe, Cu, Co, Ni) nanospheres can be obtained *via* a modified polyol process.[260]

## 5. Metal oxide nanoparticles

A nonaqueous sol–gel approach has been employed by Koziej *et al.*[261] to prepare $TiO_2$ (anatase) NPs by mixing $Ti(OiPr)_4$ and benzyl alcohol. Wu *et al.*[262] report a nonaqueous synthesis of nanostructured $TiO_2$ by condensation of titanium isopropoxide in the presence of benzyl alcohol and OAm at room temperature (Fig. 17). Liao *et al.*[263] have synthesized $TiO_2$ NP colloids from the peroxotitanate complex solution, starting from titanium(IV) sulfate and $H_2O_2$ in a water–alcohol media, under solvothermal conditions. Dai *et al.*[264] obtain $TiO_2$ (anatase) NPs with chemically active facets, by digesting electrospun nanofibers consisting of amorphous $TiO_2$ and PVP in aqueous acetic acid solution, followed by hydrothermal treatment at 150 °C for 20 h. $TiO_2$ (brookite) NPs are obtained by aminolysis of titanium carboxylate complexes.[265]

Tenne *et al.*[266] report the synthesis of hollow $V_2O_5$ NPs by laser ablation of $V_2O_5$. The as-prepared $V_2O_5$ NPs are crystalline with a low-energy surface {010} pointing outward and with some defects between the facets. Different sized $VO_x$ hierarchical hollow spheres aggregated by small NPs can be obtained by a template-free method involving the decomposition of V $(acac)_2$.[267]

Yu *et al.*[268] report the synthesis of uniform $Mn_3O_4$ NPs by the low-temperature decomposition of a mixture of manganese acetate, carboxylic acid and OAm in xylene. The size and shape of the NPs could be controlled by varying the experimental conditions such as precursors, surfactants and injection temperature. $Mn_3O_4$ NPs were obtained by Pike *et al.*[269] by mixing aqueous solutions of manganese nitrate and hexamethylenetetramine at a temperature between 20 and 80 °C. Bruce *et al.*[270] report the formation of ordered three-dimensional array of monodispersed 8 nm $Mn_3O_4$ NPs with a core–shell structure, by the thermal decomposition of $Mn(acac)_2$ in OAm. The self-assembled three-dimensional array exhibits a behavior similar to a spin-glass rather than superparamagnetic particles. The as-prepared $MnO_2$ NPs are able to remove organic pollutants from waste water. Nanoplates with a thickness of 5 nm and side dimensions of 9, 15 and 22 nm are obtained by using OAm as the surfactant where as by using a carboxylic acid as the cosurfactant along with OAm, spherical NPs are obtained in the size ranges of 5.5, 6.2, 7.2, 8.5 and 15 nm.[268] Rusakova *et al.*[271] report the conversion of hexagonal-, square- and cross-shaped MnO NPs into MnO–$Mn_3O_4$ NPs. Octahedral MnO NPs were synthesized by Xie *et al.*[272] by the thermal decomposition of $Mn(OOC_2H_5)_2$ in a mixed organic solvent containing trioctylamine and OA. MnO NPs with average diameters in the 6–14 nm range are produced by the decomposition of manganese cupferronate in the presence of TOPO, under solvothermal conditions.[273] Jiang *et al.*[274] report the synthesis of single-crystalline $Mn_3O_4$ nanooctahedra from the ethylenediaminetetraacetic acid disodium salt using the hydrothermal route. Fei *et al.*[275] demonstrate that hierarchical hollow $MnO_2$ can be synthesized by $KMnO_4$ oxidation of micrometer-scale manganese carbonate structures at room temperature, followed by the selective removal of $MnCO_3$ with HCl.

γ-$Fe_2O_3$ and $Fe_3O_4$ are synthesized by microemulsions as well as by other methods.[276–280] Coprecipitation from the iron(II) chloride and iron(III) chloride solutions is a convenient method to synthesize iron oxides, by using a base under inert atmosphere at room temperature or at an elevated temperature.[278] One of the most widely used methods is the thermal decomposition of organic soluble precursors such as $Fe(CO)_5$ and $Fe(acac)_3$ for the synthesis of monodisperse iron oxide NPs.[281] Sun *et al.*[282] report the synthesis of spherical, truncated octahedral $Fe_3O_4$ NPs by the high-temperature thermal decomposition of $Fe(acac)_3$ in the presence of OA and OAm. Another method is the nonaqueous solvothermal decomposition of iron acetylacetonate.[18] The iron NPs are prepared by the thermal decomposition of $Fe(CO)_5$ in the presence of OAm and the Fe NPs upon air oxidation, followed by heating gives rise to hollow $Fe_3O_4$ NPs (Fig. 18).[283] Spherical and cubic iron oxide NPs are obtained by the decomposition of iron oleate complex by using OA and OAm as capping agents in ODE solvent.[284] Magnetite NPs have been prepared by Sun *et al.*[285] from $Fe(acac)_3$ by using OAm as both reducing agent and stabilizer. Superparamagnetic iron oxide NPs are synthesized by coprecipitation from a ferrous + ferric salts solution by concentrated sodium hydroxide solution inside the aqueous cores of AOT–n-hexane reverse micelles.[159] Glucose and gluconic acid-coated magnetite NPs can be prepared by hydrothermal synthesis from $FeCl_3$ in the presence of sucrose.[286]

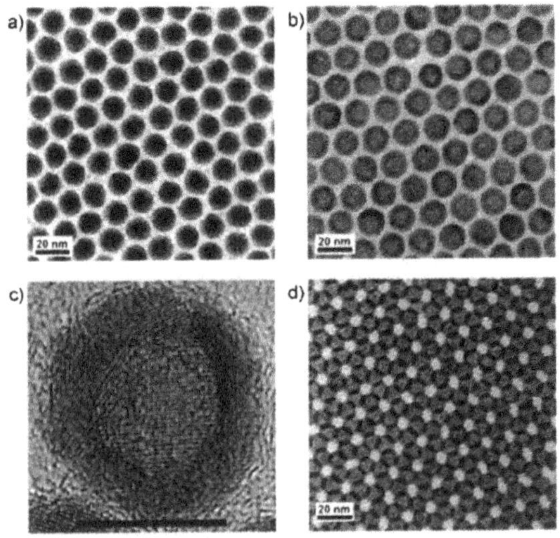


**Fig. 18** TEM images: (a) 13 nm Fe–$Fe_3O_4$ nanoparticle seeds, (b) 16 nm hollow $Fe_3O_4$ NPs, (c) a single hollow $Fe_3O_4$ nanoparticle (scale bar 10 nm), and (d) a superlattice array of the hollow $Fe_3O_4$ NPs. Reproduced with permission from Peng *et al.*, *Angew. Chem., Int. Ed.*, 2007, **46**, 4155. ©2007 Wiley-VCH Verlag GmbH & Co. KGaA.

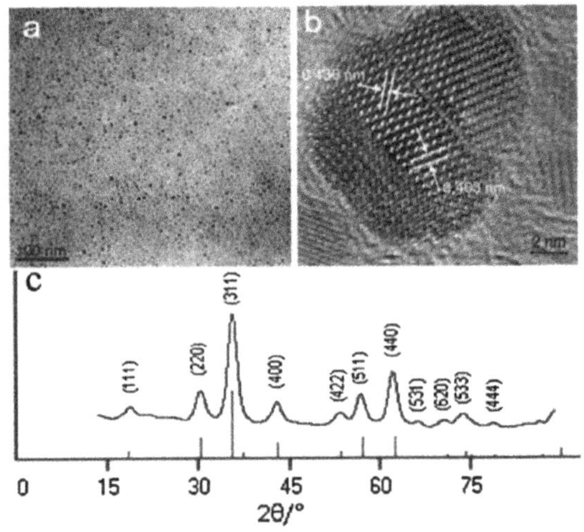


**Fig. 19** (a) TEM image, (b) HRTEM image, and (c) XRD pattern of $Fe_3O_4$ NPs synthesized by typical process and standard. Reproduced with permission from Hu *et al.*, *Chem. Commun.*, 2010, **46**, 3866–3868. ©2010 Royal Society of Chemistry.

Synthesis and characterization of iron oxide-derivatized mutant cowpea mosaic virus hybrid NPs have been reported and the NPs were found to be good candidates for MRI imaging.[287] Peng and Sun[288] report a facile solution-phase synthesis of monodisperse hollow $Fe_3O_4$ NPs by the controlled oxidation of Fe–$Fe_3O_4$ NPs.

Abou Hassan *et al.*[289] show that aqueous dispersions of 6 nm superparamagnetic $Fe_3O_4$ NPs can be prepared on a chip in a millichannel reactor by the use of tetramethylammonium hydroxide as the alkaline reagent. Hu *et al.*[290] use ionic liquid mediated microwave heating for the synthesis of magnetite NPs by the reaction of $Fe(acac)_3$ and HDD in dibenzyl ether in the presence of OA and [bmim][$BF_4$] at 250 °C. Here, [bmim][$BF_4$] acts as both stabilizing agent and a microwave-absorbing medium (Fig. 19). Cysteine-capped amorphous magnetite is prepared by the sonochemical decomposition of $Fe(CO)_5$ in the presence of cysteine.[291] Synthesis of organic soluble monodisperse iron oxide NPs has been carried out by the digestion of submicrometer hematite powder in the presence of OA at high temperatures.[292] Salazar-Alvarez *et al.*[293] have synthesized cubic and spherical shaped $Fe_3O_4$ NPs by the decomposition of iron oleate and shown that the spherical ones exhibit larger blocking temperatures ($T_B$) than the cubic ones. Thermal decomposition of FeOOH has also been employed to prepare spherical and cubic iron oxide NPs.[294] The difference in $T_B$ and other magnetic properties of spherical and cubic forms appears to be due to the change in the percent content of FeO. Feyen *et al.*[295] obtain mushroom nanostructures of iron oxides by partial coating of $Fe_3O_4$ or $Fe_2O_3$ with polymer spheres followed by attachment to silica hemispheres. Octahedral $Fe_3O_4$ NPs can be synthesized from iron(III)oleate in tetracosane medium.[296] Cuboid iron-manganese oxide NPs with dimensions as large as 33 nm are obtained by surfactant-mediated synthesis.[297]

Monodisperse fluorous amine capped γ-$Fe_2O_3$ have been prepared by the thermal decomposition of $Fe(CO)_5$ (Fig. 16(b)).[53] Different morphological single-crystal magnetites with a nanoporous structure, exhibiting excellent magnetic properties, have been obtained by a polyol process using $FeSO_4$ and EG.[298] Tan *et al.*[299] report room-temperature synthesis of magnetite NPs by the reduction of $Fe(acac)_3$ with NMP or pyrrole in the presence of surfactants OA and OAm in air. Qu *et al.*[300] used hydrothermal growth for the synthesis magnetite nanooctahedra coated with aspartic acid by using the precursor $FeCl_3{\cdot}6H_2O$.

Synthesis of various-shaped (*i.e.*, capsule, nanorings) NPs by the decomposition of $Fe(acac)_3$ has been accomplished by varying the chain length of the carboxylic acid or by the use of hydrazine.[301] Initially capsule-shaped α-$Fe_2O_3$ NPs are formed, which on preferential dissolution along the long dimension of the elongated NPs (*c*-axis of α-$Fe_2O_3$) form nanorings. The shape of the nanorings is regulated by the adsorption of phosphate ions on the faces parallel to the *c*-axis. The hollow structure of the NPs arises from the preferential dissolution of the α-$Fe_2O_3$ along the *c*-axis due to the strong coordination of the sulfate ions. α-$Fe_2O_3$ nanorings can be converted to magnetite and maghemite by reduction or reduction–oxidation, while preserving the morphology. Hu *et al.*[302] report microwave-assisted hydrothermal synthesis of α-$Fe_2O_3$ nanorings by using $FeCl_3{\cdot}6H_2O$ dissolved in aqueous $NH_4H_2PO_4$ solution. Jia *et al.*[303] have obtained single-crystal α-$Fe_2O_3$ nanorings by using the double anion-assisted hydrothermal method starting with $FeCl_3$, $NaH_2PO_4$, and $Na_2SO_4$. Kim *et al.*[304] report the synthesis of iron oxide nanocubes in the size range of 20 nm to 160 nm by the thermal decomposition of $Fe(acac)_3$ in a mixture of OA and benzyl ether. They show that the nanocubes are formed as a result of fast growth along a direction, with the surfaces of the nanocubes corresponding to the {100} planes. Tetrakaidecahedral and oblique parallelepiped iron oxide NPs with high index facets are obtained in high yields with the assistance of the sodium carboxymethyl cellulose.[305] Liz-Marzán *et al.*[306] report on the shape-controlled synthesis of iron oxide NPs from TOAB and iron oleate.

Table 2 presents some of the methods of synthesis of iron oxide NPs.

Hu *et al.*[322] report controllable synthesis of $Co_3O_4$ NPs with different shapes *via* hydrothermal processing of cobalt hydroxide and subsequent thermal decomposition. The high index {112} crystal planes of $Co_3O_4$ NPs appear to be more reactive for $CH_4$ combustion than the {001} and {011} planes. $Co_3O_4$ nanopolyhedra have been synthesized at 80 °C in aqueous solution by using $Co(NO_3)_2·6H_2O$ and hexamethylenetetramine.[323] NPs of cobalt hydroxide, cobalt oxyhydroxide, and cobalt oxide are prepared by a soft chemical route by using $Co(NO_3)_2$ solution at 60 °C under the basic conditions.[324] Single-crystalline $Co_3O_4$ hollow octahedral cages with tunable surface aperture are obtained by the carbothermal method, by using colloidal carbon spheres as templates and cobalt chloride as the reactant.[325] Zhang *et al.*[326] report the synthesis of cubic CoO NPs with different morphologies and sizes by the decomposition of cobalt(II) oleate at 280–320 °C in noncoordinating solvents such as ODE containing dodecanol–OA. Synthesis of rocksalt and wurtzite phases of CoO through thermodynamically and kinetically controlled reactions of $Co(acac)_3$ has been reported.[327] NPs of NiO with average diameters in the 3–24 nm range have been prepared by the decomposition of nickel cupferronate or acetate under solvothermal conditions.[273]

$Cu_2O$ NPs have also been prepared by the thermal decomposition of copper(I) acetate in trioctylamine in the presence of tetradecylphosphonic acid and further oxidation.[328] The $Cu_2O$ seed NPs are first prepared with SDS as the capping surfactant and sodium ascorbate as the reducing agent. Addition of NaOH forms $Cu(OH)_4^{2-}$, which subsequently gets reduced to form $Cu_2O$ seed particles. The seed particles aggregate and undergo surface reconstruction to evolve cubic morphology. Polyhedral $Cu_2O$ particles are prepared by adding a weak reducing agent such as glucose to a solution of copper citrate with PVP as capping agent.[329] $Cu_2O$ nanoframes and nanocages are obtained *in situ* by oxidative etching at room temperature. $Cu_2O$ nanocages and nanoframes possessing truncated rhombic dodecahedral structures are obtained from aqueous solution containing $CuCl_2$, SDS, $NH_2OH·HCl$ and NaOH.[330] Yao *et al.*[331] report the synthesis of monodisperse $Cu_2O$ NPs by the decomposition of Cu(II) acetate in presence of hexadecyamine. The polyol method can be employed for the synthesis of $Cu_2O$ NPs using $Cu(NO_3)_2$ and PVP as reagents and EG as the solvent.[332] Synthesis of monodisperse truncated cubic, cuboctahedral, truncated octahedral and octahedral $Cu_2O$ NPs has been carried out in aqueous solution starting with $CuCl_2$, SDS, hydroxylamine reductant, and NaOH, followed by varying the volume of hydroxylamine added to the reaction mixture.[333] Differently shaped $Cu_2O$ NPs are prepared by reducing copper citrate with glucose.[334] Park *et al.*[335] have reported gram-scale synthesis of uniform $Cu_2O$ nanocubes by a one-pot polyol process wherein $Cu(acac)_2$ is dissolved in hot PVP. Monodisperse $Cu_2O$ nanocubes with average sizes of 40, 65, 100, 230 and 420 nm are formed from seed-mediated synthesis in aqueous medium.[336] $Cu_2O$ cubes are converted to polycrystalline CuO hollow nanostructures in air through a sequential dissolution–precipitation process by adding aqueous ammonia solution.

**Table 2** Various synthetic conditions for the synthesis of iron oxide nanoparticles

| Method | Compound | Reactants and reaction conditions | Reference |
|---|---|---|---|
| Thermal decomposition | $Fe_3O_4$ | Iron oleate in the presence of OA and OAm at 320 °C | 307 |
| | | Iron oleate is decomposed in a sealed tube at low pressures and at 563 K | 308 |
| | | by decomposition of Fe $(acac)_3$ in benzyl ether using large excess of OA (15 : 1 OA–Fe) | 309 |
| | | By thermal decomposition of Fe $(acac)_3$ in a liquid mixture of triethylene glycol and triethanolamine at the refluxing temperature (245–280 °C) under the argon flow for 1 h | 310 |
| | | By decomposition of Fe $(acac)_3$ in imidazolium based ionic liquids by using oleylamine as the reducing and surface modifier agent at 150–250 °C | 311 |
| | γ-$Fe_2O_3$ | Decomposition of Fe $(acac)_3$ in the presence of poly(*N*-isopropyl-co-*t*-butylacrylamide) | 312 |
| | | By decomposition of Fe $(Cupf)_3$ complex in octylamine at 250–300 °C | 313 |
| Solvothermal–Hydrothermal | $Fe_3O_4$ | Mixture of anhydrous $FeCl_3$ and anhydrous $CH_3COONa$ at high temperature | 314 |
| | | $FeCl_2·4H_2O$ in basic aqueous solution at 138 °C | 315 |
| | | Using ultra colloidal iron hydroxide at 150 °C for 4 h | 316 |
| | | Using $Fe(acac)_3$ as the iron source, n-octylamine as the reductant, and n-octanol as the solvent 240 °C for 2 h | 317 |
| | α-$Fe_2O_3$ | $FeCl_3·6H_2O$,5-sulfosalicylic acid dehydrate and ammonia at 150 °C | 318 |
| | | By dissolving $FeCl_3·6H_2O$ in DMF at 180 °C | 319 |
| Coprecipitation | $Fe_3O_4$ | Hydrolysis of $FeCl_3$ by NaOH at a temperature above 200 °C in DEG in the presence of PAA | 320 |
| Aerosol pyrolysis | α-$Fe_2O_3$ | Ferrous acetate and ferric nitrate as precursors with hydrazine at 300–500 °C | 321 |

NPs of ZnO are usually prepared by a variety of physical and chemical methods.[44,195,337,338] Precipitation through acid-catalyzed esterification of zinc acetate dihydrate with 1-pentanol at 130 °C yields ZnO NPs.[339] The sol–gel process has also been employed by several workers to prepare ZnO NPs.[340]

Blue-emitting ZnO NPs synthesized by various means are of considerable interest. The blue emission is attributed to surface states created by OA. Investigation of the blue luminescence of ZnO NPs has been carried out by Zeng *et al.*[341] by laser ablation of zinc metal in an aqueous medium. Kaluza *et al.*[342] report high surface area ZnO NPs synthesized by a continuous precipitation method using a micromixer coupled directly to a spray dryer. Highly luminescent ZnO NPs have been synthesized by using a semiconductive polyalkyloxythiophene template.[343] ZnO NPs exhibiting various colors, including blue, green, yellow and orange, are obtained through an ethanol-based precipitation method by adjusting the pH of the solution.[344]

ZnO hollow spheres (400–500 nm in diameter) consisting of ZnO NPs of approximately 15 nm have been prepared by a rapid sonochemical process.[345] Uncapped ZnO NPs can be obtained from mixing an ethanol solution containing $Zn^{2+}$ ions with a solution of alkali metal oxides and the method can also be used for the synthesis of organic amine capped ZnO NPs.[346] A novel method to synthesize organic-capped and colloidal ZnO NPs has been developed by Omata *et al.*[347] The method involves hydrolysis of zinc-di-*t*-butoxide and zinc-di-n-butoxide in hydrophilic benzylamine followed by polymerization to ZnO by dehydration and/or elimination of butanol. Growth of ZnO NPs and exchange of the surface-capping ligand from the hydroxyl group or benzylamine to OAm is achieved by heating the colloidal solution at 100–180 °C after addition of OAm. Formation of ZnO hollow spheres is attributed to oriented attachment and subsequent Ostwald ripening. Alivisatos *et al.*[348] report anion exchange in ZnO colloidal NPs with the sulfur to form hollow ZnS NPs, whose shape is dictated by the initial ZnO structure. Crystallographic and elemental analyses provide insight into the mechanism of anion exchange which is accompanied by nanoscale Kirkendall effect where in thin layers of ZnS grow epitaxially onto the ZnO core. The surface anion exchange, generates a highly strained interface between the ZnO core and the ZnS shell. To release this interface energy, the ZnO core spontaneously diffuses into the ZnS shell, which is finally exchanged with the sulfur precursor at the outer shell surface, forming fully converted ZnS hollow NPs (Fig. 20).

Table 3 summarizes the typical methods used for the synthesis of ZnO NPs.

Monodisperse zirconia (tetragonal: T-$ZrO_2$; monoclinic: M-$ZrO_2$) NPs are prepared by the thermal decomposition of zirconium oleate by using ethanol or octane as a solvent in presence of OAm and OA as capping agents.[353] Guiot *et al.*[354] have synthesized yttria-stabilized zirconia NPs by hydrothermal synthesis, using $ZrO(NO_3)_2{\cdot}xH_2O$ and $(Y(NO_3))_3{\cdot}6H_2O$ in the presence of acetylacetone. In another method, Ninjbadgar *et al.*[355] have prepared $Eu^{3+}$ doped $ZrO_2$ NPs by using a zirconium(IV) isopropoxide–isopropanol complex, europium(III) acetate hydrate and benzyl alcohol. NPs of Mn- and Cr-doped $ZrO_2$ and $HfO_2$ have been obtained by nonaqueous sol–gel decomposition of zirconium(IV) isopropoxide and hafnium(IV)-tetrabutoxide.[356] $WO_3$ NPs in the 5–30 nm range are prepared by solvothermal treatment of tungsten chloride in benzyl alcohol up to 210 °C, followed by annealing in air.[357] Flame spray pyrolysis of tungsten chloride and tungsten hexacarbonyl is reported to yield 6–10 nm NPs of $WO_3$.[358,359]

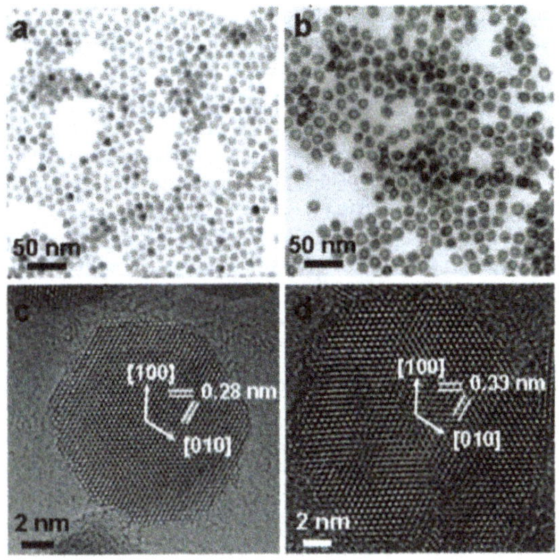


**Fig. 20** TEM and HRTEM images of initial ZnO NPs (a, c) and ZnS hollow NPs obtained after anion exchange (b, d). Reprinted with permission from J. Park, H. Zheng, Y.-W. Jun and A. P. Alivisatos, *J. Am. Chem. Soc.*, 2009, **131**, 13943. ©2009 American Chemical Society.

**Table 3** Synthesis of zinc oxide nanoparticles

| Method | Reactants and reaction conditions | Reference |
|---|---|---|
| Solvothermal | Zinc cupferronate in toluene at 180 °C | 338 |
| | Zinc(II) acetylacetonate in acetonitrile at 100 °C | 349 |
| Coprecipitation | Di-ethyl zinc into TOPO at 200 °C | 350 |
| Thermal decomposition | Di-ethyl zinc into TOPO at 200 °C | 44 |
| Oxidation | $[Zn(c\text{-}C_6H_{11})_2]$ in tetrahydrofuran at room temperature | 351 |
| Stirring | $Zn(NiBu_2)_2]_2$ with hexylamine and diethylether and stirred at room temperature with particular humidity | 352 |

$In_2O_3$ NPs with a narrow size distribution have been prepared by one-pot solvothermal synthesis using indium oleate in 1-hexanol.[360] Caruntu *et al.*[361] have employed the decomposition of indium precursors such as indium(III) acetate, indium(III) acetylacetonate and indium isopropoxide to obtain differently shaped $In_2O_3$ NPs. $In_2O_3$ NPs are obtained in solution phase by peroxide-mediated route at relatively low temperatures (120–180 °C) using indium acetylacetonate as the precursor in the presence of lauric acid and OAm.[362]

Han *et al.*[363] report the hydrothermal synthesis of high energy faceted $SnO_2$ nanooctahedra by mixing $SnCl_4{\cdot}5H_2O$ and PVP in water–ethanol mixture where the octahedral $SnO_2$ NPs are enclosed result from the coordinative effect of $H^+$ and $Cl^-$. Microwave synthesis of $SnO_2$ NPs involving decomposition of tin(IV) acetate in the presence of OA and OAm has been reported.[364] Gilstrap *et al.*[365] report the formation of organic soluble ~5 nm indium tin oxide (ITO) particles with a narrow size distribution by the thermal decomposition of indium acetate and tin acetate in ODE, by using ODA as the capping agent. Ionic liquid-mediated synthesis of conductive ITO NPs is reported by Bühler *et al.*[366] by using $SnCl_4 \cdot 5H_2O$ and $InCl_3 \cdot 4H_2O$ as precursors. Sasaki *et al.*[367] have developed a one-step solvothermal method for the synthesis of cube shaped ITO NPs by mixing $SnCl_4 \cdot 5H_2O$, $InCl_3 \cdot 4H_2O$ and EG in 1-butanol. A one-pot synthesis of colloidal, monodisperse ITO NPs with various sizes have been prepared by Choi *et al.*[368] by using In $(acac)_3$ and $Sn(acac)_2Cl$, with OAm as the capping agent. Spray pyrolysis can be employed for the synthesis of tellurium dioxide NPs by using telluric acid ($Te(OH)_2$) in deionized water as the precursor.[369] The NPs are amorphous with mean particle diameters ranging from 10 to 40 nm. Monodisperse $CeO_2$ NPs and super lattice-like colloidal particles have been successfully synthesized in ethanol–water by a one-pot procedure using icosahedral $(NH_4)_2Ce(NO_3)_6$ as the precursor.[370] Dicarboxylic acid-modified $CeO_2$ NPs are obtained under hydrothermal conditions from $Ce(OH)_4$.[371] The morphology and displayed crystallite plane of the $CeO_2$ NPs is determined by the length of the dicarboxylic acid, long dicarboxylic acids producing cubooctahedral or cubic $CeO_2$ NPs. TOPO capped $CeO_{2-\delta}$ and CeOCl NPs are prepared using nonhydrolytic condensation of cerium alkoxides with cerium halides. The NPs could be solubilized in nonpolar solvents.[47] Arita *et al.*[372] report the formation of uniform sized colloidal $CeO_2$ NPs stabilized with dodecanoic acid and decanoic acid in cyclohexane (Fig. 21). Mesoporous $CeO_2$ NPs have been prepared by Masson *et al.*[42] by the reverse micelle sol–gel technique using cerium isopropoxide in the presence of the surfactant Triton-X 100 in cyclohexane. Yang *et al.*[373] have synthesized $CeO_2$ hollow spheres by assembling $CeO_2$ nanooctahedra which are in turn prepared by hydrothermal synthesis by mixing $CeCl_3 \cdot 7H_2O$ and PVP. Template-free synthesis of single-crystalline $CeO_2$ nanooctahedra has been carried out by a hydrothermal process by using $Ce(NO_3)_3 \cdot 6H_2O$ as the cerium source and $Na_3PO_4 \cdot 6H_2O$ as the mineralizer.[374] Xu *et al.*[375] report the synthesis of $SnO_2$ quantum dots in the diameter range of ~0.5–2.5 nm from $SnCl_4 \cdot 5H_2O$, OA, and OAm under solvothermal conditions.

Monodisperse $CoFe_2O_4$ NPs are prepared by controlled thermal decomposition of $[(h_5\text{-}C_5H_5)CoFe_2(CO)_9]$.[376] $MnFe_2O_4$ NPs are obtained by the thermal decomposition of metal–surfactant complexes which in turn are prepared from $Fe(CO)_5$, $Mn_2(CO)_{10}$ and OA followed by mild chemical oxidation.[377] The particle size varies from 5 to 13 nm depending on the experimental conditions. Sun *et al.*[48] have synthesized monodisperse NPs of $CoFe_2O_4$ and $MnFe_2O_4$ from the high-temperature reaction of $Fe(acac)_3$ and $Co(acac)_2$ or $Mn(acac)_2$ with HDD. The particle diameter can be varied from 3 to 20 nm by varying the reaction conditions or by seed mediated-growth. Ferrite NPs ($MFe_2O_4$, M=Co, Ni, Mn, and Fe) can be prepared by decomposing the mixed metal–oleate complexes dissolved in ODE at 300 °C under a nitrogen atmosphere.[378] The mechanism of formation of different shapes from NPs to nanostars by controlling the nucleation and growth dynamics and through changes in the heating rate, temperature, and precursor concentration has been discussed.[379] Oxidation of $CoFe_2$ also yields $CoFe_2O_4$ NPs.[380] $MnFe_2O_4$ NPs modified with *o*-alkene-terminated ligands are successfully prepared in a one-pot reaction and covalently bound by UV irradiation to the surface of hydrogen-terminated silicon, creating stable, ordered monolayer arrays of magnetic NPs.[381] $MnFe_2O_4$ NPs are also prepared by a coprecipitation method by reacting manganous salts mixed with either ferric or ferrous salts with sodium hydroxide and subsequent digestion to form 5 to 25 nm particles.[382] Spherical and cubic $CoFe_2O_4$ NPs can be realized with greater shape control by controlling the growth rate.[383] Monodisperse NPs of $CoFe_2O_4$ covered by a bifunctional molecule containing free vinyl end groups able to link onto a silicon substrate, have been prepared from the metal–undecanoate complexes.[384]

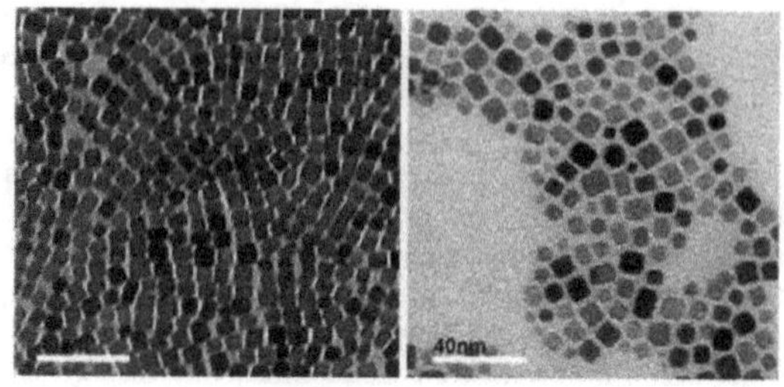


**Fig. 21** TEM images of colloidal $CeO_2$ NPs stabilized with (a) dodecanonic acid (b) decanoic acid. Reproduced with permission from T. Arita, J. Yoo, Y. Ueda and T. Adschiri, *Nanoscale*, 2010, **2**, 689. ©2010 Royal Society of Chemistry.

Monodisperse $BaTiO_3$ NPs with a diameter of 6 to 12 nm have been prepared by injection hydrolysis of barium titanium ethyl hexano-isopropoxide in a mixture of diphenyl ether and OA at 140 °C in an argon atmosphere.[385] A nonhydrolytic and halide-free procedure is used to synthesize nanocrystalline $BaTiO_3$, $SrTiO_3$, and $(Ba,Sr)TiO_3$ particles with diameters ranging from 4 to 10 nm by using the titanium isopropoxide and barium or strontium metal in benzyl alcohol as precursors.[386] Wang *et al.*[387] report a strategy to synthesize a variety of NPs based on the phase transfer and separation mechanism occurring at the interfaces of the liquid–solid-solutions by using NaOH, Ba $(NO_3)_2$ and $TiCl_3$ as starting materials. Hydrophobic $BaTiO_3$ NPs with an average crystallite size of 3.7 nm are prepared in oil–water (typical oil phases used are hexadecene, toluene, and heptane) solvothermal method using barium acetate and titanium (IV)-n-butoxide.[388] Cube-like $BaTiO_3$ NPs have been synthesized by the thermal decomposition of metal-fatty salts under hydrothermal conditions at temperatures as low as 135 °C.[389] Hydrothermal microwave synthesis of tetragonal $BaTiO_3$ NPs is reported with the NPs showing photoluminiscence.[390] $CaTiO_3$ nanostructures are prepared through an *in situ* hydrothermal method on the titanium surface.[391] Titanium citrate, lead acetate, EG, KOH and surfactants are used as reactants to form $PbTiO_3$ NPs under hydrothermal conditions in the temperature range of 90–200 °C.[392] Lead acetate, anhydrous benzyl alcohol and titanium isopropoxide are stirred for 30 min to form a uniform solution along with OA and the mixture kept at 200 °C for 2 h to yield yellow powder lead titanate of 4 nm particle size.[393,394]

Nanoparticulate perovskite oxides, $LnBO_3$ (Ln = La, Pr, Nd, Sm, Eu, Gd, Tb, Dy, Ho, Er, Tm, Yb, Lu; B = Cr, Fe), have been prepared by polyol-mediated synthesis.[395] Nanosized heterobimetallic precursors prepared from Ln and Fe oleates from their corresponding metal nitrates and sodium oleate are used for the synthesis of nanocrystalline rare-earth orthoferrites $LnFeO_3$ (Ln = La, Pr, Nd, Sm, Eu, Gd) at 500 °C.[396] Sol–gel method based on the glycol–gel reaction is used for the synthesis of single-crystalline $BiFeO_3$ NPs.[397] Uniform microcrystals of $BiFeO_3$ with various morphologies such as microspheres and microcubes are obtained by reacting $Fe(NO_3)_3{\cdot}9H_2O$ (0.01 mol), $Bi(NO_3)_3{\cdot}5H_2O$ (0.01 mol) and PEG at 200 °C under hydrothermal conditions.[398]

A hydrothermal method to prepare colloidal rare earth orthovanadate NPs, $LnVO_4$ (Ln = Y and lanthanides), by the reaction of OA, and $Na_3VO_4$ and $Ln(NO_3)_3{\cdot}xH_2O$ at 140 °C has been reported.[399] $YVO_4$ NPs doped with rare earth and phosphorous ions are obtained in aqueous media by using PVP, $Na_3VO_4$, $Na_3PO_4$ and $LnCl_3$ as reactants and heating at 180 °C for 2 h.[400] One-dimensional $YVO_4$:Ln and Y(V, P)$O_4$:Ln nanofibers and quasi-one-dimensional $YVO_4$:Ln microbelts (Ln = $Eu^{3+}$ $Sm^3$, $Dy^{3+}$ 3+) are prepared by a combination of sol–gel processing and electrospinning.[401]

Nanostructures of $Fe_2(MoO_4)_3$, $ZnMoO_4$, $MnMoO_4$ and $CoMoO_4$ have been synthesized by heating a mixture of the corresponding metal nitrates with ammonium heptamolybdate at 400 °C.[402] NPs of molybdate hydrates, $MMoO_4{\cdot}nH_2O$ (M=Co, Ni, Mn, $n$ = 0, 3 : 4, 1) are prepared hydrothermally from $(NH_4)_6Mo_7O_{24}$, $M(NO_3)_2$ or $MCl_2$ (M=Co, Ni, Mn) and $NH_3{\cdot}H_2O$.[403]

## 6. Metal chalcogenide nanoparticles

Semiconductor NPs are often referred to as quantum dots, whose wave functions are confined in all the three spatial directions.[404] Since Murray *et al.*[405] reported the pyrolysis method to synthesize CdSe NPs, there have been significant advances in the synthesis of chalcogenide NPs. Similar strategies have been employed for the synthesis of NPs of CdTe, CdS, PbS, PbSe, PbTe, ZnS and ZnSe.[406,407] NPs of metal chalcogenides are also prepared by the reaction of metal salts with an appropriate sulfiding or seleniding agent under solvothermal or thermolysis conditions. Thus, toluene-soluble CdSe NPs with a diameter of 3 nm are obtained solvothermally by reacting cadmium stearate with elemental Se in toluene in the presence of tetralin.[408] Synthesis of CdS NCs of 5–10 nm has been carried out by the reaction between cadmium stearate with sulfur in the presence of tetralin.[409] Decomposition of single molecular precursors provides a convenient and effective method for the synthesis of metal chalcogenide NPs.[410,411] Metal chalcogenide NPs of different shapes such as CdTe and CdSe tetrapods,[412] CdSe–CdTe nanobarbells,[413] PbTe cubes and CdSe hyperbranched structures have been prepared.[414] A fluorous chemical procedure for the one-step synthesis of metal chalcogenides (*e.g.*, CdSe, PbSe, ZnSe) using the fact that perfluorohydrocarbon, and hydrocarbon solvents become miscible at high temperatures has been reported (Fig. 22).[53] Growth of capped CdSe and CdS NPs formed by the reaction of selenium or sulfur with cadmium stearate in toluene solution in the presence of dodecanethiol or trioctylphosphine oxide and tetralin, has been investigated by a variety of techniques. These studies suggest that the growth of the NPs is controlled by diffusion-limited Ostwald ripening.[415] Structural transformation of hydrophobic CdSe and CdTe NPs to nanopencils and hexagonal structures is achieved through an Ostwald ripening process induced by $Cl^-$ and surfactants (Fig. 23).[416]

Ma *et al.*[417] report a one-pot synthesis of nucleic acid-functionalized chalcogenide NPs. This procedure allows the simultaneous synthesis and functionalization of CdTe NPs using DNA as the ligands. DNA-passivated NPs exhibit specific binding to nucleic acids, proteins and cell targets. The synthesis of CdSe QDs in P3HT solution without surfactants has been carried out for use in photovoltaic solar cells.[418] Choi *et al.*[419] have developed a general method for the synthesis of metal sulfide (CdS, ZnS, MnS, and PbS) NPs by the solution-phase thermal decomposition of metal–oleate complexes in alkanethiol. The size of the NPs can be tuned from 7 to 20 nm by varying the reaction conditions. Cabot *et al.*[420] report the sulfidation of spherical cadmium crystals to form hollow CdS *via* the Kirkendall effect. NPs of CdS, CdSe, ZnS, ZnSe and PbS have been obtained by using ionic liquids as solvents.[421] Changing the anion of the ionic liquid and of the addition of surfactants and capping agent's results in the change of size and shape of the nanostructures. A series of metal-chalcogenide NPs with controllable shape and size are prepared *via* a facile wet chemical route involving metal nitrates and sulfur or selenium powder and octadecylamine as solvent.[422]

Synthesis of optically active, water-soluble chiral CdS QDs, has been achieved by using *rac*, D- and L- enantiomeric forms of penicillamine as stabilizers.[423] CdS QDs prepared in ferritin show left-handed circularly polarized luminescence from the both direct transition and from surface-trapping sites with relatively large anisotropy factors.[424] Linear and dendritic peptides are used as capping agents in the synthesis of CdS nanostructures. The composition of the peptides affects the size, aggregation, and optical properties of the NPs.[425] Molecular vapor–solvent co-condensation method which involves vaporization of bulk materials followed by controlled nucleation of the molecular vapor species in cold solvent has been used to prepare metal sulfide, selenide, and telluride nanomaterials.[426] The microemulsion of water in $CO_2$ is used as a nanoreactor for the preparation of CdS and ZnS semiconductor NPs in supercritical fluid $CO_2$ by mixing microemulsions containing the appropriate ionic species in the water core.[427] Porous aerogels of thiol-capped CdS, CdSe, ZnS, PbS NPs are prepared through gelation, controlled loss of surface-groups and supercritical $CO_2$ drying.[428] Weller *et al.*[429] synthesized the thiol-capped CdTe NPs in aqueous solutions. Water soluble CdTe quantum dots have also been prepared using $(NH_4)_2Te$ as the tellurium source.[430] CdTe quantum dots are obtained by using hydrothermal[431] and microwave conditions as well.[432] Amphiphilic CdTe NPs are prepared by using thiol-modified PEG oligomers.[433]

ZnS and ZnSe crystallize in both the zinc blende (ZB) and wurtzite (WZ) structures, of which ZB is the more stable structure. The synthetic conditions employed favor one or the other structure for ZnSe NPs. ZnSe NPs in WZ structure with different morphologies have been successfully obtained by the tetraethylenepentamine directed solvothermal method.[434] Synthetic strategies for shape and phase control of ZnS NPs from a lamellar molecular precursor $ZnS(NH_2CH_2CH_2NH_2)_{0.5}$, are reported in

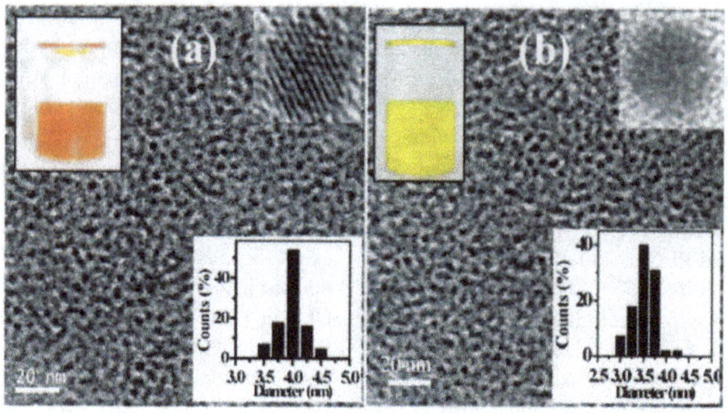


**Fig. 22** TEM images of fluorous thiol capped (a) 4 nm CdSe and (b) 3.5 nm CdS NPs with size distribution histograms as inset. Photographs of the dispersions of the NPs in PFC are also given as insets. Reproduced with permission from R. Voggu, A. Shireen and C. N. R. Rao, *Dalton Trans.*, 2010, **39**, 6021. ©2010 Royal Society of Chemistry.

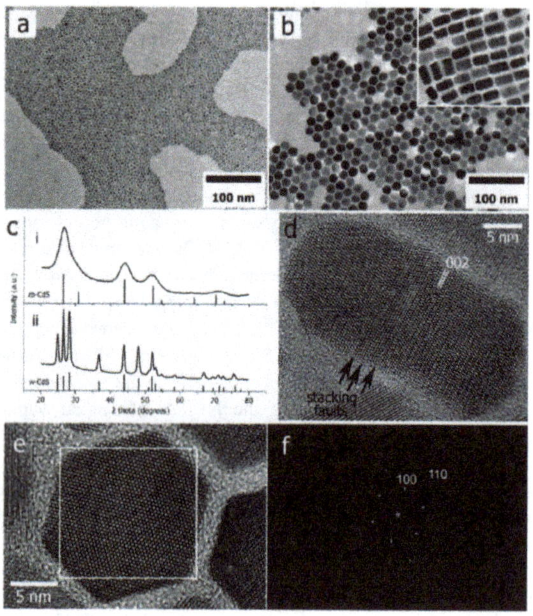


**Fig. 23** TEM images of (a) CdS seed NPs and (b) CdS nanopencils after reaction at 260 °C for 20 min. The inset in (b) shows a side view of the nanopencils. (c) XRD patterns of (i) CdS seed NPs and (ii) CdS nanopencils. (d) HRTEM image of a single nanopencil showing stacking faults. (e) HRTEM image of single nanopencil taken from the [001] zone axis. (f) Fast Fourier transform pattern of the rectangular region in (e). Reproduced with permission from M. Saruyama, M. Kanehara and T. Teranishi, *J. Am. Chem. Soc.*, 2010, **132**, 3280. ©2010 American Chemical Society.

ethylenediamine medium.[435] ZnS (WZ) NPs are formed at 150 °C in a polyol medium.[436] ZnS NPs with ZB structure with diameters less than 10 nm are synthesized from zinc diethyldithiocarbamate as the single-source precursor using spray pyrolysis method.[437] ZnO NPs, with the WZ structure, are used as templates for the formation of WZ-type ZnS and ZnSe.[438] Lactobacillus streptococcus thermophilus (Str. thermophilus), bacteria have been employed as templates to form ZnS hollow spheres.[439]

CuS NPs and their assembly to form wires and tubes by an amine-assisted hydrothermal process at low temperatures (90–110 °C) from CuCl and thiourea as reactants have been reported.[440] CuS flakes and nanodisks are obtained by using the hydrothermal microemulsion method.[441] $Cu_2S$ NPs with disk-like morphologies are synthesized by solventless thermolysis of a copper alkylthiolate molecular precursor.[442] $Cu_2S$ NPs can also be prepared by the reaction between $Cu(acac)_2$ and ammonium diethyldithiocarbamate in a mixed solvent containing dodecanethiol and OA.[443] Single-crystalline CuS hexagonal nanoplates are produced through the facile chemical vapor reaction wherein sulfur vapor is reacted with copper films in a vacuum chamber around 450 °C for about 7 h. $Cu_{2-x}Se$ NPs are prepared by the solvothermal method from CuI and Se at low temperatures.[444] Xie *et al.*[445] report nonstoichiometric and stoichiometric CuSe by a sonochemical procedure at room temperature. Phosphine-free synthesis of CuSe NPs is reported using ODE and OAm (Fig. 24).[446]

SnS and SnSe adopt an orthorhombic layered crystal structure that can be considered as a distorted NaCl structure. They are optically active in the near-infrared and infrared and are of interest for their applications in photovoltaics, near-infrared detectors, and biomedical applications. Eychmuller *et al.*[447] used organometallic $Sn[N(SiCH_3)_2]_2$ as a tin precursor for the synthesis of hydrophobic SnS NPs. Hydrophilic SnS NPs are prepared starting with $SnBr_2$ and $Na_2S$ in the presence of various ethanolamines in EG.[448] SnS NPs with shape control have been synthesized using simple tin oxide hydroxide ($Sn_6O_4(OH)_4$) and thioacetamide dissolved in OA and OAm as precursors.[449] 2D $SnS_2$ nanoplates are synthesized through the thermal decomposition of $Sn(S_2CNEt_2)_4$ in an organic solvent at elevated temperatures.[450] $SnS_2$ NPs are also obtained by heating elemental tin metal powder and $CS_2$ in a horizontal tube furnace at 800–1000 °C.[451] SnSe NPs are obtained by a method involving di-*tert*-butyl diselenide, tin(II) chloride in DDA and dodecanethiol.[452] SnSe NPs with size control are prepared by the reaction of selenium and bis[bis(trimethylsilyl)amino]tin(II).[453]

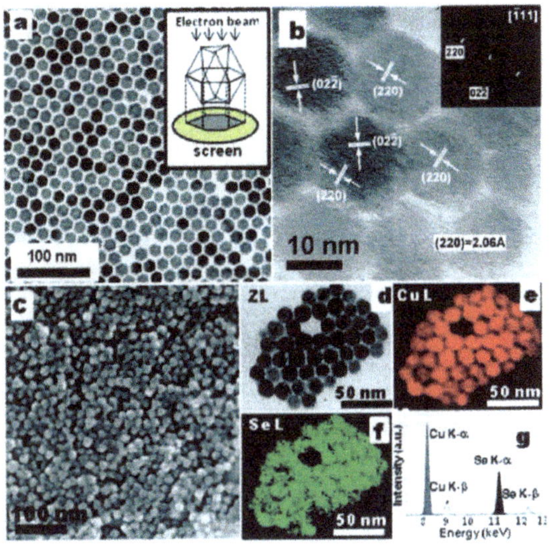


**Fig. 24** (a) TEM image of size-selected $Cu_{2-x}Se$ NPs grow for 15 min at 300 °C, having an average size of 16 nm (the size estimated by XRD was 18 nm). The inset shows a sketch of the hexagonal projection of a cuboctahedron shape. (b) HRTEM image of $Cu_{2-x}Se$ NPs. Most of the displayed NPs are seen under their [111] zone axis. The inset shows their two-dimensional fast Fourier transform. (c) Scanning electron microscopy image of $Cu_{2-x}Se$ NPs drop-casted from solution onto a glass substrate. (d) Elastic-filtered (ZL) image of several NPs. (e, f) Cu and Se elemental maps from the same group obtained by filtering the Cu L edge (at 931 eV) and the Se L edge (at 1436 eV). (g) Elemental quantification of a group of NPs by EDS. Reproduced with permission from S. Deka *et al.*, *J. Am. Chem. Soc.*, 2010, **132**, 8912. ©2010 American Chemical Society.

$Au_2S$ NP superstructures are synthesized by a hydrothermal procedure at 175 °C using $HAuCl_4$ and $Na_2S$ as reagents and CTAB as the surfactant.[454] Shape-controlled synthesis of AgS NPs is carried out by using anhydrous silver nitrate, OAm and elemental sulfur at 100 °C.[455] Cubic GdS NPs are made in solution from $GdCl_3$ and $Na_2S$ in the presence of dextrose, under continuous stirring for 2 h at 70 °C.[456] $ReS_2$ NPs are obtained by a high-temperature MOCVD process starting from $Re_2(CO)_{10}$ and elemental sulfur.[457] Monodisperse α-MnS NPs have been synthesized by decomposing manganese oleate and elemental sulfur in ODE at 250–320 °C.[458] Chen and Fan[459] have prepared a series of metal sulfides $MS_2$ (M = Ni, Co, Fe, Ni, Mo) and selenides $MSe_2$ (M = Ni, Mo) hydrothermally using metal chlorides and $Na_2S_2O_3$ or $Na_2SeSO_3$ as starting materials. Nanoplates of single-crystal antimony telluride are obtained by hydrothermal treatment of $SbCl_3$, tartaric acid, ammonia, $K_2TeO_3$ and hydrazine hydrate at 180 °C for 5 hours.[460]

## 7. Metal pnictide nanoparticles

Large GaN NPs (32 nm) were prepared by Xie *et al.*[461] by treating $GaCl_3$ with $Li_3N$ in benzene under solvothermal conditions. GaN NPs of various sizes have been prepared under solvothermal conditions, by employing gallium cupferronate or chloride as the gallium source and 1,1,1,3,3,3-hexamethyldisilazane

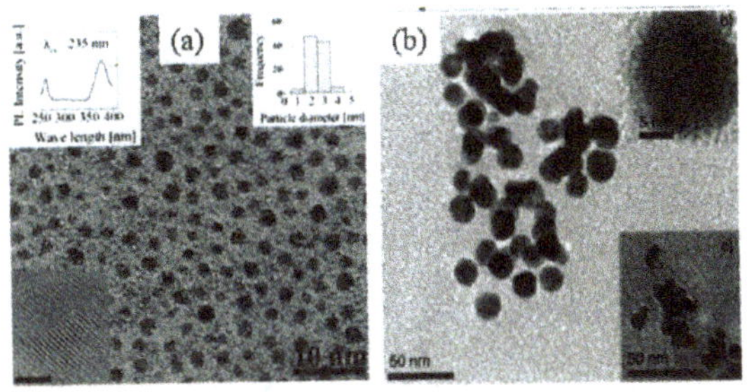


**Fig. 25** (a) TEM image of CTAB-capped 2.5 nm GaN NPs prepared starting with Ga cupferron. The upper right inset shows the size distribution. The lower inset shows a HREM image (scale bar is 2 nm) and upper inset also shows the PL spectrum of CTAB capped 2.5 nm GaN NPs; (b) TEM image of InN NPs of 15 nm average diameter prepared starting with In cupferron. The upper inset shows a HREM image of a single nanocrystal. The lower inset shows a TEM image of InN NPs prepared starting with $InCl_3$. Reproduced with permission from K. Sardar and C. N. R. Rao, *Adv. Mater.*, 2004, **6**, 425 and K. Sardar, F. L. Deepak, A. Govindaraj, M. M. Seikh and C. N. R. Rao, *Small*, 2005, **1**, 91. ©2004 and 2005 Wiley-VCH Verlag GmbH & Co. KGaA.

(HMDS) as the nitriding agent and toluene as solvent.[29] Manz *et al.*[462] have obtained NPs of hexagonal GaN of varying diameters. With $GaCl_3$ as the precursor, the reaction is, $GaCl_3$ + $(CH_3)_3SiNHSi(CH_3)_3$ → GaN + $Si(CH_3)_3Cl$ + HCl. This method has been applied for the synthesis of AlN and InN NPs as well (Fig. 25).[27,28,463] GaN NPs have also been obtained by the thermal decomposition of precursor compounds such as $(C_2H_5)N{\cdot}Ga(N_3)_3$ and polymeric $[Ga(NH)_{3/2}]_n$.[464–466] Hexagonal NPs of GaN, AlN, and InN have been obtained by refluxing the urea-complexes prepared from metal chlorides in tri-n-octylamine.[30] Nanocrystalline InN powder is obtained by the metathesis reaction of $InBr_3$ and $NaN_3$ in superheated toluene and refluxing hexadecane around 553 K[467] and from the reaction of $NaNH_2$ and $In_2S_3$ around 463 K.[468] Three-dimensional urchin-like nanostructures of AlN are formed by the DC arc-plasma reaction between Al vapor and $N_2$ gas in the absence of any catalyst or template.[469] Arrays of gallium–indium oxide NPs are heated in ammonia in a CVD reactor at temperatures in the range 600–850 °C to obtain monodisperse GaN and InN NPs with PS-P4VP as the block copolymer template.[470]

Boron nitride has been a subject of extensive investigations because of its useful properties. BN NPs have been prepared using urea as the N-source and boric acid as the source of boron.[471–473] Spray pyrolysis of poly(borazinylamine) dissolved in liquid ammonia has been used to prepare spherical BN particles.[474] Spherical BN NPs are also prepared by the chemical vapor deposition of trimethoxyborane with ammonia.[475] Nanosheets and nanopans of BN are obtained from the vapour phase reaction between $BBr_3$ and $NH_3$.[476] A solid-state reaction involving ammonium dicyanamide and the corresponding metal oxides is used to prepare NPs of BN, NbN, WN and $Mo_2N$.[477] By heating mixtures of $H_3BO_3$, $TiCl_4$, and $NbCl_5$ with urea in 1 : 6 molar ratio in the 900–1000 °C range produces NPs of BN, TiN, and NbN respectively.[471] BN with hollow sphere, onion, and peanut structures are obtained by the liquid–solid reaction between $BBr_3$ and $NaNH_2$, $NH_4Cl$ at 400–450 °C for 6–12 h.[478]

On reaction of cyanamide or urea with metal oxide NPs, the later are transformed into nanocrystalline metal nitrides.[479] Metal nitride and metal carbide NPs have been prepared by the urea route using polymeric or gel-like metal-urea complexes as the precursors.[480,481] TiN NPs are obtained by the reaction of $TiCl_4$ and $NaNH_2$.[482,483] NbN NPs are prepared by the reaction of $NbCl_5$ with $NaN_3$.[484] The benzene-thermal route has also been employed to synthesize NbN.[485] Lengauer[486] has investigated the formation of γ-$Mo_2N$ and δ-MoN by the reaction between $MoCl_5$ and ammonia. Bull *et al.*[487] obtained phase-pure δ-MoN by this reaction carried out at 933 K. γ-$Mo_2N$ and δ-MoN with superconducting $T_c$ values of 3.8 K and 7.5 K respectively, have been prepared by heating hydroxylamine complexes of molybdenum in $NH_3$.[488] Gomathi *et al.*[489] have prepared γ-$Mo_2N$ nanoparticles of ~4 nm ($T_c$ = 6.5 K) diameter by a simple procedure involving the reaction of $MoCl_5$ with urea at 873 K. The γ-$Mo_2N$ nanoparticles are transformed to nanoparticles of δ-MoN with a slightly larger diameter on heating in a $NH_3$ atmosphere at 573 K.

NPs of cobalt phosphides have been synthesized by the reaction of sodium phosphide with the metal salts in benzene or toluene in a Teflon-lined autoclave at 150 °C.[490] This route has been successfully applied to prepare $Ni_2P$[491] and FeP.[492] FeP NPs are also formed by the reaction between $Fe(acac)_3$ with P $(SiMe_3)$ in TOPO at 280 °C.[493] Decomposition of organometallic species to generate transition metal species for the reaction with alkyl phosphines like TOP, has proven to be a general and less expensive route to transition metal phosphide NPs.[494,495] White phosphorous ($P_4$) in a solution of ammonium hydroxide or ethylene diamine has been used to obtain single phase particles of $Co_2P$, $Ni_2P$, and $Cu_3P$.[496,497] Henkes *et al.*[498] transformed a variety of transition metal NPs into their respective phosphide phases by reacting them with TOP (Fig. 26). TOP can be used to convert not only metal NPs into metal phosphides but also bulk-scale metal powders, foils, wires, films, supported NPs, and patterned substrates.[499] A surfactant-aided solvothermal route has been utilized to synthesize $Ni_2P$ aggregates.[500] Yang *et al.*[501] have synthesized $Ni_2P$ with high surface area by the reduction of the phosphate in the presence of a surfactant, while Park *et al.*[502] have utilized a continuous-injection method to synthesize $Ni_2P$ nanorods by the thermal decomposition of a Ni-alkylphosphine complex. A strategy based on the transformation of Ni NPs into hollow and dense $Ni_2P$ NPs by reaction with TOP has been reported, exploiting the Kirkendall effect for hollow NP formation.[498,503] $Ni_2P$ NPs are also obtained using the solid phase reaction between nickel hypophosphite and ammonium hypophosphite in $N_2$.[504]

InP–GaP NPs with sizes ranging from 2 to 6.5 nm can be prepared from the reaction of chloroindium oxalate–chlorogallium oxalate complex and $P[Si(CH_3)_3]_3$ in a mixture of TOPO and TOP at 270 °C for 3 days.[505,506] The growth of InP and InAs NPs is facilitated by using the noncoordinating solvent ODE with precursors such as indium acetate and tris(trimethylsilyl)-phosphine.[507] Xu *et al.*[508] have prepared monodisperse InP NPs from $(CH_3)_3In$ and $P[Si(CH_3)_3]_3$ in weakly coordinating solvents such as methyl myristate or dibutyl sebacate. MnAs NPs are prepared by the reaction of triphenylarsine oxide ($Ph_3AsO$) and dimanganesedecacarbonyl ($Mn_2CO_{10}$) at temperatures ranging from 250 to 330 °C in the presence of TOPO.[509]

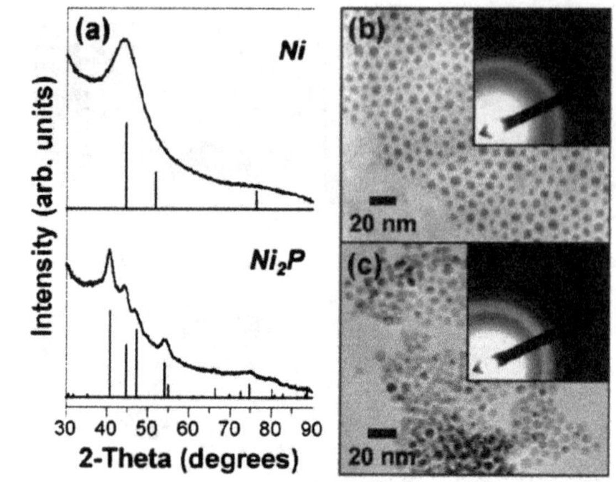


**Fig. 26** (a) Powder XRD data for Ni NPs and $Ni_2P$ formed after their reaction with TOP; TEM/SAED data for (b) Ni NPs and (c) $Ni_2P$ formed by reacting the particles in (b) with TOPA Reprinted with permission from A. E. Henkes, Y. Vasquez and R. E. Schaak, *J. Am. Chem. Soc.*, 2007, **129**, 1896. ©2007 American Chemical Society.

## 8. Core–shell nanoparticles

Core–shell nanoparticles exhibit unique properties with several possible applications. In the case of fluorescent semiconductor NPs, core–shell NPs help to increase the robustness and enhance the photoluminescence quantum yield as well as the probability of radio-active recombination. NPs with magnetic, plasmonic and semiconducting properties can be used as cores or shells for manipulating the properties of these hybrid structures. Properties of each component within the hybrids can be modulated through a conjugating component or interface.

Au@Ag core–shell nanostructures are prepared at room temperature through the selective reduction of $Ag^+$ ions onto the surface of Au nanorods, using hydroquinone as the reducing agent.[510] A modified two-step seed-mediated growth procedure for synthesizing Au@Pd nanocubes, has been reported.[511] Here Au NPs of about 3 nm in diameter are used as the seeds for growing about 30 nm Au nanooctahedra as the cores (Fig. 27). Uniform Au@Pd nanocubes have been generated by reducing $H_2PdCl_4$ on octahedral Au cores with ascorbic acid in the presence of CTAB. Feng *et al.*[512] carried out the synthesis of Au@Pt nanostructures by Au nanorod seed mediated growth wherein Pt nanodots are decorated on Au nanorods by reducing $K_2PtCl_4$ with ascorbic acid. NPs composed of magnetic cores with continuous Au shell layers have been synthesized. The procedure converts hydrophobic iron oxide NPs to water-soluble Au coated bifunctional NPs.[513] Zhang *et al.*[514] report nonepitaxial growth of hybrid Au@CdS core–shell nanostructures with large lattice mismatch by controlling the soft acid–base coordination reaction between the molecular complexes and the colloidal nanostructures (Fig. 28).

Pd–Pt core–shell nanoplates with hexagonal and triangular shapes have been synthesized through the heterogeneous, epitaxial growth of Pt on Pd nanoplates.[515] One-dimensional Ni–$Ni_3C$ core–shell nanoball chains with an average diameter of around 30 nm have been prepared in one step using TOPO as a soft

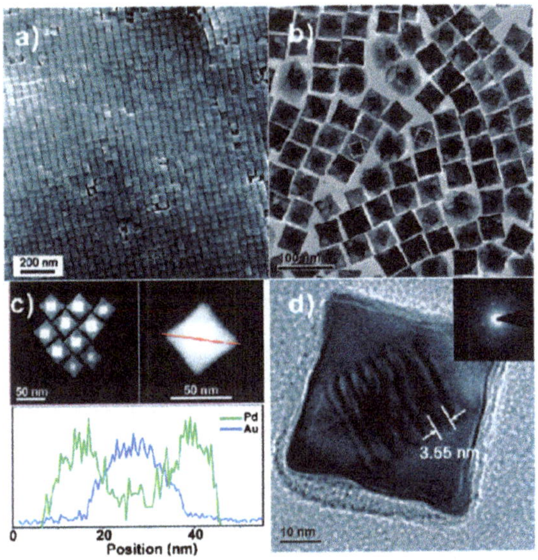


**Fig. 27** (a, b) SEM and TEM images of the overall morphology of Au@Pd nanocubes self-assembled on the Si wafer and Cu grid, respectively. The dashed frames indicate the core area of particles. (c) STEM images of the octahedral Au seed within a cubic Pd shell and cross-sectional compositional line profiles of a Au@Pd nanocube along the diagonal (indicated by a red line). (d) TEM image of a Au@Pd nanocube at high magnification. The inset is the SAED pattern taken from individual nanocube. Fan *et al.*, *J. Am. Chem. Soc.*, 2008, **130**, 6949. ©2008 American Chemical Society.

template at low temperatures.[516] Temperature-responsive γ-$Fe_2O_3$-core–Au-shell NPs are obtained from smart diblock copolymers as templates.[517] Rh@Pt core–shell, NPs can be prepared using polyol reduction in EG solvent using Pt and Rh salts with PVP as the stabilizer.[518] Catalytic studies show that the Rh@Pt core–shell NPs are superior to alloy and monometallic mixtures. $Ru_{core}Pt_{shell}$ bimetallic NPs can be prepared by the reduction of Ru NPs in TEG followed by the addition of $Pt^{2+}$ ions in EG solution.[519] By taking advantage of the different kinetics of hydrogenation of the organometallic precursors, Rh $(allyl)_3$ and $Fe[N(Si(CH_3)_3)_2]_2$, core–shell Rh@Fe NPs are prepared.[520]

Core–shell NPs based on metallic $ReO_3$ NPs have been prepared with the metallic shell $ReO_3$@Au, $ReO_3$@Ag, by the reduction of metal salts over $ReO_3$ NP seeds as well as $ReO_3$@SiO2, $ReO_3$@$TiO_2$ are prepared by the hydrolysis of the organometallic precursors over the $ReO_3$ NPs.[521] Cu–$Cu_2O$ core–shell NPs of different shapes are obtained on a H-terminated Si(100) substrate using $CuSO_4{\cdot}5H_2O$ and $NaClO_4$ as electrolytes.[522]

Peng *et al.*[523] report the synthesis of a series of CdSe–CdS NPs with core diameters ranging from 2.3 to 3.9 nm with shell thicknesses of up to 3 monolayers. Their results indicate that in the excited state the hole is confined to the core and the electron is delocalized throughout the entire structure. ZnSe–CdS core–shell NPs, are synthesized *via* the traditional pyrolysis of organometallic precursors.[524] The two-step synthesis involved fabrication of 4.5–6 nm ZnSe seeds followed by a subsequent

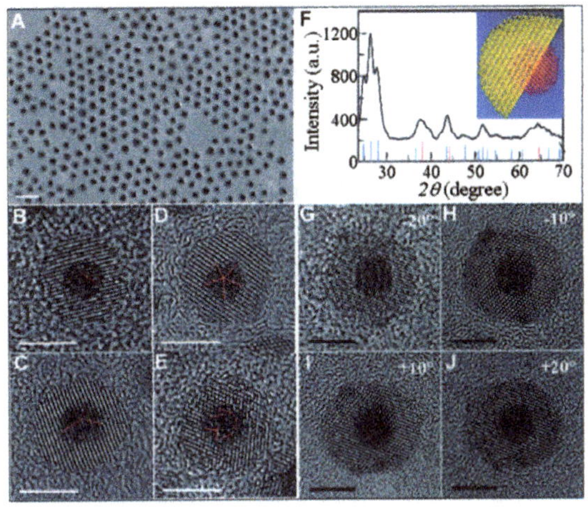


**Fig. 28** Au–CdS core–shell nanostructures with monocrystalline shell. (A) Typical TEM image showing uniform core–shell nanostructures. Scale bar, 20 nm. (B to E) High-resolution TEM images of core–shell nanostructures from (A). Whereas Au core NPs can manifest monocrystalline (B), single-fold twin (C), fivefold twin (D), and multiple-twin (E) lattice structures, all CdS shells are monocrystalline. The red lines highlight the lattice orientations within the Au core NPs. Scale bar, 5 nm. (F) XRD pattern of Au–CdS sample shown in (A). Bulk Au [red solid lines] and wurtzite CdS (blue solid lines) are also provided for reference and comparison. (Inset) A ball-and-stick molecular model of Au–CdS, illustrating a cubic core and wurtzite shell. (G to J) Angle-dependent high resolution TEM characterization. The sample depicted has a larger shell thickness than the one in (A) to emphasize the extremely high-quality crystallinity of the shell. The CdS shell shows perfect monocrystalline features without detectable structural defects under a different viewing angle. Scale bar, 5 nm. Reprinted with permission from J. Zhang, Y. Tang, K. Lee and M. Ouyang, *Science*, 2010, **327**, 1634. ©2010 AAAS. http://www.sciencemag.org.

deposition of the CdS shell. Lifshitz and coworkers[525] report the synthesis of air-stable PbSe–PbS and PbSe–$PbSe_xS_{1-x}$ core–shell NPs with lead(II) acetate, TOPSe and TOPS as precursors, OA as the stabilizer, and diphenyl ether as the solvent. Single-step synthesis of InP–ZnS core–shell NPs without precursor injection has been reported.[526] Based on the different reactivity of the core and shell precursors, the reagents are mixed at room temperature and subsequently heated to 250–300 °C. Xie *et al.*[527] report the synthesis of ZnTe–CdTe core–shell nanostructures by the addition of cadmium oleate, TOPTe, TOPSe, and sulfur dissolved in ODE to a dispersion of ZnTe core NPs.

Core–shell structures of FePt–CdSe, NiPt–CdSe, and Au–CdSe are prepared starting with a mixture of noble metal (alloy) NPs, cadmium stearate and OAm, HDA and HDD at 290–300 °C and then injecting the Se precursor solution before cooling to room temperature.[528] Hybrid NPs are prepared by decomposition of $Fe(acac)_3$ on the surfaces of Au NPs (prepared by the two-phase Brust method) in high-boiling-point solvents in the presence of OA and OAm. Octyl ether produced peanut-like particles, benzyl ether and phenyl ether gave core–shell particles and ODE yielded a mixture of peanut-like and core–shell structures.[529] Nanostructures of FePt and PbS or PbSe in form of core–shells or nanodumbbells have been obtained and the morphology-directing role of amine is examined.[530] $Fe_3O_4$ microspheres are synthesized and added to the mixture of thioacetamide and $Zn(Ac)_2$ in regular intervals, the mixture is irradiated by 40 kHz ultrasonic waves and maintained the temperature at 45 °C gives nanocomposites of $Fe_3O_4$–MS (M = Zn, Cd, Hg, Pb, Co, and Ni).[531]

Noble metal–metal oxide dumbbell shaped NPs have been synthesized based on seed-mediated growth. Metal oxides are grown over the presynthesized noble metal seeds by the thermal decomposition of the metal carbonyl followed by oxidation in air. They show enhanced catalytic activity toward CO oxidation in comparison with their counterparts.[532] Heterostructured $Cu_2S$–$In_2S_3$ with various shapes and compositions can be obtained by a high-temperature precursor-injection method in which $Cu_{1.94}S$ is used as the catalyst for the nucleation and growth of $In_2S_3$ NPs.[533]

$CeO_2$@$Ce_{1-x}Zr_xO_2$ nanocages have been prepared by using colloidal ceria as both the chemical precursor and physical template *via* Kirkendall effect.[534] A colloidal seeded-growth strategy is used to synthesize bimagnetic hybrid NPs that consist of a tetrapod-shaped ferrimagnetic iron oxide functionalized with spherical ferromagnetic Co clusters giving rise to exchange coupling between them.[535] Multiferroic and magnetoelectric properties of core–shell $CoFe_2O_4$@$BaTiO_3$ nanocomposites which are prepared by a combination of solution processing and high temperature calcinations have been explored.[536]

## 9. Conclusions

The discussion in the previous sections should clearly indicate the great interest evinced in the last few years in the synthesis of inorganic NPs. In spite of this progress, we still need well-established protocols for the synthesis of many of the important inorganic NPs of desired sizes and shapes. In this context, it is necessary to develop methods, which are simple and do not use expensive and poisonous chemicals. A green chemical approach to NP synthesis is therefore important. We believe that there will be continued effort to prepare nanoparticles of various inorganic materials of different sizes and shapes in the years to come. Functionalization and chemical manipulation of inorganic nanoparticles are of considerable importance in using these materials in the preparation of composites and such applications. We have not dealt with these aspects in this perspective, since we had to limit the scope only to synthesis. We do hope that the article will serve as a ready guide to all those pursuing synthesis of nanoparticles.

## References


1 P. J. Thomas and P. O'Brien, in *Nanomaterials Chemistry*, ed. C. N. R. Rao, A. Muller and A. K. Cheetham, Wiley-VCH, Weinhiem, 2007, pp. 1–36.
2 C. N. R. Rao, P. J. Thomas and G. U. Kulkarni, *Nanocrystals: Synthesis, Properties and Applications*, Springer-Verlag, Berlin, Heidelberg, 2007.
3 C. N. R. Rao, S. R. C. Vivekchand, K. Biswas and A. Govindaraj, *Dalton Trans.*, 2007, 3728–3749.
4 C. Burda, X. Chen, R. Narayanan and M. A. El-Sayed, *Chem. Rev.*, 2005, **105**, 1025–1102.
5 A. K. Ganguli, A. Ganguly and S. Vaidya, *Chem. Soc. Rev.*, 2010, **39**, 474–485.
6 A. B. Smetana, J. S. Wang, J. Boeckl, G. J. Brown and C. M. Wai, *Langmuir*, 2007, **23**, 10429–10432.
7 C. Petit, P. Lixon and M. P. Pileni, *J. Phys. Chem.*, 1993, **97**, 12974–12983.
8 I. Lisiecki, F. Billoudet and M. P. Pileni, *J. Phys. Chem.*, 1996, **100**, 4160–4166.
9 M. Chen, Y.-G. Feng, L.-Y. Wang, L. Zhang and J.-Y. Zhang, *Colloids Surf., A*, 2006, **281**, 119–124.
10 D.-H. Chen and C.-J. Chen, *J. Mater. Chem.*, 2002, **12**, 1557–1562.
11 A. K. Ganguli, T. Ahmad, S. Vaidya and J. Ahmed, *Pure Appl. Chem.*, 2008, **80**, 2451–2477.
12 N. Moumen and M. P. Pileni, *Chem. Mater.*, 1996, **8**, 1128–1134.
13 S. K. Haram, A. R. Mahadeshwar and S. G. Dixit, *J. Phys. Chem.*, 1996, **100**, 5868–5873.
14 S. G. Kwon and T. Hyeon, *Acc. Chem. Res.*, 2008, **41**, 1696–1709.
15 S. W. Kim, J. Park, Y. Jang, Y. Chung, S. Hwang, T. Hyeon and Y. W. Kim, *Nano Lett.*, 2003, **3**, 1289–1291.
16 H. T. Yang, C. M. Shen, Y. G. Wang, Y. K. Su, T. Z. Yang and H. J. Gao, *Nanotechnology*, 2004, **15**, 70–74.
17 A. C. S. Samia, J. A. Schlueter, J. S. Jiang, S. D. Bader, C. J. Qin and X. M. Lin, *Chem. Mater.*, 2006, **18**, 5203–5212.
18 N. Pinna, G. Garnweitner, M. Antonietti and M. Niederberger, *J. Am. Chem. Soc.*, 2005, **127**, 5608–5612.
19 M. Diab, B. Moshofsky, I. Jen-La Plante and T. Mokari, *J. Mater. Chem.*, 2011, **21**, 11626–11630.
20 J. C. Bruce, N. Revaprasadu and K. R. Koch, *New J. Chem.*, 2007, **31**, 1647–1653.
21 R. L. Wells, M. F. Self, A. T. McPhail, S. R. Aubuchon, R. C. Woudenberg and J. P. Jasinski, *Organometallics*, 1993, **12**, 2832–2834.
22 M. Green and P. O'Brien, *J. Mater. Chem.*, 2004, **14**, 629–636.
23 J. P. Carpenter, C. M. Lukehart, S. B. Milne, S. R. Stock, J. E. Wittig, B. D. Jones, R. Glosser and J. G. Zhu, *J. Organomet. Chem.*, 1998, **557**, 121–130.
24 L. M. Bronstein, J. E. Atkinson, A. G. Malyutin, F. Kidwai, B. D. Stein, D. G. Morgan, J. M. Perry and J. A. Karty, *Langmuir*, 2011, **27**, 3044–3050.
25 *Chemistry of Nanocrystalline Oxide Materials*, ed. K. C. Patil, M. S. Hegde, T. Rattan and S. T. Aruna, World Scientific Publishers, 2008.
26 V. L. Calero-DdelC, A. M. Gonzalez and C. Rinaldi, *J. Manuf. Sci. Eng.-Trans. ASME*, 2010, **132**, 7.
27 K. Sardar and C. N. R. Rao, *Solid State Sci.*, 2005, **7**, 217–220.
28 K. Sardar, F. L. Deepak, A. Govindaraj, M. M. Seikh and C. N. R. Rao, *Small*, 2005, **1**, 91–94.
29 K. Sardar and C. N. R. Rao, *Adv. Mater.*, 2004, **16**, 425–429.
30 K. Sardar, M. Dan, B. Schwenzer and C. N. R. Rao, *J. Mater. Chem.*, 2005, **15**, 2175–2177.
31 C.-C. Chang, H.-L. Wu, C.-H. Kuo and M. H. Huang, *Chem. Mater.*, 2008, **20**, 7570–7574.
32 S. Li, G. Ye and G. Chen, *J. Phys. Chem. C*, 2009, **113**, 4031–4037.
33 B. Liu and H. C. Zeng, *J. Am. Chem. Soc.*, 2003, **125**, 4430–4431.
34 K. Biswas and C. N. R. Rao, *J. Phys. Chem. B*, 2005, **110**, 842–845.
35 D. Berhanu, K. Govender, D. Smyth-Boyle, M. Archbold, D. P. Halliday and P. O'Brien, *Chem. Commun.*, 2006, 4709–4711.
36 U. K. Gautam, M. Rajamathi, F. Meldrum, P. Morgan and R. Seshadri, *Chem. Commun.*, 2001, 629–630.
37 S. Thimmaiah, M. Rajamathi, N. Singh, P. Bera, F. Meldrum, N. Chandrasekhar and R. Seshadri, *J. Mater. Chem.*, 2001, **11**, 3215–3221.
38 L. L. Hench and J. K. West, *Chem. Rev.*, 1990, **90**, 33–72.
39 C. J. Brinker and G. W. Scherer, *Sol–Gel Science: The Physics and Chemistry of Sol–Gel Processing*, Academic Press, Boston, 1990.
40 M. Niederberger, *Acc. Chem. Res.*, 2007, **40**, 793–800.
41 F. del Monte, M. P. Morales, D. Levy, A. Fernandez, M. Ocaña, A. Roig, E. Molins, K. O'Grady and C. J. Serna, *Langmuir*, 1997, **13**, 3627–3634.
42 S. Masson, P. Holliman, M. Kalaji and P. Kluson, *J. Mater. Chem.*, 2009, **19**, 3517–3522.
43 T. J. Trentler, T. E. Denler, J. F. Bertone, A. Agrawal and V. L. Colvin, *J. Am. Chem. Soc.*, 1999, **121**, 1613–1614.
44 M. Shim and P. Guyot-Sionnest, *J. Am. Chem. Soc.*, 2001, **123**, 11651–11654.
45 J. Joo, T. Yu, Y. W. Kim, H. M. Park, F. Wu, J. Z. Zhang and T. Hyeon, *J. Am. Chem. Soc.*, 2003, **125**, 6553–6557.

46 F. X. Redl, C. T. Black, G. C. Papaefthymiou, R. L. Sandstrom, M. Yin, H. Zeng, C. B. Murray and S. P. O'Brien, *J. Am. Chem. Soc.*, 2004, **126**, 14583–14599.
47 S. W. Depner, K. R. Kort, C. Jaye, D. A. Fischer and S. Banerjee, *J. Phys. Chem. C*, 2009, **113**, 14126–14134.
48 S. Sun, H. Zeng, D. B. Robinson, S. Raoux, P. M. Rice, S. X. Wang and G. Li, *J. Am. Chem. Soc.*, 2003, **126**, 273–279.
49 M. Brust, M. Walker, D. Bethell, D. J. Schiffrin and R. Whyman, *J. Chem. Soc., Chem. Commun.*, 1994, 801.
50 J. Yang, E. Sargent, S. Kelley and J. Y. Ying, *Nat. Mater.*, 2009, **8**, 683–689.
51 K. Vijaya Sarathy, G. U. Kulkarni and C. N. R. Rao, *Chem. Commun.*, 1997, 537–538.
52 R. Voggu, K. Biswas, A. Govindaraj and C. N. R. Rao, *J. Phys. Chem. B*, 2006, **110**, 20752–20755.
53 R. Voggu, A. Shireen and C. N. R. Rao, *Dalton Trans.*, 2010, **39**, 6021–6023.
54 I. Bilecka and M. Niederberger, *Nanoscale*, 2010, **2**, 1358–1374.
55 M. B. Mohamed, K. M. AbouZeid, V. Abdelsayed, A. A. Aljarash and M. S. El-Shall, *ACS Nano*, 2010, **4**, 2766–2772.
56 J. Zhu, O. Palchik, S. Chen and A. Gedanken, *J. Phys. Chem. B*, 2000, **104**, 7344–7347.
57 R. Harpeness and A. Gedanken, *Langmuir*, 2004, **20**, 3431–3434.
58 Y. Kitamoto, R. Minami, Y. Shibata, T. Chikata and S. Kat, *IEEE Trans. Magn.*, 2005, **41**, 3880.
59 X. Liao, J. Zhu, W. Zhong and H.-Y. Chen, *Mater. Lett.*, 2001, **50**, 341–346.
60 C. N. R. Rao and K. P. Kalyanikutty, *Acc. Chem. Res.*, 2008, **41**, 489–499.
61 C. N. R. Rao, V. V. Agrawal, K. Biswas, U. K. Gautam, M. Ghosh, A. Govindaraj, G. U. Kulkarni, K. R. Kalyanikutty, K. Sardar and S. R. C. Vivekchandi, *Pure Appl. Chem.*, 2006, **78**, 1619–1650.
62 C. N. R. Rao, G. U. Kulkarni, V. V. Agrawal, U. K. Gautam, M. Ghosh and U. Tumkurkar, *J. Colloid Interface Sci.*, 2005, **289**, 305–318.
63 C. N. R. Rao, G. U. Kulkarni, P. J. Thomas, V. V. Agrawal and P. Saravanan, *J. Phys. Chem. B*, 2003, **107**, 7391–7395.
64 V. V. Agrawal, P. Mahalakshmi, G. U. Kulkarni and C. N. R. Rao, *Langmuir*, 2005, **22**, 1846–1851.
65 N. Zaitseva, Z. R. Dai, C. D. Grant, J. Harper and C. Saw, *Chem. Mater.*, 2007, **19**, 5174–5178.
66 U. K. Gautam, M. Ghosh and C. N. R. Rao, *Langmuir*, 2004, **20**, 10775–10778.
67 M. Antonietti, D. Kuang, B. Smarsly and Y. Zhou, *Angew. Chem., Int. Ed.*, 2004, **43**, 4988–4992.
68 Y. Zhou, *Curr. Nanosci.*, 2005, **1**, 35.
69 H. Itoh, K. Naka and Y. Chujo, *J. Am. Chem. Soc.*, 2004, **126**, 3026–3027.
70 G. S. Fonseca, A. P. Umpierre, P. F. P. Fichtner, S. R. Teixeira and J. Dupont, *Chem.–Eur. J.*, 2003, **9**, 3263–3269.
71 L. M. Rossi, J. Dupont, G. Machado, P. F. P. Fichtner, C. D. Radtke, I. J. R. Baumvol and S. R. Teixeira, *J. Braz. Chem. Soc.*, 2004, **15**, 901–910.
72 K. Biswas and C. N. R. Rao, *Chem.–Eur. J.*, 2007, **13**, 6123–6129.
73 M. Faraday, *Philos. Trans. R. Soc. London*, 1857, **147**, 145–181.
74 B. V. Enüstün and J. Turkevich, *J. Am. Chem. Soc.*, 1963, **85**, 3317–3328.
75 Y. Xia, Y. Xiong, B. Lim and S. E. Skrabalak, *Angew. Chem., Int. Ed.*, 2009, **48**, 60–103.
76 P. Mukherjee, A. Ahmad, D. Mandal, S. Senapati, S. R. Sainkar, M. I. Khan, R. Parishcha, P. V. Ajaykumar, M. Alam, R. Kumar and M. Sastry, *Nano Lett.*, 2001, **1**, 515–519.
77 E. A. Hauser and J. E. Lynn, *Experiments in Colloid Chemistry*, MacGraw-Hill, New York, 1940.
78 J. Turkevich, P. C. Stevenson and J. Hillier, *Discuss. Faraday Soc.*, 1951, **11**, 55.
79 D. G. Duff, A. Baiker and P. P. Edwards, *Langmuir*, 1993, **9**, 2301–2309.
80 X. Ji, X. Song, J. Li, Y. Bai, W. Yang and X. Peng, *J. Am. Chem. Soc.*, 2007, **129**, 13939–13948.
81 M. Kanehara, J.-i. Sakurai, H. Sugimura and T. Teranishi, *J. Am. Chem. Soc.*, 2009, **131**, 1630–1631.
82 J. Ohyama, Y. Hitomi, Y. Higuchi, M. Shinagawa, H. Mukai, M. Kodera, K. Teramura, T. Shishido and T. Tanaka, *Chem. Commun.*, 2008, 6300–6302.
83 E. Redel, M. Walter, R. Thomann, C. Vollmer, L. Hussein, H. Scherer, M. Krüger and C. Janiak, *Chem.–Eur. J.*, 2009, **15**, 10047–10059.
84 J. Han, Y. Liu and R. Guo, *Adv. Funct. Mater.*, 2009, **19**, 1112–1117.
85 P. J. G. Goulet and R. B. Lennox, *J. Am. Chem. Soc.*, 2010, **132**, 9582–9584.
86 K. L. McGilvray, M. R. Decan, D. Wang and J. C. Scaiano, *J. Am. Chem. Soc.*, 2006, **128**, 15980–15981.
87 P. Christian and M. Bromfield, *J. Mater. Chem.*, 2010, **20**, 1135–1139.
88 K. Biswas, N. Varghese and C. N. R. Rao, *Small*, 2008, **4**, 649–655.
89 S. Egusa, P. L. Redmond and N. F. Scherer, *J. Phys. Chem. C*, 2007, **111**, 17993–17996.
90 M. L. Personick, M. R. Langille, J. Zhang and C. A. Mirkin, *Nano Lett.*, 2011, **11**, 3394–3398.
91 W. Niu, S. Zheng, D. Wang, X. Liu, H. Li, S. Han, J. Chen, Z. Tang and G. Xu, *J. Am. Chem. Soc.*, 2008, **131**, 697–703.
92 S. W. Lu, V. Frain and M. Arbab, *J. Phys. Chem. C*, 2010, **114**, 12850–12854.
93 M. S. Yavuz, W. Li and Y. Xia, *Chem.–Eur. J.*, 2009, **15**, 13181–13187.
94 D. Seo, C. I. Yoo, I. S. Chung, S. M. Park, S. Ryu and H. Song, *J. Phys. Chem. C*, 2008, **112**, 2469–2475.
95 Y. Ma, J. Zeng, W. Li, M. McKiernan, Z. Xie and Y. Xia, *Adv. Mater.*, 2010, **22**, 1930–1934.
96 J. Xu, S. Li, J. Weng, X. Wang, Z. Zhou, K. Yang, M. Liu, X. Chen, Q. Cui, M. Cao and Q. Zhang, *Adv. Funct. Mater.*, 2008, **18**, 277–284.
97 C. Li, K. L. Shuford, Q. H. Park, W. Cai, Y. Li, E. J. Lee and S. O. Cho, *Angew. Chem., Int. Ed.*, 2007, **46**, 3264–3268.
98 Y. Ma, Q. Kuang, Z. Jiang, Z. Xie, R. Huang and L. Zheng, *Angew. Chem., Int. Ed.*, 2008, **47**, 8901–8904.
99 Y. Huang, W. Wang, H. Liang and H. Xu, *Cryst. Growth Des.*, 2008, **9**, 858–862.
100 S. Guo, L. Wang and E. Wang, *Chem. Commun.*, 2007, 3163–3165.
101 H.-G. Liao, Y.-X. Jiang, Z.-Y. Zhou, S.-P. Chen and S.-G. Sun, *Angew. Chem., Int. Ed.*, 2008, **47**, 9100–9103.
102 I. Pastoriza-Santos and L. M. Liz-Marzán, *Adv. Funct. Mater.*, 2009, **19**, 679–688.
103 K. Park, H. Koerner and R. A. Vaia, *Nano Lett.*, 2010, **10**, 1433–1439.
104 M. H. Rashid and T. K. Mandal, *Adv. Funct. Mater.*, 2008, **18**, 2261–2271.
105 B. Keita, R. N. Biboum, I. M. Mbomekalle, S. Floquet, C. Simonnet-Jegat, E. Cadot, F. Miserque, P. Berthet and L. Nadjo, *J. Mater. Chem.*, 2008, **18**, 3196–3199.
106 S. Guo and E. Wang, *Inorg. Chem.*, 2007, **46**, 6740–6743.
107 J. Xie, J. Y. Lee and D. I. C. Wang, *Chem. Mater.*, 2007, **19**, 2823–2830.
108 Z. Li, A. Friedrich and A. Taubert, *J. Mater. Chem.*, 2008, **18**, 1008–1014.
109 S. D. Perrault and W. C. W. Chan, *J. Am. Chem. Soc.*, 2009, **131**, 17042–17043.
110 B. Chen, X. Jiao and D. Chen, *Cryst. Growth Des.*, 2010, **10**, 3378–3386.
111 M. Tsuji, Y. Maeda, S. Hikino, H. Kumagae, M. Matsunaga, X.-L. Tang, R. Matsuo, M. Ogino and P. Jiang, *Cryst. Growth Des.*, 2009, **9**, 4700–4705.
112 V. N. Richards, N. P. Rath and W. E. Buhro, *Chem. Mater.*, 2010, **22**, 3556–3567.
113 X. Dong, X. Ji, H. Wu, L. Zhao, J. Li and W. Yang, *J. Phys. Chem. C*, 2009, **113**, 6573–6576.
114 S. Kundu, K. Wang and H. Liang, *J. Phys. Chem. C*, 2008, **113**, 134–141.
115 H. Liang, W. Wang, Y. Huang, S. Zhang, H. Wei and H. Xu, *J. Phys. Chem. C*, 2010, **114**, 7427–7431.
116 T. U. B. Rao and T. Pradeep, *Angew. Chem., Int. Ed.*, 2010, **49**, 3925–3929.
117 C. Kind, R. Popescu, E. Muller, D. Gerthsen and C. Feldmann, *Nanoscale*, 2010, **2**, 2223–2229.
118 H. Xu and K. S. Suslick, *ACS Nano*, 2010, **4**, 3209–3214.
119 W. Li, P. H. C. Camargo, X. Lu and Y. Xia, *Nano Lett.*, 2008, **9**, 485–490.
120 Q. Zhang, W. Li, C. Moran, J. Zeng, J. Chen, L.-P. Wen and Y. Xia, *J. Am. Chem. Soc.*, 2010, **132**, 11372–11378.
121 C. Xue, J. E. Millstone, S. Li and C. A. Mirkin, *Angew. Chem., Int. Ed.*, 2007, **46**, 8436–8439.
122 B. Wiley, T. Herricks, Y. Sun and Y. Xia, *Nano Lett.*, 2004, **4**, 1733–1739.

123 B. J. Wiley, Y. Xiong, Z.-Y. Li, Y. Yin and Y. Xia, *Nano Lett.*, 2006, **6**, 765–768.
124 B. Wiley, Y. Sun and Y. Xia, *Langmuir*, 2005, **21**, 8077–8080.
125 Y. Xiong, *Chem. Commun.*, 2011, **47**, 1580–1582.
126 B. J. Wiley, Y. Chen, J. M. McLellan, Y. Xiong, Z.-Y. Li, D. Ginger and Y. Xia, *Nano Lett.*, 2007, **7**, 1032–1036.
127 J. Zeng, Y. Zheng, M. Rycenga, J. Tao, Z.-Y. Li, Q. Zhang, Y. Zhu and Y. Xia, *J. Am. Chem. Soc.*, 2010, **132**, 8552–8553.
128 A. Courty, A. I. Henry, N. Goubet and M. P. Pileni, *Nat. Mater.*, 2007, **6**, 900–907.
129 J. Zhang, S. Li, J. Wu, G. C. Schatz and C. A. Mirkin, *Angew. Chem., Int. Ed.*, 2009, **48**, 7787–7791.
130 S. E. Skrabalak, B. J. Wiley, M. Kim, E. V. Formo and Y. Xia, *Nano Lett.*, 2008, **8**, 2077–2081.
131 H. Liang, H. Yang, W. Wang, J. Li and H. Xu, *J. Am. Chem. Soc.*, 2009, **131**, 6068–6069.
132 Q. Zhang, C. Cobley, L. Au, M. McKiernan, A. Schwartz, L.-P. Wen, J. Chen and Y. Xia, *ACS Appl. Mater. Interfaces*, 2009, **1**, 2044–2048.
133 A. Tao, P. Sinsermsuksakul and P. Yang, *Angew. Chem., Int. Ed.*, 2006, **45**, 4597–4601.
134 Y. Xiong, I. Washio, J. Chen, H. Cai, Z.-Y. Li and Y. Xia, *Langmuir*, 2006, **22**, 8563–8570.
135 X. Liu, R. Huang and J. Zhu, *Chem. Mater.*, 2007, **20**, 192–197.
136 B. Pietrobon and V. Kitaev, *Chem. Mater.*, 2008, **20**, 5186–5190.
137 S. Giuffrida, G. Ventimiglia and S. Sortino, *Chem. Commun.*, 2009, 4055–4057.
138 J. An, B. Tang, X. Ning, J. Zhou, B. Zhao, W. Xu, C. Corredor and J. R. Lombardi, *J. Phys. Chem. C*, 2007, **111**, 18055–18059.
139 B. K. Jena, B. K. Mishra and S. Bohidar, *J. Phys. Chem. C*, 2009, **113**, 14753–14758.
140 T. Teranishi and M. Miyake, *Chem. Mater.*, 1998, **10**, 594–600.
141 T. Teranishi, M. Hosoe, T. Tanaka and M. Miyake, *J. Phys. Chem. B*, 1999, **103**, 3818–3827.
142 S. Chen and K. Kimura, *J. Phys. Chem. B*, 2001, **105**, 5397–5403.
143 H. Wakayama, N. Setoyama and Y. Fukushima, *Adv. Mater.*, 2003, **15**, 742–745.
144 Y. Xiong, J. M. McLellan, Y. Yin and Y. Xia, *Angew. Chem.*, 2007, **119**, 804–808.
145 T. S. Ahmadi, Z. L. Wang, T. C. Green, A. Henglein and M. A. El-Sayed, *Science*, 1996, **272**, 1924–1925.
146 K. V. Sarathy, G. Raina, R. T. Yadav, G. U. Kulkarni and C. N. R. Rao, *J. Phys. Chem. B*, 1997, **101**, 9876–9880.
147 S.-Y. Zhao, S.-H. Chen, S.-Y. Wang, D.-G. Li and H.-Y. Ma, *Langmuir*, 2002, **18**, 3315–3318.
148 K. Kurihara, J. n. H. Fendler, I. Ravet and J. n. B. Nagy, *J. Mol. Catal.*, 1986, **34**, 325–335.
149 S.-W. Kim, J. Park, Y. Jang, Y. Chung, S. Hwang, T. Hyeon and Y. W. Kim, *Nano Lett.*, 2003, **3**, 1289–1291.
150 Y. Xiong, J. Chen, B. Wiley, Y. Xia, Y. Yin and Z.-Y. Li, *Nano Lett.*, 2005, **5**, 1237–1242.
151 Y. Piao, Y. Jang, M. Shokouhimehr, I. S. Lee and T. Hyeon, *Small*, 2007, **3**, 255–260.
152 V. Mazumder and S. Sun, *J. Am. Chem. Soc.*, 2009, **131**, 4588–4589.
153 Y. Sun, L. Zhang, H. Zhou, Y. Zhu, E. Sutter, Y. Ji, M. H. Rafailovich and J. C. Sokolov, *Chem. Mater.*, 2007, **19**, 2065–2070.
154 F.-R. Fan, A. Attia, U. K. Sur, J.-B. Chen, Z.-X. Xie, J.-F. Li, B. Ren and Z.-Q. Tian, *Cryst. Growth Des.*, 2009, **9**, 2335–2340.
155 L. Bisson, C. Boissiere, L. Nicole, D. Grosso, J. P. Jolivet, C. c. Thomazeau, D. Uzio, G. Berhault and C. m. Sanchez, *Chem. Mater.*, 2009, **21**, 2668–2678.
156 E. Coronado, A. Ribera, J. Garcia-Martinez, N. Linares and L. M. Liz-Marzán, *J. Mater. Chem.*, 2008, **18**, 5682–5688.
157 Y. Xiong, J. Chen, B. Wiley, Y. Xia, S. Aloni and Y. Yin, *J. Am. Chem. Soc.*, 2005, **127**, 7332–7333.
158 Y. Xiong, J. M. McLellan, Y. Yin and Y. Xia, *Angew. Chem., Int. Ed.*, 2007, **46**, 790–794.
159 J. Chen, T. Herricks and Y. Xia, *Angew. Chem., Int. Ed.*, 2005, **44**, 2589–2592.
160 N. Tian, Z.-Y. Zhou, S.-G. Sun, Y. Ding and Z. L. Wang, *Science*, 2007, **316**, 732–735.
161 B. Lim, X. Lu, M. Jiang, P. H. C. Camargo, E. C. Cho, E. P. Lee and Y. Xia, *Nano Lett.*, 2008, **8**, 4043–4047.
162 L. Wang, H. Wang, Y. Nemoto and Y. Yamauchi, *Chem. Mater.*, 2010, **22**, 2835–2841.
163 E. Ramirez, L. Eradès, K. Philippot, P. Lecante and B. Chaudret, *Adv. Funct. Mater.*, 2007, **17**, 2219–2228.
164 J. N. Tiwari, R. N. Tiwari and K.-L. Lin, *ACS Appl. Mater. Interfaces*, 2010, **2**, 2231–2237.
165 Z.-Y. Zhou, Z.-Z. Huang, D.-J. Chen, Q. Wang, N. Tian and S.-G. Sun, *Angew. Chem., Int. Ed.*, 2010, **49**, 411–414.
166 L. Wang and Y. Yamauchi, *J. Am. Chem. Soc.*, 2009, **131**, 9152–9153.
167 M. Nogami, R. Koike, R. Jalem, G. Kawamura, Y. Yang and Y. Sasaki, *J. Phys. Chem. Lett.*, 2010, **1**, 568–571.
168 N. C. Bigall, T. Hätling, M. Klose, P. Simon, L. M. Eng and A. Eychmüller, *Nano Lett.*, 2008, **8**, 4588–4592.
169 G. Surendran, L. Ramos, B. Pansu, E. Prouzet, P. Beaunier, F. Audonnet and H. Remita, *Chem. Mater.*, 2007, **19**, 5045–5048.
170 A. Demortière, P. Launois, N. Goubet, P. A. Albouy and C. Petit, *J. Phys. Chem. B*, 2008, **112**, 14583–14592.
171 H. Zheng, R. K. Smith, Y.-w. Jun, C. Kisielowski, U. Dahmen and A. P. Alivisatos, *Science*, 2009, **324**, 1309–1312.
172 A. Mohanty, N. Garg and R. Jin, *Angew. Chem., Int. Ed.*, 2010, **49**, 4962–4966.
173 S. Cheong, J. D. Watt and R. D. Tilley, *Nanoscale*, 2010, **2**, 2045–2053.
174 N. Varghese and C. N. R. Rao, *J. Colloid Interface Sci.*, 2012, **365**, 117–121.
175 H. Kura, M. Takahashi and T. Ogawa, *J. Phys. Chem. C*, 2010, **114**, 5835–5838.
176 T. Ling, J. Zhu, H. Yu and L. Xie, *J. Phys. Chem. C*, 2009, **113**, 9450–9453.
177 T. Ling, L. Xie, J. Zhu, H. Yu, H. Ye, R. Yu, Z. Cheng, L. Liu, G. Yang, Z. Cheng, Y. Wang and X. Ma, *Nano Lett.*, 2009, **9**, 1572–1576.
178 D. Kim, J. Park, K. An, N.-K. Yang, J.-G. Park and T. Hyeon, *J. Am. Chem. Soc.*, 2007, **129**, 5812–5813.
179 M. Scariot, D. O. Silva, J. D. Scholten, G. Machado, S. R. Teixeira, M. A. Novak, G. Ebeling and J. Dupont, *Angew. Chem., Int. Ed.*, 2008, **47**, 9075–9078.
180 R. M. de Silva, V. Palshin, K. M. N. de Silva, L. L. Henry and C. S. S. R. Kumar, *J. Mater. Chem.*, 2008, **18**, 738–747.
181 L. T. Lu, L. D. Tung, I. Robinson, D. Ung, B. Tan, J. Long, A. I. Cooper, D. G. Fernig and N. T. K. Thanh, *J. Mater. Chem.*, 2008, **18**, 2453–2458.
182 I. Robinson, M. Volk, L. D. Tung, G. Caruntu, N. Kay and N. T. K. Thanh, *J. Phys. Chem. C*, 2009, **113**, 9497–9501.
183 K. M. Nam, J. H. Shim, H. Ki, S.-I. Choi, G. Lee, J. K. Jang, Y. Jo, M.-H. Jung, H. Song and J. T. Park, *Angew. Chem., Int. Ed.*, 2008, **47**, 9504–9508.
184 X. Wang, H. Fu, A. Peng, T. Zhai, Y. Ma, F. Yuan and J. Yao, *Adv. Mater.*, 2009, **21**, 1636–1640.
185 O. Metin, V. Mazumder, S. Ozkar and S. Sun, *J. Am. Chem. Soc.*, 2010, **132**, 1468–1469.
186 D. Zhang, G. Li and J. C. Yu, *Cryst. Growth Des.*, 2009, **9**, 2812–2815.
187 M. Richard-Plouet, M. Guillot, S. Vilminot, C. d. Leuvrey, C. Estournès and M. Kurmoo, *Chem. Mater.*, 2007, **19**, 865–871.
188 S. Carenco, C. d. Boissière, L. Nicole, C. m. Sanchez, P. Le Floch and N. Mézailles, *Chem. Mater.*, 2010, **22**, 1340–1349.
189 H. Wang, X. Jiao and D. Chen, *J. Phys. Chem. C*, 2008, **112**, 18793–18797.
190 H. Winnischofer, T. C. R. Rocha, W. C. Nunes, L. M. Socolovsky, M. Knobel and D. Zanchet, *ACS Nano*, 2008, **2**, 1313–1319.
191 S. B. Kalidindi, U. Sanyal and B. R. Jagirdar, *Inorg. Chem.*, 2010, **49**, 3965–3967.
192 V. Engels, F. Benaskar, D. A. Jefferson, B. F. G. Johnson and A. E. H. Wheatley, *Dalton Trans.*, 2010, **39**, 6496–6502.
193 I. C. Chang, T.-K. Huang, H.-K. Lin, Y.-F. Tzeng, C.-W. Peng, F.-M. Pan, C.-Y. Lee and H.-T. Chiu, *ACS Appl. Mater. Interfaces*, 2009, **1**, 1375–1378.
194 I. Haas, S. Shanmugam and A. Gedanken, *Chem.–Eur. J.*, 2008, **14**, 4696–4703.
195 Z. L. Wang, *ACS Nano*, 2008, **2**, 1987–1992.
196 C.-H. Huang, H. P. Wang, J.-E. Chang and E. M. Eyring, *Chem. Commun.*, 2009, 4663–4665.
197 J. F. Bondi, K. D. Oyler, X. Ke, P. Schiffer and R. E. Schaak, *J. Am. Chem. Soc.*, 2009, **131**, 9144–9145.
198 E. Redel, R. Thomann and C. Janiak, *Chem. Commun.*, 2008, 1789–1791.
199 N. Zettsu, J. M. McLellan, B. Wiley, Y. Yin, Z.-Y. Li and Y. Xia, *Angew. Chem., Int. Ed.*, 2006, **45**, 1288.

200 S. Ghosh, M. Ghosh and C. N. R. Rao, *J. Cluster Sci.*, 2007, **18**, 97–111.
201 H. Zhang, X. Xia, W. Li, J. Zeng, Y. Dai, D. Yang and Y. Xia, *Angew. Chem., Int. Ed.*, 2010, **49**, 5296–5300.
202 M. H. G. Prechtl, M. Scariot, J. D. Scholten, G. Machado, S. r. R. Teixeira and J. Dupont, *Inorg. Chem.*, 2008, **47**, 8995–9001.
203 C. Vollmer, E. Redel, K. Abu-Shandi, R. Thomann, H. Manyar, C. Hardacre and C. Janiak, *Chem.–Eur. J.*, 2010, **16**, 3849–3858.
204 Q. Yuan, Z. Zhou, J. Zhuang and X. Wang, *Inorg. Chem.*, 2010, **49**, 5515–5521.
205 Y. Zhang, H. Zhang, Y. Zhang, Y. Ma, H. Zhong and H. Ma, *Chem. Commun.*, 2009, 6589–6591.
206 J. Hu, Z. Chen, N. Wang, Y. Song, H. Jiang and Y. Sun, *Chem. Commun.*, 2009, 4503–4505.
207 S. B. Kalidindi and B. R. Jagirdar, *Inorg. Chem.*, 2009, **48**, 4524–4529.
208 M. J. Meziani, C. E. Bunker, F. Lu, H. Li, W. Wang, E. A. Guliants, R. A. Quinn and Y.-P. Sun, *ACS Appl. Mater. Interfaces*, 2009, **1**, 703–709.
209 H. Yu, P. C. Gibbons, K. F. Kelton and W. E. Buhro, *J. Am. Chem. Soc.*, 2001, **123**, 9198–9199.
210 T. H. Lim, B. Ingham, K. H. Kamarudin, P. G. Etchegoin and R. D. Tilley, *Cryst. Growth Des.*, 2010, **10**, 3854–3858.
211 N. H. Chou, X. Ke, P. Schiffer and R. E. Schaak, *J. Am. Chem. Soc.*, 2008, **130**, 8140–8141.
212 N. H. Chou, K. D. Oyler, N. E. Motl and R. E. Schaak, *Chem. Mater.*, 2009, **21**, 4105–4107.
213 S. Prabakar, A. Shiohara, S. Hanada, K. Fujioka, K. Yamamoto and R. D. Tilley, *Chem. Mater.*, 2009, **22**, 482–486.
214 L. L. A. Adams, W. R. Sweeney and H. M. Jaeger, *J. Phys. Chem. C*, 2010, **114**, 6304–6310.
215 I. Resa, H. l. n. Moreira, B. Bresson, B. Mahler, B. Dubertret and H. Aubin, *J. Phys. Chem. C*, 2009, **113**, 7120–7122.
216 J. Hu, Y. Sun and Z. Chen, *J. Phys. Chem. C*, 2009, **113**, 5201–5205.
217 Y. Wang, J. Chen, L. Chen, Y.-B. Chen and L.-M. Wu, *Cryst. Growth Des.*, 2010, **10**, 1578–1584.
218 K. Gschneidner, A. Russell, A. Pecharsky, J. Morris, Z. Zhang, T. Lograsso, D. Hsu, C. H. Chester Lo, Y. Ye, A. Slager and D. Kesse, *Nat. Mater.*, 2003, **2**, 587–591.
219 T. Krenke, E. Duman, M. Acet, E. F. Wassermann, X. Moya, L. Manosa and A. Planes, *Nat. Mater.*, 2005, **4**, 450–454.
220 D. Aherne, M. Gara, J. M. Kelly and Y. K. Gun'ko, *Adv. Funct. Mater.*, 2010, **20**, 1329–1338.
221 X. Lu, L. Au, J. McLellan, Z.-Y. Li, M. Marquez and Y. Xia, *Nano Lett.*, 2007, **7**, 1764–1769.
222 S. Anandan, F. Grieser and M. Ashokkumar, *J. Phys. Chem. C*, 2008, **112**, 15102–15105.
223 Y. Shin, I.-T. Bae, B. W. Arey and G. J. Exarhos, *J. Phys. Chem. C*, 2008, **112**, 4844–4848.
224 C. Wang, H. Yin, R. Chan, S. Peng, S. Dai and S. Sun, *Chem. Mater.*, 2009, **21**, 433–435.
225 Y. Liu and A. R. H. Walker, *Angew. Chem., Int. Ed.*, 2010, **49**, 6781–6785.
226 W. Chen, R. Yu, L. Li, A. Wang, Q. Peng and Y. Li, *Angew. Chem., Int. Ed.*, 2010, **122**, 2979–2983.
227 P. H. C. Camargo, Y. Xiong, L. Ji, J. M. Zuo and Y. Xia, *J. Am. Chem. Soc.*, 2007, **129**, 15452–15453.
228 D. Ferrer, A. Torres-Castro, X. Gao, S. Sepúlveda-Guzmán, U. Ortiz-Méndez and M. José-Yacamán, *Nano Lett.*, 2007, **7**, 1701–1705.
229 M. Friedrich and M. Armbrüster, *Chem. Mater.*, 2009, **21**, 5886–5891.
230 C. Femoni, M. C. Iapalucci, G. Longoni, C. Tiozzo and S. Zacchini, *Angew. Chem., Int. Ed.*, 2008, **47**, 6666–6669.
231 Y. Vasquez, Z. Luo and R. E. Schaak, *J. Am. Chem. Soc.*, 2008, **130**, 11866–11867.
232 Z. Xu, C. E. Carlton, L. F. Allard, Y. Shao-Horn and K. Hamad-Schifferli, *J. Phys. Chem. Lett.*, 2010, **1**, 2514–2518.
233 C. Wang, W. Tian, Y. Ding, Y.-q. Ma, Z. L. Wang, N. M. Markovic, V. R. Stamenkovic, H. Daimon and S. Sun, *J. Am. Chem. Soc.*, 2010, **132**, 6524–6529.
234 J. Wang, D. F. Thomas and A. Chen, *Chem. Commun.*, 2008, 5010–5012.
235 X.-W. Wei, G.-X. Zhu, Y.-J. Liu, Y.-H. Ni, Y. Song and Z. Xu, *Chem. Mater.*, 2008, **20**, 6248–6253.
236 B. Lim, J. Wang, P. H. C. Camargo, C. M. Cobley, M. J. Kim and Y. Xia, *Angew. Chem., Int. Ed.*, 2009, **48**, 6304–6308.
237 B. Lim, M. Jiang, P. H. C. Camargo, E. C. Cho, J. Tao, X. Lu, Y. Zhu and Y. Xia, *Science*, 2009, **324**, 1302–1305.
238 J. Zhang, H. Yang, J. Fang and S. Zou, *Nano Lett.*, 2010, **10**, 638–644.
239 J. Wang, R. M. Asmussen, B. Adams, D. F. Thomas and A. Chen, *Chem. Mater.*, 2009, **21**, 1716–1724.
240 S. Yang, Z. Peng and H. Yang, *Adv. Funct. Mater.*, 2008, **18**, 2745–2753.
241 Z. Liu, G. S. Jackson and B. W. Eichhorn, *Angew. Chem., Int. Ed.*, 2010, **49**, 3173–3176.
242 M. Boualleg, D. Baudouin, J.-M. Basset, F. Bayard, J.-P. Candy, J.-C. Jumas, L. Veyre and C. Thieuleux, *Chem. Commun.*, 2010, **46**, 4722–4724.
243 S. T. Christensen, H. Feng, J. L. Libera, N. Guo, J. T. Miller, P. C. Stair and J. W. Elam, *Nano Lett.*, 2010, **10**, 3047–3051.
244 S. Sun, C. B. Murray, D. Weller, L. Folks and A. Moser, *Science*, 2000, **287**, 1989–1992.
245 E. V. Shevchenko, D. V. Talapin, A. L. Rogach, A. Kornowski, M. Haase and H. Weller, *J. Am. Chem. Soc.*, 2002, **124**, 11480–11485.
246 I. Robinson, S. Zacchini, L. D. Tung, S. Maenosono and N. T. K. Thanh, *Chem. Mater.*, 2009, **21**, 3021–3026.
247 M. S. Wellons, W. H. Morris, Z. Gai, J. Shen, J. Bentley, J. E. Wittig and C. M. Lukehart, *Chem. Mater.*, 2007, **19**, 2483–2488.
248 H. M. Song, W. S. Kim, Y. B. Lee, J. H. Hong, H. G. Lee and N. H. Hur, *J. Mater. Chem.*, 2009, **19**, 3677–3681.
249 H. L. Nguyen, L. E. M. Howard, G. W. Stinton, S. R. Giblin, B. K. Tanner, I. Terry, A. K. Hughes, I. M. Ross, A. Serres and J. S. O. Evans, *Chem. Mater.*, 2006, **18**, 6414–6424.
250 Y. Ofir, B. Samanta, P. Arumugam and V. M. Rotello, *Adv. Mater.*, 2007, **19**, 4075–4079.
251 Y. Zhai, J. Zhai and S. Dong, *Chem. Commun.*, 2010, **46**, 1500–1502.
252 K. Ahrenstorf, H. Heller, A. Kornowski, J. A. C. Broekaert and H. Weller, *Adv. Funct. Mater.*, 2008, **18**, 3850–3856.
253 J. Wu, J. Zhang, Z. Peng, S. Yang, F. T. Wagner and H. Yang, *J. Am. Chem. Soc.*, 2010, **132**, 4984–4985.
254 C. Desvaux, P. Lecante, M. Respaud and B. Chaudret, *J. Mater. Chem.*, 2010, **20**, 103–109.
255 L. Wang, J. Liu, Y. Zhou, Y. Song, J. He and D. G. Evans, *Chem. Commun.*, 2010, **46**, 3911–3913.
256 G. S. Chaubey, C. Barcena, N. Poudyal, C. Rong, J. Gao, S. Sun and J. P. Liu, *J. Am. Chem. Soc.*, 2007, **129**, 7214–7215.
257 R. E. Cable and R. E. Schaak, *Chem. Mater.*, 2007, **19**, 4098–4104.
258 Y. Zhao and C. Burda, *ACS Appl. Mater. Interfaces*, 2009, **1**, 1259–1263.
259 N. Dahal and V. Chikan, *Chem. Mater.*, 2010, **22**, 2892–2897.
260 X.-L. Wang, W.-Q. Han, J. Chen and J. Graetz, *ACS Appl. Mater. Interfaces*, 2010, **2**, 1548–1551.
261 D. Koziej, F. Fischer, N. KraÃànzlin, W. R. Caseri and M. Niederberger, *ACS Appl. Mater. Interfaces*, 2009, **1**, 1097–1104.
262 B. Wu, C. Guo, N. Zheng, Z. Xie and G. D. Stucky, *J. Am. Chem. Soc.*, 2008, **130**, 17563–17567.
263 J. Liao, L. Shi, S. Yuan, Y. Zhao and J. Fang, *J. Phys. Chem. C*, 2009, **113**, 18778–18783.
264 Y. Dai, C. M. Cobley, J. Zeng, Y. Sun and Y. Xia, *Nano Lett.*, 2009, **9**, 2455–2459.
265 R. Buonsanti, V. Grillo, E. Carlino, C. Giannini, T. Kipp, R. Cingolani and P. D. Cozzoli, *J. Am. Chem. Soc.*, 2008, **130**, 11223–11233.
266 R. Levi, M. Bar-Sadan, A. Albu-Yaron, R. Popovitz-Biro, L. Houben, C. Shahar, A. Enyashin, G. Seifert, Y. Prior and R. Tenne, *J. Am. Chem. Soc.*, 2010, **132**, 11214–11222.
267 L.-j. Mao, C.-y. Liu and J. Li, *J. Mater. Chem.*, 2008, **18**, 1640–1643.
268 T. Yu, J. Moon, J. Park, Y. I. Park, H. B. Na, B. H. Kim, I. C. Song, W. K. Moon and T. Hyeon, *Chem. Mater.*, 2009, **21**, 2272–2279.
269 J. Pike, J. Hanson, L. Zhang and S.-W. Chan, *Chem. Mater.*, 2007, **19**, 5609–5616.
270 F. Jiao, A. Harrison and P. G. Bruce, *Angew. Chem.*, 2007, **119**, 4020–4024.
271 I. Rusakova, T. Ould-Ely, C. Hofmann, D. Prieto-Centurion, C. S. Levin, N. J. Halas, A. Lüttge and K. H. Whitmire, *Chem. Mater.*, 2007, **19**, 1369–1375.
272 S. Xie, X. Zhou, X. Han, Q. Kuang, M. Jin, Y. Jiang, Z. Xie and L. Zheng, *J. Phys. Chem. C*, 2009, **113**, 19107–19111.
273 M. Ghosh, K. Biswas, A. Sundaresan and C. N. R. Rao, *J. Mater. Chem.*, 2006, **16**, 106–111.
274 H. Jiang, T. Zhao, C. Yan, J. Ma and C. Li, *Nanoscale*, 2010, **2**, 2195–2198.
275 J. B. Fei, Y. Cui, X. H. Yan, W. Qi, Y. Yang, K. W. Wang, Q. He and J. B. Li, *Adv. Mater.*, 2008, **20**, 452–456.


 

276 G. R. Patzke, Y. Zhou, R. Kontic and F. Conrad, *Angew. Chem., Int. Ed.*, 2011, **50**, 826–859.
277 A. K. Gupta and M. Gupta, *Biomaterials*, 2005, **26**, 3995–4021.
278 A.-H. Lu, E. L. Salabas and F. Schüth, *Angew. Chem., Int. Ed.*, 2007, **46**, 1222–1244.
279 Y. S. Kang, S. Risbud, J. F. Rabolt and P. Stroeve, *Chem. Mater.*, 1996, **8**, 2209–2211.
280 S. Laurent, D. Forge, M. Port, A. Roch, C. Robic, L. Vander Elst and R. N. Muller, *Chem. Rev.*, 2008, **108**, 2064–2110.
281 T. Hyeon, S. S. Lee, J. Park, Y. Chung and H. B. Na, *J. Am. Chem. Soc.*, 2001, **123**, 12798–12801.
282 Y. Hou, Z. Xu and S. Sun, *Angew. Chem., Int. Ed.*, 2007, **46**, 6329–6332.
283 S. Peng and S. Sun, *Angew. Chem., Int. Ed.*, 2007, **46**, 4155–4158.
284 G. Zhen, B. W. Muir, B. A. Moffat, P. Harbour, K. S. Murray, B. Moubaraki, K. Suzuki, I. Madsen, N. Agron-Olshina, L. Waddington, P. Mulvaney and P. G. Hartley, *J. Phys. Chem. C*, 2010, **115**, 327–334.
285 Z. Xu, C. Shen, Y. Hou, H. Gao and S. Sun, *Chem. Mater.*, 2009, **21**, 1778–1780.
286 X. Sun, C. Zheng, F. Zhang, Y. Yang, G. Wu, A. Yu and N. Guan, *J. Phys. Chem. C*, 2009, **113**, 16002–16008.
287 A. A. Martinez-Morales, N. G. Portney, Y. Zhang, G. Destito, G. Budak, E. Ozbay, M. Manchester, C. S. Ozkan and M. Ozkan, *Adv. Mater.*, 2008, **20**, 4816–4820.
288 S. Peng and S. Sun, *Angew. Chem.*, 2007, **119**, 4233–4236.
289 A. Abou Hassan, O. Sandre, V. Cabuil and P. Tabeling, *Chem. Commun.*, 2008, 1783–1785.
290 H. Hu, H. Yang, P. Huang, D. Cui, Y. Peng, J. Zhang, F. Lu, J. Lian and D. Shi, *Chem. Commun.*, 2010, **46**, 3866–3868.
291 H. Cohen, A. Gedanken and Z. Zhong, *J. Phys. Chem. C*, 2008, **112**, 15429–15438.
292 C.-J. Cheng, C.-C. Lin, R.-K. Chiang, C.-R. Lin, I. S. Lyubutin, E. A. Alkaev and H.-Y. Lai, *Cryst. Growth Des.*, 2008, **8**, 877–883.
293 G. Salazar-Alvarez, J. Qin, V. Sepelak, I. Bergmann, M. Vasilakaki, K. N. Trohidou, J. D. Ardisson, W. A. A. Macedo, M. Mikhaylova, M. Muhammed, M. D. Baro and J. Nogues, *J. Am. Chem. Soc.*, 2008, **130**, 13234–13239.
294 R. Chalasani and S. Vasudevan, *J. Phys. Chem. C*, 2011, **115**, 18088–18093.
295 M. Feyen, C. Weidenthaler, F. Schuth and A.-H. Lu, *J. Am. Chem. Soc.*, 2010, **132**, 6791–6799.
296 L. Zhang, J. Wu, H. Liao, Y. Hou and S. Gao, *Chem. Commun.*, 2009, 4378–4380.
297 G. Leem, S. Sarangi, S. Zhang, I. Rusakova, A. Brazdeikis, D. Litvinov and T. R. Lee, *Cryst. Growth Des.*, 2008, **9**, 32–34.
298 L. Zhao, H. Zhang, Y. Xing, S. Song, S. Yu, W. Shi, X. Guo, J. Yang, Y. Lei and F. Cao, *Chem. Mater.*, 2007, **20**, 198–204.
299 Y. Tan, Z. Zhuang, Q. Peng and Y. Li, *Chem. Mater.*, 2008, **20**, 5029–5034.
300 X.-F. Qu, G.-T. Zhou, Q.-Z. Yao and S.-Q. Fu, *J. Phys. Chem. C*, 2009, **114**, 284–289.
301 P. Guardia, A. Labarta and X. Batlle, *J. Phys. Chem. C*, 2010, **115**, 390–396.
302 X. Hu, J. C. Yu, J. Gong, Q. Li and G. Li, *Adv. Mater.*, 2007, **19**, 2324–2329.
303 C.-J. Jia, L.-D. Sun, F. Luo, X.-D. Han, L. J. Heyderman, Z.-G. Yan, C.-H. Yan, K. Zheng, Z. Zhang, M. Takano, N. Hayashi, M. Eltschka, M. Klaui, U. Rudiger, T. Kasama, L. Cervera-Gontard, R. E. Dunin-Borkowski, G. Tzvetkov and J. Raabe, *J. Am. Chem. Soc.*, 2008, **130**, 16968–16977.
304 D. Kim, N. Lee, M. Park, B. H. Kim, K. An and T. Hyeon, *J. Am. Chem. Soc.*, 2008, **131**, 454–455.
305 J. Yin, Z. Yu, F. Gao, J. Wang, H. Pang and Q. Lu, *Angew. Chem., Int. Ed.*, 2010, **49**, 6328–6332.
306 A. Shavel, B. Rodríguez-González, J. Pacifico, M. Spasova, M. Farle and L. M. Liz-Marzán, *Chem. Mater.*, 2009, **21**, 1326–1332.
307 J. Park, K. An, Y. Hwang, J.-G. Park, H.-J. Noh, J.-Y. Kim, J.-H. Park, N.-M. Hwang and T. Hyeon, *Nat. Mater.*, 2004, **3**, 891–895.
308 H. G. Cha, D. K. Lee, Y. H. Kim, C. W. Kim, C. S. Lee and Y. S. Kang, *Inorg. Chem.*, 2007, **47**, 121–127.
309 Z. Xu, C. Shen, Y. Hou, H. Gao and S. Sun, *Chem. Mater.*, 2009, **21**, 1778–1780.
310 D. Maity, P. Chandrasekharan, P. Pradhan, K.-H. Chuang, J.-M. Xue, S.-S. Feng and J. Ding, *J. Mater. Chem.*, 2011, **21**, 14717–14724.
311 F. C. C. Oliveira, F. B. Effenberger, M. H. Sousa, R. F. Jardim, P. K. Kiyohara, J. Dupont, J. C. Rubim and L. M. Rossi, *Phys. Chem. Chem. Phys.*, 2011, **13**, 13558–13564.
312 I. Robinson, C. Alexander, L. T. Lu, L. D. Tung, D. G. Fernig and N. T. K. Thanh, *Chem. Commun.*, 2007, 4602–4604.
313 J. Rockenberger, E. C. Scher and A. P. Alivisatos, *J. Am. Chem. Soc.*, 1999, **121**, 11595–11596.
314 X. Jia, D. Chen, X. Jiao and S. Zhai, *Chem. Commun.*, 2009, 968–970.
315 S. Ge, X. Shi, K. Sun, C. Li, C. Uher, J. R. Baker, M. M. Banaszak Holl and B. G. Orr, *J. Phys. Chem. C*, 2009, **113**, 13593–13599.
316 L. Xiao, J. Li, D. F. Brougham, E. K. Fox, N. Feliu, A. Bushmelev, A. Schmidt, N. Mertens, F. Kiessling, M. Valldor, B. Fadeel and S. Mathur, *ACS Nano*, 2011, **5**, 6315–6324.
317 Y. Tian, B. Yu, X. Li and K. Li, *J. Mater. Chem.*, 2011, **21**, 2476–2481.
318 W. Yin, X. Chen, M. Cao, C. Hu and B. Wei, *J. Phys. Chem. C*, 2009, **113**, 15897–15903.
319 X. Wang, L. Zhang, Y. Ni, J. Hong and X. Cao, *J. Phys. Chem. C*, 2009, **113**, 7003–7008.
320 J. Ge, Y. Hu, M. Biasini, C. Dong, J. Guo, W. P. Beyermann and Y. Yin, *Chem.–Eur. J.*, 2007, **13**, 7153–7161.
321 S. Basak, K. S. Rane and P. Biswas, *Chem. Mater.*, 2008, **20**, 4906–4914.
322 L. Hu, Q. Peng and Y. Li, *J. Am. Chem. Soc.*, 2008, **130**, 16136–16137.
323 H. Liang, J. M. Raitano, L. Zhang and S.-W. Chan, *Chem. Commun.*, 2009, 7569–7571.
324 J. Yang, H. Liu, W. N. Martens and R. L. Frost, *J. Phys. Chem. C*, 2009, **114**, 111–119.
325 X. Wang, L. Yu, X.-L. Wu, F. Yuan, Y.-G. Guo, Y. Ma and J. Yao, *J. Phys. Chem. C*, 2009, **113**, 15553–15558.
326 Y. Zhang, J. Zhu, X. Song and X. Zhong, *J. Phys. Chem. C*, 2008, **112**, 5322–5327.
327 K. M. Nam, J. H. Shim, D. W. Han, H. S. Kwon, Y. M. Kang, Y. Li, H. Song, W. S. Seo and J. T. Park, *Chem. Mater.*, 2010, **22**, 4446–4454.
328 L.-I. Hung, C.-K. Tsung, W. Huang and P. Yang, *Adv. Mater.*, 2010, **22**, 1910–1914.
329 Y. Sui, W. Fu, Y. Zeng, H. Yang, Y. Zhang, H. Chen, Y. Li, M. Li and G. Zou, *Angew. Chem., Int. Ed.*, 2010, **49**, 4282–4285.
330 C.-H. Kuo and M. H. Huang, *J. Am. Chem. Soc.*, 2008, **130**, 12815–12820.
331 K. X. Yao, X. M. Yin, T. H. Wang and H. C. Zeng, *J. Am. Chem. Soc.*, 2010, **132**, 6131–6144.
332 M. H. Kim, B. Lim, E. P. Lee and Y. Xia, *J. Mater. Chem.*, 2008, **18**, 4069–4073.
333 J.-Y. Ho and M. H. Huang, *J. Phys. Chem. C*, 2009, **113**, 14159.
334 Y. Sui, W. Fu, H. Yang, Y. Zeng, Y. Zhang, Q. Zhao, Y. Li, X. Zhou, Y. Leng, M. Li and G. Zou, *Cryst. Growth Des.*, 2009, **10**, 99–108.
335 J. C. Park, J. Kim, H. Kwon and H. Song, *Adv. Mater.*, 2009, **21**, 803–807.
336 C. H. Kuo, C. H. Chen and M. H. Huang, *Adv. Funct. Mater.*, 2007, **17**, 3773–3780.
337 H. Usui, Y. Shimizu, T. Sasaki and N. Koshizaki, *J. Phys. Chem. B*, 2004, **109**, 120–124.
338 M. Ghosh, R. Seshadri and C. N. R. Rao, *J. Nanosci. Nanotechnol.*, 2004, **4**, 136.
339 M. M. Demir, R. Muñoz-Espí, I. Lieberwirth and G. Wegner, *J. Mater. Chem.*, 2006, **16**, 2940.
340 L. Yeghiazarian, S. Mahajan, C. Montemagno, C. Cohen and U. Wiesner, *Adv. Mater.*, 2005, **17**, 1869–1873.
341 H. Zeng, G. Duan, Y. Li, S. Yang, X. Xu and W. Cai, *Adv. Funct. Mater.*, 2010, **20**, 561–572.
342 S. Kaluza, M. K. Schröter, R. Naumann d'Alnoncourt, T. Reinecke and M. Muhler, *Adv. Funct. Mater.*, 2008, **18**, 3670–3677.
343 C.-H. Lai, W.-F. Lee, I. C. Wu, C.-C. Kang, D.-Y. Chen, L.-J. Chen and P.-T. Chou, *J. Mater. Chem.*, 2009, **19**, 7284–7289.
344 X. Tang, E. S. G. Choo, L. Li, J. Ding and J. Xue, *Chem. Mater.*, 2010, **22**, 3383–3388.
345 C.-X. He, B.-X. Lei, Y.-F. Wang, C.-Y. Su, Y.-P. Fang and D.-B. Kuang, *Chem.–Eur. J.*, 2010, **16**, 8757–8761.
346 L. Chen, J. Xu, J. D. Holmes and M. A. Morris, *J. Phys. Chem. C*, 2010, **114**, 2003–2011.
347 T. Omata, K. Takahashi, S. Hashimoto, Y. Maeda, K. Nose, S. Otsuka-Yao-Matsuo and K. Kanaori, *J. Colloid Interface Sci.*, 2011, **355**, 274–281.
348 J. Park, H. Zheng, Y.-w. Jun and A. P. Alivisatos, *J. Am. Chem. Soc.*, 2009, **131**, 13943–13945.

349 J. Buha, I. Djerdj and M. Niederberger, *Cryst. Growth Des.*, 2007, **7**, 133.
350 U. Koch, A. Fojtik, H. Weller and A. Henglein, *Chem. Phys. Lett.*, 1985, **122**, 507.
351 M. Monge, M. L. Kahn, A. Maisonnat and B. Chaudret, *Angew. Chem., Int. Ed.*, 2003, **42**, 5321–5324.
352 B. Luo, J. E. Rossini and W. L. Gladfelter, *Langmuir*, 2009, **25**, 13133–13141.
353 X. Xu and X. Wang, *Nano Res.*, 2009, **2**, 891–902.
354 C. Guiot, S. Grandjean, S. Lemonnier, J.-P. Jolivet and P. Batail, *Cryst. Growth Des.*, 2009, **9**, 3548–3550.
355 T. Ninjbadgar, G. Garnweitner, A. Börger, L. M. Goldenberg, O. V. Sakhno and J. Stumpe, *Adv. Funct. Mater.*, 2009, **19**, 1819–1825.
356 A. Pucci, G. Clavel, M.-G. Willinger, D. Zitoun and N. Pinna, *J. Phys. Chem. C*, 2009, **113**, 12048–12058.
357 N. Le Houx, G. Pourroy, F. Camerel, M. Comet and D. Spitzer, *J. Phys. Chem. C*, 2009, **114**, 155–161.
358 S. Pokhrel, J. Birkenstock, M. Schowalter, A. Rosenauer and L. MaÃädler, *Cryst. Growth Des.*, 2009, **10**, 632–639.
359 S. Pokhrel, C. E. Simion, V. S. Teodorescu, N. Barsan and U. Weimar, *Adv. Funct. Mater.*, 2009, **19**, 1767–1774.
360 C. Chen, D. Chen, X. Jiao and S. Chen, *J. Phys. Chem. C*, 2007, **111**, 18039–18043.
361 D. Caruntu, K. Yao, Z. Zhang, T. Austin, W. Zhou and C. J. O'Connor, *J. Phys. Chem. C*, 2010, **114**, 4875–4886.
362 M. A. Franzman, V. Pérez and R. L. Brutchey, *J. Phys. Chem. C*, 2008, **113**, 630–636.
363 X. Han, M. Jin, S. Xie, Q. Kuang, Z. Jiang, Y. Jiang, Z. Xie and L. Zheng, *Angew. Chem., Int. Ed.*, 2009, **48**, 9180–9183.
364 A. Kar, S. Kundu and A. Patra, *J. Phys. Chem. C*, 2010, **115**, 118–124.
365 R. A. Gilstrap, C. J. Capozzi, C. G. Carson, R. A. Gerhardt and C. J. Summers, *Adv. Mater.*, 2008, **20**, 4163–4166.
366 G. Bühler, D. Thölmann and C. Feldmann, *Adv. Mater.*, 2007, **19**, 2224–2227.
367 T. Sasaki, Y. Endo, M. Nakaya, K. Kanie, A. Nagatomi, K. Tanoue, R. Nakamura and A. Muramatsu, *J. Mater. Chem.*, 2010, **20**, 8153–8157.
368 S.-I. Choi, K. M. Nam, B. K. Park, W. S. Seo and J. T. Park, *Chem. Mater.*, 2008, **20**, 2609–2611.
369 H. W. Zhang and M. T. Swihart, *Chem. Mater.*, 2007, **19**, 1290–1301.
370 Z. Huo, C. Chen, X. Liu, D. Chu, H. Li, Q. Peng and Y. Li, *Chem. Commun.*, 2008, 3741–3743.
371 M. Taguchi, S. Takami, T. Naka and T. Adschiri, *Cryst. Growth Des.*, 2009, **9**, 5297–5303.
372 T. Arita, J. Yoo, Y. Ueda and T. Adschiri, *Nanoscale*, 2010, **2**, 689–693.
373 Z. Yang, D. Han, D. Ma, H. Ling, L. Liu and Y. Yang, *Cryst. Growth Des.*, 2010, **10**, 291.
374 L. Yan, Y. Ranbo, J. Chen and X. Xing, *Cryst. Growth Des.*, 2008, **8**, 1474.
375 X. X. Xu, J. Zhuang and X. Wang, *J. Am. Chem. Soc.*, 2008, **130**, 12527–12535.
376 T. Hyeon, Y. Chung, J. Park, S. S. Lee, Y.-W. Kim and B. H. Park, *J. Phys. Chem. B*, 2002, **106**, 6831–6833.
377 E. Kang, J. Park, Y. Hwang, M. Kang, J.-G. Park and T. Hyeon, *J. Phys. Chem. B*, 2004, **108**, 13932–13935.
378 N. Bao, L. Shen, Y. Wang, P. Padhan and A. Gupta, *J. Am. Chem. Soc.*, 2007, **129**, 12374–12375.
379 N. Bao, L. Shen, W. An, P. Padhan, C. Heath Turner and A. Gupta, *Chem. Mater.*, 2009, **21**, 3458–3468.
380 K. E. Mooney, J. A. Nelson and M. J. Wagner, *Chem. Mater.*, 2004, **16**, 3155–3161.
381 G. Leem, A. C. Jamison, S. Zhang, D. Litvinov and T. R. Lee, *Chem. Commun.*, 2008, 4989–4991.
382 Z. X. Tang, C. M. Sorensen, K. J. Klabunde and G. C. Hadjipanayis, *J. Colloid Interface Sci.*, 1991, **146**, 38–52.
383 Q. Song and Z. J. Zhang, *J. Am. Chem. Soc.*, 2004, **126**, 6164–6168.
384 C. Altavilla, E. Ciliberto, A. Aiello, C. Sangregorio and D. Gatteschi, *Chem. Mater.*, 2007, **19**, 5980–5985.
385 S. O'Brien, L. Brus and C. B. Murray, *J. Am. Chem. Soc.*, 2001, **123**, 12085–12086.
386 M. Niederberger, G. Garnweitner, N. Pinna and M. Antonietti, *J. Am. Chem. Soc.*, 2004, **126**, 9120–9126.
387 X. Wang, J. Zhuang, Q. Peng and Y. Li, *Nature*, 2005, **437**, 121–124.
388 H.C. Du, S. Wohlrab, M. Weiss and S. Kaskel, *J. Mater. Chem.*, 2007, **17**, 4605–4610.
389 S. Adireddy, C. Lin, B. Cao, W. Zhou and G. Caruntu, *Chem. Mater.*, 2010, **22**, 1946–1948.
390 M. L. Moreira, G. P. Mambrini, D. P. Volanti, E. R. Leite, M. O. Orlandi, P. S. Pizani, V. R. Mastelaro, C. O. Paiva-Santos, E. Longo and J. A. Varela, *Chem. Mater.*, 2008, **20**, 5381–5387.
391 D. Wang, Z. Guo, Y. Chen, J. Hao and W. Liu, *Inorg. Chem.*, 2007, **46**, 7707–7709.
392 G. Wang, P. M. Rævik, A. T. J. van Helvoort, R. Holmestad, T. Grande and M.-A. Einarsrud, *Chem. Mater.*, 2007, **19**, 2213–2221.
393 J. Wang, X. Pang, M. Akinc and Z. Lin, *J. Mater. Chem.*, 2010, **20**, 5945–5949.
394 S. Xuan, F. Wang, Y.-X. J. Wang, J. C. Yu and K. C.-F. Leung, *J. Mater. Chem.*, 2010, **20**, 5086–5094.
395 M. Siemons, A. Leifert and U. Simon, *Adv. Funct. Mater.*, 2007, **17**, 2189–2197.
396 H. Xu, X. Hu and L. Zhang, *Cryst. Growth Des.*, 2008, **8**, 2061–2065.
397 T.-J. Park, G. C. Papaefthymiou, A. J. Viescas, A. R. Moodenbaugh and S. S. Wong, *Nano Lett.*, 2007, **7**, 766–772.
398 S. Li, Y.-H. Lin, B.-P. Zhang, Y. Wang and C.-W. Nan, *J. Phys. Chem. C*, 2010, **114**, 2903–2908.
399 J. Liu and Y. Li, *J. Mater. Chem.*, 2007, **17**, 1797–1803.
400 F. Wang, X. Xue and X. Liu, *Angew. Chem., Int. Ed.*, 2008, **47**, 906–909.
401 Z. Hou, P. Yang, C. Li, L. Wang, H. Lian, Z. Quan and J. Lin, *Chem. Mater.*, 2008, **20**, 6686–6696.
402 C. Peng, L. Gao, S. Yang and J. Sun, *Chem. Commun.*, 2008, 5601–5603.
403 Y. Ding, Y. Wan, Y.-L. Min, W. Zhang and S.-H. Yu, *Inorg. Chem.*, 2008, **47**, 7813–7823.
404 F. o. H. Julien and A. Alexandrou, *Science*, 1998, **282**, 1429–1430.
405 C. B. Murray, D. J. Norris and M. G. Bawendi, *J. Am. Chem. Soc.*, 1993, **115**, 8706–8715.
406 B. O. Dabbousi, J. Rodriguez-Viejo, F. V. Mikulec, J. R. Heine, H. Mattoussi, R. Ober, K. F. Jensen and M. G. Bawendi, *J. Phys. Chem. B*, 1997, **101**, 9463–9475.
407 M. A. Hines and P. Guyot-Sionnest, *J. Phys. Chem.*, 1996, **100**, 468–471.
408 U. K. Gautam, M. Rajamathi, F. Meldrum, P. Morgan and R. Seshadri, *Chem. Commun.*, 2001, 629.
409 U. K. Gautam, R. Seshadri and C. N. R. Rao, *Chem. Phys. Lett.*, 2003, **375**, 560–564.
410 B. Ludolph, M. A. Malik, P. O'Brien and N. Revaprasadu, *Chem. Commun.*, 1998, 1849.
411 N. Pradhan and S. Efrima, *J. Am. Chem. Soc.*, 2003, **125**, 2050–2051.
412 L. Manna, D. J. Milliron, A. Meisel, E. C. Scher and A. P. Alivisatos, *Nat. Mater.*, 2003, **2**, 382–385.
413 J. E. Halpert, V. J. Porter, J. P. Zimmer and M. G. Bawendi, *J. Am. Chem. Soc.*, 2006, **128**, 12590–12591.
414 A. G. Kanaras, C. Sönnichsen, H. Liu and A. P. Alivisatos, *Nano Lett.*, 2005, **5**, 2164–2167.
415 N. Varghese, K. Biswas and C. N. R. Rao, *Chem.–Asian J.*, 2008, **3**, 1435–1442.
416 M. Saruyama, M. Kanehara and T. Teranishi, *J. Am. Chem. Soc.*, 2010, **132**, 3280–3282.
417 N. Ma, E. H. Sargent and S. O. Kelley, *Nat. Nanotechnol.*, 2009, **4**, 121–125.
418 S. Dayal, N. Kopidakis, D. C. Olson, D. S. Ginley and G. Rumbles, *J. Am. Chem. Soc.*, 2009, **131**, 17726–17727.
419 S.-H. Choi, K. An, E.-G. Kim, J. H. Yu, J. H. Kim and T. Hyeon, *Adv. Funct. Mater.*, 2009, **19**, 1645–1649.
420 A. Cabot, R. K. Smith, Y. Yin, H. Zheng, B. M. Reinhard, H. Liu and A. P. Alivisatos, *ACS Nano*, 2008, **2**, 1452–1458.
421 K. Biswas and C. N. R. Rao, *Chem.–Eur. J.*, 2007, **13**, 6123–6129.
422 D.-S. Wang, W. Zheng, C.-H. Hao, Q. Peng and Y.-D. Li, *Chem.–Eur. J.*, 2009, **15**, 1870–1875.
423 M. P. Moloney, Y. K. Gun'ko and J. M. Kelly, *Chem. Commun.*, 2007, 3900–3902.
424 M. Naito, K. Iwahori, A. Miura, M. Yamane and I. Yamashita, *Angew. Chem., Int. Ed.*, 2010, **49**, 7006–7009.
425 E. D. Spoerke and J. A. Voigt, *Adv. Funct. Mater.*, 2007, **17**, 2031–2037.
426 D. Heroux, A. Ponce, S. Cingarapu and K. J. Klabunde, *Adv. Funct. Mater.*, 2007, **17**, 3562–3568.
427 H. Ohde, M. Ohde, F. Bailey, H. Kim and C. M. Wai, *Nano Lett.*, 2002, **2**, 721–724.

428 J. L. Mohanan, I. U. Arachchige and S. L. Brock, *Science*, 2005, **307**, 397–400.
429 A. L. Rogach, L. Katsikas, A. Kornowski, D. S. Su, A. Eychmuller and H. Weller, *Ber. Bunsen-Ges. Phys. Chem. Chem. Phys.*, 1996, **100**, 1772–1778.
430 M. Green, H. Harwood, C. Barrowman, P. Rahman, A. Eggeman, F. Festry, P. Dobson and T. Ng, *J. Mater. Chem.*, 2007, **17**, 1989–1994.
431 H. Zhang, L. Wang, H. Xiong, L. Hu, B. Yang and W. Li, *Adv. Mater.*, 2003, **15**, 1712–1715.
432 L. Li, H. Qian and J. Ren, *Chem. Commun.*, 2005, 528–530.
433 A. Dubavik, V. Lesnyak, W. Thiessen, N. Gaponik, T. Wolff and A. Eychmüller, *J. Phys. Chem. C*, 2009, **113**, 4748–4750.
434 B. Xi, S. Xiong, D. Xu, J. Li, H. Zhou, J. Pan, J. Li and Y. Qian, *Chem.–Eur. J.*, 2008, **14**, 9786–9791.
435 S. H. Yu and M. Yoshimura, *Adv. Mater.*, 2002, **14**, 296–300.
436 Y. Zhao, Y. Zhang, H. Zhu, G. C. Hadjipanayis and J. Q. Xiao, *J. Am. Chem. Soc.*, 2004, **126**, 6874–6875.
437 L. Sha, H. Zhang and M. T. Swihart, *Nanotechnology*, 2009, **20**, 235603.
438 F. Dawood and R. E. Schaak, *J. Am. Chem. Soc.*, 2008, **131**, 424–425.
439 H. Zhou, T. Fan, D. Zhang, Q. Guo and H. Ogawa, *Chem. Mater.*, 2007, **19**, 2144–2146.
440 Q. Lu, F. Gao and D. Zhao, *Nano Lett.*, 2002, **2**, 725–728.
441 P. Zhang and L. Gao, *J. Mater. Chem.*, 2003, **13**, 2007–2010.
442 M. B. Sigman, A. Ghezelbash, T. Hanrath, A. E. Saunders, F. Lee and B. A. Korgel, *J. Am. Chem. Soc.*, 2003, **125**, 16050–16057.
443 Y. Wu, C. Wadia, W. Ma, B. Sadtler and A. P. Alivisatos, *Nano Lett.*, 2008, **8**, 2551–2555.
444 W. Wang, P. Yan, F. Liu, Y. Xie, Y. Geng and Y. Qian, *J. Mater. Chem.*, 1998, **8**, 2321–2322.
445 Y. Xie, X. Zheng, X. Jiang, J. Lu and L. Zhu, *Inorg. Chem.*, 2001, **41**, 387–392.
446 S. Deka, A. Genovese, Y. Zhang, K. Miszta, G. Bertoni, R. Krahne, C. Giannini and L. Manna, *J. Am. Chem. Soc.*, 2010, **132**, 8912–8914.
447 S. G. Hickey, C. Waurisch, B. Rellinghaus and A. Eychmuller, *J. Am. Chem. Soc.*, 2008, **130**, 14978–14980.
448 Y. Xu, N. Al-Salim, C. W. Bumby and R. D. Tilley, *J. Am. Chem. Soc.*, 2009, **131**, 15990–15991.
449 J. Ning, K. Men, G. Xiao, L. Wang, Q. Dai, B. Zou, B. Liu and G. Zou, *Nanoscale*, 2010, **2**, 1699–1703.
450 J.-w. Seo, J.-t. Jang, S.-w. Park, C. Kim, B. Park and J. Cheon, *Adv. Mater.*, 2008, **20**, 4269–4273.
451 A. Yella, E. Mugnaioli, H. A. Therese, M. Panthoİfer, U. Kolb and W. Tremel, *Chem. Mater.*, 2009, **21**, 2474–2481.
452 M. A. Franzman, C. W. Schlenker, M. E. Thompson and R. L. Brutchey, *J. Am. Chem. Soc.*, 2010, **132**, 4060–4061.
453 W. J. Baumgardner, J. J. Choi, Y.-F. Lim and T. Hanrath, *J. Am. Chem. Soc.*, 2010, **132**, 9519–9521.
454 C.-L. Kuo and M. H. Huang, *J. Phys. Chem. C*, 2008, **112**, 11661–11666.
455 K. Jang, S. Y. Kim, K. H. Park, E. Jang, S. Jun and S. U. Son, *Chem. Commun.*, 2007, 4474–4476.
456 V. Parashar, S. K. Pandey and A. C. Pandey, *Chem. Commun.*, 2010, **46**, 3143–3145.
457 A. Yella, H. A. Therese, N. Zink, M. Panthoİfer and W. Tremel, *Chem. Mater.*, 2008, **20**, 3587–3593.
458 A. Puglisi, S. Mondini, S. Cenedese, A. M. Ferretti, N. Santo and A. Ponti, *Chem. Mater.*, 2010, **22**, 2804–2813.
459 X. Chen and R. Fan, *Chem. Mater.*, 2001, **13**, 802–805.
460 W. Shi, L. Zhou, S. Song, J. Yang and H. Zhang, *Adv. Mater.*, 2008, **20**, 1892–1897.
461 Y. Xie, Y. Qian, W. Wang, S. Zhang and Y. Zhang, *Science*, 1996, **272**, 1926–1927.
462 A. Manz, A. Birkner, M. Kolbe and R. A. Fischer, *Adv. Mater.*, 2000, **12**, 569–573.
463 U. K. Gautam, K. Sardar, F. L. Deepak and C. N. R. Rao, *Pramana*, 2005, **65**, 549–564.
464 O. I. Micic, S. P. Ahrenkiel, D. Bertram and A. J. Nozik, *Appl. Phys. Lett.*, 1999, **75**, 478–480.
465 J. L. Coffer, M. A. Johnson, L. Zhang, R. L. Wells and J. F. Janik, *Chem. Mater.*, 1997, **9**, 2671–2673.
466 A. C. Frank, F. Stowasser, H. Sussek, H. Pritzkow, C. R. Miskys, O. Ambacher, M. Giersig and R. A. Fischer, *J. Am. Chem. Soc.*, 1998, **120**, 3512–3513.
467 J. Choi and E. G. Gillan, *J. Mater. Chem.*, 2006, **16**, 3774–3784.
468 J. P. Xiao, Y. Xie, R. Tang and W. Luo, *Inorg. Chem.*, 2003, **42**, 107–111.
469 W. Lei, D. Liu, J. Zhang, P. Zhu, Q. Cui and G. Zou, *Cryst. Growth Des.*, 2009, **9**, 1489–1493.
470 S. Bhaviripudi, J. Qi, E. L. Hu and A. M. Belcher, *Nano Lett.*, 2007, **7**, 3512–3517.
471 A. Gomathi and C. N. R. Rao, *Mater. Res. Bull.*, 2006, **41**, 941–947.
472 A. Nag, K. Raidongia, K. P. S. S. Hembram, R. Datta, U. V. Waghmare and C. N. R. Rao, *ACS Nano*, 2010, **4**, 1539–1544.
473 C. N. R. Rao and A. Nag, *Eur. J. Inorg. Chem.*, 2010, **2010**, 4244–4250.
474 D. A. Lindquist, T. T. Kodas, D. M. Smith, X. M. Xiu, S. L. Hietala and R. T. Paine, *J. Am. Chem. Soc.*, 1991, **74**, 3126–3128.
475 C. Tang, Y. Bando, Y. Huang, C. Zhi and D. Golberg, *Adv. Funct. Mater.*, 2008, **18**, 3653–3661.
476 K. Raidongia, A. Gomathi and C. N. R. Rao, *Isr. J. Chem.*, 2010, **50**, 399–404.
477 M. Lei, H. Z. Zhao, H. Yang, B. Song, P. G. Li and W. H. Tang, *Mater. Lett.*, 2008, **62**, 1539–1542.
478 F. Xu, Y. Xie, X. Zhang, S. Zhang, X. Liu and X. Tian, *Inorg. Chem.*, 2003, **43**, 822–829.
479 J. Buha, I. Djerdj, M. Antonietti and M. Niederberger, *Chem. Mater.*, 2007, **19**, 3499–3505.
480 C. Giordano, C. Erpen, W. Yao and M. Antonietti, *Nano Lett.*, 2008, **8**, 4659–4663.
481 C. Giordano, C. Erpen, W. Yao, B. Milke and M. Antonietti, *Chem. Mater.*, 2009, **21**, 5136–5144.
482 G. Qixun, X. Yi, W. Xinjun, L. Shichang, H. Tao and B. Chuannan, *J. Am. Ceram. Soc.*, 2005, **88**, 249–251.
483 X. Feng, Y.-J. Bai, B. Lu, C.-G. Wang, Y.-X. Qi, Y.-X. Liu, G.-L. Geng and Li, *Inorg. Chem.*, 2004, **43**, 3558–3560.
484 L. Shi, Y. Gu, L. Chen, Z. Yang, J. Ma and Y. Qian, *J. Nanosci. Nanotechnol.*, 2005, **5**, 296–299.
485 J. Ma, Y. Du and Y. Qian, *J. Alloys Compd.*, 2005, **389**, 296–298.
486 W. Lengauer, *J. Cryst. Growth*, 1988, **87**, 295–298.
487 C. L. Bull, P. F. McMillan, E. Soignard and K. Leinenweber, *J. Solid State Chem.*, 2004, **177**, 1488.
488 N. S. Gajbhiye and R. S. Ningthoujam, *Phys. Status Solidi C*, 2004, **1**, 3449–3454.
489 A. Gomathi, A. Sundaresan and C. N. R. Rao, *J. Solid State Chem.*, 2007, **180**, 291–295.
490 X. F. Qian, Y. Xie, Y. T. Qian, X. M. Zhang, W. Z. Wang and L. Yang, *Mater. Sci. Eng., B-Solid State Mater. Adv. Technol.*, 1997, **49**, 135–137.
491 X. F. Qian, X. M. Zhang, C. Wang, W. Z. Wang and Y. T. Qian, *Mater. Res. Bull.*, 1998, **33**, 669–672.
492 G. Yunle, G. Fan, Q. Yitai, Z. Huagui and Y. Ziping, *Mater. Res. Bull.*, 2002, **37**, 1101–1105.
493 S. C. Perera, P. S. Fodor, G. M. Tsoi, L. E. Wenger and S. L. Brock, *Chem. Mater.*, 2003, **15**, 4034–4038.
494 S. L. Brock and K. Senevirathne, *J. Solid State Chem.*, 2008, **181**, 1552–1559.
495 S. L. Brock, S. C. Perera and K. L. Stamm, *Chem.–Eur. J.*, 2004, **10**, 3364–3371.
496 Y. Xie, H. L. Su, X. F. Qian, X. M. Liu and Y. T. Qian, *J. Solid State Chem.*, 2000, **149**, 88–91.
497 H. L. Su, Y. Xie, B. Li, X. M. Liu and Y. T. Qian, *Solid State Ionics*, 1999, **122**, 157–160.
498 A. E. Henkes, Y. Vasquez and R. E. Schaak, *J. Am. Chem. Soc.*, 2007, **129**, 1896–1897.
499 A. E. Henkes and R. E. Schaak, *Chem. Mater.*, 2007, **19**, 4234–4242.
500 J. Liu, X. Chen, M. Shao, C. An, W. Yu and Y. Qian, *J. Cryst. Growth*, 2003, **252**, 297–301.
501 S. Yang, C. Liang and R. Prins, *J. Catal.*, 2006, **241**, 465–469.
502 J. Park, B. Koo, K. Y. Yoon, Y. Hwang, M. Kang, J.-G. Park and T. Hyeon, *J. Am. Chem. Soc.*, 2005, **127**, 8433–8440.
503 R.-K. Chiang and R.-T. Chiang, *Inorg. Chem.*, 2006, **46**, 369–371.
504 G. Shi and J. Shen, *J. Mater. Chem.*, 2009, **19**, 2295–2297.
505 O. I. Micic, C. J. Curtis, K. M. Jones, J. R. Sprague and A. J. Nozik, *J. Phys. Chem.*, 1994, **98**, 4966–4969.
506 O. I. Micic, J. R. Sprague, C. J. Curtis, K. M. Jones, J. L. Machol, A. J. Nozik, H. Giessen, B. Fluegel, G. Mohs and N. Peyghambarian, *J. Phys. Chem.*, 1995, **99**, 7754–7759.
507 D. Battaglia and X. Peng, *Nano Lett.*, 2002, **2**, 1027–1030.
508 S. Xu, S. Kumar and T. Nann, *J. Am. Chem. Soc.*, 2006, **128**, 1054–1055.

509 K. Senevirathne, R. Tackett, P. R. Kharel, G. Lawes, K. Somaskandan and S. L. Brock, *ACS Nano*, 2009, **3**, 1129–1138.
510 A. Sánchez-Iglesias, E. Carbó-Argibay, A. Glaria, B. Rodríguez-González, J. Pérez-Juste, I. Pastoriza-Santos and L. M. Liz-Marzán, *Chem.–Eur. J.*, 2010, **16**, 5558–5563.
511 F.-R. Fan, D.-Y. Liu, Y.-F. Wu, S. Duan, Z.-X. Xie, Z.-Y. Jiang and Z.-Q. Tian, *J. Am. Chem. Soc.*, 2008, **130**, 6949–6951.
512 L. Feng, X. Wu, L. Ren, Y. Xiang, W. He, K. Zhang, W. Zhou and S. Xie, *Chem.–Eur. J.*, 2008, **14**, 9764–9771.
513 C. S. Levin, C. Hofmann, T. A. Ali, A. T. Kelly, E. Morosan, P. Nordlander, K. H. Whitmire and N. J. Halas, *ACS Nano*, 2009, **3**, 1379–1388.
514 J. Zhang, Y. Tang, K. Lee and M. Ouyang, *Science*, 2010, **327**, 1634–1638.
515 B. Lim, J. Wang, P. H. C. Camargo, M. Jiang, M. J. Kim and Y. Xia, *Nano Lett.*, 2008, **8**, 2535–2540.
516 W. Zhou, K. Zheng, L. He, R. Wang, L. Guo, C. Chen, X. Han and Z. Zhang, *Nano Lett.*, 2008, **8**, 1147–1152.
517 M. A. Nash, J. J. Lai, A. S. Hoffman, P. Yager and P. S. Stayton, *Nano Lett.*, 2009, **10**, 85–91.
518 S. Alayoglu and B. Eichhorn, *J. Am. Chem. Soc.*, 2008, **130**, 17479–17486.
519 B.-J. Hwang, L. S. Sarma, C.-H. Chen, C. Bock, F.-J. Lai, S.-H. Chang, S.-C. Yen, D.-G. Liu, H.-S. Sheu and J.-F. Lee, *J. Phys. Chem. C*, 2008, **112**, 19922–19929.
520 D. Ciuculescu, C. Amiens, M. Respaud, A. Falqui, P. Lecante, R. E. Benfield, L. Jiang, K. Fauth and B. Chaudret, *Chem. Mater.*, 2007, **19**, 4624–4626.
521 S. Ghosh, K. Biswas and C. N. R. Rao, *J. Mater. Chem.*, 2007, **17**, 2412–2417.
522 A. Radi, D. Pradhan, Y. Sohn and K. T. Leung, *ACS Nano*, 2010, **4**, 1553–1560.
523 X. Peng, M. C. Schlamp, A. V. Kadavanich and A. P. Alivisatos, *J. Am. Chem. Soc.*, 1997, **119**, 7019–7029.
524 A. Nemchinov, M. Kirsanova, N. N. Hewa-Kasakarage and M. Zamkov, *J. Phys. Chem. C*, 2008, **112**, 9301–9307.
525 E. Lifshitz, M. Brumer, A. Kigel, A. Sashchiuk, M. Bashouti, M. Sirota, E. Galun, Z. Burshtein, A. Q. Le Quang, I. Ledoux-Rak and J. Zyss, *J. Phys. Chem. B*, 2006, **110**, 25356–25365.
526 L. Li and P. Reiss, *J. Am. Chem. Soc.*, 2008, **130**, 11588–11589.
527 R. Xie, X. Zhong and T. Basché, *Adv. Mater.*, 2005, **17**, 2741–2745.
528 Z.-Q. Tian, Z.-L. Zhang, P. Jiang, M.-X. Zhang, H.-Y. Xie and D.-W. Pang, *Chem. Mater.*, 2009, **21**, 3039–3041.
529 W. Shi, H. Zeng, Y. Sahoo, T. Y. Ohulchanskyy, Y. Ding, Z. L. Wang, M. Swihart and P. N. Prasad, *Nano Lett.*, 2006, **6**, 875–881.
530 J.-S. Lee, M. I. Bodnarchuk, E. V. Shevchenko and D. V. Talapin, *J. Am. Chem. Soc.*, 2010, **132**, 6382–6391.
531 X. Liu, Q. Hu, X. Zhang, Z. Fang and Q. Wang, *J. Phys. Chem. C*, 2008, **112**, 12728–12735.
532 C. Wang, H. Yin, S. Dai and S. Sun, *Chem. Mater.*, 2010, **22**, 3277–3282.
533 W. Han, L. Yi, N. Zhao, A. Tang, M. Gao and Z. Tang, *J. Am. Chem. Soc.*, 2008, **130**, 13152–13161.
534 X. Liang, X. Wang, Y. Zhuang, B. Xu, S. Kuang and Y. Li, *J. Am. Chem. Soc.*, 2008, **130**, 2736–2737.
535 M. Casavola, A. Falqui, M. A. Garcila, M. Garcila-Hernandez, C. Giannini, R. Cingolani and P. D. Cozzoli, *Nano Lett.*, 2008, **9**, 366–376.
536 K. Raidongia, A. Nag, A. Sundaresan and C. N. R. Rao, *Appl. Phys. Lett.*, 2010, **97**, 062904.

ACCOUNTS
of chemical research

# The Liquid–Liquid Interface as a Medium To Generate Nanocrystalline Films of Inorganic Materials

C. N. R. RAO* AND K. P. KALYANIKUTTY

*Chemistry and Physics of Materials Unit, DST Nanoscience Unit, and CSIR Centre of Excellence in Chemistry, Jawaharlal Nehru Centre for Advanced Scientific Research, Jakkur P.O., Bangalore 560 064, India*

RECEIVED ON AUGUST 31, 2007

## CONSPECTUS

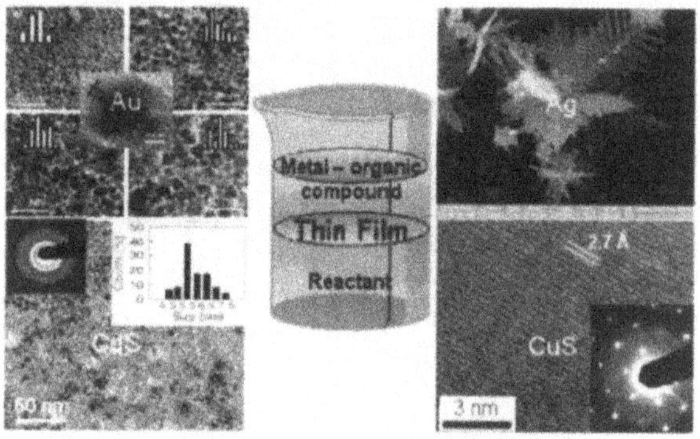


Unlike the air–water interface, the organic–aqueous (liquid–liquid) interface has not been exploited sufficiently for materials synthesis. In this Account, we demonstrate how ultrathin nanocrystalline films of metals such as gold and silver as well as of inorganic materials such as semiconducting metal chalcogenides (e.g., CdS, CuS, CdSe) and oxides are readily generated at the liquid–liquid interface. What is particularly noteworthy is that single-crystalline films of certain metal chalcogenides are also obtained by this method. The as-prepared gold films at the toluene–water interface comprise fairly monodisperse nanocrystals that are closely packed, the nature and properties of the films being influenced by various reaction parameters such as reaction temperature, time, reactant concentrations, mechanical vibrations, and the viscosity of the medium. The surface plasmon band of gold is markedly red-shifted in the films due to electronic coupling between the particles. The shift of the surface plasmon band of the Au film toward higher wavelengths with an accompanying increase in intensity as a function of reaction time marks the growth of the film. Depending on the reaction temperature, the Au films show interesting electrical transport properties. Films of metals such as gold are disintegrated by the addition of alkanethiols, the effectiveness depending on the alkane chain length, clearly evidenced by shifts of the surface plasmon bands. A time evolution study of the polycrystalline Au and CdS films as well as the single-crystalline CuS films is carried out by employing atomic force microscopy. X-ray reflectivity studies reveal the formation of a monolayer of capped clusters having 13 gold atoms each, arranged in a hexagonal manner at the toluene–water interface. The measurements also reveal an extremely small value of the interfacial tension. Besides describing features of such nanocrystalline films and their mode of formation, their rheological properties have been examined. Interfacial rheological studies show that the nanocrystalline film of Ag nanoparticles, the single-crystalline CuS film, and the multilayered CdS film exhibit a viscoelastic behavior strongly reminiscent of soft-glassy systems. Though both CuS and CdS films exhibit a finite yield stress under steady shear, the CdS films are found to rupture at high shear rates. An important advantage of the study of materials formed at the liquid–liquid interface is that it provides a means to investigate the interface itself. In addition, it enables one to obtain substrate-free single-crystalline films of materials.

Published on the Web 03/12/2008 www.pubs.acs.org/acr
10.1021/ar700192d CCC: $40.75 

## Introduction

Self-assembly, the process of formation of ordered aggregates, is a powerful strategy for creating novel structures of great academic interest as well as of technological value. Interfaces are an important means to generate two-dimensional self-assemblies of nanocrystals, providing a constrained environment for organized assembly. The air–water interface has been exploited for the preparation of films of metals and semiconductors, which have potential applications in nanodevices. For example, nanocrystal assemblies of gold have been prepared at the air–water interface by employing Langmuir–Blodgett (LB) technique.[1,2] Using a chloroform solution of $Au_{55}(PPh_3)_{12}Cl_6$, Schmid et al.[3] have self-assembled $Au_{55}$ nanocrystals into monolayers in a LB trough. Somorjai and co-workers[4] have employed a LB trough to fabricate monolayers of monodisperse Rh nanocrystals on Si wafers as model 2D catalysts. Catalysts for the oxidation of formic acid have been prepared in the form of Langmuir layers of $Fe_{20}Pt_{80}$ nanoparticles.[5] Bawendi and co-workers[6] have prepared monolayers of monodisperse trioctylphosphine oxide capped CdSe quantum dots by the LB method. By exposing an LB film of lead stearate to $H_2S$, nanoparticles of PbS have been generated in the form of films.[7] LB films of iron oxide nanoparticles have been obtained by spreading a hexane suspension of oleic acid capped $\gamma$-$Fe_2O_3$ nanoparticles at the air–water interface.[8] Properties of thin films obtained by the LB technique are determined by the nature of the substrate as well as the reaction conditions, and the films are generally polycrystalline.

Unlike the air–water interface, the liquid–liquid (organic–aqueous) interface has not been investigated sufficiently, and it is only recently that there have been concerted efforts to understand the structure of the liquid–liquid interface. The liquid–liquid surface possesses unique thermodynamic properties such as viscosity and density. A liquid–liquid interface is a nonhomogeneous region having a thickness on the order of a few nanometers. The interface is not sharp, since there is always a little solubility of one phase in the other. One of the problems that has been studied in detail at the liquid–liquid interface relates to interfacial charge transfer reactions and dynamics.[9] Distribution of ions and solvent molecules, which determines the structure of a liquid–liquid interface, has recently been investigated by Schlossman and co-workers.[10] Using X-ray reflectivity and molecular dynamics simulations, these workers find that ion sizes and ion–solvent interactions affect the ion distributions near the interface. The relevance of the liquid–liquid interface has been noted in some other areas such as environmental chemistry, cell biology, and catalysis. The interface between two immiscible liquids offers an important alternative path for the self-assembly and chemical manipulation of nanocrystals.[11] Nanoparticles are highly mobile at the interface and rapidly achieve an equilibrium assembly by reduction in interfacial energy. The three parameters that have been found to influence the energy of the assembly process at the liquid–liquid interface are (i) the nature of the interface, (ii) surface modification of the nanoparticles at the interface, and (iii) the effective radius of the nanoparticles, smaller nanoparticles adsorbing more weakly to the interface than larger ones. Binks and Clint[12] have theoretically treated the wetting of silica particles in terms of the surface energies at the oil–water interface to interpret the interactions between the solid and the liquid phases and to predict the oil–water contact angles for a solid of given hydrophobicity. Only if the contact angle is exactly 90° will the particle be located at the middle of the oil–water interface.

Russel and co-workers[13] have investigated the assembly of phosphine oxide functionalized CdSe nanoparticles of two different diameters by competitive adsorption at the toluene–water interface by employing fluorescence spectroscopy. Benkoski et al.[14] have developed the so-called fossilized liquid assembly for the creation of 2-D assemblies from nanoscale building blocks. By investigating the interactions of a variety of particles such as uncharged, charged, functionalized, and nonfunctionalized, deposited at dodecanediol dimethacrylate/water interface, they have shown that nanoparticles aggregate into a wide variety of complex morphologies. These results provide evidence for the importance of asymmetric dipole interaction in generating the complex morphologies. There are a few assorted reports in the literature where the liquid–liquid interface or a mixture of immiscible liquids has been used for the synthesis or crystallization of nanostructures and other materials. The Brust method,[15] which has been widely employed for the preparation of Au nanocrystals by the reduction of $AuCl_4^-$ by $NaBH_4$ in the presence of an alkanethiol, is carried out in a water–toluene mixture in the presence of a phase-transfer reagent such as tetraoctylammonium bromide. Stucky and co-workers[16] have prepared mesoporous fibers of silica by treating the silica precursor dissolved in an organic phase such as hexane, toluene, or $CCl_4$ with surfactant molecules dissolved in the aqueous phase. CdS nanoparticles in the form of LB films have been prepared by reacting an aqueous $CdCO_3$ solution with $CS_2$ in $CCl_4$.[17] Monodisperse, luminescent nanocrystals of CdS have been prepared by mixing a solution of cadmium–myristic acid and *n*-triphenylphosphine oxide in toluene with an aqueous solution of thiourea, followed by heating under stirring.[18] Langmuir films of silver nanoparticles have been prepared at the water–dichloromethane interface.[19] With Pickering emulsions as a template, dodecanethiol-capped Ag nanoparticles have been self-assembled at the trichloroethylene–water interface.[20] Song et al.[21] have employed the butanol–water interface for the crystallization of Se nanorods. Amorphous Se nanoparticles were first prepared in the aqueous medium and then transported to the butanol–water interface using polyvinylpyrrolidone (PVP), to obtain crystalline nanorods of Se. Cheetham and co-workers[22] have employed a cyclohexanol–water mixture to prepare single crystals of copper adipate. Although the assembly of pregenerated nanoparticles at liquid–liquid interfaces has been examined to some extent, this interface has not been exploited for the synthesis of nanoparticles and their assemblies.

In this Account, we describe a simple but elegant means of generating ultrathin nanocrystalline films of various materials at the organic–aqueous interface. The method primarily involves taking a metal organic compound in the organic layer and a reducing, a sulfiding, or an oxidizing agent in the aqueous layer. The reaction occurs at the interface giving rise to a film at the interface with several interesting features. We describe nanocrystalline films of gold in detail to show how the various reaction parameters affect the films formed at the interface and how alkanethiols bring about the disintegration of the films. We also demonstrate the power of the method in generating ultrathin polycrystalline, as well as single-crystalline, films of some metal chalcogenides at the interface. The formation of single-crystalline films is indeed a noteworthy feature. We believe that this method can be adopted not only for generating nanocrystalline films of various materials but also to study processes occurring at the liquid–liquid interface.

## General Experimental Procedure

Use of the liquid–liquid interface for preparing materials in the form of nanocrystalline films is simple and straightforward.[1a,23] The procedure to prepare nanocrystalline films of metals involves taking a metal organic precursor in the organic layer and then injecting an appropriate reducing agent into the aqueous layer to obtain a film comprising metal nanocrystals. In order to prepare nanocrystalline films of metal sulfides, $Na_2S$ is used as the source of sulfur, while $Na_2Se$ or *N,N*-dimethylselenourea is used as the source of selenium to prepare films of metal selenides.

We shall illustrate the method of preparation of nanocrystalline films with two examples. To prepare nanocrystalline

**FIGURE 1.** Nanocrystalline films of (a) Au and (b) CdS formed at the toluene–water interface.

gold films, $Au(PPh_3)Cl$ (Ph = phenyl) is found to be a good precursor. In a typical preparation, 10 mL of a 1.5 mM solution of $Au(PPh_3)Cl$ in toluene is allowed to stand in contact with 16 mL of a 3.25 mM solution of NaOH in water in a 100 mL beaker at 300 K. Tetrakishydroxymethylphosphonium chloride (THPC; 300 $\mu$L of 50 mM), which acts as the reducing agent,[23a] is slowly injected to the aqueous layer with minimal disturbance to the organic layer. A slight pink coloration of the interface indicates the onset of reduction of the gold salt. As the reaction proceeds, the color of the interface intensifies, finally resulting in a robust film at the interface as shown in Figure 1a. For preparing a nanocrystalline film of CdS, 0.0045 g of $Na_2S$ is dissolved in 30 mL of water (2 mM) in a 100 mL beaker, and 0.0125 g of cadmium cupferronate $[Cd(cup)_2]$ is dissolved in 30 mL of toluene (1 mM) by ultrasonication. A few drops of *n*-octylamine are added to the $Cd(cup)_2$ solution in order to make it completely soluble. The toluene solution is slowly added to the aqueous $Na_2S$ solution in a 100 mL beaker at 30 °C. The interface attains a yellow color within a few minutes, and a distinct film is formed after 10 h. In Figure 1b, we show a nanocrystalline film of CdS, so formed.

## Gold and Other Metals

We shall discuss ultrathin films of gold nanocrystals in some detail to illustrate the features of the method as well as the interesting features of the materials obtained by this method. As-prepared Au films, obtained at 30 °C after 24 h of the reaction at the toluene–water interface comprise fairly monodisperse nanoparticles with a mean diameter of ~7 nm (see the TEM image given in Figure 2a). The nanocrystals are closely packed with a typical interparticle distance of ~1 nm. Furthermore, they are single-crystalline as revealed by the high-resolution electron microscope (HREM) image given as an inset in Figure 2. The image shows the (111) planes of gold, sepa-

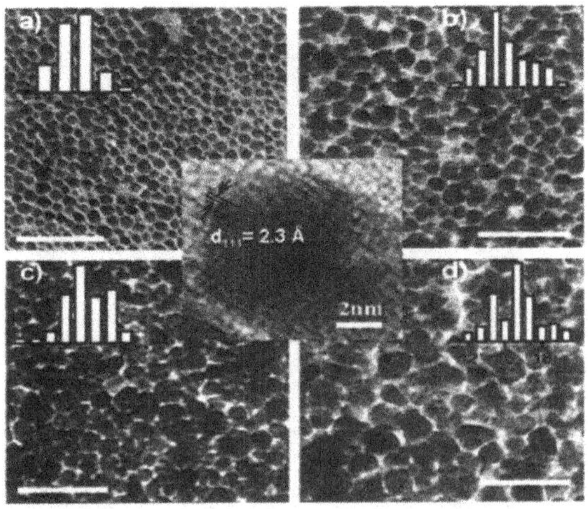


**FIGURE 2.** TEM images of the ultrathin nanocrystalline Au films obtained at the liquid–liquid interface after 24 h at (a) 30, (b) 45, (c) 60, and (d) 75 °C. Histograms of particle size distribution are shown as insets. The scale bars correspond to 50 nm. A high-resolution image of an individual particle is shown at the center. Reproduced from ref 24. Copyright 2005 American Chemical Society.

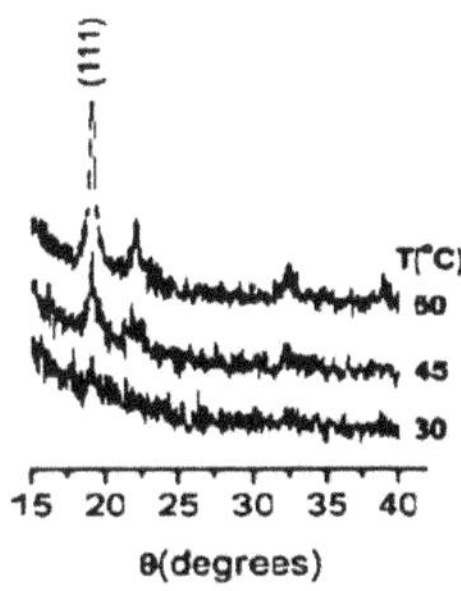


**FIGURE 3.** X-ray diffraction patterns of nanocrystalline Au films obtained at different temperatures. Reproduced from ref 24. Copyright 2005 American Chemical Society.

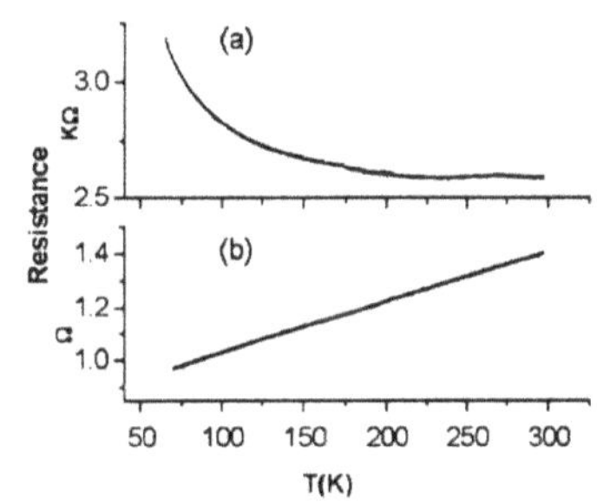


**FIGURE 4.** Temperature variation of the electrical resistance of the Au films prepared at (a) 45 and (b) 60 °C (current used, 10 mA). Modified from ref 24. Copyright 2005 American Chemical Society.

rated by a distance of ~2.3 Å. Such nanocrystalline films could also be obtained by using other solvent pairs such as $CCl_4$–water and butanol–water. Reaction parameters such as the temperature, reaction time, concentrations of the metal precursor and the reducing agent, and the viscosity of the aqueous layer affect the nature and properties of the nanocrystalline films.

The effect of temperature on the size distribution of the Au nanocrystals can be readily seen from the TEM images in Figure 2. The mean diameters of the nanocrystals formed at 30, 45, 60, and 75 °C are 7, 10, 12, and 15 nm, respectively, but the interparticle separation remains nearly the same at ~1 nm. X-ray diffraction measurements show that with increase in temperature, the crystallinity of the film increases (Figure 3). The films obtained at 45 and 60 °C exhibit prominent (111) peaks ($d$ = 2.33 Å), while those obtained at 30 °C show weak reflections, probably due to the small particle size. The growth of the (111) peak with temperature indicates an increase in the particle size.

Increasing the concentration of the metal precursor yields nanocrystalline films with a larger number of particles, but the size distribution is essentially unaffected. The thickness of the film also increases with the increase in the metal precursor concentration. The use of high concentrations of the reducing agent results in less uniform films with altered distributions in the nanoparticle diameter. A slight increase in the size of the Au nanoparticles was observed when the viscosity of the aqueous layer was increased by the addition of glycerol. Reactions at the interface carried out on a vibration-free table yielded nanocrystalline films with reduced roughness, comprising particles of smaller size.

The Au films formed at the interface show interesting electrical transport properties that are dependent on the reaction temperature (see Figure 4).[24] Four-probe electrical resistance measurements on the nanocrystalline films show a metal to insulator transition, metallic behavior being shown by the films formed at high temperatures (>45 °C). The films formed at lower temperatures (≤45 °C) show insulating behavior.

Atomic force microscopy (AFM) shows the thickness of the films to be in the 40–140 nm range. The contact mode AFM image in Figure 5a shows the boundary of a Au film on a mica substrate. The height profile of the film in Figure 5b gives an estimate of the thickness to be ~60 nm. AFM images covering a few micrometers yield a root-mean-square roughness in the range 30–35 nm.

The growth of Au films could be followed as a function of reaction time by UV–visible absorption spectroscopy (Figure 6). The Au plasmon band gets red-shifted due to the increase in the electronic coupling between the particles, accompanied by an increase in the intensity. The absorption band shows

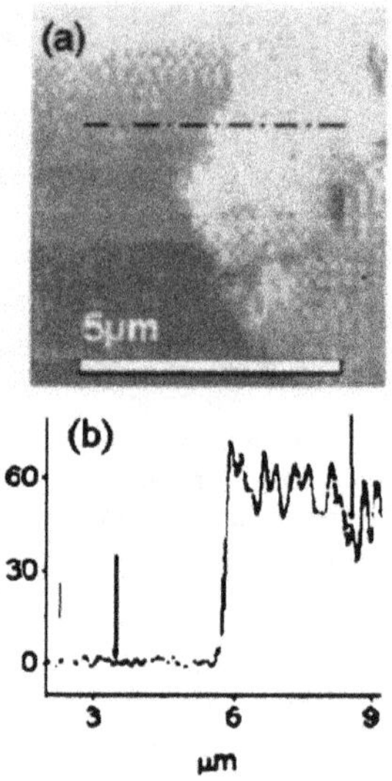


FIGURE 5. (a) Contact-mode AFM image showing the boundary of Au film on a mica substrate and (b) the *z*-profile. Reproduced from ref 24. Copyright 2005 American Chemical Society.

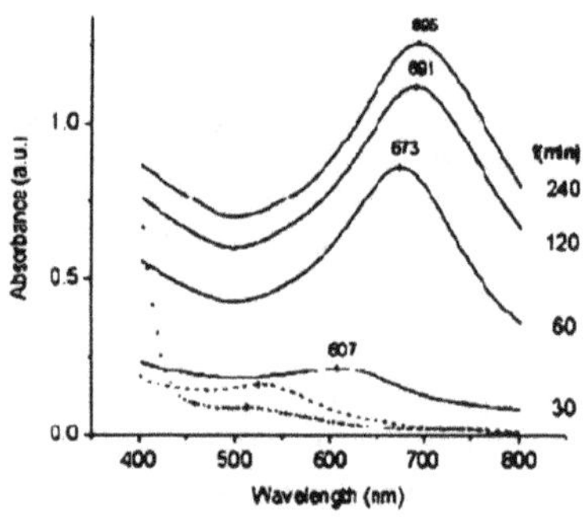


FIGURE 6. Evolution of the electronic absorption spectra with the growth of the nanocrystalline Au film at the interface at 45 °C (full curves) and spectra of octylamine-capped Au nanocrystals in toluene (broken line) and mercaptoundecanoic acid-capped nanocrystals in water (dotted line). Modified from ref 24. Copyright 2005 American Chemical Society.

negligible change after 120 min. At this stage, the films presumably consist of well-packed nanocrystals.

The effect of surfactants such as tetraoctylammonium bromide (TOAB) and cetyltrimethylammonium bromide (CTAB) on the nanostructures formed when the gold ions present in the organic phase are reduced at the interface by hydrazine (in the aqueous phase) has been investigated.[25] The surfactants give rise to extended fractal networks with a fractal dimension of 1.7 at the interface (Figure 7a). The fractals themselves comprise cauliflower-like spherical units (see top inset in Figure 7a) formed of pentagonal nanorods.

Till now, we have been examining the formation of uniform and robust nanocrystalline films of Au. The films, once

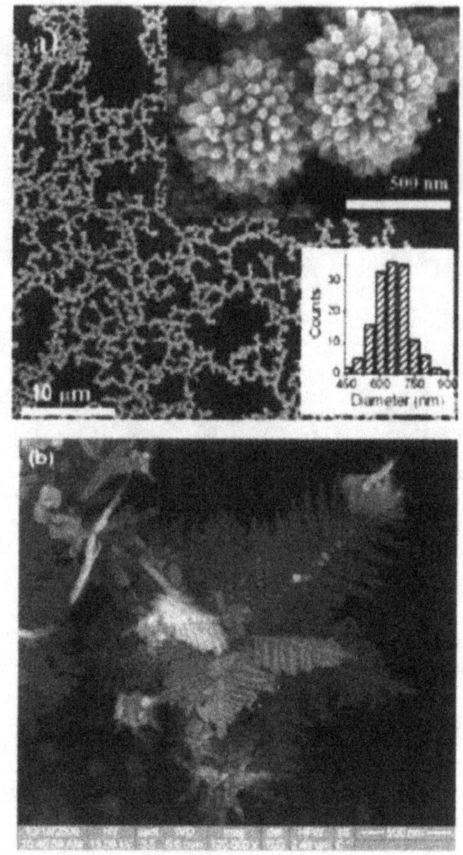


FIGURE 7. Panel a shows an SEM image of the fractal network formed by cauliflower-like gold structures by employing TOAB. The inset on the top right corner shows a high-resolution image of the cauliflower-like structures. The inset at the bottom shows the histogram of the size distribution of cauliflower-like structures. Panel b shows a dendritic nanostructure of Ag. Reproduced with permission from ref 25. Copyright 2008 Elsevier.

formed, can be disintegrated or disordered by the addition of *n*-alkanethiols.[24,26] Addition of alkanethiols is accompanied by a progressive blue shift of the plasmon absorption band of the film, suggesting that the electronic coupling between the nanocrystals gets reduced due to the increased separation between the nanocrystals. Figure 8a shows the variation in the shift of the absorption band brought about by the addition of hexadecanethiol. Interaction with thiols also brings about a change in the morphology of the film as shown in the adjoining AFM images. By varying the chain length of the alkanethiol, one is actually varying the distance between the nanoparticles, thereby giving rise to a blue shift proportional to the length of the alkane chain (Figure 8b). The rate of disordering or disintegration of the Au films is also affected by the chain length of the alkanethiols; the longer the chain

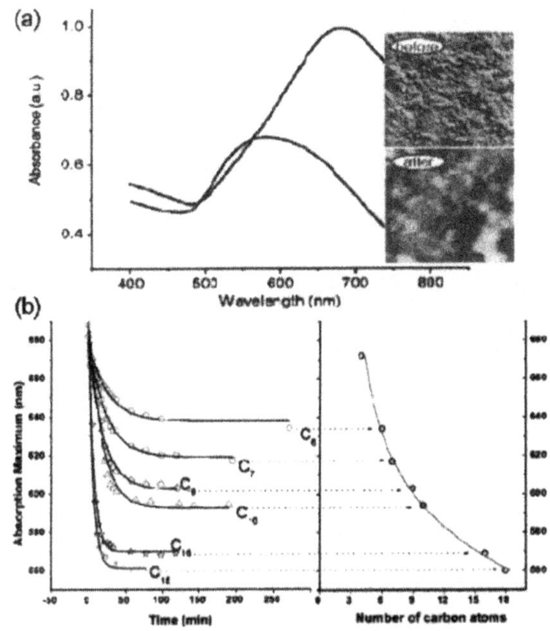


FIGURE 8. Panel a shows the electronic absorption spectra of a Au film before and after treatment with hexadecanethiol solution for 6 h. Tapping mode AFM images are shown alongside (scan area 500 nm × 500 nm). Reproduced from ref 24. Copyright 2005 American Chemical Society. (b) Time variation of the absorption maximum of Au nanocrystalline films on interaction with alkanethiols of different chain lengths (left) and variation in the limiting absorption maximum with chain length of the thiol adsorbed (right). Reproduced from ref 26. Copyright 2008 American Chemical Society.

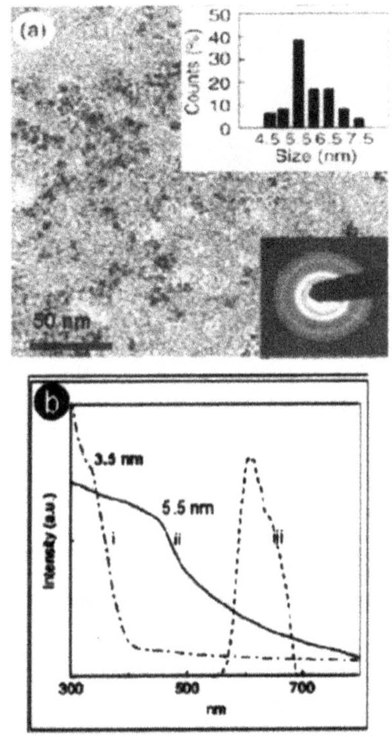


FIGURE 9. Panel a shows the TEM image of 5.5 nm CdS nanocrystals obtained at room temperature at the interface. Insets give the SAED pattern and the particle size distribution. Panel b shows the UV–visible absorption spectra of (i) 3.4 nm and (ii) 5.5 nm CdS nanocrystals and (iii) the photoluminescence spectrum of 5.5 nm CdS nanocrystals. Modified with permission from ref 28. Copyright 2003 Elsevier.

length, the faster is the rate. Calorimetric measurements also lend evidence for such a process.

It has also been possible to obtain nanocrystalline films of other metals such as Ag, Pd, and Cu at the toluene–water interface by taking $Ag_2(PPh_3)_4$, palladium acetate, and $Cu(PPh_3)Cl$, respectively, in the organic layer.[23] By carrying out the reaction in the presence of TOAB, we have obtained dendritic structures of Ag (Figure 7b).[25] By taking mixtures of the corresponding metal precursors in the organic layer, we have prepared nanocrystalline films of binary Au–Ag and Au–Cu alloys and ternary Au–Ag–Cu alloys.[27]

## Metal Chalcogenides

Ultrathin films of metal sulfides such as CdS, CuS, ZnS, and PbS are obtained at the organic–aqueous interface by the reaction of $Na_2S$ in the aqueous layer with the corresponding metal cupferronate in the organic layer.[28–30] We show typical results in the case of CdS. TEM images reveal the films to be composed of nanocrystals of 5.5 nm diameter (Figure 9a). The powder XRD pattern of the film shows that the CdS nanocrystals crystallize in the rock-salt structure rather than in the wurtzite structure. Increasing the reaction temperature and the concentration of the reactants yields bigger nanocrystals. When the viscosity of the aqueous medium is doubled by the addition of glycerol, the size of the nanocrystals is reduced to 3.5 nm. The UV–vis absorption spectrum of the nanocrystals (Figure 9b) shows a broad absorption maximum, which is blue-shifted compared with the bulk CdS due to quantum confinement. The PL spectrum of the 5.5 nm particles shows a peak at 610 nm.

Polycrystalline thin films of CdSe have been prepared at the organic–water interface by reacting cadmium cupferronate in the toluene layer with dimethylselenourea in the aqueous layer.[31] XRD measurements confirm the formation of cubic CdSe at the interface. TEM images reveal the films to be made up of nanocrystals with diameters ranging from 8 to 20 nm (Figure 10a). Time-dependent growth of the CdSe film at 20 °C has been examined by UV–vis absorption spectroscopy. All

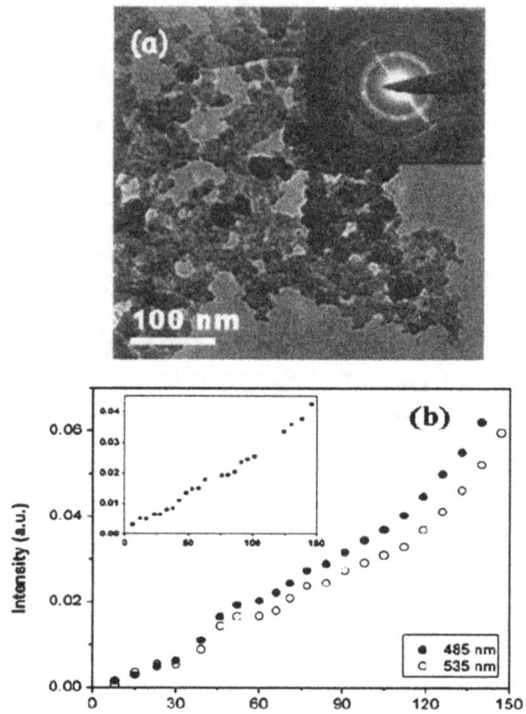


**FIGURE 10.** (a) TEM image of a CdSe film showing particles obtained by reacting aqueous solution of DMSU at 30 °C and (b) a plot showing the time-dependent growth of the absorption bands at two different wavelengths. Reproduced with permission from ref 31. Copyright 2007 American Scientific Publishers.

the films, including the one obtained after 3 min of the reaction, showed an absorption onset around 700 nm corresponding to the bulk band gap of CdSe and two higher order absorption bands at 485 and 535 nm. With increase in reaction time, the spectra indeed grew in intensity, without any appreciable shift in the positions of the band edge or the absorption bands, suggesting that CdSe nanoparticles possessing a size beyond the quantum confinement effects are formed at the interface within the first few minutes. Accordingly, the intensity versus time plots for the 485 and 535 nm bands, in Figure 10b, show a monotonic increase.

What is truly noteworthy is that we have been able to get single-crystalline films of some of the metal chalcogenides at the interface. In the case of CuS, we obtain continuous films, extending over wide areas (Figure 11a). A HREM image showing the (006) planes of hexagonal CuS and the corresponding SAED pattern given in Figure 11b reveal the single-crystalline and essentially defect-free nature of the CuS films. The thickness of the film was estimated to be ~50 nm from AFM and ellipsometric studies. In Figure 11c, we show a HREM

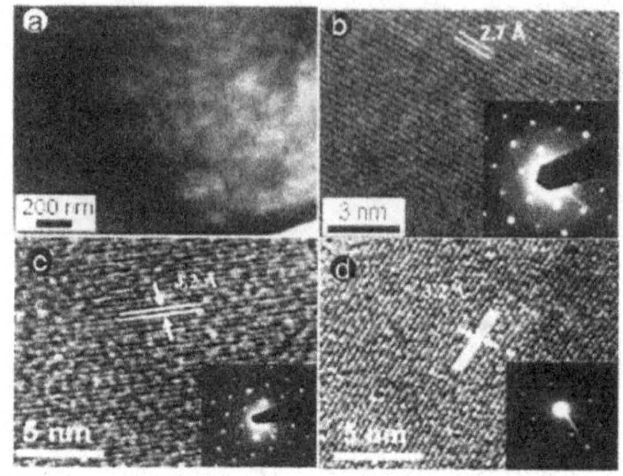


**FIGURE 11.** (a) TEM image of a CuS film, (b) HREM and SAED pattern of the film shown in panel a (Reproduced from ref 29. Copyright 2004 American Chemical Society), and HREM images and SAED patterns of (c) ZnS film and (d) PbS film (Reproduced with permission from ref 30. Copyright 2006 Elsevier).

image of a single-crystalline ZnS film. The (100) planes, separated by a distance of 3.2 Å, as seen in the HREM image correspond to hexagonal ZnS. The Bragg spots in the SAED pattern conform to the 0001 zone axis of hexagonal ZnS. In Figure 11d, we show the HREM image and the SAED pattern of a PbS film generated at the interface to reveal the single-crystalline nature. The HREM image gives a separation of 3.2 Å, corresponding to the separation between the (100) planes of cubic PbS. Single-crystalline CuSe thin films have also been obtained at the interface. Apart from single-crystalline films of metal sulfides, we could obtain metal sulfide bilayers such as CuS–CdS, CuS–PbS, and CdS–PbS at the toluene–water interface by employing the respective metal cupferronates.

## Metal Oxides

By reacting $Cu(cup)_2$ in the organic layer with an aqueous NaOH solution, one obtains single-crystalline films of monoclinic CuO at the organic–aqueous interface at 70 °C.[29] It has been possible to obtain crystalline films of ZnO by the reaction of $Zn(cup)_2$ in toluene with an aqueous solution of NaOH at 25 °C.

## Mode of Growth of the Films at the Interface

A time evolution study of Au films by AFM shows that just after 10 min of the reaction, a film made up of nanocrystals of size ~5–7 nm is formed. Although the size of the nanoparticles remains nearly the same even after 180 min of the reaction, we have found some evidence for the aggregation of nanoparticles in the films. The organic capped nanocrys-

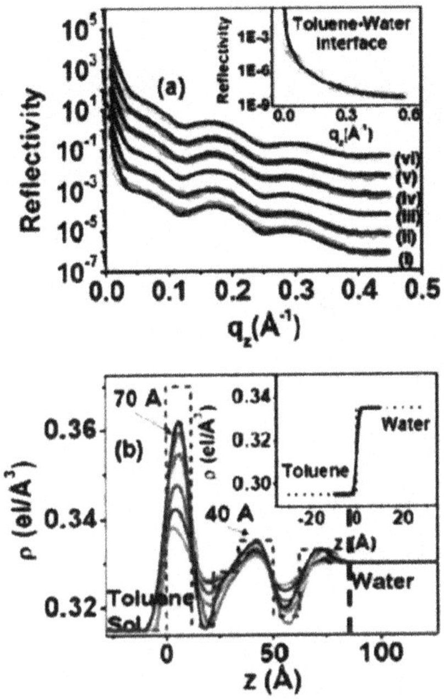


**FIGURE 12.** Reflectivity curves collected after initiation of the reaction (i, 194; ii, 224; iii, 253; iv, 283; v, 312; vi, 364 min) and the fits (solid line) and (b) extracted electron density profiles for the six reflectivity curves (i–vi) given in panel a as a function of depth. A simple model without (dashed line) and with roughness convolution (solid line) is also shown. The insets in panels a and b show the reflectivity data and the EDPs, respectively, of the bare toluene–water interface and the corresponding fit (solid line). Taken from ref 32.

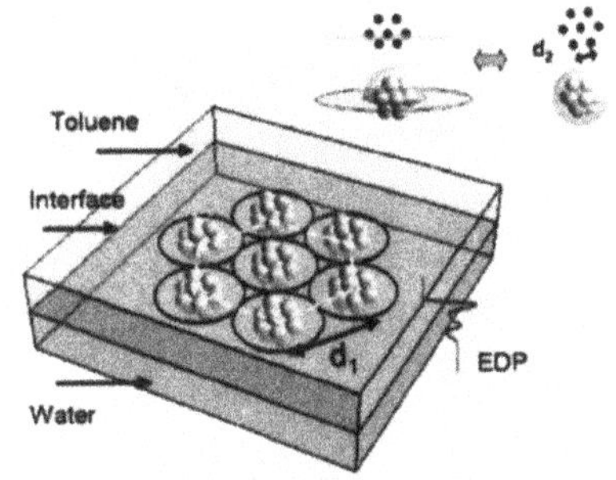


**FIGURE 13.** Schematic showing the hexagonal arrangement of organic-capped Au nanocrystals at the toluene–water interface. Taken from ref 32.

tals form an ordered arrangement giving rise to large features in the AFM images. The appearance of such large features makes it difficult to carry out a proper AFM study of the Au films. However, the AFM study shows that the thickness of the films increases progressively with time.

The formation of Au nanoparticle films has been investigated by X-ray reflectivity and diffuse scattering employing synchrotron radiation.[32] The oscillations in the reflectivity curves, collected after the initiation of the reaction (see Figure 12a), indicate the presence of a thin film at the toluene–water interface. The extracted final electron density profiles (EDPs; Figure 12b) from the reflectivity curves show three layers of gold clusters at 70 Å, 40 Å, and just above the water surface with central electron density values of 0.37, 0.33, and 0.33 electrons Å$^{-3}$. The layers have a vertical separation of 30 Å, the electron density between these layers corresponding to that of a typical organic material. A preliminary study shows that small Au nanoparticles covered by organic coatings form hexagonal clusters at the toluene–water interface (Figure 13). Each cluster consists of 13 nanoparticles of diameter ~12 Å with an organic capping of ~11 Å. Diffuse scattering measurements give a small value of interfacial tension, indicating an enhancement in the interfacial roughness caused by the presence of an organic layer at the interface.

The growth of the CdS polycrystalline thin films at the organic–aqueous interface has been studied as a function of time by employing AFM. The CdS film obtained after 30 min is made up of nanoparticles of diameter in the range ~5–7 nm. Nanoparticle aggregates of size in the range 80–200 nm are seen on the surface of the film. After 60 min of the reaction, the vertical growth of the aggregates saturates. The nanoparticles constituting the aggregates spread out laterally, thus reducing the overall thickness of the sample. With time, the aggregates gradually self-assemble to form a close-packed layer.

The growth of single-crystalline CuS films has also been examined as a function of reaction time by employing AFM. Nanoparticles of diameter in the range 7–8 nm are found to constitute the films. The amplitude image of the 30 min sample (Figure 14a) shows the presence of nanoparticle aggregates of size mainly in the range 50–100 nm. As the reaction proceeds, the initially formed nanoparticle aggregates come closer to each other and pack themselves at the interface to give flakes or uneven pieces of CuS within 50 min of reaction (denoted by arrows in Figure 14b). In the next 15 min of the reaction, the flakes coalesce to give a continuous and extended film at the interface. The process of formation of flakes and their coalescence repeats, giving rise to a thick film at the interface at the end of 180 min of the reaction (Figure 14c). Here we could hardly observe any individual nanoparticles. This is in contrast to the Au and CdS films, where the individual nanoparticles remained.

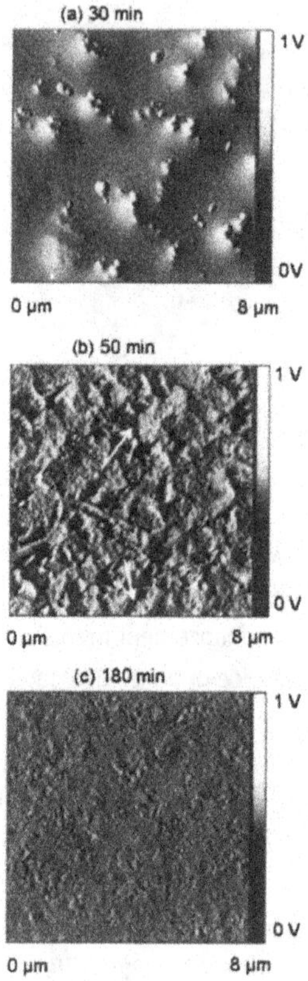


**FIGURE 14.** Tapping mode AFM images (amplitude channel) of CuS films grown at the toluene–water interface for (a) 30 min and (b) 50 min and (c) contact mode image obtained from the deflection channel of the film grown for 180 min.

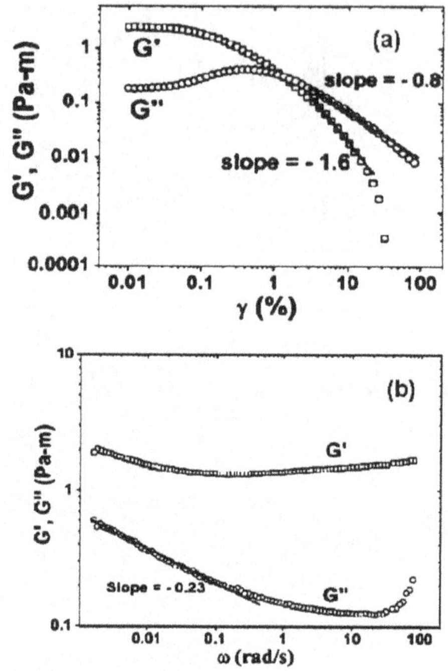


**FIGURE 15.** Panel a shows the strain amplitude sweep experiment on the 2D film of Ag nanoparticles. The storage modulus, $G'$ (□), is higher than the loss modulus, $G''$ (○) at low strain amplitudes. Panel b shows the frequency dependence of interfacial storage, $G'$ (□), and loss, $G''$ (○), moduli of the film. Reproduced from ref 33. Copyright 2007 American Chemical Society.

## Rheological Properties of the Films

Interfacial properties of Ag nanoparticles formed at the toluene–water interface have been investigated by using a biconal bob interfacial rheometer.[33] Strain amplitude measurements carried out on the film reveal a shear-thickening peak in the loss moduli ($G''$) at large amplitudes, followed by a power law decay of storage ($G'$) and loss moduli with exponents in the ratio 2:1 (see Figure 15a). In the frequency sweep measurements carried out at low frequencies, $G'$ remains nearly independent of the frequency over the range of frequencies probed, whereas $G''$ shows a power law dependence with a negative slope (Figure 15b). Such a low-frequency response of the 2D film of metal nanoparticles is reminiscent of a soft glassy system with long structural relaxation times. Steady shear measurements carried out on the film revealing a finite yield stress as the shear rate goes to zero, along with a significant deviation from the Cox–Merz rule, confirm that the monolayer of Ag nanoparticles at the interface forms a soft two-dimensional colloidal glass.

Interfacial rheological measurements carried out on nanocrystalline films of CdS, as well as on single-crystalline films of CuS, indicate a distinct nonlinear viscoelastic behavior for the films under oscillatory shear.[34] A smooth multilayered CdS film formed at higher concentrations and the single-crystalline CuS film exhibit distinct peaks in the loss modulus above a critical strain amplitude, followed by a power law decay of $G'$ and $G''$ at higher strain amplitudes, with the decay exponents in the ratio 2:1 (Figure 16). The frequency sweep response of both films, exhibiting a solid-like behavior over the range of angular frequencies probed, is similar to that of the Ag film, a characteristic feature of soft glassy systems. However, the mesoporous CdS film formed at low concentrations, where the pore boundaries form a surface fractal with a fractal dimension of 2.5, exhibits a monotonic decay of storage and loss modulus at large strain amplitudes. Under steady shear, compared with the CuS film, the finite yield stress exhibited by the CdS film is a magnitude lower, and the film is found to rupture under steady shear or at high strain amplitudes.

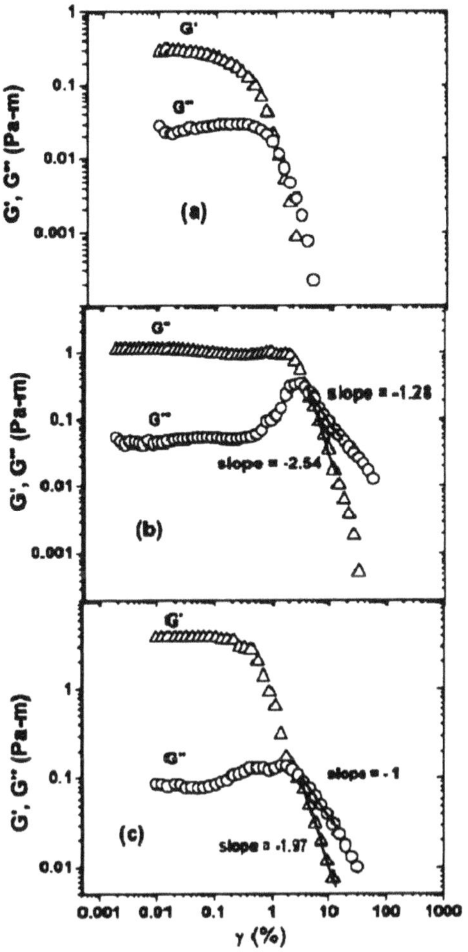


**FIGURE 16.** Strain amplitude sweep measurements on (a) CdS (low reactant concentrations), (b) CdS (high reactant concentrations), and (c) nanofilms formed at the toluene–water interface at 22 °C. Reproduced from ref 34. Copyright 2008 Elsevier.

## Conclusions

This Account should clearly bring out how the liquid–liquid interface can be exploited to prepare nanocrystalline films of various materials with wide ranging properties. What is noteworthy is that all the films are ultrathin with the thickness ranging from 40 to 100 nm. Furthermore, the method enables one to obtain substrate-free films, some of which are single-crystalline. It must be noted that obtaining single-crystalline films is difficult by any other physical or chemical method. The films prepared at the interface can be readily transferred to any substrate. By employing the organic–aqueous interface, one can prepare interesting films of inorganic materials with hydrophilic and hydrophobic coatings on opposite sides. It should also be possible to obtain bilayers of materials at the interface. Preliminary experiments show that CuS–CdS, CuS–PbS, and CdS–PbS bilayers can be prepared at the interface. The study of films of materials at the liquid–liquid interface will also provide a means to understand the nature of the interface itself.

*The authors thank their colleagues, Prof. G. U. Kulkarni, Dr. U. K. Gautam, Dr. M. Ghosh, and V. V. Agrawal, who have collaborated on some aspects of the films at the liquid–liquid interface.*

BIOGRAPHICAL INFORMATION

**C. N. R. Rao** is National Research Professor of India, Linus Pauling Research Professor at the Jawaharlal Nehru Centre for Advanced Scientific Research (JNCASR), and Honorary Professor at the Indian Institute of Science, Bangalore. He is a member of many academies including the Royal Society and the U.S. National Academy of Sciences. His main research interests are in solid state and materials chemistry, in which area he has published over 1000 research papers and 30 books. He is the recipient of the Dan David Prize for Materials Research.

**K. P. Kalyanikutty** received her B.Sc. degree from the University of Calicut and M.S. degree from JNCASR. She is continuing work for her Ph.D. degree in the area of nanomaterials.

FOOTNOTES

*Corresponding author. Fax: 00 91 80 2208 2766. E-mail: cnrrao@jncasr.ac.in.

REFERENCES


1 (a) Petty, M. C. *Langmuir-Blodgett films: An introduction*; Cambridge University Press: Cambridge, U.K., 1996. (b) Gelbart, W. M.; Sear, R. P.; Heath, J. R.; Chaney, S. Array Formation in Nano-colloids: Theory and Experiment in 2D. *Faraday Discuss.* **1999**, *112*, 299–307. (c) Rao, C. N. R.; Agrawal, V. V.; Biswas, K.; Gautam, U. K.; Ghosh, M.; Govindaraj, A.; Kulkarni, G. U.; Kalyanikutty, K. P.; Sardar, K.; Vivekchand, S. R.C. Soft Chemical Approaches to Inorganic Nanostructures. *Pure Appl. Chem.* **2006**, *78*, 1619–1650.

2 Mayya, K. S.; Patil, V.; Sastry, M. Lamellar Multilayer Gold Cluster Films Deposited by the Langmuir–Blodgett Technique. *Langmuir* **1997**, *13*, 2575–2577.

3 Schmid, G.; Baumle, M.; Beyer, N. Ordered Two-Dimensional Monolayers of $Au_{55}$ Clusters. *Angew. Chem., Int. Ed.* **2000**, *39*, 181–183.

4 Zhang, Y.; Grass, M. E.; Habas, S. E.; Tao, F.; Zhang, T.; Yang, P.; Somorjai, G. A. One-step Polyol Synthesis and Langmuir–Blodgett Monolayer Formation of Size-Tunable Monodisperse Rhodium Nanocrystals with Catalytically Active (111) Surface Structures. *J. Phys. Chem. C* **2007**, *111*, 12243–12253.

5 Chen, W.; Kim, J.; Xu, L-P.; Sun, S.; Chen, S. Langmuir–Blodgett Thin Films of $Fe_{20}Pt_{80}$ Nanoparticles for the Electrocatalytic Oxidation of Formic Acid. *J. Phys. Chem. C* **2007**, *111*, 13452–13459.

6 Dabbousi, B. O.; Murray, C. B.; Rubner, M. F.; Bawendi, M. G. Langmuir–Blodgett Manipulation of Size-Selected CdSe Nanocrystallites. *Chem. Mater.* **1994**, *6*, 216–219.

7 Zhu, R.; Min, G.; Wei, Y.; Schmitt, H. J. Scanning Tunneling Microscopy and UV–Visible Spectroscopy Studies of Lead Sulfide Ultrafine Particles Synthesized in Langmuir–Blodgett Films. *J. Phys. Chem.* **1992**, *96*, 8210–8211.

8 Guo, Q.; Teng, X.; Rahman, S.; Yang, H. Patterned Langmuir–Blodgett Films of Monodisperse Nanoparticles of Iron Oxide Using Soft Lithography. *J. Am. Chem. Soc.* **2003**, *125*, 630–631.

9 Benjamin, I. Chemical Reactions and Solvation at Liquid Interfaces: A Microscopic Perspective. *Chem. Rev.* **1996**, *96*, 1449–1475.

10 Luo, G.; Malkova, S.; Yoon, J.; Schultz, D. G.; Lin, B.; Meron, M.; Benjamin, I.; Vanysek, P.; Schlossman, M. L. Ion-Distributions Near a Liquid-Liquid Interface. *Science* **2006**, *311*, 216–218.

11 Binder, W. H. Supramolecular Assembly of Nanoparticles at Liquid-Liquid Interfaces. *Angew. Chem., Int. Ed.* **2005**, *44*, 2–5.

12 Binks, B. P.; Clint, J. H. Solid Wettability from Surface Energy Components: Relevance to Pickering Emulsions. *Langmuir* **2002**, *18*, 1270–1273.

13 Lin, Y.; Skaff, H.; Emrick, T.; Dinsmore, A. D.; Russel, T. P. Nanoparticle Assembly and Transport at Liquid-Liquid Interfaces. *Science* **2003**, *299*, 226–229.

14 Benkoski, J.; Jones, R. L.; Doughlas, J. F.; Karim, A. Photocurable Oil/Water Interfaces as a Universal Platform for 2-D Self-Assembly. *Langmuir* **2007**, *23*, 3530–3537.

15 Brust, M.; Walker, M.; Bethell, D.; Schiffrin, D. J.; Whyman, R. Synthesis of Thiol-Derivatised Gold Nanoparticles in a Two-Phase Liquid-Liquid System. *J. Chem. Soc., Chem. Commun.* **1994**, 801–802.

16 Kleitz, F.; Marlow, F.; Stucky, G. D.; Schuth, F. Mesoporous Silica Fibers: Synthesis, Internal Structure, and Growth Kinetics. *Chem. Mater.* **2001**, *13*, 3587–3595.

17 Sathaye, S. D.; Patil, K. R.; Paranjape, D. V.; Mitra, A.; Awate, S. V.; Mandale, A. B. Preparation of Q-Cadmium Sulfide Ultrathin Films by a New Liquid–Liquid Interface Reaction Technique (LLRT). *Langmuir* **2000**, *16*, 3487–3490.

18 Pan, D.; Jiang, S.; An, L.; Jiang, B. Controllable Synthesis of Highly Luminescent and Monodisperse CdS Nanocrystals by a Two-Phase Approach under Mild Conditions. *Adv. Mater.* **2004**, *16*, 982–985.

19 Schwartz, H.; Harel, Y.; Efrima, S. Surface Behavior and Buckling of Silver Interfacial Colloid Films. *Langmuir* **2001**, *17*, 3884–3892.

20 Dai, L. L.; Sharma, R.; Wu, C. Self-Assembled Structure of Nanoparticles at a Liquid–Liquid Interface. *Langmuir* **2005**, *21*, 2641–2643.

21 Song, J. M.; Zhu, J. H.; Yu, S. H. Crystallization and Shape Evolution of Single-Crystalline Selenium Nanorods at Liquid–Liquid Interface: From Monodisperse Amorphous Se Nanospheres toward Se Nanorods. *J. Phys. Chem. B* **2006**, *110*, 23790–23795.

22 Foster, P. M.; Thomas, P. M.; Cheetham, A. K. Biphasic Solvothermal Synthesis: A New Approach for Hybrid Inorganic–Organic Materials. *Chem. Mater.* **2002**, *14*, 17–20.

23 (a) Rao, C. N. R.; Kulkarni, G. U.; Thomas, P. J.; Agrawal, V. V.; Saravanan, P. Films of Metal Nanocrystals Formed at Aqueous–Organic Interfaces. *J. Phys. Chem. B* **2003**, *107*, 7391–7395. (b) Rao, C. N. R.; Kulkarni, G. U.; Agrawal, V. V.; Gautam, U. K.; Ghosh, M.; Tumkurkar, U. Use of the Liquid-Liquid Interface for Generating Ultrathin Nanocrystalline Films of Metals, Chalcogenides, and Oxides. *J. Colloid Interface Sci.* **2005**, *289*, 305–318.

24 Agrawal, V. V.; Kulkarni, G. U.; Rao, C. N. R. Nature and Properties of Ultrathin Nanocrystalline Gold Films Formed at the Organic–Aqueous Interface. *J. Phys. Chem. B* **2005**, *109*, 7300–7305.

25 Agrawal, V. V.; Kulkarni, G. U.; Rao, C. N. R. Surfactant-Promoted Formation of Fractal and Dendritic Nanostructures of Gold and Silver at the Organic-Aqueous Interface. *J. Colloid Interface Sci.* **2008**, *318*, 501–506.

26 Agrawal, V. V.; Varghese, N.; Kulkarni, G. U.; Rao, C. N. R. Effects of the Change in Interparticle Distance Induced by Alkanethiols on the Optical Spectra and Other Properties of Nanocrystalline Gold Films. *Langmuir*, in press).

27 Agrawal, V. V.; Mahalakshmi, P.; Kulkarni, G. U.; Rao, C. N. R. Nanocrystalline Films of Au–Ag, Au–Cu, and Au–Ag–Cu Alloys Formed at the Organic–Aqueous Interface. *Langmuir* **2006**, *22*, 1846–1851.

28 Gautam, U. K.; Ghosh, M.; Rao, C. N. R. A Strategy for the Synthesis of Nanocrystal Films of Metal Chalcogenides and Oxides by Employing the Liquid-Liquid Interface. *Chem. Phys. Lett.* **2003**, *381*, 1–6.

29 Gautam, U. K.; Ghosh, M.; Rao, C. N. R. Template-Free Chemical Route to Ultrathin Single-Crystalline Films of CuS and CuO Employing the Liquid–Liquid Interface. *Langmuir* **2004**, *20*, 10776–10778.

30 Kalyanikutty, K. P.; Gautam, U. K.; Rao, C. N. R. Ultra-Thin Crystalline Films of ZnS and PbS Formed at the Organic–Aqueous Interface. *Solid State Sci.* **2006**, *8*, 296–302.

31 Kalyanikutty, K. P.; Gautam, U. K.; Rao, C. N. R. Ultra-Thin Crystalline Films of CdSe and CuSe Formed at the Organic–Aqueous Interface. *J. Nanosci. Nanotechnol.* **2007**, *7*, 1916–1922.

32 Sanyal, M. K.; Agrawal, V. V.; Bera, M. K.; Kalyanikutty, K. P.; Daillant, J.; Blot, C.; Kubowicz, S.; Konovalov, O.; Rao, C. N. R. Ordering of Gold Nanoparticles Formed at Liquid-Liquid Interface. *J. Phys. Chem. C*, in press.

33 Krishnaswamy, R.; Majumdar, S.; Ganapathy, R.; Agrawal, V. V.; Sood, A. K.; Rao, C. N. R. Interfacial Rheology of an Ultrathin Nanocrystalline Film Formed at the Liquid/Liquid Interface. *Langmuir* **2007**, *23*, 3084–3087.

34 Krishnaswamy, R.; Kalyanikutty, K. P.; Kanishka, B.; Sood, A. K.; Rao, C. N. R. Nonlinear Viscoelasticity of Ultrathin Nanocrystalline Semiconducting Films of CdS and CuS at Liquid-Liquid Interfaces. *J. Colloid Interface Sci.*, (submitted for publication).

Nano Today (2009) 4, 96–106

available at www.sciencedirect.com

ScienceDirect

journal homepage: www.elsevier.com/locate/nanotoday

REVIEW

# Ferromagnetism as a universal feature of inorganic nanoparticles

A. Sundaresan*, C.N.R. Rao*

*Chemistry and Physics of Materials Unit, and Department of Science and Technology Unit on Nanoscience, Jawaharlal Nehru Centre for Advanced Scientific Research, Jakkur P.O., Bangalore 560 064, India*

Received 15 August 2008; received in revised form 16 September 2008; accepted 3 October 2008

**KEYWORDS**
Nanoparticles;
Ferromagnetism;
Defects;
Nonmagnetic

**Summary** Room-temperature ferromagnetism is exhibited by nanoparticles of a variety of inorganic materials although they are intrinsically non-magnetic. Typical of such nanomaterials are the oxides, $CeO_2$, $TiO_2$, $Al_2O_3$, and MgO. Nanoparticles of nitrides such as GaN and chalcogenides such as CdS and CdSe also exhibit ferromagnetism. Ferromagnetism of the nanoparticles is confined to the surface. This phenomenon has been utilized to render the classic ferroelectric $BaTiO_3$ to be multiferroic wherein surface ferromagnetism coexists with bulk ferroelectricity. Interestingly, nanoparticles of superconducting $YBa_2Cu_3O_7$ show surface ferromagnetism. It is possible that surface ferromagnetism of inorganic nanoparticles can be usefully exploited.


Magnetism in solid materials such as iron, $Fe_3O_4$ and $CrO_2$ arises from the collective interaction of magnetic moments associated with the atoms or ions constituting the material. The magnetic atoms or ions in these materials are arranged in a periodic lattice and their moments interact through a molecular exchange field to give rise to long-range magnetic ordering. Several interesting phenomena have been discovered in recent years based on studies of nanoparticles of magnetic and non-magnetic materials [1–3]. Magnetic nanoparticles are used as active components of ferrofluids, in recording tapes, and in biomedical applications [4–7]. It has been known for some time that nanoparticles of magnetic materials exhibit properties which are significantly different from those of the bulk materials, due to finite size effects [8]. As early as 1961, Néel predicted that nanoparticles of antiferromagnetic materials will have a small net magnetic moment due to the unequal number of spins on the two sublattices as a result of finite size [9]. Recent experiments have indeed shown large magnetic moments at room temperature, low-temperature hysteresis, and loop shifts up to 10 kOe in NiO and CoO nanoparticles [8,10,11]. Above a blocking temperature, which is characterized by a peak in the magnetization-temperature curve, where there is divergence between the zero-field cooled (ZFC) and field cooled (FC) magnetization, these particles display giant moments whose dynamics are described by superparamagnetism. In the case of ferromagnetic metallic nanoparticles, superparamagnetism as well as an increase in magnetic anisotropy energy due to surface anisotropy have been observed for particles with an average diameter of 2 nm [12]. Nanoparticles of $\gamma$-$Fe_2O_3$ are known to

* Corresponding authors.
*E-mail addresses:* sundaresan@jncasr.ac.in (A. Sundaresan), cnrrao@jncasr.ac.in (C.N.R. Rao).

1748-0132/$ — see front matter © 2008 Elsevier Ltd. All rights reserved.
doi:10.1016/j.nantod.2008.10.002

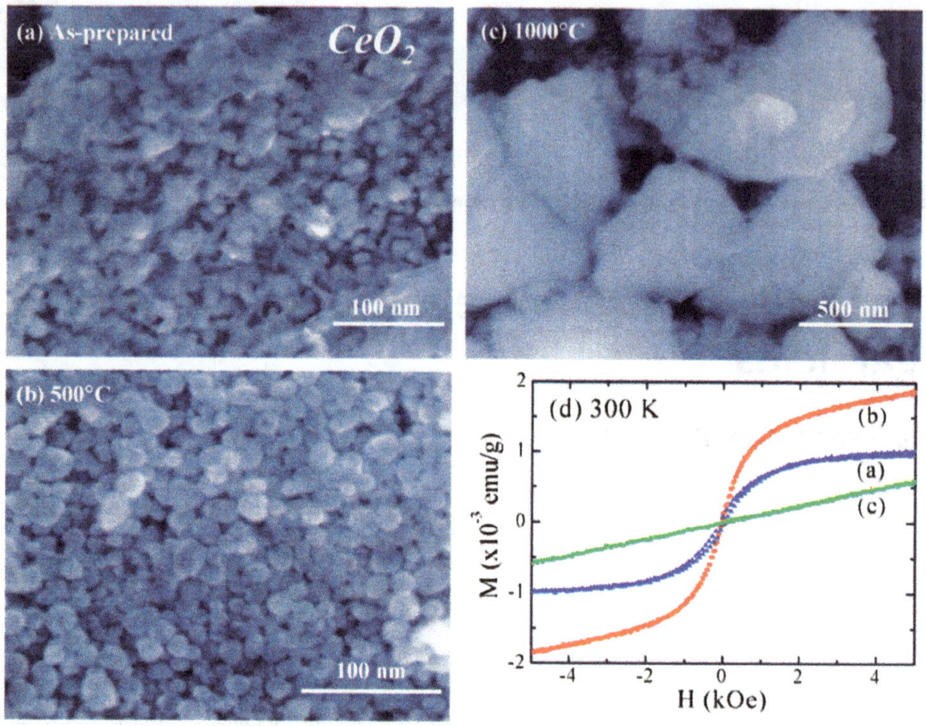


**Figure 1** FESEM images of $CeO_2$ particles: (a) as-prepared (7 nm) nanoparticles, (b) nanoparticles (15 nm) heated to 500 °C, (c) particles (500 nm) heated at 1000 °C (bulk sample) and their (d) magnetization data at room temperature. Note the absence of ferromagnetism in the bigger particles in contrast to smaller particles (a) and (b) (from Ref. **[13]**).

be superparamagnetic and show spin-canting due to surface effects.

In this article, our main concern relates to a new type of ferromagnetism observed by us in nanoparticles of a large variety of inorganic materials such as non-magnetic $Al_2O_3$, $TiO_2$ and $CeO_2$, superconducting cuprates, metal nitrides and chalcogenides **[13–15]**. We propose ferromagnetism arising from surface defects is a universal feature of inorganic nanoparticles.

## Ferromagnetism in nanoparticles of nonmagnetic metal oxides

Magnetic properties of nanoparticles of several non-magnetic oxides such as $CeO_2$, $Al_2O_3$, ZnO, $In_2O_3$, $SnO_2$, prepared by chemical methods **[16–18]**, have been investigated after proper characterization by the various physical methods. In what follows, we briefly describe the results of this study.

Nanoparticles of $CeO_2$ were prepared by the addition of hexamethylenetetramine to a solution of cerium nitrate under constant stirring. The particles were filtered, dried and heated to 500 °C in flowing oxygen to make the nanoparticles free from organic matter. In order to make micron-sized (bulk) $CeO_2$ particles, these nanoparticles were pressed into pellets and sintered in air at 1400 °C. XRD patterns showed that both the nanoparticles and the bulk $CeO_2$ were single phase with the cubic fluorite structure. The lattice parameter ($a$ = 5.424(3) Å) of the nanoparticles (7 nm) was higher than that of the bulk sample ($a$ = 5.413(1) Å) consistent with an earlier report **[19]**. Field emission scanning electron microscope (FESEM) images of the as-prepared nanoparticles and the particles heated at 500 °C as well as of the bulk sample are shown in **Fig. 1**(a), (b) and (c), respectively. The average particle size of the as-prepared sample is 7 nm whereas the one heated at 500 °C, that is free from organic moieties, is 15 nm. Bulk $CeO_2$ has an average particle size of 500 nm. Magnetization data of these samples recorded at room temperature are shown in **Fig. 1**(d). The nanoparticles (a) and (b) show ferromagnetic behavior with a coercive field of around 100 Oe. Furthermore, the saturation magnetization of the nanoparticles (15 nm) heated to 500 °C (b) which is free from organic impurities is higher than that of (a) the as-prepared particles (7 nm) which is covered with organic matter, indicating the importance of the surface of the particles to the observed ferromagnetism. As the mass of the organic matter was negligibly small, the magnitude of magnetization in the as-prepared sample refers to emu/g of sample. At magnetic fields greater than ~3000 Oe, the magnetization shows a paramagnetic behavior suggests that it originates from the core of the particles while the ferromagnetism is confined to the surface of the nanoparticles as discussed below. The bulk sample shows a weak paramagnetism while normally diamagnetic behavior is expected for $CeO_2$ with the $4f^0$ electronic configuration of $Ce^{IV}$. The appearance of ferromagnetism in nanoparticles of $CeO_2$ is intriguing as the bulk sample is diamagnetic. The observed

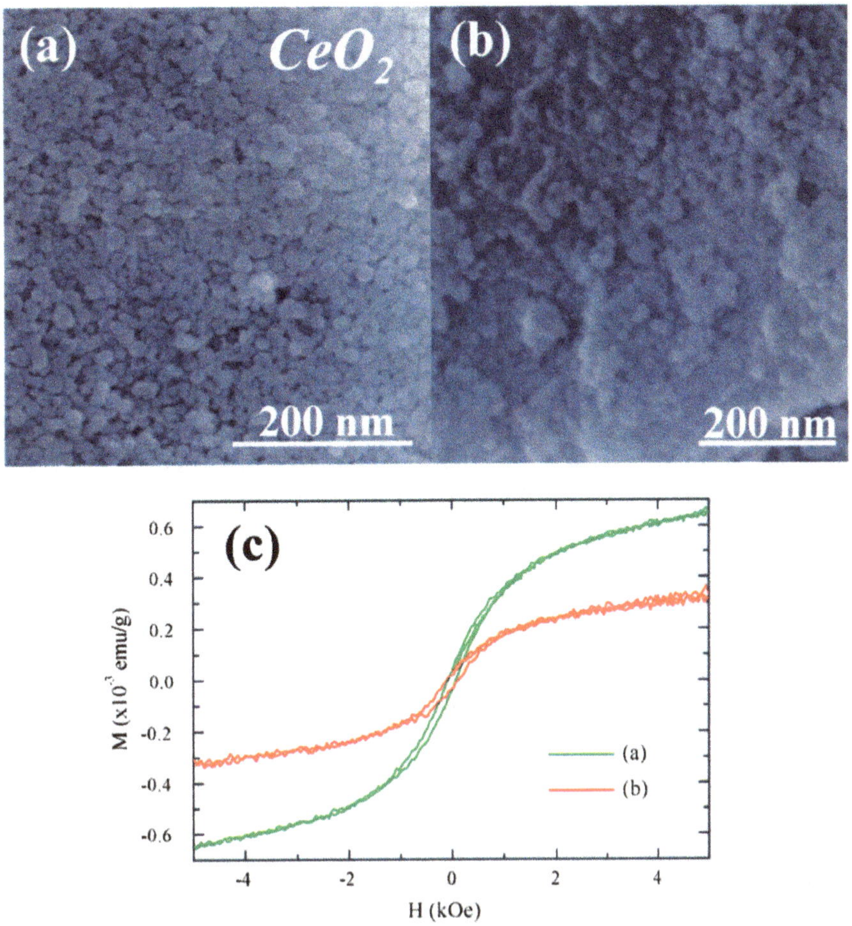


**Figure 2** FESEM images of $CeO_2$ (a) powder and (b) pelletized sample, heated to 500 °C showing more agglomeration of the particles in (b), and their magnetization curves at room temperature. The saturation magnetization of (b) is lower than that of (a) demonstrating the surface origin of ferromagnetism.

ferromagnetism is certainly intrinsic to the nanoparticles as there were no magnetic elements present in the sample. The source of ferromagnetism in these nanoparticles is likely to have a origin similar to that in thin films of $HfO_2$, $TiO_2$, ZnO, and $TiO_2$ where the oxygen deficiencies are believed to be responsible for creating magnetic moments [20–22]. Unlike thin films, where the contamination of films from extraneous sources can vitiate the results, ferromagnetism in the nanoparticles is robust. We, therefore, suggest that oxygen vacancies at the surface of the nanoparticles are likely to be responsible for the ferromagnetism of the nanoparticles. As cerium can have variable valence states ($Ce^{4+}/Ce^{3+}$) and oxygen vacancies on the surface of the $CeO_2$ nanoparticles [23], it is possible that oxygen vacancies can create magnetic moments on the neighboring Ce-ions, as shown by X-ray spectroscopic and electron paramagnetic resonance studies [24,25]. The fact that the nanoparticles show ferromagnetism suggests that there is a sufficient concentration of oxygen vacancies at the surface to provide magnetic percolation. The paramagnetic behavior of bulk $CeO_2$ suggests that in this sample the concentration of oxygen vacancies is less than the critical value required for magnetic percolation. In order to further verify that the magnetism is confined to the surface, the nanoparticles were pressed into pellets and heated at the same temperature (500 °C). The FESEM images (**Fig. 2**) shows more agglomerated particles in the palletized sample (b) than in the powder (a). If magnetism were associated with the surface, one would expect a relatively lower saturation moment in the agglomerated particles. This is what we observe in the magnetization measurements (**Fig. 2**(c)). It may be mentioned here that the results shown in **Fig. 1**(d) (curve b) and **Fig. 2** are from different batches of sample and therefore their magnitude of magnetization should not be compared. Ferromagnetism in nanoparticles of $CeO_2$ has been noticed recently by other workers as well [26].

An even better example of surface ferromagnetism is that of $Al_2O_3$ nanoparticles (3 nm size) where the valence of Al can only be 3+ (**Fig. 3**). Ferromagnetism in this oxide could be seen even at 390 K, indicating that the material has a Curie temperature ($T_C$) much higher than room-temperature. Since bulk $Al_2O_3$ is diamagnetic, the fer-

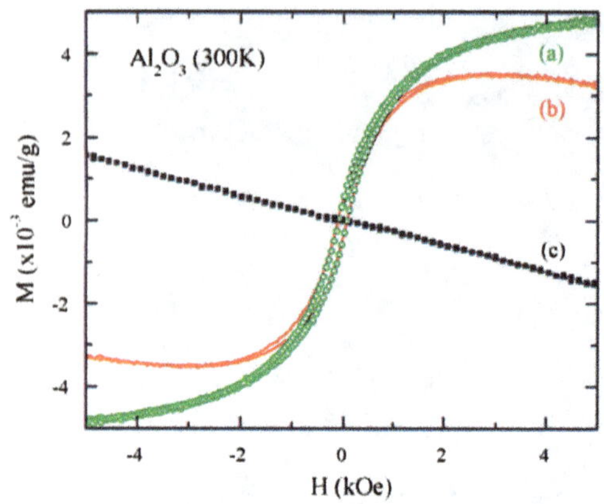


**Figure 3** Room-temperature magnetic hysteresis curves of $Al_2O_3$ nanoparticles: magnetization (a) corrected and (b) uncorrected for core diamagnetism. Note the bulk $Al_2O_3$ exhibits diamagnetic behavior (c).

romagnetism of the nanoparticles must have its origin at the surface. This suggestion is consistent with the observation that the magnetization above ~3000 Oe is that of a diamagnetic material as shown by the curve (a) in **Fig. 3**. It is seen from curve (b) that after subtracting the diamagnetic contribution from the core of the particles the magnetization is close to saturation at 5000 Oe. Although we suggested oxygen defects as a common source of magnetic moment at the surface **[13]**, theoretical studies have shown that cation vacancies could give rise to magnetic moments in simple sp oxides **[27]**. We have observed room-temperature ferromagnetism in nanoparticles of other oxides such as ZnO, $In_2O_3$, $TiO_2$, MgO and $SnO_2$ while their corresponding bulk samples are diamagnetic **[13]**. Since oxygen and cation vacancies are commonly encountered in oxides, we suggest that ferromagnetism is a universal characteristic of metal oxide nanoparticles.

## Ferromagnetism in nanoparticles of non-magnetic metal nitrides and chalcogenides

Like metal oxides, metal nitrides and chalcogenides can also have native surface defects. This motivated a study of the magnetic properties of nanoparticles of GaN and CdS. GaN is an important wide-gap semiconductor for micro- and optoelectronic applications such a blue light emitting diodes and laser **[28]**. Moreover, the prediction of room-temperature ferromagnetism in transition metal doped GaN intensified the research on this material[22]. Among the various forms, thin films of Mn-doped GaN have been investigated because of the potential applications in spintronics **[29,30]**. While thin films of Mn-doped GaN prepared by some groups have been reported to be ferromagnetic, there are reports of paramagnetic behavior in this material **[31—33]**. It has been suggested that the observed ferromagnetism may arise from possible impurity phases such as Ga—Mn or Mn—N alloys. Meanwhile, other nanoforms such as nanowires and nanocrystals of Mn-doped GaN have been reported to exhibit ferromagnetism **[34]**. We have found that nanoparticles of undoped GaN synthesized from the metal trichloride—urea complex exhibit room-temperature ferromagnetism while the sintered sample shows diamagnetic behavior **[35]**. In the synthetic method employed, the $GaCl_3$—urea complex on thermal decomposition above 800 C in a $N_2$ or $NH_3$ atmosphere yields GaN nanoparticles. In order to obtain bulk GaN, the powder heated at 850 °C was pressed into rectangular bar and sintered at 950 C in a $NH_3$ atmosphere for 12—16 h. A FESEM image of GaN particles heated at 850 °C, as seen in **Fig. 4**(a), shows distribution of particle size in the 15—20 nm range whereas the one heated at 950 °C has particles in the (25—30 nm) diameter range as shown in **Fig. 4**(b). As most of the nitride materials are difficult to sinter without sintering aids, the particle size of the bulk GaN sample remains in the range 150—350 nm (**Fig. 4**(c)) but the particles are agglomerated. The particles shown in **Fig. 4**(a) and b exhibit ferromagnetic hysteresis at room temperature whereas the bigger particles show diamagnetism. The saturation moment of the 850 °C heated particles is ~0.0012 emu/g with a coercivity of around 200 Oe. This behavior is similar to that of non-magnetic oxide nanoparticles **[13]**, suggesting that the ferromagnetism in GaN is essentially confined to the surface of the nanoparticles due to possible defects such as Ga or N vacancies. Both Ga and N vacancies are suggested to be present in the as-grown GaN depending on the growth conditions **[36]**. Calculations of the energies of formation of Ga and N vacancies in bulk GaN have been reported in the literature **[37]**. The calculations show that nitrogen vacancies have relatively lower energy and give rise to paramagnetism. On the other hand, Ga vacancies introduce magnetic moments on neighboring N atoms that are spin polarized due to Hund's rule **[38,39]**. Such defect-induced ferromagnetism has not been observed in GaN due to large formation energy of defects in the bulk GaN. However, the formation energy at the surface may be significantly lower than that in the bulk. Since the nanoparticles have a large surface area, there can be a sufficient number of defects at the surface giving rise to ferromagnetism. As the size of the particle increases, the defect concentration decreases and magnetism disappears.

CdS is an important material that exhibits interesting nonlinear optical and luminescence properties **[40]**. We prepared CdS nanoparticles by a simple sol-gel method where mercaptoethanol was used as the capping agent. To vary the particle size, the concentration of mercaptoethanol was varied (30, 15 and 3.75 mM). In order to obtain bigger particles, a sample was prepared without mercaptoethanol. TEM image of the sample prepared with 15 mM mercaptoethanol shows (**Fig. 5**(a)) the average particle size to be 18 nm whereas the sample prepared without mercaptoethanol (**Fig. 5**(b)) has larger particles (25 nm). The particle size increases with decrease in concentration of mercaptoethanol, consistent with the UV absorption spectra (**Fig. 5**(c)) where the absorption band shifts towards longer wave length with increasing particle size. The insets in the figure display the electron diffraction patterns which reveal the polycrystalline nature of the particles. While all the samples show ferromagnetic behavior with the coercive field in the range 100—150 Oe, the smallest particles show the

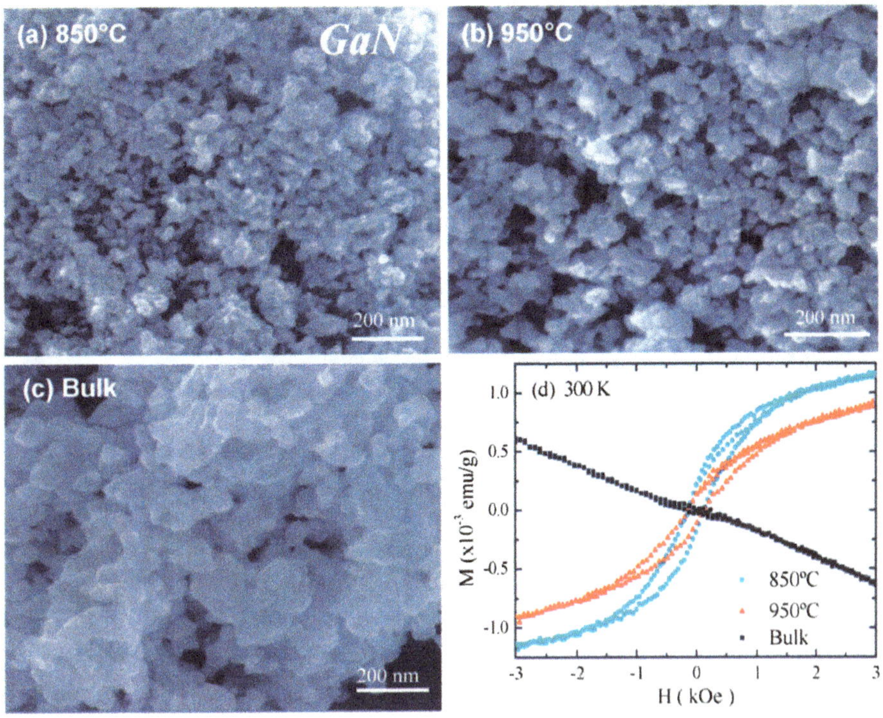


**Figure 4** FESEM images of GaN particles heated to (a) 850 C, and (b) 950 °C (c) pelletized sample heated at 950 °C. (d) Magnetization curves. Note that the small particles (a) and (b) show magnetic hysteresis whereas the bigger particles show diamagnetism (from Ref. [15]).

highest magnetization (**Fig. 5**(d)). We have also observed room-temperature ferromagnetism in CdSe nanoparticles. These results suggest that ferromagnetism occurs not only in nanoparticles of metal oxides but also of metal nitrides and chalcogenides.

## Ferromagnetism in nanoparticles of superconductors

Magnetism and superconductivity are often thought to be incompatible with each other due to the suppression of superconductivity by the applied magnetic field. However, a recent study has shown that a magnetic nano-dot can actually enhance superconductivity instead of destroying it **[41]**. In our study, we have combined surface ferromagnetism induced by the point defects with the core superconductivity in nanoparticles of $YBa_2Cu_3O_{7-y}$ (YBCO). The nanoparticles of YBCO were prepared by the citrate gel method reported earlier **[14]**. The FESEM image shows (**Fig. 6**(a)) that the particles are irregular in shape with an average size of 150 nm. The zero-field cooled (ZFC) and field-cooled (FC) susceptibilities measured at 20 Oe (**Fig. 6**(b)) show the diamagnetic superconducting onset temperature of 91 K. The nature of magnetic hysteresis at 5 K (see inset **Fig. 6**(b)) and the absence of irreversibility between the ZFC and FC curves down to 40 K suggests that inter-vortex spacing is larger than the particles size **[42,43]**. It is important and interesting to note the ferromagnetic hysteresis, with a coercivity of ~200 Oe, at room temperature (**Fig. 6**(c)). Ferromagnetic hysteresis is seen down to the superconducting onset temperature (91 K) below which the magnetization is dominated by diamagnetic hysteresis as seen in the inset of **Fig. 6**(d). Bulk YBCO, obtained by sintering the nanoparticles at 940 °C, shows paramagnetic behavior (**Fig. 6**(c)) as expected, confirming that nanoparticles of cuprate superconductors behave like those of other metal oxides with regard to the defects **[44]**. The coexistence of surface magnetism and bulk superconductivity, however, needs to be established. The ferromagnetic Curie temperature determined from temperature-dependent magnetization measurements (**Fig. 7**), where the ferromagnetic hysteresis disappears is around 800 K.

Nanoparticles of NbN (30 nm diameter), prepared by heating $NbCl_5$ with urea in 1:6 molar ratio at ~1000 °C **[45]**, show a superconducting transition of around 6 K as can be seen from the bottom inset in **Fig. 8**(a). More importantly, they also exhibit room-temperature ferromagnetism with a coercivity of ~500 Oe. Magnetic hysteresis in the superconducting state shown in the inset (top) of **Fig. 8**(a) is typical of fine particles. Bulk NbN (~300 nm diameter) prepared by sintering the nanoparticles at 920 °C for a prolonged time in ammonia, shows linear magnetization at room temperature (**Fig. 8**(b)). The diamagnetic susceptibility and the hysteretic behavior in the superconducting state are both shown as insets in **Fig. 8**. The behavior of NbN is indeed similar to that of GaN. Similarly, room-temperature ferromagnetism is observed in nanoparticles of δ-MoN **[14]**.

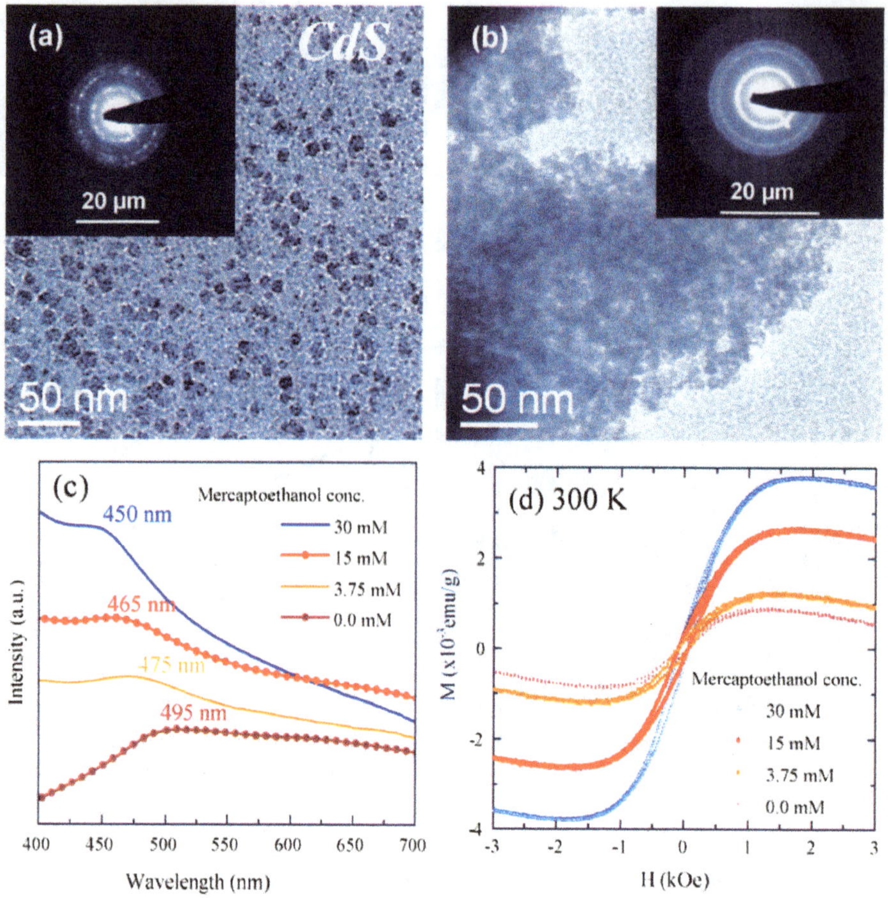


**Figure 5** TEM images of CdS nanoparticles (15–20 nm) (a) capped with mercaptoethanol and (b) without capping showing agglomeration of particles. (c) UV absorption spectra of samples showing shifts of the band depending on the particles size. Note the decrease of saturation of magnetic moment with increasing particle size (from Ref. [10]).

## Multiferroic properties of nanocrystalline $BaTiO_3$

Multiferroics are materials possessing two or all the three ferroic properties, ferroelectricity, ferroelasticity and ferromagnetism [46]. Recent interest has been on magnetoelectric multiferroics materials exhibiting both ferroelectricity and ferromagnetism, with coupling between the two [47,48]. Magnetoelectric multiferroics are expected to be very useful for device applications. $BaTiO_3$ is a classic ferroelectric where the ferroelectricity arises from the $d^0$ configuration of $Ti^{4+}$ ions which are distorted away from its ideal position in cubic symmetry, giving rise to a spontaneous polarization. This mechanism of ferroelectricity contraindicates the origin of ferromagnetism which requires occupancy of the d-orbitals. We have exploited the occurrence of surface ferromagnetism in nanoparticles with the ferroelectricity of nanocrystalline $BaTiO_3$ to obtain a new kind of multiferroic material [49].

Nanocrystalline $BaTiO_3$ samples were prepared by taking the stoichiometric amount of titanium (IV) isopropoxide in a mixture of ethanol and acetic acid, and adding to it a citric acid solution of $BaCO_3$ and polyvinyl alcohol. The mixture was stirred continuously till the formation of a white gel. The dried gel was heated at 700 °C to remove organic matter and the resulting powder pressed into a pellet and heated at 1000 °C for 1 h. The FESEM image shows in **Fig. 9** the average particle size of the 700 °C sample to be 300 nm while the size of the bulk sample is 2 μm. Magnetization measurements at room temperature confirmed (**Fig. 10**(a)) the ferromagnetic behavior of the nanocrystalline sample and the diamagnetic behavior of the bulk sample. The ferroelectric transitions of $BaTiO_3$ are seen in the nanocrystalline sample (**Fig. 10**(b)). The transition at 300 K is due to the tetragonal–orthorhombic transition and the one at 225 K is associated with the rhombohedral transition. Ferroelectric switching is seen in P–E hysteresis curve (**Fig. 10**(c)) which shows a maximum polarization of 8.42 $\mu C/cm^2$, a remnant polarization of 2.04 $\mu C/cm^2$ and a coercive field of 10 kV/cm. The coupling of the surface ferromagnetism with ferroelectricity is manifested in the positive magnetocapacitance (MC),

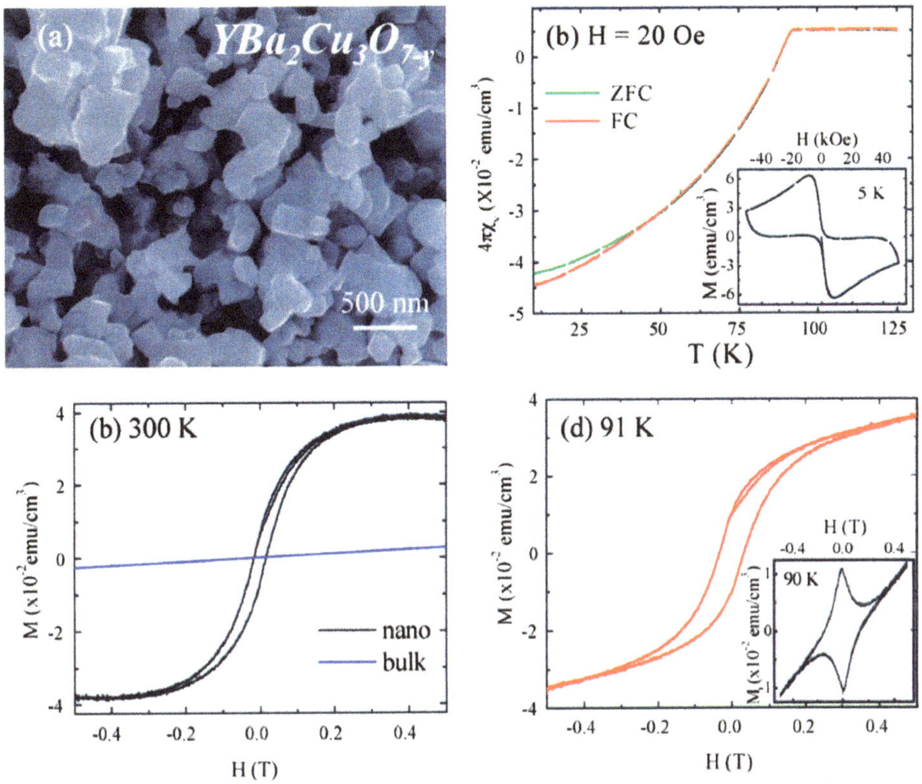


**Figure 6** (a) FESEM image of $YBa_2Cu_3O_{7-y}$ particles with an average size of 150 nm. (b) ZFC and FC susceptibility showing the diamagnetic superconducting onset at 91 K. Inset shows the magnetic hysteresis at 5 K. (c) Magnetic hysteresis of YBCO particles at 300 K. Note the paramagnetic behavior of bulk YBCO. (d) Magnetic hysteresis at $T_C = 91$ K. The inset shows hysteresis at 90 K.

$\Delta\varepsilon(H)/\varepsilon_0 = [\varepsilon(H) - \varepsilon(0)]/\varepsilon_0$, as shown in **Fig. 10**(d) near room temperature.

First-principles calculations on nano-sized (1.2 nm) $BaTiO_3$ have shown that the energy is lower when the oxygen vacancies are located at the surface **[49]**. These vacancies result in the occupation of d-states of neighboring Ti-ions. The relatively extended nature of the occupied d-states enables magnetic exchange interaction leading to the ferromagnetic ground state. On the other hand, ferroelectricity survives in the nano-slab with relatively large polarization in the interior of the slab. The calculations confirm the presence of coupling between the spins and the surface phonon mode and thus explain the observed magnetocapacitance effect.

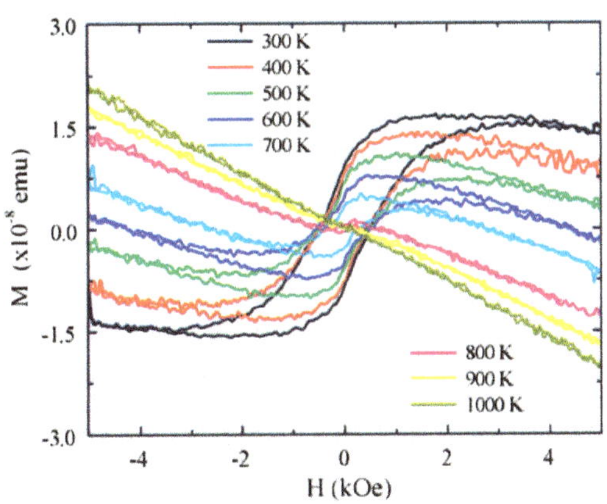


**Figure 7** Magnetization curves of nanocrystalline YBCO at various temperatures showing the disappearance of ferromagnetic ordering at 800 K.

## Ferromagnetism in metallic nanoparticles

We must recall here that nanoparticles of nonmagnetic metals such as thiol-capped gold nanoparticles have been reported to show room-temperature ferromagnetism **[50]**. The ferromagnetic spin polarization in gold nanoparticles has been observed experimentally by element-specific X-ray magnetic circular dichroism (XMCD) spectroscopy **[51]**. Coupling between the capping agent and gold surface seems to be a prerequisite for the observation of ferromagnetism. There have been several explanations of the ferromagnetism in gold nanoparticles. These include, the Fermi hole effect, Hund's rule electron filling and giant magnetic anisotropy. Ferromagnetic behavior has also been observed experimentally in Pd and Pt nanoparticles prepared by different chemical methods **[52,53]**.

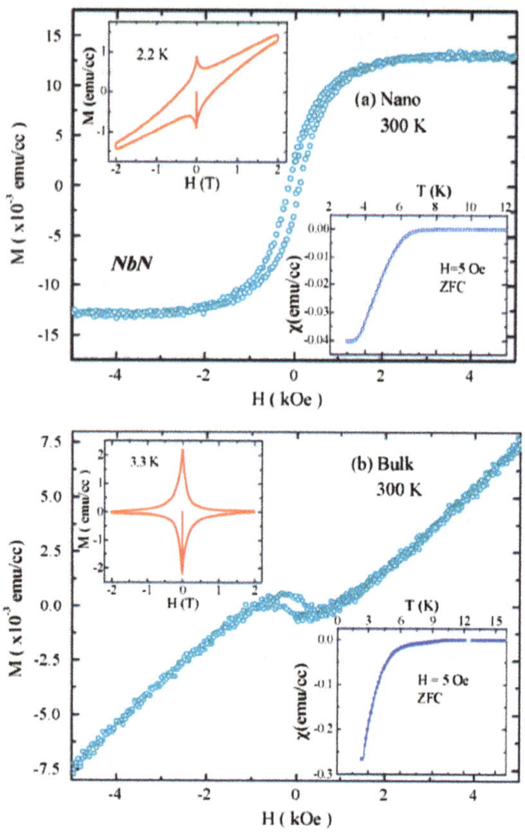


**Figure 8** (a) Room-temperature ferromagnetic hysteresis of NbN particles. Bottom inset shows the superconducting transition at 6 K. Top inset shows the magnetic hysteresis in the superconducting state. (b) Magnetization curve of bulk (200 nm) NbN at 300 K. Bottom inset shows the diamagnetic susceptibility and the top one shows the hysteresis at 3.3 K (from Ref. [14]).

## Dilute magnetic semiconductors

Nearly a decade ago, collective ferromagnetism was reported in nonmagnetic semiconductors substituted with a small amount (<10%) of magnetic ions. These dilute magnetic semiconductors, exemplified by (In,Mn)As **[54]** and (Ga,Mn)As **[55]**, exhibit low ferromagnetic ordering temperatures (61 and 173 K, respectively). The origin of ferromagnetism in these materials was explained by Dietl et al. **[29]** on the basis of the Zener model **[56,57]** of ferromagnetism where the exchange interaction between the localized moments is mediated by charge carriers. Based on this mechanism, it was predicted that wurtzite semiconductors, ZnO and GaN substituted with 5 at.% of Mn will be ferromagnetic with $T_C$ above room-temperature **[29]**. This created considerable interest in the field of semiconductors and magnetism to look for new semiconductors that exhibit room-temperature ferromagnetism since such materials can have functionalities coming from spin-polarized charge carriers with applications in new semiconductor devices. Both GaN and ZnO substituted with a few percent of transition metals were reported to be ferromagnetic with critical temperatures above room temperature **[58,59]**. Following this report, ferromagnetism was reported in other wide band gap semiconductors such as $TiO_2$ **[60]**. There have been many studies of Mn and Co substituted ZnO and some of them report ferromagnetism whereas others do not observe ferromagnetism **[61]**. A recent review article analyzed the experimental situation on the occurrence or absence ferromagnetism in transition-metal substituted ZnO and concluded the absence of ferromagnetism in well-characterized samples with low substitution **[62]**. Furthermore, the reported ferromagnetism seems to be weak ($10^{-4}$ emu/g) which casts doubts on the origin of magnetism that also include the contamination of samples while handling them with stainless steel tweezers **[63]**. It is possible that the origin of magnetism in some of these materials may be related to the surface ferromagnetism of nanoparticles associated with point defects.

## Origin of the magnetism in nanoparticles

It has been known for some time that point defects in insulators can create localized electronic levels within the band gap that can have different electron occupancies and thus different charge states and magnetic moments **[64]**. The well-known examples are CaO and MgO where it has been shown theoretically **[65]** and experimentally that the neutral charge state of cation vacancies have magnetic moments **[66,67]**. In the case of CaO, theoretical studies show that a minimum vacancy concentration of $1.8 \times 10^{21}$ cm$^{-3}$ or 4.9% is required to achieve magnetic percolation with a ferromagnetic interaction range limited to four neighbors **[68]**. However, a high vacancy formation energy even under the most favorable growth conditions limits the vacancy concentration to be three orders lower than that required for magnetism, emphasizing the need for a non-equilibrium growth processes to achieve the critical vacancy concentration. This seems to be consistent with the recent discovery of room-temperature ferromagnetism in thin films of $HfO_2$ **[20]**. Observation of ferromagnetism in thin films of non-magnetic oxides such as $TiO_2$, $In_2O_3$, and ZnO further indicate that defect induced ferromagnetism is possible in undoped non-magnetic oxide thin films **[21,22]**. Although the absence of magnetism in a $HfO_2$ thin film has been reported **[63]**, *Ab initio* electronic structure calculations using density functional theory have shown that charge neutral Hf vacancies can have partially occupied electronic levels within the band gap. The vacancies thus carry magnetic moments which couple ferromagnetically with a short range magnetic interaction **[69]**. The Hf vacancy concentration, under the most favorable growth conditions, is much lower than that required to achieve wall-to-wall percolation in the $HfO_2$ lattice. It is, therefore, concluded that equilibrium growth cannot lead to ferromagnetism and that ferromagnetism can be attained by nonequilibrium growth conditions that gives the required concentration of vacancies. At this point, it is instructive to note the energies involved in creating such point defects in the bulk and at the surface of a material. The defect formation energy at the surface is significantly different from that of the bulk due to structural and electronic effects **[70]**. *Ab initio* studies on the formation of cation vacancies in MgO have shown that the energy required to create vacancy at

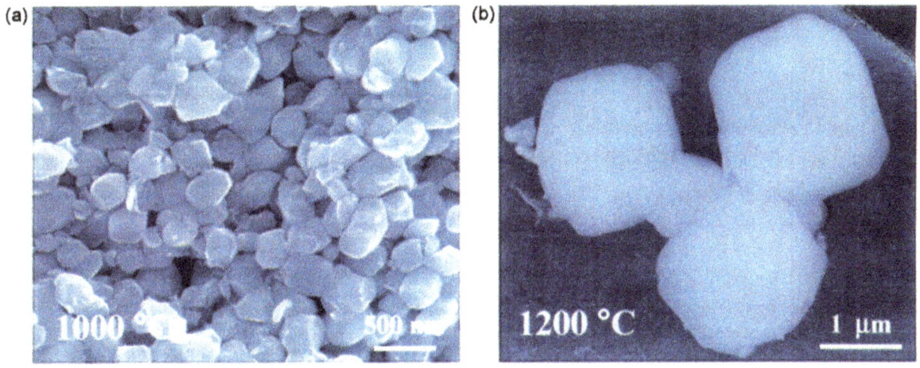


**Figure 9** FESEM images of (a) nanocrystalline and (b) bulk $BaTiO_3$.

the surface is lower than at the bulk of the material, indicating that the migration of defects to the surface should be thermodynamically favored **[71]**.

## Outlook

The observation that room-temperature ferromagnetism occurs universally in nanoparticles of inorganic materials is of great interest and has important implications. This result is relevant to understand the origin of magnetism in dilute magnetic semiconductors. Furthermore, it opens up a new way of introducing magnetism in non-magnetic functional materials. Multiferroic and magnetoelectric properties of nanocrystalline $BaTiO_3$ may be useful for device applications. Surface ferromagnetism in nanocrystalline superconductors may provide a means to study the magnetism in relation to superconductivity of inorganic nanoparticles. It would be worthwhile to further investigate magnetism of nanoparticles of diverse materials to fully understand this phenomenon.

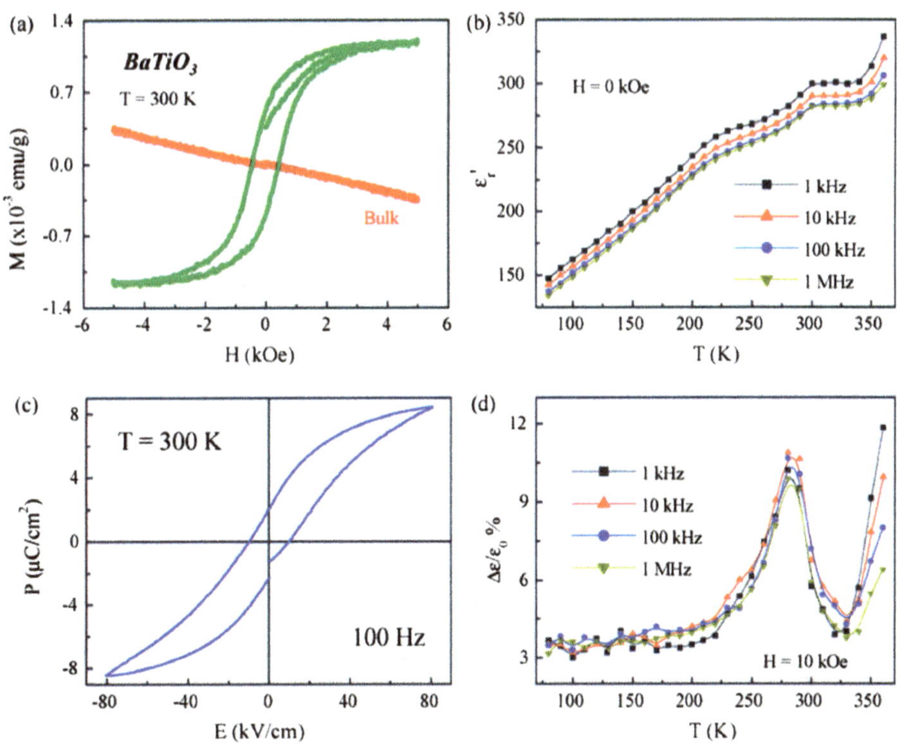


**Figure 10** (a) Magnetization curves of nanocrystalline and bulk $BaTiO_3$ (BTO) showing ferromagnetic and diamagnetic behavior respectively. (b) Dielectric constant of nanocrystalline BTO showing anomalies indicating the ferroelectric phase transitions. (c) P–E hysteresis curve of nanocrystalline BTO. (d) Magnetocapacitance of nanocrystalline BTO showing coupling of surface ferromagnetism with ferroelectricity (from Ref. **[49]**).

## Acknowledgements

We acknowledge Department of Science and Technology, India for financial support and thank R.V.K. Mangalam, Shipra, A. Gomathi and C. Madhu for their valuable help. We also thank Neil Dilley (Quantum Design) for high temperature magnetization measurements.

## References

[1] R.F. Ziolo, E.P. Giannelis, B.A. Weinstein, M.P. O'Horo, B.N. Ganguly, V. Mehrotra, et al., Science 257 (1992) 219.
[2] S. Sun, C.B. Murray, D. Weller, L. Folks, A. Moser, Science 287 (2000) 1989.
[3] T. Teranishi, H. Hori, M. Miyake, J. Phys. Chem. B 101 (1997) 5774.
[4] R.W. Chantrell, K. O'Grady, in: R. Gerber, C.D. Wright, G. Asti (Eds.), Applied Magnetism, Kluwer Academic Publishers, Dordrecht, 1994, p. 113.
[5] M. Zahn, K.E. Shenton, IEEE Trans. Magn. MAG-16 (1980) 387.
[6] L. Gunther, Phys. World 2 (1990) 28.
[7] H. Gu, P.L. Ho, K.W.T. Tsang, L. Wang, B. Xu, J. Am. Chem. Soc. 125 (2003) 15702.
[8] R.H. Kodama, S.A. Makhlouf, A.E. Berkowitz, Phys. Rev. Lett. 79 (1997) 1393.
[9] L. Néel, in: C. Dewitt, C. DeWitt-Morette, B. Dreyfus, P.D. de Gennes (Eds.), Low Temperature Physics: Lectures Delivered at Les Houches During the 1961 Session of the Summer School of Theoretical Physics, University of Grenoble, Gordon and Beach, New York, 1962, p. 433.
[10] M. Ghosh, K. Biswas, A. Sundaresan, C.N.R. Rao, J. Mater. Chem. 16 (2006) 106.
[11] M. Ghosh, E.V. Sampathkumaran, C.N.R. Rao, Chem. Mater. 17 (2005) 2348.
[12] F. Bødker, S. Mørup, S. Linderoth, Phys. Rev. Lett. 72 (1994) 282.
[13] A. Sundaresan, R. Bhargavi, N. Rangarajan, U. Siddesh, C.N.R. Rao, Phys. Rev. B 74 (2006) 161306.
[14] Shipra, A. Gomathi, A. Sundaresan, C.N.R. Rao, Solid State Commun. 142 (2007) 685.
[15] C. Madhu, A. Sundaresan, C.N.R. Rao, Phys. Rev. B 77 (2008) 201306.
[16] F. Li, X. Yu, H. Pan, M. Wang, X. Xin, Solid State Sci. 2 (2000) 767.
[17] X.-l. Song, P. Qu, H.-p. Yang, X. He, G.-z. Qiu, J. Cent. South Univ. Technol. 12 (2005) 536.
[18] C.S. Rout, A.R. Raju, A. Govindaraj, C.N.R. Rao, Solid State Commun. 138 (2006) 136.
[19] F. Zhang, S.-W. Chan, J.E. Spanier, E. Apak, Q. Jin, R.D. Robinson, et al., Appl. Phys. Lett. 80 (2002) 127.
[20] M. Venkatesan, C.B. Fitzgerald, J.M.D. Coey, Nature 430 (2004) 630.
[21] N.H. Hong, J. Sakai, N. Poirot, V. Brize, Phys. Rev. B 73 (2006) 132404.
[22] N.H. Hong, J. Sakai, V. Brize, J. Phys.: Condens. Matter 19 (2007) 036219.
[23] F. Esch, S. Fabris, L. Zhou, T. Montini, C. Africh, P. Fornasiero, et al., Science 309 (2005) 752.
[24] F. Zhang, P. Wang, J. Koberstein, S. Khalid, S.-W. Chan, Surf. Sci. 563 (2004) 74.
[25] P. Dutta, S. Pal, M.S. Seehra, Y. Shi, E.M. Eyring, R.D. Ernst, Chem. Mater. 18 (2006) 5144.
[26] Y. Liu, Z. Lockman, A. Aziz, J. MacManus-Driscoll, J. Phys.: Condens. Matter 20 (2008) 165201.
[27] A.M. Stoneham, A.P. Pathak, R.H. Bartram, J. Phys. C: Solid State Phys. 9 (1976) 73.
[28] F.A. Ponce, D.P. Bour, Nature 386 (1997) 351.
[29] T. Dietl, H. Ohno, F. Matsukura, J. Cibert, D. Ferrand, Science 287 (2000) 1019.
[30] C. Xu, J. Chun, K. Rho, D.E. Kim, B.J. Kim, S. Yoon, et al., J. Appl. Phys. 99 (2006) 064312.
[31] P.P. Chen, H. Makino, J.J. Kim, T. Yao, J. Cryst. Growth 251 (2003) 331.
[32] M. Zajac, R. Doradzinski, J. Gosk, J. Szczytko, M. Lefeld-Sosnowska, M. Kaminska, et al., Appl. Phys. Lett. 78 (2001) 1276.
[33] A. Bonanni, M. Kiecana, C. Simbrunner, T. Li, M. Sawicki, M. Wegscheider, et al., Phys. Rev. B 75 (2007) 125210.
[34] K. Biswas, K. Sardar, C.N.R. Rao, Appl. Phys. Lett. 89 (2006) 132503.
[35] K. Sardar, M. Dan, B. Schwenzer, C.N.R. Rao, J. Mater. Chem. 15 (2005) 2175.
[36] C.G. Van de Walle, J. Neugebauer, J. Appl. Phys. 95 (2004) 3851.
[37] S. Limpijumnong, C.G. Van de Walle, Phys. Rev. B 69 (2004) 035207.
[38] P. Mahadevan, S. Mahalakshmi, Phys. Rev. B 73 (2006)153201.
[39] P. Larson, S. Satpathy, Phys. Rev. B 76 (2007) 245205.
[40] W. Wang, I. Germanenko, M.S. El-Shall, Chem. Mater. 14 (2002) 3028.
[41] M. Lange, M.J.V. Bael, Y. Bruynseraede, V.V. Moshchalkov, Phys. Rev. Lett. 90 (2003) 197006.
[42] J.R. Thompson, D.K. Christen, H.R. Kerchner, L.A. Boatner, B.C. Sales, Y.R. Sun, in: R.A. Hein, T.L. Francavilla, D.H. Liebenberg (Eds.), Magnetic Susceptibility of Superconductors and Other Spin Systems, Plenum Press, New York, 1992, p. 157.
[43] U. Atzmony, R.D. Shull, C.K. Chiang, L.J. Swartzendruber, L.H. Bennett, R.E. Watson, J. Appl. Phys. 63 (1988) 4179.
[44] A.M. Stoneham, L.W. Smith, J. Phys.: Condens. Matter 3 (1991) 225.
[45] A. Gomathi, C.N.R. Rao, Mater. Res. Bull. 41 (2006) 941.
[46] H. Schmid, Ferroelectrics 162 (1994) 317.
[47] C. Ederer, N.A. Spaldin, Nat. Mater. 3 (2004) 849.
[48] C.N.R. Rao, C.R. Serrao, J. Mater. Chem. 17 (2007) 4931.
[49] R.V.K. Mangalam, N. Ray, U.V. Waghmare, A. Sundaresan, C.N.R. Rao, Solid State Commun. (2008), in press.
[50] P. Crespo, R. Litrán, T.C. Rojas, M. Multigner, J.M. de la Fuente, J.C. Sánchez-López, et al., Phys. Rev. Lett. 93 (2004)087204.
[51] Y. Yamamoto, H. Hori, Rev. Adv. Mater. Sci. 12 (2006) 23.
[52] T. Taniyama, E. Ohta, T. Sato, EPL (Europhys. Lett.) 38 (1997) 195.
[53] A.H.B. Sampedro, Phys. Status Solidi (c) 1 (2004) 3670.
[54] H. Ohno, H. Munekata, T. Penney, S. von Molnár, L.L. Chang, Phys. Rev. Lett. 68 (1992) 2664.
[55] H. Ohno, A. Shen, F. Matsukura, A. Oiwa, A. Endo, S. Katsumoto, et al., Appl. Phys. Lett. 69 (1996) 363.
[56] C. Zener, Phys. Rev. 81 (1951) 440.
[57] C. Zener, Phys. Rev. 83 (1951) 299.
[58] S. Sonoda, S. Shimizu, T. Sasaki, Y. Yamamoto, H. Hori, J. Cryst. Growth 237–239 (2002) 1358.
[59] K. Ueda, H. Tabata, T. Kawai, Appl. Phys. Lett. 79 (2001) 988.
[60] Y. Matsumoto, M. Murakami, T. Shono, T. Hasegawa, T. Fukumura, M. Kawasaki, et al., Science 291 (2001) 854.
[61] C.N.R. Rao, F.L. Deepak, J. Mater. Chem. 15 (2005) 573.
[62] R. Seshadri, Curr. Opin. Solid State Mater. Sci. 9 (2005) 1.
[63] D.W. Abraham, M.M. Frank, S. Guha, Appl. Phys. Lett. 87 (2005) 252502.
[64] A.M. Stoneham, Theory of Defects in Solids: Electronic Structure of Defects in Insulators and Semiconductors, Oxford University Press, Oxford, 2001.
[65] I.S. Elfimov, S. Yunoki, G.A. Sawatzky, Phys. Rev. Lett. 89 (2002) 216403.
[66] L.E. Halliburton, D.L. Cowan, W.B.J. Blake, J.E. Wertz, Phys. Rev. B 8 (1973) 1610.

[67] B.H. Rose, L.E. Halliburton, J. Phys. C: Solid State Phys. 7 (1974) 3981.
[68] J. Osorio-Guillen, S. Lany, S.V. Barabash, A. Zunger, Phys. Rev. Lett. 96 (2006) 107203.
[69] C.D. Pemmaraju, S. Sanvito, Phys. Rev. Lett. 94 (2005) 217205.
[70] G. Schwarz, A. Kley, J. Neugebauer, M. Scheffler, Phys. Rev. B 58 (1998) 1392.
[71] P. Baranek, G. Pinarello, C. Pisani, R. Dovesi, PCCP 2 (2000) 3893.

**A. Sundaresan** is an Associate Professor at Jawaharlal Nehru Centre for Advanced Scientific Research (JNCASR). He received his PhD degree in Chemistry from Indian Institute of Technology, Bombay in 1994. He was a post doctoral fellow at Laboratoire CRISMAT, France during 1995–1997. He was an Invited Researcher at LEPES, CNRS, Grenoble and Researcher of CREST-JST at AIST, Tsukuba before joining JNCASR as an Assistant Professor in 2004. His current research interests are in chemistry and physics of oxide materials with particular emphasis on superconductivity, multiferroism and magnetism in bulk and nanomaterials.

**C.N.R. Rao** obtained his PhD degree from Purdue University and DSc degree from the University of Mysore. He is the National Research Professor of India, Linus Pauling Research Professor at the Jawaharlal Nehru Centre for Advanced Scientific Research and Honorary Professor at the Indian Institute of Science. His research interests are in the chemistry of materials. He has authored nearly 1400 research papers and edited or written 40 books in chemistry. A member of several academies including the Royal Society and the US National Academy of Sciences, French Academy and the Pontifical Academy of Sciences. He is the recipient of the Einstein Gold Medal of UNESCO, Hughes Medal of the Royal Society, and the Somiya Award of the International Union of Materials Research Societies (IUMRS). In 2005, he received the Dan David Prize for materials research from Israel and the first India Science Prize.

2404 *J. Phys. Chem. C* **2008,** *112,* 2404–2411

# Growth Kinetics of ZnO Nanorods: Capping-Dependent Mechanism and Other Interesting Features

**Kanishka Biswas, Barun Das, and C. N. R. Rao***

*Chemistry and Physics of Materials Unit, DST nanoscience unit and CSIR Centre of Excellence in Chemistry, Jawaharlal Nehru Centre for Advanced Scientific Research, Jakkur P. O., Bangalore-560064, India, and Solid State and Structural Chemistry Unit, Indian Institute of Science, Bangalore-560012, India*

*Received: September 18, 2007; In Final Form: November 6, 2007*

Although the growth of nanocrystals has been investigated by several workers, investigations of the growth of 1-D nanostructures have been limited. We have investigated the growth kinetics of both uncapped and poly(vinyl pyrollidone) (PVP)-capped ZnO nanorods carefully by a combined use of transmission electron microscopy (TEM) and small-angle X-ray scattering (SAXS) which provide direct information on size and shape and compensate for the deficiency of each other. Values of average length and diameter of the ZnO nanorods obtained by TEM and SAXS are comparable. In the presence of the capping agent, the length of the nanorods grows faster while the diameter becomes narrower. The length distribution shows periodic changes in the width in the case of the uncapped nanorods, a feature absent in the case of the capped nanorods. In the absence of the capping agent, we observe the presence of small nanocrystals next to the nanorods after a lapse of time. The occurrence of small nanocrystals as well as the periodic focusing and defocusing of the width of the length distribution lend support to the diffusion-limited growth model for the growth of uncapped ZnO nanorods. Accordingly, the time dependence of the length of uncapped nanorods follows the $L^3$ law as required for diffusion-limited Ostwald ripening, while the PVP-capped nanorods show a time dependence which is best described by a combination of diffusion and surface reaction with a $L^3 + L^2$ type behavior. Collapse of all distribution curves obtained at different times of the reaction into a single universal Gaussian in the case of the PVP-capped nanorods also shows that the growth mechanism is more complex than Ostwald ripening.

## Introduction

Nanorods and nanowires constitute an important class of nanomaterials[1,2] and chemical routes have proved to be more useful for the synthesis of these one-dimensional materials, just as in the case of nanocrystals.[3,4] The growth mechanism of nanocrystals has attracted the attention of a few workers recently. One of the popular mechanisms employed to explain the growth kinetics of nanocrystals is the diffusion-limited Ostwald ripening process following the Lifshitz–Slyozov–Wagner (LSW) theory.[5,6] An early study of the growth kinetics of ZnO nanocrystals in the absence of capping agents revealed the diffusion-limited growth mechanism to be valid.[7] A more recent study of ZnO nanocrystals in the absence of NaOH has also shown that the variation of particle size follows the LSW model for diffusion-limited coarsening.[8] Alivisatos and co-workers[9] examined the growth of CdSe and InAs nanocrystals by employing UV–visible absorption spectroscopy to determine the size of nanocrystals and by employing the band widths of the photoluminescence spectra to determine their size distribution. These workers observed a focusing and defocusing effect of the size distribution similar to that expected in Ostwald ripening. Qu et al.[10] have shown that the growth of CdSe nanocrystals involves a prolonged formation of relatively small particles (nucleation) followed by focusing and then defocusing of the size distribution because of the growth of the bigger particles and disappearance of the small ones. These workers also found the particle size distribution to be asymmetric, indicating a diffusion-limited mechanism in the last stage of the growth. In the case of $TiO_2$ nanocrystals, the average particle radius cubed is reported to increase linearly with time consistent with the LSW model of coarsening.[11] On the basis of an in situ transmission electron microscope (TEM) investigation, El-Sayed and co-workers[12] report a diffusion-controlled growth of small gold nanoclusters.

There are a few reports in the literature where the growth kinetics of nanocrystals are found to deviate from the simple diffusion-limited Ostwald ripening model. Thus, Seshadri et al.[13] propose the growth of gold nanoparticles to be essentially stochastic wherein the nucleation and growth steps are well-separated. Furthermore, these workers observe that the average diameter and the standard deviation of the size distribution exhibit the same time dependence. Theoretical considerations suggest that the growth of the nanocrystals could be either controlled by the diffusion of particles, by the reaction at the surface, or by both factors.[14] Thus, Viswanatha et al.[15] observe the growth of the ZnO nanocrystals in water to follow a growth mechanism intermediate between diffusion-control and surface reaction-control. These workers also report the growth kinetics of the ZnO nanocrystals in the absence of any capping agent to be slower than that predicted by diffusion-controlled Ostwald ripening and to be dependent on the concentration of $OH^-$ in a nonmonotonic manner as well as on the temperature.[16]

Growth kinetics of nanocrystals in the presence of capping agents is determined by several complex factors, and signatures of either the diffusion or the reaction-controlled regimes are

* To whom correspondence should be addressed. E-mail: cnrrao@jncasr.ac.in. Fax: +91 80 22082760.

10.1021/jp077506p CCC: $40.75 
Published on Web 01/29/2008

seen. The effect of capping agents on the growth of nanocrystals have been examined by a few workers.[9,10,17–23] For example, the growth of ZnO nanocrystals in the presence of poly(vinyl pyrollidone) (PVP) is reported to show deviations from Ostwald ripening.[17] To our knowledge, information of the mechanism of the growth of nanorods, especially in the presence of capping agents, is limited. A useful study in this context is that of Peng and Peng[24,25] who examined the growth kinetics of CdSe nanorods by UV–vis spectroscopy and TEM images and found the diffusion-controlled model to be valid when the monomer concentration was sufficiently high. At low monomer concentrations, the aspect ratio of the rod decreases because of intraparticle diffusion on the surface of the nanocrystal. Thoma et al.[26] have observed anisotropic growth of CdSe in the presence of a surfactant. On the basis of TEM studies, Pascholski et al.[27] proposed that small ZnO nanoparticles are converted to rods by the oriented attachment mechanism assisted by Ostwald ripening. On the basis of the estimation of nanorod lengths from XRD peak broadening, Zhu et al.[28] have found the growth kinetics of oleic acid-capped ZnO nanorods to deviate from Ostwald ripening.

In the present study, we have carried out a detailed investigation of the growth of ZnO nanorods prepared solvothermally in the presence and absence of a capping agent by employing TEM and SAXS. Employing two such independent techniques is important because of the limitations of the techniques themselves. While TEM is the most direct probe to observe the size, shape, and the size distribution of nanostructures, it is not possible to carry out in situ measurements. Furthermore, the sampling size in TEM is rather small. SAXS, on the other hand, provides a direct probe to determine the size and shape of nanomaterials, and the sampling size is much larger than what can be used in TEM. The determination of size and shape in SAXS, however, is dependent on the model employed for fitting the data. A combined use of TEM and SAXS provides a satisfactory means to study the growth of one-dimensional (1D) nanorods. We should note that UV–vis, photoluminescence, and such spectroscopic methods are indirect and are strongly affected by the change in the electronic structure of the nanomaterials. Furthermore, the spectroscopic methods are not as suitable for the study of nanorods as for nanocrystals, since the size of the former is determined both by length and by radius. Widths of X-ray diffraction peaks provide only average values of size which are not reliable in the case of nanorods.

The present study of the growth of ZnO nanorods in the presence and absence of a capping agent (PVP) has enabled us not only to determine the growth mechanism of ZnO nanorods but also to determine the effect of the capping agent. The length of the nanorods increases preferentially in the presence of PVP. We find that the growth mechanism of the ZnO nanorods in the absence of PVP follows the diffusion-limited Ostwald ripening mechanism, with evidence of focusing and defocusing of the length distribution. In the presence of PVP, the growth mechanism deviates from the diffusion-limited LSW model requiring an additional contribution from a surface process.

## Experimental Section

**Synthesis of ZnO Nanorods.** In order to carry out the growth study in the absence of any capping agent, ZnO nanorods were prepared by the reaction of zinc acetate dihydrate ($Zn(CH_3COO)_2{\cdot}2H_2O$) and sodium hydroxide in ethanol at 100 °C under solvothermal conditions. The reaction was stopped at different times (1, 2, 3, 6, 12, 18, and 24 h), and the products were analyzed by TEM and SAXS. In a typical synthesis, $Zn(CH_3COO)_2{\cdot}2H_2O$ (0.05 g. 0.23 mmol) was dissolved in 12 mL ethanol, and NaOH (0.228 g, 5.69 mmol) was added under stirring. The reaction mixture was sealed in a Teflon-lined autoclave of 20 mL capacity (60% filling fraction) and maintained at 100 °C in a hot air oven. The solid products obtained at different times after the reaction were thoroughly washed with ethanol and distilled water. To study the effect of the capping agent on the growth process, we prepared the ZnO nanorods in the presence of PVP (0.25 g, MW ~ 55 000) by maintaining the other reaction parameters the same as in the synthesis of the uncapped ZnO nanorods. The samples were taken out after different reaction times (1, 2, 3, 6, 12, 18, and 24 h) for investigation.

**TEM Characterization.** The solid products obtained after different reaction periods were dispersed in ethanol by sonication, and the dispersions were taken on holey carbon-coated Cu grids for TEM investigations with a JEOL (JEM3010) microscope operating with an accelerating voltage of 300 kV. The length and diameter distribution were obtained from magnified micrographs by using DigitalMicrograph 3.4 software. Typically, 150–200 well-separated nanorods from three or four micrographs of the same sample were used to arrive at the size distribution.

**SAXS Characterization.** The average length and diameter of the ZnO nanorods could be readily obtained by SAXS.[29–31] We performed SAXS experiments with a Bruker-AXS NanoSTAR instrument modified and optimized for solution scattering. The instrument is equipped with a X-ray tube (Cu Kα radiation, operated at 45 kV/35 mA), cross-coupled Göbel mirrors, three-pinhole collimation, evacuated beam path, and a two-dimentional (2D) gas-detector (HI-STAR).[31] The modulus of the scattering vector is $q = 4\pi \sin \theta/\lambda$, where $2\theta$ is the scattering angle and $\lambda$ is the X-ray wavelength. We recorded the SAXS data in the $q = 0.007$ to 2.2 Å$^{-1}$ range, that is, $2\theta = 0.1$ to 3°. Solutions of the ZnO nanorods in ethanol (approximately 0.1 w/v % concentration) obtained after lapse of different reaction times were used for SAXS measurements. The ethanol solutions of nanorods were taken in quartz capillaries (diameter of about 2 mm) for the measurements. A capillary filled with only ethanol was used for background correction. The concentration of the nanorods was sufficiently low to neglect interparticle interference effects. The experimental SAXS data were fitted by Bruker-AXS DIFFRAC$^{plus}$ NANOFIT software by using a solid cylinder model. The form factor of the cylinder used in this software is given by Fournet.[32,33]

## Results and Discussion

In order to prepare the ZnO nanorods, we employed a high monomer concentration with a $Zn^{2+}/OH^-$ ratio of 1:25. High monomer concentrations favor the growth of the ZnO nanorods[27,34] because the chemical potential of an elongated structure is generally higher than a dot-shaped nanocrystal because of the lower surface energy of anisotropic structure.[25] As a result, the growth of such an anisotropic structure requires a relatively high chemical potential environment, that is, high monomer concentration in the solution. We could obtain sufficient concentrations of the nanorods with the reactant concentrations and the reaction conditions employed by us. In what follows, we discuss the results of TEM and SAXS studies of the nanorods obtain after different reaction times.

Figure 1a,b shows typical TEM images of the uncapped product after 3 and 12 h of reaction. The inset in Figure 1a indicates the presence of a small nanocrystal on the side wall of an uncapped nanorod in the process of dissolution. The inset

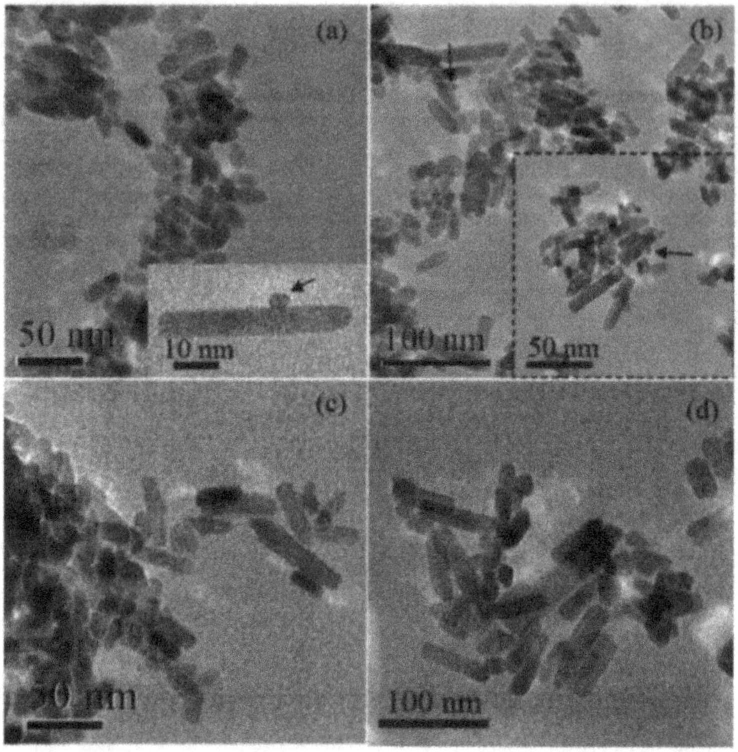


**Figure 1.** (a,b) TEM images of uncapped ZnO nanorods obtained after 3 and 12 h of reaction. The inset in (a) shows how a small ZnO nanoparticle attached to the side wall of an uncapped ZnO nanorod. The inset in (b) is a TEM image of uncapped ZnO nanorods obtained after 18 h of reaction. The arrow indicates the existence of small nanocrystal along with the nanorods. Several such nanocrystals can be seen in the image. (c,d) TEM images of PVP-capped ZnO nanorods obtained after 3 and 12 h of reaction. Note the absence of nanocrystals sticking to the nanorods.

in Figure 1b shows a TEM image of the uncapped ZnO nanorods after 18 h of reaction. The presence of small nanocrystals with the nanorods is indicated by black arrows in Figure 1b. The existence of small nanocrystals along with the uncapped nanorods after a sufficiently long period of reaction (e.g., 12 or 18 h) suggests that the diffusion-limited Ostwald ripening may be operative, wherein large particles grow at the cost of smaller ones. At the early stage of the growth, the monomer concentration is sufficiently high for the nanorods to grow, and the growth process is fast. After longer times, the monomer concentration gets depleted below a certain critical level, rendering the smaller nanocrystals to dissolve. El-Sayed and co-workers[12] have reported the existence of smaller gold nanocrystals located near gold nanorods and observed shrinkage of the nanocrystals with increasing temperature.

From the TEM studies, we have estimated the distributions of the length as well as the diameter of the ZnO nanorods after different times of the reaction. Figure 2 shows the time evolution of the length distribution of the uncapped ZnO nanorods (in blue). We have fitted each distribution curve by a Gaussian distribution function (solid blue curve). We see from Figure 2, length distribution becomes sharper and broader in a periodic fashion. Such focusing and defocusing of the diameter distribution is a phenomenon noticed earlier in the case of CdSe nanoparticles growth.[9] Since the monomer concentration is high at the beginning of the reaction, focusing of the length distribution (e.g., $t = 3$ h) occurs because of a faster growth rate of the small nanorods. When the monomer concentration gets depleted because of the faster growth of the nanorods, the smaller nanorods start to shrink while the longer ones keep growing. The size distribution, therefore, becomes broader (e.g., $t = 6$ h). Dissolution of the small nanorods again enriches the monomer concentration in the solution, with the growth of the longer rods continuing through the diffusion of the monomer from solution to the nanorod surface, giving rise to focusing of the length distribution (e.g., $t = 9$, 12 h). Periodic focusing and defocusing of the length distribution is thus dependent on the periodic variation of the monomer concentration in the solution. Focusing and defocusing of the length distribution gives a clean qualitative indication of diffusion-limited growth of uncapped ZnO nanorods.

In Figure 1c,d, we show typical TEM images of PVP-capped ZnO nanorods after 3 and 12 h of reaction. We do not see the presence of small nanocrystals along with the nanorods as found earlier in the case of the uncapped nanorods. We have estimated the length distribution of the PVP-capped ZnO nanorods from the TEM images. In Figure 2, we show the histograms (in black) with Gaussian distribution fittings (solid black curve). In the presence of PVP, we do not observe periodic changes in the width of the distribution curves. Instead, we find the width of the length distribution to become broader continuously with reaction time. This reflects the growth of ZnO nanorods in the presence of PVP to be more complex than simple diffusion-limited growth. Seshadri et al.[13] had observed a similar increase in the width of the distribution with time in the case of gold nanocrystals.

We now examine the results of SAXS measurements at the same reaction time intervals as in the case of TEM studies. Figure 3a,b shows typical plots of intensity versus scattering vector in the logarithmic scale for different times of the reaction

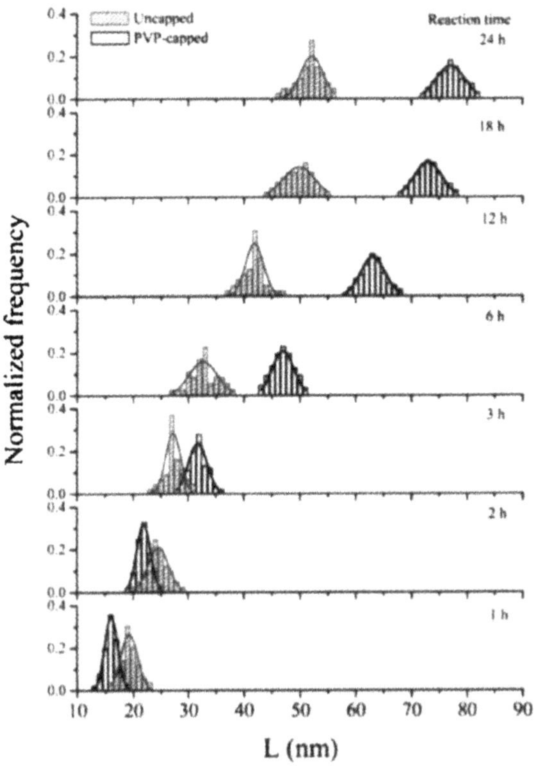


**Figure 2.** Time evolution of the length distributions of the uncapped (blue) and the PVP-capped (black) ZnO nanorods obtained from TEM images.

in the case of uncapped and PVP−capped ZnO respectively. Under the reaction conditions employed, the length of the nanorods varies significantly with time, while the diameter varies only slightly. On the basis of this observation, we assume the cylindrical shape for the nanorods to analyze the SAXS data. A cylinder is a close approximation of a hexagonal rod. To date, the hexagonal rod model has not been employed in the literature to analyze SAXS data. In order to estimate the average length and diameter of ZnO nanorods, we have fitted the experimental SAXS data to the cylinder model of Bruker-AXS DIFFRAC$^{plus}$ NANOFIT software. The form factor of the cylinder is that due to Fournet:[32,33]

$$P(q,R,L) = \int_0^{\pi/2} \left[\frac{2J_1(qR \sin \alpha)}{qR \sin \alpha}\,\frac{\sin(qL \cos \alpha/2)}{qL \cos \alpha/2}\right]^2 \sin \alpha \, d\alpha$$

Here, $q$ is the scattering vector, $R$ is the radius of the cylinder, $L$ is the length of the cylinder, $\alpha$ is the orientation dependent parameter (angle between scattering vector and cylinder long axis), and $J_1(x)$ is the first-order Bessel function of the first kind. We have not introduced the length and diameter distribution in the scattering cross-section equation because the form factor amplitude of a cylinder is a more complicated function of length, diameter, and the orientation of the cylinder than the form factor amplitude of a spherical model where the diameter is the only variable. Such a cylindrical model fitting of simulated SAXS data have been reported in the literature.[33,35] The solid lines in Figure 3 are the cylinder model fits of the experimental SAXS data. In the case of the uncapped ZnO nanorods, the theoretical fits are good, but the fits are not as good in the case of PVP-capped ZnO nanorods. We have constructed our model only considering the cylindrical ZnO nanorods present in the solution. The PVP molecules present on the side walls of the nanorods, may be responsible for not obtaining very good fits of the experimental data.

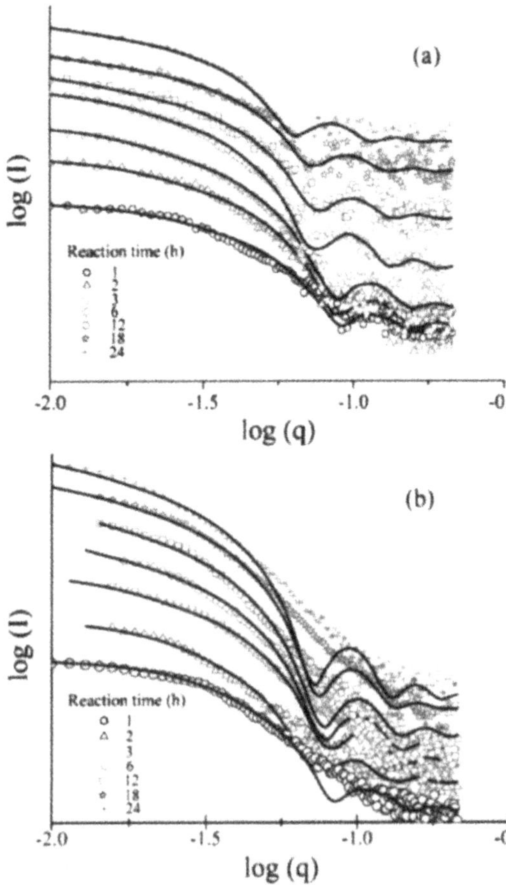


**Figure 3.** SAXS data of (a) uncapped and (b) PVP-capped ZnO nanorods obtained after different times of reaction. Solid lines are the model fits to the experimental data.

Figure 4a,b shows the time evolution of the average length calculated from TEM images (filled circles) and SAXS (open circle) of uncapped and PVP-capped ZnO nanorods, respectively. In the absence of PVP, the length of the nanorods increases from 19 to 52 nm after 24 h of reaction. In the presence of PVP, the growth is greater, and the length increases up to 77 nm after 24 h of reaction. In Figure 5a,b, we show the time evolution of the average diameter and aspect ratio calculated from TEM (filled circles) and SAXS (open circles) of uncapped ZnO nanorods, respectively. The diameter of the uncapped nanorods increases slightly starting from 8.3 to 12.3 nm. The aspect ratio increases from 2.3 to 4.3 after 24 h of reaction. Figure 6a,b shows the time evolution of the average diameter and aspect ratio obtained from TEM images (filled circles) and SAXS (open circles) of PVP-capped ZnO nanorods, respectively. In the presence of PVP, we observe an even smaller variation in the diameter (9.1 to 10.4 nm) compared with the capped ones. The aspect ratio, therefore, increases from 1.8 to 7.4 after 24 h of reaction. It is noteworthy that the estimates of lengths and diameters from TEM and SAXS agree closely.

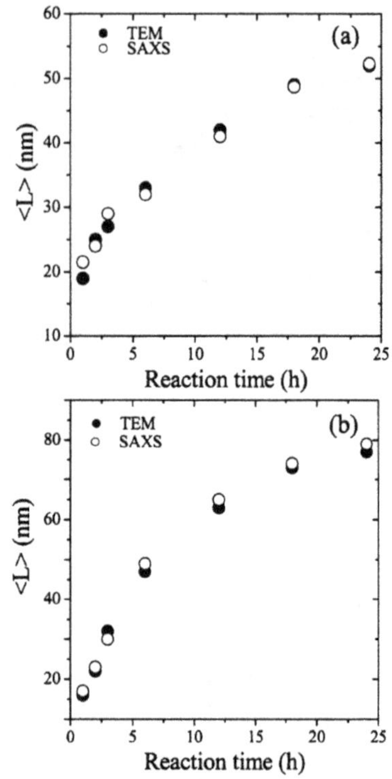


**Figure 4.** Time evolution of the average length (⟨*L*⟩) of (a) uncapped and (b) PVP-capped ZnO nanorods obtained from TEM and SAXS.

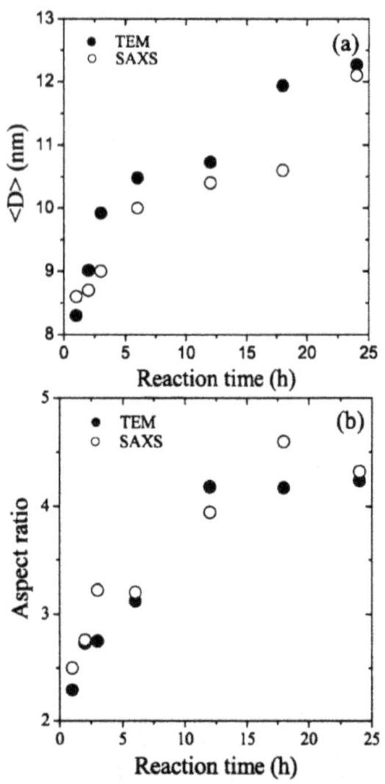


**Figure 5.** Time evolution of the (a) average diameter (⟨*D*⟩) and (b) aspect ratio of the uncapped ZnO nanorods obtained from TEM and SAXS.

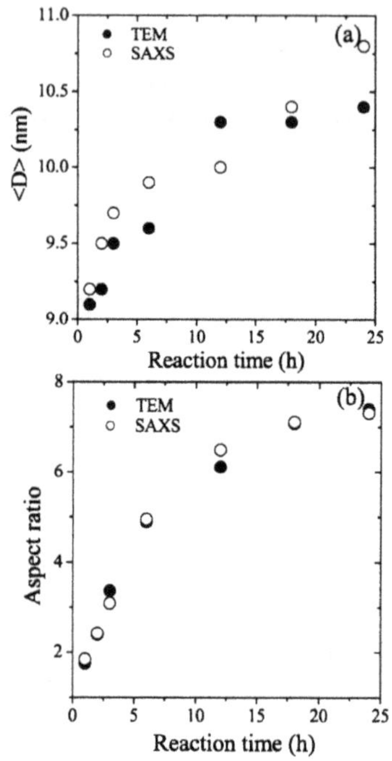


**Figure 6.** Time evolution of the (a) average diameter (⟨*D*⟩) and (b) aspect ratio of the PVP-capped ZnO nanorods obtained from TEM and SAXS.

ZnO with a polar hexagonal wurzite structure can be described as being due to the hexagonal close packing of the oxygen and zinc atoms, in point group 3*m* and space group $P6_3mc$, the zinc atoms being present in the tetrahedral sites. Thus, the crystal habit of wurzite ZnO exhibits well-defined crystallographic faces; that is, polar-terminated (0001) planes and six side facet are generally bound by the (10$\bar{1}$0) family of planes. The growth rates of different family of planes follow the sequence (0001) > (10$\bar{1}$1) > (10$\bar{1}$0).[36] ZnO anisotropic structures are normally bound by six (10$\bar{1}$0) facets grown along the (0001) direction, that is, along the *c* axis of the rods. Powder X-ray diffraction patterns of PVP-capped and uncapped ZnO nanorods could be indexed on the hexagonal space group $P6_3mc$ of ZnO. High-resolution electron microscope (HREM) images along with electron diffraction patterns also support the hexagonal crystal structure as well as the preferential *c*-directional growth of nanorods. Electron diffraction patterns confirm the single crystallinity of nanorods. PVP appears to selectively cap the side facet of the ZnO nanorods, allowing the growth to occur selectively along the *c* direction compared with the uncapped ZnO nanorods. As a result, we find the length to grow to a greater extent in the presence of PVP compared with the uncapped situation. When PVP is not present in the solution, there is a possibility of diffusion of the monomer to the side walls of the nanorods, rendering the nanorods thicker compared with the capped ones. The inset in Figure 1a gives a typical TEM image showing the presence of a small nanocrystal on the side wall of an uncapped nanorod in the process of dissolution which supports the proposed mechanism. We do not observe the presence of such small nanocrystal on the side of PVP-capped nanorods in the TEM images.

Figure 7a,b shows the time evolution of the standard deviation, $\sigma$, of the length calculated from TEM images for uncapped and PVP-capped ZnO nanorods, respectively. In the lower insets of Figure 7, we show the time evolution of the $\sigma$ of the diameter obtained from TEM images for uncapped and

J. Phys. Chem. C, Vol. 112, No. 7, 2008 **2409**

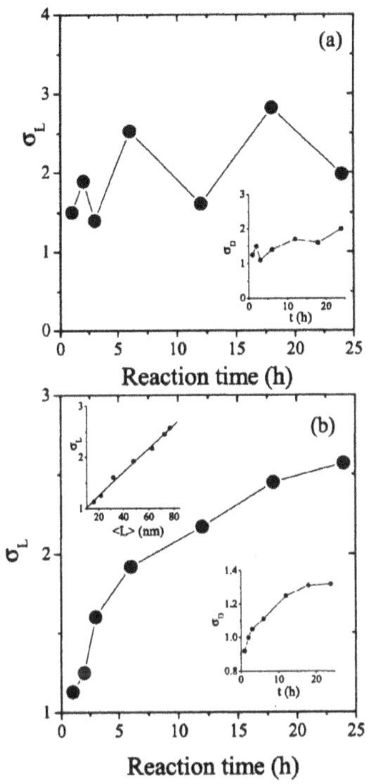


**Figure 7.** Time evolution of the standard deviation of the length ($\sigma_L$) of (a) uncapped and (b) PVP-capped ZnO nanorods. Lower insets in (a) and (b) show the time evolution of the standard deviation of the diameter ($\sigma_D$). Upper inset in (b) shows the plot of mean rod length, $\langle L \rangle$, against the $\sigma_L$.

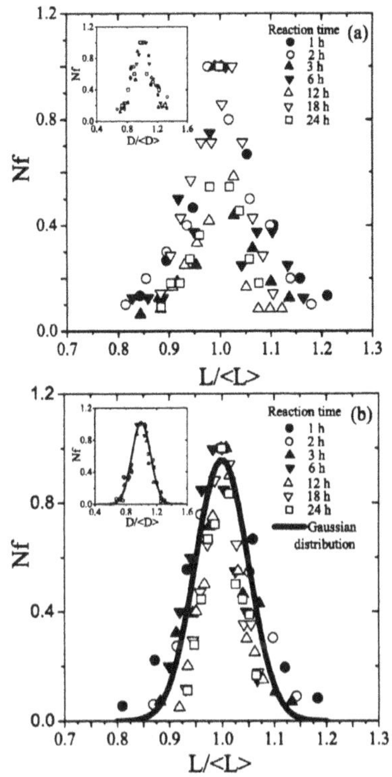


**Figure 8.** Universal length distribution curves of (a) uncapped and (b) PVP-capped ZnO nanorods obtained by scaling mean lengths at different times. Insets in (a) and (b) show the universal diameter distribution curves of (a) uncapped and (b) PVP-capped ZnO nanorods. (Nf = normalized frequency).

PVP-capped ZnO nanorods, respectively. We observe that $\sigma$ of the length ($\sigma_L$) and the diameter ($\sigma_D$) vary periodically with time in the absence of PVP as in Figure 7a. This indicates periodic fluctuation of the monomer concentration in the solution during the growth process of uncapped ZnO nanorods. In the presence of PVP, both $\sigma_L$ and $\sigma_D$ increase continuously with time. By plotting the average length as a function of $\sigma_L$ for different times in the case of PVP-capped ZnO nanorods, we find that the mean and the $\sigma_L$ bear a linear relationship, with a correlation coefficient ($R$) of 0.996. This means that both $\sigma_L$ and average length have the same time dependence (see upper inset in Figure 7b). We have observed the same linear relationship ($R = 0.988$) between $\sigma_D$ and the average diameter of the PVP-capped ZnO nanorods for different times. This observation suggests that it should be possible to represent the length or the diameter distribution by means of a variable scaled by the mean length or diameter. We do not observe such a relationship between $\sigma$ and average length or average diameter in the case of uncapped ZnO nanorods.

Figure 8a,b shows plots of normalized frequency (normalized by the maximum counts) against $L/\langle L \rangle$ in the case of uncapped and PVP-capped nanorods, respectively. In the insets in Figure 8a,b, we show the plots of normalized frequency (normalized by the maximum counts) against $D/\langle D \rangle$ in the case of uncapped and PVP-capped nanorods, respectively. In the case of the uncapped nanorods (Figure 8a), the distribution is somewhat asymmetric, and we do not obtain as good a universal curve. The asymmetric nature of universal length as well as diameter distribution curves supports a diffusion-limited Ostwald ripening growth mechanism in the case of the uncapped ZnO nanorods. In the case of PVP-capped nanorods, however, all of the data collapse to a single curve that can be represented by the Gaussian distribution function (solid line) as shown in Figure 8b. Ostwald ripening does not appear to be the sole factor determining growth in the case of the PVP-capped nanorods.

We have attempted to obtain information regarding the growth mechanisms of ZnO nanorods in the presence and the absence of the capping agent based on the time evolution of the average length of nanorods. If the diffusion-limited Ostwald ripening according to LSW theory[5,6] were to be the sole contributor of the growth mechanism, then the rate law would be given by

$$L^3 - L^3_0 = Kt$$

where, $L$ is the average length at time $t$ and $L_0$ is average initial length of the nanorods. The rate constant $K$ is given by $K = 8\gamma D V_m^2 C_\infty / 9RT$, where $D$ is the diffusion constant at temperature $T$, $V_m$ is the molar volume, $\gamma$ is the surface energy, and $C_\alpha$ is the equilibrium concentration at flat surface. We have tried to fit the $L(t)$ data obtained from TEM and SAXS to the Ostwald ripening model in the case of the uncapped (solid curve in Figure 9a) and the PVP-capped ZnO nanorods (broken curve in Figure 9b). The fit is reasonably good for the uncapped nanorods (reduced $\chi^2$ of the fit = 0.71 and coefficient of determination, $R^2 = 0.996$) as can be seen from Figure 9a. We have not found it possible to fit the experimental $L(t)$ data of the PVP-capped ZnO nanorods to the diffusion-limited model ($\chi^2 = 58.59$ and $R^2 = 0.903$). The growth process appears to deviate from diffusion-limited Ostwald ripening. We have also tried to fit the $L(t)$ data of the PVP-capped nanorods to the surface-limited reaction model (i.e., $L^2 \propto t$) or by varying the value of the

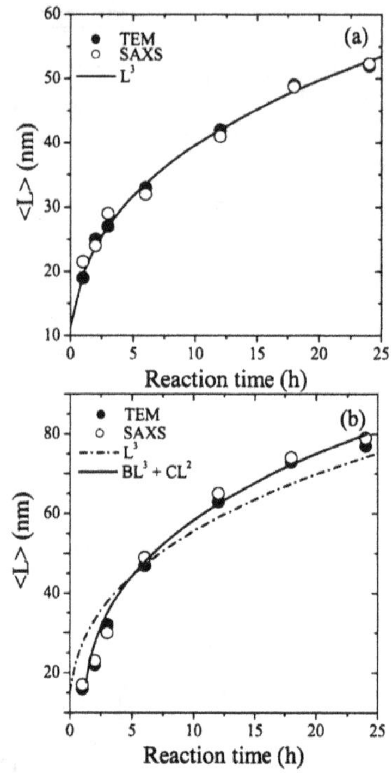


**Figure 9.** (a) Fits of $L(t)$ data (solid circles from TEM and open circles from SAXS) of uncapped ZnO nanorods to the Ostwald ripening ($L^3$) model (solid curve). (b) Fits of $L(t)$ data (solid circles from TEM and open circles from SAXS) of PVP-capped ZnO nanorods to the Ostwald ripening ($L^3$) model (broken curve) and mixed diffusion-surface control growth ($L^3 + L^2$) model (solid curve).

exponent ($n$ in $L^n \propto t$,). We have found the fit of the data either to the surface reaction model ($\chi^2 = 16.97$ and $R^2 = 0.972$) or to a variable $n$ model to be unsatisfactory. In order to fit the experimental $L(t)$ data of the PVP-capped ZnO nanorods, we have, therefore, used a model which contains both the diffusion-limited and surface-limited growth,[15]

$$BL^3 + CL^2 + \text{const} = t$$

where, $B = AT/\exp(-E_a/k_BT)$, $A \propto 1/(D_0\gamma V_m^2C_\infty)$, $C \propto T/(k_d\gamma V_m^2C_\infty)$, and $k_d$ is the rate constant of surface reaction. The remarkable goodness of fits ($\chi^2 = 0.95$ and $R^2 = 0.989$) over the entire range of experimental data by the mixed diffusion-reaction control model is shown by a thick solid curve in Figure 9b. Thus, the growth of the PVP-capped ZnO nanorods deviates sufficiently from the diffusion-limited Ostwald ripening model and follows a mechanism involving both diffusion-control and surface reaction control.

The growth of the nanorods mainly occurs by the diffusion of monomers from solution to the nanorod surface or by the reaction at the surface where units of the diffusing particles get assimilated into the growing nanorods. Diffusion and surface reaction are two limiting cases in the growth of nanorods. In the absence of a capping agent, the nanorod growth is essentially controlled by diffusion as suggested by TEM, as well as the goodness of fits of $L(t)$ data to diffusion-limited growth model. Presence of PVP gives rise to a barrier to diffusion. As a result, the contribution of the surface reaction becomes more prominent. The growth of PVP-capped nanorods occurs through a combination of diffusion and surface reaction processes. There is no barrier present to limit the diffusion process in the case of uncapped nanorods.

## Conclusions

Several major conclusions can be drawn of the growth mechanism of ZnO nanorods based on the present study. The values of the average length and diameter of the nanorods obtained by TEM and SAXS techniques are close to each other in the cases of both uncapped and PVP-capped ZnO nanorods. The average length of the PVP-capped ZnO nanorods grows at a higher rate compared with the uncapped ones, while the average diameter of the capped nanorods grows at a slower rate. PVP appears to selectively cap the side walls of the ZnO nanorods giving rise to such a preferential increase in the aspect ratio. The observation of small nanocrystals along with the nanorods in TEM images lends support to a growth mechanism based on the diffusion-limited Ostwald ripening process in the case of the uncapped nanorods. The uncapped nanorods exhibit periodic changes in the width of the length distribution curves because of changes in the monomer concentration. The periodic focusing and defocusing as well as the somewhat asymmetric nature of the distribution curves (Figure 8) are also the consequence of the diffusion-limited growth of uncapped ZnO nanorods. Good fits of the $L(t)$ data of uncapped ZnO nanorods with $L^3$ model confirms the growth mechanism to be mainly diffusion-controlled. In the presence of PVP, however, the nonexistence of small nanocrystals with the nanorods in the TEM images and the continuous broadening of the width of the size distributions with time reveal the growth mechanism to be more complex than a diffusion-limited process. Fits of the $L(t)$ data to $L^3 + L^2$ model suggest that the growth mechanism of the PVP-capped ZnO nanorods involves both diffusion and surface reactions. A similar time dependence of standard deviation $\sigma_L$ (or $\sigma_D$) and the average length (or diameter) as well as the collapse of all distributions onto a single Gaussian also suggest that the basic process contributing to the growth of PVP-capped nanorods to be distinct from Ostwald ripening alone.

## References and Notes


(1) Rao, C. N. R.; Govindaraj, A. *Nanotubes and Nanowires*, RSC series on Nanoscience, Royal Society of Chemistry, London, 2006.

(2) Rao, C. N. R., Muller, A., Cheetham, A. K., Eds. *The Chemistry of Nanomaterials*; Wiley-VCH: Weinheim, 2004; Vols. 1,2.

(3) Rao, C. N. R.; Thomas P. J.; Kulkarni, G. U. *Nanocrystals: Synthesis, Properties and Applications*; Springer: New York, 2007.

(4) Burda, C.; Chen, X.; Narayanan, R.; El-Sayed, M. A. *Chem. Rev.* **2005**, *105*, 1025.

(5) Lifshitz, I. M.; Slyozov, V. V. *J. Phys. Chem. Solids* **1961**, *19*, 35.

(6) Wagner, C. *Z. Elektrochem.* **1961**, *65*, 581.

(7) Wong, E. M.; Bonevich, J. E.; Searson, P. C. *J. Phys. Chem. B* **1998**, *102*, 7770.

(8) Hu, Z.; Ramírez, D. J. E.; Cervera, B. E. H.; Oskam, G.; and Searson, P. C. *J. Phys. Chem. B* **2005**, *109*, 11209.

(9) Peng, X.; Wickham, J.; Alivisatos, A. P. *J. Am. Chem. Soc.* **1998**, *120*, 5343.

(10) Qu, L.; Yu, W. W.; Peng, X. *Nano Lett.* **2004**, *4*, 465.

(11) Oskam, G.; Nellore, A.; Penn, R. L.; Searson, P. C. *J. Phys. Chem. B* **2003**, *107*, 1734.

(12) Mohamed, M. B.; Wang, Z. L.; El-Sayed, M. A. *J. Phys. Chem. A* **1999**, *103*, 10255.

(13) Seshadri, R.; Subbanna, G. N.; Vijayakrishnan, V.; Kulkarni, G. U.; Ananthakrishna, G.; Rao, C. N. R. *J. Phys. Chem.* **1995**, *99*, 5639.

(14) Talapin, D. V.; Rogach, A. L.; Haase, M.; Weller, H. *J. Phys. Chem.* **2001**, *105*, 12278.

(15) Viswanatha, R.; Santra, P. K.; Dasgupta, C.; Sarma, D. D. *Phys. Rev. Lett.* **2007**, *98*, 255501.

(16) Viswanatha, R.; Amenitsch, H.; Sarma, D. D. *J. Am. Chem. Soc.* **2007**, *129*, 4470.

(17) Viswanatha, R.; Sarma, D. D. *Chem. Eur. J.* **2006**, *12*, 180.

(18) Pesika, N. S.; Hu, Z.; Stebe, K. J.; Searson, P. C. *J. Phys. Chem.* **2002**, *106*, 6985.
(19) Bullen, C. R.; Mulvaney, P. *Nano Lett.* **2004**, *4*, 2303.
(20) Yu, W. W.; Peng, X. *Angew. Chem., Int. Ed.* **2002**, *41*, 2368.
(21) Qu, L.; Peng, X. *J. Am. Chem. Soc.* **2002**, *124*, 2049.
(22) Dai, Q.; Li, D.; Chen, H.; Kan, S.; Li, H.; Gao, S.; Hou, Y.; Liu, B.; Zou, G. *J. Phys. Chem. B* **2006**, *110*, 16508.
(23) Wong, E. M.; Hoertz, P. G.; Liang, C. J.; Shi, B.; Meyer, G. J.; Searson, P. C. *Langmuir* **2001**, *17*, 8362.
(24) Peng, Z. A.; Peng, X. *J. Am. Chem. Soc.* **2001**, *123*, 1389.
(25) Peng, Z. A.; Peng, X. *J. Am. Chem. Soc.* **2002**, *124*, 3343.
(26) Thoma, S. G.; Sanchez, A.; Provencio, P. P.; Abrams. B. L.; Wilcoxon, J. P. *J. Am. Chem. Soc.* **2005**, *127*, 7611.
(27) Pacholski, C.; Kornowski, A.; Weller, H. *Angew. Chem., Int. Ed.* **2002**, *41*, 1188.
(28) Zhu, Z.; Andelman, T.; Yin, M.; Chen, T.-L.; Ehrlich, S. N.; O'Brien, S. P.; Osgood, R. M. *J. Mater. Res.* **2005**, *20*, 1033.
(29) Pedersen, J. S. *Adv. Colloid Interface Sci.* **1997**, *70*, 171.
(30) Pedersen, J. S. In *Neutrons, X-rays and Light: Scattering Methods Applied to Soft Condensed Matter*; Lindner, P., Zemb, Th., Eds.; North-Holland: New York, 2002, 391.
(31) Pedersen, J. S. *J. Appl. Crystallogr.* **2004**, *37*, 369.
(32) Fournet, G. *Bull. Soc. Franc. Minéral. Crist.* **1951**, *74*, 39.
(33) Pedersen, J. S. *J. Appl. Crystallogr.* **2000**, *33*, 637.
(34) Cao, H. L.; Qian, X. F.; Gong, Q.; Du, W. M.; Ma, X. D.; Zhu, Z. K. *Nanotechnology* **2006**, *17*, 3632.
(35) Kaya, H. *J. Appl. Crystallogr.* **2004**, *37*, 223.
(36) Kar, S.; Dev, A.; Chaudhuri, S. *J. Phys. Chem. B* **2006**, *110*, 17848.

Acc. Chem. Res. **2002,** *35,* 998–1007

# *ARTICLES*

# Carbon Nanotubes from Organometallic Precursors

C. N. R. RAO* AND A. GOVINDARAJ

*Chemistry and Physics of Materials Unit and CSIR Centre of Excellence in Chemistry, Jawaharlal Nehru Centre for Advanced Scientific Research, Jakkur P. O., Bangalore 560 064, India*

Received January 7, 2002

**ABSTRACT**

Multiwalled as well as single-walled carbon nanotubes are conveniently prepared by the pyrolysis of organometallic precursors such as metallocenes and phthalocyanines in a reducing atmosphere. More importantly, pyrolysis of organometallics alone or in mixture with hydrocarbons yields aligned nanotube bundles with useful field emission and hydrogen storage properties. By pyrolysis of organometallics in the presence of thiophene, Y-junction nanotubes are obtained in large quantities. The Y-junction tubes have a good potential in nanoelectronics. Carbon nanotubes prepared from organometallics are useful to prepare nanowires and nanotubes of other materials such as BN, GaN, SiC, and $Si_3N_4$.

## Introduction

Carbon nanotubes were discovered as a microscopic miracle in the cathode deposits obtained in the arc evaporation of graphite.[1] The arc method has since been improved and modified to obtain good yields of both multiwalled and single-walled nanotubes. Carbon nanotubes are conveniently obtained by carrying out the pyrolysis of hydrocarbons such as ethylene and acetylene over nanoparticles of iron, cobalt, and other metals dispersed over a solid substrate.[2–4] The presence of nanoparticles is essential not only to form nanotubes but also to control the diameter of the nanotube to some extent.[5] Since a carbon source as well as metal nanoparticles is necessary for producing carbon nanotubes by the pyrolysis of hydrocarbons, the strategy of employing an appropriate organometallic precursor which can serve as a dual source of both the carbon and the metal nanoparticles was explored by us. The very first experiments carried out on the pyrolysis of organometallic precursors such as metallocenes and iron pentacarbonyl were successful in producing multiwalled nanotubes.[6,7] We have employed this method not only to produce multiwalled and single-walled nanotubes but also to make aligned nanotube bundles and Y-junction nanotubes. In this Account, we present the salient features of the various types of nanotubes obtained by us in Bangalore by employing the organometallic route and also examine some of the properties of the nanotubes so produced. Aligned nanotube bundles are expected to have applications in electronic displays and hydrogen storage while Y-junction nanotubes could be useful as building blocks in nanoelectronics. The nanotubes produced from organometallics are also usefully employed as starting materials to prepare other types of nanostructures.

## Multiwalled and Single-Walled Nanotubes

Sen et al.[6] prepared multiwalled carbon nanotubes (MWNTs)and metal-filled onionlike structures by the pyrolysis of metallocenes such as ferrocene, cobaltocene, and nickelocene under reducing conditions, wherein the precursor acts as the source of the metal as well as carbon. The pyrolysis setup consists of stainless steel gas flow lines and a two-stage furnace system fitted with a quartz tube (25 mm i.d.) as shown in Figure 1a, the flow rate of the gases being controlled by the use of mass flow controllers. In a typical preparation, a known quantity (100 mg) of the metallocene (presublimed 99.99% purity) is taken in a quartz boat and placed at the center of the first furnace, and a mixture of Ar and $H_2$ of the desired composition is passed through the quartz tube. The metallocene is sublimed by raising the temperature of the first furnace to 200 °C at a controlled heating rate (20 °C/min). The metallocene vapor so generated is carried by the Ar–$H_2$ gas stream into the second furnace, maintained at 900 °C, where the pyrolysis occurs. The main variables in the experiment are the heating rate of ferrocene, the flow rate of Ar gas, and the pyrolysis temperature. Ferrocene vapor carried by a 75% Ar + 25% $H_2$ mixture at a flow rate of 900 sccm (sccm = standard cubic centimeter per minute) into the furnace yields large quantities of carbon deposits, mainly containing carbon nanotubes. These deposits are examined by a scanning electron microscope (SEM) and transmission electron microscope (TEM). In Figure 2a, we show a TEM image of nanotubes produced by the pyrolysis of ferrocene. To increase the yield of the multiwalled carbon nanotubes with metallocene, vapors of an additional hydrocarbon source were mixed along with the metallocene vapor in the first furnace. Thus, pyrolysis of benzene in the presence of ferrocene or $Fe(CO)_5$ gives high yields of multiwalled nanotubes, the wall thickness of the nanotubes depending on the proportion of the carbon

C. N. R. Rao obtained his Ph.D. degree from Purdue University and D.Sc. degree from the University of Mysore. He is Linus Pauling Research Professor at the JNCASR and Honorary Professor at the Indian Institute of Science, Bangalore, India. He is a member of several academies including the Royal Society (London), U.S. National Academy of Sciences, French Academy of Sciences, and Japan Academy. His main research interests are in solid state and materials chemistry and nanomaterials. The most recent award received by him is the Hughes medal for physical sciences from the Royal Society.

A. Govindaraj obtained his Ph.D. degree from the University of Mysore and is a Scientific Officer at the Indian Institute of Science, Bangalore, India, and Honorary Senior Research Officer at the JNCASR. His main research interests are nanotubes and fullerenes. He is a recipient of the MRS (India) medal.

* To whom correspondence should be addressed. Fax: 91-80-8462760. E-mail: cnrrao@jncasr.ac.in.



10.1021/ar0101584 CCC: $22.00  Published on Web 05/18/2002

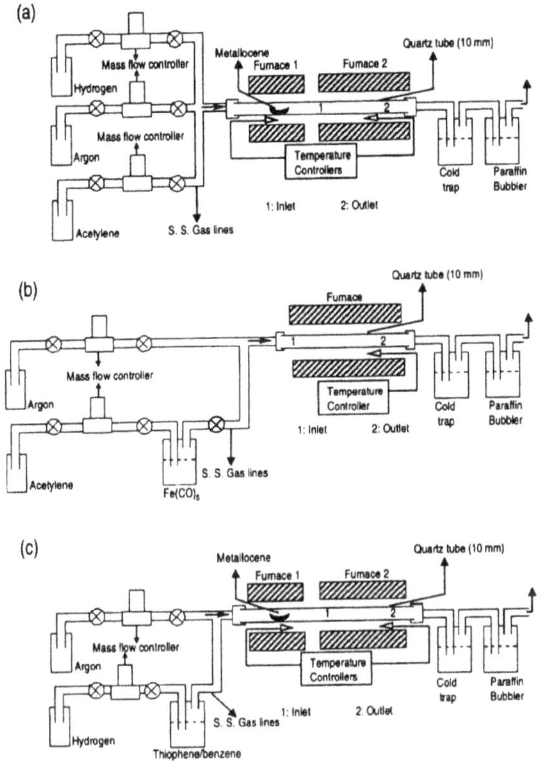


**FIGURE 1.** Pyrolysis apparatus employed for the synthesis of carbon nanotubes by pyrolysis of mixtures of (a) metallocene + $C_2H_2$, (b) $Fe(CO)_5$ + $C_2H_2$, and (c) metallocene + benzene or thiophene. The numbers 1 and 2 indicated in the figure represents inlet and outlet, respectively.[6]

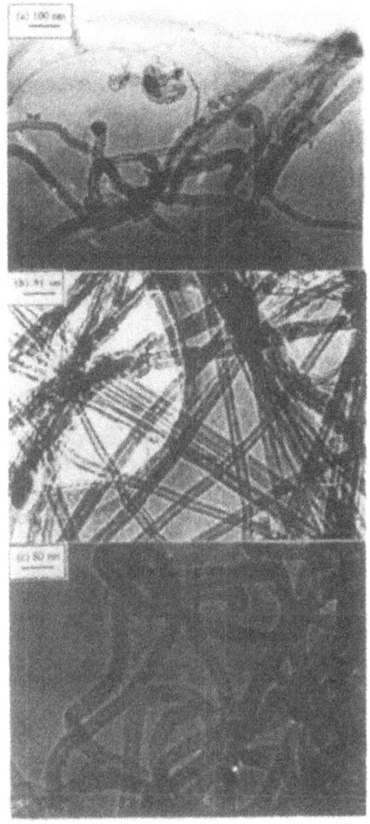

**FIGURE 2.** TEM image of a multiwalled carbon nanotube obtained by pyrolysis of (a) ferrocene at 900 °C in a mixture of 75% Ar/25% $H_2$ at a flow rate of 900 sccm, (b) a mixture of $C_2H_2$ (25 sccm) and ferrocene at 1100 °C at 1000 sccm Ar flow, and (c) a mixture of nickelocene and benzene at 900 °C in 85% Ar and 15% $H_2$ mixture at a flow rate of 1000 sccm.[6,7]

source and the metal precursor.[7] In Figure 2b, we show a TEM image of nanotubes obtained by the pyrolysis of a mixture of $C_2H_2$ (25 sccm) and ferrocene at 1100 °C in a Ar flow rate of 1000 sccm. The image clearly reveals that the addition of hydrocarbon not only increases the yield of hollow MWNTs but also reduces the amount of carbon-coated metal nanoparticles. In Figure 2c we show a TEM image of MWNTs obtained by the pyrolysis of a mixture of nickelocene and benzene at 900 °C in 85% Ar and 15% $H_2$ mixture at a flow rate of 1000 sccm. Besides metallocenes, one can also employ metal phthalocyanines as precursors to prepare MWNTs.[8,9]

To prepare single-walled nanotubes (SWNTs), alternate synthetic strategies have been explored. Under controlled conditions of pyrolysis, dilute hydrocarbon–organometallic mixtures yield SWNTs.[10,11] High-resolution TEM images of SWNTs, obtained by the pyrolysis of a nickelocene–acetylene mixture at 1100 °C,[11] are shown in the Figure 3a,b. The diameter of the SWNT in Figure 3a is 1.4 nm. It may be recalled that the pyrolysis of nickelocene in admixture with benzene under similar conditions primarily yields MWNTs. Acetylene appears to be a better carbon source for the preparation of SWNTs, since it contains a smaller number of carbon atoms. The bottom portion of the SWNT in Figure 3a shows an amorphous carbon coating around the tube, commonly observed in many of the preparations. This can be avoided by reducing the proportion of the hydrocarbon and mixing hydrogen in the Ar stream. Pyrolysis of cobaltocene and acetylene under similar conditions gives rise to isolated SWNTs. In Figure 3c,d, we show the high-resolution electron microscope (HREM) images of the SWNTs obtained by the pyrolysis of ferrocene with methane at 1100 °C. Surprisingly, the pyrolysis of nickelocene or cobaltocene in admixture with methane under similar conditions did not give SWNTs in good yield. Pyrolysis of binary mixtures (1:1 by weight) of the metallocenes along with acetylene gives good yields of SWNTs, due to the beneficial effect of binary alloys.[12]

SWNTs are obtained in good yields by the pyrolysis of acetylene in mixture with $Fe(CO)_5$ at 1100 °C employing a setup of the type shown in Figure 1b. We show TEM images of the SWNTs so obtained in Figure 4. Pyrolysis of ferrocene–thiophene mixtures yield SWNTs, but the yield is somewhat low. Pyrolysis of a mixture of benzene and thiophene along with ferrocene (Figure. 1c), however, gives a high yield of SWNTs.[13]

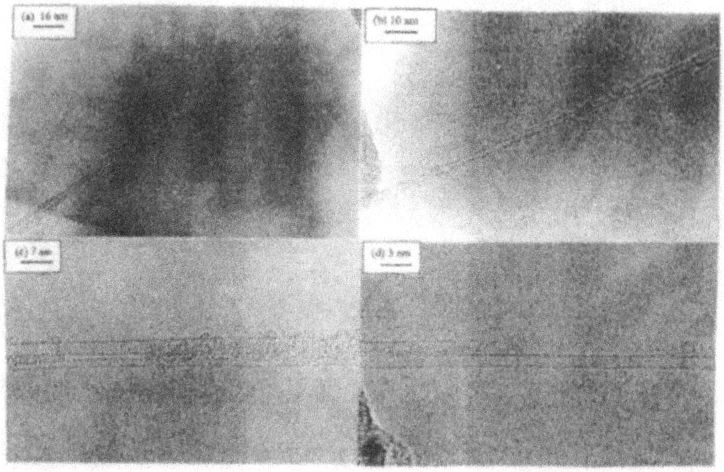

**FIGURE 3.** (a, b) HREM images of SWNTs obtained by the pyrolysis of nickelocene and $C_2H_2$ at 1100 °C in a flow of Ar (1000 sccm) with $C_2H_2$ flow rate of 50 sccm. (c, d) HREM images of SWNTs obtained by the pyrolysis of ferrocene and $CH_4$ at 1100 °C in a flow of Ar (990 sccm) with $CH_4$ flow rate of 10 sccm.[11]

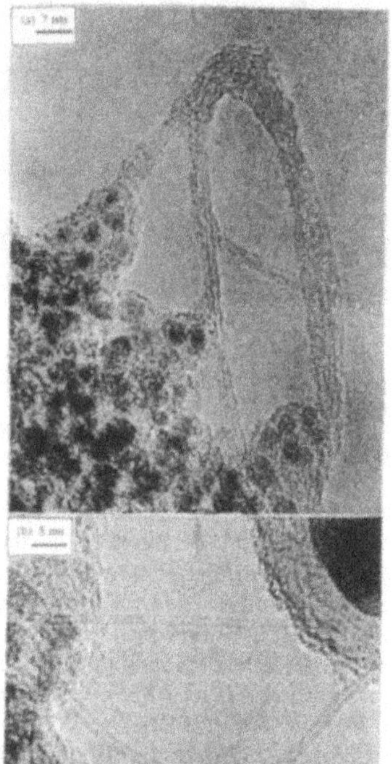

**FIGURE 4.** (a, b) TEM images of SWNTs obtained by the pyrolysis of $Fe(CO)_5$ and $C_2H_2$ (flow rate = 50 sccm) at 1100 °C in Ar (flow rate = 1000 sccm) flow.[10,11]

TEM examination of the various carbonaceous products obtained from the pyrolysis of hydrocarbons and organometallic precursors indicates that the size of the catalyst particle plays an important role with regard to the nature of the product. In the case of organometallic precursors, it seems that metal nanoclusters of ~1 nm diameter are produced under controlled conditions. When the concentration of the organometallic precursor is high, MWNTs are formed around the metal particles of 5–20 nm diameter. This is true of carbon nanotubes obtained by the metallocene route.[6,7] In the higher size range of ≥50 nm, graphite-covered metal particles are formed predominantly.[15]

## Aligned Carbon Nanotube Bundles

Since the pyrolysis of mixtures of organometallic precursors and hydrocarbons yields good quantities of multi-walled nanotubes,[6,7] we considered the possibility of obtaining aligned nanotube bundles under appropriate conditions. For this purpose, we carried out pyrolysis of metallocenes along with other hydrocarbon sources, in the apparatus shown in Figure 1a.[10,15,16] To obtain aligned nanotube bundles, a typical heating rate 50 °C/min of the first furnace and an Ar flow rate of 1000 sccm have been employed. Compact aligned nanotube bundles could be obtained by introducing $C_2H_2$ (50–100 sccm) during the sublimation of ferrocene. Scanning electron microscope images of aligned nanotubes obtained by the pyrolysis of ferrocene are shown in Figure 5. The image in Figure 5a shows the top view of the aligned nanotubes, wherein the nanotube tips are seen. The image in Figure 5b shows the side view. Aligned nanotubes obtained by the pyrolysis of ferrocene with different alkanes are shown in the SEM images in Figure 6. The average length of the nanotubes is generally around 60 $\mu$m with methane and acetylene. In the case of methane, the nanotubes are aligned, but the packing density is not high. The nanotube bundles obtained with ferrocene + acetylene mixtures appear to have a packing density greater than that obtained with Fe/silica catalysts.[17] SEM images of large bundles of the aligned nanotubes are shown in Figure 7. A small proportion of graphite-covered metal nanoparticles is often present along with the nanotubes. In Table 1, we sum-

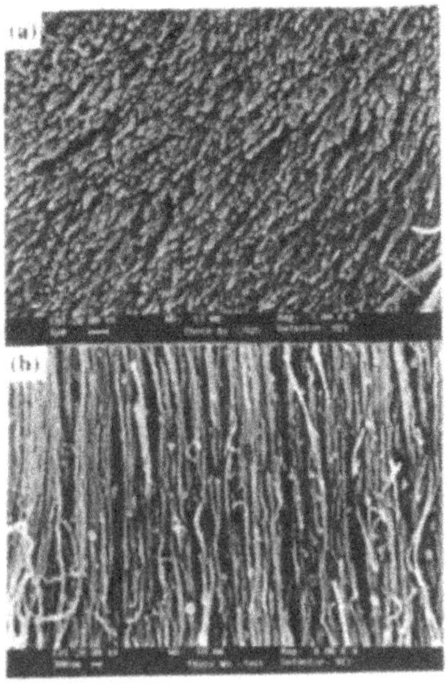

**FIGURE 5.** SEM images of aligned carbon nanotubes obtained by the pyrolysis of ferrocene. (a) and (b) show views of the aligned nanotubes along and perpendicular to the axis of the nanotubes, respectively.[16]

marize the products obtained by the pyrolysis of hydrocarbon + organometallic mixtures at 1100 °C in a stream of Ar + $H_2$.

Andrews et al.[18] have carried out the pyrolysis of ferrocene–xylene mixtures to obtain aligned carbon nanotubes. Pyrolysis of Fe(II) phthalocyanine also yields aligned nanotubes.[19] Chen et al.[20] have grown three-dimensional micropatterns of well-aligned carbon nanotubes on photolithographically prepatterned substrates, by the pyrolysis of iron(II) phthalocyanine in an Ar/$H_2$ atmosphere around 950 °C. They achieved the photopatterning by photolithographic cross-linking of a chemically amplified photoresist layer spin-cast on a quartz plate or a silicon wafer coupled with solution development. Owing to the appropriate surface characteristics, the patterned photoresist layer supports aligned nanotube growth. The difference in the chemical nature between the surfaces covered and uncovered by the photoresist film causes a region-specific growth of nanotubes with different tubular lengths and packing densities leading to the formation of three-dimensional aligned nanotube patterns.

Considering all aspects, we feel that the pyrolysis of organometallic precursors is the most convenient means of preparing aligned MWNTs. The other methods reported

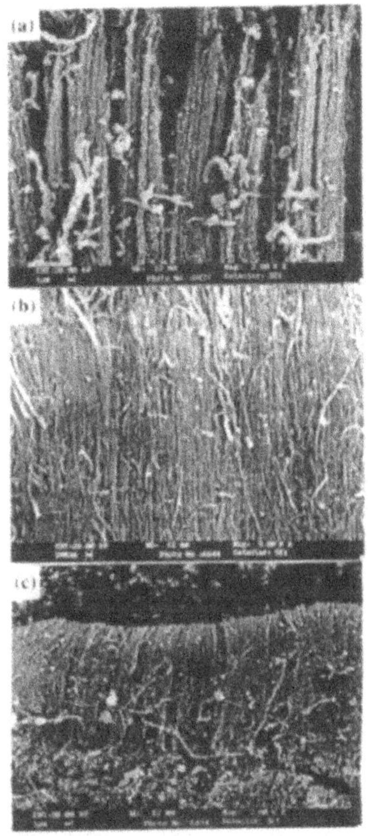

**FIGURE 6.** SEM images showing the bundles of aligned nanotubes obtained by the pyrolysis of ferrocene along with (a) methane, (b) acetylene, and (c) butane (hydrocarbon at 50 sccm) at 1100 °C in a Ar flow of 950 sccm.[15]

in the literature, such as hydrocarbon pyrolysis on patterned metal films,[21] are more difficult and are not amenable for large-scale synthesis. The advantage of the precursor method is that the aligned bundles are produced in one step, at a relatively low cost, without prior preparation of substrates. It is possible that the self-assembly of carbon nanotubes from precursor pyrolysis is influenced by the transition metal particles which take part in the nucleation and growth of the nanotubes. The metal often gets encapsulated to form nanorods or nanoparticles inside the carbon nanotubes. Ferromagnetism of these metal particles is also a property of significance. The iron nanorods encapsulated in the nanotubes exhibit a complex behavior with respect to magnetization reversal and could be useful as probes in magnetic force micros-

**Table 1. Products Obtained by the Pyrolysis of Hydrocarbon–Organometallic Mixtures at 1100 °C**

| organometallic precursor | hydrocarbon | inlet | outlet |
|---|---|---|---|
| ferrocene | acetylene | aligned MWNTs, nanorods | metal nanoparticles* |
| nickelocene | acetylene | MWNTs | SWNTs |
| cobaltocene | acetylene | MWNTs | SWNTs |
| ferrocene + nickelocene | acetylene | aligned MWNTs | SWNTs |
| ferrocene + cobaltocene | acetylene | aligned MWNTs | SWNTs |
| nickelocene + cobaltocene | acetylene | MWNTs | SWNTs |
| $Fe(CO)_5$ | acetylene | MWNTs | SWNTs (bundles) |

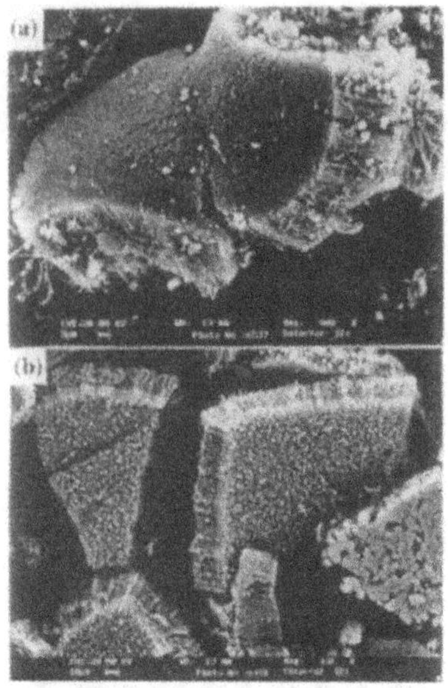

**FIGURE 7.** SEM images showing the bundles of aligned nanotubes obtained by pyrolyzing ferrocene along with (a) acetylene and (b) butane.[15]

copy. Aligned carbon nanotubes are potential candidates for use as field emitters,[14] and the easy synthesis from organometallic precursors is therefore of importance.

## Y-Junction Carbon Nanotubes

Device miniaturization in semiconductor technology is expected to reach its limits due to the inherent quantum effects as one goes toward smaller size. In such a scenario, an alternative would be nanoelectronics based on molecules. The possible use of carbon nanotubes in nanoelectronics has aroused considerable interest. For such applications it is important to be able to connect the nanotubes of different diameters and chirality.[22–24] Complex three-point nanotube junctions have been proposed as the building blocks of nanoelectronics, and in this regard Y- and T-junctions have been considered as prototypes.[25,26] While we would expect an equal number of five- and seven-membered rings to create nanotube junctions, it appears that they can be created with an equal number of five- and eight-membered rings as well.[26] To date, there have been no practical devices made of real three-point nanotube junctions. However, junctions consisting of crossed nanotubes have been fabricated to study their transport characteristics.[27] Y-junction nanotubes have been produced by using Y-shaped nanochannel alumina as templates.[28] We have prepared Y-junction nanotubes in large quanties by carrying out the pyrolysis of a mixture of a metallocene with thiophene.[29,30]

The experimental setup employed by us for the synthesis of the Y-junction nanotubes employed a two-stage furnace system similar to that described earlier[29] (Figure 1c). A known quantity of metallocene was sublimed in the first furnace and carried along with a flow of argon (Ar) gas to the pyrolysis zone in the second furnace. Simultaneously hydrogen was bubbled through thiophene and was mixed with the argon–metallocene vapors at the inlet of the furnace and carried to the pyrolysis zone. Pyrolyzing the mixed vapors at 1000 °C yielded Y-junction nanotubes in plenty. Pyrolysis of Ni/Fe phthalocyanine in mixture with thiophene was carried out in a similar manner taking phthalocyanine in place of the metallocenes to obtain good Y-junction nanotubes. Pyrolysis of various organometallics with sulfur-containing compounds has shown that pyrolysis of thiophene with nickelocene, ferrocene, and cobaltocene yields excellent Y-junction nanotubes. A TEM image of a Y-junction nanotube obtained by the pyrolysis of nickelocene/thiophene mixture is shown in Figure 8a. A TEM image revealing the presence of several Y-junction carbon nanotubes is shown in Figure 8b. Many of the nanotubes show multiple Y-junctions.

Pyrolysis of thiophene with Fe or Ni phthalocyanine or iron pentacarbonyl also yields Y-junction nanotubes. In Figure 9a, we show few Y-junctions, where as in Figure 9b we show the multiple junctions formed continuously by the pyrolysis of thiophene with Ni phthalocyanine. While the pyrolysis of nickelocene with $CS_2$ yields similar junctions, the yield and quality of the nanotubes is not satisfactory. At higher flow rates of $CS_2$, with ferrocene it pyrolyses and gives Y-junction carbon fibers. In Figure 9c we show the Y-junctions obtained by the pyrolysis of thiophene with Fe phthalocyanine. Pyrolysis of thiophene with $Fe(CO)_5$ carried out by bubbling $H_2$ (50–100 sccm) through the pentacarbonyl along with Ar (150–200 sccm) bubbled through thiophene showed the presence of interesting junction structures as revealed in Figure 9d. The pyrolysis of thiophene over a $Ni(Fe)/SiO_2$ catalyst provides an alternative procedure to the metallocene route, but the yield of Y-junction tubes is rather low. The availability of large quantities of Y-junctions should render them useful for exploitation in nanoelectronics. Compositional analysis at the Y-junction has shown absence of sulfur indicating that the junction is formed by the curvature caused by different carbon rings. The metal nanoparticles abstract the sulfur from thiophene forming a sulfide, the remaining carbon fragment probably being involved in ring formation.

HREM images show that the graphitic layers bend parallely with respect to the junction in many of these nanotubes. Our studies in collaboration with Torsteen Seeger and Manfred Ruhle have shown that the Y-juction tubes are entirely composed of carbon with no sulfur impurity. These observations suggest the presence of equal numbers of five- and seven-/eight-membered rings at the junctions. The metal particles formed in the pyrolysis contain both sulfur and carbon.

Scanning tunneling spectroscopic studies of Y-junction carbon nanotubes show interesting diodelike device characteristics at the junctions. A typical $I-V$ curve obtained from positioning the tip atop a Y-junction (the point of contact between the three arms) as well as on the individual arms of the Y-junction is shown in Figure 10. The $I-V$ plot at the junction is asymmetric (Figure 10b)

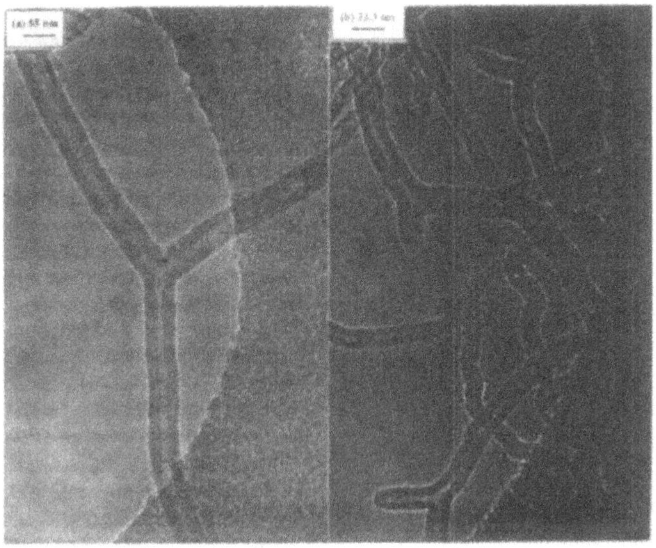

**FIGURE 8.** (a, b) TEM images of Y-junction carbon nanotubes obtained by the pyrolysis of nickelocene and thiophene at 1000 °C.[29]

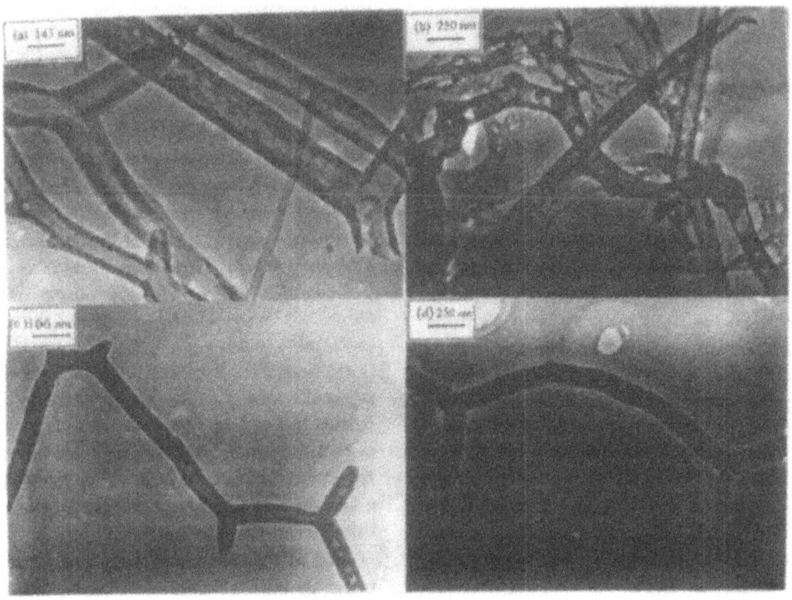

**FIGURE 9.** TEM images of Y-junction nanotubes: (a, b) obtained by the pyrolysis of thiophene along with Ni phthalocyanine at 1000 °C; (c) obtained by the pyrolysis of thiophene along with Fe phthalocyanine at 1000 °C; (d) obtained by the pyrolysis of an $Fe(CO)_5$–thiophene mixture at 1000 °C.[30]

with respect to bias polarity, unlike that along the arm (Figure 10a). The insets in Figure 10 gives a plot of differential conductance vs bias with respect to zero bias which is symmetric in Figure 10a and asymmetric in Figure 10b. Such asymmetry is characteristic of a junction diode, and this in turn indicates the existence of intramolecular junctions in the carbon nanotubes. The findings discussed above open up the possibility of assembling carbon nanotubes possessing novel devicelike properties[29,31-33] into multifunctional circuits and ultimately toward the realization of a carbon nanotube based computer chip. Rueckes et al.[34] have described the concept of carbon-nanotube-based nonvolatile random access memory for molecular computing. The viability of the concept has been demonstrated.

## Nanorods, Nanowires, and Nanotubes of Other Materials

Preparation of metal nanorods covered by carbon has been reported in the recent literature.[10,35-37] The organometallic precursor route provides a means of preparing

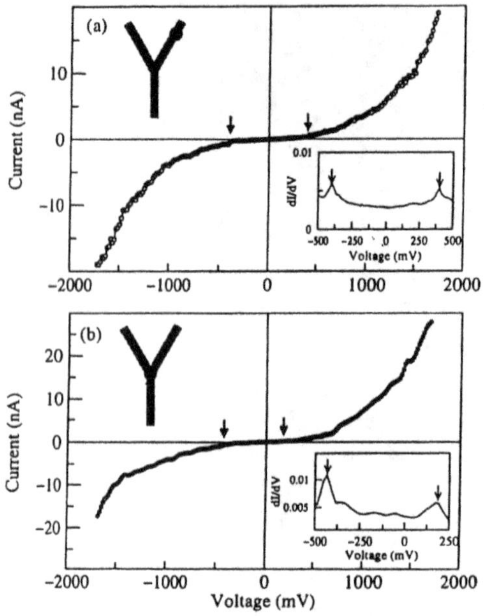


**FIGURE 10.** *I*−*V* curves collected from different points in a Y-junction carbon nanotube (a) from one of the arms away from the junction and (b) from the junction of the three arms. Inset show plots of d*I*/d*V* vs bias voltage. The observed gaps are indicated by arrows.[29]

metal nanorods. The pyrolysis of ferrocene + hydrocarbon mixtures or of ferrocene alone yields iron nanorods encapsulated inside the carbon nanotubes as evidenced from TEM, the proportion of the nanorods depending on the proportion of ferrocene. Typical TEM images of such nanorods are shown in Figure 11a,b. The selected area electron diffraction (SAED) pattern of the nanorods (inset in Figure 11c) shows spots due to (010) and (011) planes of α-Fe. The HREM image of the iron nanorod in Figure 11c shows well-resolved (011) planes of α-Fe in single-crystalline form. X-ray diffraction patterns show the presence of α-Fe with a small portion of $Fe_3C$ as the minor phase. In addition to the nanorods, iron nanoparticles (20−40 nm diameter) encapsulated inside the graphite layers are also obtained in many of the preparations. The iron nanorods and nanoparticles are well protected against oxidation by the graphitic layers. Ni/Fe phthalocyanines can be used in the place of metallocenes to prepare metal nanowires. However, the yields are not as good as in the case of metallocenes.

There are several reports on the preparation of SiC nanowires in the literature but fewer on the preparation of $Si_3N_4$ nanowires.[38,39] The methods employed for the synthesis of SiC nanowires have been varied. Since both SiC and $Si_3N_4$ are products of the carbothermal reduction of $SiO_2$, it should be possible to establish conditions wherein one set of specific conditions favor one over the other. We have been able to prepare $Si_3N_4$ nanowires,[40] by reacting multiwalled carbon nanotubes produced by ferrocene pyrolysis with ammonia and silica gel at 1360 °C. The MWNTs were used as a carbon source in the carbothermal reduction because they have higher thermal stability compared to activated carbon. The reaction of MWNTs with silica gel and $NH_3$ at 1360 °C yields a mixture of α- and β-$Si_3N_4$. SEM images of the product obtained by this reaction showed the nanowires to have large diameters (5−7 μm), with lengths of the order of hundreds of micrometers (Figure 12a). By addition of catalytic iron particles (0.5 at. %), we obtain β-SiC nanowires under the same conditions[40] (Figure 12b).

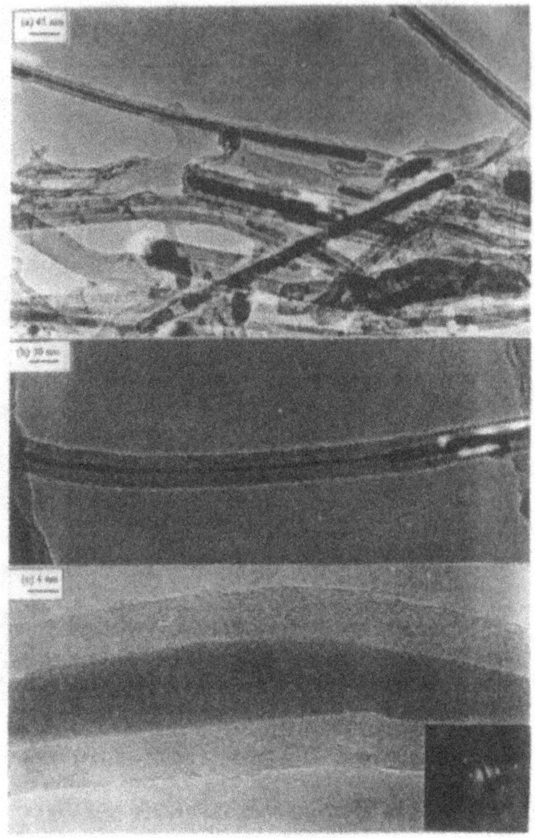

**FIGURE 11.** (a, b) TEM images of the iron nanorods encapsulated inside the carbon nanotubes from aligned nanotube bundles. (c) HREM image of a single-crystal iron nanorod encapsulated inside a carbon nanotube. The inset in (c) represents the selected area electron diffraction (SAED) pattern of a iron nanorod.[16,35]

Several chemical methods of preparing boron nitride nanotubes and nanowires have been investigated.[41] The general methods involve reacting boric acid with ammonia in the presence of MWNTs. Good yields of clean BN nanotubes are obtained with MWNTs. Aligned BN nanotubes are obtained by heating aligned bundles of MWNTs with boric acid in the presence of ammonia at the 1000°−1300 °C range. The SEM images of aligned BN nanotube bundles shown in Figure 13 a,b clearly reveal BN nanotubes aligned in two different orientations. The carbon MWNTs appear to not only take part in the reaction but also serve as templates. The average outer diameter of the

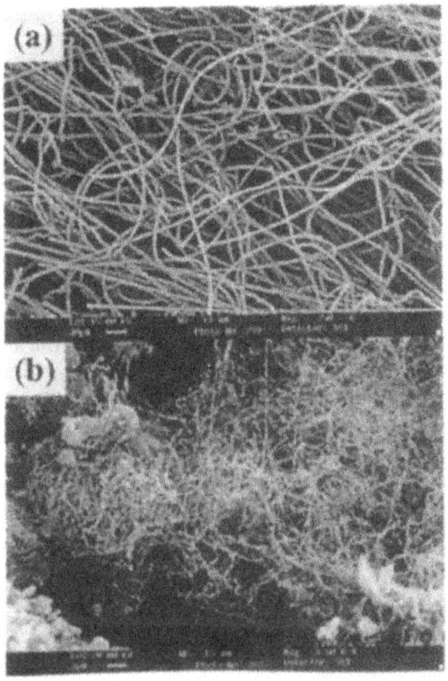

FIGURE 12. (a) SEM image of $Si_3N_4$ nanowires obtained by the reaction of aligned multiwalled nanotubes (produced by metallocene route) with silica gel at 1360 °C. (b) SiC nanowires.[40]

FIGURE 13. (a, b) SEM images of BN nanotubes obtained by heating aligned bundles of MWNTs with boric acid in the presence of ammonia at 1300 °C. (c) TEM image of the BN nanotubes obtained by the same procedure as in (a) and (b).[41]

aligned BN nanotubes varies from 15 to 40 nm as revealed by the TEM image in Figure 13c. This suggests that, during the formation of BN nanotubes, the carbon MWNTs not only take part in the reaction but also serve as templates. It is noteworthy that the BN nanotubes can be produced at a temperature as low as 1000 °C by this procedure. On the basis of elemental analysis and X-ray diffraction, it is found that the carbon content of the BN nanotubes becomes marginal if the initial proportion of carbon nanotubes is kept low.

Gallium nitride nanowires have been prepared by us by employing several procedures involving the use of carbon nanotube templates.[42] We have employed gallium acetylacetonate as the precursor for the in-situ production of small particles of the oxide ($GaO_x$) which then react with $NH_3$ vapor around 900 °C in the presence of nanotubes. When multiwalled carbon nanotubes prepared by arc discharge are used as templates, the yield of the GaN nanowires is excellent and the diameter of the majority of the nanowires is in the 35–100 nm range. The length of the nanowires extends to a few micrometers. The linear nanowires are generally single crystalline, showing a layer spacing of 0.276 nm corresponding to the [100] planes. This variation in the diameter of the nanowires occurs because of the nonuniformity in the diameter of the nanotubes. The diameter of the nanowires could be reduced to 20 nm by using single-walled nanotubes in place of multiwalled nanotubes. The growth direction of the nanowires is nearly perpendicular to the [100] planes. The nanowires show satisfactory photoluminescence characteristics similar to those of bulk GaN. The non-carbon nanowires and nanotubes have potential applications. For example, GaN nanowires, suitably doped, can have uses in nanoelectronics and in optical devices. BN nanotubes and $Si_3N_4$ nanowires are useful ceramic materials.

## Properties

Field emission properties of carbon nanotubes have direct applications in vacuum microelectronic devices.[43–46] We have found that carbon nanotubes produced by the pyrolysis of ferrocene on a pointed tungsten tip exhibit high emission current densities with good performance characteristics.[47] In Figure 14a we show a typical $I-V$ plot for the carbon nanotube covered tungsten tip for currents ranging from 0.1 nA to 1 mA. The applied voltage was 4.3 kV for a total current of 1 $\mu$A and 16.5 kV for 1000 $\mu$A. The Fowler–Nordheim (F–N) plot shown in Figure 14b has two distinct regions. The behavior is metal-like in the low-field region, while it saturates at higher fields as the voltage is increased. We have obtained a field emission current density of 1.5 A cm$^{-2}$ at a field of 290 V/mm, a value considerably higher than that found with planar cathodes.

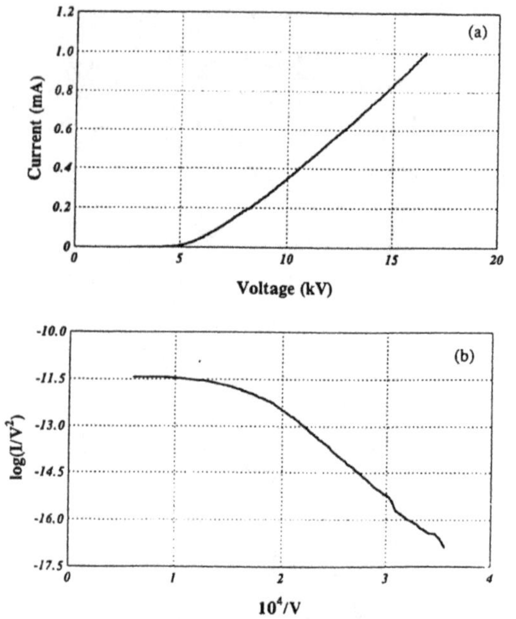


**FIGURE 14.** *I*–*V* characteristics showing field emission currents in the range of 0.1 nA to 1 mA. (b) Fowler–Nordheim plot corresponding to the data in (a).[47]

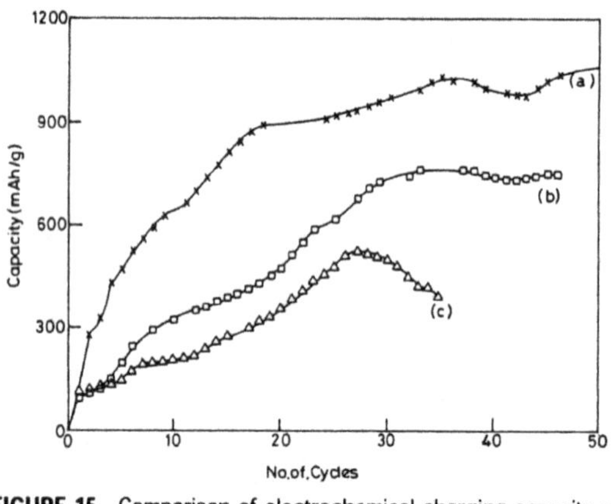


**FIGURE 15.** Comparison of electrochemical charging capacity of hydrogen of (a) aligned MWNT bundles, (b) SWNTs, and (c) MWNTs (arc-generated).[49]

Accordingly, the field enhancement factor calculated from the slope of the F–N plot in the low-field region is also large. The field emission micrographs reveal the lobe structure symmetries typical of carbon nanotube bundles. The emission current is remarkably stable over an operating period of more than 3 h for various current values in the 10–500 mA range. The relative fluctuations decrease with increasing current level, and the emitter can be operated continuously at the high current levels for at least 3 h without any degradation in the current. It appears that aligned carbon nanotubes from ferrocene pyrolysis hold promise as good field electron emission sources.

Carbon nanotubes are considered to be good hosts for hydrogen storage, although there is some controversy about the magnitude of the hydrogen uptake.[35,48] High-pressure adsorption experiments carried out by us with G. Gundiah show that the storage capacity of compact aligned nanotubes prepared by the pyrolysis of ferrocene–hydrocarbon mixtures is between 3 and 4 wt % (143 bar, 27 °C). Electrochemical hydrogen storage in these nanotubes is also substantial. In Figure 15 a–c we show (for comparison) the plots of electrochemical charging capacity of aligned nanotubes, SWNTs, and MWNTs (arc-generated), respectively.[49] Electrodes made out of aligned MWNTs clearly demonstrate higher electrochemical charging capacities up to 1100 mA h $g^{-1}$ which correspond to a hydrogen storage capacity of 3.75 wt %. SWNTs and MWNTs (arc-generated), however, show capacity in the range of 2–3 wt %.

## Concluding Remarks

Although arc evaporation of graphite has traditionally been found to yield both single-walled and multiwalled carbon nanotubes, the pyrolysis of organometallic precursors with or with out the presence of additional carbon sources seems to provide a direct and effective method of producing nanotubes of various kinds. A particularly important finding is the one-step synthesis of aligned carbon nanotubes and Y-junction nanotubes which cannot be made by arc evaporation or other methods easily. It is also noteworthy that nanotubes produced by organometallic precursors may also find applications in field emission and hydrogen storage. The successful synthesis of nanowires of gallium nitride and silicon nitride and of boron nitride nanotubes by using carbon nanotubes produced from organometallic precursors is also of interest.

*The authors thank the Department of Science and Technology, Government of India, and the DRDO (India) for supporting this research. They acknowledge productive collaboration with Dr. B. C. Satishkumar, Dr. R. Sen, Mr. F. L. Deepak, and Mr. G. Gundiah.*

## References

(1) Iijima, S. Helical microtubules of graphitic carbon. *Nature* **1991**, *354*, 56.
(2) Jose-Yacaman, M.; Miki-Yoshida, M.; Rendon, L.; Santiesteban, T. G. Catalytic growth of carbon microtubles with fullerene structure. *Appl. Phys. Lett.* **1993**, *62*, 202.
(3) Ivanov, V.; Nagy, J. B.; Lambin, Ph.; Lucas, A.; Zhang, X. B.; Zhang, X. F.; Bernaerts, D.; Van Tendeloo, G.; Amelinckx, S.; Van Landuyt, J. The study of carbon nanotubules produced by catalytic method. *Chem. Phys. Lett.* **1994**, *223*, 329.
(4) Hernadi, K.; Fonseca, A.; Nagy, J. B.; Bernaerts, D.; Riga, J.; Lucas, A. Catalytic synthesis and purification of carbon nanotubes. *Synth. Met.* **1996**, *77*, 31.
(5) Rodriguez, N. M. A review of catalytically grown carbon nanofibers. *J. Mater. Res.* **1993**, *8*, 3233.
(6) Sen, R.; Govindaraj, A.; Rao, C. N. R. Carbon nanotubes by metallocene route. *Chem. Phys. Lett.* **1997**, *267*, 276.
(7) Sen, R.; Govindaraj, A.; Rao, C. N. R. Metal-filled and hollow carbon nanotubes obtained by the decomposition of metal containing free precursor molecules. *Chem. Mater.* **1997**, *9*, 2078.
(8) Yudasaka, M.; Kikuchi, R.; Ohki, Y.; Yoshimura, S. Nitrogen-containing carbon nanotube growth from Ni-phthalocyanine by chemical vapor deposition *Carbon* **1997**, *35*, 195.
(9) Fan, S.; Chapline, M. C.; Franklin, N. R.; Tombler, T. W.; Cassel, A. M.; Dai, H. Self-oriented regular arrays of carbon nanotubes and their field emission devices. *Science* **1999**, *283*, 512.

(10) Rao, C. N. R.; Govindaraj, A.; Sen, R.; Satishkumar, B. C. Synthesis of multiwalled and single-walled nanotubes, aligned bundles and nanorods by employing organometallic precursors. *Mater. Res. Innovations* **1998** *2*, 128.
(11) Satishkumar, B. C.; Govindaraj, A.; Sen, R.; Rao, C. N. R. Single-walled nanotubes by the pyrolysis of acetylene-organometallic precursors. *Chem. Phys. Lett.* **1998** *293*, 47.
(12) Seraphin, S.; Zhou, D. Single-walled carbon nanotubes produced at high yield by mixed catalysts. *Appl. Phys. Lett.* **1994** *64*, 2087.
(13) Cheng, H. M.; Li, F.; Su, G.; Pan, H. Y.; He, L. L.; Sun, X.; Dresselhaus, M. S. Large-scale and low-cost synthesis of single-walled carbon nanotubes by the catalytic pyrolysis of hydrocarbons. *Appl. Phys. Lett.* **1998** *72*, 3282.
(14) de Heer, W. A.; Bonard, J. M.; Fauth, K.; Châtelain, A.; Forro, L.; Ugarte, D. Electron field emitters based on carbon nanotube films. *Adv. Mater.* **1997** *9*, 87.
(15) Satishkumar, B. C.; Govindaraj, A.; Rao, C. N. R. Bundles of aligned carbon nanotubes obtained by the pyrolysis of ferrocene-hydrocarbon mixtures: role of the metal nanoparticles produced in situ. *Chem. Phys. Lett.* **1999** *307*, 158.
(16) Rao, C. N. R.; Sen, R.; Satishkumar, B. C.; Govindaraj, A. Aligned nanotube bundles from ferrocene pyrolysis. *Chem. Commun.* **1998** 1525.
(17) Pan, Z. W.; Xie, S. S.; Chang, B. H.; Sun, L. F.; Zhou, W. Y.; Wang, G. Direct growth of aligned open carbon nanotubes by chemical vapor deposition. *Chem. Phys. Lett.* **1999** *299*, 97.
(18) Andrews, R.; Jacques, D.; Rao, A. M.; Derbyshire, F.; Qian, D.; Fan, X.; Dickey, E. C.; Chen, J. Continuous production of aligned carbon nanotubes: a step closer to commercial realization. *Chem. Phys. Lett.* **1999** *303*, 467.
(19) Huang, S.; Mau, A. W. H.; Turney, T. W.; White, P. A.; Dai, L. Patterned growth of well-aligned carbon nanotubes: A soft-lithographic approach *J. Phys. Chem. B* **2000** *104*, 2193.
(20) Chen. Q.; Dai, L. Three-dimensional micropatterns of well-aligned carbon nanotubes produced by photolithography. *J. Nanosci. Nanotechnol.* **2001** *1*, 43.
(21) Terrones, M.; Grobert, N.; Zhang, J. P.; Terrones, H.; Olivares, J.; Hsu, H. K.; Hare, J. P.; Cheetham, A. K.; Kroto, H. W.; Walton, D. R. M. Preparation of aligned carbon nanotubes catalysed by laser-etched cobalt thin films. *Chem. Phys. Lett.* **1998** *285*, 299.
(22) Chico, L.; Crespi, V. H.; Benedict, L. X.; Louie, S. G.; Cohen, M. L. Pure carbon nanoscale devices: Nanotube heterojunctions. *Phys. Rev. Lett.* **1996** *76*, 971.
(23) Kouwenhoven, L. Single-molecule transistors. *Science* **1997** *275*, 1896.
(24) McEuen, P. L. Nanotechnology: Carbon-based electronics. *Nature (London)* **1998** *393*, 15.
(25) Menon, M.; Srivastava, D. Carbon nanotube "T-junctions": Nanoscale metal-semiconductor-metal contact devices. *Phys. Rev. Lett.* **1997** *79*, 4453.
(26) Menon, M.; Srivastava, D. Carbon nanotube based molecular electronic devices. *J. Mater. Res.* **1998** *13*, 2357.
(27) Fuhrer, M. S.; Nygard, J.; Shih, L.; Forero, M.; Yoon, Y. G.; Mazzoni, M. S. C.; Choi, H. J.; Ihm, J.; Louie, S. G.; Zettl, A.; McEuen, P. L. Crossed nanotube junctions. *Science* **2000** *288*, 494.
(28) Li, J.; Papadopoulos, C.; Xu, J. Nanoelectronics: Growing Y-junction carbon nanotubes. *Nature (London)* **1999** *402*, 253.
(29) Satishkumar, B. C.; Thomas, P. J.; Govindaraj, A.; Rao, C. N. R. Y-junction carbon nanotubes. *Appl. Phys. Lett.* **2000** *77*, 2530.
(30) Deepak, F. L.; Govindaraj, A.; Rao, C. N. R. Synthetic strategies for Y-junction carbon nanotubes. *Chem. Phys. Lett.* **2001** *345*, 5.
(31) Tans, S. J.; Verschueren, A. R. M.; Dekker, C. Room-temperature transistor based on a single carbon nanotube *Nature* **1998** *393*, 49.
(32) Martel, R.; Schmidt, T.; Shea, H. R.; Hertel, T.; Avouris, Ph. Single- and multiwall carbon nanotube field-effect transistors. *Appl. Phys. Lett.* **1998** *73*, 2447.
(33) Yao, Z.; Postma, H. W. Ch.; Balents, L.; Dekker: C. Carbon nanotube intramolecular junctions. *Nature* **1999** *402*, 273.
(34) Rueckes, T.; Kim, K.; Joseluich, E.; Tsang, G. Y.; Cheung, C. L.; Leiber, C. M. Carbon nanotube-based nonvolatile random access memory for molecular computing. *Science* **2000** *289*, 94.
(35) Rao, C. N. R.; Satishkumar, B. C.; Govindaraj, A.; Nath, M. Nanotubes. *Chem. Phys. Chem.* **2001** *2*, 78.
(36) Morales, A. M.; Leiber, C. M. A laser ablation method for the synthesis of crystalline semiconductor nanowires. *Science* **1998** *279*, 208.
(37) Zhang, Y.; Suenaga, K.; Colliex, C.; Iijima, S. Coaxial nanocable: Silicon carbide and silicon oxide sheathed with boron nitride and carbon. *Science* **1998** *281*, 973.
(38) Han, W.; Fan, S.; Li, Q.; Gu, B.; Zhang, X.; Yu, D. Synthesis of silicon nitride nanorods using carbon nanotube as a template. *Appl. Phys. Lett.* **1997** *71*, 2271.
(39) Wang, M. J.; Wada, H. Synthesis and characterization of silicon nitride whiskers. *J. Mater. Sci.* **1990** *25*, 1690. Wu, X. C.; Song, W. H.; Zhao, B.; Huang, W. D.; Pu, M. H.; Sun, Y. P.; Du, J. J. Synthesis of coaxial nanowires of silicon nitride sheathed with silicon and silicon oxide. *Solid State Commun.* **2000** *115*, 683.
(40) Gundiah, G.; Madhav, G. V.; Govindaraj, A.; Rao, C. N. R. Synthesis and characterization of silicon carbide, silicon oxynitride and silicon nitride nanowires. *J. Mater. Chem.* **2002** *12*, 1606–1611.
(41) Deepak, F. L.; Mukhopadhyay, K.; Vinod, C. P.; Govindaraj, A.; Rao, C. N. R. Boron nitride nanotubes and nanowires. *Chem. Phys. Lett.* **2002** *353*, 345.
(42) Deepak, F. L.; Govindaraj, A.; Rao, C. N. R. Single-crystal GaN nanowires. *J. Nanosci. Nanotechnol.* **2001** *1*, 303.
(43) de Heer, W. A.; Chatelain, A.; Ugarte, D. A carbon nanotube field-emission electron source. *Science* **1995** *270*, 1179.
(44) Wang, Q. H.; Corrigan, T. D.; Dai, J. Y.; Chang, R. P. H.; Krauss, A. R. Field emission from nanotube bundle emitters at low fields. *Appl. Phys. Lett.* **1997** *70*, 3308.
(45) Bonard, J.-M.; Maier, F.; Stockli, T.; Chatelain, A.; de Heer, W. A.; Salvetat, J.-P.; Ferro, L. Field emission properties of multiwalled carbon nanotubes. *Ultramicroscopy* **1998** *73*, 7.
(46) Saito, Y.; Hamaguchi, K.; Hata, K.; Tohji, K.; Kasuya, A.; Nishina, Y.; Uchida, K.; Tasaka, Y.; Ikazaki, F.; Yumura, M. Field emission from carbon nanotubes; Purified single-walled and multiwalled tubes. *Ultramicroscopy* **1998** *73*, 1.
(47) Sharma, R. B.; Tondare, V. N.; Joag, D. S.; Govindaraj, A.; Rao, C. N. R. Field emission from carbon nanotubes grown on a tungsten tip. *Chem. Phys. Lett.* **2001** *344*, 283.
(48) Dresselhaus, M. S.; Williams, K. A.; Eklund, P. C. Hydrogen adsorption in carbon materials. *MRS Bull.* **1999** *24*, 45.
(49) Rajalakshmi, N.; Dhathathreyan, K. S.; Govindaraj. A.; Rao, C. N. R. To be submitted for publication.


AR0101584

ARTICLE

# Nitrogen- and Boron-Doped Double-Walled Carbon Nanotubes

L. S. Panchakarla, A. Govindaraj, and C. N. R. Rao*

Chemistry and Physics of Materials Unit, CSIR Center of Excellence in Chemistry and DST Unit on Nanoscience, Jawaharlal Nehru Centre for Advanced Scientific Research, Jakkur P.O., Bangalore 560 064, India

**ABSTRACT** Double-walled carbon nanotubes (DWNTs) doped with nitrogen and boron have been prepared by the decomposition of a $CH_4$ + Ar mixture along with pyridine (or $NH_3$) and diborane, respectively, over a $Mo_{0.1}Fe_{0.9}Mg_{13}O$ catalyst, prepared by the combustion route. The doped DWNTs bave been characterized by transmission electron microscopy (TEM), X-ray photoelectron spectroscopy, electron energy loss spectroscopy, and Raman spectroscopy. The dopant concentration is around 1 atom % for both boron and nitrogen. The radial breathing modes in the Raman spectra have been employed along with TEM to obtain the inner and outer diameters of the DWNTs. The diameter ranges for the undoped, N-doped (pyridine), N-doped ($NH_3$), and B-doped DWNTs are 0.73–2.20, 0.74–2.30, 0.73–2.32, and 0.74–2.36 nm, respectively, the boron-doped DWNTs giving rise to a high proportion of the large diameter DWNTs. Besides affecting the G-band in the Raman spectra, N- and B-doping affect the proportion of semiconducting nanotubes.

**KEYWORDS:** carbon nanotubes · doped nanotubes · double-walled nanotubes · Raman spectroscopy · transmission electron microscopy

Double-walled carbon nanotubes (DWNTs), first observed in 1996, constitute a unique family of carbon nanotubes (CNTs).[1,2] DWNTs occupy a position between the single-walled carbon nanotubes (SWNTs) and the multiwalled carbon nanotubes (MWNTs), as they consist of two concentric cylinders of rolled graphene. DWNTs possess useful electrical and mechanical properties with potential applications. Thus, DWNTs and SWNTs have similar threshold voltages in field electron emission, but the DWNTs exhibit longer lifetimes.[3] Unlike SWNTs, which get modified structurally and electronically upon functionalization, chemical functionalization of DWNTs surfaces would lead to novel carbon nanotube materials where the inner tubes are intact. The stability of DWNTs is controlled by the spacing of the inner and outer layers but not by the chirality of the tubes;[4] therefore, one obtains a mixture of DWNTs with varying diameters and chirality indices of the inner and outer tubes. DWNTs have been prepared by several techniques, such as arc discharge[5] and chemical vapor deposition (CVD) using a mixture of ferrocene with a hydrocarbon or alcohol (typical hydrocarbons are methane, alcohol, *n*-hexane, and benzene).[6−10] DWNTs have also been prepared by a sulfur-assisted CVD method using methane as the carbon source.[11,12]

Applications of CNTs based on their electrical properties strongly depend on the diameter and helicity as well as parity.[2,13] Doping of CNTs by boron and nitrogen renders them p-type and n-type, respectively. MWNTs and SWNTs doped with nitrogen[14−17] and boron[18,19] have been reported. Boron-doped carbon nanotubes appear to exhibit enhanced electron field emission due to the presence of the boron atom at the nanotube edges.[20,21] N-doped CNTs show n-type behavior regardless of tube chirality.[22]

We were interested in the synthesis and characterization of boron- and nitrogen-doped DWNTs in view of their potential applications. We have focused on the low-doping regime (~1 atom %), where the fundamental band structure is expected to be unchanged relative to the all-carbon model. To our knowledge, except for a report on the preparation of nitrogen-doped DWNTs by using a mixture of methane, ammonia, and argon over an iron−molybdenum catalyst,[23] there has been no detailed study of these materials. We have prepared nitrogen-doped DWNTs by the thermal decomposition of a $CH_4$ + $NH_3$ + Ar mixture as well as a $CH_4$ + pyridine + Ar mixture over a $Mo_{0.1}Fe_{0.9}Mg_{13}O$ catalyst, prepared by a new procedure. It may be noted that pyridine has been found to be a good nitrogen source to prepare N-doped MWNTs.[16] We have prepared boron-doped DWNTs by the thermal decomposition of a $CH_4$ + $B_2H_6$ + Ar mixture over the $Mo_{0.1}Fe_{0.9}Mg_{13}O$ catalyst at 950 °C.

*Address correspondence to cnrrao@jncasr.ac.in.

Received for review September 13, 2007 and accepted November 01, 2007.

**Published online November 29, 2007.**
10.1021/nn700230n CCC: $37.00

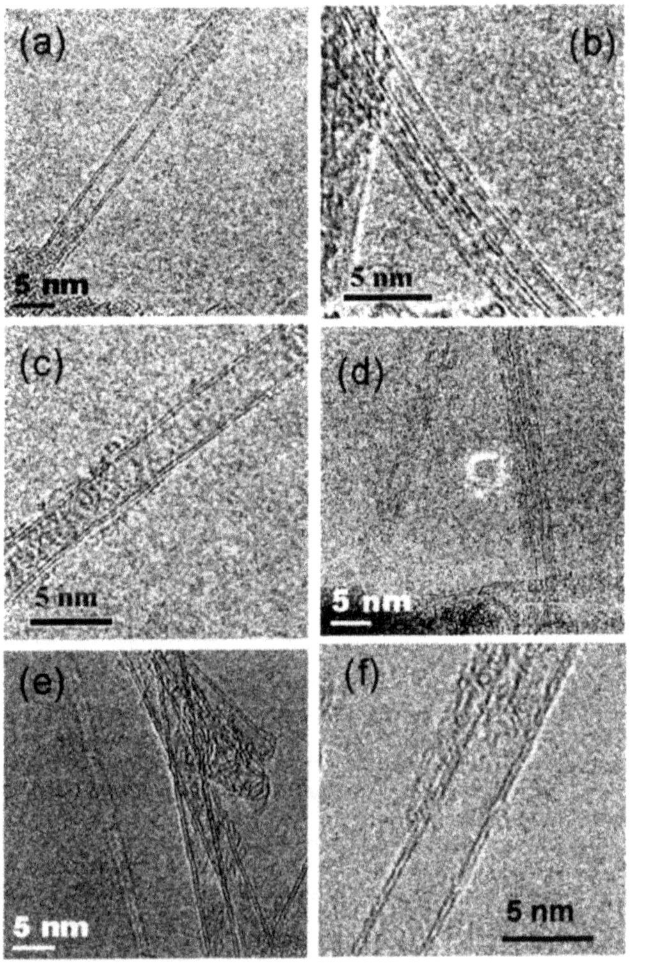


**Figure 1. HRTEM images of (a,b) undoped DWNTs, (c) N-doped DWNTs using ammonia, (d) N-doped DWNTs using pyridine, and (e,f) of B-doped DWNTs.**

The various DWNTs have been characterized with respect to composition and structure. In particular, the effect of B- and N-doping on the dimensions of the nanotubes has been examined by Raman spectroscopy.

## RESULTS AND DISCUSSION

While the decomposition of the $CH_4$ + Ar over the $Mo_{0.1}Fe_{0.9}Mg_{13}O$ catalyst at 950 °C yielded undoped DWNTs, decomposition of the $CH_4$ + $NH_3$ + Ar and $CH_4$ + pyridine + Ar mixtures gave nitrogen-doped DWNTs. Decomposition of the $CH_4$ + $BH_3$ + Ar mixture over the catalyst at 950 °C yielded boron-doped DWNTs. The N- and B-doped DWNTs could not be produced at temperatures lower than 950 °C. The combustion method employed for the preparation of the catalyst seems to help in producing DWNTs exclusively with only a very small or negligible proportion of SWNTs.

We have used both transmission electron microscopy (TEM) and Raman spectroscopy to characterize the different DWNTs samples. TEM allows direct imaging of the DWNTs and gives indications for the presence of other species along with the DWNTs. In the TEM images, we seldom encountered SWNTs or MWNTs. Besides providing information on the nature and dimensions of DWNTs, Raman spectroscopy helps to characterize the purity and quality of the DWNTs. Electron energy loss spectroscopy (EELS), carried out in a high-resolution electron microscope, and X-ray photoelectron microscopy have been employed to determine the elemental composition of the DWNTs.

Undoped DWNTs obtained by us generally had outer tube diameters of 2.2−2.8 nm and inner tube diameters of 1.4−2.1 nm, as shown by the high-resolution TEM (HRTEM) images in Figure 1a,b. In Figure 1c,d, we show typical HRTEM images of purified nitrogen-doped DWNTs synthesized by using $NH_3$ and pyridine as the nitrogen source, respectively. In Figure 1e,f we show HRTEM images of the boron-doped DWNTs. The HRTEM images indicate that the purified samples of the DWNTs have well-resolved walls and that most of the amorphous carbon was eliminated from the surface during the purification process. HRTEM images reveal that the outer tube diameters of the N-doped DWNTs prepared by using $NH_3$ as the nitrogen source are in the 1.7–3.2 nm range and the inner tube diameters are in the 1–2.4 nm range. The interlayer spacing is around 0.38 nm. In the case of N-doped DWNTs prepared by using pyridine as the nitrogen source, the outer tube diameters are generally in the 1.6–2.6 nm range, while the inner tube diameters are in the 0.9–1.8 nm range. The interlayer spacing ranges from 0.34 to 0.41 nm. From the HRTEM studies, we surmise that the diameters of the N-doped DWNTs obtained by using pyri-

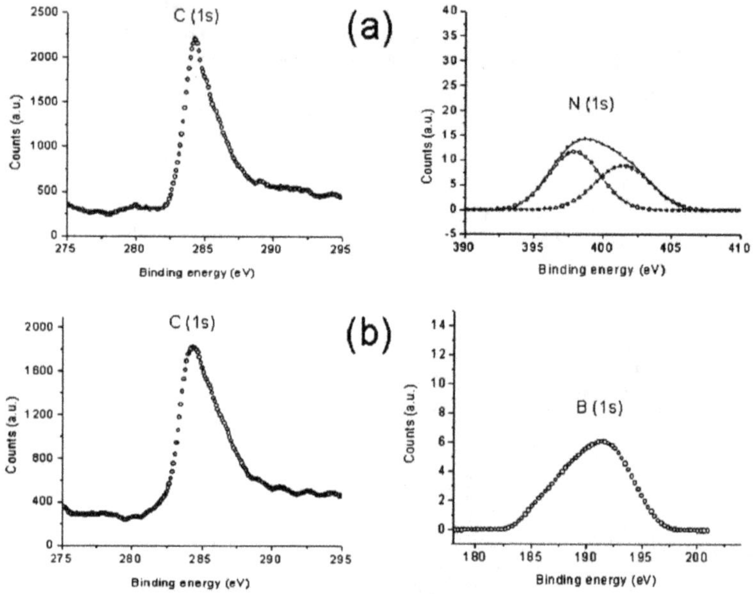


**Figure 2. (a) C 1s and N 1s XPS signals of N-doped DWNTs prepared using ammonia. (b) C 1s and B 1s signals of B-doped DWNTs.**

dine are smaller than those obtained with $NH_3$. Thus, the diameters of the N-doped DWNTs appear to depend on the nitrogen source and the reaction conditions employed.

HRTEM images of the boron-doped DWNTs show that they possess larger diameters than the undoped DWNTs as well as the N-doped DWNTs (Figure 1e,f). The outer tube diameters of the B-doped DWNTs range from 2.5 to 4.7 nm, and the inner tube diameters are in the 1.8–3.9 nm range. The interlayer spacing ranges from 0.35 to 0.41 nm. Figure 1f shows the HRTEM image of a large diameter B-doped DWNT with an outer diameter of 4.7 nm and an inner diameter of 3.9 nm.

It has been reported in the literature that boron and nitrogen are incorporated to SWNTs to a smaller extent than in MWNTs.[14–19] We have estimated the compositions of the N- and B-doped DWNTs prepared by us by employing X-ray photoelectron spectroscopy. A core-level X-ray photoelectron spectrumf the N-doped DWNTs obtained by using $NH_3$ as the N-source is shown in Figure 2a. The C 1s signal is at 284.3 eV, while the N 1s signal is centered at 399.6 eV, indicating of nitrogen substitution in the graphene sheet. It is possible that there is a small amount of amorphous carbon, as suggested by the asymmetry of the C 1s signal, although most of it gets removed on treatment with hydrogen (see Experimental Section). The asymmetric shape of the N 1s peak indicates the existence of at least two components and could be deconvoluted into two peaks at 398 and 401.3 eV. The 398 eV feature is characteristic of pyridinic nitrogen ($sp^2$ hybridization), while the peak centered at 401.3 eV is due to nitrogen present in graphene sheets.[15] The areas of the two bands bear a ratio of 1:1. On the basis of the total N 1s and C 1s intensities, the nitrogen-to-carbon ratio in the nanotubes samples was calculated by taking the photoionization cross sections of the 1s levels into account. The average composition was thus found to correspond to 1.3 atom % nitrogen. This value is lower than that réported in the literature for DWNTs (~2.9 atom %)[13] and MWNTs (3–10 atom %).[16] The N 1s spectrum of N-doped DWNTs obtained by using pyridine as the nitrogen source shows mainly the band at 398 eV, the intensity of the 401.3 eV band being very small. Thus, there is an intrinsic difference in the nature of N-substitution between the N-doped DWNTs prepared by using $NH_3$ and pyridine.

Figure 2b shows the core-level spectra of the B-doped DWNTs. The B 1s feature is at 191.4 eV, and the C 1s signal is at 284.3 eV. The shift of the B 1s signal toward higher binding energy compared to that of pure boron (188 eV) indicates that boron is in the $sp^2$ carbon network. The slight asymmetry of the B 1s signal would, however, suggest the presence of another possible mode of substitution. The boron content works out to be 1 atom %. Around 3 atom % B-doped MWNTs have been reported.[18] EELS measurements in a high-resolution electron microscope confirmed the presence of nitrogen as well as boron in the respective doped DWNTs. The %B and %N were found to be small (~1 atom %), consistent with the XPS data.

The resonance Raman spectrum of DWNTs shows three main features: the G band, the D band, and bands due to the radial breathing modes (RBMs). The tangential stretch G-band modes are in the 1550–1600 $cm^{-1}$ range. The disorder-induced D-band is observed between 1200 and 1450 $cm^{-1}$. The D-band is activated in the first-order scattering process by the presence of in-plane substitutional heteroatoms, vacancies, grain boundaries, or other defects and by finite size effects, all of which lower the crystalline symmetry of the quasi-infinite lattice.[24] RBM frequencies provide information about the nanotube diameter in the case of SWNTS and DWNTs. We have recorded the Raman spectra of the undoped as well as the N- and B-doped DWNTs by using 632.8 nm excitation using a He–Ne laser. The spectra were collected in a backscattering geometry at room

ARTICLE

**TABLE 1. RBM Frequencies (cm$^{-1}$) and Diameter (in Parentheses, nm) of Undoped and N- and B-Doped DWNTs**

| DWNTs[a] | N-DWNTs (Py)[b] | N-DWNTs ($NH_3$)[c] | B-DWNTs[d] |
|---|---|---|---|
| 338 (0.73) | 337 (0.74) | 334 (0.73) | 336 (0.74) |
| 252 (0.98) | 241 (1.03) | 308 (0.81) | 217 (1.14)* |
| 213 (1.16)* | 220 (1.13)* | 281 (0.88) | 201 (1.23)** |
| 199 (1.25) | 214 (1.16)* | 247 (1.0) | 189 (1.31)** |
| 189 (1.31)* | 191 (1.30) | 213 (1.16)** | 177 (1.40) |
| 175 (1.42) | 160 (1.55)** | 199 (1.25) | 163 (1.52) |
| 171 (1.45) | 142 (1.75)** | 191 (1.30) | 151 (1.64)* |
| 158 (1.57)** | 134 (1.85) | 158 (1.57)* | 130 (1.91)** |
| 143 (1.73)** | 125 (1.98) | 152 (1.63)** | 118 (2.10)* |
| 132 (1.88)** | 118 (2.10) | 147 (1.69)** | 105 (2.36) |
| 113 (2.20) | 108 (2.30) | 134 (1.85) | |
| | | 107 (2.32) | |

[a]Possible ($n$,$m$) values for the intense bands are as follows: 213 [(7,10)], 189 [(3,15); (6,13)], 158 [(14,9); (19,2)], 143 [(17,8)], 132 [(18,9); (21,5)]. [b]Possible ($n$,$m$) values for the intense bands are as follows: 220 [(4,12)], 214 [(7,10)], 160 [(18,3); (6,16)], 142 [(20,4); (7,18)]. [c]Possible ($n$,$m$) values for the intense bands are as follows: 213 [(7,10)], 158 [(7,10)], 152 [(13,11); (12,12)], 147 [(21,1); (6,18)]. [d]Possible ($n$,$m$) values for the intense bands are as follows: 217 [(1,14)], 201 [(3,14); (7,11)], 189 [(3,15)], 151 [(18,5); (15,9)], 130 [(16,12)], 118 [(5,24); (15,16)]. *Highest intensity RBM frequency. **Medium intensity RBM frequency.

temperature. Figure 3a shows the G-bands of the pure as well as doped DWNTs. The G-band of the N-doped DWNTs (Py) appears at a lower frequency (1574 cm$^{-1}$) compared to that of undoped DWNTs (1575 cm$^{-1}$), whereas the G-band of the B-doped DWNTs appears at a higher frequency (1579 cm$^{-1}$). The G-band of the N-doped DWNTs ($NH_3$) also appears at a lower frequency (1571 cm$^{-1}$). Thus, the shifts of the G-band are opposite for n- and p-doping of the DWNTs. Such shifts of the G-band have been reported for B- and N-doped SWNTs by Yang *et al.*[25] and McGuire *et al.*[26] The small-intensity shoulder around 1540 cm$^{-1}$ seen in the spectra of undoped DWNTs shows a decrease in intensity in the N-doped DWNTs and is negligible in the case of B-doped DWNTs. This band is related to the metallic nature of the nanotubes,[27] and its near absence in N- and B-doped DWNTs suggests a greater prevalence of semiconducting nanotubes. The inset in Figure 3a shows that the intensity of the D-band is high in the case of the B-doped DWNTs and low in the case of the N-doped DWNTs. The $I$(D)/$I$(G) ratios are 0.04, 0.06, and 0.16 for undoped, N-doped (py), and B-doped DWNTs respectively.

We observe several RBM bands in the DWNTs (Figure 3b), resulting from various sizes of the nanotubes, just as in earlier reports.[28] By using the relation $\omega = 248/d$, where $\omega$ is the RBM frequency in cm$^{-1}$ and $d$ is the nanotube diameter in nm, we have obtained the diameters of the DWNTs.[28] The RBM frequencies and the corresponding diameters are tabulated in Table 1 for undoped as well as N- and B-doped DWNTs, along with ($n$,$m$) indices for the intense features. From the table, we see that the diameter distribution of the nanotubes is markedly affected by N- and B-doping. The undoped DWNTs show the highest intensity RBM bands centered at 213 and 189 cm$^{-1}$, corresponding to diameters of 1.16 and 1.31 nm, respectively. The slightly lower intensity or medium-intensity RBM bands are at 158, 143, and 132 cm$^{-1}$, corresponding respectively to diameters of 1.57, 1.73, and 1.88 nm. The N-doped DWNTs (Py) show the highest intensity RBM bands centered at 220 and 214 cm$^{-1}$, corresponding to diameters of 1.13 and 1.16 nm, respectively. The slightly lower intensity bands centered at 160 and 142 cm$^{-1}$ correspond to 1.55 and 1.75 nm diameters, respectively. The diameters of the N-doped DWNTs (Py) are somewhat smaller compared to those of the undoped DWNTs. This is, however, not the case with N-doped DWNTs prepared using $NH_3$ as the nitrogen source. The DWNTs ($NH_3$) show the highest intensity RBM band at 158 cm$^{-1}$, corresponding to a diameter of 1.57 nm. The slightly lower intensity bands centered at 213, 152, and 147 cm$^{-1}$ correspond to diameters of 1.16, 1.63, and 1.69 nm, respectively. Since the nature of

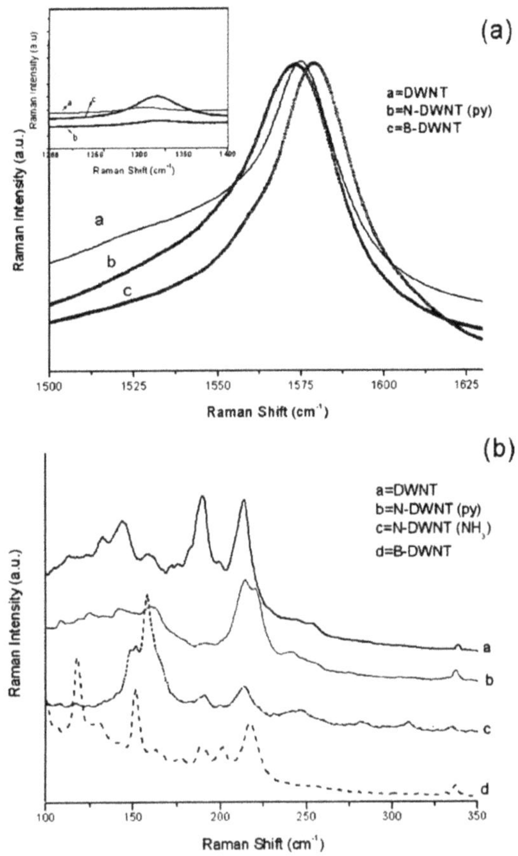


**Figure 3. (a) G-bands in the Raman spectra of undoped and doped DWNTs. Inset shows the D-bands of the same. (b) RBM bands of undoped and doped DWNTs.**

N-substitution as well as the nature of the defects is different in the N-doped DWNTs prepared by using $NH_3$ and pyridine, the difference in diameters is not entirely surprising.

The B-doped DWNTs exhibit a high proportion of large-diameter DWNTs compared to the undoped or N-doped DWNTs. The most intense RBM bands of the B-doped DWNTs are at 217, 151, and 118 $cm^{-1}$, corresponding to diameters of 1.14, 1.64, and 2.1 nm, respectively. The slightly lower intensity bands centered at 201, 189, and 130 $cm^{-1}$ correspond to diameters of 1.23, 1.31, and 1.91 nm, respectively. Due to a cutoff filter, the peaks below 100 $cm^{-1}$ were not detected.

The diameters of the various DWNTs calculated from the RBM modes are comparable with those obtained from the TEM images, but the larger diameter nanotubes seen in the TEM images are not registered in the Raman spectra since the RBM modes below 100 $cm^{-1}$ could not be recorded by us. We can identify DWNT pairs by taking the difference between the inner and outer diameters to be around 0.7 nm. The frequencies ($cm^{-1}$) of such pairs of the RBM bands in the case of undoped DWNTs are (252,143), (213,132), and (189,113). The metallic (m) and semiconducting (s) natures of these pairs are respectively (s,m), (m,m) or (m,s), and (m,s) or (s,s). In the N-doped DWNTs (py), the pairs are (241,142), (214,125), (220,134), and (160,108), and they are (s,s), (m,s), (s,s), and (m,m) or (m,s), respectively. For N-doped DWNTs ($NH_3$), the pairs are (281,158), (247,147), (213,134), and (152,107), and these pairs are respectively (s,s), (s,s) or (s,m), (m,s), and (m,m) or (m,s) or (s,m) or (s,s). In the B-doped DWNTs, the pairs are (217,130), (177,118), (189,118), and (151,105), and these pairs are (s,s), (m,s) or (s,s), (m,s), and (m,m) or (m,s) or (s,m) or (s,s), respectively. Taking the semiconducting and metallic nature of all the RBM bands, the ratio of semiconductor to metallic nanotubes in the case of undoped DWNTs works out to be 2:1, while it is 2:1, 2.2:1, and 2:1, respectively, in the case of N-doped DWNTs (py), N-doped DWNTs ($NH_3$), and B-doped DWNTs. Thus, the RBM modes predict a greater proportion of semiconducting nanotubes in the doped DWNTs as well.

The electronic absorption spectra of undoped as well as doped DWNTs show bands in the 900–1200 nm region due overlapping $E_{22}^{s}$ (s = semiconductor) features of the outer tubes and $E_{11}^{S}$ of inner tubes.[29] The absorption bands in the 1600–2400 nm regions are due to $E_{11}^{S}$ of the outer tubes. The absorption bands due to $E_{11}^{m}$ (m = metal) of the outer tubes are found in the 400–600 nm region. The metallic feature seems to prominent in the undoped DWNTs. Accordingly, the 1540 $cm^{-1}$ $G^{+}$-band in the Raman spectrum is less prominent in the doped nanotubes.

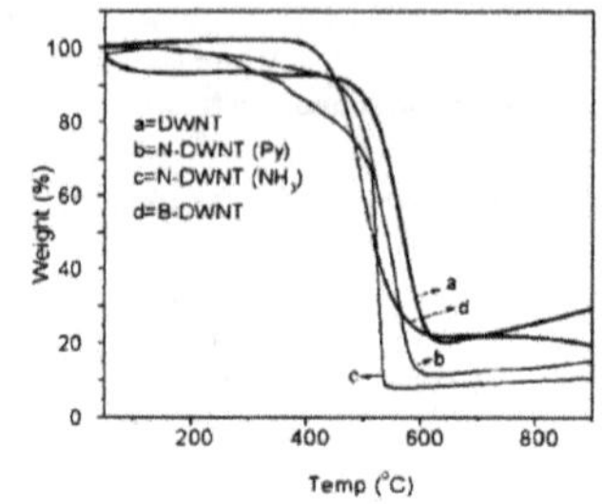


**Figure 4. TGA curves of undoped and doped DWNTs.**

The smaller diameter carbon nanotubes are known to be less stable than their larger diameter counterparts and tend to oxidize at lower temperatures. Amorphous carbon and carbon nanotubes with defects undergo combustion at lower temperatures. In Figure 4, we show the thermogravimetric analysis (TGA) curves of undoped as well as N- and B-doped DWNTs. The decomposition temperatures of all these doped DWNTs are comparable to but slightly lower that the decomposition temperature of pure DWNTs. Derivative TGA curves also shows the same trend. The slight increase in mass at high temperature may be due to the small metallic impurity.

## CONCLUSION

In conclusion, the $Mo_{0.1}Fe_{0.9}Mg_{13}O$ catalyst prepared by the combustion route preferentially yields DWNTs, the proportion of SWNTs being very small or negligible. The use of this catalyst has enabled the synthesis of 1 atom % N- and B-doped DWNTs. The diameters of the nanotubes obtained from the Raman RBM modes and transmission electron microscopy are comparable. The N-doped nanotubes show the G-band in the Raman spectrum at a lower frequency than the undoped ones, while the B-doped nanotubes show an increase in the frequency. The proportion of the metallic nanotubes appears to decrease on N- or B-doping, but the average diameter is substantially larger in the B-doped DWNTs.

## EXPERIMENTAL SECTION

Synthesis of DWNTs was carried out in a quartz tube reactor. For each synthesis, 200 mg of the supported Fe–Mo catalyst (Fe–Mo/MgO) was placed in a quartz boat by spreading it uniformly. The quartz boat was inserted into the center of the quartz tube (25 mm i.d. and 1 m long) mounted inside an electrical tube furnace. Subsequently, the furnace was heated to 950 °C in an argon atmosphere at a heating rate of 3 °C/min. A mixture of methane and Ar gas was introduced into the reactor. The flow rates of methane and Ar were maintained at 50 and 150 sccm (standard cubic centimeters per minute), respectively. After 20 min, the reactor was cooled to room temperature in an Ar atmosphere. The resulting black dense mat contained a homogeneous dispersion of carbon nanotubes around the oxide grains. This crude material was carefully collected from the boat and subjected to purification.

The oxide precursors required to prepare the catalyst for the synthesis of DWNTs were prepared by the combustion route.[30,31] The required amount of $(NH_4)_6Mo_7O_{24}\cdot 4H_2O$ was added to an aqueous solution containing ferric nitrate ($Fe(NO_3)_3\cdot 9H_2O$) and magnesium nitrate ($Mg(NO_3)_2\cdot 6H_2O$) in a Pyrex dish, keeping the molar ratio of Mo:Fe:MgO at 0.1:0.9:13. To this mixture was added an appropriate amount of urea (three times the stoichiometric ratio), which acts as the fuel in the combustion process. The mixture was dissolved by using a minimum amount of distilled water and kept in an oven at 70 °C for 12 h. The Pyrex dish containing the solution was placed in a furnace preheated at 550 °C. The thick orange-red solution immediately started boiling and underwent dehydration. The resulting thick paste frothed and blazed with a white flame, with the production of a light material which then swelled to the capacity of the Pyrex dish. The total combustion process was over in 10 min. The combustion product was baked at 550 °C for 3 h and ground to a fine powder. Preparation of the catalyst by conventional methods, such as wet impigration or coprecipitation, yields inhomogeneous catalyst particles, whereas the combustion route employed here gives homogeneous catalyst particles.

For obtaining nitrogen-doped DWNTs, the procedure was similar to that used for undoped DWNTs, expect that ammonia or pyridine vapor was taken in mixture with $CH_4$.[16] For doping nitrogen by using ammonia, the supported Fe–Mo catalyst (200 mg) was placed in a quartz boat at the center of the quartz reactor tube. The quartz tube was heated to 950 °C in an Ar atmosphere. Subsequently, $CH_4$ (50 sccm), $NH_3$ (5 sccm), and Ar (150 sccm) were mixed and introduced at the inlet of the reactor tube. After 20 min, the reactor was cooled to room temperature in an Ar atmosphere. For N-doping using pyridine, the supported Fe–Mo catalyst (200 mg) was placed in a quartz boat at the center of the quartz reactor tube. The quartz tube was heated to 950 °C in an Ar atmosphere. Subsequently, 40 sccm of $CH_4$ was passed through a bubbler containing pyridine, which carries the pyridine vapor to the furnace. These vapors were mixed with 150 sccm of Ar and passed over the MgO-supported catalyst, maintained at 950 °C for 20 min.

For the synthesis of boron-doped DWNTs, diborane ($B_2H_6$) was used as the boron source, the rest of the procedure being similar to that for undoped DWNts. $B_2H_6$ vapor was generated by the addition of $BF_3$–diethyl etherate to sodium borohydride in tetraglyme.[18] Fifty sccm of $CH_4$ was mixed and passed along with $B_2H_6$ vapors. These vapors were further mixed along with 150 sccm of Ar and passed over the MgO-supported catalyst powder, maintained at 950 °C for 20 min.

The percentage of DWNTs in all our preparations was 90% or higher. Such high preferential yields of DWNTs have been reported in the literature.[9,10] In order to dissolve the metal nanoparticles in the DWNTs, the as-prepared nanotubes were treated with concentrated HCl at 60 °C for 24 h. The product was washed with distilled water, dried, dispersed in ethanol under sonication, and filtered using Millipore (0.2 μm) filter paper. The filtered product was dried in an oven at 100 °C for 2 h and heated to 850 °C in a furnace at a rate of 3 °C per minute in flowing hydrogen at 100 sccm and held at that temperature for 6 h to remove the amorphous carbon present on the nanotube walls.[32] The resulting sample was again stirred in concentrated HCl at 60 °C for 3 h and heated in a furnace at 850 °C for 6 h in flowing hydrogen (100 sccm). The same procedure was employed to purify doped DWNTs, except that dilute HCl was used instead of concentrated HCl. It must be noted that hydrogen treatment at high temperatures has been found to be a very good method to eliminate amorphous carbon present on the carbon nanotubes.[32]

We have characterized the undoped and doped DWNTs by various techniques. Field emission scanning electron microscope (FESEM) images were recorded with a FEI NOVA NANOSEM 600. UV–vis absorption spectra of the nanotubes were recorded using a Perkin-Elmer Lambda 900 UV/vis/NIR spectrometer. Thermogravimetric analysis was carried out using a Mettler Toledo TGA 850 instrument. Raman spectra were recorded with a LabRAM HR high-resolution Raman spectrometer (Horiba Jobin Yvon) using a He–Ne laser (λ = 630 nm). Transmission electron microscope images were obtained with a JEOL JEM 3010 instrument. X-ray photoelectron spectroscopy was recorded using a VG scientific ESCA Laboratory V spectrometer. EELS were recorded with a transsmision electron microscope (FEI, TECNAI F30) equipped with an energy filter for EELS operating at 300 kV.

*Acknowledgment.* L.S.P. acknowledges CSIR, New Delhi, for a junior research fellowship.

## REFERENCES AND NOTES


1. Dai, H.; Rinzler, A.; Nikolaev, P.; Thess, A.; Colbert, D.; Smalley, R. Single-Wall Nanotubes Produced by Metal-Catalyzed Disproportionation of Carbon Monoxide. *Chem. Phys. Lett.* **1996**, *260*, 471–475.
2. Rao, C. N. R.; Govindaraj, A. *Nanotubes and Nanowires*; RSC Nanoscience & Nanotechnology Series; RSC Publishing: Cambridge, UK, 2005.
3. Kurachi, H.; Uemura, S.; Yotani, J.; Nagasako, T.; Yamada, H.; Ezaki, T.; Maesoba, T.; Loutfy, R.; Moravsky, A.; Nakazawa, T.; Saito, Y. Proceedings of the 21st International Display Research Conference in conjunction with the 8th International Display Workshops, 2001; pp 1237–1240.
4. Saito, R.; Matsuo, R.; Kimura, T.; Dresselhaus, G.; Dresselhaus, M. S. Anomalous Potential Barrier of Double-Wall Carbon Nanotube. *Chem. Phys. Lett.* **2001**, *348*, 187–193.
5. Sugai, T.; Yoshida, H.; Shimada, T.; Okazaki, T.; Shinohara, H. New Synthesis of High-Quality Double-Walled Carbon Nanotubes by High-Temperature Pulsed Arc Discharge. *Nano Lett.* **2003**, *3*, 769–773.
6. Lee, Y. D.; Lee, H. J.; Han, J. H.; Yoo, J. E.; Lee, Y.-H.; Kim, J. K.; Nahm, S.; Ju, B.-K. Synthesis of Double-Walled Carbon Nanotubes by Catalytic Chemical Vapor Deposition and Their Field Emission Properties. *J. Phys. Chem. B* **2006**, *110*, 5310–5314.
7. Liu, B. C.; Lyu, S. C.; Lee, T. J.; Choi, S. K.; Eum, S. J.; Yang, C. W.; Park, C. Y.; Lee, C. J. Synthesis of Single- and Double-Walled Carbon Nanotubes by Catalytic Decomposition of Methane. *Chem. Phys. Lett.* **2003**, *373*, 475–479.
8. Lyu, S. C.; Lee, T. J.; Yang, C. W.; Lee, C. J. Synthesis and Characterization of High-Quality Double-Walled Carbon Nanotubes by Catalytic Decomposition of Alcohol. *Chem. Commun.* **2003**, *1404*, 1405.
9. Lyu, S. C.; Liu, B. C.; Lee, S. H.; Park, C. Y.; Kang, H. K.; Yang, C.-W; Lee, C. J. Large-Scale Synthesis of High-Quality Double-Walled Carbon Nanotubes by Catalytic Decomposition of n-Hexane. *J. Phys. Chem. B* **2004**, *108*, 2192–2194.
10. Lyu, S. C.; Liu, B. C.; Lee, C. J.; Kang, H. K.; Yang, C-W; Park, C. Y. High-Quality Double-Walled Carbon Nanotubes Produced by Catalytic Decomposition of Benzene. *Chem. Mater.* **2003**, *15*, 3951–3954.
11. Wei, J.; Ci, L.; Jiang, B.; Li, Y.; Zhang, X.; Zhu, H.; Xua, C.; Wua, D. Preparation of Highly Pure Double-Walled Carbon Nanotubes. *J. Mater. Chem.* **2003**, *13*, 1340–1344.
12. Ren, W.; Cheng, H.-M. Aligned Double-Walled Carbon Nanotube Long Ropes with a Narrow Diameter Distribution. *J. Phys. Chem. B* **2005**, *109*, 7169–7173.
13. Hassanien, A.; Tokumoto, M.; Kumazawa, Y.; Kataura, H.; Maniwa, Y.; Suzuki, S.; Achida, Y. Atomic Structure and Electronic Properties of Single-Wall Carbon Nanotubes Probed by Scanning Tunneling Microscope at Room Temperature. *Appl. Phys. Lett.* **1998**, *73*, 3839–3841.
14. Sen, R.; Satishkumar, B. C.; Govindaraj, A.; Harikumar, K. R.; Renganathan, M. K.; Rao, C. N. R. Nitrogen-Containing Carbon Nanotubes. *J. Mater. Chem.* **1997**, *7*, 2335–2337.
15. Sen, R.; Satishkumar, B. C.; Govindaraj, A.; Harikumar, K. R.; Raina, G.; Zhang, J.-P.; Cheetham, A. K.; Rao, C. N. R. B-C-N, C-N and B-N Nanotubes Produced by the Pyrolysis of Precursor Molecules over Co Catalysts. *Chem. Phys. Lett.* **1998**, *287*, 671–676.
16. Nath, M.; Satishkumar, B. C.; Govindaraj, A.; Vinod, C. P.; Rao, C. N. R. Production of Bundles of Aligned Carbon and Carbon-Nitrogen Nanotubes by the Pyrolysis of

Precursors on Silica-Supported Iron and Cobalt Catalysts. *Chem. Phys. Lett.* **2000**, *322*, 333–340.
17 Villalpando-Paez, F.; Zamudio, A.; Elias, A. L.; Son, H.; Barros, E. B.; Chou, S. G.; Kim, Y. A.; Muramatsu, H.; Hayashi, T.; Kong, J.; *et al.* Synthesis and Characterization of Long Strands of Nitrogen-Doped Single-Walled Carbon Nanotubes. *Chem. Phys. Lett.* **2006**, *424*, 345–352.
18 Satishkumar, B. C.; Govindaraj, A.; Harikumar, K. R.; Zhang, J.-P; Cheetham, A. K.; Rao, C. N. R. Boron-Carbon Nanotubes from the Pyrolysis of $C_2H_2$-$B_2H_6$ Mixtures. *Chem. Phys. Lett.* **1999**, *300*, 473–477.
19 McGuire, K.; Gothard, N.; Gai, P. L.; Dresselhaus, M. S.; Sumanasekera, G.; Rao, A. M. Synthesis and Raman Characterization of Boron-Doped Single-Walled Carbon Nanotubes. *Carbon* **2005**, *43*, 219–227.
20 Charlier, J.-C.; Terrones, M.; Baxendale, M.; Meunier, V.; Zacharia, T.; Rupesinghe, N. L.; Hsu, W. K.; Grobert, N.; Terrones, H.; Amaratunga, G. A. J. Enhanced Electron Field Emission in B-doped Carbon Nanotubes. *Nano Lett.* **2002**, *2*, 1191–1195.
21 Sharma, R. B.; Late, D. J.; Joag, D. S.; Govindaraj, A.; Rao, C. N. R. Field Emission Properties of Boron and Nitrogen Doped Carbon Nanotubes. *Chem. Phys. Lett.* **2006**, *428*, 102–108.
22 Czerw, R.; Terrones, M.; Charlier, J.-C.; Blase, X.; Foley, B.; Kamalakaran, R.; Grobert, N.; Terrones, H.; Tekleab, D.; Ajayan, P. M.; *et al.* Identification of Electron Donor States in N-Doped Carbon Nanotubes. *Nano Lett.* **2001**, *1*, 457–460.
23 Kim, S. Y.; Lee, J.; Na, C. W.; Park, J.; Seo, K.; Kim, B. N-Doped Double-Walled Carbon Nanotubes Synthesized by Chemical Vapor Deposition. *Chem. Phys. Lett.* **2005**, *413*, 300–305.
24 Dresselhaus, M. S.; Eklund, P. C. Phonons in Carbon Nanotubes. *Adv. Phys.* **2000**, *49*, 705–814.
25 Yang, Q. H.; Hou, P. X.; Unno, M.; Yamauchi, S.; Saito, R.; Kyotani, T. Dual Raman Features of Double Coaxial Carbon Nanotubes with N-Doped and B-Doped Multiwalls. *Nano Lett.* **2005**, *5*, 2465–2469.
26 McGuire, K.; Gothard, N.; Gai, P. L.; Dresselhaus, M. S.; Sumanasekera, G.; Rao, A. M. Synthesis and Raman Characterization of Boron-Doped Single-Walled Carbon Nanotubes. *Carbon* **2005**, *43*, 219–227.
27 Das, A.; Sood, A. K.; Govindaraj, A.; Saitta, A. M.; Lazzeri, M.; Mauri, F.; Rao, C. N. R. Doping in Carbon Nanotubes Probed by Raman and Transport Measurements. *Phys. Rev. Lett.* **2007**, *99*, 136803 (1–4).
28 Li, F.; Chou, S. G.; Ren, W.; Gardecki, J. A.; Swan, A. K.; Unlu, M. S.; Goldbeg, B. B.; Cheng, H.-M.; Dresselhaus, M. S. Identification of the Constituents of Double-Walled Carbon Nanotubes using Raman Spectra Taken with Different Laser-Excitation Energies. *J. Mater. Res.* **2003**, *18*, 1251–1258.
29 Kishi, N.; Kikuchi, S.; Ramesh, P.; Sugai, T.; Watanabe, Y.; Shinohara, H. Enhanced Photoluminescence from Very Thin Double-Wall Carbon Nanotubes Synthesized by the Zeolite-CCVD Method. *J. Phys. Chem. B* **2006**, *110*, 24816–24821.
30 Patil, K. C. Advanced Ceramics: Combustion Synthesis and Properties. *Bull. Mater. Sci.* **1993**, *16*, 533–541.
31 Flahaut, E.; Peigney, A.; Bacsa, W. S.; Bacsa, R. R.; Laurent, C. CCVD Synthesis of Carbon Nanotubes from (Mg,Co,Mo)O Catalysts: Influence of the Proportions of Cobalt and Molybdenum. *J. Mater. Chem.* **2004**, *14*, 646–653.
32 Vivekchand, S. R. C.; Govindaraj, A.; Motin Seikh, Md.; Rao, C. N. R. New Method of Purification of Carbon Nanotubes Based on Hydrogen Treatment. *J. Phys. Chem. B* **2004**, *108*, 6935–6937.

# Charge-transfer with graphene and nanotubes

**Charge-transfer between electron–donor and –acceptor molecules is a widely studied subject of great chemical interest. Some of the charge-transfer compounds in solid state exhibit novel electronic properties. In the last two to three years, occurrence of molecular charge-transfer involving single-walled carbon nanotubes (SWNTs) and graphene has been demonstrated. This interaction gives rise to significant changes in the electronic properties of these nanocarbons. We examine charge-transfer phenomenon in graphene and SWNTs in this article in view of its potential utility in device applications.**

**C. N. R. Rao* and Rakesh Voggu**

*International Centre for Materials Science, Chemistry and Physics of Materials Unit and CSIR Centre of Excellence in Chemistry Jawaharlal Nehru Centre for Advanced Scientific Research, Jakkur P. O., Bangalore 560064, India.*

**E-mail: cnrrao@jncasr.ac.in*

**Three types of nanocarbons have emerged in the last 25 years. These are fullerenes, carbon nanotubes and graphene. Fullerenes are zero-dimensional molecular compounds while carbon nanotubes are one-dimensional materials. Graphene is a two-dimensional network of carbon atoms with fascinating properties. All the three nanocarbons contain $sp^2$ carbon atoms and one would expect certain similarities in the properties of these nanocarbons. One of the interesting characteristics that is worthy of comparison is molecular charge-transfer involving these nanocarbons. $C_{60}$ is known to exhibit charge-transfer interaction with electron donating molecules such as organic amines both in the ground and excited states. What is more interesting is that both graphene and single-walled carbon nanotubes (SWNTs) interact with electron -donor and -acceptor molecules giving rise to significant changes in the electronic structure.**

$C_{60}$ loves electrons and its salts exhibit fascinating magnetic and superconducinting properties[1]. Recent investigations reveal that charge-transfer occurs between SWNTs and electron –donor and –acceptor molecules[2,3], causing marked changes in the Raman and electronic spectra. Molecular charge-transfer is accompanied by the transformation of semiconducting nanotubes to metallic species and vice versa[2,4]. Graphene also undergoes charge-transfer interaction with electron -donor and -acceptor molecules[5]. Raman spectroscopy is eminently effective in probing charge-transfer interactions with graphene and SWNTs. In this article, we discuss molecular charge-transfer doping of graphene as well as of SWNTs and compare the effects with those found by chemical or electrochemical doping.

## Graphene

Graphene has become an exciting two-dimensional material with wondrous properties[6-8]. Some of the significant properties include



ISSN:1369 7021 © Elsevier Ltd 2010

ballistic electron transport[9] and anomalous integer quantum Hall effect at room temperature[10,11]. Raman spectroscopy has emerged as an effective probe to characterize graphene samples in terms of the number of layers and their quality. Single-layer graphene shows the well-known G-band characteristic of the $sp^2$ carbon network around 1580 $cm^{-1}$. The D band around 1350 $cm^{-1}$ and D' band around 1620 $cm^{-1}$ are both defect-induced. The 2D band at ~ 2680 $cm^{-1}$ differs in single and few-layer graphene and can be understood on the basis of the double resonance Raman process involving different electronic dispersions[12]. The 2D band can be employed to determine the number of layers in few-layer graphene. By combining Raman experiments with in-situ transport measurements of graphene in field-effect transistor geometry, it has been shown that the G-modes of single and bi -layer graphenes blue shift on doping with electrons as well as holes[13,14]. On the other hand, the 2D band blue-shifts on hole doping whereas it red shifts on doping with electrons. The relative intensity of the 2D band is quite sensitive to doping. Theoretical calculations based on time-dependent perturbation theory have been employed to explain the observed shifts of the G-band. Comparison between theory and experiment, however, is not entirely satisfactory at high doping levels (> $1 \times 10^{13}/cm^2$) and the disagreement is greater for the 2D band. In the case of bilayer graphene[14], the blue-shift of the G-band with doping has contributions from phonon-induced inter-band and intra-band electronic transitions, thereby giving an experimental measure of the overlap integral between A and B atoms in the two layers. Furthermore, the in-plane vibration in bilayer graphene splits into a symmetric Raman active mode ($E_g$) and an anti-symmetric infrared active mode ($E_u$). Doping dependence of these modes has been examined by Raman scattering[15] and infrared reflectivity measurements[16]. The latter show a drastic enhancement of intensity and a softening of the mode as a function of doping, along with a Fano-like asymmetric line shape due to a strong coupling of the $E_u$ mode to inter-band transitions.

There has been progress in the study of charge-transfer interactions of graphene with various electron donors and acceptors. The G-band softens progressively with the increasing concentration of electron-donor molecules such as aniline and tetrathiafulvalene (TTF) while the band stiffens with increasing concentration of electron-withdrawing molecules such as nitrobenzene and tetracyanoethylene (TCNE) as can be seen from Figs. 1 and 2[5,17]. Both electron-donors and -acceptors broaden the G-band. The full-width at half maximum (FWHM) of the G band increases on interaction with these molecules (Fig. 3a). The intensity of the 2D-band decreases markedly with the concentration of either donor or acceptor molecule. The ratio of intensities of the 2D and G bands, I(2D)/I(G), is a sensitive probe to examine doping of graphene by electron-donor and -acceptor molecules (Fig. 3b). Dong *et al.*[18] have observed similar effects with single layer graphene on adsorption of various aromatic molecules while Brus *et al.*[19] have examined the effect of charge-transfer doping of graphene

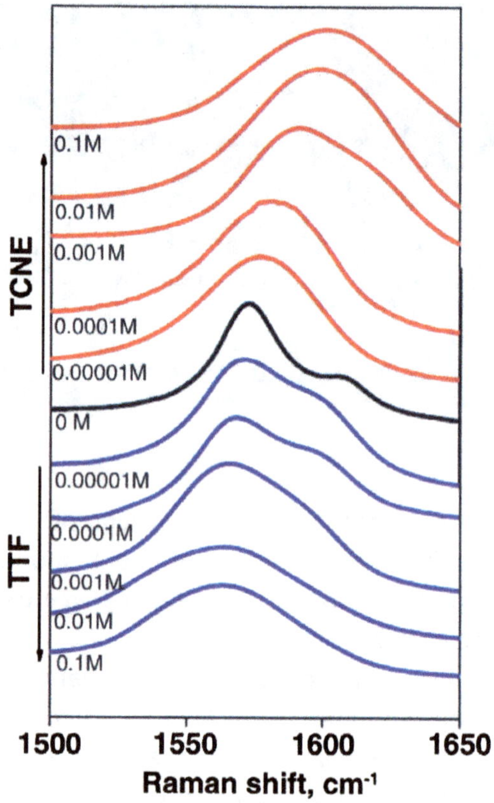


*Fig. 1 Shifts of the Raman G-band of graphene caused by interaction with varying concentrations of TTF and TCNE[17].*

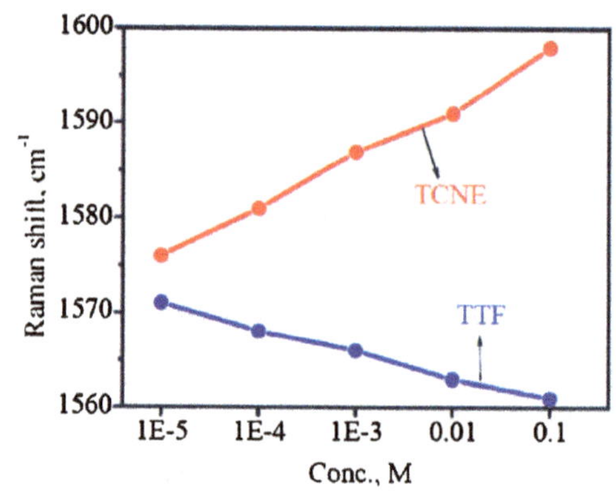


*Fig. 2 Variation in the Raman G-band position of graphene on interaction with varying concentrations of electron-donor (TTF) and electron-acceptor (TCNE) molecules[17].*

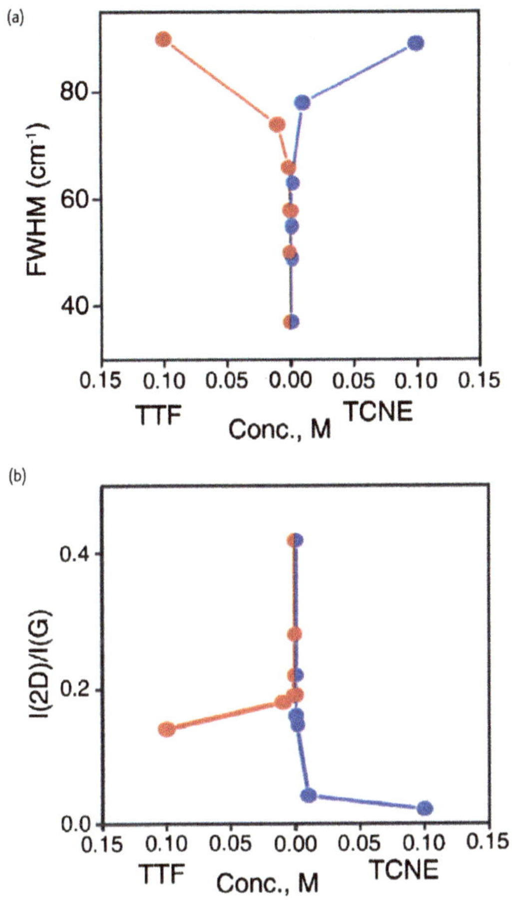


*Fig. 3 Variation in the (a) FWHM and the (b) 2D/G intensity ratio with the concentration of TTF and TCNE*[17].

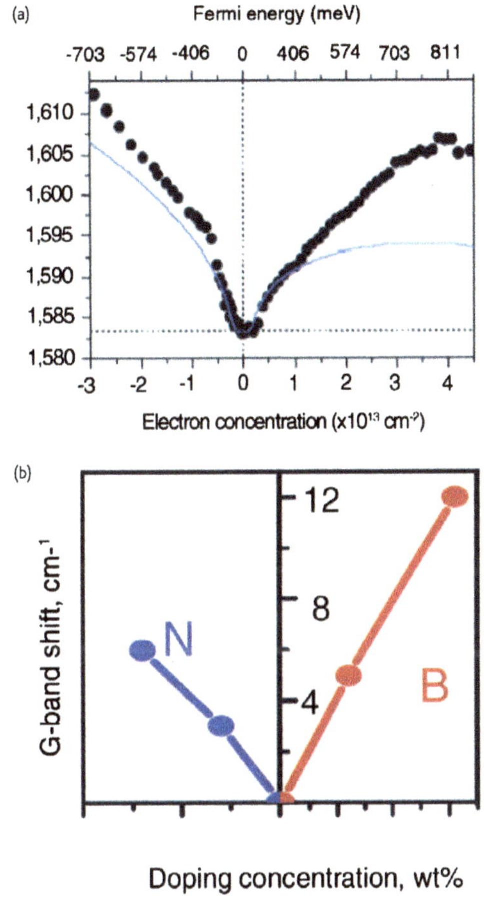


*Fig. 4 (a) Position of the Raman G-band as a function of electron and hole doping*[13]*. (b) Shifts of the G-band caused by nitrogen and boron doping*[26].

with bromine and iodine. They show that charge-transfer effects are greater on single and bi–layer graphenes compared to three and four–layer graphenes. Evidence for molecular charge-transfer involving graphene is also evidenced by the observation of charge-transfer bands in the electronic absorption spectra. Electrical resistivity of graphene varies in opposite directions on interaction with electron-donors and –acceptor molecules. The magnitude of interaction between graphene and donor/acceptor molecules seems to depend on the surface area of the graphene sample[20]. Isothermal titration calorimetry (ITC) measurements show that the interaction energies of graphene with electron-acceptor molecules are higher than those with -donor molecules[21]. DFT calculations confirm the occurrence of charge-transfer induced changes in graphene giving rise to mid-gap molecular levels with tuning of band gap region near the Dirac point and show how they are different from the effects of electrochemical doping[22,23]. It has been shown that n-type and p-type graphenes result from charge-transfer interaction of graphene with donor and acceptor molecules respectively. It is also predicted that the extent doping depends on the coverage of organic molecules. Synchrotron-based high-resolution photoemission spectroscopy studies reveal that charge transfer from graphene to adsorbed $F_4$-TCNQ is responsible for the p-type doping of graphene[24]. Recent studies of the core-level spectra of the dopant molecules (TTF and TCNE) provide direct evidence for charge transfer involving graphene[25].

It is interesting to compare the effects of doping graphene by gating[13] or chemical doping by boron and nitrogen[26] with those caused by molecular charge-transfer[17]. The G-band is shifted to higher frequencies when an electron acceptor molecule is adsorbed

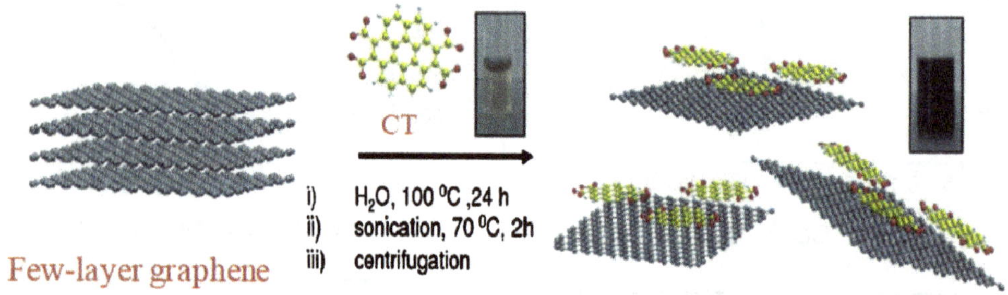


*Fig. 5 Schematic illustration of exfoliation of few-layer graphene with CT to yield monolayer graphene-CT composites*[27].

on graphene, while it is shifted to lower frequencies when an electron donor molecule is adsorbed. This is in contrast to the gate-induced or chemical doping where the G-band increases in frequency for both the electron and hole doping (Compare Fig. 2 and 4). The difference may partly be due to local distortions caused in molecular interaction and this aspect needs further study. It is also possible that some of the graphene samples may have accidentally got doped (p or n type) giving rise to such differences. It would be instructive to investigate the effects of TTF and TCNE on B- and N- doped graphenes.

Organic molecules containing aromatic π systems can be used to solubilize and modify the electronic structure of graphene. Charge-transfer with coronene tetracarboxylate (CT) has been exploited recently to solubilize graphene sheets[27]. It was shown that the CT molecules help to exfoliate few-layer graphene and selectively solubilize single- and double-layer graphenes (Fig. 5). Graphene quenches the fluorescence of aromatic molecules, probably due to the electron transfer, a feature of possible use in photovoltaics. Charge-transfer from fluorescent molecules to graphene has been utilized in visualization of graphene sheets by fluorescence microscopy[28] and in the use of graphene as a substrate for resonance Raman spectroscopy[29]. Molecular charge-transfer affects the magnetic properties of graphene[30]. Magnetization of graphene decreases on adsorption of TTF and TCNE, interaction with TTF having a greater effect than with TCNE. It is difficult to know the exact cause of such differences, but effects related to molecular planarity and area as well as mechanical compression would be relevant factor.

Charge-transfer effects would be expected to be observed with graphene nanoribbons (GNRs) as well. Graphene nanoribbons are quasi one-dimensional materials with novel electronic, magnetic, optical and conduction properties[31]. Two different edge geometries, namely zigzag and armchair are possible, due to the finite termination of graphene which control the electronic properties of graphene nanoribbons. The periodic zigzag graphene nanoribbons (ZGNRs) show interesting localized electronic states at the edges[32-34]. These edge states are ferromagnetically ordered but antiferromagnetically coupled to each other. Theoretical investigations have shown that that ZGNRs become half-metallic when an external transverse electric field is applied but very high critical electric fields are required to achieve half-metallicity, suggesting that the realization of half-metallicity is practically difficult[35,36]. Chemical modification of edges either by passivation with functional groups such as H, $NH_2$, NO, and $CH_3$[37] or the replacement of edge carbon atoms with boron (B) or nitrogen (N) atoms is suggested as an alternative way to induce half-metallicity in ZGNRs[38,39]. Molecular charge-transfer doping by the adsorption of various electron-donors and –acceptor molecules might lead to interesting changes in the electronic, magnetic and other properties of GNRs.

## Single-walled carbon nanotubes

Carbon nanotubes are one-dimensional (1D) materials with different chiralities and diameters[40-42]. SWNTs are formed by rolling two-dimensional graphene sheets into cylinders along a (n, m) lattice vector ($C_h$) in the graphene plane, where n and m are integers. We can obtain the diameter and chiral angle from the (n, m) values. Nanotubes with chiral numbers n=m are metallic, and quasi metallic if n-m is divisible by 3. The other nanotubes are semiconducting. As-prepared SWNTs generally contain a mixture of metallic and semiconducting species. In conventional synthetic processes employed for SWNTs, metallic nanotubes constitute 33%, the remaining being semiconducting nanotubes. Metallic and semiconducting SWNTs are readily characterized by Raman and electronic spectra[40,41,43,44]. The radial breathing mode (RBM) in the Raman spectrum of SWNTs is useful for determining the diameter and the (n, m) values of the nanotubes. Electronic properties of the nanotubes can be predicted by using the Kataura plots based on the RBM bands. The Raman band of SWNTs centered around 1580 $cm^{-1}$ (G-band) exhibits a feature around ~1540 $cm^{-1}$ which is characteristic of metallic SWNTs[43]. The G-band can be deconvoluted to get the relative proportions of metallic and semiconducting species. Absorption spectra of SWNTs in the visible and near IR regions contain characteristic bands due to the metallic and semiconducting species.

Ability to tune the electronic properties of SWNTs is important for many applications. A control of the carrier type and concentration has been achieved by electrochemical doping or chemical doping. Through electrochemical top-gating, it is possible to achieve a high level of doping[45]. The electronic structure and phonon frequencies of SWNTs are affected on doping with electrons or holes. Changing the electronic properties of SWNTs by chemical means has attracted considerable attention. Chemical approaches employ covalent or noncovalent functionalization. Covalent functionalization methods include direct addition of fluorine atoms, cycloaddition reactions, radical and nucleophilic additions to side walls of nanotubes[41,46]. Another method is to introduce carboxylic groups on the side walls of SWNTs by oxidation. The carboxylic acid groups can be further transformed to amide or ester linkages. Attachment of various molecules also enables modification of the electronic structure of SWNTs[41,46]. The main problem with covalent functionalization is that it causes a change in the hybridization of carbon from $sp^2$ to $sp^3$, and hence in the electronic structure. Excessive covalent functionalization completely destroys the electronic structure of SWNTs.

Noncovalent functionalization offers non-invasive approaches to modify SWNTs properties. Noncovalent modification can be brought by adsorption of aromatic compounds, surfactants or polymers and also through $\pi$–$\pi$ stacking and hydrophobic interaction. Earlier studies of noncovalent modification focused on the solubilization and exfoliation of SWNT bundles. Another aspect which has come into light in recent years is to modify the electronic structure of SWNTs by interaction with electron-donor and -acceptor molecules. This involves the adsorption of the electron-donor or -acceptor molecules on the surface of SWNTs. The advantage of molecular doping over other means is that one has an electronic system with less charged impurities. Furthermore, this effect of doping is reversible.

Doping of electrons and holes into SWNTs through molecular charge-transfer is of special interest[2]. Interaction of electron withdrawing molecules such as nitrobenzene and TCNE stiffen the Raman of SWNTs G-band while donating molecules such as TTF soften the G-band (Fig. 6a). The feature due to the metallic species around 1540 $cm^{-1}$ in the Raman spectrum nearly disappears on interaction of SWNTs with electron acceptors where as interaction of electron donors with SWNTs increases the metallic feature. It appears that electron donor molecules interact selectively with semiconducting SWNTs where as electron acceptors molecules interact with metallic SWNTs giving rise to possible metal ↔ semiconductor transitions. The changes in the Raman spectra are accompanied by changes in the electrical resistivity. The I-V curves become more nonlinear as one goes from aniline to nitrobenzene (Fig. 6b). The slope of the I-V curve also increases going from nitrobenzene to aniline, probably due to the presence of a higher proportion of metallic nanotubes in the presence of aniline. The above experiments were performed on mixture of metallic and semiconducting nanotubes. The experiments with pure

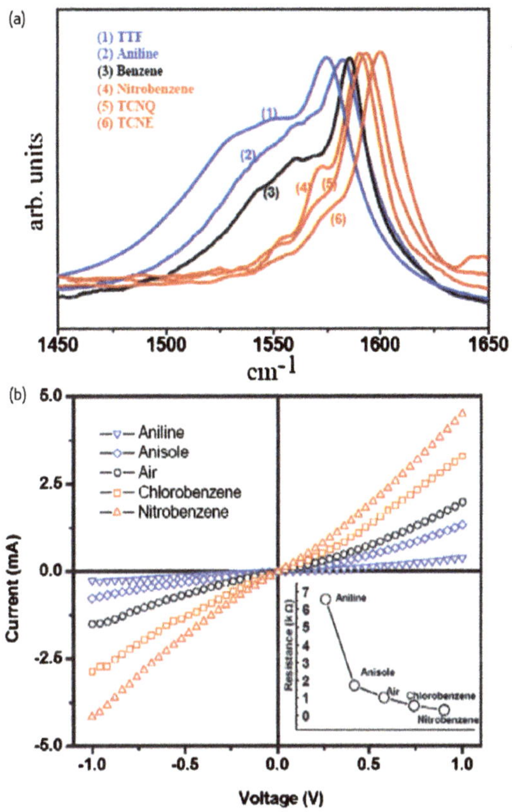


*Fig. 6 (a) Effects of interaction of SWNTs with electron donor and acceptor molecules on G-band in the Raman spectra. (b) I-V characteristics of the SWNTs in air and in the presence of different benzene derivatives[2].*

metallic nanotubes separately show that on interaction with TCNE, the 1540 $cm^{-1}$ feature in Raman spectrum due to the metallic species disappears due to the change in the Fermi level of the nanotubes. TTF had no effect on the Raman spectrum of metallic SWNTs. Interaction of TTF with semiconducting carbon nanotubes, on the other hand, increases the intensity of the 1540 $cm^{-1}$ feature. This remarkable change in the electronic structure of SWNTs is entirely reversible. Electrochemical top gating of SWNTs leads to blue-shift in the G-band of SWNTs accompanied by a decrease in the line width with both electron and hole doping[45].

ITC experiments on the interaction of SWNTs with molecules provide an insight into the affinities of the different molecules[21]. Interaction energies of electron acceptor molecules (eg., TCNE) with SWNTs are higher than those of electron donor molecules (e.g., TTF). Metallic SWNTs interact reversibly with electron acceptor molecules such as TCNE, the interaction energy being higher than

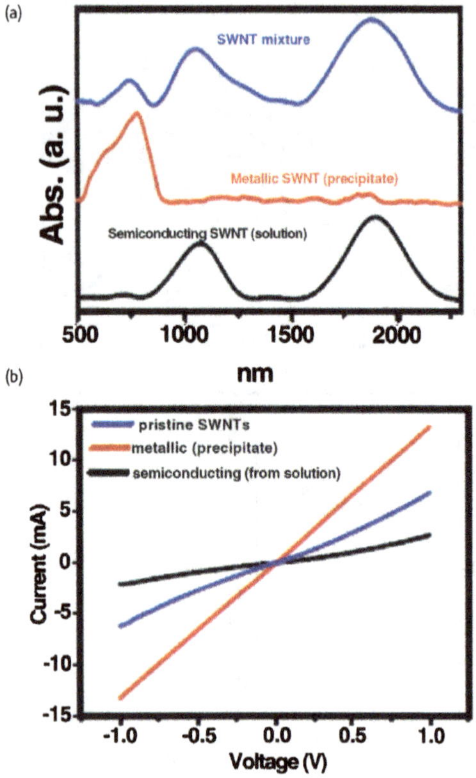


*Fig. 7 (a) Optical absorption spectra and (b) I-V characteristics of pristine SWNTs (blue), precipitate (red) and SWNTs from solution (black)*[4].

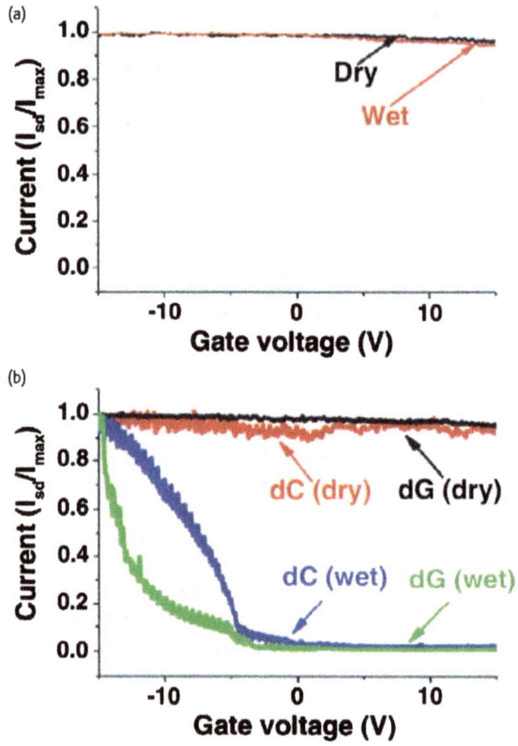


*Fig. 8 Dependence of current of the ssDNA–SWNT hybrid on gate voltage. (a) A metallic behavior with SWNTs without DNA. (b) A p-type semiconductor behavior with the ssDNA–SWNT hybrid*[39].

with as-prepared SWNTs (containing a mixture of metallic and semiconducting species). The interaction energy of metallic nanotubes with a donor molecule such as TTF is negligible and could not be measured by ITC. ITC measurements clearly show that metallic nanotubes specifically interact with electron-withdrawing molecules. The interaction energy with acceptor molecules varies with the electron affinity as well as with the charge-transfer transition energy for different aromatics. Density functional theory calculations have shown that the nature of interaction is in physisorption regime and mainly governed by Coulombic forces[47]. The large band gap of semiconducting (8,0) SWNTs can be tuned through adsorption of selective organic molecules which gives mid gap localized molecular levels near the Fermi energy with tuning the band gap region. Metallic (5,5) SWNTs, and semiconducting (8,0) SWNTs turn into semiconducting and metallic nanotubes respectively in the presence of adsorbed molecules.

Molecular charge-transfer between SWNTs and an appropriate π-system can be exploited for the effective separation of metallic and semiconducting nanotubes, since π–π interaction with aromatic molecules enables the solubilization of SWNTs[4]. Thus, the potassium salt of coronene tetracarboxylic acid (CS) which has a large π skeleton attached to four electron-withdrawing groups, exhibits charge-transfer interaction with SWNTs causing solubilization in aqueous medium. In this process, CS debundles the SWNTs and precipitates metallic SWNTs leaving semiconducting SWNTs in the solution. Fig. 7a shows the optical absorption spectra of SWNTs before and after separation. Pristine SWNTs show a nonlinear I-V curve while the metallic nanotubes show linear behavior with conductivities of 92.5 of and 1538.5 mS/cm respectively at 300 K where as Semiconducting nanotubes exhibit a low conductivity of 53.5 mS/cm and a nonlinear I-V curve (Fig. 7b).

Nanoparticles of gold and platinum deposited on SWNTs also transform the semiconducting species to metallic ones due to Columbic charge-transfer[48]. Metal to semiconductor transition in SWNTs has been induced by helical wrapping of DNA[49]. Water appears to be critical to this reversible transition which accompanied by hybrid formation with DNA. It is predicted that a band gap can open up in metallic SWNTs wrapped with ssDNA in the presence of water molecules, due to charge-transfer (Fig. 8). Kim *et al.*[50] have shown that

**Instrument Citation**

- PerkinElmer instruments Lambda UV/VIS/NIR spectrometer.
- HORIBA JOBIN YVON LabRAM Raman spectrometer fitted with Raman mapping and heating and cooling stages.
- HORIBA JOBIN YVON Fluorolog spectrofluorometer fitted with single photon counting controller and iHR 320.

adsorption of $AuCl_3$ to SWNTs results in high level of p-doping due to strong charge-transfer from the SWNTs to $AuCl_3$ and they have shown that sheet resistance was systematically reduced with the increasing doping concentration. Ehli *et al.*[51] have shown that charge-transfer between SWNTs and perylene dyes leads to individualized nanotubes and they have observed radical ion-pair state is formed in the excited state. There are reports of charge-transfer interaction of SWNTs with $I_2$ and $Br_2$[52,53]. Doping of double-walled carbon nanotubes with bromine and iodine has been investigated[54,55]. Charge-transfer doping of DWNTs has been used to distinguish between the behavior of the S/M and M/S outer/inner semiconducting (S) and metallic (M) tube configurations. The binding of electron-accepting molecules ($F_4TCNQ$ and $NO_2$) to SWNTs leads to a threshold voltage shift toward positive gate voltages, while the binding of electron donating molecules ($NH_3$ and PEI) leads to a shift toward negative gate voltages[56-58]. Field-effect transistor devices made of semiconducting SWNTs have been used to obtain quantitative information on charge-transfer with aromatic compounds[59]. Stoddart *et al.*[60] have fabricated SWNT/ FET devices to investigate the electron/charge-transfer with the donor-acceptor SWNT hybrids. A SWNT/FET device, functionalized noncovaletly with a zinc porphyrin derivative shows that SWNTs act as electron donors and that the porphyrin molecules act as the electron acceptors.

## Conclusions and outlook

The discussion in the previous sections should make it clear how charge-transfer interaction of SWNTs with electron–donor and –acceptor molecules causes major changes in the Raman and electronic spectra of nanotubes. It is noteworthy that the transformation of semiconducting to metallic and vice versa is possible through charge-transfer interaction. Changes in the electronic and Raman spectra of graphene brought about by electron–donor and –acceptor molecules is equally fascinating. It should be possible to exploit the changes brought about by the molecular charge-transfer in these nanocarbons for device applications. mt

**REFERENCES**

1. Haddon, R. C., *Acc Chem Res* (1992) **25**, 127.
2. Voggu, R., *et al.*, *J Phys Chem C* (2008) **112**, 13053.
3. Shin, H.-J., *et al.*, *J Am Chem Soc* (2008) **130**, 2062.
4. Voggu, R., *et al.*, *J Am Chem Soc* (2010) **132**, 5560.
5. Das, B., *et al.*, *Chem Commun* (2008), 5155.
6. Geim, A. K., *Science* (2009) **324**, 1530.
7. Rao, C. N. R., *et al.*, *Angew Chem Int Ed* (2009) **48**, 7752.
8. Rao, C. N. R., *et al.*, *J Phys Chem Lett* (2010) **1**, 572.
9. Novoselov, K. S., *et al.*, *Science* (2004) **306**, 666.
10. Novoselov, K. S., *et al.*, *Nature* (2005) **438**, 197.
11. Zhang, Y., *et al.*, *Nature* (2005) **438**, 201.
12. Ferrari, A. C., *et al.*, *Phys Rev Lett* (2006) **97**, 187401.
13. Das, A., *et al.*, *Nat Nanotechnol* (2008) **3**, 210.
14. Das, A., *et al.*, *Phys Rev B* (2009) **79**, 155417.
15. Malard, L. M., *et al.*, *Phys Rev Lett* (2008) **101**, 257401.
16. Kuzmenko, A. B., *et al.*, *Phys Rev Lett* (2009) **103**, 116804.
17. Voggu, R., *et al.*, *J Phys Cond Mat* (2008) **20**, 472204.
18. Dong, X., *et al.*, *Small* (2009) **5**, 1422.
19. Jung, N., *et al.*, *Nano Lett* (2009) **9**, 4133.
20. Subrahmanyam, K. S., *et al.*, *Chem Phys Lett* (2009) **472**, 96.
21. Varghese, N., *et al.*, *J Phys Chem C* (2009) **113**, 16855.
22. Manna, A. K., and Pati, S. K., *Chem Asian J* (2009) **4**, 855.
23. Saha, S. K., *et al.*, *Phys Rev B* (2009) **80**, 155414.
24. Chen, W., *et al.*, *J Am Chem Soc* (2007) **129**, 10418.
25. Choudhury, D., *et al.*, *To be published*, .
26. Panchakarla, L. S., *et al.*, *Adv Mat* (2009) **21**, 4726.
27. Ghosh, A., *et al.*, *Chem European J* (2010) **16**, 2700.
28. Kim, J., *et al.*, *J Am Chem Soc* (2009) **132**, 260.
29. Xie, L., *et al.*, *J Am Chem Soc* (2009) **131**, 9890.
30. Matte, H. S. S. R., *et al.*, *J Phys Chem C* (2009) **113**, 9982.
31. Dutta, S., and Pati, S. K., *J Mater Chem* (2010), doi: 10.1039/c0jm00261e.
32. Pisani, L., *et al.*, *Phys Rev B* (2007) **75** (6), 064418.
33. Rudberg, E., *et al.*, *Nano Lett* (2007) **7** (8), 2211.
34. Dutta, S., *et al.*, *Phys Rev B* (2008) **77** (7), 073412.
35. Son, Y.-W., *et al.*, *Nature* (2006) **444** (7117), 347.
36. Kan, E.-J., *et al.*, *App Phys Lett* (2007) **91** (24), 243116.
37. Li, Z., *et al.*, *J Am Chem Soc* (2008) **130** (13), 4224.
38. Dutta, S., *et al.*, *Phys Rev Lett* (2009) **102** (9), 096601.
39. Ci, L., *et al.*, *Nat Mater* **9** (5), 430.
40. Saito, R., *et al.*, *Physical Properties of Carbon Nanotubes*. Imperial College Press: London, 1998.
41. Rao, C. N. R., and Govindaraj, A., *Nanotubes and Nanowires: RSC. Nanoscience & Nanotechnology series*. Royal Society of Chemistry: Cambridge, 2005.
42. Rao, C. N. R., *et al.*, *Nanoscale* (2009) **1**, 96.
43. Dresselhaus, M. S., *et al.*, *J Phys Chem C* (2007) **111**, 17887.
44. Strano, M. S., *J Am Chem Soc* (2003) **125**, 16148.
45. Das, A., *et al.*, *Phys Rev Lett* (2007) **99**, 136803.
46. Tasis, D., *et al.*, *Chem Rev* (2006) **106**, 1105.
47. Manna, A. K., and Pati, S. K., *Nanoscale* (2010) 2, 1190 .
48. Voggu, R., *et al.*, *J Phys Cond Mat* (2008) **20**, 215211.
49. Cha, M., *et al.*, *Nano Lett* (2009) **9**, 1345.
50. Kim, K. K., *et al.*, *J Am Chem Soc* (2008) **130**, 12757.
51. Ehli, C., *et al.*, *Nat Chem* (2009) **1**, 243.
52. Rao, A. M., *et al.*, *Nature* (1997) **388**, 257.
53. Chen, J., *et al.*, *Science* (1998) **282**, 95.
54. do Nascimento, G. M., *et al.*, *Nano Lett* (2008) **8**, 4168.
55. do Nascimento, G. M., *et al.*, *J Phys Chem C* (2009) **113**, 3934.
56. Shim, M., *et al.*, *J Am Chem Soc* (2001) **123**, 11512.
57. Kong, J., *et al.*, *Science* (2000) **287**, 622.
58. Takenobu, T., *et al*, *Adv Mat* (2005) **17**, 2430.
59. Star, A., *et al.*, *Nano Lett* (2003) **3**, 1421.
60. Hecht, D. S., *et al.*, *Nano Lett* (2006) **6**, 2031.

REVIEW

ADVANCED MATERIALS www.advmat.de

# Synthesis of Inorganic Nanotubes

By *C. N. R. Rao** and *Achutharao Govindaraj*

**Nanotubes constitute an exciting class of one-dimensional nanomaterials of which carbon nanotubes are recognized widely as materials of importance. The possibility of having inorganic nanotubes was recognized early in the 1990s, accompanied by the report of nanotubes of $MoS_2$ and $WS_2$. Since then, nanotubes of several inorganic materials have been prepared and characterized. While nanotubes of metal chalcogenides and oxides form a high proportion of the inorganic nanotubes investigated hither to, nanotubes of many other materials have also been prepared and characterized. Several synthetic strategies including both physical and chemical methods have been employed, of which the use of templates, precursors, and hydro- or solvothermal methods are prominent. In this article, we shall present a brief account of the present status of the synthesis of nanotubes of elemental materials as well as binary and complex metal oxides, chalcogenides, pnictides and carbides.**

## 1. Introduction

Zero-dimensional nanoparticles and one-dimensional (1D) nanowires and nanotubes are important classes of nanomaterials.[1,2] The first family of nanotubes is that of carbon nanotubes described by Iijima.[3] Nanotubes are, however, no longer confined to carbon but encompass a variety of inorganic materials,[4,5] and peptides.[6] In this article, our concern is with nanotubes of inorganic materials excluding carbon.

The early examples of inorganic nanotubes synthesized in the laboratory are those of molybdenum and tungsten sulfides by Tenne and coworkers.[7] These layered sulfides form fullerene-type structures and hence also nanotubes. Several methods to prepare nanotubes of Mo and W sulfides and of the analogous selenides have been reported in the last few years.[1,2] The synthesis of BN nanotubes has also received considerable attention because of the similarity of the structure of BN to graphite. In the last few years, nanotubes of several inorganic materials including binary oxides, nitrides, halides as well as metals and non-metallic elemental materials have been prepared and characterized.[1,2,4,5] Besides nanotubes of binary compounds, those of complex materials such as perovskite titanates and spinels have also been reported. Composites that involve nanotubes, nanowires, and nanoparticles are also known. In this article, we shall provide a status report on the synthesis of inorganic nanotubes. In doing so, we shall cover the synthetic strategies employed for the different classes of inorganic nanotubes. In view of the vast literature that has emerged in the last 2–3 years, we were unable to cite all the papers in this area and have restricted ourselves to representative ones. We apologize for any oversight or error in judgement.

[*] Prof. C. N. R. Rao, Dr. A. Govindaraj
International Center for Materials Science, New Chemistry Unit and CSIR Centre of Excellence in Chemistry
Jawaharlal Nehru Centre for Advanced Scientific Research
Jakkur P. O., Bangalore, 560 064 (India)
E-mail: cnrrao@jncasr.ac.in
Prof. C. N. R. Rao, Dr. A. Govindaraj
Solid State and Structural Chemistry Unit
Indian Institute of Science
Bangalore, 560 012 (India)

DOI: 10.1002/adma.200803720

## 2. General Synthetic Strategies

Several strategies have been employed for the synthesis of inorganic nanotubes. In the case of molybdenum and tungsten sulfides and such layered chalcogenides, decomposition of precursor compounds such as the trisulfides (e.g., $MoS_3$ or $WS_3$) and ammonium thiometallate or selenometallate has been successful. An important method, used particularly in the case of oxide nanotubes, is the hydrothermal and solvothermal route, carried out in the presence of surfactants or other additives in certain instances. Electric arcing and laser ablation have been used to synthesize nanotubes of BN and other materials. Sol–gel chemistry is useful for the synthesis of nanotubes, especially of oxides. Chemical vapor deposition (CVD) is commonly used for the synthesis of some of the nanotubes.

A popular method of synthesis in the last few years has employed templates. The templates can be porous membranes of alumina or polycarbonate. The pores are used to deposit the relevant materials or these precursors, followed by annealing and removal of the template. Deposition of the material in the porous channels is carried out by the sol–gel method or by an electrochemical procedure. Electrochemical anodization is commonly used for the synthesis of nanotubes of $TiO_2$, ZnO, and such oxides. The porous membrane method has emerged to be a general means of preparing a large variety of inorganic nanotubes and nanowires. Carbon nanotubes, surfactants, polymer gels, and liquid crystals have all been used as templates, wherein the precursor material is covered over the templates, followed by annealing and removal (burning or dissolution) of the template.

In what follows, we shall discuss the synthesis of various inorganic nanomaterials where we will indicate the method and give the most essential aspects of the procedure. In order for the reader to obtain greater details, we have provided a large list of references.

## 3. Nanotubes of Metals and other Elemental Materials

Synthesis of gold nanotubes was reported by Martin and co-workers[8,9] in the 1990s. They prepared Au tubules with lengths of up to a few micrometers and diameters of a few hundred nanometers by electrochemically depositing gold into the pores of a microporous alumina membrane (AM). To obtain gold tubules, initially the AM pore walls were chemically derivatized by attaching a molecular anchor such as a cyanosilane, so that the electrodeposited metal preferentially deposits on the pore wall, which leads to tubule formation.

Electrodeposition in membrane pores is an important method for the synthesis of metal nanotubes. Thus, Au nanotube arrays have been prepared by direct electrodeposition in the nanochannels of alumina templates. The nanochannel alumina templates with pore diameters of about 105 and 45 nm were used to synthesize nanotubes with average outer diameters respectively of 105 and 45 nm and a wall thickness of 15 nm. The lengths went up to several micrometers. The alumina templates are readily removed by treatment with NaOH. The nanotube arrays so obtained have a well-controlled microstructure and are polycrystalline with a face-centered cubic structure.[10] Au nanotubes have also been prepared by electroless deposition in the pores of track-etched polycarbonate membranes that contain 10 μm thick and 220 nm diameter pores.[11] The inner surfaces of the polycarbonate membranes were first sensitized with a $Sn^{2+}$ salt and then activated by forming a layer of Ag, before depositing Au for a period of 2 h. The gold nanotubes were cleaned with 25% $HNO_3$ solution for 15 h. Hydrophobic or hydrophilic self-assembled monolayers on gold nanotubes can be formed by rinsing the samples in ethanol for 20 min, followed by immersion in a solution of ethanol that contains $HS(CH_2)_{15}CH_3$ or $HS(CH_2)_{15}COOH$. The Au nanotubule are polycrystalline and have lengths of up to 6 μm and inner diameters of ~1 nm. By controlling the Au deposition time, Au nanotubules of effective inside diameters of molecular dimensions (< 1 nm) can be prepared. They found electroless deposition allows for more uniform gold deposition in a short duration of time. Since the electroless plating method used to deposit Au nanotubes in polymeric templates does not work in AMs, Martin and coworkers[12] have developed a modified electroless plating strategy that can be used to deposit high-quality Au nanotubes within the pores of alumina templates.

Three-dimensional (3D) Au nanotube arrays with smooth as well as nanoporous walls have been obtained by using anodic alumina and conducting polyaniline nanorod templates.[13] In this procedure, polyaniline nanorods were predeposited electrochemically in the interior of a porous alumina membrane, which was then used as a template for the formation of vertically aligned Au nanotubes. For the synthesis of Au nanotube arrays with nanoporous walls, gold/silver alloy nanowalls were electrodeposited from cyanide solutions that contain gold/silver ions (mole ratio, $Au^+/Ag^+ = 1:3$). The nanoporous walls were generated by de-alloying (selective dissolution of the less noble metal) the gold/silver alloy shells with concentrated nitric acid, which also dissolves the polyaniline nanorods. The porous architecture is formed because of an intrinsic dynamic pattern formation process, in which the more noble metal (Au) atoms tend to aggregate into two-dimensional clusters through a phase separation process at the solid–acid interface. The length, average inner diameter, and wall thickness of the Au nanotubes with smooth walls was ~4 μm, ~196 nm, and 62 (± 18 nm), respectively. The Au nanotubes with smooth nanoporous walls (nanopore diameter of ~8 nm) had similar physical dimensions as the nanotubes.

Gold nanotubes embedded within the pores of the polycarbonate template membranes were subjected to reactive ion etching (RIE) using an oxygen plasma to selectively etch approximately 2.3 mm of the polycarbonate, leaving the polycrystalline gold nanotubes intact. Figure 1 shows field-emission scanning electron microscopy (FESEM) images of the top surface of a template membrane after electroless deposition of gold followed by RIE.[14] Single crystalline and bamboo-like Au nanotube arrays growing in the [111] direction, with a diameter of 100–150 nm and a length of 10 μm, and standing perpendicular to the Ti metal foil substrates are obtained by using a

C.N.R. Rao obtained his Ph.D. from Purdue University and D.Sc. from the University of Mysore. He is the National Research Professor and Linus Pauling Research Professor at the Jawaharlal Nehru Centre for Advanced Scientific Research as well as Honorary Professor at the Indian Institute of Science (both at Bangalore). His research interests are in the chemistry of materials. He has authored over 1000 research papers and edited or written 30 books in materials chemistry.

A. Govindaraj obtained his Ph.D. from the University of Mysore and is a Senior Scientific Officer at the Indian Institute of Science, and Honorary Faculty Fellow at the Jawaharlal Nehru Centre for Advanced Scientific Research. He works on various nanomaterials. He has authored more than 100 research papers and co-authored a book on nanotubes and nanowires.

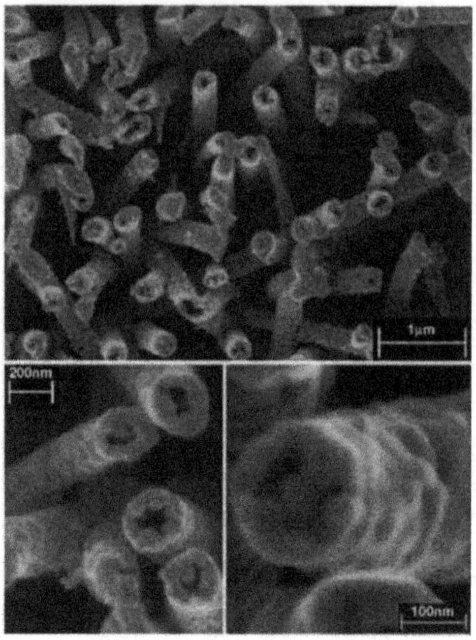


**Figure 1.** Field-emission SEM images of gold nanotubes at different magnifications. Micrographs were taken on the etched side of the membrane. Reproduced with permission from [14]. Copyright 2004, Wiley-VCH.

radiation track-etched hydrophilic polycarbonate membrane.[15] By using water-dissolvable $Na_2SO_4$ nanowires as templates, Au nanoparticle tubes have been obtained by the self-assembly of Au nanoparticles.[16] The as-synthesized Au nanoparticle tubes were further calcined at 300 °C (5 °C $min^{-1}$) for 30 min to transform them into polycrystalline Au nanotubes. At this temperature, the Au nanoparticles (3–5 nm diameter) melt and form nanotubes of ~1 μm length and ~80 nm diameter. Using an electroless deposition procedure, continuous, polycrystalline Au nanotubes with controllable shape, size (tens of nanometer in diameter), shell thickness (~5 nm), and length (up to 5 μm) can be grown by using Co nanoparticles as sacrificial templates.[17] Here, the alignment of Co nanoparticles into a 1D structure is induced by manipulation of the magnetic field. In this reaction, $Au^{3+}$ is reduced to $Au^0$ by the $Co^0$ nanoparticles as given by the following reaction:

$$3Co^0 + 2Au^{3+} \rightarrow 3Co^{2+} + 2Au^0 \quad (1)$$

Goethite (FeOOH) nanorods have been used as templates to grow Au nanotubes with a length of a few hundred nanometers and an aspect ratio between 3 and 4.[18] The uniform growth of gold nanoshells on goethite nanorods was achieved by $SiO_2$-mediated assembly/attachment of Au nanoparticles/seeds on these rods, followed by a one-step seeded growth by the catalyzed reduction of $HAuCl_4$ using formaldehyde. The successful attachment of small Au seeds on goethite nanorods requires modification of the nanorod surface, which was carried out by depositing a thin layer of silica using tetraethoxysilane (TEOS), followed by silanization using (3-aminopropyl) trimethoxysilane (APS). Gold nanoparticles (~3 nm in diameter) are prepared by the reduction with tetrakishydroxymethylphosphonium chloride (THPC), which were used for the self-assembly of Au seeds on the surface of goethite rods. The growth of Au shells on the goethite surface was accomplished by the selective reduction of $HAuCl_4$ with a weak reducing agent such as formaldehyde, catalyzed by the Au seeds. In a typical synthesis, 100 mL of the dispersion of goethite was added to 48 g of poly(vinyl pyrrolidone) (PVP) and the mixture stirred for 24 h before transferring to ethanol (100 mL). Twelve millilitres of the goethite–PVP dispersion was diluted with 200 mL of ethanol that contained 20 mL of ammonium hydroxide and 0.75 mL of TEOS, under mechanical stirring. The goethite nanorods coated with silica were washed several times with ethanol. Silanization was carried out by adding 0.6 mL of pure APS to 10 mL of the goethite@silica dispersion under magnetic stirring, followed by additional washing with ethanol to remove excess APS. For the assembly of Au nanoparticles on the nanorod surface, 5 mL of the solution of silanized goethite particles was centrifuged and re-dispersed in 5 mL of the Au colloid solution in an ultrasound bath for 1 to 2 min. This step was repeated until no further adsorption of gold on the surface of the particles was observed. Further growth of the Au coating on the nanorods was carried out as follows. A mixture of 0.425 mL of 0.01 M $HAuCl_4$ and 10 mL of $1.8 \times 10^{-3}$ M $K_2CO_3$ was aged in the dark for a day so that $Au^{III}$ ions were reduced to $Au^{I}$ ions. This solution (5 mL) was mixed with 0.01–0.03 mL of the Au-covered goethite solution and 0.01 mL of formaldehyde solution. The thickness and surface roughness of the obtained shells could be adjusted by simply varying the concentration ratio between the seeds (modified goethite rods) and the growth reagents ($HAuCl_4$ and formaldehyde).

Platinum and PtPd alloy nanotubes with polycrystalline walls and a face-centered cubic structure (50 nm in diameter, 5–20 μm long, and 4–7 nm wall thickness) can be synthesized by the galvanic replacement reaction of Ag nanowires.[19] In this procedure, Ag nanowires are refluxed for 10 min with platinum acetate in aqueous solution. When an aqueous platinum acetate solution is mixed with a dispersion of Ag nanowires, the galvanic replacement reaction generates a tubular sheath whose morphology is complementary to that of the Ag nanowire (step 1 in Scheme 1). Following the same scheme for multiple-walled nanoshells, coaxial nanotubes with more than two walls can be prepared. Mesoporous Pt nanotubes have been prepared by incorporating lyotropic liquid crystals in the pores of an AM before depositing the metal (Fig. 2).[20] By using a mixed non-ionic–cationic liquid crystalline surfactant (nonaethylene glycol monododecyl ether ($C_{12}EO_9$), polyoxyethylene (20) sorbitan monostearate (Tween 60) and water), polycrystalline Pt nanotubes (with a face-centered cubic structure) with small inner (3–4 nm) and outer (6–7 nm) diameters have been obtained.[21] In a typical synthesis, the liquid crystalline phase that contains hexachloroplatinic acid ($H_2PtCl_6$), $C_{12}EO_9$, Tween 60, and water at a molar

   

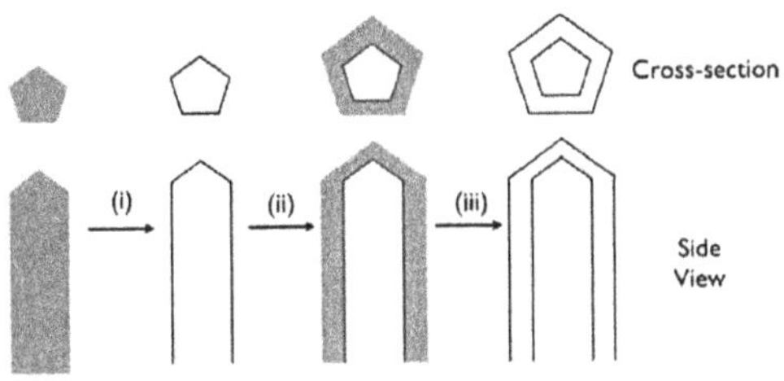


**Scheme 1.** Schematic representation of the fabrication of multiple-walled nanotubes composed of Pt/Ag alloys.

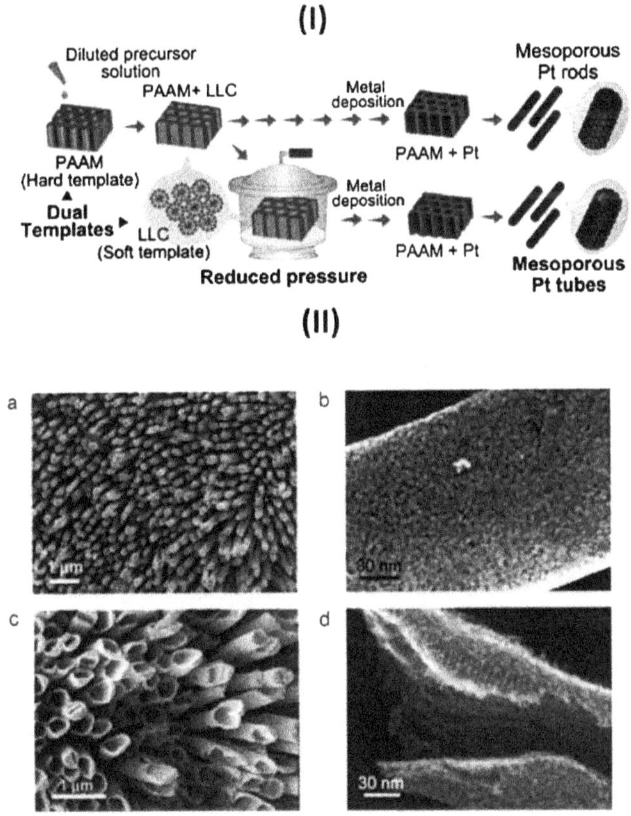


**Figure 2.** I) Schematic view of the preparative procedure for mesoporous Pt nanorods and nanotubes. II) SEM images of: a,b) mesoporous Pt nanorods, and c,d) mesoporous Pt nanotubes. Figures (b) and (d) are highly magnified images of (a) and (c), respectively. Reproduced with permission from [20]. Copyright 2008, RSC.

ratio of 1: 1: 1: 60 was treated with hydrazine to cause the reduction of the metal salts confined in the lyotropic mixed crystals of the two different surfactants to yield the metal nanotubes. The resulting solid was separated, and washed with water and ethanol prior to drying in air. The same procedure has been used to prepare nanotubes of other metals such as Pd and Ag.

Hierarchical assemblies of hollow Pd nanostructures have been grown by using Co nanoparticles as self-sacrificial templates.[22] Assemblies of hollow Pd nanostructures (from 80 nm Pd nanoparticles) are obtained with raspberry-like or nanotube-like geometries, by the replacement reaction between $H_2PdCl_4$ and Co nanoparticles, which occurs rapidly, wherein the reduced Pd atoms nucleate and grow into very small particles, and eventually evolve into a thin shell around the cobalt nanoparticles. The replacement reaction is given by:

$$Co + PdCl_4^{2-} \rightarrow Pd + Co^{2+} + 4Cl^{-} \quad (2)$$

The polycrystalline 1D Pd nanostructures have a face-centered cubic structure with lengths that run to several micrometers and diameters up to ~60 nm.

Ni and Co microtubules were prepared by Han et al.[23] by the pyrolysis of composite fibers consisting of a poly(ethylenetetraphthalate) (PET) core fiber with the electroless-plated metal at the exterior. Ni microtubules prepared by this method were single-crystalline, but the Cu microtubules were polycrystalline. Electrodeposition in the pores of AMs has been carried out by Bao et al.[24] to obtain ordered arrays of Ni nanotubules. The pore walls were modified by these workers with an organic amine to assist the formation of nanotubes (Fig. 3). In the absence of the amine, nickel nanowires were obtained. Nickel, when electrodeposited in the pores, binds preferentially to the pore walls because of its strong affinity towards the amine. The AM is removed by treatment with NaOH. The top-view SEM image in Figure 3b shows open-ends of the Ni nanotubules after the removal of the top layer of the AM with NaOH solution. The electron diffraction pattern of the Ni nanotubules in the inset of Figure 3c shows diffuse rings, which indicates that the Ni nanotubules are polycrystalline and have a face-centered cubic structure. The transmission electron microscopy (TEM) image of the Ni nanotubules in Figure 3c shows that the diameter of the Ni nanotubules are uniform with average outer diameter of 160 ± 20 nm. The deposition follows a bottom-up approach. Thus, the growth of nanotubules starts at the Au cathode at the bottom of the pores. The length and the wall thickness of the Ni nanotubules depend on experimental conditions, such as pore-wall modifying agent, and electrodeposition parameters. A low current density appears to be a key factor. For instance, with a current density of 0.3 mA $cm^{-2}$ and an electrodeposition time of 24 h, Ni nanotubules of 20 μm in length and 30 nm in wall

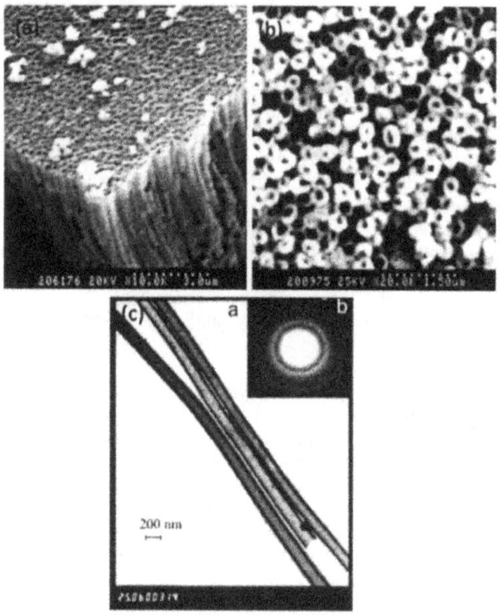


Figure 3. SEM images of a) Ni nanotubules after dissolution of the alumina template, b) showing the top-view of an array of Ni nanotubules after partial removal of the template, and c) transmission electron microscopy (TEM) image of the Ni tubules after dissolution of the alumina template. Inset shows electron diffraction pattern of the Ni nanotubules. Reproduced with permission from [24]. Copyright 2001, Wiley-VCH.

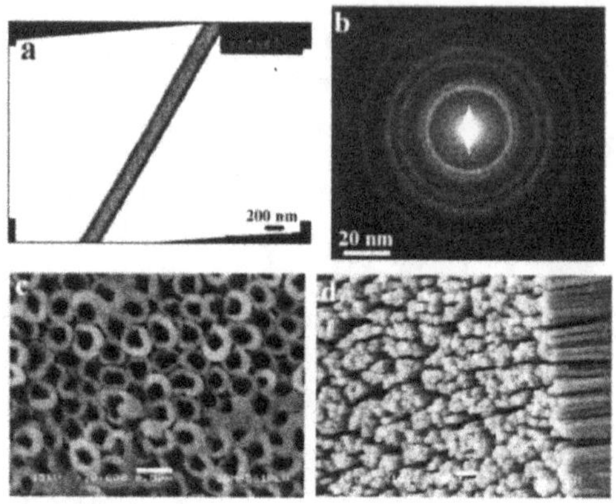


Figure 4. a) TEM image and b) SAED pattern of a Ni nanotube after removing the alumina template. c) Top-view and d) side-view SEM images of an ordered array of Ni nanotubes after partial removal of the template. The experimental parameters are: current density, 0.13 mA $cm^{-2}$, concentration of P123 of 37 g $L^{-1}$, and deposition time of 48 h. Reproduced with permission from [25]. Copyright 2006, Wiley-VCH.

thickness were obtained, while with electrodeposition for 48 h, Ni nanotubules grow up to 35 μm in length and 60 nm in wall thickness. Ordered magnetic Ni nanotubes have been prepared by employing electrodeposition by adding an amphiphilic triblock co-polymer (Pluronic P123) to the electrodeposition solution.[25] By adjusting experimental parameters such as current density and electrodeposition time, the wall thickness and length of the nanotube are controlled (Fig. 4). The Ni nanotubes prepared at a current density of 0.13 mA $cm^{-2}$ with an electrodeposition solution that contains 37 g $L^{-1}$ of P123 and a deposition time of 48 h yielded uniform tubes of about 50 nm wall thickness, 60 μm length, and an outer diameter of ~250 nm, which corresponds to the pore diameter of the alumina template. Weak, diffuse rings in the selected-area electron diffraction (SAED) pattern of the nanotubes showed they were polycrystalline.

The preparation and growth mechanism of Ni, Co, and Fe nanotubes have been reported by Cao et al.[26] wherein the nanotubes are actually constructed from non-layered materials of the metal, and the tubular growth is directed by the current in the template-based electrodeposition process (Fig. 5). Typically, the length of the as-synthesized tubular structures can reach about 60 μm, which corresponds to the thickness of the AM. Most of them have an outer diameter of 50–100 nm, which corresponds to the pore diameter of the AM, and an inner diameter of about 30–50 nm. The electron diffraction patterns reveal that the nanotubes are single-crystalline and have a body-centered cubic structure for iron, a face-centered cubic structure for nickel, and a hexagonal close packed structure for cobalt. Employing AMs, aligned Fe nanotubes have been grown electrochemically.[27] The wall thickness of the nanotubes could be controlled by changing the deposition parameters. Co nanotubes have also been grown by electrodeposition in an AM[28] Electrodeposition in a rotating electric field produces dense arrays of single crystalline Cu nanotubes.[29] The applied rotating field makes the ions graze the surface of the pores in helical paths and makes the deposition occur selectively in the region near the wall of the nanopores. The wall thickness of the metal nanotubes so obtained are in the range of 15–20 nm and can be controlled by changing the amplitude of the rotating field. The nanotubes had a diameter of ~230 nm with lengths going up to several micrometers.

Electrochemical deposition methods have also been used for the synthesis of nanotubes of many other metals including Zn, Sn, and Ag having polycrystalline structures.[30] In order to fabricate these metal nanotube arrays, a thin layer of Au was deposited on the bottom surface of the porous AM, for use as a working electrode. The circular step edges of the Au nanorings so formed at the bottom of the AM serve as the preferential sites for the deposition of metal ions. Nanotube arrays of Zn and Sn were then obtained within the porous AM by preferentially plating at circular step edges of the Au nanorings. This was accompanied by the evolution of hydrogen gas, which is critical for nanotube formation. It is well known that circular step edges have the ability to catalyze electron transfer to metal ions in solutions. Since the reductive potential of hydrogen is higher than that of Zn and Sn, evolution of hydrogen gas accompanies metal deposition. Hydrogen gas evolves continuously from the cathode through the central region of the nanopores while the metallic elements become attached to the pore walls, which results in the formation

  

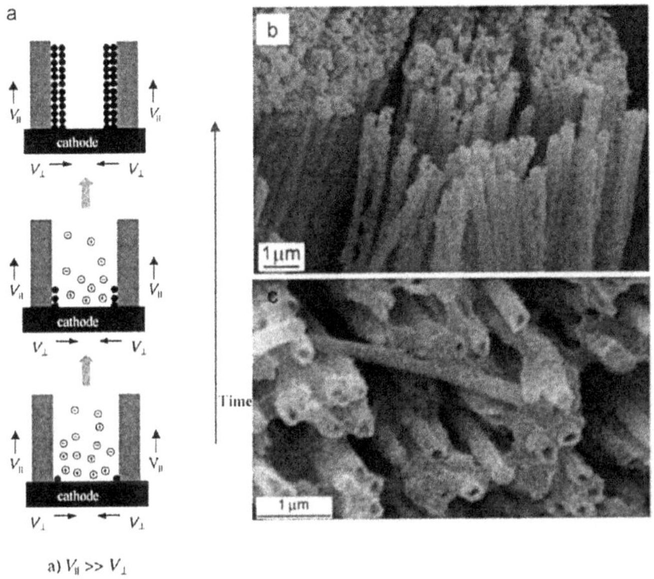


**Figure 5.** a) Schematic diagram of the steps in the growth of metal nanotubes by template-based electrochemical deposition method. b,c) SEM images of cobalt and iron nanotube arrays, respectively. Reproduced with permission from [26]. Copyright 2006, Wiley-VCH.

of nanotubes. The nanotubes have open ends on top and are arranged in a well-ordered manner. The outer diameter of the nanotubes is in the range 90–110 nm, which corresponds to the pore diameter of the membrane. The inner diameter of the nanotubes is in the range of 40–60 nm. The length of the nanotubes goes up to several micrometers. The compactness of the nanotubes is quite high, about $1 \times 10^9 cm^{-2}$, which corresponds to the pore density of the membrane.

Metallic indium nanotubes can be prepared by direct thermal evaporation of the metal.[31] A metal source is heated in an Ar atmosphere between 900–1100 °C and the vapor is passed over a catalyst such as a Au-coated silicon substrate to obtain crystalline nanotubes. The length of the nanotubes increases with temperature. The diameter of the head portion (100–300 nm) is approximately three times larger than the tail diameter. The diameter of the tail portion remains uniform throughout the nanostructure.

Electrodeposition in AMs has been employed to prepare polycrystalline nanotubes of alloys such as BiSb,[32] FeCo,[33] FeNi,[34] and CoCu.[35]

Lyotropic liquid crystal templates along with surfactants are used to synthesize metal–boron nanotubes (M–B (M = Fe, Co, and Ni)).[36] The non-ionic surfactants used are Tween 40 (polyoxyethylene sorbitan monopalmitate), Tween 60, and Tween 80 (polyoxyethylene sorbitan monooleate), and the anionic surfactant is camphorsulfonic acid. In a typical synthesis, $FeCl_3 \cdot 6H_2O$ was dissolved in water that contained (1S)-(+)-10-camphorsulfonic acid (CSA) and Tween 40 at 60 °C and the mixture cooled to 20 °C. To this a mixture of 4 M $NaBH_4$ and 0.1 M NaOH was added and the mixture kept for 48 h in an inert atmosphere. The resulting solid was collected, washed with distilled water and ethanol, and dried in flowing nitrogen. Non-crystalline Co–B and Ni–B nanotubes were prepared from $CoCl_2 \cdot 6H_2O$ or $NiCl_2 \cdot 6H_2O$ and Tween 60 in place of $FeCl_3 \cdot 6H_2O$ and Tween 40 under similar conditions. This method provides a route for the synthesis of metal–boron nanotubes and can also be extended to other tubular materials of non-crystalline alloys. Fe–B nanotubes prepared using Tween 40 and CSA are several micrometers in length and have inner and outer diameters of around 50–55 and 60–65 nm, respectively. The continuous broad halo rings in the SAED pattern further suggest a non-crystalline nature of the Fe–B nanotubes.

One-dimensional nanostructures of Se and Te were reported some time ago by Gautam et al.[37,38] Recently, *t*-Se nanotubes have been grown hydrothermally in the absence of a surfactant or a polymer.[39] In this procedure, an aqueous solution of sodium selenite ($Na_2SeO_3$), NaOH, and sodium formate ($NaCHO_2$) is reacted in a hydrothermal bomb at 100 °C for 25 h. Zhang et al.[40] have reported the fabrication of *t*-Se nanotubes by a hydrothermal–ultrasonic route. Ma et al.[41] synthesized *t*-Se nanotubes in micelles of a non-ionic surfactant, while Li et al.[42] synthesized them by a sonochemical process. Single-crystalline Te nanotubes have been prepared by a solvothermal method using PVP to modulate the reaction time, the reactants being $TeO_2$ and ethanolamine.[43]

## 4. Metal Chalcogenide Nanotubes

$MoS_2$ and $WS_2$ are the first chalcogenides whose fullerene type structures and nanotubes were prepared in the laboratory.[44,45] The method involved heating of metal oxide nanorods in $H_2S$. A similar strategy was also employed to prepare the corresponding selenide nanotubes. Recognizing that amorphous $MoS_3$ and $WS_3$ are the likely intermediates in the formation of the disulfides, the trisulfides have been directly decomposed in a $H_2$ atmosphere to obtain the disulfide nanotubes.[46] Diselenide nanotubes were similarly obtained by the decomposition of metal triselenides.[47] The trisulfide route provides a general route for the synthesis of the nanotubes of many metal disulfides such as $NbS_2$ and $HfS_2$ that have a single crystalline nature.[48,49] The decomposition of precursor ammonium salts $(NH_4)_2MX_4$ (X = S, Se; M = Mo, W), is even better, all the products, except the dichalcogenides, being gases.[46,47] Metal trichalcogenides are intermediates in the decomposition of the ammonium salts as well. These nanotubes have a hexagonal structure with a layer spacing of ~0.6 nm, which corresponds to a *d*-spacing of (002), with an external diameter of ~25 nm, a wall thickness of ~10 nm, and lengths of up to several micrometers. Employing trisulfides and triselenides as starting

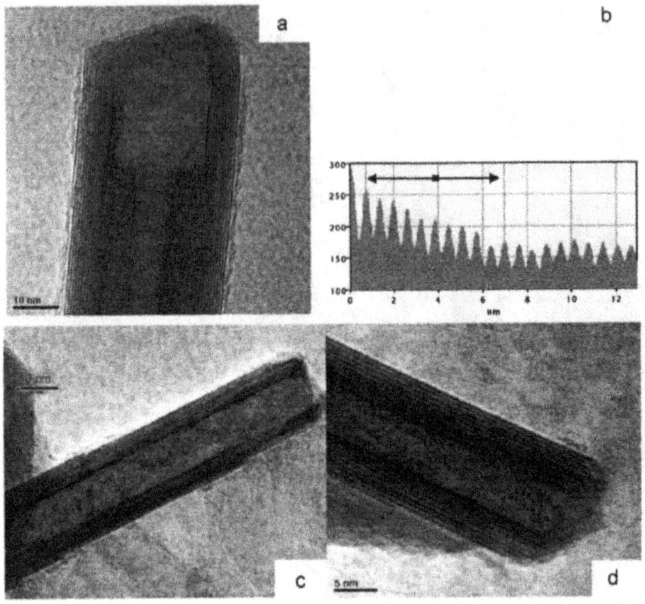

Figure 6. a) HRTEM image of the $MoS_2$ nanotube, b) the line profile of the boxed area in (a) gives an interlayer spacing of 6.2 Å. c,d) TEM images of $WS_2$ nanotubes obtained in the reaction between $WCl_5$ and $H_2S$ in the vertical reactor. Images (a,b) reproduced with permission from [51a]. Copyright 2006, World Scientific Publishing. Images (c,d) reproduced with permission from [51b]. Copyright 2008, IOP.

materials, nanotubes of $TiS_2$, $HfS_2$, $NbS_2$, $NbSe_2$ and related layered metal chalcogenide nanotubes have been prepared.[50] Recently, $MoS_2$ and $WS_2$ have been made by using gas phase reactions using metal chlorides and carbonyls (Fig. 6).[51] Solar ablation can also be used to generate $MoS_2$ nanotubes.[52] Nanotubes and onions of GaS and GaSe have been generated through laser and thermally induced exfoliation of the bulk powders.[53]

CdS nanotubes and related structures have been prepared by the thermal evaporation of CdS powder (Fig. 7).[54] Nanotubes of CdS and CdSe had earlier been prepared by using surfactants as templates.[55] CdS occurs in the hexagonal structure in the nanotubes. The diameter and length are in the ranges of 40–160 nm and 3–4 μm, respectively. The CdSe nanotubes are generally long, with lengths up to 5 mm. The outer diameter of the nanotubes is in the 15–20 nm range while the diameter of the central tubule is in the 10–15 nm range. These nanotubes are polycrystalline and form through the oriental attachment of nanoparticles. Recently, CdS, ZnS, and CuS nanotubes have been made by the hydrogel-assisted route (Fig. 8).[56] The TEM image of CdS in Figure 8a, obtained after the removal of the hydrogel template, demonstrates the hollow nature of the nanotubes. The lengths of the nanotubes extend to a few hundred nanometers, while the diameter of the inner tubule is ~2–3 nm, the outer diameter being in the 20–25 nm range. A SAED pattern of a single nanotube is given in the bottom inset of Figure 8a. The diffuse rings correspond to the (100) and (110) Bragg planes of hexagonal CdS showing the nanotubes to be generally polycrystalline. Clearly, the tripodal cholamide hydrogel fibers act as templates, on which the CdS particles get deposited, giving rise to the nanotubes. The low magnification TEM image in Figure 8b suggests a possible assembly or attachment of the initially formed shorter nanotubes to form linear chains (indicated by arrows in the figure). The hydrogel might be responsible for such an attachment, the hydrogel playing a dual role of being a template to produce hollow nanotubes as well as favoring the attachment or assembly of the nanotubes. The ZnS nanotubes prepared similarly (Figs. 8c and d) have an inner diameter in the ~4–6 nm range, with lengths going up to a micrometer. The TEM image shows tiny nanocrystals of ZnS (Fig. 8c) making up the walls of the nanotubes. The electron diffraction patterns also reveal the nanotubes to be polycrystalline and of hexagonal structure. The CuS nanotubes prepared by this route have an inner diameter of ~5 nm with lengths extending to a few hundreds of nanometers. The outer diameter is in the 20–30 nm range. The nanotubes are polycrystalline as found from electron diffraction as well as the TEM images. The polycrystalline nanotubes are formed through the oriented attachment of nanoparticles. CuS nanotubes have also been made by the solution reaction of Cu nanowires in ethylene glycol with a suitable sulfur source such as thiourea and thioacetamide at 80 °C.[57] The CuS nanotubes so-produced in large quantities possess a hexagonal structure and have an inner diameter of 30–90 nm, a wall thickness of 20–50 nm, and a length of more than 40 μm. The straight nanotubes are made up of nanoparticles of around 30 nm diameter. This study also shows that sulfur sources such as thiourea and thiacetamide, which release ionic sulfur rather than molecular sulfur at their decomposition temperature, are favorable for the formation of CuS nanotubes, compared with sulfur powder.

CdSe nanotubes have been prepared by a sono-electrochemical method,[58] while CuSe nanotubes have been prepared by using trigonal Se nanotubes as templates.[59] The CdSe nanostructures (with a cubic structure) were formed by electrodeposition onto the sonic probe cathode.[58] The deposit could be spheroidal or have a 2D structure. Subsequent sonic shock waves remove the deposit from the probe surface. The sonochemical treatment provides the required energy to roll the nanosheets to form tubular nanoscrolls. After the 2D nanosheets are ejected from the probe, because of the high surface energy of the ends of the nanosheets, the flexible and unstable nanosheets easily roll-up in the sonochemical process. It is well known that collapsing bubbles produced in a liquid solution during sonication can instantaneously generate local spots of high temperature, pressure, and cooling rates. The worm-like morphology of the CdSe prepared by this method involves a tubular structure and the nanotubes (diameter: 80 nm, wall thickness: 10 nm) are entangled with each other. The nanotube with a straight part reveals a round open tip, which is

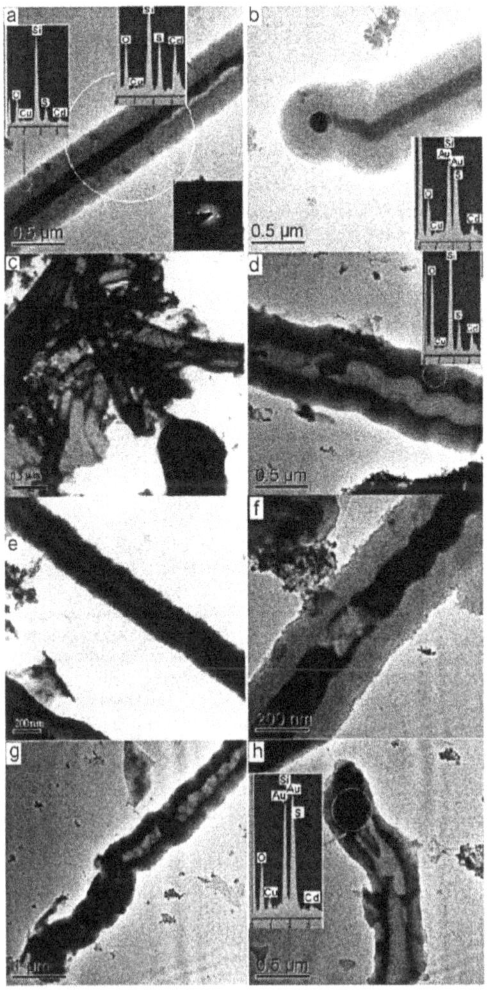


**Figure 7.** TEM images of the 1D CdS nanostructures: a) a core–sheath nanowire with insets that contain the energy dispersive X-ray analysis (EDX) of the core and the sheath and SAED of the core, b) the top of a core–sheath nanowire with an inset of the EDX of the catalyst, c,d) CdS nanotubes, e) tube–wire nanojunction, f) wire–tube–wire nanojunction, g) top part of a nanotube with nanoparticles with the channel, and (h) top part of one nanotube with the catalyst with the inset showing the EDX of the particle. Reproduced with permission from [54]. Copyright 2008, ACS.

not completely seamed. This indicates that the nanotubes were probably formed through a roll-up process. The CdSe nanotubes show many stacking faults and missing layers in the nanotube walls. The appearance of defects is mainly a result of the high stress and strain present in the nanotubes.

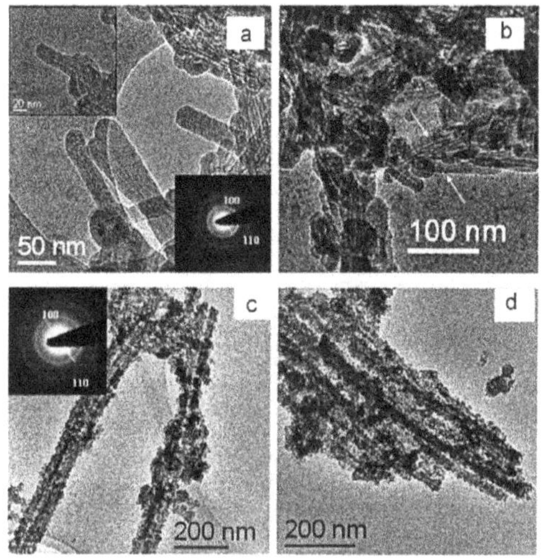


**Figure 8.** a) TEM image of CdS nanotubes obtained after the removal of the hydrogel template. Top inset is a high-magnification image of a single nanotube. Bottom inset is the SAED pattern of the nanotubes. b) TEM image showing a bunch of nanotubes assembled spontaneously, indicated by the arrows. c,d) TEM images showing nanotubes of ZnS obtained using 0.02 mmol of $Zn(OAc)_2$. Reproduced with permission from [56]. Copyright 2006, Elsevier.

CuSe with a tubular nanostructure is formed by the template mechanism wherein diffusion of Cu atoms occur into t-Se nanotubes.[59] The synthesis process involves the t-Se nanotubes acting as templates and reactants were converted into crystalline nanotubes of CuSe by reacting with Cu nanoparticles freshly produced from an aqueous $CuSO_4$ solution. Apart from CuSe nanotubes, 1D nanocrystallites of $Cu_3Se_2$, $Cu_{2-x}Se$, and $Cu_2Se$ were also obtained by changing the atom ratio of Cu and Se in the precursors. The tubular nanostructures have the hexagonal structure of CuSe. The wall thickness and diameter of the CuSe nanotubes were around 80 and 300 nm, respectively.

CdTe nanotubes of controlled diameter are prepared by first reacting $CdCl_2$ with thioglycolic acid (TGA) to obtain 1D Cd–TGA nanowires (thickness ~8 nm), the aqueous dispersion of which is then used as a sacrificial template to generate long CdTe nanotubes by reaction with NaHTe.[60] The length of the nanotubes is typically hundreds of micrometers, similar to the initial length of the precursor. The inner and outer diameters of the nanotubes are in the ranges of 12–20 nm and 30–50 nm, respectively. The CdTe nanotubes are polycrystalline and adopt a cubic structure. The average diameter of the nanowires of the 1D Cd–TGA precursors obtained in the presence of poly(acrylic acid) (PAA), could substantially be increased as a function of the amount of PAA. During the formation of CdTe, 1D Cd–TGA precursors are gradually consumed to finally lead to hollow structures, preserving the original shape, size, and morphology of the precursor.

Photochemical decomposition of $CSe_2$ adsorbed on Ag nanowires yields $Ag_2Se$ nanotubes.[61] The evolution of Ag nanowires to core–shell structures and finally to hollow $Ag_2Se$ nanotubes was studied in detail by TEM analysis. Upon irradiation for 15 min, the TEM image of the nanowires began to show evidence of core–shell nanowires with mean diameters of ~85 nm and ~45 nm cores. The shell grew thicker at the expense of the core with increasing irradiation time, and voids were observed to grow from both ends of the nanowires along the longitudinal axis, ultimately merging to form hollow nanotubes of mean diameters of ~90 nm with ~45 nm voids. The nanotubes are polycrystalline with a structure corresponding to the orthorhombic phase of R-$Ag_2Se$. Trigonal Se nanotubes can be used as templates to prepare $Ag_2Se$ nanotubes.[62] $Ag_2Te$ nanotubes have been generated by the reaction of $AgNO_3$ with sodium tellurate ($Na_2TeO_3$) in the presence of hydrazine and ammonia by a hydrothermal process in the absence of a template or a surfactant.[63] All these nanotubes are bent and curled, with diameters of 80–250 nm and several tens of micrometers in length. They show the characteristics of tubular structures with open-ended and uncovered hollow interiors. The nanotubes are single-crystalline, and free of dislocations and stacking faults. A structural phase transition of the as-prepared $Ag_2Te$ nanotubes from the low-temperature monoclinic structure ($\beta$-$Ag_2Te$) to the high-temperature face-centered cubic structure ($\alpha$-$Ag_2Te$) has been observed.

$Bi_2S_3$ nanotubes and nanorods are prepared solvothermally at a low temperature of 120 °C, using a mixed solvent (acetone–water, methanol–water, ethanol–water, water, ethylene glycol–water, or glycerol–water) as the reaction medium and urea as the mineralizer.[64] In a typical synthesis, $Bi(NO_3)_3 \cdot 5H_2O$ is dissolved in the mixed solvent and aqueous $Na_2S \cdot 9H_2O$ (S/Bi = 3: 1) solution added drop by drop into the solution under vigorous stirring. The mixture of precursors and urea is transferred into a Teflon-lined autoclave and heated solvothermally. The gray-black powder so obtained is washed with distilled water and ethanol several times and dried. A mixture of nanorods and nanotubes is found in the final product synthesized in methanol–water mixtures. The diameter of the nanotubes was more than that of the nanorods (in the 80–100 nm range) with lengths up to a micrometer. The powders synthesized in water are nanotubes, which are polycrystalline with a diameter of 200 nm and a length of about 1 μm. $Bi_2S_3$ nanotubes synthesized in ethylene glycol–water mixtures are single crystalline, with a diameter in the 200–500 nm range and the lengths up to several micrometers. The inner diameter of the single hollow nanotube is about 100 nm and the walls of the tube are around 100 nm thick. $Bi_2S_3$ microtubes synthesized in the mixed solvent of glycerol–water have a diameter of about 1 μm and a length of about 7 μm. During the process of solvothermal synthesis, there is a dynamic equilibrium between the $Bi_2S_3$ solid particles or nuclei and the $Bi^{3+}$ and $S^{2-}$ ions in solution ($2Bi^{3+} + 3S^{2-} = Bi_2S_3$). In such an equilibrium, $Bi^{3+}$ and $S^{2-}$ ions tend to dissolve from the small particles into the solution and precipitate onto the surfaces of large particles so that the total energy of the interface between the particles and the solution is decreased. $Bi_2S_3$ has a layered structure and the weak bonds between the layers give rise to an anisotropic growth of $Bi_2S_3$ particles during the solvothermal synthesis. At the low reaction temperature, the rate of crystal growth is greater than that of crystal nucleation. The growth of $Bi_2S_3$ nanosheets is favoured at higher viscosity and surface tension of the reaction medium. The powders synthesized in the mixed solvent of water, ethylene glycol–water, and glycerol–water possess nanosheet structures and the nanosheet nanostructures self-roll to form tube-like structures. The interface energy between bismuth sulfide and the mixed solvent of water, ethylene glycol–water, and glycerol–water with a high surface tension appears to be higher than that between bismuth sulfide and the mixed solvent of acetone–water, ethanol–water, and methanol–water. A solvent with higher viscosity and surface tension favours the formation of tube-like structures and a solvent with lower viscosity and surface tension favours the formation of a rod-like structure during the synthesis. Hence the morphology of the nanostructure depends on the viscosity and surface tension of the mixed solvent used in the solvothermal synthesis. $Bi_2S_3$ nanotubes have also been prepared by the conventional evaporation method,[65] as well as by employing the micelle-template method at 115 °C.[66]

$Bi_2Se_3$ nanotubes can be prepared solvothermally starting with ammonium bismuth citrate and elemental Se in dimethylformamide solution.[67] However, the diameters of the nanotubes are not uniform and vary in the range of 15–150 nm with the wall thickness in the 5–20 nm range. These polycrystalline nanotubes have a rhombohedral structure. $Bi_2Te_3$ nanotubes have been obtained by the galvanic displacement of nickel nanowires in nitric acid that contains $Bi^{3+}$and $HTeO_2^+$ ions (Fig. 9).[68] When Ni nanowires are immersed in an nitric acid solution that contains $Bi^{3+}$ and $HTeO_2^+$ ions, the Ni nanowires are galvanically displaced to form $Bi_2Te_3$, because of the difference in the reduction potentials. The nanotubes have diameters in the 100–130 nm range, a wall thickness of approximately 20 nm, and lengths running to several micrometers. The nanotubes are formed out of highly crystalline rhombohedral $Bi_2Te_3$ crystals without obvious preferential orientation. The composition of $Bi_2Te_3$ nanotubes was precisely tuned by adjusting the $[Bi^{3+}]/[HTeO^{2+}]$ ratio. The galvanic displacement reaction can be written as:

$$2Bi^{3+}(aq) + 3HTeO_2^+(aq) + 9Ni^0(s) + 9H^+(aq) \rightarrow Bi_2Te_3(s) + 9Ni^{2+}(aq) + 6H_2O(aq) \quad (3)$$

Nanotubular $Bi_2Te_3$, as well as its alloys with Se and Sb, are obtained by electrodeposition in the nanochannels of alumina templates.[69] $Sb_2S_3$ nanotubes with thin walls (1.5 nm–2 nm) have been made by the solvothermal reaction of $SbCl_3$ with sulfur in oleylamine solution at a relatively low temperature (175 °C).[70] The average diameter of the $Sb_2S_3$ nanotubes (orthorhombic structure) was 10.4 nm, and the length was in the 100 to 300 nm range with an atomic ratio of Sb to S of 1: 1.51. By changing the concentration of sulfur, the aspect ratios of the initially formed $Sb_2S_3$ nanoribbons were delicately controlled. Reducing the number of equivalents of sulfur results in a decrease in the length of the nanoribbons with a simultaneous increase in the width. In addition to the wider nanoribbons, a significant amount of nanotubes with 6.7 nm average width were observed. Further

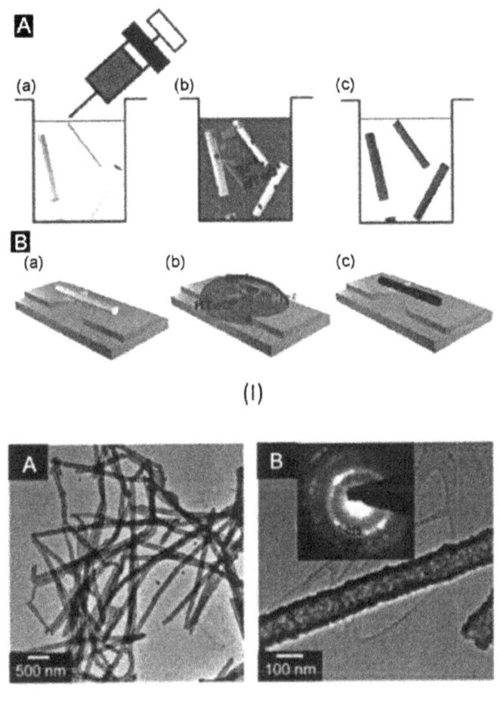

Figure 9. I) Schematic illustrations of $Bi_2Te_3$ nanotube synthesis (A) and individual $Bi_2Te_3$ nanotube laid across electrodes (B). TEM images and SAED pattern of high aspect ratio $Bi_2Te_3$ nanotubes (A) synthesized from nickel nanowire (~100 nm in diameter). Tube thickness was approximately 20 nm (B). Reproduced with permission from [68]. Copyright 2007, ACS.

reduction in sulfur yields pure small aspect ratio nanotubes. Clearly, the nanotubes are formed by the rolling of the nanoribbons.

Nanotubes of lead chalcogenides are prepared by the reaction of $Pb(NO_3)_2$ with cysteine in ethanolamine solution followed by the subsequent reaction of the nanowire product with the chalcogenide source solution at room temperature.[71] The nanowires are formed by the self-assembly of nanocrystals (10 nm diameter) and the nanowires act as templates for the subsequent formation of lead chalcogenide nanotubes. The nanotubes are polycrystalline, have a typical diameter of about 200 nm and are several micrometers long. The lead chalcogenide nanotubes show face-centered cubic phases with a disorderly aggregation of the nanocrystals. The lead chalcogenide nanocrystals are formed on the surface of the precursor nanowires by an anion-replacement reaction that is the result of the lower solubility of lead chalcogenide in solution. An ion exchange solvothermal reaction between $Na_4P_2S_6$ and $MnCl_2$ gives rise to nanotubes of $Mn_2P_2S_6$.[72] These nanotubes are single-crystalline (monoclinic) and have uniform outer diameters of 40–50 nm and lengths that range from 110 to 170 nm.

## 5. Metal Oxide Nanotubes

Metal oxide nanotubes have been investigated widely in the last 2–3 years because of the potential applications of some of these materials. They have been prepared by employing several methods including sol–gel chemistry, hydrothermal reactions and use of templates. The template method is used particularly widely in combination with electrodeposition. Besides the popular porous alumina template, carbon nanotubes and other materials are used in the preparation of oxide nanotubes. The template-directed synthesis of oxide nanotubes has been reviewed by Bae et al.[73]

### 5.1. $SiO_2$ Nanotubes

$SiO_2$ nanotubes are readily prepared by using a variety of templates wherein the templates are coated with the oxide precursor followed by hydrolysis and heat treatment. Besides carbon nanotubes, carbon nanofibres have been used as templates to obtain $SiO_2$ nanotubes. The same method is also applicable for the synthesis of $ZrO_2$ and $Al_2O_3$ nanotubes.[74] Peptide amphiphile (PA) nanofibres have been successfully used as templates. The catalytic activities of PAs that contain lysine, histidine, or glutamic acid have been compared and only the PAs that contain lysine or histidine found to be good as catalytic templates. Depending on the reaction conditions and the size of the PA assembler, the nanotube wall thickness can be varied between 5–9 nm.[75] Self-assembled peptidic lipids that form tubular structures are also used as templates to prepare silica nanotubes.[76] The use of self-assembled structures as templates for preparing silica nanotubes and other nanostructured oxides generally involves the coating of the super-structure with metal alkoxides, sol–gel condensation, and the removal of the templates. This template method has been used to prepare nanotubes of silica and other materials using organogelators including those that are fluorinated.[77]

Thermal evaporation of SiO yields $SiO_2$ nanotubes.[78] This in-situ template-like process (carried out in the presence of GaN and ZnS) in a vertical induction furnace gives amorphous $SiO_2$ nanotubes wherein both the diameters and lengths can be tuned by changing the reaction temperature. The $SiO_2$ nanotubes are amorphous, with a diameter of 30 nm and length of several hundred micrometers at 1450 °C, change to a diameter of 100 nm and length of 2–10 micrometers at 1300 °C. At high reaction temperatures, GaN decomposes to give Ga and the newly formed Ga is in the form of small-sized liquid clusters. The Ga clusters are transported by the Ar gas to the low temperature region, where they deposit as liquid droplets on the inner walls of the graphite crucible. The Ga droplets are the favored sites for the adsorption of ZnS vapor generated in the system. On supersaturation, ZnS segregates and gives rise to ZnS nanowires. With an increase in temperature, the decomposition of the SiO results in the formation of Si vapor, which is oxidized by the residual $O_2$

  

to $SiO_2$. This is transported by the Ar gas and is deposited on the surface of the ZnS nanowires, which results in $ZnS/SiO_2$ hetero-nanowires. Because of the high temperature, the inner ZnS nanowires evaporate, and hollow $SiO_2$ nanotubes remain. This method can also be used for nanotubes of other inorganic nanomaterials such as ZnS and GaN. CdSe nanocrystals are used as seeds in a one-step thermal evaporation process to prepare long, high density $SiO_2$ nanotubes.[79] $SiO_2$ nanotubes generally have smooth surfaces, uniform thickness, and round cross sections of both the interiors and exteriors which are amorphous in nature. The nanotubes are open-ended and have diameters of about 80–110 nm and a length of several hundreds of micrometers. High purity $SiO_2$ nanotubes are obtained by using track-etched membrane templates that involve $O_2$ plasma treatment. Oxygen plasma pre-treatment ensures pore-filling of the precursor solution and covalent bonding between template and precursor, while pyrolysis of the template–nanostructure composite completely removes organics and produces inorganic nanostructures.[80] These nanotubes are also amorphous and have diameters of 100 nm and lengths of 2–6 micrometers.

Hybrid $SiO_2$ nanotubes with walls that contain chiral aromatic rings are obtained by using self-assemblies of appropriate amphiphiles.[81] Mesoporous $SiO_2$ nanotubes with helical channels have been prepared by the self-assembly of surfactants in the presence of chiral molecules.[82] These are formed by the self-assembly of the achiral surfactant sodium dodecyl sulfate (SDS) in the presence of (*R*)-(+)- and (*S*)-(−)-2-amino-3-phenylpropan-1-ol (APP) (Fig. 10). TEM combined with computer simulations confirm the presence of ordered chiral channels winding around the central axis of the tubes with an inner diameter of 100 nm.

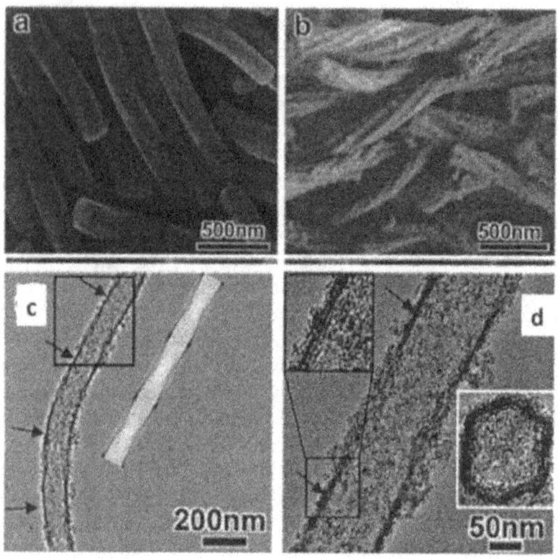


**Figure 10.** a,b) SEM images of calcined silica nanotubes with chiral mesoporous wall structure synthesized with different (*R*)-(+)-APP/SDS molar ratios of 0 and 0.2, respectively. These materials were synthesized at 30 °C for 6 h and then allowed to age for 1 day at 90 °C. c) Low and d) high magnification TEM images of calcined samples. Reproduced with permission from [82]. Copyright 2007, ACS.

Double-walled silica nanotubes have been obtained by biomimetic synthesis under mild conditions. Pouget et al.[83] use the peptide lanreotide in which the silica phase and the lanreotide nanotube grow synergistically in a concerted manner by mutually neutralizing their charges (positive on the lanreotide and negative on on the silica) (Fig. 11). This requires kinetic coupling of the two chemical processes. The presence of an X-ray diffraction peak at $0.35\,Å^{-1}$, characteristic of the $\beta$-sheet organization in the lanreotide wall surface, gives a mean diameter of 24.6 nm. The walls of the tubes are separated by 2 nm, which corresponds to the thickness of the lanreotide tube. The walls of the silica tubes are thin (1.4 nm) with lengths up to 3 μm, in agreement with the TEM data shown in Figure 11. Calcination at 600 °C converts hybrid organic–inorganic nanostructures into pure silica double-walled nanotubes. A dynamic template mechanism can explain these results. Silica deposition occurs on both sides of the lanreotide molecule, and stops immediately after neutralization of the surface charge. This is a well-controlled process and yields a hierarchical structure from the nanometer-scale to the macroscopic level. The double-walled silica tubes form bundles 1–3 μm in length, and bundles of closely packed aligned nanotubes as much as a centimetre long. Shape-coded $SiO_2$ nanotubes are obtained by using porous alumina along with the sol–gel method.[84] The template synthesis of shape-coded nanotubes begins with the fabrication of a porous alumina film that contains well-defined cylindrical pores with two or more different diameter segments created by multistep anodization of the aluminum substrate. The nanotubes are fabricated with a surface sol–gel method that controls the wall thickness at a single-nanometer level. Amorphous silica nanotubes seeded by copper sulfide nanoparticles have been synthesized by using a supercritical organic solvent.[85] Addition of copper sulfide nanocrystals, monophenylsilane, and small amounts of water and oxygen to supercritical toluene at 500 °C at 10.3 MPa yields silica nanotubes.

### 5.2. $TiO_2$ Nanotubes

Amongst the nanotubes of various metal oxides, those of $TiO_2$ have been investigated most widely. $TiO_2$ nanotubes can be prepared hydrothermally, by anodization of titanium by template-assisted growth as well as seeded growth. Porous alumina templates are specially useful for fabricating dense, uniform, aligned arrays of $TiO_2$ nanotubes on substrates such as glass, silicon, and polymers. Free-standing porous alumina templates have been employed for atomic layer deposition (ALD) of ordered $TiO_2$ nanotube arrays on various substrates (Fig. 12).[86] The diameter and length of the nanotubes, as well as the distance between two neighboring nanotubes, can be controlled by varying the dimensions of the template and the anodization conditions. Typically, hexagonally packed pores with a diameter of ∼65 nm and interpore distance of ∼110 nm are used. The synthesis of highly ordered arrays of $TiO_2$ nanotubes by potentiostatic anodization of Ti has been reviewed by Grimes.[87] This appears to be an excellent method wherein anodic oxidation is carried out in a dimethyl sulphoxide (DMSO) medium that contains hydrofluoric acid, potassium fluoride, or ammonium fluoride

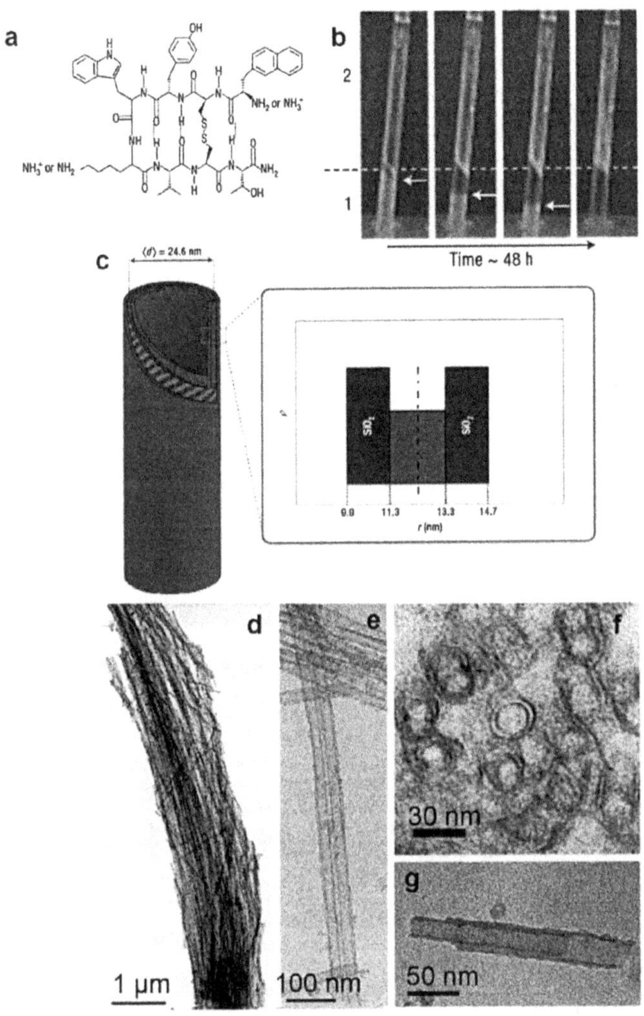


**Figure 11.** Silica mineralization of lanreotide: a) Structure of the lanreotide octapeptide showing the two charged amine sites. b) Time-lapsed pictures of the capillaries during the mineralization process. Initially, a volume of lanreotide gel at 5% (w/w) poured into the bottom of the capillary (region 1) and is covered with the same volume of a 30% (w/w) TEOS/water mixture (region 2) (tube 1). On ageing, the lanreotide turbid gel recedes (white arrows), to yield a transparent region, while long white fibres appear in region 2, starting from the interface (tubes 2 and 3). After 48 h, the lanreotide gel has completely disappeared and only two phases separated by a white ring are observed: the lower clear one containing a dilute lanreotide solution and the upper one filled with mineralized fibres (tube 4). c) Schematic representation of a multi-scale organization of a silica–lanreotide nanotube with an inner and an outer 1.4 nm thick silica shell and a central 2.0 nm thick lanreotide tube as deduced from radial density ($\rho$) profiles. d) TEM image of double-walled silica nanotube replica obtained after the calcination of silica–lanreotide nanotubes. e) TEM image of a 7 μm long bundle of dried mineralized nanotubes. f) Cross-sectional TEM image of dried fibres, revealing several concentric circles attributable to the double-wall structure. g) TEM image of a fragment of dried nanotube, showing that the internal and external cylinders are independent and free to slide. Reproduced with permission from [83]. Copyright 2007, Nature Publishing Group.

as the electrolyte.[88] Self-aligned, hexagonally close packed $TiO_2$ nanotube arrays, 1000 μm in length with high aspect ratios (~10 000) are obtained by the anodization of titanium. These are polycrystalline in the anatase structure after annealing in oxygen at 280 °C for 1 h. Such nanotubes can be transformed into self-standing membranes.[89]

The formation of $TiO_2$ membranes by the anodization of Ti foil in fluorine-containing ethylene glycol has been described.[90] This method yields self-organized, free-standing $TiO_2$ nanotube arrays with ultra-high aspect ratio of the diameter/length (~1500) by simply using solvent-evaporation-induced delamination of the $TiO_2$ barrier layer formed between the $TiO_2$ membrane and Ti foil during anodization. The resulting membrane consists of highly ordered, vertically aligned, one-side open $TiO_2$ nanotube arrays with pore diameter, wall thickness, and length of around 90 nm, 15 nm, and 135 μm, respectively. The as-grown $TiO_2$ nanotubes are amorphous and transform into the anatase structure after annealing at high temperature in air. Aligned $TiO_2$ nanotubes with novel morphologies, such as bamboo-type reinforced nanotubes and 2D nanolace sheets, obtained by an anodization process carried out under alternating-voltage conditions in fluoride-containing electrolytes, have been reported.[91] The experiment was carried out under constant voltage conditions, and after 2 h of anodization at 120 V in an electrolyte that consists of 0.2 mol $L^{-1}$ HF in ethylene glycol, yielded a regular layer of aligned, individual $TiO_2$ nanotubes with thickness of about 10 μm and diameter of 150 nm. If the voltage is lowered to 40 V, the growth of the nanotubes slows down and may even stop. A bamboo-type structure can be grown under certain conditions (i.e., when the voltage is alternated between 120 and 40 V). The spacing between the bamboo rings can be altered by means of changing the time for which the sample is held at 120 V, and spacing is reduced from 200 to 70 nm by reducing the holding time. If anodization takes place for a long time at a low voltage, nanotubular features with a reduced diameter start to grow. This can be exploited to grow a double-layer structure. In this case, branching of the main tube with a diameter of 150 nm into several (typically 2–3) smaller tubes of about 50 nm in diameter occurs. The structures can be transformed to the anatase structure without losing structural integrity by annealing in air at 450 °C. Anodization under constant-voltage conditions leads to an ordered layer that consists of smooth tubes with a defined cylindrical or hexagonal cross section.[92,93] Fluoride-free aqueous HCl elec-

ADVANCED MATERIALS www.advmat.de

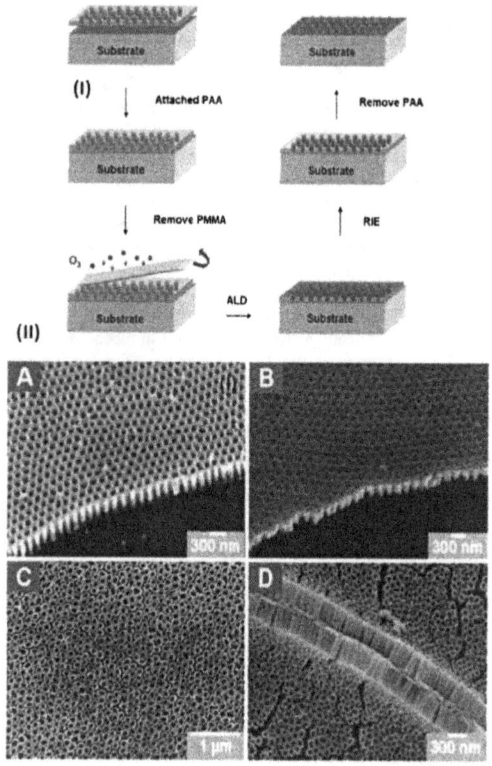


Figure 12. I) Schematic of the process to fabricate highly ordered $TiO_2$ nanotube arrays on substrates using ALD on a free-standing porous alumina template. Free-standing AM supported with a poly(methyl methacrylate) (PMMA) layer was first attached onto the substrate. Substrates were alternatively exposed to $TiCl_4$ and water vapor for ALD at a pressure of $1 \times 10^{-3}$ Torr. The PMMA was then removed for ALD on the AM template. The $TiO_2$ overlayer was etched by RIE, and finally highly ordered ALD $TiO_2$ nanotube arrays were released from the AM template. II) Dense, uniform, highly ordered, and well-aligned ALD $TiO_2$ nanotube arrays on Si and a flexible polyimide film. SEM images of a highly ordered free-standing AM template (~300 nm) attached on Si before (A) and after (B) 150-cycle $TiO_2$ ALD. C) Highly ordered $TiO_2$ nanotube arrays released from the template on the Si substrate. D) Cross-section of well-aligned $TiO_2$ nanotube arrays on a bending polyimide film. Reproduced with permission from [86]. Copyright 2008, ACS.

trolyte is also used to obtain vertically oriented $TiO_2$ nanotube arrays.[94] These nanotube arrays, obtained by using a 3 M HCl aqueous electrolyte with an anodization potential of 20 V, have an inner pore diameter of 15 nm, wall thickness of 10 nm, and length up to 600 nm. The nanotubes are polycrystalline and have an anatase structure, with a rutile barrier layer separating the tubes from the underlying metal foil.

Sulfur-doped $TiO_2$ nanotubular arrays are obtained by potentiostatic anodization of Ti foils followed by the annealing of $TiO_2$ tubular arrays in a flow of $H_2S$ at 380 °C.[95] Ordered, vertically oriented B-doped $TiO_2$ nanotube arrays are prepared by forming a nanotube-like $TiO_2$ film in the anodization process on a Ti sheet followed by CVD treatment with trimethylborate vapor.[96] A double-template-assisted sol–gel method has been used to prepare $TiO_2$ nanotube arrays with nanopores on their walls.[97] In this method, poly(ethylene glycol) dissolved in a $TiO_2$ sol is used as a soft template to form nanopores on the walls of $TiO_2$ nanotube arrays, which were templated from ZnO nanorod hard templates by the dip-coating technique. The microstructure of nanoporous $TiO_2$ nanotube arrays can change from end-opened to end-closed by increasing the number of dip-coating cycles.

Gas-phase ALD of metal oxides in combination with a micro-contact printing (μ-CP) technique has been used to attain precise atomic-level control over the dimensions (wall thickness) of nanotubes as well as the one-step fabrication of the free-standing oxide nanotubes.[98] In this procedure, octadecyltrichlorosilane (OTS) molecules were transferred onto both sides of the surfaces of the template by the μ-CP technique and self-assembled monolayers (SAMs) formed on the surfaces, thus exposing chemically inert methyl surfaces. The ALD process allows atomic-level control over the thickness of the wall of the nanotubes, and the OTS self-assembled mono-layers function as resistant layers to materials deposition, thus allowing free-standing cylindrical nanotubes to be collected after dissolution of the template without an additional polishing step. This method has been used to obtain high aspect ratio (~300) nanotubes of $TiO_2$, $ZrO_2$, and $Al_2O_3$. Crystalline and homogeneous $CeO_2$ naoparticles have been used as seeds to grow $TiO_2$ nanotubes from the hydrolysis of aqueous titanium sulfate solution.[99] Layer-by-layer deposition of a water-soluble titania precursor (titanium(IV) bis(ammonium lactato) dihydroxide, along with the oppositely charged poly(ethylenimine) gives rise to multilayer films. The tubular structure is obtained by depositing inside the cylindrical pores of a polycarbonate membrane followed by calcination.[100] Nanotubular $TiO_2$ can be obtained by using uncharged or negatively charged L-lysine-based organic gelators as templates.[101] By heating nanotubes of titanic acid (obtained by heating P25-$TiO_2$ (Degussa) with an NaOH solution at 110 °C) in ammonia, N-doped $TiO_2$ nanotubes with an anatase structure are obtained.[102]

The use of carbon nanotubes to obtain oxide nanotubes is well documented.[103,104] This method has been used to obtain pure rutile nanotubes.[105] The method involves coating of the carbon nanotubes with amorphous titania through a sol–gel process followed by heating in air to convert titania into anatase. Further heating in nitrogen transforms the anatase predominantly to rutile and finally heating in air again at a higher temperature removes the carbon nanotubes and transforms the remaining anatase to rutile. Brookite-type $TiO_2$ nanotubes are obtained by heating titanate nanotubes obtained by the reaction of $TiO_2$ powder with NaOH solution, followed by hydrothermal treatment.[106] Transparent thin films of titanate nanotube arrays with super-hydrophilic characteristics are grown on sapphire substrates by the hydrothermal reaction of sputter-deposited Ti films in an aqueous NaOH solution.[107]

### 5.3. ZnO, CdO, and $Al_2O_3$ Nanotubes

ZnO nanotubes have been prepared by CVD and thermal evaporation, as well as by hydrothermal and solution methods.

  Adv. Mater. 2009, 21, 4208–4233

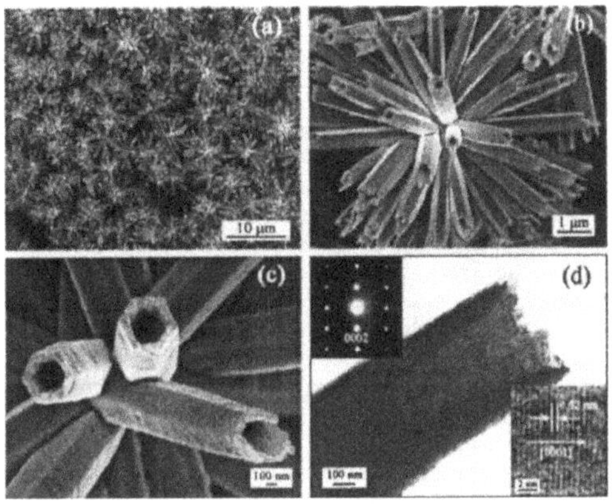


**Figure 13.** Morphological and structural characterization of ZnO nanotube bundles: a,b) Low-magnification FE-SEM images. c) High-magnification FE-SEM image. d) TEM image. The upper left and lower right inset images of (d) are the SAED pattern and the high-resolution transmission electron microscopy (HRTEM) image of a single nanotube, respectively. Reproduced with permission from [108]. Copyright 2007, ACS.

For example, large-scale ZnO nanotube bundles have been synthesized by a simple wet chemical approach (Fig. 13).[108] In this method, an aqueous solution of $Zn(NO_3)_2$ and hexamethylenetetramine is stirred over long periods and heated at 90 °C. This method yields nanotubes with an inner diameter of ~350 nm and a wall thickness of ~60 nm, and forms radiating structures. These ZnO nanotubes are single crystalline in nature, having the wurtzite structure, and preferentially grow along the [0001] direction. High aspect ratio nanotubes are obtained by a three-step low temperature process that involves ionic layer absorption, deposition of the ZnO seed layer, followed by hydrothermal annealing of the seed layer and deposition of the 1D ZnO nanostructures.[109] The uniform ZnO nanotubes have a single-crystalline wurtzite structure with lengths that exceed 10 μm and diameters of around 27 nm. By hydrothermal annealing, ZnO nanotubes grown along the < 001> direction have been obtained.[109] Nanotube-based paint-brush structures of ZnO have been prepared on Zn foils by solvothermal means by adjusting the pH of the solution.[110]

Electrochemical deposition from aqueous solution into porous alumina membranes can be employed to prepare ZnO nanotube arrays.[111] Single-crystalline ZnO nanotube arrays with an hexagonal wurtzite structure have been generated on glass substrates by a two-step solution approach.[112] The method involves the electrodeposition of oriented ZnO nanorods and subsequent coordination-assisted selective dissolution along the *c*-axis to form tubular structures caused by the preferential adsorption of ethylenediamine (EDA) and $OH^-$ on different crystal faces. After dissolution in aqueous EDA solution for 10–15 h, the inner/outer wall surfaces of the ZnO nanotubes become smooth with a wall thickness of ~10–30 nm.

Reaction of $Zn(NO_3)_2$ with methenamine in aqueous medium under hydrothermal conditions gives rise to ZnO nanotubes.[113] These nanotubes are hollow with rough surfaces, which indicates a layer-stack structure. The ZnO nanotubes have an hexagonal wurtzite structure with lengths in the range of 1–3 μm and wall thickness in the 50–100 nm range. The cross section is hexagonally faceted, which provides strong evidence that the single nanotube grows along the *c*-axis direction. ZnO nanotube structures have been obtained by the hydrothermal self-assembly of zinc ions at the interface of surfactants, such as cetyltrimethylammonium bromide (CTAB) and the triblock copolymer of poly(ethylene oxide)–poly(propylene oxide)–poly(ethylene oxide) (P123).[114] These molecules are suitable for the assembly of layered zinc species–surfactant hybrids that can be exfoliated into intermediate single sheets that roll to form tubular ZnO nanostructures because of the heat stress and the crystallization of ZnO sheets. The size of the tubular ZnO is determined by the type of surfactants, the concentration of the surfactant, and the Zn species/surfactant molar ratio. Tubular products can be obtained at a relatively high $Zn^{2+}$/surfactant molar ratio as the concentrations of CTAB and P123 are higher than their critical micelle concentrations. The hydrophobic side of each surfactant layer is connected through a weak bond, which gives rise to an organic layer, while the hydrophilic side of each surfactant layer adsorbs metal ions or hydrates, to form an inorganic layer. Dehydration and crystallization of amorphous inorganic layers cause the volume to shrink and results in stress in the inorganic layers, and breaks the equilibrium between the organic and the inorganic layers. Crystallography of these ZnO nanotubes is determined by how the nanosheets exfoliate and roll. The large cylindrical ZnO nanotubes have diameters ranging from 200 to 400 nm (wall thickness, 80 nm) with lengths going up to 14 μm. The nanotubes are single-crystalline and possess a wurtzite structure. With an increase in the amount of $Zn(CH_3COO)_2 \cdot 2H_2O$, middle-size ZnO nanotubes with diameters in the range from 40 to 70 nm and lengths up to 550 nm are obtained. The inner diameter of these nanotubes varies from 14 to 30 nm.

ZnO nanotubes with nanometer-scaled holes on the side-walls are obtained by a low-temperature hydrothermal procedure based on a preferential etching strategy.[115] The procedure is as follows. By a two-step heating of a solution mixture of $ZnCl_2$ and ammonia hydrothermally, nanotubes of nearly homogeneous size with ~250 nm diameter, 40 nm wall thickness, and 500 nm length are formed. The first stage involves the transformation of the precursor $Zn(NH_3)_4^{2+}$ to ZnO through hydrothermal decomposition at 95 °C, which leads to the formation of single crystal ZnO nanorods. At the same time, ZnO dissolves as the equilibrium moves to the left. The growth of ZnO rods becomes dominant since the high concentration of $Zn(NH_3)_4^{2+}$ favors the precipitation of ZnO. As the temperature goes down (95 to 75 °C), along with the partial consumption of the $Zn(NH_3)_4^{2+}$, etching of the ZnO nanorods occurs. Formation of porous ZnO nanotubes with nanoholes, which are single crystalline with a wurtzite structure, result from the preferential etching along the *c*-axis and slow etching along the radial directions. Nanoholes are created on the side-walls of the tubular structure. ZnO nanotubes have been assembled into microsphere superstructures by employing a mixture of poly(ethylene glycol) (PEG), ethanol, and water.[116] Addition of metallic zinc species into the solution mixture leads to aggregation of the PEG polymer coils to $Zn^{II}$/PEG globules with a diameter of ~500 nm. The globules turn into tube-like-structured

ADVANCED MATERIALS www.advmat.de

REVIEW

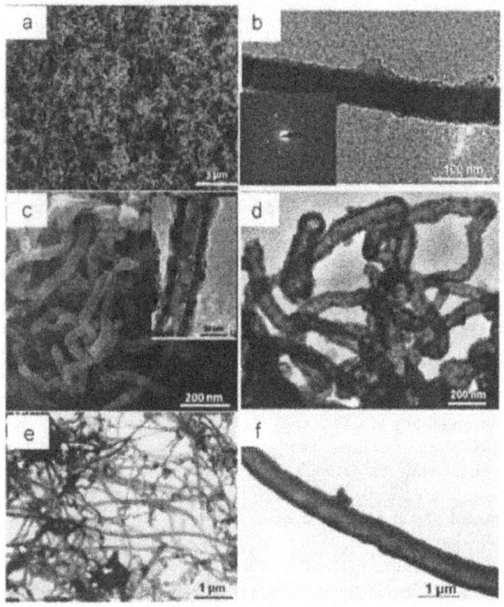


**Figure 14.** Nanowire–nanotube transformation as a result of the Kirkendall effect: a) Low-magnification FESEM images of Zn nanowires. b) TEM image of a Zn nanowire (the inset is the SAED pattern). c) FESEM image of ZnO nanotubes (inset shows the TEM image of a ZnO nanotube). d) TEM image of ZnS nanotubes. e,f) Scanning transmission electron microscope images of Si nanowires and $SiO_2$ nanotubes, respectively. Reproduced with permission from [117]. Copyright 2008, ACS.

ZnO–PEG microsphere assemblies (~2 mm) after ultrasonic pretreatment.

ZnO nanotubes are also produced by the oxidation of Zn nanowires because of the Kirkendall effect. This effect has also been used to produce nanotubes of $SiO_2$, $Co_3O_4$, and $ZnCr_2O_4$ as well as of ZnS, CdS, and CdSe (Fig. 14).[117] It is observed that there is a distribution in the diameters of the nanotubes just as in the case of the diameters of the starting nanowires. The Zn nanowires possess a smooth surface with an average diameter of 50 nm and lengths of several tens of micrometers with a zig-zag morphology. The ZnO nanotubes formed from the thermal oxidation of the Zn nanowires have an outer diameter that goes up to 90 nm with lengths that vary from 400 nm to a few micrometers. The ZnO nanotubes have a wall thickness of around 20 nm and an inner diameter close to the diameter of the starting metal nanowires.

CdO nanotubes with a mean diameter of 50 nm have been obtained by the thermal evaporation of Cd powder without using any catalyst or template.[118] The CdO nanotubes are single-crystalline with a cubic structure and have diameters of around 40–65 nm, a wall thickness of around 15 nm, and lengths of over a few tens of micrometers. Some of the nanotubes appear straight in morphology, while others are twisted with some straight parts.

Well-defined uniform $Al_2O_3$ nanotubes are obtained by the pulse anodization of aluminium in $H_2SO_4$ solution.[119] Periodic galvanic pulses were employed to achieve mild and hard anodization (HA) conditions, where the pulse duration for HA determines the length of nanotubes. By properly choosing the pulse parameters, continuous tailoring of the pore structure of the resulting nanoporous anodic alumina (i.e., periodic modulation of pore diameters along the pore axis) is achieved. This also enables weakening of the junction strength between cells, thereby helping in the separation of individual alumina nanotubes from the porous anodic alumina. The average inner and outer diameter of the nanotubes is estimated to be 80 and 95 nm, respectively, with lengths of up to several tens of micrometers. The inner diameter of the alumina nanotubes can be controlled by varying the etching time of the pore walls by using an appropriate etching solution (e.g., $H_3PO_4$). Aluminium oxide nanotubes have been fabricated by using tris(8-hydroxyquinoline) gallium organic nanowires ($GaQ_3$) as soft templates to coat alumina using ALD.[120] By dissolving the template in toluene or by heat treatment, alumina nanotubes are prepared, where the alumina shell thickness is controlled by the number of precursor/purge cycles (Fig. 15).

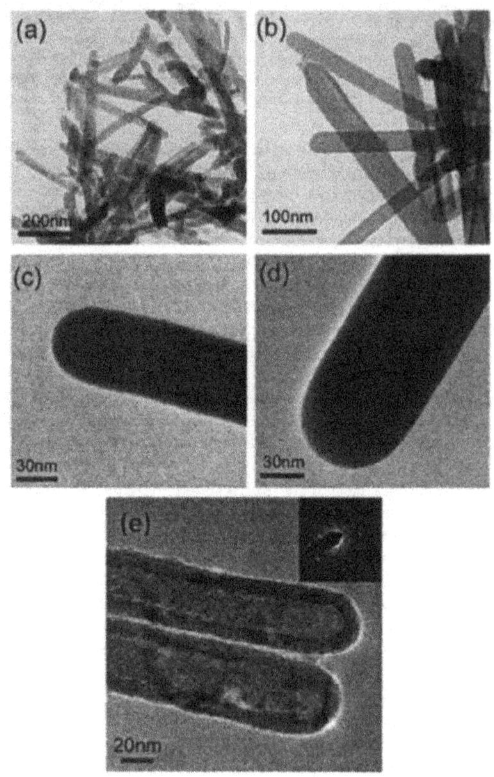


**Figure 15.** TEM images of a) $GaQ_3$ nanowires prepared by thermal evaporation. b,c) $GaQ_3$–$Al_2O_3$ core–shell nanowires fabricated by 100 cycles of ALD, and (d) same as (c), after 200 cycles of ALD. e) TEM image of the alumina nanotubes after heat treatment at 900 °C for 1 h. The inset shows the electron diffraction pattern of alumina. Reproduced with permission from [120]. Copyright 2007, ACS.

### 5.4. Nanotubes of Vanadium and Niobium Oxides

Vanadium oxides have received considerable attention because of their structural flexibility and useful catalytic, electrochemical, and other properties. The structure of $V_2O_5$ permits the intercalation of various cationic species in the interlamellar space.[121] The cations include alkylammonium ions, which are readily intercalated between the layers under hydrothermal conditions.[122] This system is thus analogous to graphite and layered dichalcogenides. Nesper and coworkers[123,124] synthesized nanotubules of alkylammonium intercalated $VO_x$ by hydrothermal means. The vanadium alkoxide precursor was hydrolyzed in the presence of hexadecylamine and the hydrolysis product (lamellar structured composite of the surfactant and the vanadium oxide) yielded $VO_x$ nanotubes along with the intercalated amine under hydrothermal conditions. Vanadium in these materials is in the mixed-valent state, and is redox active. The template cannot be removed by calcination as the structural stability is lost above 523 K. It is possible to partially extract the surfactant under mild acidic conditions. Nesper et al. have shown that the alkylamine intercalated in the intertubular space could be exchanged with other alkylamines of varying chain lengths as well as $\alpha,\omega$-diamines.[124] Such mixed valent $VO_x$ nanotubes are also obtained under hydrothermal conditions using $V_2O_5$ and 3-phenylpropylamine.[125] Most of the $VO_x$ nanotubes obtained by the hydrothermal method are open ended. Very few closed tubes had flat or pointed conical tips. Cross-sectional TEM images of the nanotubular phases show that instead of concentric cylinders (i.e., layers that fold and close within themselves), the tubes are made up of single or double layer scrolls that provide a serpentine-like morphology.[124,126] Non-symmetric fringe patterns in the tube walls exemplify that most of the nanotubes are not rotationally symmetric and carry depressions and holes in the walls. Diamine-intercalated $VO_x$ nanotubes are multilayer scrolls with narrow cores and thick walls, composed of vanadium oxide layers. Diamine-containing $VO_x$ nanotubes show a smaller number of holes in the wall structure and the tubes are well ordered with uniform distances throughout the tube length.[124] The scroll-like structures of $VO_x$ are not real nanotubes of the type formed by carbon or metal dichalcogenides. Vanadium oxide nanotubes that contain primary monoamines with long alkyl chains have been prepared by employing non-alkoxide vanadium precursors such as $VOCl_3$ and $V_2O_5$. The amine complexes of the vanadium precursors are then hydrolyzed. Hydrothermal treatment of the precursors gives good yields of $VO_x$ nanotubes that incorporate the amines.[127] In general the inner diameter varies between 15 and 50 nm in the $VO_x$ nanotubes, independent of the precursor. The outer diameters are similar, generally between 50 and 150 nm, with a range of tube lengths. Starting from the alkoxides, the tube length varies from 1 to 15 μm. In the samples obtained starting with $VOCl_3$ and $V_2O_5$, the average length of the tubes is shorter (1–3 μm). The distribution of the number of layers is similar in all the nanotubes, generally between 6 and 15 layers. The layers are preformed in the lamellar phase and roll up during the autoclave treatment, because the tetragonal structure of the layers gives rise to a more or less square-shaped growth. However, since many scrolls consist of multiple independent layers and the maximum length approaches 15 μm, growth along the tube axis may occur as well. Mn–V oxides have been prepared by mixing $V_2O_5$ with dodecylamine in the presence of ethanol and water. The amine templates are easily substituted or ion-exchanged with ions like $Mn^{2+}$ in an aqueous alcohol solution to yield Mn–V–O nanotubes.[128] Most of the nanotubes had open ends, while some of them had closed ends, with the side of the tubes wrapped around the end to close it. The $Mn^{2+}$ ions replace the organic cations in the structures and hence are intercalated in between the layers.

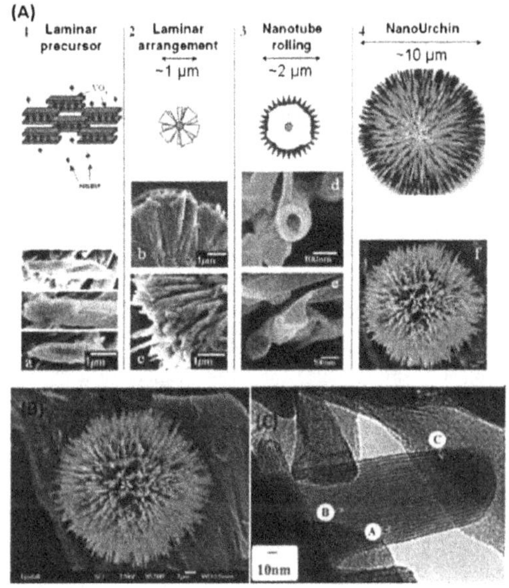


**Figure 16.** Schematic summary of the stages of growth of $VO_x$ nano-urchins. FESEM image of an individual nano-urchin. This fully grown nano-urchin is 12 μm in diameter and is covered in $VO_x$ nanotubes with a volumetric density of ~40 sr$^{-1}$. HRTEM image of an early stage nanotube. Lattice planes are resolved (A) and have a measured lattice spacing $a = 2.85$ nm. The hollow center (B) extends to the tip of the nanotube. A lattice plane termination dislocation is also observed (C). Reproduced with permission from [130]. Copyright 2006, ACS.

Arrays of $V_2O_5 \cdot nH_2O$ nanotubes with diameters of 200 nm and 5 μm length have been fabricated by template-based physical wetting of $V_2O_5$ sols.[129] Urchine-like nanostructures (Fig. 16) that consist of high-density spherical nanotube radial arrays of vanadium oxide nanocomposites, were successfully synthesized by a simple chemical route using an ethanolic solution of vanadium tri-isopropoxide and alkylamine hexadecylamine for 7 days at 180 °C.[130] $M-Nb_2O_5$ nanotube arrays have been prepared starting from $H-Nb_2O_5$ nanorods, taking advantage of the phase transformation accompanied by the formation of voids.[131]

### 5.5. Nanotubes of other Transition Metal Oxides

A hydrothermal reaction of $KMnO_4$ in HCl gives rise to $\alpha$-$MnO_2$ nanotubes.[132] These $MnO_2$ nanotubes are single-crystalline and possess a tetragonal structure, with an average outer diameter of

ADVANCED MATERIALS www.advmat.de

REVIEW

around 100 nm. The wall thickness is 30 nm and the length goes up to several micrometers. The nanotubes show nearly perfect tetragonal cross sections, consistent with the crystal structure.

By the reaction of $FeCl_3$ with water in the presence of polyisobutylene bis-succinimide (L113B) or the surfactant span80 as well as butanol under solvothermal conditions, $Fe_2O_3$ nanotubes are produced.[133] $Fe_2O_3$ nanotubes (rhombohedral structure) prepared from the surfactant (span80 as the template) were tube-like with diameters of 18–29 nm, wall thickness of 3–7 nm, and lengths of 110–360 nm. They have a multi-walled structure, with an interlayer spacing of 2.76 Å. $\alpha$-$Fe_2O_3$ nanotubes are also obtained by the deposition of a metal salt solution and $NaOH/NH_3$ in the pores of templates with the initial formation of an insoluble metal hydroxide precursor, and its subsequent transformation by dehydration and crystallization to metal oxide nanotubular structures.[134] The $\alpha$-$Fe_2O_3$ nanotubes so obtained are polycrystalline and have diameters and lengths of $\sim 260 \pm 60$ nm and $6 \pm 3$ μm, respectively. These nanotubes possess a rhombohedral structure and consist of small mis-oriented single-crystalline nanocrystalline domains. Ordered $Fe_2O_3$ nanotube arrays can also be prepared by ALD in an AM.[135] In this method, the thermal decomposition of a homoleptic dinuclear iron(III) *tert*-butoxide complex ($Fe_2(OtBu)_6$) is carried out in the presence of water inside a self-ordered porous anodic AM to yield arrays of nanocrystalline $Fe_2O_3$ tubes with aspect ratios up to 100. Polycarbonate membranes are used to obtain nanoparticle–nanotube arrays of $Fe_2O_3$ by employing electrodeposition followed by calcination.[136] The arrays have lengths in the 5–6 μm range and a diameter of ~200 nm, which corresponds closely to the pore dimension and pore diameter, respectively, of the membrane. The open ends of the arrays demonstrate the hollow structure of the product. The nanotubes have a hexagonal structure.

Ferromagnetic $\gamma$-$Fe_2O_3$ nanotubes have been synthesized by a template process with the aid of a high magnetic field.[137] These are polycrystalline, with a cubic spinel structure and lengths of about 30 μm, a wall thickness of 20 nm, and a diameter in the 300–400 nm range. $\alpha$-$Fe_2O_3$ is first formed at 500 °C by the thermal decomposition of $Fe(NO_3)_3$ inside the template and $\alpha$-$Fe_2O_3$ is then transformed into $\gamma$-$Fe_2O_3$ in the presence of the high magnetic field. Porous alumina templates are also used to obtain $Fe_3O_4$ nanotubes.[138] The hydrothermal reaction of $Fe(NO_3)_2$ in ethanol at pH 12 gives rise to $\alpha$-FeOOH nanotubes.[139] The so obtained $\alpha$-FeOOH nanotubes were ~10 nm in outer diameter and ~6 nm in inner diameter. Electron microscopic images of the nanotubes clearly show the resolved interplanar spacing of about 4.18 Å, which corresponds to the spacing between the (110) planes of orthorhombic $\alpha$-FeOOH.

Needle-like $Co_3O_4$ nanotubes have been prepared by employing a one-step self-supported topotactic transformation of nanoneedles of $\beta$-$Co(OH)_2$ (Fig. 17).[140] The nanotubes can be as long as 10 μm with a variable diameter in the range of 150–400 nm. The nanotubes are cylindrical and constructed from $Co_3O_4$ building blocks of less than 100 nm. The high-resolution (HR) TEM images in Figures 17d–f correspond to spots 1 to 3, respectively, of an individual nanotube (Fig. 17c). All show (111) lattice fringes perpendicular to the tube axis with an interplane spacing of 0.47 nm, while the two other sets of lattice fringes seen in Figure 17e are (220) and (311) planes, which correspond to interplanar spacings of 0.28 and 0.25 nm, respectively. On the basis of structural analysis, it is found that the tube axis is along the [111] direction with possible small mis-orientations for individual $Co_3O_4$ nanocrystals. The quasi-single-crystallinity of the cubic $Co_3O_4$ nanotubes is also confirmed by the SAED pattern (Fig. 17h) and the corresponding TEM image in Figure 17g. $Co_3O_4$ nanotubes are also prepared in the pores of AM by CVD, starting with cobaltacetylacetonate.[141] These nanotubes are highly ordered with a uniform diameter in the range of 100–300 nm and lengths of up to tens of micrometers. The nanotubes are composed of cubic polycrystalline $Co_3O_4$. Porous AMs have been used to obtain $CoFe_2O_4$ nanotube arrays by employing a sol–gel procedure.[142] These nanotubes are several micrometers in length with a mean outer diameter of 50 nm, which corresponds to the diameter of the alumina pore. The wall thickness is 15 nm.

NiO nanotubes can be fabricated through a MOCVD route using an AM as the template and $Ni(tta)_2tmeda$ (Htta = 2-thenoyl-trifluoroacetone, tmeda = tetramethylendiamine)

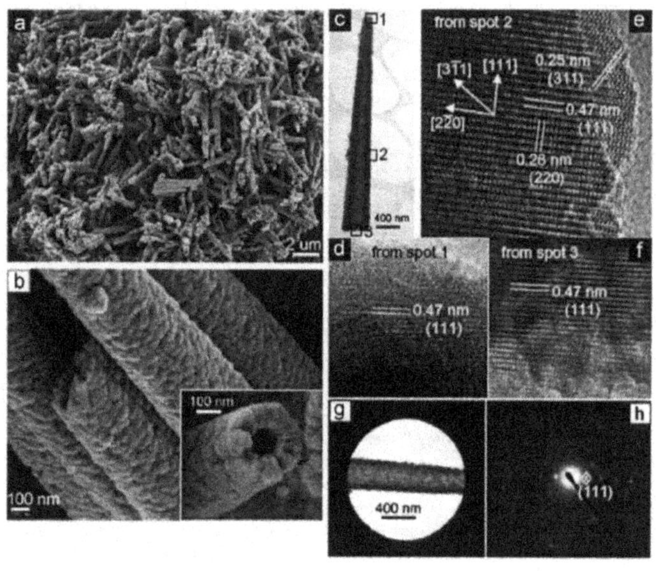


**Figure 17.** a,b) FE-SEM images of needle-like $Co_3O_4$ nanotubes. The inset in (b) shows a cross-sectional view of a nanotube. c) Low-magnification TEM image of an individual needlelike $Co_3O_4$ nanotube. d–f) HREM images taken from spots 1–3, respectively, of the nanotube shown in (c). g) TEM image showing a selected area of a $Co_3O_4$ nanotube for electron diffraction and h) the corresponding SAED pattern. Reproduced with permission from [140]. Copyright 2008, Wiley-VCH.

as the Ni source.[143] The nanotubes are polycrystalline and have the bunsenite structure of NiO. The length is around 1 μm, with an outer diameter of ~200 nm with a wall thickness of 20 nm. The nickel–ammine complex ($Ni(NH_3)_x^{2+}$) has also been used as a precursor to produce NiO nanotubes in an AM.[144] These NiO nanotubes are about 60 μm in length, with an outer diameter of about 200 nm and a wall thickness of 60 nm. They are polycrystalline with a face-centered cubic structure.

Synthesis of $WO_3$ nanotubes has been carried out by various methods, in particular starting from tungstic acid hydrate nanotubes without using any templates.[145] $WO_3$ nanotubes are obtained on slow calcination at 450 °C. The tungstic acid hydrate nanotubes obtained from the solvothermal reaction have outer diameters of 300–1000 nm and lengths of 2–20 μm. The nanotubes have a nearly rectangular pore of around 250 nm and the surface of the nanotube is not smooth. The polycrystalline nanotubes when calcined at 450 °C for 3 h, retain the open ended tubular structure and the overall dimensions of their precursor nanotubes. The nanotubes have a triclinic structure. The calcined nanotubes are free-standing and show no signs of aggregation. The annealed nanotube walls consist of individual nanoparticles arranged one-dimensionally with numerous self-supported pores, which are formed by the incomplete aggregation of nanoparticles. The sidewalls of the nanotubes are, therefore, porous. Although the diameter of the nanoparticles is approximately 40–80 nm, the nanotubes are 2–20 μm long. HR-TEM images of $WO_3$ nanotubes show that the lattice fringes of the nanocrystals have a spacing of 0.309 nm, which corresponds to the interplanar distance of the (112) plane of triclinic tungsten trioxide.

### 5.6. Nanotubes of other Binary Oxides

MgO nanotubes can be synthesized by introducing Sn as a catalyst in the CVD process. The nanotubes are formed by the vapor–liquid–solid (VLS) mechanism.[146] MgO nanotubes can also be formed by the thermal evaporation of a mixture of Zn and Mg powders.[147] $In_2O_3$ nanotubes are prepared by annealing InOOH nanotubes,[148] having obtained the latter under solvothermal conditions starting with $InCl_3$ in the presence of a surfactant and formamide in anhydrous ethanol. $In_2O_3$ tubular nanotubes are also generated by CVD starting with nanoporous structures of InP.[149] The $In_2O_3$ nanotubes are single crystalline with a cubic structure and have outer diameters of around 200–300 nm and lengths of around 2–6 μm and retain the size and square shape of the pores. Porous $In_2O_3$ nanotubes can also be prepared by layer-by-layer assembly/deposition on carbon nanotube templates followed by calcination. The layer-by-layer assembly is used to form a polyelectrolyte on the surface of pristine nanotubes, which enhances the adsorption of the metal-complex species on the surface of the carbon nanotube as a result of electrostatic attraction between the charged species.[150] In a typical procedure, the layer-by-layer assembly is formed with a polyelectrolyte such as sodium poly(styrene sulfonate) (PSS) and poly(diallyldimethylammonium chloride) (PDDA) on the surface of pristine carbon nanotubes. A solution mixture of $InCl_3$ and citric acid is added into the solution of the polyelectrolyte-modified carbon nanotubes. A solution of $NaBH_4$

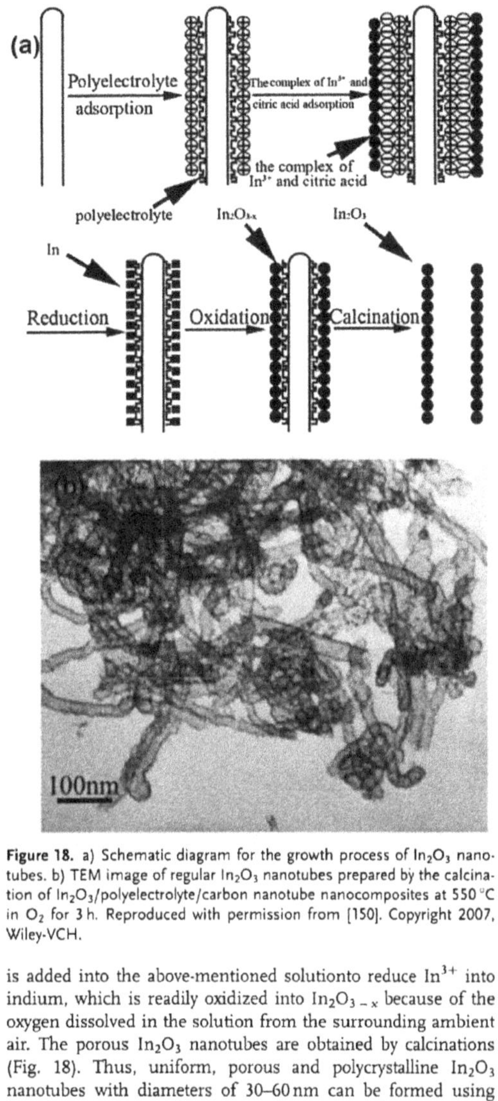


Figure 18. a) Schematic diagram for the growth process of $In_2O_3$ nanotubes. b) TEM image of regular $In_2O_3$ nanotubes prepared by the calcination of $In_2O_3$/polyelectrolyte/carbon nanotube nanocomposites at 550 °C in $O_2$ for 3 h. Reproduced with permission from [150]. Copyright 2007, Wiley-VCH.

is added into the above-mentioned solutionto reduce $In^{3+}$ into indium, which is readily oxidized into $In_2O_{3-x}$ because of the oxygen dissolved in the solution from the surrounding ambient air. The porous $In_2O_3$ nanotubes are obtained by calcinations (Fig. 18). Thus, uniform, porous and polycrystalline $In_2O_3$ nanotubes with diameters of 30–60 nm can be formed using layer-by-layer assembly on carbon nanotube templates, followed by calcination. $In_2O_3$ nanotubes have a cubic structure and comprize nanoparticles of about 5 nm. The wall thickness is about 9 nm. This technique can also be used for the preparation of nanotubes of other oxides such NiO, $SnO_2$, $Fe_2O_3$, and CuO.[150]

$Y_2O_3$ nanotubes are prepared by a non-aqueous electrochemical method involving oxide transfer to $Y^{III}$ precursors.[151] These horn-shaped nanotubes (nanohorns) with a narrow size distribution coexist with small nodular deposits. The horn-shaped structures are hollow. The diameter of these structures tapers from approximately 900 nm at the base to less than 100 nm at the tip. The nanohorns have lengths that range from 3.9 to 16.5 μm. The thickness of the walls of the cylinder is approximately 250 nm, and the inner diameter is approximately 400 nm. These open cap structures have diameters similar to the nanohorn base diameters and are commonly observed at the initial stages of growth. A hydrothermal procedure has also been employed starting with $Y(OH)_3$ nanotubes, followed by calcination of the hydroxide.[152]

The one-pot synthesis of $SnO_2$ nanotubes has been accomplished at room temperature starting from Sn nanorods.[153] The method involves the Kirkendall effect. The $SnO_2$ nanotubes are polycrystalline, tetragonal with diameters that range from 50 to 60 nm, and lengths from 300 to 500 nm. A 10–20 nm increase in the diameter of $SnO_2$ nanotubes compared with that of the starting Sn nanorods is seen as a result of the outward flow of Sn during the oxidation. A solution phase synthesis of $SnO_2$ nanotubes using surfactant-assisted micelles as templates has been reported.[154] In this method, a mixture of $SnCl_4$ with PVP and dimethyl sulfoxide (DMSO) is refluxed in the presence of $Na_2S$. The nanotubes consist of nanocrystalline particles that have a face-centered cubic structure, a diameter of ~200 nm, and lengths up to a few micrometers. Electrosynthesis of $SnO_2$ nanotubes employing the track-etched polymer polycarbonate membranes has been carried out.[155] A gold electrode modified with a porous polycarbonate membrane is immersed in an aqueous tin chloride solution. Electrochemistry is employed to control the local pH within the pores and drive the precipitation reaction. Removal of the gold and dissolution of the polymer yields 1D polycrystalline tin oxide particles. The crystallinity of the material is enhanced by annealing at 650 °C. The particles are hollow, with a wall thickness of approximately 10 nm. Nanotubes result from the continuous side-wall precipitation along a reaction front, which are polycrystalline $SnO_2$ (rutile structure), with a diameter in the 100 nm range and a length between 0.4 and 1.4 μm. Nanotubular indium-tin oxide (ITO) has been prepared by the sol–gel process using cellulose paper as a template.[156]

$ZrO_2$ nanotube arrays with diameters of about 130 nm and lengths up to 190 μm are obtained by anodizing zirconium foil in a mixture of formamide and glycerol that contains $NH_4F$ and 3 wt % water.[157] The as-prepared $ZrO_2$ nanotube arrays are amorphous, with the coexistence of monoclinic and tetragonal phases when annealed from 400 to 600 °C. Monoclinic zirconia nanotubes were obtained at 800 °C with retention of shape. By employing a sol–gel method and AM, yttria-stabilized $ZrO_2$ nanotubes have been prepared.[158] The length and the diameter of these nanotubes are 50 μm and 200 nm, respectively, in agreement with the dimensions of the template pores. The wall thickness of the nanotubes depends on the impregnation time. The nanotubes after sintering at 800 °C are polycrystalline with a cubic structure.

Formation of $CeO_2$ nanotubes has been reported by a few workers. $CeO_2$ nanotubes are produced by a two-step procedure involving precipitation at 100 °C and ageing at 0 °C for 45 days.[159] $CeO_{2-x}$ nanotubes are crystalline, having a cubic fluorite structure. They tend to align the (111) planes parallel to the axis direction. The diameter of the nanotubes ranges from 5 to 30 nm, while the length goes up to several micrometers. The thickness of the wall of the nanotubes is about 5.5 nm. $CeO_2$ nanotubes are also obtained by annealing layered $Ce(OH)_3$ nanotubes by a simple oxidation–coordination–assisted dissolution process of the $Ce(OH)_3$ nanotubes/nanorods. One-dimensional structures of $Ce(OH)_3$ are synthesized by the hydrothermal treatment of $Ce_2(SO_4)\cdot 9H_2O$ with a 10 M NaOH solution at 130 °C. Starting from $Ce(OH)_3$ nanorods (as well as from narrow cavity nanotubes), it is possible to obtain $CeO_2$ nanotubes with large cavities by a simple oxidation and dissolution process.[160] These nanotubes with open ends have a cubic structure and display an outer diameter of about 15–25 nm and a length of about 100 nm. Nanoparticles of about 8 nm are attached to the walls and the thickness of the wall is about 5–7 nm. The inner diameter of the nanotubes is about 10–15 nm. In this method, freshly prepared 1D $Ce(OH)_3$ is exposed to air at room temperature for 24 h. The partially oxidized $Ce(OH)_3$ nanotubes with narrow cavities are dispersed in distilled water and subjected to ultrasonication for 2 h after adding 15% $H_2O_2$. Partial oxidation of the $Ce(OH)_3$ is essential to form the ceria tubular structures. Electrosynthesis of $CeO_2$ nanotubes using AMs has been reported by employing a non-aqueous electrolyte.[161] $ThO_2$ nanotubes are reported by the sol–gel method by using a porous alumina template.[162]

### 5.7. Nanotubes of Titanates and other Complex Oxides

The hydrothermal method has been employed to synthesize nanotubes of monocrystalline $BaTiO_3$,[163] and tubular $PbTiO_3$.[164] Pulsed laser ablation also yields $PbTiO_3$ nanotubes within the pores of an AM,[165] while sol–gel electrophoretic deposition of an acetic acid-based highly stabilized lead zirconate titanate (PZT) sol in AMs gives rise to PZT nanotubes.[166] The PZT sol with a near-morphotropic phase boundary composition and no polymeric addition was prepared using lead acetate trihydrate, zirconium, and titanium tetra-butoxides. By anodization of a Ti–Zr alloy, $ZrTiO_3$ nanotubes are obtained.[167]

Mallouk et al.[168] and Peng et al.[169] have reported the synthesis of niobate-based nanotubes at low temperatures by the exfoliation of acid-exchanged $K_4Nb_6O_{17}$ with tetra(*n*-butyl)-ammonium hydroxide ($TBA^+OH^-$) in aqueous solution. In the presence of excess acid, around 80% of the potassium ions in $K_4Nb_6O_{17}$ are replaced by protons. The remaining potassium ions appear to reside in the slowly exchanging interlayer galleries that alternate along the stacking axis.[168] The bilamellar colloid initially formed by reacting this material with $TBA^+OH^-$ is unstable relative to the formation of unilamellar sheets and tubules. Exfoliation initially produces a bilayer colloid and then transforms (depending on conditions) irreversibly into tubules, unilamellar sheets, or a mixture of the two. The concentration of the colloid and the pH are important factors in controlling the coiling equilibrium. Both $H^+$ and alkali ions help to precipitate the aggregated tubules. The individual tubules are formed by the rolling of sheets of exfoliated $K_{4-x}H_xNb_6O_{17}$ ($x \approx 3.2$). The tubules are 0.1–1 μm in length, with outer diameters that range from 15 to 30 nm. The colloids and the precipitated tubules are both highly stable.

Potassium hexaniobate nanotubes have also been fabricated from polycrystalline $K_4Nb_6O_{17}$ at room temperature using the intercalating and exfoliating methods.[169] These are multilayer crystalline nanotubes with interlayer spacings from 0.83 to 3.6 nm, depending on the intercalating molecules such as tetra($n$-butyl)ammonium hydroxide ($TBA^+OH^-$) and alkylamines ($C_nH_{2n+1}NH_2$). The number of layers in the wall is in the range of 3 to 8. The outer diameter varies between 20 and 90 nm for the nanotubes obtained with different alkylamines, and the length of the nanotubes ranges from a few hundred nanometers to several micrometers. When a single-layer $(-Nb_6O_{17}-)_n$ sheet rolls up into a nanotube, $C_nH_{2n+1}NH_2$ ($n \neq 1$) molecules are already adsorbed on both sides of the sheet and then reside in the interlayer spaces of the nanotube. A model of the spiral structural growth of these nanotubes has been proposed and the tube axis found to be parallel to the [100] direction of the $K_4Nb_6O_{17}$ crystal. Spiral nanotubes of potassium niobate are obtained by introduction of a polyfluorinated cationic azobenzene derivative, *trans*-[2-(2,2,3,3,4,4,4-heptafluorobutylamino)ethyl] {2-[4-(4-hexyphenylazo)phenoxy]ethyl} dimethylammonium (abbreviated as $C_3F_7$-Azo+), into the layered niobate interlayer by a two-step guest–guest exchange method, with methyl viologen ($MV^{2+}$)-$K_4Nb_6O_{17}$ as the precursor.[170] When $MV^{2+}$-intercalated niobate was used as the precursor, the polyfluorinated $C_3F_7$-Azo+ results in the quantitative formation of spiral nanotubes from exfoliated nanosheets of the niobate, by rolling along the sandwiched microstructure. Nanotubes of $FePO_4$ have been prepared under solvothermal conditions in the presence of a sodium dodecyl sulfate (SDS) surfactant.[171] Iron phosphate nanotubes have mesoporous walls and diameters of 50–400 nm and lengths of several micrometers. The walls of the nanotubes range from 20 to 40 nm in thickness. The removal of the surfactant by acetate exchange and heat treatment results in amorphous mesoporous nanotubes of $FePO_4$. Mesoporous $NiPO_4$ nanotubes have been prepared in the presence of a cationic surfactant and different bases by the sol-gel method.[172] The solution–liquid–solid (SLS) method has been exploited to obtain tin-filled $In(OH)_3$ nanotubes using liquid droplets of an In–Sn mixture.[173]

Nanotubes and how they are formed of other complex oxides reported are: $In_2Ge_2O_7$ by thermal evaporation,[174] $InVO_4$ nanotubes using templates,[175] $WO_3-H_2O$ nanotubes with the aid of intercalated polyaniline,[176] chrysotile nanotubes by the hydrothermal method,[177] aluminogermanate nanotubes by a simple solution procedure,[178] $\alpha$-FeOOH nanotubes by employing reverse micelles,[179] boehmite nanotubes by a hydrothermal procedure,[180] and hydroxyapatite nanotubes by a sol-gel procedure in an AM.[181]

$La_{0.5}Sr_{0.5}CoO_3$ nanotubes have been prepared using an AM template by the sol-gel method,[182] while $ZnAl_2O_4$ spinel nanotubes have been obtained by making use of the Kirkendall effect.[183] $ZrCr_2O_4$ has also been prepared by this method.[117] Nanotubes of $SrAl_2O_4$,[184] as well as $AgIn(WO_4)_2$,[185] are produced under hydrothermal conditions. By controlling the pH, single-walled aluminogermanate nanotubes have been synthesized by the reaction of tetraethylorthogermanate and $AlCl_3$ in NaOH solution.[186] Nickel hexacyanoferrate nanotubes have been fabricated in an AM using the electrokinetic method.[187] Single-crystal calcium sulfate nanotubes can be generated by the reverse micelle method at room temperature.[188] Self-assembled supramolecular $C_{32}H_{70}N_2ZnSO_4$ nanotubes with high thermal stability have been prepared.[189] Bismuth subcarbonate nanotubes are prepared by a simple reflux of bismuth citrate and urea in ethylene glycol.[190]

## 6. Pnictide Nanotubes

Several methods to synthesize BN nanotubes including CVD and electrical discharge, as well as templating have been described in the literature.[4] Thin BN tubes of less than 200 nm diameter were first obtained by arc discharge with hollow tungsten electrodes filled with h-BN powder. Following this initial report, a variety of methods have been employed to prepare BN nanotubes. The other methods of synthesis of BN nanotubes include those that are far from equilibrium, such as the electrical arc method,[191,192] arcing between h-BN and Ta rods in a $N_2$ atmosphere,[193] laser ablation of h-BN,[194] and continuous laser heating of BN.[195] The last method produces long ropes of BN nanotubes with thin walls. Single-walled BN nanotubes are formed in some cases.[191,196] Single-walled nanotubes of BN have been deposited on polycrystalline W substrates by using electron-cyclotron resonance nitrogen and electron beam boron sources.[197]

Bando and coworkers,[198,199] have carried out extensive work on BN nanotubes. They prepared the nanotubes by the reaction of MgO and B in the presence of ammonia at 1300 °C.[198] $B_2O_2$ (obtained by heating B and MgO at 1300 °C) was heated in the presence of $NH_3$ in a long BN boat, which was placed into a graphite susceptor that was heated from the outside by an RF furnace to obtain BN nanotubes. The BN nanotubes exhibit a 1D nanostructure and have diameters that range from several nanometers to about 70 nm, and lengths that go up to 10 micrometers. They show the presence of mixed hexagonal and rhombohedral BN phases. The nanotubes of diameter around 10 nm exhibit a perfectly cylindrical structure, although edge dislocations are occasionally observed from the rhombohedrally stacking, ordered nanotubes. The concentric tube structures do not contain internal wall closures or internal cap structures. Bando and coworkers,[199] have also prepared BN nanotubes by the reaction of MgO, FeO, and B in the presence of $NH_3$ at 1400 °C. The nanotubes are in the pure h-BN phase and have diameters of ~50 nm and lengths of up to several tens of micrometers. The reaction of boric acid or $B_2O_3$ with $N_2$ or $NH_3$ at high temperatures in the presence of carbon or catalytic metal particles has also been employed in the preparation of BN nanotubes.[200] BN nanotubes can be grown directly on substrates at 873 K by a plasma-enhanced laser-deposition technique.[201]

$B_2O_3$-coated multi-walled carbon nanotubes, on reaction with $NH_3$ at high temperatures, yield BN nanotubes.[202] Bando and coworkers have reviewed various aspects of BN nanotubes and cite methods involving arc-discharge, laser ablation, and the plasma jet method (Fig. 19).[203] Straight BN nanotubes have been grown by heating a mixture of $Mg(BO_2)_2 \cdot H_2O$, $NH_4Cl$, $NaN_3$, and Mg powder in an autoclave at 600 °C for 20–60 h. These BN nanotubes had diameters mainly in the range of 30–300 nm and lengths of up to ~5 μm, and a majority of them had at least one closed end.[204] Multi-walled BN nanotubes have been obtained by a reduction–nitridation route wherein the reaction of boron trifluoride etherate and sodium azide was carried out in the presence of Fe–Ni powder in a sealed autoclave at 600 °C for 12 h. The BN nanotubes so obtained have an average outer diameter of 60 nm, an inner diameter of 30 nm, and lengths up to 300 nm.[205]

ADVANCED MATERIALS
www.advmat.de
REVIEW

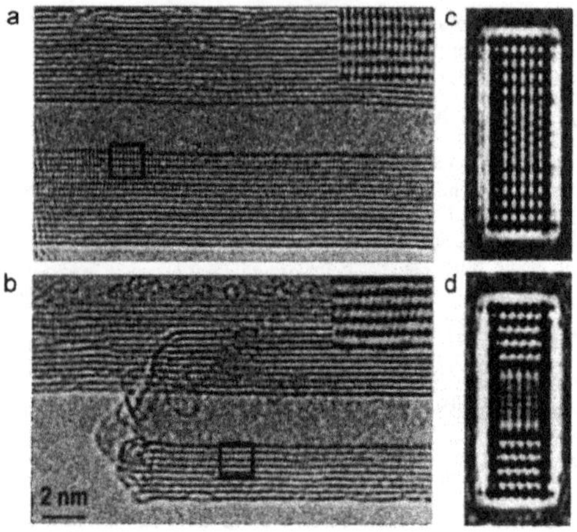


**Figure 19.** a,b) HRTEM images of two zig-zag BN MWNTs. A different stacking order is apparent in the marked areas in (a) and (b), as highlighted in the insets. Hexagonal- (a) and rhombohedral-type (b) stacking (in 12.5° fringe inclinations with respect to the tube axis) are verified by the corresponding computer-simulated HRTEM images (c and d, respectively) for BN MWNTs having the axes strictly parallel to the [10–10] orientations ('zig-zag' NTs). Note a characteristically open tip-end of a BNNT in (b). This feature is frequently seen in BNNTs obtained from high-temperature chemical syntheses. Reproduced with permission from [203b]. Copyright 2000, American Institute of Physics.

By annealing an Fe film evaporated on boron pellets at 1000 °C in $N_2$ gives rise to BN nanotubes.[206] Microwave plasma-enhanced CVD has been employed to obtain BN nanotube arrays.[207] In this method, AM are used (for confinement) along with borane and ammonia or nitrogen. BN nanotubes that are 1 mm long are obtained by an optimized ball-milling and annealing method.[208] The annealing temperature of 1100 °C is crucial for the growth of the long BN nanotubes because at this temperature there is a fast nitrogen dissolution rate in Fe and the B/N ratio in Fe is 1.

By combining polymer thermolysis and the use of templates, ordered arrays of BN nanotubes have been obtained.[209] In this method, liquid borazine in AM is thermolyzed at high temperatures in a nitrogen atmosphere. Double-walled BN nanotubes are obtained by the reaction of a flowing stream of borazine and ammonia with a nickelocene catalyst vapor in the hot zone of a furnace maintained at 1200 °C.[210] The mechanism of growth of single-walled BN nanotubes by laser vaporization has been examined in detail.[211] It appears that the root-growth model that involves a droplet of B is applicable for this growth. It may be noted that laser vaporization is a unique route for the synthesis of single-walled BN nanotubes.

Single-walled BCN nanotubes are obtained by bias-assisted hot-filament CVD.[212] In this method, a Fe–Mo catalyst supported on MgO powder is used for the decomposition of a mixture of $CH_4$, $B_2H_6$, and ethylenediamine. The as-grown SWNTs have clean and smooth surfaces with diameters in the range of 0.8–2.5 nm and lengths up to micrometers. Homogeneous $BC_4N$ nanotube brushes have been obtained recently by the reaction of amorphous carbon nanotubes with boric acid and urea (Fig. 20).[213] The BCN nanotubes have open ends with a diameter of 170 nm, a length of 15 mm, and thickness of around 50 nm. The SAED pattern shows faint rings and a few spots. The X-ray diffraction (XRD) pattern of the BCN nanotube brushes shows broad reflections with *d*-spacings of 3.43 Å and 2.13 Å, which correspond to the (002) and (100) planes, respectively, similar to the pattern reported for $BC_3N$. The broad reflections in the XRD pattern and the diffuse rings in the electron diffraction pattern suggest the turbo static nature of the nanotubes.

GaN nanotubes have been grown in AMs by employing MO CVD. The reactants used are trimethylgallium and $NH_3$.[214] The diameter of the GaN nanotubes is approximately 200–250 nm and the wall thickness is about 40–50 nm. Nanotubular GaN consists of numerous fine GaN particulates with a size range of 15–30 nm. XRD and TEM analyses indicate that the grains in the GaN nanotubular material is nanocrystalline with a hexagonal structure. By the reaction of $GaCl_3$ with excess $NaN_3$ in dry benzene, GaN nanotubes with branched tubules have been grown in mesoporous MCM 48.[215] These GaN nanotubes are polycrystalline with a wurtzite structure and an outer diameter of 50–150 nm. The tube wall thickness is ~10 nm and the lengths go up to several micrometers. Amorphous carbon nanotube brushes, on reaction with $GaCl_3$ followed by reaction with $NH_3$, give rise to GaN nanotube brushes (Fig. 21).[216] The close packing of 1D GaN nanotube-nanobristles have diameters from 100 to 230 nm and lengths in the 7–10 μm range. The tubes are open at one end and their surface appears to be smooth. The nanotube brushes have a wurtzite structure. The TEM images in Figures 21c–e shows a nanotube with an outer diameter of 200 nm and the wall thickness is in the range of 10–15 nm. The HRTEM image of the nanotube wall in Figure 21f shows interplanar spacing of 0.244 nm, which indicates the single crystalline nature of the GaN nanotube.

$Si_3N_4$ nanotubes can be prepared by the sol–gel route in which tetraethoxysilane (TEOS) and phenolic resin are used to prepare a carbonaceous silica xerogel, and ferric nitrate is employed as an additive. The xero-gel is heated at 1300 °C for 10 h in nitrogen.[217] Molybdenum nitride nanotubes have been obtained by depositing a nitride film on anodic alumina and etching away the template with NaOH.[218]

Single crystalline nanotubes of $Cd_3P_2$ and $Zn_3P_2$ have been obtained by a template process in which Cd and Zn nanorods formed in-situ act as self-sacrificing templates.[219] A mixture of reactants that contain CdS (or ZnS), P, and $Mn_3P_2$ is heated to 1350 °C in an induction furnace to obtain nanotubes. Single-crystalline, single-walled, $SbPS_{4-x}Se$ nanotubes with a tunable band-gap have been prepared by taking a mixture of stoichiometric amounts of the elements Sb, P, S, for $SbPS_4$, and the corresponding amounts for $SbPS_3Se$, $SbPS_2Se_2$, and $SbPSSe_3$ in fused silica tubes sealed under vacuum, and heated.[220]

## 7. Nanotubes of Carbides and other Materials

SiC nanowires, $SiC/SiO_2$ core-shell nanocables, and SiC nanotubes have been synthesized simultaneously by directly heating Si powder and multiwall carbon nanotubes

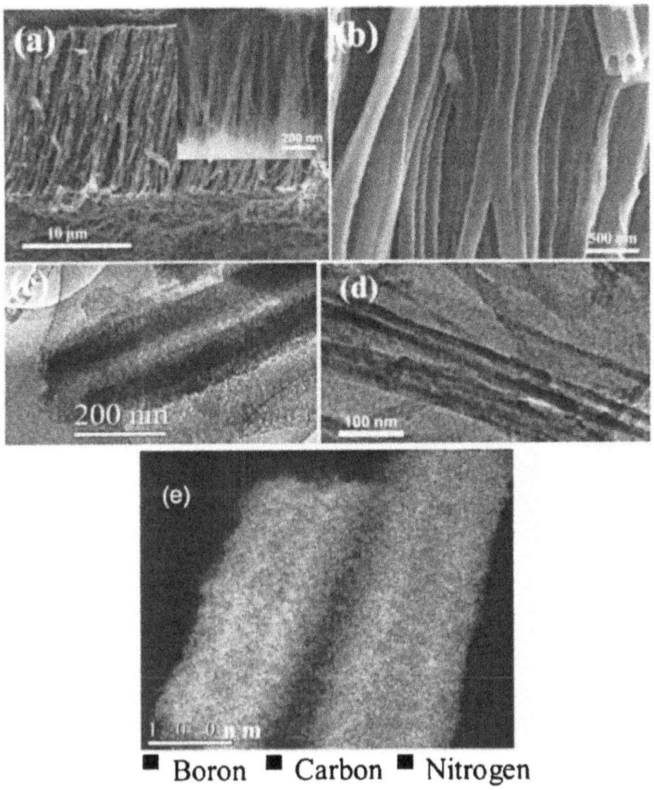


Figure 20. a) FESEM images of BCN nanotube brushes with the average diameter of a single tube being around 170 nm. The inset shows a FESEM image of BCN nanotube brushes of 40 nm diameter. b) Higher magnification FESEM images of BCN nanotube brushes. TEM images of a BCN nanotube: c) 170 nm diameter and d) 40 nm diameter. e) Elemental mapping of the boron, carbon, and nitrogen of BCN nanotubes obtained from EELS. Reproduced with permission from [213]. Copyright 2008, RSC.

(MWCNTs).[221] Single-phase TiC nanotubes are prepared by the reaction of carbon nanotubes with Ti powder at 1300 °C for 30 h.[222]

Silicon oxycarbide ceramic nanotubes can be obtained by the pyrolysis of polysilicone nanotubes using a sacrificial AM as a template.[223] In a typical synthesis, the AM templates were immersed in a mixture of low-molecular-weight silicone starting materials, such as $\{CH_3(CH{=}CH)SiO\}_n$ ($n = 3$–$7$), named VC4, and a cyclic-type silicone bearing vinyl groups, and $(CH_3)_3SiO\{CH_3(H)SiO\}_m$ ($m = 20$), named KF-99 and heated in an oven at two different temperatures, 50 and 200 °C. Large-scale aligned silicon carbonitride nanotube arrays have been synthesized by microwave-plasma-assisted CVD using $SiH_4$, $CH_4$, and $N_2$ as precursors. The nanotubes are 6–7 μm in length and 100–200 nm in diameter.[224]

Transition metal halides such as $NiCl_2$ crystallize in the $CdCl_2$ structure, with the metal halide layers held together by weak van der Waals forces. $NiCl_2$ has been shown to form closed cage structures and nanotubes.[225] These were prepared by heating $NiCl_2 \cdot 6H_2O$ initially in air to remove the water of crystallization, and heated further at 960 °C under Ar. Zeng et al.[226] have carried out UV-light-induced fabrication of $CdCl_2$ nanotubes starting from CdSe solid nanocrystals through a Kirkendall effect. In this procedure, $CdCl_2$ nanotubes are obtained starting with a solution of CdSe nanocrystals in *o*-dichlorobenzene (which generates chlorine free radicals on photolysis).

## 8. Complex Inorganic Nanostructures Based on Nanotubes

In addition to the simple nanotube structures, more complex nanotubular structures of various inorganic materials such as coaxial cables, heterojunctions, and nanotube-nanowire or nanotube-nanoparticle composites have been prepared and characterized by several workers. Typical examples of such nanostructures are the following: Integration of ZnO nanotubes with ordered nanorods,[227] conversion of ZnO nanorod arrays into ZnO–ZnS nanocable and ZnS nanotube arrays,[228] core–sheath heterostructure $CdS{-}TiO_2$ nanotube arrays,[229] ZnS nanotube–In nanowire core–shell heterostructures,[230] ZnO–ZnS core–shall nanotube arrays,[231] Cu nanotube–Bi nanowire heterojunctions,[232] carbon nanotubes in $TiO_2$ nanotubes,[233] $TiO_2{-}Pt$ coaxial nanotube arrays,[234] Sn nanowires on $TiO_2$ nanotubes,[235] $Fe_2O_3{-}TiO_2$ nanorod–nanotube arrays,[236] $SiO_2{-}Ta_2O_5$ core–shell nanowires and nanotubes,[237] multi-walled BCN–carbon nanotube junctions,[238] and BN nanotubes with periodic iron nanoparticles.[239]

## 9. Outlook

The preceding presentation should suffice to demonstrate the great progress made in the synthesis of inorganic nanotubes. It would appear that today we are able to synthesize almost any inorganic material in tubular form. Not only do we have an arsenal of inorganic nanotubes but also several strategies to synthesize them. There are still many synthetic challenges. One of them relates to the need to synthesize single-walled single-crystalline nanotubes of inorganic materials. There are hardly one or two reports of single-walled inorganic nanotubes at present. With the availability of inorganic nanotubes in sufficient quantities, there is much scope to measure their properties and investigate phenomina associated with them.

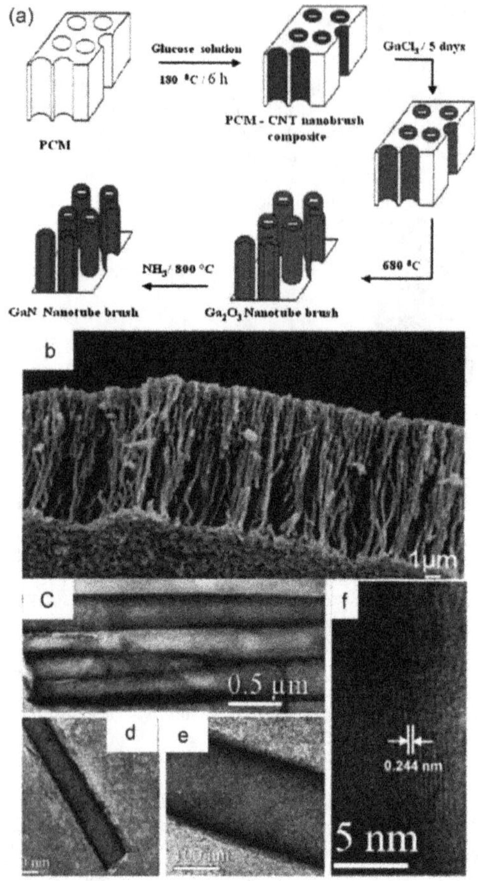


**Figure 21.** a) Schematic showing the formation of nanotube brushes of carbon, $Ga_2O_3$, and GaN. b) SEM image of a GaN nanotube brush. c–e) TEM images of GaN nanotubes and f) HRTEM image of the wall of a GaN nanotube. Reproduced with permission from [216]. Copyright 2007, ACS.

Received: December 16, 2008
Revised: March 18, 2009
Published online: August 3, 2009

[1] *Nanomaterials Chemistry: Recent Developments* (Eds: C. N. R. Rao, A. Muller, A. K. Cheetham), Wiley-VCH, Weinheim **2007**.
[2] C. N. R. Rao, A. Govindaraj, *Nanotubes and Nanowires*, RSC series on Nanoscience, Royal Society of Chemistry, London **2005**.
[3] S. Iijima, *Nature* **1991**, *354*, 56.
[4] C. N. R. Rao, M. Nath, *Dalton Trans.* **2003**, 1.
[5] R. Tenne, C. N. R. Rao, *Phil. Trans. Royal. Soc. (London)* **2004**, *362*, 2099.
[6] a) M. R. Ghadiri, J. R. Granja, R. A. Milligan, D. E. McRee, N. Kazanovich, *Nature* **1993**, *366*, 324. b) E. Gazit, *Chem. Soc. Rev.* **2007**, *36*, 1263.
[7] a) R. Tenne, *Chem. Eur. J* **2002**, *8*, 5296 and the references there in. b) F. L. Deepak, R. Tenne, *Cent. Eur. Chem.* **2008**, *6*, 373.
[8] C. J. Brumlik, C. R. Martin, *J. Am. Chem. Soc.* **1991**, *113*, 3174.
[9] J. C. Hulteen, C. R. Martin, *J. Mater. Chem.* **1997**, *7*, 1075.
[10] X. Zhang, H. Wang, L. Bourgeois, R. Pen, D. Zhao, P. A. Webly, *J. Mater. Chem.* **2008**, *18*, 463.
[11] a) C. R. Martin, M. Nishizawa, K. Jirage, M. Kang, *J. Phys. Chem. B* **2001**, *105*, 1925. b) M. Wirtz, C. R. Martin, *Adv. Mater.* **2003**, *15*, 455.
[12] P. Kohli, J. E. Wharton, O. Braide, C. R. Martin, *J. Nanosci. Nanotechnol.* **2004**, *4*, 605.
[13] T.-Y. Shin, S.-H. Yoo, S. Park, *Chem. Mater.* **2008**, *20*, 5682.
[14] M. A. Sachez-Castillo, C. Couto, W. B. Kim, J. A. Dumesic, *Angew. Chem. Int. Ed.* **2004**, *43*, 1140.
[15] H.-W. Wang, C.-F. Sheih, H.-Y. Chen, W.-C. Shiu, B. Russo, G. Cao, *Nanotechnology* **2006**, *17*, 2689.
[16] Y.-C. Pu, J. R. Hwu, W.-C. Su, D.-B. Shieh, Y. Tzeng, C.-S. Yeh, *J. Am. Chem. Soc.* **2006**, *128*, 11 606.
[17] A. M. Schwartzberg, T. Y. Olson, C. E. Talley, J. Z. Zhang, *J. Phys. Chem. C* **2007**, *111*, 16 080.
[18] M. S. Calvar, J. Pacifico, J. P. Juste, L. M. Liz-Marzan, *Langmiur* **2008**, *24*, 9675.
[19] Z. Chen, M. Waje, W. Li, Y. Yan, *Angew. Chem. Int. Ed.* **2007**, *46*, 4060.
[20] A. Takai, Y. Yamauchi, K. Kuroda, *Chem. Commun.* **2008**, 4171.
[21] T. Kijima, T. Yoshimura, M. Uota, T. Ikeda, D. Fujikawa, S. Mouri, S. Uoyama, *Angew. Chem. Int. Ed.* **2004**, *43*, 228.
[22] H.-P. Liang, N. S. Lawrence, L.-J. Wan, L. Jiang, W.-G. Song, T. G. J. Jones, *J. Phys. Chem. C* **2008**, *112*, 338.
[23] C. C. Han, M. Y. Bai, J. T. Lee, *Chem. Mater.* **2001**, *13*, 4260.
[24] J. Bao, C. Tie, Z. Xu, Q. Zhou, D. Shen, Q. Ma, *Adv. Mater.* **2001**, *13*, 1631.
[25] F. Tao, M. Guan, Y. Jiang, J. Zhu, Z. Xu, Z. Xue, *Adv. Mater.* **2006**, *18*, 2161.
[26] H. Cao, L. Wang, Y. Qiu, Q. Wu, G. Wang, L. Zhang, X. Liu, *ChemPhysChem* **2006**, *7*, 1500.
[27] X. J. Xu, S. F. Yu, S. P. Lau, L. Li, B. C. Zhao, *J. Phys. Chem. C* **2008**, *112*, 4168.
[28] T. N. Narayanan, M. M. Shaijumon, P. M. Ajayan, M. R. Anantharaman, *J. Phys. Chem. C* **2008**, *112*, 14 281.
[29] M. V. Kamalakar, A. K. Raychaudhuri, *Adv. Mater.* **2008**, *20*, 149.
[30] C.-L. Cheng, J.-S. Lin, Y.-F. Chen, *Mater. Lett.* **2008**, *62*, 1666.
[31] S. Kar, S. Santra, S. Chaudhuri, *Cryst. Growth Des.* **2008**, *8*, 344.
[32] X. Dou, G. Li, X. Huang, L. Li, *J. Phys. Chem. C* **2008**, *112*, 8167.
[33] F. S. Li, D. Zhou, T. Wang, Y. Wang, L. J. Song, C. T. Xu, *J. Appl. Phys.* **2007**, *101*, 014 309.
[34] D. Zhou, L.-H. Cai, F.-S. Wen, F.-S. Li, *Chi. J. Chem. Phys.* **2007**, *20*, 821.
[35] L. Liu, W. Zhou, S. Xie, L. Song, S. Luo, D. Liu, J. Shen, Z. Zhang, Y. Xiang, W. Ma, Y. Ren, C. Wang, G. Wang, *J. Phys. Chem. C* **2008**, *112*, 2256.
[36] Y. Zhu, F. Liu, W. Ding, X. Guo, Y. Chin, *Angew. Chem. Int. Ed.* **2006**, *45*, 7211.
[37] U. K. Gautam, M. Nath, C. N. R. Rao, *J. Mater. Chem.* **2003**, *13*, 2845.
[38] U. K. Gautam, C. N. R. Rao, *J. Mater. Chem.* **2004**, *14*, 2530.
[39] G. Xi, K. Xiong, Q. Zhao, R. Zhang, H. Zhang, Y. Qian, *Cryst. Growth Des.* **2006**, *6*, 577.
[40] H. Zhang, D. Yang, Y. J. Ji, X. Y. Ma, J. Xu, D. L. Que, *J. Phys. Chem. B* **2004**, *108*, 1179.
[41] Y. R. Ma, L. M. Qi, J. M. Ma, H. M. Cheng, *Adv. Mater.* **2004**, *16*, 1023.
[42] X. M. Li, S. Q. Li, W. W. Zhou, H. B. Chu, W. Chen, I. L. Li, Z. K. Tang, *Cryst. Growth Des.* **2005**, *5*, 911.
[43] B. Xi, S. Xiong, H. Fan, X. Wang, Y. Qian, *Cryst. Growth Des.* **2007**, *7*, 1185.
[44] R. Tenne, L. Margulis, M. Genut, G. Hodes, *Nature* **1992**, *360*, 444.
[45] Y. Feldman, E. Wasserman, D. J. Srolovitch, R. Tenne, *Science* **1995**, *267*, 222.
[46] M. Nath, A. Govindaraj, C. N. R. Rao, *Adv. Mater.* **2001**, *13*, 283.
[47] M. Nath, C. N. R. Rao, *Chem. Commun.* **2001**, 2336.

[48] M. Nath, C. N. R. Rao, *J. Am. Chem. Soc.* **2001**, *123*, 4841.
[49] M. Nath, C. N. R. Rao, *Angew. Chem. Int. Ed.* **2002**, *41*, 3451.
[50] M. Nath, C. N. R. Rao, *Pure Appl. Chem.* **2002**, *74*, 1545.
[51] a) F. L. Deepak, A. Margolin, I. Wiesel, M. Bar-Sadan, R. Popovitz-Biro, R. Tenne, *Nano* **2006**, *1*, 167. b) A. Margolin, F. L. Deepak, R. Popovitz-Biro, M. Bar-Sadan, Y. Feldman, R. Tenne, *Nanotechnology* **2008**, *19*, 095 601.
[52] J. M. Gordon, E. A. Katz, D. Feuermann, A. A. Yaron, M. Levy, R. Tenne, *J. Mater. Res.* **2008**, *18*, 458.
[53] U. K. Gautam, S. R. C. Vivekchand, A. Govindaraj, G. U. Kulkarni, N. R. Selvi, C. N. R. Rao, *J. Am. Chem. Soc.* **2005**, *127*, 3658.
[54] H. Pan, C. K. Poh, Y. Zhu, G. Xing, K. C. Chin, Y. P. Feng, J. Lin, C. H. Sow, W. Ji, A. T. S. Wee, *J. Phys. Chem. C.* **2008**, *112*, 11 227.
[55] C. N. R. Rao, A. Govindaraj, F. L. Deepak, N. A. Gunari, M. Nath, *Appl. Phys. Lett.* **2001**, *78*, 1853.
[56] K. P. Kalyanikutty, M. Nikhila, U. Maitra, C. N. R. Rao, *Chem. Phys. Lett.* **2006**, *432*, 190.
[57] C. Wu, S.-H. Yu, S. Chen, G. Liu, B. Liu, *J. Mater. Chem.* **2006**, *16*, 3326.
[58] Q. Shen, L. Jiang, J. Miao, W. Hou, J.-J. Zhu, *Chem. Commun.* **2008**, 1683.
[59] S.-Y. Zhang, C.-X. Fang, Y.-P. Tian, K.-R. Zhu, B.-K. Jin, Y.-H. Shen, J.-X. Yang, *Cryst. Growth Des.* **2006**, *6*, 2089.
[60] H. Niu, M. Gao, *Angew. Chem. Int. Ed.* **2006**, *45*, 6462.
[61] C. H. B. Ng, H. Tan, W. Y. Fan, *Langmuir* **2006**, *22*, 9712.
[62] S.-Y. Zhang, C.-X. Fang, W. Wei, B.-K. Jin, Y.-P. Tian, Y.-H. Shen, J.-X. Yang, H.-W. Gao, *J. Phys. Chem. C* **2007**, *111*, 4168.
[63] A. Qin, Y. Fang, P. Tao, J. Zhang, C. Su, *Inorg. Chem.* **2007**, *46*, 7403.
[64] G. Zhu, P. Liu, J. Zhou, X. Bian, X. Wang, J. Li, B. Chen, *Mater. Lett.* **2008**, *62*, 2335.
[65] C. H. Ye, G. W. Meng, Z. Jiang, Y. H. Wang, G. Z. Wang, L. D. Zhang, *J. Am. Chem. Soc.* **2002**, *124*, 15 180.
[66] R. O. Jyoti, K. S. Suneel, *Nanotechnology* **2005**, *16*, 2415.
[67] S. K. Batabyal, C. Basu, A. R. Das, G. S. Sanyal, *Mater. Lett.* **2006**, *60*, 2582.
[68] F. Xiao, B. Yoo, K. H. Lee, N. V. Myung, *J. Am. Chem. Soc.* **2007**, *129*, 10068.
[69] X.-H. Li, B. Zhou, L. Pu, J.-J. Zhu, *Cryst. Growth Des.* **2008**, *8*, 771.
[70] K. H. Park, J. Choi, H. J. Kim, J. B. Lee, S. Uk Son, *Chem. Mater.* **2007**, *19*, 3861.
[71] H. Tong, Y.-J. Zhu, L.-X. Yang, L. Li, L. Zhang, *Angew. Chem. Int. Ed.* **2006**, *45*, 7739.
[72] C. Li, X. Wang, Q. Peng, Y. Li, *Inorg. Chem.* **2005**, *44*, 6641.
[73] C. Bae, H. Yoo, S. Kim, K. Lee, J. Kim, M. M. Sung, H. Shin, *Chem. Mater.* **2008**, *20*, 756.
[74] H. Ogihara, M. Sadakane, Y. Nodasaka, W. Ueda, *Chem. Mater.* **2006**, *18*, 4981.
[75] V. M. Yuwono, J. D. Hartgerink, *Langmuir* **2007**, *23*, 5033.
[76] Q. Ji, R. Iwaura, T. Shimizu, *Chem. Mater.* **2007**, *19*, 1329.
[77] M. Yamanaka, Y. Miyake, S. Akita, K. Nakano, *Chem. Mater.* **2008**, *20*, 2072.
[78] G. Shen, Y. Bando, D. Golberg, *J. Phys. Chem. B* **2006**, *110*, 23 170.
[79] T. Zhai, Z. Gu, Y. Dong, H. Zhong, Y. Ma, H. Fu, Y. Li, J. Yao, *J. Phys. Chem. C* **2007**, *111*, 11 604.
[80] H. Chen, Y. A. Elabd, G. R. Palmese, *J. Mater. Chem.* **2007**, *17*, 1593.
[81] Y. Chen, B. Li, X. Wu, X. Zhu, M. Suzuki, K. Hanabusa, Y. Yang, *Chem. Commun.* **2008**, 4948.
[82] X. Wu, J. Ruan, T. Ohsuna, O. Terasaki, S. Che, *Chem. Mater.* **2007**, *19*, 1577.
[83] E. Pouget, E. Dujardin, A. Cavalier, A. Moreac, C. Valéry, V. Marchi-Artzner, T. Weiss, A. Renault, M. Paternostre, F. Artzner, *Nature Mater.* **2007**, *6*, 434.
[84] B. He, S. J. Son, S. B. Lee, *Langmuir* **2006**, *22*, 8263.
[85] H.-Y. Tuan, A. Ghezelbash, B. A. Korgel, *Chem. Mater.* **2008**, *20*, 2306.
[86] L. K. Tan, M. A. S. Chong, H. Gao, *J. Phys. Chem. C* **2008**, *112*, 69.
[87] C. A. Grimes, *J. Mater. Chem.* **2007**, *17*, 1451.
[88] S. Yoriya, M. Paulose, O. K. Varghese, G. K. Mor, C. A. Grimes, *J. Phys. Chem. C* **2007**, *111*, 13 770.
[89] M. Paulose, H. E. Prakasam, O. K. Varghese, L. Peng, K. C. Popat, *J. Phys. Chem. C* **2007**, *111*, 14 992.
[90] J. Wang, Z. Lin, *Chem. Mater.* **2008**, *20*, 1257.
[91] S. P. Albu, D. Kim, P. Schmuki, *Angew. Chem. Int. Ed.* **2008**, *47*, 1916.
[92] S. P. Albu, A. Ghicov, J. M. Macak, P. Schmuki, *Phys. Status Solidi RRL* **2007**, *1*, R65.
[93] J. M. Macak, S. P. Albu, P. Schmuki, *Phys. Status Solidi RRL* **2007**, *1*, 181.
[94] N. K. Allam, C. A. Grimes, *J. Phys. Chem. C* **2007**, *111*, 13 028.
[95] X. Tang, D. Li, *J. Phys. Chem. C* **2008**, *112*, 5405.
[96] N. Lu, X. Quan, J. Y. Li, S. Chen, H. T. Yu, G. H. Chen, *J. Phys. Chem. C* **2007**, *111*, 11 836.
[97] J. Qiu, W. Yu, X. Gao, X. Li, *Nanotechnology* **2007**, *18*, 295 604.
[98] C. Bae, S. Kim, B. Ahn, J. Kim, M. M. Sung, H. Shin, *J. Mater. Chem.* **2008**, *18*, 1362.
[99] L. Yue, W. Gao, D. Zhang, X. Guo, W. Ding, Y. Chen, *J. Am. Chem. Soc.* **2006**, *128*, 11 042.
[100] A. Yu, G. Q. M. Lu, J. Drennan, I. R. Gentle, *Adv. Funct. Mater.* **2007**, *17*, 2600.
[101] M. Suzuki, Y. Nakajima, T. Sato, H. Shirai, K. Hanabusa, *Chem. Commun.* **2006**, 377.
[102] C. Feng, Y. Wang, Z. Jin, J. Zhang, S. Zhang, Z. Wu, Z. Zhang, *New J. Chem.* **2008**, *32*, 1038.
[103] B. C. Satishkumar, A. Govindaraj, E. M. Vogl, L. Basumallick, C. N. R. Rao, *J. Mater. Res.* **1997**, *12*, 604.
[104] B. C. Satishkumar, A. Govindaraj, M. Nath, C. N. R. Rao, *J. Mater. Chem.* **2000**, *10*, 2115.
[105] D. Eder, I. A. Kinloch, A. H. Windle, *Chem. Commun.* **2006**, 1448.
[106] Q. Deng, M. Wei, X. Ding, L. Jiang, B. Ye, K. Wei, *Chem. Commun.* **2008**, 3657.
[107] M. Miyauchi, H. Tokudome, *J. Mater. Chem.* **2007**, *17*, 2095.
[108] Q. Yu, W. Fu, C. Yu, H. Yang, R. Wei, M. Li, S. Liu, Y. Sui, Z. Liu, M. Yuan, G. Zou, G. Wang, C. Shao, Y. Liu, *J. Phys. Chem. C* **2007**, *111*, 17 521.
[109] C.-H. Ku, J.-J. Wu, *J. Phys. Chem. B* **2006**, *110*, 12 981.
[110] S. Kar, S. Santra, *J. Phys. Chem. C* **2008**, *112*, 8144.
[111] L. Li, S. Pan, X. Dou, Y. Zhu, X. Huang, Y. Yang, G. Li, L. Zhang, *J. Phys. Chem. C* **2007**, *111*, 7288.
[112] L. Xu, Q. Liao, J. Zhang, X. Ai, D. Xu, *J. Phys. Chem. C* **2007**, *111*, 4549.
[113] Y. Tong, Y. Liu, C. Shao, Y. Liu, C. Xu, J. Zhang, Y. Lu, D. Shen, X. Fan, *J. Phys. Chem. B* **2006**, *110*, 14 714.
[114] L. Shen, N. Bao, K. Yanagisawa, K. Domen, C. A. Grimes, A. Gupta, *J. Phys. Chem. C* **2007**, *111*, 7280.
[115] H. Wang, G. Li, L. Jia, G. Wang, C. Tang, *J. Phys. Chem. C* **2008**, *112*, 11 738.
[116] X. F. Zhou, Z. L. Hu, Y. Chen, H. Y. Shang, *Mater. Res. Bull.* **2008**, *43*, 2790.
[117] K. Raidongia, C. N. R. Rao, *J. Phys. Chem. C* **2008**, *112*, 13 366.
[118] H. B. Lu, L. Liao, H. Li, Y. Tian, D. F. Wang, J. C. Li, Q. Fu, B. P. Zhu, Y. Wu, *Mater. Lett.* **2008**, *62*, 3928.
[119] W. Lee, R. Scholz, U. Gosele, *Nano Lett.* **2008**, *8*, 2155.
[120] C.-C. Wang, C.-C. Kei, Y.-W. Yu, T.-P. Perng, *Nano Lett.* **2007**, *7*, 1566.
[121] J. Galy, *J. Solid State Chem.* **1992**, *100*, 229.
[122] M. S. Whittingham, J. Guo, R. Chen, T. Chirayil, G. Janauer, P. Y. Zavalji, *Solid State Ionics* **1995**, *75*, 257.
[123] M. E. Spahr, P. Bitterli, R. Nesper, M. Müller, F. Krumeich, H. U. Nissen, *Angew. Chem. Int. Ed.* **1998**, *37*, 1263. *Angew. Chem.* **1998**, *110*, 1339.
[124] F. Krumeich, H.-J. Muhr, M. Niederberger, F. Bieri, B. Schnyder, R. Nesper, *J. Am. Chem. Soc.* **1999**, *121*, 8324.
[125] F. Sediri, F. Touati, N. Gharbi, *Mater. Lett.* **2007**, *61*, 1946.
[126] H.-J. Muhr, F. Krumeich, U. P. Schönholzer, F. Bieri, M. Niederberger, L. J. Gauckler, R. Nesper, *Adv. Mater.* **2000**, *12*, 231.
[127] M. Niederberger, H.-J. Muhr, F. Krumeich, F. Bieri, D. Günther, R. Nesper, *Chem. Mater.* **2000**, *12*, 1995.
[128] A. Dobley, K. Ngala, S. Yang, P. Y. Zavalji, M. S. Whittingham, *Chem. Mater.* **2001**, *13*, 4382.

[129] C. Zhou, L. Mai, Y. Liu, Y. Qi, Y. Dai, W. Chen, *J. Phys. Chem. C* **2007**, *111*, 8202.
[130] C. O'Dwyer, D. Navas, V. Lavayen, E. Benavente, M. A. Santa Ana, G. González, S. B. Newcomb, C. M. S. Torres, *Chem. Mater.* **2006**, *18*, 3016.
[131] C. Yan, D. Xue, *Adv. Mater.* **2008**, *20*, 1055.
[132] J. Luo, H. T. Zhu, H. M. Fan, J. K. Liang, H. L. Shi, G. H. Rao, J. B. Li, Z. M. Du, Z. X. Shen, *J. Phys. Chem. C* **2008**, *112*, 12 594.
[133] L. Liu, H.-Z. Kou, W. Mo, H. Liu, Y. Wang, *J. Phys. Chem. B* **2006**, *110*, 15 218.
[134] H. Zhou, S. S. Wong, *ACS Nano* **2008**, *2*, 944.
[135] J. Bachmann, J. Jing, M. Knez, S. Barth, H. Shen, S. Mathur, U. Gosele, K. Nielsch, *J. Am. Chem. Soc.* **2007**, *129*, 9554.
[136] X. Yu, C. Cao, X. An, *Chem. Mater.* **2008**, *20*, 1936.
[137] J. Wang, Y. Ma, K. Watanabe, *Chem. Mater.* **2008**, *20*, 20.
[138] T. Wang, Y. Wang, F. Li, C. Xu, D. Zhou, *J. Phys. Condens. Matter* **2006**, *18*, 10 545.
[139] F. Geng, Z. Zhao, J. Geng, H. Cong, H.-M. Cheng, *Mater. Lett.* **2007**, *61*, 4794.
[140] X. W. Lou, D. Deng, J. Y. Lee, J. Feng, L. A. Archer, *Adv. Mater.* **2008**, *20*, 258.
[141] X.-P. Shen, H.-J. Miao, H. Zhao, Z. Xu, *Appl. Phys. A* **2008**, *91*, 47.
[142] Y. Xu, J. Wei, J. Yao, J. Fu, D. Xue, *Mater. Lett.* **2008**, *62*, 1403.
[143] G. Malandrino, L. M. S. Perdicaro, I. L. Fragala, R. Lo Nigro, M. Losurdo, G. Bruno, *J. Phys. Chem. C* **2007**, *111*, 3211.
[144] C. Shi, G. Wang, N. Zhao, X. Du, J. Li, *Chem. Phys. Lett.* **2008**, *454*, 75.
[145] Z.-G. Zhao, M. Miyauchi, *Angew. Chem. Int. Ed.* **2008**, *47*, 7051.
[146] Y. Yan, L. Zhou, J. Zhang, H. Zeng, Y. Zhang, L. Zhang, *J. Phys. Chem. C* **2008**, *112*, 10 412.
[147] H. B. Lu, L. Liao, H. Li, Y. Tian, J. C. Li, D. F. Wang, B. P. Zhu, *J. Phys. Chem. C* **2007**, *111*, 10 273.
[148] C. Chen, D. Chen, X. Jiao, C. Wang, *Chem. Commun.* **2006**, 4632.
[149] M. Zhong, M. Zheng, L. Ma, Y. Li, *Nanotechnology* **2007**, *18*, 465 605.
[150] N. Du, H. Zhang, B. Chen, X. Ma, Z. Liu, J. Wu, D. Yang, *Adv. Mater.* **2007**, *19*, 1641.
[151] V. V. Rajasekharan, D. A. Buttry, *Chem. Mater.* **2006**, *18*, 4541.
[152] Y. Mao, J. Y. Huang, R. Ostroumov, K. L. Wang, J. P. Chang, *J. Phys. Chem. C* **2008**, *112*, 2278.
[153] N. Du, H. Zhang, B. Chen, X. Ma, D. Yang, *Chem. Commun.* **2008**, 3028.
[154] N. Wang, X. Cao, L. Guo, *J. Phys. Chem. C* **2008**, *112*, 12 616.
[155] M. Lai, J. A. G. Martinez, M. Gratzel, D. J. Riley, *J. Mater. Chem.* **2006**, *16*, 2843.
[156] Y. Aoki, J. Huang, T. Kunitake, *J. Mater. Chem.* **2006**, *16*, 292.
[157] J. Zhao, X. Wang, R. Xu, F. Meng, L. Guo, Y. Li, *Mater. Lett.* **2008**, *62*, 4428.
[158] X. Meng, X. Tan, B. Meng, N. Yang, Z.-F. Ma, *Mater. Chem. Phys.* **2008**, *111*, 275.
[159] W. Q. Han, L. J. Wu, Y. M. Zhu, *J. Am. Chem. Soc.* **2005**, *127*, 12 814.
[160] K. Zhou, Z. Yang, S. Yang, *Chem. Mater.* **2007**, *19*, 1215.
[161] R. Inguanta, S. Piazza, C. Sunseri, *Nanotechnology* **2007**, *18*, 485 605.
[162] Z.-W. Lin, Q. Kuang, W. Lian, Z.-Y. Jiang, Z.-X. Xie, R.-B. Huang, L.-S. Zheng, *J. Phys. Chem. B* **2006**, *110*, 23 007.
[163] N. Bao, L. Shen, G. Srinivasan, K. Yanagisawa, A. Gupta, *J. Phys. Chem. C* **2008**, *112*, 8634.
[164] Y. Yang, X. Wang, C. Zhong, C. Sun, L. Li, *Appl. Phys. Lett.* **2008**, *92*, 122 907.
[165] S. Singh, S. B. Krupanidhi, *Appl. Phys. A* **2007**, *87*, 27.
[166] A. Nourmohammadi, M. A. Bahrevar, M. Hietschold, *Mater. Lett.* **2008**, *62*, 3349.
[167] K. Yasuda, P. Schmuki, *Adv. Mater.* **2007**, *19*, 1757.
[168] G. B. Saupe, C. C. Waraksa, H. N. Kim, Y. J. Han, D. M. Kaschak, D. M. Skinner, T. E. Mallouk, *Chem. Mater.* **2000**, *12*, 1556.
[169] G. Du, Q. Chen, Y. Yu, S. Zhang, W. Zhu, L. M. Peng, *J. Mater. Chem.* **2004**, *14*, 1437.
[170] Z. Tong, S. Takagi, T. Shimada, H. Tachibana, H. Inoue, *J. Am. Chem. Soc.* **2006**, *128*, 684.
[171] D. Yu, J. Qian, N. Xue, D. Zhang, C. Wang, X. Guo, W. Ding, Y. Chen, *Langmuir* **2007**, *23*, 382.
[172] J. Yu, A. Wang, J. Tan, X. Li, J. A. V. Bokhoven, Y. Hu, *J. Mater. Chem.* **2008**, *18*, 3601.
[173] Y. Fang, X. Wen, S. Yang, *Angew. Chem. Int. Ed.* **2006**, *45*, 4655.
[174] C. Yan, T. Zhang, P. S. Lee, *Cryst. Growth Des.* **2008**, *8*, 3144.
[175] Y. Wang, G. Cao, *J. Mater. Chem.* **2007**, *17*, 894.
[176] Z. Wang, S. Zhou, L. Wu, *Adv. Funct. Mater.* **2007**, *17*, 1790.
[177] B. G. Olson, J. J. Decker, S. Nazarenko, V. E. Yudin, J. U. Otaigbe, E. N. Korytkova, V. V. Gusarov, *J. Phys. Chem. C* **2008**, *112*, 12 943.
[178] C. Levard, J. Rose, A. Masion, E. Doelsch, D. Borschneck, L. Olivi, C. Dominici, O. Grauby, J. C. Woicik, J.-Y. Bottero, *J. Am. Chem. Soc.* **2008**, *130*, 5862.
[179] T. Yu, J. Park, J. Moon, K. An, Y. Piao, T. Hyeon, *J. Am. Chem. Soc.* **2007**, *129*, 14 558.
[180] Y. Zhao, R. L. Frost, W. N. Martens, H. Y. Zhu, *Langmuir* **2007**, *23*, 9850.
[181] Y. Yuan, C. Liu, Y. Zhang, X. Shan, *Mater. Chem. Phys.* **2008**, *112*, 275.
[182] W. Liu, S. Wang, Y. Chen, G. Fang, M. Li, X.-Z. Zhao, *Sensors Actuators B* **2008**, *134*, 62.
[183] H. J. Fan, M. Knez, R. Scholz, K. Nielsch, E. Pippel, D. Hesse, M. Zacharias, U. Gosele, *Nat. Mater.* **2006**, *5*, 627. .
[184] C. Ye, Y. Bando, G. Shen, D. Golberg, *Angew. Chem. Int. Ed.* **2006**, *45*, 4922.
[185] S. Song, Y. Zhang, Y. Xing, C. Wang, J. Feng, W. Shi, G. Zheng, H. Zhang, *Adv. Funct. Mater.* **2008**, *18*, 2328.
[186] S. Mukherjee, K. Kim, S. Nair, *J. Am. Chem. Soc.* **2007**, *129*, 6820.
[187] W. Chen, X.-H. Xia, *Adv. Funct. Mater.* **2007**, *17*, 2943.
[188] Y. Chen, Q. Wu, *Colloids Surf. A* **2008**, *325*, 33.
[189] K.-W. Huang, J.-H. Wang, H.-C. Chen, H.-C. Hsu, Y.-C. Chang, M.-Y. Lu, C.-Y. Lee, L.-J. Chen, *J. Mater. Chem.* **2007**, *17*, 2307.
[190] R. Chen, M. H. So, J. Yang, F. Deng, C.-M. Che, H. Sun, *Chem. Commun.* **2006**, 2265.
[191] A. Loiseau, F. Williame, N. Demonecy, G. Hug, H. Pascard, *Phys. Rev. Lett.* **1996**, *76*, 4737.
[192] N. G. Chopra, R. G. Luyken, K. Cherrey, V. H. Crespi, M. L. Cohen, S. G. Louie, A. Zettl, *Science* **1995**, *269*, 966.
[193] M. Terrones, W. K. Hsu, H. Terrones, J. P. Zhang, S. Ramos, J. P. Hare, R. Castillo, K. Prassides, A. K. Cheetham, H. W. Kroto, D. R. M. Walton, *Chem. Phys. Lett.* **1996**, *259*, 568.
[194] a) D. P. Yu, X. S. Sun, C. S. Lee, I. Bello, S. T. Lee, H. D. Gu, K. M. Leung, G. W. Zhou, Z. F. Dong, Z. Zhang, *Appl. Phys. Lett.* **1998**, *72*, 1966. b) G. W. Zhou, Z. Zhang, Z. G. Bai, D. P. Yu, *Solid State Commun.* **1999**, *109*, 555. c) M. Terauchi, M. Tanaka, H. Matsuda, M. Takeda, K. Kimura, *J. Electron Microsc.* **1997**, *1*, 75.
[195] T. Laude, Y. Matsui, A. Marraud, B. Jouffrey, *Appl. Phys. Lett.* **2000**, *76*, 3239.
[196] a) D. Golberg, Y. Bando, W. Han, K. Kurashima, T. Sato, *Chem. Phys. Lett.* **1999**, *308*, 337. b) P. W. Fowler, K. M. Rogers, G. Seifert, M. Terrones, H. Terrones, *Chem. Phys. Lett.* **1999**, *299*, 359.
[197] E. Bengu, L. D. Marks, *Phys. Rev. Lett.* **2001**, *86*, 2385.
[198] C. C. Tang, Y. Bando, T. Sato, K. Kurashima, *Chem. Commun.* **2002**, 1290.
[199] C. Y. Zhi, Y. Bando, C. Tang, D. Golberg, *Solid State Commun.* **2005**, *135*, 67.
[200] F. L. Deepak, C. P. Vinod, K. Mukhopadhyay, A. Govindaraj, C. N. R. Rao, *Chem. Phys. Lett.* **2002**, *353*, 345.
[201] J. Wang, V. K. Kayastha, Y. K. Yap, Z. Fan, J. G. Lu, Z. Pan, I. N. Ivanov, A. A. Puretzky, D. B. Geohegan, *Nano Lett.* **2005**, *5*, 2528.
[202] S. Pal, S. R. C. Vivekchand, A. Govindaraj, C. N. R. Rao, *J. Mater. Chem.* **2007**, *17*, 450.
[203] a) D. Golberg, Y. Bando, C. Tang, C. Zhi, *Adv. Mater.* **2007**, *19*, 2413. b) D. Golberg, Y. Bando, L. Bourgeois, K. Kurashima, T. Sato, *Appl. Phys. Lett.* **2000**, *77*, 1979.

[204] J. Dai, L. Xu, Z. Fang, D. Sheng, Q. Guo, Z. Ren, K. Wang, Y. Qian, *Chem. Phys. Lett.* **2007**, *440*, 253.
[205] X. Chen, X. Wang, J. Liu, Z. Wang, Y. Qian, *Appl. Phys. A* **2005**, *81*, 1035.
[206] T. Oku, N. Koi, K. Suganuma, *Diamond Relat. Mater.* **2008**, *17*, 1805.
[207] X. Z. Wang, Q. Wu, Z. Hu, Y. Chen, *Electrochem. Acta* **2007**, *52*, 2841.
[208] H. Chen, Y. Chen, Y. Liu, L. Fu, C. Huang, D. Llewellyn, *Chem. Phys. Lett.* **2008**, *463*, 130.
[209] M. Bechelany, S. Bernard, A. Brioude, D. Cornu, P. Stadelmann, C. Charcosset, K. Fiaty, P. Miele, *J. Phys. Chem. C* **2007**, *111*, 13 378.
[210] M. J. Kim, S. Chatterjee, S. M. Kim, E. A. Stach, M. G. Bradley, M. J. Pender, L. G. Sneddon, B. Maruyama, *Nano Lett.* **2008**, *8*, 3298.
[211] R. Arenal, O. Stephan, J.-L. Cochon, A. Loiseau, *J. Am. Chem. Soc.* **2007**, *129*, 16 183.
[212] W. L. Wang, X. D. Bai, K. H. Liu, Z. Xu, D. Golberg, Y. Bando, E. G. Wang, *J. Am. Chem. Soc.* **2006**, *128*, 6530.
[213] K. Raidongia, D. Jagadeesan, M. Upadhyay-Kahaly, U. V. Waghmare, S. K. Pati, M. Eswaramoorthy, C. N. R. Rao, *J. Mater. Chem.* **2008**, *18*, 83.
[214] W.-G. Jung, S.-H. Jung, P. Kung, M. Razeghi, *Nanotechnology* **2006**, *17*, 54.
[215] L. Gai, H. Jiang, W. Ma, D. Cui, N. Lun, Q. Wang, *J. Phys. Chem. C* **2007**, *111*, 2386.
[216] J. Dinesh, M. Eswaramoorthy, C. N. R. Rao, *J. Phys. Chem. C* **2007**, *111*, 510.
[217] F. Wang, G.-Q. Jin, X.-Y. Guo, *Mater. Lett.* **2006**, *60*, 330.
[218] V. Miikkulainen, M. Suvanto, T. A. Pakkanen, *Thin Solid Films* **2008**, *516*, 6041.
[219] G. Shen, Y. Bando, C. Ye, X. Yuan, T. Sekiguchi, D. Golberg, *Angew. Chem. Int. Ed.* **2006**, *45*, 7568.
[220] C. D. Malliakas, M. G. Kanatzidis, *J. Am. Chem. Soc.* **2006**, *128*, 6538.
[221] B. Li, R. Wu, Y. Pan, L. Wu, G. Yang, J. Chen, Q. Zhu, *J. Alloys Compounds* **2008**, *462*, 446.
[222] T. Taguchi, H. Yamamoto, *J. Phys. Chem. C* **2007**, *111*, 18 888.
[223] C. Wan, G. Guo, Q. Zhang, *Mater. Lett.* **2008**, *62*, 2776.
[224] L. Liao, Z. Xu, K. H. Liu, W. L. Wang, S. Liu, X. D. Bai, E. G. Wang, J. C. Li, C. Liu, *J. Appl. Phys.* **2007**, *101*, 114 306.
[225] Y. R. Hacohen, E. Grunbaum, R. Tenne, J. Sloan, J. L. Hutchison, *Nature* **1998**, *395*, 336.
[226] J. Zeng, C. Liu, J. Huang, X. Wang, S. Zhang, G. Li, J. Hou, *Nano Lett.* **2008**, *8*, 1318.
[227] D. H. Fan, W. Z. Shen, M. J. Zheng, Y. F. Zhu, J. J. Lu, *J. Phys. Chem. C* **2007**, *111*, 9116.
[228] C. Yan, D. Xue, *J. Phys. Chem. B* **2006**, *110*, 25 850.
[229] Y. Yin, Z. Jin, F. Hou, *Nanotechnology* **2007**, *18*, 495 608.
[230] U. K. Gautam, X. Fang, Y. Bando, J. Zhan, D. Golberg, *ACS Nano* **2008**, *2*, 1015.
[231] H.-C. Liao, P.-C. Kuo, C.-C. Lin, S.-Y. Chen, *J. Vac. Sci. Technol.* **2006**, *24*, 2198.
[232] D. Yang, G. Meng, S. Zhang, Y. Hao, X. An, Q. Wei, M. Ye, L. Zhang, *Chem. Commun.* **2007**, 1733.
[233] L. Yang, S. Luo, S. Liu, Q. Cai, *J. Phys. Chem. C* **2008**, *112*, 8939.
[234] H. Chen, S. Chen, X. Quan, H. Yu, H. Zhao, Y. Zhang, *J. Phys. Chem. C* **2008**, *112*, 9285.
[235] T. Djenizian, I. Hanzu, Y. D. Premchand, *Nanotechnology* **2008**, *19*, 205 601.
[236] S. K. Mohapatra, S. Banerjee, M. Misra, *Nanotechnology* **2008**, *19*, 315 601.
[237] Y.-L. Chueh, L.-J. Chou, Z. L. Wang, *Angew. Chem. Int. Ed.* **2006**, *45*, 7773.
[238] L. Liao, K. Liu, W. Wang, X. Bai, E. Wang, Y. Liu, J. Li, C. Liu, *J. Am. Chem. Soc.* **2007**, *129*, 9562.
[239] Z.-G. Chen, J. Zou, F. Li, G. Liu, D.-M. Tang, D. Li, C. Liu, X. Ma, H.-M. Cheng, G. Q. Lu, Z. Zhang, *Adv. Funct. Mater.* **2007**, *17*, 3371.

ACCOUNTS
of chemical research

# Synthesis and Selected Properties of Graphene and Graphene Mimics

C. N. R. RAO,* H. S. S. RAMAKRISHNA MATTE, AND
K. S. SUBRAHMANYAM

*Chemistry and Physics of Materials Unit, CSIR Centre of Excellence in Chemistry and International Centre for Materials Science, Jawaharlal Nehru Centre for Advanced Scientific Research, Jakkur P.O., Bangalore 560 064, India*

RECEIVED ON JANUARY 29, 2012

## CONSPECTUS

Graphene has generated great excitement in the last few years because of its novel properties with potential applications. Graphene exhibits an ambipolar electric field effect, ballistic conduction of charge carriers, and the quantum Hall effect at room temperature. Some of the other interesting characteristics of graphene include high transparency toward visible light, high elasticity and thermal conductivity, unusual magnetic properties, and charge transfer interactions with molecules.

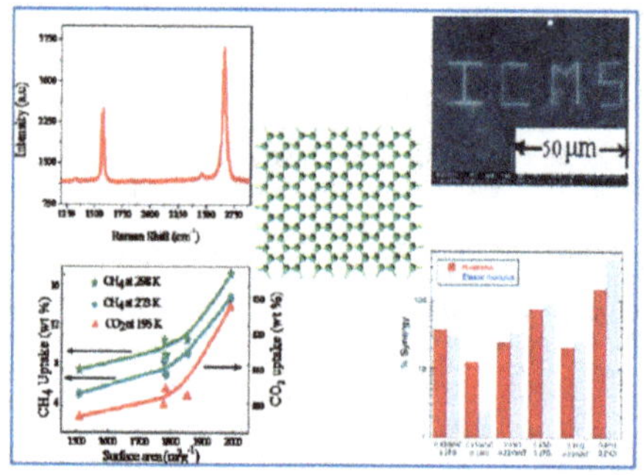


In this Account, we present the highlights of some of our research on the synthesis of graphene and its properties. Since the isolation and characterization of graphene by micromechanical cleavage from graphite, several strategies have been developed for the synthesis of graphene with either a single or just a few layers. The most significant contribution from our laboratory is the synthesis of two to four layer graphene by arc-discharge of graphite in a hydrogen atmosphere. Besides providing clean graphene surfaces, this method allows for doping with boron and nitrogen. UV and laser irradiation of graphene oxide provides fairly good graphene samples, and laser unzipping of nanotubes produces graphene nanoribbons. We have exploited Raman spectroscopy to investigate the charge-transfer interactions of graphene with electron-donor and -acceptor molecules, as well as with nanoparticles of noble metals. Graphene quenches the fluorescence of aromatics because of electron transfer or energy transfer.

Notable potential applications of the properties of graphene are low turn-on field emission and radiation detection. High-temperature ferromagnetism is another intriguing feature of graphene. Although incorporation of graphene improves the mechanical properties of polymers, its incorporation with nanodiamond or carbon nanotubes exhibits extraordinary synergy. The potential of graphene and its analogues as adsorbents and chemical storage materials for $H_2$ and $CO_2$ is noteworthy.

## Introduction

Graphene has generated great excitement in the last few years owing to its novel properties with potential applications.[1] Graphene exhibits ambipolar electric field effect, ballistic conduction of charge carriers, and quantum Hall effect at room temperature.[2] Some of the other interesting aspects of graphene include high transparency toward visible light, high values of elasticity and thermal conductivity, unusual magnetic properties, and charge transfer interaction with molecules.[3,4] While graphene research is of recent origin, synthesis and modification of graphite-related materials have received the attention of chemistry for some decades.[4] Thus, early theoretical studies predicted some unusual electronic properties of graphene.[5] Procedures employed to generate single-layer graphene (SLG) in the early years were not successful.[6] It is only in 2004 that Novoselov et al.[1,2] successfully isolated single- and few-layer graphenes from highly oriented pyrolytic graphite (HOPG) by micromechanical cleavage and reported their unusual electronic characteristics. This technique has since been extended to other layered materials such as metal dichalcogenides.[7] In the past few years, several other strategies have been developed for the synthesis of graphene. A variety of microscopic and spectroscopic techniques have been

Published on the Web 06/27/2012 www.pubs.acs.org/accounts
10.1021/ar300033m 

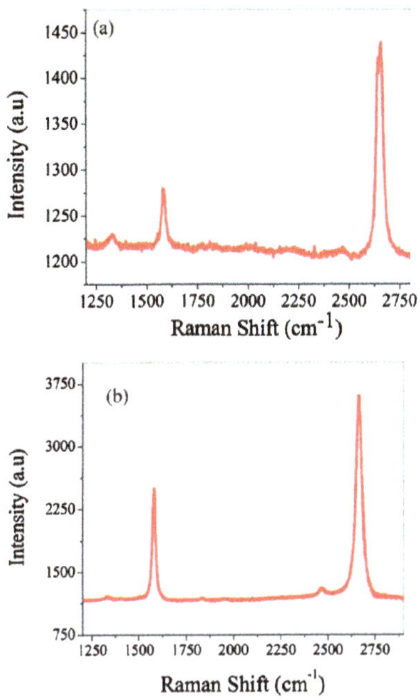


**FIGURE 1.** Raman spectra of graphene prepared by the thermal decomposition of (a) methane (70 sccm) and (b) benzene (argon passed through benzene with flow rate of 200 sccm) at 1000 °C on a nickel sheet. From ref 11.

employed to determine the purity and the number of layers in graphene samples.[8−10] Raman spectroscopy is an invaluable tool to characterize graphene in terms of both the number of layers and the electronic structure. We started research on graphene in 2006 and have since explored the synthesis and several of its properties. In this Account, we present the highlights of some of the results obtained by us on various aspects of graphene.

## Synthesis

SLG has been synthesized by several methods such as epitaxial growth, chemical vapor deposition (CVD), solution phase exfoliation of graphite, and reduction of single-layer graphene oxide (SGO).[3] We have carried out CVD using various hydrocarbons on different metal catalysts[11] and found the growth of graphene films to depend on the hydrocarbon source and reaction conditions. On a nickel foil, CVD of methane (60−70 sccm) or ethylene (4−8 sccm) diluted with hydrogen (500 sccm) at 1000 °C for 5−10 min yielded SLG. In the case of benzene, vapors diluted with argon and hydrogen were decomposed at 1000 °C for 5 min. On a cobalt foil, acetylene (4 sccm) and methane (65 sccm)

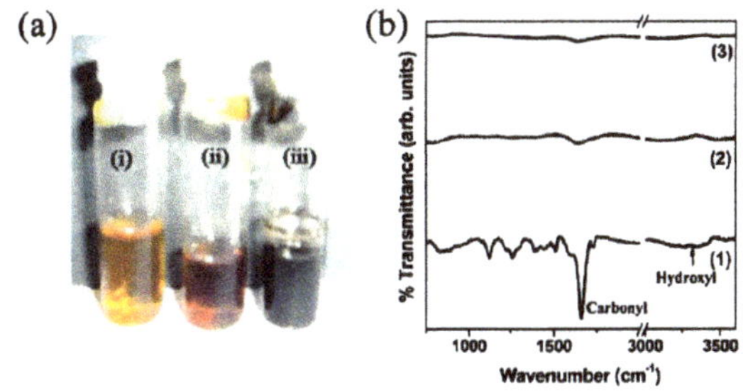


**FIGURE 2.** (a) Photographs of SGO (i) before and (ii) after irradiation with an excimer laser (300 mJ, 1000 shots) and (iii) after 18000 shots. (b) IR spectra of (1) SGO and SGO (2) after ultraviolet light exposure for 2 h and (3) after excimer laser treatment for 40 min. From ref 14.

diluted with hydrogen (500 sccm) were decomposed at 800 and 1000 °C, respectively. Figure 1 shows the Raman spectra of graphene sheets obtained by CVD on a nickel foil. The narrow line width (30−40 $cm^{-1}$) and relatively high intensity of the 2D band (~2670 $cm^{-1}$) indicate that these graphene samples predominantly contain single layers.[12]

A standard method employed for the synthesis of SLG involves the exfoliation of graphite oxide, followed by reduction.[13] The product obtained by this method is reduced graphene oxide (RGO), rather than graphene, since the surface of the graphene contains considerable oxygen functionalities. Reduction of SGO is generally carried out by using hydrazine hydrate as the reducing agent in dimethylformamide medium. We have found that microwave irradiation during reduction of SGO is effective, and it takes only 10 min for the reduction. RGO formed by this method has fewer defects as revealed by Raman spectroscopy. Oxygen functionalities in SGO can be minimized by irradiation with ultraviolet radiation or excimer laser.[14,15] In order to do so, a SGO solution is irradiated with a mercury lamp (254 nm, 25 W, 90 $\mu W/cm^2$). In the IR spectrum, bands due to carbonyl stretching and other oxygen-containing functional groups decrease in intensity markedly after ultraviolet irradiation for 2 h. Some changes are observed in the FESEM images, and more graphene edges become visible.[14,16] For laser irradiation, a KrF excimer laser (248 nm, 5 Hz) is generally employed. Figure 2a shows how the yellow colored solution of graphene oxide turns black on irradiation with the laser. Infrared spectra in Figure 2b show similar effects by irradiation with ultraviolet light (for ~2 h) and by excimer laser radiation (30−40 min). This technique has been usefully employed for patterning of graphene oxide (Figure 3a).[16] We have also carried out e-beam patterning of graphene oxide films to obtain patterns of RGO as thin as 250 nm

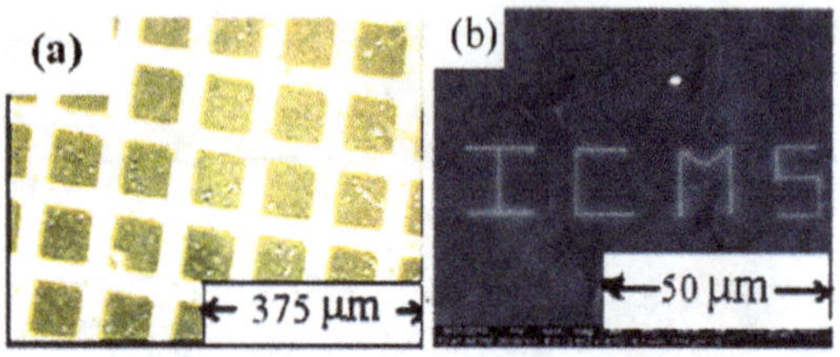


**FIGURE 3.** (a) Optical microscope image of the pattern achieved after excimer laser reduction of graphene oxide and (b) electron-beam pattern with 250 nm wide lines of RGO on GO films. From ref 16.

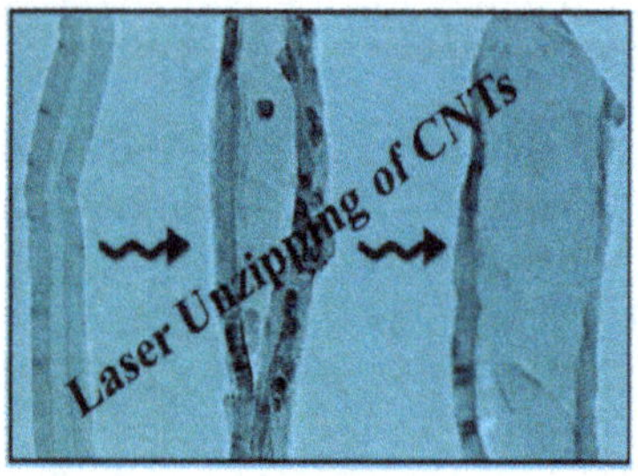


**FIGURE 4.** The process of unzipping of carbon nanotubes on irradiation with laser. From ref 17.

(Figure 3b). Prior to the irradiation, the material is insulating, and it becomes conducting after irradiation. Another important use of laser irradiation is to unzip carbon nanotubes to obtain graphene nanoribbons (GNRs).[17] Figure 4 shows the process of laser unzipping of carbon nanotubes.

The most common method to produce few-layer graphene is by exfoliation of graphite oxide at high temperatures.[18] We designate the few-layer graphene prepared by this procedure as exfoliated graphene (EG) and show typical TEM and AFM images in Figure 5a,b, respectively. Another approach to prepare few-layer graphene (DG) is thermal conversion of nanodiamond in the temperature range of 1650–2200 °C in helium atmosphere.[11,18,19] We have developed a method to prepare graphene samples with 2–4 layers by carrying out arc-evaporation of graphite in a hydrogen atmosphere (HG).[20] The presence of hydrogen during the arc-discharge process terminates the dangling carbon bonds preventing the formation of closed structures. In a typical experiment, a graphite rod (6 mm in diameter and 50 mm long) was used as the anode, and another graphite rod (13 mm in diameter and 60 mm in length) was used as the cathode. Figure 5c,d shows TEM and AFM images, respectively, of graphene sheets with two to three layers.

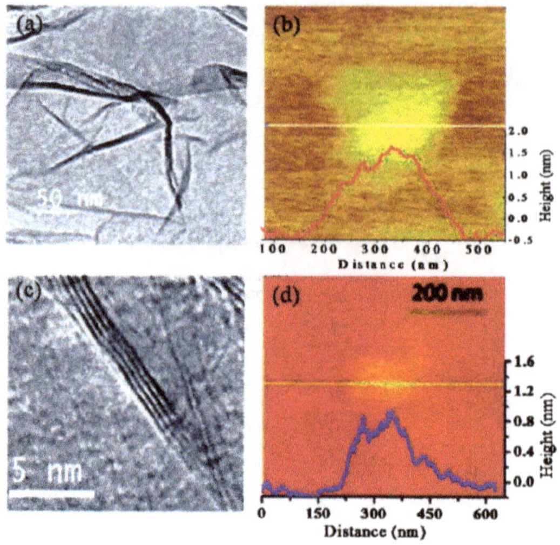


**FIGURE 5.** TEM images of (a) EG and (c) HG and AFM images of (b) EG and (d) HG. From ref 22.

## Surface Area and Gas Adsorption

Brunauer–Emmet–Teller (BET) surface areas of few-layer graphenes are in the range of 270–1550 $m^2/g$, with the trend EG > DG > RGO > HG.[21,22] In Figure 6a, we show the $N_2$ adsorption data of EG with a surface area of 1250 $m^2/g$ at 77 K. Uptake of $CO_2$ by graphene samples varies between 5 and 45 wt % (at 195 K and 1 atm), with EG exhibiting the highest uptake. The uptake of $CO_2$ by EG at 298 K and 50 bar is 51 wt %.[22] In Figure 6b, we show typical $CO_2$ adsorption data of an EG sample at 195 K, 1 atm. $H_2$ adsorption measurements show that EG, DG, and HG absorb 1.7, 1.2, and 1.0 wt % of $H_2$ at 1 atm and 77 K and 3.1, 2.5, and 2.0 wt % at at 100 bar and 300 K. The uptake of $CH_4$ by graphene samples varies between 0 and 3 wt %, EG showing the highest value.[22]

## Charge-Transfer Interaction with Donor and Acceptor Molecules

Being an extended $\pi$-system, graphene exhibits charge-transfer interactions with electron-donor and electron-acceptor molecules. This aspect has been studied in detail by us by employing Raman spectroscopy and calorimetric experiments.[23–26] On interaction with electron-donor molecules such as tetrathiafulvalene (TTF), the Raman G-band softens progressively with the increase in concentration while the band stiffens with increasing concentration of electron-withdrawing molecules such as tetracyanoethylene (TCNE) (Figure 7a). The full-width at half-maximum (fwhm) of the G-band increases (Figure 7b) and the intensity

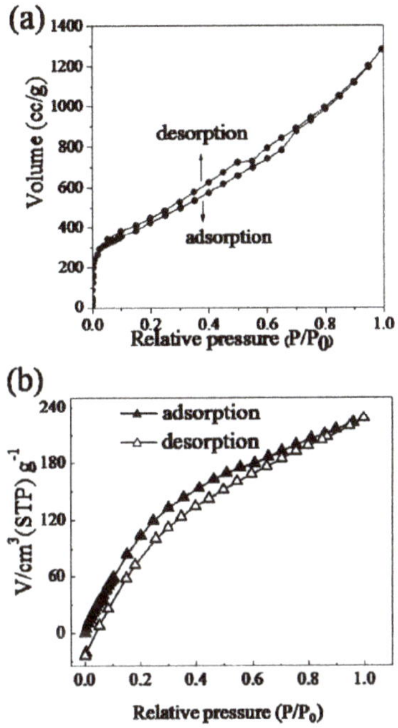


**FIGURE 6.** (a) Nitrogen and (b) carbon dioxide adsorption and desorption isotherms of EG. From ref 22.

of the 2D-band decreases markedly on interaction with these molecules. The ratio of intensities of the 2D and G bands, $I_{2D}/I_G$, is a sensitive probe to examine the effect of the electron-donor and -acceptor molecules on the electronic structure of graphene. The $I_{2D}/I_G$ ratio decreases markedly with the concentration of both electron-donor and electron-acceptor molecules (Figure 7c). Electronic absorption spectra also provide the evidence of molecular charge transfer. The shifts in the G-band caused by molecular charge transfer differ from those due to electrochemical doping where the G-band increases in frequency for both the electron and hole doping.[27,28] Isothermal titration calorimetry (ITC) experiments show that interaction energies of the graphene with electron-acceptor molecules are higher than those with electron-donor molecules and increase progressively with the electron affinities of the acceptors.[26] DFT calculations show that interaction with molecules such as TCNE open a gap in graphene. The magnitude of interaction between graphene and donor/acceptor molecules is dependent on the surface area of the graphene.[29] Charge transfer of graphene with coronene tetracarboxylate (CS) has been exploited for solubilization and exfoliation (Figure 8).[30] Such charge-transfer also provides the basis for sensor properties of graphene for gases like $NO_2$.[31]

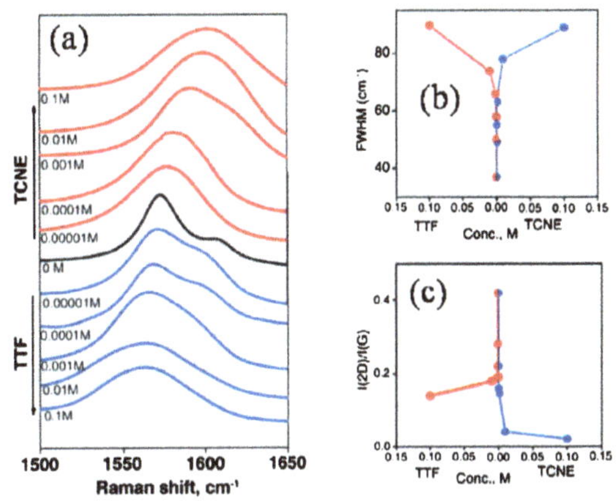


**FIGURE 7.** Variation in (a) position and (b) fwhm of Raman G-band and (c) $I_{2D}/I_G$ intensity ratio of graphene with the concentrations of TTF and TCNE. From ref 25.

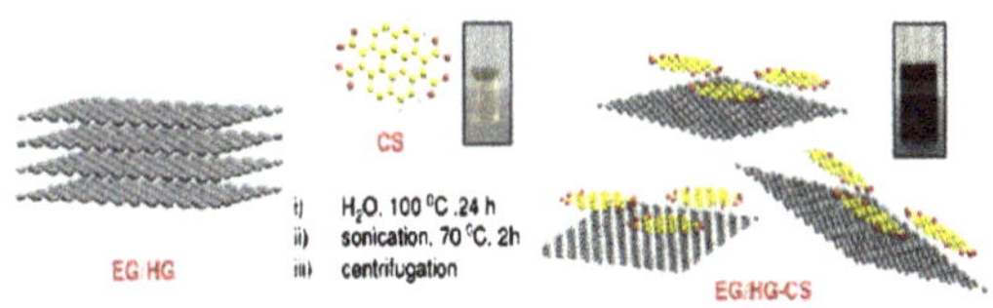


**FIGURE 8.** Illustration of the exfoliation of few-layer graphene with CS to yield monolayer graphene–CS composites. From ref 30.

## Decoration with Metal Nanoparticles

Decoration of graphene with metal nanoparticles causes changes in the electronic structure of graphene through Columbic charge transfer. Such hybrid materials may be of use in catalysis, nanoelectronics, optics, and nanobiotechnology.[32−35] We have employed a simple, green approach for the decoration of noble metal nanoparticles on the graphene surface, using ethylene glycol as the solvent and reducing agent in a domestic microwave oven. On interaction with nanoparticles of metals such as Ag, Au, Pt, and Pd, the Raman G-band shifts to higher frequencies, accompanied by a variation in the relative intensities of D and 2D bands.[36] The magnitude of the G-band shift decreases with increase in ionization energy (Au > Pt > Pd > Ag) of the metal. In addition, nanoparticles of ZnO and $TiO_2$, as well as magnetic nanoparticles of oxides such as $Fe_3O_4$ and $CoFe_2O_4$, interact with graphene giving rise to changes in the Raman spectra and magnetic properties.[37]

## Quenching of Fluorescence of Aromatics

Graphene possesses many interesting optical properties. One of them is the blue emission exhibited by RGO and other oxidized graphene samples.[38] White light has been

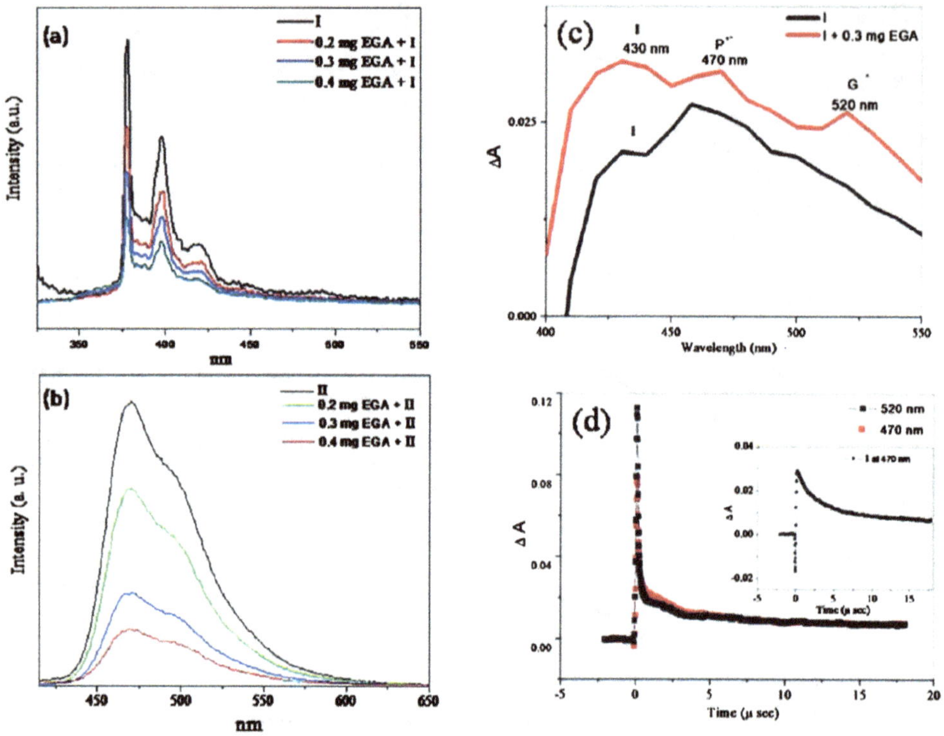


**FIGURE 9.** Fluorescence spectra of (a) PyBS (I, $10^{-5}$ M in DMF) and (b) OPV ester (II, $10^{-5}$ M in chloroform) with increasing concentrations of graphene (EGA). (c) Effect of addition of EGA on the transient absorption spectrum of PyBS, (I, $\lambda_{exc}$=355 nm) after 500 ns and (d) lifetime decay of transient species of PyBS + EGA recorded at 470 and 520 nm. Inset shows the decay of pure PyBS at 470 nm. From ref 39.

produced in combination with a yellow emitter. Interaction with amide-functionalized graphene (EGA) with the electron donors, pyrene-butanaoic acid succinimidyl ester (PyBS, I) and oligo(*p*-phenylene vinylene) methyl ester (OPV-ester, II), causes a marked decrease in the intensity of the fluorescence bands of the donors as illustrated in Figure 9a,b.[39] Transient absorption measurements show a broad band in the 450–530 nm range due to the triplet state and new bands around 470 and 520 nm in the transient absorption spectrum at 500 ns (Figure 9c) due to the pyrenyl radical cation and the graphene radical anion, respectively. The decay of the radical cation is fast as evidenced from the appearance of a short-lived component (900 ns) in the decay profile (Figure 9d). Kamat et al.[40] invoke electron transfer to explain interaction of porphyrin with graphene. Preliminary results show that electron acceptor molecules are likely to interact with graphene via energy transfer.

## Doping with Nitrogen and Boron

Efforts have been made to dope graphene with boron and nitrogen.[41−43] In order to dope graphene without affecting its surface, the arc-evaporation technique is useful.[43] To dope with boron, arc discharge of graphite is carried out in the presence of $H_2$ + diborane or by using boron-stuffed carbon electrodes. Doping with nitrogen is accomplished by carrying out discharge in an atmosphere of $NH_3$ or pyridine + $H_2$ mixture. The Raman G-band shifts to higher frequencies in comparison with pristine samples on B and N doping just as in the case of electrochemical doping.[26] In Figure 10, we show the shifts of the G-band caused by nitrogen and boron doping. The ratio of intensities of the D and G bands, $I_{(D)}/I_{(G)}$, increases where as the ratio of intensities of the 2D and G bands, $I_{(2D)}/I_{(G)}$, decreases in the doped samples. N- and B-doped graphenes show different photodegradation kinetics with rhodamine 6G and methylene blue.[44]

## Chemical Storage of Hydrogen and Halogens

Since the uptake of hydrogen by graphene samples by adsorption does not exceed ~3 wt %, we have examined chemical storage of hydrogen through hydrogenation. Hydrogenation of graphene by cold hydrogen plasma and

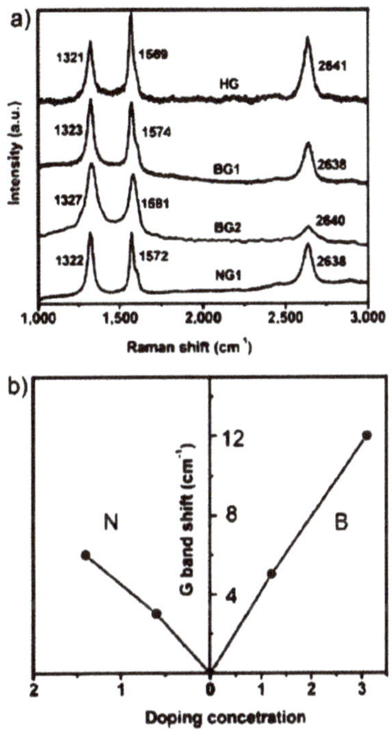


FIGURE 10. (a) Raman Spectra of undoped (HG) and doped (BG and NG) graphene samples. (b) Shifts of the G-band caused by electron (N) doping and hole (B) doping. From ref 43.

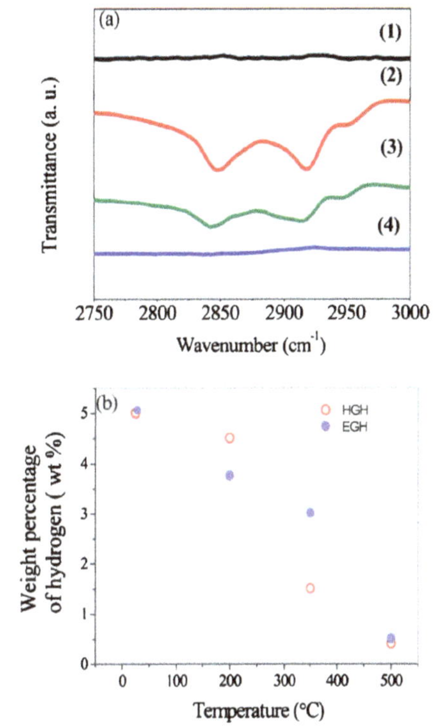


FIGURE 11. (a) IR spectra of (1) EG, (2) EGH, and EGH heated for 4 h (3) at 200 °C and (4) at 500 °C. (b) Change in hydrogen content of EGH and HGH with temperature. From ref 47.

by using radio frequency catalytic CVD has been reported.[45,46] We have also carried out the plasma hydrogenation of graphene but only obtained a hydrogen content of up to 1.8 wt %. We have, therefore, carried out hydrogenation of various graphene samples by Birch reduction.[47] Infrared (IR) spectra of the hydrogenated samples (EGH, HGH) show aliphatic C–H stretching bands (Figure 11a), and elemental analysis revealed the hydrogen content to be around 5 wt % (maximum possible value is ~7.7 wt %). Birch reduction of graphene nanoribbons (GNR) prepared by oxidative unzipping of carbon nanotubes showed a $H_2$ uptake of 3 wt %. The hydrogenated samples were stable at room temperature and get dehydrogenated on annealing at 500 °C (Figure 11b). Dehydrogenation could also be carried out by UV or excimer laser irradiation. This transformation can be exploited for patterning.[16]

Chlorination of graphene by irradiation with UV light in liquid chlorine yields up to 56 wt % chlorine in the sample.[48] Interestingly, chlorination too is reversible. Reversible bromination of graphene up to 25 wt % has also been achieved.

## Magnetic Properties

Room-temperature ferromagnetism in graphene and graphite-based materials has attracted considerable interest. Yazyev and Helm[49] showed that magnetism in graphene can be induced by defects. Zig-zag edges have been considered to be responsible for the magnetism of graphene by many workers.[50] Wang et al.[51] reported room-temperature ferromagnetism in a graphene sample prepared by reduction of graphene oxide. We have measured the magnetic properties of a variety of graphene samples.[52,53] All the samples show divergence between the field-cooled (FC) and zero-field-cooled (ZFC) data, starting around 300 K (Figure 12a). The divergence nearly disappears on application of 1 T. Temperature-dependent magnetization data of graphenes are comparable to those of magnetically frustrated systems, wherein different types of magnetic states coexist. The graphene samples show magnetic hysteresis at room temperature (Figure 12b) and the saturation magnetization ($M_s$) increases with decrease in temperature, the value being highest in the case of HG. EPR measurements in the

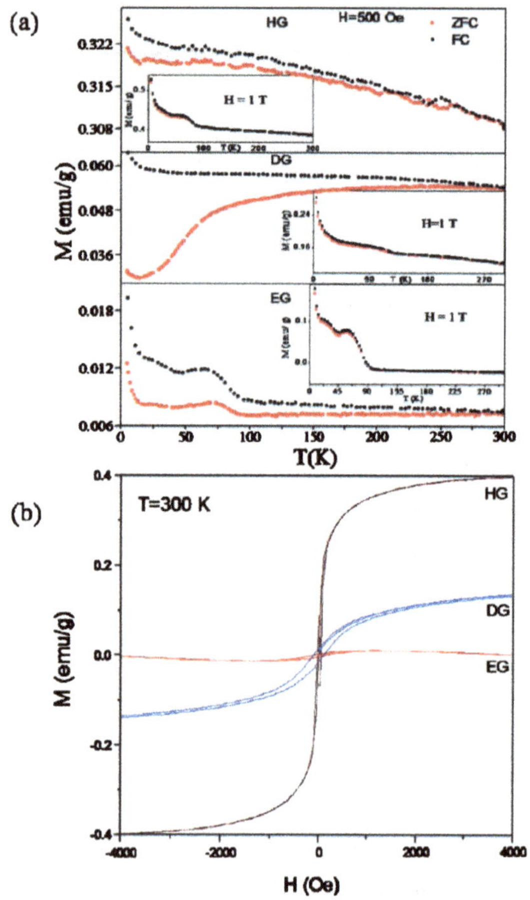


**FIGURE 12.** (a) Temperature variation of magnetization of EG, DG, and HG at 500 Oe. (Inset shows data at 1 T). (b) Magnetic hysteresis in EG, DG, and HG at 300 K. Reproduced from ref 52. Copyright 2009 American Chemical Society.

2.5–300 K range show a signal with a line width of 0.7–2.9 mT and a *g*-value in the 2.006–2.013 range, suggesting that spins do not originate from transition-metal impurities but from the spin species in the graphenes.

Adsorption of the benzene solutions of tetrathiafulvalane (TTF) and tetracyanoethylene (TCNE) on graphene decreases the magnetization value while still showing room-temperature hysteresis.[52,53] Hydrogenated graphenes show an increase in the magnetic moment with increase in the hydrogen content. On dehydrogenation at 500 °C for 4 h, the samples exhibit a decrease in the magnetic moment.[53]

## Field-Emission Properties

Graphene films prepared by electrophoretic deposition have been demonstrated to show good field-emission properties

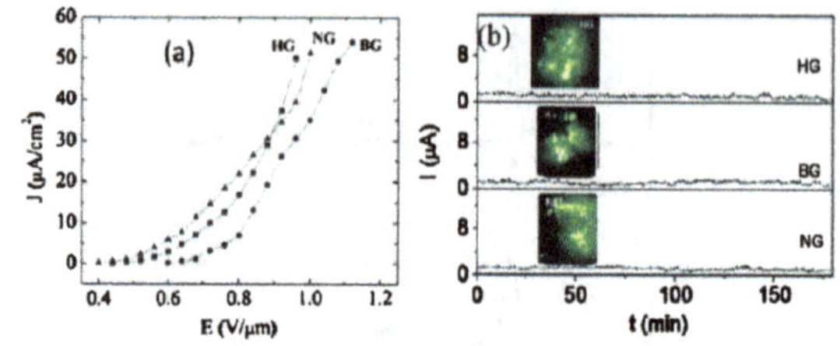


**FIGURE 13.** (a) Current density (*J*) of HG, BG, and NG as a function of electric field. (b) Current stability of HG, BG, and NG (Inset shows field emission pattern corresponding to HG, BG, and NG). From ref 55.

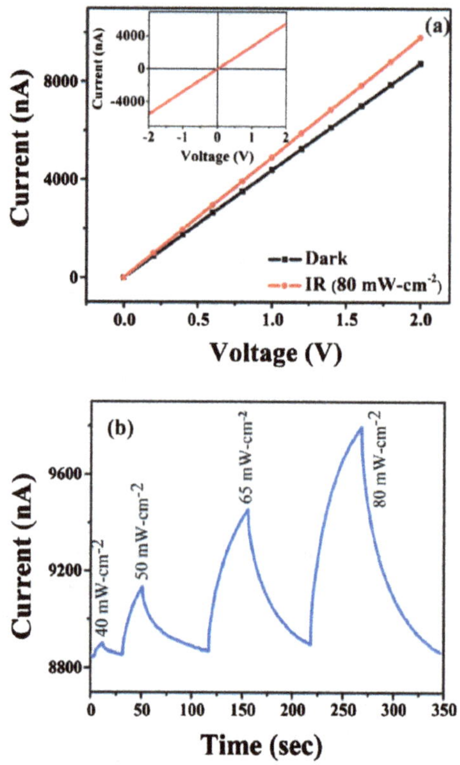


**FIGURE 14.** (a) Typical *I*–*V* characteristics of RGO in the dark and under illumination at 1550 nm at 80 mW cm$^{-2}$. The inset shows metal–RGO–metal contact. (b) Photocurrent as a function of time with different IR intensities at 2 V for RGO. From ref 59.

with a low turn-on electric field ($E_{to}$ = 2.3 V $\mu$m$^{-1}$ at 10 $\mu$A cm$^{-2}$), a large field-enhancement factor (3000), and good emission stability, better than those of CNTs.[54] In Figure 13a, we show the variation in current density (*J*) of the HG, as well as boron and nitrogen doped HG, samples (BG and NG, respectively) with electric field (*E*).[55] The turn-on electric field (at 10 $\mu$A cm$^{-2}$) is 0.7, 0.8, and 0.6 V $\mu$m$^{-1}$ for HG, BG, and NG, respectively. The low turn-on electric field[56] for NG

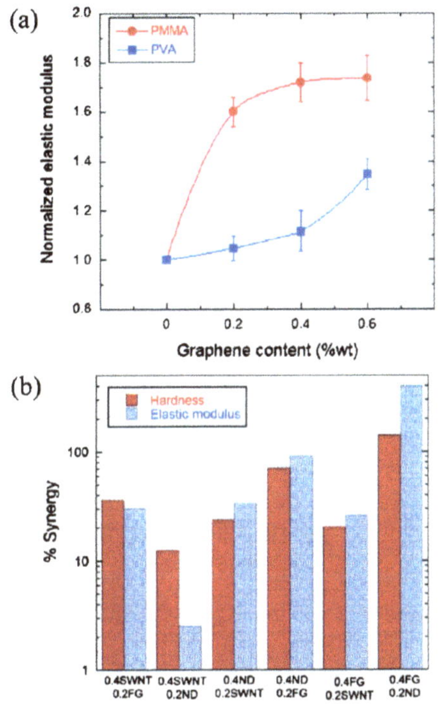


**FIGURE 15.** (a) Normalized elastic modulus plotted with graphene content for PVA and PMMA composites. (b) Percentage synergy in hardness and elastic modulus for different binary composites. From ref 63.

might be due to the upshift of Fermi level of graphene. The emission currents are stable over a period of more than 3 h (Figure 13b), revealing that the field-emission characteristics of the graphene samples are superior to many other nanomaterials.

## Photodetectors

Graphene FETs have been studied widely. Our own studies have shown that FETs fabricated from RGO exhibit ambipolar characteristics.[57] N- and B-doped graphenes show n-type and p-type transfer characteristics, respectively. Graphene also exhibits good infrared (IR) detection properties.[58] We have fabricated IR detectors of graphene prepared by chemical methods.[59] Electrical conductivities of RGO and graphene nanoribbons (GNRs) increase with IR laser radiation and can even sense the IR radiation emitted from a human body. In Figure 14, we show the change in $I-V$ characteristics of RGO with the illumination of IR radiation (1550 nm at 80 mW cm $^{-2}$) and the effect of intensity of IR radiation. The inset in the Figure 14a shows metal–RGO–metal contact. The current

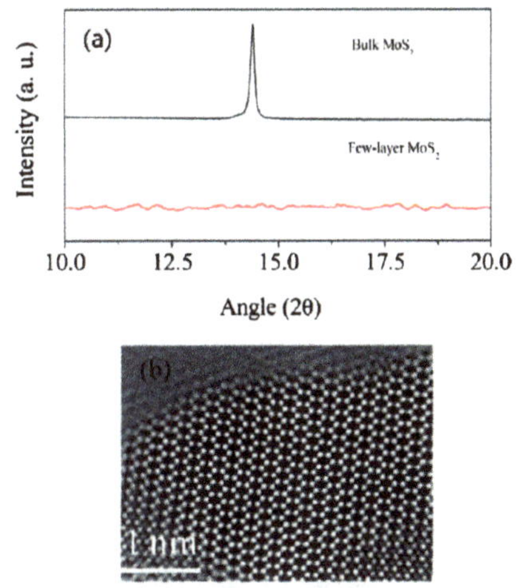


**FIGURE 16.** (a) XRD patterns of bulk $MoS_2$ and few-layered $MoS_2$. (b) High-resolution TEM image of $MoSe_2$. From refs 66 and 67.

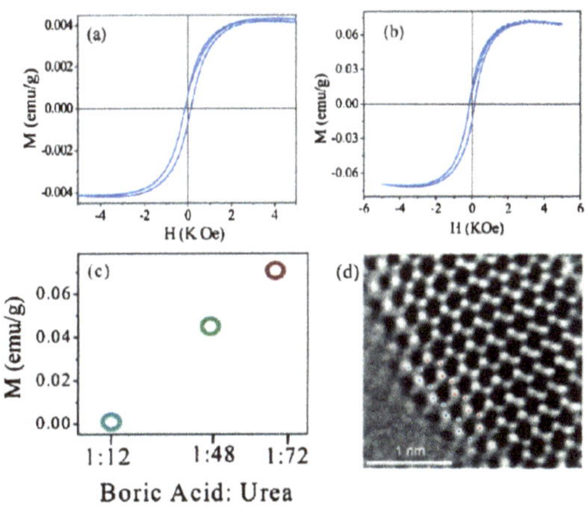


**FIGURE 17.** Magnetic hysteresis exhibited by graphene-like (a) $MoS_2$ and (b) BN at 300 K. (c) shows layer dependence of magnetism of BN. (d) High-resolution TEM image showing zigzag edges in graphene-like $MoS_2$. From ref 53.

responsivity ($R_\lambda$) and the external quantum efficiency (EQE) for RGO is 4 mA/W and 0.3%, respectively, while for GNR these values are higher, being 1 A/W and 80%, respectively, for an incident wavelength of 1550 nm at 2 V. Graphene is also a good UV absorber.[60] UV-detectors based on RGO show a responsivity of 0.12 A/W with an EQE of 40%.[61]

## Mechanical Properties

Ramanathan et al.[62] reported that incorporation of ~1 wt % of functionalized graphene in a poly(methyl methacrylate)

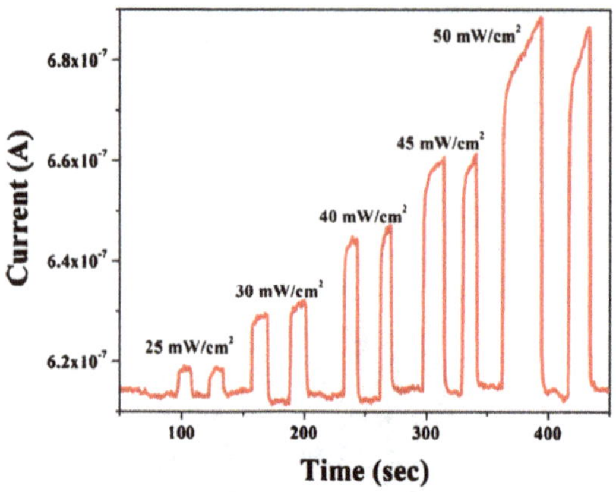


**FIGURE 18.** Photocurrent as a function of time with different IR intensities at 0.1 V for few-layer $MoSe_2$. From ref 69.

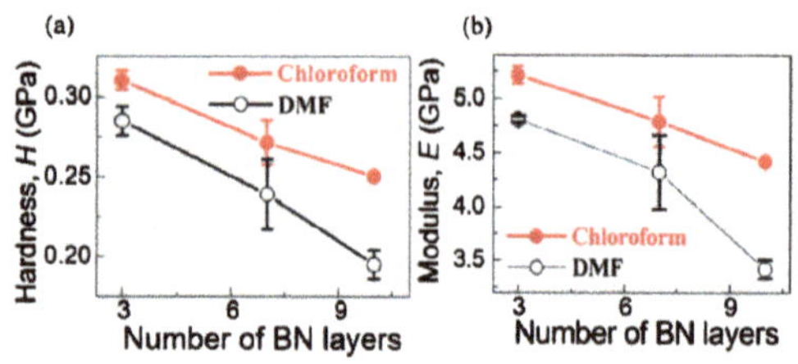


**FIGURE 19.** Variation of (a) hardness and (b) elastic modulus of BN–PMMA composites prepared in chloroform and dimethylformamide (DMF) as a function of number of BN layers. From ref 70.

(PMMA) matrix causes an increase of 80% in elastic modulus and 20% in ultimate tensile strength. We have investigated the performance of functionalized EG as a filler material in poly(vinyl alcohol) (PVA) and PMMA matrices by employing the nanoindentation technique. The addition of 0.6 wt % of EG results a significant increase in both the elastic modulus and hardness (Figure 15a). The observed improvement in the mechanical properties of the polymers is attributed to the interaction between the polymer and graphene, which in turn provides better load-transfer between the matrix and the fiber. We have studied the effect of binary combinations of nanodiamond (ND), few-layer graphene (EG), and single-walled nanotubes (SWNT) as fillers in PVA matrices.[63] There is an extraordinary synergy with improvement by as much as 400% in stiffness and hardness compared with those obtained with single nanocarbon reinforcements. The synergistic effect was dramatic in the case of ND–EG composite with 4- and 1.5-fold increases in elastic modulus and hardness, respectively (Figure 15b).

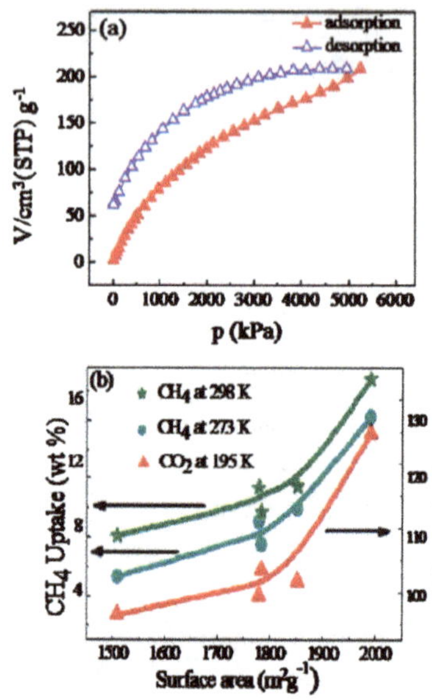


**FIGURE 20.** (a) $CH_4$ uptake curves of BCN at 298 K and 50 bar. (b) Plot of the BET surface area and the weight percentage of methane uptake (at 50 bar of pressure, 273 and 298 K) and carbon dioxide uptake (at 1 bar and 195 K) of BCN. From ref 22.

## Graphene Mimics

Inorganic analogues of zero-dimensional fullerenes and one-dimensional carbon nanotubes formed by various layered materials have been prepared and characterized.[64] There have been efforts to prepare graphene-like $MoS_2$, BN, and other layered materials. Single- and few-layer materials of these have been prepared by micromechanical cleavage or ultrasonication of dispersions in liquids.[65] Chemical methods have been successfully applied for generating these graphene analogues. In the case of $MoS_2$, $WS_2$, $MoSe_2$, and $WSe_2$, intercalation of the bulk material with lithium followed by sonication in water, reaction of molybdic or tungstic acid with excess thiourea or selenourea at 773 K in a nitrogen atmosphere, or reaction of KSCN or Se with $H_2MoO_4$ under hydrothermal conditions have been employed.[66,67] The XRD patterns provide evidence for the formation of the graphene analogues, through the absence of the (002) reflection (Figure 16a) while the Raman spectra show shifts and broadening of the bands due to the $A_{1g}$ and $E_{2g}$ modes. In Figure 16b, we show a high-resolution TEM image of $MoSe_2$. We have prepared BN sheets by the reaction of boric acid and urea at high temperatures.[68]

Graphene analogues of $MoS_2$, $WS_2$, $MoSe_2$, $WSe_2$, and BN exhibit room-temperature ferromagnetism due to edge effects (Figure 17a,b). There is an increase in the magnetic moment with the decreasing number of layers (see Figure 17c). We show the presence of edges in graphene-like $MoS_2$ in Figure 17d. Few-layer $MoS_2$ and $MoSe_2$ exhibit good IR radiation detection properties. The responsivity and EQE of the $MoSe_2$ device are specially good (Figure 18), the values being 50 mA/W and 6%, respectively, at an incident wavelength of 1064 nm.[69] Mechanical properties of PMMA–BN composites with different numbers of BN layers show that the composite with the least number of BN layers has the highest hardness and elastic modulus (Figure 19).[70]

Besides graphene analogues of known materials, it is of value to generate new graphene-like materials. One success in this direction is the synthesis of high-surface area $B_xC_yN_z$ containing a few layers, with an average composition between BCN and $BC_2N$, prepared by the reaction of activated charcoal, boric acid, and urea.[22] They exhibit high uptake of both $CO_2$ and $CH_4$, the highest value of $CO_2$ uptake being 128 wt % at 195 K and 1 atm and 64% at room temperature and 50 bar pressure. The $CH_4$ uptake value varies from 7.5 to 17 wt % at 273 K. At 298 K, the highest uptake is 15 wt % (see Figure 20).

## Conclusions

Our research on graphene has been exciting and rewarding. Some of the properties such as the synergy in mechanical properties of polymer composites found when graphene is incorporated along with another nanocarbon and the radiation detection characteristics are impressive. It is possible that graphene can be employed in flatbed displays due to the favorable field-emission characteristics. Possible use of graphene in organic photovoltaics is suggested by its excited-state interactions with fluorescent molecules. An early study in this laboratory indicated the possible use of graphene in supercapacitors.[71] Recent studies have indeed shown this to be a potential application. Graphene has also been used in lithium ion batteries and other energy related devices. While the 5 wt % reversible storage of hydrogen in graphene could possibly be useful in energy-related applications, the remarkable adsorptive properties of graphene-like BCN for $CO_2$ and $CH_4$ should certainly be explored for further use. Properties and applications of inorganic graphene mimics are worthy of exploration.

BIOGRAPHICAL INFORMATION

**C. N. R. Rao** obtained his Ph.D. degree from Purdue University (1958) and D.Sc. degree from the University of Mysore (1961). He is the National Research Professor and Linus Pauling Research Professor at the Jawaharlal Nehru Centre for Advanced Scientific Research (JNCASR) and honorary professor at the Indian Institute of Science (both at Bangalore). His research interests are mainly in the chemistry of materials. He is a member of many of the science academies and is the recipient of the Hughes and Royal Medals of the Royal Society, the August Wilhelm von Hofmann medal of the German Chemical Society, and the Dan David Prize for materials research.

**H. S. S. Ramakrishna Matte** obtained his M.S. degree in chemical science in 2009 from JNCASR and is pursuing Ph.D. studies.

**K. S. Subrahmanyam** received his M.S. (Engg) and Ph.D. degrees from JNCASR in 2008 and 2012, respectively.

FOOTNOTES

*Corresponding author. Fax: 00 91 80 2208 2766. E-mail: cnrrao@jncasr.ac.in.
The authors declare no competing financial interest.

REFERENCES


1 Novoselov, K. S.; Geim, A. K.; Morozov, S. V.; Jiang, D.; Zhang, Y.; Dubonos, S. V.; Grigorieva, I. V.; Firsov, A. A. Electric Field Effect in Atomically Thin Carbon Films. *Science* **2004**, *306*, 666–669.

2 Geim, A. K.; Novoselov, K. S. The Rise of Graphene. *Nat. Mater.* **2007**, *6*, 183–191.

3 Rao, C. N. R.; Sood, A. K.; Subrahmanyam, K. S.; Govindaraj, A. Graphene: The New Two-Dimensional Nanomaterial. *Angew. Chem., Int. Ed.* **2009**, *48*, 7752–7777.

4 Pati, S. K., Enoki, T., Rao, C. N. R., Eds. *Graphene and Its Fascinating Attributes*; World Scientific: Singapore, 2011.

5 Wallace, P. R. The Band Theory of Graphite. *Phys. Rev.* **1947**, *71*, 622–634.

6 Dreyer, D. R.; Ruoff, R. S.; Bielawski, C. W. From Conception to Realization: An Historial Account of Graphene and Some Perspectives for Its Future. *Angew. Chem., Int. Ed.* **2010**, *49*, 9336–9344.

7 Novoselov, K. S.; Jiang, D.; Schedin, F.; Booth, T. J.; Khotkevich, V. V.; Morozov, S. V.; Geim, A. K. Two-Dimensional Atomic Crystals. *Proc. Natl. Acad. Sci. U.S.A.* **2005**, *102*, 10451–10453.

8 Abergel, D. S.; Russell, A.; Falko, V. I. Visibility of Graphene Flakes on a Dielectric Substrate. *Appl. Phys. Lett.* **2007**, *91*, No. 063125.

9 Meyer, J. C.; Kisielowski, C.; Erni, R.; Rossell, M. D.; Crommie, M. F.; Zettl, A. Direct Imaging of Lattice Atoms and Topological Defects in Graphene Membranes. *Nano Lett.* **2008**, *8*, 3582–3586.

10 Hernandez, Y.; Nicolosi, V.; Lotya, M.; Blighe, F. M.; Sun, Z.; De, S.; McGovern, I. T.; Holland, B.; Byrne, M.; Gun'Ko, Y. K.; Boland, J. J.; Niraj, P.; Duesberg, G.; Krishnamurthy, S.; Goodhue, R.; Hutchison, J.; Scardaci, V.; Ferrari, A. C.; Coleman, J. N. High-Yield Production of Graphene by Liquid-Phase Exfoliation of Graphite. *Nat. Nanotechnol.* **2008**, *3*, 563–568.

11 Rao, C. N. R.; Subrahmanyam, K. S.; Matte, H. S. S. R.; Abdulhakeem, B.; Govindaraj, A.; Das, B.; Kumar, P.; Ghosh, A.; Late, D. J. A Study of the Synthetic Methods and Properties of Graphenes. *Sci. Technol. Adv. Mater.* **2010**, *11*, No. 054502.

12 Reina, A.; Jia, X. T.; Ho, J.; Nezich, D.; Son, H. B.; Bulovic, V.; Dresselhaus, M. S.; Kong, J. Large Area, Few-Layer Graphene Films on Arbitrary Substrates by Chemical Vapor Deposition. *Nano Lett.* **2009**, *9*, 30–35.

13 Dreyer, D. R.; Park, S.; Bielawski, C. W.; Ruoff, R. S. The Chemistry of Graphene Oxide. *Chem. Soc. Rev.* **2010**, *39*, 228–240.

14 Kumar, P.; Subrahmanyam, K. S.; Rao, C. N. R. Graphene Produced by Radiation-Induced Reduction of Graphene Oxide. *Int. J. Nanosci.* **2011**, *10*, 559–566.

15 Abdelsayed, V.; Moussa, S.; Hassan, H. M.; Aluri, H. S.; Collinson, M. M.; El-Shall, M. S. Photothermal Deoxygenation of Graphite Oxide with Laser Excitation in Solution and Graphene-Aided Increase in Water Temperature. *J. Phys. Chem. Lett.* **2010**, *1*, 2804–2809.

16 Kumar, P.; Subrahmanyam, K. S.; Rao, C. N. R. Graphene Patterning and Lithography Employing Laser/Electron-Beam Reduced Graphene Oxide and Hydrogenated Graphene. *Mater. Express* **2011**, *1*, 252–256.

17 Kumar, P.; Panchakarla, L. S.; Rao, C. N. R. Laser-Induced Unzipping of Carbon Nanotubes to Yield Graphene Nanoribbons. *Nanoscale* **2011**, *3*, 2127–2129.

18 Subrahmanyam, K. S.; Vivekchand, S. R. C.; Govindaraj, A.; Rao, C. N. R. A Study of Graphenes Prepared by Different Methods: Characterization, Properties and Solubilization. *J. Mater. Chem.* **2008**, *18*, 1517–1523.

19 Prasad, B. L. V.; Sato, H.; Enoki, T.; Hishiyama, Y.; Kaburagi, Y.; Rao, A. M.; Eklund, P. C.; Oshida, K.; Endo, M. Heat-Treatment Effect on the Nanosized Graphite Pi-Electron System during Diamond to Graphite Conversion. *Phys. Rev. B* **2000**, *62*, 11209–11218.

20 Subrahmanyam, K. S.; Panchakarla, L. S.; Govindaraj, A.; Rao, C. N. R. Simple Method of Preparing Graphene Flakes by an Arc-Discharge Method. *J. Phys. Chem. C* **2009**, *113*, 4257–4259.

21 Ghosh, A.; Subrahmanyam, K. S.; Krishna, K. S.; Datta, S.; Govindaraj, A.; Pati, S. K.; Rao, C. N. R. Uptake of $H_2$ and $CO_2$ by Graphene. *J. Phys. Chem. C* **2008**, *112*, 15704–15707.

22 Kumar, N.; Subrahmanyam, K. S.; Chaturbedy, P.; Raidongia, K.; Govindaraj, A.; Hembram, K. P. S. S.; Mishra, A. K.; Waghmare, U. V.; Rao, C. N. R. Remarkable Uptake of $CO_2$ and $CH_4$ by Graphene-Like Borocarbonitrides, BxCyNz. *ChemSusChem* **2011**, *4*, 1662–1670.

23 Das, B.; Voggu, R.; Rout, C. S.; Rao, C. N. R. Changes in the Electronic Structure and Properties of Graphene Induced by Molecular Charge-Transfer. *Chem. Commun.* **2008**, 5155–5157.

24 Voggu, R.; Das, B.; Rout, C. S.; Rao, C. N. R. Effects of Charge Transfer Interaction of Graphene with Electron Donor and Acceptor Molecules Examined Using Raman Spectroscopy and Cognate Techniques. *J. Phys.: Condens. Matter* **2008**, *20*, No. 472204.

25 Rao, C. N. R.; Voggu, R. Charge-Transfer with Graphene and Nanotubes. *Mater. Today* **2010**, *13*, 34–40.

26 Varghese, N.; Ghosh, A.; Voggu, R.; Ghosh, S.; Rao, C. N. R. Selectivity in the Interaction of Electron Donor and Acceptor Molecules with Graphene and Single-Walled Carbon Nanotubes. *J. Phys. Chem. C* **2009**, *113*, 16855–16859.

27 Das, A.; Pisana, S.; Chakraborty, B.; Piscanec, S.; Saha, S. K.; Waghmare, U. V.; Novoselov, K. S.; Krishnamurthy, H. R.; Geim, A. K.; Ferrari, A. C.; Sood, A. K. Monitoring Dopants by Raman Scattering in an Electrochemically Top-Gated Graphene Transistor. *Nat. Nanotechnol.* **2008**, *3*, 210–215.

28 Manna, A. K.; Pati, S. K. Tuning the Electronic Structure of Graphene by Molecular Charge Transfer: A Computational Study. *Chem.—Asian J.* **2009**, *4*, 855–860.

29 Subrahmanyam, K. S.; Voggu, R.; Govindaraj, A.; Rao, C. N. R. A Comparative Raman Study of the Interaction of Electron Donor and Acceptor Molecules with Graphene Prepared by Different Methods. *Chem. Phys. Lett.* **2009**, *472*, 96–98.

30 Ghosh, A.; Rao, K. V.; George, S. J.; Rao, C. N. R. Noncovalent Functionalization, Exfoliation, and Solubilization of Graphene in Water by Employing a Fluorescent Coronene Carboxylate. *Chem.—Eur. J.* **2010**, *16*, 2700–2704.

31 Ghosh, A.; Late, D. J.; Panchakarla, L. S.; Govindaraj, A.; Rao, C. N. R. $NO_2$ and Humidity Sensing Characteristics of Few-Layer Graphenes. *J. Exp. Nanosci.* **2009**, *4*, 313.

32 Hassan, H. M. A.; Abdelsayed, V.; Khder, A. E. R. S.; AbouZeid, K. M.; Terner, J.; El-Shall, M. S.; Al-Resayes, S. I.; El-Azhary, A. A. Microwave Synthesis of Graphene Sheets Supporting Metal Nanocrystals in Aqueous and Organic Media. *J. Mater. Chem* **2009**, *19*, 3832–3837.

33 Siamaki, A. R.; Khder, A. E. R. S.; Abdelsayed, V.; El-Shall, M. S.; Gupton, B. F. Microwave-Assisted Synthesis of Palladium Nanoparticles Supported on Graphene: A Highly Active and Recyclable Catalyst for Carbon-Carbon Cross-Coupling Reactions. *J. Catal.* **2011**, *279*, 1–11.

34 Moussa, S.; Abdelsayed, V.; Samy El-Shall, M. Laser Synthesis of Pt, Pd, CoO and Pd-CoO Nanoparticle Catalysts Supported on Graphene. *Chem. Phys. Lett.* **2011**, *510*, 179–184.

35 Moussa, S.; Siamaki, A. R.; Gupton, B. F.; El-Shall, M. S. Pd-Partially Reduced Graphene Oxide Catalysts (Pd/PRGO): Laser Synthesis of Pd Nanoparticles Supported on PRGO Nanosheets for Carbon-Carbon Cross Coupling Reactions. *ACS Catal.* **2012**, *2*, 145–154.

36 Subrahmanyam, K. S.; Manna, A. K.; Pati, S. K.; Rao, C. N. R. A Study of Graphene Decorated with Metal Nanoparticles. *Chem. Phys. Lett.* **2010**, *497*, 70–75.

37 Das, B.; Choudhury, B.; Gomathi, A.; Manna, A. K.; Pati, S. K.; Rao, C. N. R. Interaction of Inorganic Nanoparticles with Graphene. *Chem. Phys. Chem.* **2011**, *12*, 937–943.

38 Subrahmanyam, K. S.; Kumar, P.; Nag, A.; Rao, C. N. R. Blue Light Emitting Graphene-Based Materials and Their Use in Generating White Light. *Solid State Commun.* **2010**, *150*, 1774–1777.

39 Matte, H. S. S. R.; Subrahmanyam, K. S.; Venkata Rao, K.; George, S. J.; Rao, C. N. R. Quenching of Fluorescence of Aromatic Molecules by Graphene Due to Electron Transfer. *Chem. Phys. Lett.* **2011**, *506*, 260–264.

40 Wojcik, A.; Kamat, P. V. Reduced Graphene Oxide and Porphyrin. An Interactive Affair in 2-D. *ACS Nano* **2010**, *4*, 6697–6706.

41 Wang, X.; Li, X.; Zhang, L.; Yoon, Y.; Weber, P. K.; Wang, H.; Guo, J.; Dai, H. N-Doping of Graphene through Electrothermal Reactions with Ammonia. *Science* **2009**, *324*, 768–771.

42 Li, X. L.; Wang, H. L.; Robinson, J. T.; Sanchez, H.; Diankov, G.; Dai, H. J. Simultaneous Nitrogen Doping and Reduction of Graphene Oxide. *J. Am. Chem. Soc.* **2009**, *131*, 15939–15944.

43 Panchakarla, L. S.; Subrahmanyam, K. S.; Saha, S. K.; Govindaraj, A.; Krishnamurthy, H. R.; Waghmare, U. V.; Rao, C. N. R. Synthesis, Structure, and Properties of Boron- and Nitrogen-Doped Graphene. *Adv. Mater.* **2009**, *21*, 4726–4730.

44 Gopalakrishnan, K.; Joshi, H. M.; Kumar, P.; Panchakarla, L. S.; Rao, C. N. R. Selectivity in the Photocatalytic Properties of the Composites of $TiO_2$ Nanoparticles with B- and N-Doped Graphenes. *Chem. Phys. Lett.* **2011**, *511*, 304–308.

45 Elias, D. C.; Nair, R. R.; Mohiuddin, T. M. G.; Morozov, S. V.; Blake, P.; Halsall, M. P.; Ferrari, A. C.; Boukhvalov, D. W.; Katsnelson, M. I.; Geim, A. K.; Novoselov, K. S. Control of Graphene's Properties by Reversible Hydrogenation: Evidence for Graphane. *Science* **2009**, *323*, 610–613.

46 Zheng, L.; Li, Z.; Bourdo, S.; Watanabe, F.; Ryerson, C. C.; Biris, A. S. Catalytic Hydrogenation of Graphene Films. *Chem. Commun.* **2010**, *47*, 1213–1215.

47 Subrahmanyam, K. S.; Kumar, P.; Maitra, U.; Govindaraj, A.; Hembram, K.; Waghmare, U. V.; Rao, C. N. R. Chemical Storage of Hydrogen in Few-Layer Graphene. *Proc. Natl. Acad. Sci. U.S.A.* **2011**, *108*, 2674–2677.

48 Gopalakrishnan, K.; Subrahmanyam, K. S.; Kumar, P.; Govindaraj, A.; Rao, C. N. R. Reversible Chemical Storage of Halogens in Few-Layer Graphene. *RSC Adv.* **2012**, *2*, 1605–1608.

49 Yazyev, O. V.; Helm, L. Defect-Induced Magnetism in Graphene. *Phys. Rev. B* **2007**, *75*, No. 125408.

50 Bhowmick, S.; Shenoy, V. B. Edge State Magnetism of Single Layer Graphene Nanostructures. *J. Chem. Phys.* **2008**, *128*, No. 244717.

51 Wang, Y.; Huang, Y.; Song, Y.; Zhang, X. Y.; Ma, Y. F.; Liang, J. J.; Chen, Y. S. Room-Temperature Ferromagnetism of Graphene. *Nano Lett.* **2009**, *9*, 220–224.

52 Matte, H. S. S. R.; Subrahmanyam, K. S.; Rao, C. N. R. Novel Magnetic Properties of Graphene: Presence of Both Ferromagnetic and Antiferromagnetic Features and Other Aspects. *J. Phys. Chem. C* **2009**, *113*, 9982–9985.

53 Rao, C. N. R.; Matte, H. S. S. R.; Subrahmanyam, K. S.; Maitra, U. Unusual Magnetic Properties of Graphene and Related Materials. *Chem. Sci.* **2012**, *3*, 45–52.

54 Wu, Z.-S.; Pei, S.; Ren, W.; Tang, D.; Gao, L.; Liu, B.; Li, F.; Liu, C.; Cheng, H.-M. Field Emission of Single-Layer Graphene Films Prepared by Electrophoretic Deposition. *Adv. Mater.* **2009**, *21*, 1756–1760.

55 Palnitkar, U. A.; Kashid, R. V.; More, M. A.; Joag, D. S.; Panchakarla, L. S.; Rao, C. N. R. Remarkably Low Turn-on Field Emission in Undoped, Nitrogen-Doped, And Boron-Doped Graphene. *Appl. Phys. Lett.* **2010**, *97*, 063102.

56 Ye, D.; Moussa, S.; Ferguson, J. D.; Baski, A. A.; El-Shall, M. S. Highly Efficient Electron Field Emission from Graphene Oxide Sheets Supported by Nickel Nanotip Arrays. *Nano Lett.* **2012**, *12*, 1265–1268.

57 Late, D. J.; Ghosh, A.; Subrahmanyam, K. S.; Panchakarla, L. S.; Krupanidhi, S. B.; Rao, C. N. R. Characteristics of Field-Effect Transistors Based on Undoped and B- and N-Doped Few-Layer Graphenes. *Solid State Commun.* **2010**, *150*, 734–738.

58 Xia, F.; Mueller, T.; Lin, Y.-m.; Valdes-Garcia, A.; Avouris, P. Ultrafast Graphene Photodetector. *Nat. Nanotechnol.* **2009**, *4*, 839–843.

59 Chitara, B.; Panchakarla, L. S.; Krupanidhi, S. B.; Rao, C. N. R. Infrared Photodetectors Based on Reduced Graphene Oxide and Graphene Nanoribbons. *Adv. Mater.* **2011**, *23*, 5419–5424.

60 Lee, C.; Kim, J. Y.; Bae, S.; Kim, K. S.; Hong, B. H.; Choi, E. J. Optical Response of Large Scale Single Layer Graphene. *Appl. Phys. Lett.* **2011**, *98*, No. 071905.

61 Chitara, B.; Krupanidhi, S. B.; Rao, C. N. R. Solution Processed Reduced Graphene Oxide Ultraviolet Detector. *Appl. Phys. Lett.* **2011**, *99*, No. 113114.

62 Ramanathan, T.; Abdala, A. A.; Stankovich, S.; Dikin, D. A.; Herrera Alonso, M.; Piner, R. D.; Adamson, D. H.; Schniepp, H. C.; ChenX; Ruoff, R. S.; Nguyen, S. T.; Aksay, I. A.; Prud'Homme, R. K.; Brinson, L. C. Functionalized Graphene Sheets for Polymer Nanocomposites. *Nat. Nanotechnol.* **2008**, *3*, 327–331.

63 Prasad, K. E.; Das, B.; Maitra, U.; Ramamurty, U.; Rao, C. N. R. Extraordinary Synergy in the Mechanical Properties of Polymer Matrix Composites Reinforced with 2 Nanocarbons. *Proc. Natl. Acad. Sci. U.S.A.* **2009**, *106*, 13186–13189.

64 Tenne, R. Inorganic Nanotubes and Fullerene-Like Nanoparticles. *Nat. Nanotechnol.* **2006**, *1*, 103–111.

65 Coleman, J. N.; Lotya, M.; O'Neill, A.; Bergin, S. D.; King, P. J.; Khan, U.; Young, K.; Gaucher, A.; De, S.; Smith, R. J.; Shvets, I. V.; Arora, S. K.; Stanton, G.; Kim, H.-Y.; Lee, K.; Kim, G. T.; Duesberg, G. S.; Hallam, T.; Boland, J. J.; Wang, J. J.; Donegan, J. F.; Grunlan, J. C.; Moriarty, G.; Shmeliov, A.; Nicholls, R. J.; Perkins, J. M.; Grieveson, E. M.; Theuwissen, K.; McComb, D. W.; Nellist, P. D.; Nicolosi, V. Two-Dimensional Nanosheets Produced by Liquid Exfoliation of Layered Materials. *Science* **2011**, *331*, 568–571.

66 Matte, H. S. S. R.; Gomathi, A.; Manna, A. K.; Late, D. J.; Datta, R.; Pati, S. K.; Rao, C. N. R. $MoS_2$ and $WS_2$ Analogues of Graphene. *Angew. Chem., Int. Ed.* **2010**, *49*, 4059–4062.

67 Matte, H. S. S. R.; Plowman, B.; Datta, R.; Rao, C. N. R. Graphene Analogues of Layered Metal Selenides. *Dalton Trans.* **2011**, *40*, 10322–10325.

68 Nag, A.; Raidongia, K.; Hembram, K. P. S. S.; Datta, R.; Waghmare, U. V.; Rao, C. N. R. Graphene Analogues of BN: Novel Synthesis and Properties. *ACS Nano* **2010**, *4*, 1539–1544.

69 Matte, H. S. S. R.; Chitara, B.; Maitra, U.; Sharmila, N. S.; Krupanidhi, S. B.; Waghmare, U. V.; Rao, C. N. R. Superior Infrared Detection Properties of Few-layer $MoSe_2$. *ACS Nano.* **2012**, submitted for publication.

70 Kiran, M. S. R. N.; Raidongia, K.; Ramamurty, U.; Rao, C. N. R. Improved Mechanical Properties of Polymer Nanocomposites Incorporating Graphene-Like BN: Dependence on the Number of BN Layers. *Scr. Mater.* **2010**, *64*, 592–595.

71 Vivekchand, S. R. C.; Rout, C. S.; Subrahmanyam, K. S.; Govindaraj, A.; Rao, C. N. R. Graphene-Based Electrochemical Supercapacitors. *J. Chem. Sci.* **2008**, *120*, 9–13.

MICROREVIEW

DOI: 10.1002/ejic.201000408

# Inorganic Analogues of Graphene

C. N. R. Rao*[a,b] and Angshuman Nag[a,b]

**Keywords:** Graphene / Nanostructres / Molybdenum / Tungsten / Sulfides / Boron / Nitrides

The discovery of graphene has aroused great interest in the properties and phenomena exhibited by two-dimensional inorganic materials, especially when they comprise only a single, two or a few layers. Graphene-like $MoS_2$ and $WS_2$ have been prepared by chemical methods, and the materials have been characterized by electron microscopy, atomic force microscopy (AFM) and other methods. Boron nitride analogues of graphene have been obtained by a simple chemical procedure starting with boric acid and urea and have been characterized by various techniques that include surface area measurements. A new layered material with the composition BCN possessing a few layers and a large surface area discovered recently exhibits a large uptake of $CO_2$.

## Introduction

The discovery of fullerenes[1] in 1985 prompted the discovery of fullerene-like structures of other layered inorganic materials, such as the dichalcogenides of tungsten and molybdenum by Tenne et al.,[2–4] which was soon followed by the fullerenes of BN and other nitrides.[5] Similarly, soon after the characterization of carbon nanotubes by Iijima et al.,[6] inorganic nanotubes of $WS_2$, $MoS_2$ and BN were prepared and characterized.[7–12] Graphene, which is a two-dimensional network of $sp^2$-hybridized carbon atoms has created a great sensation in the last few years because of the unique properties and phenomena exhibited by it. Graphene is fascinating not only academically but also promises numerous applications.[13–16] A natural extension of the study of graphene would be to synthesize graphene analogues of other layered inorganic materials, the obvious candidates being $WS_2$, $MoS_2$ and BN. These materials possess graphite-like layered structures in which the layers are held together by weak van der Waals forces. In this article, we present recent developments in the synthesis, characterization and properties of inorganic analogues of graphene, such as $MoS_2$, $WS_2$ and BN.[17,18] We also report the synthesis of an exciting new material with the composition BCN, which has graphene-like features.[19]

[a] Chemistry and Physics of Materials Unit, International Centre for Materials Science, CSIR Centre of Excellence in Chemistry, Jawaharlal Nehru Centre for Advanced Scientific Research, Jakkur P. O., Bangalore 560064, India
E-mail: cnrrao@jncasr.ac.in
[b] Solid State and Structural Chemistry Unit, Indian Institute of Science,
Bangalore 560012, India

## $MoS_2$ and $WS_2$

$MoS_2$ and $WS_2$ are quasi-two-dimensional compounds in which the atoms in the layer are bound by strong covalent forces, while van der Waals forces hold the layers together just as in graphite (Figure 1a). Intercalation of al-

*C. N. R. Rao obtained his Ph.D. degree from Purdue University (1958) and D.Sc. degree from the University of Mysore (1961). He is the National Research Professor and Linus Pauling Research Professor at the Jawaharlal Nehru Centre for Advanced Scientific Research and Honorary Professor at the Indian Institute of Science (both at Bangalore). His research interests are mainly in the chemistry of materials. He is a member of all the major science academies and the recipient of the Einstein Gold Medal of UNESCO, the Hughes Medal as well as the Royal Medal of the Royal Society, London. In 2005, he received the Dan David Prize for materials research and the first India Science Prize.*

*Angshuman Nag received his M.Sc. (Chemistry) degree from Indian Institute of Technology, Guwahati, in 2003. He obtained his Ph.D. degree from the Indian Institute of Science, Bangalore, in 2009. He is currently a centenary postdoctoral fellow at the Indian Institute of Science, Bangalore. He works on nanomaterials.*

kali metals in layered metal dichalcogenides was reported sometime ago, but the products of exfoliation have not been studied adequately.[20] A graphene-like two-dimensional structure of $WS_2$ seems to have been obtained[21,22] by lithium intercalation and exfoliation, but the sample was characterized only by using X-ray diffraction (XRD). Such layered sulfides containing five layers or less do not exhibit the (002) reflection in the XRD pattern, and XRD is therefore not sufficient to estimate the number of layers. Preparation of graphene-analogues of $MoS_2$ was also attempted by lithium intercalation and exfoliation, but the characterization was only based on XRD.[23] $MoS_2$ samples prepared by lithium intercalation followed by exfoliation have been examined by employing scanning force microscopy and X-ray absorption fine structure spectroscopy,[24,25] but little information is available on the number of layers resulting from these preparations. Frindt at al.[26] reported that exfoliated $MoS_2$ forms aqueous suspensions of single layers wherein S atoms are bonded to Mo in an octahedral arrangement. Suspensions of layered chalcogenides have been used to prepare inclusion compounds with organic molecules[27,28] and to fabricate white-light-emitting diodes.[29,30] Optical properties of $MoS_2$ containing several layers prepared by micromechanical cleavage have been examined.[31,32] Rao et al.[18] have successfully carried out the reported the synthesis of few-layer $MoS_2$ and $WS_2$ by employing multiple chemical routes followed by unambiguous characterization involving different experimental techniques, including atomic force microscopy (AFM) and high-resolution transmission electron microscopy.

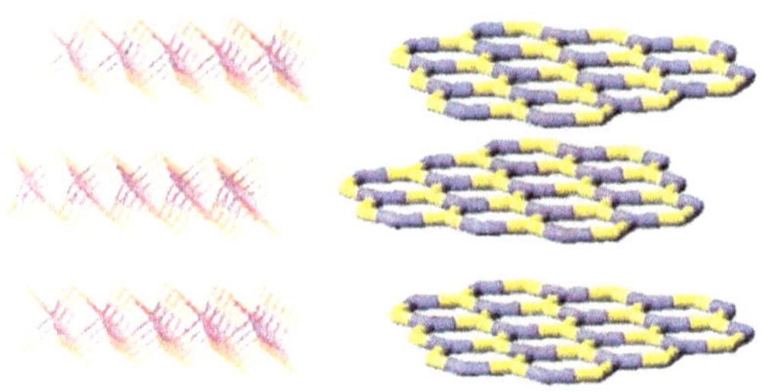

Figure 1. Layered structures of (a) $WS_2$/$Mos_2$ and (b) BN. In $WS_2$/$Mos_2$, the metal is in an octahedral coordination. Adapted from ref.[9]

Graphene analogues of $MoS_2$ and $WS_2$ can be synthesized by employing different chemical methods.[18] While methods (1) and (2) are common for both few-layer $WS_2$ and $MoS_2$, method (3) is applicable for only few-layer $MoS_2$. In method 1, bulk $WS_2$ or $MoS_2$ is intercalated with lithium followed by exfoliation of the layers by ultra sonication to form graphene analogues of $MoS_2$ and $WS_2$. In method 2, molybdic acid or tungstic acid is treated with excess thiourea. Graphene analogues of $MoS_2$ are produced by the reaction between $MoO_3$ and KSCN under hydrothermal conditions, as shown in the method 3; few-layer $MoS_2$ is formed.

$$MoS_2/WS_2 \xrightarrow[\text{Reflux in hexane, 3 d}]{n\text{BuLi}} Li\text{-}MoS_2/Li\text{-}WS_2 \xrightarrow[\text{in water}]{\text{Exfoliation}} \text{Few-layer } MoS_2/WS_2 \quad (1)$$

$$H_2MoO_4/H_2WO_4 + 48\ NH_2CSNH_2 \xrightarrow[3\ h]{500\ ^\circ C} \text{Few-layer } MoS_2/WS_2 \quad (2)$$

$$MoO_3 + 2\ KSCN \xrightarrow[180\ ^\circ C,\ 1\ d]{H_2O} \text{Few-layer } MoS_2 \quad (3)$$

The products obtained from the above reactions were characterized by employing TEM, AFM, Raman spectroscopy and XRD. The XRD patterns of few-layer $WS_2$ and $MoS_2$ along with the patterns of their bulk counterpart are shown in Figure 2a and b, respectively. $WS_2$ samples synthesized by methods 1 and 2 do not exhibit the (002) reflection, in contrast to the pattern of bulk $WS_2$ (see Figure 2a). The absence of the (002) reflection indicates that the material contains five or less layers of $WS_2$. The XRD patterns of the $MoS_2$ samples synthesized by all the three methods also do not exhibit the (002) reflection, which indicates the formation of few-layer $MoS_2$. Energy dispersive X-ray analysis (EDAX) shows the products to be stoichiometric.

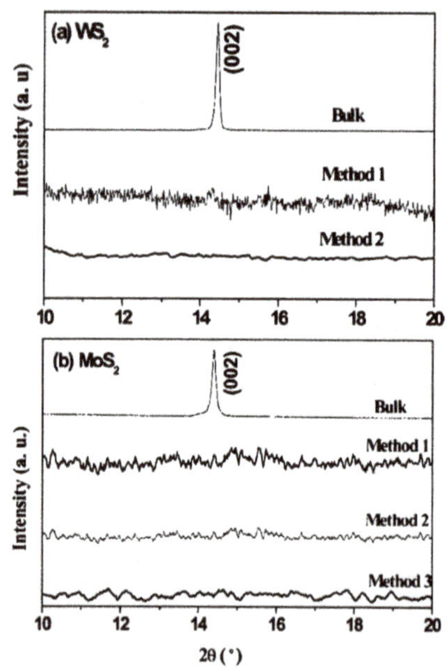


Figure 2. XRD patterns of (a) bulk $WS_2$ and few-layer $WS_2$ obtained by methods 1 and 2, and (b) bulk $MoS_2$ and few-layer $MoS_2$ obtained by methods 1, 2 and 3. Adapted from ref.[18]

The TEM images of the $WS_2$ and $MoS_2$ samples synthesized by the different methods are shown in Figure 3, which reveals that $WS_2$ obtained from both methods 1 and 2 mostly consists of single or bilayers. The interlayer separation between the $WS_2$ layers obtained from method 1 is in the range 0.65–0.70 nm (see inset of Figure 3a), and is slightly larger than the interplanar distance for the (002) plane of the bulk $WS_2$. However, $WS_2$ obtained by the thiourea method shows somewhat broken layers with an interlayer spacing of about 0.9 nm, and further studies are needed to understand the nature of these layers. The hexagonal atomic arrangement in a $WS_2$ layer is shown by the high resolution TEM image in the inset of Figure 3b. The $MoS_2$ samples synthesized by method 2 exhibit graphene-like layers (Figure 3c) with an interlayer separation of ca. 0.68 nm, which is slightly larger than the interplanar distance (0.62 nm) for the (002) plane of bulk $MoS_2$. The high resolution TEM image in Figure 3d shows the hexagonal atomic arrangement formed by Mo and S atoms with a Mo–S distance of 2.30 Å, which is in agreement with the known Mo–S distance in bulk $MoS_2$. It is to be noted that the $MoS_2$ samples prepared by all the three methods exhibit graphene-like layers, when examined by using high resolution TEM.

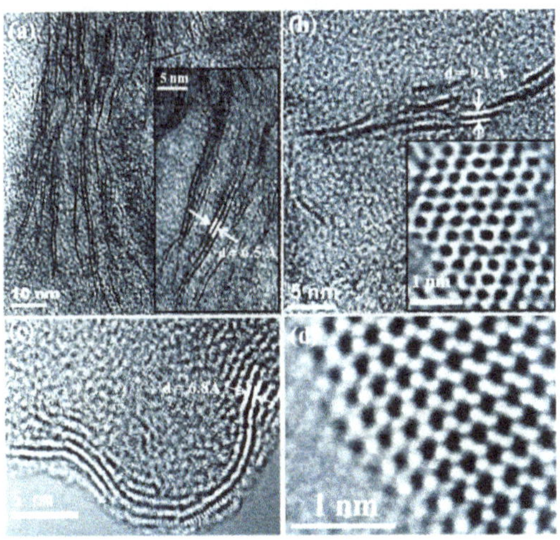


Figure 3. TEM images of $WS_2$ layers obtained from (a) method 1 and (b) method 2. (c) The TEM image of $MoS_2$ layers obtained from method 2; (d) the high-resolution TEM image of layered $MoS_2$ obtained from method 3. Adapted from ref.[18]

In addition to the TEM images, all the prepared $WS_2$ and $MoS_2$ samples were investigated by using AFM. Representative AFM images and the corresponding height profiles of the $WS_2$ sample prepared by method 1 are shown in Figure 4a and b, respectively. The AFM results confirm the formation of bilayer $WS_2$ with an average thickness of ca. 1.3 nm. AFM images of the $MoS_2$ samples similarly show graphene-like layers.

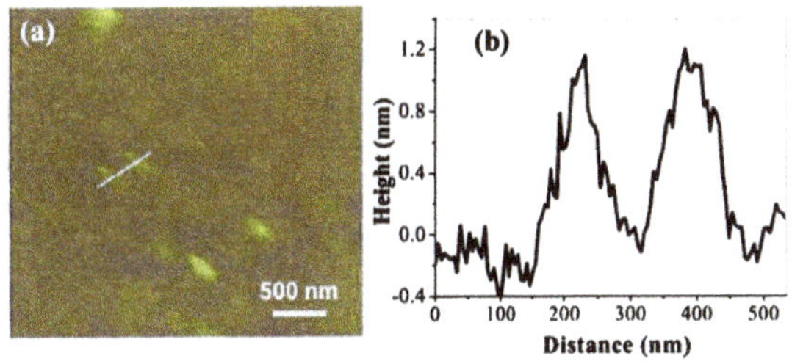


Figure 4. (a) AFM image and the corresponding (b) height profiles of $WS_2$ layers obtained by method 1. Adapted from ref.[18]

We compare the Raman spectra of graphene-like $WS_2$ and $MoS_2$ with those of their bulk counterpart. The spectrum for bulk $WS_2$ shows bands at 351 and 420 $cm^{-1}$ arising from the $E_{2g}$ and $A_{1g}$ modes with FWHM values of about 7.8 and 2.4 $cm^{-1}$, respectively. However, the spectrum of $WS_2$ samples obtained from method 1 shows bands at 350 and 415 $cm^{-1}$ with FWHM values of 13.7 and 8.4 $cm^{-1}$. A similar softening of the $A_{1g}$ band and an increase in FWHM has been observed for few-layer $WS_2$ synthesized by method 2. Furthermore, the spectral width also increases in the spectrum of graphene-like $WS_2$. Figure 5 clearly shows that both the $A_{1g}$ and $E_{2g}$ modes soften in the graphene analogues of $MoS_2$ along with a spectral broadening compared to that seen for bulk $MoS_2$. The broadening of the Raman spectra occurs because of phonon confinement. We have prepared graphene-like $MoS_2$ by micromechanical cleavage of a $MoS_2$ single crystal by using the scotch-tape technique. Raman spectra of these samples exhibit a progressive softening of the $A_{1g}$ and $E_{2g}$ bands with a decrease in the number of layers. Thus, softening and broadening of Raman bands seem to be characteristic of few-layer $MoS_2$ and $WS_2$.

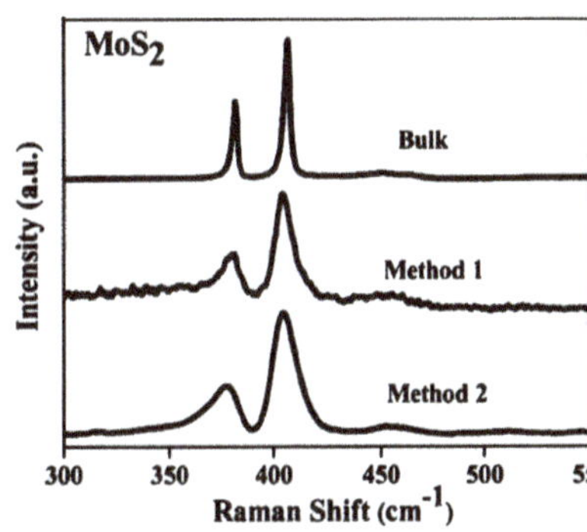


Figure 5. Raman spectra of bulk $MoS_2$ and few-layer $MoS_2$ obtained by methods 1 and 2. Adapted from ref.[18]

## Boron Nitride

BN which is a layered material similar to graphite (Figure 1b) has a wide band gap. This material has received continuous scientific attention because of its remarkable mechanical, optical and electronic properties, with resistance to oxidation over a wide range of temperatures. A pair of boron and nitrogen is isoelectronic with a pair of carbon, and hexagonal boron nitride (h-BN) with a layered structure provides the motivation to prepare a graphene-analogue of BN, namely, mono- and few-layer BN. Mono- and few-layer BN has been prepared by peeling off BN layers from a bulk BN crystal by micromechanical cleavage,[33] by sonication[34] of BN particles, or by using a high-energy electron beam[35,36] inside a transmission electron microscope. We have been able to carry out the controlled chemical synthesis of graphene analogues of BN possessing a desired number of layers that can be used for the large-scale production of the material for the preparation of composites and for other applications.[17] In the course of this study, single-layer BN has been identified.

Graphene analogues of BN could be synthesized by employing the reaction of boric acid with urea at high temperatures.[37–39] Boric acid/urea mixtures with different molar ratios (1:6, 1:12, 1:24, 1:48 and 1:72) were used to obtain the desired number of BN layers.[17] The formation of BN is given by the following reactions (4), (5) and (6).[17]

$$2\,B(OH)_3 \xrightarrow{\Delta} B_2O_3 + 3\,H_2O \quad (4)$$

$$NH_2CONH_2 \xrightarrow{\Delta} NH_3 + HNCO \quad (5)$$

$$B_2O_3 + 2\,NH_3 \xrightarrow{\Delta} 2\,BN + 3\,H_2O \quad (6)$$

Characterization of few-layer BN was carried out by XRD, Raman spectroscopy, Fourier transform infrared (FTIR) spectroscopy, TEM and AFM. The XRD patterns of the BN samples prepared with different boric acid to urea ratio exhibit two broad reflections with *d* spacings of ca. 3.51 Å and ca. 2.12 Å, which correspond to the (002) and (100) planes of h-BN, respectively. Interestingly, the width of the (002) reflection increases systematically with an increase in the proportion of urea in the starting reaction mixture. Also, the *d* spacing of the (002) plane increases slightly when the boric acid to urea ratio is 1:48, which suggests lower sheet-to-sheet registry in the sample. However, the Raman spectra of these samples remain unaltered with a band centred at 1373 cm$^{-1}$ arising from the $E_{2g}$ tangential mode.[40] The FTIR spectra show peaks at 1384 and 800 cm$^{-1}$ assigned to the in-plane B–N transverse optical mode and out-of-plane B–N–B bending mode, respectively, along with the signature of the N–H stretching band from the amino groups on the surface.[41]

Electron energy loss spectroscopy (EELS) carried out in the transmission electron microscope reveals the π* and σ* peaks of the sp$^2$-hybridized B and N[42] and also that the composition of the product is stoichiometric. The TEM images in Figure 6 show the presence of single- as well as few-layer BN in the samples synthesized with different boric acid to urea ratios. The interlayer spacing is ca. 0.35 nm, which is slightly larger than that for the (002) planes of bulk h-BN (0.33 nm). The samples prepared with boric acid to urea ratios of 1:12 (Figure 6a) and 1:24 show the presence of eight and six layers of BN, respectively. Significantly, Figure 6b shows the presence of single layers along with two layers in the sample prepared with a ratio of 1:48 for boric acid/urea. The inset in Figure 6b shows the hexagonal atomic arrangement of the BN crystal. While the B–N bond length has been measured to be 1.44 Å, the separation between two nearest B atoms (or two nearest N atoms) is found to be 2.49 Å, both these values are in good agreement with the corresponding values for bulk h-BN.[34–36]

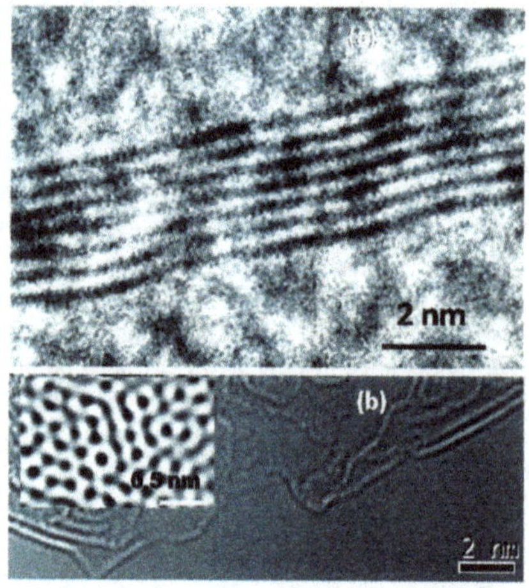


Figure 6. TEM images of few-layer BN prepared with boric acid/urea mixture with a ratio of (a) 1:12 and (b) 1:48. The inset in Figure 6b shows the HRTEM image of few-layer BN prepared with a 1:48 ratio for boric acid/urea. Adapted from ref.[17]

More reliable information on the number of layers in the BN samples was obtained by the AFM analysis, as shown in Figure 7. The AFM image of a BN flake prepared with a reactant ratio of 1:24 (Figure 7a) shows a steplike increase in the height from 0.4 to 0.8 nm, which corresponds to a single layer and bilayers. Figure 7b shows single-layer BN

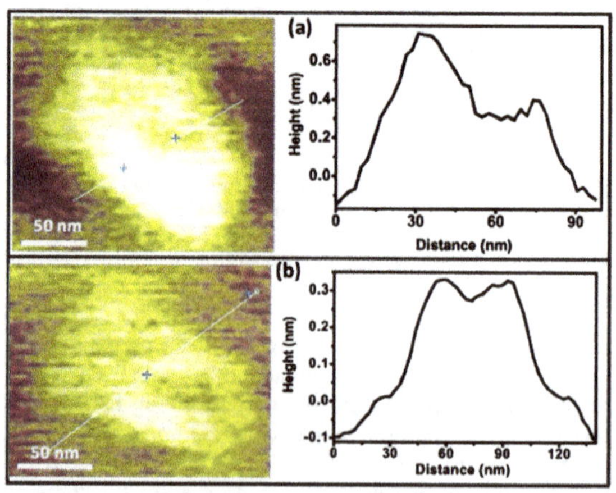


Figure 7. AFM images and corresponding height profiles of few-layer BN prepared with a (a) 1:24 and (b) 1:48 ratio for boric acid/urea. Adapted from ref.[17]

obtained from the sample prepared with a boric acid to urea ratio of 1:48. We prepared histograms of the thickness of the layer by analyzing AFM images of ca. 100 flakes for each sample synthesized with different urea content in the reaction mixtures, and the extracted average number of layers for different reactant ratios is plotted in Figure 8a. Clearly, the number of BN layers decreases with an increase in the proportion of urea in the starting reaction mixture. Thus, it is not surprising that the width of the (002) reflection in XRD increases with the proportion of urea in the reaction mixture; this graph is shown in the inset of the Figure 8a. A further increase in the boric acid/urea ratio of up to 1:72 does not result in thinner BN flakes relative to those obtained with a boric acid/urea ratio of 1:48. Recently, we have been able to synthesize few-layer BN using chemical vapour deposition techniques starting with a mixture of $BBr_3$ and $NH_3$.

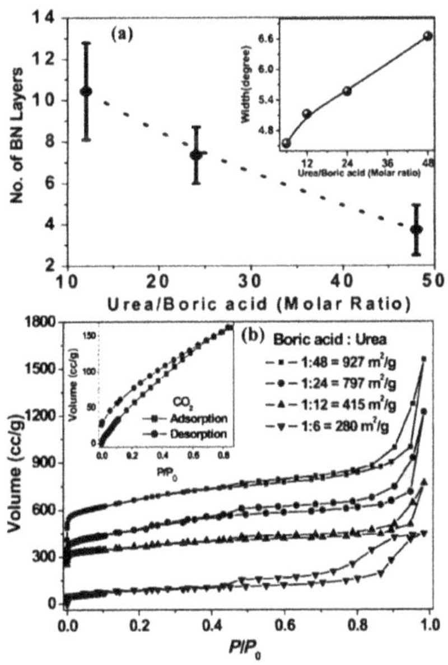


Figure 8. (a) Variation in the average number of BN layers obtained from AFM analysis vs. the urea/boric acid ratio. The inset shows the variation in the width of the (002) diffraction, obtained from XRD patterns, vs. the urea/boric acid precursor ratio. (b) Nitrogen adsorption–desorption isotherms of few-layer BN prepared with different boric acid/urea ratios in the reaction mixture. For a clear representation, plots for the ratios of 1:12, 1:24 and 1:48 have been moved vertically upward by 150, 270 and 400 units. The inset shows the $CO_2$ adsorption–desorption isotherms at 195 K and 1 atm for few-layer BN prepared with a boric acid/urea ratio of 1:48.

Figure 8b shows the $N_2$ adsorption–desorption isotherms of few-layer BN samples synthesized with different boric acid/urea ratios. Brunaure–Emmett–Teller (BET) surface areas of these samples calculated from the isotherms (Figure 8b) show an increase in the surface area with an increase in the proportion of urea in the reaction mixture. This systematic increase in the surface area is because of the decrease in the thickness of the layer with increasing urea content in the reaction mixture. To the best of our knowledge, the surface area of 927 $m^2/g$ obtained with a boric acid to urea ratio of 1:48 is the highest value reported for BN so far. Remarkably, BN with the highest surface area adsorbs 32 wt.-% of $CO_2$ at 195 K and $p/p_0 = 0.85$, as shown in the inset of Figure 8b.

Solubilization of these inorganic graphenes is required for possible applications in composites and elsewhere. We have solubilized few-layer BN by interaction with Lewis bases.[43] The inherent electron deficiency of boron allows us to form Lewis acid/Lewis base adducts, by interacting few-layer BN with Lewis bases. Thus, adducts with trioctylamine (TOA) and trioctylphosphane (TOP) are soluble in nonpolar solvents such as toluene, while the as-prepared BN samples are insoluble in both polar and nonpolar solvents.

## BCN – A New Graphene-Like Material

Similarly to h-BN and graphite, materials containing boron, carbon and nitrogen ($B_xC_yN_z$) also possess layered structures.[44–46] There are rare examples[47,48] of materials with compositions close to BCN, but their compositions and properties have not been investigated fully. We have recently been able to synthesize a graphite-like material with the composition BCN.[19] The material mainly consists of 2–3 layers of BCN, thus establishing a new graphene analogue.

Few-layer BCN could be synthesized by employing a reaction similar to that employed for few-layer BN. Activated charcoal with a high-surface area (1256 $m^2/g$) was used as a precursor for carbon. Addition of activated charcoal modifies reaction (7) to give BCN.[19]

$$B_2O_3 + 2\,NH_3 + 2\,C \xrightarrow{\Delta} 2\,BCN + 3\,H_2O \qquad (7)$$

Characterization of few-layer BCN was carried out by using XRD, Raman spectroscopy, TEM, AFM, X-ray photoelectron spectroscopy (XPS), EELS and thermogravimetric analysis (TGA). XRD and selected area electron diffraction (SAED) patterns show the (002) and (100) reflections of the graphite-like material. The obtained interlayer distance in BCN (3.46 Å) is slightly larger than that in both graphite (3.33 Å) and h-BN (3.36 Å). The Raman spectrum of BCN is similar to that of graphitic materials with a D band at 1324 $cm^{-1}$ and G band at 1592 $cm^{-1}$.[49]

The unusual composition of BCN requires confirmation and verification by several means, in contrast to that of few-layer $WS_2$, $MoS_2$ and BN. TGA analysis carried out in air shows that BCN loses about 57% of weight at ca. 600 °C to leave a residue that is stable up to 1000 °C, while activated charcoal completely oxidizes before 600 °C. The XRD pattern of the BCN residue shows the presence of BN and $C_3N_4$, which suggests the presence of B, C and N in the final product. XPS core level spectra of few-layer $B_xC_yN_z$ products show broad signals for B 1s, C 1s and N 1s. The C 1s signal could be deconvoluted into three component spectra with binding energies of 283.1, 285.4 and 288.2 eV, which correspond to carbon bonded to boron, carbon and nitrogen, respectively.[42] The N 1s signal was deconvoluted into two bands with binding energies of 397.1 and 399.8 eV, which correspond to nitrogen bonded to boron and carbon, respectively.[42] Similarly, the B 1s signal was deconvoluted into two bands corresponding to boron bonded to carbon and nitrogen. The XPS spectra provide evidence for B–C, B–N, N–C and C–C bonds in the final product but also give an estimate of the composition to be $BC_{1.5}N_{1.1}$.

To have a more accurate estimation of the elemental composition and its homogeneity throughout the flake, we carried out EELS measurements on the K-edge absorption for B, C and N, as shown in Figure 9. K-shell ionization edges (Figure 9) at 188, 284 and 401 eV correspond to B, C and N respectively. The two peaks at each core edge suggest that all three elements are $sp^2$ hybridized, as expected in a graphite-like layered structure.[42] EELS measurements reveal the average chemical composition of the product to be $BC_{1.2}N$, which is in reasonable agreement with the XPS results. We conclude the composition to be similar to BCN. EEL mapping of boron, carbon and nitrogen showed that the composition is homogeneous throughout a BCN flake.

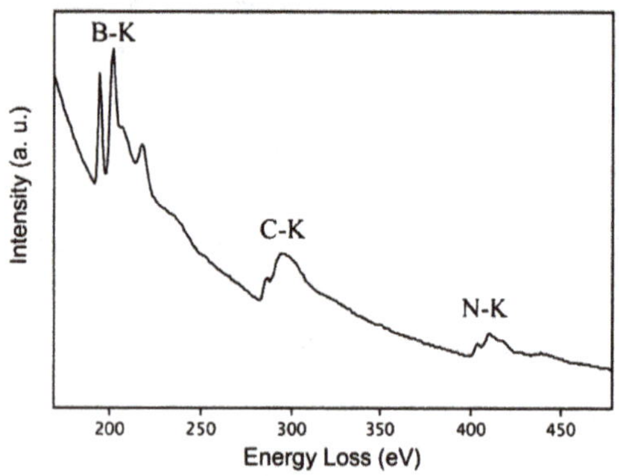


Figure 9. EEL spectrum showing the K-edges of boron, carbon and nitrogen of BCN. Adapted from ref.[19]

TEM and FESEM images of BCN show a flaky morphology with ripples and entanglement, similar to that of graphene. High-resolution TEM images show that the majority of the flakes consist of 2 or 3 layers with an interplanar distance of 3.43 Å. AFM images and the corresponding height profiles in Figure 10a and b exhibit flakes with thicknesses of 0.65 and 1.1 nm, which correspond to two and three layers, respectively. AFM analysis of over 100 BCN flakes has shown that they generally comprise two to three layers, with an occasional presence of single layers.

Few-layer BCN exhibits interesting gas adsorption properties. The BET surface area of the BCN sample was calculated to be 2911 $m^2/g$, by using the $N_2$ adsorption–desorption isotherms shown in Figure 11. The obtained surface area of BCN is higher than that for any composition of the $B_xC_yN_z$ materials reported so far. This high-surface-area BCN exhibits a remarkably high $CO_2$ uptake of 100 wt.-% at 195 K and 1 atm [inset (i) in Figure 11]. However, at room temperature and 40 bar, the $CO_2$ uptake is 44 wt.-%. Materials with high $H_2$ uptake are important for clean-energy alternatives. $H_2$ adsorption on BCN at 77 K and 1 atm is found to be 2.6 wt.-% [see inset (ii) in Figure 11]. $H_2$ uptake on BCN remains the same even after increasing the gas pressure. The BCN sample also adsorbs methanol vapour (about 26 wt.-% at room temperature and atmospheric pressure).

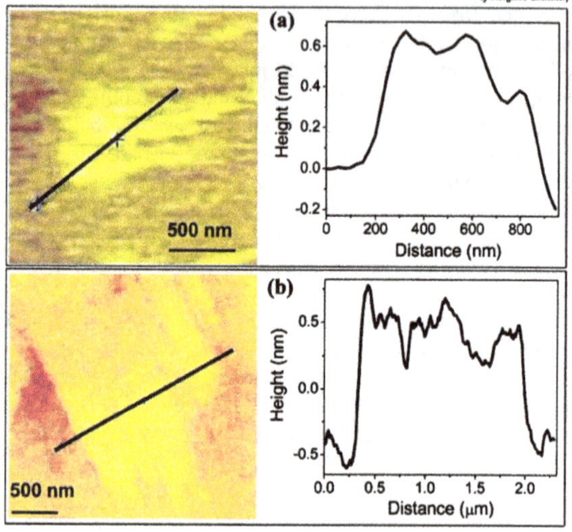


Figure 10. AFM images and their corresponding height profiles for (a) two-layered and (b) three-layered BCN. Adapted from ref.[19]

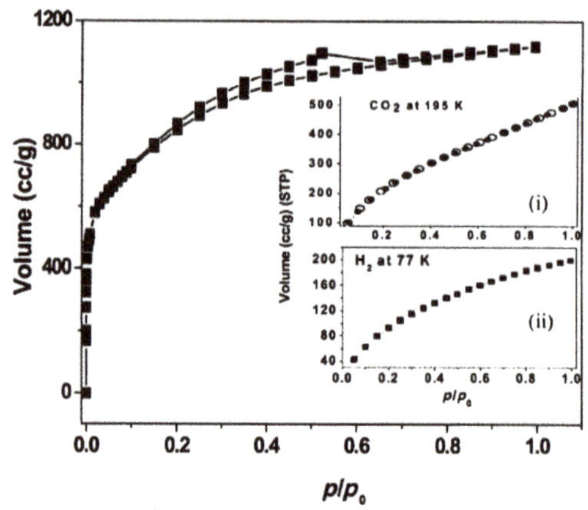


Figure 11. Nitrogen adsorption–desorption isotherms of few-layer BCN. Insets (i) and (ii) show the $CO_2$ and $H_2$ adsorption–desorption isotherms of the BCN sample at 195 K and 77 K, respectively and at 1 atm. Adapted from ref.[19]

The magnetic properties of BCN with 2–3 layers were compared with those of graphene. The temperature-dependent magnetic properties of few-layer BCN at 500 Oe show a divergence between field-cooled (FC) and zero-field-cooled (ZFC) data below 300 K. The divergence disappears at higher fields. Such a divergence between FC and ZFC at lower fields is somewhat similar to that seen in magnetically frustrated systems or phase-separated oxides. The Curie–Weiss temperature is –350 K, which indicates antiferromagnetic interactions. However, the material shows weak ferromagnetic hysteresis at 300 K. The presence of both ferromagnetic and antiferromagnetic interactions in BCN is similar to those in graphene,[50] and such magnetism could arise from the contribution of edge effects and defects.

## Conclusions

It is noteworthy that we have been able to prepare graphene analogues of $WS_2$, $MoS_2$, BN and BCN using bottom-up chemical methods. These are rare examples of inorganic graphene analogues, which constitute a new family of two-dimensional nanostructures. All these layered materials have been characterized unambiguously by using a plethora of techniques to establish the formation of single- and bi- or few-layer material. Inorganic graphenes are expected to exhibit unique properties. For example, few-layer BCN exhibits high surface areas as well as high $CO_2$ uptake. Few-layer BN shows a systematic increase in the surface area with decreasing thickness of the layer. The interesting gas adsorption properties of these materials arise because of their graphene-like morphology. Few-layer BN can be used to prepare novel polymer composites. Few-layer $MoS_2$ could be useful in solid-state lubrication. Clearly, there is much to be explored with this new class of two-dimensional inorganic nanomaterials.

## Acknowledgments

Angshuman Nag acknowledges the Indian Institute of Science for the centenary postdoctoral fellowship.

[1] H. W. Kroto, J. R. Heath, S. C. O'Brien, R. F. Curl, R. E. Smalley, *Nature* **1985**, *318*, 162–163.
[2] R. Tenne, L. Margulis, M. Genut, G. Hodes, *Nature* **1992**, *360*, 444–446.
[3] L. Margulis, G. Salitra, R. Tenne, M. Talianker, *Nature* **1993**, *365*, 113–114.
[4] R. Rosentsveig, A. Margolin, A. Gorodnev, R. P. Biro, L. Rapaport, Y. Novema, G. Naveh, R. Tenne, *J. Mater. Chem.* **2009**, *19*, 4368–4374.
[5] D. Golberg, Y. Bando, O. Stephan, K. Kurashima, *Appl. Phys. Lett.* **1998**, *73*, 2441.
[6] S. Iijima, *Nature* **1991**, *354*, 56–58.
[7] N. G. Chopra, R. J. Luyken, K. C. Cherrey, V. H. Crespi, M. L. Cohen, S. G. Louie, A. Zettl, *Science* **1995**, *269*, 966–967.
[8] A. Rothschild, J. Sloan, R. Tenne, *J. Am. Chem. Soc.* **2000**, *122*, 5169–5179.
[9] C. N. R. Rao, M. Nath, *Dalton Trans.* **2003**, 1–24.
[10] R. Tenne, C. N. R. Rao, *Phil. Trans. R. Soc. London A* **2004**, *362*, 2099–2125.
[11] C. Y. Zhi, Y. Bando, C. Tang, Q. Huang, D. Golberg, *J. Mater. Chem.* **2008**, *18*, 3900–3908.
[12] C. N. R. Rao, A. Govindaraj, *Adv. Mater.* **2009**, *21*, 4208–4233.
[13] K. S. Novoselov, A. K. Geim, S. V. Morozov, D. Jiang, Y. Zhang, S. V. Dubonos, I. V. Grigorieva, A. A. Firsov, *Science* **2004**, *306*, 666–669.
[14] A. K. Geim, K. S. Novoselov, *Nat. Mater.* **2007**, *6*, 183–191.
[15] C. N. R. Rao, A. K. Sood, K. S. Subrahmanyam, A. Govindaraj, *Angew. Chem. Int. Ed.* **2009**, *48*, 7752–7777.
[16] A. K. Geim, *Science* **2009**, *324*, 1530–1534.
[17] A. Nag, K. Raidongia, K. P. S. S. Hembram, R. Datta, U. V. Waghmare, C. N. R. Rao, *ACS Nano* **2010**, *4*, 1539–1544.
[18] H. S. S. R. Matte, A. Gomathi, A. K. Manna, D. J. Late, R. Datta, S. K. Pati, C. N. R. Rao, *Angew. Chem. Int. Ed.* **2010**, *49*, 4059–4062.
[19] K. Raidongia, A. Nag, K. P. S. S. Hembram, U. V. Waghmare, R. Datta, C. N. R. Rao, *Chem. Eur. J.* **2010**, *16*, 149–157.
[20] S. M. Zahurak, D. W. Murphy, *J. Solid State Chem.* **1987**, *70*, 137–140.
[21] B. K. Miremadi, S. R. Morrison, *J. Appl. Phys.* **1988**, *63*, 4970–4974.
[22] D. Yang, R. F. Frindt, *J. Phys. Chem. Solids* **1996**, *57*, 1113–1116.
[23] P. Joensen, R. F. Frindt, S. R. Morrison, *Mater. Res. Bull.* **1986**, *21*, 457–461.
[24] A. Schumacher, L. Scandella, N. Kruse, R. Prins, *Sur. Sci. Lett.* **1993**, *289*, 595–598.
[25] R. A. Gordon, D. Yang, E. D. Crozier, D. T. Jiang, R. F. Frindt, *Phys. Rev. B* **2002**, *65*, 125407–125415.
[26] D. Yang, S. J. Sandoval, W. M. R. Divigalpitiya, J. C. Irwin, R. F. Frindt, *Phys. Rev. B* **1991**, *15*, 12053–1205.
[27] M. A. Gee, R. F. Frindt, P. Joensen, S. R. Morrison, *Mater. Res. Bull.* **1986**, *21*, 543–549.
[28] W. M. R. Divigalpitiya, R. F. Frindt, S. R. Morrison, *Science* **1989**, *246*, 369–371.
[29] G. L. Frey, K. J. Reynolds, R. H. Friend, H. Cohen, Y. Feldman, *J. Am. Chem. Soc.* **2003**, *125*, 5998–6007.
[30] G. L. Frey, K. J. Reynolds, R. H. Friend, *Adv. Mater.* **2002**, *14*, 265–268.
[31] R. F. Frindt, *J. Appl. Phys.* **1996**, *37*, 1928–1929.
[32] R. F. Frindt, A. D. Yoffe, *Proc. R. Soc.* **1963**, *273*, 69–83.
[33] W. Q. Han, L. J. Wu, Y. M. Zhu, K. Watanabe, T. Takiguchi, *Appl. Phys. Lett.* **2008**, *93*, 223103.
[34] C. Zhi, Y. Bando, C. Tang, H. Kuwahara, D. Golberg, *Adv. Mater.* **2009**, *21*, 2889–2893.
[35] J. C. Meyer, A. Chuvilin, G. Algara-Siller, J. Biskupek, U. Kaiser, *Nano Lett.* **2009**, *9*, 2683–2689.
[36] C. Jin, F. Lin, K. Suenaga, S. Iijima, *Phys. Rev. Lett.* **2009**, *102*, 195505.
[37] A. Gomathi, A. Sundaresan, C. N. R. Rao, *J. Solid State Chem.* **2007**, *180*, 291–295.
[38] A. Gomathi, M. R. Harika, C. N. R. Rao, *Mater. Sci. Eng.* **2008**, *476*, 29–33.
[39] A. Gomathi, S. Reshma, C. N. R. Rao, *J. Solid State Chem.* **2009**, *182*, 72–76.
[40] S. Saha, D. V. S. Muthu, D. Goldberg, C. Tang, C. Zhi, Y. Bando, A. K. Sood, *Chem. Phys. Lett.* **2006**, *421*, 86–90.
[41] C. Zhi, Y. Bando, C. Tang, D. Golberg, *J. Am. Chem. Soc.* **2005**, *127*, 17144–17145.
[42] S. Y. Kim, J. Park, H. Chul Choi, J. P. Ahn, J. Q. Hou, H. S. Kang, *J. Am. Chem. Soc.* **2007**, *129*, 1705–1716.
[43] S. Pal, S. R. C. Vivekchand, A. Govindaraj, C. N. R. Rao, *J. Mater. Chem.* **2007**, *17*, 450–452.
[44] M. Kawaguchi, *Adv. Mater.* **1997**, *9*, 615–625.
[45] K. Raidongia, D. Jagadeesan, M. U. Kahaly, U. V. Waghmare, S. K. Pati, M. Eswaramoorthy, C. N. R. Rao, *J. Mater. Chem.* **2008**, *18*, 83–90.
[46] K. Raidongia, K. P. S. S. Hembram, U. V. Waghmare, M. Eswaramoorthy, C. N. R. Rao, *Z. Anorg. Allg. Chem.* **2010**, *636*, 30–35.
[47] R. B. Kaner, J. Kouvetakis, C. E. Warble, M. L. Sattler, N. Bartlett, *Mater. Res. Bull.* **1987**, *22*, 399–404.
[48] A. Vinu, M. Terrones, D. Golberg, S. Hishita, K. Ariga, T. Mori, *Chem. Mater.* **2005**, *17*, 5887–5890.
[49] C. N. R. Rao, K. Biswas, K. S. Subrahmanyam, A. Govindaraj, *J. Mater. Chem.* **2009**, *19*, 2457.
[50] H. S. S. R. Matte, K. S. Subrahmanyam, C. N. R. Rao, *J. Phys. Chem. C* **2009**, *113*, 9982–9985.

Received: April 14, 2010
Published Online: June 23, 2010

  

DOI: 10.1002/zaac.201200283

# Synthesis, Characterization, and Properties of Few-layer Metal Dichalcogenides and their Nanocomposites with Noble Metal Particles, Polyaniline, and Reduced Graphene Oxide

H. S. S. Ramakrishna Matte,[a] Urmimala Maitra,[a] Prashant Kumar,[a] Bolla Govinda Rao,[a] K. Pramoda,[a] and C. N. R. Rao*[a]

**Keywords:** Graphene; Nanosheets; $MoS_2$; $WS_2$; Dichalcogenides; Nanoparticles; Nanocomposites

**Abstract.** Few-layer graphene-like sheets of $MoS_2$, $WS_2$, $MoSe_2$ and $WSe_2$ were prepared by employing two methods. The first method uses microwave synthesis in ethylene glycol or water medium and the second method involves laser irradiation of the bulk samples dispersed in dimethylformamide. The few-layer dichalcogenide samples were characterized by various microscopic and spectroscopic techniques. These graphene mimics show room-temperature ferromagnetism and exchange bias besides magnetoresistance at room temperature. We have deposited nanoparticles of gold, silver, and platinum on few-layer $WS_2$ and characterized the nanocomposites. Nanocomposites of $MoS_2$ with reduced graphene oxide show increase in resistivity with increasing amount of $MoS_2$. The polyaniline $MoS_2$ composite prepared by *in situ* polymerization of aniline in the presence of few-layer $MoS_2$ shows evidence of polaronic character.

## Introduction

Graphene has been investigated widely in the last few years and several interesting properties have been studied.[1] Graphene-like species of $MoS_2$ and other layered inorganic materials have been prepared by both physical and chemical methods.[2] The chemical methods involve hydrothermal synthesis and alkali metal intercalation followed by exfoliation.[3] Microwave synthesis is considered to be an effective method for the synthesis of inorganic materials because of the small reaction time and high product yield compared to conventional methods.[4] There has been some effort to prepare different morphologies of molybdenum and tungsten chalacogenides using microwaves.[5] We have found it possible to synthesize few-layer sulfides and selenides of molybdenum under microwave heating using water and ethylene glycol as solvents. We have also developed another method for synthesis of few-layer chalcogenides based on laser irradiation of the bulk samples dispersed in dimethylformamide (DMF) medium. The uniqueness of this single-step method lies in that it does not involve intercalation or any chemical reactions. The few-layered chalcogenides prepared by the above methods were characterized by X-ray diffraction (XRD), field emission scanning electron microscopy (FESEM), transmission electron microscopy (TEM), atomic force microscopy (AFM), and Raman spectroscopy. We have measured magnetic properties of the few-layered metal chalcogenides, to surprisingly find room-temperature ferromagnetism. We have decorated and characterized few-layer $WS_2$ with nanoparticles of gold, silver, and platinum. Such nanocomposites have generated considerable interest in the recent years.[6] We have also prepared nanocomposites of few-layered $MoS_2$ with polyaniline (PANI) reduced graphene oxide (RGO) and we measured their electrical properties.

* Prof. Dr. C. N. R. Rao
E-Mail: cnrrao@jncasr.ac.in
[a] Chemistry and Physics of Materials Unit
New Chemistry Unit, and Sheik Saqr Laboratory
International Centre for Materials Science
Jawaharlal Nehru Centre for Advanced Scientific Research
Jakkur P. O.
Bangalore 560 064, India

## Results and Discussion

### *Synthesis*

In Figure 1(a) and (b), we show the XRD patterns of graphene-like layered structures of $MoS_2$ and $MoSe_2$ prepared by microwave synthesis using ethylene glycol (EG) as the medium. Absence of the (002) reflection in the XRD patterns of the nanosheets indicates that the products formed in the reactions contain only a few layers. In Figure 1(c) we compare the Raman spectrum of 3–4 layered $MoS_2$ with that of the bulk sample. The bulk sample gives bands at 406.5 and 381.2 $cm^{-1}$ due to the $A_{1g}$ and $E_{2g}$ modes. The few-layered $MoS_2$ prepared by the microwave method exhibits the corresponding bands at 403.7 and 376.7 $cm^{-1}$ and the bands are much broader. There is a clear softening of both the $A_{1g}$ and $E_{2g}$ modes in the graphene analogues of $MoS_2$. The Raman spectrum of few-layer $MoSe_2$ shown in Figure 1(d) gives Raman bands at 168.1 and 236.4 $cm^{-1}$ due to the $E_{1g}$ and $A_{1g}$ modes, respectively, compared to 168.8 and 242 $cm^{-1}$ of the bulk sample. There is also softening of both the $A_{1g}$ and $E_{1g}$ modes in few-layer samples. The FWHM values of the few-layer samples are much larger than those of the bulk sample. The broadening of the Raman bands is considered to be due to phonon confinement.[2b,2c,7]

Special Issue

  

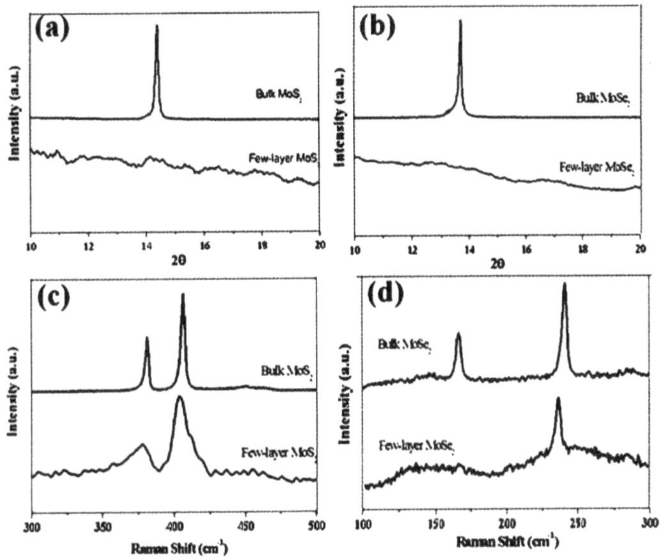


**Figure 1.** XRD patterns of bulk samples and few-layer (a) $MoS_2$ and (b) $MoSe_2$, Raman spectra of bulk and few-layered (c) $MoS_2$ and (d) $MoSe_2$ prepared by microwave treatment.



TEM studies show the formation of predominantly 3–4 layered $MoS_2$ by the microwave method, with an interlayer spacing in the 0.6–0.9 nm range as can be seen from Figure 2(a). AFM images and the corresponding height profiles in Figure 2(b) confirm the presence of three to four layers with a height profile of 2.5 nm and lateral dimensions varying from 300 nm to 500 nm in accordance with the TEM observations. FESEM images also reveal the layered morphologies of the $MoS_2$ samples obtained by microwave treatment.

A typical TEM image of a few-layer $MoSe_2$ sample obtained by the microwave method consists of 3–5 layers (Figure 2(c)). To further confirm the number of layers, we have measured the height profiles using AFM and found it to contain 2–3 layers, the height being around 1.7 nm (Figure 2(d)). By using water as a solvent instead of EG during the microwave synthesis, we were also able to obtain graphene analogues of $MoS_2$ and $MoSe_2$.

By laser irradiation of bulk $MoS_2$, we could obtain nanosheets with lateral dimensions in the range of 150–600 nm as can be seen from the FESEM image in Figure 3(a). Figure 3(b) shows a TEM image of few-layered $MoS_2$. The inset in Figure 3(b) represents the electron diffraction (ED) pattern showing the hexagonal pattern indicating crystallinity of the sample. The laser irradiated samples consist of 3–5 layers as confirmed by the AFM studies shown in Figure 3(c) and (d). We could also obtain two-dimensional sheets of $MoSe_2$ by laser irradiation of the bulk sample. In Figure 4(a) and (b) we show the FESEM and TEM images of the $MoSe_2$ sample obtained by this means. The lateral dimensions of the $MoSe_2$ sheets are in the range 400 nm to 1 μm consisting of 3–6 layers. The inset in Figure 4(b) shows the electron diffraction pattern characteristic of hexagonal $MoSe_2$. Nanosheets of $WS_2$ and $WSe_2$ could also be prepared by laser radiation as can be seen from the FESEM images in Figure 4(c) and (d), respectively. All the

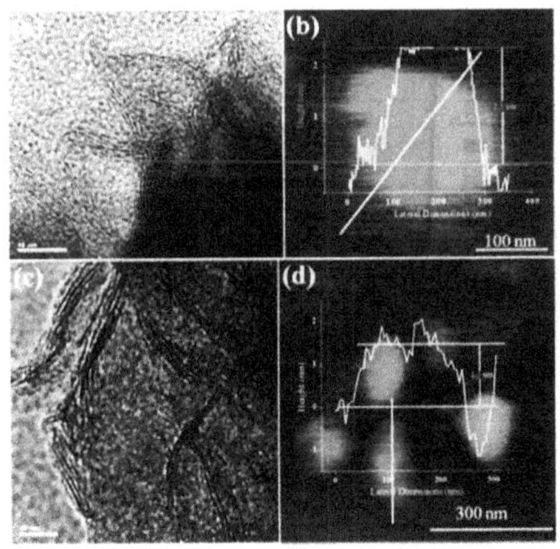


**Figure 2.** (a) TEM and (b) AFM images of few-layered $MoS_2$, height profile is shown in (b). (c) TEM and (d) AFM images of few-layered $MoSe_2$, height profile is shown in (d).

  

nanosheets of the chalcogenides prepared were are highly dispersible in DMF.

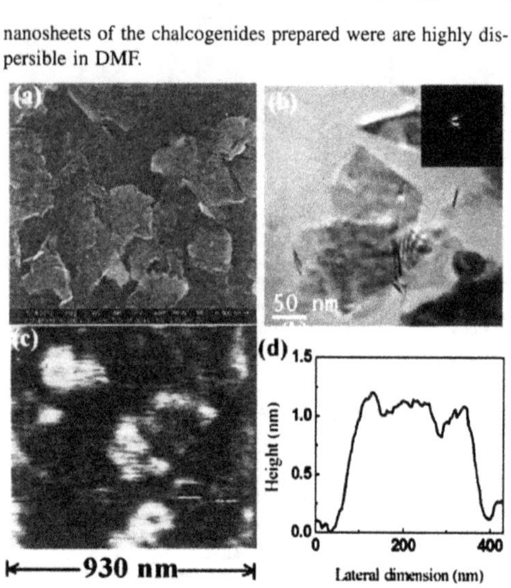


**Figure 3.** (a) FESEM, (b) TEM, (c) AFM images, and (d) height profile for laser exfoliated $MoS_2$ sheets. Inset in (b) shows the electron diffraction pattern for the sheets.

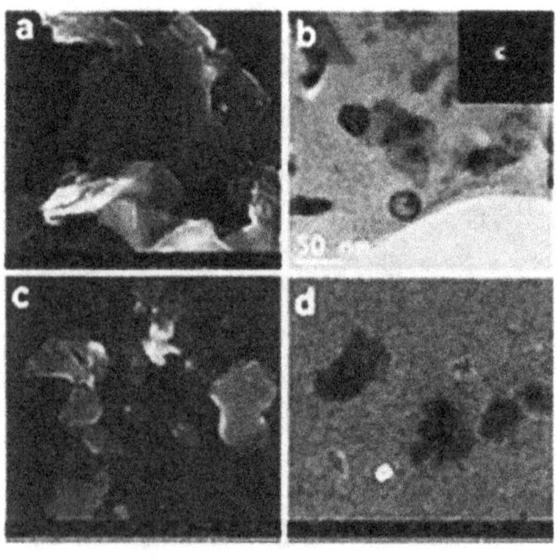


**Figure 4.** Laser exfoliated sheets of $MoSe_2$ imaged by (a) FESEM, (b) TEM (inset shows the electron diffraction pattern), FESEM images of laser exfoliated sheets of (c) $WS_2$ and (d) $WSe_2$.

### *Magnetic Properties*

We have carried out magnetic investigations of few-layer molybdenum and tungsten sulfides and selenides. Temperature-dependent magnetization data of the 3–6 layered $MoS_2$ sample at 500 Oe and 1 T are shown in Figure 5(a). We see divergence between the field-cooled (FC) and zero-field-cooled (ZFC) data starting around 300 K, a behavior common to all the layered samples prepared by us. The divergence nearly disappears on application of higher fields. This type of behavior is generally observed in magnetically frustrated systems such as spin glasses, where usually ferromagnetic and antiferromagnetic bonds are randomly distributed. A similar behavior has been observed in few-layer graphene as well.[8] We also observe a low temperature upturn in the magnetization showing a Curie-like tail. Such a Curie-like tail is known to arise at low temperatures, where the paramagnetic component takes over ferromagnetic or antiferromagnetic interactions. This behavior is found in several graphitic materials.[9] Interestingly, we observe a hump around 220 K in the temperature-dependent magnetization data at 1 T. In the case of $WS_2$ a stronger hump is noticed around 270 K at a much lower field of 500 Oe, whereas in the case of $MoSe_2$ (inset in Figure 5(a)) and $WSe_2$ no distinguishable humps can be seen. Such magnetic anomalies are often found in graphene samples.[8]

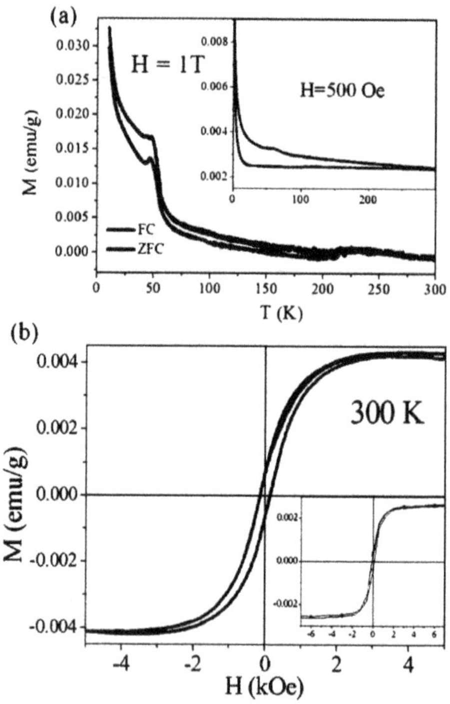


**Figure 5.** (a) Temperature dependent magnetization (ZFC and FC) of few-layer $MoS_2$ under 1 T field. Inset shows the temperature dependence of magnetization of few-layer $MoSe_2$ under 500 Oe field. (b) Hysteresis of few-layer $MoS_2$ at 300 K. Inset shows hysteresis of few-layer $MoSe_2$ at 300 K.

In Figure 5(b) we have shown magnetic hysteresis of few-layer $MoS_2$ at room temperature, which is a characteristic of

Special Issue

  

**Table 1.** Magnetic properties of few-layer metal disulfides and diselenides.

| Samples | No. of layers | $M_S$ /× $10^{-2}$ emu·g$^{-1}$ | | $M_R$ /× $10^{-2}$emu·g$^{-1}$ | | $H_C$ /Oe | |
|---|---|---|---|---|---|---|---|
| | | 10 K | 300 K | 10 K | 300 K | 10 K | 300 K |
| $MoS_2$ | 3–5 | 1.9 | 0.43 | 0.14 | 0.06 | 517 | 146 |
| $WS_2$ | 3–5 | 9.8 | 0.90 | 0.4 | 0.09 | 295 | 130 |
| $MoSe_2$ | 3–6 | 1.3 | 0.26 | 0.09 | 0.04 | 435 | 40 |
| $WSe_2$ | 3–5 | 7.8 | 0.73 | 0.1 | 0.09 | 578 | 200 |

ferromagnetic materials. The saturation magnetization ($M_S$), remnant magnetization ($M_R$), and coercive field ($H_C$) are listed in Table 1. All the other layered metal sulfides and selenides show a similar ferromagnetic behavior at room temperature. Room temperature magnetization of $MoSe_2$ is shown as inset in Figure 5(b). $WS_2$ exhibits the maximum saturation magnetization value while $WSe_2$ exhibits the maximum coercive field. The saturation magnetization increases by almost an order of magnitude at 10 K as compared to that at 300 K (see Table 1). It may be noted that ferromagnetism is a general feature of all inorganic nanomaterials.[10] Interestingly, few-layer $MoS_2$ also exhibits magnetoresistance of 0.5–2% at 300 K.

The temperature-dependent magnetization together with room temperature hysteresis of few-layer $MoS_2$ and other dichalcogenides suggest the presence of both ferromagnetic and antiferromagnetic domains. In order to verify this, we carried out isothermal magnetization measurements at different temperatures by cooling the sample under an applied field to observe the exchange bias behavior. Exchange bias is a phenomenon known to occur in magnetic materials, where ferromagnetic and antiferromagnetic layers or domains are strongly coupled to each other. In such strongly coupled systems, the ferromagnetic spins get pinned by the more anisotropic antiferromagnetic spins. Reversal of the ferromagnetic moment will therefore have an energy cost corresponding to the amount of energy required to reverse the pinned moments. This manifests as a shift in the hysteresis. Such phenomenon has been observed in nanoribbons of graphene and is considered to be due to the presence of ferromagnetic clusters along with antiferromagnetic regions.[11] Figure 6(a) shows the exchange bias behavior in 3–6 layered $MoS_2$ samples under different fields. A negative exchange bias of 15.5 Oe is observed with an applied field of 1 T. The exchange bias decreases with the decrease in the applied field and reverses by applying bias field of –1 T to –15.5 Oe as shown in Figure 6(b). Metal selenides also show similar behavior with an exchange bias field of 19 Oe on application of 1 T.

Layered materials like graphene, graphene nanoribbons,[12] and $MoS_2$ are known to exhibit edge states.[13] In the case of graphene nanoribbons some of these edge states are magnetic.[14] Graphene prepared by different methods is predominantly ferromagnetic at room temperature along with some antiferromagnetic interactions. *Matte* et al.[8] and *Enoki* et al.[15] have shown that edge states as well as adsorbed species affect the magnetic properties of graphene. Density functional calculations on inorganic analogue of graphite $MoS_2$ reveal that edge states are magnetic.[16] It appears that magnetism originates at the sulfur terminated edges due to the splitting of metallic edge states at the Fermi level. *Zhang* et al.[17] have prepared $MoS_2$ samples with high density of prismatic edges and showed them to be ferromagnetic at room temperature, the magnetism arising from nonstoichiometry of the unsaturated molybdenum and sulfur atoms at the edge. In an earlier work, $MoS_2$ and $WS_2$ clusters ($Mo_6S_{12}$ and $W_6S_{12}$) were shown to be magnetic, the magnetism arising from the unsaturated central metal atom due to partially filled d orbitals.[18]

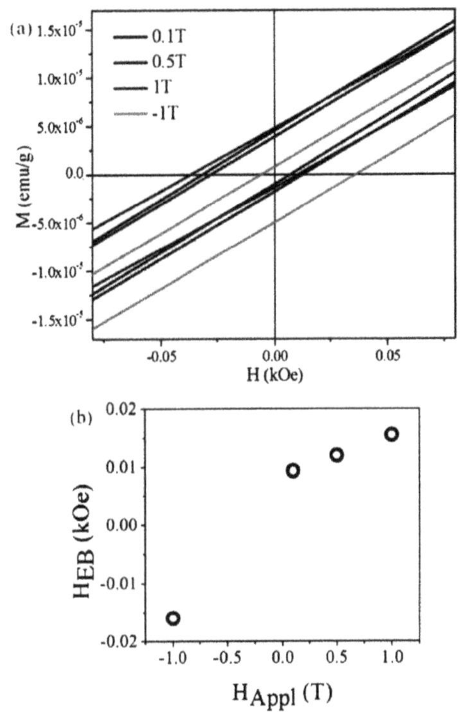


**Figure 6.** (a) Plot of magnetzation vs. field of few-layer $MoS_2$ at 2 K cooled under different applied fields. (b) Dependence of exchange bias on the applied field.

In order to understand the origin of magnetism in few-layer $MoS_2$ and $MoSe_2$, we have carried out high-resolution electron microscopic studies. The samples, which were 3–5 layers thick, show the presence of a high density of edges as can be seen from the images in Figure 7(a) and (b). The bends in the layers shown in the inset of Figure 7(a) may arise from defects. Such defect centers and edges would be associated with the metal atoms, which are undercoordinated, resulting in partially

Special Issue

  

filled d orbitals. This would render the spin-polarized state lower in energy compared to the spin-unpolarized state. A high concentration of such edges and defects in our samples could be one of the possible reasons for the observation of ferromagnetism. High resolution TEM images of $MoS_2$ and $MoSe_2$, shown in Figure 7(c) and (d), respectively, reveal a hexagonal arrangement of atoms with zigzag edges. First principles calculations on arm-chair and zigzag $MoS_2$ nanoribbons show that zigzag nanoribbons have the spin-polarized state as the ground state, while for the arm-chair ribbons the ground state is spin-unpolarized. It is argued that zigzag $MoS_2$ nanoribbons with a ribbon width in the nanoscale will be magnetic.[19]

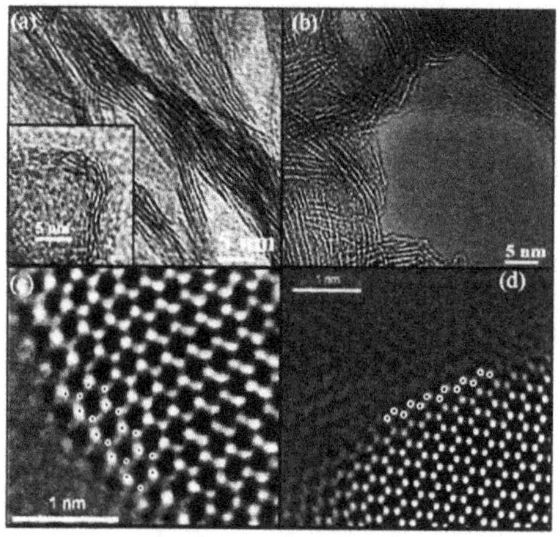


**Figure 7.** TEM images of few-layer (a) $MoS_2$, (b) $MoSe_2$. Fourier transform high resolution TEM images of (c) $MoS_2$ and (d) $MoSe_2$. Bends in the edges can be seen in the inset of (a).

### *Decoration with Metal Nanoparticles*

We have prepared nanocomposites of noble metal nanoparticles with few-layer $WS_2$ and $WSe_2$. In Figure 8, we show the XRD patterns of the nanocomposites formed with few-layer $WS_2$. The XRD patterns confirm the presence of the metal nanoparticles along with $WS_2$. TEM images of the metal-decorated samples are shown in Figure 9. Figure 9(a) shows gold-decorated $WS_2$ sheets with the gold nanoparticles in the range of 19–20 nm. The high resolution TEM image (see inset in Figure 9(a)) shows that the gold nanoparticles decorate the surface as well as at the edges of few-layer $WS_2$. A fairly uniform decoration by platinum and silver nanoparticles is seen in the TEM images of silver- and platinum-decorated $WS_2$ sheets (Figure 9(b) and (c)). The average diameter of the silver nanoparticles is ca. 5 nm and that of platinum nanoparticles is 6 nm. Few-layer $WSe_2$ was also decorated with gold, silver, and platinum nanoparticles. Figure 9(d) shows gold-decorated $WSe_2$ nanosheets with the gold particle size of around 10 nm. Clear evidence of composite formation is evident from the HRTEM images (see inset in Figure 9(d)), which shows nanoparticles attached to edges of the sheets. Raman studies were carried out to examine the interaction between the metal nanoparticles with few-layered $WS_2$. We observe around 2 $cm^{-1}$ shift in the $E_{2g}$ and $A_{1g}$ modes of $WS_2$ in the metal-decorated sheets indicating the presence of mild interaction. We could not observe surface plasmon absorbance of the metal nanoparticles in the UV/Vis spectra because of the overlap with $WS_2$ absorption.

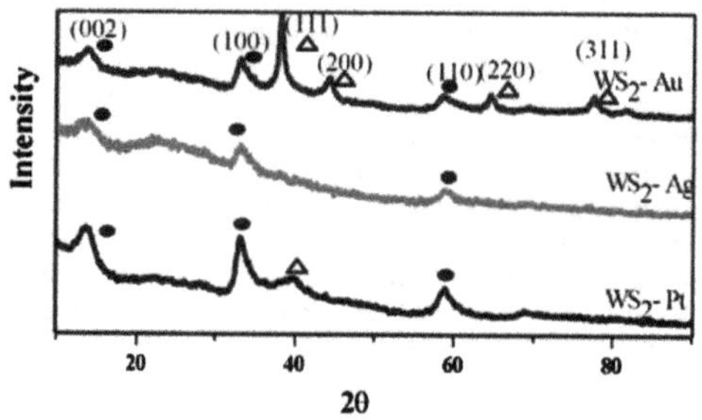


**Figure 8.** XRD patterns of metal nanoparticle decorated on few-layer $WS_2$. Circle represents the reflections corresponding to $WS_2$ and triangle represents that due to metal nanoparticles.

### *Composite with Reduced Graphene Oxide*

We have prepared nanocomposites of the few-layer metal chalcogenides with reduced graphene oxide (RGO). The XRD pattern of these composites show a broad hump at $2\theta = 26°$ arising from RGO along with peaks corresponding to $MoS_2$. Raman studies show that with increasing proportion of $MoS_2$, the intensity of the graphene D band at ca. 1350 $cm^{-1}$ increases, while that of the G band remains essentially constant. The $I_D/I_G$ ratio shows only a slight increase with the increasing proportion of $MoS_2$ in the samples. We do not observe a significant shift in the position of the G band (1595 $cm^{-1}$) in the composites. Figure 10(a) and (b) shows the FESEM images of the 1:1 (wt/wt) RGO/$MoS_2$ composites along with the EDAX pattern indicating the presence of carbon, molybdenum, and sulfur. The TEM image in Figure 10(c) shows the presence of a RGO sheet, in which smaller sheets of $MoS_2$ are embedded. This is confirmed by the HRTEM image (Figure 10(d)) of this composite, which shows two types of interlayer spacing: 0.35 nm corresponding to RGO and 0.68 nm due to $MoS_2$. Electrical resistivity measurements (Figure 11) show that the resistivity of the nanocomposite increases with lowering temperature and follows a trend similar to that of RGO. With the increasing proportion of RGO in the composite, the resistivity decreases progressively.

### *Composite with Polyaniline*

We have prepared a composite of few-layer $MoS_2$ with polyaniline (PANI) containing 30 wt-% of $MoS_2$. The composite

Special Issue

  

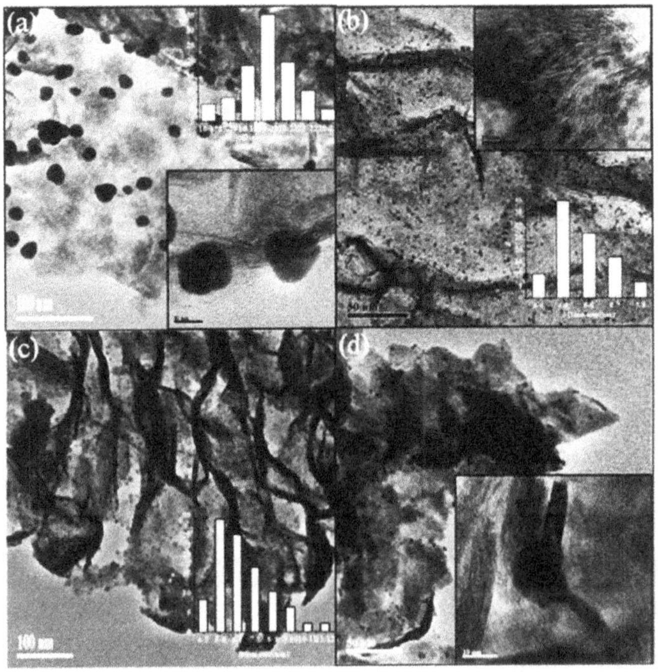

**Figure 9.** TEM images of few-layer $WS_2$ decorated with (a) Au, (b) Ag, (c) Pt nanoparticles (d) Au-decorated $WSe_2$. HRTEM images and particle size distributions of the metal nanoparticles are shown in insets.

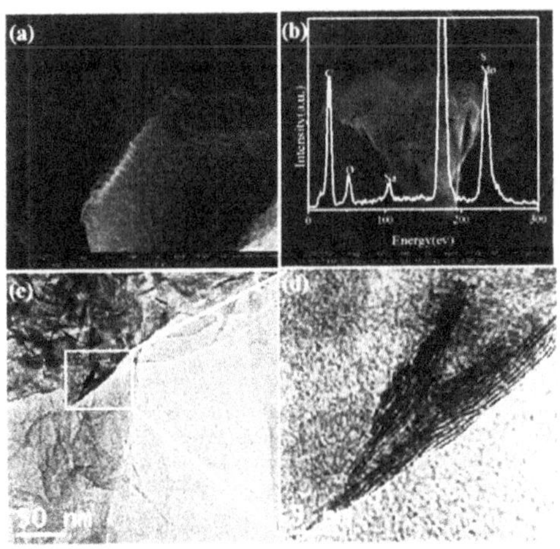

**Figure 10.** (a) FESEM image, (b) EDAX pattern, and FESEM image, (c) TEM image, and (d) high resolution TEM image of 1:1 RGO-$MoS_2$ composite.

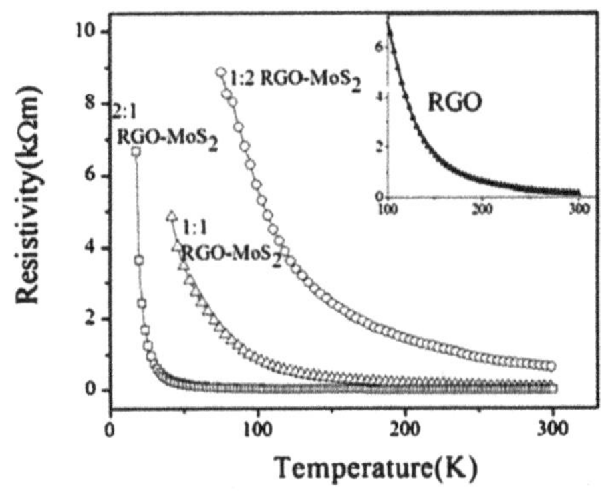


**Figure 11.** Resistivity vs. temperature plots for RGO-$MoS_2$ composites. Inset shows the temperature dependent resistivity for RGO.

was synthesized by *in situ* polymerization of a mixture of few-layer $MoS_2$ and aniline with ammonium persulfate in diluted HCl. For the purpose of comparison pure PANI was prepared under similar conditions. The XRD pattern of the composite shows reflections corresponding to both PANI and few-layer $MoS_2$. Figure 12(a) shows the FESEM image of the $MoS_2$–PANI composite. A TEM image of the composite is shown in Figure 12(b). Homogenous distribution of small flakes of few-layered $MoS_2$ in the PANI matrix can be seen. The Zeta potential of few-layer $MoS_2$ and PANI are −30 mV and +20 mV, respectively, thereby enabling the positively charged PANI to completely wrap around the negatively charged few-layered $MoS_2$ during the *in situ* polymerization. Few-layer $MoS_2$ with an interlayer spacing of 0.68 nm can be clearly seen embedded in the PANI matrix in the HREM image shown as inset in Figure 12(b). Figure 13 shows the Raman spectra of few-layer $MoS_2$, PANI, and the $MoS_2$-PANI composite. The Raman spectrum of few-layer $MoS_2$ displays two bands at 380 and 403 corresponding to the $A_{1g}$ and $E_{2g}$ modes. PANI gives bands at 1590, 1560, 1323, 1215, and 1160 $cm^{-1}$ corresponding to benzenoid C–C stretching, quininoid C=N stretching, $C–N^+$ stretching, benzenoid and quininoid C–H bending, respectively, indicating the formation of emeraldine state of PANI[20]. The composite shows bands corresponding to both PANI and few-layer $MoS_2$ with slight increase in the intensity of the band corresponding to the benzenoid C–H bending relative to that of quinoid in PANI.[21] This indicates that the polaronic structure gets stabilized in the few-layer $MoS_2$-PANI composite.

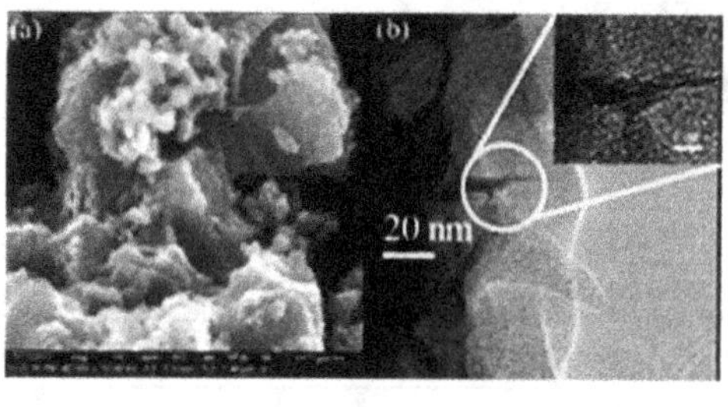


**Figure 12.** (a) FESEM and (b) TEM image of 30 wt-% $MoS_2$-PANI composite. Inset in (b) shows HRTEM image of the same.

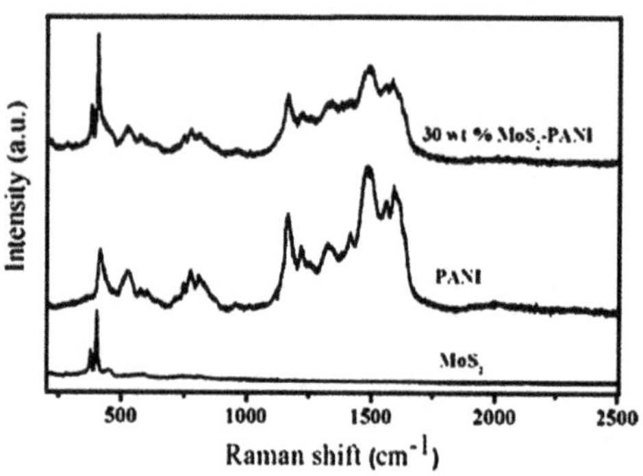


**Figure 13.** Raman spectra of few-layered $MoS_2$, PANI and 30 wt-% $MoS_2$-PANI composite.

## Conclusions

We have successfully synthesized few-layer of sulfides and selenides of molybdenum using microwave irradiation. Laser irradiation of the bulk samples dispersed in DMF also yields few-layered samples. Both of these methods are simple and provide easy methods for the preparation of few-layered metal dichalcogenides. Ferromagnetism exhibited by these few-layer chalcogenides is a unique feature and can be related to the defects especially arising from edges. Nanocomposites of metal nanoparticles with dichalcogenides might find use in applications such as catalysis. In particular, cobalt and nickel particles deposited on few-layer $MoS_2$ are good catalysts in hydrodesulfurization. By addition of the few-layered dichalcogenides to reduced graphene oxide one can progressively increase the resistivity. We have prepared a few-layer $MoS_2$–PANI composite showing polaronic character.

## Experimental Section

**Synthesis of Layered Chalcogenides:** Few-layer $MoS_2$ was prepared starting with a mixture of molybdic acid (1 mM, $H_2MoO_4$) and potassium thiocyanate (3 mM, KSCN) in ethylene glycol (10 ml, EG) taken in a Teflon-coated autoclave (30 mL). The mixture was heated at 450 W for 15 min in a domestic microwave oven operating at 2.45 GHz. Few-layer $MoSe_2$ was prepared by microwave treatment of a mixture of molybdic acid (1 mM, $H_2MoO_4$), with selenium powder (2 mM, Se) and sodium borohydride (3 mM, $NaBH_4$) in EG (10 mL) taken in a 30 mL capacity of Teflon-coated autoclave. The products were washed with ethanol, dried at 60 °C, and used for further characterization.

KrF excimer laser (248 nm wavelength, 30 ns pulse width) was employed to exfoliate bulk powders of sulfides and selenides of molybdenum and tungsten. Irradiation was carried out for 1 h of dispersions of these materials in a quartz vessel in DMF medium as solvent. The laser fluency and repetition rate used for the purpose were 1.5 $J·cm^{-2}$ and 5 Hz, respectively. During laser treatment the dispersions were stirred with a magnetic stirrer. After the laser exfoliation experiment, the resultant solutions were kept for 24 h and additionally centrifuged at 2500 RPM for 5 min. The materials used for microscopy were from the top half of the centrifuged solution. After centrifugation, the supernatant fluid was quite stable for a long period.

**Decoration of Few-layer $WS_2$ and $WSe_2$ with Metal Nanoparticles:** For the decoration of noble metal particles on graphene analogues, we have prepared few-layer $WS_2$ and $WSe_2$ by thiourea/selenourea method respectively.[2b,2c] Decoration with gold nanoparticles was accomplished starting with a mixture of few-layer $WS_2$/$WSe_2$ (15 mg), hexachloroauricacid (2 mL, 4.78 mM), and ethylene glycol (6 mL) (as the reducing agent and solvent) in a 50 ml autoclave. The bomb was kept in a microwave oven at 300 W for 30 min. The product of the reaction was washed with ethanol and dried at 40 °C for 12 h. In order to decorate with silver nanoparticles, to a mixture of few-layer $WS_2$/$WSe_2$ (15 mg) in silver nitrate solution (2 mL, 10 mM) in a 50 mL autoclave, ethylene glycol (2 mL) was added. The bomb was kept in the microwave oven at 300 W for 20 min. The product was washed with ethanol and dried at 40 °C for 12 h. Platinum nanoparticle decorated samples were obtained by starting with a mixture of few-layer $WS_2$/$WSe_2$ (15 mg) and hexachloroplatinic acid (1 ml, 11 mM) in a 50 mL autoclave bomb, to which ethylene glycol (2 mL) was added. The autoclave was kept in a microwave oven at 300 W for 30 min. Later the product was washed with ethanol and dried at 40 °C for 12 h.

**Nanocomposites of $MoS_2$ with RGO:** To prepare nanocomposites of $MoS_2$ with RGO, we first dispersed the required quantities graphene

Special Issue

  

oxide and $MoS_2$ in water (5 mL) and DMF (5 mL), respectively. The dispersions were mixed and sonicated together for 30 min and finally heated to reflux with hydrazine hydrate (500 μL) for 20 h. In a typical experiment to prepare 1:1 composite the starting quantities of GO and $MoS_2$ were 15 mg each.

**Nanocomposite of $MoS_2$ with PANI:** Aniline (300 μL) was added to few-layer $MoS_2$ (90 mg), and to this micture HCl (10 mL, 1 M) was added. The resulting mixture was sonicated for 1 h. The *in situ* chemical polymerization was carried out by dropwise addition of ammonium persulfate in HCl (1 M). The resulting suspension was kept in ice-cold temperature with vigorous stirring for 2 h.[22] The product was filtered and washed with methanol and characterized by XRD, Raman spectroscopy, and electron microscopy.

**Characterization:** XRD patterns were recorded with Cu-$K_\alpha$ radiation with a Rich-Siefert XRD -300-TT diffractometer. Raman spectra of samples were recorded with 514.5 nm argon laser with a Jobin Yvon LabRam HR spectrometer. TEM images were obtained with a JEOL JEM3010 instrument fitted with a Gatan CCD camera operating at an accelerating voltage of 300 KV. FESEM images were recorded with a FEI NOVA NANOSEM 600. Samples for magnetoresistance measurements were prepared using electron beam induced deposition of Pt in FEI NOVA NANOSEM 600 at dosage of $35mC{\cdot}cm^{-2}$ on pre-patterned 300 nm $Si/SiO_2$ substrate with 6 μm gap between two gold electrodes. AFM measurements were carried out with Vecco digital instruments (di Innova). Samples for AFM measurements were prepared by spin coating the $MoS_2$ and $MoSe_2$ solutions on $Si/SiO_2$ substrate. Electronic absorption spectra were recorded with a Perkin-Elmer Lambda 900 UV/Vis/NIR Spectrometer. UV/Vis spectra were recorded in a 1 cm path length quartz cuvette. While preparing samples for magnetic measurement care was taken to avoid contact with materials containing magnetic elements thereby ruling out possibility of magnetic impurities. Magnetic measurements were carried out with vibrating sample magnetometer in Physical Properties Measurement System (PPMS, Quantum Design, USA). For carrying out resistivity measurements the composites were deposited as a film on cellulose nitrate membrane by filtration of the dispersion of the composite through the membrane. Four probe measurements were carried out in PPMS.

Special Issue

## References


[1] a) A. K. Geim, *Science* **2009**, *324*, 1530–1534; b) C. N. R. Rao, A. K. Sood, K. S. Subrahmanyam, A. Govindaraj, *Angew. Chem. Int. Ed.* **2009**, *48*, 7752.

[2] a) J. N. Coleman, M. Lotya, A. O'Neill, S. D. Bergin, P. J. King, U. Khan, K. Young, A. Gaucher, S. De, R. J. Smith, I. V. Shvets, S. K. Arora, G. Stanton, H.-Y. Kim, K. Lee, G. T. Kim, G. S. Duesberg, T. Hallam, J. J. Boland, J. J. Wang, J. F. Donegan, J. C. Grunlan, G. Moriarty, A. Shmeliov, R. J. Nicholls, J. M. Perkins, E. M. Grieveson, K. Theuwissen, D. W. McComb, P. D. Nellist, V. Nicolosi, *Science* **2011**, *331*, 568–571; b) H. S. S. R. Matte, B. Plowman, R. Datta, C. N. R. Rao, *Dalton Trans.* **2011**, *40*, 10322; c) H. S. S. Ramakrishna Matte, A. Gomathi, A. K. Manna, D. J. Late, R. Datta, S. K. Pati, C. N. R. Rao, *Angew. Chem. Int. Ed.* **2010**, *49*, 4059.

[3] a) H. Huang, Y. Xia, X. Tao, J. Du, J. Fang, Y. Gan, W. Zhang, *J. Mater. Chem.* **2012**, *22*, 10452–10456; b) B. K. Miremadi, S. R. Morrison, *J. Appl. Phys.* **1988**, *63*, 4970–4974.

[4] a) K. J. Rao, B. Vaidhyanathan, M. Ganguli, P. A. Ramakrishnan, *Chem. Mater.* **1999**, *11*, 882–895; b) I. Bilecka, M. Niederberger, *Nanoscale* **2010**, *2*, 1358–1374; c) C. N. R. Rao, H. S. S. Ramakrishna Matte, R. Voggu, A. Govindaraj, *Dalton Trans.* **2012**, *41*, 5089–5120.

[5] a) D. Vollath, D. V. Szabo, *Mater. Lett.* **1998**, *35*, 236–244; b) R. Harpeness, A. Gedanken, A. M. Weiss, M. A. Slifkin, *J. Mater. Chem.* **2003**, *13*, 2603–2606; c) P. Pravas Kumar, P. Amita, *Sci. Technol. Adv. Mater.* **2008**, *9*, 045008.

[6] a) K. S. Subrahmanyam, A. K. Manna, S. K. Pati, C. N. R. Rao, *Chem. Phys. Lett.* **2010**, *497*, 70–75; b) P. V. Kamat, *J. Phys. Chem. Lett.* **2009**, *1*, 520–527; c) H. M. A. Hassan, V. Abdelsayed, A. E. R. S. Khder, K. M. AbouZeid, J. Terner, M. S. El-Shall, S. I. Al-Resayes, A. A. El-Azhary, *J. Mater. Chem.* **2009**, *19*, 3832–3837; d) B. Das, B. Choudhury, A. Gomathi, A. K. Manna, S. K. Pati, C. N. R. Rao, *ChemPhysChem* **2011**, *12*, 937–943; e) e) B. Govinda Rao, H. S. S. R. Matte, C. N. R. Rao, *J. Cluster Sci.* **2012**, *23*, 929–937.

[7] G. L. Frey, R. Tenne, M. J. Matthews, M. S. Dresselhaus, G. Dresselhaus, *Phys. Rev. B* **1999**, *60*, 2883–2892.

[8] a) H. S. S. R. Matte, K. S. Subrahmanyam, C. N. R. Rao, *J. Phys. Chem. C* **2009**, *113*, 9982–9985; b) C. N. R. Rao, H. S. S. R. Matte, K. S. Subrahmanyam, U. Maitra, *Chem. Sci.* **2012**, *3*, 45–52.

[9] T. L. Makarova, K. H. Han, P. Esquinazi, R. R. da Silva, Y. Kopelevich, I. B. Zakharova, B. Sundqvist, *Carbon* **2003**, *41*, 1575–1584.

[10] A. Sundaresan, C. N. R. Rao, *Nano Today* **2009**, *12–14*, 96–106.

[11] S. S. Rao, S. N. Jammalamadaka, A. Stesmans, V. V. Moshchalkov, J. v. Tol, D. V. Kosynkin, A. Higginbotham-Duque, J. M. Tour, *Nano Lett.* **2012**, *12*, 1210–1217.

[12] a) M. Fujita, K. Wakabayashi, K. Nakada, K. Kusakabe, *J. Phys. Soc. Jpn.* **1996**, *65*, 1920–1923; b) K. Nakada, M. Fujita, G. Dresselhaus, M. S. Dresselhaus, *Phys. Rev. B* **1996**, *54*, 17954; c) P. L. Giunta, S. P. Kelty, *J. Chem. Phys.* **2001**, *114*, 1807–1812.

[13] a) M. V. Bollinger, J. V. Lauritsen, K. W. Jacobsen, J. K. Nørskov, S. Helveg, F. Besenbacher, *Phys. Rev. Lett.* **2001**, *87*, 196803; b) S. Helveg, J. V. Lauritsen, E. Lægsgaard, I. Stensgaard, J. K. Nørskov, B. S. Clausen, H. Topsøe, F. Besenbacher, *Phys. Rev. Lett.* **2000**, *84*, 951–954; c) M. V. Bollinger, K. W. Jacobsen, J. K. Nørskov, *Phys. Rev. B* **2003**, *67*, 085410.

[14] a) K. Wakabayashi, M. Fujita, H. Ajiki, M. Sigrist, *Phys. Rev. B* **1999**, *59*, 8271; b) T. Enoki, N. Kawatsu, Y. Shibayama, H. Sato, R. Kobori, S. Maruyama, K. Kaneko, *Polyhedron* **2001**, *20*, 1311–1315.

[15] a) T. Enoki, Y. Kobayashi, *J. Mater. Chem.* **2005**, *15*, 3999–4002; b) T. Enoki, K. Takai, *Dalton Trans.* **2008**, 3773–3781.

[16] A. Vojvodic, B. Hinnemann, J. K. Nørskov, *Phys. Rev. B* **2009**, *80*, 125416.

[17] J. Zhang, J. M. Soon, K. P. Loh, J. Yin, J. Ding, M. B. Sullivian, P. Wu, *Nano Lett.* **2007**, *7*, 2370–2376.

[18] P. Murugan, V. Kumar, Y. Kawazoe, N. Ota, *Phys. Rev. A* **2005**, *71*, 063203.

[19] Y. Li, Z. Zhou, S. Zhang, Z. Chen, *J. Am. Chem. Soc.* **2008**, *130*, 16739–16744.

[20] H. Wang, Q. Hao, X. Yang, L. Lu, X. Wang, *ACS Appl. Mater. Interfaces* **2010**, *2*, 821–828.

[21] R. V. Salvatierra, M. M. Oliveira, A. J. G. Zarbin, *Chem. Mater.* **2010**, *22*, 5222–5234.

[22] C. Fite, Y. Cao, A. J. Heeger, *Solid State Commun.* **1989**, *70*, 245–247.


Received: June 16, 2012
Published Online: November 9, 2012

  

ChemComm

RSCPublishing

COMMUNICATION

View Article Online
View Journal | View Issue

# Hybrid nanocomposites of ZIF-8 with graphene oxide exhibiting tunable morphology, significant $CO_2$ uptake and other novel properties†

Cite this: Chem. Commun., 2013, 49, 4947

Received 7th January 2013, Accepted 8th April 2013

DOI: 10.1039/c3cc00136a

www.rsc.org/chemcomm

Ram Kumar,[a] Kolleboyina Jayaramulu,[a] Tapas Kumar Maji*[a] and C. N. R. Rao*[ab]

**Hybrid nanocomposites of graphene oxide (GO) with ZIF-8 exhibit tunable nanoscale morphology and porosity, both determined by the GO content, coordination modulation being responsible for such properties. These materials also give rise to high $CO_2$ storage capability and can be used as precursors to prepare GO@ZnS nanocomposites.**

Metal–organic frameworks (MOFs) or porous coordination polymers (PCPs) have attracted great attention in recent years because of their possible use in gas storage, catalysis, and optoelectronic devices.[1–3] The modifiable pore surface of MOFs props them to the forefront of clean energy and drug delivery research.[4,5] Scaling down these materials to the nanoscale retaining the same periodic structure of the bulk state can give rise to novel applications in bioimaging, bio-sensing and other areas.[6] Furthermore nanoscale MOFs enable solution-based processability and increased gas uptake.[7] Methodologies such as solvent induced precipitation, solvo(hydro)thermal or microwave treatment, coordination modulation techniques and reverse micro-emulsion have been employed for the synthesis of nanoscale MOFs.[6,7] In the last few years, graphene oxide (GO) has become a material of vital importance because of its unique properties.[8] GO has a functionalized basal plane with anchoring groups like hydroxy, epoxy or carboxyl groups which are useful for forming graphene-based soft materials and hybrid composites.[9] Hybrid composites based on MOFs and graphene enable integration of the unique properties of these two fascinating materials, thus allowing one to design materials possessing properties not possessed by either component.[10,11] Bandosz *et al.* have recently reported GO–MOF composites for the adsorption of ammonia and nitric oxide[10b] while Loh *et al.* have described conducting nanowires comprising MOF-5 and benzoic acid functionalized graphene wherein MOF-5 is intercalated between graphene sheets.[11a] Growth of MOF nanocrystals of different morphologies on graphene sheets has not been explored adequately. One can envisage GO as a potential platform and as a structure-directing agent for the growth and stabilization of MOF nanoparticles, where coordination modulation occurs through the different functional groups on the surface.[9b] In this context, making use of zeolitic imidazolate frameworks (ZIFs), a new class of zeolite-like MOFs, possessing a high degree of microporosity and exceptional thermal and chemical stabilities to prepare composites with GO, appeared to be attractive.[12] We have therefore made use of the prototypical ZIF-8 ($Zn(MeIM)_2$, MeIM = 2-methylimidazolate) with a sodalite zeolite structure, containing large cavities (~11.7 Å of diameter) interconnected by narrow windows (~3.40 Å) for this purpose (Scheme 1a).[12a] It may be noted that ZIF-8 has been exploited for developing thin films, membranes for catalysis and separation of gases.[12,13] In this communication we report, for the first time, a simple route to the preparation of hybrid nanocomposites of graphene oxide and ZIF-8, where the growth of the ZIF-8 nanocrystals as well as their sizes and morphologies are tuned by the different concentration of GO. The obtained GO@ZIF-8 (ZG) hybrid nanocomposites have tunable surface areas and pore volumes depending on the proportion of GO. Interestingly the hybrid nanocomposites exhibit characteristics by nature of both GO and ZIF-8. The composites exhibit remarkable $CO_2$ uptake characteristics with increasing GO concentration. Furthermore, we demonstrate the generation of ZnS nanoparticles on the GO sheets by using an external sulphur source, where GO acts as a template

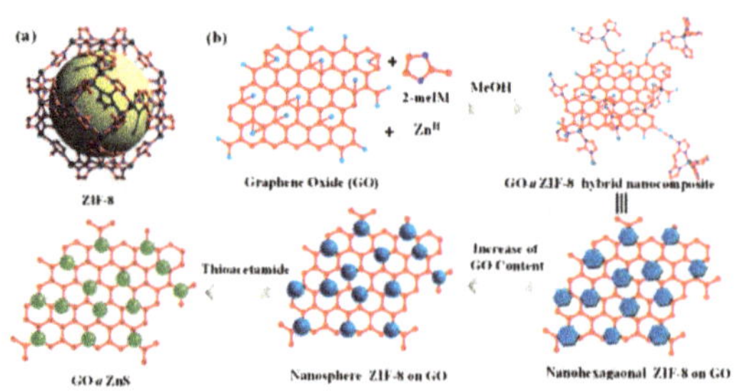


**Scheme 1** (a) View of the ZIF-8 framework showing large cavities with narrow windows; (b) stepwise synthesis of GO@ZIF-8 hybrid nanocomposites with tunable morphology, nanospheres from hexagonal ZIF-8 nanocrystals and *in situ* stabilization of ZnS nanoparticles on the GO sheet.

[a] *International Centre for Material Science, Chemistry and Physics of Materials Unit and New Chemistry Unit, Jawaharal Nehru Centre for Advanced Scientific Research, Jakkur, Bangalore 560064, India*

[b] *CSIR Centre of Excellence in Chemistry, Jawaharal Nehru Centre for Advanced Scientific Research, Jakkur, Bangalore 560064, India. E-mail: cnrrao@jncasr.ac.in, tmaji@jncasr.ac.in*

† Electronic supplementary information (ESI) available. See DOI: 10.1039/c3cc00136a

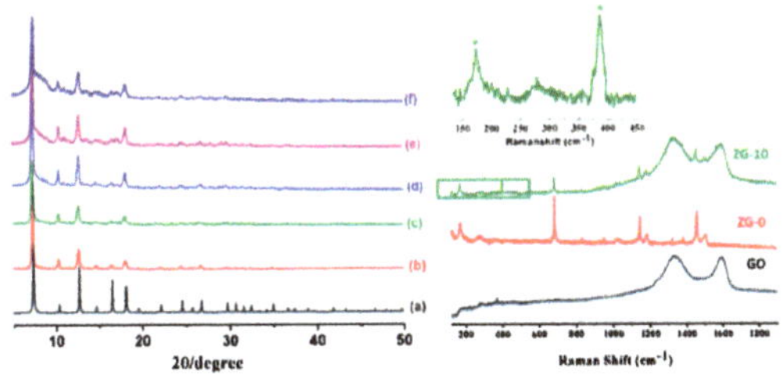

**Fig. 1** (Left) PXRD patterns of ZIF-8 and GO@ZIF-8 hybrid nanocomposites: (a) ZG-0 (ZIF-8), (b) ZG-1, (c) ZG-2, (d) ZG-4, (e) ZG-10, and (f) ZG- 20; (right) Raman spectra of GO, ZIF-8 and ZG-10, the inset shows bands corresponding to Zn–O bonding.

and ZIF-8 as the sacrificial Zn source. In Scheme 1b we show the method for synthesis of GO@ZIF-8 hybrid nanocomposites with different GO contents (ZG-0, ZG-1, ZG-2, ZG-4, ZG-10 and ZG-20 corresponding to 0, 1, 2, 4, 10 and 20 wt% GO, respectively). The phase purity of the nanocomposite was established through elemental analysis and EDAX (see Fig. S1, ESI†). Powder X-ray diffraction patterns of the composites with different GO contents show the reflections of ZIF-8 (cubic space group) (Fig. 1). With an increase in the GO content, a new reflection appears at $2\theta = 8.6°$ close to the 110 peak of ZIF-8, which would be attributed to GO. Such low angle shift compared to that of pristine GO was observed indicating intercalation of ZIF-8 nanocrystals between the GO sheets.

The infrared spectra of the ZG hybrid nanocomposites show bands due to ZIF-8 as well as GO, the frequencies being shifted to slightly lower frequencies (Fig S2, ESI†). ZIF-8 shows bands in the Raman spectrum corresponding to the 2-methylimidazole linker while GO shows bands at 1333 $cm^{-1}$ and 1594 $cm^{-1}$ due to the D and G bands respectively. The Raman spectra of composites show bands corresponding to ZIF-8 as well as the D and G bands (Fig. 1 and Fig. S3, ESI†). With increasing GO concentration, the intensities of the D and G bands increase. Furthermore, peaks at 175 $cm^{-1}$ and 390 $cm^{-1}$ corresponding to Zn–O and Zn–O–Zn vibrational bands are also seen due to bonding between –OH and –COOH functional groups decorated on the GO with Zn(II) (see inset of Fig. 1). Thermogravimetric (TG) analysis of ZIF-8 and the composites shows high thermal stability as shown in Fig. S4 (ESI†).

The morphologies of ZIF-8 and the ZG composites were characterized using transmission electron microscopy (TEM). We show typical TEM images in Fig. 2 and Fig. S5 (ESI†). The as-synthesized ZIF-8 (without any GO) shows hexagonal morphology with the particle size in the range of ~100 to 150 nm as shown in Fig. 2a. Similar hexagonal morphology and particle size are also found for ZG-1 (1 wt% of GO), with the ZIF-8 particles being stabilized within the dense GO sheets (Fig. 2b and Fig. S5, ESI†). With increasing GO content, the particle size of ZIF-8 decreases. In the case of ZG-2, irregular hexagons with a few squares or rectangles of 50–100 nm size are seen and covered between the GO sheets (Fig. 2c and Fig. S5, ESI†). Upon increasing the GO content to 4 wt% in (ZG-4), irregular particles transform to spherical shape, the size being 40–50 nm (Fig. 2d). In ZG-10 (10 wt% GO) spheres of 20 nm diameter (a few with bigger size of 40–50 nm) are formed (Fig. 2e and Fig. S5, ESI†). In the case of ZG-20 (20 wt% GO), the size of the nanocrystals gets significantly reduced and we see a

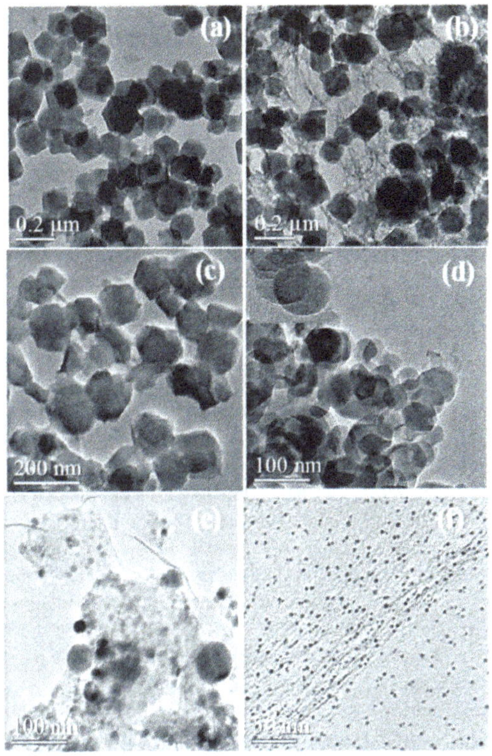

**Fig. 2** TEM images of ZG samples: (a) ZG-0, (b) ZG-1, (c) ZG-2, (d) ZG-4, (e) ZG-10 and (f) ZG-20.

homogeneous distribution of uniform nanospheres of ~4 nm diameter in Fig. 2f and Fig. S5 (ESI†).

Clearly, the size and morphology of ZIF-8 are controlled by the percentage of GO in the composite. The GO sheets appear to provide a platform for the nucleation and growth of nanocrystals. The shapes and sizes (hexagonal to spherical) are probably controlled by –OH and –COOH groups through coordination modulation which inhibits the growth of large crystals. Bonding between $Zn^{II}$ metal ions of ZIF-8 with the GO sheets is evidenced from the Raman spectra (inset, Fig. 1). We therefore suggest that the functionalized GO sheets act as both structure-directing and size-controlling agents for ZIF-8 nanocrystals.

Textural parameters such as surface area, pore volume and pore size of the ZG hybrid nanocomposites were obtained from $N_2$ adsorption–desorption measurements at 77 K (Fig. 3, left). The BET surface area of the GO sample is around ~15 $m^2\ g^{-1}$ (Fig S6, ESI†). On the other hand, $N_2$ adsorption of the nanoscale ZIF-8 (ZG-0) shows typical type-I profile with a surface area of about 1120 $m^2\ g^{-1}$ comparable with the literature value.[10] The textural parameters of the ZG composites are given in Table S1 (ESI†). The BET surface area of the composites decreases relative to ZG-0 (ZIF-8) with increasing GO content. Thus, the surface areas of ZG-1, ZG-2, ZG-4, ZG-10 and ZG-20 are 819, 675, 450, 368, 289 $m^2\ g^{-1}$ respectively. The decrease in surface area can arise from an increasing proportion of nonporous GO in the composites. To our surprise,

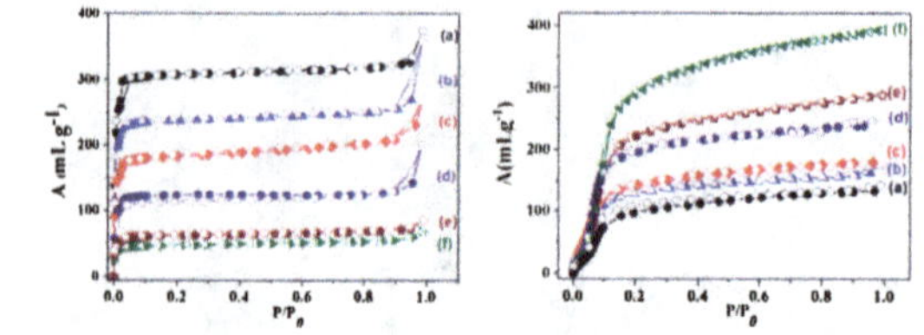

**Fig. 3** (Left) Nitrogen (at 77 K), (right) carbon dioxide (at 195 K) adsorption–desorption isotherms for ZG samples: (a) ZG-0; (b) ZG-1; (c) ZG-2; (d) ZG-4 (e) ZG-10 and (f) ZG-20.

the ZG composites exhibit good $CO_2$ storage capacity and $CO_2$ uptake at 195 K increasing with the increasing GO content (Fig. 3 (right) and Table S1, ESI†). Note that pristine ZIF-8 (ZG-0) and GO show 27.2 wt% and 33 wt% of $CO_2$ uptake, respectively (Fig. S6, ESI†). ZG-4 exhibits $CO_2$ uptake of about 49 wt% at 195 K (1 atm) which increases further to 72 wt% in ZG-20. This unusual $CO_2$ uptake can be attributed to the cumulative effect of ZIF-8 and GO as the latter provides specific interaction sites based on different polar functional groups for $CO_2$. This is supported by the high heat of $CO_2$ sorption ($q_{st,\phi}$) 31.9 kJ $mol^{-1}$ in the case of ZG-20 obtained using the Dubinin–Radushkevich (DR) equation.[14] To the best of our knowledge, this is the first report of GO@ZIF-8 hybrid nanocomposites with tunable porosity and significant $CO_2$ storage capacity.

Zinc sulphide (ZnS) shows fascinating electrical and optical properties, which can be further modified on the nanoscale or in hybridization with other materials like graphene oxide (GO) as a hybrid composite.[15] We have made use of the GO@ZIF-8 composite, ZG-20, for the synthesis of the GO@ZnS composite. Upon reacting ZG-20 with thioacetamide, the ZIF-8 in the composite acts as the sacrificial Zn source and reacts with the sulphur produced using thioacetamide. Fig. 4a shows the powder X-ray diffraction pattern of the GO@ZnS nanocomposite with broad reflections corresponding to the Wurtzite structure of ZnS (JCPDS No. 39-1363). The reflection at 12.21° corresponds to GO and the slight shift to a higher angle corroborates a decrease in the interlayer separation between two successive graphene sheets. Fig. 4b–c and Fig. S7 (ESI†) show TEM images of the GO@ ZnS, revealing that the ZnS nanoparticles of ~3 nm are stabilized on the GO sheet. This particle size is consistent with the XRD data and this hybrid also shows $CO_2$ uptake of about 9 wt% as shown in Fig. S8 (ESI†).

In conclusion, we have demonstrated a simple procedure to prepare hybrid GO@ZIF-8 hybrid nanocomposites. ZIF-8 is stabilized on the GO surfaces through functional groups and the nanoscale morphology of the composite (from hexagonal to spherical) can be tuned by the concentration of GO. GO acts as a structure-directing agent for the growth of nanocrystals of ZIF-8 through coordination modulation. The composites show remarkable $CO_2$ storage capacity compared to the parent ZIF-8. To the best of our knowledge, this is the first account of the growth and stabilization of nanoscale MOFs on a graphene surface, with tunable morphology and gas storage characteristics. The GO–ZIF composite can be used as a precursor to obtain the GO@ZnS composite by reaction with thioacetamide, where ZIF-8 acts as the sacrificial source of Zn(II).

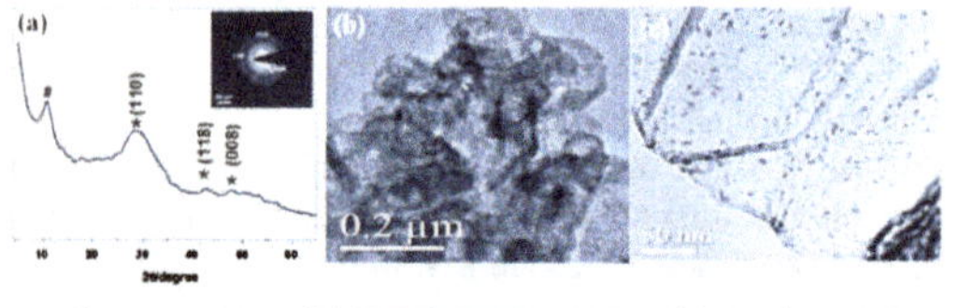

**Fig. 4** Characterization of GO@ZnS nanocomposites: (a) powder XRD pattern (the ED pattern inset shows three planes of the ZnS corresponding Wurtzite structure); (b and c) TEM images showing ZnS nanoparticles on the GO sheet.

## Notes and references

1 (*a*) L. J. Murray, M. Dinca and J. R. Long, *Chem. Soc. Rev.*, 2009, **38**, 1294; (*b*) A. K. Cheetham and C. N. R. Rao, *Science*, 2007, **318**, 58; (*c*) S. Kitagawa, R. Kitaura and S. I. Noro, *Angew. Chem., Int. Ed.*, 2004, **43**, 2334; (*d*) D. Bradshaw, J. B. Claridge, E. J. Cussen, T. J. Prior and M. J. Rosseinsky, *Acc. Chem. Res.*, 2005, **38**, 273.

2 (*a*) J. Lee, O. K. Farha, J. Roberts, K. A. Scheidt, S. T. Nguyen and J. T. Hupp, *Chem. Soc. Rev.*, 2009, **38**, 1450; (*b*) A. Dhakshinamoorthy, M. Alvaro, A. Corma and H. Garcia, *Dalton Trans.*, 2011, **40**, 6344.

3 (*a*) C. H. Hendon, D. Tiana and A. Walsh, *Phys. Chem. Chem. Phys.*, 2012, **14**, 13120; (*b*) K. Jayaramulu, P. Kanoo, S. J. George and T. K. Maji, *Chem. Commun.*, 2010, **46**, 7906.

4 J. R. Li, R. J. Kuppler and H. C. Zhou, *Chem. Soc. Rev.*, 2009, **38**, 1477.

5 P. Horcajada, T. Chalati, C. Serre, B. Gillet, C. Sebrie, T. Baati, J. F. Eubank, D. Heurtaux, P. Clayette, C. Kreuz, J.-S. Chang, Y. K. Hwang, V. Marsaud, P.-N. Bories, L. Cynober, S. Gil, G. Ferey, P. Couvreur and R. Gref, *Nat. Mater.*, 2010, **9**, 172.

6 (*a*) W. Lin, W. J. Rieter and K. M. L. Taylor, *Angew. Chem., Int. Ed.*, 2009, **48**, 650; (*b*) M. Spokoyny, D. Kim, A. Sumrein and C. A. Mirkin, *Chem. Soc. Rev.*, 2009, **38**, 1218; (*c*) A. Carne, C. Carbonell, I. Imaz and D. Maspoch, *Chem. Soc. Rev.*, 2011, **40**, 291.

7 (*a*) H. J. Lee, W. Cho, S. Jung and M. Oh, *Adv. Mater.*, 2009, **21**, 674; (*b*) O. K. Farha, A. M. Spokoyny, K. L. Mulfort, S. Galli, J. T. Hupp and C. A. Mirkin, *Small*, 2009, **5**, 1727; (*c*) T. Tsuruoka, S. Furukawa, Y. Takashima, K. Yoshida, S. Isoda and S. Kitagawa, *Angew. Chem., Int. Ed.*, 2009, **48**, 4739.

8 (*a*) C. N. R. Rao, A. K. Sood, K. S. Subrahmanyam and A. Govindaraj, *Angew. Chem., Int. Ed.*, 2009, **48**, 7752; (*b*) D. Chen, H. Feng and J. Li, *Chem. Soc. Rev.*, 2012, **112**, 6027.

9 (*a*) K. Yang, L. Feng, X. Shi and Z. Liu, *Chem. Soc. Rev.*, 2013, **42**, 530; (*b*) X. Li, W. Qi, D. Mei, M. L. Sushko, I. Aksay and J. Liu, *Adv. Mater.*, 2012, **24**, 5136; (*c*) J. Kim, L. J. Cote and J. Huang, *Acc. Chem. Res.*, 2012, **45**, 1356.

10 (*a*) Z.-H. Huang, G. Liu and F. Kang, *ACS Appl. Mater. Interfaces*, 2012, **4**, 4942; (*b*) C. Petit and T. J. Bandosz, *Adv. Mater.*, 2009, **21**, 4753; (*c*) L. D. O'Neill, H. Zhang and D. Bradshaw, *J. Mater. Chem.*, 2010, **20**, 5720; (*d*) Y. Guo, Y. Han, S. Shuang and C. Dong, *J. Mater. Chem.*, 2012, **22**, 13166.

11 (*a*) M. Jahan, Q. Bao, J.-X. Yang and K. P. Loh, *J. Am. Chem. Soc.*, 2010, **132**, 14487; (*b*) D. Li, L. Qiu, K. Wang, Y. Zeng, T. Williams, Y. Huang, M. Tsapatsis and H. Wang, *Chem. Commun.*, 2012, **48**, 2249; (*c*) H. Guo, Y. Zhu, S. Qiu, J. A. Lercher and H. Zhang, *Adv. Mater.*, 2010, **22**, 4190.

12 (*a*) K. S. Park, Z. Ni, A. P. Côté, J. Y. Choi, R. Huang, F. J. Uribe-Romo, H. K. Chae, M. O'Keeffe and O. M. Yaghi, *Proc. Natl. Acad. Sci. U. S. A.*, 2006, **103**, 10186; (*b*) Y. Pan, Y. Liu, G. Zeng, L. Zhao and Z. Lai, *Chem. Commun.*, 2011, **47**, 2071; (*c*) Z. Jiang, H. Sun, Z. Qin, X. Jiao and D. Chen, *Chem. Commun.*, 2012, **48**, 3620; (*d*) J. Cravillon, S. Münzer, S.-J. Lohmeier, A. Feldhoff, K. Huber and M. Wiebcke, *Chem. Mater.*, 2009, **21**, 1410.

13 (*a*) Q. Song, S. K. Nataraj, M. V. Roussenova, J. C. Tan, D. J. Hughes, W. Li, P. Bourgoin, M. A. Alam, A. K. Cheetham, S. A. Al-Muhtaseb and E. Sivaniah, *Energy Environ. Sci.*, 2012, **5**, 8359; (*b*) G. Lu, O. K. Farha, W. Zhang, F. Huo and J. T. Hupp, *Adv. Mater.*, 2012, **24**, 3970.

14 M. M. Dubinin, *Chem. Rev.*, 1960, **60**, 235.

15 (*a*) T. Zhai, L. Li, X. Wang and X. Fang, *J. Phys. Chem. Lett.*, 2009, **1**, 520; (*b*) S. Pan and X. Liu, *J. Solid State Chem.*, 2012, **191**, 50.

www.ingramcontent.com/pod-product-compliance
Lightning Source LLC
LaVergne TN
LVHW080356011026
849161LV00009B/2324
* 9 7 8 9 8 1 4 5 3 1 6 8 9 *